Springer Optimization and Its Applications

Volume 179

Aims and Scope

Optimization has continued to expand in all directions at an astonishing rate. New algorithmic and theoretical techniques are continually developing and the diffusion into other disciplines is proceeding at a rapid pace, with a spot light on machine learning, artificial intelligence, and quantum computing. Our knowledge of all aspects of the field has grown even more profound. At the same time, one of the most striking trends in optimization is the constantly increasing emphasis on the interdisciplinary nature of the field. Optimization has been a basic tool in areas not limited to applied mathematics, engineering, medicine, economics, computer science, operations research, and other sciences.

The series **Springer Optimization and Its Applications (SOIA)** aims to publish state-of-the-art expository works (monographs, contributed volumes, textbooks, handbooks) that focus on theory, methods, and applications of optimization. Topics covered include, but are not limited to, nonlinear optimization, combinatorial optimization, continuous optimization, stochastic optimization, Bayesian optimization, optimal control, discrete optimization, multi-objective optimization, and more. New to the series portfolio include Works at the intersection of optimization and machine learning, artificial intelligence, and quantum computing.

Volumes from this series are indexed by Web of Science, zbMATH, Mathematical Reviews, and SCOPUS.

More information about this series at https://link.springer.com/bookseries/7393

Ioannis N. Parasidis • Efthimios Providas
Themistocles M. Rassias
Editors

Mathematical Analysis in Interdisciplinary Research

Editors
Ioannis N. Parasidis
Department of Environmental Sciences
University of Thessaly
Larissa, Greece

Efthimios Providas
Department of Environmental Sciences
University of Thessaly
Larissa, Greece

Themistocles M. Rassias
Department of Mathematics
Zografou Campus
National Technical University of Athens
Athens, Greece

ISSN 1931-6828 ISSN 1931-6836 (electronic)
Springer Optimization and Its Applications
ISBN 978-3-030-84723-4 ISBN 978-3-030-84721-0 (eBook)
https://doi.org/10.1007/978-3-030-84721-0

Mathematics Subject Classification: 31XX, 32XX, 33XX, 34XX, 35XX, 37XX, 39XX, 41XX, 42XX, 44XX, 45XX, 46XX, 47XX, 49XX, 58XX, 70XX, 74XX, 76XX, 78XX, 80XX, 81XX, 83XX, 90XX, 91XX, 92XX, 93XX

This Springer imprint is published by the registered company Springer Nature Switzerland AG
The registered company address is: Gewerbestrasse 11, 6330 Cham, Switzerland

Preface

Mathematical Analysis in Interdisciplinary Research provides an extensive account of research as well as research-expository articles in a broad domain of analysis and its various applications in a plethora of fields.

The book focuses to the study of several essential subjects, including optimal control problems, optimal maintenance of communication networks, optimal emergency evacuation with uncertainty, cooperative and noncooperative partial differential systems, variational inequalities and general equilibrium models, anisotropic elasticity and harmonic functions, nonlinear stochastic differential equations, operator equations, max-product operators of Kantorovich type, perturbations of operators, integral operators, dynamical systems involving maximal monotone operators, the three-body problem, deceptive systems, hyperbolic equations, strongly generalized preinvex functions, Dirichlet characters, probability distribution functions, applied statistics, integral inequalities, generalized convexity, global hyperbolicity of spacetimes, Douglas-Rachford methods, fixed point problems, the general Rodrigues problem, Banach algebras, affine group, Gibbs semigroup, relator spaces, sparse data representation, Meier-Keeler sequential contractions, hybrid contractions, and polynomial equations.

This collective effort, which ranges over the abovementioned broad spectrum of topics, is hoped to be useful to both graduate students and researchers who wish to be informed about the latest developments in the corresponding problems treated. The works published within this book will be of particular value for both theoretical and applicable interdisciplinary research.

We would like to express our gratitude to the authors who contributed their valuable papers in this volume. Last but not least, we wish to extend our sincere thanks to the staff of Springer for their valuable assistance throughout the preparation of this book.

Larissa, Greece Ioannis N. Parasidis

Larissa, Greece Efthimios Providas

Athens, Greece Themistocles M. Rassias

Contents

Quasilinear Operator Equation at Resonance 1
A. R. Abdullaev and E. A. Skachkova

A Control Problem for a System of ODE with Nonseparated Multipoint and Integral Conditions 11
V. M. Abdullayev

On Generalized Convexity and Superquadracity 33
Shoshana Abramovich

Well-Posedness of Nonsmooth Lurie Dynamical Systems Involving Maximal Monotone Operators 47
S. Adly, D. Goeleven, and R. Oujja

Numerical Method for Calculation of Unsteady Fluid Flow Regimes in Hydraulic Networks of Complex Structure 69
K. R. Aida-zade and Y. R. Ashrafova

Numerical Solution to Inverse Problems of Recovering Special-Type Source of a Parabolic Equation 85
K. R. Aida-zade and A. B. Rahimov

Using an Integrating Factor to Transform a Second Order BVP to a Fixed Point Problem 101
Richard I. Avery, Douglas R. Anderson, and Johnny Henderson

Volterra Relatively Compact Perturbations of the Laplace Operator ... 109
Bazarkan Biyarov

Computational Aspects of the General Rodrigues Problem 119
Oana-Liliana Chender

Approximation by Max-Product Operators of Kantorovich Type 135
Lucian Coroianu and Sorin G. Gal

Variational Inequalities and General Equilibrium Models 169
Maria Bernadette Donato, Antonino Maugeri, Monica Milasi, and Antonio Villanacci

The Strong Convergence of Douglas-Rachford Methods for the Split Feasibility Problem 213
Qiao-Li Dong, Lulu Liu, and Themistocles M. Rassias

Some Triple Integral Inequalities for Functions Defined on Three-Dimensional Bodies Via Gauss-Ostrogradsky Identity 235
Silvestru Sever Dragomir

Optimal Emergency Evacuation with Uncertainty 261
Georgia Fargetta and Laura Scrimali

On Global Hyperbolicity of Spacetimes: Some Recent Advances and Open Problems 281
Felix Finster, Albert Much, and Kyriakos Papadopoulos

Spectrum Perturbations of Linear Operators in a Banach Space 297
Michael Gil'

Perturbations of Operator Functions: A Survey 335
Michael Gil'

Representation Variety for the Rank One Affine Group 381
Ángel González-Prieto, Marina Logares, and Vicente Muñoz

A Regularized Stochastic Subgradient Projection Method for an Optimal Control Problem in a Stochastic Partial Differential Equation 417
Baasansuren Jadamba, Akhtar A. Khan, and Miguel Sama

A Survey on Interpolative and Hybrid Contractions 431
Erdal Karapınar

Identifying the Computational Problem in Applied Statistics 477
Christos P. Kitsos and C. S. A. Nisiotis

Fractional Integral Operators in Linear Spaces 499
Jichang Kuang

Anisotropic Elasticity and Harmonic Functions in Cartesian Geometry 523
D. Labropoulou, P. Vafeas, and G. Dassios

Hyers–Ulam Stability of Symmetric Biderivations on Banach Algebras 555
Jung Rye Lee, Choonkil Park, and Themistocles M. Rassias

Some New Classes of Higher Order Strongly Generalized Preinvex Functions 573
Muhammad Aslam Noor and Khalida Inayat Noor

Existence of Global Solutions and Stability Results for a Nonlinear Wave Problem in Unbounded Domains 589
P. Papadopoulos, N. L. Matiadou, S. Fatouros, and G. Xerogiannakis

Congestion Control and Optimal Maintenance of Communication Networks with Stochastic Cost Functions: A Variational Formulation 599
Mauro Passacantando and Fabio Raciti

Nonlocal Problems for Hyperbolic Equations 619
L. S. Pulkina

On the Solution of Boundary Value Problems for Loaded Ordinary Differential Equations 641
E. Providas and I. N. Parasidis

Set-Theoretic Properties of Generalized Topologically Open Sets in Relator Spaces 661
Themistocles M. Rassias, Muwafaq M. Salih, and Árpád Száz

On Degenerate Boundary Conditions and Finiteness of the Spectrum of Boundary Value Problems 731
Victor A. Sadovnichii, Yaudat T. Sultanaev, and Azamat M. Akhtyamov

Deceptive Systems of Differential Equations 781
Martin Schechter

Some Certain Classes of Combinatorial Numbers and Polynomials Attached to Dirichlet Characters: Their Construction by p-Adic Integration and Applications to Probability Distribution Functions 795
Yilmaz Simsek and Irem Kucukoglu

Pathwise Stability and Positivity of Semi-Discrete Approximations of the Solution of Nonlinear Stochastic Differential Equations 859
Ioannis S. Stamatiou

Solution of Polynomial Equations 875
N. Tsirivas

Meir–Keeler Sequential Contractions and Pata Fixed Point Results 935
Mihai Turinici

Existence and Stability of Equilibrium Points Under the Influence of Poynting–Robertson and Stokes Drags in the Restricted Three-Body Problem 987
Aguda Ekele Vincent and Angela E. Perdiou

Nearest Neighbor Forecasting Using Sparse Data Representation 1003
Dimitrios Vlachos and Dimitrios Thomakos

On Two Kinds of the Hardy-Type Integral Inequalities in the Whole Plane with the Equivalent Forms 1025
Bicheng Yang, Dorin Andrica, Ovidiu Bagdasar, and Michael Th. Rassias

Product Formulae for Non-Autonomous Gibbs Semigroups 1049
Valentin A. Zagrebnov

Quasilinear Operator Equation at Resonance

A. R. Abdullaev and E. A. Skachkova

Abstract This work considers a quasilinear operator equation at resonance. We obtained solvability theorems and formulated corollaries. The current approach is based on a special generalization of the classical Schauder Fixed Point Theorem.

1 Introduction

Let us consider the following equation:

$$Lx = Fx, \tag{1}$$

with a linear bounded operator $L : X \to Y$ and a continuous (generally speaking, a nonlinear) operator $F : X \to Y$, where X and Y are Banach spaces. If the operator L is non-invertible, Eq. (1) is called a resonance case. In particular, periodic problems for systems of ordinary differential equations can be considered as forms of Eq. (1) with non-invertible operator.

In 1970s, an approach, aimed at studying the Eq. (1) at resonance, was introduced. The approach is based on the Lyapunov-Schmidt method, and it reduces the problem of solvability of Eq. (1) to the problem of the existence of a fixed point of some auxiliary operator. In the literature, statements obtained in a similar manner are called "results of the Landesman-Laser theorem type." However, when applying this methodology to specific classes of resonant boundary value problems, certain

A. R. Abdullaev
Department of Higher Mathematics, Perm National Research Polytechnic University, Perm, Russia
e-mail: h.m@pstu.ru

E. A. Skachkova (✉)
Mechanics and Mathematics Faculty, Perm State University, Perm, Russia
e-mail: skachkovaea@gmail.com

I. N. Parasidis et al. (eds.), *Mathematical Analysis in Interdisciplinary Research*, Springer Optimization and Its Applications 179,
https://doi.org/10.1007/978-3-030-84721-0_1

complications arise. Therefore, the question of the effective solvability conditions for Eq. (1) remains relevant.

By now a large number of papers studying Eq. (1) with a non-invertible operator has been accumulated in the scientific literature. Some intuition about the problems and development trends in this area of research can be obtained from the following sources [1–5].

In this paper we propose an approach to study Eq. (1) at resonance based on a special generalization of the classical Schauder Fixed Point Theorem (Theorem 1). Further, the current paper is structured as follows. In Sect. 2, we explain the notation that we use in our paper. Further in the same section, we provide information related to operators L and F. In Sect. 3, we formulate existence theorems for solutions of Eq. (1). In Sect. 4, we consider applying our approach to the periodic problem for an ordinary second-order differential equation.

2 Preliminaries

The following notations and terminology will be used in the rest of the paper. Let X and Y be Banach spaces. The equality $X = X_1 \bigoplus X_2$ denotes that X is a direct sum of bounded subspaces X_1 and X_2. We denote the kernel and the image of the linear operator $L : X \to Y$ as $\ker L$ and $R(L)$, respectively. Let $P : X \to X$ be a linear bounded projection on $\ker L$, $P^c = I - P$ be an additional projection operator, $Q : Y \to Y$ be a projection on $R(L)$, and Q^c be an additional projection operator. Let $L : X \to Y$ be a Fredholm operator. Then $X = X_0 \bigoplus \ker L$, and $Y = Y_0 \bigoplus R(L)$.

A restriction of operator L is regarded as operator $L_0 : X \to R(L)$, such that $L_0x = Lx$, for all $x \in X$.

By $U(r)$ we denote a closed ball of radius $r > 0$, centered at the zero element of X or Y. For $R(L) \neq Y$, we assume

$$U_{R(L)}(r)\{y \mid y \in R(L),\ \|y\| \leq r\}.$$

The surjectivity coefficient [6, 7] of an operator $L : X \to Y$ is a non-negative number, defined by

$$q(L) = \inf_{\omega \neq \theta} \frac{\|L^*\omega\|}{\|\omega\|},$$

where $L^* : Y^* \to X^*$ is the adjoint operator of L.

If $q(L) > 0$, then the operator $L : X \to Y$ is surjective, that is $R(L) = Y$. If at the same time $\dim \ker L < \infty$, then it holds that $U(q(L)r) \subset L(U(r))$, for all $r > 0$.

If $R(L) \neq Y$, then the following characteristic of the linear operator turns out to be beneficial for studying Eq. (1).

Assume $L_0 : X \to R(L)$ is a contraction of $L : X \to Y$. The relative coefficient of surjectivity of the operator L is the number $q_0(L)$ determined by the equality [8], for all $z \in (R(L))^*$:

$$q_0(L) = \inf_{z \neq \theta} \frac{\|L_0^* z\|}{\|z\|}.$$

For the Fredholm operator $L : X \to Y$, it holds that $U_{R(L)}(q_0(L)r) \subset L(U(r))$, for all $r > 0$.

If the calculation of the precise meaning of the relative surjectivity coefficient is difficult, then its lower estimation is applied. To do this we consider the following approach.

Operator $K_p : R(L) \to X$ is called generalized inverse [9] to the operator $L : X \to Y$, associated with the projection P, if $K_p L = P^c$ and $L K_p = I_0$ hold, where $I_0 : R(L) \to Y$ is an embedding operator.

The following estimation [8] holds:

$$\|K_p\|^{-1} \leq q_0(L).$$

For a continuous (generally speaking, a nonlinear) operator, we consider the following functional characteristic:

$$b_F(r) = \sup_{\|x\| \leq r} \|Fx\|,$$

for all $r \geq 0$.

If $b_F(r) < \infty$, then $F(U(r)) \subset U(b_F(r))$. For the linear operator $L : X \to Y$, it holds that $b_L(r) = \|L\| r$. If $\|Fx\| \leq a + b\|x\|$ holds for some non-negative constants a and b, then the estimation $b_F(r) \leq a + br$ holds as well.

3 Existence Theorem

The existence theorems obtained in this section are based on a special modification of the Schauder Fixed Point Theorem [10] for the Eq. (1). It is formulated as follows.

Theorem 1 *Assume the following conditions hold:*

1. $L : X \to Y$ *is a Fredholm operator;*
2. *the operator* $F : X \to Y$ *is completely continuous;*
3. *there exists a nonempty closed bounded set* $M \subset X$, *such that* $\overline{co}F(M) \subset L(M)$.

Then there exists at least one solution of the Eq. (1).

Further in the paper, it is assumed that the following conditions hold:

(A) $L : X \to Y$ is a Fredholm an operator with the index $\text{ind}L \geq 0$;
(B) an operator $F : X \to Y$ is completely continuous.

Theorem 2 *Assume the following conditions hold:*

1. *inequality* $b_F(r) \leq q_0(L)r$ *has a positive solution* r_0*;*
2. *for all* $z \in X_0$ *there exists* $u \in \ker L$ *such that* $F(X) \in R(L)$*,* $x = z + u$*, and* $\|x\| \leq r_0$.

Then Eq. (1) *has at least one solution.*

Proof Let $r_0 > 0$ be the number, the existence of which is assumed in condition 1 of Theorem 2. Let M be a set of elements of the form $x = z + u$, $\|x\| \leq r_0$, where $z \in X_0$, and u is an element of $\ker L$, which corresponds to the given z. Since for all $x \in M$ it holds that

$$F(x) \in R(L) \text{ and } \|Fx\| \leq b(r_0),$$

then

$$F(M) \subset U_{R(L)}(b(r_0)).$$

Moreover, it holds that

$$b(r_0) < q_0(L)r_0 \text{ and } U_{R(L)}(q_0(L)r_0) \subseteq L(U(r_0)).$$

Hence,

$$F(M) \subset L(U(r_0)).$$

For a closed set $C = \overline{M}$ we have that

$$F(C) \subset \overline{F}(M) \subset L(U(r_0)).$$

To this embedding, we apply the convex closure operation, taking into account the following equality

$$L(U(r_0)) = L(C),$$

we obtain

$$\overline{\text{co}}F(C) \subset L(C).$$

Now we apply Theorem 1, which guarantees the existence of at least one solution of the Eq. (1) under the conditions of Theorem 2. □

Now we provide a statement that in some situations is more efficient in practice.

Theorem 3 *Assume the following conditions hold:*

1. *For all $x \in X$, there exists $u \in \ker L$ and $u = u(x)$, such that $F(x+u) \in R(L)$, and $\|u\| \le \eta(\|x\|)$;*
2. *$b_F(r + \eta(r)) \le q_0(L)r$ has a positive solution.*

Then, Eq. (1) *has at least one solution.*

Proof Let $x \in X$, and $u \in \ker L$, such that condition 1 of the theorem holds. Then

$$\|x+u\| \le \|x\| + \eta(\|x\|).$$

If $x \in U(r)$, then

$$\|F(x+u)\| \le b(r_1),$$

where $r_1 = r + \eta(r)$. Let r_0 be a positive solution of inequality from condition 2 of the theorem. Assume

$$M = \{x+u \mid F(x+u) \in R(L),\ \|x\| \le r_0\}.$$

The following embedding holds

$$F(M) \subset U(r_1),\ r_1 = r_0 + \eta(r_0).$$

Further we follow the lines of the proof of Theorem 2. □

Corollary 1 *Assume the following conditions hold:*

1. *there exist $a, b \ge 0$, such that $\|Fx\| \le a + b\|x\|$;*
2. *there exist $c, d \ge 0$, such that for all $x \in X$ there exists $u \in \ker L$, such that $u = u(x)$, $F(x+u) \in R(L)$, and $\|u\| \le c + d\|x\|$;*
3. *it holds that $b(1+d) < q_0(L)$.*

Then Eq. (1) *has at least one solution.*

Proof We have

$$b_F(r + \eta(r)) \le a + b(r + (c + dr)).$$

If $b(1+d) < q_0(L)$, then

$$a + b(r + (c + dr)) \le q_0(L)r$$

has a positive solution. Thus, all conditions of 3 hold. □

Remark 1 Under the conditions of Corollary 1, the constant b can be replaced by the following (provided that it exists):

$$b(F) = \limsup \frac{\|Fx\|}{\|x\|}, \quad \text{where } \|x\| \to \infty.$$

Indeed, for a sufficiently small $\varepsilon > 0$, there exists $a = a(\varepsilon) \geq 0$, such that:

$$\|Fx\| \leq a + (b(F) + \varepsilon) \|x\| .$$

It is easier to check the conditions of Theorem 3 if F has sublinear growth. We consider it in more details below.

Corollary 2 *Assume the following conditions hold:*

1. *there exist $a, b \geq 0$, and $0 \leq \delta < 1$, such that for all $x \in X$ it holds that $\|Fx\| \leq a + b \|x\|^\delta$;*
2. *there exist $c, d \geq 0$, such that for all $x \in X$ there exists $u \in \ker L$, such that $u = u(x)$, $F(x + u) \in R(L)$, and $\|u\| \leq c + d \|x\|$.*

Then Eq. (1) *has at least one solution.*

Proof It is enough to observe that for $0 \leq \delta < 1$ inequality

$$a + b(r + (c + dr))^\delta \leq q_0(L)r$$

has r_0 as a positive solution. □

Certain difficulties may arise when one applies the above-mentioned statements to specific boundary value problems, which can be expressed in the form of Eq. (1) with a non-invertible operator L. To be more specific, in the said cases it might be problematic to verify the condition 2 of Theorem 2 (or similar conditions of other statements).

For the sake of simplicity, let us consider the case when $\operatorname{ind} L = 0$. Let $n = \dim \ker L$ and $J_1 : R^n \to \ker L$, and $J_2 : Y_0 \to R^n$ be fixed isomorphisms. For an arbitrary fixed $z \in X_0$ we define the mapping $\Phi_z : R^n \to R^n$ by the following equality:

$$\Phi_z(\alpha) = J_2 Q^c F(z + J_1\alpha),$$

where $Q^c = I - Q$ is an additional projection operator.

Let $B(r) = \{\alpha \mid \alpha \in R^n, |\alpha| < r\}$ be an open ball of radius $r > 0$ in R^n. We use $\deg(\Phi_z, B(r))$ to denote the Brouwer degree [11] of $\Phi_z : R^n \to R^n$ relative to a ball of radius $r > 0$ centered at zero. To verify the mentioned above condition, one may require, for example, that for all $x \in X_0$, there exists $r = r(z) > 0$, such that $\deg(\Phi_z, B(r)) \neq 0$ and $r \leq d \|z\|$.

Taking into account the number of the existing approaches to verify the condition $\deg(\Phi_z, B(r)) \neq 0$, we can talk about efficient sufficient conditions that ensure the existence of a solution to the Eq. (1).

4 Application

As an example of the application of the statements, obtained in Sect. 3, we consider a periodic boundary value problem for an ordinary second-order differential equation:

$$x''(t) = f(t, x(h(t))),\ t \in [0, \omega], \tag{2}$$

$$x(0) = x(\omega),\ x'(0) = x'(\omega), \tag{3}$$

where $f : [0, \omega] \times R^1 \to R^1$ satisfies Carathéodory's criterion, $h : [0, \omega] \to R^1$ is measurable, and $h([0, \omega]) \subset [0, \omega]$.

We consider (2)–(3) on the space $W_2 = W_2[0, \omega]$ of the functions $x : [0, \omega] \to R^1$, which have an absolutely continuous derivative and such that $x'' \in L_2[0, \omega]$. We define the norm in W_2 by

$$\|x\|_W = |x(0)| + |x'(0)| + \left\|x''\right\|_{L_2}.$$

Consider the space $X = \{x \mid x \in W_2,\ x(0) = x(\omega),\ x'(0) = x'(\omega)\}$ and define the operators $L, F : X \to Y$, $Y = L_2$ by

$$(Lx)(t) = x''(t),\ (Fx)(t) = f(t, x(h(t))).$$

Now we consider (2)–(3) as an operator Eq. (1). Obviously:

$$\ker L = \{x \mid x \in X, x(t) \equiv \text{const}\},$$

$$R(L) = \left\{y \mid y \in Y,\ \int_0^\omega y(s)\, ds = 0\right\}.$$

We define the projection operators $P : X \to X$, $Q : Y \to Y$ by

$$(Px)(t) = x(0),$$

$$(Qy)(t) = y(t) - \frac{1}{\omega}\int_0^\omega y(s)\, ds.$$

Further we will be using the following inequalities. For an arbitrary $x \in W_2$ the following estimation holds:

$$|x(t)| \le \gamma \|x\|_W,\ t \in [0, \omega],$$

where $\gamma = \max\left\{1, \omega, \omega\sqrt{\frac{\omega}{3}}\right\}$.

Lemma 1 *The relative surjectivity coefficient of* $L : X \to Y$, $Lx = x''$ *has the estimation* $\left(1 + \sqrt{\frac{\omega}{3}}^{-1}\right) \le q_0(L)$.

Proof A direct verification shows that the generalized inverse operator K_p : $R(L) \to X$, associated with the projector $Px = x(0)$ has the following form:

$$(K_p y)(t) = \int_0^t (t-s)y(s)\,ds + \frac{t}{\omega}\int_0^\omega sy(s)\,ds.$$

Hence,

$$\|K_p y\|_W \le \|y\|_{L_2} + \frac{1}{\omega}\left(\frac{\omega}{3}\right)^{\frac{1}{2}} \|y\|_{L_2} \le \left(1+\sqrt{\frac{\omega}{3}}\right)\|y\|_{L_2}.$$

Thus, $\|K_p\| \le 1+\sqrt{\frac{\omega}{3}}$. Since $\|K_p\|^{-1} \le q_0(L)$, the statement of the lemma holds. □

Theorem 4 *Let the following conditions hold:*

1. *there exist non-negative constants* a, b, *such that* $|f(t,u)| \le a + b|u|$, *and* $(t,u) \in [0,\omega] \times R^1$;
2. *there exists* $u > 0$, *such that for all* $u \in R^1$, $|u| > u^*, t \in [0,\omega]$ *it holds that* $sign(u)f(t,u) \ge 0$ $(sign(u)f(t,u) \le 0)$;
3. $b\gamma\sqrt{\omega}(1+\gamma) < \left(1+\sqrt{\frac{\omega}{3}}\right)^{-1}$.

Then, there exists at least one solution for (2)–(3).

Proof To prove the theorem, we use Corollary 1. Note that due to condition 1 of Theorem 4, and since $W_2 \subset L_2$ is completely continuous, we have that the operator $F : W_2 \to L_2$ defined by the equality $F(x)(t) = f(t, x(h(t)))$ is completely continuous.

It is not hard to verify that $|x(h(t))| \le \gamma \|x\|_W$. Due to condition 1 of Theorem 4, we have

$$|f(t, x(h(t)))| \le a + b\gamma \|x\|_W,\ x \in X,\ t \in [0,\omega].$$

Hence, $\|Fx\|_{L_2} \le (a + b\gamma \|x\|_W)\sqrt{\omega}$. Thus, condition 1 of Corollary 1 holds.

For certainty we will check condition 2 of Corollary 1 assuming that:

$$\mathrm{sign}(u)f(t,u) \ge 0,$$

for all $u \in R^1$, $|u| > u^*$, $t \in [0,\omega]$. We fix $x \in X$ and define $\Phi_x : R^1 \to^1$ by

$$\Phi_x(C) = \int_0^\omega f(s, x(h(s)) + C)\,ds.$$

Let $C_1 = u^* + \gamma \|x\|_W$. Then for all $C \ge C_1$ it holds that $x(h(t)) + C > u^*$, hence $\Phi_x(C) \ge 0$. Similarly, $\Phi_x(C) \le 0$ for all $C \le C_2 = -u^* - \gamma \|x\|_W$. Due to the continuity of Φ, there exists a constant $\widetilde{C} = C(x)$, which satisfies $|\widetilde{C}| \le$

$\max\{|C_1|, |C_2|\} \le \gamma \|x\|_W + u^*$, and such that $\Phi_x(\widetilde{C}) = 0$. Thus, condition 2 of Corollary 1 holds.

For the case $\operatorname{sign}(u) f(t, u) \le 0$, the proof follows the same lines. □

References

1. Fucik, S.: Solvability of Nonlinear Equations and Boundary Value Problems. Reidel, Dordrecht (1980)
2. Mawhin, J.: Leray-Schauder degree: A half century of extensions and applications. Topol. Methods Nonlinear Anal. **14**, 195–228 (1999) https://projecteuclid.org/euclid.tmna/1475179840.
3. Mawhin, J.: Landesman-Lazer's type problems for nonlinear equations. Confer. Sem. Mat. Univ. Bari **147** (1977)
4. Przeradzki, B.: Three methods for the study of semilinear equations at resonance. Colloq. Math. **66** 110–129 (1993)
5. Tarafdar, E., Teo, S.K.: On the existence of solutions of the equation $Lx \infty Nx$ and a coincidence degree theory. J. Austral. Math. Soc. **28**, 139–173. (1979) https://doi.org/10.1017/S1446788700015640
6. Abdullaev, A.R., Bragina, N. A.: Green operator with minimal norm. Izv. Vyssh. Uchebn. Zaved. **4**, 3–7 (2003). Russian Math. (Iz. VUZ) **47**, 1–5 (2003)
7. Pietsch, A.: Operator Ideals [Russian translation]. Mir, Moscow (1983)
8. Burmistrova, A.B.: Relative surjectivity coefficients of linear operator. Izv. vuzov. Math. **6**, 19–26 (1999)
9. Abdullaev, A.R., Burmistrova, A.B.: Elements of Theory of Topological Noetherian Operators. Chelyabinsk State Univ., Chelyabinsk (1994)
10. Abdullaev, A.R.: Questions of the Perturbation Theory of the Stable Properties of Boundary Value Problems for Functional Differential Equations. PhD thesis, Perm (1991)
11. Fučik, S., Kufner, A.: Nonlinear Differential Equations. Studies in Applied Mechanics 2. Elsevier Scientific Publ. Co., Amsterdam-New York (1980)

A Control Problem for a System of ODE with Nonseparated Multipoint and Integral Conditions

V. M. Abdullayev

Abstract Using gradient-type methods is proposed to solve a control problem with nonseparated multipoint and integral conditions. Therefore, formulas for the gradient of the objective functional are obtained in this study. For the numerical solution of nonlocal direct and conjugate boundary value problems, an approach is proposed that allows folding the integral terms into local ones and then using an analog of the transfer of conditions. As a result, the solving of nonlocal boundary value problems is reduced to the solving of specially constructed Cauchy problems and one system of linear equations. An analysis of the obtained results of computational experiments is carried out.

1 Introduction

Recently, there has been an increase in research on boundary value problems with nonlocal conditions and corresponding control problems.

Note that boundary value problems with nonlocal conditions were started in [1–3] and then continued in the studies of many authors both for equations with ordinary and partial derivatives [4–10]. Control in boundary value problems with nonlocal multipoint integral conditions also aroused great interest [11–17]. Various studies have been carried out in this area, including the necessary conditions for optimality.

For linear nonlocal boundary value problems, numerical methods based on the sweep method are proposed in [18–22]. To take into account the integral conditions, many authors propose to reduce them to problems with multipoint conditions. For this, it was required to introduce new variables and, accordingly, to increase the dimension of the system of differential equations.

V. M. Abdullayev (✉)
Azerbaijan State Oil and Industry University, Baku, Azerbaijan

Institute of Control Systems of Azerbaijan NAS, Baku, Azerbaijan

I. N. Parasidis et al. (eds.), *Mathematical Analysis in Interdisciplinary Research*, Springer Optimization and Its Applications 179,
https://doi.org/10.1007/978-3-030-84721-0_2

We have studied an approach to the numerical solution of boundary value problems with integral conditions, which does not require an increase in the dimension of the system, and its application to solving the considered control problems. Analytical formulas for the gradient of the objective functional are obtained to use gradient methods for solving optimal control problems. The results of computational experiments and their analysis are given.

2 An Analysis of the Problem under Investigation and Obtaining Basic Formulas

Let the controlled process be described by the following ODE system:

$$\dot{x}(t) = A(t,u)\,x(t) + B(t,u), \quad t \in [t_0, T]. \tag{1}$$

Here $x(t) \in E^n$ is a phase variable; piecewise continuous function $u(t) \in U \subset E^r$ is a control vector function, compact set U is admissible values of control, and n-dimensional square matrix $A(t,u) \neq const$ and the n-dimensional vector function $B(t,u)$ are continuous with respect to t and continuously differentiable with respect to u.

The following conditions are given:

$$\sum_{i=1}^{l_1} \int_{\bar{t}_{2i-1}}^{\bar{t}_{2i}} \overline{D}_i(\tau)\,x(\tau)\,d\tau + \sum_{j=1}^{l_2} \tilde{D}_j x\left(\tilde{t}_j\right) = C_0. \tag{2}$$

Here $\overline{D}_i(\tau)$, $\tilde{D}_j$ are $(n \times n)$-dimensional given matrices and $\overline{D}_i(\tau)$ is continuous; C_0 is the n-dimensional given vector; $\bar{t}_i$, $\tilde{t}_j$ are given time instances from $[t_0, T]$; $\bar{t}_{i+1} > \bar{t}_i, \tilde{t}_{j+1} > \tilde{t}_j, i = 1, \ldots, 2l_1 - 1, \quad j = 1, \ldots, l_2 - 1$; and l_1, l_2 are given.

We assume that, first,

$$\min\left(\bar{t}_1, \quad \tilde{t}_1\right) = t_0, \quad \max\left(\bar{t}_{2l_1}, \quad \tilde{t}_{l_2}\right) = T \tag{3}$$

and, second, condition

$$\tilde{t}_j \overline{\in} \left[\bar{t}_{2i-1}, \bar{t}_{2i}\right] \tag{4}$$

is satisfied for all $i = 1, \ldots, 2l_1, \quad j = 1, \ldots, l_2$.

It is required to determine the admissible control $u(t) \in U$ and the corresponding solution $x(t)$ to the nonlocal boundary value problem (1), (2), such that the pair $(x(t), u(t))$ minimizes the following objective functional:

$$J(u) = \Phi\left(\hat{x}\left(\hat{t}\right)\right) + \int_{t_0}^{T} f^0(x, u, t)\, dt \to \min_{u(t)\in U}, \tag{5}$$

where the given function Φ is continuously differentiable and $f^0(x, u, t)$ is continuously differentiable with respect to (x, u) and continuous with respect to t; $\hat{t} = \left(\hat{t}_1, \hat{t}_2, \ldots, \hat{t}_{2l_1+l_2}\right)$ is the ordered union of the sets $\tilde{t} = \left(\tilde{t}_1, \tilde{t}_2, \ldots, \tilde{t}_{l_2}\right)$ and $\bar{t} = \left(\bar{t}_1, \bar{t}_2, \ldots, \bar{t}_{2l_1}\right)$, i.e., $\hat{t}_j < \hat{t}_{j+1}$, $j = 1, \ldots, 2l_1 + l_2 - 1$ and $\hat{x}\left(\hat{t}\right) = \left(x\left(\hat{t}_1\right), x\left(\hat{t}_2\right), \ldots, x\left(\hat{t}_{2l_1+l_2}\right)\right)$.

The fundamental difference of problem statement (1)–(5) from the optimal control problems considered, for example, in [12–14] lies in nonseparated nonlocal integral and multipoint conditions (2). By introducing some new phase variables, problem (1)–(5) can be reduced to a problem involving multipoint conditions. To demonstrate this, introduce new phase vector $X(t) = \left(x^1(t), \ldots, x^{l_1+1}(t)\right)$, $x^1(t) = x(t)$, which is the solution to the following differential equations system:

$$\begin{aligned} &\dot{x}^1(t) = A(t, u)\, x^1(t) + B(t, u), \\ &\dot{x}^{i+1}(t) = \overline{D}_i(t) x^1(t), \quad t \in \left(\bar{t}_{2i-1}, \bar{t}_{2i}\right], \quad i = 1, \ldots, 2l_1, \end{aligned} \tag{6}$$

involving the following initial conditions:

$$x^{i+1}\left(\bar{t}_{2i-1}\right) = 0, \quad i = 1, \ldots, 2l_1. \tag{7}$$

Then conditions (2) takes the following form:

$$\sum_{i=1}^{l_1} x^{i+1}\left(\bar{t}_{2i}\right) + \sum_{j=1}^{l_2} \tilde{D}_j x^1\left(\tilde{t}_j\right) = C_0. \tag{8}$$

Systems (6)–(8) are obviously equivalent to (1) and (2). In systems (6) and (7), there are $(l_1 + 1)n$ differential equations with respect to the phase vector $X(t)$, and there is the same number of conditions in (7) and (8). Obviously, the drawback of boundary problem (6), (7) is its high dimension. This is an essential point for numerical methods of solution to boundary problems based, as a rule, on the methods of sweep or shift of boundary conditions [18–22]. Also, the increase of the dimension of the phase variable complicates the solution to the optimal control problem itself due to the increase of the dimension of the adjoint problem.

Note that if we use the approach proposed in [22], then at the expense of the additional increase of the dimension of the phase variable vector up to $2(l_1 + l_2 + 1)(l_1 + 1)n$, problem (6)–(8) can be reduced to a two-point problem involving nonseparated boundary conditions.

Using the technique of the works [23–26], we can obtain existence and uniqueness conditions for the solution to problem (1), (2) under every admissible control

$u \in U$, without reducing it to a problem involving multipoint conditions (8). But this kind of investigation is not the objective of the present work.

In the present work, we propose an approach to the numerical solution to both boundary problem (1), (2) and to the corresponding optimal control problem. This approach does not require increase of the order of the differential equations system and of the phase vector.

Assume that under every admissible control $u(t) \in U$, there is a unique solution to problem (1), (2). For this purpose, we assume that the parameters of problem (1), (2), after reducing it to (6)–(8), satisfy the conditions proposed in [2, 23–26] dedicated to differential equations systems involving multipoint and two-point conditions.

To apply gradient methods to solving optimal control problem (1)–(5), we obtain the formulas for the gradient of the functional.

Suppose that $(u(t), x(t; u))$ is an arbitrary admissible process and the control $u(t)$ has received an increment $\Delta u(t)$: $\tilde{u} = u + \Delta u$. Then the phase variable $\tilde{x}(t) = x(t) + \Delta x(t)$ will also receive increments, and the following takes place:

$$\Delta \dot{x}(t) = A(t, u)\,\Delta x(t) + \Delta_u A(t, u)\,x(t) + \Delta_u B(t, u), \quad t \in [t_0, T], \tag{9}$$

$$\sum_{i=1}^{l_1} \int_{\bar{t}_{2i-1}}^{\bar{t}_{2i}} \overline{D}_i(\tau)\,\Delta x(\tau)\,d\tau + \sum_{j=1}^{l_2} \tilde{D}_j \Delta x(\tilde{t}_j) = 0. \tag{10}$$

Here we use the following designations:

$$\Delta x(t) = x(t, \tilde{u}) - x(t, u), \quad \Delta_u A(t, u) = A(t, \tilde{u}) - A(t, u), \quad \Delta_u B(t, u) = B(t, \tilde{u}) - B(t, u).$$

Consider the as-yet arbitrary almost everywhere continuously differentiable vector function $\psi(t) \in R^n$ and vector $\lambda \in R^n$. To calculate the increment of the functional, taking into account (9)–(10) and (1)–(2), we have

$$J(u) = \Phi\left(\hat{x}\left(\hat{t}\right)\right) + \int_{t_0}^{T} f^0(x, u, t)\,dt + \int_{t_0}^{T} \psi^*(t)\left[\dot{x}(t) - A\Big(t, u\Big)x(t) - B(t, u)\Big)\right]dt +$$
$$+ \lambda^*\left[\sum_{i=1}^{l_1} \int_{\bar{t}_{2i-1}}^{\bar{t}_{2i}} \overline{D}_i(\tau)\,x(\tau)\,d\tau + \sum_{j=1}^{l_2} \tilde{D}_j x\left(\tilde{t}_j\right) - C_0\right],$$

$$J(u + \Delta u) = \Phi\left(\hat{x}\left(\hat{t}\right) + \Delta\hat{x}\left(\hat{t}\right)\right) + \int_{t_0}^{T} f^0(x + \Delta x, u + \Delta u, t)\,dt +$$
$$+ \int_{t_0}^{T} \psi^*(t)\left[(\dot{x} + \Delta\dot{x}) - A\Big(t, u + \Delta u\Big)(x + \Delta x) - B(t, u + \Delta u)\Big)\right]dt +$$
$$+ \lambda^*\left[\sum_{i=1}^{l_1} \int_{\bar{t}_{2i-1}}^{\bar{t}_{2i}} \overline{D}_i(\tau)\left(x(\tau) + \Delta x(\tau)\right)d\tau + \sum_{j=1}^{l_2} \tilde{D}_j\left(x\left(\tilde{t}_j\right) + \Delta x\left(\tilde{t}_j\right)\right) - C_0\right],$$

where * is the transposition operation. Using the formula for integration by parts, we obtain

$$\Delta J(u)=\int_{t_0}^{T}\left[-\dot{\psi}^*(t)-\psi^*(t)A(t,u)+\lambda^*\sum_{i=1}^{l_1}\left[\chi\left(\bar{t}_{2i}\right)-\chi\left(\bar{t}_{2i-1}\right)\right]\overline{D}_i(t)+f_x^0(x,u,t)\right]\Delta x(t)dt+$$
$$+\int_{t_0}^{T}\left\{f_u^0(x,u,t)+\psi^*(t)\left[-A_u^*(t,u)\,x(t)-B_u\left(t,u\right)\right]\right\}\Delta u(t)dt+$$
$$+\sum_{k=2}^{2l_1+l_2-1}\left[\psi^{*-}\left(\hat{t}_k\right)-\psi^{*+}\left(\hat{t}_k\right)+\frac{\partial\Phi(\hat{x}(\hat{t}))}{\partial x\left(\hat{t}_k\right)}\right]\Delta x\left(\hat{t}_k\right)$$
$$+\sum_{j=1}^{l_2}\lambda^*\tilde{D}_j\Delta x\left(\tilde{t}_j\right)+\psi^*(T)\Delta x(T)-\psi^*\left(t_0\right)\Delta x\left(t_0\right)+$$
$$+\int_{t_0}^{T}o_1\left(\|\Delta x(t)\|\right)dt+\int_{t_0}^{T}o_2\left(\|\Delta u(t)\|\right)dt+o_3\left(\left\|\Delta\hat{x}\left(\hat{t}_k\right)\right\|\right), \tag{11}$$

where $\psi^+\left(\hat{t}_k\right)=\psi\left(\hat{t}_k+0\right),\quad \psi^-\left(\hat{t}_k\right)=\psi\left(\hat{t}_k-0\right),\quad k=1,\ldots,(2l_1+l_2)$, $\chi(t)-$ is the Heaviside function.

$o_1(\|\Delta x(t)\|)$, $o_2(\|\Delta u(t)\|)$, $o_3\left(\left\|\Delta\hat{x}\left(\hat{t}_k\right)\right\|\right)$ are the quantities of less than the first order of smallness.

Henceforth, the norms of the vector functions $\|x(t)\|$ and $\|u(t)\|$ are understood (see [27]) as $\|x(t)\|_{L_2^n[t_0,T]}$ and $\|u(t)\|_{L_2^r[t_0,T]}$, respectively.

$$A_u(t,u)=\left(\left(\frac{\partial A_{ij}(t,u)}{\partial u_s}\right)\right)\quad\text{and}\quad B_u(t,u)=\left(\left(\frac{\partial B_i(t,u)}{\partial u_s}\right)\right)$$

are considered as matrices of the dimensions $(n\times n\times r)$ and $(n\times r)$, respectively. Applying the transposition operation to these matrices, we obtain the matrices $A_u^*(t,u)$ and $B_u^*(t,u)$ of the dimensions $(n\times r\times n)$ and $(r\times n)$.

We require that $\psi(t)$ be a solution to the following nonlocal boundary value problem:

$$\dot{\psi}(t)=-A^*(t,u)\,\psi(t)+\sum_{i=1}^{l_1}\left[\chi\left(\bar{t}_{2i}\right)-\chi\left(\bar{t}_{2i-1}\right)\right]\overline{D}^*(t)\lambda+f_x^{0*}(x,u,t), \tag{12}$$

$$\psi\left(t_0\right)=\begin{cases}\left(\frac{\partial\Phi(\hat{x}(\hat{t}))}{\partial x(\tilde{t}_1)}\right)^*+\tilde{D}_1^*\lambda, & for\quad t_0=\tilde{t}_1,\\ \left(\frac{\partial\Phi(\hat{x}(\hat{t}))}{\partial x(\bar{t}_1)}\right)^*, & for\quad t_0=\bar{t}_1,\end{cases} \tag{13}$$

$$\psi(T)=\begin{cases}-\left(\frac{\partial\Phi(\hat{x}(\hat{t}))}{\partial x(\tilde{t}_{l_2})}\right)^*-\tilde{D}_{l_2}^*\lambda, & for \quad \tilde{t}_{l_2}=T,\\ -\left(\frac{\partial\Phi(\hat{x}(\hat{t}))}{\partial x(\bar{t}_{2l_1})}\right)^*, & for \quad \bar{t}_{2l_1}=T,\end{cases} \tag{14}$$

$$\psi^+\left(\tilde{t}_j\right)-\psi^-\left(\tilde{t}_j\right)=\left(\frac{\partial\Phi\left(\hat{x}\left(\hat{t}\right)\right)}{\partial x\left(\tilde{t}_j\right)}\right)^*+\tilde{D}_j^*\lambda, \quad j=1,2,\dots l_2, \tag{15}$$

$$\psi^+\left(\bar{t}_i\right)-\psi^-\left(\bar{t}_i\right)=\left(\frac{\partial\Phi\left(\hat{x}\left(\hat{t}\right)\right)}{\partial x\left(\bar{t}_i\right)}\right)^*, \quad i=1,2,\dots 2l_1. \tag{16}$$

Instead of (12) and (15)–(16), we can use a differential equations system involving impulse actions:

$$\begin{aligned}\dot{\psi}(t)=&-A^*(t,u)\,\psi(t)+\sum_{i=1}^{l_1}\left[\chi\left(\bar{t}_{2i}\right)-\chi\left(\bar{t}_{2i-1}\right)\right]\overline{D}^*(t)\lambda\\ &+\sum_{j=1}^{l_2}\left[\left(\frac{\partial\Phi(\hat{x}(\hat{t}))}{\partial x(\tilde{t}_{v_1})}\right)^*+\tilde{D}_j^*\lambda\right]\delta\left(t-\tilde{t}_j\right)+\\ &+\sum_{i=1}^{l_1}\left(\frac{\partial\Phi(\hat{x}(\hat{t}))}{\partial x(\bar{t}_i)}\right)^*\delta\left(t-\bar{t}_i\right)+f_x^{0*}\left(x,u,t\right).\end{aligned} \tag{17}$$

Here $\delta(\cdot)$ is the delta function. Problems (12)–(16) and (17), (13), and (14) are equivalent. Numerical schemes of their approximation and the solution algorithms used are identical.

To obtain estimates of $o_1(\|\Delta x(t)\|)$ and $o_3\left(\left\|\Delta\hat{x}\left(\hat{t}\right)\right\|\right)$ by known methods by increasing the dimension of the system, we can reduce the considered boundary value problem (1), (2) to the Cauchy problem and obtain the estimate of the form:

$$\|\Delta x(t)\|\leq c\,\|\Delta u(t)\|, \tag{18}$$

where $c=const>0$ does not depend on $u(t)$ [15–17].

From formula (11), taking into account that the gradient of the objective functional is determined by the linear part of the functional increment, we have

$$(\nabla J(u))^*=f_u^0\left(x,u,t\right)+\psi^*(t)\left[-A_u^*\left(t,u\right)x(t)-B_u\Big(t,u\Big)\right]. \tag{19}$$

The functions $x(t)$ and $\psi(t)$ here are, for this control, the solutions to nonlocal boundary value problem (1), (2) and the conjugate boundary value problem (12)–(16).

For the numerical solution of the problem, we use methods of minimization of the gradient type, in particular the well-known method of gradient projection [27]:

$$\begin{aligned} u^{k+1}(t) &= P_U\left(u^k(t) - \alpha_k\ \nabla\ J\left(u^k(t)\right)\right), \quad k = 0, 1, \ldots, \\ \alpha_k &= \arg\min_{\alpha \geq 0} J\left(P_U\left(u^k(t) - \alpha\ \nabla J\left(u^k(t)\right)\right)\right), \end{aligned} \tag{20}$$

where $P_U(\upsilon)$ is the operator of projection of the element $\upsilon \in E^r$ on the set U.

There are two computational difficulties in calculating the gradient of the functional. They are related to the problem of solving direct nonlocal boundary value problem (1), (2) and the conjugate boundary value problem (12)–(16) with an unknown vector λ. It is clear that in system of relations (1), (2), and (12)–(1.16), for a given control $u(t)$, we need to determine $2n$ unknown components of the vector functions $x(t)$, $\psi(t)$, $2n$ their initial values, and n component of the vector λ. For this, there are $2n$ differential Eqs. (1) and (12), n condition (2), and $2n$ conditions (13) and (14).

Proposed below is an algorithm based on the use of the one developed in [13, 18, 21], an operation of shifting conditions for solving systems of ODE with boundary conditions that also include unknown parameters [28, 29]. The proposed operation of shifting intermediate conditions generalizes the well-known operation of transfer of boundary conditions and extends the results of [13, 18, 21] to this class of problems.

3 Numerical Scheme of Solution to the Problem

One of the approaches to the numerical solution to problem (1), (2) could be the reduction of (1) and (2) to a problem involving nonseparated point conditions (6)–(8) by means of introduction of some new variables.

As stated in the first paragraph, the obvious drawback of such an approach is the need for increasing the order of the system, which complicates carrying out the operations of sweep and of shift of the functional matrix of the respective order (see [13, 18, 21]).

Below, we propose and investigate the scheme of the method of reduction of conditions (2) to initial conditions, which do not require increasing the dimension of the differential equations system. For this purpose, we transform first (2) to an integral form.

Introduce the following $(n \times n)$ matrix function:

$$D(t) = \sum_{i=1}^{l_1} \left[\chi\left(\bar{t}_{2i}\right) - \chi\left(\bar{t}_{2i-1}\right)\right] \overline{\overline{D}}_i(t) + \sum_{j=1}^{l_2} \tilde{D}_j \delta\left(t - \tilde{t}_j\right). \tag{21}$$

Function $\overline{\overline{D}}_i(t)$ is as follows:

$$\overline{\overline{D}}_i(t) = \begin{cases} \overline{D}_i(t), & t \in [\bar{t}_{2i-1}, \bar{t}_{2i}], \\ 0, & t \notin [\bar{t}_{2i-1}, \bar{t}_{2i}]. \end{cases}$$

From (21), it follows that

$$D(t) \equiv 0 \quad \text{for} \quad t \notin \bigcup_{i=1}^{l_1} [\bar{t}_{2i-1}, \bar{t}_{2i}] \quad \cup \left(\bigcup_{j=1}^{l_2} \tilde{t}_j \right).$$

In view of (3) and (4), conditions (2) can be written in the equivalent matrix form:

$$\int_{t_0}^{T} D(\tau)\, x(\tau)\, d\tau = C_0, \tag{22}$$

or each of n conditions in (22) can be written separately:

$$\int_{t_0}^{T} D^{\nu}(\tau)\, x(\tau)\, d\tau = C_{0\nu}, \quad \nu = 1, \ldots, n, \tag{23}$$

where $D^{\nu}(\tau)$ is the ν th n-dimensional row of the matrix function $D(\tau)$.

Now, in order to replace integral conditions (22) with local (point) initial conditions, we use an operation that is similar to the transfer operation (sweep) of conditions, which we call a convolution operation.

Introduce $n-$ dimensional vector functions:

$$\overline{C}(t) = \int_{t_0}^{t} D(\tau)\, x(\tau)\, d\tau, \qquad \underline{C}(t) = \int_{t}^{T} D(\tau)\, x(\tau)\, d\tau, \tag{24}$$

for which, the following relations obviously take place:

$$\overline{C}(t_0) = \underline{C}(T) = 0, \quad \overline{C}(T) = \underline{C}(t_0) = C_0. \tag{25}$$

Definition Matrix functions $\overline{\alpha}(t)$, $\underline{\alpha}(t)$ of the dimension $n \times n$ and n-dimensional vector functions $\overline{\beta}(t)$, $\underline{\beta}(t)$ convolve integral conditions (22) into point conditions at the right and left ends, respectively, if for any solution $x(t)$ to system (1), there holds the following conditions:

$$\int_{t_0}^{t} D(\tau)\, x(\tau)\, d\tau = \overline{\alpha}(t) x(t) + \overline{\beta}(t), \quad t \in [t_0, T], \tag{26}$$

$$\int_t^T D(\tau)\,x(\tau)\,d\tau = \underline{\alpha}(t)x(t) + \underline{\beta}(t), \quad t \in [t_0, T]. \tag{27}$$

From (26) and (27) in view of (24) and (25), it follows that

$$\overline{\alpha}(T)x(T) + \overline{\beta}(T) = \overline{C}(T) = C_0, \tag{28}$$

$$\underline{\alpha}(t_0)\,x(t_0) + \underline{\beta}(t_0) = \underline{C}(t_0) = C_0. \tag{29}$$

Each of conditions (28) and (29) represents a local boundary condition. Pairs $\overline{\alpha}(t)$, $\overline{\beta}(t)$ and $\underline{\alpha}(t)$, $\underline{\beta}(t)$ are called functions convolving integral conditions (22) into point conditions from left to right and from right to left, respectively.

Denote by $O_{n\times n}$ a matrix of the dimension $(n \times n)$ with null elements, $I_{n\times n}$ is the identity matrix of order n, and by O_n an n-dimensional vector with null elements. The following theorem takes place.

Theorem 1 *If functions* $\overline{\alpha}(t)$, $\overline{\beta}(t)$ *are the solution to the following Cauchy problems:*

$$\dot{\overline{\alpha}}(t) = -\overline{\alpha}(t)A(t) + D(t), \quad \overline{\alpha}(t_0) = O_{n\times n}, \tag{30}$$

$$\dot{\overline{\beta}}(t) = -\overline{\alpha}(t)B(t), \qquad \overline{\beta}(t_0) = O_n, \tag{31}$$

then these functions convolve integral conditions (22) from left to right into point condition (28).

Proof Assume that there exists the dependence:

$$\overline{C}(t) = \overline{\alpha}(t)x(t) + \overline{\beta}(t), \qquad t \in [t_0, T]. \tag{32}$$

Here $\overline{\alpha}(t)$, $\overline{\beta}(t)$ are as-yet arbitrary matrix and vector functions of the dimensions $n \times n$ and n, respectively, which satisfy conditions (31). Then obviously

$$\overline{\alpha}(t_0) = O_{n\times n}, \quad \overline{\beta}(t_0) = O_n.$$

Differentiating (32) and taking (1) and (23) into account, we have

$$[\dot{\alpha}(t) + \alpha(t)A(t) - D(t)]\,x(t) + \left[\dot{\beta}(t) + \alpha(t)B(t)\right] = 0. \tag{33}$$

Taking the arbitrariness of the functions $\overline{\alpha}(t)$, $\overline{\beta}(t)$ into account, as well as the fact that (33) must be satisfied for all the solutions $x(t)$ to system (1), then it is necessary to set each of the expressions in two brackets (33) equal to 0, i.e., to satisfy conditions (30) and (31) of the lemma1.

The following theorem is proven similarly.

Theorem 2 *If functions* $\underline{\alpha}(t)$, $\underline{\beta}(t)$ *are the solution to the following Cauchy problems:*

$$\dot{\underline{\alpha}}(t) = -\underline{\alpha}(t)A(t) - D(t), \quad \underline{\alpha}(T) = O_{n\times n}, \tag{34}$$

$$\dot{\underline{\beta}}(t) = -\underline{\alpha}(t)B(t), \quad \underline{\beta}(T) = O_n, \tag{35}$$

then these functions convolve integral conditions (22) from right to left into point conditions (29).

Thus, to solve problem (1), (2), it is necessary to solve Cauchy problems (30) and (31) or (34) and (35), to obtain the n th order of linear algebraic system (28) or (29), respectively, then $x(T)$ or $x(t_0)$ is determined from (28) or (29). They can be used as initial conditions for solving the Cauchy problem with respect to the main system (1).

The choice of the convolution scheme applied to conditions (2) from right to left or vice versa depends on the properties of the matrix $A(t)$, namely, on its eigenvalues. If they are all positive, then systems (30) and (31) are steady; if they are all negative, then systems (34) and (35) are steady. If some eigenvalues of the matrix $A(t)$ are positive, and the other are negative, and their absolute values are large, then both systems have fast-increasing solutions, and therefore their numerical solution is unsteady, and it can result in low accuracy. In this case, it is recommended to use the convolving functions which are proposed in the following theorem and which have linear growth in time.

Theorem 3 *If n-dimensional vector function* $g_1^\nu(t)$ *and scalar functions* $g_2^\nu(t)$ *and* $m^\nu(t)$ *are the solution to the following nonlinear Cauchy problems:*

$$\dot{g}_1^\nu(t) = S(t)g_1^\nu(t) - A^*(t)g_1^\nu(t) + m^\nu D^{\nu^*}(t), \quad g_1^\nu(t_0) = 0_n, \tag{36}$$

$$\dot{g}_2^\nu(t) = S(t)g_2^\nu(t) - B^*(t)g_1^\nu(t),\ g_2^\nu(t_0) = 0, \tag{37}$$

$$\dot{m}^\nu(t) = S(t)m^\nu(t), \quad m^\nu(t_0) = 1, \tag{38}$$

$$S(t)=\frac{\left[\frac{1}{2(T-t_0)}+g_1^{\nu^*}(t)A(t)g_1^{\nu}(t)+m^{\nu}(t)D^{\nu}(t)g_1^{\nu}(t)-B^*(t)g_1^{\nu}(t)g_2^{\nu}(t)\right]}{\left[g_1^{\nu^*}(t)g_1^{\nu}(t)+\left(g_2^{\nu}(t)\right)^2\right]}, \tag{39}$$

then the functions $g_1^{\nu}(t)$, $g_2^{\nu}(t)$ convolve the ν th integral condition (23) from left to right, and the following relation takes place:

$$g_1{}^{\nu*}(t)g_1{}^{\nu}(t)+\left(g_2{}^{\nu}(t)\right)^2=(t-t_0)\,/\,(T-t_0)\,,\quad t\in[t_0,T]\,. \tag{40}$$

Proof Multiply the ν th equality from (32) by as-yet arbitrary function $m^{\nu}(t)$ that satisfy the following condition:

$$m^{\nu}\left(t_0\right)=1, \tag{41}$$

and obtain

$$m^{\nu}(t)\overline{C}^{\nu}(t)=m^{\nu}(t)\overline{\alpha}^{\nu}(t)x(t)+m^{\nu}(t)\overline{\beta}^{\nu}(t).$$

Introduce the notation

$$g_1{}^{\nu}(t)=m^{\nu}(t)\,\overline{\alpha}^{\nu}(t),\quad g_2{}^{\nu}(t)=m^{\nu}(t)\,\overline{\beta}^{\nu}(t). \tag{42}$$

It is clear that

$$g_1^{\nu}\left(t_0\right)=0_n,\qquad g_2^{\nu}\left(t_0\right)=0.$$

Choose the function $m^{\nu}(t)$ in such a way that condition (40) holds, i.e., we require the linear growth of the sum of squares of the convolving functions.

Differentiating (40), we obtain

$$2\left(\dot{g}_1^{\nu}(t),g_1^{\nu}(t)\right)+2\dot{g}_2^{\nu}(t)g_2^{\nu}(t)=1/\left(T-t_0\right). \tag{43}$$

Differentiating (42) and taking (30) into account, it is not difficult to obtain

$$\dot{g}_1^{\nu}(t)=\frac{\dot{m}^{\nu}(t)}{m^{\nu}(t)}g_1{}^{\nu}(t)-A^*(t)g_1^{\nu}(t)+m^{\nu}(t)D^{\nu*}(t), \tag{44}$$

$$\dot{g}_2^{\nu}(t)=\frac{\dot{m}^{\nu}(t)}{m^{\nu}(t)}g_2{}^{\nu}(t)-B^*(t)g_1{}^{\nu}(t). \tag{45}$$

Substituting the derivatives obtained into (43), we have

$$\left(\frac{\dot{m}^{\nu}(t)}{m^{\nu}(t)}g_1^{\nu}(t) - A^*(t)g_1^{\nu}(t) + m^{\nu}(t)D^{\nu *}(t), g_1^{\nu}(t)\right) +$$
$$+ \frac{\dot{m}^{\nu}(t)}{m^{\nu}(t)}\left(g_2^{\nu}(t)\right)^2 - B^*(t)g_1{}^{\nu}(t)g_2^{\nu}(t) = \frac{1}{2(T-t_0)}.$$

From here, in view of notation (39), it is not difficult to obtain Eq. (38). Substituting (38) into (45) and (46), we obtain Eqs. (36) and (37).

For simplicity of notation, we suppose that $\tilde{t}_1 = t_0$ and $\tilde{t}_{l_2} = T$, and write conjugate problem (12)–(15) in a sufficiently general form:

$$\dot{\psi}(t) = A_1(t)\psi(t) + \sum_{i=1}^{l_1}\left[\chi\left(\overline{t}_{2i}\right) - \chi\left(\overline{t}_{2i-1}\right)\right]\overline{D}_i(t)\lambda + C(t), \tag{46}$$

$$\tilde{G}_1\psi\left(t_0\right) = \tilde{K}_1 + \tilde{D}_1\lambda, \tag{47}$$

$$\psi(T) = -\tilde{K}_{l_2} - \tilde{D}_{l_2}\lambda, \tag{48}$$

$$\psi^+\left(\tilde{t}_j\right) - \psi^-\left(\tilde{t}_j\right) = \tilde{K}_j + \tilde{D}_j\lambda, \quad j = 2, 3, \ldots, l_2 - 1, \tag{49}$$

and at the points $\overline{t}_i$, for which $t_0 < \overline{t}_i < T, i = 1, 2, \ldots 2l_1$:

$$\psi^+\left(\overline{t}_i\right) - \psi^-\left(\overline{t}_i\right) = \overline{K}_i, \quad i = 1, 2, \ldots 2l_1. \tag{50}$$

Here, we used the notation $\overline{G}_1 = I_n$ – n-dimensional matrix, all elements of which are equal to 1:

$$A_1(t) = -A^*(t, u), \quad C(t) = \partial f^0(x, u, t)/\partial x, \quad \tilde{K}_j = \partial\Phi\left(\hat{x}\left(\hat{t}\right)\right)/\partial x\left(\tilde{t}_j\right),$$
$$\tilde{D}_j^* = \tilde{D}_j, \quad j = 1, \ldots, l_2, \quad \overline{D}_i^*(t) = \overline{D}_i(t), \quad i = 1, 2, \ldots, l_1,$$
$$\overline{K}_i = \partial\Phi\left(\hat{x}\left(\hat{t}\right)\right)/\partial x\left(\overline{t}_i\right), \quad i = 1, \ldots, 2l_1.$$

As indicated above, problem (46)–(51) include n differential Eqs. (46), $2n$ boundary conditions, and an unknown vector $\lambda \in R^n$. It is clear that problems (46)–(49) are a closed one.

We say that the matrix and vector functions $G_1(t)$, $D_1(t) \in R^{n\times n}$ and $K_1(t) \in R^n$ are such that

$$G_1\left(t_0\right) = G_1\left(\tilde{t}_1\right) = \tilde{G}_1, \quad K_1\left(t_0\right) = K_1\left(\tilde{t}_1\right) = \tilde{K}_1, \quad \tilde{D}_1\left(t_0\right) = \tilde{D}_1\left(\tilde{t}_1\right) = \tilde{D}_1, \tag{51}$$

shift condition (47) to the right if for the solution of (46) – $\psi(t)$ the following takes place:

$$G_1(t)\psi(t) = K_1(t) + D_1(t)\lambda, \qquad t \in \left[\tilde{t}_1, \tilde{t}_2\right). \tag{52}$$

To obtain the functions of shifting $G_1(t)$, $D_1(t)$, we use the results of [23]. Using formula (52), we shift the initial conditions (50) to the point $t = \tilde{t}_2 - 0$ and, thereby, condition (49) for the point $t = \tilde{t}_2 + 0$:

$$G_1\left(\tilde{t}_2\right)\psi\left(\tilde{t}_2 + 0\right) = \left[K_1\left(\tilde{t}_2\right) + G\left(\tilde{t}_2\right)\tilde{K}_2\right] + \left[D_1\left(\tilde{t}_2\right) + G_1\left(\tilde{t}_2\right)\tilde{D}_2\right]\lambda.$$

Denoting

$$\tilde{t}_2 = \tilde{t}_2 + 0, \quad \tilde{G}_1^1 = G_1\left(\tilde{t}_2\right), \quad \tilde{K}_1^1 = K_1\left(\tilde{t}_2\right) + G\left(\tilde{t}_2\right)\tilde{K}_2, \quad \tilde{D}_1^1 = D_1\left(\tilde{t}_2\right) + G_1\left(\tilde{t}_2\right)\tilde{D}_2,$$

at the point $\tilde{t}_2$, we obtain conditions equivalent to condition (47):

$$\tilde{G}_1^1\psi\left(\tilde{t}_2\right) = \tilde{K}_1^1 + \tilde{D}_1^1\lambda.$$

Repeating the above procedure $l_2 - 1$ times, taking into account (50), we obtain a system of $2n$ equations relative to $\psi\left(\tilde{t}_{l_2}\right) = \psi(T)$ and λ. Determining $\psi(T)$ and λ from the solution to the Cauchy problem for Eq. (46), the vector function $\psi(t)$ is determined from right to left.

Let us consider in more detail the implementation of the stages of the process of shifting condition (47). Suppose $\left[\bar{t}_1, \bar{t}_2\right] \subset \left[\tilde{t}_1, \tilde{t}_2\right)$ and $\tilde{t}_1 = t_0$. The shift of the condition is carried out in stages for the intervals $\left[\tilde{t}_1, \bar{t}_1\right)$, $\left[\bar{t}_1, \bar{t}_2\right)$, and $\left[\bar{t}_2, \tilde{t}_2\right)$, using formula (52).

1) First, at $t \in \left[\tilde{t}_1, \bar{t}_1\right)$, we shift initial condition (47) to the point $t = \bar{t}_1 - 0$. Then, taking into account jump conditions (50) for the point $t = \bar{t}_1$, we will obtain

$$G_1\left(\bar{t}_1\right)\psi\left(\bar{t}_1 + 0\right) = \left[K_1\left(\bar{t}_1\right) + G_1\left(\bar{t}_1\right)\overline{K}_1\right] + D_1\left(\bar{t}_1\right)\lambda.$$

Denoting

$$\bar{t}_1 = \bar{t}_1 + 0, \quad \tilde{G}_1^1 = G_1\left(\bar{t}_1\right), \quad \tilde{K}_1^1 = K_1\left(\bar{t}_1\right) + G_1\left(\bar{t}_1\right)\overline{K}_1, \quad \tilde{D}_1^1 = D_1\left(\bar{t}_1\right),$$

at the point $\bar{t}_1$, we obtain conditions equivalent to the initial conditions:

$$\tilde{G}_1^1\psi\left(\bar{t}_1\right) = \tilde{K}_1^1 + \tilde{D}_1^1\lambda. \tag{53}$$

2) At $t \in \left[\bar{t}_1, \bar{t}_2\right)$, we shift condition (53) to the point $t = \bar{t}_2 - 0$, and taking into account jump conditions (50) for the point $t = \bar{t}_2$, we will obtain

$$G_1\left(\bar{t}_2\right)\psi\left(\bar{t}_2 + 0\right) = \left[K_1\left(\bar{t}_2\right) + G_1\left(\bar{t}_2\right)\overline{K}_2\right] + D_1\left(\bar{t}_2\right)\lambda.$$

Denoting

$$\bar{t}_2 = \bar{t}_2 + 0, \quad \tilde{G}_1^2 = G_1(\bar{t}_2), \quad \tilde{K}_1^2 = K_1(\bar{t}_2) + G_1(\bar{t}_2)\overline{K}_2, \quad \tilde{D}_1^2 = D_1(\bar{t}_2),$$

at the point $\bar{t}_2$, we obtain conditions equivalent to the condition (53):

$$\tilde{G}_1^2 \psi(\bar{t}_2) = \tilde{K}_1^2 + \tilde{D}_1^2 \lambda. \tag{54}$$

3) At $t \in [\bar{t}_2, \tilde{t}_2)$, we shift condition (54) to the point $t = \tilde{t}_2 - 0$, and taking into account jump conditions (49) for the point $t = \tilde{t}_2$, we will obtain

$$G_1(\tilde{t}_2)\psi(\tilde{t}_2 + 0) = \left[K_1(\tilde{t}_2) + G_1(\tilde{t}_2)\overline{K}_2\right] + \left[D_1(\tilde{t}_2) + G_1(\tilde{t}_2)\tilde{D}_2\right]\lambda.$$

Denoting

$$\tilde{t}_2 = \tilde{t}_2 + 0, \quad \tilde{G}_1^3 = G_1(\tilde{t}_2), \quad \tilde{K}_1^3 = K_1(\tilde{t}_2) + G_1(\tilde{t}_2)\overline{K}_2, \quad \tilde{D}_1^3 = D_1(\tilde{t}_2) + G_1(\tilde{t}_2)\tilde{D}_2,$$

at the point $\tilde{t}_2$, we obtain conditions equivalent to the conditions (54):

$$\tilde{G}_1^3 \psi(\tilde{t}_2) = \tilde{K}_1^3 + \tilde{D}_1^3 \lambda. \tag{55}$$

Obviously, the functions $G_j(t)$, $K_j(t)$, $D_j(t)$, $j = 1, \ldots, l_2$ satisfying (51) and (52) are not uniquely defined. The following theorem includes functions that can be used to shift conditions.

Theorem 4 *Suppose the functions $G_1(t)$, $K_1(t)$, $D_1(t)$ are the solution to the following Cauchy problems:*

$$\begin{aligned}
&\dot{G}_1(t) = Q^0(t)G_1(t) - G_1(t)A_1(t), \qquad G_1(\tilde{t}_1) = \tilde{G}_1,\\
&\dot{D}_1(t) = Q^0(t)D_1(t) + G_1(t)\sum_{i=1}^{l_1}\left[\chi(\bar{t}_{2i}) - \chi(\bar{t}_{2i-1})\right]\overline{D}_i(t), \qquad D_1(\tilde{t}_1) = \tilde{D}_1,\\
&\dot{K}_1(t) = Q^0(t)K_1(t) + G_1(t)C(t), \qquad K_1(\tilde{t}_1) = \tilde{K}_1,\\
&\dot{Q}(t) = Q^0(t)Q(t), \qquad Q(\tilde{t}_1) = I_{n\times n},\\
&Q^0(t) = \left[G_1(t)A_1(t)G_1^*(t) - G_1(t)\sum_{i=1}^{l_1}\left[\chi(\bar{t}_{2i}) - \chi(\bar{t}_{2i-1})\right]\overline{D}_i(t)D_1^*(t) - G_1(t)C(t)K_1^*(t)\right]\times\\
&\times\left[G_1(t)G_1^*(t) + D_1(t)D_1^*(t) + K_1(t)K_1^*(t)\right]^{-1}.
\end{aligned} \tag{56}$$

Then relation (52) is true for these functions on the half-interval $t \in [\tilde{t}_1, \tilde{t}_2)$, and the following condition is satisfied:

$$\begin{aligned}
\|G_1(t)\|^2_{R^{n\times n}} + \|D_1(t)\|^2_{R^{n\times n}} + \|K_1(t)\|^2_{R^n} &= \left\|\tilde{G}_1\right\|^2_{R^{n\times n}} + \left\|\tilde{D}_1\right\|^2_{R^{n\times n}} + \left\|\tilde{K}_1\right\|^2_{R^n}\\
&= const, \quad t \in (\tilde{t}_1, \tilde{t}_2).
\end{aligned} \tag{57}$$

Proof Differentiating expression (52).

$\dot{G}_1(t)\psi(t) + G_1(t)\dot{\psi}(t) = \dot{K}_1(t) + \dot{D}_1(t)\lambda$,and taking (46) into account, we come to the equality.

$$\dot{G}_1(t)\psi(t) + G_1(t)\left[A_1(t)\psi(t) + \sum_{i=1}^{l_1}\left[\chi\left(\bar{t}_{2i}\right) - \chi\left(\bar{t}_{2i-1}\right)\right]\overline{D}_i(t)\lambda + C(t)\right] = \dot{K}_1(t) + \dot{D}_1(t)\lambda.$$

After grouping, we obtain the following equation:

$$\left[\dot{G}_1(t) + G_1(t)A_1(t)\right]\psi(t) + \left[-\dot{D}_1(t) + G_1(t)\sum_{i=1}^{l_1}\left[\chi\left(\bar{t}_{2i}\right) - \chi\left(\bar{t}_{2i-1}\right)\right]\overline{D}_i(t)\right]\lambda$$
$$+\left[-\dot{K}_1(t) + G_1(t)C(t)\right] = 0.$$

Setting the expressions in brackets equal to 0, we obtain

$$\dot{G}_1(t) = -G_1(t)A_1(t), \quad \dot{D}_1(t) = G_1(t)\sum_{i=1}^{l_1}\left[\chi\left(\bar{t}_{2i}\right) - \chi\left(\bar{t}_{2i-1}\right)\right]\overline{D}_i(t),$$
$$\dot{K}_1(t) = G_1(t)C(t). \tag{58}$$

The functions $G_1(t)$, $K_1(t)$, $D_1(t)$, which are the solution to Cauchy problems (58) and (51), satisfy condition (52), i.e., they shift condition (47) from the point $\tilde{t}_1 = t_0$ to the point $\tilde{t}_2$. But numerical solution to Cauchy problems (58) and (51), as is known, confronts with instability due to the presence of fast-increasing components. This is because the matrix $A(t)$ often has both positive and negative eigenvalues. That is why we try to find shifting functions that satisfy condition (2.37).

Multiply both parts of (52) by an arbitrary matrix function $Q(t)$ such that

$$Q\left(t_0\right) = I_{n\times n}, \quad rang\ \ Q(t) = n, \quad t \in \left[\tilde{t}_1, \tilde{t}_2\right),$$

and introduce the notations

$$g(t) = Q(t)G_1(t), \quad q(t) = Q(t)D_1(t), \quad r(t) = Q(t)K_1(t). \tag{59}$$

From (52), it follows that

$$g(t)\psi(t) = r(t) + q(t)\lambda. \tag{60}$$

Differentiating (59) and taking (58) into account, we obtain

$$\dot{g}(t) = \dot{Q}(t)G_1(t) + Q(t)\dot{G}_1(t) = \dot{Q}(t)Q^{-1}(t)g(t) - g(t)A_1(t), \tag{61}$$

$$\dot{q}(t)=\dot{Q}(t)D_1(t)+Q(t)\dot{D}_1(t)=\dot{Q}(t)Q^{-1}(t)q(t)+g(t)\sum_{i=1}^{l_1}\left[\chi\left(\bar{t}_{2i}\right)-\chi\left(\bar{t}_{2i-1}\right)\right]\overline{D}_i(t), \tag{62}$$

$$\dot{r}(t) = \dot{Q}(t)K_1(t) + Q(t)\dot{K}_1(t) = \dot{Q}(t)Q^{-1}(t)r(t) + g(t)C(t). \tag{63}$$

Transposing relations (61)–(63), we obtain

$$\dot{g}^*(t) = g^*(t)\left(\dot{Q}(t)Q^{-1}(t)\right)^* - A_1^*(t)g^*(t), \tag{64}$$

$$\dot{q}^*(t) = q^*(t)\left(\dot{Q}(t)Q^{-1}(t)\right)^* + \sum_{i=1}^{l_1}\left[\chi\left(\bar{t}_{2i}\right) - \chi\left(\bar{t}_{2i-1}\right)\right]\overline{D}_i^*(t)g^*(t), \tag{65}$$

$$\dot{r}^*(t) = r^*(t)\left(\dot{Q}(t)Q^{-1}(t)\right)^* + C^*(t)g^*(t). \tag{66}$$

Choose the matrix functions $Q(t)$ such that the following relation holds

$$g(t)g^*(t) + q(t)q^*(t) + r(t)r^*(t) = const.$$

Differentiating it, we obtain

$$\dot{g}(t)g^*(t) + g(t)\dot{g}^*(t) + \dot{q}(t)q^*(t) + q(t)\dot{q}^*(t) + \dot{r}(t)r^*(t) + r(t)\dot{r}^*(t) = 0. \tag{67}$$

Substituting (61)–(66) into (67), after grouping, we obtain

$$\begin{aligned}&\left[Q(t)Q^{-1}(t)\left(g(t)g^*(t) + q(t)q^*(t) + r(t)r^*(t)\right) + \right.\\&+\left.\left(-g(t)A_1(t)g^*(t) + g(t)\sum_{i=1}^{l_1}\left[\chi\left(\bar{t}_{2i}\right) - \chi\left(\bar{t}_{2i-1}\right)\right]\overline{D}_i(t)q^*(t) + g(t)C(t)r^*(t)\right)\right] +\\&+\left[Q(t)Q^{-1}(t)\left(g(t)g^*(t) + q(t)q^*(t) + r(t)r^*(t)\right) + \right.\\&+\left.\left(-g(t)A_1(t)g^*(t) + g(t)\sum_{i=1}^{l_1}\left[\chi\left(\bar{t}_{2i}\right) - \chi\left(\bar{t}_{2i-1}\right)\right]\overline{D}_i(t)q^*(t) + g(t)C(t)r^*(t)\right)\right]^* = 0.\end{aligned}$$

Assume that the expression in both square brackets equals 0:

$$\begin{aligned}&\left[Q(t)Q^{-1}(t)\left(g(t)g^*(t) + q(t)q^*(t) + r(t)r^*(t)\right) + \right.\\&+\left.\left(-g(t)A_1(t)g^*(t) + g(t)\sum_{i=1}^{l_1}\left[\chi\left(\bar{t}_{2i}\right) - \chi\left(\bar{t}_{2i-1}\right)\right]\overline{D}_i(t)q^*(t) + g(t)C(t)r^*(t)\right)\right] = 0.\end{aligned}$$

From this, it follows that

$$Q(t)Q^{-1}(t) = Q^0(t), \tag{68}$$

where

$$Q^0(t) = \left[g(t)A_1(t)g^*(t) - g(t)\sum_{i=1}^{l_1}\left[\chi\left(\bar{t}_{2i}\right) - \chi\left(\bar{t}_{2i-1}\right)\right]\overline{D}_i(t)q^*(t) - g(t)C(t)r^*(t)\right] \times,$$
$$\times\left[g(t)g^*(t) + q(t)q^*(t) + r(t)r^*(t)\right]^{-1}.$$

Substituting (68) into (61)–(63) and renaming the functions $g(t)$ in $G_1(t)$, $q(t)$ and $D_1(t)$, $r(t)$ in $K_1(t)$, we obtain the statement of the theorem.

Clearly, a similar formula can be obtained for a successive shift of condition (48) to the left.

Thus, each iteration of procedure (20) for a given control $u(t) = u^k(t)$, $t \in [t_0, T]$, $k = 0, 1, \ldots$ requires the following steps:

1. To find the phase trajectory $x(t)$, $t \in [t_0, T]$, solve the nonlocal boundary value problem (1), (2) using the given scheme for shifting the conditions of the solution to the auxiliary Cauchy problem (36)–(40).
2. To find the conjugate vector function $\psi(t)$, $t \in [t_0, T]$ and the vector of dual variables λ, solve problem (46)–(51) using procedure (52) for shifting boundary conditions.
3. Using the found values of $x(t)$, $\psi(t)$, $t \in [t_0, T]$, calculate the values of the gradient of the functional in formula (19).

Obviously, instead of gradient projection method (20), we can use other well-known first-order optimization methods ([27]).

4 Analysis of the Results of the Computational Experiments

Problem 1 Consider the following optimal control problem:

$$\begin{cases} \dot{x}_1(t) = 2tx_1(t) - 3x_2(t) + tu - 3t^2 + t + 5, \\ \dot{x}_2(t) = 5x_1(t) + tx_2(t) - t^3 - 9t + 5, \end{cases} \quad t \in [0; 1], \tag{69}$$

$$\int_0^{0.2}\begin{pmatrix} \tau & 3 \\ 2 & \tau - 1 \end{pmatrix} x(\tau)\,d\tau + \begin{pmatrix} 2 & 1 \\ 3 & 4 \end{pmatrix} x(0.5) + \int_{0.7}^{1}\begin{pmatrix} 1 & -2\tau \\ \tau & 3 \end{pmatrix} x(\tau)\,d\tau = \begin{pmatrix} 1.16335 \\ 6.2377 \end{pmatrix},$$
$$\overline{D}_1(t) = \begin{pmatrix} t & 3 \\ 2 & t-1 \end{pmatrix}, \quad \overline{D}_2(t) = \begin{pmatrix} 1 & -2t \\ t & 3 \end{pmatrix}, \quad \tilde{D}_1 = \begin{pmatrix} 2 & 1 \\ 3 & 4 \end{pmatrix}, \tag{70}$$

$$J(u) = \int_0^1 [x_1(t) - u(t) + 2]^2 dt + x_1^2(0,5) + [x_2(0,5) - 1.25]^2 + [x_1(1) - 1]^2$$
$$+ [x_2(1) - 2]^2. \tag{71}$$

The exact solution to this problem is known: $u^*(t) = 2t + 1$, $x_1^*(t) = 2t - 1$, $x_2^*(t) = t^2 + 1$, $J(u^*) = 0$.

According to formulas (12)–(16), the adjoint problem is as follows:

$$\begin{aligned}
&\dot{\psi}_1(t) = -2t\psi_1(t) - 5\psi_2(t) + (\chi(0.2) - \chi(0)) \ (t\lambda_1 + 2\lambda_2) + \\
&(\chi(1) - \chi(0.7))(\lambda_1 + t\lambda_2) + +2(x_1(t) - u(t) + 2), \\
&\dot{\psi}_2(t) = 3\psi_1(t) - t\psi_2(t) + (\chi(0.2) - \chi(0)) \ (3\lambda_1 + (t-1)\lambda_2) + \\
&(\chi(1) - \chi(0.7))(-2t\lambda_1 + 3\lambda_2), \psi_1(0) = 0, \quad \psi_2(0) = 0, \\
&\psi_1(1) = -2[x_1(1) - 1], \quad \psi_2(1) = -2[x_2(1) - 1], \\
&\psi_1^+(0.5) - \psi_1^-(0.5) = 2x_1(0.5) + 2\lambda_1 + 3\lambda_2, \\
&\psi_2^+(0.5) - \psi_2^-(0.5) = 2[x_2(0.5) - 1.25] + \lambda_1 + 4\lambda_2.
\end{aligned}$$

Formula (19) for the gradient of the functional is as the following form:

$$\nabla \ J(u) = -[x_1(t) - u(t) + 2] - t\psi_1(t).$$

For the numerical solution using procedure (20), numerical experiments were carried out for different initial controls $u^0(t)$, different values of the steps for the Runge-Kutta fourth-order method when solving the Cauchy problems.

Table 1 shows the results of solving the system of nonlocal direct problem (69), (70), the conjugate problem, and the values of the components of the normalized gradients. The gradients were calculated both by the proposed formulas (19) $\left(\nabla^{\text{norm.}}_{analyt.} J\right)$ and using the finite difference approximation of the functional $\left(\nabla^{\text{norm.}}_{approx.} J\right)$ by the formula:

$$\partial J(u)/\partial u_j \approx \left(J\left(u + \delta e_j\right) - J(u)\right)/\delta. \tag{72}$$

Here, u_j is the value of control $u = (u_1, u_2, \ldots, u_N)$ at the j th sampling instant and e_j – N-dimensional vector, all components of which are equal to 0, except the j th, which is equal to 1. The quantity δ took values 0.1 and 0.001.

Notice that instead of formula (72) it is possible to use the formulas proposed in the author's papers [30, 31]. These formulas are most effective at points with small values of the components of the gradient (in particular, in the neighborhood of the extremum).

The initial value of the functional is $J(u^0) = 56.28717$, $\lambda_1 = 0.2387$, $\lambda_2 = 0.1781$. The values of the functional obtained in the course of the iterations are as

Table 1 Initial values of the controls, of the phase variables, and of the normalized gradients calculated using both the proposed formulas and (72)

t	$u^{(0)}(t)$	$x_1^{(0)}(t)$	$x_2^{(0)}(t)$	$\psi_1^{(0)}(t)$	$\psi_2^{(0)}(t)$	$\nabla^{norm.}_{analyt.} J$	$\nabla^{norm.}_{approx.} J$ $\delta = 10^{-2}$	$\delta = 10^{-3}$
0	1.0000	1.5886	1.2034	−9.2836	−6.3653	−0.0180	−0.0138	−0.0161
20	2.0000	1.5641	2.4702	−0.5626	3.2122	−0.0107	−0.0104	−0.0107
40	3.0000	1.2382	3.5937	2.9294	14.8639	−0.0037	−0.0049	−0.0052
60	4.0000	0.6657	4.4538	−4.5081	14.2492	0.0140	0.0147	0.0145
80	5.0000	−0.0781	4.9528	−10.9008	11.4433	0.0367	0.0411	0.0410
100	6.0000	−0.8973	5.0218	−15.1806	7.0519	0.0606	0.0688	0.0667
120	7.0000	−1.6767	4.6295	−21.2792	9.1844	0.0911	0.0994	0.0994
140	8.0000	−2.2835	3.7919	−22.7128	2.1275	0.1132	0.1167	0.1168
160	9.0000	−2.5710	2.5845	−14.7216	−0.0383	0.1078	0.1037	0.1038
180	10.0000	−2.3853	1.1544	−6.8798	−0.6530	0.0940	0.0827	0.0826
200	11.0000	−1.5759	−0.2660	0.0000	−0.0000	0.0737	0.0800	0.0800

Table 2 The exact solution to the problem and the solution obtained after the sixth iteration

	Solution obtained					Exact solution		
t	$u^{(6)}(t)$	$x_1^{(6)}(t)$	$x_2^{(6)}(t)$	$\psi_1^{(6)}(t)$	$\psi_2^{(6)}(t)$	$u^*(t)$	$x_1^*(t)$	$x_2^*(t)$
0	0.9999	−1.0000	1.0011	0.0059	−0.0040	1.0000	−1.0000	1.0000
20	1.2000	−0.8003	1.0111	0.0095	0.0027	1.2000	−0.8000	1.0100
40	1.4002	−0.6005	1.0409	0.0089	0.0098	1.4000	−0.6000	1.0400
60	1.6001	−0.4006	1.0906	0.0038	0.0114	1.6000	−0.4000	1.0900
80	1.7999	−0.2007	1.1603	−0.0012	0.0114	1.8000	−0.2000	1.1600
100	1.9997	−0.0005	1.2500	−0.0049	0.0099	2.0000	0.0000	1.2500
120	2.1995	0.1997	1.3598	−0.0055	0.0044	2.2000	0.2000	1.3600
140	2.3998	0.4000	1.4897	−0.0053	0.0025	2.4000	0.4000	1.4900
160	2.6010	0.6006	1.6398	−0.0026	0.0014	2.6000	0.6000	1.6400
180	2.8024	0.8013	1.8103	−0.0008	0.0006	2.8000	0.8000	1.8100
200	3.0041	1.0022	2.0012	0.0000	−0.0000	3.0000	1.0000	2.0000

follows: $J(u^1) = 1.93187$, $J(u^2) = 0.10445$, $J(u^3) = 0.00868$, $J(u^4) = 0.00023$, $J(u^5) = 0.00004$. On the sixth iteration of conjugate gradient method, we obtain the results given in Table 2 with the minimal value of the functional $J(u^6)$ equal to 10^{-6}.

5 Conclusion

In the work, we propose the technique for numerical solution to optimal control problems for ordinary differential equations systems involving nonseparated multipoint and integral conditions. Note that a mere numerical solution to the differential systems presents certain difficulties. The adjoint problem also has a specific character which lies both in the equation itself and in the presence of an unknown vector of Lagrange coefficients in the conditions.

The formulas proposed in the work, as well as the computational schemes, make it possible to take into account all the specific characters which occur when calculating the gradient of the functional. Overall, the proposed approach allows us to use a rich arsenal of first-order optimization methods and the corresponding standard software to solve the considered optimal control problems.

References

1. *O. Nicoletti,* Sulle condizioni iniziali che determinano gli integrali delle equazioni differenziali ordinarie, **Atti R. Sci. Torino.** 33 (1897), 746–748.
2. *Ya. D. Tamarkin,* On some general problems of ordinary differential equations theory and on series expansion of arbitrary functions, Petrograd. (1917).
3. *Ch. J. Vallee-Poussin,* Sur l'quation diffrentielle linaire du second ordre. Dtermination d'une intgrale par deux valeurs assignes. Extension aux quations d'ordre n., **J. Math. Pures Appl.,** 8 (1929), 125–144.
4. *K. R. Aida-zade and V. M. Abdullaev,* On an approach to designing control of the distributed-parameter processes, **Autom. Remote Control.** (9) 73 (2012), 1443–1455.
5. *K. R. Aida-zade and V. M. Abdullayev,* Optimizing placement of the control points at synthesis of the heating process control, **Autom. Remote Control.** (9) 78 (2017), 1585–1599.
6. *A. Bouziani,* On the solvability of parabolic and hyperbolic problems with a boundary integral condition, **Intern.J. Math. Sci.** (4) 31 (2002), 202–213.
7. *I.N. Parasidis and E. Providas*, Closed-form solutions for some classes of loaded difference equations with initial and nonlocal multipoint conditions, T. Rassias and N.J. Daras, eds., Springer, 2018, pp. 363–387.
8. *I. N. Parasidis and E. Providas,* An exact solution method for a class of nonlinear loaded difference equations with multipoint boundary conditions, **Journal of Difference Equations and Applications.** (10) 24 (2018), 1649–1663.
9. *L. S. Pulkina*, Non-local problem with integral conditions for a hyperbolic equation, **Differential Equations.** (7) 40 (2004), 887–891.
10. *Y. A. Sharifov and H. F. Quliyev*, Formula for the gradient in the optimal control problem for the non-Linear system of the hyperbolic equations with non-local boundary conditions, **TWMS Journal of Pure and Applied Mathematics.** (1) 3 (2012), 111–121.
11. *V. M. Abdullayev,* Identification of the functions of response to loading for stationary systems, **Cybern. Syst. Analysis.** (3) 53 (2017), 417–425.
12. *V. M. Abdullaev and K. R. Aida-zade*, Numerical solution of optimal control problems for loaded lumped parameter systems, **Comput. Math. Math. Phys.** (9) 46 (2006), 1487–1502.
13. *V. M. Abdullayev and K. R. Aida-zade*, Optimization of loading places and load response functions for stationary systems, **Comput. Math. Math. Phys.** (4) 57 (2017),634–644.
14. *K. R. Aida-zade and V. M. Abdullaev*, Numerical solution of optimal control problems with unseparated conditions on phase state, **Appl. and Comput. Math.** (2) 4 (2005), 165–177.
15. *L. T. Aschepkov*, Optimal control of system with intermediate conditions, **Journal of Applied Mathematics and Mechanics.** (2) 45 (1981), 215–222.
16. *O. O. Vasileva and K. Mizukami*, Dynamical processes described by boundary problem: necessary optimality conditions and methods of solution, **Journal of Computer and System Sciences International (A Journal of Optimization and Control).** 1 (2000), 95–100.
17. *O. V. Vasilev and V. A. Terleckij*, Optimal control of a boundary problem, **Proceedings of the Steklov Institute of Mathematics.** 211 (1995), 221–130.

18. *V. M. Abdullaev and K. R. Aida-zade*, On the numerical solution of loaded systems of ordinary differential equations, **Comput. Math. Math. Phys.** (9) 44 (2004), 1585–1595.
19. *V. M. Abdullaev and K. R. Aida-zade*, Numerical method of solution to loaded nonlocal boundary-value problems for ordinary differential equations, **Comput. Math. Math. Phys.** (7) 54 (2014), 1096–1109.
20. *A. A. Abramov*, On the transfer of boundary conditions for systems of ordinary linear differential equations (a variant of the dispersive method), **Zh. Vychisl. Mat. Mat. Fiz.** (3) 1 (1961), 542–545.
21. *K. R. Aida-zade and V. M. Abdullaev*, On the solution of boundary-value problems with nonseparated multipoint and integral conditions, **Differential Equations.** (9) 49 (2013), 1114–1125.
22. *K. Moszynski,* A method of solving the boundary value problem for a system of linear ordinary differential equation, **Algorytmy. Varshava.** (3) 11 (1964), 25–43.
23. *A. N. Bondarev and V. N. Laptinskii*, A multipoint boundary value problem for the Lyapunov equation in the case of strong degeneration of the boundary conditions, **Differential Equations.** (6) 47 (2011), 776–784.
24. *D. S. Dzhumabaev and A. E. Imanchiev*, Well-posed solvability of linear multipoint boundary problem, **Matematicheskij zhurnal. Almaaty.** 1(15) 5 (2005), 30–38.
25. *I. T. Kiguradze*, Boundary value problems for systems of ordinary differential equations, Itogi Nauki i Tekhniki. Ser. Sovrem. **Probl. Mat. Nov. Dostizh.** 30 (1987), 3–103.
26. *A. M. Samoilenko, V. N. Laptinskii and K. K. Kenzhebaev*, Constructive Methods of Investigating Periodic and Multipoint Boundary - Value Problems, Kiev,1999.
27. *F. P. Vasilyev*, Optimization Methods, M: Faktorial, 2002.
28. *K. R. Aida-zade*, A numerical method for the reconstruction of parameters of a dynamical system, **Cybern. Syst. Analysis.** (1) 3 (2004), 101–108.
29. *K. R. Aida-zade and V. M. Abdullayev*, Solution to a class of inverse problems for system of loaded ordinary differential equations with integral conditions, **J. of Inverse and Ill-posed Problems.** (5) 24 (2016), 543–558.
30. *A.L.Karchevsky,* Numerical solution of the one-dimensional inverse problem for the elasticity system, **Doklady RAS.** 375 (2), (2000), pp. 235-238.
31. *A.L. Karchevsky,* Reconstruction of pressure velocities and boundaries of thin layers in thinlystratified layers, **J. Inv. Ill-Posed Problems.** 18 (4) (2010), pp. 371-388.

On Generalized Convexity and Superquadracity

Shoshana Abramovich

Abstract In this paper we deal with generalized ψ-uniformly convex functions and with superquadratic functions and discuss some of their similarities and differences. Using the techniques discussed here, we obtain reversed and refined Minkowski type inequalities.

1 Introduction

Convex and convex type functions and their relations to mathematical inequalities play an important role in science, see, for instance, [3] about electrical engineering and [5] about statistical applications and their references.

In this paper we deal with generalized ψ-uniformly convex functions and with superquadratic functions and discuss some of their similarities and differences.

We start quoting the definition and properties of superquadratic functions from [1] which include the functions $f(x) = x^p$, $x \geq 0$, when $p \geq 2$, the functions $f(x) = -\left(1 + x^{\frac{1}{p}}\right)^p$ when $p > 0$ and $f(x) = 1 - \left(1 + x^{\frac{1}{p}}\right)^p$ when $p \geq \frac{1}{2}$. Also, we quote from [4] the definition of generalized ψ-uniformly convex functions.

In Sect. 2 we emphasize the importance of the general definition of superquadracity appearing in [1, Definition 2.1] compared with some of its special cases and with the generalized ψ-uniformly convex functions defined in [4].

In Sect. 3, by using the results discussed in Sect. 2 we refine and reverse the well known Minkowski inequality that says

$$\left(\sum_{i=1}^{n} a_i^p\right)^{\frac{1}{p}} + \left(\sum_{i=1}^{n} b_i^p\right)^{\frac{1}{p}} \leq \left(\sum_{i=1}^{n} (a_i + b_i)^p\right)^{\frac{1}{p}}, \tag{1}$$

S. Abramovich (✉)
Department of Mathematics, University of Haifa, Haifa, Israel
e-mail: abramos@math.haifa.ac.il

I. N. Parasidis et al. (eds.), *Mathematical Analysis in Interdisciplinary Research*, Springer Optimization and Its Applications 179,
https://doi.org/10.1007/978-3-030-84721-0_3

for $0 < p < 1,\ a_i, b_i \geq 0,\ i = 1, \ldots, n.$

Definition 1 ([1, Definition 2.1]) A function $f : [0, B) \to \mathbb{R}$ is superquadratic provided that for all $x \in [0, B)$ there exists a constant $C_f(x) \in \mathbb{R}$ such that the inequality

$$f(y) \geq f(x) + C_f(x)(y - x) + f(|y - x|) \tag{2}$$

holds for all $y \in [0, B)$ (see [1, Definition 2.1], there $[0, \infty)$ instead $[0, B)$).

f is called subquadratic if $-f$ is superquadratic.

Theorem 1 ([1, Theorem 2.2]) *The inequality*

$$\int f(g(s))\, d\mu(s) \geq f\left(\int g d\mu\right) + f\left(\left|g(s) - \int g d\mu\right|\right)$$

holds for all probability measures and all non-negative, μ-integrable functions g if and only if f is superquadratic.

Corollary 1 ([1, 2]) *Suppose that f is superquadratic. Let $0 \leq x_i < B$, $i = 1, 2$ and let $0 \leq t \leq 1$. Then*

$$\begin{aligned} &tf(x_1) + (1 - t) f(x_2) - f(tx_1 + (1 - t) x_2) \\ &\geq tf((1 - t)|x_2 - x_1|) + (1 - t) f(t|x_2 - x_1|) \end{aligned} \tag{3}$$

holds.

More generally, suppose that f is superquadratic. Let $\xi_i \geq 0$, $i = 1, \ldots, m$, and let $\overline{\xi} = \sum_{i=1}^{m} p_i \xi_i$ where $p_i \geq 0,\ i = 1, \ldots, m$, and $\sum_{i=1}^{m} p_i = 1$. Then

$$\sum_{i=1}^{m} p_i f(\xi_i) - f(\overline{\xi}) \geq \sum_{i=1}^{m} p_i f(|\xi_i - \overline{\xi}|) \tag{4}$$

holds.

If f is non-negative, it is also convex and Inequality (4) refines Jensen's inequality. In particular, the functions $f(x) = x^r$, $x \geq 0$, are superquadratic and convex when $r \geq 2$, and subquadratic and convex when $1 < r < 2$. Equality holds in inequalities (3) and (4) when $r = 2$.

Lemma 1 ([1, Lemma 2.1]) *Let f be superquadratic function with $C_f(x)$ as in Definition 1. Then:*

(i) $f(0) \leq 0$,
(ii) *if $f(0) = f'(0) = 0$, then $C_f(x) = f'(x)$ whenever f is differentiable at $x > 0$,*
(iii) *if $f \geq 0$, then f is convex and $f(0) = f'(0) = 0$.*

Lemma 2 ([1, Lemma 3.1]) *Suppose $f : [0, \infty) \to \mathbb{R}$ is continuously differentiable and $f(0) \leq 0$. If f' is superadditive or $\frac{f'(x)}{x}$ is non-decreasing, then f is superquadratic and (according to its proof) $C_f(x) = f'(x)$, where $C_f(x)$ is as in Definition 1.*

Lemma 3 ([1, Lemma 4.1]) *A non-positive, non-increasing, and superadditive function is a superquadratic function and (according to its proof) satisfies $C_f(x) = 0$, where $C_f(x)$ is as in Definition 1.*

Example 1 ([1, Example 4.2]) Let

$$f_p(x) = -\left(1 + x^{\frac{1}{p}}\right)^p, \quad x \geq 0.$$

Then f_p is superquadratic for $p > 0$ with $C_{f_p}(x) = 0$ and $g = 1 + f_p$ is superquadratic for $p \geq \frac{1}{2}$ with $C_g(x) = g'(x) = f_p'(x)$.

Lemma 4 ([1, Section 3]) *Suppose that f is a differentiable function and $f(0) = f'(0) = 0$. If f is superquadratic, then $\frac{f(x)}{x^2}$ is non-decreasing.*

The definition of **generalized ψ-uniformly convex functions** as appears in [4] is the following:

Definition 2 ([4, Page 306]) Let $I = [a, b] \subset \mathbb{R}$ be an interval and $\psi : [0, b - a] \to \mathbb{R}$ be a function. A function $f : [a, b] \to \mathbb{R}$ is said to be *generalized ψ-uniformly convex* if:

$$tf(x) + (1-t) f(y) \geq f(tx + (1-t)y) + t(1-t)\psi(|x - y|)$$
$$\text{for } x, y \in I \text{ and } t \in [0, 1]. \tag{5}$$

If in addition $\psi \geq 0$, then f is said to be *ψ-uniformly convex.*

Paper [4] deals with inequalities that extend the Levin-Stečkin's theorem. The main result in [4, Theorem 1] relates to the function ψ as appears in Definition 2, and depends on the fact that $\lim_{t \to 0^+} \frac{\psi(t)}{t^2}$ is finite. We discuss this issue in Sect. 2.

In the unpublished [6] a companion inequality to Minkowski inequality is stated and proved:

Theorem 2 ([6, Th2.1]) *For $0 < p < 1$, $a_i, b_i > 0$, $i = 1, \ldots, n$ the inequality*

$$\left(\sum_{i=1}^n a_i^p\right)^{\frac{1}{p}} + \left(\sum_{i=1}^n b_i^p\right)^{\frac{1}{p}} \leq \left(\sum_{i=1}^n (a_i + b_i)^p\right)^{\frac{1}{p}}$$
$$\leq \frac{\sum_{i=1}^n a_i b_i^{p-1}}{\left(\sum_{i=1}^n b_i^p\right)^{\frac{p-1}{p}}} + \frac{\sum_{i=1}^n b_i a_i^{p-1}}{\left(\sum_{i=1}^n a_i^p\right)^{\frac{p-1}{p}}} \tag{6}$$

holds.

In Sect. 3 we refine Minkowski's inequality in four ways using generalized ψ-uniformly convexity, subquadracity and superquadracity properties of the functions discussed in Sect. 2. The proofs of Theorems 3 and 4 apply the technique employed in [6] to prove the right hand-side of (6) in Theorem 2, besides using superquadracity and subquadracity properties of the functions involved there.

2 Superquadracity and Generalized ψ-Uniformly Convexity

We start with emphasizing the importance of the definition of superquadracity as appears in [1] vis a vis its special cases. Definition 1 does not guarantee that $C_f(x) = f'(x)$. However, from Lemmas 1 and 2 we know that in the case that f is superquadratic and $f(0) = f'(0) = 0$, and in the case that the derivative of the superquadratic function is superadditive or $\frac{f'(x)}{x}$ is non-decreasing we get $C_f(x) = f'(x)$. On the other hand when the superquadratic function satisfies Lemma 3 we get that $C_f(x) = 0$.

Although the n-th derivative of $f_p(x) = -\left(1 + x^{\frac{1}{p}}\right)^p$, $x \geq 0$, $0 < p < 1$, as discussed in Example 1, is continuous on $[0, \infty)$, we get when inserting this function in Definition 1 that $C_{f_p}(x)$ satisfies $C_{f_p}(x) = 0 \neq f_p'(x) = -x^{\frac{1}{p}-1}\left(1 + x^{\frac{1}{p}}\right)^p$.

Therefore whenever

$$f(y) - f(x) \geq f'(x)(y - x) + f(|y - x|) \tag{7}$$

is used as the definition of superquadracity, it means that it deals not with the general case of superquadratic functions but it might, but not necessarily, deal with those superquadratic functions satisfying Lemma 1(ii) or Lemma 2. The following function f is an example of a superquadratic function that satisfies (7) but as proved in [1, Example 3.3] does not satisfy Lemma 2: This function is defined by $f(0) = 0$ and

$$f'(x) = \begin{cases} 0, & x \leq 1 \\ 1 + (x-2)^3, & x \geq 1. \end{cases}$$

For such superquadratic functions, Definition 1 translates into (7), but as explained above it does not hold for all superquadratic functions.

We point out now a difference between the superquadratic functions and the generalized ψ-uniformly convex functions:

According to the proof of Theorem 1 [1, Theorem 2.2] and Corollary 1 we get that inequalities (2) and (3) are equivalent. On the other hand, Inequality (5) that defines, according to [4], the generalized ψ-uniformly convex function f, when f is continuously differentiable, leads to the inequality

$$f(y) - f(x) \geq f'(x)(y-x) + \psi(|y-x|), \tag{8}$$

as proved in [4, Theorem 1], but Inequality (8) does not lead in general to Inequality (5) but to

$$\begin{aligned} & tf(x_1) + (1-t) f(x_2) - f(tx_1 + (1-t) x_2) \\ & \quad \geq t\psi((1-t)|x_2 - x_1|) + (1-t)\psi(t|x_2 - x_1|), \end{aligned} \tag{9}$$

for $0 \leq t \leq 1$.

More generally, it is easy to verify that, similarly to Inequality (4) for superquadratic functions, when f is a generalized ψ-uniformly convex function, then

$$\sum_{i=1}^{m} p_i f(\xi_i) - f(\overline{\xi}) \geq \sum_{i=1}^{m} p_i \psi(|\xi_i - \overline{\xi}|), \tag{10}$$

holds, where $\xi_i \geq 0$, $i = 1, \ldots, m$, $\overline{\xi} = \sum_{i=1}^{m} p_i \xi_i$, $p_i \geq 0$, $i = 1, \ldots, m$, and $\sum_{i=1}^{m} p_i = 1$.

In addition, if ψ is non-negative, the function f is also convex and Inequality (10) refines Jensen's inequality.

Moreover, if instead of (5) in Definition 2 we have a set of functions f which satisfies

$$\begin{aligned} & tf(x_1) + (1-t) f(x_2) - f(tx_1 + (1-t) x_2) \\ & \quad \geq G(t)\psi(|x_1 - x_2|), \qquad t \in [0,1], \end{aligned} \tag{11}$$

then (11) still leads to (8) when $\lim_{t \to 0^+} \frac{G(t)}{t} = 1$.

However, for the special case where $\psi(x) = kx^2$, when k is constant, the inequalities (5) and (9) are the same.

Remark 1 By choosing $x = y$ in (5) or in (8) we get that ψ satisfies $\psi(0) \leq 0$.

From now on till the end of this section we deal with functions satisfying inequalities (7) and (8).

A similarity between convex superquadratic functions and ψ-uniformly convex functions is shown in Remark 2 below. The set of convex superquadratic functions f satisfies $f(0) = f'(0) = 0$. Also, the set f of ψ-uniformly convex functions satisfies $\psi(0) = \psi'(0) = 0$.

For the convenience of the reader a proof of Remark 2 is presented. This can easily be obtained by following the steps of the proof in [1] of Lemma 1(iii):

Remark 2 For a function $\psi : [0, b-a] \to \mathbb{R}$ and a continuously differentiable ψ-uniformly convex function f on $[a,b] \to \mathbb{R}$, we get that $\psi(0) = \psi'(0) = 0$.

Proof If $\psi \geq 0$, then $\psi(0) = 0$ because always as mentioned in Remark 1 $\psi(0) \leq 0$. Then by choosing in (8) first $y > x$ and then $y < x$ we get that

$$\limsup_{y \to x^-} \left(\frac{f(x) - f(y)}{x - y} + \frac{\psi(x - y)}{x - y} \right) \leq f'(x) \leq \limsup_{y \to x^+} \left(\frac{f(y) - f(x)}{y - x} + \frac{\psi(y - x)}{y - x} \right),$$

and hence

$$\limsup_{x \to 0^+} \frac{\psi(x)}{x} \leq 0.$$

Since ψ is non-negative, we have

$$0 \leq \limsup_{x \to 0^-} \frac{\psi(x)}{x} \leq \limsup_{x \to 0^+} \frac{\psi(x)}{x} \leq 0,$$

and therefore the one sided derivative at zero exists and $\psi'(0) = 0$.

We deal now with the behavior of $\frac{\psi(x)}{x^2}$ when f is generalized ψ-uniformly convex function, and with $\frac{f(x)}{x^2}$ when f is superquadratic.

Besides Lemma 4 we get the following lemma which is proved in [4, Proof of Theorem 1]:

Lemma 5 *If f is twice continuously differentiable generalized ψ-uniformly convex function, then $f''(x) \geq 2 \lim_{x \to 0^+} \frac{\psi(x)}{x^2}$.*

Corollary 2 *Let $I = [a, b]$ be an interval and $\psi : [0, b - a] \to \mathbb{R}$ be a twice differentiable function on $[0, b - a]$. Let $f : [a, b] \to \mathbb{R}$ be a continuously twice differentiable ψ-uniformly convex function, that is $\psi \geq 0$. Denote $\varphi(x) = \frac{\psi(x)}{x^2}$, $x > 0$. Then $\varphi(0) = \lim_{x = 0^+} \frac{\psi(x)}{x^2}$ is finite and non-negative.*

Indeed, Remark 2 says that $\psi(0) = \psi'(0) = 0$. Therefore,

$$\lim_{x = 0^+} \frac{\psi(x)}{x^2} = \varphi(0) = \lim_{x = 0^+} \frac{\psi'(x)}{2x} = \lim_{x = 0^+} \frac{\psi''(x)}{2} = \frac{\psi''(0)}{2}.$$

Remark 3 It is shown in Remark 1 that ψ satisfies $\psi(0) \leq 0$ and therefore when $\psi(0) < 0$ we get $\lim_{x \to 0^+} \frac{\psi(x)}{x^2} = -\infty$. Also, when ψ is differentiable on $[0, b - a]$ and $\psi(0) = 0$ but $\psi'(0) < 0$ then again

$$\lim_{x \to 0^+} \frac{\psi(x)}{x^2} = \lim_{x \to 0^+} \frac{\psi'(x)}{2x} = -\infty.$$

Example 2 shows that the conditions $\psi(0) = 0$, $\psi'(0) = 0$ do not guarantee that $\lim_{x \to 0^+} \frac{\psi(x)}{x^2}$ is finite:

Example 2 The superquadratic function $f(x) = x^2 \ln x$ for $x > 0$ and $f(0) = 0$, $f'(0) = 0$ is continuously differentiable but not twice continuously differentiable at $x = 0$. Therefore we deal now with an interval $[a, b]$, $a > 0$ for $f(x) = x^2 \ln x$ which is twice differentiable and $\psi(x) = x^2 \ln x$, $0 < x \leq b - a$. These f and ψ satisfy (8). In this case $\lim_{x \to 0^+} \frac{\psi(x)}{x^2} = -\infty$.

We show here an example where $\lim_{x \to 0^+} \frac{\psi(x)}{x^2}$ is finite, but the generalized ψ-uniformly convex function g is not necessarily convex.

Example 3 Let $g(x) = f(x) - (kx)^2$ where k is a constant and f is twice differentiable convex and superquadratic function satisfying $\lim_{x \to 0^+} \frac{f(x)}{x^2} = \varphi(0)$ and $\varphi(0) \geq 0$. In such cases $\frac{g(x)}{x^2} = \varphi(x) - k^2 \underset{x \to 0^+}{\to} \varphi(0) - k^2$ and because $\varphi(0)$ is finite and non-negative, and because equality holds in (3) for the function x^2, the function g is superquadratic satisfying (7) and therefore also (8) for $\psi = f$, but is not necessarily convex.

In addition to the monotonicity of $\frac{f(x)}{x^2}$ as proved in Lemma 4 for superquadratic functions satisfying $f(0) = f'(0) = 0$, it is easy to prove:

Remark 4 If Inequality (8) when $x \geq 0$ holds for $\psi \geq 0$ and $f(0) = 0$, then f is convex and $\left(\frac{f(x)}{x}\right)' \geq \frac{\psi(x)}{x^2} \geq 0$. In the special case that f is superquadratic and convex, we get that $\left(\frac{f(x)}{x}\right)' \geq \frac{f(x)}{x^2} \geq 0$.

Indeed, from (8) we get that

$$f(0) - f(x) \geq -xf'(x) + \psi(x)$$

holds.

From this, because $f(0) = 0$ we get that

$$\frac{xf'(x) - f(x)}{x^2} = \left(\frac{f(x)}{x}\right)' \geq \frac{\psi(x)}{x^2} \geq 0.$$

In the special case that f is superquadratic we get that

$$\left(\frac{f(x)}{x}\right)' \geq \frac{f(x)}{x^2} \geq 0.$$

We finish this section demonstrating a set of continuous differentiable functions satisfying Inequality (8). As explained above, (8) holds for continuous differentiable generalized ψ-uniformly convex functions.

Example 4 The functions $f_p = -\left(1+x^{\frac{1}{p}}\right)^p$ where $\psi_t(x) = t - \left(1+x^{\frac{1}{p}}\right)^p$, $p \geq \frac{1}{2}$, $0 \leq t \leq 1$, $x \geq 0$ are generalized ψ_t-uniformly convex functions and satisfy (8). In particular, when $t = 0$, the function f_p is superquadratic and when $t = 1$ the function $f^*(x) = 1 - \left(1+x^{\frac{1}{p}}\right)^p$ where $\psi_1(x) = 1 - \left(1+x^{\frac{1}{p}}\right)^p$ is also superquadratic.

Indeed, $f^*(x) = 1 - \left(1+x^{\frac{1}{p}}\right)^p$, $p \geq \frac{1}{2}$ is superquadratic satisfying Inequality (7). Specifically as shown in Example 1 [1, Example 4.2] the inequality

$$1-\left(1+y^{\frac{1}{p}}\right)^p-\left(1-\left(1+x^{\frac{1}{p}}\right)^p\right)$$
$$\geq -\left(1+x^{-\frac{1}{p}}\right)^{p-1}(y-x)+\left(1-\left(1+|x-y|^{\frac{1}{p}}\right)^p\right)$$

holds, which is the same as Inequality (8)

$$-\left(1+y^{\frac{1}{p}}\right)^p-\left(-\left(1+x^{\frac{1}{p}}\right)^p\right)$$
$$\geq -\left(1+x^{-\frac{1}{p}}\right)^{p-1}(y-x)+\left(1-\left(1+|x-y|^{\frac{1}{p}}\right)^p\right),$$

for $f_p(x) = -\left(1+x^{\frac{1}{p}}\right)^p$ and $\psi_1(x) = 1 - \left(1+x^{\frac{1}{p}}\right)^p$.

Therefore, also

$$-\left(1+y^{\frac{1}{p}}\right)^p-\left(-\left(1+x^{\frac{1}{p}}\right)^p\right)$$
$$\geq -\left(1+x^{-\frac{1}{p}}\right)^{p-1}(y-x)+\left(t-\left(1+|x-y|^{\frac{1}{p}}\right)^p\right)$$

holds when $t \leq 1$ and Inequality (8) is satisfied by $f_p(x) = -\left(1+x^{\frac{1}{p}}\right)^p$ and $\psi_t(x) = t - \left(1+x^{\frac{1}{p}}\right)^p$.

As shown in Example 1, when $t = 0$, the function $f_p(x) = -\left(1+x^{\frac{1}{p}}\right)^p$ is also superquadratic but this time satisfying (7) with $C_f(x) = 0$, that is,

$$-\left(1+y^{\frac{1}{p}}\right)^{p}-\left(-\left(1+x^{\frac{1}{p}}\right)^{p}\right)\geq-\left(1+|x-y|^{\frac{1}{p}}\right)^{p}$$

holds.

3 Reversed and Refined Minkowski Inequality

In this section we use the properties discussed in Sect. 2 of superquadracity and of generalized ψ-uniformly convexity.

In Example 1 [1, Example 4.2] it is shown that $f_p(x)=\left(1+x^{\frac{1}{p}}\right)^{p}$ for $x\geq 0$, is subquadratic when $p>0$. Using this property and Corollary 1 together with the convexity of f_p when $p<1$ we get a refinement of Minkowski's inequality when $0<p<1$ (see also [1, Theorem 4.1]):

Lemma 6 *Let $a_i, b_i \geq 0,\ i=1,\ldots,n$. Then, when $p>0$ the inequality*

$$\sum_{i=1}^{n}(a_i+b_i)^{p}$$
$$\leq\left(\left(\sum_{i=1}^{n}a_i^{p}\right)^{\frac{1}{p}}+\left(\sum_{i=1}^{n}b_i^{p}\right)^{\frac{1}{p}}\right)^{p}+\sum_{i=1}^{n}a_i^{p}\left(1+\left|\frac{b_i^{p}}{a_i^{p}}-\frac{\sum_{j=1}^{n}b_j^{p}}{\sum_{j=1}^{n}a_j^{p}}\right|^{\frac{1}{p}}\right)^{p} \tag{12}$$

holds, and when $0<p<1$ the inequalities

$$\left(\left(\sum_{i=1}^{n}a_i^{p}\right)^{\frac{1}{p}}+\left(\sum_{i=1}^{n}b_i^{p}\right)^{\frac{1}{p}}\right)^{p}\leq\sum_{i=1}^{n}(a_i+b_i)^{p}$$
$$\leq\left(\left(\sum_{i=1}^{n}a_i^{p}\right)^{\frac{1}{p}}+\left(\sum_{i=1}^{n}b_i^{p}\right)^{\frac{1}{p}}\right)^{p}+\sum_{i=1}^{n}a_i^{p}\left(1+\left|\frac{b_i^{p}}{a_i^{p}}-\frac{\sum_{j=1}^{n}b_j^{p}}{\sum_{j=1}^{n}a_j^{p}}\right|^{\frac{1}{p}}\right)^{p} \tag{13}$$

hold.

Proof From the subquadracity of $f_p=\left(1+x^{\frac{1}{p}}\right)^{p},\ x\geq 0,\ p>0,$ according to Lemma 3 and Example 1 we get that:

$$\sum_{i=1}^{n}x_i\left(1+\left(\frac{y_i}{x_i}\right)^{\frac{1}{p}}\right)^{p}=\sum_{i=1}^{n}\left(x_i^{\frac{1}{p}}+y_i^{\frac{1}{p}}\right)^{p}$$

$$\leq \left(\left(\sum_{i=1}^{n} x_i \right)^{\frac{1}{p}} + \left(\sum_{i=1}^{n} y_i \right)^{\frac{1}{p}} \right)^p + \sum_{i=1}^{n} x_i \left(1 + \left| \frac{y_i}{x_i} - \frac{\sum_{j=1}^{n} y_j}{\sum_{j=1}^{n} x_j} \right|^{\frac{1}{p}} \right)^p \tag{14}$$

is satisfied. Substituting $x_i^{\frac{1}{p}} = a_i$ and $y_i^{\frac{1}{p}} = b_i, i = 1, \ldots, n$, we get Inequality (12), and together with the convexity of f for $0 < p < 1$ we get from (14) that (13) holds.

The next lemma uses the generalized ψ-uniformly convex functions $g_p = -\left(1 + x^{\frac{1}{p}}\right)^p$ when $\psi(x) = t - \left(1 + x^{\frac{1}{p}}\right)^p$, $0 \leq t \leq 1$ for $p \geq \frac{1}{2}$ and the convexity of $f_p(x) = \left(1 + x^{\frac{1}{p}}\right)^p$ when $0 < p < 1$ as discussed in Example 4. Similar to Lemma 6 we get:

Lemma 7 *Let $a_i, b_i > 0$, $i = 1, \ldots, n$ and $0 \leq t \leq 1$ then when $p \geq \frac{1}{2}$ the inequality:*

$$\sum_{i=1}^{n} (a_i + b_i)^p \leq \left(\left(\sum_{i=1}^{n} a_i^p \right)^{\frac{1}{p}} + \left(\sum_{i=1}^{n} b_i^p \right)^{\frac{1}{p}} \right)^p$$

$$+ \sum_{i=1}^{n} a_i^p \left(1 + \left| \frac{b_i^p}{a_i^p} - \frac{\sum_{j=1}^{n} b_j^p}{\sum_{j=1}^{n} a_j^p} \right|^{\frac{1}{p}} \right)^p - t \sum_{i=1}^{n} a_i^p$$

holds, and when $\frac{1}{2} \leq p \leq 1$, the inequalities

$$\left(\left(\sum_{i=1}^{n} a_i^p \right)^{\frac{1}{p}} + \left(\sum_{i=1}^{n} b_i^p \right)^{\frac{1}{p}} \right)^p \leq \sum_{i=1}^{n} (a_i + b_i)^p$$

$$\leq \left(\left(\sum_{i=1}^{n} a_i^p \right)^{\frac{1}{p}} + \left(\sum_{i=1}^{n} b_i^p \right)^{\frac{1}{p}} \right)^p + \sum_{i=1}^{n} a_i^p \left(1 + \left| \frac{b_i^p}{a_i^p} - \frac{\sum_{j=1}^{n} b_j^p}{\sum_{j=1}^{n} a_j^p} \right|^{\frac{1}{p}} \right)^p$$

$$- t \sum_{i=1}^{n} a_i^p$$

hold.

We finish the paper by refining Inequality (6) in Theorem 2, and we get two new Minkowski type inequalities. In the proofs we use the technique employed in [6, Theorem 2.1] and the subquadracity of $f(x) = x^{\frac{1}{p}}$, $x \geq 0$, $\frac{1}{2} < p < 1$, the superquadracity of $f(x) = x^{\frac{1}{p}}$, $x \geq 0$, $0 < p \leq \frac{1}{2}$.

Theorem 3 *Let* $0 < p < \frac{1}{2}$, $a_i, b_i \geq 0$, $i = 1, \ldots, n$. *Then, the inequalities*

$$\left(\sum_{i=1}^{n} a_i^p\right)^{\frac{1}{p}} + \left(\sum_{i=1}^{n} b_i^p\right)^{\frac{1}{p}}$$
$$\leq \left(\sum_{i=1}^{n} (a_i + b_i)^p\right)^{\frac{1}{p}}$$
$$\leq \frac{\sum_{i=1}^{n} a_i b_i^{p-1}}{\left(\sum_{i=1}^{n} b_i^p\right)^{\frac{p-1}{p}}} + \frac{\sum_{i=1}^{n} b_i a_i^{p-1}}{\left(\sum_{i=1}^{n} a_i^p\right)^{\frac{p-1}{p}}}$$
$$- \frac{\sum_{i=1}^{n} a_i^p \left|\frac{(a_i+b_i)^p}{a_i^p} - \frac{\sum_{j=1}^{n}(a_j+b_j)^p}{\sum_{j=1}^{n} a_j^p}\right|^{\frac{1}{p}}}{\left(\sum_{j=1}^{n} a_j^p\right)^{\frac{1}{p}}} - \frac{\sum_{i=1}^{n} b_i^p \left|\frac{a_i^p}{b_i^p} - \frac{\sum_{j=1}^{n} a_j^p}{\sum_{j=1}^{n} b_j^p}\right|^{\frac{1}{p}}}{\left(\sum_{j=1}^{n} b_j^p\right)^{\frac{1}{p}}} \tag{15}$$

hold. Equality holds in the right hand-side of inequality (15) when $p = \frac{1}{2}$.

Proof We use the superquadracity of $g(x) = x^{\frac{1}{p}}$, $x \geq 0$, $0 < p \leq \frac{1}{2}$ which by Corollary 1 leads to the inequality

$$\sum_{i=1}^{n} x_i \left(\frac{y_i}{x_i}\right)^{\frac{1}{p}} = \sum_{i=1}^{n} x_i^{1-\frac{1}{p}} y_i^{\frac{1}{p}}$$
$$\geq \left(\sum_{i=1}^{n} x_i\right)^{1-\frac{1}{p}} \left(\sum_{i=1}^{n} y_i\right)^{\frac{1}{p}} + \sum_{i=1}^{n} x_i \left|\frac{y_i}{x_i} - \frac{\sum_{j=1}^{n} y_j}{\sum_{j=1}^{n} x_j}\right|^{\frac{1}{p}}, \tag{16}$$

and we get from (16) that

$$\frac{\sum_{i=1}^{n} a_i b_i^{p-1}}{\left(\sum_{i=1}^{n} b_i^p\right)^{\frac{p-1}{p}}} = \frac{\sum_{i=1}^{n} \left(a_i^p\right)^{\frac{1}{p}} \left(b_i^p\right)^{1-\frac{1}{p}}}{\left(\sum_{i=1}^{n} b_i^p\right)^{\frac{p-1}{p}}}$$
$$\geq \left(\sum_{i=1}^{n} a_i^p\right)^{\frac{1}{p}} + \frac{\sum_{i=1}^{n} b_i^p}{\left(\sum_{j=1}^{n} b_j^p\right)^{\frac{p-1}{p}}} \left|\frac{a_i^p}{b_i^p} - \frac{\sum_{j=1}^{n} a_j^p}{\sum_{j=1}^{n} b_j^p}\right|^{\frac{1}{p}}. \tag{17}$$

By denoting $c_i = a_i + b_i$, $i = 1, \ldots, n$ we get also that

$$
\begin{aligned}
\frac{\sum_{i=1}^{n} b_i a_i^{p-1}}{\left(\sum_{j=1}^{n} a_j^p\right)^{\frac{p-1}{p}}} &= \frac{\sum_{i=1}^{n} (c_i - a_i) a_i^{p-1}}{\left(\sum_{j=1}^{n} a_j^p\right)^{\frac{p-1}{p}}} \\
&= \frac{\sum_{i=1}^{n} c_i a_i^{p-1}}{\left(\sum_{j=1}^{n} a_j^p\right)^{\frac{p-1}{p}}} - \frac{\sum_{i=1}^{n} a_i a_i^{p-1}}{\left(\sum_{j=1}^{n} a_j^p\right)^{\frac{p-1}{p}}} \\
&= \frac{\sum_{i=1}^{n} c_i a_i^{p-1}}{\left(\sum_{j=1}^{n} a_j^p\right)^{\frac{p-1}{p}}} - \left(\sum_{i=1}^{n} a_i^p\right)^{\frac{1}{p}} \\
&\geq \left(\sum_{i=1}^{n} c_i^p\right)^{\frac{1}{p}} + \frac{\sum_{i=1}^{n} a_i^p}{\left(\sum_{j=1}^{n} a_j^p\right)^{\frac{p-1}{p}}} \left|\frac{c_i^p}{a_i^p} - \frac{\sum_{j=1}^{n} c_j^p}{\sum_{j=1}^{n} a_j^p}\right|^{\frac{1}{p}} - \left(\sum_{i=1}^{n} a_i^p\right)^{\frac{1}{p}}. \qquad (18)
\end{aligned}
$$

Summing (17) with (18) and using $c_i = a_i + b_i, i = 1, \ldots, n$ we get that

$$
\begin{aligned}
&\frac{\sum_{i=1}^{n} a_i b_i^{p-1}}{\left(\sum_{i=1}^{n} b_i^p\right)^{\frac{p-1}{p}}} + \frac{\sum_{i=1}^{n} b_i a_i^{p-1}}{\left(\sum_{i=1}^{n} a_i^p\right)^{\frac{p-1}{p}}} \\
&\quad \geq \left(\sum_{i=1}^{n} (a_i + b_i)^p\right)^{\frac{1}{p}} + \frac{\sum_{i=1}^{n} a_i^p}{\left(\sum_{j=1}^{n} a_j^p\right)^{\frac{p-1}{p}}} \left|\frac{(a_i + b_i)^p}{a_i^p} - \frac{\sum_{j=1}^{n} (a_j + b_j)^p}{\sum_{j=1}^{n} a_j^p}\right|^{\frac{1}{p}} \\
&\quad + \frac{\sum_{i=1}^{n} b_i^p}{\left(\sum_{j=1}^{n} b_j^p\right)^{\frac{p-1}{p}}} \left|\frac{a_i^p}{b_i^p} - \frac{\sum_{j=1}^{n} a_j^p}{\sum_{j=1}^{n} b_j^p}\right|^{\frac{1}{p}}. \qquad (19)
\end{aligned}
$$

From (19) and from Minkowski inequality (1) for $0 < p < 1$ we get for $0 < p < \frac{1}{2}$ that (15) holds.

The proof is complete.

Theorem 4 *Let* $\frac{1}{2} \leq p \leq 1,\ a_i, b_i \geq 0,\ i = 1, \ldots, n$. *Then, the inequality*

$$
\begin{aligned}
\max &\left(\frac{\sum_{i=1}^{n} a_i b_i^{p-1}}{\left(\sum_{i=1}^{n} b_i^p\right)^{\frac{p-1}{p}}} + \frac{\sum_{i=1}^{n} b_i a_i^{p-1}}{\left(\sum_{i=1}^{n} a_i^p\right)^{\frac{p-1}{p}}} \right. \\
&- \frac{\sum_{i=1}^{n} a_i^p \left|\frac{(a_i+b_i)^p}{a_i^p} - \frac{\sum_{j=1}^{n} (a_j+b_j)^p}{\sum_{j=1}^{n} a_j^p}\right|^{\frac{1}{p}}}{\left(\sum_{j=1}^{n} a_j^p\right)^{\frac{1}{p}}} - \frac{\sum_{i=1}^{n} b_i^p \left|\frac{a_i^p}{b_i^p} - \frac{\sum_{j=1}^{n} a_j^p}{\sum_{j=1}^{n} b_j^p}\right|^{\frac{1}{p}}}{\left(\sum_{j=1}^{n} b_j^p\right)^{\frac{1}{p}}},
\end{aligned}
$$

$$\left(\left(\sum_{i=1}^{n} a_i^p\right)^{\frac{1}{p}} + \left(\sum_{i=1}^{n} b_i^p\right)^{\frac{1}{p}}\right)$$

$$\leq \left(\sum_{i=1}^{n} (a_i + b_i)^p\right)^{\frac{1}{p}}$$

$$\leq \frac{\sum_{i=1}^{n} a_i b_i^{p-1}}{\left(\sum_{i=1}^{n} b_i^p\right)^{\frac{p-1}{p}}} + \frac{\sum_{i=1}^{n} b_i a_i^{p-1}}{\left(\sum_{i=1}^{n} a_i^p\right)^{\frac{p-1}{p}}} \tag{20}$$

holds.

Proof The proof of the inequalities in (20) is omitted because it is similar to the proof of Theorem 3 using here the subquadracity of $f(x) = x^{\frac{1}{p}}, x > 0, \frac{1}{2} < p < 1$.

References

1. S. Abramovich, G. Jameson, and G. Sinnamon, Refining Jensen's inequality, Bull. Math. Soc. Sci. Math. Roumanie (N.S.) 47(95) (2004), no. 1-2, 3–14.
2. S. Abramovich, G. Jameson and G. Sinnamon, "Inequalities for Averages of Convex and Superquadratic Functions", Journal of Inequalities in Pure and Applied Mathematics, 5, issue 4, Article 91, (2004) 14 pages.
3. M. J. Cloud, B.C. Drachman and I. T. Lebedev, "Inequalities with applications to engineering", Springer Cham Heidelberg New York Dordrecht, London (2014).
4. M. Niezgoda, "An extension of Levin-Steckin's theorem to uniformly convex and superquadratic functions", Aequat. Math. 94 (2020), 303–321.
5. J. Pečarić, F. Proschan and Y. L. Tong, "Convex functions, Partial orderings and statistical applications", Academic press, New York (1992).
6. R. Zhou, H. Liu and J. Miao, "Matching form of Minkowski's inequality", Unpublished Manuscript.

Well-Posedness of Nonsmooth Lurie Dynamical Systems Involving Maximal Monotone Operators

S. Adly, D. Goeleven, and R. Oujja

Abstract Many physical phenomena can be modeled as a feedback connection of a linear dynamical systems combined with a nonlinear function which satisfies a sector condition. The concept of absolute stability, proposed by Lurie and Postnikov (Appl Math Mech 8(3), 1944) in the early 1940s, constitutes an important tool in the theory of control systems. Lurie dynamical systems have been studied extensively in the literature with nonlinear (but smooth) feedback functions that can be formulated as an ordinary differential equation. Many concrete applications in engineering can be modeled by a set-valued feedback law in order to take into account the nonsmooth relation between the output and the state variables. In this paper, we show the well-posedness of nonsmooth Lurie dynamical systems involving maximal monotone operators. This includes the case where the set-valued law is given by the subdifferential of a convex, proper, and lower semicontinuous function. Some existence and uniqueness results are given depending on the data of the problem and particularly the interplay between the matrix D and the set-valued map $\mathcal{F}$. We will also give some conditions ensuring that the extended resolvent $(D+\mathcal{F})^{-1}$ is single-valued and Lipschitz continuous. The main tools used are derived from convex analysis and maximal monotone theory.

1 Introduction

Control problems described by the Lurie (or Lur'e) systems consist of a linear time-invariant forward path and a feedback path with a static nonlinearity satisfying a sector condition of the form: $\dot{x} = Ax + B\phi(y)$ with the output signal $y = Cx$.

S. Adly
DMI-XLIM, University of Limoges, Limoges, France
e-mail: adly@unilim.fr

D. Goeleven (✉) · R. Oujja
PIMENT, University of La Reunion, Saint-Denis, France
e-mail: goeleven@univ-reunion.fr; oujja@univ-reunion.fr

I. N. Parasidis et al. (eds.), *Mathematical Analysis in Interdisciplinary Research*, Springer Optimization and Its Applications 179,
https://doi.org/10.1007/978-3-030-84721-0_4

Due to their importance in practical applications in control theory, this class of problems has been investigated intensively, in both continuous and discrete time cases, in the literature of control and applied mathematics. The development of this topic is closely connected with that of the absolute stability problem, which consists of studying the stability of a system with a positive real transfer function and the feedback branch containing a sector static nonlinearity (see Fig. 1). For more details, we refer to [18]. The general mathematical formalism of Lurie systems can be written as a negative feedback interconnection of an ordinary differential equation $\dot{x} = f(x, p)$, where p is one of the two slack variables, the second one being $q = g(x, p)$, connected to each other by a possibly set-valued relation of the form $p \in \mathcal{F}(q)$, or equivalently $q \in \mathcal{F}^{-1}(p)$. The main reason for extending Lurie systems to the case where the feedback nonlinearity is a set-valued map lies in the fact that many concrete problems in engineering and other field of science can be modeled by set-valued laws. This is the case, for example, for unilateral problems in mechanical systems with Coulomb friction or electrical circuits with nonregular devices such as diodes, transistors, or DC-DC power converters. More recently, Lurie dynamical systems with a set-valued static feedback part have been used and studied in [6–8, 12, 13, 16]. It is also known that other mathematical models used to study nonsmooth dynamical systems (relay systems, evolution variational inequalities, projected dynamical systems, complementarity systems...) can be formulated into Lurie dynamical systems with a set-valued feedback nonlinearity [1, 12].

We are interested in the Lurie systems which are (possibly nonlinear) time-invariant dynamical systems with static set-valued feedback. Usually, the function g has the form $g(x, p) = Cx + Dp$, with C and D two given matrices with suitable dimensions. The case $D = 0$ appears in many applications, particularly in nonregular electrical circuits, while the case $D \neq 0$ is more general but creates some difficulties when one wants to study the possibly set-valued operator $(-D+\mathcal{F}^{-1})^{-1}(C \circ \cdot)$. In [12], Brogliato and Goeleven overcome these obstacles by assuming that $\mathcal{F}$ is the subdifferential of some proper, convex, lower semicontinuous function, to enjoy the nice properties of the Fenchel transform and maximally monotone operator theory.

In this paper, we study Lurie systems involving maximally monotone set-valued laws. The problem is formulated into a first-order differential inclusion form where the set-valued right-hand side is a Lipschitz continuous perturbation of a maximal monotone operator under composition. We discuss conditions on the data such that the problem is well-posed in the sense that for every given initial condition, the Lurie system has a unique solution. Finally, some illustrative examples are presented.

2 Lurie Systems with Maximal Monotone Operators

Let $A \in \mathbb{R}^{n\times n}$, $B \in \mathbb{R}^{n\times m}$, $C \in \mathbb{R}^{m\times n}$, and $D \in \mathbb{R}^{m\times m}$ be given matrices. Let $\mathcal{F} : \mathbb{R}^m \rightrightarrows \mathbb{R}^m$ be a set-valued map. Given an initial point $x_0 \in \mathbb{R}^n$, the problem consists to find a function $x \in W^{1,\infty}_{loc}(0, +\infty)$ such that

$$(\mathcal{L}) \begin{cases} \dot{x}(t) = Ax(t) + Bp(t), \text{ a.e. } t \in [0, +\infty[, & \text{(1a)} \\ q(t) = Cx(t) + Dp(t), & \text{(1b)} \\ q(t) \in \mathcal{F}(-p(t)), \ t \geq 0, & \text{(1c)} \\ x(0) = x_0, & \text{(1d)} \end{cases}$$

where $p, q : \mathbb{R}_+ \to \mathbb{R}^m$ are two connected unknown mappings.

When the map $\mathcal{F} : \mathbb{R}^m \to \mathbb{R}^m, \ x = (x_1, x_2, \ldots, x_m) \mapsto \mathcal{F}(x)$ is single-valued and such that

$$\mathcal{F}(x) = \Big(\phi_1(x_1), \phi_2(x_2), \ldots, \phi_m(x_m)\Big),$$

where $\phi_i : \mathbb{R} \to \mathbb{R}$ are some given functions, the above problem is known in the literature as Lurie system with multiple nonlinearities. In general the nonlinear feedback function $\Phi(\cdot)$ is always assumed to satisfy a condition known in the literature as *the sector condition*. More precisely, we say that the function $\phi_i : \mathbb{R} \to \mathbb{R}$ is in sector $[l_i, u_i]$ if for all $x \in \mathbb{R}$, we have $\phi_i(x) \in [l_i x, u_i x]$ or $[u_i x, l_i x]$ (l_i and u_i are given in $\mathbb{R} \cup \{-\infty, +\infty\}$). For example, the sector $[-1, 1]$ means that $\phi_i(x) \leq |x|, \ \forall x \in \mathbb{R}$, and the sector $[0, +\infty]$ means that $\phi_i(x)x \geq 0, \ \forall x \in \mathbb{R}$ (see Fig. 1 for an illustration). Using this sector condition, it is possible to give some criteria, expressed in terms of linear matrix inequalities (LMI) for the asymptotic stability of Lurie systems. It is possible to include a perturbation with a locally integrable external force $f(\cdot)$ and/or a nonlinear Lipschitz continuous map instead of the matrix A but, for simplicity, we restrict ourselves to the system $(\mathcal{L})$. To the best of our knowledge, such a system was first introduced and analyzed in a special case in [9].

An important example in practice is given when the static set-valued nonlinearity $\mathcal{F} : \mathbb{R}^m \rightrightarrows \mathbb{R}^m, \ x = (x_1, \ldots, x_m) \mapsto \mathcal{F}(x)$ is defined componentwise as follows

$$\mathcal{F}(x) = \Big(\mathcal{F}_1(x_1), \mathcal{F}(x_2), \ldots, \mathcal{F}_m(x_m)\Big), \tag{2}$$

where for each $i = 1, 2, \ldots, m, \ \mathcal{F}_i : \mathbb{R} \rightrightarrows \mathbb{R}$ are given scalar set-valued nonlinearities.

In order to illustrate the system $(\mathcal{L})$, we give the following classical scheme in Fig. 2 (see, e.g., [12]). Here A, B, C, D denote the state, the input, the output, and the feedthrough matrices, respectively. Most of the previous works concern the case where the matrix $D = 0$ [6] or $D \neq 0$ and $\mathcal{F}$ is a maximal monotone operator [7, 8, 12, 13]. The consideration of a nonzero matrix D makes the analysis of the system more difficult. If $\mathcal{F}$ coincides with the normal cone of $\mathbb{R}^n_+$ (i.e., $\mathcal{F} = \mathrm{N}_{\mathbb{R}^n_+}$), then system $(\mathcal{L})$ reduces to the well-known linear complementarity systems largely studied in the literature [1, 16, 21]. In [12], the authors studied the well-

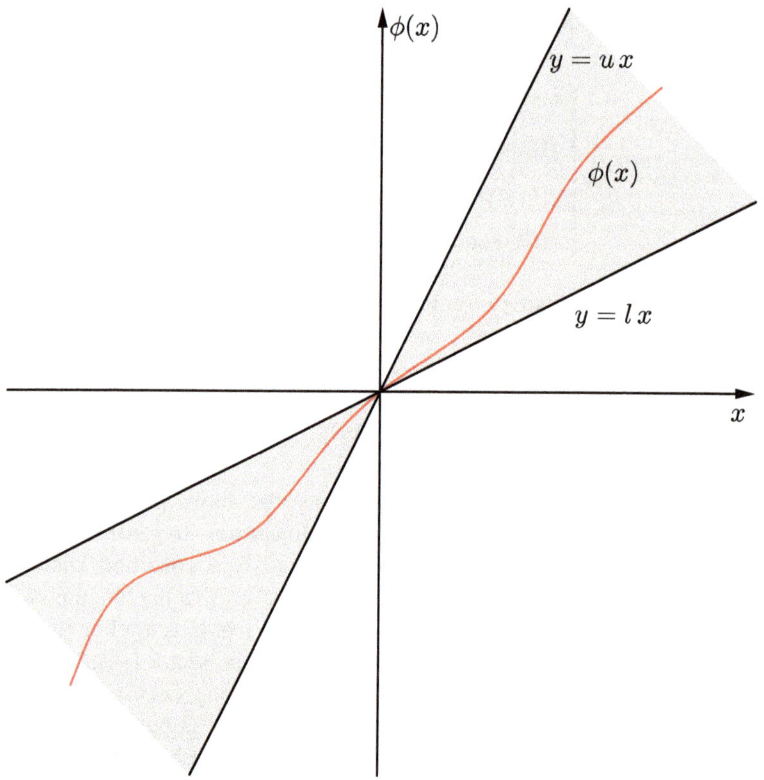

Fig. 1 Sector nonlinearities $lx \leq \phi(x) \leq ux$

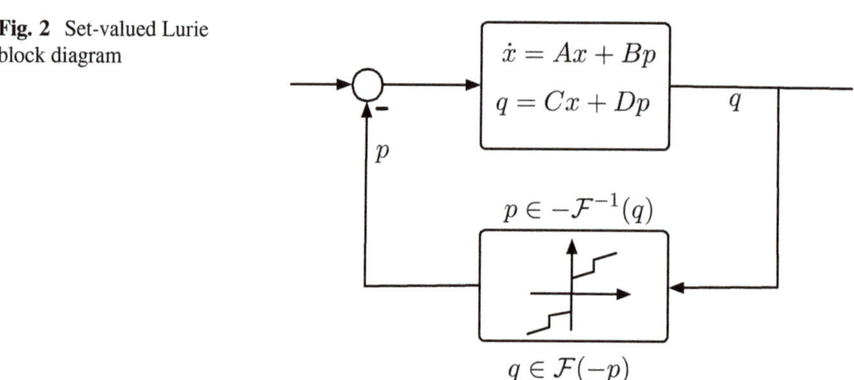

Fig. 2 Set-valued Lurie block diagram

posedness, stability, and invariance properties of system $(\mathcal{L})$ where $\mathcal{F}$ is the inverse of the subdifferential mapping of a given proper convex and lower semicontinuous function, or equivalently, the subdifferential mapping of its Fenchel conjugate. The well-posedness is improved in [13] for a general maximally monotone operator by using the passivity of the linear system. It is worth noting that in the case $D = 0$, the monotonicity of the set-valued map $\mathcal{F}$ is not necessary; in fact, only the local hypomonotonicity of $\mathcal{F}$ is sufficient (see [4] Chapter 5).

In this article, a nonzero matrix D is allowed with a maximal monotone set-valued map $\mathcal{F}$. The well-posedness of $(\mathcal{L})$ is investigated under various conditions on the data $(A, B, C, D, \mathcal{F})$.

3 Background from Convex and Set-Valued Analysis

We recall some definitions and some results about maximal monotone operators theory drawn from [11] (see also [10]). Let H be a real Hilbert space, with scalar product $\langle .,. \rangle$ and associated norm $\| \cdot \|$. For a multivalued operator $T : H \rightrightarrows H$, we denote by:

$$\mathrm{Dom}(T) := \{u \in H | \, T(u) \neq \emptyset\},$$

the *domain* of T,

$$\mathrm{Rge}(T) := \bigcup_{u \in H} T(u),$$

the *range* of T,

$$\text{Graph } (T) := \{(u, u^*) \in H \times H | \, u \in \mathrm{Dom}(T) \text{ and } u^* \in T(u)\},$$

the *graph* of T. Throughout the paper we identify operators with their graphs.

We recall that T is *monotone* if and only if for each $u, \ v \in \mathrm{Dom}(T)$ and $u^* \in T(u)$, $v^* \in T(v)$ we have

$$\langle v^* - u^*, v - u \rangle \geq 0.$$

The operator T is *maximal monotone* if it is monotone and its graph is not properly contained in the graph of any other monotone operator.

We say that a single-valued mapping A is *hemicontinuous* (following [10, p. 26]) if, for all $x, y \in H$:

$$A((1-t)x + ty) \to A(x), \ \text{ as } t \to 0.$$

Remark 3.1 A continuous map is therefore hemicontinuous. It can be shown that, if $A : \mathrm{Dom}(A) = H \to H$ is monotone and hemicontinuous, then A is maximal monotone (see [10, Proposition 2.4]). Also, if $A : H \to H$ is monotone and hemicontinuous and $B : H \rightrightarrows H$ is maximal monotone, then $A + B$ is maximal monotone [10, p. 37].

Let $T : H \rightrightarrows H$ be a monotone operator. From *Minty's theorem*, we know that:

$$T \text{ is maximal} \iff \text{Rge}(I + T) = H,$$

where I stands for the identity mapping on H.

The inverse operator T^{-1} of $T : H \rightrightarrows H$ is the operator defined by

$$v \in T^{-1}(u) \iff u \in T(v).$$

We note that

$$\text{Dom}(T^{-1}) = \text{Rge}(T). \tag{3}$$

When T is maximal monotone, the resolvent operator $\text{J}_T = (I + T)^{-1}$ is defined on the whole space H, it is single-valued and Lipschitz continuous with modulus 1; indeed, it is nonexpansive, i.e.,

$$\|\text{J}_T(x) - \text{J}_T(y)\| \leq \|x - y\|.$$

Proposition 3.1 *Let $T : H \rightrightarrows H$ be a maximal monotone operator (Fig. 3). Then the resolvent operator is well-defined, single-valued, and Lipschitz continuous with modulus* 1, *i.e.,*

$$\|\text{J}_T(x) - \text{J}_T(y)\| \leq \|x - y\|, \forall x, y \in H.$$

Proof The well-definedness is a consequence of Minty's Theorem. Let $x \in H$ and suppose that $y_1,\ y_2 \in \text{J}_T(x)$. By definition of J_T, we have

$$x - y_1 \in T(y_1) \text{ and } x - y_2 \in T(y_2).$$

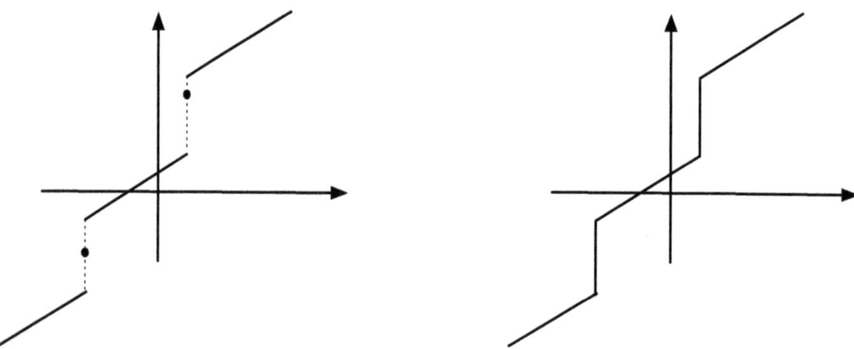

Fig. 3 Example of monotone operator but not maximal (left) and a maximal monotone operator (right)

Using the monotonicity of T, we get

$$\langle (x - y_1) - (x - y_2), y_1 - y_2 \geq 0.$$

Hence,

$$\|y_1 - y_2\|^2 \leq 0,$$

which means that $y_1 = y_2$. The resolvent operator is thus single-valued.

Let $x,\ y \in H$ and set $x^* = \mathrm{J}_T(x)$ and $y^* = \mathrm{J}_T(y)$. We have,

$$x - x^* \in T(x^*) \text{ and } y - y^* \in T(y^*).$$

Using the monotonicity of T, we get

$$\|x^* - y^*\|^2 \leq \langle x^* - y^*, x - y\rangle.$$

Hence,

$$\|x^* - y^*\| \leq \|x - y\|,$$

and the proof of the proposition is thereby completed. ■

Moreover, the operators $(\lambda I + T)^{-1}$ or $(I + \lambda T)^{-1}$ with $\lambda > 0$ are similarly well-defined, single-valued, and Lipschitz continuous.

The notation $\Gamma_0(H)$ stands for the set of all convex, lower semicontinuous, and proper extended real-valued functions $\varphi : H \to \mathbb{R} \cup \{+\infty\}$. The effective domain of $\varphi : H \to \mathbb{R} \cup \{+\infty\}$ is given by

$$\mathrm{Dom}\ (\varphi) = \{u \in H\ :\ \varphi(u) < +\infty\}.$$

For extended real-valued function $\varphi : H \to \mathbb{R} \cup \{+\infty\}$ its epigraph, denoted by $\mathrm{epi}(\varphi)$, is defined by

$$\mathrm{epi}(\varphi) = \{(u, \alpha) \in H \times \mathbb{R}\ :\ \varphi(u) \leq \alpha\}.$$

Definition 3.1 Let $\varphi \in \Gamma_0(H)$. A point $p \in H$ is called a *subgradient* of φ at the point $u \in \mathrm{Dom}\ (\varphi)$ if and only if

$$\varphi(v) \geq \varphi(u) + \langle p, v - u\rangle, \quad \forall v \in H.$$

We denote

$$\partial\varphi(u) = \{p \in H : \varphi(v) \geq \varphi(u) + \langle p, v - u\rangle, \quad \forall v \in H\}$$

and we say that $\partial\varphi(u)$ is the *subdifferential* of φ at u.

This means that elements of $\partial\varphi$ are slopes of the hyperplanes supporting the epigraph of φ at $(u, \varphi(u))$.

Let C be a subset of H. We write $I_C(x)$ for the *indicator function* of C at $x \in H$ defined by

$$I_C(x) = \begin{cases} 0 & \text{if } x \in C, \\ l + \infty & \text{if } x \notin C. \end{cases}$$

Then C is a convex set if and only if I_C is a convex function.

We define the relative interior of $C \subset H$, denoted by $\text{rint}(C)$ as the interior of C within the affine hull of C, i.e.,

$$\text{rint}(C) = \{x \in C \ : \ \exists \varepsilon > 0, \ \mathbb{B}(x, \varepsilon) \cap \text{Aff}(C) \subset C\},$$

where $\text{Aff}(C)$ is the affine hull of C and $\mathbb{B}(x, \varepsilon)$ the ball of radius ε and centered at x.

Suppose that C is a closed convex subset of H. For $\bar{x} \in C$, we define the (outward) *normal cone* at $\bar{x}$ by

$$N_C(\bar{x}) = \{p \in \mathbb{R}^n : \langle p, x - \bar{x}\rangle \leq 0, \quad \forall x \in C\}.$$

If $\bar{x}$ belongs to the interior of C, then $N_C(\bar{x}) = \{0\}$. It is easy to see that

$$N_C(x) = \partial I_C(x), \ \forall x \in H.$$

An important property of the subdifferential in Convex Analysis (established in the Hilbert setting by J.-J. Moreau [19, Proposition 12.b]) concerns its maximal monotonicity:

Theorem 3.1 (J.-J. Moreau [19]) *Let H be a real Hilbert space. Then the subdifferential $\partial\varphi$ of a proper lower semicontinuous and convex function $\varphi : H \to \mathbb{R} \cup \{+\infty\}$ is a maximal monotone operator.*

Remark 3.2

(i) The converse of Theorem 3.1 is not true in general, i.e., a maximal monotone operator is not necessarily the subdifferential of a convex, proper, and lower semicontinuous function. For example, the operator $T = \begin{bmatrix} 0 & -1 \\ 1 & 0 \end{bmatrix}$ is monotone, and hence maximal by Minty's theorem. However, there is no $\varphi \in \Gamma_0(\mathbb{R}^2)$ such that: $T = \nabla\varphi$.

(ii) The converse of Theorem 3.1 is true in dimension one. More precisely, for every maximal monotone operator $T : \mathbb{R} \rightrightarrows \mathbb{R}$, there exists $\varphi \in \Gamma_0(\mathbb{R})$ such that: $T = \partial\varphi$.

Let $y \in \mathbb{R}^n$ be given. We consider the following *variational inequality problem*: find $x \in \mathbb{R}^n$ such that

$$\langle x - y, v - x \rangle + \varphi(v) - \varphi(x) \geq 0, \forall v \in \mathbb{R}^n. \tag{4}$$

This is equivalent to $y - x \in \partial\varphi(x)$, that is, $y \in (I + \partial\varphi)(x)$, i.e., $x = (I + \partial\varphi)^{-1}(y)$. Since the subdifferential of a proper l.s.c. convex function is a maximal monotone operator, we may apply the above theory.

Thus, Problem (4) has a unique solution, that we denote by

$$\text{prox}_{\varphi}(y) = (I + \partial\varphi)^{-1}(y) = \text{J}_{\partial\varphi}(y).$$

The operator $\text{prox}_{\varphi} : \mathbb{R}^n \to \mathbb{R}^n; y \mapsto \text{prox}_{\varphi}(y)$, called the proximal operator, is well-defined on the whole of $\mathbb{R}^n$ and

$$\text{prox}_{\varphi}(\mathbb{R}^n) \subset \text{Dom}(\partial\varphi). \tag{5}$$

For instance, if $\varphi \equiv I_K$, where K is a nonempty closed convex set and I_K denotes the indicator function of K, then

$$\text{prox}_{\varphi} \equiv \text{prox}_{I_K} \equiv \text{proj}_K,$$

where proj_K denotes the projection operator onto K, which is defined by the formula:

$$\|x - \text{proj}_K x\| = \min_{w \in K} \|x - w\|.$$

We may also consider the set-valued operator $\mathcal{A}_{\varphi} : \mathbb{R}^n \rightrightarrows \mathbb{R}^n$ defined by

$$\mathcal{A}_{\varphi}(x) = \{f \in \mathbb{R}^n : \langle x - f, v - x \rangle + \varphi(v) - \varphi(x) \geq 0, \forall v \in \mathbb{R}^n\}. \tag{6}$$

We see that $f \in \mathcal{A}_{\varphi}(x) \iff f - x \in \partial\varphi(x)$ so that

$$\mathcal{A}_{\varphi}^{-1}(f) = \text{prox}_{\varphi}(f). \tag{7}$$

It is also easy to see that

$$(t\mathcal{A}_{\varphi})^{-1}(tf) = \text{prox}_{\varphi}(f), \forall t > 0. \tag{8}$$

Note that

$$\mathcal{A}_{\varphi}(x) = \partial \left\{ \frac{\|.\|^2}{2} + \varphi(.) \right\} (x), \forall x \in \mathbb{R}^n,$$

so that $\mathcal{A}_\varphi$ is a maximal monotone operator. It results that for any $t > 0$, the operator $(I_n + t\mathcal{A}_\varphi)^{-1}$ is a well-defined single-valued operator.

The basic *Linear Complementarity Problem*, LCP for short, is presented in [14] (pp. 1–32) as follows. Consider $H = \mathbb{R}^n$ with the standard scalar product and the order relation $x \geq y$ if and only if $x_i \geq y_i$ for all $1 \leq i \leq n$. This is the order associated to the cone $K = \mathbb{R}^n_+$, since $x \geq y$ if and only if $x - y \in K$. Then the LCP problem denoted by $LCP(q, M)$ requires that, given $q \in \mathbb{R}^n$ and a $n \times n$ matrix M we find z satisfying

$$z \geq 0, \quad q + Mz \geq 0, \quad z^T(q + Mz) = 0. \tag{9}$$

This is often written as

$$0 \leq z \perp q + Mz \geq 0.$$

It is easy to verify that this is equivalent to find z such that

$$-(q + Mz) \in N_K(z). \tag{10}$$

More generally, we may consider the *Nonlinear Complementarity Problem* (NCP(f)) associated in a similar manner to a function f from $\mathbb{R}^n$ into itself: find z such that

$$z \geq 0, \quad f(z) \geq 0, \quad z^T f(z) = 0. \tag{11}$$

This is equivalent to finding z such that

$$-f(z) \in N_K(z), \tag{12}$$

where again $K = \mathbb{R}^n_+$. Since $N_K(z) = \partial I_K(z)$, this is nothing but a special case of a variational inequality of first kind.

The NCP(f) may be generalized as follows. Let K be a closed convex cone in $\mathbb{R}^n$, with dual cone $K^\star$, and f be given as above. Then the *nonlinear complementarity problem over the cone* K, denoted CP(K, f), consists to find z such that

$$z \in K, \quad f(z) \in K^\star, \quad z^T f(z) = 0. \tag{13}$$

If $K = \mathbb{R}^n_+$, then it is self-dual, i. e. $K^\star = K$, and thus CP(K, f) reduces to NCP(f).

An associated problem is the following *variational inequality* $VI(K, f)$: find z such that

$$z \in K, \quad f(z)^T(y - z) \geq 0, \quad \forall\, y \in K. \tag{14}$$

It is worth mentioning that the standard existence result for the LCP is the one due to Samelson, Thrall, and Wesler, which says that (cf. [14, p. 148]):

Theorem 3.2 *The linear complementarity problem LCP*(M, q) : $0 \leq z \perp q + Mz \geq 0$ *has a unique solution for all* $q \in \mathbb{R}^n$ *if and only* M *is a P-matrix, i.e., all its principal minors are positive.*

Notice that if M is symmetric, then M is a P-matrix if and only if it is positive definite. In this situation, the LCP is equivalent to the quadratic minimization problem with bound constraint:

$$\min_{z \geq 0} \frac{1}{2}\langle Mz, z\rangle + \langle q, z\rangle.$$

Let us recall the following existence theorem for a general variational inclusion that will be useful later.

Theorem 3.3 *Let* H *be a real Hilbert space,* $T : H \rightrightarrows H$ *a maximal monotone operator and* $F : H \to H$ *be a Lipschitz continuous and strongly monotone operator, i.e.,* $\exists k > 0,\ \alpha > 0$ *such that for every* $u,\ v \in H$*, we have*

$$\|F(u) - F(v)\| \leq k\|u - v\|,$$

and

$$\langle F(u) - F(v), u - v\rangle \geq \alpha\|u - v\|^2.$$

Then for each $f \in H$*, there is exists a unique* $u \in H$ *such that:*

$$f \in F(u) + T(u), \tag{15}$$

i.e., $\mathrm{Rge}(F + T) = H$.

Proof The existence is a consequence of Corollary 32.25 [22]. The uniqueness follows immediately from the monotonicity of T and the strong monotonicity of F. In fact, for a given $f \in H$, let $u_1 \in H$ and $u_2 \in H$ be two solutions of the inclusion (15). We have $f - F(u_1) \in T(u_1)$ and $f - F(u_2) \in T(u_2)$. By the monotonicity of T, we have

$$\langle (f - F(u_1)) - (f - F(u_2), u_1 - u_2\rangle \geq 0.$$

Hence,

$$\langle F(u_2) - F(u_1), u_1 - u_2\rangle \geq 0.$$

Using the strong monotonicity of F, we get

$$\alpha\|u_1 - u_2\|^2 \leq \langle F(u_1) - F(u_2), u_1 - u_2\rangle \leq 0.$$

Hence, $u_1 = u_2$, which proves the uniqueness. ∎

The following existence and uniqueness result is essentially a consequence of Kato's theorem [17]. We refer the reader to [15, Corollary 2.2] for more details.

Theorem 3.4 *Let H be a real Hilbert space, $T : H \rightrightarrows H$ a maximal monotone operator and $F : H \to H$ be a hemicontinuous operator such that for some $\omega \geq 0$, $F + \omega I$ is monotone. Let $x_0 \in \mathrm{Dom}(T)$ be given. Then there exists a unique $x \in C^0([0, +\infty[; H)$ such that*

$$\begin{cases} \dot{x} \in L^{\infty}_{\mathrm{loc}}([0, +\infty[; H), & (16a) \\ x \text{ is right-differentiable on } [0, +\infty[, & (16b) \\ x(0) = x_0, & (16c) \\ x(t) \in \mathrm{Dom}(T),\ t \geq 0, & (16d) \\ \dot{x}(t) + F(x(t)) + T(x(t)) \ni 0,\ a.e.\ t \geq 0. & (16e) \end{cases}$$

4 Well-Posedness of Nonsmooth Lurie Systems

Let us first rewrite the nonsmooth system as a differential inclusion. We have

$$\begin{aligned} q \in \mathcal{F}(-p) &\Longleftrightarrow Cx + Dp \in \mathcal{F}(-p), \\ &\Longleftrightarrow Cx \in (D + \mathcal{F})(-p), \\ &\Longleftrightarrow -p \in (D + \mathcal{F})^{-1}(Cx). \end{aligned}$$

Hence,

$$\dot{x} \in Ax + Bp \Longleftrightarrow \dot{x} \in Ax - B(D + \mathcal{F})^{-1}(Cx).$$

Consequently problem $(\mathcal{L})$ is equivalent to the following differential inclusion: given an initial point $x_0 \in \mathbb{R}^n$, find a function $x \in W^{1,\infty}_{loc}(0, +\infty)$ such that

$$\dot{x}(t) \in Ax(t) - B(D + \mathcal{F})^{-1}(Cx(t)),\ t \geq 0. \tag{17}$$

In what follows I_p stands for the identity matrix of order $p \in \mathbb{N}^*$.

Let us suppose that the operator $\mathcal{F} : \mathbb{R}^m \rightrightarrows \mathbb{R}^m$ is a maximal monotone operator. By Proposition 3.1, the resolvent operator defined by

$$\mathrm{J}_{\mathcal{F}} : \mathbb{R}^m \to \mathbb{R}^m,\ \ \mapsto \mathrm{J}_{\mathcal{F}}(x) = (I_m + \mathcal{F})^{-1}(x),$$

is single-valued and Lipschitz continuous with modulus 1, i.e.,

$$\|\mathrm{J}_{\mathcal{F}}(x) - \mathrm{J}_{\mathcal{F}}(y)\| \leq \|x - y\|, \quad \forall x, \ y \in \mathbb{R}^m.$$

The single-valuedness of $\mathrm{J}_{\mathcal{F}}$ means the for every $y \in \mathbb{R}^m$, the following inclusion

$$x + \mathcal{F}(x) \ni y$$

has a unique solution x (which depends on y).

The main question we treat now is the following: under which conditions on the matrix $D \in \mathbb{R}^{m \times m}$, the following operator

$$\mathrm{J}_{D,\mathcal{F}} = (D + \mathcal{F})^{-1} \tag{18}$$

is single-valued and Lipschitz continuous. We give in Examples 4.1 and 4.2 some situations where the computation of the operator $\mathrm{J}_{D,\mathcal{F}}$ is possible.

Proposition 4.1 *Suppose that the matrix $D \in \mathbb{R}^{m\times m}$ is positive definite and that $\mathcal{F} : \mathbb{R}^m \rightrightarrows \mathbb{R}^m$ is a maximal monotone operator. Then the operator $\mathrm{J}_{D,\mathcal{F}} = (D + \mathcal{F})^{-1}$ is well-defined, single-valued, and Lipschitz continuous with modulus $L = \frac{2}{\lambda_1(D+D^T)}$, where $\lambda_1(D + D^T)$ is the smallest eigenvalue of the matrix $D + D^T$.*

Proof Since D is positive definite (hence strongly monotone) and $\mathcal{F}$ is maximal monotone, by Theorem 3.3 for every $y \in \mathbb{R}^m$, there exists a unique $x \in \mathbb{R}^m$ such that

$$y \in Dx + \mathcal{F}(x).$$

Hence the operator $(D + \mathcal{F})^{-1}$ is well-defined and single-valued. Let us show that $(D + \mathcal{F})^{-1}$ is Lipschitz continuous. Let $y_1, \ y_2 \in \mathbb{R}^m$ and set $x_1 = (D + \mathcal{F})^{-1}(y_1)$ and $x_2 = (D + \mathcal{F})^{-1}(y_2)$. We have

$$Dx_1 + \mathcal{F}(x_1) \ni y_1 \text{ and } Dx_2 + \mathcal{F}(x_2) \ni y_2.$$

Using the fact $\mathcal{F}$ is monotone, we get

$$\langle (y_1 - Dx_1) - (y_2 - Dx_2), x_1 - x_2\rangle \geq 0.$$

Hence,

$$\langle D(x_1 - x_2), x_1 - x_2\rangle \leq \langle y_1 - y_2, x_1 - x_2\rangle.$$

Since the matrix D is positive definite, we have

$$\langle Dx, x\rangle \geq \frac{\lambda_1(D + D^T)}{2}\|x\|^2, \ \forall x \in \mathbb{R}^m.$$

Therefore,

$$\begin{aligned}\frac{\lambda_1(D+D^T)}{2}\|x_1-x_2\|^2 &\le \langle Dx_1 - Dx_2, x_1 - x_2\rangle \\ &\le \langle y_1 - y_2, x_1 - x_2\rangle \\ &\le \|y_1 - y_2\|\,\|x_1 - x_2\|.\end{aligned}$$

Consequently,

$$\|x_1 - x_2\| \le \frac{2}{\lambda_1(D+D^T)}\|y_1 - y_2\|,$$

which means that the operator $\mathrm{J}_{D,\mathcal{F}} = (D+\mathcal{F})^{-1}$ is Lipschitz continuous with modulus $L = \frac{2}{\lambda_1(D+D^T)}$. ■

Remark 4.1

(i) If the matrix D is only positive semidefinite, then the resolvent operator $\mathrm{J}_{D,\mathcal{F}} = (D+\mathcal{F})^{-1}$ may be set-valued. Let us consider the following simple example:

$$D = \begin{bmatrix} 0 & 0 \\ 0 & 1 \end{bmatrix} \text{ and } \mathcal{F} : \mathbb{R}^2 \rightrightarrows \mathbb{R}^2,\ x = (x_1, x_2) \mapsto \mathcal{F}(x) = (\mathrm{Sign}(x_1), \mathrm{Sign}(x_2)),$$

where Sign : $\mathbb{R} \rightrightarrows \mathbb{R},\ \alpha \mapsto \mathrm{Sign}(\alpha)$ is defined by

$$\mathrm{Sign}(\alpha) = \partial|\cdot|(\alpha) = \begin{cases} 1 & \text{if } \alpha > 0, \\ [-1, 1] & \text{if } \alpha = 0, \\ -1 & \text{if } \alpha < 0. \end{cases}$$

We note that $\mathcal{F}(\cdot)$ is a maximal monotone operator defined on $\mathbb{R}^2$ and that the matrix D is positive semidefinite.

It is easy to see that the inverse (in the set-valued sense) of the operator Sign is defined by $\mathcal{S} : \mathbb{R} \rightrightarrows \mathbb{R},\ \alpha \mapsto \mathcal{S}(\alpha) := \mathrm{Sign}^{-1}(\alpha)$ (see Fig. 4):

$$\mathcal{S}(\alpha) = \begin{cases} \emptyset & \text{if } \alpha < -1, \\ ]-\infty, 0] & \text{if } \alpha = -1, \\ 0 & \text{if } \alpha \in]-1, 1[, \\ [0, +\infty[& \text{if } \alpha = 1, \\ \emptyset & \text{if } \alpha > 1. \end{cases} \tag{19}$$

The inverse of the operator (1 + Sign) is well-known as the soft thresholding operator $\mathcal{T}(\cdot)$ which is used in the FISTA method for sparse optimization (see Fig. 5):

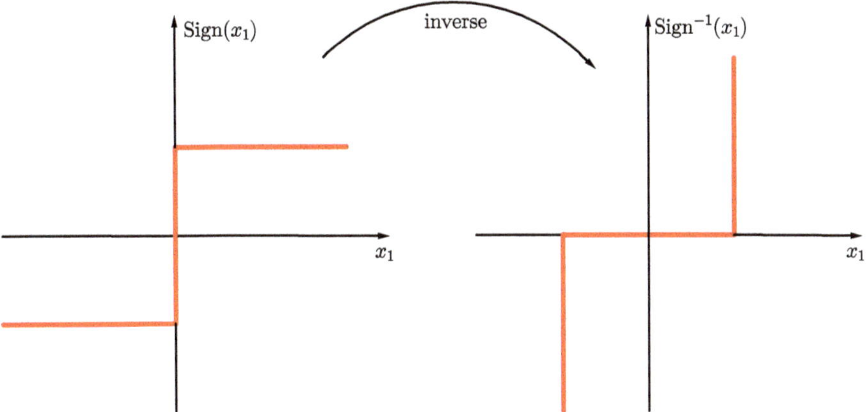

Fig. 4 Inverse operator of Sign(·)

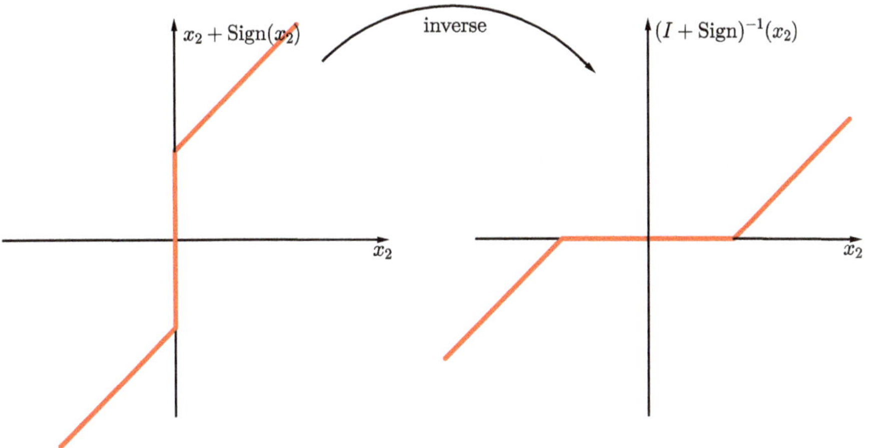

Fig. 5 Inverse operator of 1 + Sign(·)

$$\mathcal{T}(\alpha) = \text{sign}(\alpha)(|\alpha| - 1)_+ = \begin{cases} \alpha - 1 & \text{if } \alpha \geq 1, \\ 0 & \text{if } -1 \leq \alpha \leq 1, \\ \alpha + 1 & \text{if } \alpha \leq -1. \end{cases} \tag{20}$$

We note that $\mathcal{T}$ is a single-valued while $\mathcal{S}$ is set-valued operator defined on $\mathbb{R}$.

Let us compute $D + \mathcal{F}$ and its inverse. For every $x = (x_1, x_2) \in \mathbb{R}^2$, we have

$$\begin{aligned} y = (y_1, y_2) \in (D + \mathcal{F})(x) &\iff y_1 \in \text{Sign}(x_1) \text{ and } y_2 \in x_2 + \text{Sign}(x_2) \\ &\iff x_1 \in \text{Sign}^{-1}(y_1) \text{ and } x_2 = (1 + \text{Sign})^{-1}(y_2). \\ &\iff (x_1, x_2) \in (\mathcal{S}(y_1), \mathcal{T}(y_2)). \end{aligned}$$

Hence,

$$\mathrm{J}_{D,\mathcal{F}} = (D+\mathcal{F})^{-1} = (\mathcal{S}(\cdot), \mathcal{T}(\cdot)),$$

where $\mathcal{S}$ and $\mathcal{T}$ are defined, respectively, in (19) and (20). Since $\mathcal{S}$ is a set-valued operator, we conclude that $(D+\mathcal{F})^{-1}$ is also a set-valued operator.

(ii) Let us consider the following matrix D and operator $\mathcal{F} : \mathbb{R}^2 \to \mathbb{R}^2$ defined, respectively, by

$$D = \begin{bmatrix} -1 & 0 \\ 0 & 1 \end{bmatrix} \text{ and } \mathcal{F}(x_1, x_2) = \begin{bmatrix} -x_2 \\ x_1 \end{bmatrix}.$$

It is clear that $\mathcal{F}$ is a maximal monotone operator and that the inverse $D+\mathcal{F}$ does not exist.

Example 4.1 In $\mathbb{R}^m$, let us consider the following $\mathcal{F} : \mathbb{R}^m \rightrightarrows \mathbb{R}^m$, defined by

$$x = (x_1, \ldots, x_m) \mapsto \mathcal{F}(x) = \Big(\mathcal{F}_1(x_1), \ldots, \mathcal{F}_m(x_m)\Big),$$

where $\mathcal{F}_i : \mathbb{R} \rightrightarrows \mathbb{R},\ x_i \mapsto \mathcal{F}_i(x_i) = \mathrm{Sign}(x_i),\ i = 1, 2, \ldots, m$. If we take $D = I_m$, then the resolvent operator $\mathrm{J}_{D,\mathcal{F}} = \mathrm{J}_{\mathcal{F}}$ can be computed componentwise by applying the one-dimensional soft thresholding operator $\mathcal{T}$ in (20) to each component. Hence for every $x = (x_1, \ldots, x_m),\ y = (y_1, \ldots, y_m) \in \mathbb{R}^m$, we have

$$y \in (I_m + \mathcal{F})(x) \iff x_i = \mathcal{T}(y_i) = \mathrm{sign}(y_i)(|y_i| - 1)_+,\ i = 1, 2, \ldots m.$$

Hence,

$$\mathrm{J}_{\mathcal{F}}(y) = (I_m + \mathcal{F})^{-1}(y) = \big(\mathcal{T}(y_1), \ldots, \mathcal{T}(y_m)\big). \tag{21}$$

Example 4.2 Let $\varphi : \mathbb{R}^m \to \mathbb{R}$ given by $\varphi(x) = r\|x\|_2$, with $r > 0$. We have

$$\partial\varphi(x) = \begin{cases} r\frac{x}{\|x\|} & \text{if } x \neq 0, \\ \mathbb{B}(0, r) & \text{if } x = 0. \end{cases} \tag{22}$$

It is easy to show that for $D = I_m$ and $\mathcal{F} = \partial\varphi$, we have,

$$\begin{aligned} \mathrm{J}_{\partial\varphi} &= (I_m + \partial\varphi)^{-1}(x) \\ &= \Big(1 - \frac{r}{\max\{r, \|x\|\}}\Big)x \\ &= \begin{cases} 0 & \text{if } \|x\| \leq r, \\ (\|x\| - r)\frac{x}{\|x\|} & \text{if } \|x\| \geq r. \end{cases} \end{aligned} \tag{23}$$

We have the following existence and uniqueness result.

Theorem 4.1 *Assume that $\mathcal{F} : \mathbb{R}^m \rightrightarrows \mathbb{R}^m$ is a maximal monotone operator and that the matrix $D \in \mathbb{R}^{m\times m}$ is positive definite. Then for each $x_0 \in \mathbb{R}^n$, there exists a unique function $x \in W^{1,\infty}_{loc}(0, +\infty)$ such that*

$$\begin{cases} \dot{x}(t) = Ax(t) + Bp(t), \text{ a.e. } t \in [0, +\infty[, \\ q(t) = Cx(t) + Dp(t), \\ q(t) \in \mathcal{F}(-p(t)), \ t \geq 0, \\ x(0) = x_0. \end{cases}$$

Proof Since $\mathcal{F} : \mathbb{R}^m \rightrightarrows \mathbb{R}^m$ is a maximal monotone operator and that the matrix D is positive definite, by Proposition 4.1 the generalized resolvent $\mathrm{J}_{D,\mathcal{F}} = (D + \mathcal{F})^{-1}$ is well-defined, single-valued, and Lipschitz continuous. Using Eq. (17) and the notation (18), it is clear that the system $(\mathcal{L})$ is equivalent to find a trajectory satisfying the following ordinary differential equation

$$\dot{x}(t) = Ax(t) - B\mathrm{J}_{D,\mathcal{F}}(Cx(t)), \ t \geq 0. \tag{24}$$

Let us defined the following vector field $f : \mathbb{R}^n \to \mathbb{R}^n, \ x \mapsto f(x)$ associated to (24)

$$f(x) = Ax - B\mathrm{J}_{D,\mathcal{F}}(Cx).$$

It is easy to see that $f(\cdot)$ is Lipschitz continuous. Hence by the classical Cauchy-Lipschitz theorem, for each $x_0 \in \mathbb{R}^n$ there exists a unique trajectory $x(\cdot)$ satisfying (24). The proof is thereby completed. ∎

In the particular case where $\mathcal{F}$ coincides with the normal cone to $\mathbb{R}^m_+$, i.e., $\mathcal{F} : \mathbb{R}^m \rightrightarrows \mathbb{R}^m, \ x \mapsto \mathcal{F}(x) = \mathrm{N}_{\mathbb{R}^m_+}(x)$, we can relax the assumption on the matrix D.

Proposition 4.2 *Suppose that $D \in \mathbb{R}^{m\times m}$ is a P-matrix and that $\mathcal{F} = \mathrm{N}_{\mathbb{R}^m_+}$. Then the operator $\mathrm{J}_{D,\mathcal{F}} = (D + \mathrm{N}_{\mathbb{R}^m_+})^{-1}$ is well-defined, single-valued, and Lipchitz continuous.*

Proof By (10), $z = (D + \mathrm{N}_{\mathbb{R}^m_+})^{-1}(q)$ if and only if z is a solution of the linear complementarity problem LCP$(D, -q)$. Since D is a P-matrix, by Theorem 3.2, the LCP$(D, -q)$ has a unique solution for every $q \in \mathbb{R}^m$. The Lipschitz continuity property the solution map of a LCP is well-known in the literature (see, e.g., [14]). ∎

We have the following existence result for the Lurie system involving the normal cone to the nonnegative orthant $\mathbb{R}^m_+$ as a set-valued law.

Theorem 4.2 *Assume that the matrix $D \in \mathbb{R}^{m\times m}$ is a P-matrix. Then for each $x_0 \in \mathbb{R}^n$, there exists a unique function $x \in W^{1,\infty}_{loc}(0, +\infty)$ such that*

$$\begin{cases} \dot{x}(t) = Ax(t) + Bp(t) \text{ a.e. } t \in [0, +\infty[, \\ q(t) = Cx(t) + Dp(t), \\ q(t) \in \mathrm{N}_{\mathbb{R}^m_+}(-p(t)),\ t \geq 0, \\ x(0) = x_0. \end{cases}$$

Proof Use Proposition 4.2 and the same argument as in the proof of Theorem 4.1. ■

In this part, we deal with the general case, i.e., the matrix D is positive semidefinite but we will impose some other assumptions on the matrices $B \in \mathbb{R}^{n\times m}$ and $C \in \mathbb{R}^{m\times n}$. By introducing the new set-valued map $\mathcal{G} : \mathbb{R}^m \rightrightarrows \mathbb{R}^m,\ x \mapsto \mathcal{G}(x) = (D + \mathcal{F})^{-1}(x)$, the last inclusion is of the form

$$\dot{x}(t) \in Ax(t) - B\mathcal{G}(Cx(t)).$$

We will give now conditions on the data $(B, C, \mathcal{F})$ such that the set-valued map under composition $B\mathcal{G}C$ is maximal monotone. Let us recall the following theorem (see Theorem 12.43 page 556 in [20]).

Theorem 4.3 *Suppose that $T(x) = M^T S(Mx)$ for a given maximal monotone operator $S : \mathbb{R}^m \rightrightarrows \mathbb{R}^m$ and a given matrix $M \in \mathbb{R}^{m\times n}$.*

If $\mathrm{Rge}(M) \cap \mathrm{rint}(\mathrm{Dom}(S)) \neq \emptyset$, *then the operator T is maximal monotone.*

The following lemma will be useful.

Lemma 4.1 *Let $S : \mathbb{R}^m \rightrightarrows \mathbb{R}^m$ be a set-valued map and $M \in \mathbb{R}^{m\times n}$ be a matrix. Let $T : \mathbb{R}^n \rightrightarrows \mathbb{R}^n$ defined by: $x \mapsto T(x) = M^T S(Mx)$. Then*

$$\mathrm{Dom}\,(T) = M^{-1}(\mathrm{Dom}(S)),$$

where $M^{-1}(K) = \{x \in \mathbb{R}^n \ :\ Mx \in K\}$ for a given nonempty set $K \subset \mathbb{R}^m$.

Proof We have,

$$\begin{aligned} x \in \mathrm{Dom}(T) &\iff T(x) \neq \emptyset, \\ &\iff M^T S(Mx) \neq \emptyset, \\ &\iff S(Mx) \neq \emptyset, \\ &\iff Mx \in \mathrm{Dom}(S), \\ &\iff x \in M^{-1}(\mathrm{Dom}(S)). \end{aligned}$$

Hence,

$$\mathrm{Dom}\,(T) = M^{-1}(\mathrm{Dom}(S)),$$

which completes the proof of Lemma 4.1. ■

Proposition 4.3 *Assume that there exists a matrix* $P \in \mathbb{R}^{n\times n}$, $P = P^T > 0$ *such that* $PB = C^T$ *and* $\mathcal{F} : \mathbb{R}^m \rightrightarrows \mathbb{R}^m$ *is a given set-valued map. Then the dynamic* (1a)–(1c) *is equivalent to*

$$\dot{z}(t) \in RAR^{-1}z(t) - R^{-1}C^T(D+\mathcal{F})^{-1}(CR^{-1}z(t)), \tag{25}$$

with $R = P^{\frac{1}{2}}$ *and* $z(t) = Rx(t)$.

Proof Since there exists a symmetric and positive definite matrix P such that: $PB = C^T$, we set $R = P^{\frac{1}{2}}$. Hence,

$$PB = C^T \iff RB = R^{-1}C^T. \tag{26}$$

We set $z(t) = Rx(t)$, which means that $\dot{z}(t) = R\dot{x}(t)$. The inclusion (17) is equivalent to

$$\dot{z}(t) \in RAR^{-1}z(t) - R^{-1}C^T(D+\mathcal{F})^{-1}(CR^{-1}z(t)),$$

which completes the proof. ■

Remark 4.2 The existence of a matrix $P = P^T > 0$ such that $PB = C^T$ can be linked to the famous Kalman-Yakubovich-Popov Lemma (see [2]).

We have the following well-posedness result for the nonsmooth Lurie system (L) involving a maximal monotone set-valued law.

Theorem 4.4 *Assume that there exists a matrix* $R \in \mathbb{R}^{n\times n}$, $R = R^T > 0$ *such that* $R^2B = C^T$, *the matrix* $D \in \mathbb{R}^{m\times m}$ *is positive semidefinite and that* $\mathcal{F} : \mathbb{R}^m \rightrightarrows \mathbb{R}^m$ *is a maximal monotone operator. Assume also that*

$$\mathrm{Rge}(CR^{-1}) \cap \mathrm{rint}\Big(\mathrm{Rge}(D+\mathcal{F})\Big) \neq \emptyset. \tag{27}$$

Then for each $x_0 \in RC^{-1}\Big(\mathrm{Rge}(D+\mathcal{F})\Big)$, *there exists a unique function* $x \in W^{1,\infty}_{loc}(0,+\infty)$ *such that*

$$\begin{cases} \dot{x}(t) = Ax(t) + Bp(t), \text{ a.e. } t \in [0,+\infty[,\\ q(t) = Cx(t) + Dp(t),\\ q(t) \in \mathcal{F}(-p(t)),\ t \geq 0,\\ x(0) = x_0. \end{cases}$$

Proof By Proposition 4.3, the dynamic (1a)–(1c) is equivalent to

$$\dot{z}(t) \in RAR^{-1}z(t) - R^{-1}C^T(D+\mathcal{F})^{-1}(CR^{-1}z(t)),$$

with $z(t) = Rx(t)$. We use Theorem 4.3 to show that the operator $T : \mathbb{R}^n \rightrightarrows \mathbb{R}^n$ under composition defined by

$$T(z) = R^{-1}C^T(D+\mathcal{F})^{-1}(CR^{-1}z)$$

is maximal monotone. We set $M = CR^{-1}$ and $S = (D+\mathcal{F})^{-1}$. By Lemma 4.1, we have

$$\mathrm{Dom}(T) = M^{-1}(\mathrm{Dom}(S)) = RC^{-1}\Big(\mathrm{Rge}(D+\mathcal{F})\Big).$$

Since, in one hand the matrix D is positive semidefinite (hence monotone and continuous on $\mathbb{R}^m$) and the other hand $\mathcal{F} : \mathbb{R}^m \rightrightarrows \mathbb{R}^m$ is a maximal monotone operator, we deduce that the sum $(D + \mathcal{F})$ is also maximal monotone (see Remark 3.1). Consequently the inverse operator $S = (D + \mathcal{F})^{-1} : \mathbb{R}^m \rightrightarrows \mathbb{R}^m$ is maximal monotone. Using (3), we have

$$\mathrm{Dom}(S) = \mathrm{Rge}(D+\mathcal{F}).$$

Using (27), we deduce that T is a maximal monotone operator. Since the perturbation term $z \mapsto RAR^{-1}z$ is Lipschitz continuous, the existence and uniqueness of a trajectory for a given $x_0 \in RC^{-1}\Big(\mathrm{Rge}(D+\mathcal{F})\Big)$ follows from Theorem 3.4. The proof is thereby completed. ■

Remark 4.3

(i) In some applications, the matrices $B \in \mathbb{R}^{n\times m}$ and $C \in \mathbb{R}^{m\times n}$ are transpose to each other, i.e., $B = C^T$. In this case it is sufficient to set $R = I_n$ in Theorem 4.4.
(ii) If the matrix $C = B^T$ is surjective (or B is injective), then condition (27) is satisfied.

Example 4.3 We give some examples of the data A, B, C, D, and $\mathcal{F}$ satisfying conditions of Theorem 4.4.

(i) Let consider

$$A = \begin{bmatrix} 0 & -1 \\ 1 & -1 \end{bmatrix}, \quad B = \begin{bmatrix} 0 & -1 \\ 0 & 0 \end{bmatrix}, \quad C = \begin{bmatrix} 0 & 0 \\ -1 & 0 \end{bmatrix}, \quad D = \begin{bmatrix} 0 & 0 \\ 0 & 1 \end{bmatrix}.$$

Let us consider the following set-valued map $\mathcal{F} : \mathbb{R}^2 \rightrightarrows \mathbb{R}^2$, $x = (x_1, x_2) \mapsto \mathcal{F}(x) = \Big(\mathcal{F}_1(x_1), \mathcal{F}_2(x_2)\Big)$ given by

$$\mathcal{F}_1(x_1) = \begin{cases} \alpha & \text{if } x_1 < 0 \\ [\alpha, \beta] & \text{if } x_1 = 0 \\ \beta & \text{if } x_1 > 0 \end{cases} \text{ and } \mathcal{F}_2(x_2) = \partial(|\cdot|)(x_2) = \mathrm{Sign}(x_2),$$

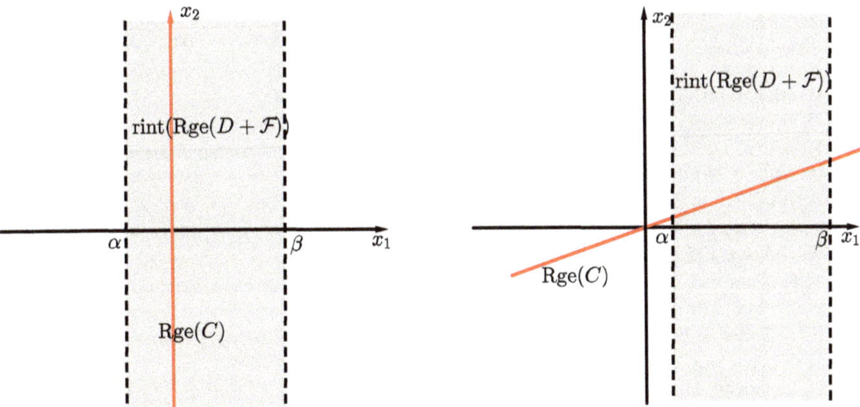

Fig. 6 Illustration of condition (27) with Example 4.3

with $\alpha,\ \beta \in \mathbb{R}$ two given parameters such that $\alpha \le \beta$.

Since $B = C^T$, according to Remark 4.3, we can take $R = I_2$. One has

$$\text{Rge}(C) = \text{span}\,\{(0, 1)\} \text{ and } \text{Rge}(D + \mathcal{F}) = [\alpha, \beta] \times \mathbb{R}.$$

It is easy to see that condition (27) is satisfied if and only if $\alpha < 0$ and $\beta > 0$.

(ii) If we take, for example, $C = \begin{bmatrix} a & 0 \\ -1 & 0 \end{bmatrix} = B^T$, with $a \in \mathbb{R} \setminus \{0\}$ and the other data as in (i). One can easily check that the condition (27) is satisfied for every $\alpha,\ \beta \in \mathbb{R}$ with $\alpha \le \beta$ (Fig. 6).

References

1. V. Acary and B. Brogliato. *Numerical Methods for Nonsmooth Dynamical Systems*. Applications in Mechanics and Electronics. Springer Verlag, LNACM 35, (2008).
2. K. Addi, S. Adly, B. Brogliato and D. Goeleven. *A method using the approach of Moreau and Panagiotopoulos for the mathematical formulation of non-regular circuits in electronics*. Nonlinear Analysis C: Hybrid Systems and Applications, **1** pp. 30–43, (2007).
3. S. Adly, D. Goeleven, *A stability theory for second-order nonsmooth dynamical systems with applications to friction problems*, Journal de Mathématiques Pures et Appliquées, **83**, 17–51, (2004).
4. S. Adly, *A Variational Approach to Nonsmooth Dynamics: Applications in Unilateral Mechanics and Electronics*. Springerbriefs (2017).
5. S. Adly. *Attractivity theory for second order non-smooth dynamical systems with application to dry friction*. J. of Math. Anal. and Appl. 322, 1055–1070, (2006).
6. S. Adly and B.K. Le. *Stability and invariance results for a class of non-monotone set-valued Lur'e dynamical systems*, Applicable Analysis, vol. 93, iss. 5, 1087–1105, (2014).
7. S. Adly, A. Hantoute and B.K. Le. *Nonsmooth Lur'e dynamical systems in Hilbert spaces*. Set-Valued Var. Anal. 24, no. 1, 13–35, (2016).

8. S. Adly, A. Hantoute and B.K. Le. *Maximal monotonicity and cyclic monotonicity arising in nonsmooth Lur'e dynamical systems*. J. Math. Anal. Appl. 448, no. 1, 691–706, (2017).
9. B. Brogliato. *Absolute stability and the Lagrange-Dirichlet theorem with monotone multivalued mappings*. Systems & Control Letters 51, 343–353, (2004).
10. H. Brezis, *Analyse fonctionnelle*, Masson, (1983).
11. H. Brezis, *Opérateurs maximaux monotones et semi groupes de contractions dans les espaces de Hilbert*, North Holland, (1973).
12. B. Brogliato and D. Goeleven. *Well-posedness, stability and invariance results for a class of multivalued Lur'e dynamical systems*, Nonlinear Analysis: Theory, Methods and Applications, vol. 74, pp. 195–212, (2011).
13. M.K. Camlibel and J.M. Schumacher. *Linear passive systems and maximal monotone mappings*, Math. Program. 157 (2016), no. 2, Ser. B, 397420.
14. R.W. Cottle, J.-S. Pang and R.E. Stone, *The linear complementarity problem*, Academic Press, (1992).
15. D. Goeleven, D. Motreanu and V. Motreanu. *On the stability of stationary solutions of first-order evolution variational inequalities*, Adv. Nonlinear Var. Inequal. 6 pp 1–30, (2003).
16. B. Brogliato, R. Lozano, B. Maschke and O. Egeland. *Dissipative Systems Analysis and Control*, Springer-Verlag London, 2nd Edition, (2007).
17. T. Kato. *Accretive Operators and Nonlinear Evolutions Equations in Banach Spaces, Part 1, Nonlinear Functional Analysis, in : Proc. Sympos. Pure Math., Amer. Math. Society,* **18** (1970).
18. A.I. Lur'e and V.N. Postnikov. *On the theory of stability of control systems*, Applied mathematics and mechanics, 8(3), (in Russian), (1944).
19. J.J. Moreau. *Proximité et dualité dans un espace hilbertien,* Bull. Soc. Math. France, 93 (1965), 273–299.
20. R.T. Rockafellar and R.B. Wets. *Variational Analysis*. Springer, Berlin, (1998).
21. J.M. Schumacher. *Complementarity systems in optimization*. Math. Program. 101, no. 1, Ser. B, 263–295, (2004).
22. E. Zeidler. *Nonlinear functional analysis and its applications. II/B. Nonlinear monotone operators*. Springer-Verlag (1990).

Numerical Method for Calculation of Unsteady Fluid Flow Regimes in Hydraulic Networks of Complex Structure

K. R. Aida-zade and Y. R. Ashrafova

Abstract The problem of calculating flow regimes of transient processes in complex hydraulic networks with loops is considered in the chapter. Fluid flow in each linear segment of pipeline network is described by a system of two linear partial differential equations of the first order. Non-separated boundary conditions are satisfied at the nodes of the network. These conditions are determined by the first Kirchhoff's law and by the continuity of flow. The scheme of numerical solution to the problem based on the application of grid method is suggested. The formulas analogous to the formulas of sweep method are derived. The obtained formulas are independent of the number of nodes, segments, and structure of the pipeline network. Numerical experiments are carried out with the use of the suggested approach, and the obtained results are analyzed.

1 Introduction

It's known that one of the most important technological processes in oil and gas sector is the transportation of hydrocarbons from deposits to the places of its processing with further delivery to customers. Pipeline transport networks of complex loopback structure are used for these purposes. The development of the networks of trunk pipelines is associated with the wide use of modern facilities of measuring, computing, and remote control systems. The most difficult situations arise in the systems of pipeline transportation of hydrocarbon raw material when it

K. R. Aida-zade (✉)
Department of Numerical decision-making methods in deterministic systems, Institute of Control Systems of Azerbaijan National Academy, Baku, Azerbaijan

Y. R. Ashrafova
Department of Numerical decision-making methods in deterministic systems, Institute of Control Systems of Azerbaijan National Academy, Baku, Azerbaijan

Department of Applied Mathematics, Baku State University, Baku, Azerbaijan

I. N. Parasidis et al. (eds.), *Mathematical Analysis in Interdisciplinary Research*, Springer Optimization and Its Applications 179,
https://doi.org/10.1007/978-3-030-84721-0_5

is necessary to change the pumping regimes which initiate transition processes in the pipelines [1–14]. Such regimes are undesirable for equipment of the pipeline systems (in particular, for pumping, compressor stations). They can result in the break of the pipelines due to the fluid hammer caused by incorrect control over transition process, when switching from the transient regime to the steady-state regime.

We study the problem of calculating flow regimes of transient processes in complex pipeline networks as opposed to many cases of the considered problem-solving exclusively for separate linear segments of the pipeline, investigated in [6–13] earlier. The number of linear sections can reach up to several tens and even hundreds of real pipeline transport networks. In the work, we propose to use the graph theory, which drastically simplifies the process of computer modeling of complex pipeline networks. We use the system of two linear partial differential equations of hyperbolic type to describe the process of fluid flow in every linear segment of the pipeline network. There are non-separated boundary conditions [15, 16] at the points of connection of linear segments, which makes it impossible to calculate the flow regimes for every individual segment separately; it must be done for all segments simultaneously. To solve the problem, we suggest to use the methods of finite difference approximation (grid method) and special scheme of sweep (transfer) method to approximate a set of systems of hyperbolic differential equations interrelated by boundary conditions. Satisfactory results of numerical experiments are given. In order to switch from the grid problem with non-separated conditions to a problem with separated boundary conditions, we obtain the dependencies between boundary values of sought-for functions for every linear section. We obtained the formulas for the procedure of obtaining such dependencies, called a sweep (transfer) method, that use the difference equations only for that section, for which is made the dependencies.

2 Statement of the Problem

We consider the problem of calculating laminar flow regimes in complex pipe networks for the sufficient general case of complex-shaped hydraulic network with loops containing M segments and N nodes. Sometimes it is difficult to decide which direction fluid will flow. The direction of flow is often obvious, but when it is not, flow direction has to be assumed. We assume that the flow directions in the segments are given a priori. If he calculated value of flow rate is greater than zero in a segment, it means that actual flow direction in this segment coincides with the given direction. If the calculated value of the flow rate is less than zero, then actual flow direction is opposite to the given one.

To simplify presentation of numerical schemes and to be specific, let us consider the pipe network, containing eight segments as shown in Fig. 1.

Numbers in brackets identify the nodes (or junctions). We denote the set of nodes by I: $I = \{k_1, \ldots, k_N\}$; where $k_i, i = \overline{1, N}$ are the nodes; $N = |I|$ is the

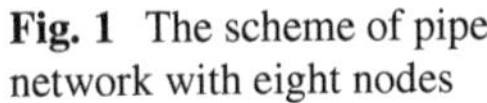

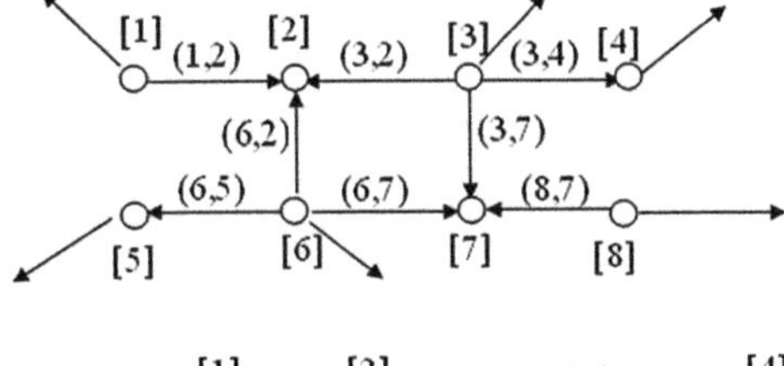

Fig. 1 The scheme of pipe network with eight nodes

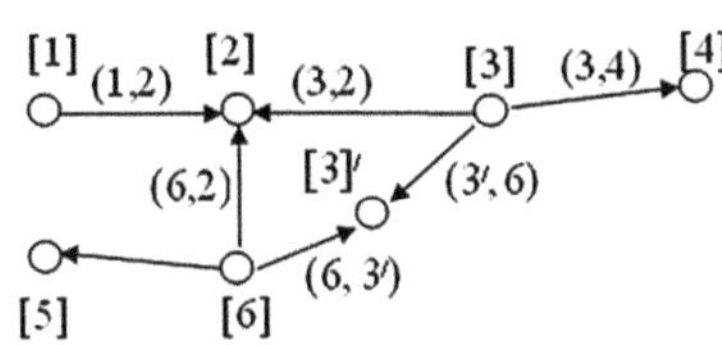

Fig. 2 The case of introduction of additional node

numbers of nodes in the network. Two numbers in parentheses identify two-index numbers of segments. The flow in these segments goes from the first index to the second (e.g., the flow in segment (1,2) is obviously from node 1 to node 2. Let J: $J = \{(k_i, k_j) : k_i, k_j \in I\}$ denote the set of segments and $M = |J|$ denote its quantity; $l_{k_i k_j}, d_{k_i k_j}, k_i, k_j \in I$ is the length and diameter of the segment (k_i, k_j), respectively.

Let I_k^+ denote the set of nodes connected with node k by segments where flow goes into the node, and let I_k^- denote the set of nodes connected with node k by segments where flow goes out of the node; $I_k = I_k^+ \cup I_k^-$ denotes the set of total nodes connected with node k and $N_k = |I_k|$, $N_{k^+} = \left|I_k^+\right|$, $N_{k^-} = \left|I_k^-\right|$, $N_k = N_{k^-} + N_{k^+}$. The arrows in Fig. 1 formally indicate the assumed (not actual) direction of the fluid flow.

Note that the arrows indicating direction of flow are selected so that each node may be only inflow or outflow. Such selection facilitates in certain extent the calculating process, but it is not principal (the reason for this will be clear further).

Such an appointment of directions can be done for any structure of the pipeline network, at the expense of artificial introduction of an additional node and partition of the whole segment into two sections in the case of necessity. For example, in the segment (6, 3), we introduced a new node [3′] as shown in Fig. 2, and the segment is divided into two sections: (6, 3′) and (3′, 6).

Besides the inflows and outflows in the segments of the network, there can be external inflows (sources) and outflows (sinks) with the rate $\tilde{q}_i(t)$ at some nodes $i \in I$ of the network. Positive and negative values of $\tilde{q}_i(t)$ indicate the existence of external inflow or outflow at the node i. However, in general case, assuming that the case $\tilde{q}_i(t) \equiv 0$ for the sources is admissible, one can consider all nodes of the network as the nodes with external inflows or outflows. Let $I^f \subset I$ denote the set of nodes $i \in I$, where i is such that the set $I_i^+ \cup I_i^-$ consists of only one segment. It means that the node i is a node of external inflow or outflow for the whole pipe network (e.g., $I^f = \{1, 4, 5, 8\}$ in Fig. 1). Let $N_f = |I^f|$ denote the number of such nodes; it is obviously that $N_f \leq N$. Let I^{int} denote the set of nodes not belonging to I^f, so $N_{\text{int}} = |I^{\text{int}}|$, i.e., $I^{\text{int}} = I/I^f$, $N_{\text{int}} = N - N_f$. In actual conditions, the pumping stations are placed, the measuring equipment is installed, and the quantitative accounting is conducted at the nodes from the set I^f.

Linearized system of differential equations for unsteady isothermal laminar flow of dripping liquid with constant density ρ in a linear pipe (k, s) of length l_{ks} and diameter d_{ks} of oil pipeline network can be written in the following form [14]:

$$\begin{cases} -\frac{\partial P^{ks}(x,t)}{\partial x} = \frac{\rho}{S^{ks}} \frac{\partial Q^{ks}(x,t)}{\partial t} + 2a^{ks} \frac{\rho}{S^{ks}} Q^{ks}(x,t), \\ -\frac{\partial P^{ks}(x,t)}{\partial t} = c^2 \frac{\rho}{S^{ks}} \frac{\partial Q^{ks}(x,t)}{\partial x}, x \in (0, l^{ks}), t \in (0, T], \end{cases} \quad s \in I_k^+, k \in I. \tag{1}$$

Here c is the sound velocity in oil; S^{ks} is the area of an internal cross section of the segment (k, s) and a^{ks} is the coefficient of dissipation (we may consider that the kinematic coefficient of viscosity γ is independent of pressure and the condition $2a^{ks} = \frac{32\gamma}{(d^{ks})^2} = const$ is quite accurate for a laminar flow). $Q^{k_i k_j}(x, t)$ and $P^{k_i, k_j}(x, t)$ are the flow rate and pressure of flow, respectively, at the time instance t in the point $x \in (0, l_{k_i, k_j})$ of the segment (k_i, k_j) of the pipe network. $P^k(t)$, $Q^k(t)$ are the pressure and flow rate at the node $k \in I$, respectively. The values $Q^{k_i k_j}(x, t)$ can be positive or negative. The positive or negative value $Q^{k_i k_j}(x, t)$ means that an actual flow in the segment (k_i, k_j) is directed from the node k_i to the node k_j or the flow direction is from k_j to k_i, respectively. It is obvious that each segment carries an inflow and outflow for some nodes and the following conditions are satisfied:

$$\begin{aligned} &Q^{k_i, k_j}(x, t) = -Q^{k_j k_i}(l^{k_i, k_j} - x, t), \\ &P^{k_i, k_j}(x, t) = P^{k_j k_i}(l^{k_i, k_j} - x, t), \quad x \in (0, l^{k_i k_j}), \quad k_i \in I, k_j \in I_{k_i}^+, \end{aligned} \tag{2}$$

where $l^{k_j k_i} = l^{k_i k_j}$.

The equation for each segment of the network appears in the system (1) only once. Indeed, the first index k takes the values of all nodes from I, and the segments (k, s) make up the set of all segments-inflows into the node s, $s \in I_k^+$. Taking into account that each segment of the network is inflow for some node, consequently, there is an equation in (1) for each segment of the network, and this equation takes place in (1) only once.

Instead of using (1), we can write the process of fluid flow in the network for the segments-outflows in the following form:

$$\begin{cases} -\frac{\partial P^{ks}(x,t)}{\partial x} = \frac{\rho}{S^{ks}} \frac{\partial Q^{ks}(x,t)}{\partial t} + 2a^{ks} \frac{\rho}{S^{ks}} Q^{ks}(x,t), \\ -\frac{\partial P^{ks}(x,t)}{\partial t} = c^2 \frac{\rho}{S^{ks}} \frac{\partial Q^{ks}(x,t)}{\partial x}, x \in (0, l^{ks}), t \in (0, T], \end{cases} \quad s \in I_k^-, k \in I,$$

because each segment is an outflow for some node.

The conditions of Kirchhoff's first law (total flow into the node must be equal to total flow out of the node) are satisfied at the nodes of the network at $t \in [0, T]$:

$$\sum_{s\in I_k^+} Q^{ks}\left(l^{ks},t\right) - \sum_{s\in I_k^-} Q^{ks}\left(0,t\right) = \tilde{q}^k(t), \quad k \in I. \tag{3}$$

Also, the following conditions of flow continuity for the nodes of the net (the equality of the values of pressures on all adjacent ends of the segments of the network) hold:

$$P^k(t) = P^{k_i k}\left(l^{k_i k},t\right) = P^{kk_j}\left(0,t\right), \quad k_i \in I_k^+, k_j \in I_k^-, \ k \in I, \tag{4}$$

where $\tilde{q}^k(t)$ is the external inflow $\left(\tilde{q}^k(t) > 0\right)$ or outflow $\left(\tilde{q}^k(t) < 0\right)$ for the node k and $P^k(t)$ is the value of the pressure in the node k.

It is evident that (3) includes $N_f + N_{\text{int}} = N$ conditions. The number of independent conditions is $(N_k - 1)$ for every node k from the set I^{int} in (4) (i.e., one less than the total number of adjacent nodes), and two boundary conditions (4) are associated with every internal segment of the network. The total number of conditions for all nodes from I^f is N_f. So, the total number of conditions in (3) and (4) is $[N_f + N_{\text{int}}] + [(2M - N_f) - N_{\text{int}}] = 2M$.

As it was noted above, the number of conditions in (3) is N, but in view of the condition of material balance $\left(\sum_{k\in I} \tilde{q}^k(t) = 0\right)$ for the whole pipeline network, we conclude that the number of linearly independent conditions is $N - 1$. So, it is necessary to add any one independent condition. As a rule the value of pressure at one of the nodes $s \in I^f$ is given for this purpose, in place of the flow rate $q^s(t)$:

$$P^s(t) = \tilde{P}^s(t). \tag{5}$$

In more general case, for every node from I^f, it is necessary to give the values of pressure ($I_p^f \subset I^f$ denotes the set of such nodes) or the values of flow rate (the set $I_q^f \subset I^f$). So, we will add the following conditions to condition (3):

$$\begin{cases} P^n(t) = P^{ns}(0,t) = \tilde{P}^n(t), & s \in I_n^+, \quad if \ I_n^- = \varnothing, \\ P^n(t) = P^{sn}(l^{sn},t) = \tilde{P}^n(t), & s \in I_n^-, \quad if \ I_n^+ = \varnothing, \end{cases} \quad n \in I_p^f, \tag{6}$$

$$\begin{cases} Q^m(t) = Q^{ms}(0,t) = \tilde{P}^m(t), & s \in I_m^+, \quad if \ I_m^- = \varnothing, \\ Q^m(t) = Q^{sm}(l^{sm},t) = \tilde{P}^m(t), & s \in I_m^-, \quad if \ I_m^+ = \varnothing. \end{cases} \quad m \in I_q^f, \tag{7}$$

Here $I^f = I_q^f \cup I_p^f$ and I_p^f must not be empty: $I_p^f \neq \varnothing$; $\tilde{Q}^m(t)$, $\tilde{P}^n(t)$, $m \in I_q^f$, and $n \in I_p^f$ are the given functions, which determine the regimes of sources' operation.

Conditions (3), (4), (6), and (7) are the boundary conditions for the system of differential Eq. (1). We must note that they have significant specific features, consisting in the fact that conditions (3) and (4) are non-separated (nonlocal) boundary conditions unlike classical cases of boundary conditions for partial differential equations.

We will assume that initial state of the process at $t = 0$ is given for all segments of the network:

$$Q^{ks}(x, 0) = \hat{Q}^{ks}(x), \quad P^{ks}(x, 0) = \hat{P}^{ks}(x), \quad x \in \left[0,\ l^{ks}\right], \quad s \in I_k^+, k \in I, \tag{8}$$

here $\hat{Q}^{ks}(x),\ \hat{P}^{ks}(x),\ (k, s) \in J$ are known functions. We also assume that the functions defining an initial state of the process and boundary conditions are continuous and satisfy the consistency conditions on the ends at initial time and all the conditions of existence and uniqueness of the solution to the corresponding boundary value problem (1) and (3)–(8) hold.

So, we are given the functions $\hat{Q}^{ks}(x)$, $\hat{P}^{ks}(x)$, $(k, s) \in J$, $\tilde{Q}^m(t)$, $m \in I_q^f$, $\tilde{P}^n(t)$, $n \in I_p^f$, and $\tilde{q}^k(t)$, $k \in I$ involved in (3)–(8) which characterize external inflows or outflows, and we need to find the functions $Q^{ks}(x, t)$, $P^{ks}(x, t)$, $x \in (0, l^{ks})$, $(k, s) \in J$, and $t \in [0, T]$ (the solutions to the system of differential eqs. (1) with non-separated boundary conditions (3)–(7) and initial conditions (8)) which determine the regimes of transient processes for all segments of the network at $t \in [0, T]$.

In general, it is impossible to build analytical solution to the considered problem due to the non-separability of boundary conditions. Thereby in this work, we suggest to use numerical approach based on application of grid method for solving boundary value problems (1) and (3)–(8). In what follows we present the implementation of the proposed numerical approach. The formulas are obtained for this approach; the algorithms based on the suggested scheme of transfer of non-separated boundary conditions are developed; numerical experiments are conducted; and an analysis of the obtained results is given.

3 Numerical Method of the Solution to the Problem

To solve numerically the problem in domain $[0, l^{ks}] \times [0, T]$, $(ks) \in J$, we introduce a uniform grid area.

$$\omega^{ks} = \left\{(x_i, t_j) : x_i = ih, t_j = j\tau, i = \overline{0, n_{ks}}, j = \overline{0, n_t}\right\}, n_{ks} = \left[l^{ks}/h\right], \tau = T/n_t, (ks) \in J, \tag{9}$$

where h, τ are the given positive numbers and $[a]$ is an integer part of the number a. We use the following notations for grid functions:

$$P_{ij}^{ks} = P^{ks}\left(x_i, t_j\right), \quad Q_{ij}^{ks} = Q^{ks}\left(x_i, t_j\right), \quad \tilde{q}_j^k = \tilde{q}^k\left(t_j\right), \quad (i, j) \in \omega^{ks}, \quad (ks) \in J, \quad k \in I. \tag{10}$$

Let's approximate the system of eqs. (1) by implicit scheme of grid method in the following form [17]:

$$\begin{cases} -\frac{P_{ij}^{ks}-P_{i-1j}^{ks}}{h} = \frac{\rho}{S^{ks}}\frac{Q_{ij}^{ks}-Q_{ij-1}^{ks}}{\tau} + 2a^{ks}\frac{\rho}{S^{ks}}Q_{ij}^{ks}, \\ -\frac{P_{ij}^{ks}-P_{ij-1}^{ks}}{\tau} = c^2\frac{\rho}{S^{ks}}\frac{Q_{ij}^{ks}-Q_{i-1j}^{ks}}{h}, \quad i=\overline{1,n_{ks}}, \quad j=\overline{1,n_t}, \end{cases} \quad s \in I_k^+, k \in I. \tag{11}$$

The approximation (11) is stable [18]. Let's approximate initial conditions (1.12):

$$Q_{i0}^{ks} = \hat{Q}^{ks}(x_i), \quad P_{i0}^{ks} = \hat{P}^{ks}(x_i), \quad i=\overline{1,n_{ks}}, \quad (ks) \in J \tag{12}$$

and boundary conditions at $j=\overline{1,n_t}$:

$$\begin{cases} Q_{0j}^{ms} = \tilde{Q}^s(t_j), \quad s \in I_m^-, \quad if \ I_n^+ = \varnothing, \\ Q_{n_{sm}j}^{sm} = \tilde{Q}^s(t_j), \quad s \in I_m^+, \quad if \ I_n^- = \varnothing, \end{cases} \quad m \in I_q^f, \tag{13}$$

$$\begin{cases} P_{0j}^{ns} = \tilde{P}^s(t_j), \quad s \in I_n^-, if \ I_n^+ = \varnothing, \\ P_{n_{sn}j}^{sn} = \tilde{P}^s(t_j), \quad s \in I_n^+, if \ I_n^- = \varnothing, \end{cases} \quad n \in I_p^f, \tag{14}$$

$$P_{n_{k_r k}j}^{k_r k} = P_{0j}^{kk_s}, \quad k_r \in I_k^+, \quad k_s \in I_k^-, \quad k \in I, \tag{15}$$

$$\sum_{s\in I_k^+} Q_{n_{sk}j}^{sk} - \sum_{s\in I_k^-} Q_{0j}^{ks} = \tilde{q}_{kj}, \quad k \in I. \tag{16}$$

Using the notations $\xi^{ks} = \left(\frac{h}{\tau} + 2a^{ks}h\right)$, $\mu = \frac{h}{\tau}$, $\eta = \frac{h}{\tau c^2}$, and $Q_{ij}^{ks} = \frac{\rho}{S^{ks}}Q_{ij}^{ks}$, we can write the expression (11) for $j=\overline{1,n_t}$ as follows:

$$\begin{cases} P_{ij}^{ks} = P_{i-1j}^{ks} - \xi^{ks}Q_{ij}^{ks} + \mu Q_{ij-1}^{ks}, \\ Q_{ij}^{ks} = Q_{i-1j}^{ks} - \eta P_{ij}^{ks} + \eta P_{ij-1}^{ks}, \quad i=\overline{1,n_{ks}}, \end{cases} \quad s \in I_k^+, k \in I. \tag{17}$$

Problem (12)–(17) is the finite-dimensional difference approximations of original problem (1) and (3)–(8). System (17) consists of M pairs of difference equations; (12)–(16) include $2M$ boundary conditions, some of them are given on the left ends and some on the right ends of the intervals $(0, l^{ks})$, $(ks) \in J$; and conditions (15) and (16) are non-separated. To solve problem (12)–(17) numerically, first of all, it is necessary to bring all of the conditions to one end: (to the left or right end) of the interval $(0, l_{ks})$ for every time layer. Since conditions (15) and (16) are non-separated, it is impossible to use immediately the traditional sweep methods [17] or the analogues of the sweep methods, introduced for non-separated boundary conditions in the problems with concentrated parameters [16].

In this work we suggest a modification of sweep method that eventually constructs the dependences between boundary values of all unknown functions (i.e., the values of pressure and flow rate for the case being considered) on the left and right ends of corresponding intervals (segments) $(0, l^{ks})$. This will afford to express the values of boundary functions on one end of each (segment) interval in terms of the values of the functions on the other end of the same (segment) interval. Consequently, all the boundary conditions will be given on the one end: on the left ends or right ends, which will allow carrying out the calculations directly by formulae (17).

Thus, for example, to transfer the conditions from the left end to the right end on the $j-$ th time layer for the segment (ks), we will obtain the following dependences, which we will call the formulas for left sweep method:

$$P_{0j}^{ks} = R\left(P_{n_{ks}j}^{ks}, Q_{n_{ks}j}^{ks}\right), \quad Q_{0j}^{ks} = G\left(P_{n_{ks}j}^{ks}, Q_{n_{ks}j}^{ks}\right), \quad (ks) \in J, \tag{18}$$

whereas to sweep the conditions from the right end to the left end, we will obtain the following dependences, which we will call the right sweep method:

$$P_{n_{ks}j}^{ks} = R\left(P_{0j}^{ks}, Q_{0j}^{ks}\right), \overline{q}_{n_{ks}j}^{ks} = G\left(P_{0j}^{ks}, Q_{0j}^{ks}\right), (ks) \in J. \tag{19}$$

For these purposes, we build the dependences for the left sweep method in the following form:

$$\begin{aligned} P_{0j}^{ks} &= \alpha_p^{ks(r)} P_{rj}^{ks} + \beta_p^{ks(r)} Q_{rj}^{ks} + \theta_p^{ks(r)}, \\ Q_{0j}^{ks} &= \alpha_q^{ks(r)} Q_{rj}^{ks} + \beta_q^{ks(r)} P_{rj}^{ks} + \theta_q^{ks(r)}, \end{aligned} \quad r = \overline{1, n_{ks}}, \tag{20}$$

whereas the analogous dependences for the right sweep method are as follows:

$$\begin{aligned} P_{n_{ks}j}^{ks} &= \alpha_p^{ks(r)} P_{rj}^{ks} + \beta_p^{ks(r)} Q_{rj}^{ks} + \theta_p^{ks(r)}, \\ Q_{n_{ks}j}^{ks} &= \alpha_q^{ks(r)} Q_{rj}^{ks} + \beta_q^{ks(r)} P_{rj}^{ks} + \theta_q^{ks(r)}, \end{aligned} \quad r = \overline{n_{ks} - 1, 0}, \tag{21}$$

where $\alpha_p^{ks(r)}$, $\beta_r^{ks(r)}$, $\theta_p^{ks(r)}$, $\alpha_q^{ks(r)}$, $\beta_q^{ks(r)}$, $\theta_q^{ks(r)}$, are the as-yet unknown sweep coefficients. Let us write (18) in the following form to obtain formulas for sweep coefficients:

$$\begin{cases} P_{i-1j}^{ks} = P_{ij}^{ks} + \xi^{ks} Q_{ij}^{ks} - \mu Q_{ij-1}^{ks}, \\ Q_{i-1j}^{ks} = Q_{ij}^{ks} + \eta P_{ij}^{ks} - \eta P_{ij-1}^{ks}, \quad i = \overline{1, n_{ks}}, \quad s \in I_k^+, k \in I. \end{cases} \tag{22}$$

and for $i = 1$ (22) will be as follows:

$$\begin{cases} P_{0j}^{ks} = P_{1j}^{ks} + \xi^{ks} Q_{1j}^{ks} - \mu Q_{1j-1}^{ks}, \\ Q_{0j}^{ks} = Q_{1j}^{ks} + \eta P_{1j}^{ks} - \eta P_{1j-1}^{ks}, \quad s \in I_k^+, k \in I. \end{cases} \tag{23}$$

We can build the recurrent relations if we use the method of mathematical induction and take into account the formulae (20) in (22):

$$\begin{aligned} &P_{0j}^{ks} = \alpha_p^{ks(r)} P_{rj}^{ks} + \beta_p^{ks(r)} Q_{rj}^{ks} + \theta_p^{ks(r)} = \\ &= \alpha_p^{ks(r)} \left(P_{r+1j}^{ks} + \xi^{ks} Q_{r+1j}^{ks} - \mu Q_{r+1j-1}^{ks} \right) + \beta_p^{ks(r)} \left(Q_{r+1j}^{ks} + \eta P_{r+1j}^{ks} - \eta P_{r+1j-1}^{ks} \right) + \\ &+ \theta_p^{ks(r)} = P_{r+1j}^{ks} \left(\alpha_p^{ks(r)} + \eta \beta_p^{ks(r)} \right) + Q_{r+1j}^{ks} \left(\alpha_p^{ks(r)} \xi^{ks} + \beta_p^{ks(r)} \right) + \theta_p^{ks(r)} - \\ &- \alpha_p^{ks(r)} \mu Q_{r+1j-1}^{ks} - \beta_p^{ks(r)} \eta P_{r+1j-1}^{ks} \\ &Q_0^{ks} = \alpha_q^{ks(r)} Q_{rj}^{ks} + \beta_q^{ks(r)} P_{rj}^{ks} + \theta_q^{ks(r)} = \\ &= \alpha_q^{ks(r)} \left(Q_{r+1j}^{ks} + \eta P_{r+1j}^{ks} - \eta P_{r+1j-1}^{ks} \right) + \beta_q^{ks(r)} \left(P_{r+1j}^{ks} + \xi^{ks} Q_{r+1j}^{ks} - \mu Q_{r+1j-1}^{ks} \right) + \\ &+ \theta_q^{ks(r)} = Q_{r+1j}^{ks} \left(\alpha_q^{ks(r)} + \xi^{ks} \beta_q^{ks(r)} \right) + P_{r+1j}^{ks} \left(\beta_q^{ks(r)} + \eta \alpha_q^{ks(r)} \right) + \theta_q^{ks(r)} - \\ &- \alpha_q^{ks(r)} \eta P_{r+1j-1}^{ks} - \beta_q^{ks(r)} \mu Q_{r+1j-1}^{ks}. \end{aligned} \tag{24}$$

On the other hand, we can write (20) for $r + 1$ in the following form:

$$\begin{aligned} P_{0j}^{ks} &= \alpha_p^{ks(r+1)} P_{r+1j}^{ks} + \beta_p^{ks(r+1)} Q_{r+1j}^{ks} + \theta_p^{ks(r+1)}, \\ Q_{0j}^{ks} &= \alpha_q^{ks(r+1)} Q_{r+1j}^{ks} + \beta_q^{ks(r+1)} P_{r+1j}^{ks} + \theta_q^{ks(r+1)}, \end{aligned} \tag{25}$$

So, if we take into account (23) and equate the right sides of (25) with (24), we obtain the formulas for finding the left sweep coefficients:

$$\begin{cases} \alpha_p^{ks(r+1)} = \alpha_p^{ks(r)} + \eta \beta_p^{ks(r)}, \quad \alpha_p^{ks(1)} = 1, \\ \beta_p^{ks(r+1)} = \alpha_p^{ks(r)} \xi^{ks} + \beta_p^{ks(r)}, \quad \beta_p^{ks(1)} = \xi^{ks}, \\ \theta_p^{ks(r+1)} = \theta_p^{ks(r)} - \alpha_p^{ks(r)} \mu Q_{r+1j-1}^{ks} - \beta_p^{ks(r)} \eta P_{r+1j-1}^{ks}, \quad \theta_p^{ks(1)} = -\mu Q_{1j-1}^{ks}, \quad r = \overline{1, n_{ks} - 1}, \end{cases}$$
$$\begin{cases} \alpha_q^{ks(r+1)} = \alpha_q^{ks(r)} + \xi^{ks} \beta_q^{ks(r)}, \quad \alpha_q^{ks(1)} = 1, \quad s \in I_k^+, k \in I, \\ \beta_q^{ks(r+1)} = \beta_q^{ks(r)} + \eta \alpha_q^{ks(r)}, \quad \beta_q^{ks(1)} = \eta, \\ \theta_q^{ks(r+1)} = \theta_q^{ks(r)} - \alpha_q^{ks(r)} \eta P_{r+1j-1}^{ks} - \beta_q^{ks(r)} \mu Q_{r+1j-1}^{ks}, \quad \theta_q^{ks(1)} = -\eta P_{1j-1}^{ks}. \end{cases} \tag{26}$$

Similarly, we derive the formulas for the sweep coefficients in the case of right sweep method. Indeed, conducting some uncomplicated transformations and including the designation $\delta^{ks} = (1 - \xi^{ks}\eta)^{-1}$, $(ks) \in J$, we can write (22) as follows:

$$\begin{cases} P_{ij}^{ks} = \delta^{ks} P_{i-1j}^{ks} - \delta^{ks} \xi^{ks} Q_{i-1j}^{ks} + \delta^{ks} \mu Q_{ij-1}^{ks} - \delta^{ks} \xi^{ks} \eta P_{ij-1}^{ks}. \\ Q_{ij}^{ks} = \delta^{ks} Q_{i-1j}^{ks} - \delta^{ks} \eta P_{i-1j}^{ks} + \delta^{ks} \eta P_{ij-1}^{ks} - \delta^{ks} \eta \mu Q_{ij-1}^{ks}, \quad i = \overline{n_{ks}, 1}, \quad s \in I_k^+, k \in I. \end{cases} \tag{27}$$

and if $i = n_{ks}$ (27) will be as follows:

$$\begin{cases} P^{ks}_{n_{ks}j} = \delta^{ks} P^{ks}_{n_{ks}-1j} - \delta^{ks}\xi^{ks} Q^{ks}_{n_{ks}-1j} + \delta^{ks}\mu Q^{ks}_{n_{ks}j-1} - \delta^{ks}\xi^{ks}\eta P^{ks}_{n_{ks}j-1}, \\ Q^{ks}_{n_{ks}j} = \delta^{ks} Q^{ks}_{n_{ks}-1j} - \delta^{ks}\eta P^{ks}_{n_{ks}-1j} + \delta^{ks}\eta P^{ks}_{n_{ks}j-1} - \delta^{ks}\eta\mu Q^{ks}_{n_{ks}j-1}, \end{cases} \quad s \in I^+_k, k \in I. \tag{28}$$

We can find the following recurrent relations if we consider (21) in (27):

$$\begin{aligned} &P^{ks}_{n_{ks}j} = \alpha_p^{ks(r)} P^{ks}_{rj} + \beta_p^{ks(r)} Q^{ks}_{rj} + \theta_p^{ks(r)} = \alpha_p^{ks(r)}\Big(\delta^{ks} P^{ks}_{r-1j} - \delta^{ks}\xi^{ks} Q^{ks}_{r-1j} + \\ &+ \delta^{ks}\mu Q^{ks}_{rj-1} - \delta^{ks}\xi^{ks}\eta P^{ks}_{rj-1}\Big) + \beta_p^{ks(s)}\left(\delta^{ks} Q^{ks}_{r-1j} - \delta^{ks}\eta P^{ks}_{r-1j} + \delta^{ks}\eta P^{ks}_{rj-1} - \delta^{ks}\eta\mu Q^{ks}_{rj-1}\right) + \\ &+ \theta_p^{ks(r)} = P^{ks}_{r-1j}\left(\delta^{ks}\alpha_p^{ks(r)} - \delta^{ks}\eta\beta_p^{ks(r)}\right) + Q^{ks}_{r-1j}\left(\delta^{ks}\beta_p^{ks(r)} - \delta^{ks}\xi^{ks}\alpha_p^{ks(r)}\right) + \\ &+ \theta_p^{ks(r)} + \alpha_p^{ks(r)}\left(\delta^{ks}\mu Q^{ks}_{rj-1} - \delta^{ks}\xi^{ks}\eta P^{ks}_{rj-1}\right) + \beta_p^{ks(r)}\left(\delta^{ks}\eta P^{ks}_{rj-1} - \delta^{ks}\eta\mu Q^{ks}_{rj-1}\right), \\ &Q^{ks}_{n_{ks}j} = \alpha_q^{ks(r)} Q^{ks}_{rj} + \beta_q^{ks(r)} P^{ks}_{rj} + \theta_q^{ks(r)} = \alpha_q^{ks(r)}\Big(\delta^{ks} Q^{ks}_{r-1j} - \delta^{ks}\eta P^{ks}_{r-1j} + \delta^{ks}\eta P^{ks}_{rj-1} - \\ &- \delta^{ks}\eta\mu Q^{ks}_{rj-1}\Big) + \beta_q^{ks(r)}\left(\delta^{ks} P^{ks}_{r-1j} - \delta^{ks}\xi^{ks} Q^{ks}_{r-1j} + \delta^{ks}\mu Q^{ks}_{rj-1} - \delta^{ks}\xi^{ks}\eta P^{ks}_{rj-1}\right) + \theta_q^{ks(r)} = \\ &= Q^{ks}_{r-1j}\left(\delta^{ks}\alpha_q^{ks(r)} - \delta^{ks}\xi^{ks}\beta_q^{ks(r)}\right) + P^{ks}_{r-1j}\left(\delta^{ks}\beta_q^{ks(r)} - \delta^{ks}\eta\alpha_q^{ks(r)}\right) + \theta_q^{ks(r)} + \\ &+ \alpha_q^{ks(r)}\left(\delta^{ks}\eta P^{ks}_{rj-1} - \delta^{ks}\eta\mu Q^{ks}_{rj-1}\right) + \beta_q^{ks(r)}\left(\delta^{ks}\mu Q^{ks}_{rj-1} - \delta^{ks}\xi^{ks}\eta P^{ks}_{rj-1}\right). \end{aligned} \tag{29}$$

On the other hand, we can write (21) for $r-1$ in the following form:

$$\begin{aligned} P^{ks}_{n_{ks}j} &= \alpha_p^{ks(r-1)} P^{ks}_{r-1j} + \beta_p^{ks(r-1)} Q^{ks}_{r-1j} + \theta_p^{ks(r-1)}, \\ Q^{ks}_{n_{ks}j} &= \alpha_q^{ks(r-1)} Q^{ks}_{r-1j} + \beta_q^{ks(r-1)} P^{ks}_{r-1j} + \theta_q^{ks(r-1)}, \end{aligned} \quad r = \overline{n_{ks}, 1}. \tag{30}$$

So, if we consider (28) and equate the right sides of (30) and (29), we obtain the formulas for finding the right sweep coefficients:

$$\begin{cases} \begin{cases} \alpha_p^{ks(r-1)} = \delta^{ks}\left(\alpha_p^{ks(r)} - \eta\beta_p^{ks(r)}\right), \quad \alpha_p^{ks(n_{ks})} = \delta^{ks}, \\ \beta_p^{ks(r-1)} = \delta^{ks}\left(\beta_p^{ks(r)} - \xi^{ks}\alpha_p^{ks(r)}\right), \quad \beta_p^{ks(n_{ks})} = -\delta^{ks}\xi^{ks}, \\ \theta_p^{ks(r-1)} = \theta_p^{ks(r)} + \delta^{ks}\left[\alpha_p^{ks(r)}\left(\mu Q^{ks}_{rj-1} - \xi^{ks}\eta P^{ks}_{rj-1}\right) + \beta_p^{ks(r)}\left(\eta P^{ks}_{rj-1} - \eta\mu Q^{ks}_{rj-1}\right)\right], \\ \theta_p^{ks(n_{ks})} = \delta^{ks}\left(\mu Q^{ks}_{n_{ks}j-1} - \xi^{ks}\eta P^{ks}_{n_{ks}j-1}\right), \end{cases} \\ \begin{cases} \alpha_q^{ks(r-1)} = \delta^{ks}\left(\alpha_q^{ks(r)} - \xi^{ks}\beta_q^{ks(r)}\right), \quad \alpha_q^{ks(n_{ks})} = \delta^{ks}, \qquad r = \overline{n_{ks}, 1}, \quad (ks) \in J, \\ \beta_q^{ks(r-1)} = \delta^{ks}\left(\beta_q^{ks(r)} - \eta\alpha_q^{ks(r)}\right), \quad \beta_q^{ks(n_{ks})} = -\delta^{ks}\eta, \\ \theta_q^{ks(r-1)} = \theta_q^{ks(r)} + \delta^{ks}\left[\alpha_q^{ks(r)}\left(\eta P^{ks}_{rj-1} - \eta\mu Q^{ks}_{rj-1}\right) + \beta_q^{ks(r)}\left(\mu Q^{ks}_{rj-1} - \xi^{ks}\eta P^{ks}_{rj-1}\right)\right., \\ \theta_q^{ks(n_{ks})} = \delta^{ks}\left(\eta P^{ks}_{n_{ks}j-1} - \eta\mu Q^{ks}_{n_{ks}j-1}\right). \end{cases} \end{cases} \tag{31}$$

Consequently, at the end of sweep method, we will find the following expressions in the case of left sweep method for $r = n_{ks}$:

$$\begin{aligned} P^{ks}_{0j} &= \alpha_p^{ks(n_{ks})} P^{ks}_{n_{ks}j} + \beta_p^{ks(n_{ks})} Q^{ks}_{n_{ks}j} + \theta_p^{ks(n_{ks})}, \\ Q^{ks}_{0j} &= \alpha_q^{ks(n_{ks})} Q^{ks}_{n_{ks}j} + \beta_q^{ks(n_{ks})} P^{ks}_{n_{ks}j} + \theta_q^{ks(n_{ks})}, \end{aligned} \tag{32}$$

and the following expressions in the case of right sweep method for $r = 0$:

$$\begin{aligned} P^{ks}_{n_{ks}j} &= \alpha_p^{ks(0)} P^{ks}_{0j} + \beta_p^{ks(0)} Q^{ks}_{0j} + \theta_p^{ks(0)}, \\ Q^{ks}_{n_{ks}j} &= \alpha_q^{ks(0)} Q^{ks}_{0j} + \beta_q^{ks(0)} P^{ks}_{0j} + \theta_q^{ks(0)}. \end{aligned} \tag{33}$$

The values of unknown functions in formulae (32), i.e., the values of pressure and flow rate on the left end (or on the right end in the case of (33)) of the segment (k, s), are expressed in terms of their values on the right end (or on the left end) of the same segment. So, after performing the operation of left sweep method and substituting obtained relations (32) in (13)–(16), we will obtain all $2M$ conditions only on the left ends of the intervals $(0, l_{ks})$, $(k, s) \in J$ for all segments:

$$\alpha_q^{ks(0)} Q^{ks}_{0j} + \beta_q^{ks(0)} P^{ks}_{0j} + \theta_q^{ks(0)} = \tilde{Q}_s\left(t_j\right), \quad s \in I_k^+, k \in I_q^f, \tag{34}$$

$$\alpha_p^{ks(0)} P^{ks}_{0j} + \beta_p^{ks(0)} Q^{ks}_{0j} + \theta_p^{ks(0)} = \tilde{P}_s\left(t_j\right), \quad s \in I_k^+, k \in I_p^f, \tag{35}$$

$$\alpha_p^{k_r k(0)} P^{k_r k}_{0j} + \beta_p^{k_r k(0)} Q^{k_r k}_{0j} + \theta_p^{k_r k(0)} = P^{kk_r}_{0j}, \ \forall k_r \in I_k^+, \ k_s \in I_k^-, \ k \in I, \tag{36}$$

$$\sum_{s \in I_k^+} \alpha_q^{ks(0)} Q^{sk}_{0j} + \beta_q^{ks(0)} P^{ks}_{0j} + \theta_q^{ks(0)} - \sum_{s \in I_k^-} Q^{ks}_{0j} = \tilde{q}_k\left(t_j\right), \quad k \in I. \tag{37}$$

Or we will obtain all $2M$ conditions only on the right ends of the intervals $(0, l_{ks})$, $(k, s) \in J$ for all segments:

$$\alpha_q^{ks(n_{ks})} Q^{ks}_{n_{ks}j} + \beta_q^{ks(n_{ks})} P^{ks}_{n_{ks}j} + \theta_q^{ks(n_{ks})} = \tilde{Q}_{sj}, \quad s \in I_k^-, \quad k \in I_q^f, \tag{38}$$

$$\alpha_p^{sk(n_{ks})} P^{ks}_{n_{ks}j} + \beta_p^{ks(n_{ks})} Q^{ks}_{n_{ks}j} + \theta_p^{ks(n_{ks})} = \tilde{P}_{sj}, \quad s \in I_k^-, \quad k \in I_p^f. \tag{39}$$

$$P^{k_r k}_{n_{k_r k}j} = \alpha_p^{kk_s(n_{kk_s})} P^{kk_s}_{n_{kk_s}j} + \beta_p^{kk_s(n_{kk_s})} Q^{kk_s}_{n_{kk_s}j} + \theta_p^{kk_s(n_{kk_s})}, \ \forall k_r \in I_k^+, \ k_s \in I_k^-, \ k \in I, \tag{40}$$

$$\sum_{s\in I_k^+} Q_{n_{sk}j}^{sk} - \sum_{s\in I_k^-} \alpha_q^{ks(n_{ks})} Q_{n_{ks}j}^{ks} + \beta_q^{ks(n_{ks})} P_{n_{ks}j}^{ks} + \theta_q^{ks(n_{ks})} = \tilde{q}^k\left(t_j\right), \quad k \in I, \tag{41}$$

if we perform the operation of right sweep method and substitute obtained relations (34) in (13)–(16).

One can write these conditions compactly in terms of the following system of linear algebraic equations of order $2M$:

$$AX = B, \tag{42}$$

here $X = (x_1, \ldots, x_{2M})^T$: $x_s = P_{n_{rs}}^{r_s}$, $s = 1, .., M$, $x_s = Q_{n_{rs}}^{r_s}$, $s = M + 1, .., 2M$ in the case of conditions (34)–(37) or $x_s = P_{0r_s}^{r_s}$, $s = 1, .., M$, $x_s = Q_{0r_s}^{r_s}$, $s = M + 1, .., 2M$ in the case of conditions (38)–(41)), A is the matrix of dimension $2M \times 2M$, B is the vector of dimension $2M$, and T is the sign of transposition. We will find the values for the pressure and flow rate on the right (or on the left) ends of all intervals (segments) $(0, l^{ks})$, $(ks) \in J$ by solving the system of (42) by any known numerical method. Consequently, problem (11)–(17) with non-separated boundary conditions is reduced to the problem with separated boundary conditions which are all linked to one end, and we will use formulae (17) to solve this problem and to find the values for the pressure and flow rate at all the points of the (segments) intervals $(0, l^{ks})$, $(ks) \in J$.

4 The Results of Numerical Experiments

We consider the following test problem for oil pipeline network consisting of five nodes, as shown in Fig. 3. Here $N = 6$, $M = 5$, $I^f = \{1, 3, 4, 6\}$, $N_f = 4$, $N_{\text{int}} = 2$. There are no external inflows and outflows inside the network. The diameter, d, of pipe segments is 530 (mm), and the lengths of segments are.

$$l^{(1,2)} = 100\ (km), l^{(5,2)} = 30\ (km), l^{(3,2)} = 70\ (km), l^{(5,4)} = 100\ (km), l^{(5,6)} = 60\ (km).$$

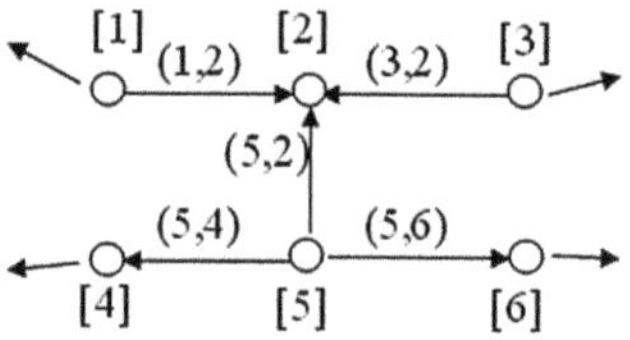

Fig. 3 The scheme of oil pipeline network with five nodes

Suppose that oil with kinematic viscosity $\nu = 1.5 \cdot 10^{-4}(m^2/s)$ and with density $\rho = 920(kg/m^3)$ is transported via the network; $2a = \frac{32\nu}{d^2} = 0.017$ for case being considered; and the sound velocity in oil is $1200(m/s)$.

There was a steady regime in the pipes at initial time instance $t = 0$ with the following values of flow rate and pressure in the pipes:

$$\hat{Q}^{1,2}(x) = 300\,(m^3/hour), \quad \hat{Q}^{5,2}(x) = 200\,(m^3/hour), \quad \hat{Q}^{3,2}(x) = 100\,(m^3/hour), \\ \hat{Q}^{5,4}(x) = 120\,(m^3/hour), \quad \hat{Q}^{5,6}(x) = 80\,(m^3/hour), \tag{43}$$

$$\hat{P}^{1,2}(x) = 2300000 - 5.8955x \text{ (Pa)}, \quad \hat{P}^{5,2}(x) = 1710451 + 1.17393x \text{ (Pa)}, \\ \hat{P}^{3,2}(x) = 1847826 - 1.37375x \text{ (Pa)}, \quad \hat{P}^{5,4}(x) = 1592058 + 2.35786x \text{ (Pa)}, \\ \hat{P}^{5,6}(x) = 1733429 + 0.94415x \text{ (Pa)}. \tag{44}$$

The pumping stations set on four ending points of oil pipeline provide current transportation regime which is defined by following values of pressure at $t > 0$:

$$\tilde{P}_0^1(t) = 2000000 \text{ (Pa)}, \quad \tilde{P}_0^3(t) = 1800000 \text{ (Pa)}, \\ \tilde{P}_l^4(t) = 1800000 \text{ (Pa)}, \quad \tilde{P}_l^6(t) = 1790000 \text{ (Pa)}. \tag{45}$$

We use the above-described scheme to obtain numerical solution to the initial boundary value problems (17) and (43)–(45), i.e., to calculate the values of fluid flow regimes in the pipe network, where the step for time variable is $h_t = 10(s)$ and the step of spatial variable is $h_x = 10(m)$ (these values were determined by the results of purpose of the conducted experiments to find the effective values of these parameters). The obtained results of the conducted numerical experiments for the solution to the test problem are given in Figs. 4 and 5. As may be seen from Figs. 4 and 5 and especially from Fig. 5, the transient process originated at $t = 0$ has finished approximately at $t = 300(sec)$, after which the fluid flow proceeds in a new stationary regime.

5 Conclusion

The mathematical statement for calculation of unsteady fluid (oil) flow regimes in the pipeline networks of complex structure is given in the work. The number of linear sections can reach up to several tens and even hundreds of real pipeline transport networks. In the work, we propose to use the graph theory, which drastically simplifies the process of computer modeling of complex pipeline networks. We use the system of two linear partial differential equations of hyperbolic type to describe the process of fluid flow in every linear segment of the pipeline network. There are

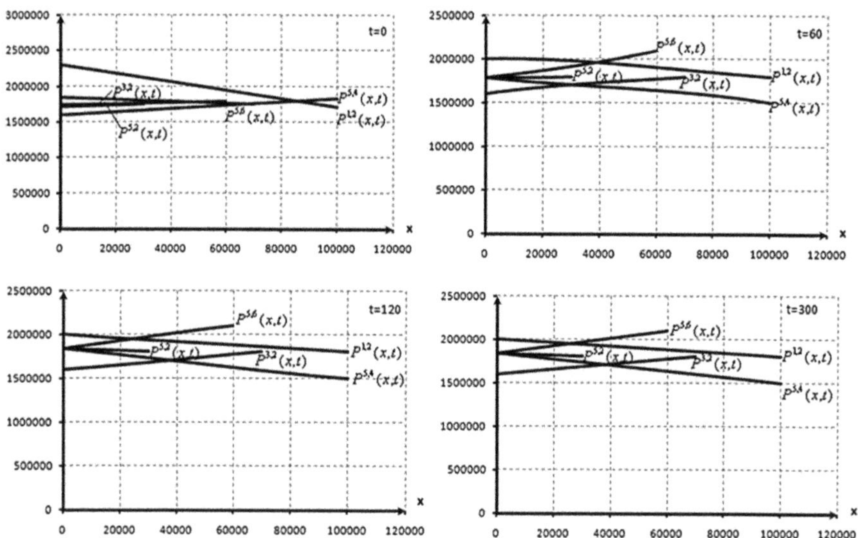

Fig. 4 The plots for pressure at segments $P^{ks}(x, t)$, $(ks) \in J$ at time instances $t = 0; 60; 120; 300$

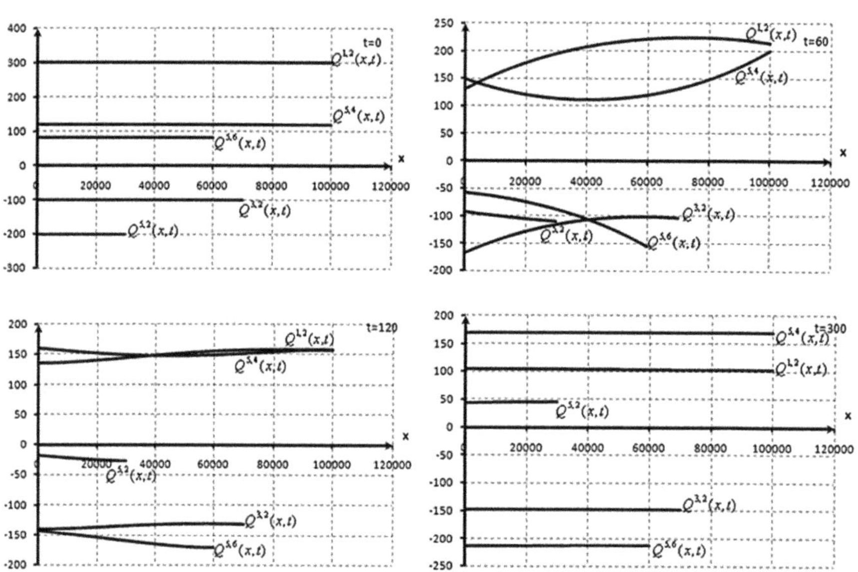

Fig. 5 The plots for flow rate at segments $Q^{ks}(x, t)$, $(ks) \in J$ at time instances $t = 0; 60; 120; 300$

non-separated boundary conditions at the points of connection of linear segments, which makes it impossible to calculate the flow regimes for every individual segment separately; it must be done for all segments simultaneously. To solve the problem, we suggest to use the methods of finite difference approximation (grid method) and special scheme of sweep (transfer) method to approximate a set of systems of hyperbolic differential equations interrelated by boundary conditions. Satisfactory results of numerical experiments are given. In order to switch from the grid problem with non-separated conditions to a problem with separated boundary conditions, we obtain the dependencies between boundary values of sought-for functions for every linear section. We obtained the formulas for the procedure of obtaining such dependencies, called a sweep (transfer) method, that use the difference equations only for that section, for which is made the dependencies.

References

1. Aida-zade K.R. Computational problems for hydraulic networks. *J.comp.math. and math. Physics*. 1989; **29**(2): 184-193. (in Russian)
2. Seok Woo Hong and Kim Ch. A new finite volume method on junction coupling and boundary treatment for flow network system analyses. *International journal for numerical methods in fluids* 2011; **65** (6): 707–742. DOI: https://doi.org/10.1002/fld.2212
3. Michael H. and Mohammed S. Simulation of transient gas flow at pipe-to-pipe intersections. *International journal for numerical methods in fluids*. 2008; **56** (5): 485–506. DOI: https://doi.org/10.1002/fld.1531
4. Freitas Rachid F. B. and Costa Mattos H. S. Modelling of pipeline integrity taking into account the fluid–structure interaction. *International journal for numerical methods in fluids*. 1998; **28** (2): 337–355. DOI: https://doi.org/10.1002/(SICI)1097-0363(19980815)28:2<337::AID-FLD724>3.0.CO;2-6
5. Ahmetzyanov A.V., Salnikov A.M., Spiridonov S.V. Multi-grid balance models of non-stationary flows in complex gas transportation systems. *Control of large systems*, 2010; Special issue 30.1 "Net models in control" M.:IPU RAN.: 230-251.
6. Szymkiewicz R. and Marek M. Numerical aspects of improvement of the unsteady pipe flow equations. *International journal for numerical methods in fluids*.2007; **55** (11):1039–1058. DOI: https://doi.org/10.1002/fld.1507
7. Aida-zade K.R. Asadova J.A. Study of Transients in Oil Pipelines. *Automation and Remote Control*. 2011; **72**, (12): 2563-2577.
8. Aliev R.A., Belousov V.D., Nemudrov A.Q. et al. Oil and gas transportation via pipeline: *Manual for students, M.:Nedra*. 1988; 368. (in Russian)
9. Smirnov M. E., Verigin A. N., and Nezamaev N. A., Computation of Unsteady Regimes of Pipeline Systems. *Russian Journal of Applied Chemistry*. 2010; **83**, (3): 572−578.
10. Changjun Li, Xia Wu and Wenlong J. Research of Heated Oil Pipeline Shutdown and Restart Process Based on VB and MATLAB. *I.J.Modern Education and Computer Science*. 2010; **2:**18-24 "http://www.mecs-press.org/"
11. Kuzmin D. A Guide to Numerical Methods for Transport Equations. *Friedrich-Alexander-Universitat Erlangen-Nürnberg*. 2010; 226.
12. Raad ISSA Simulation of intermittent flow in multiphase oil and gas pipelines Seventh International Conference on CFD in the Minerals and Process Industries CSIRO, Melbourne, Australia 9-11 December 2009.
13. Aida-zade K. R. and Ashrafova E. R. Localization of the points of leakage in an oil main pipeline under non-stationary conditions. *Journal of Engineering Physics and Thermophysics*.2012; **85**(5):1148-1156.
14. Charniy I.A. Unsteady flow of real fluid in pipelines, *M.:Nedra*, 1975; 199.

15. I. N. Parasidis & E. Providas (2018): An exact solution method for a class of nonlinear loaded difference equations with multipoint boundary conditions, *Journal of Difference Equations and Applications*, V.24, No 10, 2018. DOI: https://doi.org/10.1080/10236198.2018.1515928
16. Aida-zade K.R., Abdullaev V. M. On the Solution of Boundary value Problems with Non Separated Multipoint and Integral Conditions. *Differential Equations*. 2013; **49**,(9):1152-1162.
17. Samarskii A.A., Nikolaev E.S. Numerical methods for grid equations, *M.Nauka* 1978; 592.
18. Richtmyer R.D., Morton K.W. Difference methods for initial-value problems. *Second edition Interscience publishers, New York*, 1967.

Numerical Solution to Inverse Problems of Recovering Special-Type Source of a Parabolic Equation

K. R. Aida-zade and A. B. Rahimov

Abstract The chapter investigates inverse problems of recovering a source of a special type of parabolic equation with initial and boundary conditions. The specificity of these problems is that the identifiable parameters depend on only space or time variable and are factors of the coefficients of the right-hand side of the equation. By applying the method of lines, the problems are reduced to parametric inverse problems with respect to ordinary differential equations. A special type of representation of the solution is proposed to solve them. The most important in this work is that the proposed approach to the numerical solution to the investigated inverse problems of identifying the coefficients does not require to construct any iterative procedure. The results of numerical experiments conducted on test problems are provided.

1 Introduction

Inverse problems of mathematical physics are investigated in different directions, and the number of studies, from theoretical to specific applied problems, substantially increased in the last years [1–12]. The important class of boundary value problems with nonlocal conditions considered in the chapter leads to the coefficient inverse problems under study [1, 11–14]. The nonlocality of the conditions is due

AMS Subject Classifications 35R30; 35K20; 65N40; 65N21, 65L09, 34A55

K. R. Aida-zade
Institute of Control Systems of Azerbaijan NAS, Baku, Azerbaijan

Institute of Mathematics and Mechanics of Azerbaijan NAS, Baku, Azerbaijan

A. B. Rahimov (✉)
Institute of Control Systems of Azerbaijan NAS, Baku, Azerbaijan

I. N. Parasidis et al. (eds.), *Mathematical Analysis in Interdisciplinary Research*, Springer Optimization and Its Applications 179,
https://doi.org/10.1007/978-3-030-84721-0_6

to the practical impossibility to measure parameters of state of an object (process) immediately or at its separate points.

One of the most common approaches to the solution of inverse problems is reducing them to variation statements followed by optimization and use of optimal control methods [4–6]. Applying this approach is related, first, to problems of deriving formulas for the gradient of the functional of the variation problem and, second, to the necessity of using iterative methods of minimization of a functional.

Another approach is to construct the fundamental solution of the problem and to reduce it to an integral equation. If the functions that appear in the problem are of general form, there are some difficulties that hinder the use of such approach [3, 9–11].

Application of the finite difference method (explicit or implicit) [15] is also of interest. A shortcoming of such approach is high dimension of the obtained system of algebraic equations.

In the chapter, we consider the inverse source problems. The most important is that the proposed approach to the numerical solution of the inverse source problems under study does not use iteration algorithms. The apparatus of optimal control theory and appropriate numerical iterative methods of optimization of first order were used in [4–6] to solve such problems.

Another specific feature of the considered classes of inverse problems is that, first, the restored coefficients are for the free term and, second, they depend either on time or on space variable. This specific feature allows using the method of lines [16–18] to reduce the solution of initial problems to solution of specially constructed Cauchy problems [19, 20] with respect to the system of ordinary differential equations. In the chapter, we present the results of numerical experiments and their analysis.

2 Study of the Inverse Problem of Determination a Source Depending on a Space Variable

Let us consider the inverse source problem with respect to the parabolic equation:

$$\begin{aligned}&\frac{\partial v(x,t)}{\partial t} = a\left(x,t\right)\frac{\partial^2 v(x,t)}{\partial x^2} + a_1\left(x,t\right)\frac{\partial v(x,t)}{\partial x} + a_2\left(x,t\right)v\left(x,t\right) + f\left(x,t\right) + F\left(x,t\right),\\&(x,t) \in \Omega = \{(x,t) : 0 < x < l, \quad 0 < t \le T\},\end{aligned} \tag{1}$$

where

$$F\left(x,t\right) = \sum_{s=1}^{L} B_s\left(x,t\right)C_s(x), \tag{2}$$

under the following initial-boundary conditions and additional conditions:

$$v(x,0)=\varphi_0(x), \quad x\in[0,l], \tag{3}$$

$$v(0,t)=\psi_0(t), \quad v(l,t)=\psi_1(t), \quad t\in[0,T], \tag{4}$$

$$v\left(x,\bar{t}_s\right)=\varphi_{1s}(x), \quad x\in[0,l], \quad \bar{t}_s\in(0,T], \quad s=1,\ldots,L. \tag{5}$$

Here, $L>0$ is a given integer; $\bar{t}_s \in (0,T]$, $s=1,\ldots,L$ are given instants of time; given functions $a_0(x,t)>0$, $a_1(x,t)$, $a_2(x,t)$, $f(x,t)$, $B_s(x,t)$, $\varphi_0(x)$, $\psi_0(t)$, $\psi_1(t)$, and $\varphi_{1s}(x)$, $s=1,\ \ldots L$ are continuous with respect to x and t; $B_s(x,t)$ are linearly independent functions differentiable with respect to t and $a_2(x,t)\le 0$, $B_s(x,t)\ge 0$ and $\frac{\partial B_s(x,t)}{\partial t}\ge 0$. Functions $\varphi_0(x)$, $\varphi_{1s}(x)$, $\psi_0(t)$ and $\psi_1(t)$ satisfy the consistency conditions:

$$\varphi_0(0)=\psi_0(0),\ \ \varphi_0(l)=\psi_1(0),\ \ \varphi_{1s}(0)=\psi_0\left(\bar{t}_s\right),\ \ \varphi_{1s}(l)=\psi_1\left(\bar{t}_s\right), s=1,\ldots,L.$$

Problem (1)–(5) is to find the unknown continuous L-dimensional vector function $C(x)=(C_1(x),\ldots,C_L(x))^T$ and the respective solution of the boundary value problem $v(x,t)$, which is twice continuously differentiable with respect to x and once continuously differentiable with respect to t for $(x,t)\in\Omega$, satisfying conditions (1)–(5). Under the above assumptions, the inverse problem (1)–(5) is known to have a solution, and it is unique [7, 8].

To solve problem (1)–(5), we propose an approach based on the method of lines. Problem (1)–(5) is reduced to the system of ordinary differential equations with unknown parameters.

In the domain Ω, let us set up the lines $x_i=ih_x$, $i=0,1,\ldots,N$, $h_x=l/N$. On these lines, we define functions $v_i(t)=v(x_i,t)$, $t\in[0,T]$, $i=0,1,\ldots,N$, for which according to (3)–(5)

$$v_i(0)=\varphi_0(x_i)=\varphi_{0i}, \quad i=0,\ldots,N, \tag{6}$$

$$v_0(t)=\psi_0(t), \quad v_N(t)=\psi_1(t), \quad t\in[0,T], \tag{7}$$

$$v_i\left(\bar{t}_s\right)=\varphi_{1s}(x_i)=\varphi_{1s,i}, \quad \bar{t}_s\in(0,T], \quad s=1,\ldots,L, \quad i=0,\ldots,N. \tag{8}$$

On the lines $x=x_i$, we approximate the derivatives $\partial v/\partial x$ and $\partial^2 v/\partial x^2$ with the use of central difference schemes:

$$\left.\frac{\partial v(x,t)}{\partial x}\right|_{x=x_i}=\frac{v_{i+1}(t)-v_{i-1}(t)}{2h_x}+O\left(h_x^2\right), \quad i=1,\ldots,N-1, \tag{9}$$

$$\left.\frac{\partial^2 v(x,t)}{\partial x^2}\right|_{x=x_i} = \frac{v_{i+1}(t) - 2v_i(t) + v_{i-1}(t)}{h_x^2} + O\left(h_x^2\right), \quad i = 1, \ldots, N-1. \tag{10}$$

Then we use the notations

$$a_i(t) = a(x_i, t), \quad \overline{f}_i(t) = f(x_i, t), \quad a_{1i}(t) = a_1(x_i, t), \quad a_{2i}(t) = a_2(x_i, t),$$
$$B_{si}(t) = B_s(x_i, t), \quad C_{si} = C_s(x_i), \quad s = 1, \ldots, L, \quad i = 1, \ldots, N-1.$$

Substituting (9) and (10) into (1), we obtain the system of ordinary differential equations of $(N - 1)$ th order with unknown (identifiable) vector of parameters $C_s = (C_{s1}, \ldots, C_{s, N-1})^T$:

$$v_i'(t) = \tfrac{a_i(t)}{h_x^2}\left[v_{i+1}(t) - 2v_i(t) + v_{i-1}(t)\right] + \tfrac{a_{1i}(t)}{2h_x}\left[v_{i+1}(t) - v_{i-1}(t)\right] +$$
$$+ a_{2i}(t)v_i(t) + \overline{f}_i(t) + \sum_{s=1}^{L} B_{si}(t)C_{si}, \quad i = 1, 2, \ldots, N-1.$$

Taking into account (6) and (7), we can write this system in the vector-matrix form:

$$\dot{v}(t) = A(t)v(t) + f(t) + \sum_{s=1}^{L} E_{sB(t)}C_s, \quad t \in (0, T], \tag{11}$$

$$v(0) = \varphi_0, \tag{12}$$

$$v\left(\overline{t}_s\right) = \varphi_{1s}, \quad s = 1, \ldots, L, \tag{13}$$

where $v(t) = (v_1(t), \ldots, v_{N-1}(t))^T$, $B_s(t) = (B_{s1}(t), \ldots, B_{s, N-1}(t))^T$, $\varphi_0 = (\varphi_{01}, \ldots, \varphi_{0, N-1})^T$, $\varphi_{1s} = (\varphi_{1s, 1}, \ldots, \varphi_{1s, N-1})^T$, $E_{sB(t)}$ is the $(N - 1)$-dimensional quadratic matrix whose i th element on the principal diagonal is equal to the i th component of vector $B_s(t)$, i.e., $B_{si}(t)$; all the other elements are zero. The nonzero elements of the quadratic $(N - 1)$-dimensional three-diagonal matrix A have the form

$$\tilde{a}_{ii}(t) = \tfrac{1}{h_x^2}\left[-2a_i(t) + h_x^2 a_{2i}(t)\right], \quad i = 1, \ldots, N-1,$$
$$\tilde{a}_{i,i+1}(t) = \tfrac{1}{h_x^2}\left[a_i(t) + \tfrac{h_x}{2}a_{1i}(t)\right], \quad i = 1, \ldots, N-2,$$
$$\tilde{a}_{i,i-1}(t) = \tfrac{1}{h_x^2}\left[a_i(t) - \tfrac{h_x}{2}a_{1i}(t)\right], \quad i = 2, \ldots, N-1.$$

Vector $f(t)$ is defined as follows:

$$f(t) = \left(\overline{f}_1(t) + \left(\frac{a_1(t)}{h_x^2} - \frac{a_{11}(t)}{2h_x}\right)\psi_0(t), \ \overline{f}_2(t), \ldots, \overline{f}_{N-2}(t), \ \overline{f}_{N-1}(t) + \left(\frac{a_{N-1}(t)}{h_x^2} + \frac{a_{1,N-1}(t)}{2h_x}\right)\psi_1(t)\right)^T.$$

Problem (11)–(13) under the conditions imposed on the coefficients of Eq. (1) and functions in the initial-boundary conditions approximate problem (1)–(5) with accuracy $O\left(h_x^2\right)$ (convergence of the solution of problem (11)–(13) to the solution of problem (1)–(5) and error estimates are considered in [21]). Note that error $O\left(h_x^2\right)$ can be improved by using schemes of approximation of derivatives with respect to x of higher order [17, 18].

Denote by $0_{(N-1)\times(N-1)}$ zero $(N-1)$-dimensional quadratic matrix.

Theorem 1 Let $\alpha_s(t)$, $s = 1, \ldots, L$, which is a quadratic matrix function, and $\gamma(t)$, which is a vector function of dimension $(N-1)$, be the solution of the following Cauchy problems:

$$\dot{\alpha}_s(t) = A(t)\alpha_s(t) + E_{sB(t)}, \quad s = 1, \ldots, L, \tag{14}$$

$$\alpha_s(0) = 0_{(N-1)\times(N-1)}, \quad s = 1, \ldots, L, \tag{15}$$

$$\dot{\gamma}(t) = A(t)\gamma(t) + f(t), \tag{16}$$

$$\gamma(0) = \varphi_0. \tag{17}$$

Then for an arbitrary constant $(N-1)$-dimensional vector C_s, the solution of the Cauchy problem (11, 12) is the following vector function:

$$v(t) = \sum_{s=1}^{L} \alpha_s(t)C_s + \gamma(t). \tag{18}$$

Proof With regard to (15) and (17), it is obvious that function $v(t)$ defined from (18), for an arbitrary vector $C_s \in R^{N-1}$, $s = 1, \ldots, L$, satisfies the initial condition (12). Differentiating both sides of Eq. (18) and taking into account (14) and (16), we obtain

$$\begin{aligned}\dot{v}(t) &= \sum_{s=1}^{L} \dot{\alpha}_s(t)C_s + \dot{\gamma}(t) = \sum_{s=1}^{L} \left[A(t)\alpha_s(t) + E_{sB(t)}\right] C_s + [A(t)\gamma(t) + f(t)] = \\ &= A(t)\left(\sum_{s=1}^{L} \alpha_s(t)C_s + \gamma(t)\right) + \sum_{s=1}^{L} E_{sB(t)}C_s + f(t) = \\ &= A(t)v(t) + \sum_{s=1}^{L} E_{sB(t)}C_s + f(t).\end{aligned}$$

Hence, function $v(t)$ satisfies Eq. (11). ■.

Solving independently the Cauchy matrix problem (14, 15) to find $\alpha_s(t)$, $s = 1, \ldots, L$ and the Cauchy problem (16), (17) with respect to the vector function $\gamma(t)$ and using condition (13) and representation (18), we obtain the equality

$$v\left(\bar{t}_s\right) = \varphi_{1s} = \sum_{s=1}^{L} \alpha_s\left(\bar{t}_s\right) C_s + \gamma\left(\bar{t}_s\right), \quad s = 1, \ldots, L, \tag{19}$$

which is an algebraic system of equations of order $(N - 1)$, which can be used to find the vector C_s, $s = 1, \ldots, L$, being identified.

Then, using the values of components of vector $C_s = (C_s(x_1), \ldots, C_s(x_{N-1}))^T$, $s = 1, \ldots, L$, and applying a certain method of interpolation or approximation, we can restore the required function $C_s(x)$, $s = 1, \ldots, L$, on the given class of functions.

If it is necessary to find the solution $v(x, t)$ of the boundary value problem (1)–(5), it will suffice to solve the Cauchy problem (11), (12).

3 Study of the Inverse Problem of Determination a Source Depending on a Time Variable

Now, let us consider the inverse source problem with respect to the parabolic equation

$$\frac{\partial v(x,t)}{\partial t} = a\left(x,t\right) \frac{\partial^2 v(x,t)}{\partial x^2} + a_1\left(x,t\right) \frac{\partial v(x,t)}{\partial x} + a_2\left(x,t\right) v\left(x,t\right) + f\left(x,t\right) + F\left(x,t\right),$$
$$(x,t) \in \Omega = \{(x,t) : 0 < x < l, \quad 0 < t \le T\}, \tag{20}$$

where

$$F\left(x,t\right) = \sum_{s=1}^{L} C_s\left(x,t\right) B_s(t), \tag{21}$$

under the following initial-boundary value conditions and overdetermination conditions:

$$v\left(x,0\right) = \varphi_0(x), \quad x \in [0,l], \tag{22}$$

$$v\left(0,t\right) = \psi_0(t), \quad v\left(l,t\right) = \psi_1(t), \quad t \in [0,T], \tag{23}$$

$$v\left(\overline{x}_s, t\right) = \psi_{2s}(t), \quad \overline{x}_s \in (0,l), \quad t \in [0,T], \quad s = 1, \ldots, L. \tag{24}$$

Here, $L > 0$ is a given integer that determines the number of identified sources and overdetermination conditions; $\overline{x}_s \in (0,l)$, $s = 1, \ldots, L$ are given points; given functions $a(x,t) \ge \mu > 0$, $a_1(x,t)$, $a_2(x,t)$, $f(x,t)$, $\varphi_0(x)$, $\psi_0(t)$, $\psi_1(t)$, $\psi_{2s}(t)$, $C_s(x,t)$, $s = 1, \ldots, L$, $\mu = const > 0$; are continuous with respect to x and t;

$s = 1, \ldots, L$, are linearly independent functions, and $a_2(x,t) \le 0$, $\varphi_0(x) \in C^2([0,l])$, $\psi_{2s}(t) \in C^1([0,T])$, $|C_s(\overline{x}_k, t)| \ge \delta > 0$, $s, k = 1, \ldots, L$, $t \in [0,T]$, $\delta = const > 0$. Given points $\overline{x}_s$, $s = 1, \ldots, L, \overline{x}_i \ne \overline{x}_j$ for $i \ne j$, $i, j = 1, \ldots, L$ are observation points. Functions $\varphi_0(x)$, $\psi_0(t)$, $\psi_1(t)$ and $\psi_{2s}(t)$ satisfy the consistency conditions:

$$\varphi_0(0) = \psi_0(0), \quad \varphi_0(l) = \psi_1(0), \quad \psi_{2s}(0) = \varphi_0(\overline{x}_s), \quad s = 1, \ldots, L.$$

Problem (20)–(24) is to determine the unknown continuous L-dimensional vector function $B(t) = (B_1(t), \ldots, B_L(t))^T$ and the corresponding solution to the boundary value problem $v(x,t) \in C^{2,1}(\Omega) \cap C^{1,0}(\overline{\Omega})$ that satisfy conditions (20)–(24).

Note that under the above assumptions both the initial-boundary value problem (20), (22), (23) under the given continuous L-dimensional vector function $B(t) = (B_1(t), \ldots, B_L(t))^T$, i.e., function $F(x,t)$ [24–27], and inverse problem (20)–(24) has solutions and they are unique [1, 22, 23, 28].

Now, let us consider an approach to numerical solution of problem (20)–(24), based on the method of lines. At first, problem (20)–(24) reduces to a system of ordinary differential equations with unknown parameters.

In the domain Ω, we set up the lines

$$t_j = jh_t, \quad j = 0, 1, \ldots, N, \quad h_t = T/N.$$

On these lines, we define functions

$$v_j(x) = v(x, t_j), \quad x \in [0,l], \quad j = 0, 1, \ldots, N,$$

for which the following equalities take place on the basis of (22)–(24):

$$v_0(x) = \varphi_0(x), \quad x \in [0,l], \tag{25}$$

$$v_j(0) = \psi_0(t_j) = \psi_{0j}, \quad j = 0, \ldots, N, \tag{26}$$

$$v_j(l) = \psi_1(t_j) = \psi_{1j}, \quad j = 0, \ldots, N, \tag{27}$$

$$v_j(\overline{x}_s) = \psi_{2s}(t_j) = \psi_{2s,j}, \quad \overline{x}_s \in (0,l), \quad s = 1, \ldots, L, \quad j = 0, \ldots, N. \tag{28}$$

On the straight lines $t = t_j$, we approximate derivative $\partial v(x,t)/\partial t$ with the use of the difference scheme:

$$\left.\frac{\partial v(x,t)}{\partial t}\right|_{t=t_j} = \frac{v_j(x) - v_{j-1}(x)}{h_t} + O(h_t), \quad j = 1, \ldots, N. \tag{29}$$

Using (29) in Eq. (20), we obtain N differential equations:

$$v_j''(x) + \tilde{a}_{1j}(x)v_j'(x) + \tilde{a}_{2j}(x)v_j(x) + \tilde{f}_j(x) + \sum_{s=1}^{L} \tilde{C}_{sj}(x)B_{sj} = 0, \quad j = 1, \ldots, N, \quad x \in (0, l), \tag{30}$$

where

$$B_{sj} = B_s\left(t_j\right), \quad \tilde{f}_j(x) = \frac{v_{j-1}(x) + h_t f(x,t_j)}{a(x,t_j)h_t}, \quad \tilde{C}_{sj}(x) = \frac{C_s(x,t_j)}{a(x,t_j)},$$
$$\tilde{a}_{1j}(x) = \frac{a_1(x,t_j)}{a(x,t_j)}, \quad \tilde{a}_{2j}(x) = \frac{a_2(x,t_j)h_t - 1}{a(x,t_j)h_t}.$$

Convergence as $h_t \to 0$ and error of the method of lines in approximation of the derivatives with respect to t in Eq. (20) (in this case, in approximation of problem (20)–(24) by problem (30), (26)–(28) with an error estimated as $O(h_t)$) are analyzed in [21]. Hence, for the known B_{sj}, solution to the boundary value problem with respect to system (30) as $h_t \to 0$ converges to the solution of the initial-boundary value problem (20)–(23). From the existence and uniqueness of solution of the initial inverse problem (20)–(24), we can show the existence and uniqueness of solution of inverse problem (30), (26)–(28). Indeed, if solutions of the inverse problem (30), (26)–(28) do not exist or are nonunique, the initial problem has similar properties.

The equations of system (30) for each j can be solved independently and sequentially, beginning with $j = 1$ to N, and hence components of the vector $B_s = (B_{s1}, \ldots, B_{sN})^T$ are defined sequentially.

Theorem 2 Let functions $\alpha_j(x)$ and $\beta_{sj}(x)$, $s = 1, \ldots, L$, for $x \in [0, l]$, be the solutions of the following Cauchy problems:

$$\alpha_j''(x) + \tilde{a}_{1j}(x)\alpha_j'(x) + \tilde{a}_{2j}(x)\alpha_j(x) + \tilde{f}_j(x) = 0, \tag{31}$$

$$\alpha_j(0) = \psi_{0j}, \; \alpha_j'(0) = \psi_{2j}, \tag{32}$$

$$\beta_{sj}''(x) + \tilde{a}_{1j}(x)\beta_{sj}'(x) + \tilde{a}_{2j}(x)\beta_{sj}(x) + \tilde{C}_{sj}(x) = 0, \tag{33}$$

$$\beta_{sj}(0) = 0, \; \beta_{sj}'(0) = 0. \tag{34}$$

Then for arbitrary values of parameter B_{sj}, function

$$v_j(x) = \alpha_j(x) + \sum_{s=1}^{L} \beta_{sj}(x)B_{sj}, \quad x \in [0, l] \tag{35}$$

satisfy the system of differential Eqs. (30) and conditions (26), (27).

Proof From conditions (32), (34), it is obvious that functions $v_j(x)$ defined from (35), for arbitrary values of B_{sj}, $s = 1, \ldots, L, j = 1, \ldots, N$, satisfy conditions (26), (27). The fact that $v_j(x)$, $j = 1, \ldots, N$ satisfy the system of differential eqs. (30) can be verified by direct differentiation of (35) and substitution of $v'_j(x)$ and $v''_j(x)$ into (30) taking into account (31), (33):

$$\begin{aligned}
&v''_j(x) + \tilde{a}_{1j}(x)v'_j(x) + \tilde{a}_{2j}(x)v_j(x) + \tilde{f}_j(x) + \sum_{s=1}^{L} \tilde{C}_{sj}(x)B_{sj} = \\
&= \alpha''_j(x) + \sum_{s=1}^{L} \beta''_{sj}(x)B_{sj} + \tilde{a}_{1j}(x)\alpha'_j(x) + \tilde{a}_{1j}(x)\sum_{s=1}^{L} \beta'_{sj}(x)B_{sj} + \\
&+ \tilde{a}_{2j}(x)\alpha_j(x) + \tilde{a}_{2j}(x)\sum_{s=1}^{L} \beta_{sj}(x)B_{sj} + \tilde{f}_j(x) + \sum_{s=1}^{L} \tilde{C}_{sj}(x)B_{sj} = \\
&= \left[\alpha''_j(x) + \tilde{a}_{1j}(x)\alpha'_j(x) + \tilde{a}_{2j}(x)\alpha_j(x) + \tilde{f}_j(x)\right] + \\
&+ \sum_{s=1}^{L} \left(\beta''_{sj}(x) + \tilde{a}_{1j}(x)\beta'_{sj}(x) + \tilde{a}_{2j}(x)\beta_{sj}(x) + \tilde{C}_{sj}(x)\right) B_{sj} = 0, \\
&j = 1, \ldots, N, \quad x \in (0, l).
\end{aligned}$$

■

We can easily prove the following theorem.

Theorem 3 Representation (35) for solution of differential eq. (30) with boundary conditions (26), (27) is unique.

Let us solve separately two Cauchy problem (31)–(34), using condition (28) and representation (35). We obtain the equality

$$v_j\left(\overline{x}_s\right) = \alpha_j\left(\overline{x}_s\right) + \sum_{s=1}^{L} \beta_{sj}\left(\overline{x}_s\right) B_{sj} = \psi_{2s,j}, \quad s = 1, \ldots, L, \tag{36}$$

which is an algebraic system of equations, which can be used to find the vector B_s, $s = 1, \ldots, L$, being identified. Considering that L, which is the number of unknown functions that appear in Eq. (20), is as a rule insignificant in real problems, any well-known methods, for example, Gauss method or iterated methods, can be used to solve the algebraic system of eq. (36).

Solvability of system (36) depends on solvability of the inverse problem (30), (26)–(28) and vice versa; if system (36) has no solution, problem (30), (26)–(28) and hence initial problem (20)–(24) have no solution neither. Thus, the properties of existence and uniqueness of the solution of system (36) and of original inverse problem (20)–(24) are interrelated.

Function $v_j(x)$, $x \in [0, l]$ can be found from the solution of problem (30), (26), (27). Then procedure (31)–(36) is repeated on the line $t = t_{j+1}$, on which $v_{j+1}(x)$ is defined.

Thus, to find components of the parameter vector B_s, $s = 1, \ldots, L$, it is necessary to solve the Cauchy problem N times with respect to $(L + 1)$

independent differential equations of the second order. The calculated vector $B_s = (B_s(t_1), \ldots, B_s(t_N))^T$ with the application of the methods of interpolation or approximation can then be used to obtain the analytical form of function $B_s(t)$.

4 Results of Numerical Experiments

Let us present the solution results for the following inverse source problems.

Problem 1 Consider the problem

$$\frac{\partial v(x,t)}{\partial t} = x^2 \frac{\partial^2 v(x,t)}{\partial x^2} + e^t C(x), \quad (x,t) \in \Omega = \{(x,t) : 0 < x < 1, \ 0 < t \le 1\},$$
$$v(x,0) = x^2 \cos x, \quad v(x,1) = ex^2 \cos x, \quad x \in [0,1],$$
$$v(0,t) = 0, \quad v(1,t) = e^t \cos 1, \quad t \in [0,1].$$

The exact solution of this problem are functions

$$C(x) = x^2 \left(x^2 - 1\right) \cos x + 4x^3 \sin x, \quad v(x,t) = e^t x^2 \cos x.$$

Numerical experiments were carried out for different number N of lines $x = x_i$, $i = 1, \ldots, N$. To solve auxiliary Cauchy problems, we used the Runge-Kutta method of the fourth order for different steps h_t. We carried out calculations under the presence of random noise in function $v(x, 1)$, which was defined as follows:

$$v^\sigma(x,1) = v(x,1)(1 + \sigma \text{ rand}),$$

where σ is noise-level percentage, *rand* are the random numbers generated by means of the MATLAB function rand for uniform distribution on the interval $[-1, 1]$.

Table 1 presents the results of solution of Problem 1 for $N = 20$, $h_t = 0.001$ for the noise levels equal $\sigma = 1\%$, $\sigma = 3\%$, and $\sigma = 5\%$, as well as without noise, i.e., $\sigma = 0\%$.

Figure 1 provides the graphs of exact (analytic solution) and obtained by numerical methods (presented in Sec. 2) coefficient $C(x)$ under various noise levels based on data from Table 1 for Problem 1.

The solution accuracy of inverse problems, as one would expect, substantially depends on the number of lines N used in the method of lines for approximation of the original boundary-value problem.

In the problem of finding $C(x)$, increase in the number of lines increases the order of the system of differential equations with ordinary derivatives, equal to N^2. This substantially increases the amount of computation and hence increases the computing error. Therefore, in solving a specific problem of identification of coefficient $C(x)$ to choose the number of lines, additional numerical analysis is necessary.

Table 1 Values of the coefficient $C(x)$ for Problem 1

i	x_i	Values of $C(x)$				
			Obtained values for σ (%)			
		Exact value	$\sigma = 0.0$	$\sigma = 1.0$	$\sigma = 3.0$	$\sigma = 5.0$
1	0.05	−0.002466	−0.002459	−0.002474	−0.002504	−0.002535
2	0.10	−0.009451	−0.009427	−0.009610	−0.009978	−0.010345
3	0.15	−0.019729	−0.019675	−0.020085	−0.020904	−0.021724
4	0.20	−0.031277	−0.031182	−0.031389	−0.031803	−0.032217
5	0.25	−0.041309	−0.041165	−0.039985	−0.037625	−0.035264
6	0.30	−0.046326	−0.046126	−0.041688	−0.032811	−0.023934
7	0.35	−0.042170	−0.041910	−0.032199	−0.012779	0.006641
8	0.40	−0.024100	−0.023778	−0.007624	0.024685	0.056993
9	0.45	0.013128	0.013510	0.035243	0.078708	0.122173
10	0.50	0.075166	0.075603	0.098996	0.145783	0.192570
11	0.55	0.167971	0.168454	0.186165	0.221585	0.257006
12	0.60	0.297694	0.298211	0.300125	0.303952	0.307779
13	0.65	0.470558	0.471092	0.446190	0.396386	0.346582
14	0.70	0.692733	0.693263	0.632543	0.511105	0.389666
15	0.75	0.970201	0.970703	0.870646	0.670534	0.470421
16	0.80	1.308624	1.309067	1.174817	0.906317	0.637817
17	0.85	1.713198	1.713549	1.560817	1.255352	0.949888
18	0.90	2.188515	2.188739	2.043520	1.753082	1.462644
19	0.95	2.738424	2.738473	2.633984	2.425005	2.216027

Problem 2 Let us consider the problem

$$\frac{\partial v(x,t)}{\partial t} = a(x)\frac{\partial^2 v(x,t)}{\partial x^2} + B(t)e^{2x}\left[\frac{1}{7}x - 4a(x)\,(x+1)\right], \quad (x,t) \in \Omega = \{(x,t) : 0 < x < 1, \quad 0 < t \le 1\},$$
$$v\,(x,0) = xe^{2x}, \quad x \in [0,1]\,,$$
$$v\,(0,t) = 0, \quad v\,(1,t) = e^{\frac{t}{7}+2}, \quad \frac{\partial v(0,t)}{\partial x} = e^{\frac{t}{7}}, \quad t \in [0,1]\,,$$

where $a(x) = \frac{\cos x}{e^x}$. Exact solutions of this problem are functions

$$B(t) = e^{\frac{t}{7}}, \quad v\,(x,t) = xe^{\frac{t}{7}+2x}.$$

Numerical experiments were carried out with different number N of lines $t = t_j$, $j = 1, \ldots, N$. To solve auxiliary Cauchy problems, the fourth-order Runge-Kutta method was used at different steps h_x.

Table 2 shows the results of solving Problem 2 with the number of lines $N = 100$, 200, 500.

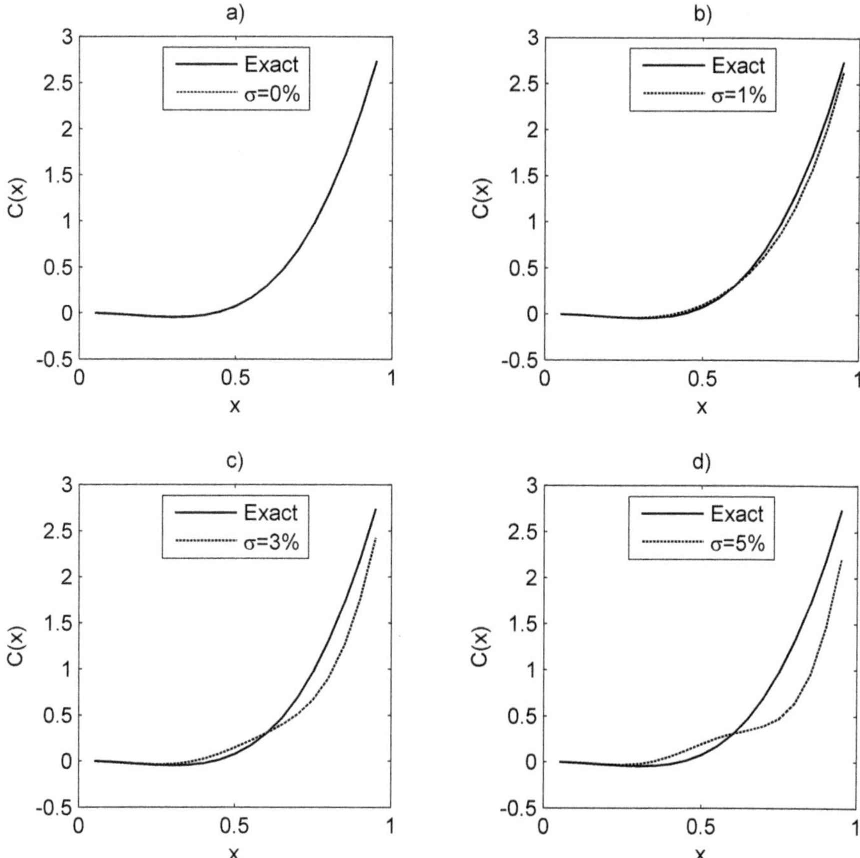

Fig. 1 Graphs of exact (Exact) and obtained by numerical methods coefficient $C(x)$ under different noise levels for Problem 1

Table 2 Obtained and exact values of $B(t)$ for Problem 2

	$N = 100$		$N = 200$		$N = 500$		
t_j	$B(t_j)$	$\lvert\Delta B(t_j)\rvert$	$B(t_j)$	$\lvert\Delta B(t_j)\rvert$	$B(t_j)$	$\lvert\Delta B(t_j)\rvert$	Exact $B(t_j)$
0.10	101.403	0.00036	101.425	0.00014	101.433	0.00006	101.439
0.20	102.856	0.00042	102.883	0.00016	102.892	0.00006	102.898
0.30	104.336	0.00043	104.363	0.00016	104.372	0.00007	104.379
0.40	105.837	0.00044	105.864	0.00016	105.874	0.00007	105.881
0.50	107.360	0.00045	107.388	0.00017	107.397	0.00007	107.404
0.60	108.904	0.00045	108.933	0.00017	108.943	0.00007	108.949
0.70	110.471	0.00046	110.500	0.00017	110.510	0.00007	110.517
0.80	112.061	0.00046	112.090	0.00017	112.100	0.00007	112.107
0.90	113.673	0.00047	113.703	0.00018	113.713	0.00007	113.720

Table 3 Values of the coefficient $B(t)$ for Problem 2

j	t_j	Values of $B(t)$				
			Obtained values for σ (%)			
		Exact value	$\sigma = 0.0$	$\sigma = 1.0$	$\sigma = 3.0$	$\sigma = 5.0$
25	0.05	1.007168	1.007117	1.004635	0.999671	0.994708
50	0.10	1.014388	1.014331	1.011494	1.005821	1.000148
75	0.15	1.021660	1.021599	1.018857	1.013373	1.007889
100	0.20	1.028984	1.028920	1.026556	1.021827	1.017098
125	0.25	1.036360	1.036295	1.034525	1.030986	1.027446
150	0.30	1.043789	1.043724	1.042709	1.040681	1.038653
175	0.35	1.051271	1.051205	1.051040	1.050710	1.050379
200	0.40	1.058807	1.058741	1.059438	1.060833	1.062228
225	0.45	1.066397	1.066330	1.067822	1.070805	1.073787
250	0.50	1.074041	1.073974	1.076113	1.080392	1.084670
275	0.55	1.081741	1.081673	1.084251	1.089406	1.094561
300	0.60	1.089495	1.089427	1.092191	1.097718	1.103246
325	0.65	1.097305	1.097236	1.099916	1.105275	1.110634
350	0.70	1.105171	1.105102	1.107435	1.112100	1.116766
375	0.75	1.113093	1.113024	1.114782	1.118297	1.121813
400	0.80	1.121072	1.121002	1.122013	1.124035	1.126056
425	0.85	1.129109	1.129038	1.129203	1.129533	1.129862
450	0.90	1.137203	1.137132	1.136434	1.135040	1.133645
475	0.95	1.145355	1.145283	1.143792	1.140810	1.137827

Calculations were carried out under the presence of random noise in function $\frac{\partial v(0,t)}{\partial x}$, caused by errors of measurement of the state $v(0, t)$ at the left end, which were defined as follows:

$$\left(\frac{\partial v\,(0,t)}{\partial x}\right)^{\sigma} = \sigma \ \text{rand}.$$

Here, σ determines the level of error in measurements, *rand* is the random number uniformly distributed on the interval $[-1, 1]$ and obtained with the use of the MATLAB function *rand*.

Table 3 shows the results of solution of Problem 2 for the number of straight lines $N = 500$, $h_x = 0.002$ for error levels $\sigma = 1\%$, $\sigma = 3\%$, and $\sigma = 5\%$ and also without noise, i.e., $\sigma = 0\%$.

Figure 2 shows graphs of exact values of the coefficient $B(t)$ (an analytical solution) and of those obtained by the numerical method proposed in Sec. 3, for different noise levels σ.

Results of a big number of numerical experiments on solving various inverse test problems on determining the coefficients $B(t)$ have shown the following. As one would expect, the accuracy of solution of inverse problems substantially depends on

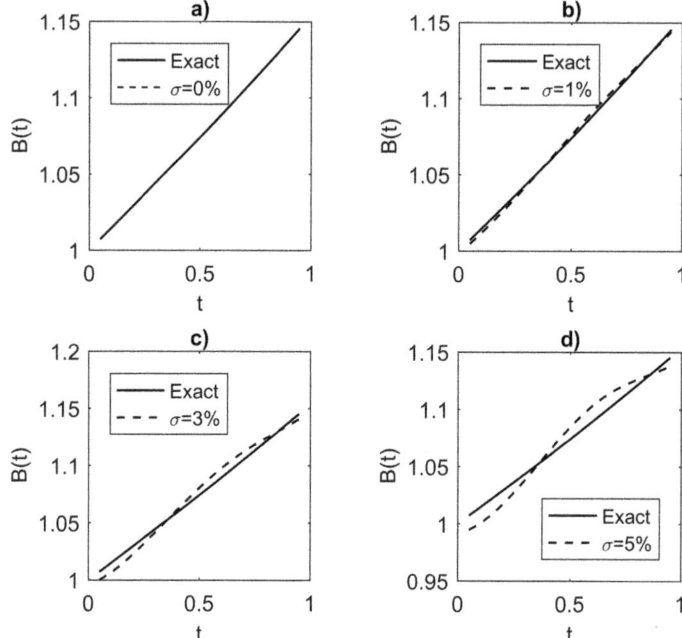

Fig. 2 Graphs of exact and numerically obtained coefficient $B(t)$ at different noise levels for Problem 2

the number of used lines N in the method of lines for approximation of the original boundary value problem.

For the problem of identification of coefficient $B(t)$, increase in the number of straight lines has no significant influence on the computing process since problem (22), (18), (19) on each line $t = t_j$ is solved independently and sequentially for $j = 1, 2, \ldots, N$. Solution of this problem is possible with almost any given accuracy with the use of well-known efficient numerical methods of solution of Cauchy problems.

5 Conclusions

The numerical methods we have proposed in the chapter to solve inverse source problems for a parabolic equation are expedient since they are reduced to the solution of auxiliary, well-analyzed Cauchy problems and do not need iterative procedures to be constructed. To this end, standard software such as MATLAB can be used. Problems with nonlocal initial and boundary conditions, which are often met in practice, are reduced to the considered classes of problems.

Noteworthy is that the proposed technique of construction of numerical methods can be used for other types of partial differential equations with other given forms of initial-boundary conditions.

References

1. A. I. Prilepko, D. G. Orlovsky, and I. A. Vasin, Methods for Solving Inverse Problems in Mathematical Physics, M. Dekker, New York (2000).
2. M. I. Ivanchov, Inverse Problems for Equations of Parabolic Type, VNTL Publications, Lviv (2003).
3. A. Farcas and D. Lesnic, "The boundary-element method for the determination of a heat source dependent on one variable," J. Eng. Math., Vol. 54, 375–388 (2006).
4. T. Johansson and D. Lesnic, "A variational method for identifying a spacewise-dependent heat source," IMA J. Appl. Math., Vol. 72, 748–760 (2007).
5. A. Hasanov, "Identification of spacewise and time dependent source terms in 1D heat conduction equation from temperature measurement at a final time," Int. J. Heat Mass Transfer, Vol. 55, 2069–2080 (2012).
6. A. Hasanov, "An inverse source problem with single Dirichlet type measured output data for a linear parabolic equation," Appl. Math. Lett., Vol. 24, 1269–1273 (2011).
7. A. I. Prilepko and A. B. Kostin, "Some inverse problems for parabolic equations with final and integral observation," Matem. Sb., Vol. 183, No. 4, 49–68 (1992).
8. E. G. Savateev, "The problem of identification of the coefficient of parabolic equation," Sib. Matem. Zhurn., Vol. 36, No. 1, 177–185 (1995).
9. L. Yan, C. L. Fu, and F. L. Yang, "The method of fundamental solutions for the inverse heat source problem," Eng. Anal. Boundary Elements, Vol. 32, 216–222 (2008).
10. M. Nili Ahmadabadi, M. Arab, and F.M. Maalek Ghaini, "The method of fundamental solutions for the inverse space-dependent heat source problem," Eng. Anal. Bound. Elem., Vol. 33, 1231–1235 (2009).
11. M. I. Ismailov, F. Kanca, and D. Lesnic, "Determination of a time-dependent heat source under nonlocal boundary and integral overdetermination conditions," Appl. Math. Comput., Vol. 218, 4138–4146 (2011).
12. V. L. Kamynin, "On the inverse problem of determining the right-hand side in the parabolic equation with the condition of integral overdetermination," Matem. Zametki, Vol. 77, No. 4, 522–534 (2005).
13. A. Mohebbia and M. Abbasia, "A fourth-order compact difference scheme for the parabolic inverse problem with an overspecification at a point," Inverse Problems in Science and Engineering, Vol. 23, No. 3, 457–478 (2015).
14. I. N. Parasidis and E. Providas, An exact solution method for a class of nonlinear loaded difference equations with multipoint boundary conditions, Journal of Difference Equations and Applications. (10) 24, 1649–1663 (2018).
15. A. A. Samarskii and E. S. Nikolaev, Methods to Solve Finite Difference Equations [in Russian], Nauka, Moscow (1978).
16. W. E. Schiesser, The Numerical Method of Lines: Integration of Partial Differential Equations, Academic Press, San Diego (1991).
17. A. V. Samusenko and S. V. Frolova, "Multipoint schemes of the longitudinal variant of highly accurate method of lines to solve some problems of mathematical physics," Vestsi NAN Belarusi, Ser. Fiz.-Mat. Navuk, No. 3, 31–39 (2009).
18. O. A. Liskovets, "The method of lines," Diff. Uravn., Vol. 1, No. 12, 1662–1678 (1965).
19. K. R. Aida-zade and A. B. Rahimov, "An approach to numerical solution of some inverse problems for parabolic equations," Inverse Problems in Science and Engineering, Vol. 22, No. 1, 96–111 (2014).

20. K. R. Aida-zade and A. B. Rahimov, "Solution to classes of inverse coefficient problems and problems with nonlocal conditions for parabolic equations," Differential Equations, Vol. 51, No. 1, 83–93 (2015).
21. E. Rothe, "Zweidimensionale parabolische Randwertaufgaben als Grenzfall eindimensionaler Randwertaufgaben," Math. Ann., Vol. 102, No. 1, 650–670 (1930).
22. A. I. Prilepko and V. V. Solov'yev, "Solvability theorems and the Rothe method in inverse problems for the equation of parabolic type. I," Diff. Uravneniya, Vol. 23, No. 10, 1791–1799 (1987).
23. V. V. Solov'yev, "Determining the source and coefficients in a parabolic equation in a multidimentional case," Diff. Uravneniya, Vol. 31, No. 6, 1060–1069 (1995).
24. V. A. Il'yin, "Solvability of mixed problems for hyperbolic and parabolic equations," Uspekhi Mat. Nauk, Vol. 15, No. 2, 97–154 (1960).
25. A. M. Il'yin, A. S. Kalashnikov, and O. A. Oleinik, "Second-order linear equations of parabolic type," Uspekhi Mat. Nauk, Vol. 17, No. 3, 3–146 (1962).
26. S. D. Eidelman, "Parabolic equations. Partial differential equations," Itogi Nauki i Tekhniki, Ser. Sovrem. Problemy Matematiki. Fundamental'nye Napravleniya, 63, VINITI, Moscow (1990), pp. 201–313.
27. V. I. Smirnov, A Course in Higher Mathematics [in Russian], Vol. IV, Pt. 2, Nauka, Moscow (1981).
28. V. V. Solov'yev, "Existence of a solution "as a whole" to the inverse problem of determining the source in a quasilinear equation of parabolic type," Differential Equations. Vol. 32, No. 4, 536– 544 (1996).

Using an Integrating Factor to Transform a Second Order BVP to a Fixed Point Problem

Richard I. Avery, Douglas R. Anderson, and Johnny Henderson

Abstract Using an integrating factor, a second order boundary value problem is transformed into a fixed point problem. We provide growth conditions for the existence of a fixed point to the associated operator for this transformation and conclude that the index of the operator applying the standard Green's function approach is zero; this does not guarantee the existence of a solution, demonstrating the value and potential for this new transformation.

1 Introduction

Converting boundary value problems to fixed point problems is a standard approach for existence of solutions arguments, and for finding solutions using iterative methods. The standard method to transform a boundary value problem to a fixed point problem is to apply Green's function techniques, see [12, 16] for a thorough treatment. Recently, other methods have been developed to bring the operator inside the nonlinear term, corresponding to a transformation of a transformation that Burton refers to as the Direct Fixed Point Mapping, see [3, 4, 7, 8] for a discussion of these transformations. There are also many different transformation results related to boundary value problems, whose transformations result in sums or products of operators, see [1, 9, 11, 18] for a discussion of these types of transformations. While others may not be fundamentally different than the standard

R. I. Avery
College of Arts and Sciences, Dakota State University, Madison, SD, USA
e-mail: rich.avery@dsu.edu

D. R. Anderson (✉)
Department of Mathematics, Concordia College, Moorhead, MN, USA
e-mail: andersod@cord.edu

J. Henderson
Department of Mathematics, Baylor University, Waco, TX, USA
e-mail: Johnny_Henderson@baylor.edu

I. N. Parasidis et al. (eds.), *Mathematical Analysis in Interdisciplinary Research*,
Springer Optimization and Its Applications 179,
https://doi.org/10.1007/978-3-030-84721-0_7

Green's function approach, the resulting transformed operator appears different due to the nature of the boundary value problem itself. For example, p-Laplacian boundary value problems, where the transformed operator is found by multiple integration steps while using the boundary conditions, see [2, 14] for some examples of transformations of this type.

Introducing an integrating factor in the transformation process brings new terms into the transformed operator, which results in more flexibility (degrees of freedom) for existence of solutions arguments, as well as iteration arguments. The Leggett–Williams generalization [17] of the Krasnoselskii fixed point theorem [5, 6, 15] altered the sets for which to apply index theory, whereas the techniques in this paper change the operator that the fixed point theorems are applied to, in order to show existence of solutions or to find solutions to boundary value problems. We will conclude with an example showing that the transformed operator is invariant in a ball in which the standard operator transforming a boundary value problem applying Green's function techniques is not invariant in a similar ball.

2 Introducing a Term to Create a New Operator

Existence of solutions of the continuous right focal boundary value problem

$$x''(t) + f(x(t)) = 0, \quad t \in (0, 1), \tag{1}$$

$$x(0) = x'(1) = 0, \tag{2}$$

will be shown via this new approach. The next result introduces the operator based on the introduction of an integrating factor into Eq. (1).

Theorem 1 *If given a function $a \in C[0, 1]$, then x is a solution of* (1), (2) *if and only if x is a fixed point of the operator H defined by*

$$Ax(t) = \int_0^t e^{-\int_s^t a(w)\,dw} \left(\int_s^1 f(x(r))\,dr + a(s)x(s) \right) ds.$$

Proof A function x is a solution of (1), (2) if and only if

$$\int_t^1 x''(r)dr = \int_t^1 -f(x(r))dr,$$

and x satisfies the boundary conditions $x(0) = 0 = x'(1)$. Thus, x is a solution of the first order boundary value problem

$$x'(t) = \int_t^1 f(x(r))dr, \quad t \in (0, 1), \tag{3}$$

$$x(0) = 0. \tag{4}$$

The equation in (3) is equivalent to

$$x'(t) + a(t)x(t) = \int_t^1 f(x(r))dr + a(t)x(t);$$

by introducing the integrating factor $e^{\int_0^t a(w)dw}$, this is equivalent to

$$\frac{d}{dt}\left(x(t)e^{\int_0^t a(w)dw}\right) = e^{\int_0^t a(w)dw}\left(\int_t^1 f(x(r))dr + a(t)x(t)\right).$$

Hence, (3) is equivalent to

$$x(t)e^{\int_0^t a(w)dw} = \int_0^t e^{\int_0^s a(w)dw}\left(\int_s^1 f(x(r))dr + a(s)x(s)\right) ds,$$

since $x(0) = 0$. Thus, we have that (3), (4) is equivalent to

$$x(t) = \int_0^t e^{-\int_s^t a(w)dw}\left(\int_s^1 f(x(r))dr + a(s)x(s)\right) ds \tag{5}$$

for $t \in [0, 1]$. Therefore, if we define the operator A by

$$Ax(t) = \int_0^t e^{-\int_s^t a(w)dw}\left(\int_s^1 f(x(r))dr + a(s)x(s)\right) ds,$$

then we have that x is a solution of (1), (2) if and only if x is a fixed point of the operator A. This ends the proof. □

3 Application of the New Fixed Point Theorem

Define the cone P of the Banach space $C[0, 1]$ by

$$P = \{x \in E \ : \ x \ \text{is non-negative and non-decreasing}\},$$

with the standard $C[0, 1]$ supnorm given by

$$\|x\| = \sup_{t\in[0,1]} |x(t)|.$$

Also, for $r > 0$, let

$$P_r = \{x \in P \;:\; \|x\| < r\}.$$

Theorem 2 *Let $\lambda < 0$, $a(t) = \lambda(1-t)$, and $f : [0,\infty) \to [0,\infty)$ be a continuous function with*

$$f(x) + \lambda x \geq 0$$

for all non-negative real numbers x. Then, $A : P \to P$.

Proof Suppose $\lambda < 0$, $a(t) = \lambda(1-t)$, and $f : [0,\infty) \to [0,\infty)$ is a continuous function with

$$f(w) + \lambda w \geq 0$$

for all non-negative real numbers w; moreover, let $x \in P$. Thus, for $t \in [0, 1]$,

$$\begin{aligned}\int_t^1 f(x(r))dr + \lambda(1-t)x(t) &\geq \int_t^1 -\lambda x(r)dr + \lambda(1-t)x(t)\\ &\geq \int_t^1 -\lambda x(t)dr + \lambda(1-t)x(t)\\ &= -(1-t)\lambda x(t) + \lambda(1-t)x(t) = 0.\end{aligned}$$

It follows that

$$Ax(t) = \int_0^t e^{-\int_s^t \lambda(1-w)dw}\left(\int_s^1 f(x(r))dr + \lambda(1-s)x(s)\right) ds \geq 0$$

and

$$\begin{aligned}(Ax)'(t) = &\int_t^1 f(x(r))dr + \lambda(1-t)x(t)\\ &+ \int_0^t \lambda(t-1)e^{\lambda\left(\frac{(t-1)^2}{2} - \frac{(s-1)^2}{2}\right)}\left(\int_s^1 f(x(r))dr + \lambda(1-s)x(s)\right) ds\\ \geq &\; 0.\end{aligned}$$

Therefore, since Ax is clearly an element of $C[0, 1]$, we have that $A : P \to P$. This ends the proof. □

In the following theorem, we will employ some elementary results from index theory, see [10, 13, 19] for a thorough treatment of index theory.

Theorem 3 *Let $\lambda < 0$, $a(t) = \lambda(1-t)$, $x_0 \equiv r > 0$, and let $f : [0,\infty) \to [0,\infty)$ be a continuous function with*

$$f(x) + \lambda x \geq 0$$

for all non-negative real numbers x such that

$$f(r) \leq \frac{-\lambda r}{1 - e^{\frac{\lambda}{2}}}.$$

Then, $A : P_r \to P_r$, and (1), (2) has at least one positive solution $x^* \in \overline{P_r}$.

Proof Let $r > 0$ and $x_0 \equiv r$. Since

$$f(r) \leq \frac{-\lambda r}{1 - e^{\frac{\lambda}{2}}},$$

we have that for all $t \in [0, 1]$

$$\begin{aligned} r &\geq \left(\frac{f(r)}{\lambda}\right)\left(e^{\frac{\lambda}{2}} - 1\right) \\ &\geq \left(\frac{f(r)}{\lambda}\right)\left(e^{\frac{-\lambda((t-1)^2-1)}{2}} - 1\right). \end{aligned}$$

As a result,

$$\left(\frac{f(r)}{\lambda}\right)\left(1 - e^{\frac{\lambda((t-1)^2-1)}{2}}\right) \leq re^{\frac{\lambda((t-1)^2-1)}{2}}.$$

Hence, for all $t \in [0, 1]$, we have

$$\begin{aligned} Ax_0(t) &= \int_0^t e^{-\int_s^t \lambda(1-w)dw}\left(\int_s^1 f(r)dz + \lambda(1-s)r\right) ds \\ &= (f(r) + \lambda r)\int_0^t e^{\lambda\left(\frac{(t-1)^2}{2} - \frac{(s-1)^2}{2}\right)}(1-s)\, ds \\ &= (f(r) + \lambda r)e^{\frac{\lambda(t-1)^2}{2}}\int_0^t e^{\frac{-\lambda(s-1)^2}{2}}(1-s)\, ds \\ &= (f(r) + \lambda r)e^{\frac{\lambda(t-1)^2}{2}}\int_{\frac{-1}{2}}^{\frac{-(t-1)^2}{2}} e^{\lambda u}\, du \end{aligned}$$

$$
\begin{aligned}
&= (f(r)+\lambda r)e^{\frac{\lambda(t-1)^2}{2}}\left(\frac{e^{\frac{-\lambda(t-1)^2}{2}}-e^{\frac{-\lambda}{2}}}{\lambda}\right)\\
&= \left(\frac{f(r)+\lambda r}{\lambda}\right)\left(1-e^{\frac{\lambda((t-1)^2-1)}{2}}\right)\\
&= \left(\frac{f(r)}{\lambda}\right)\left(1-e^{\frac{\lambda((t-1)^2-1)}{2}}\right)+r\left(1-e^{\frac{\lambda((t-1)^2-1)}{2}}\right)\\
&\le re^{\frac{\lambda((t-1)^2-1)}{2}}+r\left(1-e^{\frac{\lambda((t-1)^2-1)}{2}}\right)\\
&= r.
\end{aligned}
$$

Therefore,

$$\|Ax_0\| = \sup_{t\in[0,1]} |Ax_0(t)| \le r,$$

and by Theorem 2 we have that $A : P \to P$, hence $A : P_r \to P_r$.

If A has a fixed point in the $\partial\overline{P_r}$, we are finished, so without loss of generality, suppose that A does not have a fixed point in $\partial\overline{P_r}$. For $x \in \overline{P_r}$, define $H : [0, 1] \times \overline{P_r} \to \overline{P_r}$ by

$$H(t, x) = tAx.$$

We have assumed that $H(1, x) = Ax \ne x$ for all $x \in \partial\overline{P_r}$, and for $t \in [0, 1)$ and $x \in \partial\overline{P_r}$ we have that

$$\|H(t, x)\| = \sup_{s\in[0,1]} |tAx(s)| = t \sup_{s\in[0,1]} Ax(s) \le tr < r.$$

Since $\|x\| = r$, we have that $H(t, x) \ne x$ for all $(t, x) \in [0, 1] \times \overline{P_r}$. Thus, by the homotopy invariance property of the fixed point index, we have

$$i(A, P_r, P) = i(0, P_r, P),$$

and by the normality property of the fixed point index, we have

$$i(A, P_r, P) = i(0, P_r, P) = 1.$$

Consequently, by the solution property of the fixed point index, A has a fixed point $x^* \in P_r$.

Therefore, regardless of the case, we have that A has a fixed point $x^* \in \overline{P_r}$, which by Theorem 1 is a solution of (1), (2). This ends the proof. □

Note that the standard approach to showing the existence of a solution of (1), (2) is to show that

$$Dx(t) = \int_0^1 G(t,s)f(x(s))\,ds$$

has a fixed point, where

$$G(t,s) = \min\{t,s\}$$

is the Green's function and $x \in Q$, with Q being the cone

$$Q = \{x \in E \ : \ x \text{ is non-negative, non-decreasing, and concave}\}.$$

Since $D : Q \to Q$, the concavity condition is not restrictive. Thus, whenever $f(w) \geq -\lambda w$ and $\lambda < -3$, for any $r \in \mathbb{R}$ and any $x \in \partial\overline{Q_r}$ that satisfies

$$x(s) \geq sr$$

for all $s \in [0,1]$ by the concavity of x, hence,

$$\begin{aligned}
Dx(1) &= \int_0^1 G(1,s)f(x(s))\,ds \\
&= \int_0^1 sf(x(s))\,ds \\
&\geq -\int_0^1 s\lambda x(s)\,ds \\
&\geq -\int_0^1 s\lambda(sr)\,ds \\
&= \frac{-\lambda r}{3} > r.
\end{aligned}$$

Therefore, when $\lambda < -3$ we have that $\|Ax\| > \|x\|$ for all $x \in \partial\overline{Q_r}$, which can be used to show that $i(D, Q_r, Q) = 0$. Then, the solution property cannot be used with the operator D on sets of the form Q_r when $\lambda < -3$, which illustrates the utility of the operator A and the alternative method to convert (1), (2) to a fixed point problem utilizing an integrating factor.

Moreover, note that the foundational Krasnosleskii [15] arguments and the Leggett–Williams [17] generalization of these arguments revolve around functional wedges of the form

$$P(\beta, b) = \{x \in P \ : \ \beta(x) < b\},$$

and around being able to show that the index of sets of this form are nonzero. The transformation of Theorem 1 provides a rich opportunity to investigate existence of solution arguments not only in the choice of the function a leading to the integrating factor approach, but also in finding conditions for the index of the functional wedges to be nonzero for the transformed operator. That is, ascertaining what are the most appropriate functionals to use with Theorem 1.

In the future, we hope to use this method with more general boundary conditions, and in different applications.

References

1. D. Anderson, R. Avery, and J. Henderson, Layered compression-expansion fixed point theorem, *RFPTA*, 2018 (2018), Article ID 201825.
2. R. I. Avery and J. Henderson, Existence of Three Positive Pseudo-Symmetric Solutions for a One Dimensional p-Laplacian, *J. Math. Anal. Appl*, Vol. 277 (2003), pp. 395–404.
3. R. I. Avery and A. C. Peterson, Multiple positive solutions of a discrete second order conjugate problem, *PanAmerican Mathematical Journal*, **8.3** (1998), 1–12.
4. R. I. Avery, D. O'Regan and J. Henderson, Dual of the compression-expansion fixed point theorems, *Fixed Point Theory and Applications*, **2007** (2007), Article ID 90715, 11 pages.
5. C. Avramescu and C. Vladimirescu, Some remarks on Krasnoselskii's fixed point theorem, *Fixed Point Theory* **4** (2003), 3–13.
6. T.A. Burton, A fixed-point theorem of Krasnoselskii, *Appl. Math. Lett.*, Vol. 11 (1998), No. 1, pp. 85–88.
7. T. A. Burton, Fixed points, differential equations, and proper mappings, *Semin. Fixed Point Theory Cluj-Napoca* **3**, (2002), 19–32.
8. T. A. Burton and B. Zhang, Periodicity in delay equations by direct fixed point mapping, *Differential Equations Dynam. Systems* **6.4**, (1998), 413–424.
9. M. Cichon and M.A. Metwali, On a fixed point theorem for the product of operators, *J. Fixed Point Theory Appl.*, 18 (2016), 753–770.
10. K. Deimling, *Nonlinear Functional Analysis*, Springer-Verlag, New York, 1985.
11. B.C. Dhage, Remarks on two fixed-point theorems involving the sum and the product of two operators, *Comput Math Appl 46*, (2003), 1779–1785.
12. D. Duffy, *Green's Functions with Applications*, CRC Press, New York, 2015.
13. D. Guo and V. Lakshmikantham, *Nonlinear Problems in Abstract Cones*, Academic Press, San Diego, 1988.
14. D. Jia, Y.Yangb and W. Gea, Triple positive pseudo-symmetric solutions to a four-point boundary value problem with p-Laplacian, *Appl. Math. Lett.*, Vol. 21 (2008), 268–274.
15. M.A. Krasnoselskii, Amer. Math. Soc. Transl. **10**(2) (1958) 345–409.
16. G.S. Ladde, V. Lakshmikantham and A.S. Vatsala, *Monotone Iterative Techniques for Nonlinear Differential Equations*, Pitman, Boston, 1985.
17. R. W. Leggett and L. R. Williams, Multiple positive fixed points of nonlinear operators on ordered Banach spaces, *Indiana Univ. Math. J.* **28** (1979), 673–688.
18. D. O'Regan, Fixed-point theory for the sum of two operators, *Appl. Math. Lett.* **9** (1996), 1–8.
19. E. Zeidler, *Nonlinear Functional Analysis and its Applications I, Fixed Point Theorems*, Springer-Verlag, New York, 1986.

Volterra Relatively Compact Perturbations of the Laplace Operator

Bazarkan Biyarov

Abstract In this paper, we distinguish a class of correct restrictions and extensions with compact inverse operators which do not belong to any of the Schatten classes. Using such operators, a relatively compact Volterra correct perturbation for the Laplace operator is constructed.

1 Introduction

Let us present some definitions, notation, and terminology.

In a Hilbert space H, we consider a linear operator L with domain $D(L)$ and range $R(L)$. By the *kernel* of the operator L we mean the set

$$\operatorname{Ker} L = \{f \in D(L) : \; Lf = 0\}.$$

Definition 1 An operator L is called a *restriction* of an operator L_1, and L_1 is called an *extension* of an operator L, briefly $L \subset L_1$, if:

(1) $D(L) \subset D(L_1)$,
(2) $Lf = L_1 f$ for all f from $D(L)$.

Definition 2 A linear closed operator L_0 in a Hilbert space H is called *minimal* if there exists a bounded inverse operator L_0^{-1} on $R(L_0)$ and $R(L_0) \neq H$.

Definition 3 A linear closed operator $\widehat{L}$ in a Hilbert space H is called *maximal* if $R(\widehat{L}) = H$ and $\operatorname{Ker} \widehat{L} \neq \{0\}$.

Definition 4 A linear closed operator L in a Hilbert space H is called *correct* if there exists a bounded inverse operator L^{-1} defined on all of H.

B. Biyarov (✉)
Department of Fundamental Mathematics, L. N. Gumilyov Eurasian National University, Nur-Sultan, Kazakhstan

I. N. Parasidis et al. (eds.), *Mathematical Analysis in Interdisciplinary Research*, Springer Optimization and Its Applications 179,
https://doi.org/10.1007/978-3-030-84721-0_8

Definition 5 We say that a correct operator L in a Hilbert space H is a *correct extension* of minimal operator L_0 (*correct restriction* of maximal operator $\widehat{L}$) if $L_0 \subset L$ $(L \subset \widehat{L})$.

Definition 6 We say that a correct operator L in a Hilbert space H is a *boundary correct* extension of a minimal operator L_0 with respect to a maximal operator $\widehat{L}$ if L is simultaneously a correct restriction of the maximal operator $\widehat{L}$ and a correct extension of the minimal operator L_0, that is, $L_0 \subset L \subset \widehat{L}$.

Definition 7 A bounded operator A in a Hilbert space H is called *quasinilpotent* if its spectral radius is zero, that is, the spectrum consists of the single point zero.

Definition 8 An operator A in a Hilbert space H is called a *Volterra operator* if A is compact and quasinilpotent.

Definition 9 A correct restriction L of a maximal operator $\widehat{L}$ $(L \subset \widehat{L})$, a correct extension L of a minimal operator L_0 $(L_0 \subset L)$ or a boundary correct extension L of a minimal operator L_0 with respect to a maximal operator $\widehat{L}$ $(L_0 \subset L \subset \widehat{L})$, will be called *Volterra* if the inverse operator L^{-1} is a Volterra operator.

In a Hilbert space H, we consider a linear operator L with domain $D(L)$ and range $R(L)$. By the *kernel* of the operator L we mean the set

$$\operatorname{Ker} L = \{f \in D(L) : Lf = 0\}.$$

Let $\widehat{L}$ be a maximal linear operator in a Hilbert space H, let L be any known correct restriction of $\widehat{L}$, and let K be an arbitrary linear bounded (in H) operator satisfying the following condition:

$$R(K) \subset \operatorname{Ker} \widehat{L}. \tag{1.1}$$

Then the operator L_K^{-1} defined by the formula (see [1])

$$L_K^{-1} f = L^{-1} f + K f \tag{1.2}$$

describes the inverse operators to all possible correct restrictions L_K of $\widehat{L}$, i.e., $L_K \subset \widehat{L}$.

Let L_0 be a minimal operator in a Hilbert space H, let L be any known correct extension of L_0, and let K be a linear bounded operator in H satisfying the conditions

(a) $R(L_0) \subset \operatorname{Ker} K$,
(b) $\operatorname{Ker}(L^{-1} + K) = \{0\}$,

then the operator L_K^{-1} defined by formula (1.2) describes the inverse operators to all possible correct extensions L_K of L_0, i.e., $L_0 \subset L_K$ (see [1]).

Let L be any known boundary correct extension of L_0, i.e., $L_0 \subset L \subset \widehat{L}$. The existence of at least one boundary correct extension L was proved by Vishik in [2]. Let K be a linear bounded (in H) operator satisfying the conditions

(a) $R(L_0) \subset \operatorname{Ker} K$,
(b) $R(K) \subset \operatorname{Ker} \widehat{L}$,

then the operator L_K^{-1} defined by formula (1.2) describes the inverse operators to all possible boundary correct extensions L_K of L_0, i.e., $L_0 \subset L_K \subset \widehat{L}$ (see [1]).

From the description of (1.2) for all correct restrictions of L_K we have that $D(L_K) \subset D(\widehat{L})$. Then the operator

$$\widehat{L}u = f, \quad \text{for all } u \in D(L_K),$$

where

$$D(L_K) = \left\{u \in D(\widehat{L}) : (I - K\widehat{L})u \in D(L)\right\},$$

corresponds to the description of the direct operator L_K, here I is the identity operator in H. It is easy to see that the operator K defines the domain of $D(L_K)$, since (see [1])

$$(I - K\widehat{L})D(L_K) = D(L),$$

$$(I + K\widehat{L})D(L) = D(L_K), \quad I - K\widehat{L} = (I + K\widehat{L})^{-1}.$$

We have taken the term "boundary correct extension" in connection with the fact that for differential equations such operators are generated only by boundary conditions. And this, in turn, is due to that the minimal operator for them is usually defined using boundary conditions.

If L_0 and M_0 are minimal operators with dense domains in a Hilbert space H, and connected among themselves by the relation

$$(L_0u, v) = (u, M_0v), \quad \text{for all } u \in D(L_0), \quad \text{for all } v \in D(M_0),$$

then $\widehat{L} = M_0^*$ and $\widehat{M} = L_0^*$ are maximal operators such that $L_0 \subset \widehat{L}$ and $M_0 \subset \widehat{M}$.

Let L be some correct restriction of maximal operator $\widehat{L}$. Then the inverse operators to all correct restrictions L_K of $\widehat{L}$ are described by formula (1.2). The following is true

Assertion 1 ([1]) *The domain of correct restriction L_K is dense in H if and only if*

$$D(L^*) \cap Ker\,(I + K^*L^*) = \{0\}.$$

Note that there exists a correct restriction L_K, which the domain is not dense in H despite the fact that the range $R(L_K) = H$.

Assertion 2 ([1]) *It is obvious that any correct extension M_1 of the minimal operator M_0 is the adjoint of some correct restriction L_1 of $\widehat{L}$ with dense domain. And vice versa, that any correct restriction L_1 of $\widehat{L}$ with dense domain is the adjoint of some correct extension M_1 of the minimal operator M_0.*

In this regard, it suffices to study the correct restrictions of the maximum operator.

2 Compact Operators not in the Schatten Classes

We denote by $\mathfrak{S}_\infty(H, H_1)$ the set of all linear compact operators acting from a Hilbert space H to a Hilbert space H_1. If $T \in \mathfrak{S}_\infty(H, H_1)$, then T^*T is a non-negative self-adjoint operator in $\mathfrak{S}_\infty(H) \equiv \mathfrak{S}_\infty(H, H)$ and, moreover, there is a non-negative unique self-adjoint root $|T| = (T^*T)^{1/2}$ in $\mathfrak{S}_\infty(H)$. The eigenvalues $\lambda_n(|T|)$ numbered, taking into account their multiplicity, form a monotonically converging to zero sequence of non-negative numbers. These numbers are usually called *s-numbers* of the operator T and denoted by $s_n(T)$, $n \in \mathbb{N}$. *The Schatten class* $\mathfrak{S}_p(H, H_1)$ is the set of all compact operators $T \in \mathfrak{S}_\infty(H, H_1)$, for which

$$|T|_p^p = \sum_{j=1}^{\infty} s_j^p(T) < \infty, \quad 0 < p < \infty.$$

The following result shows the breadth of the asymptotic range of the eigenvalues of the correct constrictions with a discrete spectrum:

Theorem 3 *Let L be the fixed correct restriction of the maximal operator $\widehat{L}$ in a Hilbert space H. If L^{-1} belongs to the Schatten class $\mathfrak{S}_p(H)$ for some p $(0 < p < +\infty)$, and if $s_n(K)$, the s-numbers of the operator K from the representation* (1.2) *that are numbered in descending order (taking their multiplicities into account) satisfy the condition*

$$\lim_{n\to\infty} s_n(K) = 0, \quad \lim_{n\to\infty} s_{2n}(K)/s_n(K) = 1, \tag{2.1}$$

then the operator L_K^{-1} defined by the formula (1.2) *is compact but does not belong to any of the Schatten classes.*

Proof Denote by $\theta(n) = 1/s_n(K)$, $n = 1, 2, \ldots$. Then $\{\theta(n)\}_1^\infty$ is a positive monotonically increasing number sequence. Therefore, there exists such a monotonically increasing continuous function $f(x)$ defined on $[0, +\infty)$, $f(n) = \theta(n)$, $n = 1, 2, \ldots$.

The function $l(x)$ on $[a, +\infty)$, where $a > 0$, is called *slowly varying* (see [3]), if $l(x)$ is a positive measurable function and

$$\lim_{x\to\infty}\{l(\lambda x)/l(x)\} = 1, \quad \text{for all} \quad \lambda > 0.$$

In the future, we need the following properties of slowly varying functions $l(x)$, which proved in [3, p. 16]:

If $l(x)$ is a slowly varying function, then

$$x^{\alpha}l(x) \to \infty, \quad x^{-\alpha}l(x) \to 0, \qquad x \to \infty, \qquad \alpha > 0. \tag{2.2}$$

We give the formulation of one affirmation from the monograph of Seneta

Lemma (Seneta [4, p. 41]) *Let the function $l(x)$ be positive and monotone on $[A, \infty)$. If, for some fixed λ_0, such that $\lambda_0 > 0$, $\lambda_0 \neq 1$ takes place*

$$\lim_{x\to\infty}\{l(\lambda_0 x)/l(x)\} = 1,$$

then l is a slowly varying function.

Then it follows from the conditions (2.1) by the Seneta Lemma that the function $f(x)$ is slowly varying. Using Corollary 2.2 from the monograph of Gohberg and Krein [5, p. 49], and the representation (1.2) we get

$$s_{n+m-1}(L_K^{-1}) \leq s_n(L^{-1}) + s_m(K), \quad m, n = 1, 2, \ldots.$$

Notice that

$$\lim_{n\to\infty} f(n)s_n(L^{-1}) = 0,$$

by virtue of (2.2) and

$$\lim_{m\to\infty} f(m)s_m(K) = 1$$

by construction.

Further, according to the scheme of proof of Theorem 2.3 (K. Fan) from [5, p. 52] it is easy to obtain the validity of the property

$$\lim_{n\to\infty} s_{2n}(L_K^{-1})/s_n(L_K^{-1}) = 1.$$

Then, by virtue of (2.2) we obtain the statement of Theorem 3. Thus, Theorem 3 is proved.

Example 1 We take the complete orthonormal system of vectors $\{\varphi_i\}_1^{\infty}$ from H. Let $\{\psi_i\}_1^{\infty}$ be an orthonormal system from the infinite-dimensional subspace $\operatorname{Ker}\widehat{L}$ of H. As the operator K in the representation (1.2) we take the operator (see. [6])

$$Kf = \sum_{n=1}^{\infty} s_n(K)(\varphi_n, f)\psi_n,$$

where $s_n(K) = 1/\log(n+2)$, $n = 1, 2, \ldots$. Then, by Theorem 3 the correct restriction L_K of the maximum operator $\widehat{L}$ has the inverse operator L_K^{-1}, which is compact but does not belong to any of the Schatten classes.

3 Volterra Relatively Compact Perturbations of the Laplace Operator

In the Hilbert space $L_2(\Omega)$, where Ω is a bounded domain in $\mathbb{R}^m$ with infinitely smooth boundary $\partial\Omega$, let us consider the minimal L_0 and maximal $\widehat{L}$ operators generated by the Laplace operator

$$-\Delta u = -\left(\frac{\partial^2 u}{\partial x_1^2} + \frac{\partial^2 u}{\partial x_2^2} + \cdots + \frac{\partial^2 u}{\partial x_m^2}\right). \tag{3.1}$$

The closure L_0 in the space $L_2(\Omega)$ of the Laplace operator (3.1) with the domain $C_0^\infty(\Omega)$ is called the *minimal operator corresponding to the Laplace operator.*

The operator $\widehat{L}$, adjoint to the minimal operator L_0 corresponding to the Laplace operator is called the *maximal operator corresponding to the Laplace operator* (see [7]). Note that

$$D(\widehat{L}) = \{u \in L_2(\Omega) : \widehat{L}u = -\Delta u \in L_2(\Omega)\}.$$

Denote by L_D the operator, corresponding to the Dirichlet problem with the domain

$$D(L_D) = \{u \in W_2^2(\Omega) : u|_{\partial\Omega} = 0\}.$$

Then, by virtue of (1.2), the inverse operators L^{-1} to all possible correct restrictions of the maximal operator $\widehat{L}$, corresponding to the Laplace operator (3.1), have the following form:

$$u \equiv L^{-1}f = L_D^{-1}f + Kf, \tag{3.2}$$

where, by virtue of (1.1), K is an arbitrary linear operator bounded in $L_2(\Omega)$ with

$$R(K) \subset \operatorname{Ker}\widehat{L} = \{u \in L_2(\Omega) : -\Delta u = 0\}.$$

Then the direct operator L is determined from the following problem:

$$\widehat{L}u \equiv -\Delta u = f, \quad f \in L_2(\Omega), \tag{3.3}$$

$$D(L) = \{u \in D(\widehat{L}) : (I - K\widehat{L})u|_{\partial\Omega} = 0\}, \tag{3.4}$$

where I is the unit operator in $L_2(\Omega)$.

The operators $(L^*)^{-1}$, corresponding to the operators L^*

$$v \equiv (L^*)^{-1}g = L_D^{-1}g + K^*g,$$

describe the inverses of all possible correct extensions of the minimal operator L_0 if and only if K satisfies the condition (see [8]):

$$\text{Ker}(L_D^{-1} + K^*) = \{0\}.$$

Note that the last condition is equivalent to the following: $\overline{D(L)} = L_2(\Omega)$. If the operator K in (3.2), satisfies one more additional condition

$$KR(L_0) = \{0\},$$

then the operator L, corresponding to problem (3.3)–(3.4), will turn out to be a boundary correct extension.

Gekhtman's work (see [9]) proves the existence of a positive definite extension S of the Laplace operator in a unit disk on a plane whose spectrum is discrete and asymptotically

$$\lambda_n(S) \sim Cn^{1+\beta}, \quad -1 < \beta < 0, \quad \lambda_n(S) \sim Cn, \quad \beta > 0. \tag{3.5}$$

From the formula (3.5) it follows that, in the case of $-1 < \beta < 0$ the spectrum of the operator S is non-classical. It is easy to see that the inverse operator S^{-1} belongs to some Shatten class $\mathfrak{S}_p(L_2(D))$ $p > 1/(1+\beta)$.

It follows from Theorem 3 that, in particular, for the Laplace operator there exist correct restrictions that inverse operators are compact but do not belong to any of the Schatten classes. This is possible due to the infinite-dimensionality of the kernel of the maximal operator Ker $\widehat{L}$. In the case of a finite-dimensional kernel, this effect cannot be achieved.

The author in [8] proved a theorem for a wide class of correct restrictions of the maximal operator $\widehat{L}$ and the correct extensions of the minimal operator L_0 generated by the Laplace operator that they cannot be Volterra. The compact operators K from the Schatten class $\mathfrak{S}_p(L_2(\Omega))$ correspond to them, for any $p \le m/2$, where $m \ge 2$ is the dimension of the space $\mathbb{R}^m$.

In terms of the smoothness of the domain, this means that the correct restrictions with domain $D(L) \subset W_2^s(\Omega)$, where $2(m-1)/m < s \le 2$, cannot be Volterra. It was noted that under perturbations of the positive operator L_D^{-1} by the finite-dimensional operator K there are no Volterra restrictions L_K. At first glance, this seems strange against the background of Corollary 8.3 of [10]. However, there are no contradictions, since, in our case, the eigenvalues of the positive operator L_D^{-1} do not satisfy the condition of Matsaevs Theorem 11.5 (see [5, p. 273])

$$\lim_{n\to\infty} n^2\lambda_n(L_p^{-1}) = c < +\infty.$$

And in Corollary 8.3 of [10], the positive operator must satisfy the above condition. Therefore, Volterra perturbations for the Laplace operator (if they exist) must be sought among the infinite-dimensional operators K.

Operators generated by ordinary differential equations, equations of hyperbolic or parabolic type with Cauchy initial data, as a rule, are obtained by Volterra correct extensions of the minimal operator. But Hadamard's example shows that the Cauchy problem for the Laplace equation is not correct. At present, not a single Volterra correct restriction or extension for elliptic-type equations is known. Correct Cauchy problems for differential equations remain Volterra under perturbations with the help of lower terms. The works (see [1, 11]) contain many Volterra problems, except for the Cauchy problem, which change Volterra under disturbances by the lower terms. Guided by these guesses, we will try to construct at least some Volterra perturbation for the Laplace operator.

Theorem 4 *Let L be the correct restriction of the maximal operator $\widehat{L}$ generated by the Laplace operator in $L_2(\Omega)$, which is determined by the formulas* (3.3)–(3.4). *We take the operator K as compact positive in $L_2(\Omega)$ and its eigenvalues $\{\mu_j\}_1^\infty$ that are numbered in descending order (taking their multiplicities into account) satisfy the condition*

$$\lim_{n\to\infty} \mu_{2n}\mu_n^{-1} = 1.$$

Then there exists a compact operator S in $L_2(\Omega)$ and a relatively compact perturbation of the Laplace operator

$$\widehat{L}_S u = -\Delta u + S(-\Delta u) = f, \quad f \in L_2(\Omega), \tag{3.6}$$

with the domain

$$D(L_S) = \{u \in D(\widehat{L}) : (I - K\widehat{L})u|_{\partial\Omega} = 0\}, \tag{3.7}$$

is a Volterra boundary correct problem.

Proof A linear operator acting on a separable Hilbert space H is called complete if the system of its root vectors corresponding to nonzero eigenvalues is complete in H. By a weak perturbation of the complete compact self-adjoint operator A we mean the operator $A(I + C)$, where C is such a compact operator such that the operator $I + C$ is continuously invertible.

It follows from the conditions of Theorem 4 that the direct operator L, determined from problems (3.3) and (3.4) is positive, since its inverse operator L^{-1} of the form (3.2) is a positive and compact operator in $L_2(\Omega)$.

After applying Theorem 3, the operator L^{-1} satisfies all the conditions of the theorem of Matsaev and Mogul'skii which states:

Theorem (Matsaev, Mogul'skii [12]) *Let the eigenvalues $\{\lambda_n\}_1^\infty$ of a complete positive compact operator A numbered in descending order taking into account their multiplicities satisfy the condition*

$$\lim_{n\to\infty} \lambda_{2n}\lambda_n^{-1} = 1.$$

Then there is a weak perturbation of A which is a Volterra operator.

Using this theorem, we obtain that the operator L^{-1} has a weak perturbation $L_S^{-1} = L^{-1}(I + S_1)$, which is a Volterra operator. By the definition of a weak perturbation, the operator S_1 is compact and $I + S_1$ is continuously invertible. We denote

$$(I + S_1)^{-1} = I + S,$$

where S is the compact operator in $L_2(\Omega)$. Then the problem (3.6)–(3.7) defines the direct operator L_S. Notice that

$$D(L_S) = D(L),$$

and the action of the operator L_S is a relatively compact perturbation of the Laplace operator. Theorem 4 is proved.

As an example of K satisfying the condition of Theorem 4, we can take the operator

$$Kf = \sum_{n=1}^{\infty} \frac{1}{\log(2+n)}(\varphi_n, f)\varphi_n,$$

where $\{\varphi_n\}_1^\infty$ is an orthonormal basis in the subspace $\operatorname{Ker}\widehat{L}$, with $\widehat{L} = -\Delta$. Then $\lambda_n = 1/\log(2+n),\ n = 1, 2, \ldots$ are eigenvalues of the positive operator K, and $\{\varphi_n\}_1^\infty$ are the corresponding eigenvalues vector.

Note that Theorem 4 is not only true for the Laplace operator, it is also true in the case of an abstract operator when a maximal operator with an infinite-dimensional kernel has a positive correct restriction with a compact inverse.

References

1. B. N. Biyarov, Spectral properties of correct restrictions and extensions of the Sturm-Liouville operator, *Differ. Equations*, **30**:12, (1994), 1863–1868. (Translated from Differenisial'nye Uravneniya. **30**:12, (1994), 2027–2032.)
2. M. I. Vishik, On general boundary problems for elliptic differential equations, *Tr. Mosk. Matem. Obs.* **1**, 187–246 (1952); English transl., *Am. Math. Soc., Transl. II*, **24**, 107–172 (1963).
3. N. H. Bingham, C. M. Goldie and J. L. Teugels, *Regular variation.* Encyclopaedia of mathematics and its applications, Cambridge University press, 1987.
4. E. Seneta, *Regularly varying functions*, Berlin-Heidelberg-New-York, Springer-Verlag, 1976.
5. I. C. Gohberg and M. G. Krein, *Introduction to the theory of linear non-self adjoint operators on Hilbert space*, Nauka, Moscow, 1965 (in Russian); (English transl.: Amer. Math. Soc, Providence, R.I., 1969. p.28).
6. R. Schatten, *Norm ideals of completely continuous operators*. Berlin, Springer-Verlag, 1960.
7. L. Hörmander, *On the theory of general partial differential operators*. IL, Moscow, 1959 (in Russian).
8. B. N. Biyarov, On the spectrum of well-defined restrictions and extensions for the Laplace operator, *Math. notes*, **95**:4, 463–470 (2014) (Translated from mathematicheskie zametki, **95**:4, (2014) 507–516).
9. M. M. Gekhtman, Spectrum of some nonclassical self-adjoint extensions of the Laplace operator, *Funct. Anal. Appl.*, **4**:4, (1970), 325–326.
10. A. D. Baranov, D. V. Yakubovich, One-dimensional perturbations of unbounded self-adjoint operators with empty spectrum. *J. Math. Anal. Appl.* **1424**:2, (2015), 1404–1424
11. B. N. Biyarov, S. A. Dzhumabaev, A criterion for the Volterra property of boundary value problems for Sturm-Liouville equations, *Math. Zametki*, **56**:1, (1994), 143–146 (in Russian). English transl. in Math. Notes, **56**:1, (1994), 751–753.
12. V. I. Matsaev, and E. Z. Mogul'skii, On the possibility of a weak perturbation of a complete operator to a Volterra operator, *Dokl. Akad. Nauk SSSR*, **207**:3, 534–537 (1972), (English transl.: Soviet Math. Dokl. **13**, (1972) 1565–1568).

Computational Aspects of the General Rodrigues Problem

Oana-Liliana Chender

Abstract We discuss the general Rodrigues problem and we give explicit formulae for the coefficients when the eigenvalues of the matrix have double multiplicity (Theorem 5). An effective computation for $a_0^{(f)}(X)$ is given in the case $n = 4$.

2010 AMS Subject Classification 22Exx, 22E60, 22E70

1 Introduction

The exponential map $\exp : M_n(K) \to \mathbf{GL}(n, K)$, where $K = \mathbb{C}$ or $K = \mathbb{R}$, and $\mathbf{GL}(n, K)$ that denotes the Lie group of the invertible $n \times n$ matrices having the entrees in K is defined by (11). According to the well-known Hamilton-Cayley Theorem, it follows that every power X^k, $k \geq n$, of the matrix $X \in M_n(K)$ is a linear combination of $X^0 = I_n$, X^1, ..., X^{n-1}, hence $\exp(X)$ can be written as in (15), where the coefficients $a_0(X), \ldots, a_{n-1}(X)$ are uniquely defined and depend on X. Inspired by the classical Rodrigues formula for the special orthogonal group $SO(3)$ (see Sect. 3), we call these numbers the Rodrigues coefficients of the exponential map with respect to the matrix $X \in M_n(K)$.

The general idea of construction of matrix function generalizing the exponential map is to consider an analytic function $f(z)$ defined on an open disk containing the spectrum $\sigma(X)$ of the matrix $X \in M_n(K)$ and replace z by X. Similarly, we obtain the reduced formula (16) for $f(X)$, where $a_0^{(f)}(X), a_1^{(f)}(X), \ldots, a_{n-1}^{(f)}(X) \in K$ are the Rodrigues coefficients of f with respect to the matrix X. In this paper we discuss the Rodrigues problem, that is the problem of determining the Rodrigues coefficients.

O.-L. Chender (✉)
Faculty of Mathematics and Computer Science, "Babeş-Bolyai" University, Cluj-Napoca, Romania
e-mail: oana.broaina@pccj.ro

I. N. Parasidis et al. (eds.), *Mathematical Analysis in Interdisciplinary Research*, Springer Optimization and Its Applications 179,
https://doi.org/10.1007/978-3-030-84721-0_9

Section 2 is dealing with a brief review of matrix functions. We recall the definition of a matrix function by using the Jordan canonical form, by the Cauchy integral formula, and by the Hermite's interpolation polynomial. The connection between these definitions is given in Theorem 2.

Section 3 is devoted to the study of the general Rodrigues problem in terms of the spectrum of the matrix X and it consists in a short discussion on the case when the eigenvalues of the matrix X are pairwise distinct (Theorems 3 and 4).

The last section illustrates the importance of the Hermite interpolation polynomial in solving the general Rodrigues problem. The computation complexity of the Rodrigues problem is discussed in Sect. 4.1, the explicit formulae when the eigenvalues of X have double multiplicity are given in Theorem 5, and an effective computation for $a_0^{(f)}(X)$ and $n = 4$ is given in Sect. 4.3.

2 A Brief Review on Matrix Functions

The concept of matrix function plays an important role in many domains of mathematics with numerous applications in science and engineering, especially in control theory and in the theory of the differential equations in which $\exp(tA)$ has an important role. An example is given by the *nuclear magnetic resonance* described by the Solomon equations

$$dM/dt = -RM, M(0) = I,$$

where $M(t)$ is the matrix of intensities and R is the matrix of symmetrical relaxation.

Given a scalar function $f : D \to \mathbb{C}$, we define the matrix $f(A) \in M_n(\mathbb{C})$, formally replacing z by A, where $A \in M_n(\mathbb{C})$ and $M_n(\mathbb{C})$ is the algebra of $n \times n$ square matrices with complex entrees. This direct approach to defining a matrix function is sufficient for a wide range of functions, but does not provide a general definition. It also does not necessarily provide a correct way to numerically evaluate the matrix $f(A)$. The following four definitions are useful for the developments in this paper.

Any matrix $A \in M_n(\mathbb{C})$, not necessarily diagonalizable, can be written using the Jordan canonical form as

$$X^{-1}AX = J = \operatorname{diag}(J_1(\lambda_1), J_2(\lambda_2), \ldots, J_p(\lambda_p)), \tag{1}$$

where we have

$$J = \begin{pmatrix} J_1 & & & \\ & J_2 & & \\ & & \ddots & \\ & & & J_p \end{pmatrix}.$$

Here the transformation matrix X is non-singular, $n_1 + n_2 + \cdots + n_k = n$, and

$$J_k = J_k(\lambda_k) = \begin{pmatrix} \lambda_k & 1 & \cdots & 0 \\ 0 & \lambda_k & \ddots & 0 \\ & & \ddots & 1 \\ 0 & 0 & \cdots & \lambda_k \end{pmatrix} \in M_{n_k}(\mathbb{C}), \tag{2}$$

λ_k being the eigenvalues of A.

The Jordan matrix J is unique up to a permutation of the blocks J_k, but the transformation matrix X is not unique.

Definition 1 Let f be defined on a neighborhood of the spectrum of $A \in M_n(\mathbb{C})$. If A has the Jordan canonical form J, then

$$f(A) = Xf(J)X^{-1} = X\text{diag}(f(J_k(\lambda_k)))X^{-1}, \tag{3}$$

where

$$f(J_k) = f(J_k(\lambda_k)) = \begin{pmatrix} f(\lambda_k) & f'(\lambda_k) & \cdots & \frac{f^{(n_k-1)}(\lambda_k)}{(n_k-1)!} \\ 0 & f(\lambda_k) & \ddots & \vdots \\ & & \ddots & f'(\lambda_k) \\ 0 & 0 & \cdots & f(\lambda_k) \end{pmatrix}. \tag{4}$$

The result in the right member of the relation (3) is independent of the choice of X and J.

Any polynomial p with complex coefficients,

$$p(t) = a_0 + a_1 t + \cdots + a_{m-1} t^{m-1} + a_m t^m,\ a_m \neq 0 \tag{5}$$

defines a matrix polynomial by simply replacing t with A in (5), and obtain

$$p(A) = a_m A^m + a_{m-1} A^{m-1} + \cdots + a_0 I_n. \tag{6}$$

More general [5, p. 565, Theorem 11.2.3], if f is an analytic function defined by

$$f(t) = \sum_{i=0}^{\infty} a_i t^i$$

on an open disk containing the spectrum $\sigma(A)$ of A, then

$$f(A) = \sum_{i=0}^{\infty} a_i A^i.$$

To evaluate a polynomial matrix function if $A \in M_n(\mathbb{C})$ is diagonalizable, so we have $A = X\Lambda X^{-1}$ with $\Lambda = \mathrm{diag}(\lambda_1, \lambda_2, \ldots, \lambda_n)$, from (4) we obtain

$$p(A) = p(X\Lambda X^{-1}) = Xp(\Lambda)X^{-1},$$

thus

$$p(A) = X \begin{pmatrix} p(\lambda_1) & & 0 \\ & \ddots & \\ 0 & & p(\lambda_n) \end{pmatrix} X^{-1}, \tag{7}$$

where we have used the property $p(XAX^{-1}) = Xp(A)X^{-1}$. If A is not necessarily diagonalizable and has the Jordan canonical form $A = XJX^{-1}$ with the blocks J_k defined in (2), then

$$p(A) = p(XJX^{-1}) = Xp(J)X^{-1},$$

so

$$p(A) = X \begin{pmatrix} p(J_1(\lambda_1)) & & 0 \\ & \ddots & \\ 0 & & p(J_p(\lambda_p)) \end{pmatrix} X^{-1}. \tag{8}$$

Now, we write $J_k(\lambda) = \lambda I_{n_k} + N$, where$N \equiv J_k(0)$, and obtain

$$J_k^j(\lambda) = (\lambda I + N)^j = \sum_{i=0}^{j} \binom{j}{i} \lambda^{j\lambda - i} N^i.$$

All terms with $i \geq n_k$ are zero, because we have $N^{n_k} = O$. This takes us to the relation

$$p(J_k(\lambda)) = \sum_{i-0}^{m} \frac{1}{i!} p^{(i)}(\lambda) N^i = \sum_{i-0}^{\mu} \frac{1}{i!} p^{(i)}(\lambda) N^i,$$

where $\mu = \min\{m, n_k - 1\}$. Therefore, one obtains

$$p(J_k) = p(J_k(\lambda)) = \begin{pmatrix} p(\lambda) & p'(\lambda) & \frac{1}{2}p''(\lambda) & \cdots & \frac{p^{(n_k-1)}(\lambda)}{(n_k-1)!} \\ 0 & p(\lambda) & p'(\lambda) & \ddots & \frac{p^{(n_k-2)}(\lambda)}{(n_k-2)!} \\ 0 & 0 & p(\lambda) & \ddots & \frac{1}{2}p''(\lambda) \\ \vdots & \vdots & \cdots & \ddots & p'(\lambda) \\ 0 & 0 & 0 & \cdots & p(\lambda) \end{pmatrix}, \tag{9}$$

where all the elements i over the superdiagonal are $p^{(i)}(\lambda)/i!$. Notice that only the derivatives until the order $n_k - 1$ are necessary.

Definition 2 For a differentiable function f defined on a neighborhood of $\sigma(A)$, the numbers $f^{(j)}(\lambda_i)$, $i = 1, \ldots, s$, $j = 0, \ldots, n_i - 1$, are called the values of the function f and its derivatives on the spectrum of A. If these values exist we say that f is **defined on the spectrum of** A.

We notice that the minimal polynomial ψ_A takes the value zero on the spectrum of A.

Theorem 1 ([6, p. 5, Theorem 1.3]) *For polynomials p and q and $A \in M_n(\mathbb{C})$ we have $p(A) = q(A)$ if and only if p and q take the same values on the spectrum of A.*

The following definition of the matrix function, using the Hermite interpolation, is important for our presentation.

Definition 3 Let f be defined on the spectrum of $A \in M_n(\mathbb{C})$. Then $f(A) = r(A)$, where r is the Hermite interpolation polynomial that satisfies the interpolation conditions

$$r^{(j)}(\lambda_i) = f^{(j)}(\lambda_j), i = 1, \ldots, s, j = 0, \ldots, n_i - 1,$$

where $\lambda_1, \ldots, \lambda_s$ are the distinct eigenvalues of A with the multiplicities $n_1, \ldots n_s$, respectively.

According to Theorem 1, it follows that the polynomial r depends on A through the values of the function f on the spectrum of A.

The Cauchy's integral formula is an elegant result of complex analysis that states that under certain conditions, the value of a function can be determined using an integral. Given a function $f(z)$ we can obtain the value $f(a)$ by

$$f(a) = \frac{1}{2\pi i} \int_\Gamma \frac{f(z)}{z - a} dz,$$

where Γ is a simple closed curve around a and f is analytic on and inside Γ. This formula extends to the case of the matrices.

Definition 4 Let $\Omega \subset \mathbb{C}$ be a domain and let $f : \Omega \to \mathbb{C}$ be an analytic function. Let $A \in M_n(\mathbb{C})$ be diagonalizable so that all eigenvalues of A are in Ω. We define $f(A) \in M_n(\mathbb{C})$ by

$$f(A) = \frac{1}{2\pi i} \int_{\Gamma} f(z)(zI_n - A)^{-1} dz, \tag{10}$$

where $(zI_n - A)^{-1}$ is the resolvent of A in z and $\Gamma \subset \Omega$ is a simple closed clockwise oriented curve around the spectrum $\sigma(A)$.

Let $A \in M_n(\mathbb{C})$ be a diagonalizable matrix and f an analytic function on a domain that contains the spectrum of A. Then we have [8, p. 427, Theorem 6.2.28]

$$f(A) = Xf(\Lambda)X^{-1},$$

where $A = X\Lambda X^{-1}$, with $\Lambda = \text{diag}(\lambda_1, \lambda_2, \ldots, \lambda_n)$, and $f(\Lambda)$ is defined by the Cauchy's integral formula. In conclusion, this result shows that $f(A)$ is similar to the matrix $f(\Lambda)$.

The following result allows us to define $f(A)$ if f has a development in power series [5, p. 565, Theorem 11.2.3]. If the function f is given by

$$f(z) = \sum_{k=0}^{\infty} c_k z^k$$

on an open disk containing the spectrum of A, then

$$f(A) = \sum_{k=0}^{\infty} c_k A^k.$$

The most popular matrix function defined in this way is the matrix exponential

$$\exp(A) = \sum_{k=0}^{\infty} \frac{1}{k!} A^k, \tag{11}$$

obtained for

$$f(z) = e^z = \sum_{k=0}^{\infty} \frac{1}{k!} z^k.$$

Another important matrix function defined in this way is the Cayley transform of the special orthogonal group $SO(n)$ (see, for instance, [2]). Denoted by Cay, it is obtained from

$$f(z) = 1 + 2z + 2z^2 + \cdots = \frac{1+z}{1-z}, \; |z| < 1$$

and it is given by

$$Cay(A) = (I_n + A)(I_n - A)^{-1}.$$

The above definitions for the matrix $f(A)$, where $A \in M_n(\mathbb{C})$, are equivalent. More precisely, we have the following result (see, for example, [14]).

Theorem 2 *Let be $A \in M_n(\mathbb{C})$. Let f be an analytical function defined on a domain containing the spectrum of A. We denote by*

1. *$f_J(A)$ the matrix $f(A)$ obtained using the definition with the Jordan canonical form;*
2. *$f_H(A)$ the matrix $f(A)$ obtained using the definition with the Hermite's interpolation polynomial;*
3. *$f_C(A)$ the matrix $f(A)$ obtained using the definition with the Cauchy's integral formula.*

Then

$$f_J(A) = f_H(A) = f_C(A). \tag{12}$$

The following important two properties are satisfied by a matrix function given by any definition discussed above [10, p. 310, Theorem 1 and Theorem 2].

1. If $A, B, X \in M_n(\mathbb{C})$, where $B = XAX^{-1}$, and f is defined on the spectrum of A, then

$$f(B) = Xf(A)X^{-1}. \tag{13}$$

2. If $A \in M_n(\mathbb{C})$ is a matrix given in blocks on the principal diagonal,

$$A = \operatorname{diag}(A_1, A_2, \ldots, A_s),$$

where $A_1, A_2, \ldots, A_s$ are square matrices, then

$$f(A) = \operatorname{diag}(f(A_1), f(A_2), \ldots, f(A_s)). \tag{14}$$

3 The General Rodrigues Problem

We have seen in formula (11) the definition of the exponential matrix $X \in M_n(\mathbb{C})$.

According to the well-known Hamilton-Cayley-Frobenius theorem, it follows that every power X^k, $k \geq n$, is a linear combination of powers X^0, X^1, ..., X^{n-1}, hence we can write

$$\exp(X) = \sum_{k=0}^{n-1} a_k(X)X^k, \tag{15}$$

where the coefficients $a_0(X), \ldots, a_{n-1}(X)$ are uniquely defined and depend on the matrix X. From this formula, it follows that $\exp(X)$ is a polynomial of X with coefficients functions of X. Moreover, it is clear that the coefficients are functions of the eigenvalues of X.

The problem to find a reduced formula for $\exp(X)$ is equivalent to the determination of the coefficients $a_0(X), \ldots, a_{n-1}(X)$. We will call this problem, the *Rodrigues problem*, and the numbers $a_0(X), \ldots, a_{n-1}(X)$ the *Rodrigues coefficients* of the exponential map with respect to the matrix $X \in M_n(\mathbb{C})$.

The origin of this problem is the classical Rodriques formula

$$\exp(X) = I_3 + \frac{\sin\theta}{\theta}X + \frac{1-\cos\theta}{\theta^2}X^2$$

obtained in 1840 for the special orthogonal group $SO(3)$ with the Lie algebra $\mathfrak{so}(3)$ consisting in the set of all skew-symmetric 3×3 matrices. Here $\sqrt{2}\theta = \|X\|$, where $\|X\|$ is the Frobenius norm of the matrix X. From the numerous arguments pointing out the importance of this formula we only mention here the study of the rigid body rotation in $\mathbb{R}^3$, and the parametrization of the rotations in $\mathbb{R}^3$.

A natural way to extend this formula is to consider the analytic function f defined on the domain $D \subseteq \mathbb{C}$. If $X \in M_n(\mathbb{C})$ and $\sigma(X) \subset D$, then according to the Hamilton-Cayley-Frobenius theorem, we can write the reduced formula of the matrix $f(X)$ as

$$f(X) = \sum_{k=0}^{n-1} a_k^{(f)}(X)X^k. \tag{16}$$

We call the relation (16), the *Rodrigues formula* with respect to f. The numbers $a_0^{(f)}(X), \ldots, a_{n-1}^{(f)}(X)$ are the *Rodrigues coefficients* of f with respect to the matrix $X \in M_n(\mathbb{R})$. Clearly, the real coefficients $a_0^{(f)}(X), \ldots, a_{n-1}^{(f)}(X)$ are uniquely defined, they depend on the matrix X, and $f(X)$ is a polynomial of X.

An important property of the Rodrigues coefficients is the invariance under the matrix conjugacy (see [1]). That is, for every invertible matrix U the following relations hold

$$a_k^{(f)}(UXU^{-1}) = a_k^{(f)}(X), k = 0, \ldots, n-1. \tag{17}$$

We call the *general Rodrigues problem* the determination of the Rodrigues coefficients for a given analytic function. To understand the complexity of this problem we mention that Schwerdtfeger (see [7]) proved that for any analytic function f and for any matrix $X \in M_n(\mathbb{C})$ the following formula holds

$$f(X) = \sum_{j=1}^{\mu} X_j \sum_{k=0}^{n_j-1} \frac{1}{k!} f^{(k)}(\lambda_j)(X - \lambda_j I_n)^k, \tag{18}$$

where μ is the number of distinct eigenvalues λ_j of X, n_j is the multiplicity of λ_j and

$$X_j = \frac{1}{2\pi i} \int_{\Gamma_j} (sI_n - X)^{-1} ds = \frac{1}{2\pi i} \int_{\Gamma_j} \mathfrak{L}[e^{tX}](s) ds \tag{19}$$

are the Frobenius covariants of X. Here Γ_j is a smooth closed curve around the complex number λ_j and $\mathfrak{L}$ is the Laplace transform. A way to compute the Frobenius covariants is the use of the Penrose generalized inverse of a matrix.

An extension of the formula (18) for the matrix tX, where $t \in \mathbb{R}^*$, is given in [3].

When the eigenvalues $\lambda_1, \ldots, \lambda_n$ of the matrix $X \in M_n(\mathbb{C})$ are pairwise distinct, the general Rodrigues problem is completely solved in the paper [1]. More precisely, in this case the following result holds.

Theorem 3

(1) If the eigenvalues $\lambda_1, \ldots, \lambda_n$ of the matrix X are pairwise distinct, then the Rodrigues coefficients $a_0^{(f)}(X), \ldots, a_{n-1}^{(f)}(X)$ are given by the formulas

$$a_k^{(f)}(X) = \frac{V_{n,k}^{(f)}(\lambda_1, \ldots, \lambda_n)}{V_n(\lambda_1, \ldots, \lambda_n)}, k = 0, \ldots, n-1, \tag{20}$$

where $V_n(\lambda_1, \ldots, \lambda_n)$ is the Vandermonde determinant of order n, and $V_{n,k}^{(f)}(\lambda_1, \ldots, \lambda_n)$ is the determinant of order n obtained from $V_n(\lambda_1, \ldots, \lambda_n)$ by replacing the line $k+1$ by $f(\lambda_1), \ldots, f(\lambda_n)$.

(2) If the eigenvalues $\lambda_1, \ldots, \lambda_n$ of the matrix X are pairwise distinct, then the Rodrigues coefficients $a_0^{(f)}(X), \ldots, a_{n-1}^{(f)}(X)$ are linear combinations of $f(\lambda_1), \ldots, f(\lambda_n)$ having the coefficients rational functions of $\lambda_1, \ldots, \lambda_n$, i.e., we have

$$a_k^{(f)}(X) = b_k^{(1)}(X) f(\lambda_1) + \cdots + b_k^{(n)}(X) f(\lambda_n), k = 0, \ldots, n-1, \tag{21}$$

where $b_k^{(1)}, \ldots, b_k^{(n)} \in \mathbb{Q}[\lambda_1, \ldots, \lambda_n]$.

Expanding the determinant $V_{n,k}^{(f)}(\lambda_1, \ldots, \lambda_n)$ in (20) with respect to the line $k+1$ it follows

$$a_k^{(f)}(X) = \frac{1}{V_n} \sum_{j=1}^{n} (-1)^{k+j+1} LV_{n-1}(\lambda_1, \ldots, \widehat{\lambda_j}, \ldots, \lambda_n) f(\lambda_j), \tag{22}$$

where $LV_{n-1}(\lambda_1, \ldots, \widehat{\lambda_j}, \ldots, \lambda_n) f(\lambda_j)$ is the $(k+1)$-lacunary Vandermonde determinant in the variables $\lambda_1, \ldots, \widehat{\lambda_j}, \ldots, \lambda_n$, i.e., the determinant obtained from $V_n\,(\lambda_1, \ldots, \lambda_n)$ by cutting out the row $k+1$ and the column j. Applying the well-known formula (see the reference [15])

$$\begin{aligned} &LV_{n-1}(\lambda_1, \ldots, \widehat{\lambda_j}, \ldots, \lambda_n) \\ &\quad = s_{n-k-1}(\lambda_1, \ldots, \widehat{\lambda_j}, \ldots, \lambda_n) V_{n-1}(\lambda_1, \ldots, \widehat{\lambda_j}, \ldots, \lambda_n), \end{aligned}$$

where s_l is the l-th symmetric polynomial in the $n-1$ variables $\lambda_1, \ldots, \widehat{\lambda_j}, \ldots, \lambda_n$, where λ_j is missing, we obtain the following result which completely solves the general problem in the case when the eigenvalues $\lambda_1, \ldots, \lambda_n$ of the matrix X are pairwise distinct.

Theorem 4 *For every $k = 0, \ldots, n-1$, the following formulas hold*

$$\begin{aligned} &a_k^{(f)}(X) \\ &\quad = \sum_{j=1}^{n} (-1)^{k+j+1} \frac{V_{n-1}(\lambda_1, \ldots, \widehat{\lambda_j}, \ldots, \lambda_n) s_{n-k-1}(\lambda_1, \ldots, \widehat{\lambda_j}, \ldots, \lambda_n)}{V_n(\lambda_1, \ldots, \lambda_n)} f(\lambda_j), \end{aligned} \tag{23}$$

where s_l denotes the l-th symmetric polynomial, and $\widehat{\lambda_j}$ means that in the Vandermonde determinant V_{n-1} the variable λ_j is omitted.

Remarks

(1) In the paper [1] is given a direct proof to formulas (20) using the "trace method".
(2) The result in (20) can be obtained applying Theorem 2, therefore $f(X)$ is given by the Lagrange interpolation polynomial of f on the distinct points $\lambda_1, \ldots, \lambda_n$. That is

$$f(X) = \sum_{i=1}^{n} f(\lambda_i) \prod_{\substack{j=1 \\ j \neq i}}^{n} \frac{X - \lambda_j I_n}{\lambda_i - \lambda_j}, \tag{24}$$

and $a_k^{(f)}(X)$ is the coefficient of X^k in the algebraic form of the polynomial (24), $k = 0, \ldots, n-1$.

(3) Explicit computations when $n = 2, 3, 4$ are given in [1]. The case $n = 2$ also appears in [11, Theorem 4.7] and in the paper [4].

4 Using the Hermite Interpolation Polynomial

Assume that the function f is defined on spectrum of matrix $X \in M_n(\mathbb{C})$. Considering Theorem 2 we have $f(X) = r(X)$, where r is the Hermite interpolation polynomial that satisfies the conditions

$$r^{(j)}(\lambda_i) = f^{(j)}(\lambda_i), i = 1, \ldots, s, j = 0, \ldots, n_i - 1,$$

where $\lambda_1, \ldots, \lambda_s$ are the distinct eigenvalues of X with multiplicities $n_1, \ldots, n_s$, respectively, and $n_1 + \cdots + n_s = n$.

In this case the Rodrigues coefficients $a_0^{(f)}(X), \ldots, a_n^{(f)}(X)$ of the map f for the matrix X are the coefficients of the Hermite polynomial defined by the above interpolation conditions. This polynomial is given by

$$r(t) = \sum_{i=1}^{s} \left[\left(\sum_{j=0}^{n_i-1} \frac{1}{j!} \Phi_i^{(j)}(\lambda_i)(t - \lambda_i)^j \right) \prod_{j \neq i} (t - \lambda_j)^{n_j} \right], \tag{25}$$

where $\Phi_i(t) = f(t) / \prod_{j \neq i} (t - \lambda_j)^{n_j}$.

If the eigenvalues of the matrix X are pairwise distinct, then the Hermite polynomial r is reduced to Lagrange interpolation polynomial with conditions $r(\lambda_i) = f(\lambda_i), i = 1, \ldots, n$,

$$r(t) = \sum_{i=1}^{n} f(\lambda_i) l_i(t), \tag{26}$$

where l_i are the Lagrange fundamental polynomials defined by

$$l_i(t) = \prod_{\substack{j=1 \\ j \neq i}}^{n} \frac{t - \lambda_j}{\lambda_i - \lambda_j}, i = 1, \ldots, n. \tag{27}$$

4.1 *The Complexity of the Rodrigues Problem*

The determination of the algebraic form of the Hermite polynomial given by (25) is a problem equivalent to the problem of determining the Rodrigues coefficients of the map f when the eigenvalues of the matrix X are known. Because the effective determination of the spectrum of the matrix X is equivalent to the solving of an algebraic equation of degree n obtained after the computation of a determinant of order n (see [12] and [13] for a deep analysis), we can say that the Rodrigues problem has greater complexity than the problem of explicitly determining the coefficients of the Hermite polynomial in the general context. This is also a difficult problem if n and the multiplicities $n_1, \ldots, n_s$ are big (see [9]).

On the other hand, the Frobenius covariants X_j are polynomials in X, so we have $X_j = p_j(X)$, $j = 1, \ldots, \mu$. Developing $(X - \lambda_j I_n)^k$ in the Schwerdtfeger formula (18) we obtain

$$\sum_{k=0}^{n-1} a_k^{(f)}(X)X^k = \sum_{j=1}^{\mu} p_j(X) \sum_{k=0}^{m_j-1} \frac{1}{k!} f^{(k)}(\lambda_j) \sum_{s=0}^{k} (-1)^s \binom{k}{s} \lambda_j^s X^{k-s}.$$

Identifying the coefficient of X^k in this relation, we obtain the Rodrigues coefficients $a_k^{(f)}(X)$, for $k = 0, \ldots, n-1$. This approach provides another image of the complexity of the Rodrigues problem by reducing it to the determination of the polynomials p_j, $j = 1, \ldots, \mu$.

If the eigenvalues of the matrix X are pairwise distinct, the formulas (20) and (22) give the explicit form for the coefficients of the Lagrange polynomial that satisfies the above interpolation conditions.

4.2 *The Solution of the Rodrigues Problem When the Eigenvalues have Double Multiplicity*

In this subsection, we assume that the function f is defined on the spectrum of the matrix $X \in M_{2s}(\mathbb{C})$ and distinct eigenvalues $\lambda_1, \ldots, \lambda_s$ of X have double multiplicity, that is $n_1 = \cdots = n_s = 2$. In this case the Hermite interpolation polynomial r satisfies the conditions

$$r(\lambda_i) = f(\lambda_i), r'(\lambda_i) = f'(\lambda_i), i = 1, \ldots, s$$

and the formula (25) becomes

$$r(t) = \sum_{i=1}^{s} \left[f(\lambda_i) \left(1 - 2l_i'(\lambda_i)(t - \lambda_i)\right) + f'(\lambda_i)(t - \lambda_i) \right] l_i^2(t), \tag{28}$$

where l_i are the fundamental Lagrange polynomials defined in (27).

We notice that the formula (28) it can be written in the form

$$r(t) = \sum_{i=1}^{s} (A_i t + B_i) r_i(t), \tag{29}$$

where

$$A_i = \frac{1}{\prod\limits_{\substack{j=1 \\ j \neq i}}^{s} (\lambda_i - \lambda_j)^2} \left[f'(\lambda_i) - 2 f(\lambda_i) l_i'(\lambda_i) \right],$$

$$B_i = \frac{1}{\prod\limits_{\substack{j=1 \\ j \neq i}}^{s} (\lambda_i - \lambda_j)^2} \left[f(\lambda_i) \left(1 + 2\lambda_i l_i'(\lambda_i)\right) - \lambda_i f'(\lambda_i) \right]$$

and r_i is the polynomial $\prod\limits_{\substack{j=1 \\ j \neq i}}^{s} (t - \lambda_j)^2, i = 1, \ldots, s.$

On the other hand we have $l_i(\lambda_i) = 1$ and

$$\frac{l_i'(t)}{l_i(t)} = \sum_{\substack{j=1 \\ j \neq i}}^{s} \frac{1}{t - \lambda_j}, i = 1, \ldots, s,$$

so we get

$$l_i'(\lambda_i) = \sum_{\substack{j=1 \\ j \neq i}}^{s} \frac{1}{\lambda_i - \lambda_j}, i = 1, \ldots, s. \tag{30}$$

To obtain the algebraic form of the polynomial r_i notice that we can write

$$\begin{aligned} r_i(t) &= \prod_{\substack{j=1 \\ j \neq i}}^{s} (t - \lambda_j)^2 = \prod_{\substack{j=1 \\ j \neq i}}^{s} (t - \lambda_j)(t - \lambda_j) \\ &= t^{2s-2} - \sigma_{i,1} t^{2s-1} + \sigma_{i,2} t^{2s-2} - \cdots + \sigma_{i,2s-2}, \end{aligned} \tag{31}$$

where $\sigma_{i,k}(\lambda_1, \ldots, \lambda_s) = s_k(\lambda_1, \lambda_1, \ldots, \widehat{\lambda_i}, \widehat{\lambda_i}, \ldots, \lambda_s, \lambda_s)$ is the symmetric polynomial of order k in $2s - 2$ variable $\lambda_1, \lambda_1, \ldots, \widehat{\lambda_i}, \widehat{\lambda_i}, \ldots, \lambda_s, \lambda_s$, where λ_i is missing, for all $k = 1, \ldots, 2s - 2$.

Combining formulas (29) and (31) we obtain

$$r(t) = \sum_{i=1}^{s}(A_i t + B_i)(t^{2s-2} - \sigma_{i,1}t^{2s-1} + \sigma_{i,2}t^{2s-2} - \cdots + \sigma_{i,2s-2})$$

$$= \left(\sum_{i=1}^{s} A_i\right) t^{2s-1} + \sum_{i=1}^{s}(-A_i\sigma_{i,1} + B_i)t^{2s-2} + \cdots$$

$$+ \sum_{i=1}^{s}(A_i\sigma_{i,2} - B_i\sigma_{i,1})t^{2s-3} + \cdots + \sum_{i=1}^{s}(A_i\sigma_{i,2s-2} - B_i\sigma_{i,2s-3})t$$

$$+ \sum_{i=1}^{s} B_i\sigma_{i,2s-2}.$$

Thus we get the following result which completely solves the Rodrigues general problem if the eigenvalues $\lambda_1, \ldots, \lambda_s$ are distinct and have double multiplicity.

Theorem 5 *For any* $k = 0, 1, \ldots, n-1$, *we have*

$$a_k^{(f)}(X) = (-1)^{k+1} \sum_{i=1}^{s} \frac{1}{\prod\limits_{\substack{j=1 \\ j \neq i}}^{s} (\lambda_i - \lambda_j)^2} \left\{ \left[f'(\lambda_i) - 2f(\lambda_i) \sum_{\substack{j=1 \\ j \neq i}}^{s} \frac{1}{\lambda_i - \lambda_j} \right] \sigma_{i,2s-k-1} \right.$$

$$\left. - \left[f(\lambda_i) \left(1 + 2\lambda_i \sum_{\substack{j=1 \\ j \neq i}}^{s} \frac{1}{\lambda_i - \lambda_j} \right) - \lambda_i f'(\lambda_i) \right] \sigma_{i,2s-k-2} \right\}. \tag{32}$$

Corollary 1 *If the eigenvalues* $\lambda_1, \ldots, \lambda_s$ *of the matrix* $X \in M_n(\mathbb{C}), n = 2s$, *are pairwise distinct and have double multiplicity, than the Rodrigues coefficients* $a_0^{(f)}(X), \ldots, a_{n-1}^{(f)}(X)$ *are linear combinations of* $f(\lambda_1), \ldots, f(\lambda_s), f'(\lambda_1), \ldots, f'(\lambda_s)$ *having the coefficients rational functions of* $\lambda_1, \ldots, \lambda_s$, *that is, we have*

$$a_k^{(f)}(X) = b_k^{(1)}(X)f(\lambda_1) + \cdots + b_k^{(s)}(X)f(\lambda_s) + c_k^{(1)}(X)f'(\lambda_1) + \cdots + c_k^{(s)} f'(\lambda_s),$$

where $b_k^{(1)}, \ldots, b_k^{(s)}, c_k^{(1)}, \ldots, c_k^{(s)} \in \mathbb{Q}[\lambda_1, \ldots, \lambda_s], k = 0, \ldots, n-1$.

4.3 Example for $n = 4$ and $a_0^{(f)}(X)$

Next we apply the formulas (32) to determine the coefficient $a_0^{(f)}(X)$ in the case $n = 4$ for $\lambda_1 = \lambda_3, \lambda_2 = \lambda_4$ and $\lambda_1 \neq \lambda_2$. In this situation we have

$$\sigma_{1,1}(\lambda_1,\lambda_2)=\sigma_1(\widehat{\lambda_1},\widehat{\lambda_1},\lambda_2,\lambda_2)=2\lambda_2,\sigma_{1,2}(\lambda_1,\lambda_2)=\sigma_2(\widehat{\lambda_1},\widehat{\lambda_1},\lambda_2,\lambda_2)=\lambda_2^2,$$

$$\sigma_{2,1}(\lambda_1,\lambda_2)=\sigma_1(\lambda_1,\lambda_1,\widehat{\lambda_2},\widehat{\lambda_2})=2\lambda_1,\sigma_{2,2}(\lambda_1,\lambda_2)=\sigma_2(\lambda_1,\lambda_1,\widehat{\lambda_2},\widehat{\lambda_2})=\lambda_1^2.$$

Applying formula (32), we find the coefficient $a_0^{(f)}(X)$ in the form

$$\begin{aligned}a_0^{(f)}(X)&=\frac{1}{(\lambda_1-\lambda_2)^2}\left[f(\lambda_1)\left(\lambda_2^2+\frac{2\lambda_1\lambda_2^2}{\lambda_1-\lambda_2}\right)-\lambda_1\lambda_2^2f'(\lambda_1)\right.\\&\left.+f(\lambda_2)\left(\lambda_1^2+\frac{2\lambda_1^2\lambda_2}{\lambda_2-\lambda_1}\right)-\lambda_1^2\lambda_2f'(\lambda_2)\right]\\&=\frac{\lambda_2^2(-3\lambda_1+\lambda_2)}{(\lambda_2-\lambda_1)^3}f(\lambda_1)-\frac{\lambda_1\lambda_2^2}{(\lambda_2-\lambda_1)^2}f'(\lambda_1)\\&+\frac{-\lambda_1^2(\lambda_1-3\lambda_2)}{(\lambda_2-\lambda_1)^3}f(\lambda_2)-\frac{\lambda_1^2\lambda_2}{(\lambda_2-\lambda_1)^2}f'(\lambda_2).\end{aligned}$$

References

1. D. Andrica, O.L. Chender, *A New Way to Compute the Rodrigues Coefficients of Functions of the Lie Groups of Matrices*, in "Essays in Mathematics and its Applications" in the Honor of Vladimir Arnold, Springer, 2016, 1–24.
2. D. Andrica, O.L. Chender, *Rodrigues formula for the Cayley transform of groups $SO(n)$ and $SE(n)$*, Studia Univ. Babeş-Bolyai-Mathematica, Vol 60(2015), No. 1, 31–38.
3. O.L. Chender, *Schwerdtfeger formula for matrix functions*, The 16th International Conference on Applied Mathematics and Computer Science, July 3 rd to 6 th, 2019, Annals of the Tiberiu Popoviciu Seminar, Vol. 16, 2018, în curs de publicare.
4. O. Furdui, *Computing exponential and trigonometric functions of matrices in $M_2(\mathbb{C})$*, Gazeta Matematică, Seria A, Anul XXXVI, Nr. 1-2/2018, 1–13.
5. G.H. Golub, C.F. Van Loan, *Matrix Computations*, The Johns Hopkins University Press, 1996.
6. N.J. Higham, *Functions of Matrices Theory and Computation*, SIAM Society for Industrial and Applied Mathematics, Philadelphia, PA. USA, 2008.
7. R.A. Horn, Ch.R. Johnson, *Matrix Analysis*, Cambridge University Press, 1990.
8. R.A. Horn, Ch.R. Johnson, *Topics in Matrix Analysis*, Cambridge University Press, 1991.
9. S. Kida, E. Trimandalawati, S. Ogawa, *Matrix Expression of Hermite Interpolation Polynomials*, Computer Math. Applic., Vol. **33** (1997), No. 11, 11–13.
10. P. Lancaster, M. Tismenetsky *The Theory of Matrices*, Second Edition With Application, Academis Press, 1984.
11. V. Pop, O. Furdui, *Squares Matrices of Order 2. Theory, Applications and Problems*, Springer, 2017.
12. V. Pan, *Algebraic Complexity of Computing Polynomial Zeros*, Comput. Math. Applic. Vol. 14, No. 4, pp. 285–304, 1987.
13. V. Pan, *Solving a Polynomial Equation: Some History and Recent Progress*, SIAM REV. Vol. 39, No. 2, pp. 187–220, June 1997.
14. R.F. Rinehart, *The equivalence of Definitions of a Matrix Function*, American Mathematical Monthly, Vol. 62, 1955, 395–414.
15. R. Vein, P. Dale, *Determinants and Their Applications in Mathematical Physics*, Springer, 1999.

Approximation by Max-Product Operators of Kantorovich Type

Lucian Coroianu and Sorin G. Gal

Abstract The main goal of this survey is to describe the results of the present authors concerning approximation properties of various max-product Kantorovich operators, fulfilling thus this gap in their very recent research monograph (Bede et al., Approximation by Max-product type operators. Springer, New York, 2016). Section 1 contains a short introduction in the topic. In Sect. 2, after presenting some general results, we state approximation results including upper estimates, direct and inverse results, localization results and shape preserving results, for the max-product: Bernstein–Kantorovich operators, truncated and non-truncated Favard–Szász-Mirakjan—Kantorovich operators, truncated and non-truncated Baskakov–Kantorovich operators, Meyer–König-Zeller–Kantorovich operators, Hermite–Fejér–Kantorovich operators based on the Chebyshev knots of first kind, discrete Picard–Kantorovich operators, discrete Weierstrass–Kantorovich operators and discrete Poisson–Cauchy–Kantorovich operators. All these approximation properties are deduced directly from the approximation properties of their corresponding usual max-product operators. Section 3 presents the approximation properties with quantitative estimates in the L^p-norm, $1 \leq p \leq +\infty$, for the Kantorovich variant of the truncated max-product sampling operators based on the Fejér kernel. In Sect. 4, we introduce and study the approximation properties in L^p-spaces, $1 \leq p \leq +\infty$ for truncated max-product Kantorovich operators based on generalized type kernels depending on two functions φ and ψ satisfying a set of suitable conditions. The goal of Sect. 5 is to present approximation in L^p, $1 \leq p \leq +\infty$, by sampling max-product Kantorovich operators based on generalized kernels, not necessarily with bounded support, or generated by

L. Coroianu
Department of Mathematics and Computer Sciences, University of Oradea, Oradea, Romania
e-mail: lcoroianu@uoradea.ro

S. G. Gal (✉)
Department of Mathematics and Computer Sciences, University of Oradea, Oradea, Romania

The Academy of Romanian Scientists, Bucharest, Romania
e-mail: galso@uoradea.ro

I. N. Parasidis et al. (eds.), *Mathematical Analysis in Interdisciplinary Research*,
Springer Optimization and Its Applications 179,
https://doi.org/10.1007/978-3-030-84721-0_10

sigmoidal functions. Several types of kernels for which the theory applies and possible extensions and applications to higher dimensions are presented. Finally, some new directions for future researches are presented, including applications to learning theory.

1 Introduction

It is known that the general form of a linear discrete operator attached to $f : I \to [0, +\infty)$ can be expressed by

$$D_n(f)(x) = \sum_{k \in I_n} p_{n,k}(x) f(x_{n,k}), x \in I, n \in \mathbb{N},$$

where $p_{n,k}(x)$ are various kinds of function basis on I with $\sum_{k \in I_n} p_{n,k}(x) = 1$, I_n are finite or infinite families of indices and $\{x_{n,k}; k \in I_n\}$ represents a division of I.

With the notation $\bigvee_{k \in A} a_k = \sup_{k \in A} a_k$, by the Open Problem 5.5.4, pp. 324–326 in [17], to each $D_n(f)(x)$, was attached the so-called max-product type operator defined by

$$L_n^{(M)}(f)(x) = \frac{\bigvee_{k \in I_n} p_{n,k}(x) \cdot f(x_{n,k})}{\bigvee_{k \in I_n} p_{n,k}(x)}, x \in I, n \in \mathbb{N}. \tag{1}$$

Note that if, for example, $p_{n,k}(x)$, $n \in \mathbb{N}$, $k = 0, \ldots, n$ is a polynomial basis, then the operators $L_n^{(M)}(f)(x)$ become piecewise rational functions.

In a long series of papers, the present authors have introduced and studied approximation properties (including upper estimates, saturation, localization, inverse results, shape preservation and global smoothness preservation) of the max-product operators of the form in (1), attached as follows: to Bernstein-type operators, like the Bernstein polynomials, Favard–Szász–Mirakjan operators (truncated and non-truncated cases), Baskakov operators (truncated and non-truncated cases), Meyer–König and Zeller operators, Bleimann–Butzer–Hahn operators, to interpolation polynomials of Lagrange and Hermite–Feéjer on various special knots and to sampling operators based on various kernels, like those of Whittaker type based on sinc-type kernels and those based on Fejér-type kernels.

All these results were collected in the recent research monograph [1] co-authored by the present authors.

After the appearance of this research monograph, the study of the max-product operators of the form (1) has been continued by other authors in many papers, like, for example, [8, 12–14, 18–27].

Remark It is worth noting that the max-product operators can also be naturally called as **possibilistic operators**, since they can be obtained by analogy with the Feller probabilistic scheme used to generate positive and linear operators, by replacing the probability (σ-additive), with a maxitive set function and the classical integral with the possibilistic integral (see, e.g., [1], Chapter 10, Section 10.2).

Now, to each max-product operator $L_n^{(M)}$, we can formally attach its Kantorovich variant, defined by

$$LK_n^{(M)}(f)(x) = \frac{\bigvee_{k \in I_n} p_{n,k}(x) \cdot (1/(x_{n,k+1} - x_{n,k})) \cdot \int_{x_{n,k}}^{x_{n,k+1}} f(t)dt}{\bigvee_{k \in I_n} p_{n,k}(x)}, \quad (2)$$

with $\{x_{n,k}; k \in I_n\}$ a division of the finite or infinite interval I.

The study of these kinds of max-product operators is missing from the research monograph [1] and has been completed only in the very recent papers [3–6, 9, 11, 12, 15, 16].

The goal of this chapter is to survey the main results for various kinds of Kantorovich max-product operators, focusing mainly on those obtained by the present authors in [3–7].

2 Approximation Properties of $LK_n^{(M)}$ Deduced from Those of $L_n^{(M)}$

Keeping the notations in the formulas (1) and (2), let us denote

$$C_+(I) = \{f : I \to \mathbb{R}_+; f \text{ is continuous on } I\},$$

where I is a bounded or unbounded interval, and suppose that all $p_{n,k}(x)$ are continuous functions on I, satisfying $p_{n,k}(x) \geq 0$, for all $x \in I, n \in \mathbb{N}, k \in I_n$ and $\sum_{k \in I_n} p_{n,k}(x) = 1$, for all $x \in I, n \in \mathbb{N}$.

In this section, from the properties of the max-product operators $L_n^{(M)}$, we present the properties of their Kantorovich variants $LK_n^{(M)}$ deduced form those of $L_n^{(M)}$.

All the results in this section are from Coroianu–Gal [5].

After presenting some general results, we state here approximation results for the max-product Bernstein–Kantorovich operators, truncated and non-truncated Favard–Szász–Mirakjan–Kantorovich operators, truncated and non-truncated Baskakov–Kantorovich operators, Meyer–König–Zeller–Kantorovich operators, Hermite-Fejér–Kantorovich operators based on the Chebyshev knots of first kind, discrete Picard-Kantorovich operators, discrete Weierstrass–Kantorovich operators and discrete Poisson–Cauchy–Kantorovich operators. Notice that the moment results obtained for these max-product type operators parallel somehow those obtained for the classical positive and linear operators in, e.g., [20].

Firstly, we present the following general results.

Lemma 2.1

(i) For any $f \in C_+(I)$, $LK_n^{(M)}(f)$ is continuous on I;
(ii) If $f \leq g$, then $LK_n^{(M)}(f) \leq LK_n^{(M)}(g)$;

(iii) $LK_n^{(M)}(f+g) \le LK_n^{(M)}(f) + LK_n^{(M)}(g)$;

(iv) *If* $f \in C_+(I)$ *and* $\lambda \ge 0$, *then* $LK_n^{(M)}(\lambda f) = \lambda LK_n^{(M)}(f)$.

(v) *If* $LK_n^{(M)}(e_0) = e_0$, *where* $e_0(x) = 1$, *for all* $x \in I$, *then for any* $f \in C_+(I)$, *we have*

$$\left|LK_n^{(M)}(f)(x) - f(x)\right| \le \left[1 + \frac{1}{\delta}LK_n^{(M)}(\varphi_x)(x)\right]\omega_1(f;\delta),$$

for any $x \in I$ *and* $\delta > 0$. *Here,* $\varphi_x(t) = |t-x|$, $t \in I$, *and* $\omega_1(f;\delta) = \sup\{|f(x)-f(y)|; x, y \in I, |x-y| \le \delta\}$;

(vi) $\left|LK_n^{(M)}(f) - LK_n^{(M)}(g)\right| \le LK_n^{(M)}(|f-g|)$.

Lemma 2.2 *With the notations in (1) and (2), suppose that, in addition,* $|x_{n,k+1} - x_{n,k}| \le \frac{C}{n+1}$ *for all* $k \in I_n$, *with* $C > 0$ *an absolute constant. Then, for all* $x \in I$ *and* $n \in \mathbb{N}$, *we have*

$$LK_n^{(M)}(\varphi_x)(x) \le L_n^{(M)}(\varphi_x)(x) + \frac{C}{n+1}.$$

Corollary 2.3 *With the notations in (1) and (2) and supposing that, in addition,* $|x_{n,k+1} - x_{n,k}| \le \frac{C}{n+1}$ *for all* $k \in I_n$, *for any* $f \in C_+(I)$, *we have*

$$\left|LK_n^{(M)}(f)(x) - f(x)\right| \le 2\left[\omega_1(f; L_n^{(M)}(\varphi_x)(x)) + \omega_1(f; C/(n+1))\right], \tag{3}$$

for any $x \in I$ *and* $n \in \mathbb{N}$.

This corollary shows that knowing quantitative estimates in approximation by a given max-product operator, we can deduce a quantitative estimate for its Kantorovich variant. Also, this corollary does not worsen the orders of approximation of the original operators. Let us exemplify below for several known max-product operators.

Firstly, let us choose $p_{n,k}(x) = \binom{n}{k}x^k(1-x)^{n-k}$, $I = [0,1]$, $I_n = \{0, \ldots, n-1\}$ and $x_{n,k} = \frac{k}{n+1}$. In this case, $L_n^{(M)}$ in (1) become the Bernstein max-product operators. Let us denote by $BK_n^{(M)}$ their Kantorovich variant, given by the formula

$$BK_n^{(M)}(f)(x) = \frac{\bigvee_{k=0}^{n}\binom{n}{k}x^k(1-x)^{n-k} \cdot (n+1)\int_{k/(n+1)}^{(k+1)/(n+1)} f(t)dt}{\bigvee_{k=0}^{n}\binom{n}{k}x^k(1-x)^{n-k}}. \tag{4}$$

We can state the following result.

Theorem 2.4

(i) *If* $f \in C_+([0,1])$, *then we have*

$$|BK_n^{(M)}(f)(x) - f(x)| \le 24\omega_1(f; 1/\sqrt{n+1}) + 2\omega_1(f; 1/(n+1)), x \in [0,1], n \in \mathbb{N}.$$

(ii) If $f \in C_+([0, 1])$ *is concave on* $[0, 1]$*, then we have*

$$|BK_n^{(M)}(f)(x) - f(x)| \leq 6\omega_1(f; 1/n), x \in [0, 1], n \in \mathbb{N}.$$

(iii) If $f \in C_+([0, 1])$ *is strictly positive on* $[0, 1]$*, then we have*

$$|BK_n^{(M)}(f)(x) - f(x)| \leq 2\omega_1(f; 1/n) \cdot \left(\frac{n\omega_1(f; 1/n)}{m_f} + 4\right) + 2\omega_1(f; 1/n),$$

for all $x \in [0, 1], n \in \mathbb{N}$*, where* $m_f = \min\{f(x); x \in [0, 1]\}$.

Now, let us choose $p_{n,k}(x) = \frac{(nx)^k}{k!}$, $I = [0, +\infty)$, $I_n = \{0, \ldots, n, \ldots, \}$ and $x_{n,k} = \frac{k}{n+1}$. In this case, $L_n^{(M)}$ in (1) become the non-truncated Favard–Szász–Mirakjan max-product operators. Let us denote by $FK_n^{(M)}$ their Kantorovich variant defined by

$$FK_n^{(M)}(f)(x) = \frac{\bigvee_{k=0}^{\infty} \frac{(nx)^k}{k!} \cdot (n+1) \int_{k/(n+1)}^{(k+1)/(n+1)} f(t)dt}{\bigvee_{k=0}^{\infty} \frac{(nx)^k}{k!}}. \tag{5}$$

We can state the following result.

Theorem 2.5

(i) If $f : [0, +\infty) \to [0, +\infty)$ *is bounded and continuous on* $[0, +\infty)$*, then we have*

$$|FK_n^{(M)}(f)(x) - f(x)| \leq 16\omega_1(f; \sqrt{x}/\sqrt{n}) + 2\omega_1(f; 1/n), x \in [0, +\infty), n \in \mathbb{N}.$$

(ii) If $f : [0, +\infty) \to [0, +\infty)$ *is continuous, bounded, non-decreasing, concave function on* $[0, +\infty)$*, then we have*

$$|FK_n^{(M)}(f)(x) - f(x)| \leq 4\omega_1(f; 1/n), x \in [0, +\infty), n \in \mathbb{N}.$$

If we choose $p_{n,k}(x) = \frac{(nx)^k}{k!}$, $I = [0, 1]$, $I_n = \{0, \ldots, n\}$ and $x_{n,k} = \frac{k}{n+1}$, in this case, $L_n^{(M)}$ in (1) become the truncated Favard–Szász–Mirakjan max-product operators. Let us denote by $TK_n^{(M)}$ their Kantorovich variant given by the formula

$$TK_n^{(M)}(f)(x) = \frac{\bigvee_{k=0}^{n} \frac{(nx)^k}{k!} \cdot (n+1) \int_{k/(n+1)}^{(k+1)/(n+1)} f(t)dt}{\bigvee_{k=0}^{n} \frac{(nx)^k}{k!}}. \tag{6}$$

We have the following theorem:

Theorem 2.6

(i) If $f \in C_+([0,1])$*, then we have*

$$|TK_n^{(M)}(f)(x) - f(x)| \le 12\omega_1(f; 1/\sqrt{n}) + 2\omega_1(f; 1/n), x \in [0,1], n \in \mathbb{N}.$$

(ii) If $f \in C_+([0,1])$ *is non-decreasing, concave function on* $[0,1]$*, then we have*

$$|TK_n^{(M)}(f)(x) - f(x)| \le 4\omega_1(f; 1/n), x \in [0,1], n \in \mathbb{N}.$$

Take now $p_{n,k}(x) = \binom{n+k-1}{k} x^k/(1+x)^{n+k}$, $I = [0, +\infty)$, $I_n = \{0, \dots, n, \dots,\}$ and $x_{n,k} = \frac{k}{n+1}$. In this case, $L_n^{(M)}$ in (1) become the non-truncated Baskakov max-product operators. Let us denote by $VK_n^{(M)}$ their Kantorovich variant defined by

$$VK_n^{(M)}(f)(x) = \frac{\bigvee_{k=0}^{\infty} \binom{n+k-1}{k} \frac{x^k}{(1+x)^{n+k}} \cdot (n+1) \int_{k/(n+1)}^{(k+1)/(n+1)} f(t)dt}{\bigvee_{k=0}^{\infty} \binom{n+k-1}{k} \frac{x^k}{(1+x)^{n+k}}}. \tag{7}$$

The following result holds.

Theorem 2.7

(i) If $f : [0, +\infty) \to [0, +\infty)$ *is bounded and continuous on* $[0, +\infty)$*, then for all* $x \in [0, +\infty)$ *and* $n \ge 3$*, we have*

$$|VK_n^{(M)}(f)(x) - f(x)| \le 24\omega_1(f; \sqrt{x(x+1)}/\sqrt{n-1}) + 2\omega_1(f; 1/(n+1)).$$

(ii) If $f : [0, +\infty) \to [0, +\infty)$ *is continuous, bounded, non-decreasing, concave function on* $[0, +\infty)$*, then for* $x \in [0, +\infty)$ *and* $n \ge 3$*, we have*

$$|VK_n^{(M)}(f)(x) - f(x)| \le 4\omega_1(f; 1/n).$$

For $p_{n,k}(x) = \binom{n+k-1}{k} x^k/(1+x)^{n+k}$, $I = [0,1]$, $I_n = \{0, \dots, n\}$ and $x_{n,k} = \frac{k}{n+1}$, $L_n^{(M)}$ in (1) become the truncated Baskakov max-product operators. Let us denote by $UK_n^{(M)}$ their Kantorovich variant defined by

$$UK_n^{(M)}(f)(x) = \frac{\bigvee_{k=0}^{n} \binom{n+k-1}{k} \frac{x^k}{(1+x)^{n+k}} \cdot (n+1) \int_{k/(n+1)}^{(k+1)/(n+1)} f(t)dt}{\bigvee_{k=0}^{\infty} \binom{n+k-1}{k} \frac{x^k}{(1+x)^{n+k}}}. \tag{8}$$

The following result holds.

Theorem 2.8

(i) If $f \in C_+([0,1])$*, then we have*

$$|UK_n^{(M)}(f)(x) - f(x)| \le 48\omega_1(f; 1/\sqrt{n+1}) + 2\omega_1(f; 1/(n+1)), x \in [0,1], n \ge 2.$$

(ii) If $f \in C_+([0,1])$ *is non-decreasing, concave function on* $[0,1]$*, then we have*

$$|UK_n^{(M)}(f)(x) - f(x)| \le 6\omega_1(f; 1/n), x \in [0,1], n \in \mathbb{N}.$$

Now, let us choose $p_{n,k}(x) = \binom{n+k}{k}x^k$, $I = [0,1]$, $I_n = \{0, \ldots, n, \ldots\}$ and $x_{n,k} = \frac{k}{n+1+k}$. In this case, $L_n^{(M)}$ in (1) become the Meyer–König and Zeller max-product operators. Also, it is easy to see that $|x_{n,k+1} - x_{n,k}| \le \frac{1}{n+1}$, for all $k \in I_n$. Let us denote by $ZK_n^{(M)}$ their Kantorovich variant defined by

$$ZK_n^{(M)}(f)(x) = \frac{\bigvee_{k=0}^{\infty} \binom{n+k}{k}x^k \cdot \frac{(n+k+1)(n+k+2)}{n+1} \int_{k/(n+1+k)}^{(k+1)/(n+k+2)} f(t)dt}{\bigvee_{k=0}^{\infty} \binom{n+k}{k}x^k}. \tag{9}$$

We have the following theorem:

Theorem 2.9

(i) If $f \in C_+([0,1])$*, then for* $n \ge 4$, $x \in [0,1]$*, we have*

$$|ZK_n^{(M)}(f)(x) - f(x)| \le 36\omega_1(f; \sqrt{x}(1-x)/\sqrt{n}) + 2\omega_1(f; 1/n).$$

(ii) If $f \in C_+([0,1])$ *is non-decreasing, concave function on* $[0,1]$*, then for* $x \in [0,1]$ *and* $n \ge 2x$*, we have*

$$|ZK_n^{(M)}(f)(x) - f(x)| \le 4\omega_1(f; 1/n).$$

Now, let us choose $p_{n,k}(x) = h_{n,k}(x)$—the fundamental Hermite–Fejér interpolation polynomials based on the Chebyshev knots of first kind $x_{n,k} = \cos\left(\frac{2(n-k)+1}{2(n+1)}\pi\right)$, $I = [-1,1]$, and $I_n = \{0, \ldots, n\}$. In this case, $L_n^{(M)}$ in (1) become the Hermite–Fejér max-product operators. Also, applying the mean value theorem to cos, it is easy to see that $|x_{n,k+1} - x_{n,k}| \le \frac{4}{n+1}$, for all $k \in I_n$. Let us denote by $HK_n^{(M)}$ their Kantorovich variant defined by

$$HK_n^{(M)}(f)(x) = \frac{\bigvee_{k=0}^{n} h_{n,k}(x) \cdot \frac{1}{x_{n,k} - x_{n,k+1}} \cdot \int_{x_{n,k}}^{x_{n,k+1}} f(t)dt}{\bigvee_{k=0}^{\infty} h_{n,k}(x)}, \tag{10}$$

where $x_{n,k} = \cos\left(\frac{2(n-k)+1}{2(n+1)}\pi\right)$.

The following result holds.

Theorem 2.10 *If* $f \in C_+([-1,1])$*, then for* $n \in \mathbb{N}$, $x \in [-1,1]$*, we have*

$$|HK_n^{(M)}(f)(x) - f(x)| \le 30\omega_1(f; 1/n).$$

Now, let us consider choose $p_{n,k}(x) = e^{-|x-k/(n+1)|}$, $I = (-\infty, +\infty)$, $I_n = \mathbb{Z}$—the set of integers and $x_{n,k} = \frac{k}{n+1}$. In this case, $L_n^{(M)}$ in (1) become the Picard max-product operators. Let us denote by $\mathscr{P}K_n^{(M)}$ their Kantorovich variant defined by

$$\mathscr{P}K_n^{(M)}(f)(x) = \frac{\bigvee_{k=0}^{\infty} e^{-|x-k/(n+1)|} \cdot (n+1) \int_{k/(n+1)}^{(k+1)/(n+1)} f(t)dt}{\bigvee_{k=0}^{\infty} e^{-|x-k/(n+1)|}}. \tag{11}$$

We can state the following result.

Theorem 2.11 *If $f : \mathbb{R} \to [0, +\infty)$ is bounded and uniformly continuous on $\mathbb{R}$, then we have*

$$|\mathscr{P}K_n^{(M)}(f)(x) - f(x)| \leq 6\omega_1(f; 1/n), x \in \mathbb{R}, n \in \mathbb{N}.$$

In what follows, let us choose $p_{n,k}(x) = e^{-(x-k/(n+1))^2}$, $I = (-\infty, +\infty)$, $I_n = \mathbb{Z}$—the set of integers and $x_{n,k} = \frac{k}{n+1}$. In this case, $L_n^{(M)}$ in (1) become the Weierstrass max-product operators. Let us denote by $\mathscr{W}K_n^{(M)}$ their Kantorovich variant defined by

$$\mathscr{W}K_n^{(M)}(f)(x) = \frac{\bigvee_{k=0}^{\infty} e^{-(x-k/(n+1))^2} \cdot (n+1) \int_{k/(n+1)}^{(k+1)/(n+1)} f(t)dt}{\bigvee_{k=0}^{\infty} e^{-(x-k/(n+1))^2}}. \tag{12}$$

We have the following theorem:

Theorem 2.12 *If $f : \mathbb{R} \to [0, +\infty)$ is bounded and uniformly continuous on $\mathbb{R}$, then we have*

$$|\mathscr{W}K_n^{(M)}(f)(x) - f(x)| \leq 4\omega_1(f; 1/\sqrt{n}) + 2\omega_1(f; 1/n), x \in \mathbb{R}, n \in \mathbb{N}.$$

At the end of this subsection, let us choose $p_{n,k}(x) = \frac{1}{n^2(x-k/n)^2+1}$, $I = (-\infty, +\infty)$, $I_n = \mathbb{Z}$—the set of integers and $x_{n,k} = \frac{k}{n+1}$. In this case, $L_n^{(M)}$ in (1) become the Poisson–Cauchy max-product operators. Let us denote by $\mathscr{C}K_n^{(M)}$ their Kantorovich variant

$$\mathscr{C}K_n^{(M)}(f)(x) = \frac{\bigvee_{k=0}^{\infty} \frac{1}{n^2(x-k/(n+1))^2+1} \cdot (n+1) \int_{k/(n+1)}^{(k+1)/(n+1)} f(t)dt}{\bigvee_{k=0}^{\infty} \frac{1}{n^2(x-k/(n+1))^2+1}}. \tag{13}$$

Concerning these operators, the following result holds.

Theorem 2.13 *If $f : \mathbb{R} \to [0, +\infty)$ is bounded and uniformly continuous on $\mathbb{R}$, then we have*

$$|\mathcal{C}K_n^{(M)}(f)(x) - f(x)| \leq 6\omega_1(f; 1/n), x \in \mathbb{R}, n \in \mathbb{N}.$$

Remark 2.14 All the Kantorovich kind max-product operators $LK_n^{(M)}$ given by (2) are defined and used for approximation of positive-valued functions. But, they can be used for approximation of lower bounded functions of variable sign too, by introducing the new operators

$$N_n^{(M)}(f)(x) = LK_n^{(M)}(f + c)(x) - c,$$

where $c > 0$ is such that $f(x) + c > 0$, for all x in the domain of definition of f.

It is easy to see that the operators $N_n^{(M)}$ give the same approximation orders as $LK_n^{(M)}$.

At the end of this section, we present the shape preserving properties, direct results and localization results of the Bernstein–Kantorovich max-product operators $BK_n^{(M)}$ given by (4).

They can be deduced from the corresponding results of $B_n^{(M)}$, based on the remark that the operator $BK_n^{(M)}$ can be obtained from the operator $B_n^{(M)}$, as follows. Suppose that f is arbitrary in $C_+([0, 1])$. Let us consider

$$f_n(x) = (n + 1) \int_{nx/(n+1)}^{(nx+1)/(n+1)} f(t)dt. \tag{14}$$

It is readily seen that $B_n^{(M)}(f_n)(x) = BK_n^{(M)}(f)(x)$, for all $x \in [0, 1]$. We also notice that $f_n \in C_+([0, 1])$. What is more, if f is strictly positive then so is f_n.

The following two shape preserving results hold.

Theorem 2.15 *Let $f \in C_+([0, 1])$.*

(i) If f is non-decreasing (non-increasing) on $[0, 1]$, then for all $n \in \mathbb{N}$, $BK_n^{(M)}(f)$ is non-decreasing (non-increasing, respectively) on $[0, 1]$.
(ii) If f is quasi-convex on $[0, 1]$, then for all $n \in \mathbb{N}$, $BK_n^{(M)}(f)$ is quasi-convex on $[0, 1]$. Here, quasi-convexity on $[0, 1]$ means that $f(\lambda x + (1 - \lambda)y) \leq \max\{f(x), f(y)\}$, for all $x, y, \lambda \in [0, 1]$.

Recall that a continuous function $f : [a, b] \to \mathbb{R}$ is quasi-concave, if and only if there exists $c \in [a, b]$ such that f is non-decreasing on $[a, c]$ and non-increasing on $[c, b]$.

Theorem 2.16 *Let $f \in C_+([0, 1])$. If f is quasi-concave on $[0, 1]$, then $BK_n^{(M)}(f)$ is quasi-concave on $[0, 1]$.*

Let us return now to the functions f_n given in (14), and let us find now an upper bound for the approximation of f by f_n in terms of the uniform norm. For some $x \in [0, 1]$, using the mean value theorem, there exists $\xi_x \in \left[\frac{nx}{n+1}, \frac{nx+1}{n+1}\right]$ such that $f_n(x) = f(\xi_x)$. We also easily notice that $|\xi_x - x| \leq \frac{1}{n+1}$. It means that

$$|f(x) - f_n(x)| \leq \omega_1(f; 1/(n+1)), x \in \mathbb{R}, n \in \mathbb{N}. \tag{15}$$

In particular, if f is Lipschitz with constant C, then f_n is Lipschitz continuous with constant $3C$. These estimations are useful to prove some inverse results in the case of the operator $BK_n^{(M)}$ by using analogue results already obtained for the operator $B_n^{(M)}$.

Below we present a result that gives for the class of Lipschitz function the order of approximation $1/n$ in the approximation by the operator $BK_n^{(M)}$, hence an analogue result that holds in the case of the operator $B_n^{(M)}$.

Theorem 2.17 *Suppose that f is Lipschitz on $[0, 1]$ with Lipschitz constant C, and suppose that the lower bound of f is $m_f > 0$. Then, we have*

$$\left\| BK_n^{(M)}(f) - f \right\| \leq 2C\left(\frac{C}{m_f} + 5\right) \cdot \frac{1}{n}, n \geq 1.$$

Here, $\| \cdot \|$ denotes the uniform norm on $C[0, 1]$.

In what follows, we deal with localization properties for the operators $BK_n^{(M)}$. We firstly present a very strong localization property.

Theorem 2.18 *Let $f, g : [0, 1] \to [0, \infty)$ be both bounded on $[0, 1]$ with strictly positive lower Bounds, and suppose that there exist $a, b \in [0, 1]$, $0 < a < b < 1$ such that $f(x) = g(x)$ for all $x \in [a, b]$. Then, for all $c, d \in [a, b]$ satisfying $a < c < d < b$, there exists $\widetilde{n} \in \mathbb{N}$ depending only on f, g, a, b, c and d such that $BK_n^{(M)}(f)(x) = BK_n^{(M)}(g)(x)$ for all $x \in [c, d]$ and $n \in \mathbb{N}$ with $n \geq \widetilde{n}$.*

As an immediate consequence of the localization result in Theorem 2.18, can be deduced the following local direct approximation result.

Corollary 2.19 *Let $f : [0, 1] \to [0, \infty)$ be bounded on $[0, 1]$ with the lower bound strictly positive and $0 < a < b < 1$ be such that $f|_{[a,b]} \in Lip\,[a, b]$ with Lipschitz constant $\overline{C}$. Then, for any $c, d \in [0, 1]$ satisfying $a < c < d < b$, we have*

$$\left|BK_n^{(M)}(f)(x) - f(x)\right| \leq \frac{C}{n} \text{ for all } n \in \mathbb{N} \text{ and } x \in [c, d],$$

where the constant C depends only on f and a, b, c, d.

Previously, we presented results which show that $BK_n^{(M)}$ preserves monotonicity and more generally quasi-convexity. By the localization result in Theorem 2.18 and then applying a very similar reasoning to the one used in the proof of Corollary 2.19, one can obtain as corollaries local versions for these shape preserving properties.

Corollary 2.20 *Let $f : [0, 1] \to [0, \infty)$ be bounded on $[0, 1]$ with strictly positive lower bound, and suppose that there exist $a, b \in [0, 1]$, $0 < a < b < 1$, such that f is non-decreasing (non-increasing) on $[a, b]$. Then, for any $c, d \in [a, b]$ with*

$a < c < d < b$, there exists $\widetilde{n} \in \mathbb{N}$ depending only on a, b, c, d and f such that $BK_n^{(M)}(f)$ is non-decreasing (non-increasing) on $[c, d]$ for all $n \in \mathbb{N}$ with $n \geq \widetilde{n}$.

Corollary 2.21 *Let $f : [0, 1] \to [0, \infty)$ be a continuous and strictly positive function, and suppose that there exist $a, b \in [0, 1]$, $0 < a < b < 1$, such that f is quasi-convex on $[a, b]$. Then, for any $c, d \in [a, b]$ with $a < c < d < b$, there exists $\widetilde{n} \in \mathbb{N}$ depending only on a, b, c, d and f such that $BK_n^{(M)}(f)$ is quasi-convex on $[c, d]$ for all $n \in \mathbb{N}$ with $n \geq \widetilde{n}$.*

Corollary 2.22 *Let $f : [0, 1] \to [0, \infty)$ be a continuous and strictly positive function, and suppose that there exist $a, b \in [0, 1]$, $0 < a < b < 1$, such that f is quasi-concave on $[a, b]$. Then for any $c, d \in [a, b]$ with $a < c < d < b$, there exists $\widetilde{n} \in \mathbb{N}$ depending only on a, b, c, d and f such that $BK_n^{(M)}(f)$ is quasi-concave on $[c, d]$ for all $n \in \mathbb{N}$ with $n \geq \widetilde{n}$.*

Remark 2.23 As in the cases of Bernstein-type max-product operators studied in the research monograph [1], for the max-product Kantorovich-type operators, we can find natural interpretation as possibilistic operators, which can be deduced from the Feller scheme written in terms of the possibilistic integral. These approaches also offer new proofs for the uniform convergence, based on a Chebyshev-type inequality in the theory of possibility.

Remark 2.24 The max-product Kantorovich operators $LK_n^{(M)}$ given by the formula (2) can be generalized by replacing the classical linear integral $\int dt$, with the nonlinear Choquet integral $(C) \int d\mu(t)$ with respect to a monotone and submodular set function μ, obtaining thus the new operators

$$LK_{n,\mu}^{(M)}(f)(x) = \frac{\bigvee_{k \in I_n} p_{n,k}(x) \cdot (1/(x_{n,k+1} - x_{n,k})) \cdot (C) \int_{x_{n,k}}^{x_{n,k+1}} f(t) d\mu(t)}{\bigvee_{k \in I_n} p_{n,k}(x)}. \tag{16}$$

It is worth noting that the above max-product Kantorovich–Choquet operators are *doubly nonlinear operators*: firstly due to max and secondly due to the Choquet integral. The study of these max-product Kantorovich–Choquet operators for various particular choices of $p_{n,k}(x)$, $x_{n,k}$ and μ remains as open questions for future researches.

3 Max-Product Sampling Kantorovich Operators Based on Fejér Kernel

Sampling operators are among the best tools in the approximation of signals when we have information from the past. Starting with the seminal works of Plana, Wittaker and others, this topic gained a continuous interest in the last century. A very detailed survey on this topic can be found in, e.g., [2].

In a series of papers, all included in Chapter 8 of the book co-authored by us [1], we applied the max-product method to sampling operators, as follows.

Applying this idea to Whittaker's cardinal series, we have obtained a Jackson-type estimate in uniform approximation of f by the max-product Whittaker sampling operator given by

$$S_{W,\varphi}^{(M)}(f)(t) = \frac{\bigvee_{k=-\infty}^{\infty} \varphi(Wt-k) f\left(\frac{k}{W}\right)}{\bigvee_{k=-\infty}^{\infty} \varphi(Wt-k)}, t \in \mathbb{R}, \tag{17}$$

where $W > 0$, $f : \mathbb{R} \to \mathbb{R}_+$ and φ is a kernel given by the formula $\varphi(t) = sinc(t)$, where $sinc(t) = \frac{sin(\pi t)}{\pi t}$, for $t \neq 0$ and at $t = 0$, $sinc(t)$ is defined to be the limiting value, that is, $sinc(0) = 1$.

A similar idea and study was applied to the sampling operator based on the Fejér-type kernel $\varphi(t) = \frac{1}{2} \cdot [sinc(t/2)]^2$.

Applying the max-product idea to the truncated sampling operator based on Fejér's kernel and defined by

$$T_n(f)(x) = \sum_{k=0}^{n} \frac{\sin^2(nx - k\pi)}{(nx - k\pi)^2} \cdot f\left(\frac{k\pi}{n}\right), x \in [0, \pi],$$

we have introduced and studied the uniform approximation by the truncated max-product operator based on the Fejér kernel, given by

$$T_n^{(M)}(f)(x) = \frac{\bigvee_{k=0}^{n} \frac{\sin^2(nx-k\pi)}{(nx-k\pi)^2} \cdot f\left(\frac{k\pi}{n}\right)}{\bigvee_{k=0}^{n} \frac{\sin^2(nx-k\pi)}{(nx-k\pi)^2}}, x \in [0, \pi], \tag{18}$$

where $f : [0, \pi] \to \mathbb{R}_+$. Here, since $sinc(0) = 1$, it means that above, for every $x = k\pi/n$, $k \in \{0, 1, \ldots, n\}$, we have $\frac{\sin(nx-k\pi)}{nx-k\pi} = 1$.

In the present section, we study the approximation properties with quantitative estimates in the L^p-norm, $1 \leq p \leq \infty$, for the Kantorovich variant of the above truncated max-product sampling operators $T_n^{(M)}(f)(x)$, defined for $x \in [0, \pi]$ and $n \in \mathbb{N}$ by

$$K_n^{(M)}(f)(x) = \frac{1}{\pi} \cdot \frac{\bigvee_{k=0}^{n} \frac{\sin^2(nx-k\pi)}{(nx-k\pi)^2} \cdot \left[(n+1) \int_{k\pi/(n+1)}^{(k+1)\pi/(n+1)} f(v)\, dv\right]}{\bigvee_{k=0}^{n} \frac{\sin^2(nx-k\pi)}{(nx-k\pi)^2}}, \tag{19}$$

where $f : [0, \pi] \to \mathbb{R}_+$, $f \in L^p[0, \pi]$, $1 \leq p \leq \infty$.

All the results in this section were obtained in the paper by Coroianu–Gal [3].

Firstly, we present some properties of the operator $K_n^{(M)}$, which are useful in the proofs of the approximation results.

Lemma 3.1

(i) *For any integrable function* $f : [0, \pi] \to \mathbb{R}$, $K_n^{(M)}(f)$ *is continuous on* $[0, \pi]$*;*
(ii) *If* $f \leq g$ *then* $K_n^{(M)}(f) \leq K_n^{(M)}(g)$*;*
(iii) $K_n^{(M)}(f + g) \leq K_n^{(M)}(f) + K_n^{(M)}(g)$*;*
(iv) $\left|K_n^{(M)}(f) - K_n^{(M)}(g)\right| \leq K_n^{(M)}(|f - g|)$*;*
(v) *If, in addition,* f *is positive on* $[0, \pi]$ *and* $\lambda \geq 0$*, then* $K_n^{(M)}(\lambda f) = \lambda K_n^{(M)}(f)$*.*

For the next result, we need the first-order modulus of continuity on $[0, \pi]$ defined for $f : [0, \pi] \to \mathbb{R}$ and $\delta \geq 0$, by

$$\omega_1(f; \delta) = \max\{|f(x) - f(y)| : x, y \in [0, \pi], |x - y| \leq \delta\}.$$

Lemma 3.2 *For any continuous function* $f : [0, \pi] \to \mathbb{R}_+$*, we have*

$$\left|K_n^{(M)}(f)(x) - f(x)\right| \leq \left[1 + \frac{1}{\delta} K_n^{(M)}(\varphi_x)(x)\right] \omega_1(f; \delta), \tag{20}$$

for any $x \in [0, \pi]$ *and* $\delta > 0$*. Here,* $\varphi_x(t) = |t - x|$, $t \in [0, \pi]$*.*

Our first main result proves that $K_n^{(M)}(f)(x)$ converges to $f(x)$ at any point of continuity for f.

Theorem 3.3 *Suppose that* $f : [0, \pi] \to \mathbb{R}_+$ *is bounded on its domain and integrable on any subinterval of* $[0, \pi]$*. If* f *is continuous at* $x_0 \in [0, \pi]$*, then*

$$\lim_{n\to\infty} K_n^{(M)}(f)(x_0) = f(x_0).$$

Also, a quantitative result holds.

Theorem 3.4 *Suppose that* $f : [0, \pi] \to \mathbb{R}_+$ *is continuous on* $[0, \pi]$*. Then, for any* $n \in \mathbb{N}$, $n \geq 1$*, we have*

$$\left\|K_n^{(M)}(f) - f\right\| \leq 10\omega_1\left(f; \frac{1}{n}\right).$$

Remark 3.5 The estimate in the statement of Theorem 3.4 remains valid for functions of arbitrary sign, lower bounded. Indeed, if $c \in \mathbb{R}$ is such that $f(x) \geq c$ for all $x \in [0, \pi]$, then it is easy to see that defining the new max-product operator $\overline{K}^{(M)}(f)(x) = K_n^{(M)}(f - c)(x) + c$, we get $|f(x) - \overline{K}^{(M)}(f)(x)| \leq 10\omega_1(f; 1/n)$, for all $x \in [0, \pi]$, $n \in \mathbb{N}$.

Now, let the L^p-norm, $\|f\|_p = \left(\int_0^\pi |f(t)|^p dt\right)^{1/p}$, with $1 \leq p < +\infty$. In this section, we present approximation results by $K_n^{(M)}$ in the L^p-norm. For this purpose, firstly, we need the following Lipschitz property of the operator $K_n^{(M)}$.

Theorem 3.6 *We have*

$$\left\| K_n^{(M)}(f) - K_n^{(M)}(g) \right\|_p \leq 2^{(1-2p)/p} \pi^2 \cdot \|f - g\|_p,$$

for any $n \in \mathbb{N}$, $n \geq 1$, $f, g : [0, \pi] \to \mathbb{R}_+$, $f, g \in L^p[0, \pi]$ and $1 \leq p < \infty$.

Now, let us define

$$C_+^1[0, \pi] = \{g : [0, \pi] \to \mathbb{R}_+; g \text{ is differentiable on } [0, \pi]\},$$

$\| \cdot \|_{C[0,\pi]}$ the uniform norm of continuous functions on $[0, \pi]$ and the Petree K-functional

$$K(f; t)_p = \inf_{g \in C_+^1[0,\pi]} \{\|f - g\|_p + t\|g'\|_{C[0,\pi]}\}.$$

The second main result of this section is the following.

Theorem 3.7 *Let $1 \leq p < \infty$. For all $f : [0, \pi] \to \mathbb{R}_+$, $f \in L^p[0, \pi]$, $n \in \mathbb{N}$, we have*

$$\|f - K_n^{(M)}(f)\|_p \leq c \cdot K\left(f; \frac{a}{n}\right)_p,$$

where $c = 1 + 2^{(1-2p)/p} \cdot \pi^2$, $a = \frac{3\pi^{1+1/p}}{2c}$.

Remark 3.8 The statement of Theorem 3.7 can be restated for functions of arbitrary sign, lower bounded. Indeed, if $c \in \mathbb{R}$ is such that $f(x) \geq c$ for all $x \in [0, \pi]$, then it is easy to see that defining the slightly modified max-product operator $\overline{K}^{(M)}(f)(x) = K_n^{(M)}(f - c)(x) + c$, we get $|f(x) - \overline{K}^{(M)}(f)(x)| = |(f(x) - c) - K_n^{(M)}(f - c)(x)|$ and since we may consider here that $c < 0$, we immediately get the relations

$$\begin{aligned} K(f - m; t)_p &= \inf_{g \in C_+^1[0,\pi]} \{\|f - (g + c)\|_p + t\|g'\|_{C[0,\pi]}\} \\ &= \inf_{g \in C_+^1[0,\pi]} \{\|f - (g + c)\|_p + t\|(g + c)'\|_{C[0,\pi]}\} \\ &= \inf_{h \in C_+^1[0,\pi],\, h \geq c} \{\|f - h\|_p + t\|h'\|_{C[0,\pi]}\}. \end{aligned}$$

4 Max-Product Kantorovich Operators Based on (ϕ, ψ)-Kernels

All the results in this section are from Coroianu–Gal [4].

Suggested by the max-product sampling operators based on sinc-Fejér kernels presented in the previous section, in this section we introduce truncated max-product Kantorovich operators based on generalized type kernels depending on two functions φ and ψ satisfying a set of suitable conditions. Pointwise convergence, quantitative uniform convergence in terms of the moduli of continuity and quantitative L^p-approximation results in terms of a K-functional are obtained. Previous results in sampling and neural network approximation are recaptured and new results for many concrete examples are obtained.

In this sense, we introduce the more general Kantorovich max-product operators based on a generalized (φ, ψ)-kernel, by the formula

$$K_n^{(M)}(f;\varphi,\psi)(x) = \frac{1}{b} \cdot \frac{\bigvee_{k=0}^{n} \frac{\varphi(nx-kb)}{\psi(nx-kb)} \cdot \left[(n+1)\int_{kb/(n+1)}^{(k+1)b/(n+1)} f(v)\,dv\right]}{\bigvee_{k=0}^{n} \frac{\varphi(nx-kb)}{\psi(nx-kb)}}, \tag{21}$$

where $b > 0$, $f : [0, b] \to \mathbb{R}_+$, $f \in L^p[0, b]$, $1 \le p \le \infty$ and φ and ψ satisfy some properties specific to max-product operators and required to prove pointwise, uniform or L^p convergence, as follows:

Definition 4.1 We say that (φ, ψ) forms a generalized kernel if satisfy some (not necessary all, depending on the type of convergence intended for study) of the following properties:

(i) $\varphi, \psi : \mathbb{R} \to \mathbb{R}_+$ are continuous on $\mathbb{R}$, $\varphi(x) \neq 0$ for all $x \in (0, b/2]$ and $\psi(x) \neq 0$ for all $x \neq 0$, $\frac{\varphi(x)}{\psi(x)}$ is an even function on $\mathbb{R}$ and $\lim_{x\to 0} \frac{\varphi(x)}{\psi(x)} = \alpha \in (0, 1]$.

(ii) There exists a constant $C \in \mathbb{R}$ such that $\varphi(x) \le C \cdot \psi(x)$, for all $x \in \mathbb{R}$.

(iii) There exist the positive constants $M > 0$ and $\beta > 0$, such that $\frac{\varphi(x)}{\psi(x)} \le \frac{M}{x^\beta}$, for all $x \in (0, \infty)$.

(iv) For any $n \in \mathbb{N}$, $j \in \{0, \ldots, n\}$ and $x \in \left[\frac{jb}{n}, \frac{(j+1)b}{n}\right]$,

$$\bigvee_{k=0}^{n} \frac{\varphi(nx-kb)}{\psi(nx-kb)} = \max\left\{\frac{\varphi(nx-jb)}{\psi(nx-jb)}, \frac{\varphi(nx-(j+1)b)}{\psi(nx-(j+1)b)}\right\}.$$

(v) $\int_{-\infty}^{+\infty} \frac{\varphi(y)}{\psi(y)} dy = c$, where $c > 0$ is a positive real constant.

Remark 4.2 The properties of φ and ψ in Definition 4.1 were suggested by the methods characteristic in the proofs for various convergence results of max-product Kantorovich sinc-type operators. The use of the two functions in the generalized kernels offers a large flexibility in finding many concrete examples.

Remark 4.3 Let us note that if properties (i) and (iii) hold simultaneously, then (ii) holds too. Indeed, firstly if (i) holds, clearly that we may extend the continuity of $\frac{\varphi(x)}{\psi(x)}$ in the origin too, that is, we take $\frac{\varphi(0)}{\psi(0)} = \lim_{x\to 0}\frac{\varphi(x)}{\psi(x)}$. This means that $\frac{\varphi(x)}{\psi(x)}$ is continuous on the whole $\mathbb{R}$. Secondly, from (iii), it is readily seen that there exists a constant $a > 0$ such that $\frac{M}{x^\beta} \le 1$, for all $x \in [a, \infty)$. It means that $\varphi(x) \le \psi(x)$, for all $x \in [a, \infty)$. This fact combined with the continuity of $\frac{\varphi(x)}{\psi(x)}$ on $[-a, a]$ easily implies that (ii) holds.

Remark 4.4 Another important remark is that if (i) and (iii), $\beta > 1$ case, hold simultaneously, then (v) holds too. Indeed, since $\frac{\varphi(x)}{\psi(x)}$ is an even function on $\mathbb{R}$, it suffices to prove that $\int_0^{+\infty}\frac{\varphi(y)}{\psi(y)}dy$ is finite. From the continuity of $\frac{\varphi(x)}{\psi(x)}$, this later integral is finite if and only if $\int_1^{+\infty}\frac{\varphi(y)}{\psi(y)}dy$ is finite. Now, since $\frac{\varphi(x)}{\psi(x)} \le \frac{M}{x^\beta}$, for all $x \in [0, \infty)$, and since we easily note that $\int_1^{+\infty}\frac{M}{x^\beta}dx$ is finite, we conclude that $\int_1^{+\infty}\frac{\varphi(y)}{\psi(y)}dy$ is finite. Thus, $\int_{-\infty}^{+\infty}\frac{\varphi(y)}{\psi(y)}dy$ is finite, which means that (v) holds.

Remark 4.5 If in the pair (φ, ψ), we consider that ψ is a strictly positive constant function, then in order that (φ, ψ) be a generalized kernel satisfying all the properties (i)–(v) in Definition 1.1, it is good enough if $\varphi : \mathbb{R} \to \mathbb{R}_+$ is a continuous, even function, satisfying $\varphi(x) > 0$, for all $x \in (0, b/2)$, $\varphi(0) \ne 0$ (this implies (i)), $\varphi(x)$ is bounded on $\mathbb{R}$ (this implies (ii)), $\varphi(x) = \mathcal{O}\left(\frac{1}{x^\beta}\right)$, $x \in [0, +\infty)$, $\beta > 0$ (this implies (iii)), $\varphi(x)$ is non-increasing on $[0, +\infty)$ (this implies (iv)) and $\int_0^{+\infty}\varphi(x)dx < +\infty$ (this implies (v)). Note that this particular type of choice for the generalized kernel (φ, ψ) may cover some sampling approximation operators (see Application 5.3 in Sect. 5) and neural network operators (see Application 5.6).

Firstly, we present some properties of the operator $K_n^{(M)}(\cdot; \varphi, \psi)$, which will be useful to prove the approximation results.

Lemma 4.6 *Suppose that φ and ψ satisfy condition (i) from the end of the previous section.*

(i) *For any integrable function $f : [0, b] \to \mathbb{R}$, $K_n^{(M)}(f; \varphi, \psi)$ is continuous on $[0, b]$;*
(ii) *If $f \le g$, then $K_n^{(M)}(f; \varphi, \psi) \le K_n^{(M)}(g; \varphi, \psi)$;*
(iii) $K_n^{(M)}(f + g; \varphi, \psi) \le K_n^{(M)}(f; \varphi, \psi) + K_n^{(M)}(g; \varphi, \psi)$;
(iv) $\left|K_n^{(M)}(f; \varphi, \psi) - K_n^{(M)}(g; \varphi, \psi)\right| \le K_n^{(M)}(|f - g|; \varphi, \psi)$;
(v) *If, in addition, f is positive on $[0, b]$ and $\lambda \ge 0$, then $K_n^{(M)}(\lambda f; \varphi, \psi) = \lambda K_n^{(M)}(f; \varphi, \psi)$.*

For the next result, we need the first-order modulus of continuity on $[0, b]$ defined for $f : [0, b] \to \mathbb{R}$ and $\delta \ge 0$, by

$$\omega_1(f; \delta) = \max\{|f(x) - f(y)| : x, y \in [0, b], |x - y| \le \delta\}.$$

Lemma 4.7 *Suppose that φ and ψ satisfy condition (i) from the end of the previous section. For any continuous function $f : [0, b] \to \mathbb{R}_+$, we have*

$$\left|K_n^{(M)}(f; \varphi, \psi)(x) - f(x)\right| \leq \left[1 + \frac{1}{\delta} K_n^{(M)}(\varphi_x; \varphi, \psi)(x)\right] \omega_1(f; \delta), \tag{22}$$

for any $x \in [0, b]$ and $\delta > 0$. Here, $\varphi_x(t) = |t - x|$, $t \in [0, b]$.

Our first main result in this section proves that for $n \to \infty$, $K_n^{(M)}(f; \varphi, \psi)(x)$ converges to $f(x)$ at any point of continuity for f.

Theorem 4.8 *Suppose that $f : [0, b] \to \mathbb{R}_+$ is bounded on its domain and integrable on any subinterval of $[0, b]$. Then, suppose that φ and ψ satisfy the properties (i)–(iii). If f is continuous at $x_0 \in [0, b]$, then we have*

$$\lim_{n \to \infty} K_n^{(M)}(f; \varphi, \psi)(x_0) = f(x_0).$$

Corollary 4.9 *Suppose that $f : [0, b] \to \mathbb{R}_+$ is continuous on $[0, b]$ and that φ and ψ satisfy the properties (i)–(iii). Then, $K_n^{(M)}(f; \varphi, \psi)$ converges to f uniformly on $[0, b]$.*

We present a quantitative estimate that involves the modulus of continuity, as follows.

Theorem 4.10 *Suppose that $f : [0, b] \to \mathbb{R}_+$ is continuous on $[0, b]$ and that properties (i), (iii) and (iv) are fulfilled by φ and ψ. Then, for any $n \in \mathbb{N}$, we have*

$$\left\|K_n^{(M)}(f; \varphi, \psi) - f\right\|$$

$$\leq 2\left(b + \frac{M \cdot (2b)^{1-\beta}}{c_1} + 1\right) \max\left\{\omega_1\left(f; \frac{1}{n}\right), \omega_1\left(f; \frac{1}{n^\beta}\right)\right\}.$$

Here, c_1 denotes a constant.

Remark 4.11 The estimate in the statement of Theorem 4.10 remains valid for functions of arbitrary sign, lower bounded. Indeed, if $m \in \mathbb{R}$ is such that $f(x) \geq m$ for all $x \in [0, b]$, then it is easy to see that defining the new max-product operator $\overline{K}_n^{(M)}(f; \varphi, \psi)(x) = K_n^{(M)}(f - m; \varphi, \psi)(x) + m$, for $|f(x) - \overline{K}_n^{(M)}(f; \varphi, \psi)(x)|$, we get the same estimate as in the statement of Theorem 4.10.

Now, let the L^p-norm, $\|f\|_p = \left(\int_0^b |f(t)|^p dt\right)^{1/p}$, with $1 \leq p < +\infty$, $0 < b < +\infty$. In what follows, we present the approximation properties of $K_n^{(M)}$ in the L^p-norm. For this purpose, firstly, we need the following Lipschitz property of the operator $K_n^{(M)}$.

Theorem 4.12 *Supposing that φ and ψ satisfy (i), (ii) and (v), we have*

$$\left\| K_n^{(M)}(f;\varphi,\psi) - K_n^{(M)}(g;\varphi,\psi)\right\|_p \le \frac{1}{c_1}\cdot\left(\frac{2c}{b}\right)^{1/p}\cdot \|f-g\|_p,$$

for all $n \in \mathbb{N}$, $f, g : [0, b] \to \mathbb{R}_+$, $f, g \in L^p[0, b]$ *and* $1 \le p < \infty$.

Now, let us consider by $C_+^1[0,b]$ the space of all non-negative and continuous differentiable functions on $[0, b]$ and the K-functional

$$K(f;t)_p = \inf_{g\in C_+^1[0,b]} \{\|f-g\|_p + t\|g'\|_{C[0,b]}\}.$$

The second main result of this section is the following.

Theorem 4.13 *Suppose that* φ *and* ψ *satisfy (i) (ii) and (v), and let* $1 \le p < \infty$. *For all* $f : [0, b] \to \mathbb{R}_+$, $f \in L^p[0, b]$, $n \in \mathbb{N}$, *we have*

$$\|f - K_n^{(M)}(f;\varphi,\psi)\|_p \le d\cdot K\left(f;\frac{\Delta_{n,p}}{d}\right)_p,$$

where $\Delta_{n,p} = \|K_n^{(M)}(\varphi_x;\varphi,\psi)\|_p$, $\varphi_x(t) = |t-x|$ *and* $d = 1 + \frac{1}{c_1}\cdot\left(\frac{2c}{b}\right)^{1/p}$.

Corollary 4.14 *Let* $1 \le p < +\infty$. *Suppose now that* φ *and* ψ *satisfy all the conditions (i)–(v). For all* $f : [0, b] \to \mathbb{R}_+$, $f \in L^p[0, b]$, $n \in \mathbb{N}$, *we have*

$$\|f - K_n^{(M)}(f;\varphi,\psi)\|_p \le d\cdot K\left(f; D\cdot\max\left\{\frac{1}{n},\frac{1}{n^\beta}\right\}\right)_p,$$

where $D = \frac{b^{1/p}}{d}\cdot\left(b + \frac{M(2b)^{1-\beta}}{c_1}\right)$ *and* $d = 1 + \frac{1}{c_1}\cdot\left(\frac{2c}{b}\right)^{1/p}$.

Remark 4.15 The statements of Theorem 4.13 and Corollary 4.14 can be restated for functions of arbitrary sign, lower bounded. Indeed, if $m \in \mathbb{R}$ is such that $f(x) \ge m$ for all $x \in [0, b]$, then it is easy to see that defining the slightly modified max-product operator $\overline{K}^{(M)}(f:\varphi,\psi)(x) = K_n^{(M)}(f-m;\varphi,\psi)(x)+m$, we get $|f(x) - \overline{K}^{(M)}(f;\varphi,\psi)(x)| = |(f(x)-m) - K_n^{(M)}(f-m;\varphi,\psi)(x)|$, and since we may consider here that $m < 0$, we immediately get the relations

$$K(f-m;t)_p = \inf_{g\in C_+^1[0,b]} \{\|f-(g+m)\|_p + t\|g'\|_{C[0,b]}\}$$

$$= \inf_{g\in C_+^1[0,b]} \{\|f-(g+m)\|_p + t\|(g+m)'\|_{C[0,b]}\}$$

$$= \inf_{h\in C_+^1[0,b],\, h\ge m} \{\|f-h\|_p + t\|h'\|_{C[0,b]}\}.$$

In the next lines, we present some concrete examples of (φ, ψ)-kernels satisfying the conditions in Definition 4.1.

Application 4.16 Let us choose $\varphi(x) = \sin^{2r}(x)$, $\psi(x) = x^{2r}$, with $r \in \mathbb{N}$. In this case, $\frac{\varphi(x)}{\psi(x)}$ represents in fact the so-called generalized Jackson kernel. Now, in Definition 4.1 by taking $b = \pi$, condition (i) is evidently satisfied with $\alpha = 1$, condition (ii) is evidently satisfied with $C = 1$, condition (iii) holds with $M = 1$ and $\beta = 2r$ and condition (v) is satisfied with $c = \frac{\pi}{(2r-1)!} \cdot e_r$, where e_r is the so-called Eulerian number given by

$$e_r = \sum_{j=0}^{r}(-1)^j \binom{2r}{j}(r-j)^{2r-1}.$$

Due to the fact that $\sin^{2r}(nx - k\pi) = \sin^{2r}(nx)$, the equality in condition (iv) in Definition 4.1, one reduces to

$$\bigvee_{k=0}^{n} \frac{1}{(nx - kb)^{2r}} = \max\left\{\frac{1}{(nx - jb)^{2r}}, \frac{1}{(nx - (j+1)b)^{2r}}\right\}, \qquad (23)$$

for all $x \in \left[\frac{jb}{n}, \frac{(j+1)b}{n}\right]$, which follows by simple calculation.

Concluding, all the results in this chapter are valid for the max-product Kantorovich sampling operators based on this kernel (φ, ψ) and given by (21).

Application 4.17 Let us choose $\varphi(x) = \sin(x/2)\sin(3x/2)$, $\psi(x) = 9x^2/4$. We note that $\frac{\varphi(x)}{\psi(x)}$ represents in fact the so-called de la Vallée-Poussin kernel used in approximation by sampling operators. Similar reasonings with those in Application 4.16 easily lead to the fact that in this case too conditions (i)–(v) in Definition 4.1 hold and that the max-product Kantorovich sampling operators in (21) based on this (φ, ψ)-kernel satisfy all the results in this chapter.

Application 4.18 Let us choose as $\varphi(x)$ the B-spline of order 3 given by

$$\varphi(x) = \frac{3}{4} - x^2, \text{ if } |x| \le \frac{1}{2}, \ \varphi(x) = \frac{1}{2}\left(\frac{3}{2} - |x|\right)^2, \text{ if } \frac{1}{2} < |x| \le \frac{3}{2},$$

$$\varphi(x) = 0, \text{ if } |x| > \frac{3}{2}.$$

Choosing, for example, $\psi(x) = 1$, for all $x \in \mathbb{R}$, it is easy to see that (φ, ψ) verifies all the conditions in Definition 4.1, as follows: condition (1) with $b = \frac{1}{2}$, condition (ii) with a sufficiently large constant $C > 0$, condition (iii) with $\beta = 2$ and $M > 0$ sufficiently large, and evidently condition (iv) and condition (v). In conclusion, all the results in this chapter hold for the max-product Kantorovich operator in (21) based on this kernel (φ, ψ).

In fact, if we choose for $\varphi(x)$ any B-spline of an arbitrary order n and $\psi(x) = 1$, $x \in \mathbb{R}$, then (φ, ψ) verifies, as in the previous lines, all the conditions in Definition 4.1, which means that all the results in this chapter hold for the max-product Kantorovich operators in (21) based on this (φ, ψ)-kernel.

Application 4.19 Let us consider $\varphi(x) = 2\arctan\left(\frac{1}{x^2}\right)$, $x \neq 0$, $\varphi(0) = \pi$ and $\psi(x) = \pi$, $x \in \mathbb{R}$, where $\arctan : \mathbb{R} \to \left(-\frac{\pi}{2}, \frac{\pi}{2}\right)$ and $\lim_{y\to\infty} \arctan(y) = \frac{\pi}{2}$. We check the conditions in Definition 4.1. Indeed, it is clear that condition (i) is satisfied for $b = 1$ and with $\alpha = 1$, while since $0 \leq \arctan(y) \leq \frac{\pi}{2}$ for all $y \geq 0$, condition (ii) follows with $C = 1$. Note here that since $\arctan(y) \geq \frac{\pi}{4}$, for all $y \in [1, +\infty)$, putting $y = \frac{1}{x^2}$, we can take $b = 1$. By $2\arctan(1/x^2) \leq \frac{2}{x^2}$, for all $x > 0$, we obtain that condition (iii) holds too with $\beta = 2$ and $M = 2\pi$

Then, since $\arctan(y) \leq y$ for all $y \in [0 + \infty)$, we get

$$\int_{-\infty}^{+\infty} \frac{\varphi(x)}{\psi(x)} dx = \frac{2}{\pi} \int_{0}^{+\infty} \arctan(1/x^2) dx$$

$$= \frac{2}{\pi} \int_{0}^{1} \arctan(1/x^2) dx + \frac{2}{\pi} \int_{1}^{+\infty} \arctan(1/x^2) dx \leq 1 + \frac{2}{\pi} \int_{1}^{+\infty} \arctan(1/x^2) dx$$

$$\leq 1 + \frac{2}{\pi} \int_{1}^{+\infty} \frac{1}{x^2} dx = 1 + \frac{2}{\pi} < +\infty,$$

which shows that condition (v) holds.

Now, since for $x \in [j/n, (j+1)/n]$, we evidently have (see also the similar relation (23))

$$\frac{1}{(x - k/n)^2} \leq \frac{1}{(x - j/n)^2} \text{ and } \frac{1}{(x - k/n)^2} \leq \frac{1}{(x - (j+1)/n)^2}, \text{ for } 0 \leq k \leq n, \tag{24}$$

applying here the increasing function arctan, we immediately obtain (iv).

In conclusion, for this choice of the (φ, ψ)-kernel, all the results in this chapter remain valid for the max-product operators given by (21).

Application 4.20 Let us choose $\varphi(x) = |x|$ and $\psi(x) = e^{|x|} - 1$. We will check the conditions in Definition 4.1. Firstly, it is easy to see that condition (i) is satisfied with, e.g., $b = \ln(2)$, since by using l'Hopital's rule, we have $\lim_{x\to 0} \frac{\varphi(x)}{\psi(x)} = 1$. Then, by $|x| \leq e^{|x|} - 1$ for all $x \in \mathbb{R}$, it follows that condition (ii) holds with $C = 1$, and we can take $b = \ln(2)$). Condition (iii) obviously holds for $M = 2$ and $\beta = 1$.

Then, condition (v) also is satisfied, since

$$\int_{-\infty}^{+\infty} \frac{|x|}{e^{|x|} - 1} dx = 2 \int_{0}^{+\infty} \frac{x}{e^{x} - 1} dx$$

$$= 2\int_0^1 \frac{x}{e^x - 1}dx + 2\int_1^{+\infty} \frac{x}{e^x - 1}dx = c > 0,\ c \text{ finite},$$

since $\int_1^{+\infty} \frac{x}{e^x-1}dx \leq \int_1^{+\infty} x \cdot e^{-x/2}dx < +\infty$.

It remains to check condition (iv). Firstly, by similar reasonings to those used for the proofs of relations (23) and (24), for all $x \in [jb/n, (j+1)b/n]$, we get

$$|nx - kb| \geq |nx - jb| \text{ and } |nx - kb| \geq |nx - (j+1)b|, \text{ for all } k = 0, \ldots, n.$$

Now, denote $F(u) = \frac{u}{e^u-1}$, $u \geq 0$. If we prove that F is non-increasing on $[0, +\infty)$, then we immediately get that condition (iv) in Definition 4.1 is satisfied.

In this sense, by $F'(u) = \frac{e^u-1-ue^u}{(e^u-1)^2} = \frac{G(u)}{(e^u-1)^2}$, with $G(u) = e^u - 1 - ue^u$, since $G(0) = 0$ and $G'(u) = -ue^u \leq 0$, we immediately obtain $G(u) \leq 0$, for all $u \geq 0$, and consequently $F'(u) \leq 0$, for all $u \geq 0$.

In conclusion, in the case of this (φ, ψ)-kernel too, all the results in this chapter remain valid for the max-product operators given by (21).

Application 4.21 It is worth mentioning that if (φ_1, ψ_1) and (φ_2, ψ_2) are two kernels satisfying the conditions (i), (ii), (iii) and (v) in Definition 4.1, then the new kernel $(\varphi_1 \cdot \varphi_2, \psi_1 \cdot \psi_2)$ also satisfies these conditions. This remark is useful in order to generate new generalized kernels for which at least the convergence results in Theorem 4.8 and Corollary 4.9 still hold.

The only problem is that, condition (iv) is not, in general, satisfied by the $(\varphi_1 \cdot \varphi_2, \psi_1 \cdot \psi_2)$-kernel.

5 Max-Product Sampling Kantorovich Operators Based on Generalized Kernels

All the results in this section are from Coroianu–Costarelli–Gal–Vinti [6].

The goal of this section is to present generalized kernels, not necessarily with bounded support or generated by sigmoidal functions, such that the approximation capability and the approximation quality remain good for the sampling Kantorovich operators based on them.

It will be revealed that it suffices to consider measurable and bounded kernels, which in addition have bounded generalized absolute moment of order β, for some $\beta > 0$. These absolute moments are max-product variants of the linear counterparts by replacing the sum (or series) with the supremum. These assumptions are sufficient to obtain pointwise and uniform convergence properties of generalized sampling max-product Kantorovich operators. In addition, a fast Jackson-type estimation is obtained for the approximation of continuous functions. Also, we obtain estimations with respect to the L_p-norm and in terms of the corresponding K-functionals and the modulus of continuity. In particular, we obtain the convergence to the approximated function in the space of L_p-integrable functions. Several types of kernels for which the theory applies and possible extensions and applications to higher dimensions are presented.

In this section, firstly, we present the basic mathematical tools used, as well as the definition and some general properties of the generalized sampling max-product Kantorovich operators. We will consider functions $f : I \to \mathbb{R}$, the domain being a compact interval $I = [a, b]$ or $I = \mathbb{R}$. In the first case, $C(I)$ denotes the set of all continuous functions defined on I, and in the second case $C(I)$ denotes the set of uniformly continuous and bounded functions defined on I. Furthermore, by $C_+(I)$ we denote the subspace of $C(I)$ of the non-negative functions. We also consider the larger space $B(I)$ of all bounded real functions defined on I and denote by $B_+(I)$ the space of all bounded and non-negative functions on I. Then, for some $p \geq 1$, we denote with $L_p(I)$ the set of all L_p-integrable functions defined on I. For $f \in B(I)$, we denote $\|f\|_\infty = \sup_{x \in I} |f(x)|$, while for $f \in L_p(I)$, we denote $\|f\|_p = \left(\int |f(x)|^p \, dx\right)^{1/p}$.

Consider a set of indices J and the set of real numbers $\{x_k : k \in J\}$. We denote the supremum of this set as $\bigvee_{k \in J} x_k$. If J is finite, then $\bigvee_{k \in J} x_k = \max_{k \in J} x_k$.

Here, we will consider kernels of the form $\chi : \mathbb{R} \to \mathbb{R}$, where χ is bounded and measurable and satisfies the properties:

(χ_1)

$$\text{there exists } \beta > 0 \text{ such that } m_\beta(\chi) := \sup_{x \in \mathbb{R}} \bigvee_{k \in \mathbb{Z}} |\chi(x-k)| \cdot |x-k|^\beta < \infty$$

(we call $m_\beta(\chi)$ the generalized absolute moment of order β of χ)

(χ_2)

$$\text{we have } \inf_{x \in [-3/2, 3/2]} \chi(x) =: a_\chi > 0;$$

or

(χ_2')

$$\text{we have } \inf_{x \in [-1/2, 1/2]} \chi(x) =: a_\chi > 0;$$

Here, when we study the max-product sampling Kantorovich operators in the compact case $I = [a, b]$, we need to assume the slightly stronger assumption $(\chi 2)$ in place of (χ_2'), while in the case $I = \mathbb{R}$ assumptions (χ_2') is still enough to prove the desired results.

The goal of this chapter is to prove that these conditions will generate similar properties for the Kantorovich type max-product sampling operators. In addition, we will obtain qualitative and quantitative convergence properties with respect to the norm $\|\cdot\|_p$.

We recall here some useful results.

Lemma 5.1 *If $\chi : \mathbb{R} \to \mathbb{R}$ is bounded and such that $\chi(x) = O\left(|x|^{-\alpha}\right)$, as $|x| \to \infty$, for $\alpha > 0$, then*

$$m_\beta(\chi) < \infty, \text{for every} \, 0 \leq \beta \leq \alpha.$$

Lemma 5.2 *If $\chi : \mathbb{R} \to \mathbb{R}$ is bounded and satisfies (χ_1) for some $\beta > 0$, then*

$$m_\upsilon(\chi) < \infty, \text{for every} 0 \leq \upsilon \leq \beta.$$

In particular, we have $m_0(\chi) \leq \|\chi\|_\infty$.

Lemma 5.3 *Let $\chi : \mathbb{R} \to \mathbb{R}$ be a given function, and consider a fixed compact interval $[a, b]$.*

If χ satisfies assumption (χ_2'), for every $n \in \mathbb{N}^+$, we have

$$\bigvee_{k \in \mathbb{Z}} \chi(nx - k) \geq a_\chi, \text{for all} x \in \mathbb{R}. \tag{25}$$

If χ satisfies assumption (χ_2), for every $n \in \mathbb{N}^+$ sufficiently large, we also have

$$\bigvee_{k \in \mathscr{J}_n} \chi(nx - k) \geq a_\chi, \text{for all} x \in [a, b]. \tag{26}$$

Here, $\mathscr{J}_n := \{k \in \mathbb{Z} : k = \lceil na \rceil, \ldots, \lfloor nb \rfloor - 1\}$, where $\lceil \cdot \rceil$ and $\lfloor \cdot \rfloor$ denote, respectively, the "ceiling" and the "integral part" of a given number, a_χ is the constant from condition $\left(\chi_2'\right)$ and (χ_2), respectively.

If we keep condition (χ_1) but replace conditions (χ_2) and $\left(\chi_2'\right)$ with the weaker condition that both relations (1)–(2) hold, then the main approximation results of this chapter still hold. It means that we can obtain such results for kernels that are not necessarily bounded from below by a positive value on $\left[-\frac{1}{2}, \frac{1}{2}\right]$ or $[-3/2, 3/2]$.

Lemma 5.4 *Suppose that $\chi : \mathbb{R} \to \mathbb{R}$ is any bounded and measurable function, which satisfies (χ_1) with $\beta > 0$. Then, for every $\gamma > 0$, we have*

$$\bigvee_{k \in \mathbb{Z}: |x-k| > \gamma n} |\chi(x - k)| = \mathscr{O}\left(n^{-\beta}\right), \text{as} \, n \to \infty,$$

uniformly with respect to $x \in \mathbb{R}$.

We are now in position to introduce the sampling Kantorovich operators. In the next definition, $\mathcal{J}_n = \mathbb{Z}$, if $I = \mathbb{R}$ and $\mathcal{J}_n = \{k \in \mathbb{Z} : k = \lceil na \rceil, \ldots, \lfloor nb \rfloor - 1\}$, if $I = [a, b]$.

Definition 5.5 Let $f : I \to \mathbb{R}$ be a locally integrable function, and let $\chi : \mathbb{R} \to \mathbb{R}$ be a kernel such that $\bigvee_{k\in \mathcal{J}_n} \chi(nx - k) \neq 0$, for all $x \in I$.

The max-product Kantorovich generalized sampling operator on f based upon χ is defined as

$$K_n^{\chi}(f)(x) := \frac{\bigvee_{k\in\mathcal{J}_n} \chi(nx-k)\, n \int_{k/n}^{(k+1)/n} f(t)\, dt}{\bigvee_{k\in\mathcal{J}_n} \chi(nx-k)}, \quad x \in I.$$

Obviously, if χ satisfies (χ_2) or (χ_2') (if $I = [a, b]$ or $I = \mathbb{R}$, respectively), we always have $\bigvee_{k\in\mathcal{J}_n} \chi(nx - k) \neq 0$, for all $x \in I$.

We observe that $K_n^{\chi}(f)$ is well defined, for instance, if f is bounded. Indeed, it is easily seen that

$$\left|K_n^{\chi}(f)(x)\right| \leq \frac{m_0(\chi)}{a_\chi} \|f\|_\infty < \infty.$$

The purpose of this section is to study convergence properties for $K_n^{\chi} f$ without needing kernels with compact support.

As for other types of max-product operators, we can prove some important properties for K_n^{χ}.

Lemma 5.6 *Let χ be a kernel satisfying the conditions (χ_1) and (χ_2) or (χ_2') for $I = [a, b]$ and $I = \mathbb{R}$, respectively, and let $f, g \in B_+(I)$ be locally integrable functions on I. For all $n \in \mathbb{N}^+$, we have*

(i) *if $f \leq g$, then $K_n^{\chi}(f) \leq K_n^{\chi}(g)$;*
(ii) *$K_n^{\chi}(f+g) \leq K_n^{\chi}(f) + K_n^{\chi}(g)$ (i.e., K_n^{χ} is subadditive);*
(iii) *$\left|K_n^{\chi}(f) - K_n^{\chi}(g)\right| \leq K_n^{\chi}(|f - g|)$;*
(iv) *$K_n^{\chi}(\lambda f) = \lambda K_n^{\chi}(f)$, for each $\lambda \geq 0$ (i.e., K_n^{χ} is positive homogeneous).*

We are now in position to present qualitative pointwise and uniform convergence properties for K_n^{χ}.

In all that follows in this section, if otherwise not stated, the kernel χ from the definition of K_n^{χ} will satisfy properties (χ_1) and (χ_2) or (χ_2') for $I = [a, b]$ and $I = \mathbb{R}$, respectively.

Theorem 5.7 *Let χ be a given kernel. Let $f : I \to [0, \infty)$ be a non-negative and bounded function. Then,*

$$\lim_{n\to\infty} K_n^{\chi}(f)(x) = f(x),$$

at any point $x \in I$ where f is continuous. In addition, if $f \in C_+(I)$, then

$$\lim_{n\to\infty} \left\| K_n^{\chi}(f) - f \right\|_{\infty} = 0.$$

Remark 5.8 If we look carefully over the proof of Theorem 5.7, we notice that both conclusions hold if instead of assuming that χ satisfies (χ_1) and (χ_2), we only assume the weaker hypotheses that χ satisfies (χ_2) and that $\lim_{t\to\infty} \chi(t) = 0$ and $\lim_{t\to-\infty} \chi(t) = 0$. First of all, let us notice that without any loss of generality, we may assume that χ is non-negative because $K_n^{\chi} = K_n^{\chi_+}$, where $\chi_+(t) = \max\{\chi(t), 0\}$, for all $t \in \mathbb{R}$. It means that for any $\varepsilon > 0$ and $\gamma > 0$, there exists $N \in \mathbb{N}$ such that $\chi(nx - k) < \varepsilon/(2\|f\|_{\infty})$, whenever $|nx - k| \geq n\gamma$ and $n \geq N$. On the other hand, by the same reasoning as in the proof of Theorem 5, for any $\varepsilon > 0$, there exists $\gamma > 0$ such $n \int_{k/n}^{(k+1)/n} |f(t) - f(x)|\, dt < \varepsilon$ whenever $|nx - k| < n\gamma$. To sum up, repeating the same steps to estimate I_1 and I_2 introduced in the proof of Theorem 5.7, we conclude that for any $\varepsilon > 0$, there exists $N \in \mathbb{N}$ such that

$$\begin{aligned} \left|K_n^{\chi}(f)(x) - f(x)\right| &\leq K_n^{\chi}(|f - f_x|)(x) \\ &\leq \max\left\{\varepsilon, \frac{\varepsilon}{a_{\chi}}\right\}, \end{aligned}$$

for all $n \geq N$. From here we easily obtain the same conclusions as in the statement of Theorem 5.7.

In what follows, we will present a quantitative estimate in terms of the modulus of continuity of a function $f \in C(I)$ denoted by $\omega(f, \cdot)$, where

$$\omega(f, \delta) := \sup\{|f(x) - f(y)| : x, y \in I, |x - y| \leq \delta\}.$$

It is well known that $\omega(f, \cdot)$ is non-decreasing, continuous, subadditive and

$$\lim_{\delta\to 0^+} \omega(f, \delta) = 0.$$

Theorem 5.9 *Let χ be a given kernel which satisfies (χ_1) with $\beta \geq 1$. Then, for any $f \in C_+(I)$ and sufficiently large $n \in \mathbb{N}^+$, we have*

$$\left\|K_n^{\chi}(f) - f\right\|_{\infty} \leq \frac{2m_0(\chi) + m_1(\chi)}{a_{\chi}}\, \omega(f, 1/n).$$

Remark 5.10 As in the case of other max-product operators, the results of this section can be extended for functions that are bounded from below, hence not necessarily positive. Indeed, under the previous assumptions, for $f : I \to \mathbb{R}$

bounded from the below, and by considering the operators $\left(T_n^{\chi}(f)\right)_{n\geq 1}$, where $T_n^{\chi}(f)(x) = K_n^{\chi}(f-c)(x) + c$, $c = \inf_{x\in I} f(x)$, we obtain the same convergence results and estimations on f by $\left(T_n^{\chi}(f)\right)_{n\geq 1}$.

Remark 5.11 We also notice that all the results of this section hold too, if instead of assuming that the kernel χ satisfies condition (χ_2) or (χ_2'), we only assume that $\bigvee_{k\in\mathcal{J}_n} \chi(nx-k) \geq a_\chi$, *for all* $n \in \mathbb{N}_0$ and $x \in \mathbb{R}$.

In all that follows in this section, if otherwise not stated, the kernel χ from the definition of K_n^{χ} will satisfy properties (χ_1) and (χ_2) or (χ_2') for $I = [a, b]$ and $I = \mathbb{R}$, respectively.

Here, in the special case when $\chi \in L_1(\mathbb{R})$ and $f \in L_p(\mathbb{R})$, we prove a Lipschitz property for K_n^{χ}, which will imply a convergence property of K_n^{χ} in the norm $\|\cdot\|_p$.

Theorem 5.12 *Suppose that a given kernel $\chi \in L_1(\mathbb{R})$, and let $1 \leq p < \infty$ be fixed. Then, for any non-negative and L_p-integrable on I functions f and g, we have*

$$\left\|K_n^{\chi}(f) - K_n^{\chi}(g)\right\|_p \leq \frac{1}{a_\chi}\left((m_0(\chi))^{p-1}\|\chi\|_1\right)^{1/p} \cdot \|f-g\|_p.$$

In the statement of Theorem 5.12, we believe it is natural to assume that $\chi \in L_1(\mathbb{R})$ because in this way the conclusion is valid for any $p \geq 1$ and any non-negative functions $f, g \in L_p(I)$. However, considering a fixed $p \geq 1$, we can also obtain a Lipschitz estimation in the case when we assume that $\chi \in L_p(\mathbb{R})$.

Theorem 5.13 *Let $1 \leq p < \infty$ be fixed. If $\chi \in L_p(\mathbb{R})$, then for any non-negative and L_p-integrable on I functions f and g, we have*

$$\left\|K_n^{\chi}(f) - K_n^{\chi}(g)\right\|_p \leq \frac{\|\chi\|_p}{a_\chi} \cdot \|f-g\|_p.$$

Now, in order to obtain the L_p-convergence for the max-product sampling Kantorovich operators in case of L_p-integrable functions on I, we firstly need to test the L_p convergence in case of functions $f \in C_+(I)$, when $I = [a, b]$, and in case of $C_c^+(I)$ when $I = \mathbb{R}$, where $C_c^+(I)$ denotes the space of the non-negative function with compact support on I. We can prove what follows.

Theorem 5.14 *Let $\chi \in L_1(\mathbb{R})$ be a fixed kernel. Let $f \in C_+(I)$, $I = [a, b]$, be fixed. Then,*

$$\lim_{n\to+\infty} \|K_n^{\chi}(f) - f\|_p = 0,$$

for $1 \leq p < \infty$. Moreover, if $f \in C_c^+(I)$, with $I = \mathbb{R}$, we also have

$$\lim_{n\to+\infty} \|K_n^{\chi}(f) - f\|_p = 0,$$

for $1 \leq p < \infty$.

The following L_p-convergence theorem.

Theorem 5.15 *Let* $\chi \in L_1(\mathbb{R})$ *be a given kernel. For any non-negative* L_p-*integrable function* f *on* $\mathbb{R}$, *there holds*

$$\lim_{n\to+\infty} \|K_n^\chi(f) - f\|_p = 0,$$

for $1 \leq p < \infty$.

Remark 5.16 Obviously, in view of what has been established in Theorem 5.13, it is also possible to prove the convergence results of Theorems 5.14 and 5.15 under the assumption $\chi \in L^p(\mathbb{R})$.

The next lines present quantitative L_p-estimates in terms of a K-functional and of a modulus of continuity.

Let us define

$$C_+^1(I) = \{g : I \to \mathbb{R}_+;\ g \text{ is differentiable with } g' \text{ bounded on } I\},$$

and the Petree K-functional

$$K\,(f;t)_p = \inf_{g\in C_+^1(I)} \{\|f-g\|_p + t\|g'\|_\infty\}.$$

Firstly, we can prove an estimate in terms of the above K-functional, for approximation by $K_n^\chi(f)$ of f, in the $\|\cdot\|_p$ norm.

Theorem 5.17 *Suppose that* $\chi \in L_1(\mathbb{R})$ *satisfies* (χ_1) *for* $\beta \geq 1$, *and let* $1 \leq p < \infty$ *be fixed. Also, denote by* $\varphi_x(t) = |t-x|$, $x, t \in I$. *Then, for any non-negative and* L_p-*integrable function* f *on* I, *we have*

$$\left\|K_n^\chi(f) - f\right\|_p \leq c \cdot K(f; \Delta_{n,p}/c)_p,$$

with $c = 1 + \frac{1}{a_\chi}\left((m_0(\chi))^{p-1}\,\|\chi\|_1\right)^{1/p}$ *and* $\Delta_{n,p} = \|K_n^\chi(\varphi_x)\|_p$.

As a consequence of Theorem 5.17, we immediately get the following.

Corollary 5.18 *Let* $\chi \in L_1(\mathbb{R})$ *be a given kernel which satisfies* (χ_1) *with* $\beta \geq 1$, *and let* $1 \leq p < \infty$ *be fixed. Then for any non-negative and* L_p-*integrable function* f *on* $I = [a, b]$, *we have*

$$\|f - K_n^\chi(f)\|_p \leq c \cdot K\left(f; \frac{c_p}{n}\right)_p,$$

where $c_p = \frac{(b-a)^{1/p}}{c}\left(\frac{m_0(\chi)}{a_\chi} + \frac{m_1(\chi)}{a_\chi}\right)$ *and* c *is the constant given by the statement of Theorem 5.17.*

Note that the proof of Corollary 5.18 holds only in the case when $I = [a, b]$.

In the case when $I = \mathbb{R}$, we can present a Jackson-type estimate in terms of the L_p-modulus of continuity defined by

$$\omega_p(f, \delta) := \sup_{0<|h|\le\delta} \|f(\cdot + h) - f(\cdot)\|_p = \sup_{0<|h|\le\delta} \left(\int_{\mathbb{R}} |f(u+h) - f(u)|^p \, du\right)^{1/p},$$

with $\delta > 0$, $f \in L_p(\mathbb{R})$. It is well known that the above modulus of continuity satisfies the following inequality:

$$\omega_p(f, \lambda\,\delta) \le (1+\lambda)\,\omega_p(f, \delta), \qquad \lambda,\, \delta > 0, \tag{27}$$

for every $f \in L_p(\mathbb{R})$.

We can prove the following.

Theorem 5.19 *Let $1 \le p < \infty$ and χ be a kernel satisfying (χ_1) and (χ_2). Suppose in addition that*

$$M_p(\chi) := \left(\int_{\mathbb{R}} |x|^p \, |\chi(x)| \, dx\right)^{1/p} < +\infty. \tag{28}$$

Then, for any non-negative and L_p-integrable function f on $\mathbb{R}$, there holds

$$\|K_n^{\chi}(f) - f\|_p \le a_{\chi}^{-1} \left(2\|\chi\|_{\infty}^{(p-1)/p} \cdot \left[\|\chi\|_1 + M_p(\chi)^p\right]^{1/p} + \|\chi\|_{\infty}\right) \omega_p(f,\, 1/n),$$

for every $n \in \mathbb{N}^+$.

Remark 5.20 The modulus of continuity in Theorem 5.19 is equivalent to the functional given by

$$K^*(f;t)_p = \inf_{g' \in AC_{loc}(\mathbb{R})} \{\|f - g\|_{L_p(\mathbb{R})} + t\|g'\|_{L_p(\mathbb{R})}\},$$

where $AC_{loc}(\mathbb{R})$ denotes the space of all locally absolutely continuous functions on $\mathbb{R}$.

Since $C_+^1(\mathbb{R}) \subset AC_{loc}(\mathbb{R})$, between $K(f;t)_p$ defined at the beginning of this section and K^*, we evidently have the inequality

$$K^*(f;t)_p \le K(f;t)_p, \quad \text{for all } t \ge 0.$$

Note that for $f \in L_p([a,b])$ non-negative, we have that $\lim_{t\to 0} K(f;t)_p = 0$. Indeed, there exists a sequence of non-negative polynomials $(P_n)_{n\in\mathbb{N}}$, such that $\|f - P_n\|_p \to 0$ as $n \to \infty$. For arbitrary $\varepsilon > 0$, let P_m be such that $\|f - P_m\|_p < \varepsilon/2$. Then, for all $t \in (0,\, \varepsilon/(2\|P_m'\|_{\infty}))$, we get

$$K(f;t)_p \le \|f - P_m\|_p + t\|P_m'\|_{\infty} < \varepsilon/2 + \varepsilon/2 = \varepsilon,$$

which proves our assertion.

At the end of this section, we present several types of kernels which satisfy conditions (χ_1) and (χ_2) or (χ_2').

The first example consists in the well known sinc-function, that is,

$$\operatorname{sinc}(x) = \begin{cases} \frac{\sin \pi x}{\pi x}, & x \neq 0, \\ 1, & x = 0. \end{cases}$$

As $\operatorname{sinc}(x) = \mathcal{O}\left(|x|^{-1}\right)$ as $|x| \to \infty$, we get that (χ_1) is satisfied with $\beta = 1$ by Lemma 5.1. Moreover, (χ_2') is satisfied with $a_{\text{sinc}} = \operatorname{sinc}(1/2) \approx 0.6366$, while (χ_2) is not fulfilled. Moreover, it can also be noticed that the *sinc* function does not belong to $L_1(\mathbb{R})$ and then cannot be used as kernel in order to achieve L_p-approximation results.

Now, consider the non-negative Fejér kernel, $F(x) = \frac{1}{2}\operatorname{sinc}^2(x/2)$, $x \in \mathbb{R}$. It is well known that $F(x) = \mathcal{O}\left(|x|^{-2}\right)$ as $|x| \to \infty$, which also implies that $F \in L_1(\mathbb{R})$ and that (χ_1) is satisfied with $\beta = 2$. Then, it is easy to check that (χ_2) is satisfied with $a_F = F(3/2)$, and then also (χ_2') holds. However, in case of the Fejér Kernel, the continuous moment $M_1(F)$ of assumption (28) is not finite, and then the p-estimate of Theorem 5.19 cannot be achieved for the max-product sampling Kantorovich operators based upon F, but only the L_p-convergence and Corollary 5.18 occur.

Let us now consider the Jackson-type kernels of order $k \in \mathbb{N}_0$, where

$$J_k(x) = c_k \operatorname{sinc}^{2k}\left(\frac{x}{2k\pi\alpha}\right), x \in \mathbb{R},$$

with $\alpha \geq 1$ and c_k the normalization coefficient such that

$$c_k = \left[\int_{\mathbb{R}} \operatorname{sinc}^{2k}\left(\frac{x}{2k\pi\alpha}\right) du\right]^{-1}.$$

In the above cases, all the assumptions of the previous sections are satisfied, where assumption (28) holds for suitable $1 \leq p < +\infty$.

We may also consider radial kernels such as the well-known Bochner–Riesz kernels

$$b_\gamma(x) = \frac{2^\gamma}{\sqrt{2\pi}} \Gamma(\gamma + 1)\,(|x|)^{-\frac{1}{2}-\gamma} J_{\frac{1}{2}+\gamma}(|x|), \gamma > 0,$$

where J_λ is the Bessel function of order λ and $\Gamma(x)$ is Euler's gamma function. As in the case of the Jackson-type kernels, also the Bochner–Riesz kernels satisfy all the assumptions made in the previous sections, where assumption (28) holds again for suitable $1 \leq p < +\infty$.

Other useful kernels can be obtained by the so-called sigmoidal functions, that is, functions $\sigma : \mathbb{R} \to \mathbb{R}$, such that

$$\lim_{x\to-\infty} \sigma(x) = 0 \quad \text{and} \quad \lim_{x\to\infty} \sigma(x) = 1.$$

If in addition σ is such that
$(\Sigma 1)$ $\sigma(x) - 1/2$ is an odd function;
$(\Sigma 2)$ $\sigma \in C^2(\mathbb{R})$ is concave for $x \geq 0$;
$(\Sigma 3)$ $\sigma(x) = \mathcal{O}(|x|^{-\alpha})$ as $x \to -\infty$, for some $\alpha > 0$, then

$$\phi_\sigma(x) = \frac{1}{2}\left[\sigma(x+1) - \sigma(x-1)\right], \quad x \in \mathbb{R},$$

satisfies (χ_1) for every $0 \leq \beta \leq \alpha$, and it also satisfies (χ_2) with $a_{\phi_\sigma} = \phi_\sigma(3/2)$.

Other important kernels can be obtained from the well-known central B-spline of order $s \in \mathbb{N}_0$, given by

$$M_s(x) = \frac{1}{(s-1)!}\sum_{i=0}^{s}(-1)^i\binom{s}{i}\left(\frac{s}{2} + x - i\right)_+^{s-1},$$

where $(x)_+ = \max\{x, 0\}$. The support of M_s is the interval $[-s/2, s/2]$. It means that for $s \geq 2$, these central B-splines can be considered as kernels for the Kantorovich max-product sampling operator, and moreover they satisfy (χ_2') and, if $s \geq 4$, they satisfy also (χ_2).

Note that we can use as kernels shifted B-splines too. Although they do not satisfy in general (χ_2) or (χ_2'), they satisfy the property mentioned in Remark 5.11.

The kernels presented so far are continuous, but one can also construct kernels having discontinuities without affecting the main results. For example, we can take $\chi(x) = 1$ if $x \in [-2, 2]$ and $\chi(x) = 0$ otherwise.

Another example can be the so-called de la Valée-Poussin kernel χ, defined by

$$\chi(x) = \frac{\sin(x/2)\,\sin(3x/2)}{9x^2/4}, \quad x \neq 0,$$

and $\chi(0) = \frac{1}{3}$. We observe that $\chi(x) = \mathcal{O}\left(|x|^{-2}\right)$ as $|x| \to \infty$, which means that (χ_1) is satisfied with $\beta = 2$. Then, one can easily prove that (χ_2) is satisfied with $a_\chi = 4/3\pi^2$.

Finally, consider $\chi(x) = \frac{2}{\pi}\arctan\left(\frac{1}{x^2}\right)$, if $x \neq 0$ and $\chi(0) = 1$. Again, $\chi(x) = \mathcal{O}\left(|x|^{-2}\right)$ as $|x| \to \infty$, and hence, (χ_1) is satisfied with $\beta = 2$. Then, (χ_2) is satisfied with $a_\chi = \frac{2}{\pi}\arctan\left(\frac{1}{4}\right)$.

6 Future Researches

We present here on short some extensions for future researches and applications.

It is known the fact that the approximation results for the multidimensional linear Bernstein–Durrmeyer operators (polynomials) of degree d with respect to an arbitrary Borel probability measure, given by the formula

$$D_d(f)(x) = \sum_{|k|\le d} \frac{\int_X f(t)P_{d,k}(t)d\mu_X(t)}{\int_X P_{d,k}(t)d\mu_X(t)} \cdot P_{d,k}(x),$$

where X denotes the simplex $\{x = (x_1, \ldots, x_n); x_i \ge 0, i = 1, \ldots, n, 1-|x| \ge 0\}$, $k = (k_1, \ldots, k_n)$, $|x| = \sum_{i=1}^{n} |x_i|$, $P_{d,k}(x) = \binom{d}{k} x^k (1-|x|)^{d-|k|}$, $|k| \le d$, $x^k = x_1^{k_1} \cdot \ldots \cdot x_n^{k_n}$, $k! = k_1! \cdot \ldots \cdot k_n!$, $\binom{d}{k} = \frac{d!}{k!(d-|k|)!}$ and μ_X is a Borel probability measure on X, have important applications in learning theory, see Li [25].

Similar problems with applications in learning theory could be considered for the multidimensional Kantorovich linear operators (polynomials) of degree d, given by

$$K_d(f)(x) = \sum_{|k|\le d} P_{d,k}(x) \cdot \int_X f\left(\frac{k+t}{d}\right) d\mu_X(t).$$

Furthermore, one can consider for future researches extensions of the approximation properties and applications to learning theory for the max-product nonlinear operators corresponding to the above defined multidimensional Bernstein–Durrmeyer linear operators and Kantorovich linear operators, given by the formulas

$$D_d^{(M)}(f)(x) = \frac{\bigvee_{|k|\le d} P_{d,k}(x) \cdot \left[\int_X f(t)P_{d,k}(t)d\mu_X(t) / \int_X P_{d,k}(t)d\mu_X(t)\right]}{\bigvee_{|k|\le d} P_{d,k}(x)},$$

$$K_d^{(M)}(f)(x) = \frac{\bigvee_{|k|\le d} P_{d,k}(x) \cdot \int_X f\left(\frac{k+t}{d}\right) d\mu_X(t)}{\bigvee_{|k|\le d} P_{d,k}(x)},$$

or, more general and to be in accordance with the max-product Kantorovich type operators studied in the previous sections, for

$$D_d^{\chi,(M)}(f)(x)$$

$$= \frac{\bigvee_{k\in \mathcal{J}_d} \chi(d\cdot x - k) \cdot \left[\int_X f(t)\chi(d\cdot t - k)d\mu_X(t) / \int_X \chi(d\cdot t - k)d\mu_X(t)\right]}{\bigvee_{k\in \mathcal{J}_d} \chi(d\cdot x - k)}$$

and

$$K_d^{\chi,(M)}(f)(x) = \frac{\bigvee_{k\in \mathcal{J}_d} \chi(d\cdot x - k)\cdot \int_X f\left(\frac{k+t}{d}\right) d\mu_X(t)}{\bigvee_{k\in \mathcal{J}_d} \chi(d\cdot x - k)},$$

where $x \in \mathbb{R}^n$, $k = (k_1, \ldots, k_n)$ is a multi-index, $\mathcal{J}_d$ is a family of multi-indices and $\chi(d\cdot x - k)$ is a multidimensional kernel.

Notice that in the very recent paper by Coroianu–Costarelli–Gal–Vinti [7], the above problems were treated for another variant of the max-product multidimensional Kantorovich operators, defined on short below.

Let $f : I^n \to \mathbb{R}$ be a locally integrable function with respect to a Borel probability measure on I^n, $\mu_{I^n}(t) = \mu_1(t_1)\mu_2(t_2)\cdot\ldots\cdot\mu_n(t_n), t = (t_1, t_2, \ldots, t_n) \in I^n$, where each μ_j, $j = 1, \ldots, n$, is a Borel probability measure on I.

For $\chi : \mathbb{R}^n \to \mathbb{R}$ a kernel such that $\bigvee_{k\in \mathcal{J}_n} \chi(dx - k) \neq 0$, *for all* $x \in I^n$, the multidimensional max-product Kantorovich sampling operator on f based upon χ is defined as

$$K_{d,\mu_{I^n}}^{\chi,(M)}(f)(x)$$
$$:= \frac{\bigvee_{k\in \mathcal{J}_{d,n}} \chi(dx-k) \int_{[k/d,(k+1)/d]} f(t)\, d\mu_{I^n}(t)/\mu_{I^n}([k/d,(k+1)/d])}{\bigvee_{k\in \mathcal{J}_{d,n}} \chi(dx-k)}, x \in I^n,$$

where for $k = (k_1, k_2, \ldots, k_n) \in \mathcal{J}_{d,n}$,

$$[k/d, (k+1)/d] = [k_1/d, (k_1+1)/d] \times [k_2/d, (k_2+1)/d] \times \ldots \times [k_n/d, (k_n+1)/d]$$

and

$$\int_{[k/d,(k+1)/d]} f(t)\, d\mu_{I^n}(t)$$
$$= \int_{[k_1/d,(k_1+1)/d]} \cdots \int_{[k_n/d,(k_n+1)/d]} f(t_1, \ldots, t_n) d\mu_1(t_1)\ldots d\mu_n(t_n).$$

Remark 6.1 It is worth mentioning that the max-product neural networks of Kantorovich-type have been studied in the recent papers [9–11, 15, 16]. We do not enter into details, since this topic does not belong to the present authors.

Acknowledgments The work of the authors was supported by a grant of Ministry of Research and Innovation, CNCS-UEFISCDI, project number PN-III-P1-1.1-PD-2016-1416, within PNCDI III.

References

1. B. Bede, L. Coroianu, S.G. Gal, *Approximation by Max-Product Type Operators*, Springer, New York, 2016.
2. P.L. Butzer, *A survey of the Whittaker-Shannon sampling theorem and some of its extensions*, J. Math. Res. Expos., **3** (1983), 185–212.
3. L. Coroianu, S.G. Gal, *L^p-approximation by truncated max-product sampling operators of Kantorovich-type based on Fejer kernel*, J. Integral Equations Appl., **29** (2017), no. 2, 349–364.
4. L. Coroianu, S.G. Gal, *Approximation by truncated max-product operators of Kantorovich-type based on generalized (ϕ, ψ) - kernels*, Math. Methods. Appl. Sci, **41** (2018), no. 17, 7971–7984.
5. L. Coroianu, S.G. Gal, *Approximation by max-product operators of Kantorovich type*, Stud. Univ. Babe-Bolyai Math., **64** (2019), no. 2, 207–223.
6. L. Coroianu, D. Costarelli, S.G. Gal, G. Vinti, *Approximation by max-product sampling Kantorovich operators with generalized kernels*, Analysis and Applications, online access, https://doi.org/10.1142/S0219530519500155.
7. L. Coroianu, D. Costarelli, S.G. Gal, G. Vinti, *Approximation by multivariate max-product Kantorovich-type operators and learning rates of least-squares regularized regression*, to appear in Communications on Pure and Applied Analysis, **19** (2020), no. 8, 4213–4225. doi:10.3934/cpaa.2020189
8. L. Coroianu, D. Costarelli, S.G. Gal, G. Vinti, *The max-product generalized sampling operators: convergence and quantitative estimates*, Appl. Math. Comput., **355** (2019), 173–183.
9. D. Costarelli, A. R. Sambucini, *Approximation results in Orlicz spaces for sequences of Kantorovich max-product neural network operators*, Results in Mathematics, **73** (1) (2018), Art. 15. DOI: 10.1007/s00025-018-0799-4.
10. D. Costarelli, A.R. Sambucini, G. Vinti, *Convergence in Orlicz spaces by means of the multivariate max-product neural network operators of the Kantorovich type and applications*, in print in: Neural Computing & Applications (2019). DOI: 10.1007/s00521-018-03998-6.
11. D. Costarelli, G. Vinti, *Approximation by max-product neural network operators of Kantorovich type*, Results Math., **69** (1-2) (2016), 505–519.
12. D. Costarelli, G. Vinti, *Max-product neural network and quasi-interpolation operators activated by sigmoidal functions*, J. Approx. Theory, **209** (2016), 1–22.
13. D. Costarelli, G. Vinti, *Saturation classes for max-product neural network operators activated by sigmoidal functions*, Results in Mathematics, **72** (3) (2017), 1555–1569.
14. D. Costarelli, G. Vinti, *Pointwise and uniform approximation by multivariate neural network operators of the max-product type*, Neural Networks, **81** (2016), 81–90.
15. D. Costarelli, G. Vinti, *Convergence results for a family of Kantorovich max-product neural network operators in a multivariate setting*, Math. Slovaca, **67** (6) (2017), 1469–1480.
16. D. Costarelli, G. Vinti, *Estimates for the neural network operators of the max-product type with continuous and p-integrable functions*, Results in Mathematics, **73** (1) (2018), Art. 12. DOI: 10.1007/s00025-018-0790-0.
17. S.G. Gal, *Shape-Preserving Approximation by Real and Complex Polynomials*, XIV + 352 pp., Birkhäuser, Boston, Basel, Berlin, 2008.
18. T.Y. Göker, D. Oktay, *Summation process by max-product operators*, Computational analysis, 59–67, Springer Proc. Math. Stat., **155**, Springer, Cham, 2016.
19. S. Y. Güngör, N. Ispir, *Approximation by Bernstein-Chlodowsky operators of max-product kind*, Mathematical Communic., **23** (2018), 205–225.
20. V. Gupta, M.Th. Rassias, *Moments of Linear Positive Operators and Approximation*, SpringerBriefs in Mathematics, Springer, Cham, VIII+96 pp., 2019.
21. A. Holhos, *Approximation of functions by some exponential operators of max-product type*, Studia Sci. Math. Hungar., **56** (1) (2019), 94–102.

22. A. Holhos, *Weighted approximation of functions by Favard operators of max-product type*, Period. Math. Hungar., **77** (2) (2018), 340–346.
23. A. Holhos, *Weighted Approximation of functions by Meyer-König and Zeller operators of max-product type*, Numer. Funct. Anal. Optim., **39** (6) (2018), 689–703.
24. A. Holhos, *Approximation of functions by Favard-Szász-Mirakyan operators of max-product type in weighted spaces*, Filomat, **32** (7) (2018), 2567–2576.
25. B.-Z. Li, *Approximation by multivariate Bernstein-Durrmeyer operators and learning rates of least-squares regularized regression with multivariate polynomial kernels*, J. Approx. Theory, **173** (2013), 33–55.
26. R.-H. Shen, L.-Y. Wei, *Convexity of functions produced by Bernstein operators of max-product kind*, Results Math. **74** (3) (2019), Art. 92.
27. T. Yurdakadim, E. Taş, *Some results for max-product operators via power series method*, Acta Math. Univ. Comenian., (N.S.), **87** (2) (2018), 191–198.

Variational Inequalities and General Equilibrium Models

Maria Bernadette Donato, Antonino Maugeri, Monica Milasi, and Antonio Villanacci

1 Introduction

We deal with the study of several general equilibrium models by using the variational inequality theory. The theory of variational inequalities was introduced in the sixties of the past century by Fichera (1964) [27], and Lions and Stampacchia (1965) [28], as an innovative and effective method to solve equilibrium problems arising in mathematical physics. Afterward this theory turned out as a powerful tool, and it was used to analyze different kinds of equilibrium problems. We mention the equilibrium problems of the oligopoly, of the market, of the traffic, of Nash, see, e.g., [9, 12, 26, 29].

Here is the content of this chapter. We first introduce some general equilibrium economic models; then, after recalling some basic notions on variational inequality theory, we make use of such a tool to analyze some proposed models. We consider the exchange economy model, subsequently we study the models with nominal and numeraire assets, and, finally, we consider the case of restricted participation.

M. B. Donato · M. Milasi (✉)
Department of Economics, University of Messina, Messina, Italy
e-mail: mbdonato@unime.it; mmilasi@unime.it

A. Maugeri
Department of Mathematics and Computer Sciences, University of Catania, Catania, Italy
e-mail: maugeri@dmi.unict.it

A. Villanacci
Department of Economics and Management, Universitá degli Studi di Firenze, Firenze, Italy
e-mail: antonio.villanacci@unifi.it

I. N. Parasidis et al. (eds.), *Mathematical Analysis in Interdisciplinary Research*,
Springer Optimization and Its Applications 179,
https://doi.org/10.1007/978-3-030-84721-0_11

2 General Equilibrium Economic Models

General Economic Equilibrium (GEE) models analyze *equilibria* in different markets, i.e., economic situations described by prices and consumption/production vectors that are consistent with households' and firms' maximizing behaviors and with demand of goods being smaller than or equal to supply. Standard references for the economic literature on GEE are Debreu [13] and Arrow and Hahn [4].

One main virtue of the GEE models is the fact that they analyze very complex frameworks, but they can be described by means of few and simple elements we list below.

A verbal description of the functioning of the economic environment under analysis is the starting point of the analysis. A formalization of the above environment is then provided presenting three main ingredients: a list of exogenous variables, or parameters, defining the economy; a description of the behavior of economic agents who are supposed to maximize a goal or objective function under some physical, institutional, and economic constraints; and an aggregate consistency condition for agents' behavior in terms of market clearing, rationing rules, and/or expectation fulfillment.

The simplest GEE model is the so-called exchange economy model. The economic environment under analysis is a set of individuals owning goods they want to exchange on a market in order to maximize their well-being: just think about the Sunday market square of a village. The exogenous objects are the characteristics of the main actors in the market, i.e., the consumers or households. $\mathcal{H} = \{1, \ldots, H\}$ is the set of households and each household $h \in \mathcal{H}$ is described by the endowment vector $e_h \in \mathbb{R}^C$ of the quantities of C commodities he or she owns and by the utility function u_h defined on a consumption set $X_h \subseteq \mathbb{R}^C$. The consumption set X_h describes what can be potentially consumed by household h and it is often assumed to $\mathbb{R}^C_+$. The utility function u_h represents the preferences of households in choosing two consumption vectors x_h and y_h in X_h. We also define $e = (e_h)_{h \in \mathcal{H}}$ and, then, an economy is a pair $(e, u) \in \prod_{h \in \mathcal{H}} (X_h \times \mathcal{U}_h)$, where $\mathcal{U}_h$ is the set of all utility functions on X_h.

To describe households' behavior, we introduce prices $p := (p^c)_{c \in \mathcal{C}} \in \mathbb{R}^C_+$, where for any $c \in \mathcal{C} := \{1, \ldots, C\}$, p^c is the price of good or commodity c, i.e., the number of units of account (say euros or dollars) needed to purchase one unit of good c. Moreover, define the budget constraint set of household h, as follows:

$$B_h(p) := \{x_h \in X_h : \ \langle p, x_h - e_h \rangle_C \leq 0\}.$$

Observe that the condition $\langle p, x_h \rangle_C \leq \langle p, e_h \rangle_C$ imposes that the value of the expenditure must not exceed the value of household's wealth.

The assumption about household's behavior is formalized as follows. For given $(\widetilde{p}, e_h, u_h) \in \mathbb{R}^C_+ \times X_h \times \mathcal{U}_h$, $\widetilde{x}_h$ solves the household h's maximization problem

$$\max_{x_h \in B_h(\widetilde{p})} u_h(x_h). \tag{1}$$

In the above simple framework, the aggregate consistency condition is just a market clearing condition, which, in its simplest form, requires that the aggregate demand does not exceeds the aggregate supply, i.e.,

$$\sum_{h \in \mathcal{H}} \widetilde{x}_h \leq \sum_{h \in \mathcal{H}} e_h. \tag{2}$$

Given the above structure, we can then present the definition that is central in any GEE model.

Definition 1 The pair $(\tilde{x}, \tilde{p}) \in \left(\prod_{h \in \mathcal{H}} X_h\right) \times \mathbb{R}^C_+$ is an allocation-price *equilibrium* for an economy ϵ if

(i) for any $h \in \mathcal{H}$, $\tilde{x}_h$ solves the problem

$$\max_{x_h \in B(\tilde{p})} u_h(x_h) = u_h(\tilde{x}_h), \tag{3}$$

(ii) $\tilde{x}$ satisfies the market clearing conditions, for all $c \in \mathcal{C}$

$$\sum_{h \in \mathcal{H}} (\tilde{x}^c_h - e^c_h) \leq 0 \text{ if } \tilde{p}^c = 0, \tag{4}$$

$$\sum_{h \in \mathcal{H}} (\tilde{x}^c_h - e^c_h) = 0 \text{ if } \tilde{p}^c > 0. \tag{5}$$

An extremely important generalization of the exchange economy model is the GEE model with time, uncertainty, and incomplete financial markets. The description of that framework is as follows. We assume that there are 2 periods of time, say today and tomorrow: the state of the world today is known to individuals, and it is called state 0; in the following period, S, with $S > 1$, states of the world are possible. We label each state of the world, or spot by s, where $s = 0$ corresponds to the first period, and we set $\mathcal{S}^0 := \{0\} \cup \mathcal{S}$ and $\mathcal{S} := \{1, 2, \ldots, S\}$. In this framework, a commodity may be defined in terms not only of its physical or chemical characteristics but also in terms of the period or the state of nature in which it is available. In other words, presume we consider bananas today and apples tomorrow, if it rains. Spot commodity markets open in the first and second period, and there are C, with $C > 1$, commodities in each spot, labelled by $c \in \mathcal{C} := \{1, 2, \ldots, C\}$ and the total number of commodities available in the economy is $G := (S + 1)C$. There are H households, $H > 1$, labelled by $h \in \mathcal{H} := \{1, 2, \ldots, H\}$, and A assets, $A \geq 1$, labelled by $a \in \mathcal{A} := \{1, \ldots, a, \ldots, A\}$. An asset is an $S + 1$ dimensional vector whose first component is the price of the asset, and the other S components are the returns of that asset in each state—see below for a more precise definition. The time structure of the model is the following one. In the first period, commodities and assets are exchanged and first period consumption takes place. Then uncertainty is resolved, households fulfill their financial commitments, and, finally, they exchange and consume second period commodities. Following the standard notation, we have that x_h^{sc} is the consumption of commodity c in state s by household h, e_h^{sc} is the endowment of commodity c in state s owned by household h, and p^{sc} denotes the price of commodity c in state s. Moreover, X_h^s is the consumption set of household $h \in \mathcal{H}$ in state $s \in \mathcal{S}^0$, $X_h := \prod_{s\in\mathcal{S}^0} X_h^s$, $X := \prod_{h\in\mathcal{H}} X_h$. Moreover,

$$x_h^s := (x_h^{sc})_{c\in\mathcal{C}} \in X_h^s, \quad x_h := (x_h^s)_{s\in\mathcal{S}^0} \in X_h, \quad x := (x_h)_{h\in\mathcal{H}} \in X,$$

$$e_h^s := (e_h^{sc})_{c\in\mathcal{C}} \in X_h^s, \quad e_h := (e_h^s)_{s\in\mathcal{S}^0} \in X_h, \quad e := (e_h)_{h\in\mathcal{H}} \in X,$$

$$p^s := (p^{sc})_{c\in\mathcal{C}} \in \mathbb{R}_+^C, \quad p^{\mathbf{1}} := (p^s)_{s\in\mathcal{S}} \in \mathbb{R}_+^{SC} \; p := (p^s)_{s\in\mathcal{S}^0} \in \mathbb{R}_+^G.$$

Household h's preferences are represented by the utility function $u_h : X_h \to \mathbb{R}$.

The description of the so-called financial side of the economy is as follows. q^a is the price of asset $a \in \mathcal{A}$ and $q := (q^a)_{a\in\mathcal{A}} \in \mathbb{R}^A$, b_h^a is the demand of asset a by household h, with $b_h := \left(b_h^a\right)_{a\in\mathcal{A}}$ and $b := (b_h)_{h\in\mathcal{H}}$. Moreover,

$$R : \mathbb{R}_{++}^G \to \mathcal{M}_{S,A}, \quad p \mapsto R(p) = \begin{bmatrix} r^{11}(p) \cdots r^{1a}(p) \cdots r^{1A}(p) \\ \cdots \\ r^{s1}(p) \cdots r^{sa}(p) \cdots r^{sA}(p) \\ \cdots \\ r^{S1}(p) \cdots r^{Sa}(p) \cdots r^{SA}(p) \end{bmatrix}$$

is the return matrix function,[1] where $r^{sa}(p)$ is the return of asset $a \in \mathcal{A}$ in state $s \in \mathcal{S}$. Denote by $R^s(p)$ the s-th row of matrix $R(p)$. Different results about equilibria are obtained for different specification of the return matrix function $R(p)$. In the literature, authors distinguish among financial models with nominal, real, and numeraire assets. *Nominal* assets promise to deliver units of account, *real* assets a *vector* of goods, and *numeraire* assets just some amount of a *given good*, the so-called numeraire good. In general, y^{sa} denotes the number of units of economics objects (or yields) delivered by the specific asset under consideration. More formally, we have what follows.

Assets are nominal if for any $(s, a) \in \mathcal{S} \times \mathcal{A}$, and any $p \in P$, $r^{sa}(p) = y^{sa}$, where $y^{si} \in \mathbb{R}$ is the number of units of account that asset i promises to pay in state s.

Assets are real if $r^{sa}(p) = p^s y^{sa}$ and in each state $s \in \mathcal{S}$, asset a pays a vector $y^{sa} \equiv (y^{sac})_{c=1}^C \in \mathbb{R}^C$ of goods, i.e., $y^{sac} \in \mathbb{R}$ is the number of units of good c delivered in state s.

[1] Recall that $\mathcal{M}_{S,A}$ is the set of all $S \times A$ dimensional matrices of real numbers.

Assets are numeraire if they pay in units of the numeraire commodity, say gold, only. y^{sa} is the number of units of the numeraire commodity that asset a promises to pay in state s. Taking the numeraire commodity as commodity C, the return in state s is $r^{sa}(p) = p^{sC} y^{sa}$.

An economy in a financial economy model with numeraire assets is an element $\mathcal{E} := (e, R, u) \in \mathbb{R}^{GH}_{++} \times \mathcal{R} \times \mathcal{U}$, where $\mathcal{R}$ is the set of return functions. Each household chooses the set of most preferred consumption vectors under the constraints that in period 0 expenditure for goods and assets is smaller than the value of wealth in that period and, similarly, in each state in the future, expenditure for consumption is smaller than wealth increased by the value of the asset's yields.

We can then define a consumption, portfolio holding, commodity, and asset price vector as an equilibrium vector associated with a given economy described by commodity endowments, household's preferences, and financial structure if at those prices and economies, households maximize, and market clears, i.e., commodities' demand is smaller than or equal to commodities' supply and assets' demand is equal to zero.

The so-called budget set is defined as follows:

$$B_h(p, q) := \{ (x_h, b_h) \in X_h \times \mathbb{R}^A : \langle p^0, x_h^0 - e_h^0 \rangle_C + \langle q, b_h \rangle_A \leq 0$$

$$\text{and for any } s \in \mathcal{S}, \langle p^s, x_h^s - e_h^s \rangle_C - \langle R^s(p^1, y), b_h \rangle_A \leq 0 \}.$$

We can finally give the formal definition of equilibrium.

Definition 2 The vector $(\widetilde{p}, \widetilde{q}, \widetilde{x}, \widetilde{b}) \in \mathbb{R}^G_+ \times \mathbb{R}^A \times X \times \mathbb{R}^{AH}$ is an *equilibrium* vector for the economy $\mathcal{E}$ if

1. for any $h \in \mathcal{H}$, $(\widetilde{x}_h, \widetilde{b}_h)$ solves problem

$$\max_{(x_h, b_h) \in B_h(\widetilde{p}, \widetilde{q})} u_h(x_h) = u_h(\widetilde{x}_h);$$

2. for any $s \in \mathcal{S}^0$ and $c \in \mathcal{C}$,

$$\sum_{h \in \mathcal{H}} \widetilde{x}_h^{sc} \leq \sum_{h \in \mathcal{H}} e_h^{sc} \text{ if } \widetilde{p}^{sc} = 0,$$

$$\sum_{h \in \mathcal{H}} \widetilde{x}_h^{sc} = \sum_{h \in \mathcal{H}} e_h^{sc} \text{ if } \widetilde{p}^{sc} > 0;$$

3. for any $a \in \mathcal{A}$,

$$\sum_{h \in \mathcal{H}} \widetilde{b}_h^a = 0.$$

From now on, we assume that for any $h \in \mathcal{H}$, $X_h = \mathbb{R}^C_+$ and $X_h = \mathbb{R}^G_+$.

An important assumption to economic theories is the concavity of the utility functions. We conclude this section to recall some basic definitions. Let X be a nonempty, convex set of $\mathbb{R}^n$ and $u : X \to \mathbb{R}$ a function. We say that u is:

- *strongly concave* if: there exists $\tau > 0$ such that for any $x, y \in X$ and $\lambda \in [0, 1]$ one has

$$u(\lambda x + (1-\lambda)y) \geq \lambda u(x) + (1-\lambda)u(y) - \frac{\tau}{2}\lambda(1-\lambda)\|x-y\|^2 \,;$$

- *strictly concave* if: for any $x, y \in X$ and $\lambda \in (0, 1)$ one has

$$u(\lambda x + (1-\lambda)y) > \lambda u(x) + (1-\lambda)u(y) \,;$$

- *quasiconcave* if: for any $x, y \in X$ and $\lambda \in [0, 1]$ one has

$$u(\lambda x + (1-\lambda)y) \geq \min\{u(x), u(y)\} \,;$$

- *semistrictly quasiconcave* if: for any $x, y \in X$ such that $u(x) \neq u(y)$ one has

$$u(\lambda x + (1-\lambda)y) > \min\{u(x), u(y)\}, \quad \forall \lambda \in (0, 1) \,.$$

3 Different Variational Inequality Problems

This section is devoted to recall the definitions and some results about variational problems. Let us consider a nonempty, closed, and convex subset C of $\mathbb{R}^n$.

(GQVI) Let the set-valued maps $S : C \to 2^{\mathbb{R}^n}$ and $\Phi : C \to 2^{\mathbb{R}^n}$ be given. A *Generalized Quasi-Variational Inequality* associated with C, S, Φ, denoted by GQVI, is the following problem:

$$\text{“Find } \widetilde{x} \in S(\widetilde{x}) \,,\ \varphi \in \Phi(\widetilde{x}) \text{ such that } \langle \varphi, x - \widetilde{x} \rangle_n \geq 0 \qquad \forall x \in S(\widetilde{x}) \,.\text{”} \tag{6}$$

(GVI) Let the set-valued map $\Phi : C \to 2^{\mathbb{R}^n}$ be given. A *Generalized Variational Inequality* associated with C, Φ, denoted by GQVI, is the following problem:

$$\text{“Find } \widetilde{x} \in C \,,\ \varphi \in \Phi(\widetilde{x}) \text{ such that } \langle \varphi, x - \widetilde{x} \rangle_n \geq 0 \qquad \forall x \in C \,.\text{”} \tag{7}$$

(QVI) Let be given the set-valued map $S : C \to 2^{\mathbb{R}^n}$ and the function $\phi : C \to \mathbb{R}^n$. A *Quasi-Variational Inequality* associated with C, S, ϕ, denoted by QVI, is the following problem:

$$\text{“Find } \widetilde{x} \in S(\widetilde{x}) \text{ such that } \langle \phi(\widetilde{x}), x - \widetilde{x} \rangle_n \geq 0 \qquad \forall x \in S(\widetilde{x}) \,.\text{”} \tag{8}$$

(VI) Let the function $\phi : C \to \mathbb{R}^n$ be given. A *(Classical) Variational Inequality* associated with C, ϕ, denoted by VI, is the following problem:

$$\text{“Find } \widetilde{x} \in C \text{ such that } \langle \phi(\widetilde{x}), x - \widetilde{x} \rangle_n \geq 0 \qquad \forall x \in C\text{.”} \tag{9}$$

We now present some results about the existence of solution to variational problems.

Theorem 1 (Existence Under Compactness and Continuity, Section 4 of [31]) *If C is a compact set and the function $\phi : C \to \mathbb{R}^n$ is continuous, then the VI (9) admits at least a solution.*

Theorem 2 (Existence Under Coercivity, Section 4 of [31]) *Let ϕ satisfy*

$$\lim_{\|x\| \to +\infty} \frac{\langle \phi(x) - \phi(x_0), x - x_0 \rangle}{|x - x_0|} = +\infty \tag{10}$$

for some $x_0 \in C$. Then there exists a solution to (9).

Theorem 3 (Uniqueness, Section 4 of [31]) *Let ϕ be strictly monotone:*

$$\langle \phi(x) - \phi(x'), x - x' \rangle_n > 0 \qquad \forall x, x' \in C,\ x \neq x'. \tag{11}$$

Then, if there exists the solution to (9), it is unique.

From Theorems 2 and 3, it follows the following.

Theorem 4 (Existence and Uniqueness) *Let $\phi : C \to \mathbb{R}^n$ be strongly monotone; that is, there exists $\nu > 0$ such that:*

$$\langle \phi(x) - \phi(x'), x - x' \rangle_n \geq \nu \|x - x'\|^2 \qquad \forall x, x' \in C,\ x \neq x'. \tag{12}$$

Then the VI (9) admits a unique solution.

Theorem 5 (Corollary 3.1 of [11]) *Let C be a compact set and ϕ a set-valued map usc, with compact and convex values. Then there exists the solution to GVI (7).*

Theorem 6 ([32]) *Let Φ and S be two set-valued maps satisfying the following properties:*

(i) Φ is upper semicontinuous with nonempty, convex, and compact values;
(ii) S is closed, lower semicontinuous and with nonempty, convex, and compact values.

Then, the GQVI problem (6) admits at least a solution.

An important ingredient of our framework is the analysis of the connection between a well-chosen variational inequality problem and a standard maximization problem. Let $u : X \subseteq \mathbb{R}^n \to \mathbb{R}$ be a function and $C \subset \mathbb{R}^n$ be a closed and convex set; let us consider the following maximization problem:

$$\max_{x \in C} u(x)\,. \tag{13}$$

When the objective function u is *continuous differentiable*, if $\tilde{x}$ is a solution to the maximization problem (13), then $\tilde{x}$ is solution to the following VI:

$$\langle \nabla u(\tilde{x}), x - \tilde{x} \rangle_n \leq 0 \qquad \forall x \in C. \tag{14}$$

When u is also *concave*, VI (14) represents a necessary and sufficient optimality condition to be $\tilde{x}$ a solution to (13) (see, e.g., [31]). Whereas, under more general assumptions, if the function u is *concave and continuous*, without differentiability assumption, in (14) the operator ∇u is replaced by a supergradient of u in $\tilde{x}$. Hence, the problem (13) is equivalent to the following generalized variational inequality:

$$\text{“Find}\, \tilde{x} \in C \,\text{such that}\, \exists\; k \in \partial u(\tilde{x}) \;\; \text{with}\; \langle k, x - \tilde{x} \rangle_n \geq 0 \qquad \forall x \in C\text{''}$$

where $\partial u(x) = \{l \in \mathbb{R}^n : \;\; u(y) \leq u(x) + \langle l, y - x \rangle_n \quad \forall y \in \mathbb{R}^n\}$ is the superdifferential of the concave function u and $k \in \partial u(\tilde{x})$ is a supergradient of u in $\tilde{x}$. If the function u is not concave, then the supergradient is not suitable to characterize a maximum problem. To this aim, in [5, 6, 8] authors introduced a new operator to obtain necessary and sufficient conditions when the objective function is quasiconcave. For any $\alpha \in \mathbb{R}$, let us denote by $U_\alpha^>(u)$ the strict upper level set associated with u and α, i.e.,

$$U_\alpha^>(u) := \{x \in \mathbb{R}_+^n : u(x) > \alpha\},$$

and, for all $x \in X$, define

$$N^>(x) := \{h \in \mathbb{R}^n : \; \langle h, z - x \rangle_n \leq 0 \quad \forall\, z \in U_{u(x)}^>\}.$$

We recall that a function u is quasiconcave if and only if, for any $\alpha \in \mathbb{R}$, the strict upper level set $U_\alpha^>(u)$ is a convex set. Then $N^>(x)$ represents the normal cone to $U_{u(x)}^>$. Hence, one has the following.

Theorem 7 (See Proposition 4.1 in [7]) *Let u be continuous and semistrictly quasiconcave and C nonempty convex. Then $\tilde{x} \in C$ is a solution to the maximization problem (13) if and only if $\tilde{x}$ is solution to GVI:*

$$\text{“Find}\, \tilde{x} \in C \text{ such that}\, \exists\; g \in N^>(\tilde{x}) \setminus \{0\} \,\text{with}\, \langle g, x - \tilde{x} \rangle_n \geq 0 \qquad \forall x \in C\text{''}. \tag{15}$$

4 Exchange Economy Model

This section deals with the analysis of equilibria in the pure exchange economy model. The below listed results have been published in [1, 2, 16, 17, 23]. Such results have been generalized to a model with consumption and production in [10, 25]; a further generalization considers the case in which the market evolves in a time interval $[0, T]$; the last model is studied in the Hilbert space $L^2([0, T])$ (see, e.g., [15, 21, 22, 24]).

In this section, for all $h \in \mathcal{H}$, we make the following.

Assumptions on Utility Function $u_h : \mathbb{R}^C_+ \to \mathbb{R}$:

- *Survival assumption*: $e_h >> 0$ for any $h \in \mathcal{H}$;
- u_h is *non-satiated*: for any $x \in \mathbb{R}^C_+$ there exists $y \in \mathbb{R}^C_+$ such that $u_h(y) > u_h(x)$.

Moreover, without loss of generality, we can consider the prices in the simplex set:

$$P := \{p \in \mathbb{R}^C_+ : \sum_{c \in C} p^c = 1\}.$$

4.1 Characterization by Means of a Variational Problem

The goal of this subsection is to reformulate the equilibrium problem by means of suitable variational inequalities, by considering different assumptions on utility functions.

Theorem 8 *Let $u_h \in C^1(\mathbb{R}^C_+)$ be non-satiated and concave for all $h \in \mathcal{H}$. The pair $(\tilde{x}, \tilde{p}) \in \mathbb{R}^{CH}_+ \times P$ is a competitive equilibrium of a pure exchange economy if and only if $(\tilde{x}, \tilde{p}) \in B(p) \times P$ is such that:*

$$\sum_{h \in \mathcal{H}} \langle \nabla u_h(\tilde{x}_h), x_h - \tilde{x}_h \rangle_C + \left\langle \sum_{h \in \mathcal{H}} (\tilde{x}_h - e_h), p - \tilde{p} \right\rangle_C \leq 0 \qquad \forall\, (x, p) \in B(\tilde{p}) \times P\,. \tag{16}$$

Remark 1 The vector $(\tilde{x}, \tilde{p})$ is a solution to the QVI (16) if and only if

$$\text{for all } h \in \mathcal{H} \qquad \langle \nabla u_h(\tilde{x}_h), x_h - \tilde{x}_h \rangle_C \leq 0 \qquad \forall x_h \in B_h(\tilde{p}), \tag{17}$$

$$\left\langle \sum_{h \in \mathcal{H}} (\tilde{x}_h - e_h), p - \tilde{p} \right\rangle_C \leq 0 \qquad \forall p \in P\,. \tag{18}$$

Proof of Theorem 8 Firstly, one has:

(i) Since u_h is non-satiated, if $\tilde{x}_h$ is a solution to problem (3), then $\langle \tilde{p}, \tilde{x}_h - e_h \rangle_C = 0$. Hence, if, for all $h \in \mathcal{H}$, $\tilde{x}_h$ is a maximum, one has

$$\left\langle \tilde{p}, \sum_{h \in \mathcal{H}} (\tilde{x}_h - e_h) \right\rangle_C = 0 . \tag{19}$$

(ii) Since $u_h \in C^1(\mathbb{R}^C_+)$ and u_h is concave, then the maximization problem (3) is equivalent to (17).

Let $(\tilde{x}, \tilde{p})$ be a competitive equilibrium. From items (i) and (ii), conditions (19) and (17) hold. From (19), (4), and (5), one has

$$\left\langle \sum_{h \in \mathcal{H}} (\tilde{x}_h - e_h), p - \tilde{p} \right\rangle_C = \left\langle \sum_{h \in \mathcal{H}} (\tilde{x}_h - e_h), p \right\rangle_C \leq 0 \qquad \forall p \in P .$$

Hence, condition (18) holds, and from the Remark 1, $(\tilde{x}, \tilde{p})$ is a solution to (16).

Vice versa, let $(\tilde{x}, \tilde{p})$ be a solution to (16); that is, (17) and (18) hold. From item (ii), $\tilde{x}$ is a solution to maximization problem (3) and then from (i), the condition (19) holds. Then,

$$\left\langle \sum_{h \in \mathcal{H}} (\tilde{x}_h - e_h), p \right\rangle_C \leq \left\langle \sum_{h \in \mathcal{H}} (\tilde{x}_h - e_h), \tilde{p} \right\rangle_C = 0 .$$

Then, selecting $p = (0, \ldots, 0, 1, 0, \ldots, 0)$, with 1 at the c-th position, we get $\sum_{h \in \mathcal{H}} (\tilde{x}_h^c - e_h^c) \leq 0$ for all $c \in \mathcal{C}$. From last inequality and from (19), it follows that conditions (4) and (5) hold. Hence, $(\tilde{x}, \tilde{p})$ is a competitive equilibrium. □

For all $h \in \mathcal{H}$, we introduce the map $G_h : \mathbb{R}^C_+ \to 2^{\mathbb{R}^l}$ such that

$$G_h(x_h) := conv\Big(N^{>}(x_h) \cap S(0, 1)\Big),$$

where $\overline{B}(0, 1)$ and $S(0, 1)$ are, respectively, the closed unit ball and the unit sphere of $\mathbb{R}^C$. One has the following.

Theorem 9 *Let u_h be non-satiated and continuous for all $h \in \mathcal{H}$. The pair $(\tilde{x}, \tilde{p}) \in \mathbb{R}^{CH}_+ \times P$ is a competitive equilibrium of a pure exchange economy if and only if*

(i) *if u_h are concave, there exist $k_h \in \partial u_h(\tilde{x}_h)$ such that*

$$\sum_{h \in \mathcal{H}} \langle k_h, x_h - \tilde{x}_h \rangle_C + \left\langle \sum_{h \in \mathcal{H}} (\tilde{x}_h - e_h), p - \tilde{p} \right\rangle_C \leq 0 \qquad \forall\, (x, p) \in B(\tilde{p}) \times P ; \tag{20}$$

(ii) *if u_h are semistrictly quasiconcave, there exist $g_h \in G_h$ such that*

$$\sum_{h\in\mathcal{H}} \langle g_h, x_h - \tilde{x}_h\rangle_C + \left\langle \sum_{h\in\mathcal{H}} (\tilde{x}_h - e_h), p - \tilde{p} \right\rangle_C \geq 0 \qquad \forall\,(x,p) \in B(\tilde{p}) \times P\,. \tag{21}$$

Proof

(i) The desired result follows from Remark 1 and adapting the Proof of Theorem 8.

(ii) We suppose that $\tilde{x}_h \in B_h(\tilde{p})$ with $g_h \in G_h(\tilde{x}_h)$ is a solution to

$$\langle g_h, x_h - \tilde{x}_h\rangle_C \geq 0 \qquad \forall\, x_h \in B_h(\tilde{p})\,. \tag{22}$$

Since u_h is non-satiated, $\tilde{x}_h \notin argmax_{\mathbb{R}^C}\, u_h$ and it follows that $0 \notin G_h(\tilde{x}_h)$; hence, $g_h \in N^{>}(\tilde{x}_h) \setminus \{0\}$. Clearly, if $\tilde{x}_h \in B_h(\tilde{p})$ with $g_h \in N^{>}(\tilde{x}_h) \setminus \{0\}$ is a solution to (22), one has $\frac{g_h}{\|g_h\|} \in G_h(\tilde{x}_h)$. Hence, also in this case the desired result follows from Remark 1, Theorem 7 and adapting the Proof of Theorem 8.

□

4.2 Existence of Equilibria

This subsection deals with the existence of equilibrium by means of the variational problems, thanks to the characterization theorems proven in Sect. 4.1. To prove the existence, a key role is played by the properties of set-valued map $B_h : P \to \mathbb{R}^C$.

Proposition 1 *For all $h \in \mathcal{H}$, the set-valued map B_h is*

(i) lower semicontinuous*: for any $p \in P$, for any sequence $\{p_n\}_{n\in\mathbb{N}} \subset P$, $p_n \to p$, and for any $x_h \in B_h(p)$, there exists a sequence of elements $\{x_{hn}\}_{n\in\mathbb{N}} \subset \mathbb{R}^C_+$, with $x_{hn} \in B_h(p_n)$ for all $n \in \mathbb{N}$ and $x_{hn} \to x_h$;*

(ii) closed*: for any sequences $\{x_{hn}\}_{n\in\mathbb{N}} \subset \mathbb{R}^C_+$, $\{p_n\}_{n\in\mathbb{N}} \subset P$, if $p_n \to p$, $x_{hn} \in B_h(p_n)$ and $x_{hn} \to x_h$, then $x_h \in B_h(p)$.*

Proof (i) : *B_h is lower semicontinuous.*

Let $p \in P$ and $\{p_n\}_{n\in\mathbb{N}} \subset P$ be such that $p_n \to p \in P$. Let $x_h \in B_h(p)$. Let us pose $I = \{c \in C : x_h^c > 0\}$. We consider the following sequence:

$$x_{hn} := x_h - \eta_n \qquad \forall n \in \mathbb{N}, \tag{23}$$

where the sequence $\{\eta_n\}_{n\in\mathbb{N}} \subset \mathbb{R}^C$ is such that $\eta_n^c = 0$ if $c \notin I$ and $\eta_n^c = \eta_n$ if $c \in I$ with $\{\eta_n\}_{n\in\mathbb{N}}$ converging to zero and, if $\sum_{c\in I} p_n^c > 0$, satisfying

$$\frac{\langle p_n - p, x_h - e_h\rangle_C}{\sum_{c\in I} p_n^c} < \eta_n < \min_{c\in I}\{x_h^c\}.$$

Let us verify that $x_{hn} \in B_h(p_n)$ for all $n \in \mathbb{N}$.

- If $\sum_{c\in I} p_n^c = 0$: since $p_n^c \geq 0$, one has $p_n^c = 0$ for all $c \in I$. Then,

$$\langle p_n, x_{hn} - e_h\rangle_C = \sum_{c\in I} p_n^c(x_{hn}^c - e_h^c) + \sum_{c\notin I} p_n^c(x_{hn}^c - e_h^c) = \sum_{c\notin I} p_n^c(-e_h^c) \leq 0.$$

- If $\sum_{c\in I} p_n^c > 0$: by the choice of η_n and since $x_h \in B_h(p)$, it results

$$\langle p_n, x_{hn} - e_h\rangle_C = \langle p_n, x_h - e_h\rangle_C - \eta_n \sum_{c\in C} p_n^c < \langle p, x_h - e_h\rangle_C \leq 0\,.$$

Then, $x_{hn} \in B_h(p_n)$ for all $n \in \mathbb{N}$. Moreover, we have $x_{hn} = (x_h - \eta_n) \to x_h$. Hence, B_h is lower semicontinuous.

(ii) : *B_h is closed.*

Let $p \in P$ and $\{p_n\}_{n\in\mathbb{N}} \subset P$ be such that $p_n \to p \in P$. Let $\{x_{hn}\}_{n\in\mathbb{N}}$ be a sequence such that $x_{hn} \in B_a(p_n)$ for all $n \in \mathbb{N}$ and $x_{hn} \to x_h$. Since $\langle p_n, x_{hn} - e_h^c\rangle_C \leq 0$ and $x_{hn} \geq 0 \quad \forall n \in \mathbb{N}$, we get $\langle p, x_h - e_h^c\rangle_C \leq 0$ and $x_h \geq 0$; that is, $x_h \in B_h(p)$. □

Theorem 10 *If u_h is strongly concave for all $h \in \mathcal{H}$, then there exists a competitive equilibrium.*

Proof We consider the quasi-variational inequality (16), and we prove that there exists at least one solution. First, for all $h \in \mathcal{H}$ and $p \in P$ we consider the parametric variational inequality:

$$\text{“Find } \tilde{x}_h \in B_h(p) \text{ such that:}$$
$$\langle \nabla u_h(\tilde{x}_h), x_h - \tilde{x}_h\rangle_C \leq 0 \qquad \forall x_h \in B_h(p).\text{”} \tag{24}$$

From strong concavity of utility function, it follows that the operator $-\nabla u_h(x_h)$ is strongly monotone; hence, from Theorem 4 there exists a unique solution to (24). We can introduce the function of solutions:

$$\tilde{x}_h : P \to \mathbb{R}_+^C,$$

such that, for all $p \in P$, $\tilde{x}_h(p)$ is the unique solution to (24). We prove that the function $\tilde{x}_h$ is continuous on P. Let $p \in P$ and $\{p_n\}_{n\in\mathbb{N}} \subset P$ be such that $p_n \to p$. We consider the sequence $\{\tilde{x}_h(p_n)\}_{n\in\mathbb{N}}$ such that, for all $n \in \mathbb{N}$, $\tilde{x}(p_n)$ is the unique solution to variational inequality:

$$\langle \nabla u_h(\tilde{x}_h(p_n)), x_{hn} - \tilde{x}_h(p_n)\rangle_C \leq 0, \qquad \forall x_{hn} \in B_h(p_n)\,. \tag{25}$$

We prove that $\tilde{x}_h(p_n) \to \tilde{x}_h(p)$. Since B_h is lower semicontinuous, there exists a sequence $\{y_{hn}\}_{n\in\mathbb{N}} \subseteq \mathbb{R}^C_+$ such that:

$$y_{hn} \in B_h(p_n) \quad \forall n \in \mathbb{N}, \qquad \text{and} \qquad \lim_{n\to+\infty} y_{hn} = \tilde{x}_h(p). \tag{26}$$

Then, from the condition of strong monotonicity of $-\nabla u_h$, with $\tilde{x}_{hn}$ and y_{hn}, one has

$$\langle -\nabla u_h(\tilde{x}_h(p_n)) + \nabla u_h(y_{hn}), \tilde{x}_h(p_n) - y_{hn}\rangle_C \geq \nu \|\tilde{x}_h(p_n) - y_{hn}\|^2 .$$

From the last inequality and from (25), with $x_{hn} = y_{hn}$, we have

$$\nu \|\tilde{x}_h(p_n) - y_{hn}\|^2 \leq \langle -\nabla u_h(\tilde{x}_h(p_n)) + \nabla u_h(y_{hn}), \tilde{x}_h(p_n) - y_{hn}\rangle_C =$$

$$= \langle \nabla u_h(\tilde{x}_h(p_n)), y_{hn} - \tilde{x}_h(p_n)\rangle_C + \langle \nabla u_h(y_{hn}), \tilde{x}_h(p_n) - y_{hn}\rangle_C \leq$$

$$\leq \|\nabla u_h(y_{hn})\| \cdot \|\tilde{x}_h(p_n) - y_{hn})\|,$$

namely, $\|\tilde{x}_h(p_n) - y_{hn}\| \leq \dfrac{\|\nabla u_h(y_{hn})\|}{\nu}$. Then, we have

$$\|\tilde{x}_h(p_n)\| \leq \|\tilde{x}_h(p_n) - y_{hn}\| + \|y_{hn}\| \leq \frac{\|\nabla u_h(y_{hn})\|}{\nu} + \|y_{hn}\| . \tag{27}$$

Since $y_{hn} \to \tilde{x}_h(p)$ and, being $u_h \in C^1(\mathbb{R}^l_+)$, one has $\nabla u_h(y_{hn}) \to \nabla u_h(\tilde{x}_h(p))$; then, there exist $s, k \in \mathbb{R}_+$ such that $\|\nabla u_h(y_{hn})\| \leq h$ and $\|y_{hn}\| \leq k \quad \forall n \in \mathbb{N}$. So, from (27), it follows that $\|\tilde{x}_h(p_n)\|\| \leq \frac{h}{\nu} + k$, for all $n \in \mathbb{N}$, where the constant $\frac{s}{\nu} + k$ does not depend on n. Hence, there exists a subsequence $\{\tilde{x}_h(p_{k_n})\}$ of $\{\tilde{x}_h(p_n)\}$ converging to an element $y_h \in \mathbb{R}^C$: $\lim_{n\to+\infty} \tilde{x}_h(p_{k_n}) = y_h$. Since B_h is a closed map, it follows that $y_h \in B_h(p)$. We prove that $y_h = \tilde{x}_h(p)$. For all $x_h \in B_h(p)$, since B_h is lower semicontinuous, there exists $\{x_{hn}\}_{n\in\mathbb{N}}$ such that $x_{hn} \in B_a(p_n)$ for all $n \in \mathbb{N}$ and $x_{hn} \to x_h$. From variational inequality (25), with x_{hk_n}, one has

$$\langle \nabla u_h(\tilde{x}_h(p_{k_n})), x_{hk_n} - \tilde{x}_h(p_{k_n})\rangle_C \leq 0 ,$$

and passing to the limit as $n \to +\infty$, it follows that

$$\langle \nabla u_h(y_h), x_h - y_h\rangle_C \leq 0 .$$

Namely, we have $y_h = \tilde{x}_h(p)$. Hence, since $\tilde{x}_h(p_n) \to \tilde{x}_h(p)$, we can conclude that the solutions map $\tilde{x}_h$ is continuous on P.

Now, we consider the variational inequality:

"Find $\tilde{p} \in P$ such that:

$$\left\langle -\sum_{h\in\mathcal{H}}(\tilde{x}_h(p)-e_h), p-\tilde{p}\right\rangle_C \geq 0, \qquad \forall p \in P." \tag{28}$$

Since P is a compact and convex set and $\tilde{x}_a(p)$ is a continuous function, by Theorem 1, there exists at least one $\tilde{p}$ solution to (28).

Finally, we consider the pair $(\tilde{x}, \tilde{p}) \in P \times B(\tilde{p})$, where $\tilde{p}$ is a solution to (28) and $\tilde{x}$ is such that for all $h \in \mathcal{H}$, $\tilde{x}_h$ is the solution to (24) with parameter $\tilde{p}$. It follows that $(\tilde{x}, \tilde{p})$ is a solution to quasi-variational inequality (16). □

Theorem 11 *Let $u_h \in C^1(\mathbb{R}^C_+)$ be non-satiated and strictly concave for all $h \in \mathcal{H}$, then there exists the equilibrium.*

Proof First, we consider the variational inequality:

"Find $(\tilde{x}, \tilde{p}) \in K(\tilde{p}) \times P$:

$$\sum_{h\in\mathcal{H}}\langle\nabla u_h(\tilde{x}_h), x_h-\tilde{x}_h\rangle_C+\left\langle\sum_{h\in\mathcal{H}}(\tilde{x}_h-e_h), p-\tilde{p}\right\rangle_C \leq 0 \qquad \forall(x,p)\in K(\tilde{p})\times P.", \tag{29}$$

where $K(p) := \prod_{h\in\mathcal{H}} K_h(p)$ is the bounded convex set, with $K_h(p) := B_h(p) \cap \prod_{c\in\mathcal{C}}[0, \sum_{h\in\mathcal{H}} e^c_h]$. In order to prove the existence, we can proceed as in the proof of Theorem 10. Since $u_h \in C^1(\mathbb{R}^C_+)$ and u_h is strictly concave, from Theorem 3 there exists a unique solution to the parametric variational inequality (24), hence we can introduce the function of solutions $\tilde{x}_h(\cdot)$. For all $\{p_n\}_{n\in\mathbb{N}}$ converging to p, we consider $\{\tilde{x}_h(p_n)\}_{n\in\mathbb{N}}$; since the sequence is in the bounded set $\prod_{c\in\mathcal{C}}[0, \sum_{h\in\mathcal{H}} e^c_h]$, there exists a subsequence converging to y. Being $u_h \in C^1$, $\tilde{x}_h(p_n) \to y$ and $\nabla u_h(\tilde{x}_h(p_n)) \to \nabla u_h(y)$, and for properties of the set-valued map B_h, it follows that $y = \tilde{x}_h$. Hence, the function of solution $\tilde{x}$ is continuous and there exists $\tilde{p}$ solution to (18). Then, there exists $(\tilde{x}, \tilde{p})$ solution to (29).

Now, it remains to prove that if $\tilde{x}_h$ is a solution to the variational inequality:

$$\langle\nabla u_h(\tilde{x}_h), x_h - \tilde{x}_h\rangle_C \leq 0 \qquad \forall x_h \in K_h(\tilde{p}), \tag{30}$$

then $\tilde{x}_h$ is a solution to the variational inequality in the unbounded set

$$\langle\nabla u_h(\tilde{x}_h), x_h - \tilde{x}_h\rangle_C \leq 0 \qquad \forall x_h \in B_h(\tilde{p}). \tag{31}$$

We suppose that there exists $x' \in B_h(\tilde{p})$ such that $\langle\nabla u_h(\tilde{x}_h), x' - \tilde{x}_h\rangle_C > 0$. Since $\tilde{x}_h \in K_h(\tilde{p}) \subseteq B_h(\tilde{p})$, $x' \in B_h(\tilde{p})$, and $B_h(\tilde{p})$ is a convex set, then for all $\lambda \in (0, 1)$, $\bar{x} = \lambda x' + (1-\lambda)\tilde{x}_h \in B_h(\tilde{p})$. Then,

$$\langle\nabla u_h(\tilde{x}_h), \bar{x}-\tilde{x}_h\rangle_C = \langle\nabla u_h(\tilde{x}_h), \lambda x'+(1-\lambda)\tilde{x}_h-\tilde{x}_h\rangle_C = \lambda\langle\nabla u_h(\tilde{x}_h), x'-\tilde{x}_h\rangle > 0.$$

We choose $\lambda \in (0, 1)$ such that

$$\lambda < \min\left\{1, -\frac{\tilde{x}_h^c - \sum_{h\in\mathcal{H}} e_h^c}{(x')^c - \tilde{x}_h^c}, \text{ with } c \in \mathcal{C} \text{ such that } (x')^c - \tilde{x}_h^c > 0\right\},$$

for all $c \in \mathcal{C}$ it results

$$\bar{x}^c - \sum_{h\in\mathcal{H}} e_h^c = \lambda(x')^c + (1-\lambda)\tilde{x}_h^c - \sum_{h\in\mathcal{H}} e_h^c = \lambda[(x')^c - \tilde{x}_h^c] + \tilde{x}_h^c - \sum_{h\in\mathcal{H}} e_h^c.$$

Observe that

- if $(x')^c - \tilde{x}_h^c = 0$, then $\bar{x}^c - \sum_{h\in\mathcal{H}} e_h^c = \tilde{x}_h^c - \sum_{h\in\mathcal{H}} e_h^c \le 0$;
- if $(x')^c - \tilde{x}_h^c < 0$, then $\bar{x}^c - \sum_{i\in\mathcal{H}} e_h^c < 0$;
- if $(x')^c - \tilde{x}_h^c > 0$, by the choice of λ, we have

$$\bar{x}^c - \sum_{h\in\mathcal{H}} e_h^c < -\frac{\tilde{x}_h^c - \sum_{h\in\mathcal{H}} e_h^c}{(x')^c - \tilde{x}_h^c}[(x')^c - \tilde{x}_h^c] + \tilde{x}_h^c - \sum_{h\in\mathcal{H}} e_h^c = 0\,.$$

Hence, $\bar{x} \in \prod_{c\in\tilde{\mathcal{C}}}[0, \sum_{h\in\mathcal{H}} e_h^c]$. Then, there exists $\bar{x} \in K_h(\tilde{p})$ such that $\langle \nabla u_h(\tilde{x}_h), \bar{x} - \tilde{x}_h\rangle < 0$, but this contradicts $\tilde{x}_h$ solution to (30). Then, $\tilde{x}$ is a solution to (31) and we can conclude that there exists $(\tilde{x}, \tilde{p})$ solution to (16). □

Theorem 12 *Let u_h be non-satiated, continuous, and semistrictly quasiconcave for all $h \in \mathcal{H}$, then there exists the equilibrium.*

Proof As proven in Theorem 11, also under above assumptions, it is sufficient to prove the existence of a solution to the variational problem in the bounded set:

"Find $(\tilde{x}, \tilde{p}) \in K((\tilde{p})) \times P$ and $g = \{g_h\}_{h\in\mathcal{H}}$, with $g_h \in G_h(\tilde{x}_h)\ \forall h \in \mathcal{H}$:

$$\sum_{h\in\mathcal{H}} \langle g_h, x_h - \tilde{x}_h\rangle_C + \left\langle \sum_{h\in\mathcal{H}} (\tilde{x}_h - e_h), p - \tilde{p}\right\rangle_C \le 0 \qquad \forall (x, p) \in K(\tilde{p}) \times P.\text{"} \quad (32)$$

from Remark 1, condition (18) and the following hold:

$$\text{for all } h \in \mathcal{H} \qquad \langle g_h, x_h - \tilde{x}_h\rangle \le 0 \qquad \forall x_h \in K_h(\tilde{p}) \quad (33)$$

For all $h \in \mathcal{H}$ and for all $p \in P$, we consider the parametric generalized variational inequality:

$$\langle g_h, x_h - \tilde{x}_h\rangle \le 0 \qquad \forall x_h \in K_h(p)\,. \quad (34)$$

Since u_h is continuous and semistrictly quasiconcave for all $h \in \mathcal{H}$, the set-valued map G_h is usc with convex and compact values. Then, thanks to Theorem 5, there exists a solution to (34). So, we can define the set-valued map of solutions:

$$\Theta_h : P \rightrightarrows \mathbb{R}_+^C$$

such that for all $p \in P$, $\quad \Theta_h(p) = \{\tilde{x}_h : \tilde{x}_h \text{ isasolutionto(34)}\} \neq \emptyset$.

The map Θ_h enjoys the following properties.

Θ_h is a closed map:

let $\{p_n\} \subseteq P$ and $\{\tilde{x}_{h,n}\} \subseteq \mathbb{R}_+^C$ be two sequences with $\tilde{x}_{h,n} \in \Theta_h(p_n)$ and such that $p_n \to p$ and $\tilde{x}_{h,n} \to \widetilde{x}_h$. We have to prove that $\widetilde{x}_h \in \Theta_h(p)$.

First, we observe that $\widetilde{x}_h \in K_h(p)$, indeed:

$$\tilde{x}_{h,n} \in K_h(p_n) \quad \Rightarrow \quad 0 \leq \tilde{x}_{h,n}^c \leq \sum_{h \in \mathcal{H}} e_h^c \quad \forall c \in \mathcal{C}, \quad \langle p_n, \tilde{x}_{h,n} - e_h \rangle_C \leq 0 \quad \forall n \in \mathbb{N};$$

passing to the limit, it results

$$0 \leq \widetilde{x}_h^c \leq \sum_{h \in \mathcal{H}} e_h^c \quad \forall c \in \mathcal{C}, \quad \langle p, \widetilde{x}_h - e_h \rangle_C \leq 0 \quad \Rightarrow \quad \widetilde{x}_h \in K_h(p).$$

Since $\tilde{x}_{h,n}$ is a solution to (34) in $K_h(p_n)$, for all $n \in \mathbb{N}$, there exists $g_{hn} \in G_h(\tilde{x}_{h,n})$ such that

$$\langle g_{hn}, x_{hn} - \tilde{x}_{hn} \rangle \leq 0 \qquad \forall x_{hn} \in K_h(p_n). \tag{35}$$

Since the map G_h is a closed set-valued map, the sequence $\{g_{hn}\}_{n \in \mathbb{N}}$ converges to $g_h \in G_h(\tilde{x})$. From the lower semicontinuity of K_h for all $y_h \in K_h(p)$, there exists $\{y_{hn}\}_{n \in \mathbb{N}}$ converging to y_h such that $y_n \in K_h(p_n)$ for all $n \in \mathbb{N}$. Hence, replacing y_{hn} in (35) and passing to the limit, it follows that $\langle g_h, y_h - \widetilde{x}_h \rangle \leq 0$; that is, $\widetilde{x}_h \in \Theta_h(p)$. We can conclude that Θ_h is closed.

Θ_h is with compact values because, for all $p \in P$, the set $K_h(p)$ is compact.

Θ_h is usc:

since $\Theta_h(p) \subseteq \prod_{c \in \mathcal{C}} [0, \sum_{h \in \mathcal{H}} e_h^c]$. Hence, $\overline{\Theta_h(P)}$ is compact; that is, Θ_h is compact. Namely, since Θ_h is closed and compact, it follows that Θ_h is usc.

Θ is with convex values.

For all $p \in P$, let $\tilde{x}_h, \ \tilde{y}_h \in \Theta_h(p)$. Being u_h semistrictly quasiconcave, it follows that, for all $\lambda \in (0, 1)$, $z = \lambda \tilde{x}_h + (1 - \lambda)\tilde{y}_h \in K(\tilde{p})$ and $z \in \Theta_h(p)$.

Now, we can consider the following generalized variational inequality:

$$\text{Find } \tilde{p} \in P \text{ such that there exist } \varphi_h \in \Theta_h(\tilde{p}), \text{ with } h \in \mathcal{H}:$$
$$\left\langle \sum_{h \in \mathcal{H}} (\varphi_h - e_h), p - \tilde{p} \right\rangle \leq 0 \qquad \forall p \in P. \tag{36}$$

From properties of $\Theta_h(\cdot)$, it follows that the operator of variational problem (36) is usc with compact and convex values. Then, since P is a compact set, from Theorem 5 there exists $\tilde{p} \in P$ and $\varphi_h \in \Theta_h(\tilde{p})$ solutions to (36).

Then, the pair $(\tilde{x}, \tilde{p}) \in K(\tilde{p}) \times P$ such that $\tilde{p}$ is a solution to (36) and $\tilde{x}_h \in \Theta_h(\tilde{p})$, for $h \in \mathcal{H}$, is a solution to variational problem (21) and, then, $(\tilde{x}, \tilde{p})$ is an equilibrium. □

4.3 Remark

In this section we analyzed the exchange economy model, by means of a quasi-variational inequality problem. The characteristic of the quasi-variational inequality is that the convex set depends on the solution of the problem; and the latter fact makes the problem more complicated. We gave a procedure that allowed us to consider $H + 1$ variational inequalities, instead of a single quasi-variational inequality. Indeed, it is certainly easier to solve variational problems where the convex set does not depend on the solution. This method represents a new and useful methodology not only to provide existence results but also to provide an efficient computational procedure for the calculus of solutions.

In order to obtain the existence of equilibrium, we firstly considered the economic model under strong assumptions on utility functions: strong concavity and continuous differentiability. Even if the convex set $B_h(p)$ might be unbounded, under those assumptions, there exists a unique solution to the parametric variational inequality (24); hence, we introduced the demand functions $\tilde{x}(p)$ and we proved that this function is continuous. Moreover, since a solution of the quasi-variational problem is in a compact set, we solved our problem in a bounded set where the utility functions are strictly concave instead of strongly concave.

Finally, we considered the utility functions being semistrictly quasiconcave. Under such assumptions, since the solution of the parametric variational inequality is not unique, the map of the solution is set-valued. In this more general case, we present the same procedure by using tools of set-valued analysis and generalized variational inequalities.

5 Model with Nominal Assets

In this section, we investigate the existence of an equilibrium vector for a financial economy with incomplete markets and nominal assets. More precisely, we characterize households' maximization problems and market clearing conditions in terms of well-chosen Variational Inequality problems. We then introduce a sequence of "artificial" price sets and associated Generalized Quasi-Variational Problems whose solutions have nice properties. We finally show that the sequence admits a subsequence converging to an equilibrium.

For this model, we make the following Assumptions.

Assumption 1

1. *For any $h \in \mathcal{H}$, the utility function $u_h : \mathbb{R}^G_+ \to \mathbb{R}$ is continuously differentiable and concave.*
2. *For any $h \in \mathcal{H}$ and for any $s \in \mathcal{S}^0$, there exists $c \in \mathcal{C}$ such that u_h is strictly increasing in x_h^{sc}, i.e.,*

$$\text{if } \overline{x}_h \geq^2 \overline{\overline{x}}_h \text{ and } \exists\, c \in \mathcal{C} : \ \overline{x}_h{}^{sc} > \overline{\overline{x}}_h{}^{sc}, \text{ then } u_h(\overline{x}_h) > u_h\left(\overline{\overline{x}}_h\right).$$

3. *For any $(s, c) \in \mathcal{S}^0 \times \mathcal{C}$, there exists $h \in \mathcal{H}$ such that u_h is strictly increasing in x_h^{sc}.*

Assumption 2

For every $h \in \mathcal{H}$, $e_h >> 0$, i.e., each household is endowed with each commodity c in each state s.

Assumption 3

1. *$Y \geq 0$, i.e., $\forall s \in \mathcal{S}$ and $a \in A$, $y^{sa} \geq 0$.*
2. *For $s \in \mathcal{S}$, $y^s := (Row\ s\ of\ Y) > 0$; that is, $\exists a \in \mathcal{A}$ such that $y^{sa} > 0$, i.e., in each state at least one asset delivers something.*
3. *rank $Y = A < S$, i.e., markets are incomplete.*

By using Assumption 1.2, we prove the following result.

Proposition 2 *If $(\widetilde{x}_h, \widetilde{b}_h) \in B_h(\widetilde{p}, \widetilde{q})$ is a solution to household h's maximization problem, $\max_{(x_h, b_h) \in B_h(\widetilde{p}, \widetilde{q})} u_h(x_h)$, then the following statements hold true:*

1. *For any $s \in \mathcal{S}^0$, there exists c^* such that $\widetilde{p}^{sc^*} > 0$.*
2. $\widetilde{q} \in Q := \left\{ q' \in \mathbb{R}^A : \text{ there is no } b \in \mathbb{R}^A \text{such that} \begin{bmatrix} -q \\ Y \end{bmatrix} b > 0 \right\} =$

$$= \left\{ q' \in \mathbb{R}^A : \exists \nu = \left(\nu^s\right)_{s \in \mathcal{S}} \in \mathbb{R}^S_{++} \text{such that} q' = \nu Y \right\}, \tag{37}$$

where the set Q is open and it is called the set of no-abitrage prices.

Proof See pp. 1356–1357 in [18]. □

[2] Following standard notation, for vectors $y := (y_i)_{i=1}^n$, $z := (z_i)_{i=1}^n \in \mathbb{R}^n$; $y \geq z$ means that for $i \in \{1, \ldots, n\}$, $y_i \geq z_i$; $y >> z$ means that for $i \in \{1, \ldots, n\}$, $y_i > z_i$, and $y > z$ means that $y \geq z$ but $y \neq z$.

5.1 Variational Inequality Approach

From Proposition 2, without loss of generality, we can restrict prices to belong to the following sets:

$$\Delta_0 := \left\{ (\nu, p^0) \in \mathbb{R}^S_+ \times \mathbb{R}^C_+ : \sum_{s \in \mathcal{S}} \nu^s + \sum_{c \in \mathcal{C}} p^{0c} = S + 1 \right\},$$

$$\Delta_C := \left\{ p^s \in \mathbb{R}^C_+ : \sum_{c \in \mathcal{C}} p^{sc} = 1 \right\}, \qquad \Delta := \prod_{s \in \mathcal{S}} \Delta_C.$$

We consider the set-valued map

$$B_h^* : \Delta_0 \times \Delta \rightrightarrows \mathbb{R}^G_+ \times \mathbb{R}^A,$$

$$B_h^*\Big((\nu, p^0), p^1\Big) := \{(x_h, b_h) \in \mathbb{R}^G_+ \times \mathbb{R}^A : \langle (\nu, p^0), (Yb_h, x_h^0 - e_h^0) \rangle_{S+C} \leq 0,$$
$$\langle p^s, x_h^s - e_h^s \rangle_C - \langle y^s, b_h \rangle_A \leq 0, \ \forall s \in \mathcal{S}\}.$$

We define $B^*\Big((\nu, p^0), p^{\mathbf{1}}\Big) := \prod_{h \in \mathcal{H}} B_h^*\Big((\nu, p^0), p^{\mathbf{1}}\Big)$.

To study equilibria using the variational approach, we introduce a non-zero lower bound on prices. For each positive number $\varepsilon \leq \dfrac{1}{C}$, we define the following prices sets:

$$\Delta_0^\varepsilon := \Big\{ (\nu, p^0) \in \Delta_0 : \ p^{0c} \geq \varepsilon, \ \nu^s \geq \varepsilon, \ \forall c \in \mathcal{C}, \ \forall s \in \mathcal{S} \Big\};$$

$$\Delta_C^\varepsilon := \{ p^s \in \Delta_C : p^{sc} \geq \varepsilon \, , \forall c \in \mathcal{C} \}; \qquad \Delta^\varepsilon := \prod_{s \in \mathcal{S}} \Delta_C^\varepsilon.$$

In this subsection, we drop ε in the variables for a lighter notation. Now, we present some properties of the set-valued map B_h.

Proposition 3 *Under Assumptions 2 and 3, for any $\varepsilon \in \left(0, \frac{1}{C}\right]$ and any $h \in \mathcal{H}$, the set-valued map B_h is nonempty, convex, and compact valued, and l.s.c. and closed in $\Delta_0^\varepsilon \times \Delta^\varepsilon$.*

Proof See p. 1360 in [18]. □

We introduce the following problem:

" *Find* $\Big((\widetilde{x}, \widetilde{b}), (\widetilde{\nu}, \widetilde{p}^{\,0}), \widetilde{p}^{\mathbf{1}}\Big) \in B^*((\widetilde{\nu}, \widetilde{p}^{\,0}), \widetilde{p}^{\mathbf{1}}) \times \Delta_0^\varepsilon \times \Delta^\varepsilon$ *such that*

$$\sum_{h\in\mathcal{H}}\langle\nabla u_h(\widetilde{x}_h), x_h - \widetilde{x}_h\rangle_G + \langle(Y\sum_{h\in\mathcal{H}}\widetilde{b}_h, \sum_{h\in\mathcal{H}}(\widetilde{x}^0_h - e^0_h)), (\nu, p^0) - (\widetilde{\nu}, \widetilde{p}^0)\rangle_{S+C}$$

$$+\sum_{s\in\mathcal{S}}\langle\sum_{h\in\mathcal{H}}(\widetilde{x}^s_h - e^s_h), p^s - \widetilde{p}^s\rangle_C \leq 0 \tag{38}$$

$$\forall\Big((x, b), (\nu, p^0), p^{\mathbf{1}}\Big) \in B^*((\widetilde{\nu}, \widetilde{p}^0), \widetilde{p}^{\mathbf{1}}) \times \Delta^\varepsilon_0 \times \Delta^\varepsilon."$$

Remark 2 It is important to observe that solving QVI (38) is equivalent to solving simultaneously $(H + 1 + S)$ variational inequalities. Indeed, $\Big((\widetilde{x}, \widetilde{b}), (\widetilde{\nu}, \widetilde{p}^0), \widetilde{p}^{\mathbf{1}}\Big)$ is a solution to QVI (38) if and only if

- for all $h \in \mathcal{H}$, $(\widetilde{x}_h, \widetilde{b}_h)$ is a solution to VI:

$$\langle\nabla u_h(\widetilde{x}_h), x_h - \widetilde{x}_h\rangle_G \leq 0 \qquad \forall(x_h, b_h) \in B^*_h((\widetilde{\nu}, \widetilde{p}^0), \widetilde{p}^{\mathbf{1}}) \tag{39}$$

- $(\widetilde{\nu}, \widetilde{p}^0)$ is a solution to VI:

$$\left\langle\left(Y\sum_{h\in\mathcal{H}}\widetilde{b}_h, \sum_{h\in\mathcal{H}}(\widetilde{x}^0_h - e^0_h)\right), (\nu, p^0) - (\widetilde{\nu}, \widetilde{p}^0)\right\rangle_{S+C} \leq 0 \qquad \forall(\nu, p^0) \in \Delta^\varepsilon_0 \tag{40}$$

- for all $s \in \mathcal{S}$, $\widetilde{p}^s$ is a solution to VI:

$$\left\langle\sum_{h\in\mathcal{H}}(\widetilde{x}^s_h - e^s_h), p^s - \widetilde{p}^s\right\rangle_C \leq 0 \qquad \forall p^s \in \Delta^\varepsilon_C. \tag{41}$$

Clearly, $(\widetilde{\nu}, \widetilde{p}^0)$ is solution to VI (40) and for all $s \in \mathcal{S}$, $\widetilde{p}^s$ is solution to VI (41) if and only if $\big((\widetilde{\nu}, \widetilde{p}^0), \widetilde{p}_{\mathbf{1}}\big)$ is solution to VI:

$$\left\langle\left(Y\sum_{h\in\mathcal{H}}\widetilde{b}_h, \sum_{h\in\mathcal{H}}(\widetilde{x}^0_h - e^0_h)\right), (\nu, p^0) - (\widetilde{\nu}, \widetilde{p}^0)\right\rangle_{S+C} + \sum_{s\in\mathcal{S}}\left\langle\sum_{h\in\mathcal{H}}(\widetilde{x}^s_h - e^s_h), p^s - \widetilde{p}^s\right\rangle_C \leq 0 \tag{42}$$

$$\forall((\nu, p^0), p^{\mathbf{1}}) \in \Delta^\varepsilon_0 \times \Delta^\varepsilon.$$

Theorem 13 *Under Assumptions 1, 2, and 3, for all* $\varepsilon \in \left(0, \frac{1}{C}\right]$, *QVI (38) admits at least one solution.*

Proof From Assumptions 1, 2, and 3 and from Proposition 3, it follows that all Assumptions of Theorem 6 are satisfied. Then, there exists $\left((\widetilde{x}, \widetilde{b}), (\widetilde{\nu}, \widetilde{p}^0), \widetilde{p}^{\mathbf{1}}\right) \in B^*((\widetilde{\nu}, \widetilde{p}^0), \widetilde{p}^{\mathbf{1}}) \times \Delta_0^{\varepsilon} \times \Delta^{\varepsilon}$, which is solution to QVI (38). □

Theorem 14 *For any* $\varepsilon \in \left(0, \frac{1}{C}\right]$, *let* $\left((\widetilde{x}, \widetilde{b}), (\widetilde{\nu}, \widetilde{p}^0), \widetilde{p}^{\mathbf{1}}\right) \in B^*((\nu, p^0), p^{\mathbf{1}}) \times \Delta_0^{\varepsilon} \times \Delta^{\varepsilon}$ *be a solution to QVI (38). Then, under Assumptions 1, 2, and 3 one has:*

(i) *for any* $h \in \mathcal{H}$:

$$\max_{(x_h, b_h) \in B_h^*((\widetilde{\nu}, \widetilde{p}^0), \widetilde{p}_1)} u_h(x_h) = u_h(\widetilde{x}_h); \tag{43}$$

(ii) *for any* $h \in \mathcal{H}$:

$$\langle (\widetilde{\nu}, \widetilde{p}^0), (Y\widetilde{b}_h, \widetilde{x}_h^0 - e_h^0) \rangle_{S+C} = 0, \tag{44}$$

$$\langle \widetilde{p}^s, \widetilde{x}_h^s - e_h^s \rangle_C - \langle y^s, \widetilde{b}_h \rangle_A = 0 \qquad \forall s \in \mathcal{S}; \tag{45}$$

(iii) *the following equations hold true:*

$$\left\langle (\widetilde{\nu}, \widetilde{p}^0), \left(Y \sum_{h \in \mathcal{H}} \widetilde{b}_h, \sum_{h \in \mathcal{H}} (\widetilde{x}_h^0 - e_h^0) \right) \right\rangle_{S+C} = 0, \tag{46}$$

$$\left\langle \widetilde{p}^s, \sum_{h \in \mathcal{H}} (\widetilde{x}_h^s - e_h^s) \right\rangle_C - \left\langle y^s, \sum_{h \in \mathcal{H}} \widetilde{b}_h \right\rangle_A = 0 \qquad \forall s \in \mathcal{S}. \tag{47}$$

Moreover,

$$\left\langle (\widetilde{\nu} - \mathbf{1}, \widetilde{p}^0), \left(Y \sum_{h \in \mathcal{H}} \widetilde{b}_h, \sum_{h \in \mathcal{H}} (\widetilde{x}_h^0 - e_h^0) \right) \right\rangle_{S+C} + \sum_{s \in \mathcal{S}} \left\langle \widetilde{p}^s, \sum_{h \in \mathcal{H}} (\widetilde{x}_h^s - e_h^s) \right\rangle_C = 0, \tag{48}$$

where $\mathbf{1} = (1, \ldots, 1) \in \mathbb{R}^S$.

(iv) *For any* $h \in \mathcal{H}$, $c \in \mathcal{C}$, *and* $s \in \mathcal{S}$,

$$0 \le \widetilde{x}_h^{sc} \le \sum_{s \in \mathcal{S}^0} \sum_{c \in \mathcal{C}} \sum_{h \in \mathcal{H}} \widetilde{x}_h^{sc} \le \sum_{s \in \mathcal{S}^0} \sum_{c \in \mathcal{C}} \sum_{h \in \mathcal{H}} e_h^{sc}. \tag{49}$$

5.2 Existence of Equilibrium

We need now to recall an important result usually called "boundary condition" in the general equilibrium literature - for a proof, see, for example, Werner, Lemma 2 in [34].

Proposition 4 *If the sequence* (p_n, q_n) *of strictly positive price vectors converges to* (p, q) *that is not strictly positive and such that, for some* $h \in H$, *either conditions below holds:*

1. $\langle p^0, e_h^0 \rangle_C > 0,$
2. $q \neq 0$ *and for any* $s \in \mathcal{S}$, $\langle p^s, e_h^s \rangle_C > 0,$

then,

$$\inf \left\{ \|x_{h,n}\| \in \mathbb{R}_+ : \left(x_{h,n}, b_{h,n}\right) \in \underset{B(p_n, q_n)}{argmax}\; u_h(x_{h,n}) \text{ for some } b_{h,n} \in \mathbb{R}^A \right\} \xrightarrow{n} +\infty.$$

Theorem 15 *Under Assumptions 1, 2, and 3, there exists an equilibrium vector for any financial economy with incomplete markets and nominal assets.*

Proof Let $\{\varepsilon_n\}_{n \in \mathbb{N}}$ be a sequence of positive real numbers such that

$$\lim_{n \to +\infty} \varepsilon_n = 0. \tag{50}$$

By Theorem 13, for all $n \in \mathbb{N}$, there exists $\left((\widetilde{x}_n, \widetilde{b}_n), (\widetilde{\nu}_n, \widetilde{p}_n^{\,0}), \widetilde{p}^{\mathbf{1},n}\right) \in B^*((\widetilde{\nu}_n, \widetilde{p}_n^0), \widetilde{p}^{\mathbf{1},n}) \times \Delta_0^{\varepsilon_n} \times \Delta^{\varepsilon_n}$ solution to QVI (38); that is, from Remark 2 one has:

- for all $h \in \mathcal{H}$, $(\widetilde{x}_{h,n}, \widetilde{b}_{h,n})$ is a solution to VI

$$\langle \nabla u_h(\widetilde{x}_{h,n}), x_{h,n} - \widetilde{x}_{h,n} \rangle_G \leq 0 \qquad \forall (x_{h,n}, b_{h,n}) \in B_h^*((\widetilde{\nu}_n, \widetilde{p}_n{}^0), \widetilde{p}_{\mathbf{1},n}) \tag{51}$$

- $(\widetilde{\nu}_n, \widetilde{p}_n{}^0)$ is a solution to VI

$$\left\langle \left(Y \sum_{h \in \mathcal{H}} \widetilde{b}_{h,n}, \sum_{h \in \mathcal{H}} (\widetilde{x}^{\,0}_{h,n} - e_h^0) \right), (\nu_n, p_n^0) - (\widetilde{\nu}_n, \widetilde{p}_n^0) \right\rangle_{S+C} \leq 0 \qquad \forall (\nu_n, p_n^0) \in \Delta_0^{\varepsilon_n} \tag{52}$$

- and for all $s \in \mathcal{S}$, $\widetilde{p}_n^{\,s}$ is a solution to VI

$$\left\langle \sum_{h \in \mathcal{H}} (\widetilde{x}^{\,s}_{h,n} - e_h^s), p_n^s - \widetilde{p}_n^{\,s} \right\rangle_C \leq 0 \qquad \forall p_n^s \in \Delta_C^{\varepsilon_n}. \tag{53}$$

Moreover, from Theorem 14, one has:

(i) for any $h \in \mathcal{H}$:

$$\max_{(x_{h,n},b_{h,n})\in B_h^*((\widetilde{v}_n,\widetilde{p}_n^0),\widetilde{p}_n^1)} u_h(x_{h,n}) = u_h(\widetilde{x}_{h,n}); \tag{54}$$

(ii) for any $h \in \mathcal{H}$,

$$\langle(\widetilde{v}_n, \widetilde{p}_n^0), (Y\widetilde{b}_h, \widetilde{x}_{h,n}^0 - e_h^0)\rangle_{S+C} = 0, \tag{55}$$

$$\langle \widetilde{p}_n{}^s, \widetilde{x}_{h,n}^s - e_h^s\rangle_C - \langle y^s, \widetilde{b}_{h,n}\rangle_A = 0 \qquad \forall s \in \mathcal{S}; \tag{56}$$

(iv)

$$0 \le \widetilde{x}_{h,n}^{sc} \le \sum_{s\in\mathcal{S}^0}\sum_{c\in\mathcal{C}}\sum_{h\in\mathcal{H}} \widetilde{x}_{h,n}^{sc} \le \sum_{s\in\mathcal{S}^0}\sum_{c\in\mathcal{C}}\sum_{h\in\mathcal{H}} e_h^{sc}. \tag{57}$$

We consider the sequence $\left\{\left((\widetilde{x}_n, \widetilde{b}_n), (\widetilde{v}_n, \widetilde{p}_n^0), \widetilde{p}_n^{\mathbf{1}}\right)\right\}_{n\in\mathbb{N}}$ and we prove that this sequence converges to an equilibrium. Since

$$\{(\widetilde{v}_n, \widetilde{p}_n^0)\}_{n\in\mathbb{N}} \subseteq \Delta_0, \qquad \{\widetilde{p}_n^s\}_{n\in\mathbb{N}} \subseteq \Delta_C, \qquad 0 \le \widetilde{x}_{h,n}^{sc} \le \sum_{s\in\mathcal{S}^0}\sum_{c\in\mathcal{C}}\sum_{h\in\mathcal{H}} e_h^{sc}$$

with Δ_0 and Δ_C compact sets, without loss of generality, for all $s \in \mathcal{S}^0$, $c \in \mathcal{C}$, and $h \in \mathcal{H}$, it follows that

$$\lim_{n\to+\infty}(\widetilde{v}_n, \widetilde{p}_n^0) = (\widetilde{v}, \widetilde{p}^0) \in \Delta_0, \qquad \lim_{n\to+\infty}\widetilde{p}_n^s = \widetilde{p}^s \in \Delta_C, \qquad \lim_{n\to+\infty}\widetilde{x}_{h,n}^{sc} = \widetilde{x}_h^{sc}. \tag{58}$$

From (56), we have $\langle Y, \widetilde{b}_{h,n}\rangle_A = \left(\langle \widetilde{p}_n{}^s, \widetilde{x}_{h,n}^s - e_h^s\rangle_C\right)_{s\in\mathcal{S}}$ and taking limits, we get

$$\langle Y, \lim_{n\to+\infty}\widetilde{b}_{h,n}\rangle_A = \left(\langle \widetilde{p}^s, \widetilde{x}_h^s - e_h^s\rangle_C\right)_{s\in\mathcal{S}}.$$

Since $rank\, Y = A$, for all $a \in \mathcal{A}$ and $h \in \mathcal{H}$, we can conclude that

$$\lim_{n\to+\infty}\widetilde{b}^a{}_{h,n} = \widetilde{b}_h^a.$$

Since B_h^* is a closed set-valued map, then $(\widetilde{x}_h, \widetilde{b}_h) \in B_h^*((\widetilde{v}, \widetilde{p}^0), \widetilde{p}_n^{\mathbf{1}})$.
Now, we prove that the limit point $\left((\widetilde{x}, \widetilde{b}), \left((\widetilde{v}, \widetilde{p}^0), \widetilde{p}_n^{\mathbf{1}}\right)\right)$ is an equilibrium vector.

Condition 1 of Definition 3
We show that

$$\left((\widetilde{\nu}, \widetilde{p}^{\,0}), \widetilde{p}^{\mathbf{1}}\right) >> 0 \text{ and } \widetilde{q} = \widetilde{\nu}Y \in Q. \tag{59}$$

To get the first result, we first show for any $h \in \mathcal{H}$, either condition below holds true:

(a) $\langle \widetilde{p}^{\,0}, e_h^0 \rangle_C > 0$,
(b) $\widetilde{q} = \widetilde{\nu}Y \neq 0_{\mathbb{R}^A}$ and for any $s \in \mathcal{S}$, $\langle \widetilde{p}^{\,s}, e_h^s \rangle_C > 0$.

Observe that:

(a) If $\widetilde{p}^{\,0c} \neq 0$ for some $c \in \mathcal{C}$, from Assumptions 2, we have $\langle \widetilde{p}^{\,0}, e_h^0 \rangle_C > 0$.
(b) If $\widetilde{p}^{\,0c} = 0$ for any $c \in \mathcal{C}$, since $(\widetilde{\nu}, \widetilde{p}^{\,0}) \in \Delta^0$, then there exists $s^* \in \mathcal{S}$ such that $\nu^{s^*} > 0$. Therefore,

$$\widetilde{\nu}Y = \left(\sum_{s \neq s^*} \overset{(\geq 0)}{\widetilde{\nu}^{\,s}} \cdot \overset{(>0)}{y^s}\right) + \overset{(>0)}{\widetilde{\nu}^{\,s^*}} \cdot \overset{(>0)}{y^{s^*}} > 0\,.$$

Moreover, since $\widetilde{p}^{\,s} \in \Delta_C$, then there exists $c^* \in \mathcal{C}$ such that $\widetilde{p}^{sc^*} > 0$ and from Assumptions 2, we have

$$\forall s \in \mathcal{S},\ \langle \widetilde{p}^s, e_h^s \rangle_C = \left(\sum_{c \neq c^*} \overset{(\geq 0)}{\widetilde{p}^{sc}} \cdot \overset{(>0)}{e_h^{sc}}\right) + \overset{(>0)}{\widetilde{p}^{sc^*}} \cdot \overset{(>0)}{e_h^{sc^*}} > 0.$$

By Proposition 4, it follows that $((\widetilde{\nu}, \widetilde{p}^0), \widetilde{p}^{\mathbf{1}})$ must be strictly positive; otherwise, $\{\widetilde{x}_{h,n}\}$ would be unbounded.

Define $\widetilde{q} := \widetilde{\nu}Y \in Q$. Since B^* is l.s.c., then for any $(x_h, b_h) \in B_h(\widetilde{p}, \widetilde{q})$, there exists a sequence $\{(x_{h,n}, b_{h,n})\}_{n \in \mathbb{N}}$ converging to (x_h, b_h) and such that $(x_{h,n}, b_{h,n}) \in B_h^*((\widetilde{\nu}_n, \widetilde{p}_n{}^0), \widetilde{p}_n^{\mathbf{1}})$. Then, from (54), one has that $u_h(x_{h,n}) \leq u_h(\widetilde{x}_{h,n})$. Hence, taking limits, it follows that $u_h(x_h) \leq u_h(\widetilde{x}_h)$. We can therefore conclude that, for any $h \in \mathcal{H}$:

$$\max_{(x_h, b_h) \in B_h(\widetilde{p}, \widetilde{q})} u_h(x_h) = u_h(\widetilde{x}_h)\,. \tag{60}$$

Condition 2 (a) of Definition 3
We have to prove that for all $s \in \mathcal{S}^0$ and $c \in \mathcal{C}$,

$$\widetilde{Z}^{sc} := \sum_{h \in \mathcal{H}} (\widetilde{x}_h^{\,sc} - e_h^{sc}) \leq 0. \tag{61}$$

We suppose that there exist $s^* \in \mathcal{S}^0$ and $c^* \in \mathcal{C}$ such that $\widetilde{Z}^{s^*c^*} > 0$; that is, from (58) there exists $k_1 \in \mathbb{N}$ such that, for all $n > k_1$,

$$\widetilde{Z}_n^{s^*c^*} = \textstyle\sum_{h \in \mathcal{H}} (\widetilde{x}_{h,n}^{\,s^*c^*} - e_h^{s^*c^*}) > 0. \tag{62}$$

Observe that, since $((\widetilde{\nu}, \widetilde{p}^0), \widetilde{p}^1) >> 0$, there exists $\widetilde{\varepsilon} = \min_{s\in\mathcal{S}^0 c\in\mathcal{C}}\{\widetilde{\nu}^s, \widetilde{p}^{sc}\} > 0$, then there exists $k_2 \in \mathbb{N}$ such that, for all $n > k_2$, $\widetilde{\nu}_n^s > \widetilde{\varepsilon}$ and $\widetilde{p}_n{}^{sc} > \widetilde{\varepsilon}$. Moreover, since $\{\varepsilon_n\}_{n\in\mathbb{N}}$ converges to zero, there exists $k_3 \in \mathbb{N}$ such that $\varepsilon_n < \widetilde{\varepsilon}$ for all $n > k_3$. Let $k := \max\{k_1, k_2, k_3\}$ and we fix $n > k$. For any $s \in \mathcal{S}^0$, such that $\widetilde{Z}^{sc} > 0$ for some $c \in \mathcal{C}$, define the sets

$$\mathcal{C}_s^- := \{c \in \mathcal{C} : \widetilde{Z}_n^{sc} \le 0\}, \qquad \mathcal{C}_s^+ := \{c \in \mathcal{C} : \widetilde{Z}_n^{sc} > 0\}.$$

We assume that there exists $s^* \in \mathcal{S}^0$ such that $\mathcal{C}_{s^*}^+ \neq \emptyset$. Then, fixed $s^* \in \mathcal{S}^0$, we need to analyze some cases.

1. $\mathcal{C}_{s^*}^- \neq \emptyset$.
   Take $((\widehat{\nu}_n, \widehat{p}_n{}^0), \widehat{p}_n^1) \in \Delta_0^{\varepsilon_n} \times \Delta^{\varepsilon_n}$ such that $\widehat{\nu}_n := \widetilde{\nu}_n$, $\widehat{p}_n^s := \widetilde{p}_n^s$ for all $s \neq s^*$ and

$$\widehat{p}_n^{s^*c} := \begin{cases} \widetilde{p}_n^{s^*c} + K & \text{if } \ c \in \mathcal{C}_{s^*}^+, \\ \widetilde{p}_n^{s^*c} - K\dfrac{|\mathcal{C}_{s^*}^+|}{|\mathcal{C}_{s^*}^-|} & \text{if } \ c \in \mathcal{C}_{s^*}^-, \end{cases}$$

   with $0 < K \le \dfrac{|\mathcal{C}_{s^*}^-|}{|\mathcal{C}_{s^*}^+|}(\widetilde{p}_n^{s^*c} - \varepsilon_n)$ for all $c \in \mathcal{C}$. One has

$$\left\langle \left(Y \sum_{h\in\mathcal{H}} \widetilde{b}_{h,n}, \sum_{h\in\mathcal{H}} (\widetilde{x}_{h,n}^0 - e_h^0) \right), (\widehat{\nu}_n, \widehat{p}_n^0) - (\widetilde{\nu}_n, \widetilde{p}_n^0) \right\rangle_{S+C} + \sum_{s\in\mathcal{S}} \left\langle \sum_{h\in\mathcal{H}} (\widetilde{x}_{h,n}^s - e_h^s), \widehat{p}_n^s - \widetilde{p}_n^s \right\rangle_C > 0,$$

   contradicting VI (42).
2. $\mathcal{C}_{s^*}^- = \emptyset$ and there exist $c, c' \in \mathcal{C}$ such that $\widetilde{Z}_n^{s^*c} \neq \widetilde{Z}_n^{s^*c'}$.
   Let $c^* \in \mathcal{C}$ be such that $\widetilde{Z}_n^{s^*c^*} := \max\left\{\widetilde{Z}_n^{s^*c} : c \in \mathcal{C}\right\} > 0$. Take $((\widehat{\nu}_n, \widehat{p}_n{}^0), \widehat{p}_{1,n}) \in \Delta_0^{\varepsilon_n} \times \Delta^{\varepsilon_n}$ such that $\widehat{\nu}_n := \widetilde{\nu}_n$, $\widehat{p}_n{}^s := \widetilde{p}_n{}^s$ for all $s \neq s^*$ and

$$\widehat{p}_n{}^{s^*c} := \begin{cases} \widetilde{p}_n^{s^*c^*} + M & \text{if } \ c = c^*, \\ \widetilde{p}_n^{s^*c} - \frac{M}{C-1} & \text{if } \ c \neq c^*, \end{cases}$$

   with $0 < M \le (C-1)(\widetilde{p}_n^{s^*c} - \varepsilon_n)$ for all $c \in \mathcal{C}$. One has

$$\left\langle \left(Y \sum_{h\in\mathcal{H}} \widetilde{b}_{h,n}, \sum_{h\in\mathcal{H}} (\widetilde{x}_{h,n}^0 - e_h^0) \right), (\widehat{\nu}_n, \widehat{p}_n^0) - (\widetilde{\nu}_n, \widetilde{p}_n^0) \right\rangle_{S+C} + \sum_{s\in\mathcal{S}} \left\langle \sum_{h\in\mathcal{H}} (\widetilde{x}_{h,n}^s - e_h^s), \widehat{p}_n^s - \widetilde{p}_n^s \right\rangle_C > 0,$$

   contradicting VI (42).

3. $\mathcal{C}_{s^*}^- = \emptyset$ and $\widetilde{Z}_n^{s^*c} = \gamma_n^{s^*} > 0$ for all $c \in \mathcal{C}$.

a. $s^* = 0$. From (46), there exists $s' \in \mathcal{S}$ such that $\langle y^{s'}, \sum_{h\in\mathcal{H}} \widetilde{b}_{h,n}\rangle_A < 0$. Take $(\widehat{\nu}_n, \widehat{p}\,_n^0) \in \Delta_0^{\varepsilon_n}$ such that

$$\widehat{\nu}^s = \begin{cases} \widetilde{\nu}^{s'} - D & \text{if} \quad s = s' \\ \\ \widetilde{\nu}^s & \text{if} \quad s \neq s' \end{cases}, \qquad \widehat{p}_n^{0c} = \widetilde{p}_n^{0c} + \frac{1}{C} D \ \ \forall c \in \mathcal{C}$$

with $0 < D < \widetilde{\nu}^{s'} - \varepsilon_n$. One has

$$\left\langle \left(Y \sum_{h\in\mathcal{H}} \widetilde{b}_{h,n}, \sum_{h\in\mathcal{H}} (\widetilde{x}\,^0_{h,n} - e_h^0) \right), (\widehat{\nu}_n, \widehat{p}\,_n^0) - (\widetilde{\nu}_n, \widetilde{p}\,_n^0) \right\rangle_{S+C} = -D \left\langle y^{s'}, \sum_{h\in\mathcal{H}} \widetilde{b}_{h,n} \right\rangle_A + \gamma_n^0 D > 0,$$

contradicting VI (52).

b. $s^* \neq 0$. From (47), one has that $\langle y^{s^*}, \sum_{h\in\mathcal{H}} \widetilde{b}_{h,n}\rangle_A > 0$. Take $(\widehat{\nu}_n, \widehat{p}\,_n^0) \in \Delta_0^{\varepsilon_n}$ such that

$$\widehat{\nu}_n{}^s := \begin{cases} \widetilde{\nu}_n^{s^*} + B & \text{if} \quad s = s^* \\ \\ \widetilde{\nu}_n{}^s & \text{if} \quad s \neq s^* \end{cases}, \qquad \widehat{p}_n{}^{0c} := \widetilde{p}_n^{oc} - \frac{1}{C} B \ \ \forall c \in \mathcal{C}$$

with $0 < B \leq C(\widetilde{p}_n^{0c} - \varepsilon_n)$ for all $c \in \mathcal{C}$. One has

$$\left\langle \left(Y \sum_{h\in\mathcal{H}} \widetilde{b}_{h,n}, \sum_{h\in\mathcal{H}} (\widetilde{x}\,^0_{h,n} - e_h^0) \right), (\widehat{\nu}_n, \widehat{p}\,_n^0) - (\widetilde{\nu}_n, \widetilde{p}\,_n^0) \right\rangle_{S+C} > 0,$$

contradicting VI (52).

Condition 2 (b) of Definition 3

From Eq. (47), we get

$$\left\langle \widetilde{\nu}^s y^s, \sum_{h\in\mathcal{H}} \widetilde{b}_h \right\rangle_A = \left\langle \widetilde{\nu}^s \widetilde{p}\,^s, \sum_{h\in\mathcal{H}} (\widetilde{x}_h^s - e_h^s) \right\rangle_C \leq 0 \qquad \forall s \in \mathcal{S},$$

where the inequality follows from (61) and the fact that $\widetilde{\nu} \in \mathbb{R}^S_{++}$ and $\widetilde{p} \in \mathbb{R}^G_{++}$. Therefore,

$$\left\langle \widetilde{\nu} Y, \sum_{h\in\mathcal{H}} \widetilde{b}_h \right\rangle_A \leq 0, \tag{63}$$

and from (61), we get

$$\left\langle \widetilde{p}^0, \sum_{h \in \mathcal{H}} (\widetilde{x}_h^0 - e_h^0) \right\rangle_C \leq 0. \tag{64}$$

From (63), (64), and (46), we get

$$\left\langle \widetilde{\nu} Y, \sum_{h \in \mathcal{H}} \widetilde{b}_h \right\rangle_A = 0. \tag{65}$$

From (45), (61), and the fact that for any $s \in \mathcal{S}^0$, $\widetilde{p}^s \in \mathbb{R}^C_{++}$, we have

$$Y \sum_{h \in \mathcal{H}} \widetilde{b}_h = \left(\langle \widetilde{x}_h^s - e_h^s, \widetilde{p}^s \rangle_C \right)_{s \in \mathcal{S}} \leq 0. \tag{66}$$

Then, from (65), (66), and the fact that $\nu >> 0$, we have $Y \sum_{h \in \mathcal{H}} \widetilde{b}_h = 0$. Since Y has full rank A from Assumption 3, we obtain then for all $a \in \mathcal{A}$:

$$\sum_{h \in \mathcal{H}} \widetilde{b}_h^a = 0. \tag{67}$$

Hence, from (60), (61), and (67), we can conclude that $(\widetilde{x}, \widetilde{b}, \widetilde{p}, \widetilde{q})$ is an equilibrium vector.

□

Remark 3 Observe that since good prices are strictly positive, then (61) holds in the form of equalities, i.e., for any $s \in \mathcal{S}^0$ and any $c \in \mathcal{C}$

$$\sum_{h \in \mathcal{H}} \left(\widetilde{x}_h^{sc} - e_h^{sc} \right) = 0\,.$$

Indeed, from (ii) in Theorem 14 and (67), for any $s \in \mathcal{S}^0$ and any $c \in \mathcal{C}$, we have that

$$\widetilde{p}^{sc} \sum_{h \in \mathcal{H}} \left(\widetilde{x}_h^{sc} - e_h^{sc} \right) = 0\,. \tag{68}$$

Since $\widetilde{p} >> 0$ and from (61), $\sum_{h \in \mathcal{H}} \left(\widetilde{x}_h^{sc} - e_h^{sc} \right) \leq 0$, we get the desired result.

6 Model with Numeraire Assets

In this section, we deal with the existence of an equilibrium vector for a financial economy with incomplete markets and numeraire assets.

In the previous section, following a more standard variational inequality approach, we studied the case of differentiable and concave utility functions. Consistently with the more demanding assumptions usually made in the economic literature, in the model presented below, we assume instead that those functions are continuous and semistrictly quasi-concave. Moreover, the numeraire asset case is both more economically appealing and more mathematically general than the nominal one. Indeed, in the latter case, while assets pay in units of "money," the model itself describes a barter economy in which there is no role for what is usually defined as money. In the numeraire case, instead, in a natural and simple manner, yields are denominated in units of a specific good, which is the unit of measure of the value, the so-called numeraire good. Moreover, existence of equilibria in the numeraire case implies existence in the nominal case, but not vice versa. Indeed, typically in the space of economies, while the cardinality of the equilibrium set is finite in the numeraire case, in the nominal one, that cardinality is infinite.

For this model, we make the following Assumptions:

Assumption 4

1. *For any* $h \in \mathcal{H}$, $e_h >> 0$.

Assumption 5

1. *For any* $h \in \mathcal{H}$:

   a. *the utility function* u_h *is continuous and quasi-concave;*
   b. u_h *is strictly increasing in the numeraire good* sC, *for every* $s \in \mathcal{S}^0$.

2. *For every* $s \in \mathcal{S}^0$ *and* $c \in \mathcal{C}$, *there exists* $h' \in \mathcal{H}$ *such that* $u_{h'}$ *is strictly increasing in* sc.

Assumption 6

1. $\operatorname{rank} Y = A < S$.
2. $\mathbf{1} := (1, \ldots, 1) \in col \operatorname{span} Y$, *i.e., for any* $h \in \mathcal{H}$, $\exists\, b_h^* \in \mathbb{R}^A$ *such that* $Yb_h^* = \mathbf{1}$.

Assumption 5.1b says that the chosen numeraire good is "highly evaluated" by each household. It is used to show that if household's maximization problem has a solution, then the price of the numeraire good is strictly positive and budget inequalities are satisfied as equalities, i.e., Walras' laws do hold true —see Proposition 5 below.

Assumption 5.2 allows to get strictly positive prices of each good.

Assumption 6.1 means that markets are incomplete.

Assumption 6.2 means that households can store the numeraire good without any change in its quantity in the second period. In other words, if gold is the numeraire good, you can keep it as it is in a safe from today until tomorrow.

As shown in the following Proposition, solutions to the household's maximization problem satisfy some properties we are going to use in the remainder of the paper.

Proposition 5 *Let Assumptions 5.1b and 6.1 be satisfied. If* $(\widetilde{x}_h, \widetilde{b}_h)$ *is a solution to household h's maximization problem at prices* $\widetilde{p}, \widetilde{q}$*, then*

1. $(\widetilde{p}^{sC})_{s\in\mathcal{S}^0} >> 0.$
2. $\widetilde{q} \in Q$*, where*

$$Q := \left\{q \in \mathbb{R}^A : \text{ there is no} b_h \in \mathbb{R}^A \text{such that} \begin{bmatrix} -q \\ P^{\cdot}Y \end{bmatrix} b_h > 0\right\}$$

$$= \left\{q \in \mathbb{R}^A : \exists \nu \in \mathbb{R}^S \text{such that } \nu >> 0 \text{ and } q = \nu P^{\cdot} Y\right\}$$

and $P^{\cdot} := (p^{sj})_{s,j\in\mathcal{S}}$ *is an* $S \times S$ *matrix such that* $p^{sj} = 0$ *for all* $s \neq j$ *and* $p^{ss} = p^{sC}$ *for* $s \in \mathcal{S}$.

3. *For any* $h \in \mathcal{H}$, $\langle \widetilde{p}^0, \widetilde{x}_h^0 - e_h^0\rangle_C + \langle \widetilde{q}, \widetilde{b}_h\rangle_A = 0\,,$

$$\langle \widetilde{p}^s, \widetilde{x}_h^s - e_h^s\rangle_C - \widetilde{p}^{sC}\langle y^s, \widetilde{b}_h\rangle_A = 0 \qquad \forall s \in \mathcal{S}\,.$$

4. *The following so-called* $S + 1$ *Walras laws hold true*

$$\left\langle \widetilde{p}^0, \sum_{h\in\mathcal{H}} (\widetilde{x}_h^0 - e_h^0)\right\rangle_C + \left\langle \widetilde{q}, \sum_{h\in\mathcal{H}} \widetilde{b}_h\right\rangle_A = 0,$$

$$\left\langle \widetilde{p}^s, \sum_{h\in\mathcal{H}} (\widetilde{x}_h^s - e_h^s)\right\rangle_C - \widetilde{p}^{sC}\left\langle y^s, \sum_{h\in\mathcal{H}} \widetilde{b}_h\right\rangle_A = 0 \qquad \forall s \in \mathcal{S}\,.$$

From claim 1 of Proposition 5, without loss of generality, we can consider the same price sets, introduced in the previous model, see p. 17. In particular here, we use Δ_s instead of Δ_C on page 17.

It is useful to consider the budget constraints set as the set-valued map $B_h : \Delta_0 \times \Delta \rightrightarrows \mathbb{R}_+^G \times \mathbb{R}^A$ such that

$$B_h((\nu, p^0), p^1) := \{(x_h, b_h) \in \mathbb{R}_+^G \times \mathbb{R}^A : \langle p^0, x_h^0 - e_h^0\rangle_C + \langle \nu P^{\cdot} Y, b_h\rangle_A \leq 0,$$

$$\langle p^s, x_h^s - e_h^s\rangle_C - p^{sC}\langle y^s, b_h\rangle_A \leq 0 \;\forall s \in \mathcal{S}\}\,.$$

Moreover, we define $B((\nu, p^0), p^1) := \prod_{h\in\mathcal{H}} B_h((\nu, p^0), p^1)$.

Now, we introduce a suitable variational inequality problem, which allows us to prove the existence of the equilibrium. To this aim, we consider a non-zero lower bound on prices: for any $n \in \mathbb{N}$ such that $n \geq C$, we define the prices sets

$$\Delta_0^n := \left\{(\nu, p^0) \in \Delta_0 : p^0 \geq \frac{1}{n}\mathbf{1},\ \nu \geq \frac{1}{n}\mathbf{1}\right\}, \quad \Delta_s^n := \left\{p^s \in \Delta_s : p^s \geq \frac{1}{n}\mathbf{1}\right\},$$

and $\Delta^n := \prod_{s\in\mathcal{S}} \Delta^n_s$. To relate the consumer h's maximization problem with a variational problem, we use the same operator introduced in Sect. 4, $\mathcal{G}_h : \mathbb{R}^G \rightrightarrows \mathbb{R}^G$ such that

$$\forall x_h \in \mathbb{R}^G \qquad \mathcal{G}_h(x_h) := conv\Big(N_h^{>}(x_h) \cap S(0,1)\Big).$$

We introduce the following Generalized Quasi-Variational Inequality GQVI$_n$:

"Find $\Big((\widetilde{x}_n, \widetilde{b}_n), (\widetilde{\nu}_n, \widetilde{p}^{\,0}_n), \widetilde{p}^{\,\mathbf{1}}_n\Big) \in B((\widetilde{\nu}_n, \widetilde{p}^{\,0}_n), \widetilde{p}^{\,\mathbf{1}}_n) \times \Delta^n_0 \times \Delta^n$ such that

$$\text{there exists } g_n = (g_{h,n})_{h\in\mathcal{H}} \in \prod_{h\in\mathcal{H}} \mathcal{G}_h(\widetilde{x}_{h,n}) \text{ and}$$

$$\langle -g_n, x_n - \widetilde{x}_n\rangle_{GH} + \langle (\widetilde{P}^{\cdot}_n Y \sum_{h\in\mathcal{H}} \widetilde{b}_{h,n}, \sum_{h\in\mathcal{H}} (\widetilde{x}^{\,0}_{h,n} - e^0_h)), (\nu_n, p^0_n) - (\widetilde{\nu}_n, \widetilde{p}^{\,0}_n)\rangle_{S+C}$$

$$+\langle \sum_{h\in\mathcal{H}} (\widetilde{x}_{h,n} - e_h), p^{\mathbf{1}}_n - \widetilde{p}^{\,\mathbf{1}}_n\rangle_{CS} \leq 0$$

$$\forall\Big((x_n, b_n), (\nu_n, p^0_n), p^{\mathbf{1}}_n\Big) \in B((\widetilde{\nu}_n, \widetilde{p}^{\,0}_n), \widetilde{p}^{\,\mathbf{1}}_n) \times \Delta^n_0 \times \Delta^n \text{."} \tag{69}$$

Under our assumptions, the set-valued maps $B_h : \Delta_0 \times \Delta \rightrightarrows \mathbb{R}^G_+ \times \mathbb{R}^A$ satisfy the following nice properties.

Proposition 6 *Let Assumptions 4, 5.2, and 6 hold true. For any $n \in \mathbb{N}$ with $n \geq C$ and for any $h \in \mathcal{H}$, the set-valued map B_h is lower semicontinuous and closed with nonempty, convex, compact values in $\Delta^n_0 \times \Delta^n$. Moreover B_h is lower semicontinuous on $\Delta^+ := \Big\{\Big((\nu, p^0), p^1\Big) \in \Delta_0 \times \Delta : \big(p^{sC}\big)_{s\in\mathcal{S}^0} > 0\Big\}$.*

Proof See pp. 432–433 in [19]. □

Remark 4 Also in this case, it is important to observe that $\Big((\widetilde{x}_n, \widetilde{b}_n), (\widetilde{\nu}_n, \widetilde{p}^{\,0}_n), \widetilde{p}^{\,\mathbf{1}}_n\Big)$ is a solution to GQVI$_n$ (69) if and only if

- for all $h \in \mathcal{H}$, $(\widetilde{x}_{h,n}, \widetilde{b}_{h,n})$ is a solution to GVI$_n$

$$\langle -g_{h,n}, x_{h,n} - \widetilde{x}_{h,n}\rangle_G \leq 0 \qquad \forall (x_{h,n}, b_{h,n}) \in B_h((\widetilde{\nu}_n, \widetilde{p}_n{}^0), \widetilde{p}^{\,\mathbf{1}}_n), \tag{70}$$

  where $g_{h,n} \in \mathcal{G}_h(\widetilde{x}_{h,n})$
- $(\widetilde{\nu}_n, \widetilde{p}_n{}^0)$ is a solution to VI$_n$

$$\left\langle\left(\widetilde{P}_{n}^{\cdot}Y\sum_{h\in\mathcal{H}}\widetilde{b}_{h,n},\sum_{h\in\mathcal{H}}(\widetilde{x}_{h,n}^{0}-e_{h}^{0})\right),(\nu_{n},p_{n}^{0})-(\widetilde{\nu}_{n},\widetilde{p}_{n}^{0})\right\rangle_{S+C}\leq 0$$

$$\forall(\nu_{n},p_{n}^{0})\in\Delta_{0}^{n}, \tag{71}$$

- for all $s\in\mathcal{S}$, $\widetilde{p}_{n}^{s}$ is a solution to VI_n

$$\left\langle\sum_{h\in\mathcal{H}}(\widetilde{x}_{h,n}^{s}-e_{h}^{s}),p_{n}^{s}-\widetilde{p}_{n}^{s}\right\rangle_{C}\leq 0\qquad\forall p_{n}^{s}\in\Delta_{s}^{n}. \tag{72}$$

For any $n\in\mathbb{N}$, with $n\geq C$, the solution to GQVI_n (69) verifies following properties.

Theorem 16 *Let Assumptions 5 be satisfied. For any $n\in\mathbb{N}$ such that $n\geq C$, let $\left((\widetilde{x}_n,\widetilde{b}_n),(\widetilde{\nu}_n,\widetilde{p}_{n}^{0}),\widetilde{p}_{n}^{1}\right)$ be a solution to GQVI_n (69). Then, one has:*

(i) for any $h\in\mathcal{H}$, $(\widetilde{x}_{h,n},\widetilde{b}_{h,n})$ is a solution to maximization problem

$$\max_{(x_{h,n},b_{h,n})\in B_{h}((\widetilde{\nu}_{n},\widetilde{p}_{n}^{0}),\widetilde{p}_{n}^{1})}u_{h}(x_{h,n}); \tag{73}$$

(ii) for any $h\in\mathcal{H}$,

$$\langle\widetilde{p}_{n}^{0},\widetilde{x}_{h,n}^{0}-e_{h}^{0}\rangle_{C}+\langle\widetilde{\nu}_{n}\widetilde{P}_{n}^{\cdot}Y,\widetilde{b}_{h,n}\rangle_{A}=0, \tag{74}$$

$$(\langle\widetilde{p}_{n}^{s},\widetilde{x}_{h,n}^{s}-e_{h}^{s}\rangle_{C})_{s\in\mathcal{S}}=\widetilde{P}_{n}^{\cdot}Y\widetilde{b}_{h,n}; \tag{75}$$

(iii)

$$\left\langle\widetilde{p}_{n}^{0},\sum_{h\in\mathcal{H}}(\widetilde{x}_{h,n}^{0}-e_{h}^{0})\right\rangle_{C}+\left\langle\widetilde{\nu}_{n}\widetilde{P}_{n}^{\cdot}Y,\sum_{h\in\mathcal{H}}\widetilde{b}_{h,n}\right\rangle_{A}=0, \tag{76}$$

$$\left(\left\langle\widetilde{p}_{n}^{s},\sum_{h\in\mathcal{H}}(\widetilde{x}_{h,n}^{s}-e_{h}^{s})\right\rangle_{C}\right)_{s\in\mathcal{S}}=\widetilde{P}_{n}^{\cdot}Y\sum_{h\in\mathcal{H}}\widetilde{b}_{h,n}; \tag{77}$$

(iv) for any $h\in\mathcal{H}$, $c\in\mathcal{C}$ and $s\in\mathcal{S}^0$,

$$0\leq\widetilde{x}_{h,n}^{sc}\leq\sum_{s\in\mathcal{S}^{0}}\sum_{c\in\mathcal{C}}\sum_{h\in\mathcal{H}}e_{h}^{sc}. \tag{78}$$

Proof See pp. 436–437 in [19]. □

Proposition 7 *Let Assumptions 4, 5, and 6 be satisfied. Let* $\left\{\left((\widetilde{x}_n, \widetilde{b}_n), (\widetilde{\nu}_n, \widetilde{p}^{\,0}_n), \widetilde{p}^{\,\mathbf{1}}_n\right)\right\}_{n\in\mathbb{N}}$ *be the sequence such that, for all* $n \in \mathbb{N}$ *with* $n \geq C$, $\left((\widetilde{x}_n, \widetilde{b}_n), (\widetilde{\nu}_n, \widetilde{p}^{\,0}_n), \widetilde{p}^{\,\mathbf{1}}_n\right)$ *is a solution to* GQVI$_n$ (69). *Then,*

$$\lim_{n\to+\infty}\left((\widetilde{x}_n, \widetilde{b}_n), (\widetilde{\nu}_n, \widetilde{p}^{\,0}_n), \widetilde{p}^{\,\mathbf{1}}_n\right) = \left((\widetilde{x}, \widetilde{b}), (\widetilde{\nu}, \widetilde{p}^{\,0}), \widetilde{p}^{\,\mathbf{1}}\right)$$

with $\widetilde{p}^{\,0} >> 0$ *and* $\widetilde{p}^{\,\mathbf{1}} >> 0$.

Proof See pp. 337–341 in [19]. □

Theorem 17 *Let* (e, Y, u) *be a financial economy with incomplete markets and numeraire assets, which satisfies all Assumptions 4, 5, and 6. Then,* $(\widetilde{x}, \widetilde{b}, \widetilde{p}, \widetilde{q} = \widetilde{\nu}\widetilde{P}^{\cdot}Y)$ *is an equilibrium associated with the economy* (e, Y, u).

Proof See pp. 441–444 in [19]. □

7 Model with Restricted Participation

In a model with restricted participation, households are allowed to choose portfolios in a personalized subset B_h of $\mathbb{R}^A$; B_h is the financial constrained set of household h; define $B = \prod_{h\in\mathcal{H}} B_h$.

An economy in a financial economy model with numeraire assets and restricted participation is an element $\Sigma := (e, u, Y, B) \in \mathbb{R}^{GH}_{++} \times \mathcal{U} \times \mathcal{M}_{S,A} \times \mathcal{B}$, where $\mathcal{M}_{S,A}$ is the set of $S \times A$ dimensional matrices, $\mathcal{U}$ is the set of functions $u_h : \mathbb{R}^G \to \mathbb{R}$, and $\mathcal{B}$ is the family of all financial constrained sets of households. Each household maximizes his or her utility under the constraints that in period 0 expenditure for goods and assets is smaller than the value of wealth in that period and, similarly, in each state in the future, expenditure for consumption is smaller than wealth increased by the value of the assets yields. For any $h \in \mathcal{H}$, we define the budget set of h at prices $(q, p^0, p^{\mathbf{1}})$ as follows:

$$\Gamma_h(q, p^0, p^{\mathbf{1}}) := \{(x_h, b_h) \in \mathbb{R}^G_+ \times B_h : \langle p^0, x^0_h - e^0_h\rangle_C + \langle q, b_h\rangle_A \leq 0,$$

$$\langle p^s, x^s_h - e^s_h\rangle_C - p^{sC}\langle y^s, b_h\rangle_A \leq 0 \quad \forall s \in \mathcal{S}\}$$

and $\Gamma(q, p^0, p^{\mathbf{1}}) := \prod_{h\in\mathcal{H}} \Gamma(q, p^0, p^{\mathbf{1}})$.

Household h choice variables are his or her consumption vector $x_h \in \mathbb{R}^G$ and his or her constrained portfolio $b_h \in B_h$. We then say that a consumption, portfolio holding, commodity, and asset price vector is an equilibrium vector for the economy Σ if at those prices, households maximize their utility functions and market clears, i.e., commodities demand is smaller than or equal to commodities supply and assets demand is equal to zero.

The formal definition of equilibrium is presented below.

Definition 3 The vector $(\widetilde{x}, \widetilde{b}, \widetilde{q}, \widetilde{p}) \in \mathbb{R}^{GH} \times \mathbb{R}^{AH} \times \mathbb{R}^{A} \times \mathbb{R}^{G}_{+}$ is an *equilibrium* vector for the economy Σ if

1. for any $h \in \mathcal{H}$,

$$\max\ u_h\ (x_h) = u_h(\widetilde{x}_h)$$

$$\text{s.t. } (x_h, b_h) \in \Gamma_h(\widetilde{q}, \widetilde{p}^0, \widetilde{p}^1); \tag{79}$$

2. for any $s \in \mathcal{S}^0$ and $c \in \mathcal{C}$,

$$\sum_{h \in \mathcal{H}} \widetilde{x}_h^{sc} \leq \sum_{h \in \mathcal{H}} e_h^{sc} \quad \text{if } \widetilde{p}^{\,sc} = 0,$$

$$\sum_{h \in \mathcal{H}} \widetilde{x}_h^{sc} = \sum_{h \in \mathcal{H}} e_h^{sc} \quad \text{if } \widetilde{p}^{\,sc} > 0;$$

3. for any $a \in \mathcal{A}$,

$$\sum_{h \in \mathcal{H}} \widetilde{b}_h^a = 0.$$

The description of the set of no free lunch good prices and no arbitrage assets prices is a convenient preliminary step in the process of proving existence of equilibrium prices: prices outside that set cannot be equilibrium prices. In the case of unrestricted financial participation and numeraire asset, the set of no-arbitrage asset prices[3] for household h is given by[4]

$$\begin{aligned} Q^u(P^{\cdot}, Y) &:= \left\{ q \in \mathbb{R}^A : \text{ there is no } b_h \in \mathbb{R}^A \text{such that} \begin{bmatrix} -q \\ P^{\cdot} Y \end{bmatrix} b_h > 0 \right\} \\ &= \left\{ q \in \mathbb{R}^A : \ \forall b_h \in \mathbb{R}^A \text{such that} P^{\cdot} Y b_h > 0 \text{ we have } \langle q, b_h \rangle_A > 0 \right\}. \end{aligned}$$

By using, a form of the Alternative Lemma (for details, see Lemma 14, page 297, in [33]), one has that

[3] In the symbol Q^u, the superscript u stays for “unrestricted.”

[4] For vectors $y, z \in \mathbb{R}^n$, $y \geq z$ means that for $i = 1, \ldots, n$, $y_i \geq z_i$; $y >> z$ means that for $i = 1, \ldots, n$, $y_i > z_i$, and $y > z$ means that $y \geq z$ but $y \neq z$.

$$Q^u(P^{\cdot}, Y) = \left\{q \in \mathbb{R}^A : \exists \nu \in \mathbb{R}^S_{++} \text{such that } q = \nu P^{\cdot} Y\right\}.$$

Observe that if $\left(p^{sC}\right)_{s\in\mathcal{S}} >> 0$, then

$$Q^u\left(P^{\cdot}, Y\right) = Q^u(Y) := \left\{q \in \mathbb{R}^A : \exists \nu \in \mathbb{R}^S_{++} \text{such that } q = \nu Y\right\}.$$

In the case of restricted participation, it may be that there is $b_h^* \in \mathbb{R}^A$ such that $\begin{bmatrix} -q \\ P^{\cdot}Y \end{bmatrix} b_h^* > 0$, but if B_h is bounded in the direction of b_h^*, then household h is not allowed to demand an unbounded amount of that portfolio. Therefore, in the case of presence of financial restriction, for given $\left(p^{sC}\right)_{s\in\mathcal{S}} \in \mathbb{R}^S_+$, $Y \in \mathcal{M}_{S,A}$, and $B \in \mathcal{B}$, we define the set[5] of no-arbitrage asset prices for household h as

$$Q_h(P^{\cdot}, Y, B_h) := \left\{q \in \mathbb{R}^A : \text{ there is no } b_h \in \text{rec } B_h \text{ such that} \begin{bmatrix} -q \\ P^{\cdot}Y \end{bmatrix} b_h > 0\right\}$$

$$= \left\{q \in \mathbb{R}^A : \forall b_h \in \text{rec } B_h \text{such that } P^{\cdot}Yb_h > 0 \text{ we have } \langle q, b_h\rangle_A > 0\right\},$$

and the set of no-arbitrage asset prices as

$$Q(P^{\cdot}, Y, B) := \bigcap_{h\in\mathcal{H}} Q_h(P^{\cdot}, Y, B_h)$$

$$= \left\{q \in \mathbb{R}^A : \forall b \in \cup_{h\in\mathcal{H}} \text{rec } B_h \text{such that } P^{\cdot}Yb > 0 \text{ we have } \langle q, b\rangle_A > 0\right\}.$$

From an economic viewpoint, prices in Q_h are such that if there exists a portfolio b_h that gives a positive return in some state and non-negative return in each state tomorrow, i.e., such that $P^{\cdot}Yb_h > 0$, and that can be bought in an unbounded amount by household h, i.e., $b_h \in \text{rec } B_h$, then that portfolio must cost a positive amount today, i.e., $\langle q, b_h\rangle_A > 0$. Moreover, define

$$Q_h(Y, B_h) := \left\{q \in \mathbb{R}^A : \text{ there is no } b_h \in \text{rec } B_h \text{ such that} \begin{bmatrix} -q \\ Y \end{bmatrix} b_h > 0\right\},$$

$$Q(Y, B) := \bigcap_{h\in\mathcal{H}} Q_h(Y, B_h).$$

[5] rec B_h is the recession cone of B_h, which is defined as follows $\text{rec}B_h = \left\{y \in \mathbb{R}^A : \forall x^0 \in B_h, \forall\lambda \geq 0, x^0 + \lambda y \in B_h\right\}$.

Remark 5 For any $Y \in \mathcal{M}_{S,A}$, $B \in \mathcal{B}$, and $\left(p^{sC}\right)_{s \in \mathcal{S}} \in \mathbb{R}^{S}_{++}$, one has that $Q_h(Y, B_h) = Q_h(P^{\cdot}, Y, B_h)$.

From now on, we make the following Assumptions.

Assumption 7 *For any* $h \in \mathcal{H}$, $e_h >> 0$.

Assumption 8 *For any* $h \in \mathcal{H}$, *the utility function* u_h *is*

1. *continuous and quasi-concave;*
2. *strictly increasing in the numeraire good* sC, *for every* $s \in \mathcal{S}$, *i.e.,*

$$\forall \widehat{x}_h, \widehat{\widehat{x}}_h \in \mathbb{R}^G_+ : \widehat{x}_h \geq \widehat{\widehat{x}}_h \;\; with \;\; \widehat{x}^{sC}_h > \widehat{\widehat{x}}^{sC}_h \quad \Rightarrow \quad u_h(\widehat{x}_h) > u_h(\widehat{\widehat{x}}_h);$$

3. *locally non-satiated in state* 0, *i.e.,*

$$\forall x_h = (x^0_h, x^1_h, \ldots, x^S_h) \in \mathbb{R}^G_+ \;\; and \;\; \forall \varepsilon > 0, \;\; \exists \widehat{x}_h = (\widehat{x}^0_h, x^1_h, \ldots, x^S_h) \in \mathbb{R}^G_+$$

$$such \; that \;\; \|\widehat{x}^0_h - x^0_h\| < \varepsilon \; and \;\; u_h(\widehat{x}_h) > u_h(x_h)\,.$$

Assumption 9 *For every* $s \in \mathcal{S}^0$ *and* $c \in \mathcal{C}$, $\exists h' \in \mathcal{H}$ *such that* $u_{h'}$ *is strictly increasing in* sc.

Assumption 10 *For any* $h \in \mathcal{H}$,

1. B_h *is a convex and closed subset of* $\mathbb{R}^A$ *and* $0_A \in B_h$*;*
2. $\mathrm{Ker}Y \cap rec\ B_h = \{0_A\}$*;*
3. *for any* $p \in \mathbb{R}^G_+$ *and for any* $q \in \mathrm{Cl}(Q_h(P^{\cdot}, Y, B)) \setminus \{0_A\}$, *there exists* $b_h \in B_h$ *such that* $\langle -q, b_h \rangle_A > 0$.

Assumption 7 is a survival assumption on the commodity side of the economy: it helps ensuring households are able to buy, and consume, some good in each state of the world.

Assumption 8.1 is relatively general and standard in the general equilibrium literature.

Assumption 8.2 is based on the fact that, by construction of the model, households agreed upon choosing the numeraire good as the unit of measure of asset yields and therefore "they strongly like that good."

Assumption 8.3 simply says that households care about consumption in period zero.

Assumption 9 stresses the fact that each good is appreciated at least by one household.

Assumption 10.1 is quite general and implies that households are allowed to stay out of the financial market.

Assumption 10.2 is crucial in several steps in the proofs below and it is implied by any of the following conditions: B_h is bounded (which implies that rec $B_h = \{0_A\}$); there are no redundant assets, i.e., rank $Y = A$ (which implies that $\mathrm{Ker}Y = \{0_A\}$).

Assumption 10.3 is a survival assumption on the financial side of the economy: it ensures that even if the available endowment at time zero has no value, then there exists an admissible portfolio that generates positive wealth in state zero itself.

Several reasonable conditions are indeed sufficient for Assumption 10.3 (see Proposition 2, page 776, in [3]). For example, it is enough that for any households there is a lower bond on some asset demand, or the origin of $\mathbb{R}^A$ is an interior point of the portfolio set.

We now define $\rho := \sum_{s\in\mathcal{S}}\sum_{a\in\mathcal{A}} y^{sa}$. The proposition below says that there is no loss of generality in assuming $\rho \geq 0$, a condition that is crucial in the arguments below.

Let $\mathcal{U}^*$ be the set of utility functions satisfying Assumptions 8 and 9; let $\mathcal{F}$[6] be the family of pairs $(Y, B) \in \mathcal{M}_{S,A} \times \mathcal{B}$ satisfying Assumptions 10 and define $\mathcal{E} = \mathbb{R}^{GH}_{++} \times \mathcal{U}^* \times \mathcal{F}$ to be the set of economies satisfying all our maintained assumptions.

Proposition 8 *For any* $\alpha \in \mathbb{R}\setminus\{0\}$,

1. $\Sigma = (e, u, Y, B) \in \mathcal{E} \Leftrightarrow \Sigma_\alpha := \left(e, u, \alpha Y, \left(\frac{B}{\alpha}\right)\right) \in \mathcal{E}$;
2. $(\widetilde{x}, \widetilde{b}, \widetilde{q}, \widetilde{p})$ *is an equilibrium for* Σ *if and only if* $\left(\widetilde{x}, \frac{\widetilde{b}}{\alpha}, \alpha\widetilde{q}, \widetilde{p}\right)$ *is an equilibrium for* Σ_α.

Proof 1. Economy Σ satisfies Assumptions 7, 8, 9, and 10.1 and 10.2 if and only if Σ_α satisfies Assumptions 7, 8, and 9 and 10.1 and 10.2. About Assumption 10.3, observe what follows. We assume that economy Σ satisfies 10.3. Taken $\widehat{q} \in \mathrm{Cl}\left(Q_h(P^\cdot, \alpha Y, \frac{B_h}{\alpha})\right)\setminus\{0_A\}$, we have that $\frac{\widehat{q}}{\alpha} \in \mathrm{Cl}\,(Q_h(P^\cdot, Y, B_h))\setminus\{0_A\}$. Then, by Assumption 10.3, there exists $b_h \in B_h$ such that $\langle\frac{\widehat{q}}{\alpha}, b_h\rangle_A < 0$, that is equivalent to have there exists $\widehat{b}_h = \frac{b_h}{\alpha} \in \frac{B_h}{\alpha}$ such that $\langle\widehat{q}, \widehat{b}_h\rangle_A < 0$, as desired. The proof of the opposite implication is symmetric to the above one.

2. Since $(\widetilde{x}_h, \widetilde{b}_h) \in \Gamma_h(\widetilde{q}, \widetilde{p}^0, \widetilde{p}^1)$ if and only if $\left(\widetilde{x}_h, \frac{\widetilde{b}_h}{\alpha}\right) \in \Gamma_h(\alpha\widetilde{q}, \widetilde{p}^0, \widetilde{p}^1)$, then the desired result holds true. □

Remark 6 Thanks to Proposition 8, in order to prove existence of equilibria, we can assume that $\rho \geq 0$. Indeed, let $\Sigma = (e, Y, B, u)$ be an economy with associated ρ being strictly negative and consider the economy $\Sigma_{-1} := (e, -Y, -B, u)$, whose associated ρ is strictly positive. Then, from Proposition 8, if $(\widetilde{x}, \widetilde{b}, \widetilde{q}, \widetilde{p})$ is an equilibrium for Σ_{-1}, then $(\widetilde{x}, -\widetilde{b}, -\widetilde{q}, \widetilde{p})$ is an equilibrium for the original economy Σ.

The following proposition gives some preliminary properties of solutions to the household h's maximization problem.

Proposition 9 *Let Assumption 8 be satisfied. If for any* $h \in \mathcal{H}$, $(\widetilde{x}_h, \widetilde{b}_h)$ *is a solution to maximization problem (79) at prices* $(\widetilde{q}, \widetilde{p}) \in \mathbb{R}^A \times \mathbb{R}^G_+$, *then*

[6] $\mathcal{F}$ stays for financial structure.

1. $\widetilde{p}^0 > 0, \ (\widetilde{p}^{sC})_{s\in\mathcal{S}} >> 0$;
2. for any $h \in \mathcal{H}$,

$$\langle \widetilde{p}^0, \widetilde{x}_h^0 - e_h^0 \rangle_C + \langle \widetilde{q}, \widetilde{b}_h \rangle_A = 0\,,$$

$$\langle \widetilde{p}^s, \widetilde{x}_h^s - e_h^s \rangle_C - \widetilde{p}^{sC} \langle y^s, \widetilde{b}_h \rangle_A = 0, \qquad \forall s \in \mathcal{S};$$

3. the following so-called $S+1$ Walras laws hold true

$$\left\langle \widetilde{p}^0, \sum_{h\in\mathcal{H}} (\widetilde{x}_h^0 - e_h^0) \right\rangle_C + \left\langle \widetilde{q}, \sum_{h\in\mathcal{H}} \widetilde{b}_h \right\rangle_A = 0\,,$$

$$\left\langle \widetilde{p}^s, \sum_{h\in\mathcal{H}} (\widetilde{x}_h^s - e_h^s) \right\rangle_C - \widetilde{p}^{sC} \left\langle y^s, \sum_{h\in\mathcal{H}} \widetilde{b}_h \right\rangle_A = 0, \qquad \forall s \in \mathcal{S};$$

4. $\widetilde{q} \in Q(P^{\cdot}, Y, B) = Q(Y, B)$.

We consider the following price sets:

$$\Delta_0 := \left\{ (q, p^0) \in \mathrm{Cl}\,(Q\,(Y, B)) \times \mathbb{R}_+^C : \sum_{c\in\mathcal{C}} p^{0c} + \sum_{a\in\mathcal{A}} q^a = 1 + \rho \right\},$$

$$\Delta_s := \left\{ p^s \in \mathbb{R}_+^C : \sum_{c\in\mathcal{C}} p^{sc} = 1 \right\}, \qquad \Delta_1 := \prod_{s\in\mathcal{S}} \Delta_s, \qquad \Delta := \prod_{s\in\mathcal{S}^0} \Delta_s\ .$$

The following proposition describes useful properties of the budget set-valued function $\Gamma_h : \Delta \rightrightarrows \mathbb{R}_+^G \times B_h$.

Proposition 10 *Let Assumptions 10.1 and 10.2 be satisfied. Then, for any $h \in \mathcal{H}$, the set-valued function Γ_h is*

1. *nonempty and convex valued;*
2. *closed;*
3. *compact valued for any $(q, p^0, p^1) \in \Delta$ such that $p \in \mathbb{R}_{++}^G$ and $q \in\in Q(Y, B)$;*
4. *lower semicontinuous for any $(q, p^0, p^1) \in \Delta$ such that $(p^{sC})_{s\in\mathcal{S}^0} >> 0$;*
5. *upper semicontinuous for any $(q, p^0, p^1) \in \Delta$ such that $p \in \mathbb{R}_{++}^G$ and $q \in Q(Y, B)$.*[7]

[7] Recall that, from Remark 5, if $(p^{sC})_{s\in\mathcal{S}} \in \mathbb{R}_{++}^S$, then $Q_h(Y, B_h) = Q_h(P^{\cdot}, Y, B_h)$.

First of all, we observe that, given our assumptions on the utility functions, the no-free lunch good price set is $\mathbb{R}^G_{++}$. Now, we consider a non-zero lower bound on prices and to this aim we define the following price sets:

$$\Delta_0^n := \left\{(q, p^0) \in \Delta_0 : p^0 \geq \frac{1}{n}\mathbf{1}_C,\ q \in \left\{\frac{1}{n}\mathbf{1}_S Y\right\} + \mathrm{Cl}\,(\in Q(Y, B)),\ q \geq -n\mathbf{1}_A\right\},$$

$$\Delta_s^n := \left\{p^s \in \Delta_s : p^s \geq \frac{1}{n}\mathbf{1}_C\right\}, \qquad \Delta_{\mathbf{1}}^n := \prod_{s \in \mathcal{S}} \Delta_s^n, \qquad \text{and} \qquad \Delta^n := \prod_{s \in \mathcal{S}^0} \Delta_s.$$

Proposition 11 *Let* $\left(q, p^0, p^{\mathbf{1}}\right) \in \Delta^n$ *be given; the following properties hold true:*

1. $Q(P^{\cdot}, Y, B) = Q(Y, B) := Q$ *and* $Q^u(P^{\cdot}, Y) = Q^u(Y) := Q^u$*;*
2. *for any* $n \in \mathbb{N}$*, one has*

$$\frac{1}{n}\mathbf{1}_S Y \in Q^u \subseteq Q \subseteq \mathrm{Cl}\,(Q), \quad \left\{\frac{1}{n}\mathbf{1}_S Y\right\} + \mathrm{Cl}\,(Q) \subseteq \mathrm{Cl}\,(Q)$$

   and $\quad \mathbf{1}_S Y \in \left\{\frac{1}{n}\mathbf{1}_S Y\right\} + \mathrm{Cl}\,(Q)$;
3. *let* $\{q_n\}_{n \in \mathbb{N}}$ *be a sequence such that* $\lim_{n \to +\infty} q_n = q$ *and* $q \in Q$*. Then, there exists* $\nu \in \mathbb{N}$ *such that for all* $n > \nu$ *one has* $q_n \in \left\{\frac{1}{n}\mathbf{1}_S Y\right\} + \mathrm{Cl}\,(Q)$*;*
4. $q \in Q$ *and* $p^0 \in \mathbb{R}^S_{++}$.

Proof See p. 17 in [20]. □

Proposition 12 *For any* $n \geq C$ *and* $n^2 > \max_{a \in \mathcal{A}} \left\{-\sum_{s \in \mathcal{S}} y^{sa}\right\}$*, one has:*

1. Δ_0^n *is nonempty, convex, compact;*
2. Δ_s^n *is nonempty, convex, compact for any* $s \in \mathcal{S}$*;*
3. Δ^n *is nonempty, convex, compact.*

Proof See pp. 18–19 in [20]. □

Now, for any $h \in \mathcal{H}$, we consider $\mathcal{G}_h : \mathbb{R}^G \rightrightarrows \mathbb{R}^G$ such that

$$\mathcal{G}_h(x_h) = conv\left(N_h^{>}(x_h) \cap S(0, 1)\right) \qquad \forall x_h \in \mathbb{R}^G.$$

Let f_C^C be the element of the canonical base of $\mathbb{R}^C$ with 1 in the component C. Now, we introduce the following GQVI$_n$:

Find $\left((\widetilde{x}_n, \widetilde{b}_n), (\widetilde{q}_n, \widetilde{p}_n^{\,0}, \widetilde{p}_n^{\,\mathbf{1}})\right) \in \Gamma(\widetilde{q}_n, \widetilde{p}_n^{\,0}, \widetilde{p}_n^{\,\mathbf{1}}) \times \Delta^n$ such that there exists $g_n = (g_{h,n})_{h \in \mathcal{H}} \in \prod_{h \in \mathcal{H}} \mathcal{G}_h(\widetilde{x}_{h,n})$ with

$$\langle -g_n, x_n - \widetilde{x}_n \rangle_{GH} + \left\langle \sum_{h \in \mathcal{H}} \widetilde{b}_{h,n}, q_n - \widetilde{q}_n \right\rangle_A + \left\langle \sum_{h \in \mathcal{H}} (\widetilde{x}^0_{h,n} - e^0_h), p^0_n - \widetilde{p}^0_n \right\rangle_C$$

$$+ \sum_{s \in \mathcal{S}} \left\langle \sum_{h \in \mathcal{H}} (\widetilde{x}^s_{h,n} - e^s_h - f^C_C \langle y^s, \widetilde{b}_h \rangle_A), p^s_n - \widetilde{p}^s_n \right\rangle_C \leq 0,$$

$$\forall \Big((x_n, b_n), (q_n, p^0_n, p^{\mathbf{1}}_n) \Big) \in \Gamma(\widetilde{q}_n, \widetilde{p}^0_n, \widetilde{p}^{\mathbf{1}}_n) \times \Delta^n. \tag{80}$$

Remark 7 $\Big((\widetilde{x}_n, \widetilde{b}_n), (q_n, \widetilde{p}^0_n, \widetilde{p}^{\mathbf{1}}_n) \Big)$ is a solution to GQVI_n (80) if and only if, simultaneously we have,

for any $h \in \mathcal{H}$,

$$\langle -g_{h,n}, x_{h,n} - \widetilde{x}_{h,n} \rangle_G \leq 0, \quad \forall (x_{h,n}, b_{h,n}) \in \Gamma_h(\widetilde{q}_n, \widetilde{p}^0_n, \widetilde{p}^{\mathbf{1}}_n); \tag{81}$$

$$\left\langle \sum_{h \in \mathcal{H}} \widetilde{b}_{h,n}, q_n - \widetilde{q}_n \right\rangle_A + \left\langle \sum_{h \in \mathcal{H}} (\widetilde{x}^0_{h,n} - e^0_h), p^0_n - \widetilde{p}^0_n \right\rangle_C \leq 0 \quad \forall (q_n, p^0_n) \in \Delta^n_0; \tag{82}$$

for any $s \in \mathcal{S}$,

$$\left\langle \sum_{h \in \mathcal{H}} (\widetilde{x}^s_{h,n} - e^s_h - f^C_C \langle y^s, \widetilde{b}_h \rangle_A), p^s_n - \widetilde{p}^s_n \right\rangle_C \leq 0, \quad \forall p^s_n \in \Delta^n_s. \tag{83}$$

Theorem 18 *Let Assumptions 7, 8, and 10 hold true. For any $n \in \mathbb{N}$, with $n \geq C$ and $n^2 > \max_{a \in \mathcal{A}} \{-\sum_{s \in \mathcal{S}} y^{sa}\}$, GQVI_n (80) admits at least one solution.*

Proof To get the desired result, we apply Theorem 6. Observe as the variational inequality in (80) has a simple structure, made up of two parts. A first one, involving (x, b), relates to the households' maximization problems, and the second one, involving (q, p^0, p^1) to market clearing conditions.

Consistently with Definition 6, the variational problem (80) represents a generalized quasi-variational inequality associated with

$$C := conv\left(\Gamma\left(\Delta^n\right)\right) \times \Delta^n$$

and, for any $\Big((x_n, b_n), (q_n, p^0_n, p^1_n) \Big) \in \Gamma(q_n, p^0_n, p^1_n) \times \Delta^n$,

$$S\Big((x_n, b_n), (q_n, p_n^0, p_n^1)\Big) := \Gamma\left(q_n, p_n^0, p^1\right) \times \Delta^n$$

$$\Phi\Big((x_n, b_n), (q_n, p_n^0, p_n^1)\Big) :=$$

$$-\Big(-\prod_{h\in\mathcal{H}} \mathcal{G}_h(x_{h,n}), \sum_{h\in\mathcal{H}} b_{h,n}, \sum_{h\in\mathcal{H}} (x_{h,n}^0 - e_h^0), (\sum_{h\in\mathcal{H}} (x_{h,n}^s - e_h^s - f_C^C \langle y^s, \widetilde{b}_h\rangle_A))_{s\in\mathcal{S}}\Big).$$

Under our Assumptions, we have that all Assumptions of Theorem 6 are satisfied (see proof p. 20 in [20]), then GQVI_n (80) associated with C, S, and Φ admits at least a solution. □

Theorem 19 *Let Assumption 8 be satisfied. For any $n \in \mathbb{N}$ such that $n \geq C$, $n > \max_{a\in\mathcal{A}}\{-\sum_{s\in\mathcal{S}} y^{sa}\}$, and $n^2 > \max_{a\in\mathcal{A}}\left\{-\sum_{s\in\mathcal{S}} y^{sa}\right\}$, let $\left((\widetilde{x}_n, \widetilde{b}_n), (\widetilde{q}_n, \widetilde{p}_n^0, \widetilde{p}_n^1)\right)$ be a solution to GQVI_n (80). Then, one has*

(i) for any $h \in \mathcal{H}$, $(\widetilde{x}_{h,n}, \widetilde{b}_{h,n})$ is a solution to maximization problem

$$\max_{(x_{h,n}, b_{h,n})\in\Gamma_h((\widetilde{q}_n, \widetilde{p}_n^0), \widetilde{p}_n^1)} u_h(x_{h,n}); \tag{84}$$

(ii) for any $h \in \mathcal{H}$ and $s \in \mathcal{S}$,

$$\langle \widetilde{p}_n^0, \widetilde{x}_{h,n}^0 - e_h^0\rangle_C + \langle \widetilde{q}_n, \widetilde{b}_{h,n}\rangle_A = 0, \tag{85}$$

$$\langle \widetilde{p}_n^s, \widetilde{x}_{h,n}^s - e_h^s\rangle_C = \widetilde{p}_n^{sC} \langle y^s, \widetilde{b}_{h,n}\rangle; \tag{86}$$

(iii) for any $s \in \mathcal{S}$,

$$\left\langle \widetilde{p}_n^0, \sum_{h\in\mathcal{H}} (\widetilde{x}_{h,n}^0 - e_h^0)\right\rangle_C + \left\langle \widetilde{q}_n, \sum_{h\in\mathcal{H}} \widetilde{b}_{h,n}\right\rangle_A = 0, \tag{87}$$

$$\left\langle \widetilde{p}_n^s, \sum_{h\in\mathcal{H}} (\widetilde{x}_{h,n}^s - e_h^s)\right\rangle_C = \widetilde{p}_n^{sC} \left\langle y^s, \sum_{h\in\mathcal{H}} \widetilde{b}_{h,n}\right\rangle; \tag{88}$$

(iv) for any $h \in \mathcal{H}$, $c \in \mathcal{C}$ and $s \in \mathcal{S}^0$,

$$0 \leq \widetilde{x}_{h,n}^{sc} \leq \sum_{c\in\mathcal{C}} \sum_{h\in\mathcal{H}} e_h^{0c} + C \sum_{s\in\mathcal{S}} \sum_{c\in\mathcal{C}} \sum_{h\in\mathcal{H}} e_h^{sc}. \tag{89}$$

Proof Thanks to Remark 7, for any $h \in \mathcal{H}$, $(\widetilde{x}_{h,n}, \widetilde{b}_{h,n})$ is a solution to GVI_n (81), $(\widetilde{q}_n, \widetilde{p}_n^0)$ is a solution to (82), and for all $s \in S$, $\widetilde{p}_n^s$ is a solution to (83).

(i). $(\widetilde{x}_{h,n}, \widetilde{b}_{h,n})$ is a solution to maximization problem (84).

(ii) and (iii). Both statements follow from the fact that $(\widetilde{x}_{h,n}, \widetilde{b}_{h,n})$ is a solution to maximization problem (84) and from Proposition 9.

(iv). From (87), inequality (82) becomes

$$\left\langle \sum_{h\in\mathcal{H}} \widetilde{b}_{h,n}, q_n \right\rangle_A + \left\langle \sum_{h\in\mathcal{H}} (\widetilde{x}^{\,0}_{h,n} - e^0_h), p^0_n \right\rangle_C \leq 0 \qquad \forall (q_n, p^0_n) \in \Delta^n_0; \tag{90}$$

and, from (86), for any $s \in \mathcal{S}$, inequality (83) becomes

$$\left\langle \sum_{h\in\mathcal{H}} (\widetilde{x}^s_{h,n} - e^s_h), p^{\,s}_n \right\rangle_C - p^{sC}_n \left\langle y^s, \sum_{h\in\mathcal{H}} \widetilde{b}_{h,n} \right\rangle_A \leq 0 \qquad \forall p^s_n \in \Delta^n_s,$$

that is,

$$\left\langle \sum_{h\in\mathcal{H}} (\widetilde{x}^s_{h,n} - e^s_h), \frac{p^{\,s}_n}{p^{sC}_n} \right\rangle_C - \left\langle y^s, \sum_{h\in\mathcal{H}} \widetilde{b}_{h,n} \right\rangle_A \leq 0, \quad \forall p^s_n \in \Delta^n_s. \tag{91}$$

Summing up (90) and (91), for any $(q_n, p^0_n, p^{\,\mathbf{1}}_n) \in \Delta^n$, one has

$$\left\langle \sum_{h\in\mathcal{H}} \widetilde{b}_{h,n}, q_n \right\rangle_A + \left\langle \sum_{h\in\mathcal{H}} (\widetilde{x}^{\,0}_{h,n} - e^0_h), p^0_n \right\rangle_C + \sum_{s\in\mathcal{S}} \left\langle \sum_{h\in\mathcal{H}} (\widetilde{x}^s_{h,n} - e^s_h), \frac{p^{\,s}_n}{p^{sC}_n} \right\rangle_C$$

$$- \left\langle \sum_{s\in\mathcal{S}} y^s, \sum_{h\in\mathcal{H}} \widetilde{b}_{h,n} \right\rangle_A \leq 0. \tag{92}$$

Now, choose $q_n = \mathbf{1}_S Y$ and $\widehat{p}^{\,sc} = \dfrac{1}{C}$[8] for any $c \in \mathcal{C}$ and $s \in \mathcal{S}^0$. From Proposition 11.2, $\mathbf{1}_S Y \in \{\frac{1}{n}\mathbf{1}_S Y\} + \mathrm{Cl}(Q)$; since $n > \max_{a\in\mathcal{A}} \left\{-\sum_{s\in\mathcal{S}} y^{sa}\right\}$, $q_n = \sum_{s\in\mathcal{S}} y^s \geq -n\mathbf{1}_A$, and $\sum_{a\in\mathcal{A}} q^a_n = \rho$. Hence, $(q_n, \widehat{p}^{\,0}_n, \widehat{p}^{\,\mathbf{1}}_n) \in \Delta^n$ and replacing $(q_n, p^0_n, p^{\mathbf{1}}_n)$ with $(\mathbf{1}_S Y, \widehat{p}^{\,0}_n, \widehat{p}^{\,\mathbf{1}}_n)$ in (92),

$$\frac{1}{C} \sum_{c\in\mathcal{C}} \sum_{h\in\mathcal{H}} (\widetilde{x}^{0c}_{h,n} - e^{0c}_h) + \sum_{s\in\mathcal{S}} \sum_{c\in\mathcal{C}} \sum_{h\in\mathcal{H}} (\widetilde{x}^{sc}_{h,n} - e^{sc}_h) \leq 0.$$

Then,

[8] Observe that $\frac{1}{C} \geq \frac{1}{n}$ since by assumption $n \geq C$.

$$\frac{1}{C}\sum_{c\in\mathcal{C}}\sum_{h\in\mathcal{H}}\widetilde{x}^{0c}_{h,n}+\sum_{s\in\mathcal{S}}\sum_{c\in\mathcal{C}}\sum_{h\in\mathcal{H}}\widetilde{x}^{sc}_{h,n}\leq\frac{1}{C}\sum_{c\in\mathcal{C}}\sum_{h\in\mathcal{H}}e^{0c}_{h}+\sum_{s\in\mathcal{S}}\sum_{c\in\mathcal{C}}\sum_{h\in\mathcal{H}}e^{sc}_{h}.$$

Hence, being $C > 1$, for any $s \in \mathcal{S}^0$, $c \in \mathcal{C}$, and $h \in \mathcal{H}$, we have

$$0\leq\widetilde{x}^{sc}_{h,n}\leq\sum_{s\in\mathcal{S}^0}\sum_{c\in\mathcal{C}}\sum_{h\in\mathcal{H}}\widetilde{x}^{sc}_{h,n}\leq\sum_{c\in\mathcal{C}}\sum_{h\in\mathcal{H}}e^{0c}_{h}++C\sum_{s\in\mathcal{S}}\sum_{c\in\mathcal{C}}\sum_{h\in\mathcal{H}}e^{sc}_{h}.$$

□

Proposition 13 *Let Assumptions 7, 8, 9, and 10.3 be satisfied.*
Let $\left\{\left((\widetilde{x}_n,\widetilde{b}_n),(\widetilde{q}_n,\widetilde{p}^{\,0}_{n},\widetilde{p}^{\,\mathbf{1}}_{n})\right)\right\}_{n\in\mathbb{N}}$ *be the sequence such that, for any* $n \in \mathbb{N}$ *with* $n \geq C$ *and* $n^2 > \max_{a\in\mathcal{A}}\left\{-\sum_{s\in\mathcal{S}}y^{sa}\right\}$, $\left((\widetilde{x}_n,\widetilde{b}_n),(\widetilde{q}_n,\widetilde{p}^{\,0}_{n},\widetilde{p}^{\,\mathbf{1}}_{n})\right)$ *is a solution to* GQVI$_n$ *(80). Then, there exists a subsequence converging to* $((\widetilde{x},\widetilde{b}),(\widetilde{q},\widetilde{p}^{\,0},\widetilde{p}^{\,\mathbf{1}}))$ *such that* $(\widetilde{q},\widetilde{p}^{\,0},\widetilde{p}^{\,\mathbf{1}})\in\Delta$ *with* $\widetilde{p}>>0$, $\widetilde{q}\in Q$ *and* $(\widetilde{x},\widetilde{b})\in\Gamma(\widetilde{q},\widetilde{p}^{\,0},\widetilde{p}^{\,\mathbf{1}})$.

Proof See pp. 22–26 in [20]. □

Theorem 20 *Let Assumptions 7, 8, 9, and 10 be satisfied. Then, for any financial economy* $\Sigma\in\mathcal{E}$, $(\widetilde{x},\widetilde{b},\widetilde{q},\widetilde{p})$ *is an equilibrium vector with restricted participation and numeraire assets.*

Proof See pp. 26–29 in [20]. □

References

1. G. Anello, M.B. Donato, M. Milasi, "A quasi-variational approach to a competitive economic equilibrium problem without strong monotonicity assumption", *Journal of Global Optimization*, **48** n. 2, pp. 279–287(2010).
2. G. Anello, M.B. Donato, M. Milasi, "Variational methods for equilibrium problems involving quasi-concave utility functions", *Optimization and Engineering*, **13** n. 2, pp. 169–179 (2012).
3. Z. Aouani, B. Cornet, "Existence of financial equilibria with restricted participation", *Journal of Mathematical Economics*, **45**, pp. 772–786, (2009).
4. K.J. Arrow, F.H. Hahn, General Competitive Analysis. Holden-Day, Inc., San Francisco (1971).
5. D. Aussel and N. Hadjisavvas, *Adjusted sublevel sets, normal operator and quasiconvex programming*, SIAM Journal of Optimization, **16**, 358–367 (2005).
6. D. Aussel, J. Dutta, *Generalized Nash equilibrium problem, variational inequality and quasiconvexity*. Operation Research Letters, **36**, 461–464 (2008).
7. D. Aussel, J. Cotrina, *Quasimonotone Quasivariational Inequalities: Existence Results and Applications*, Journal of Optimization Theory and Applications, **158**, n. 3, 637–652 (2013).
8. D. Aussel and J. Cotrina, *Stability of Quasimonotone Variational Inequality Under Sign-Continuity*, Journal of Optimization Theory and Applications, **158**, 653–667 (2013).
9. A. Barbagallo, P. Daniele, S. Giuffrè, A. Maugeri, *Variational approach for a general financial equilibrium problem: The Deficit Formula, the Balance Law and the Liability Formula. A path to the economy recovery*, European Journal of Operational Research, **237**, n. 1, 231–244 (2014).

10. I. Benedetti, M.B. Donato, M. Milasi, “Existence for Competitive Equilibrium by Means of Generalized Quasivariational Inequalities”, *Abstract and Applied Analysis*, article n. 648986 (2013).
11. D. Chan, J.S. Pang (1982). The generalized quasi-variational inequality problem. Math. Oper. Res. 7: 211–222.
12. P. Daniele, L. Scrimali, Strong Nash equilibria for cybersecurity investments with nonlinear budget constraints. New trends in emerging complex real life problems, 199–207, AIRO Springer Ser., 1, Springer, Cham, 2018.
13. G. Debreu, Theory of Value, An Axiomatic Analysis of Economic Equilibrium. New York: Wiley (1959).
14. M.B. Donato, A. Maugeri, M. Milasi, C. Vitanza, “Duality theory for a dynamic Walrasian pure exchange economy”, *Pacific Journal of Optimization*, **4** n. 3, pp. 537–547 (2008).
15. M.B. Donato, M. Milasi, “Lagrangean variables in infinite dimensional spaces for a dynamic economic equilibrium problem”, *Nonlinear Analysis-Theory Methods and Applications*, **74** n. 15, pp. 5048–5056 (2011).
16. M.B. Donato, M. Milasi, C. Vitanza, “An existence result of a quasi-variational inequality associated to an equilibrium problem”, *Journal of Global Optimization*, **40** n. 1–3, pp. 87–97 (2008).
17. M.B. Donato, M. Milasi, C. Vitanza, “Quasi-variational approach of a competitive economic equilibrium problem with utility function: existence of equilibrium”, *Mathematical Models and Methods in Applied Sciences*, **18**, n. 3, pp. 351–367 (2008).
18. M.B. Donato, M. Milasi, A. Villanacci, “Incomplete financial markets model with nominal assets: variational approach”, *Journal of Mathematical Analysis and Applications*, **457**, pp. 1353–1369 (2018).
19. M.B. Donato, M. Milasi, A. Villanacci, “Variational formulation of a general equilibrium model with incomplete financial markets and numeraire assets: existence”, *Journal of Optimization Theory and Application*, **179**(2), pp. 425–451 (2018).
20. M.B. Donato, M. Milasi, A. Villanacci, “Restricted Participation on Financial Markets: A General Equilibrium Approach Using Variational Inequality Methods”, *Netw Spat Econ*, (2020). https://doi.org/10.1007/s11067-019-09491-4
21. M.B. Donato, M. Milasi, C. Vitanza, “Quasivariational inequalities for a dynamic competitive economic equilibrium problem”, *Journal of Inequalities and Applications*, pp.1–17, article number 519623 (2009).
22. M.B. Donato, M. Milasi, C. Vitanza, “A new contribution to a dynamic competitive equilibrium problem”, *Applied Mathematics Letters*, **23** n. 2, pp. 148–151 (2010).
23. M.B. Donato, M.Milasi, C. Vitanza, “Variational problem, generalized convexity and application to an equilibrium problem”, *Numerical Functional Analysis and Optimization*, **35** n. 7–9, pp. 962–983 (2014).
24. M.B. Donato, M. Milasi, C. Vitanza, “Evolutionary quasi-variational inequality for a production economy”, *Nonlinear Analysis: Real World Applications*, pp. 328–336 (2018).
25. M.B. Donato, M. Milasi, C. Vitanza, “Generalized variational inequality and general equilibrium problem”, *Journal of Convex Analysis*, 25 n. 2, pp. 515–527 (2018).
26. M.B. Donato, M. Milasi, C. Vitanza, “Quasivariational inequalities for a dynamic competitive economic equilibrium problem”, *Journal of Inequality and Applications*, 519623 (2009).
27. G. Fichera, Problemi elastostatici con vincoli unilaterali: il problema di signorini con ambigue condizioni al contorno, Mem. Accad. Naz. Lincei (1964) 91–140.
28. J. Lions, G. Stampacchia, Variational inequalities, Comm. Pure Appl. Math. (1967) 493–519.
29. M. Milasi, C. Vitanza, Variational inequality and evolutionary market disequilibria: the case of quantity formulation. Variational analysis and applications, 681–696, Nonconvex Optim. Appl., 79, Springer, New York, 2005.
30. G. Stampacchia, Variational Inequalities. In: Ghizzetti, A. (ed.): Theory and Applications of Monotone Operators, pp. 101–191. Edizioni Oderisi, Gubbio (1969)
31. D. Kinderlehrer, G. Stampacchia, An introduction to variational inequaities and their applications, Academic Press, (1980)

32. N.X. Tan, (1985). Quasi-variational inequality in topological linear locally convex Hausdorff spaces. *Math. Nachr.* 122, 231–245.
33. A. Villanacci, L. Carosi, P. Benevieri, A. Battinelli, Differential Topology and General Equilibrium with Complete and Incomplete Markets, Kluwer Academic Publishers (2002).
34. J. Werner, "Equilibrium in economies with incomplete financial markets", *Journal of Economic Theory*, **36**, pp. 110–119 (1985).

The Strong Convergence of Douglas-Rachford Methods for the Split Feasibility Problem

Qiao-Li Dong, Lulu Liu, and Themistocles M. Rassias

Abstract In this article, we introduce several Douglas-Rachford method to solve the split feasibility problems (SFP). Firstly, we propose a new iterative method by combining Douglas-Rachford method and Halpern iteration. The stepsize is determined dynamically which does not need any prior information about the operator norm. A relaxed version is presented for the SFP where the two closed convex sets are both level sets of convex functions. The strong convergence of two proposed methods is established under standard assumptions. We also propose an iterative method by combining Douglas-Rachford method with Haugazeau algorithm, and show its strong convergence. The numerical examples are presented to illustrate the advantage of our methods by comparing with other methods.

1 Introduction

In this article, we consider the split feasibility problem (SFP) which is formulated as finding a point x^* with the property

$$x^* \in C \quad \text{and} \quad Ax^* \in Q, \tag{1}$$

where C and Q are the nonempty closed convex subsets of the real Hilbert spaces $\mathscr{H}_1$ and $\mathscr{H}_2$, respectively, and $A : \mathscr{H}_1 \to \mathscr{H}_2$ is a bounded linear operator. The SFP was introduced by Censor and Elfving [5] for inverse problems which arise from phase retrievals and medical image reconstruction [3]. Recently, it was extended to systems biology [25] and electricity production [26].

Q.-L. Dong (✉) · L. Liu
College of Science, Civil Aviation University of China, Tianjin, China
e-mail: dongql@lsec.cc.ac.cn

Th. M. Rassias
Department of Mathematics, Zografou Campus, National Technical University of Athens, Athens, Greece
e-mail: trassias@math.ntua.gr

I. N. Parasidis et al. (eds.), *Mathematical Analysis in Interdisciplinary Research*, Springer Optimization and Its Applications 179,
https://doi.org/10.1007/978-3-030-84721-0_12

Throughout this article, we assume that SFP (1) is consistent, i.e., its solution set, denoted by

$$\Gamma = \{x \mid x \in C \quad \text{and} \quad Ax \in Q\},$$

is nonempty. It is easy to see that the SFP equals to the following constrained optimization

$$\min_{x \in C} f(x), \tag{2}$$

where

$$f(x) = \frac{1}{2}\|(I - P_Q)Ax\|^2.$$

Recall that the objective function f is convex, differentiable and has a Lipschitz gradient given by

$$\nabla f(x) = A^*(I - P_Q)Ax,$$

whose Lipschitz constant is $\|A\|^2$.

One of the simplest and popular methods for the problem (2) is the gradient-projection method

$$x^{k+1} = P_C(x^k - \lambda_k \nabla f(x^k)), \tag{3}$$

where λ_k is the stepsize. There are three ways to determine the stepsize λ_k in the algorithm (3). The first one is to take the stepsize $\lambda_k \in (0, \frac{2}{\|A\|^2})$ which depends on the operator (matrix) norm $\|A\|$ (see [3, 4]). The second one is to select the stepsize $\lambda_k > 0$ self-adaptively by adopting Armijo-like (see [28]) searches:

$$\lambda_k \|\nabla f(x^k) - \nabla f(y^k)\| \leq \mu \|x^k - y^k\|, \quad \forall \mu \in (0, 1).$$

The third one introduced by López et al. [19] is to dynamically determine the stepsize by

$$\lambda_k = \rho_k \frac{f(x^k)}{\|\nabla f(x^k)\|^2}, \tag{4}$$

where $\rho_k \in (0, 4)$.

The algorithm (3) was firstly introduced to solve the SFP by Byrne [3, 4] and referred as CQ algorithm since it involves the (metric) projections onto the sets C and Q. The CQ algorithm has received a great deal of attention by many authors, who improved it in various ways; see, e.g., [6, 9, 12, 15, 20, 21, 27].

Under the simple assumptions, it is easy to show that the CQ algorithm with different stepsizes converges weakly to a solution of the SFP. However, the strong convergence is often much more desirable than the weak convergence in many problems that arise in infinite dimensional spaces (see [2] and the references therein). So, attempts have been made to modify CQ algorithm so that the strong convergence is guaranteed. By combining Haugazeau's method [13] and CQ algorithm, and using the stepsize given in (4), López et al. [19] introduced the following modification of CQ algorithm:

$$\begin{cases} y^k = P_C(x^k - \lambda_k \nabla f(x^k)), \\ S_k = \left\{ z \in \mathscr{H}_1 : \|y^k - z\|^2 \leq \|x^k - z\|^2 - \rho_k(4 - \rho_k)\dfrac{f^2(x^k)}{\|\nabla f(x^k)\|^2} \right\}, \\ T_k = \left\{ z \in \mathscr{H}_1 : \langle x^0 - x^k, z - x^k \rangle \leq 0 \right\}, \\ x^{k+1} = P_{S_k \cap T_k}(x^0). \end{cases}$$

and showed that the sequence $\{x^k\}_{k\in\mathbb{N}}$ converges strongly to $P_\Gamma x^0$. This type of modification may cost much computation since it involves the projection onto the intersection of two half-spaces S_k and T_k. Another type of modifications in [19] is combining CQ algorithm with Halpern iteration as following

$$x^{k+1} = \beta_k u + (1 - \beta_k) P_C(x^k - \lambda_k \nabla f(x^k)), \tag{5}$$

where $\{\beta_k\}_{k\in\mathbb{N}} \subset [0, 1]$ and the stepsize λ_k is given in (4). The algorithm (5) strongly converges to $P_\Gamma u$ provided that $\{\beta_k\}_{k\in\mathbb{N}}$ satisfies $\lim_{k\to\infty} \beta_k = 0$ and $\sum_{k=1}^{\infty} \beta_k = \infty$ and ρ_k is far away from 0 and 4.

By combining Polyak's gradient method and Haugazeau's method, Wang [24] proposed the following algorithm:

$$\begin{cases} y^k = x^k - \lambda_k \left[(x^k - P_C x^k) + A^*(I - P_Q) A x^k \right], \\ S_k = \left\{ z \in \mathscr{H}_1 : \langle y^k - z, x^k - y^k \rangle \geq 0 \right\}, \\ T_k = \left\{ z \in \mathscr{H}_1 : \langle x^k - x^0, z - x^k \rangle \geq 0 \right\}, \\ x^{k+1} = P_{S_k \cap T_k}(x^0), \end{cases} \tag{6}$$

where $\lambda_k \in (0, \frac{1}{1+\|A\|^2})$ or $\lambda_k = \dfrac{\|x^k - P_C x^k\|^2 + \|(I - P_Q) A x^k\|^2}{2\|(x^k - P_C x^k) + A^*(I - P_Q) A x^k\|^2}$ and showed that the sequence $\{x^k\}_{k\in\mathbb{N}}$ converges strongly to $P_\Gamma x^0$. By putting Polyak's gradient method and Halpern iteration together, Wang [23] introduced an algorithm as follows:

$$x^{k+1} = \beta_k u + (1-\beta_k)\left[x^k - \lambda_k((x^k - P_{C_k}x^k) + A^*(I - P_{Q_k})Ax^k)\right], \quad (7)$$

where $\lambda_k = \rho_k \frac{\|x^k - P_{C_k}x^k\|^2 + \|(I-P_{Q_k})Ax^k\|^2}{2\|(x^k - P_{C_k}x^k) + A^*(I-P_{Q_k})Ax^k\|^2}$, $\rho_k \in (0,4)$ and C_k and Q_k are half-spaces including C and Q, respectively. The algorithm (7) strongly converges to $P_\Gamma u$ under the similar assumptions with the algorithm (5) for $\{\beta_k\}_{k\in\mathbb{N}}$ and $\{\rho_k\}_{k\in\mathbb{N}}$.

The algorithms (5) and (7) are Halpern-type iteration and their convergence is generally slow due to the strict conditions on the parameters β_k. This results in seldom applications of Halpern iteration method in actual computation. Recently, He et al. [14] presented two optimal choices of the parameters β_k and showed that Halpern iteration method with their choices of β_k highly improves the convergence rate of Halpern iteration method with the general choice through numerical examples. The results in [8] also support this conclusion.

Douglas-Rachford method was originally introduced in [10] to solve nonlinear heat flow problems and later Lions and Mercier [18] extended it to two closed convex sets with nonempty intersection. It is regarded as one of the most classical algorithms for finding zeroes of sums of maximally monotone operators (see, e.g., [17]). Very recently, Douglas-Rachford method was used to solve the SFP by transforming the SFP into an optimization problem and linearizing the minimization problem of the regularization for $f(x)$ (see [7]) as follows:

$$\begin{cases} y^{k+1} := x^k - \lambda_k \nabla f(x^k), \\ z^{k+1} := P_C(2y^{k+1} - x^k), \\ x^{k+1} := x^k + \alpha(z^{k+1} - y^{k+1}), \end{cases} \quad (8)$$

where

$$\lambda_k := \begin{cases} \gamma \frac{f(x^k)}{\|\nabla f(x^k)\|^2}, & \text{if } \nabla f(x^k) \neq 0, \\ 0, & \text{otherwise}, \end{cases} \quad (9)$$

and $\gamma \in (0,2)$. Douglas-Rachford method was shown to have good numerical performance comparing with other methods.

The aim of this paper is to present some Douglas-Rachford type algorithms with strong convergence. Firstly, we introduce a new algorithm by combining Douglas-Rachford method and Halpern iteration, where the stepsize does not involves L. Secondly, a relaxed variant of the proposed algorithm is presented and the stepsize is selected by a similar way. The strong convergence of the algorithms is shown under the standard conditions. Thirdly, we propose an iterative algorithm by combining Douglas-Rachford method and Haugazeau algorithm. Finally, a numerical example is provided to illustrate that the proposed algorithms outperform the algorithms (5) and (7).

The paper is organized as follows: In Sect. 2, we recall some concepts and lemmas which will be used in the proof of main results and, in Sect. 3, we present Halpern-type method and prove its strong convergence. The relaxed version of Halpern-type method is introduced and shown to strongly converge to a solution of the SFP. In Sect. 4, we present a Haugazeau-type algorithm and show its convergence. Finally, in Sect. 5, a preliminary numerical experiment is provided to illustrate the behavior of the proposed methods.

2 Preliminaries

Let $\mathcal{H}$ be a Hilbert space and D be a nonempty closed convex subset of $\mathcal{H}$. We use the notation:

- $\rightharpoonup$ for weak convergence and $\to$ for strong convergence;
- $\omega_w(x^k) = \{x : \exists x^{k_l} \rightharpoonup x\}$ denotes the weak ω-limit set of $\{x^k\}_{k\in\mathbb{N}}$.

The following inequalities will be used for the main results:

$$\|\alpha x + (1-\alpha)y\|^2 = \alpha\|x\|^2 + (1-\alpha)\|y\|^2 - \alpha(1-\alpha)\|x-y\|^2, \tag{10}$$

for all $x, y \in \mathcal{H}$.

For each point $x \in \mathcal{H}$, there exists a unique nearest point in D, denoted by $P_D(x)$. That is,

$$\|x - P_D(x)\| \le \|x - y\| \text{ for all } y \in D. \tag{11}$$

The mapping $P_D : \mathcal{H} \to D$ is called the metric projection of $\mathcal{H}$ onto D.

Next two lemmas give the fundamental properties of the metric projection.

Lemma 1 *For any $x \in \mathcal{H}$ and $z \in D$, then $z = P_D x$ if and only if*

$$\langle x - z, y - z\rangle \le 0, \quad \forall y \in D.$$

Lemma 2 *For any $x, y \in \mathcal{H}$ and $z \in D$, the following hold:*

(i) $\|P_D(x) - P_D(y)\|^2 \le \langle P_D(x) - P_D(y), x - y\rangle$;
(ii) $\|P_D(x) - z\|^2 \le \|x - z\|^2 - \|P_D(x) - x\|^2$;
(iii) $\langle (I - P_D)x - (I - P_D)y, x - y\rangle \ge \|(I - P_D)x - (I - P_D)y\|^2$.

It follows from Lemma 2 (iii) that

$$\langle x - P_D x, x - z\rangle \ge \|x - P_D x\|^2, \quad \forall x \in \mathcal{H},\ \forall z \in D. \tag{12}$$

Recall that a mapping $T : \mathcal{H} \to \mathcal{H}$ is called to be *nonexpansive* if

$$\|Tx - Ty\| \le \|x - y\|, \quad \forall x, y \in \mathcal{H},$$

and *firmly nonexpansive*

$$\|Tx - Ty\|^2 \leq \langle Tx - Ty, x - y\rangle, \quad \forall x, y \in \mathscr{H}.$$

It is obvious that a firmly nonexpansive mapping is nonexpansive. Lemma 2 (i) implies that P_D and $I - P_D$ are firmly nonexpansive.

Next lemma shows that the nonexpansive mappings are demiclosed at 0.

Lemma 3 ([1, Theorem 4.27]) *Let D be a nonempty closed convex subset of $\mathscr{H}$ and $T : D \to \mathscr{H}$ be a nonexpansive mapping. Let $\{x^k\}_{k\in\mathbb{N}}$ be a sequence in D and $x \in \mathscr{H}$ such that $x^k \rightharpoonup x$ and $Tx^k - x^k \to 0$ as $k \to +\infty$. Then $x \in \mathrm{Fix}(T)$.*

3 Halpern-Type Algorithm

In this section, we introduce a Halpern-type algorithm by combining Douglas-Rachford method with Halpern iteration and establish the strong convergence of the iterative sequence generated by the proposed method.

Before presenting the method, we assume the sequence of parameters $\{\beta_k\} \subseteq [0, 1]$ satisfying

$$\text{(H1)} \quad \lim_{k\to\infty} \beta_k = 0 \quad \text{and} \quad \text{(H2)} \quad \sum_{k=1}^{\infty} \beta_k = \infty.$$

Now we present the first iterative algorithm.

Algorithm 1

Step 0. Input $k := 0$, $x^0 \in \mathscr{H}_1$ and $\alpha \in (0, 2)$.
Step 1. Generate x^{k+1} by

$$\begin{cases} y^{k+1} := x^k - \lambda_k \nabla f(x^k), \\ z^{k+1} := P_C(2y^{k+1} - x^k), \\ w^{k+1} := x^k + \alpha(z^{k+1} - y^{k+1}), \\ x^{k+1} := \beta_k u + (1 - \beta_k) w^{k+1}, \end{cases} \tag{13}$$

where

$$\lambda_k := \gamma \frac{f(x^k)}{\|\nabla f(x^k)\|^2}, \tag{14}$$

and $\gamma \in (0, 2)$.
Step 2. If $\nabla f(x^k) = 0$ and $y^{k+1} = z^{k+1}$, then terminate. Otherwise, set $k := k + 1$ and go to Step 1.

The iterative scheme (13) can be rewritten as following:

$$\begin{aligned}x^{k+1} =&\beta_k u + (1-\beta_k)\left[x^k + \alpha(z^{k+1} - y^{k+1})\right]\\ =&\beta_k u + (1-\beta_k)\left[\left(1-\frac{\alpha}{2}\right)x^k + \frac{\alpha}{2}(2\lambda_k \nabla f(x^k) - x^k)\right.\\ &\left.+\frac{\alpha}{2}2P_C(x^k - 2\lambda_k \nabla f(x^k))\right]\\ =&\beta_k u + (1-\beta_k)\left[\left(1-\frac{\alpha}{2}\right)x^k + \frac{\alpha}{2}(2P_C(v^k) - v^k)\right],\end{aligned} \tag{15}$$

where

$$v^k = x^k - 2\lambda_k \nabla f(x^k). \tag{16}$$

In Algorithm 1, we assume that the projections P_C and P_Q are easily calculated. However, in some cases, it is impossible or needs too much work to compute the projection. To deal with this situation, we introduce the relaxed method, in which the projections onto the approximated half-spaces are adopted in place of the projections onto C and Q (see, for example, [11]).

In this section, we consider a general case of the SFP (1), where C and Q are given by level sets of convex functions. Throughout this section, we assume that each $c : \mathscr{H}_1 \to \mathbb{R}$ and $q : \mathscr{H}_2 \to \mathbb{R}$ are convex functions and the sets C and Q are given, respectively, by

$$C = \{x \in \mathscr{H}_1 : c(x) \le 0\}, \quad \text{and} \quad Q = \{y \in \mathscr{H}_2 : q(y) \le 0\}.$$

We assume that ∂c and ∂q are bounded operators (i.e., bounded on any bounded set).

Define the sets C_k and Q_k by the following half-spaces:

$$C_k = \left\{x \in \mathscr{H}_1 : c(x^k) + \langle \xi^k, x - x^k\rangle \le 0\right\},$$

where $\xi^k \in \partial c(x^k)$, and

$$Q_k = \left\{y \in \mathscr{H}_2 : q(Ax^k) + \langle \eta^k, y - Ax^k\rangle \le 0\right\},$$

where $\eta^k \in \partial q(Ax^k)$.

By the definition of the subgradient, it is clear that $C \subseteq C_k$ and $Q \subseteq Q_k$. The projections onto C_k and Q_k are easy to compute since C_k and Q_k are half-spaces.

Define $f_k(x) := \frac{1}{2}\|(I - P_{Q_k})A(x)\|^2$ and then $\nabla f_k(x) := A^T(I - P_{Q_k})A(x)$. Below we introduce the relaxed method.

Algorithm 2

Step 0. Input $k := 0$, $x^0 \in \mathscr{H}_1$ and $\alpha \in (0, 2)$.
Step 1. Generate x^{k+1} by

$$\begin{cases} y^{k+1} := x^k - \lambda_k \nabla f_k(x^k), \\ z^{k+1} := P_{C_k}(2y^{k+1} - x^k), \\ w^{k+1} := x^k + \alpha(z^{k+1} - y^{k+1}), \\ x^{k+1} := \beta_k u + (1 - \beta_k) w^{k+1}, \end{cases} \tag{17}$$

where

$$\lambda_k := \gamma \frac{f_k(x^k)}{\|\nabla f_k(x^k)\|^2}, \tag{18}$$

and $\gamma \in (0, 2)$.
Step 2. If $\nabla f_k(x^k) = 0$ and $y^{k+1} = z^{k+1}$, then terminate. Otherwise, set $k := k + 1$ and go to Step 1.

Before establishing convergence of Algorithm 1, we need to ascertain the validity of the stopping criterion used in Step 2.

Lemma 4 *If $\nabla f(x^k) = 0$ and $y^{k+1} = z^{k+1}$ for some k, then x^k is a solution of the SFP* (1).

Proof Suppose $\nabla f(x^k) = 0$. Then $y^{k+1} = x^k$ and $z^{k+1} = P_C(x^k)$. Using $y^{k+1} = z^{k+1}$, we get $x^k = P_C(x^k)$ and therefore $x^k \in C$. Since $I - P_Q$ is firmly nonexpansive, then we have

$$\begin{aligned} \langle \nabla f(x^k), x^k - x^* \rangle =& \langle (I - P_Q)Ax^k, Ax^k - Ax^* \rangle \\ =& \langle (I - P_Q)Ax^k - (I - P_Q)Ax^*, Ax^k - Ax^* \rangle \\ \geq& \|(I - P_Q)Ax^k\|^2 = 2f(x^k). \end{aligned} \tag{19}$$

By (19) and $\nabla f(x^k) = 0$, we get $f(x^k) = 0$. Then $Ax^k = P_Q(Ax^k)$, i.e., $Ax^k \in Q$. Therefore, x^k is a solution of the SFP (1).

Next we present two lemmas which play important role in the proof of the main results.

Lemma 5 *The sequence $\{x^k\}_{k \in \mathbb{N}}$ generated by Algorithm 1 is bounded.*

Proof Let $x^* = P_\Gamma u$. Using (12), we have

$$\begin{aligned} \langle v^k - P_C(v^k), v^k - x^* \rangle =& \langle (v^k - P_C(v^k)) - (x^* - P_C(x^*)), v^k - x^* \rangle \\ \geq& \|v^k - P_C(v^k)\|^2. \end{aligned} \tag{20}$$

From (20), it follows

$$\begin{aligned}
&\|2P_C(v^k) - v^k - x^*\|^2 \\
&\quad = \|2(P_C(v^k) - v^k) + (v^k - x^*)\|^2 \\
&\quad = 4\Big[\|P_C(v^k) - v^k\|^2 - \langle v^k - P_C(v^k), v^k - x^*\rangle\Big] + \|v^k - x^*\|^2 \\
&\quad \le \|v^k - x^*\|^2.
\end{aligned} \tag{21}$$

By the definition of v^k, we get

$$\begin{aligned}
\|v^k - x^*\|^2 &= \|(x^k - x^*) - 2\lambda_k \nabla f(x^k)\|^2 \\
&= \|x^k - x^*\|^2 + 4\lambda_k^2 \|\nabla f(x^k)\|^2 - 4\lambda_k \langle \nabla f(x^k), x^k - x^*\rangle \\
&\le \|x^k - x^*\|^2 + 4\lambda_k^2 \|\nabla f(x^k)\|^2 - 8\lambda_k f(x^k) \\
&= \|x^k - x^*\|^2 - 4\gamma(2-\gamma)\frac{f^2(x^k)}{\|\nabla f(x^k)\|^2},
\end{aligned} \tag{22}$$

where the inequality comes from (19) and the last equality originates from (14). By (15),

$$\begin{aligned}
\|x^{k+1} - x^*\| \le &\beta_k \|u - x^*\| + (1-\beta_k)\left\| \frac{\alpha}{2}(2P_C(v^k) - v^k) + \left(1 - \frac{\alpha}{2}\right)x^k - x^*\right\| \\
\le &\beta_k \|u - x^*\| + (1-\beta_k)\frac{\alpha}{2}\left\|2P_C(v^k) - v^k - x^*\right\| \\
&+ (1-\beta_k)\left(1 - \frac{\alpha}{2}\right)\|x^k - x^*\|.
\end{aligned} \tag{23}$$

Combining (21)–(23), we obtain

$$\begin{aligned}
\|x^{k+1} - x^*\| \le &\beta_k \|u - x^*\| + (1-\beta_k)\|x^k - x^*\| \\
\le &\max\{\|u - x^*\|, \|x^k - x^*\|\}.
\end{aligned}$$

Furthermore, we know

$$\|x^{k+1} - x^*\| \le \max\{\|u - x^*\|, \|x^0 - x^*\|\}.$$

Thus, $\{x^k\}_{k\in\mathbb{N}}$ is bounded.

Lemma 6 *Let the sequence $\{x^k\}_{k\in\mathbb{N}}$ be generated by Algorithm 1. If*

$$\lim_{k\to\infty} \frac{f^2(x^k)}{\|\nabla f(x^k)\|^2} = 0 \tag{24}$$

and

$$\lim_{k\to\infty} \|2P_C(v^k) - v^k - x^k\| = 0, \tag{25}$$

then $\omega_w(x^k) \subseteq \Gamma$.

Proof Since

$$\|\nabla f(x^k)\|^2 \le 2\|A\|^2 f(x^k),$$

from (24), we get

$$\lim_{k\to\infty} f(x^k) = 0, \tag{26}$$

and consequently

$$\lim_{k\to\infty} \|\nabla f(x^k)\| = 0.$$

From (24) and the definition of λ_k, we get

$$\lim_{k\to\infty} \lambda_k^2 \|\nabla f(x^k)\|^2 = 0. \tag{27}$$

Combining (16) and (27) above, we get

$$\lim_{k\to\infty} \|v^k - x^k\| = 0. \tag{28}$$

Since $\|P_C(v^k) - v^k\| \le \frac{1}{2}\left(\|2P_C(v^k) - v^k - x^k\| + \|x^k - v^k\|\right)$, using (25) and (28), we have

$$\lim_{k\to\infty} \|P_C(v^k) - v^k\| = 0. \tag{29}$$

Since P_C is nonexpansive, by (28) and (29), we get

$$\lim_{k\to\infty} \|P_C(x^k) - x^k\| = 0. \tag{30}$$

Because $\{x^k\}_{k\in\mathbb{N}}$ is bounded, we can take $\hat{x} \in \omega_w(x^k)$ and let $\{x^{k_l}\}_{l\in\mathbb{N}}$ be a subsequence of $\{x^k\}_{k\in\mathbb{N}}$ weakly converging to $\hat{x}$. By Lemma 3 and (30), we get $\hat{x} \in C$. From (26) and the weak lower semicontinuity of f, it follows that

$$0 \le f(\hat{x}) \le \liminf_{l\to\infty} f(x^{k_l}) = \lim_{k\to\infty} f(x^k) = 0. \tag{31}$$

Hence $f(\hat{x}) = 0$, i.e., $A\hat{x} \in Q$. Therefore, $\omega_w(x^k) \subseteq \Gamma$.

In the proof of the main results, we also need the following technical lemma from [16].

Lemma 7 *Assume $\{s^k\}_{k\in\mathbb{N}}$ is a sequence of nonnegative real numbers such that*

$$\begin{aligned} s^{k+1} &\le (1-\beta_k)s^k + \beta_k\delta^k, \quad k \ge 0, \\ s^{k+1} &\le s^k - \eta^k + \gamma^k, \quad k \ge 0, \end{aligned}$$

where $\{\beta_k\}_{k\in\mathbb{N}}$ is a sequence in $(0, 1)$, $\{\eta^k\}_{k\in\mathbb{N}}$ is a sequence of nonnegative real numbers, and $\{\delta^k\}_{k\in\mathbb{N}}$ and $\{\gamma^k\}_{k\in\mathbb{N}}$ are two sequences in $\mathbb{R}$ such that

(i) $\sum_{k=0}^{\infty} \beta_k = \infty$,
(ii) $\lim_{k\to\infty} \gamma^k = 0$,
(iii) $\lim_{l\to\infty} \eta^{k_l} = 0$ *implies* $\limsup_{l\to\infty} \delta^{k_l} \le 0$ *for any subsequence* $\{k_l\}_{l\in\mathbb{N}} \subseteq \{k\}_{k\in\mathbb{N}}$.

Then $\lim_{k\to\infty} s^k = 0$.

Theorem 1 *Let $\{x^k\}_{k\in\mathbb{N}}$ be the sequence generated by Algorithm 1. Then $\{x^k\}_{k\in\mathbb{N}}$ strongly converges to $x^* = P_\Gamma u$.*

Proof By (15), we have

$$\begin{aligned} &\|x^{k+1} - x^*\|^2 \\ &= \left\| \beta_k(u - x^*) + (1-\beta_k)\left[\frac{\alpha}{2}(2P_C(v^k) - v^k - x^*) + \left(1 - \frac{\alpha}{2}\right)(x^k - x^*)\right]\right\|^2 \\ &= \beta_k^2\|u - x^*\|^2 + (1-\beta_k)^2\left\|\frac{\alpha}{2}(2P_C(v^k) - v^k - x^*) + \left(1 - \frac{\alpha}{2}\right)(x^k - x^*)\right\|^2 \\ &\quad + 2\beta_k(1-\beta_k)\left\langle u - x^*, \frac{\alpha}{2}(2P_C(v^k) - v^k - x^*) + \left(1 - \frac{\alpha}{2}\right)(x^k - x^*)\right\rangle \\ &\le \beta_k^2\|u - x^*\|^2 + (1-\beta_k)^2\left[\frac{\alpha}{2}\|2P_C(v^k) - v^k - x^*\|^2 + \left(1 - \frac{\alpha}{2}\right)\|x^k - x^*\|^2\right] \\ &\quad + 2\beta_k(1-\beta_k)\left\langle u - x^*, \frac{\alpha}{2}(2P_C(v^k) - v^k - x^*) + \left(1 - \frac{\alpha}{2}\right)(x^k - x^*)\right\rangle \\ &\le (1-\beta_k)\|x^k - x^*\|^2 + \beta_k\big[\beta_k\|u - x^*\|^2 + \alpha(1-\beta_k)\langle u - x^*, 2P_C(v^k) - v^k - x^*\rangle \\ &\quad + (1-\beta_k)(2-\alpha)\langle u - x^*, x^k - x^*\rangle\big], \end{aligned} \tag{32}$$

where the first inequality comes from (10) and the second inequality comes from (21) and (22). On the other hand, from (10) and (15), we get

$$\begin{aligned}
&\|x^{k+1}-x^*\|^2\\
&=\left\|\beta_k(u-x^*)+(1-\beta_k)\left[\frac{\alpha}{2}(2P_C(v^k)-v^k-x^*)+\left(1-\frac{\alpha}{2}\right)(x^k-x^*)\right]\right\|^2\\
&\le \beta_k\|u-x^*\|^2+(1-\beta_k)\left\|\frac{\alpha}{2}(2P_C(v^k)-v^k-x^*)+\left(1-\frac{\alpha}{2}\right)(x^k-x^*)\right\|^2\\
&\le \beta_k\|u-x^*\|^2+\left[\frac{\alpha}{2}\|2P_C(v^k)-v^k-x^*\|^2+\left(1-\frac{\alpha}{2}\right)\|x^k-x^*\|^2\right.\\
&\quad\left.-\frac{\alpha}{2}\left(1-\frac{\alpha}{2}\right)\|2P_C(v^k)-v^k-x^k\|^2\right]\\
&\le \|x^k-x^*\|^2+\beta_k\|u-x^*\|^2-2\alpha\gamma(2-\gamma)\frac{f^2(x^k)}{\|\nabla f(x^k)\|^2}\\
&\quad-\frac{\alpha}{2}\left(1-\frac{\alpha}{2}\right)\|2P_C(v^k)-v^k-x^k\|^2, \qquad (33)
\end{aligned}$$

where the last inequality comes from (21) and (22). Set

$$\begin{aligned}
s^k&=\|x^k-x^*\|^2,\\
\gamma^k&=\beta_k\|u-x^*\|^2,\\
\eta^k&=2\alpha\gamma(2-\gamma)\frac{f^2(x^k)}{\|\nabla f(x^k)\|^2}+\frac{\alpha}{2}\left(1-\frac{\alpha}{2}\right)\|2P_C(v^k)-v^k-x^k\|^2,\\
\delta^k&=\beta_k\|u-x^*\|^2+\alpha(1-\beta_k)\langle u-x^*,2P_C(v^k)-v^k-x^*\rangle\\
&\quad+(1-\beta_k)(2-\alpha)\langle u-x^*,x^k-x^*\rangle.
\end{aligned}$$

From (32) and (33), we derive the inequalities as follows

$$\begin{aligned}
s^{k+1}&\le(1-\beta_k)s^k+\beta_k\delta^k, \quad k\ge 0,\\
s^{k+1}&\le s^k-\eta^k+\gamma^k, \quad k\ge 0.
\end{aligned}$$

Since β_k satisfies assumption (H1), we obtain $\lim_{k\to\infty}\gamma^k=0$. To use Lemma 7, it suffices to verify that, for any subsequence $\{k_l\}_{l\in\mathbb{N}}\subseteq\{k\}_{k\in\mathbb{N}}$, $\lim_{l\to\infty}\eta^{k_l}=0$ implies

$$\limsup_{l\to\infty}\delta^{k_l}\le 0.$$

From $\lim_{l\to\infty}\eta^{k_l}=0$, we have

$$\lim_{l\to\infty}\frac{f^2(x^{k_l})}{\|\nabla f(x^{k_l})\|^2}=0, \tag{34}$$

and

$$\lim_{l\to\infty}\|2P_C(v^{k_l})-v^{k_l}-x^{k_l}\|=0. \tag{35}$$

According to (35) and (H2), in order to prove $\limsup_{l\to\infty}\delta^{k_l}\le 0$, we only need to prove

$$\limsup_{l\to\infty}\langle u-x^*, x^{k_l}-x^*\rangle\le 0.$$

Since $\{x^{k_l}\}_{l\in\mathbb{N}}$ is bounded by Lemma 5, it is easy to choose a subsequence $\{x^{k_{l_j}}\}_{j\in\mathbb{N}}$ which weakly to $\hat{x}$ and such that

$$\limsup_{l\to\infty}\langle u-x^*, x^{k_l}-x^*\rangle=\lim_{j\to\infty}\langle u-x^*, x^{k_{l_j}}-x^*\rangle=\langle u-x^*, \hat{x}-x^*\rangle.$$

From (34), (35) and Lemma 6, we get $\hat{x}\in\omega_w(x^{k_l})\subseteq\Gamma$. By Lemma 1, $\langle u-x^*, \hat{x}-x^*\rangle\le 0$. Therefore $\limsup_{l\to\infty}\langle u-x^*, x^{k_l}-x^*\rangle\le 0$ and then we get

$$\limsup_{l\to\infty}\delta^{k_l}\le 0.$$

From Lemma 7, we conclude that $x^k\to x^*$. The proof is complete.

Now we will show the convergence of Algorithm 2. It is easy to extend Lemma 5 to Algorithm 2.

Lemma 8 *The sequence $\{x^k\}_{k\in\mathbb{N}}$ generated by Algorithm 2 is bounded.*

Lemma 9 *Let the sequence $\{x^k\}_{k\in\mathbb{N}}$ be generated by Algorithm 2. If*

$$\lim_{k\to\infty}\frac{f_k^2(x^k)}{\|\nabla f_k(x^k)\|^2}=0 \tag{36}$$

and

$$\lim_{k\to\infty}\|2P_{C_k}(v^k)-v^k-x^k\|=0, \tag{37}$$

then $\omega_w(x^k)\subseteq\Gamma$.

Proof From (36), (37) and Lemma 6, we get

$$\lim_{k\to\infty} f_k(x^k) = 0, \tag{38}$$

$$\lim_{k\to\infty} \lambda_k \|\nabla f_k(x^k)\| = 0, \tag{39}$$

and

$$\lim_{k\to\infty} \|P_{C_k}(x^k) - x^k\| = 0. \tag{40}$$

By the definition of z^{k+1},

$$\begin{aligned} \|z^{k+1} - x^k\| &= \|P_{C_k}(2y^{k+1} - x^k) - x^k\| \\ &\le \|P_{C_k}(2y^{k+1} - x^k) - P_{C_k}(x^k)\| + \|P_{C_k}(x^k) - x^k\| \\ &\le 2\|y^{k+1} - x^k\| + \|P_{C_k}(x^k) - x^k\| \\ &= 2\lambda_k \|\nabla f_k(x^k)\| + \|P_{C_k}(x^k) - x^k\|. \end{aligned} \tag{41}$$

Putting (39) and (40) into (41), we obtain

$$\lim_{k\to\infty} \|z^{k+1} - x^k\| = 0. \tag{42}$$

By Lemma 8, we know that there exists a subsequence $\{x^{k_l}\}_{l\in\mathbb{N}}$ of $\{x^k\}_{k\in\mathbb{N}}$ converging to $\hat{x}$. Next, we show that $\hat{x} \in \Gamma$. In fact, since $z^{k_l+1} \in C_{k_l}$, by the definition of C_{k_l}, we have

$$c(x^{k_l}) + \langle \xi^{k_l}, z^{k_l+1} - x^{k_l}\rangle \le 0,$$

where $\xi^{k_l} \in \partial c(x^{k_l})$. By the assumption that ξ^{k_l} is bounded and (42), we have

$$c(x^{k_l}) \le -\langle \xi^{k_l}, z^{k_l+1} - x^{k_l}\rangle \le \|\xi^{k_l}\|\|z^{k_l+1} - x^{k_l}\| \to 0, \quad l \to \infty,$$

which implies $c(\hat{x}) \le 0$, i.e., $\hat{x} \in C$. Since $P_{Q_{k_l}}(Ay^{k_l}) \in Q_{k_l}$, we have

$$q(Ay^{k_l}) + \langle \eta^{k_l}, P_{Q_{k_l}}(Ay^{k_l}) - Ay^{k_l}\rangle \le 0,$$

where $\eta^{k_l} \in \partial q(Ay^{k_l})$. From the boundedness of $\{\eta^{k_l}\}$ and (38), it follows that

$$q(Ay^{k_l}) \le \|\eta^{k_l}\|\|P_{Q_{k_l}}(Ay^{k_l}) - Ay^{k_l}\| \to 0, \quad l \to \infty.$$

Similarly, we can obtain that $q(A\hat{x}) \le 0$, i.e., $A\hat{x} \in Q$ and $\omega_w(x^k) \subseteq \Gamma$.

Theorem 2 *Let $\{x^k\}_{k\in\mathbb{N}}$ be generated by Algorithm 2, then $\{x^k\}_{k\in\mathbb{N}}$ converges to a solution of the SFP* (1).

Proof Using similar arguments in the proof of Theorem 1, we have

$$\lim_{l\to\infty} \frac{f_{k_l}^2(x^{k_l})}{\|\nabla f_{k_l}(x^{k_l})\|^2} = 0, \tag{43}$$

and

$$\lim_{l\to\infty} \|2P_{C_{k_l}}(v^{k_l}) - v^{k_l} - x^{k_l}\| = 0. \tag{44}$$

From (43), (44) and Lemma 9, we get $\hat{x} \in \omega_w(x^{k_l}) \subseteq \Gamma$. Following the rest of the proof of Theorem 1, we get that $\{x^k\}_{k\in\mathbb{N}}$ converges to a solution of the SFP (1). This completes the proof.

4 Haugazeau-Type Algorithm

By combining Haugazeau method and Douglas-Rachford method, we introduce a new algorithm with strong convergence.

Algorithm 3

Step 0. Input $k := 0$, $x^0 \in \mathscr{H}_1$ and $\alpha \in (0, 2)$.
Step 1. Generate x^{k+1} by

$$\begin{cases} y^{k+1} := x^k - \lambda_k \nabla f(x^k), \\ z^{k+1} := P_C(2y^{k+1} - x^k), \\ w^{k+1} := x^k + \alpha(z^{k+1} - y^{k+1}), \\ v^k = x^k - 2\lambda_k \nabla f(x^k) \\ S_k = \Big\{z \in \mathscr{H}_1 : \|w^{k+1} - z\|^2 \le \|x^k - z\|^2 - 2\alpha\gamma(2-\gamma)\dfrac{f^2(x^k)}{\|\nabla f(x^k)\|^2} \\ \qquad - \left(1 - \dfrac{\alpha}{2}\right)\dfrac{\alpha}{2}\|2P_C(v^k) - v^k - x^k\|^2\Big\}, \\ T_k = \Big\{z \in \mathscr{H}_1 : \langle x^0 - x^k, z - x^k\rangle \le 0\Big\}, \\ x^{k+1} = P_{S_k\cap T_k}(x^0). \end{cases} \tag{45}$$

where λ_k is given by (14).
Step 2. If $\nabla f(x^k) = 0$ and $x^{k+1} = x^k$, then terminate. Otherwise, set $k := k + 1$ and go to Step 1.

Now we ascertain the validity of the stopping criterion used in Step 2.

Lemma 10 *If $\nabla f(x^k) = 0$ and $x^{k+1} = x^k$ for some k, then x^k is a solution of the SFP* (1).

Proof Suppose $\nabla f(x^k) = 0$. Then $y^{k+1} = x^k$ and $z^{k+1} = P_C(x^k)$. Observe that $x^{k+1} = x^k$ implies $x^k \in S_k$. By the definition of S_k, $w^{k+1} = x^k$. So we have $x^k = P_C(x^k)$, i.e., $x^k \in C$. By (19), we get $f(x^k) = 0$. Then $Ax^k = P_Q(Ax^k)$, i.e., $Ax^k \in Q$. Thus, x^k is a solution of the SFP (1).

Lemma 11 *It holds*

$$\Gamma \subseteq S_k \cap T_k. \tag{46}$$

Proof Let $z \in \Gamma$. Similar to the proof of equality (15), from (13) we can get

$$w^{k+1} = \left(1 - \frac{\alpha}{2}\right)x^k + \frac{\alpha}{2}(2P_C(v^k) - v^k). \tag{47}$$

Combining (21), (22) and (47), we get

$$\begin{aligned}
\|w^{k+1} - z\|^2 &= \left(1 - \frac{\alpha}{2}\right)\|x^k - z\|^2 + \frac{\alpha}{2}\|2P_C(v^k) - v^k - z\|^2 \\
&\quad - \left(1 - \frac{\alpha}{2}\right)\frac{\alpha}{2}\|2P_C(v^k) - v^k - x^k\|^2 \\
&\le \|x^k - z\|^2 - 2\alpha\gamma(2-\gamma)\frac{f^2(x^k)}{\|\nabla f(x^k)\|^2} \\
&\quad - \left(1 - \frac{\alpha}{2}\right)\frac{\alpha}{2}\|2P_C(v^k) - v^k - x^k\|^2,
\end{aligned}$$

which means that $\Gamma \subset S_k$. Next, we use induction to prove that (46) holds for all $k \ge 0$. It is easy to check that $\Gamma \subset T_0 = \mathscr{H}_1$. Assume now (46) holds for $k = n$. It then turns out that $x^{n+1} = P_{S_n \cap T_n}(x^0)$ is well defined. By Lemma 1,

$$\langle x^{n+1} - z, x^{n+1} - x^0 \rangle \le 0.$$

This implies that $z \in T_{n+1}$ and hence $\Gamma \subset T_{n+1}$. Hence, (46) holds for $k = n + 1$, and thus for all $k \ge 0$.

Theorem 3 *Let $\{x^k\}_{k\in\mathbb{N}}$ be generated by Algorithm 3. Then the following hold:*

(i) $\lim_{k\to\infty} \|x^{k+1} - x^k\| = 0$,
(ii) $\omega_w(x^k) \subseteq \Gamma$,
(iii) $x^k \to x^* \in \Gamma$, *where* $x^* = P_\Gamma x^0$.

Proof

(i) From the definition of T_k, we know that $x^k = P_{T_k}x^0$. Then, we obtain

$$\|x^k - x^0\| = \| P_{T_k} x^0 - x^0 \| \leq \| P_\Gamma x^0 - x^0 \|, \tag{48}$$

which means that $\{\|x^k - x^0\|\}_{k \in \mathbb{N}}$ is increasing and $\{x^k\}_{k \in \mathbb{N}}$ is bounded. Therefore, $\lim_{k \to \infty} \|x^k - x^0\|$ exists. Using $x^k = P_{T_k} x^0$ and $x^{k+1} \in T_k$, we have $\langle x^{k+1} - x^k, x^k - x^0 \rangle \geq 0$. Hence, we get

$$\begin{aligned} \|x^{k+1} - x^0\|^2 - \|x^k - x^0\|^2 &= \|x^{k+1} - x^k\|^2 + 2\langle x^{k+1} - x^k, x^k - x^0 \rangle \\ &\geq \|x^{k+1} - x^k\|^2. \end{aligned} \tag{49}$$

Consequently, from (48) and (49), we have

$$\sum_{l=0}^{k} \|x^{l+1} - x^l\|^2 \leq \|P_\Gamma x^0 - x^0\|^2.$$

Therefore,

$$\sum_{k=0}^{\infty} \|x^{k+1} - x^k\|^2 < \infty,$$

and

$$\lim_{k \to \infty} \|x^{k+1} - x^k\| = 0.$$

(ii) Since $x^{k+1} \in S_k$, by the definition of S_k, we get $\|x^{k+1} - w^{k+1}\| \leq \|x^{k+1} - x^k\| \to 0$ and also that

$$\frac{f^2(x^k)}{\|\nabla f(x^k)\|^2} \to 0, \tag{50}$$

and

$$\|2P_C(v^k) - v^k - x^k\| \to 0.$$

By Lemma 6, we get $\omega_w(x^k) \subseteq \Gamma$.

(iii) Take $x^* \in \omega_w(x^k)$ and let $\{x^{k_l}\}_{l \in \mathbb{N}}$ be a subsequence of $\{x^k\}_{k \in \mathbb{N}}$ weakly converging to x^*. From (48), it follows

$$\begin{aligned} \|x^{k_l} - P_\Gamma x^0\|^2 &= \|x^{k_l} - x^0\|^2 + \|x^0 - P_\Gamma x^0\|^2 + 2\langle x^{k_l} - x^0, x^0 - P_\Gamma x^0 \rangle \\ &\leq 2\|x^0 - P_\Gamma x^0\|^2 + 2\langle x^{k_l} - x^0, x^0 - P_\Gamma x^0 \rangle \\ &= 2\langle x^{k_l} - P_\Gamma x^0, x^0 - P_\Gamma x^0 \rangle. \end{aligned}$$

This implies that

$$\limsup_{l\to\infty} \|x^{k_l} - P_\Gamma x^0\|^2 \le \langle x^* - P_\Gamma x^0, x^0 - P_\Gamma x^0\rangle \le 0.$$

Therefore, $x^{k_l} \to P_\Gamma x^0$, i.e., $x^k \to P_\Gamma x^0$. The proof is complete.

Remark 1 It is easy to present a relaxed version of Algorithm 3 and show its convergence by combining Algorithms 2 and 3.

5 Example Results

In this section, we provide computational experiment and compare our Algorithm 2 with Algorithm 5.1 in [19] and Algorithm 2 in [23]. In addition, we compare our Algorithm 3 with Algorithm 3.7 in [19]. In the numerical results listed in the following table, "Iter." and "CPU time" denote the number of iterations and CPU times in seconds, respectively.

Example 1 Consider the following LASSO problem [22]:

$$\min\left\{\frac{1}{2}\|Ax-b\|_2^2 \,:\, x\in\mathbb{R}^n,\ \|x\|_1\le\tau\right\}, \tag{51}$$

where $A \in \mathbb{R}^{m\times n}$, $m < n$, $b \in \mathbb{R}^n$ and $\tau > 0$. We generate the system matrix A from a standard normal distribution with mean zero and unit variance. The true sparse signal x^* is generated from uniformly distribution in the interval $[-2, 2]$ with random K position nonzero while the rest is kept zero. The sample data $b = Ax^*$.

In this example, we apply the proposed Algorithm 2 to solve the LASSO problem, which aims to finding a sparse solution of an underdetermined linear system.

Under certain conditions on matrix A, the solution of the minimization problem (51) is equivalent to the ℓ_0-norm solution of the underdetermined linear system. For the considered SFP (1), we define $C = \{x \mid \|x\|_1 \le \tau\}$ and $Q \doteq \{b\}$. Since the projection onto the closed convex C does not have a closed form solution and so we make use of the subgradient projection. Define a convex function $c(x) = \|x\|_1 - \tau$ and denote the level set C_k by:

$$C_k = \{x \,:\, c(x^k) + \langle \xi^k, x - x^k\rangle \le 0\},$$

where $\xi^k \in \partial c(x^k)$. Then the orthogonal projection onto C_k can be calculated by the following:

$$P_{C_k}(x) = \begin{cases} x, & \text{if } c(x^k) + \langle \xi^k, x - x^k \rangle \leq 0, \\ x - \dfrac{c(x^k) + \langle \xi^k, x - x^k \rangle}{\|\xi^k\|^2} \xi^k, & \text{otherwise.} \end{cases}$$

It is worth noting that the subdifferential ∂c at x^k is

$$\partial c(x^k) = \begin{cases} 1, & \text{if } x^k > 0, \\ [-1, 1], & \text{if } x^k = 0, \\ -1, & \text{if } x^k < 0. \end{cases}$$

We initialize the algorithms at the origin and terminate it when

$$\frac{\|x^k - x^*\|}{\max\{1, \|x^k\|\}} < 10^{-3}.$$

We took $\alpha = 0.8$ and $\gamma = 1.9$ in the Algorithm 2, $\rho_n = 1$ in Algorithm 5.1 of [19] and $\lambda_n = 2$ in Algorithm 2 of [23]. In addition, the parameter β_k is taken as (3.18) in [14].

The corresponding results are reported in Table 1, where "Max" means that the number of iterations hits 10,000. We can see from Table 1 that the Algorithm 2 performs better than the other algorithms from the iteration numbers and CPU time.

We took $\alpha = 1.9$ and $\gamma = 0.3$ in the Algorithm 3 and $\rho_n = 0.5$ in Algorithm 3.7 of [19]. We initialize the algorithms at the origin and terminate it when

$$\frac{\|x^k - x^*\|}{\max\{1, \|x^k\|\}} < 10^{-2}.$$

The corresponding results are reported in Table 2. We can see from Table 2 that the Algorithm 3 is better than Algorithm 3.7 of [19] from the iteration numbers and

Table 1 Computational results of three algorithms for solving SFP (1)

Problem size			Iter			CPU time		
m	n	K	Alg 2	Alg 5.1 in [19]	Alg 2 in [23]	Alg 2	Alg 5.1 in [19]	Alg 2 in [23]
240	1024	30	795	1135	Max	2.1564	2.9900	26.8720
480	2048	60	818	1050	Max	7.4100	9.0140	84.0586
720	3072	90	921	1156	Max	16.6737	20.1963	174.7703
960	4096	120	735	1001	Max	23.2229	30.3517	301.0742
1200	5120	150	800	1110	Max	36.8285	50.1296	454.7727
1440	6144	180	823	1073	Max	54.7822	71.1720	637.0593
1680	7168	210	859	1244	Max	75.6329	104.4183	840.7035
1920	8192	240	795	1055	Max	85.2829	113.5449	1077.3

Table 2 Computational results for solving the SFP (1) by the Algorithm 3 and Algorithm 3.7 in [19]

Problem size			Iter		CPU time	
m	n	K	Alg 3	Alg 3.7 in [19]	Alg 3	Alg 3.7 in [19]
240	1024	30	7196	12,555	14.2062	23.2478
480	2048	60	7121	12,845	43.7493	73.3263
720	3072	90	5214	8632	60.4976	99.9253
960	4096	120	6408	11,280	125.4982	213.7182
1200	5120	150	5581	9285	160.8265	262.7904
1440	6144	180	5787	9693	231.0650	383.5354
1680	7168	210	6585	11,102	349.7393	579.9234
1920	8192	240	6144	10,596	416.8434	712.2142

CPU time. We did not compare Algorithm 3 with the algorithm (6) since its error decreases very slowly comparing Algorithm and Algorithm 3.7 of [19].

Comparing Tables 1 and 2, it concludes that Haugazeau-type algorithm performs better than Halpern-type algorithm.

References

1. Bauschke, H.H., Combettes, P.L., *Convex Analysis and Monotone Operator Theory in Hilbert Spaces*, Second Edition, Springer, 2017.
2. H.H. Bauschke, P.L. Combettes, A weak-to-strong convergence principle for Fejer-monotone methods in Hilbert spaces, Math. Oper. Res. 26(2) (2001) 248–264.
3. Byrne, C.L.: Iterative oblique projection onto convex sets and the split feasibility problem, *Inverse Probl.* **18**, 441–453 (2002)
4. Byrne, C.L.: A unified treatment of some iterative algorithms in signal processing and image reconstruction, *Inverse Probl.* **20**, 103–120 (2004)
5. Censor, Y., Elfving, T.: A multiprojection algorithm using Bregman projections in a product space, *Numer. Algorithms* **8**, 221–239 (1994)
6. Dang, Y., Sun, J., Zhang, S., Double projection algorithms for solving the split feasibility problems, *J. Ind. Manag. Optim.* **15**, 2023–2034 (2019)
7. Dong, Q.L., He, S., Rassias, M. Th.: Douglas-Rachford splitting methods with linearization for the split feasibility problem, J. Global Optim. **79**, 813–836 (2021).
8. Dong, Q.L., Li, X.H., He, S.: Outer perturbations of a projection method and two approximation methods for the split equality problem, *Optimization*, **67**, 1429C1446 (2018)
9. Dong, Q.L., Yao, Y., He, S.: Weak convergence theorems of the modified relaxed projection algorithms for the split feasibility problem in Hilbert spaces, *Optim. Lett.* **8**, 1031–1046 (2014)
10. Douglas, J., Rachford, H.H.: On the numerical solution of heat conduction problems in two or three space variables, *Trans. Am. Math. Soc.* **82**, 421–439 (1956)
11. Fukushima, M.A.: relaxed projection method for variational inequalities, *Math. Program.* **35**, 58–70, (1986)
12. Gibali, A., Liu, L., Tang, Y.C.: Note on the modified relaxation CQ algorithm for the split feasibility problem, *Optim. Lett.* **12**, 817–830 (2018)
13. Haugazeau, Y.: Sur les inéquations variationnelles et la minimisation de fonctionnelles convexes. Paris: Thèse Université de Paris, 1968.

14. He, S., Wu, T., Cho, Y.J., Rassias, Th.M.: Optimal parameter selections for a general Halpern iteration, *Numer Algorithms* **82(7)**, 1171–1188 (2019).
15. He, S., Xu, H.K.: The selective projection method for convex feasibility and split feasibility problems, *J. Nonlinear Sci. Appl.* **19(7)**, 1199–1215 (2018)
16. He, S., Yang, C.: Solving the variational inequality problem defined on intersection of finite level sets, *Abstr. Appl. Anal.* (2013) 8 p; Article ID 942315
17. Lindstrom, S.B., Sims, B., Survey: Sixty Years of Douglas Ratchford. J. Austral. Math. Soc. **110**, 333–370 (2021)
18. Lions, P.L., Mercier, B.: Splitting algorithms for the sum of two nonlinear operators, *SIAM J. Numer. Anal.* **16**, 964–979 (1979)
19. López, G., Martín-Márquez, V., Wang, F., Xu, H.K.: Solving the split feasibility problem without prior knowledge of matrix norms, *Inverse Probl.* **27**, 085004 (2012)
20. Qu, B., Wang, C., Xiu, N.: Analysis on Newton projection method for the split feasibility problem, *Comput. Optim. Appl.* **67**, 175–199 (2017)
21. Shehu, Y., Iyiola, O.S.: Nonlinear iteration method for proximal split feasibility problems, *Math. Method Appl. Sci.* **41**, 781–802 (2018)
22. Tibshirani, R.: Regression shrinkage and selection via the lasso, *J.R. Stat. Soc. Ser. B. Stat. Methodol.* **58**, 267–288 (1996)
23. Wang, F.: Polyak's gradient method for split feasibility problem constrained by level sets, *Numer. Algorithm* **77**, 925–938 (2018)
24. Wang, F.: Strong convergence of two algorithms for the split feasibility problem in Banach spaces, *Optimization* **67(10)**, 1649–1660 (2018)
25. Wang, J.H., Hu, Y.H., Li, C., Yao, J.C.: Linear convergence of CQ algorithms and applications in gene regulatory network inference. *Inverse Probl.* **33**, 055017 (2017)
26. Yen, L.H., Muu, L.D., Huyen, N.T.T.: An algorithm for a class of split feasibility problems: application to a model in electricity production, *Math. Meth. Oper. Res.* **84**, 549–565 (2016)
27. Zhao, J., Zong, H.: Iterative algorithms for solving the split feasibility problem in Hilbert spaces, *J. Fix. Point Theory A.* **20**, 11 (2018)
28. Zhao, J., Yang, Q.: Self-adaptive projection methods for the multiple-sets split feasibility problem, *Inverse Probl.* **27**, 035009 (2011)

Some Triple Integral Inequalities for Functions Defined on Three-Dimensional Bodies Via Gauss-Ostrogradsky Identity

Silvestru Sever Dragomir

Abstract In this paper, by the use of Gauss-Ostrogradsky identity, we establish some inequalities for functions of three variables defined on closed and bounded bodies of the Euclidean space $\mathbb{R}^3$. Some examples for three-dimensional balls are also provided.

1991 Mathematics Subject Classification 26D15

1 Introduction

Recall the following inequalities of Hermite-Hadamard's type for convex functions defined on a ball $B(C, R)$, where $C = (a, b, c) \in \mathbb{R}^3$, $R > 0$ and

$$B(C, R) := \left\{ (x, y, z) \in \mathbb{R}^3 \middle| (x - a)^2 + (y - b)^2 + (z - c)^2 \le R^2 \right\}.$$

The following theorem holds [10].

Theorem 1 *Let $f : B(C, R) \to \mathbb{R}$ be a convex function on the ball $B(C, R)$. Then we have the inequality:*

$$f(a, b, c) \le \frac{1}{V(B(C, R))} \iiint_{B(C,R)} f(x, y, z)\, dxdydz$$

S. S. Dragomir (✉)
Mathematics, College of Engineering & Science, Victoria University, Melbourne City, MC, Australia

DST-NRF Centre of Excellence in the Mathematical and Statistical Sciences, School of Computer Science & Applied Mathematics, University of the Witwatersrand, Johannesburg, South Africa
e-mail: sever.dragomir@vu.edu.au
http://rgmia.org/dragomir

I. N. Parasidis et al. (eds.), *Mathematical Analysis in Interdisciplinary Research*, Springer Optimization and Its Applications 179,
https://doi.org/10.1007/978-3-030-84721-0_13

$$\leq \frac{1}{\sigma(B(C,R))} \iint_{S(C,R)} f(x,y,z)\, dS, \tag{1.1}$$

where

$$S(C,R) := \left\{ (x,y,z) \in \mathbb{R}^3 \middle| (x-a)^2 + (y-b)^2 + (z-c)^2 = R^2 \right\}$$

and

$$V(B(C,R)) = \frac{4\pi R^3}{3}, \sigma(B(C,R)) = 4\pi R^2.$$

If the assumption of convexity is dropped, then one can prove the following Ostrowski type inequality for the center of the ball as well, see [11].

Theorem 2 *Assume that $f : B(C,R) \to \mathbb{C}$ is differentiable on $B(C,R)$. Then*

$$\left| f(a,b,c) - \frac{1}{V(B(C,R))} \iiint_{B(C,R)} f(x,y,z)\, dxdydz \right|$$
$$\leq \frac{3}{8} R \left[\left\| \frac{\partial f}{\partial x} \right\|_{B(C,R),\infty} + \left\| \frac{\partial f}{\partial y} \right\|_{B(C,R),\infty} + \left\| \frac{\partial f}{\partial z} \right\|_{B(C,R),\infty} \right], \tag{1.2}$$

provided

$$\left\| \frac{\partial f}{\partial x} \right\|_{B(C,R),\infty} := \sup_{(x,y,z)\in B(C,R)} \left| \frac{\partial f(x,y,z)}{\partial x} \right| < \infty,$$

$$\left\| \frac{\partial f}{\partial y} \right\|_{B(C,R),\infty} := \sup_{(x,y,z)\in B(C,R)} \left| \frac{\partial f(x,y,z)}{\partial y} \right| < \infty$$

and

$$\left\| \frac{\partial f}{\partial z} \right\|_{B(C,R),\infty} := \sup_{(x,y,z)\in B(C,R)} \left| \frac{\partial f(x,y,z)}{\partial y} \right| < \infty.$$

This fact can be furthermore generalized to the following Ostrowski type inequality for any point in a convex body $B \subset \mathbb{R}^3$, see [11].

Theorem 3 *Assume that $f : B \to \mathbb{C}$ is differentiable on the convex body B and $(u, v, w) \in B$. If $V(B)$ is the volume of B,, then*

$$\begin{aligned}
&\left| f(u, v, w) - \frac{1}{V(B)} \iiint_B f(x, y, z)\, dxdydz \right| \\
&\leq \frac{1}{V(B)} \iiint_B |x - u| \left(\int_0^1 \left| \frac{\partial f}{\partial x} [t(x, y, z) + (1 - t)(u, v, w)] \right| dt \right) dxdydz \\
&\quad + \frac{1}{V(B)} \iiint_B |y - v| \left(\int_0^1 \left| \frac{\partial f}{\partial y} [t(x, y, z) + (1 - t)(u, v, w)] \right| dt \right) dxdydz \\
&\quad + \frac{1}{V(B)} \iiint_B |z - w| \left(\int_0^1 \left| \frac{\partial f}{\partial y} [t(x, y, z) + (1 - t)(u, v, w)] \right| dt \right) dxdydz \\
&\leq \left\| \frac{\partial f}{\partial x} \right\|_{B,\infty} \frac{1}{V(B)} \iiint_B |x - u|\, dxdydz \\
&\quad + \left\| \frac{\partial f}{\partial y} \right\|_{B,\infty} \frac{1}{V(B)} \iiint_B |y - v|\, dxdydz \\
&\quad + \left\| \frac{\partial f}{\partial z} \right\|_{B,\infty} \frac{1}{V(B)} \iiint_B |z - w|\, dxdydz
\end{aligned} \tag{1.3}$$

provided

$$\left\| \frac{\partial f}{\partial x} \right\|_{B,\infty}, \left\| \frac{\partial f}{\partial y} \right\|_{B,\infty}, \left\| \frac{\partial f}{\partial z} \right\|_{B,\infty} < \infty.$$

In particular,

$$\begin{aligned}
&\left| f(\overline{x_B}, \overline{y_B}, \overline{z_B}) - \frac{1}{V(B)} \iiint_B f(x, y, z)\, dxdydz \right| \\
&\leq \frac{1}{V(B)} \iiint_B |x - \overline{x_B}| \left(\int_0^1 \left| \frac{\partial f}{\partial x} [t(x, y, z) + (1 - t)(\overline{x_B}, \overline{y_B}, \overline{z_B})] \right| dt \right) dxdydz \\
&\quad + \frac{1}{V(B)} \iiint_B |y - \overline{y_B}| \left(\int_0^1 \left| \frac{\partial f}{\partial y} [t(x, y, z) + (1 - t)(\overline{x_B}, \overline{y_B}, \overline{z_B})] \right| dt \right) dxdydz \\
&\quad + \frac{1}{V(B)} \iiint_B |z - \overline{z_B}| \left(\int_0^1 \left| \frac{\partial f}{\partial y} [t(x, y, z) + (1 - t)(\overline{x_B}, \overline{y_B}, \overline{z_B})] \right| dt \right) dxdydz \\
&\leq \left\| \frac{\partial f}{\partial x} \right\|_{B,\infty} \frac{1}{V(B)} \iiint_B |x - \overline{x_B}|\, dxdydz \\
&\quad + \left\| \frac{\partial f}{\partial y} \right\|_{B,\infty} \frac{1}{V(B)} \iiint_B |y - \overline{y_B}|\, dxdydz \\
&\quad + \left\| \frac{\partial f}{\partial z} \right\|_{B,\infty} \frac{1}{V(B)} \iiint_B |z - \overline{z_B}|\, dxdydz,
\end{aligned} \tag{1.4}$$

where

$$\begin{aligned}
\overline{x_B} &:= \frac{1}{V(B)} \iiint_B x dxdydz, \overline{y_B} = \frac{1}{V(B)} \iiint_B y dxdydz, \\
\overline{z_B} &= \frac{1}{V(B)} \iiint_B z dxdydz
\end{aligned}$$

are the center of gravity coordinates for the convex body B.

For some Hermite-Hadamard type inequalities for multiple integrals see [2, 6, 8–10, 17–20, 25–27]. For some Ostrowski type inequalities see [3–5, 7, 11–16, 21–24].

In this paper we establish some error bounds in approximating the triple integral

$$\frac{1}{V(B)}\iiint_B f(x,y,z)\,dxdydz$$

by either the surface integrals

$$\frac{1}{3}\left[\int\int_S (x-\alpha) f(x,y,z)\,dy\wedge dz + \int\int_S (y-\beta) f(x,y,z)\,dz\wedge dx \right. \\ \left. + \int\int_S (z-\gamma) f(x,y,z)\,dx\wedge dy\right] \tag{1.5}$$

or by, the possibly simpler, triple integrals

$$\frac{1}{3}\iiint_B \left[(\alpha-x)\frac{\partial f(x,y,z)}{\partial x} + (\beta-y)\frac{\partial f(x,y,z)}{\partial y} \right. \\ \left. + (\gamma-z)\frac{\partial f(x,y,z)}{\partial z}\right] dxdydz \tag{1.6}$$

for some α, β, and γ complex numbers.

Examples for functions defined on a ball $B(C,R)$ centered in $C=(a,b,c)\in\mathbb{R}^3$ and with the radius $R>0$ are also provided.

2 Some Preliminary Facts

Following Apostol [1], consider a surface described by the vector equation

$$r(u,v) = x(u,v)\overrightarrow{i} + y(u,v)\overrightarrow{j} + z(u,v)\overrightarrow{k}, \tag{2.1}$$

where $(u,v)\in[a,b]\times[c,d]$.

If $x,\ y,\ z$ are differentiable on $[a,b]\times[c,d]$ we consider the two vectors

$$\frac{\partial r}{\partial u} = \frac{\partial x}{\partial u}\overrightarrow{i} + \frac{\partial y}{\partial u}\overrightarrow{j} + \frac{\partial z}{\partial u}\overrightarrow{k}$$

and

$$\frac{\partial r}{\partial v} = \frac{\partial x}{\partial v}\overrightarrow{i} + \frac{\partial y}{\partial v}\overrightarrow{j} + \frac{\partial z}{\partial v}\overrightarrow{k}.$$

The *cross product* of these two vectors $\frac{\partial r}{\partial u} \times \frac{\partial r}{\partial v}$ will be referred to as the fundamental vector product of the representation r. Its components can be expressed as *Jacobian determinants*. In fact, we have [1, p. 420]

$$\frac{\partial r}{\partial u} \times \frac{\partial r}{\partial v} = \begin{vmatrix} \frac{\partial y}{\partial u} & \frac{\partial z}{\partial u} \\ \frac{\partial y}{\partial v} & \frac{\partial z}{\partial v} \end{vmatrix} \overrightarrow{i} + \begin{vmatrix} \frac{\partial z}{\partial u} & \frac{\partial x}{\partial u} \\ \frac{\partial z}{\partial v} & \frac{\partial x}{\partial v} \end{vmatrix} \overrightarrow{j} + \begin{vmatrix} \frac{\partial x}{\partial u} & \frac{\partial y}{\partial u} \\ \frac{\partial x}{\partial v} & \frac{\partial y}{\partial v} \end{vmatrix} \overrightarrow{k} \tag{2.2}$$

$$= \frac{\partial (y, z)}{\partial (u, v)} \overrightarrow{i} + \frac{\partial (z, x)}{\partial (u, v)} \overrightarrow{j} + \frac{\partial (x, y)}{\partial (u, v)} \overrightarrow{k} .$$

Let $S = r(T)$ be a parametric surface described by a vector-valued function r defined on the box $T = [a, b] \times [c, d]$. The area of S denoted A_S is defined by the double integral [1, pp. 424–425]

$$A_S = \int_a^b \int_c^d \left\| \frac{\partial r}{\partial u} \times \frac{\partial r}{\partial v} \right\| dudv \tag{2.3}$$

$$= \int_a^b \int_c^d \sqrt{\left(\frac{\partial (y, z)}{\partial (u, v)} \right)^2 + \left(\frac{\partial (z, x)}{\partial (u, v)} \right)^2 + \left(\frac{\partial (x, y)}{\partial (u, v)} \right)^2} dudv.$$

We define surface integrals in terms of a parametric representation for the surface. One can prove that under certain general conditions the value of the integral is independent of the representation.

Let $S = r(T)$ be a parametric surface described by a vector-valued differentiable function r defined on the box $T = [a, b] \times [c, d]$ and let $f : S \to \mathbb{C}$ defined and bounded on S. The surface integral of f over S is defined by Apostol [1, p. 430]

$$\int \int_S f dS = \int_a^b \int_c^d f(x, y, z) \left\| \frac{\partial r}{\partial u} \times \frac{\partial r}{\partial v} \right\| dudv \tag{2.4}$$

$$= \int_a^b \int_c^d f(x(u, v), y(u, v), z(u, v))$$

$$\times \sqrt{\left(\frac{\partial (y, z)}{\partial (u, v)} \right)^2 + \left(\frac{\partial (z, x)}{\partial (u, v)} \right)^2 + \left(\frac{\partial (x, y)}{\partial (u, v)} \right)^2} dudv.$$

If $S = r(T)$ is a parametric surface, the fundamental vector product $N = \frac{\partial r}{\partial u} \times \frac{\partial r}{\partial v}$ is normal to S at each regular point of the surface. At each such point there are two unit normals, a unit normal n_1, which has the same direction as N, and a unit normal n_2 which has the opposite direction. Thus

$$n_1 = \frac{N}{\|N\|} \text{ and } n_2 = -n_1.$$

Let n be one of the two normals n_1 or n_2. Let also F be a vector field defined on S and assume that the surface integral,

$$\int\int_S (F \cdot n)\, dS,$$

called *the flux surface integral,* exists. Here $F \cdot n$ is the dot or inner product.

We can write [1, p. 434]

$$\int\int_S (F \cdot n)\, dS = \pm \int_a^b \int_c^d F(r(u,v)) \cdot \left(\frac{\partial r}{\partial u} \times \frac{\partial r}{\partial v} \right) dudv,$$

where the sign "+" is used if $n = n_1$ and the "−" sign is used if $n = n_2$.

If

$$F(x,y,z) = P(x,y,z)\,\overrightarrow{i} + Q(x,y,z)\,\overrightarrow{j} + R(x,y,z)\,\overrightarrow{k}$$

and

$$r(u,v) = x(u,v)\,\overrightarrow{i} + y(u,v)\,\overrightarrow{j} + z(u,v)\,\overrightarrow{k} \text{ where } (u,v) \in [a,b] \times [c,d],$$

then the flux surface integral for $n = n_1$ can be explicitly calculated as [1, p. 435]

$$\begin{aligned} \int\int_S (F \cdot n)\, dS &= \int_a^b \int_c^d P(x(u,v), y(u,v), z(u,v)) \frac{\partial(y,z)}{\partial(u,v)} dudv \\ &+ \int_a^b \int_c^d Q(x(u,v), y(u,v), z(u,v)) \frac{\partial(z,x)}{\partial(u,v)} dudv \\ &+ \int_a^b \int_c^d R(x(u,v), y(u,v), z(u,v)) \frac{\partial(x,y)}{\partial(u,v)} dudv. \end{aligned} \tag{2.5}$$

The sum of the double integrals on the right is often written more briefly as [1, p. 435]

$$\int\int_S P(x,y,z)\, dy \wedge dz + \int\int_S Q(x,y,z)\, dz \wedge dx + \int\int_S R(x,y,z)\, dx \wedge dy.$$

Let $B \subset \mathbb{R}^3$ be a solid in 3-space bounded by an orientable closed surface S, and let n be the unit outer normal to S. If F is a continuously differentiable vector field defined on B, we have the *Gauss-Ostrogradsky identity*

$$\iiint_B (\text{div} F) \, dV = \int\int_S (F \cdot n) \, dS. \tag{GO}$$

If we express

$$F(x, y, z) = P(x, y, z) \overrightarrow{i} + Q(x, y, z) \overrightarrow{j} + R(x, y, z) \overrightarrow{k},$$

then (GO) can be written as

$$\begin{aligned} &\iiint_B \left(\frac{\partial P(x, y, z)}{\partial x} + \frac{\partial Q(x, y, z)}{\partial y} + \frac{\partial R(x, y, z)}{\partial z} \right) dxdydz \\ &= \int\int_S P(x, y, z) \, dy \wedge dz + \int\int_S Q(x, y, z) \, dz \wedge dx \\ &+ \int\int_S R(x, y, z) \, dx \wedge dy. \end{aligned} \tag{2.6}$$

By taking the real and imaginary part, we can extend the above inequality for complex valued functions P, Q, R defined on B.

3 Identities of Interest

We have:

Lemma 1 *Let B be a solid in the three-dimensional space $\mathbb{R}^3$ bounded by an orientable closed surface S. If $f : B \to \mathbb{C}$ is a continuously differentiable function defined on an open set containing B, then we have the equality*

$$\begin{aligned} &\iiint_B f(x, y, z) \, dxdydz \\ &= \frac{1}{3} \iiint_B \left[(\alpha - x) \frac{\partial f(x, y, z)}{\partial x} + (\beta - y) \frac{\partial f(x, y, z)}{\partial y} \right. \\ &\left. + (\gamma - z) \frac{\partial f(x, y, z)}{\partial z} \right] dxdydz \\ &+ \frac{1}{3} \left[\int\int_S (x - \alpha) f(x, y, z) \, dy \wedge dz + \int\int_S (y - \beta) f(x, y, z) \, dz \wedge dx \right. \\ &\left. + \int\int_S (z - \gamma) f(x, y, z) \, dx \wedge dy \right] \end{aligned} \tag{3.1}$$

for all α, β, and γ complex numbers.

In particular, we have

$$\iiint_B f(x,y,z)\,dxdydz$$
$$= \frac{1}{3}\iiint_B \left[(\overline{x_B} - x)\frac{\partial f(x,y,z)}{\partial x} + (\overline{y_B} - y)\frac{\partial f(x,y,z)}{\partial y} + (\overline{z_B} - z)\frac{\partial f(x,y,z)}{\partial z} \right] dxdydz$$
$$+ \frac{1}{3}\left[\int\int_S (x - \overline{x_B}) f(x,y,z)\,dy \wedge dz + \int\int_S (y - \overline{y_B}) f(x,y,z)\,dz \wedge dx + \int\int_S (z - \overline{z_B}) f(x,y,z)\,dx \wedge dy \right]. \quad (3.2)$$

Proof We have

$$\frac{\partial[(x-\alpha) f(x,y,z)]}{\partial x} = f(x,y,z) + (x-\alpha)\frac{\partial f(x,y,z)}{\partial x},$$
$$\frac{\partial[(y-\beta) f(x,y,z)]}{\partial y} = f(x,y,z) + (y-\beta)\frac{\partial f(x,y,z)}{\partial y}$$

and

$$\frac{\partial[(z-\gamma) f(x,y,z)]}{\partial z} = f(x,y,z) + (z-\gamma)\frac{\partial f(x,y,z)}{\partial z}.$$

By adding these three equalities we get

$$\frac{\partial[(x-\alpha) f(x,y,z)]}{\partial x} + \frac{\partial[(y-\beta) f(x,y,z)]}{\partial y} + \frac{\partial[(z-\gamma) f(x,y,z)]}{\partial z}$$
$$= 3f(x,y,z) + (x-\alpha)\frac{\partial f(x,y,z)}{\partial x} + (y-\beta)\frac{\partial f(x,y,z)}{\partial y} + (z-\gamma)\frac{\partial f(x,y,z)}{\partial z} \quad (3.3)$$

for all $(x,y,z) \in B$.

Integrating this equality on B we get

$$\iiint_B \left(\frac{\partial[(x-\alpha) f(x,y,z)]}{\partial x} + \frac{\partial[(y-\beta) f(x,y,z)]}{\partial y} + \frac{\partial[(z-\gamma) f(x,y,z)]}{\partial z} \right) dxdydz$$

$$= 3 \iiint_B f(x, y, z)\, dxdydz$$
$$+ \iiint_B \left[(x - \alpha) \frac{\partial f(x, y, z)}{\partial x} + (y - \beta) \frac{\partial f(x, y, z)}{\partial y} \right.$$
$$\left. + (z - \gamma) \frac{\partial f(x, y, z)}{\partial z} \right] dxdydz. \tag{3.4}$$

Applying the *Gauss-Ostrogradsky identity (2.6)* for the functions

$$P(x, y, z) = (x - \alpha) f(x, y, z),\ Q(x, y, z) = (y - \beta) f(x, y, z)$$

and

$$R(x, y, z) = (z - \gamma) f(x, y, z)$$

we obtain

$$\iiint_B \left(\frac{\partial [(x - \alpha) f(x, y, z)]}{\partial x} + \frac{\partial [(y - \beta) f(x, y, z)]}{\partial y} \right.$$
$$\left. + \frac{\partial [(z - \gamma) f(x, y, z)]}{\partial z} \right) dxdydz$$
$$= \int \int_S (x - \alpha) f(x, y, z)\, dy \wedge dz + \int \int_S (y - \beta) f(x, y, z)\, dz \wedge dx$$
$$+ \int \int_S (z - \gamma) f(x, y, z)\, dx \wedge dy. \tag{3.5}$$

By (3.4) and (3.5) we get

$$3 \iiint_B f(x, y, z)\, dxdydz$$
$$+ \iiint_B \left[(x - \alpha) \frac{\partial f(x, y, z)}{\partial x} + (y - \beta) \frac{\partial f(x, y, z)}{\partial y} \right.$$
$$\left. + (z - \gamma) \frac{\partial f(x, y, z)}{\partial z} \right] dxdydz$$
$$= \int \int_S (x - \alpha) f(x, y, z)\, dy \wedge dz + \int \int_S (y - \beta) f(x, y, z)\, dz \wedge dx$$
$$+ \int \int_S (z - \gamma) f(x, y, z)\, dx \wedge dy,$$

which is equivalent to the desired result (3.1). □

Remark 1 For a function f as in Lemma 1 above, we define the points

$$x_{B,\partial f} := \frac{\iiint_B x \frac{\partial f(x,y,z)}{\partial x} dxdydz}{\iiint_B \frac{\partial f(x,y,z)}{\partial x} dxdydz}, \quad y_{B,\partial f} := \frac{\iiint_B y \frac{\partial f(x,y,z)}{\partial y} dxdydz}{\iiint_B \frac{\partial f(x,y,z)}{\partial y} dxdydz},$$

and

$$z_{B,\partial f} := \frac{\iiint_B z \frac{\partial f(x,y,z)}{\partial z} dxdydz}{\iiint_B \frac{\partial f(x,y,z)}{\partial z} dxdydz}$$

provided the denominators are not zero.

If we take $\alpha = x_{B,\partial f}$, $\beta = y_{B,\partial f}$ and $\gamma = z_{B,\partial f}$ in (3.1), then we get

$$\begin{aligned} &\iiint_B f(x,y,z)\,dxdydz \\ &= \frac{1}{3}\left[\int\int_S \left(x - x_{B,\partial f}\right) f(x,y,z)\,dy \wedge dz \right. \\ &\quad + \int\int_S \left(y - \beta y_{B,\partial f}\right) f(x,y,z)\,dz \wedge dx \\ &\quad \left. + \int\int_S \left(z - z_{B,\partial f}\right) f(x,y,z)\,dx \wedge dy\right], \end{aligned} \tag{3.6}$$

since, obviously,

$$\begin{aligned} &\iiint_B \left[\left(x_{B,\partial f} - x\right) \frac{\partial f(x,y,z)}{\partial x} + \left(y_{B,\partial f} - y\right) \frac{\partial f(x,y,z)}{\partial y} \right. \\ &\left. + \left(z_{B,\partial f} - z\right) \frac{\partial f(x,y,z)}{\partial z}\right] dxdydz = 0. \end{aligned}$$

We also have the following dual approach:

Remark 2 For a function f as in Lemma 1 above, we define the points

$$x_{S,f} := \frac{\int\int_S x f(x,y,z)\,dy \wedge dz}{\int\int_S f(x,y,z)\,dy \wedge dz}, \quad y_{S,f} := \frac{\int\int_S y f(x,y,z)\,dz \wedge dx}{\int\int_S f(x,y,z)\,dz \wedge dx}$$

and

$$z_{S,f} := \frac{\int\int_S z f(x,y,z)\,dx \wedge dy}{\int\int_S f(x,y,z)\,dx \wedge dy}$$

provided the denominators are not zero.

If we take $\alpha = x_{S,f}$, $\beta = y_{S,f}$ and $\gamma = z_{S,f}$ in (3.1), then we get

$$\begin{aligned}
&\iiint_B f(x,y,z)\,dxdydz \\
&= \frac{1}{3}\iiint_B \left[\left(x_{S,f} - x\right) \frac{\partial f(x,y,z)}{\partial x} + \left(y_{S,f} - y\right) \frac{\partial f(x,y,z)}{\partial y} \right. \\
&\quad \left. + \left(z_{S,f} - z\right) \frac{\partial f(x,y,z)}{\partial z} \right] dxdydz \qquad (3.7)
\end{aligned}$$

since, obviously,

$$\begin{aligned}
&\int\int_S \left(x - x_{S,f}\right) f(x,y,z)\,dy \wedge dz + \int\int_S \left(y - y_{S,f}\right) f(x,y,z)\,dz \wedge dx \\
&+ \int\int_S \left(z - z_{S,f}\right) f(x,y,z)\,dx \wedge dy = 0.
\end{aligned}$$

4 Integral Inequalities

For a measurable function $g : B \to \mathbb{C}$ we define the *Lebesgue norms*

$$\|g\|_{B,p} := \left(\iiint_B |g(x,y,z)|^p \,dxdydz \right)^{1/p} < \infty$$

for $p \geq 1$ and

$$\|g\|_{B,\infty} := \sup_{(x,y,z)\in B} |g(x,y,z)| < \infty$$

for $p = \infty$.

We have:

Theorem 4 *Let B be a solid in the three-dimensional space $\mathbb{R}^3$ bounded by an orientable closed surface S. If $f : B \to \mathbb{C}$ is a continuously differentiable function defined on an open set containing B, then for all α, β, γ complex numbers we have the inequality*

$$\begin{aligned}
&\left| \iiint_B f(x,y,z)\,dxdydz - \frac{1}{3}\left[\int\int_S (x-\alpha) f(x,y,z)\,dy \wedge dz \right.\right. \\
&\left.\left. + \int\int_S (y-\beta) f(x,y,z)\,dz \wedge dx + \int\int_S (z-\gamma) f(x,y,z)\,dx \wedge dy \right] \right|
\end{aligned}$$

$$\leq \frac{1}{3} \iiint_B \left[|\alpha - x| \left| \frac{\partial f(x,y,z)}{\partial x} \right| + |\beta - y| \left| \frac{\partial f(x,y,z)}{\partial y} \right| \right.$$
$$\left. + |\gamma - z| \left| \frac{\partial f(x,y,z)}{\partial z} \right| \right] dxdydz =: M(\alpha, \beta, \gamma; f). \tag{4.1}$$

Moreover, we have the bounds

$$M(\alpha, \beta, \gamma; f)$$
$$\leq \frac{1}{3} \begin{cases} \left\| \frac{\partial f}{\partial x} \right\|_{B,\infty} \iiint_B |\alpha - x| \, dxdydz + \left\| \frac{\partial f}{\partial y} \right\|_{B,\infty} \iiint_B |\beta - y| \, dxdydz \\ + \left\| \frac{\partial f}{\partial z} \right\|_{B,\infty} \iiint_B |\gamma - z| \, dxdydz; \\ \left\| \frac{\partial f}{\partial x} \right\|_{B,p} \left(\iiint_B |\alpha - x|^q \, dxdydz \right)^{1/q} + \left\| \frac{\partial f}{\partial y} \right\|_{B,p} \left(\iiint_B |\beta - y|^q \, dxdydz \right)^{1/q} \\ + \left\| \frac{\partial f}{\partial z} \right\|_{B,p} \left(\iiint_B |\gamma - z| \, dxdydz \right)^{1/q}, \ p, q > 1, \ \frac{1}{p} + \frac{1}{q} = 1; \\ \sup_{(x,y,z) \in B} |\alpha - x| \left\| \frac{\partial f}{\partial x} \right\|_{B,1} + \sup_{(x,y,z) \in B} |\beta - y| \left\| \frac{\partial f}{\partial y} \right\|_{B,1} \\ + \sup_{(x,y,z) \in B} |\gamma - z| \left\| \frac{\partial f}{\partial z} \right\|_{B,1}. \end{cases} \tag{4.2}$$

Proof From the identity (3.1) we have

$$\left| \iiint_B f(x,y,z) \, dxdydz - \frac{1}{3} \left[\iint_S (x - \alpha) f(x,y,z) \, dy \wedge dz \right. \right.$$
$$\left. \left. + \iint_S (y - \beta) f(x,y,z) \, dz \wedge dx + \iint_S (z - \gamma) f(x,y,z) \, dx \wedge dy \right] \right|$$
$$= \frac{1}{3} \left| \iiint_B \left[(\alpha - x) \frac{\partial f(x,y,z)}{\partial x} + (\beta - y) \frac{\partial f(x,y,z)}{\partial y} \right. \right.$$
$$\left. \left. + (\gamma - z) \frac{\partial f(x,y,z)}{\partial z} \right] dxdydz \right|$$
$$\leq \frac{1}{3} \iiint_B \left| \left[(\alpha - x) \frac{\partial f(x,y,z)}{\partial x} + (\beta - y) \frac{\partial f(x,y,z)}{\partial y} \right. \right.$$
$$\left. \left. + (\gamma - z) \frac{\partial f(x,y,z)}{\partial z} \right] \right| dxdydz$$
$$\leq \frac{1}{3} \iiint_B \left[\left| (\alpha - x) \frac{\partial f(x,y,z)}{\partial x} \right| + \left| (\beta - y) \frac{\partial f(x,y,z)}{\partial y} \right| \right.$$
$$\left. + \left| (\gamma - z) \frac{\partial f(x,y,z)}{\partial z} \right| \right] dxdydz = M(\alpha, \beta, \gamma; f),$$

which proves the inequality (4.1).

By Hölder's multiple integral inequality we also have

$$\iiint_B \left| (\alpha - x) \frac{\partial f(x,y,z)}{\partial x} \right| dxdydz$$
$$\leq \begin{cases} \left\| \frac{\partial f}{\partial x} \right\|_{B,\infty} \iiint_B |\alpha - x| \, dxdydz; \\ \\ \left\| \frac{\partial f}{\partial x} \right\|_{B,p} \left(\iiint_B |\alpha - x|^q \, dxdydz \right)^{1/q}, \ p,q > 1, \ \frac{1}{p} + \frac{1}{q} = 1; \\ \\ \sup_{(x,y,z) \in} |\alpha - x| \left\| \frac{\partial f}{\partial x} \right\|_{B,1} \end{cases}$$

and the other two similar inequalities for the partial derivatives $\frac{\partial f}{\partial y}$ and $\frac{\partial f}{\partial z}$, which, by addition, provide the bound from (4.2). □

Corollary 1 *With the assumptions of Theorem 4 we have the inequalities*

$$\left| \iiint_B f(x,y,z) \, dxdydz - \frac{1}{3} \left[\int\int_S (x - \overline{x_B}) f(x,y,z) \, dy \wedge dz \right. \right.$$
$$\left. \left. + \int\int_S (\overline{y_B} - \beta) f(x,y,z) \, dz \wedge dx + \int\int_S (z - \overline{z_B}) f(x,y,z) \, dx \wedge dy \right] \right|$$
$$\leq \frac{1}{3} \iiint_B \left[|\overline{x_B} - x| \left| \frac{\partial f(x,y,z)}{\partial x} \right| + |\overline{y_B} - y| \left| \frac{\partial f(x,y,z)}{\partial y} \right| \right.$$
$$\left. + |\overline{z_B} - z| \left| \frac{\partial f(x,y,z)}{\partial z} \right| \right] dxdydz =: M(\overline{x_B}, \overline{y_B}, \overline{z_B}; f) \tag{4.3}$$

with

$$M(\overline{x_B}, \overline{y_B}, \overline{z_B}; f)$$
$$\leq \frac{1}{3} \begin{cases} \left\| \frac{\partial f}{\partial x} \right\|_{B,\infty} \iiint_B |\overline{x_B} - x| \, dxdydz + \left\| \frac{\partial f}{\partial y} \right\|_{B,\infty} \iiint_B |\overline{y_B} - y| \, dxdydz \\ + \left\| \frac{\partial f}{\partial z} \right\|_{B,\infty} \iiint_B |\overline{z_B} - z| \, dxdydz; \\ \\ \left\| \frac{\partial f}{\partial x} \right\|_{B,p} \left(\iiint_B |\overline{x_B} - x|^q \, dxdydz \right)^{1/q} + \left\| \frac{\partial f}{\partial y} \right\|_{B,p} \left(\iiint_B |\overline{y_B} - y|^q \, dxdydz \right)^{1/q} \\ + \left\| \frac{\partial f}{\partial z} \right\|_{B,p} \left(\iiint_B |\overline{z_B} - z| \, dxdydz \right)^{1/q}, \ p,q > 1, \ \frac{1}{p} + \frac{1}{q} = 1; \\ \\ \sup_{(x,y,z) \in} |\overline{x_B} - x| \left\| \frac{\partial f}{\partial x} \right\|_{B,1} + \sup_{(x,y,z) \in} |\overline{y_B} - y| \left\| \frac{\partial f}{\partial y} \right\|_{B,1} \\ + \sup_{(x,y,z) \in} |\overline{z_B} - z| \left\| \frac{\partial f}{\partial z} \right\|_{B,1} . \end{cases} \tag{4.4}$$

We also have

$$\begin{aligned}&\left|\iiint_B f(x,y,z)\,dxdydz\right|\\&\le \frac{1}{3}\iiint_B\left[\left|x_{S,f}-x\right|\left|\frac{\partial f(x,y,z)}{\partial x}\right|+\left|y_{S,f}-y\right|\left|\frac{\partial f(x,y,z)}{\partial y}\right|\right.\\&\quad\left.+\left|z_{S,f}-z\right|\left|\frac{\partial f(x,y,z)}{\partial z}\right|\right]dxdydz=:M\left(x_{S,f},y_{S,f},z_{S,f};f\right)\end{aligned}\tag{4.5}$$

with

$$M\left(x_{S,f},y_{S,f},z_{S,f};f\right)\le\frac{1}{3}\begin{cases}\left\|\frac{\partial f}{\partial x}\right\|_{B,\infty}\iiint_B\left|x_{S,f}-x\right|dxdydz+\left\|\frac{\partial f}{\partial y}\right\|_{B,\infty}\iiint_B\left|y_{S,f}-y\right|dxdydz\\+\left\|\frac{\partial f}{\partial z}\right\|_{B,\infty}\iiint_B\left|z_{S,f}-z\right|dxdydz;\\ \\ \left\|\frac{\partial f}{\partial x}\right\|_{B,p}\left(\iiint_B\left|x_{S,f}-x\right|^q dxdydz\right)^{1/q}+\left\|\frac{\partial f}{\partial y}\right\|_{B,p}\left(\iiint_B\left|y_{S,f}-y\right|^q dxdydz\right)^{1/q}\\+\left\|\frac{\partial f}{\partial z}\right\|_{B,p}\left(\iiint_B\left|z_{S,f}-z\right|dxdydz\right)^{1/q},\ p,q>1,\ \frac{1}{p}+\frac{1}{q}=1;\\ \\ \sup_{(x,y,z)\in}\left|x_{S,f}-x\right|\left\|\frac{\partial f}{\partial x}\right\|_{B,1}+\sup_{(x,y,z)\in}\left|y_{S,f}-y\right|\left\|\frac{\partial f}{\partial y}\right\|_{B,1}\\+\sup_{(x,y,z)\in}\left|z_{S,f}-z\right|\left\|\frac{\partial f}{\partial z}\right\|_{B,1}.\end{cases}\tag{4.6}$$

Remark 3 Using the discrete Hölder's inequality we have

$$\begin{aligned}&|\alpha-x|\left|\frac{\partial f(x,y,z)}{\partial x}\right|+|\beta-y|\left|\frac{\partial f(x,y,z)}{\partial y}\right|+|\gamma-z|\left|\frac{\partial f(x,y,z)}{\partial z}\right|\\&\le\begin{cases}\max\{|\alpha-x|,|\beta-y|,|\gamma-z|\}\left[\left|\frac{\partial f(x,y,z)}{\partial x}\right|+\left|\frac{\partial f(x,y,z)}{\partial y}\right|+\left|\frac{\partial f(x,y,z)}{\partial z}\right|\right];\\ \\ \left(|\alpha-x|^q+|\beta-y|^q+|\gamma-z|^q\right)^{1/q}\left[\left|\frac{\partial f(x,y,z)}{\partial x}\right|^p+\left|\frac{\partial f(x,y,z)}{\partial y}\right|^p+\left|\frac{\partial f(x,y,z)}{\partial z}\right|^p\right]^{1/p}\\ \text{for}p,\ q>1,\ \frac{1}{p}+\frac{1}{q}=1;\\ \\ \max\left\{\left|\frac{\partial f(x,y,z)}{\partial x}\right|,\left|\frac{\partial f(x,y,z)}{\partial y}\right|,\left|\frac{\partial f(x,y,z)}{\partial z}\right|\right\}[|\alpha-x|+|\beta-y|+|\gamma-z|]\end{cases}\end{aligned}$$

for all $(x,y,z)\in B$ and all $\alpha,\ \beta,\ \gamma$ complex numbers.

By taking the integral we get

$$M(\alpha,\beta,\gamma;f)$$

$$
\leq \frac{1}{3}\begin{cases}
\iiint_B \max\{|\alpha - x|, |\beta - y|, |\gamma - z|\} \\
\times \left[\left|\frac{\partial f(x,y,z)}{\partial x}\right| + \left|\frac{\partial f(x,y,z)}{\partial y}\right| + \left|\frac{\partial f(x,y,z)}{\partial z}\right|\right] dxdydz; \\
\\
\iiint_B \left(|\alpha - x|^q + |\beta - y|^q + |\gamma - z|^q\right)^{1/q} \\
\times \left[\left|\frac{\partial f(x,y,z)}{\partial x}\right|^p + \left|\frac{\partial f(x,y,z)}{\partial y}\right|^p + \left|\frac{\partial f(x,y,z)}{\partial z}\right|^p\right]^{1/p} dxdydz \\
\text{for} p,\ q > 1,\ \frac{1}{p} + \frac{1}{q} = 1; \\
\\
\iiint_B \max\left\{\left|\frac{\partial f(x,y,z)}{\partial x}\right|, \left|\frac{\partial f(x,y,z)}{\partial y}\right|, \left|\frac{\partial f(x,y,z)}{\partial z}\right|\right\} \\
\times \left[|\alpha - x| + |\beta - y| + |\gamma - z|\right] dxdydz
\end{cases}
$$

for all α, β, γ complex numbers.

One can separate the factors in the above inequality by using Hölder's integral inequality. For instance, we have

$$
\begin{aligned}
&\iiint_B \left(|\alpha - x|^q + |\beta - y|^q + |\gamma - z|^q\right)^{1/q} \\
&\times \left[\left|\frac{\partial f(x, y, z)}{\partial x}\right|^p + \left|\frac{\partial f(x, y, z)}{\partial y}\right|^p + \left|\frac{\partial f(x, y, z)}{\partial z}\right|^p\right]^{1/p} dxdydz \\
&\leq \left(\iiint_B \left[\left(|\alpha - x|^q + |\beta - y|^q + |\gamma - z|^q\right)^{1/q}\right]^q dxdydz\right)^{1/q} \\
&\times \left(\iiint_B \left(\left[\left|\frac{\partial f(x, y, z)}{\partial x}\right|^p + \left|\frac{\partial f(x, y, z)}{\partial y}\right|^p + \left|\frac{\partial f(x, y, z)}{\partial z}\right|^p\right]^{1/p}\right)^p dxdydz\right)^{1/p} \\
&= \left(\iiint_B \left(|\alpha - x|^q + |\beta - y|^q + |\gamma - z|^q\right) dxdydz\right)^{1/q} \\
&\times \left(\iiint_B \left[\left|\frac{\partial f(x, y, z)}{\partial x}\right|^p + \left|\frac{\partial f(x, y, z)}{\partial y}\right|^p + \left|\frac{\partial f(x, y, z)}{\partial z}\right|^p\right] dxdydz\right)^{1/p},
\end{aligned}
$$

which gives

$$
\begin{aligned}
M(\alpha, \beta, \gamma; f) &\leq \frac{1}{3}\left(\iiint_B \left(|\alpha - x|^q + |\beta - y|^q + |\gamma - z|^q\right) dxdydz\right)^{1/q} \\
&\times \left(\iiint_B \left[\left|\frac{\partial f(x, y, z)}{\partial x}\right|^p + \left|\frac{\partial f(x, y, z)}{\partial y}\right|^p + \left|\frac{\partial f(x, y, z)}{\partial z}\right|^p\right] dxdydz\right)^{1/p}, \quad (4.7)
\end{aligned}
$$

for $p, q > 1$, $\frac{1}{p} + \frac{1}{q} = 1$.

We also have:

Theorem 5 *Let B be a solid in the three-dimensional space $\mathbb{R}^3$ bounded by an orientable closed surface S described by the vector equation*

$$r(u,v) = x(u,v)\vec{i} + y(u,v)\vec{j} + z(u,v)\vec{k}, \ (u,v) \in [a,b] \times [c,d],$$

where $x(u,v)$, $y(u,v)$, $z(u,v)$ are differentiable. If $f : B \to \mathbb{C}$ is a continuously differentiable function defined on an open set containing B, then for all α, β, γ complex numbers we have the inequality

$$\begin{aligned}
&\left| \iiint_B f(x,y,z)\,dxdydz - \frac{1}{3}\iiint_B \left[(\alpha - x)\frac{\partial f(x,y,z)}{\partial x} \right.\right. \\
&\quad \left.\left. + (\beta - y)\frac{\partial f(x,y,z)}{\partial y} + (\gamma - z)\frac{\partial f(x,y,z)}{\partial z} \right] dxdydz \right| \\
&\quad \le \frac{1}{3}\left[\int_a^b \int_c^d |f(x(u,v), y(u,v), z(u,v))|\,|x(u,v) - \alpha| \left| \frac{\partial(y,z)}{\partial(u,v)} \right| dudv \right. \\
&\quad + \int_a^b \int_c^d |f(x(u,v), y(u,v), z(u,v))|\,|y(u,v) - \beta| \left| \frac{\partial(z,x)}{\partial(u,v)} \right| dudv \\
&\quad \left. + \int_a^b \int_c^d |f(x(u,v), y(u,v), z(u,v))|\,|z(u,v) - \gamma| \left| \frac{\partial(x,y)}{\partial(u,v)} \right| dudv \right] \\
&\quad =: N(\alpha, \beta, \gamma; f).
\end{aligned} \tag{4.8}$$

Moreover, if we put $\square := [a,b] \times [c,d]$, then we have the bounds

$$\begin{aligned}
N(\alpha,\beta,\gamma;f) &\le \frac{1}{3}\|f\|_{S,\infty} \left[\int_a^b \int_c^d |x(u,v) - \alpha| \left| \frac{\partial(y,z)}{\partial(u,v)} \right| \right. \\
&\quad \left. + |y(u,v) - \beta| \left| \frac{\partial(z,x)}{\partial(u,v)} \right| + |z(u,v) - \gamma| \left| \frac{\partial(x,y)}{\partial(u,v)} \right| dudv \right] \\
&\le \frac{1}{3}\|f\|_{S,\infty} \\
&\quad \times \begin{cases}
\left\| \frac{\partial(y,z)}{\partial(\cdot,\cdot)} \right\|_{\square,\infty} \|x - \alpha\|_{\square,1} + \left\| \frac{\partial(z,x)}{\partial(\cdot,\cdot)} \right\|_{\square,\infty} \|y - \beta\|_{\square,1} \\
\qquad + \left\| \frac{\partial(x,y)}{\partial(\cdot,\cdot)} \right\|_{\square,\infty} \|z - \gamma\|_{\square,1}, \\
\\
\left\| \frac{\partial(y,z)}{\partial(\cdot,\cdot)} \right\|_{\square,p} \|x - \alpha\|_{\square,q} + \left\| \frac{\partial(z,x)}{\partial(\cdot,\cdot)} \right\|_{\square,p} \|y - \beta\|_{\square,q} \\
\qquad + \left\| \frac{\partial(x,y)}{\partial(\cdot,\cdot)} \right\|_{\square,p} \|z - \gamma\|_{\square,q}, \\
\\
\left\| \frac{\partial(y,z)}{\partial(\cdot,\cdot)} \right\|_{\square,1} \|x - \alpha\|_{\square,\infty} + \left\| \frac{\partial(z,x)}{\partial(\cdot,\cdot)} \right\|_{\square,p} \|y - \beta\|_{\square,\infty} \\
\qquad + \left\| \frac{\partial(x,y)}{\partial(\cdot,\cdot)} \right\|_{\square,1} \|z - \gamma\|_{\square,\infty}.
\end{cases}
\end{aligned} \tag{4.9}$$

Proof From the identity (3.1) we get

$$\begin{aligned}
&\iiint_B f(x,y,z)\,dxdydz \\
&\quad - \frac{1}{3}\iiint_B \left[(\alpha - x)\frac{\partial f(x,y,z)}{\partial x} + (\beta - y)\frac{\partial f(x,y,z)}{\partial y}\right. \\
&\quad \left. + (\gamma - z)\frac{\partial f(x,y,z)}{\partial z}\right] dxdydz \\
&\quad = \frac{1}{3}\left[\int_a^b\int_c^d (x(u,v) - \alpha)\, f(x(u,v), y(u,v), z(u,v))\frac{\partial(y,z)}{\partial(u,v)}dudv\right. \\
&\qquad + \int_a^b\int_c^d (y(u,v) - \beta)\, f(x(u,v), y(u,v), z(u,v))\frac{\partial(z,x)}{\partial(u,v)}dudv \\
&\qquad \left. + \int_a^b\int_c^d (z(u,v) - \gamma)\, f(x(u,v), y(u,v), z(u,v))\frac{\partial(x,y)}{\partial(u,v)}dudv\right]
\end{aligned} \tag{4.10}$$

for all α, β, γ complex numbers.

By taking the modulus in (4.10) we get

$$\begin{aligned}
&\left|\iiint_B f(x,y,z)\,dxdydz - \frac{1}{3}\iiint_B \left[(\alpha - x)\frac{\partial f(x,y,z)}{\partial x}\right.\right. \\
&\quad \left.\left. + (\beta - y)\frac{\partial f(x,y,z)}{\partial y} + (\gamma - z)\frac{\partial f(x,y,z)}{\partial z}\right] dxdydz\right| \\
&\quad \le \frac{1}{3}\left[\int_a^b\int_c^d \left|(x(u,v) - \alpha)\, f(x(u,v), y(u,v), z(u,v))\frac{\partial(y,z)}{\partial(u,v)}\right| dudv\right. \\
&\qquad + \int_a^b\int_c^d \left|(y(u,v) - \beta)\, f(x(u,v), y(u,v), z(u,v))\frac{\partial(z,x)}{\partial(u,v)}\right| dudv \\
&\qquad \left. + \int_a^b\int_c^d \left|(z(u,v) - \gamma)\, f(x(u,v), y(u,v), z(u,v))\frac{\partial(x,y)}{\partial(u,v)}\right| dudv\right] \\
&\quad = \frac{1}{3}\left[\int_a^b\int_c^d |x(u,v) - \alpha|\,|f(x(u,v), y(u,v), z(u,v))|\left|\frac{\partial(y,z)}{\partial(u,v)}\right| dudv\right. \\
&\qquad + \int_a^b\int_c^d |y(u,v) - \beta|\,|f(x(u,v), y(u,v), z(u,v))|\left|\frac{\partial(z,x)}{\partial(u,v)}\right| dudv \\
&\qquad \left. + \int_a^b\int_c^d |z(u,v) - \gamma|\,|f(x(u,v), y(u,v), z(u,v))|\left|\frac{\partial(x,y)}{\partial(u,v)}\right| dudv\right] \\
&\quad = N(\alpha, \beta, \gamma; f),
\end{aligned}$$

which proves the first inequality in (4.8).

We have

$$N(\alpha,\beta,\gamma;f) \le \frac{1}{3}\|f\|_{S,\infty}\left[\int_a^b\int_c^d |x(u,v)-\alpha|\left|\frac{\partial(y,z)}{\partial(u,v)}\right|dudv\right.$$
$$+\int_a^b\int_c^d |y(u,v)-\beta|\left|\frac{\partial(z,x)}{\partial(u,v)}\right|dudv$$
$$\left.+\int_a^b\int_c^d |z(u,v)-\gamma|\left|\frac{\partial(x,y)}{\partial(u,v)}\right|dudv\right]$$

and by Hölder's inequality for each integral we get the last part of (4.9). □

Corollary 2 *With the assumptions of Theorem 5 we have*

$$N(\alpha,\beta,\gamma;f) \le \frac{1}{3}\left(\int\int_S |f(x,y,z)|^2 dS\right)^{1/2}$$
$$\times\left(\int\int_S\left(|x-\alpha|^2+|y-\beta|^2+|z-\gamma|^2\right)dS\right)^{1/2} \quad (4.11)$$

for all α, β, γ complex numbers.

Proof We have, by Cauchy-Bunyakovsky-Schwarz (CBS) discrete inequality, that

$$|x(u,v)-\alpha|\left|\frac{\partial(y,z)}{\partial(u,v)}\right|+|y(u,v)-\beta|\left|\frac{\partial(z,x)}{\partial(u,v)}\right|+|z(u,v)-\gamma|\left|\frac{\partial(x,y)}{\partial(u,v)}\right|$$
$$\le\left(|x(u,v)-\alpha|^2+|y(u,v)-\beta|^2+|z(u,v)-\gamma|^2\right)^{1/2}$$
$$\times\left(\left|\frac{\partial(y,z)}{\partial(u,v)}\right|^2+\left|\frac{\partial(z,x)}{\partial(u,v)}\right|^2+\left|\frac{\partial(x,y)}{\partial(u,v)}\right|^2\right)^{1/2}$$

for $(u,v)\in[a,b]\times[c,d]$.

Therefore we get

$$N(\alpha,\beta,\gamma;f) \le \int_a^b\int_c^d |f(x(u,v),y(u,v),z(u,v))|$$
$$\times\left(|x(u,v)-\alpha|^2+|y(u,v)-\beta|^2+|z(u,v)-\gamma|^2\right)^{1/2}$$
$$\times\left(\left|\frac{\partial(y,z)}{\partial(u,v)}\right|^2+\left|\frac{\partial(z,x)}{\partial(u,v)}\right|^2+\left|\frac{\partial(x,y)}{\partial(u,v)}\right|^2\right)^{1/2}dudv$$
$$=: P(\alpha,\beta,\gamma;f).$$

By using CBS weighted integral inequality we get

$$\begin{aligned}
&P(\alpha,\beta,\gamma;f)\\
&\leq\left(\int_a^b\int_c^d|f(x(u,v),y(u,v),z(u,v))|^2\right.\\
&\quad\times\left.\left(\left|\frac{\partial(y,z)}{\partial(u,v)}\right|^2+\left|\frac{\partial(z,x)}{\partial(u,v)}\right|^2+\left|\frac{\partial(x,y)}{\partial(u,v)}\right|^2\right)^{1/2}dudv\right)^{1/2}\\
&\quad\times\left(\int_a^b\int_c^d\left(|x(u,v)-\alpha|^2+|y(u,v)-\beta|^2+|z(u,v)-\gamma|^2\right)\right.\\
&\quad\times\left.\left(\left|\frac{\partial(y,z)}{\partial(u,v)}\right|^2+\left|\frac{\partial(z,x)}{\partial(u,v)}\right|^2+\left|\frac{\partial(x,y)}{\partial(u,v)}\right|^2\right)^{1/2}dudv\right)^{1/2}\\
&=\left(\int\int_S|f(x,y,z)|^2dS\right)^{1/2}\left(\int\int_S\left(|x-\alpha|^2+|y-\beta|^2+|z-\gamma|^2\right)dS\right)^{1/2},
\end{aligned}$$

which proves the desired result (4.11). □

Remark 4 From (4.8) we get

$$\begin{aligned}
&\left|\iiint_B f(x,y,z)\,dxdydz-\frac{1}{3}\iiint_B\left[(\overline{x_B}-x)\frac{\partial f(x,y,z)}{\partial x}\right.\right.\\
&\quad\left.\left.+(\overline{y_B}-y)\frac{\partial f(x,y,z)}{\partial y}+(\overline{z_B}-z)\frac{\partial f(x,y,z)}{\partial z}\right]dxdydz\right|\\
&\leq\frac{1}{3}\left[\int_a^b\int_c^d|f(x(u,v),y(u,v),z(u,v))|\,|x(u,v)-\overline{x_B}|\left|\frac{\partial(y,z)}{\partial(u,v)}\right|dudv\right.\\
&\quad+\int_a^b\int_c^d|f(x(u,v),y(u,v),z(u,v))|\,|y(u,v)-\overline{y_B}|\left|\frac{\partial(z,x)}{\partial(u,v)}\right|dudv\\
&\quad\left.+\int_a^b\int_c^d|f(x(u,v),y(u,v),z(u,v))|\,|z(u,v)-\overline{z_B}|\left|\frac{\partial(x,y)}{\partial(u,v)}\right|dudv\right]\\
&=:N(\alpha,\beta,\gamma;f).
\end{aligned}\tag{4.12}$$

Moreover, if we put $\square:=[a,b]\times[c,d]$, then we have the bounds

$$\begin{aligned}
N(\alpha,\beta,\gamma;f)&\leq\frac{1}{3}\|f\|_{S,\infty}\left[\int_a^b\int_c^d|x(u,v)-\overline{x_B}|\left|\frac{\partial(y,z)}{\partial(u,v)}\right|\right.\\
&\quad\left.+|y(u,v)-\overline{y_B}|\left|\frac{\partial(z,x)}{\partial(u,v)}\right|+|z(u,v)-\overline{z_B}|\left|\frac{\partial(x,y)}{\partial(u,v)}\right|dudv\right]
\end{aligned}$$

$$\leq \frac{1}{3} \|f\|_{S,\infty}$$
$$\times \begin{cases} \left\|\frac{\partial(y,z)}{\partial(\cdot,\cdot)}\right\|_{\square,\infty} \|x - \overline{x_B}\|_{\square,1} + \left\|\frac{\partial(z,x)}{\partial(\cdot,\cdot)}\right\|_{\square,\infty} \|y - \overline{y_B}\|_{\square,1} \\ \qquad + \left\|\frac{\partial(x,y)}{\partial(\cdot,\cdot)}\right\|_{\square,\infty} \|z - \overline{z_B}\|_{\square,1}, \\ \\ \left\|\frac{\partial(y,z)}{\partial(\cdot,\cdot)}\right\|_{\square,p} \|x - \overline{x_B}\|_{\square,q} + \left\|\frac{\partial(z,x)}{\partial(\cdot,\cdot)}\right\|_{\square,p} \|y - \overline{y_B}\|_{\square,q} \\ \qquad + \left\|\frac{\partial(x,y)}{\partial(\cdot,\cdot)}\right\|_{\square,p} \|z - \overline{z_B}\|_{\square,q}, \\ \\ \left\|\frac{\partial(y,z)}{\partial(\cdot,\cdot)}\right\|_{\square,1} \|x - \overline{x_B}\|_{\square,\infty} + \left\|\frac{\partial(z,x)}{\partial(\cdot,\cdot)}\right\|_{\square,p} \|y - \overline{y_B}\|_{\square,\infty} \\ \qquad + \left\|\frac{\partial(x,y)}{\partial(\cdot,\cdot)}\right\|_{\square,1} \|z - \overline{z_B}\|_{\square,\infty}. \end{cases} \tag{4.13}$$

We also observe that under the assumptions of Theorem 5 we have

$$\left|\iiint_B f(x,y,z)\, dxdydz\right|$$
$$\leq \frac{1}{3}\left[\int_a^b \int_c^d |f(x(u,v), y(u,v), z(u,v))| \left|x(u,v) - x_{B,\partial f}\right| \left|\frac{\partial(y,z)}{\partial(u,v)}\right| dudv\right.$$
$$+ \int_a^b \int_c^d |f(x(u,v), y(u,v), z(u,v))| \left|y(u,v) - y_{B,\partial f}\right| \left|\frac{\partial(z,x)}{\partial(u,v)}\right| dudv$$
$$\left. + \int_a^b \int_c^d |f(x(u,v), y(u,v), z(u,v))| \left|z(u,v) - z_{B,\partial f}\right| \left|\frac{\partial(x,y)}{\partial(u,v)}\right| dudv\right]$$
$$=: N\left(x_{B,\partial f}, y_{B,\partial f}, z_{B,\partial f}; f\right). \tag{4.14}$$

Moreover, we have the bounds

$$N\left(x_{B,\partial f}, y_{B,\partial f}, z_{B,\partial f}; f\right) \leq \frac{1}{3}\|f\|_{S,\infty}\left[\int_a^b \int_c^d \left|x(u,v) - x_{B,\partial f}\right| \left|\frac{\partial(y,z)}{\partial(u,v)}\right|\right.$$
$$\left. + \left|y(u,v) - y_{B,\partial f}\right| \left|\frac{\partial(z,x)}{\partial(u,v)}\right| + \left|z(u,v) - z_{B,\partial f}\right| \left|\frac{\partial(x,y)}{\partial(u,v)}\right| dudv\right]$$
$$\leq \frac{1}{3}\|f\|_{S,\infty}$$

$$\times \begin{cases} \left\| \frac{\partial(y,z)}{\partial(\cdot,\cdot)} \right\|_{\square,\infty} \left\| x - x_{B,\partial f} \right\|_{\square,1} + \left\| \frac{\partial(z,x)}{\partial(\cdot,\cdot)} \right\|_{\square,\infty} \left\| y - y_{B,\partial f} \right\|_{\square,1} \\ \quad + \left\| \frac{\partial(x,y)}{\partial(\cdot,\cdot)} \right\|_{\square,\infty} \left\| z - z_{B,\partial f} \right\|_{\square,1}, \\ \\ \left\| \frac{\partial(y,z)}{\partial(\cdot,\cdot)} \right\|_{\square,p} \left\| x - x_{B,\partial f} \right\|_{\square,q} + \left\| \frac{\partial(z,x)}{\partial(\cdot,\cdot)} \right\|_{\square,p} \left\| y - y_{B,\partial f} \right\|_{\square,q} \\ \quad + \left\| \frac{\partial(x,y)}{\partial(\cdot,\cdot)} \right\|_{\square,p} \left\| z - z_{B,\partial f} \right\|_{\square,q}, \\ \\ \left\| \frac{\partial(y,z)}{\partial(\cdot,\cdot)} \right\|_{\square,1} \left\| x - x_{B,\partial f} \right\|_{\square,\infty} + \left\| \frac{\partial(z,x)}{\partial(\cdot,\cdot)} \right\|_{\square,p} \left\| y - y_{B,\partial f} \right\|_{\square,\infty} \\ \quad + \left\| \frac{\partial(x,y)}{\partial(\cdot,\cdot)} \right\|_{\square,1} \left\| z - z_{B,\partial f} \right\|_{\square,\infty}. \end{cases} \tag{4.15}$$

5 Applications for Three-Dimensional Balls

Now, let us compute the surface integral

$$K\left(S\left(C,R\right),f\right) := \iint_{S(C,R)} f\left(x,y,z\right) dS,$$

where

$$S\left(C,R\right) := \left\{ (x,y,z) \in \mathbb{R}^3 \middle| \, (x-a)^2 + (y-b)^2 + (z-c)^2 = R^2 \right\}.$$

If we consider the parametrization of $S\left(C,R\right)$ given by:

$$S\left(C,R\right) : \begin{cases} x = R\cos\psi\cos\varphi + a \\ y = R\cos\psi\sin\varphi + b \\ z = R\sin\psi + c \end{cases} ; \; (\psi,\varphi) \in \left[-\frac{\pi}{2},\frac{\pi}{2}\right] \times [0,2\pi]$$

and putting

$$A := \begin{vmatrix} \frac{\partial y}{\partial \psi} & \frac{\partial z}{\partial \psi} \\ \frac{\partial y}{\partial \varphi} & \frac{\partial z}{\partial \varphi} \end{vmatrix} = -R^2\cos^2\psi\cos\varphi,$$

$$B := \begin{vmatrix} \frac{\partial x}{\partial \psi} & \frac{\partial z}{\partial \psi} \\ \frac{\partial x}{\partial \varphi} & \frac{\partial z}{\partial \varphi} \end{vmatrix} = R^2\cos^2\psi\sin\varphi,$$

and

$$C := \begin{vmatrix} \frac{\partial x}{\partial \psi} & \frac{\partial y}{\partial \psi} \\ \frac{\partial x}{\partial \varphi} & \frac{\partial y}{\partial \varphi} \end{vmatrix} = -R^2 \sin \psi \cos \psi,$$

we have that

$$A^2 + B^2 + C^2 = R^4 \cos^2 \psi \text{forall} (\psi, \varphi) \in \left[-\frac{\pi}{2}, \frac{\pi}{2}\right] \times [0, 2\pi].$$

Thus,

$$\begin{aligned} K(S(C,R), f) &= \iint_{S(C,R)} f(x,y,z)\, dS \\ &= \int_{-\frac{\pi}{2}}^{\frac{\pi}{2}} \int_0^{2\pi} \Big[f(R\cos\psi\cos\varphi + a, R\cos\psi\sin\varphi + b, R\sin\psi + c) \\ &\quad \times \sqrt{A^2 + B^2 + C^2} \Big] d\psi d\varphi \\ &= R^2 \int_{-\frac{\pi}{2}}^{\frac{\pi}{2}} \int_0^{2\pi} \cos\psi f(R\cos\psi\cos\varphi + a, R\cos\psi\sin\varphi \\ &\quad + b, R\sin\psi + c)\, d\psi d\varphi. \end{aligned} \tag{5.1}$$

We also have

$$\begin{aligned} L(S(C,R), f) &:= \int\int_{S(C,R)} (x-a) f(x,y,z)\, dy \wedge dz \\ &\quad + \int\int_{S(C,R)} (y-b) f(x,y,z)\, dz \wedge dx \\ &\quad + \int\int_{S(C,R)} (z-c) f(x,y,z)\, dx \wedge dy \\ &= -R^3 \int_{-\frac{\pi}{2}}^{\frac{\pi}{2}} \int_0^{2\pi} \cos^3\psi \cos^2\varphi \\ &\quad \times f(R\cos\psi\cos\varphi + a, R\cos\psi\sin\varphi + b, R\sin\psi + c)\, d\psi d\varphi \\ &\quad + R^3 \int_{-\frac{\pi}{2}}^{\frac{\pi}{2}} \int_0^{2\pi} \cos^3\psi \sin^2\varphi \\ &\quad \times f(R\cos\psi\cos\varphi + a, R\cos\psi\sin\varphi + b, R\sin\psi + c)\, d\psi d\varphi \\ &\quad - R^3 \int\int_S \sin^2\psi \cos\psi f(R\cos\psi\cos\varphi + a, R\cos\psi\sin\varphi \\ &\quad + b, R\sin\psi + c)\, d\psi d\varphi. \end{aligned} \tag{5.2}$$

Let us consider the transformation $T_2 : \mathbb{R}^3 \to \mathbb{R}^3$ given by:

$$T_2(r, \psi, \varphi) := (r\cos\psi\cos\varphi + a, r\cos\psi\sin\varphi + b, r\sin\psi + c).$$

It is well known that the Jacobian of T_2 is

$$J(T_2) = r^2\cos\psi$$

and T_2 is a one-to-one mapping defined on the interval of $\mathbb{R}^3$, $[0, R] \times \left[-\frac{\pi}{2}, \frac{\pi}{2}\right] \times [0, 2\pi]$, with values in the ball $B(C, R)$ from $\mathbb{R}^3$. Thus we have the change of variable:

$$\begin{aligned} I(B(C,R), f) &:= \iiint_{B(C,R)} f(x, y, z)\,dxdydz \\ &= \int_0^R \int_{-\frac{\pi}{2}}^{\frac{\pi}{2}} \int_0^{2\pi} f(r\cos\psi\cos\varphi + a, r\cos\psi\sin\varphi \\ &\quad + b, r\sin\psi + c)\, r^2\cos\psi\, dr d\psi d\varphi. \end{aligned} \tag{5.3}$$

We also have

$$\iiint_{B(C,R)} |a - x|\,dxdydz = \int_0^R \int_{-\frac{\pi}{2}}^{\frac{\pi}{2}} \int_0^{2\pi} r^3\cos^2\psi\,|\cos\varphi|\,dr d\psi d\varphi = \frac{\pi}{2}R^4$$

and, similarly

$$\iiint_{B(C,R)} |b - y|\,dxdydz = \iiint_{B(C,R)} |c - z|\,dxdydz = \frac{\pi}{2}R^4.$$

Therefore

$$\begin{aligned} &\left\|\frac{\partial f}{\partial x}\right\|_{B(C,R),\infty} \iiint_{B(C,R)} |\overline{x_B} - x|\,dxdydz \\ &\quad + \left\|\frac{\partial f}{\partial y}\right\|_{B(C,R),\infty} \iiint_{B(C,R)} |\overline{y_B} - y|\,dxdydz \\ &\quad + \left\|\frac{\partial f}{\partial z}\right\|_{B(C,R),\infty} \iiint_{B(C,R)} |\overline{z_B} - z|\,dxdydz \\ &\quad = \frac{\pi}{2}R^4 \left(\left\|\frac{\partial f}{\partial x}\right\|_{B(C,R),\infty} + \left\|\frac{\partial f}{\partial y}\right\|_{B(C,R),\infty} + \left\|\frac{\partial f}{\partial z}\right\|_{B(C,R),\infty}\right) \end{aligned}$$

and by the inequalities (4.3) and (4.4) we get

$$\left|I(B(C,R),f)-\frac{1}{3}L(S(C,R),f)\right| \le \frac{\pi}{6}R^4\left(\left\|\frac{\partial f}{\partial x}\right\|_{B(C,R),\infty}+\left\|\frac{\partial f}{\partial y}\right\|_{B(C,R),\infty}+\left\|\frac{\partial f}{\partial z}\right\|_{B(C,R),\infty}\right) \tag{5.4}$$

provided $f : B(C,R) \to \mathbb{C}$ is a continuously differentiable function defined on an open set containing $B(C,R)$.

We also consider

$$\begin{aligned} T(B(C,R),f) &:= \iiint_B \Big[(\overline{x_B}-x)\frac{\partial f(x,y,z)}{\partial x} \\ &+ (\overline{y_B}-y)\frac{\partial f(x,y,z)}{\partial y} + (\overline{z_B}-z)\frac{\partial f(x,y,z)}{\partial z}\Big] dxdydz \\ &= -\int_0^R\int_{-\frac{\pi}{2}}^{\frac{\pi}{2}}\int_0^{2\pi}\left(r^3\cos^2\psi\cos\varphi\right) \\ &\times \frac{\partial f(r\cos\psi\cos\varphi+a, r\cos\psi\sin\varphi+b, r\sin\psi+c)}{\partial x} drd\psi d\varphi \\ &- \int_0^R\int_{-\frac{\pi}{2}}^{\frac{\pi}{2}}\int_0^{2\pi}\left(r^3\cos^2\psi\sin\varphi\right) \\ &\times \frac{\partial f(r\cos\psi\cos\varphi+a, r\cos\psi\sin\varphi+b, r\sin\psi+c)}{\partial y} drd\psi d\varphi \\ &- \int_0^R\int_{-\frac{\pi}{2}}^{\frac{\pi}{2}}\int_0^{2\pi} r^3\sin\psi\cos\psi \\ &\times \frac{\partial f(r\cos\psi\cos\varphi+a, r\cos\psi\sin\varphi+b, r\sin\psi+c)}{\partial z} drd\psi d\varphi. \end{aligned} \tag{5.5}$$

By the inequality (4.11) we obtain

$$\begin{aligned} N(a,b,c;f) &\le \frac{1}{3}\left(\int\int_{S(C,R)} |f(x,y,z)|^2 dS\right)^{1/2} \\ &\times \left(\int\int_{S(C,R)} \left(|x-a|^2+|y-b|^2+|z-c|^2\right) dS\right)^{1/2} \\ &= \frac{2}{3}\sqrt{\pi}R^2\left[K\left(S(C,R),|f|^2\right)\right]^{1/2} \end{aligned}$$

and by utilizing (4.12) we also get

$$\left|I(B(C,R),f)-\frac{1}{3}T(B(C,R),f)\right| \le \frac{2}{3}\sqrt{\pi}R^2\left[K\left(S(C,R),|f|^2\right)\right]^{1/2}. \tag{5.6}$$

References

1. Apostol, T. M.; *Calculus Volume II, Multi Variable Calculus and Linear Algebra, with Applications to Differential Equations and Probability*, Second Edition, John Wiley & Sons, New York, London, Sydney, Toronto, 1969.
2. Barani, A.; Hermite–Hadamard and Ostrowski type inequalities on hemispheres, *Mediterr. J. Math.* **13** (2016), 4253–4263
3. Barnett, N. S.; Cîrstea, F. C. and Dragomir, S. S. Some inequalities for the integral mean of Hölder continuous functions defined on disks in a plane, in *Inequality Theory and Applications*, Vol. **2** (Chinju/Masan, 2001), 7–18, Nova Sci. Publ., Hauppauge, NY, 2003. Preprint *RGMIA Res. Rep. Coll.* **5** (2002), Nr. 1, Art. 7, 10 pp. https://rgmia.org/papers/v5n1/BCD.pdf.
4. Barnett, N. S.; Dragomir, S. S. An Ostrowski type inequality for double integrals and applications for cubature formulae. Soochow J. Math. 27 (2001), no. 1, 1–10.
5. Barnett, N. S.; Dragomir, S. S.; Pearce, C. E. M. A quasi-trapezoid inequality for double integrals. ANZIAM J. 44 (2003), no. 3, 355–364.
6. Bessenyei, M.; The Hermite–Hadamard inequality on simplices, *Amer. Math. Monthly* **115** (2008), 339–345.
7. Budak, Hüseyin; Sarıkaya, Mehmet Zeki An inequality of Ostrowski-Grüss type for double integrals. Stud. Univ. Babeş-Bolyai Math. 62 (2017), no. 2, 163–173.
8. Cal, J. de la and Cárcamo, J.; Multidimensional Hermite-Hadamard inequalities and the convex order, *Journal of Mathematical Analysis and Applications*, vol. **324**, no. 1, pp. 248–261, 2006.
9. Dragomir S. S.; On Hadamard's inequality on a disk, *Journal of Ineq. Pure & Appl. Math.*, **1** (2000), No. 1, Article 2. https://www.emis.de/journals/JIPAM/article95.html?sid=95.
10. Dragomir S. S.; On Hadamard's inequality for the convex mappings defined on a ball in the space and applications, *Math. Ineq. & Appl.*, **3** (2) (2000), 177–187.
11. Dragomir, S. S.; Ostrowski type integral inequalities for multiple integral on general convex bodies, Preprint *RGMIA Res. Rep. Coll.* **22** (2019), Art. 50, 13 pp. http://rgmia.org/papers/v22/v22a50.pdf.
12. Dragomir, S. S.; Cerone, P.; Barnett, N. S.; Roumeliotis, J. An inequality of the Ostrowski type for double integrals and applications for cubature formulae. Tamsui Oxf. J. Math. Sci. 16 (2000), no. 1, 1–16.
13. Erden, Samet; Sarikaya, Mehmet Zeki on exponential Pompeiu's type inequalities for double integrals with applications to Ostrowski's inequality. New Trends Math. Sci. 4 (2016), no. 1, 256–267.
14. Hanna, G.; Some results for double integrals based on an Ostrowski type inequality. Ostrowski type inequalities and applications in numerical integration, 331–371, Kluwer Acad. Publ., Dordrecht, 2002.
15. Hanna, G.; Dragomir, S. S.; Cerone, P. A general Ostrowski type inequality for double integrals. Tamkang J. Math. 33 (2002), no. 4, 319–333.
16. Liu, Z.; A sharp general Ostrowski type inequality for double integrals. Tamsui Oxf. J. Inf. Math. Sci. 28 (2012), no. 2, 217–226.
17. Matłoka, M.; On Hadamard's inequality for h-convex function on a disk, *Applied Mathematics and Computation* 235 (2014) 118–123
18. Mitroi, F.-C. and Symeonidis, E.; The converse of the Hermite-Hadamard inequality on simplices. *Expo. Math.* **30** (2012), 389–396.
19. Neuman, E.; Inequalities involving multivariate convex functions II, *Proc. Amer. Math. Soc.* **109** (1990), 965–974.
20. Neuman, E. and Pečarić, J.; Inequalities involving multivariate convex functions, *J. Math. Anal. Appl.* **137** (1989), 541–549.
21. Özdemir, M. Emin; Akdemir, Ahmet Ocak; Set, Erhan A new Ostrowski-type inequality for double integrals. J. Inequal. Spec. Funct. 2 (2011), no. 1, 27–34.
22. Pachpatte, B. G.; A new Ostrowski type inequality for double integrals. Soochow J. Math. 32 (2006), no. 2, 317–322.

23. Sarikaya, Mehmet Zeki on the Ostrowski type integral inequality for double integrals. Demonstratio Math. 45 (2012), no. 3, 533–540.
24. Sarikaya, Mehmet Zeki; Ogunmez, Hasan On the weighted Ostrowski-type integral inequality for double integrals. Arab. J. Sci. Eng. 36 (2011), no. 6, 1153–1160.
25. Wasowicz, S. and Witkowski, A.; On some inequality of Hermite–Hadamard type. *Opusc. Math.* **32** (3)(2012), 591–600
26. Wang, F.-L.;The generalizations to several-dimensions of the classical Hadamard's inequality, *Mathematics in Practice and Theory*, vol. **36**, no. 9, pp. 370–373, 2006 (Chinese).
27. Wang, F.-L.; A family of mappings associated with Hadamard's inequality on a hypercube, *International Scholarly Research Network ISRN Mathematical Analysis* Volume **2011**, Article ID 594758, 9 pages https://doi.org/10.5402/2011/594758.

Optimal Emergency Evacuation with Uncertainty

Georgia Fargetta and Laura Scrimali

Abstract Emergency management after crises or natural disasters is a very important issue shared by many countries. In this chapter, we focus on evacuation planning which is a complex and challenging process able to predict or evaluate different disaster scenarios. In particular, we present an evacuation model where a population has to be evacuated from crisis areas to shelters and propose an optimization formulation for minimizing a combination of the transportation cost and the transportation time. In addition, we admit uncertainty in the size of the population to be evacuated and provide a two-stage stochastic programming model. In order to illustrate the modeling framework, we present a numerical example.

1 Introduction

Natural disasters (earthquakes, hurricanes, landslides, etc.) as well as unnatural ones (wars, terrorist attacks, etc.) are a serious threat for the humankind. Evacuation of the disaster region is the most used strategy to save people affected by a disaster. Generally, disasters cannot be predicted, and it is extremely difficult to estimate their intensity and damages; hence, evacuation planning must be done under uncertainty. For this reason, it is generally formulated as a stochastic programming problem (see [16]). Incomplete information may regard different factors, such as evacuation demand, link capacity, disruption in the road network, or how much infrastructures may be impacted by disasters.

In this chapter, we propose a scenario-based evacuation planning model that provides the optimal flows of evacuees from crisis areas to shelters, in order to minimize both the transportation cost and the transportation time, under uncertainty on the evacuation demand and the link capacities. Inspired by [18], we admit real-time information availability, which makes the evacuation process be divided into

G. Fargetta · L. Scrimali (✉)
Department of Mathematics and Computer Science, University of Catania, Catania, Italy
e-mail: georgia.fargetta@phd.unict.it; laura.scrimali@unict.it

I. N. Parasidis et al. (eds.), *Mathematical Analysis in Interdisciplinary Research*,
Springer Optimization and Its Applications 179,
https://doi.org/10.1007/978-3-030-84721-0_14

two stages. In the first stage, people at risk receive the early warning information about the disaster and escape from the crisis areas; however, they cannot obtain the exact information of disaster intensity. After a certain time period, accurate real-time information is observed and the process reaches the second stage, where the decision relies on the first-stage solution and on the observed scenario. Moreover, we introduce a penalty to the unmet demand of evacuation which will affect the second-stage decision.

The importance of an efficient approach to emergency management and evacuation planning has been emphasized in several papers.

In [5], the authors model an escape situation in a labyrinth, where people are agents that act as two different kinds of ant colonies. Payoff values in both the competitive and the cooperative framework are studied, merging a game theoretical approach and Ant Colony Optimization.

In [1], the authors present a two-stage stochastic programming model to plan the transportation of vital first-aid commodities to disaster-affected areas during emergency response. A multi-commodity, multi-modal network flow formulation is then developed. Since it is difficult to predict the timing and magnitude of any disaster, uncertainty and information asymmetry naturally arise. The authors introduce randomness as a finite sample of scenarios for capacity, supply, and demand.

In [2], the authors propose a scenario-based two-stage stochastic evacuation planning model that optimally chooses shelter sites and assigns evacuees to nearest shelters within a tolerance degree to minimize the expected total evacuation time. The model takes into account the uncertainty in the evacuation demand and the disruption in the road network and shelter sites.

In [12], the authors develop a stochastic optimization model that determines the order in which patients should be evacuated over time, based on the evolution of the storm by considering a weighted sum of the expected risk and the expected cost of evacuation.

In [17], a bi-objective optimization model is proposed, which studies critical management before and after the disaster. The first level investigates the locations of shelters and warehouses before the disaster and maximizes the weights of the sites selected for construction of shelters. The second level minimizes the distances from warehouses to the shelters and the distances from crisis areas to the shelters.

In [18], the authors study the regional emergency resources storage and, in particular, the region division. A two-stage stochastic programming model is proposed to solve the region division problem.

Recently, two-stage stochastic variational inequalities were introduced, where one seeks a decision vector before the stochastic variables are known and a decision vector after the scenario has been realized. In [13], Rockafellar and Wets propose the multistage stochastic variational inequality. In [14], the authors develop progressive hedging methods for solving multistage convex stochastic programming, see also [15]. In [4], the authors formulate the two-stage stochastic variational inequality as a two-stage stochastic programming problem with recourse.

In this chapter, we present a two-stage stochastic programming problem for the evacuation planning and give an equivalent formulation as a two-stage stochastic variational inequality, using the Lagrangian relaxation approach. We also discuss the qualitative properties of the two-stage stochastic variational inequality.

The structure of this chapter is as follows. In Sect. 2, we present the deterministic evacuation model and derive an equivalent variational inequality formulation. In Sect. 3, we present the two-stage stochastic model. In Sect. 4, we propose an equivalent two-stage variational inequality formulation. In Sect. 5, we provide a numerical example, and, finally, we present our conclusions in Sect. 6.

2 The Deterministic Model

We assume that a population of N individuals is located in some crisis areas and must be evacuated to some shelters. Different modes of transportation are considered to enhance node accessibility. We denote by A the set of crisis areas, with typical area denoted by i, by S the set of shelters, with typical shelter denoted by j, and by M the set of transportation modes, with typical mode denoted by m. We consider a network representation as in Fig. 1. The links between the levels of the network represent all the possible connections between the crisis areas and the shelters. Multiple links between each area and each shelter depict the possibility of alternative modes of transportation.

Let d_i^m be the demand of crisis area i for evacuation with mode m, namely, the number of people to be evacuated from area i with mode m, where $\sum_{i \in A} \sum_{m \in M} d_i^m \leq N$. We denote by $d_i = \sum_{m \in M} d_i^m$ the demand of area i on all modes. Moreover, let x_{ij} be the flows of evacuees from area i to shelter j. We also assume that K_j is the maximum number of people that can be hosted in shelter j, and k_j is the minimum number of people required to open shelter j. Thus, the following conditions have to be satisfied:

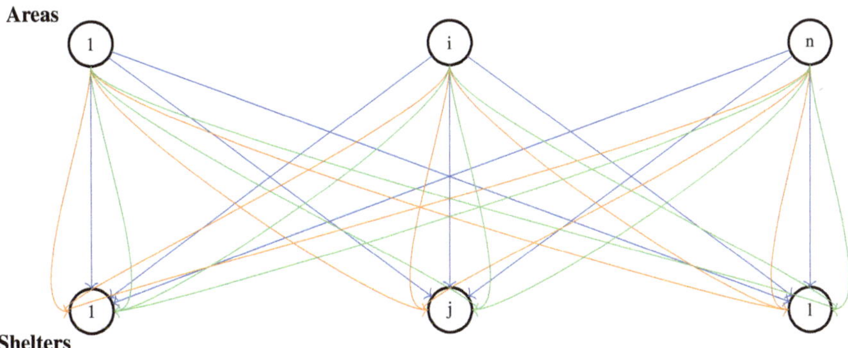

Fig. 1 The network representation

Table 1 The notation for the deterministic model

Symbols	Definitions
A	Set of crisis areas, with typical area denoted by i, $card(A) = n$
S	Set of shelters, with typical shelter denoted by j, $card(S) = l$
M	Set of transportation modes, with typical mode denoted by m, $card(M) = \bar{m}$
d_i^m	Demand of crisis area i for evacuation with mode m
$d_i = \sum_{m \in M} d_i^m$	The demand of area i on all modes
K_j	Maximum number of people that can be hosted in shelter j
k_j	Recommended number of people to open shelter j
w	Weight in [0, 1]
ε	Positive balance tolerance
μ_{ij}	Flow capacity on link (i, j)
$\mu = (\mu_{ij})_{i,j}$	Total flow capacity
x_{ij}	Flow of evacuees from area i to shelter j
$x = (x_{ij})_{i,j}$	Total flow of evacuees
$c_{ij}^m(x_{ij})$	Transportation cost from area i to shelter j with mode m
$t_{ij}^m(x_{ij})$	Transportation time from area i to shelter j with mode m

$$\sum_{i \in A} x_{ij} \leq K_j, \forall j \in S, \quad \sum_{i \in A} x_{ij} \geq k_j, \forall j \in S.$$

We group the flows x_{ij} into a column vector $x \in \mathbb{R}^{nl}$. In addition, we introduce the transportation cost c_{ij}^m from area i to shelter j with mode m and assume that it depends on the flow x_{ij}, namely, $c_{ij}^m = c_{ij}^m(x_{ij})$. Analogously, we define the transportation time t_{ij}^m from area i to shelter j with mode m and assume that it depends on the flow x_{ij}, namely, $t_{ij}^m = t_{ij}^m(x_{ij})$.

We summarize the relevant notations used in the mathematical formulation in Table 1.

We introduce the total evacuation cost C_{ij}^m given by

$$C_{ij}^m(x_{ij}) = w\, c_{ij}^m(x_{ij}) + (1 - w)\, t_{ij}^m(x_{ij}), \quad \forall i \in A\,, j \in S\,, m \in M.$$

Our aim is to minimize the total evacuation costs; hence, we seek to solve the following optimization problem:

$$\min \sum_{i \in A} \sum_{j \in S} \sum_{m \in M} C_{ij}^m(x_{ij}) \tag{1}$$

$$\sum_{i \in A} x_{ij} \leq K_j, \; \forall j \in S, \tag{2}$$

$$\sum_{i \in A} x_{ij} \geq k_j, \ \forall j \in S, \tag{3}$$

$$x_{ij} \leq \mu_{ij}, \ \forall i \in A, \ j \in S, \tag{4}$$

$$\frac{\sum_{i \in A} x_{ij}}{K_j} - \frac{\sum_{i \in A} x_{ij'}}{K_j'} \leq \varepsilon, \ \forall j, \ j' \in S, \tag{5}$$

$$\sum_{j \in S} x_{ij} \leq d_i, \ \forall i \in A, \tag{6}$$

$$x_{ij} \geq 0, \ \forall i \in A, j \in S. \tag{7}$$

The objective function (1) is the weighted sum of transportation cost and transportation time, where $w \in [0, 1]$. Constraint (2) ensures that the capacity of each shelter j is not exceeded. Constraint (3) guarantees that each shelter j is used. Constraint (4) requires that the flow on link (i, j) must satisfy the flow capacity on that link. Constraint (5) states that the number of evacuees is balanced among the shelters. Constraint (6) establishes that for each area i, the evacuation demand d_i on all modes is satisfied, and, finally, (7) represents the non-negativity requirement on flows.

We now introduce the set of feasible flows

$$X = \Big\{ x \in \mathbb{R}^{nl} : x_{ij} \geq 0, \forall i, j \, ; \sum_{i \in A} x_{ij} \leq K_j, \ \forall j; \sum_{i \in A} x_{ij} \geq k_j, \ \forall j;$$

$$x_{ij} \leq \mu_{ij}, \ \forall i, j, \ ; \ \frac{\sum_{i \in A} x_{ij}}{K_j} - \frac{\sum_{i \in A} x_{ij'}}{K_j'} \leq \varepsilon, \ \forall j, \ j'; \ \sum_{j \in S} x_{ij} \leq d_i, \ \forall i \Big\}.$$

We assume that the transportation cost and the transportation time functions are continuously differentiable and convex. In addition, since the set X is closed, bounded, and convex, we can apply the classical theory on variational inequalities (see, for instance, [6, 9], or [11]) and formulate problem (1)–(7) as the following variational inequality:

$$\text{Find} x^* \in X : \quad \sum_{i \in A} \sum_{j \in S} \sum_{m \in M} \left[w \frac{\partial c_{ij}^m(x_{ij}^*)}{\partial x_{ij}} + (1 - w) \frac{\partial t_{ij}^m(x_{ij}^*)}{\partial x_{ij}} \right] \times \left(x_{ij} - x_{ij}^* \right), \ \forall x \in X.$$

The above variational inequality can be put in standard form as follows (see [11]):

$$\text{Find } x^* \in X \text{ such that } \langle F(x^*), x - x^* \rangle \geq 0, \quad \forall x \in X, \tag{8}$$

where $\langle \cdot, \cdot \rangle$ denotes the inner product in the (nl)-dimensional Euclidean space and

$$F(x^*) = \left[w \frac{\partial c_{ij}^m(x_{ij}^*)}{\partial x_{ij}} + (1-w) \frac{\partial t_{ij}^m(x_{ij}^*)}{\partial x_{ij}} \right]_{\substack{i \in A \\ j \in S \\ m \in M}} .$$

Under the assumptions on the feasible set X and the objective function, we can ensure the existence of at least one solution to (8) (see [9]). Moreover, if the function $F(x)$ in (8) is strictly monotone on X, namely,

$$\langle F(x^1) - F(x^2), x^1 - x^2 \rangle > 0 \quad \forall x^1, x^2 \in X, \ x^1 \neq x^2,$$

then variational inequality (8) admits a unique solution.

3 Two-Stage Stochastic Model

In this section, we present a scenario-based stochastic optimization model to represent the evacuation process after an earthquake. At the occurrence of the disaster event, affected people receive the early warning information and escape from the crisis areas. After the event, accurate information is available, for instance, due to technological communication tools. Since the initial response will depend on a number of disaster scenarios, we propose a two-stage stochastic programming model with recourse, where both the first stage and the second stage arise in different time phases in the same evacuation network. We remark that the first-stage decision is taken before the realization of the disaster scenario is observed. After that this information is accessible, the decision process reaches the second stage, where the decision depends on the first-stage solution and on the observed scenario.

As it is very hard to estimate exactly the impact of a natural disaster, we allow for uncertainty in the modeling assumptions and introduce some random parameters. In particular, since the effects of disasters naturally randomize the number of people who survive, we consider random evacuation demands at the emergency sites. Moreover, we take into account possible disruptions on network links and deal with random capacities on links.

We note that, in the first stage, the evacuation demand and the link capacities are known only from a probabilistic point of view. Thus, in the first stage, people at risk must be evacuated from crisis areas before observing the real data. Instead, in the second stage, the evacuation plan will be obtained with the realization of the evacuation demand and the link capacities.

Our aim is to formulate the random evacuation problem as a two-stage optimization problem. In the first stage, we seek for the optimal flows to minimize the penalty cost generated by the decisions taken before the acquisition of the information plus the recourse cost for the disaster scenario. In the second stage, another optimization problem is solved, based on a given realization of each scenario.

Table 2 The notation for the two-stage stochastic model

Symbols	Definitions
μ_{ij}	Flow capacity on link (i, j) in stage one
$\mu_{ij}(\omega)$	Random flow capacity in stage two
d_i^m	Evacuation demand of crisis area i with mode m in stage one
$d_i = \sum_{m \in M} d_i^m$	Demand of area i on all modes in stage one
$d_i^m(\omega)$	Evacuation demand of crisis area i with mode m in stage two
$d_i(\omega) = \sum_{m \in M} d_i^m(\omega)$	Demand of area i on all modes in stage two
$\xi(\omega)$	Random parameters in stage two resulting from decisions made in stage one according to ω
x_{ij}	Flow of evacuees from area i to shelter j in stage one
$y_{ij}(\omega)$	Flow of evacuees from area i to shelter j in stage two under scenario ω
$\pi_{ij}(x_{ij})$	The penalty cost on link (i, j)

Let $(\Omega, \mathcal{F}, P)$ be a probability space, where the random parameter $\omega \in \Omega$ represents the typical disaster scenario. For each $\omega \in \Omega$, we denote by $\xi : \Omega \to \mathbb{R}^K$ a finite-dimensional random vector and by $\mathbb{E}_\xi$ the mathematical expectation with respect to ξ.

In order to formulate the two-stage stochastic evacuation model, we introduce two types of decision variables. In the first stage, the decision variable x_{ij} is used to represent the flow of evacuees from area i to shelter j in stage one. The second-stage decision variable $y_{ij}(\omega)$ represents the flow of evacuees from area i to shelter j in stage two under scenario ω. From the perspective of the entire system, the decision planner chooses x_{ij} before a realization of ξ is revealed and later selects $y(\omega)$ with known realization.

Table 2 summarizes the relevant notations used in the model formulation.

We denote by $\pi_{ij} = \pi_{ij}(x_{ij})$ the penalty cost generated by the decisions taken before obtaining the information. In order to minimize the penalty for the prior evacuation plan and the expected total evacuation expenses (time and cost) for each scenario, we formulate the following two-stage evacuation problem:

$$\min \sum_{i \in A} \sum_{j \in S} \sum_{m \in M} \pi_{ij}(x_{ij}) + \mathbb{E}_\xi(\Phi(x, \xi(\omega))) \tag{9}$$

subject to

$$\sum_{i \in A} x_{ij} \leq K_j, \ \forall j \in S, \tag{10}$$

$$\sum_{i \in A} x_{ij} \geq k_j, \ \forall j \in S, \tag{11}$$

$$x_{ij} \leq \mu_{ij}, \ \forall i \in A, \ j \in S, \tag{12}$$

$$\frac{\sum_{i\in A} x_{ij}}{K_j} - \frac{\sum_{i\in A} x_{ij'}}{K'_j} \leq \varepsilon, \; \forall j, \; j' \in S, \tag{13}$$

$$\sum_{j\in S} x_{ij} \leq d_i, \; \forall i \in A, \tag{14}$$

$$x_{ij} \geq 0, \; \forall i \in A, \; j \in S, \tag{15}$$

where

$$\Phi(x, \xi(\omega)) = \min \sum_{i\in A} \sum_{j\in S} \Big(w \, c^m_{ij}(y_{ij}(\omega)) + (1-w) \, t^m_{ij}(y_{ij}(\omega)) \Big) \tag{16}$$

subject to

$$\sum_{i\in A} y_{ij}(\omega) \leq K_j, \; \forall j \in S, \; P\text{-}a.s., \tag{17}$$

$$\sum_{i\in A} y^m_{ij}(\omega) \geq k_j, \; \forall j \in S, \; P\text{-}a.s., \tag{18}$$

$$y_{ij}(\omega) \leq \mu_{ij}(\omega), \; \forall i \in A, \; j \in S, \; P\text{-}a.s., \tag{19}$$

$$\frac{\sum_{i\in A} y_{ij}(\omega)}{K_j} - \frac{\sum_{i\in A} y_{ij'}(\omega)}{K'_j} \leq \varepsilon, \; \forall j, \; j' \in S, \; P\text{-}a.s., \tag{20}$$

$$\sum_{j\in S} y_{ij}(\omega) + \big(d_i - \sum_{j} x_{ij}\big) \leq d_i(\omega), \; \forall i \in A, \; P\text{-}a.s., \tag{21}$$

$$y_{ij}(\omega) \geq 0, \; \forall i \in A, \; j \in S, \; P\text{-}a.s. \tag{22}$$

Problem (9)–(15) is the first-stage problem. The objective function (9) minimizes the sum of the penalty for early evacuation plan and the recourse cost $\Phi(x, \xi(\omega))$. Constraints (10)–(11) ensure that the capacity of each shelter j is satisfied. Constraint (12) is the link flow capacity, constraint (13) is the balance constraint among the shelters, and constraint (14) states that for each area i, the number of evacuees cannot exceed the number of affected people. Finally, (15) is the non-negativity requirements on flows.

For a given realization $\omega \in \Omega$, $\Phi(x, \xi(\omega))$ is the optimal value of the second-stage problem (16)–(22), where the constraints hold almost surely (P-a.s.). The objective function (16) minimizes the total evacuation cost and the transportation time in the second stage. Constraints (17)–(18) are the shelter capacities, constraint (19) is the flow capacity, and constraint (20) is the balance constraint among the shelters. Constraint (21) establishes that the number of evacuees at the second stage plus the unmet demand of evacuation at the first stage, given by $d_i - \sum_{j\in S} x_{ij}$, cannot exceed the two-stage demand of people at risk $d_i(\omega)$. We emphasize that the connections between stage-wise decision variables x and y are captured by constraint (21). It is the linking factor between the first and second stages and communicates the first-stage decisions to the second one. Finally, (22) is the non-negativity constraint.

We assume that

1. $\pi_{ij}(\cdot)$ is continuously differentiable and convex for all i, j;
2. $c_{ij}^m(\cdot, \omega)$ and $t_{ij}^m(\cdot, \omega)$, a.e. in Ω, are continuously differentiable and convex for all i, j;
3. for each $u \in \mathbb{R}^{pq}$, $c_{ij}^m(u, \cdot)$ and $t_{ij}^m(u, \cdot)$ are measurable with respect to the random parameter in Ω for all i, j;
4. $y_{ij} : \Omega \to \mathbb{R}$ and $\mu_{ij} : \Omega \to \mathbb{R}$ are measurable mappings for all i, j;
5. $d_i^m : \Omega \to \mathbb{R}$ is a measurable mapping for all i and all m.

Now, we set

$$Y = \Bigg\{ y(\omega) \in \mathbb{R}^{nl} : y_{ij}(\omega) \geq 0, \forall i,\ j\,; \sum_{i \in A} y_{ij}(\omega) \leq K_j,\ \forall j; \sum_{i \in A} y_{ij}(\omega) \geq k_j,\ \forall j;$$

$$y_{ij}(\omega) \leq \mu_{ij}(\omega),\ \forall i,\ j;\ \frac{\sum_{i \in A} y_{ij}(\omega)}{K_j} - \frac{\sum_{i \in A} y_{ij'}(\omega)}{K_j'} \leq \varepsilon,\ \forall j,\ j',\ P-a.s. \Bigg\},$$

which is a closed, convex, and bounded subset of $\mathbb{R}^{pq}$.

Thus, the two-stage stochastic problem can be stated in a more compact form as

$$\min_{x \in X} \sum_{i \in A} \sum_{j \in S} \pi_{ij}(x_{ij}) + \mathbb{E}_\xi(\Phi(x, \xi(\omega))), \tag{23}$$

$$\Phi(x, \xi(\omega)) = \min_{y(\omega) \in Y} \sum_{i \in A} \sum_{j \in S} \sum_{m \in M} \left(w\, c_{ij}^m(y_{ij}(\omega)) + (1 - w)\, t_{ij}^m(y_{ij}(\omega)) \right) \tag{24}$$

$$\sum_{j \in S} y_{ij}(\omega) + \left(d_i - \sum_j x_{ij} \right) \leq d_i(\omega),\ \forall i \in A,\ P\text{-}a.s. \tag{25}$$

If the random parameter $\omega \in \Omega$ follows a discrete distribution with finite support $\Omega = \{\omega_1, \ldots, \omega_{\bar{r}}\}$ and probabilities $p(\omega_1), \ldots, p(\omega_{\bar{r}})$ associated with each realization $\omega_1, \ldots, \omega_{\bar{r}}$, then the two-stage problem can be formulated as the unique large-scale problem

$$\min \pi_{ij}(x_{ij}) + \sum_{r \in R} p(\omega_r) \sum_{i \in A} \sum_{j \in S} \sum_{m \in M} \left(w\, c_{ij}^m(y_{ij}(\omega_r)) + (1 - w)\, t_{ij}^m(y_{ij}(\omega_r)) \right)$$

$$\sum_{i \in A} x_{ij} \leq K_j, \forall j \in S,$$

$$\sum_{i \in A} x_{ij} \geq k_j, \forall j \in S,$$

$$x_{ij} \leq \mu_{ij}, \forall i \in A,\ j \in S,$$

$$\frac{\sum_{i\in A} x_{ij}}{K_j} - \frac{\sum_{i\in A} x_{ij'}}{K'_j} \leq \varepsilon,\ \forall j,\ j' \in S,$$

$$\sum_{j\in S} x_{ijr} \leq d_i,\ \forall i \in A,$$

$$\sum_{i\in A} y_{ij}(\omega_r) \leq K_j, \forall j \in S,\ r \in R,$$

$$\sum_{i\in A} y_{ij}(\omega_r) \geq k_j, \forall j \in S,\ r \in R,$$

$$y_{ij}(\omega_r) \leq \mu_{ij}(\omega_r), \forall i \in A,\ j \in S,\ r \in R,$$

$$\frac{\sum_{i\in A} y_{ij}(\omega_r)}{K_j} - \frac{\sum_{i\in A} y_{ij'}(\omega_r)}{K'_j} \leq \varepsilon,\ \forall j,\ j' \in S,\ r \in R,$$

$$\sum_{j\in S} y_{ij}(\omega_r) + d_i - \sum_{j\in S} x_{ij} \leq d_i(\omega_t),\ \forall i \in A,\ r \in R,$$

$$x_{ij} \geq 0, \forall i \in A,\ j \in S,$$

$$y_{ij}(\omega_r) \geq 0, \forall i \in A,\ j \in S,\ r \in R,$$

where $R = \{1, \ldots, \bar{r}\}$. If the number of scenarios is not excessive, then a possible approach is to solve directly the linear programming problem using a solver such as CPLEX. In the next section, we suggest an alternative approach that decomposes the original problem into two variational inequality subproblems.

4 Two-Stage Variational Inequality Formulation

In this section, we propose an equivalent two-stage variational inequality formulation, using the Lagrangian relaxation approach.

We note that the second-stage problem (24)–(25), due to constraint (25), contains the variable x that is not yet known at that stage. For this reason, the problem is not easy to solve. Thus, we suggest to relax (25) into the objective function by the Lagrangian relaxation approach (see [3, 8, 10]). As a consequence, we decompose the original problem into two subproblems, which can be easily solved, and provide a lower bound of the optimal value of the initial model (9)–(22).

Now, we focus on the second-stage problem and give its Lagrangian formulation. We introduce the Lagrange multiplier vector $\lambda : \Omega \to \mathbb{R}^n$, with $\lambda_i(\omega) \geq 0$, i.e., $\omega \in \Omega$, for all $i \in A$, and consider the relaxed constraints

$$\sum_{i \in A} \lambda_i(\omega)\Big(\sum_{j \in S} y_{ij}(\omega) + d_i - \sum_{j \in S} x_{ij} - d_i(\omega)\Big).$$

The Lagrange multiplier $\lambda_i(\omega)$ represents the price or disutility deriving from the unmet demand at the first stage. Therefore, the Lagrangian of the second-stage problem with general probability distribution is

$$L(x, y(\omega), \lambda(\omega), \omega) = \sum_{i \in A} \sum_{j \in S} \sum_{m \in M} \Big(w\, c_{ij}^m(y_{ij}(\omega)) + (1-w)\, t_{ij}^m(y_{ij}(\omega))\Big)$$
$$+ \sum_{i \in A} \lambda_i(\omega)\Big(\sum_{j \in S} y_{ij}(\omega) + d_i - \sum_{j \in S} x_{ij} - d_i(\omega)\Big).$$

We have

$$\inf_{y \in Y} L(x, y(\omega), \lambda(\omega), \omega) = \sum_{i \in A} \lambda_i(\omega)\Big(d_i - \sum_{j \in S} x_{ij} - d_i(\omega)\Big)$$
$$+ \inf_{y \in Y} \Big(\sum_{i \in A} \sum_{j \in S} \sum_{m \in M} \Big(w\, c_{ij}^m(y_{ij}(\omega)) + (1-w)\, t_{ij}^m(y_{ij}(\omega))\Big) + \sum_{i \in A} \lambda_i(\omega) \sum_{j \in S} y_{ij}(\omega)\Big).$$

Then, the dual problem is

$$\max_{\lambda \geq 0} \Big(\sum_{i \in A} \lambda_i(\omega)\Big(d_i - \sum_{j \in S} x_{ij} - d_i(\omega)\Big) \tag{26}$$
$$+ \inf_{y \in Y} \Big(\sum_{i \in A} \sum_{j \in S} \sum_{m \in M} \Big(w\, c_{ij}^m(y_{ij}(\omega)) + (1-w)\, t_{ij}^m(y_{ij}(\omega))\Big) + \sum_{i \in A} \lambda_i(\omega) \sum_{j \in S} y_{ij}(\omega)\Big)\Big). \tag{27}$$

Thus, the two-stage problem becomes

$$\min_{x \in X} \sum_{i \in A} \sum_{j \in S} \sum_{m \in M} \pi_{ij}^m(x_{ij}) + \mathbb{E}_{\xi}(\Phi^1(x, \xi(\omega))),$$
$$\Phi^1(x, \xi(\omega)) = \max_{\lambda \geq 0} \inf_{y \in Y} L(x, y(\omega), \lambda(\omega), \omega).$$

Under the assumption of discrete probability space, we find

$$\mathbb{E}_{\xi}(\Phi^1(x, \xi(\omega))) = \sum_{r \in R} p(\omega_r) \Phi^1(x, \xi(\omega_r)),$$
$$\nabla_x \mathbb{E}_{\xi}(\Phi^1(x, \xi(\omega))) = \mathbb{E}_{\xi}(-\lambda(\omega)) = -\sum_{r \in R} p(\omega_r) \lambda(\omega_r).$$

Theorem 1 *The pair* $(x^*, y^*(\omega))$*, where* $x^* \in \mathbb{R}^{nl}$ *and* $y^* : \Omega \to \mathbb{R}^{nl}$ *is a measurable map, is an optimal solution of the two-stage problem if and only if there exists* $\lambda^* : \Omega \to \mathbb{R}^n$ *measurable such that*

1. x^* *is a solution of the variational inequality*

$$\sum_{i\in A}\sum_{j\in S}\Big(\sum_{m\in M}\frac{\partial \pi_{ij}^m(x_{ij}^*)}{\partial x_{ij}} - \sum_{r\in R} p(\omega_r)\lambda_i^*(\omega_r)\Big) \times (x_{ij} - x_{ij}^*) \geq 0, \quad \forall x \in X. \tag{28}$$

2. $(y^*(\omega_r), \lambda^*(\omega_r))$ *is a solution of the parametric variational inequality*

$$\begin{aligned}
&\sum_{r\in R}\sum_{i\in A}\sum_{j\in S} p(\omega_r)\Big(w\sum_{m\in M}\frac{\partial c_{ij}^m(y_{ij}^*(\omega_r))}{\partial y_{ij}} + (1-w)\sum_{m\in M}\frac{\partial t_{ij}^m(y_{ij}^*(\omega_r))}{\partial y_{ij}} + \lambda_i(\omega_r)\Big)\\
&\times (y_{ij}(\omega_r) - y_{ij}^*(\omega_r))\\
&+\sum_{r\in R} p(\omega_r)\sum_{i\in A}\Big(\sum_{j\in S} x_{ij}^* - \sum_{j\in S} y_{ij}^*(\omega_r) - d_i + d_i(\omega_r)\Big)\\
&\times (\lambda_i(\omega_r) - \lambda_i^*(\omega_r)) \geq 0, \quad \forall y(\omega_r) \in Y, \forall \lambda_i(\omega_r) \geq 0.
\end{aligned} \tag{29}$$

Proof Function $\Phi^1(x, \xi(\omega_r))$ is linear w.r.t. x and, hence, is convex for all $\omega_r \in \Omega$. This implies the convexity of the expectation function $\mathbb{E}_\xi(\Phi^1(x, \xi(\omega_r)))$. Since Ω is finite, for any $x_0 \in \cap_{r\in R}\Phi^1(x, \xi(\omega_r))$, the expectation function is differentiable at x_0. Then, by interchangeability of the gradient and the expectation operators and by classical variational inequality theory, we conclude that the first-stage problem (23) is equivalent to variational inequality (28). Finally, from the optimality conditions of the dual problem (27), it is easy to see that $\lambda^*(\omega_r)$ implies the existence of $y^*(\omega_r)$ such that $(y^*(\omega_r), \lambda^*(\omega_r))$ satisfies (29) .

We observe that problem (28)–(29) can be put in the standard form variational inequality,

$$\langle G(z^*), z - z^*\rangle \geq 0, \quad \forall z \in X \times Y \times \mathbb{R}_+^n, \tag{30}$$

where

$$z = \begin{bmatrix} x \\ y \\ \lambda \end{bmatrix}, \; G(z) = \begin{bmatrix} \sum_{m\in M}\frac{\partial \pi_{ij}^m(x_{ij})}{\partial x_{ij}} - \sum_{r\in R} p(\omega_r)\lambda_i(\omega_r) \\ \sum_{r\in R} p(\omega_r)\Big(w\sum_{m\in M}\frac{\partial c_{ij}^m(y_{ij}(\omega_r))}{\partial y_{ij}} + (1-w)\sum_{m\in M}\frac{\partial t_{ij}^m(y_{ij}(\omega_r))}{\partial y_{ij}} + \lambda_i(\omega_r)\Big) \\ p(\omega_r)\Big(\sum_{j\in S} x_{ij} - \sum_{j\in S} y_{ij}(\omega_r) - d_i + d_i(\omega_r)\Big)\end{bmatrix}_{\substack{r\in R\\ i\in A\\ j\in S}}.$$

In virtue of Theorem 1, variational inequality (30) represents the optimality conditions of the decision planner that is faced with the two-stage stochastic optimization problem (23)–(25). We now highlight the economic interpretation of these conditions. From variational inequality (28), we can infer that, if there is a positive evacuation flow between area i and shelter j in the first stage, then the expected price $p(\omega_r)\lambda_r(\omega_r)$ is equal to the marginal penalty cost of the unmet demand at the first stage. From the first term in inequality (29), we have that, if there is a positive evacuation flow between area i and shelter j in the second stage, then the marginal evacuation costs plus the price $\lambda(\omega_r)$ must be null. From the second term, we also note that if no flow is positive, then the sum of the marginal evacuation costs plus the price $\lambda(\omega_r)$ can be positive. Finally, from the second term in inequality (29), we see that the price $\lambda(\omega_r)$ serves as the price to balance the system.

We now discuss some qualitative properties of (30). Since the feasible set underlying the variational inequality problem is not compact, we cannot derive the existence of a solution simply from the assumption of continuity of the functions. Instead, we should require some coercivity conditions (see, for instance, [6]). It is well known that the uniqueness of the solution to the above variational inequality is ensured by the strict monotonicity of mapping $G(z)$. The theorem below presents the sufficient conditions for the uniqueness.

Theorem 2 *Let us assume that functions $\pi_{ij}^m(x_{ij})$ are strictly convex in x_{ij}, and $c_{ij}^m(y_{ij})$ and $t_{ij}^m(y_{ij})$ are strictly convex in y_{ij}. Then, the vector function G involved in the variational inequality* (30) *is strictly monotone, that is,*

$$\langle G(\bar{z}) - G(\tilde{z}), \bar{z} - \tilde{z}\rangle > 0, \forall \bar{z}, \tilde{z} \in X \times Y \times \mathbb{R}_+^n, \bar{z} \neq \tilde{z}.$$

Proof For any $\bar{z} = (\bar{x}^T, \bar{y}^T, \bar{\lambda}^T)^T, \tilde{z} = (\tilde{x}^T, \tilde{y}^T, \tilde{\lambda}^T)^T \in X \times Y \times \mathbb{R}_+^n$, and using the linearity of constraint (21) (see [10], Theorem 2), direct computations lead to

$$\langle (G(\bar{z}) - G(\tilde{z})), \bar{z} - \tilde{z}\rangle = \sum_{i\in A}\sum_{j\in S}\sum_{m\in M}\left(\frac{\partial \pi_{ij}^m(\bar{x}_{ij})}{\partial x_{ij}} - \frac{\partial \pi_{ij}^m(\tilde{x}_{ij})}{\partial x_{ij}}\right)(\bar{x}_{ij} - \tilde{x}_{ij})$$

$$+\sum_{r\in R}\sum_{i\in A}\sum_{j\in S} p(\omega_r)\Bigg(w\sum_{m\in M}\left(\frac{\partial c_{ij}^m(\bar{y}_{ij}(\omega_r))}{\partial y_{ij}} - \frac{\partial c_{ij}^m(\tilde{y}_{ij}(\omega_r))}{\partial y_{ij}}\right)$$

$$+(1-w)\sum_{m\in M}\left(\frac{\partial t_{ij}^m(\bar{y}_{ij}(\omega_r))}{\partial y_{ij}} - \frac{\partial t_{ij}^m(\tilde{y}_{ij}(\omega_r))}{\partial y_{ij}}\right)\Bigg)(\bar{y}_{ij} - \tilde{y}_{ij}).$$

Since π_{ij}^m, c_{ij}^m, and t_{ij}^m are strictly convex functions, the matrices of the second derivatives of those functions are positive definite, and the functions are strictly monotone. Thus, $G(z)$ is strict monotone if and only if $\bar{z} \neq \tilde{z}$.

5 Numerical Results

In this section, we introduce a numerical example for the aim of validating our approach.

For simplicity, we consider only a single mode of transportation between each crisis area and each shelter ($m = 1$). Moreover, we set $w = 0.6$, $\varepsilon = 0.01$, $\bar{r} = 100$. Then, we choose the random parameter $\omega_r \in [0, 1]$ and fix the following parameters for $i = 1, 2,\ j = 1, 2$, and $r = 1, \ldots, 100$, as follows:

- the probabilities $p(\omega_r) = \frac{1}{100}$ associated with each realization ω_r, randomly taken in $[0, 1]$;
- the recommended number and the maximum number of people that can be hosted in shelter:

$$k_1 = 4, \qquad k_2 = 3,$$
$$K_1 = 40, \qquad K_2 = 40;$$

- capacity of flows in stage one and random flow capacity in stage two:

$$\mu_{ij} = 50 \quad \forall i = 1, 2,\ j = 1, 2;$$

- evacuation demand in stage one and random evacuation demand in stage two:

$$d_1 = 50, \qquad d_1(\omega_r) = 50\omega_r;$$
$$d_2 = 60, \qquad d_2(\omega_r) = 60\omega_r.$$

We now describe the two-stage variational inequality formulation, based on the Lagrangian relaxation approach. The procedure is structured in two steps:

1. we first solve the second-stage parametric variational inequality and find the solution $(y^*(x, \omega), \lambda^*(\omega))$;
2. we write the operator $\sum_{m\in M} \sum_{i\in A} \sum_{j\in S} \pi_{ij}^m (x_{ij}^*) + \mathbb{E}_\xi(F(x^*, \xi(\omega)))$, solve the first-stage variational inequality, and find x^*. We obtain the solution $(x^*, y^*(\omega), \lambda^*(\omega))$. We remark that $\sum_{m\in M} \sum_{i\in A} \sum_{j\in S} \pi_{ij}^m x_{ij}^* + \mathbb{E}_\xi(F(x^*, \xi(\omega)))$ is a lower bound of the optimal value of the original problem.

We apply the extragradient method (see [7]) to compute solutions to our numerical problem and implement it as M-script files of MatLab.

The first step of this approach consists in solving the second-stage parametric variational inequality, using the Lagrangian approach. The profit functions F_{ij} for each area $i = 1, 2$ and shelter $j = 1, 2$ are given by

$$F_{11} = \sum_{r=1}^{100} p(\omega_r)\Big(w(20y_{11}(\omega_r)^2 - 90y_{11}(\omega_r) + 200) + (1-w)(20y_{11}(\omega_r)^2 - 180y_{11}(\omega_r) + 600) +$$

$$+\lambda_1(y_{11}(\omega_r)+y_{12}(\omega_r)+50-x_{11}-x_{12}-50\omega_r)\Big)+\frac{2x_{11}^2+x_{12}^2-5x_{11}}{2},$$

$$F_{12}=\sum_{r=1}^{100}p(\omega_r)\Big(w(45y_{12}(\omega_r)^2-105y_{12}(\omega_r)+225)+(1-w)(60y_{12}(\omega_r)^2-180y_{12}(\omega_r)+900)+$$

$$+\lambda_1(y_{11}(\omega_r)+y_{12}(\omega_r)+50-x_{11}-x_{12}-50\omega_r)\Big)+\frac{x_{11}^2+1.5x_{12}^2-5.4x_{12}}{2},$$

$$F_{21}=\sum_{r=1}^{100}p(\omega_r)\Big(w(80y_{21}(\omega_r)^2-200y_{21}(\omega_r)+300)+(1-w)(45y_{21}(\omega_r)^2-120y_{21}(\omega_r)+450)+$$

$$+\lambda_2(y_{21}(\omega_r)+y_{22}(\omega_r)+60-x_{21}-x_{22}-60\omega_r)\Big)+\frac{x_{22}^2+3x_{21}^2-7.4x_{21}}{2},$$

$$F_{22}=\sum_{r=1}^{100}p(\omega_r)\Big(w(30y_{22}(\omega_r)^2-110y_{22}(\omega_r)+350)+(1-w)(54y_{22}(\omega_r)^2-168y_{22}(\omega_r)+675)+$$

$$+\lambda_2(y_{21}(\omega_r)+y_{22}(\omega_r)+60-x_{21}-x_{22}-60\omega_r)\Big)+\frac{2x_{22}^2+x_{21}^2-6.2x_{22}}{2}.$$

We obtain the following solutions:

$$y_{11}=-35.9789+35.9155\omega_r+0.71831x_{11}+0.71831x_{12},$$
$$y_{12}=-14.0211+14.0845\omega_r+0.28169x_{11}+0.28169x_{12},$$
$$y_{21}=-22.375+22.5\omega_r+0.375x_{21}+0.375x_{22},$$
$$y_{22}=-37.625+37.5\omega_r+0.625x_{21}+0.625x_{22},$$
$$\lambda_1=156515-143662\omega_r-2873.24x_{11}-2873.24x_{12},$$
$$\lambda_2=312150-297000\omega_r-4950x_{21}-4950x_{22}.$$

For all i and j, each flow y_{ij} and the corresponding Lagrange multipliers λ_i depend on x_{ij} and ω_r, for $r=1,\ldots,100$.
The second step consists in calculating x_{ij} using the profit function of the first stage F_{ij}; hence, we obtain the flows of the first and the second stages and the Lagrange multipliers of the second stage, which depend on ω_r. We find

$$x_{11}(\omega_r)=32.9972-30.8684\omega_r,$$
$$x_{12}(\omega_r)=18.3999-16.1403\omega_r,$$
$$x_{21}(\omega_r)=17.6953-16.3862\omega_r,$$
$$x_{22}(\omega_r)=42.7048-40.9655\omega_r.$$

$$y_{11}(\omega_r)=0.940151+2.14868\omega_r,$$
$$y_{12}(\omega_r)=0.456949+0.842619\omega_r,$$

Table 3 Average of flows x_{ij}, y_{ij}, and multipliers λ_{ij} in each scenarios and the total profit function F_{ij}

(i, j)	(1, 1)	(1, 2)	(2, 1)	(2, 2)
$\bar{x}_{ij}$	17.6921	10.3973	9.5707	22.3934
$\bar{\lambda}_{ij}$	4577.4	4577.4	6669.8	6669.8
$\bar{y}_{ij}$	2.0055	0.8747	0.7674	0.9457

$$y_{21}(\omega_r) = 0.275037 + 0.993113\omega_r,$$
$$y_{22}(\omega_r) = 0.125063 + 1.65519\omega_r,$$
$$\lambda_1(\omega_r) = 8838.8 - 8594.72\omega_r,$$
$$\lambda_2(\omega_r) = 13169.5 - 13109.1\omega_r.$$

Thus, the average values of profit functions F_{ij}, for $i = 1, 2$, $j = 1, 2$ and $\omega_1, \ldots, \omega_{100}$, are

$$F_{11} = 39602, \qquad F_{12} = 25789,$$
$$F_{21} = 43715, \qquad F_{22} = 60095.$$

In Table 3, we present the average value of flows x_{ij}, y_{ij}, and the multipliers λ_{ij}.

In order to verify the effectiveness of the proposed model, we compare the deterministic model and the Lagrange relaxation approach.

In the deterministic case, using the same data, we find the solutions:

$$x_{11} = 2.82246, \qquad x_{12} = 1.47495,$$
$$x_{21} = 1.177549, \qquad x_{22} = 2.12505.$$

The values of the deterministic profit functions F_{ij}, $i = 1, 2$ and $j = 1, 2$, are then

$$F_{11} = 163.696, \qquad F_{12} = 406.831,$$
$$F_{21} = 253.689, \qquad F_{22} = 357.92.$$

We note that the values of the deterministic profit functions F_{ij} are greater than the respective values of the Lagrange relaxation approach. This observation implies that the stochastic framework and the real-time updating of information allow one to evaluate more precisely the situation and to lower evacuation costs. Of course, for small dimensional models, the two-stage stochastic evacuation problem can be directly solved.

Thus, we consider the two-stage evacuation model without the Lagrange relaxation. We define F_{ij} for $i = 1, 2$ and $j = 1, 2$ as

$$F_{11} = \sum_{r=1}^{100} p(\omega_r)\Big(w(20y_{11}(\omega_r)^2 - 90y_{11}(\omega_r) + 200) + (1-w)(20y_{11}(\omega_r)^2 - 180y_{11}(\omega_r) + 600)\Big) +$$
$$+ \frac{2x_{11}^2 + x_{12}^2 - 5x_{11}}{2},$$

$$F_{12} = \sum_{r=1}^{100} p(\omega_r)\Big(w(45y_{12}(\omega_r)^2 - 105y_{12}(\omega_r) + 225) + (1-w)(60y_{12}(\omega_r)^2 - 180y_{12}(\omega_r) + 900)\Big) +$$
$$+ \frac{x_{11}^2 + 1.5x_{12}^2 - 5.4x_{12}}{2},$$

$$F_{21} = \sum_{r=1}^{100} p(\omega_r)\Big(w(80y_{21}(\omega_r)^2 - 200y_{21}(\omega_r) + 300) + (1-w)(45y_{21}(\omega_r)^2 - 120y_{21}(\omega_r) + 450)\Big) +$$
$$+ \frac{x_{22}^2 + 3x_{21}^2 - 7.4x_{21}}{2},$$

$$F_{22} = \sum_{r=1}^{100} p(\omega_r)\Big(w(30y_{22}(\omega_r)^2 - 110y_{22}(\omega_r) + 350) + (1-w)(54y_{22}(\omega_r)^2 - 168y_{22}(\omega_r) + 675)\Big) +$$
$$+ \frac{2x_{22}^2 + x_{21}^2 - 6.2x_{22}}{2}.$$

We obtain the following average of the flows of the first and second stages over $r = 1, \ldots, 100$, respectively:

$$\bar{x}_{11} = 13.69769, \qquad \bar{y}_{11} = 3.15,$$
$$\bar{x}_{12} = 13.59372, \qquad \bar{y}_{12} = 1.32,$$
$$\bar{x}_{21} = 12.85349, \qquad \bar{y}_{21} = 1.27,$$
$$\bar{x}_{22} = 16.96938, \qquad \bar{y}_{22} = 1.78.$$

As a consequence, the values of profit functions F_{ij}, for $i = 1, 2,\ j = 1, 2$, and $\omega_1, \ldots, \omega_{100}$, are

$$F_{11} = 43343, \qquad F_{12} = 62724,$$
$$F_{21} = 72244, \qquad F_{22} = 74621.$$

As expected, the values of the deterministic profit functions F_{ij}, for all i, j, are greater than the respective values of the two-stage evacuation model without the Lagrange relaxation. This confirms the efficiency of the stochastic approach.

6 Conclusions

In this chapter, we introduced a two-stage stochastic programming model for the emergency evacuation problem. We proposed a scenario-based evacuation planning model able to provide the optimal flows of evacuees from crisis areas to shelters, in order to minimize both the transportation cost and the transportation time, and under uncertainty on the evacuation demand and the link capacities.

We then proposed a variational inequality formulation of the model; in particular, we reduced our problem to a two-stage stochastic variational inequality, using the Lagrangian relaxation approach. We also discussed the qualitative properties of the two-stage stochastic variational inequality. In addition, we analyzed the role of Lagrange multipliers associated with the relaxed constraints. Finally, in order to show the applicability and effectiveness of our model, we provided a numerical example.

Future research may include extending this framework to multistage stochastic models.

Acknowledgments The research of the authors was partially supported by the research project "Problemi di equilibrio: metodi variazionali e teoria dei giochi" GNAMPA-INdAM and by the research project PON SCN 00451 CLARA—CLoud plAtform and smart underground imaging for natural Risk Assessment, Smart Cities and Communities, and Social Innovation. These supports are gratefully acknowledged.

References

1. G. Barbarosoğlu and Y. Arda, A Two-Stage Stochastic Programming Framework for Transportation Planning in DisasterResponse, The Journal of the Operational Research Society, Vol. 55, No. 1 (Jan., 2004), pp. 43–53
2. V. Bayram, H. Yaman, A Stochastic Programming Approach for Shelter Location and Evacuation Planning
3. P. Beraldi, M.E. Bruni, D. Conforti. A Solution Approach for Two-Stage Stochastic Nonlinear Mixed Integer Programs. Algorithmic Operations Research Vol.4 (2009) 76–85.
4. X. Chen, T.K. Pong, R.J.-B. Wets. Two-stage stochastic variational inequalities: an ERM-solution procedure. Math. Program. 165, 1–41 (2017).
5. C. Crespi, G. Fargetta, M. Pavone, R.A. Scollo, L. Scrimali. A Game Theory Approach for Crowd Evacuation Modelling. In: Filipič B., Minisci E., Vasile M. (eds) Bioinspired Optimization Methods and Their Applications. BIOMA 2020. Lecture Notes in Computer Science, vol 12438 (2020). Springer, Cham. https://doi.org/10.1007/978-3-030-63710-1_18
6. F. Facchinei, J.S. Pang, *Finite-Dimensional variational inequalities and complementarity problems*, Vol. I. Springer, New York (2003).
7. G.M. Korpelevich, *The extragradient method for finding saddle points and other problems*, Matekon **13**, 1977, 35–49.
8. J. Jiang, Y. Shi, X. Wang, X. Chen. Regularized Two-Stage Stochastic Variational Inequalities for Cournot-Nash Equilibrium Under Uncertainty. J. Comp. Math., 37 (2019), 813–842.
9. D. Kinderlehrer, G. Stampacchia, An introduction to variational inequalities and their applications, New York: Academic Press, 1980.
10. M. Li, C. Zhang. Two-Stage Stochastic variational inequality arising from stochastic programming. Journal of Optimization Theory and Applications (2020) 186:324–343
11. A. Nagurney, Network economics: A variational inequality approach (2nd ed. (revised)), Boston, Massachusetts: Kluwer Academic Publishers, 1999.
12. T. Rambha, L. K. Nozick, R. Davidson, W.Yi, K. Yang, A Stochastic Optimization Model for Staged Hospital Evacuation during Hurricanes

13. R.T. Rockafellar, R.J.-B. Wets. Stochastic variational inequalities: single-stage to multistage. Math. Program. 165, 1–30 (2016)
14. Rockafellar, R.T., Sun, J.: Solving monotone stochastic variational inequalities and complementarity problems by progressive hedging. Math. Program. 174, 453–471 (2019)
15. Rockafellar, R.T., Sun, J.: Solving Lagrangian variational inequalities with applications to stochastic programming. Math. Program. (2020). https://doi.org/10.1007/s10107-019-01458-0
16. Shapiro, A., Dentcheva, D., Ruszczynski, A.: Lectures on Stochastic Programming: Modeling and Theory. SIAM, Philadelphia (2009)
17. H. Seraji, R. Tavakkoli-Moghaddam, R. Soltani, A two-stage mathematical model for evacuation planning and relief logistics in a response phase. Journal of Industrial and Systems Engineering Vol. 12, No. 1, pp. 129–146 (2019)
18. L. Wang. A two-stage stochastic programming framework for evacuation planning in disaster responses. Computers & Industrial Engineering 145 (2020) Article 106458

On Global Hyperbolicity of Spacetimes: Some Recent Advances and Open Problems

Felix Finster, Albert Much, and Kyriakos Papadopoulos

Abstract This chapter is an up-to-date account of results on globally hyperbolic spacetimes and serves as a multitool; we start the exposition of results from a foundational level, where the main tools are order-theory and general topology, we continue with results of a more geometric nature, and we finally reach results that are connected to the most recent advances in theoretical physics. In each case, we list a number of open questions and we finally introduce a conjecture, on sliced spaces.

1 Introduction to the Terrain

While in Riemannian geometry it is natural to consider geodesic completeness, in Lorentzian geometry—and in particular in spacetime geometry—it is more reasonable to consider global hyperbolicity as a condition that two events, that are chronologically related, can be joined by a maximal timelike geodesic. This is due to the validity of the Hopf–Rinow theorem, whose naive analogue fails to exist in Lorentzian geometry (see [9]).

The advantage of global hyperbolicity is its multifaceted perspectives: for example, it can be examined from an order-theoretic and topological perspective, as the strongest of the causal conditions on a spacetime (the causal diamonds are compact—see [35]) or from a purely geometrical view (existence of a Cauchy surface—see [12]). The equivalence of such seemingly different statements is what supplies mathematical physics with mathematical tools and physical insights. The

F. Finster
Fakultät für Mathematik, Universität Regensburg, Regensburg, Germany

A. Much
Institut für Theoretische Physik, Universität Leipzig, Leipzig, Germany

K. Papadopoulos (✉)
Department of Mathematics, Kuwait University, Safat, Kuwait

I. N. Parasidis et al. (eds.), *Mathematical Analysis in Interdisciplinary Research*, Springer Optimization and Its Applications 179,
https://doi.org/10.1007/978-3-030-84721-0_15

importance of such an interdisciplinary work cannot be overstated; in mathematical physics, this is significant in order to construct quantum field theories in curved spacetimes in a mathematically rigorous fashion. In particular, the input that comes from a topological direction is fruitful towards a more abstract formulation. From a mathematical point of view, it is important to see the applications of general theorems and in particular examine and push the scope of their validity.

It is worth mentioning that there is a theoretical background behind the notion of globally hyperbolic spacetimes that uses tools from domain theory and interconnects mathematical physics with the foundations of mathematics. A recent example is reference [16], where the authors showed that globally hyperbolic spacetimes belong to a category,[1] which is equivalent to integral domains. Integral domains are partially ordered sets that carry, intrinsically, notions of completeness and approximation and were used in theoretical computer science (see [37]). The equivalence of causality between events to an order on regions of a spacetime suggests that questions about spacetime can be translated to questions on domain theory. This is of a great interest, since the type of domains called ω-continuous are the ideal completions of countable abstract bases, so that a spacetime can be reconstructed (in a purely order-theoretical manner) from a dense discrete set. In particular, one can claim that a globally hyperbolic spacetime is linked to something discrete.

Strongly causal spacetimes, and thus globally hyperbolic ones that are strongly causal by definition, are conformal to spacetimes in which all null geodesics are complete; that is, their null geodesics can be extended to infinite values of their affine parameters, as the following theorem suggests (see [4]).

Theorem 1.1 (Clarke) *If M is a strongly causal spacetime with metric g, then there is a C^∞-function Ω, such that null geodesics with respect to $\Omega^2 g$ are complete.*

As for timelike geodesic completeness, even if there is no obvious relation to global hyperbolicity, it has been shown that under particular sectional or Ricci curvature conditions, if a spacelike hypersurface is future-timelike geodesically complete, then it is globally hyperbolic; see [9]. Next we turn to the spacelike geodesic completeness of the Cauchy surface of a globally hyperbolic manifold. An interesting result connecting the spacelike geodesic completeness to global hyperbolicity was given in [18, Proposition 5.3]. In particular, the author proved that an ultra-static spacetime (M, g) is globally hyperbolic if and only if the global Cauchy surface is geodesically complete. The physical advantage of working in globally hyperbolic manifolds is that the wave equation (in what follows referred to as the Klein–Gordon equation), i.e. the differential equation describing the dynamics of a spin zero particle, is well-posed. In particular, given smooth initial data on the Cauchy hypersurface, there exists a corresponding global smooth solution. For these manifolds, it is in general possible to rewrite the second order differential equation as a first order system of equations; this is referred to as the Hamiltonian

[1] In the frame of the field of mathematics called Category Theory; see, for example, [25].

formulation. The generator of the time evolution can then be written as a 2×2 matrix differential operator. In order to define a corresponding quantum field theory using the Hamiltonian formulation a necessary requirement is the essentially self-adjointness of weighted Laplace–Beltrami operators stemming from the Klein–Gordon equation. The essential self-adjointness of these operators depends on the spacelike geodesic completeness of the Cauchy surfaces. In what follows we will expand more on the obtained results and a still open *conjecture*.

2 Some Preliminaries

2.1 A Spell of Domain Theory and General Topology

In this section we will list some definitions, which will be needed in Sect. 3, where we will discuss about a link between a certain type of partially ordered sets (posets) with globally hyperbolic spacetimes.

Throughout the text, the *power set* of a set X will be denoted by $\mathcal{P}(X)$ and it will be considered as the set of all subsets of X.

Definition 2.1 If $(A, \prec)$ is a poset, then:

1. a nonempty set $D \in \mathcal{P}(A)$ is called *directed*, if for every $x, y \in D$, there exists $z \in D$, such that both $x \prec z$ and $y \prec z$.
2. (dually to (1)) a nonempty set $F \in \mathcal{P}(A)$ is called *filtered*, if for every $x, y \in F$, there exists $z \in F$, such that $z \prec x$ and $z \prec y$.
3. a nonempty set $L \in \mathcal{P}(A)$ is called *lower set*, if for every $x \in A$ and for every $y \in L$, $x \prec y$ implies that $x \in L$.
4. $(A, \prec)$ is a *dcpo* (directed, complete poset), if every directed set has a supremum; that is, if $D \in \mathcal{P}(A)$ is directed, then D has a least upper bound.
5. $(A, \prec)$ is *continuous*, if there exists $B \in \mathcal{P}(A)$, such that $B \cap \downarrow x$ contains a directed set with supremum, for all $x \in A$, where $\downarrow x = \{a \in A : a \ll x\}$ and where $\ll$ is defined as $x \ll y$, if for all directed sets D with supremum, $y \prec \sup D$ implies that there exists $d \in D$, such that $x \prec d$.
6. $(A, \prec)$ is *bicontinuous*, if it is continuous and for every $x, y \in A$, $x \ll y$ if for every filtered set $F \in \mathcal{A}$ with infimum, $\inf F \prec x$ implies that there exists $f \in F$, such that $f \prec y$ and also, for every $x \in A$, $\uparrow x$ is filtered, with infimum x (and where, dually to $\downarrow x$, $\uparrow x = \{a \in A : x \ll a\}$).
7. if $(A, \prec)$ is bicontinuous, then its *interval topology* is the topology that has a basis that consists of the intervals $(a, b) := \{x \in A : a \ll x \ll b\}$.
8. $(A, \prec)$ is *globally hyperbolic*, if it is bicontinuous and the closure of the intervals (a, b), that is $\overline{(a, b)} := [a, b] = \{x : a \preceq x \preceq b\}$, is compact in the interval topology on A.
9. A *domain* is a continuous dcpo. An example of an *interval domain* is the domain of compact intervals of the real line.

10. Let $\Uparrow x = \{a \in D : x \ll a\}$, where D is a dcpo. Then the class $\{\Uparrow x : x \in D\}$ forms a base for the *Scott topology* on a continuous poset.

2.2 Causal Relations in a Spacetime

It is standard (see [35]) in a spacetime to consider two partial orders, the *chronological* order $\ll$, which is considered to be irreflexive if one wants to avoid closed timelike curves, and the *causal order* $\prec$, which is reflexive, and are defined, respectively, as follows:

(1) $x \ll y$ if and only if $y \in C_+^T(x)$ and
(2) $x \prec y$ if and only if $y \in C_+^T(x) \cup C_+^L(x)$
where $C_+^T(x)$ denotes the future *time-cone* of an event x, $C_+^L(x)$ its future *light-cone* and $C_+^T(L) \cup C_+^L(x)$ its future *causal-cone* (for an analytical exposition of time-coordinates, space-coordinates and the spacetime "metric", that leads to the definition of these cones, see [12]; for a detailed exposition of these definitions on Minkowski space, see [2]).

In addition, the reflexive relation *horismos* $\rightarrow$ is defined as $x \rightarrow y \mathrm{iff} x \prec y \mathrm{butnot} x \ll y$.

Naturally, in a spacetime M one can define the sets $I^+(x) = \{y \in M : x \ll y\}$ and $J^+(x) = \{y \in M : x \prec y\}$; $I^-(x)$ and $J^-(x)$ are defined dually.

In Sect. 3, we will examine the relation between $\ll$ in Definition 2.1 (5) and the chronological order $\ll$ that will be defined in Definition 2.2 (1). Furthermore, the interval topology of Definition 2.1 (7), defined for bicontinuous posets, will be called the *Alexandrov topology* T_A for a spacetime M, whenever $\ll$ is the chronological order on M.

2.3 Weighted Riemannian Manifolds and All That

In the following we define a weighted manifold, [48, Chapter 3.6, Definition 3.17].

Definition 2.2 A triple (Σ, g_Σ, μ) is called a *weighted manifold*, if (Σ, g_Σ) is a Riemannian manifold and μ is a measure on Σ with a smooth and everywhere positive density function ρ, i.e. $d\mu = \rho \, d\Sigma$. A *weighted Hilbert space*, denoted by $L^2(\Sigma, \mu)$, is given as the space of all square-integrable functions on the manifold Σ with respect to the measure μ. The corresponding *weighted Laplace–Beltrami operator* (also called the Dirichlet–Laplace operator), denoted by $\Delta_{g_\Sigma,\mu}$, is given by

$$\Delta_{g_\Sigma,\mu} = \frac{1}{\rho\sqrt{|g_\Sigma|}} \partial_i (\rho \sqrt{|g_\Sigma|}\, g_\Sigma^{ij} \partial_j). \tag{2.1}$$

In regards to the upcoming discussion on essential self-adjointness, we need an essential result that was given in [39, Theorem 1].

Theorem 2.1 *Let the Riemannian manifold* (Σ, g_Σ) *be complete and* μ *be a measure on* Σ *with a smooth and everywhere positive density function* ρ *and suppose that the potential* $Y \in L^2_{loc}(\Sigma)$ *point-wise. Furthermore, let* $A \in \Lambda^1_{(1)}(\Sigma)$ *and let the operator* $L = -\Delta^A_{g_\Sigma,\mu} + Y$*, where the minimally coupled Laplace Beltrami operator is defined by*

$$-\Delta^A_{g_\Sigma,\mu} := -(\rho\sqrt{|g_\Sigma|})^{-1}(\partial - A)^*_i(\sqrt{|g_\Sigma|}\,\rho\, g^{ij}_\Sigma\,(\partial - A)_j)$$

be semi-bounded from below. Then, the operator L *is an essentially self-adjoint operator on* $C^\infty_0(\Sigma)$.

Where we write $f \in L^2_{loc}(\Sigma)$ for a local $L^2(\Sigma)$ function f that is an element of the Hilbert space $L^2(\Sigma)$ on every compact subset of the manifold Σ and we denote $\Lambda^p_{(k)}(\Sigma)$ as the set of all k-smooth (i.e. of the class C^k) complex-valued p-forms on Σ. Note that the above result is (apart from the potential term and the gauge field) a reformulation of the result of [1] for weighted Riemannian manifolds.

3 Globally Hyperbolic Spacetimes

Classically, when we talk about global hyperbolicity, in terms of the causal structure of a spacetime, we consider the definition that follows (see [35]).

Definition 3.1 A spacetime M is *globally hyperbolic*, if and only if M is strongly causal and every set $J^+(x) \cap J^-(y)$ is compact, for some $x, y \in M$.

In [46] it was proven that any globally hyperbolic spacetime admits a *smooth* foliation into Cauchy surfaces [46, Theorem 1.1]. Moreover, the induced metric of such a globally hyperbolic spacetime admits a specific form [47, Theorem 1.1].

Theorem 3.1 *Let* (M, g) *be an* $(n+1)$*-dimensional globally hyperbolic spacetime. Then, it is isometric to the smooth product manifold* $\mathbb{R} \times \Sigma$ *with a metric* g*, i.e.*

$$g = -N^2 dt^2 + h_{ij} dx^i dx^j, \tag{3.1}$$

where Σ *is a smooth* n*-manifold,* $t : \mathbb{R} \times \Sigma \mapsto \mathbb{R}$ *the natural projection,* $N : \mathbb{R} \times \Sigma \mapsto (0, \infty)$ *a smooth function, and* $\mathbf{h}$ *a 2-covariant symmetric tensor field on* $\mathbb{R} \times \Sigma$*, satisfying the following condition: Each hypersurface* $\Sigma_t \subset M$ *at constant* t *is a Cauchy surface, and the restriction* $\mathbf{h}(t)$ *of* $\mathbf{h}$ *to such a* Σ_t *is a Riemannian metric (i.e.* Σ_t *is spacelike).*

Hounnonkpe and Minguzzi have shown that the (strong-)causality condition is needless in the definition of global hyperbolicity, for a reasonable (meaning non-compact) spacetime of dimension strictly greater than 2 (see [16]).

Theorem 3.2 (Hounnonkpe–Minguzzi) *A non-compact spacetime of dimension strictly greater than 2 is globally hyperbolic, if and only if the causal diamonds are compact in the topology,* T_A.

Recently, there has been a generous step towards an understanding of the nature of globally hyperbolic spacetimes down to the most fundamental level. In [26], it has been shown that the structure of such spacetimes is equivalent to the structure of interval domains, which are purely order-theoretic objects. Let us have a closer look on the main results.

Definition 3.2 A set B equipped with a transitive relation $\ll$ is an *abstract basis*, if $\ll$ is $-$-interpolative; that is, for all $S \in \mathcal{P}(B)$, such that S is finite, if $S \ll x$, then there exists $y \in B$, such that $S \ll y \ll x$, where by $S \ll x$ one means that $y \ll x$, for all $y \in S$.

By $int(C)$ one denotes the set $\{(a, b) : a \ll b\}$, where $(a, b) \ll (c, d)$ if and only if $a \ll c$ and $d \ll b$. It is easy to see that if $(B, \ll)$ is an abstract basis, which is $-$- and $+$-interpolative (where $+$-interpolation is the dual to $-$-interpolation), then the set $(int(C), \ll)$ is an abstract basis.

Definition 3.3 An *ideal* of an abstract basis $(B, \ll)$ is a nonempty set $I \in \mathcal{P}(B)$, which is lower and directed.

The set $(\overline{B}, \subset)$ of all ideals of the abstract basis $(B, \ll)$ is a poset and is called the *ideal completion* of B.

Before introducing the main theorems of [26], we remind that since spacetimes are considered naturally as being second countable and second countability implies separability, a spacetime admits a countable dense subset. In the theorem that follows a countable set, which is equipped with a causal relation, determines the entire space.

Theorem 3.3 (Martin–Panangaden) *If C is a countable dense subset of a globally hyperbolic spacetime M, where $\ll$ denotes chronology, then* $\max IC \simeq M$*, where the maximal elements are equipped with the Scott topology.*

Theorem 3.3 can be written more generally as follows.

Theorem 3.4 *If C is a countable dense subset of a globally hyperbolic set M, then* $\max IC \simeq M$.

The proof of Theorem 3.4 is the same as that of Theorem 3.3. Since M is bicontinuous by definition, $(C, \ll)$ will be an abstract basis and so $(int(C), \ll)$ will be an abstract basis for IM. Since ideal completions for bases of IM are isomorphic to IM, the result follows immediately. The result of Theorem 3.3 follows from the fact that the relation $\ll$ of Definition 2.1 (5), which reads $x \ll y$ as "x

approximates y", coincides with the chronological partial order $\ll$ in a globally hyperbolic spacetime (see [27], Proposition 4.4).

Theorem 3.5 (Martin–Panangaden) *The category of globally hyperbolic posets is equivalent to the category of interval domains.*

The proof of Theorem 3.5 depends on several technical results that precede it in [26], but the result on its own is powerful; within a globally hyperbolic spacetime, one can convert questions of a physical meaning to questions in domain theory (and vice versa).

Question 1 *Consider a globally hyperbolic poset $(A, \prec)$. Does the compactness of the intervals $[a, b]$, under the interval topology, imply the bicontinuity of $(A, \prec)$, in a similar fashion that the compactness in the closed diamonds in a spacetime M, under the Alexandrov topology T_A, implies the (strong) causality of M (for dimensions strictly greater than 2)? This is a reasonable question, given the results in [16] as well as the main result of [26], which guarantees the equivalence (up to category theory) of interval domains and globally hyperbolic spacetimes. The only problem, towards [16], of considering a globally hyperbolic poset, is that we are only left with an order-theoretic structure; what would be, if there is such, the equivalent condition in [16] of "dimension strictly greater than 2"? Is there a possibility that the condition on the dimension applies only to globally hyperbolic spacetimes, while the proposed conjecture holds in globally hyperbolic posets without any restriction?*

In the section that follows, we consider a spacetime, and we discuss how topology can affect the way that we look at it.

4 Different Candidates for a Topology: Which Is the Most "Fruitful" One?

When one considers a spacetime manifold (M, g), the spacetime topology is traditionally, and without any doubt, taken to be the manifold topology T_M, a topology that does not incorporate the causal structure of M, a structure that is linked to the Lorentz "metric" g. This problem was first addressed by Zeeman in his papers [43] and [44], where he restricted the discussion in the Minkowski space. It was then extended to curved spacetimes by Göbel in [10] and by Hawking–King–McCarthy in [13], and the discussion was taken into a further level by several authors, including an important feedback by Low in [23] (for a recent survey on the topologization of spacetime, see [34]). In particular, Low proved that the Limit Curve Theorem (LCT) does not hold under the Path topology of Hawking–King–McCarthy and claimed that this fact gives the right for manifold topology to be called a fruitful one, against any other known candidate. Given that the LCT plays an important role in building contradictions in proofs in singularity theorems, in general relativity (see, for example, [8] and [4]), the authors of [33] posed the question on whether the singularity problem is a purely topological one, depending largely on the topology that one chooses to equip the spacetime with. For example, if one considers the space of timelike paths, the notion of convergence stays unaffected, if choosing either the Path topology $\mathcal{P}$ or the manifold topology T_M, while convergence in the space of causal paths under T_M does not imply convergence in the space of causal paths under $\mathcal{P}$ (see [23]). Consider the causal cone of an event x, in a spacetime M; that is, consider the time-cone $C^T(x)$ of x union its light-cone $C^L(x)$. Consider now an open ball $B^d_\epsilon(x)$, of radius ϵ, under the manifold topology T_M (where the distance is defined via an appropriate Riemann metric d). Consider the intersection $A = (C^T(x) \cup C^L(x)) \cap B^d_\epsilon(x)$; the sets A form a local base for a topology Z^{LT} on M, where the notion of convergence under Z^{LT} and under T_M stays unaffected in the space of causal paths (see [33]). Obviously, this assertion cannot hold if Z^{LT} is considered in the space of timelike paths.

It is clear that the Path topology $\mathcal{P}$ in [13] has several advantages against the manifold topology on a spacetime; $\mathcal{P}$ incorporates the causal, differential and smooth conformal structure of spacetime and, most importantly, the group of homeomorphisms of $\mathcal{P}$ is the conformal group. According to Göbel (see [10]), the reason for considering the Euclidean topology as a "natural" topology for the Minkowski spacetime (and, more generally, the manifold topology for a spacetime manifold) was that people were mostly concerned with Riemannian spaces and not with spaces equipped with a Lorentz metric; it really seems that the blind use of the manifold topology, while proving theorems in spacetime geometry, was due to ignorance (in the words of Göbel)!

In the frame of globally hyperbolic spacetimes, Low introduced a list of interesting topological results, including the following two (see [24]).

Proposition 4.1 (Low) *A strongly causal spacetime M is globally hyperbolic, if and ony if the space of smooth endless causal curves $\mathcal{C}$ is Hausdorff.*

Proposition 4.2 (Low) *M is globally hyperbolic, if and only if $\mathcal{C}$ is metrizable.*

The topologization of the space of smooth endless causal curves $\mathcal{C}$ is important, since it affects the topology induced on C, the space of causal geodesics, by $\mathcal{C}$. C need not be a manifold but, in particular, if the spacetime M is strongly causal, then C is a smooth manifold with boundary, with a smooth structure inherited from the homogeneous tangent bundle UM (see [24]). The canonical lift of a smooth causal curve from M to CM (where CM is the bundle of causal directions) and the lift from C to CM results into a foliation of CM; the space of leaves of this foliation is a topological space, with the quotient topology coming from UM. The topology of this space need not be Hausdorff, if M is strongly causal. In the case that C, under this topology, is non-Hausdorff, then M will be nakedly singular (a TIP lies inside a PIP; see [24] and for more general exposition see [35]).

Low suggests a topology T^0, on $\mathcal{C}$, which can be described as follows. Given a smooth endless causal curve γ in $\mathcal{C}$, let Γ be the corresponding curve in M (which can be considered as a submanifold). Consider $x \in \Gamma$ and an open set U in M, under the manifold topology T_M, such that $x \in U$. Set $\mathcal{U} \subset \mathcal{C}$ to be the set consisting of all (smooth endless causal-) curves which pass through U. T^0 will then be a topology in which the sets $\mathcal{U}$ form a basis. T^0 is obviously a topology of pointwise convergence. It follows, straightforwardly that, if a spacetime M has a closed timelike curve (CTC), then (M, T^0) cannot satisfy the T_1-separation axiom and if M is totally vicious,[2] then there exists $x \in \mathcal{C}$ such that x is dense in $\mathcal{C}$ under T^0. For the proof of Proposition 4.2, Low constructed a metric, which induces T^0, and remarked that one can find metrics that induce the topologies of convergence to any degree of smoothness (by means of the slicing of the jet bundles over the spacetime manifold M).

In regards to slicing, in [5], sliced spaces were considered to have uniformly bounded lapse, shift and spatial metric, in order to achieve the equivalence of global hyperbolicity of (M, g) with the completeness of the slice (Σ, g_Σ) (Theorem 2.1). Being motivated by this result, the authors of [22] considered global topological conditions, for showing the equivalence of the global hyperbolicity of (M, g) with a slice (Σ_t, g_{Σ_t}) being T_A-complete. Theorem 4.1, below, differs from Theorem 2.1 of [5] in that the slices in [5] are complete Riemannian manifolds (with uniformly bounded spatial metric, lapse and shift functions) while in [22] the slices are T_A-complete.

Theorem 4.1 *Let (M, g) be a sliced space, equipped with its natural product topology T_P, where $M = \mathbb{R} \times \Sigma$, Σ is an n-dimensional manifold ($n \geq 2$)and g the $n+1$-Lorentz "metric" on M. Let also T_A be the Alexandrov spacetime topology on M. Then, the following statements are equivalent:*

(1) (M, g) is globally hyperbolic.
(2) For every basic-open set $D \in T_A$, there exists a basic-open set $B \in T_P$, such that $D \subset B$.
(3) (M_t, g_t) is complete with respect to T_A.

We will talk, in more detail, about sliced spacetimes in Sect. 5.

Question 2 *As we have seen in this section, there are many different candidates for spacetime topology, other than the manifold one. Is there a fruitful[3] and physically meaningful spacetime topology that we ignore? What are the criteria for choosing it? For example, how should such a topology be related to the structural levels of a spacetime? These questions certainly need a more systematic work, since even if they can touch important problems (for example the singularity problem in relativity theory as well as Penrose's cosmic censorship conjecture), they are a bit underestimated until now.*

Question 3 *In the light of Theorem 3.5, we wonder whether the topology of a spacetime can be reconstructed, in some technical and rigorous way, from the topology of an interval domain. The spacetime topologies that we have mentioned*

[2] For the definition and properties of (non-) totally vicious spacetimes, see [28].

[3] Fruitful in the sense of [10, 44] and [13].

in this section, like for example the Zeeman Fine topology or its generalization by Göbel or the one by Hawking–King–McCarthy, are not metrizable; does this give us any insight about the appropriate candidate for a spacetime topology (other than the manifold topology), given that interval domains are order-theoretic objects?

5 The Klein–Gordon Equation in Globally Hyperbolic Manifolds

In what follows, we consider the Klein–Gordon operator on a Lorentzian manifold (M, g) minimally coupled to an electromagnetic potential A and with a scalar potential Y:

$$K\phi := \Big((\sqrt{|g|})^{-1} D^A_\mu (\sqrt{|g|} g^{\mu\nu} D^A_\nu) + Y \Big) \phi = 0, \tag{5.1}$$

where $|g| = \det(g_{\mu\nu})$ and $D^A_\mu = (i\partial - A)_\mu$. Next, we insert the form of the metric given in Theorem 3.1 and instead of working with the operator K (from (5.1)), it is more convenient to work with the operator

$$\tilde{K} := N\, K\, N,$$

where now $\tilde{K}$ can be expressed as

$$\tilde{K} = -(D_t + W^*)(D_t + W) + L, \tag{5.2}$$

with

$$W := -A_0 - \frac{1}{2}(N\, |g_\Sigma|^{1/2})^{-1} D_t\, (N\, |g_\Sigma|^{1/2})$$

and $L(t)$ is the spatial Klein–Gordon operator given by

$$L(t) = -N\, (\sqrt{|g_\Sigma|})^{-1} (D - A)^*_i (\sqrt{|g_\Sigma|}\, N g^{ij}_\Sigma\, (D - A)_j) + N^2\, Y. \tag{5.3}$$

Next, we define the operator B by

$$B(t) = \begin{pmatrix} W(t) & \mathbb{I} \\ L(t) & W(t) \end{pmatrix}.$$

Let $u_1(t) = u(t)$ and $u_2(t) = -(D_t + W(t))\, u(t)$ be the Cauchy data for u at time t, then

$$(\partial_t + iB(t)) \begin{pmatrix} u_1(t) \\ u_2(t) \end{pmatrix} = 0$$

only if u is a (weak) solution of the Klein–Gordon equation $\tilde{K}u = 0$. Then the Hamiltonian is given by the multiplication of B with the matrix Q

$$Q = \begin{pmatrix} 0 & \mathbb{I} \\ \mathbb{I} & 0 \end{pmatrix}$$

i.e. $H(t) = Q\,B(t)$. Since in this case $W(t)$ is equal to a function, the problem of proving self-adjointness of the Hamiltonian reduces to proving self-adjointness of $L(t)$ for all (fixed) t.

In order to gain coherence on the nature of this problem, let us assume that the electromagnetic potential A is equal to zero, the lapse function N is equal to one and the metric g_Σ is time independent. Let us further assume time independence and positivity for the scalar potential Y. By taking the assumptions into account, the operator L is simply the standard Laplace–Beltrami operator w.r.t. the Riemannian manifold (Σ, g_Σ). The results by [1] then state that the geodesic completeness of the Riemannian manifold implies the essential self-adjointness of Laplace–Beltrami operator L. Since the Riemannian manifold stems from a globally hyperbolic spacetime, it is a priori not clear why it should be geodesically complete. At this point, let us state [18, Proposition 5.3].

Proposition 5.1 *Let a Lorentzian manifold* (M, g) *with metric tensor*

$$g = -dt^2 + g_{\Sigma,ij}(\mathbf{x})dx^i dx^j \tag{5.4}$$

be given. Then, the manifold (M, g) *is globally hyperbolic if and only if Riemannian manifold* (Σ, g_Σ) *is complete.*

Hence back to our problem at hand, it follows from Theorem 2.1 that the simplified Laplace–Beltrami operator we considered is an essentially self-adjoint operator on $C_0^\infty(\Sigma)$.

In order to prove essential self-adjointness for the case where N is unequal to one, in the absence of the electromagnetic potential, we used techniques from weighted manifolds and reduced the problem to the following theorem, [31, Theorem 4.1].

Theorem 5.1 *Let the Riemannian manifold* $(\Sigma, N^{-2}g_\Sigma)$ *be geodesically complete and let the scaled potential* $N^2V \in L^2_{loc}(\Sigma)$ *point-wise. Furthermore, let the operator L (from Eq. (5.3)) be semi-bounded from below. Then, the operator L is essentially self-adjoint on* $C_0^\infty(\Sigma)$.

Proof For the proof, see [31]. □

Next, we generalize the former result to the case of nonvanishing electromagnetic potential and lapse function unequal to the unit.

Theorem 5.2 *Let the Riemannian manifold* $(\Sigma, \tilde{g}_\Sigma := N^{-2} g_\Sigma)$ *be geodesically complete and let the scaled potential* $N^2 V \in L^2_{loc}(\Sigma)$ *point-wise. Furthermore, let* $A \in \Lambda^1_{(1)}(\Sigma)$ *and let the operator*

$$L(t) = -N\,(\sqrt{|g_\Sigma|})^{-1}(D-A)^*_i(\sqrt{|g_\Sigma|}\,N g^{ij}_\Sigma\,(D-A)_j) + N^2\,Y$$

be semi-bounded from below. Then, the operator L (from Eq. (5.3)) is essentially self-adjoint on $C^\infty_0(\Sigma) \subset L^2(\Sigma,\,\tilde{\mu})$.

Proof We rewrite the operator L as a weighted Laplace–Beltrami operator multiplied with the Lapse function, i.e.

$$L = -N^2 \Delta^A_{g_\Sigma,\mu} + N^2 Y.$$

By redefining the metric and measure μ as follows:

$$\tilde{g}_\Sigma := N^{-2} g_\Sigma, \qquad d\tilde{\mu} = N^{-2} d\mu,$$

the operator L reads (for proof, see [31, Proposition 3.1])

$$L = -\Delta^A_{\tilde{g}_\Sigma,\tilde{\mu}} + N^2\,Y.$$

By rewriting the operator as a minimally coupled, weighted Laplace–Beltrami operator, the condition of positivity of the potential and Theorem 2.1 lead to the essential self-adjointness. □

Hence, the problem of proving essential self-adjointess of L on is reduced to proving geodesic completeness of the Riemannian manifold $(\Sigma, N^{-2} g_\Sigma)$. For all globally hyperbolic static cases, this can be done by the use of Proposition 5.1. Theorem 5.1 was as well-generalized to the case of stationary spacetimes, see [45].

In the case of nonstatic spacetimes, however, this is still an open problem. In the following, we simplify the problem. Since any globally hyperbolic spacetime (M, g) can be brought in to the following form:

$$g = -N^2(t, \mathbf{x})dt^2 + g_{\Sigma,ij}(t, \mathbf{x})dx^i dx^j,$$

and conformal transformations do not change the causal structure of the manifold, we can consider the conformally transformed globally hyperbolic spacetime $(M, \tilde{g})$

$$\begin{aligned}\tilde{g} &= -dt^2 + N^{-2} g_{\Sigma,ij}(t, \mathbf{x})dx^i dx^j \\ &= -dt^2 + \tilde{g}_{\Sigma,ij}(t, \mathbf{x})\,dx^i dx^j.\end{aligned}$$

Our strategical reason for the conformal transformation is our interest in the connection of global hyperbolicity and the geodesic completeness of the

Riemannian manifold $(\Sigma_t, \tilde{g}_{\Sigma_t})$ for all $t \in \mathbb{R}$. For three large classes of Lorentzian manifolds, the condition of global hyperbolicity is equivalent to the geodesic completeness.

- Static case. Lapse function and the spatial metric are time independent and thus Proposition 5.1 leads to the equivalence.

- The case of warped manifolds $M = \mathbb{R} \times_f \Sigma$, see [50, Theorem 3.66.] or [3, Lemma A.5.14.],

  Theorem 5.3 *Let (Σ, g_Σ) be a Riemannian manifold, and let $I = (a, b)$ with $-\infty \leq a < b \leq +\infty$ be given the negative definite metric $-dt^2$. Furthermore, let $f : I \mapsto (0, \infty)$ be a smooth function and the metric g be given by*

  $$g = -dt^2 + f(t)g_\Sigma.$$

  Then, the Lorentzian warped product $(I \times_f \Sigma, g)$ is globally hyperbolic iff (H, h) is complete.

- Sliced spaces. Assume that the metric $g_{\Sigma,ij}(\mathbf{x}, t)$ is uniformly bounded by the metric $g_{\Sigma,ij}(\mathbf{x}, 0)$ for all $t \in \mathbb{R}$ and tangent vectors $u \in T\Sigma$, i.e. that is there exist constants $A, D \in \mathbb{R} > 0$ such that

  $$A\, g_{\Sigma,ij}(\mathbf{x}, 0)u^i\, u^j \leq g_{\Sigma,ij}(\mathbf{x}, t)u^i\, u^j \leq D\, g_{\Sigma,ij}(\mathbf{x}, 0)u^i\, u^j.$$

  Then by [5, Theorem 2.1] or Theorem 4.1, the equivalence of globally hyperbolic spacetimes and geodesic completeness follows.

Due to these various cases, we are led to the following conjecture.

Conjecture 5.1 A Lorentzian manifold $(M = \mathbb{R} \times \Sigma, g)$ with metric tensor

$$g = -dt^2 + g_{\Sigma,ij}(\mathbf{x}, t)dx^i dx^j \tag{5.5}$$

is globally hyperbolic if and only if Riemannian manifold (Σ_t, g_{Σ_t}) is complete.

References

1. R. Strichartz, *Analysis of the Laplacian on the complete Riemannian manifold*, Journal of functional analysis; Vol. 52; PP. 48–79; 1983
2. W. Al-Qallaf, K. Papadopoulos, *On a Duality between Time and Space Cones*, Kuwait Journal of Science, Vol. 47 No. 2, 2020.
3. C. Bär, N. Ginoux, F. Pfäffle, *Wave Equations on Lorentzian Manifolds and Quantization*, ESI lectures in mathematics and physics, European Mathematical Society, 2007.

4. C.J.S. Clarke, *On the geodesic completeness of causal space-times*, Proc. Camb. Phil. Soc., 69, 319, 1970.
5. S. Cotsakis, *Global hyperbolicity of sliced spaces*, Gen. Rel. Grav. **36** (2004) 1183–1188.
6. J. Derezinski, D. Siemssen, *An evolution equation approach to the Klein-Gordon operator on curved spacetime*, Pure and Applied Analysis **2** (2019) 215–261
7. L. H. Ford, *Quantum field theory in curved space-time*, Particles and fields. Proceedings, 9th Jorge Andre Swieca Summer School, Campos do Jordao, Brazil, February 16–28, 1997, 1997, pp. 345–388.
8. G.J. Galloway, *Curvature, causality and completeness in space-times with causally complete spacelike slices*, Math. Proc. Camb. Phil. Soc., 99, 367, 1986.
9. G.J. Galloway, *Some connections between global hyperbolicity and geodesic completeness*, Journal of Geometry and Physics, Vol. 6, Issue 1, 127–141, 1989.
10. R. Gobel, *Zeeman Topologies on Space-Times of General Relativity Theory*, Comm. Math. Phys. 46, 289–307 (1976).
11. W. Gordon, *An Analytical Criterion for the Completeness of Riemannian Manifolds*, Proc. Amer. Math. Soc. **37** (1973) 221–221.
12. S.W. Hawking and G.F.R. Ellis, *The Large Scale Structure of Space-Time*, Cambridge University Press, 1973
13. S.W. Hawking, A.R. King, P. J. and McCarthy, *A new topology for curved space–time which incorporates the causal, differential, and conformal structures*. Journal of Mathematical Physics, 17 (2). pp. 174–181, 1976.
14. S. Hawking, W. Israel, *General Relativity: an Einstein Centenary Survey*, 2010.
15. S. Hollands, R. Wald, *Quantum fields in curved spacetime*, Phys. Rept. **574** (2015) 1–35.
16. E. Hounnonkpe, E. Minguzzi, *Globally hyperbolic spacetimes can be defined without the 'causal' condition*, Classical and Quantum Gravity, Vol. 36, No 19, 2019.
17. A. Ishibashi, R. Wald, *Dynamics in non-globally-hyperbolic static spacetimes: II. General analysis of prescriptions for dynamics*, Classical and Quantum Gravity **20** (2003), no. 16 3815–3826.
18. B. S. Kay, *Linear Spin 0 Quantum Fields in External Gravitational and Scalar Fields. 1. A One Particle Structure for the Stationary Case*, Commun. Math. Phys. **62** (1978) 55–70.
19. B. S. Kay, R. M. Wald, *Theorems on the Uniqueness and Thermal Properties of Stationary, Nonsingular, Quasifree States on Space-Times with a Bifurcate Killing Horizon*, Phys. Rept. **207** (1991) 49–136.
20. B. S. Kay, *The Principle of locality and quantum field theory on (nonglobally hyperbolic) curved space-times*, Rev. Math. Phys. **4** (1992), no. spec01 167–195.
21. B. S. Kay, *Quantum field theory in curved spacetime*, Encyclopedia of Mathematical Physics, Françoise, J.P. and Naber, G.L. and Tsou, S.T. **4** (2006) 202–214.
22. Kurt, N.; Papadopoulos, K. *On Completeness of Sliced Spaces under the Alexandrov Topology*, Mathematics, 8, 99, 2020.
23. R.J. Low, *Spaces of paths and the path topology*, Journal of Mathematical Physics, 57, 092503 (2016).
24. Low, R.J., *Spaces of Causal Paths and Naked Singularities*, Class. Quantum, Grav., Vol. 7, No. 6, 1990.
25. S. Mac Lane, *Categories for the Working Mathematician*, Springer, 1971.
26. K. Martin, P. Panangaden, *A Domain of Spacetime Intervals in General Relativity*, Commun. Math. Phys. 267, 563–586, 2006.
27. K. Martin, P. Panangaden, *Spacetime topology from causality*, arXiv:gr-qc/0407093.
28. E. Minguzzi, M. Śanchez, *The causal hierarchy of spacetimes*, in Recent developments in pseudo-Riemannian geometry, ed. by H. Baum and D. Alekseevsky, Zurich, ESI Lect. Math. Phys., EMS Pub. House, p. 299–358, 2006.
29. J.-P. Nicolas, *A nonlinear Klein–Gordon equation on Kerr metrics*, Journal de Mathématiques Pures et Appliquées **81** (2002), no. 9 885 – 914.
30. K. Nomizu, H. Ozeki, *The existence of complete Riemannian metrics*, 1961.

31. A. Much, R. Oeckl, *Self-adjointness in Klein-Gordon theory on globally hyperbolic spacetimes*, Mathematical Physics, Analysis and Geometry **24**, Springer, PP.5; 2021
32. Kyriakos Papadopoulos, Nazli Kurt, Basil K. Papadopoulos, *On Sliced Spaces; Global Hyperbolicity Revisited* Symmetry, 11(3), 304, 2019.
33. Papadopoulos, K., Papadopoulos, B.K., *Space-time Singularities vs. Topologies in the Zeeman—Göbel Class*, Gravit. Cosmol. 25, 116121, 2019.
34. Kyriakos Papadopoulos, *Natural vs. Artificial Topologies on a Relativistic Spacetime*, Chapter in Nonlinear Analysis and Global Optimization, eds Th. M. Rassias and P.M. Pardalos, Springer, accepted and to appear in 2020.
35. R. Penrose, *Techniques of Differential Topology in Relativity*, CBMS-NSF Regional Conference Series in Applied Mathematics, 1972.
36. P. Petersen, *Manifold Theory*, 2010.
37. D. Scott, *Outline of a mathematical theory of computation*, Technical Monograph PRG-2, Oxford University Computing Laboratory, November 1970.
38. I. Seggev, *Dynamics in stationary, non-globally hyperbolic spacetimes*, Classical and Quantum Gravity **21** (2004), no. 11 2651–2668.
39. M. Shubin, *Essential self-adjointness for semi-bounded magnetic Schrödinger operators on non-compact manifolds*, J. Funct. Anal. **186** (2001), no. 1 92 – 116.
40. R. Verch, J. Tolksdorf, *Quantum physics, fields and closed timelike curves: The D-CTC condition in quantum field theory*, arXiv:1609.01496 [math-ph], Commun. Math. Phys. **357** (2018) 319–351
41. R. M. Wald, *Quantum Field Theory in Curved Spacetime and Black Hole Thermodynamics*, Chicago Lectures in Physics, University of Chicago Press, 1994.
42. R. M. Wald, *Dynamics in nonglobally hyperbolic, static space-times*, J. Math. Phys **21** (1980) 2802–2805
43. E.C. Zeeman, *Causality implies the Lorentz group*, J. Math. Phys. 5 (1964), 490–493.
44. E.C. Zeeman, *The Topology of Minkowski Space*, Topology, Vol. 6, 161–170 (1967).
45. Felix Finster and Albert Much and Robert Oeckl, *Stationary spacetimes and self-adjointness in Klein–Gordon theory*, Journal of Geometry and Physics; Vol. 148; PP. 103561; 2020
46. A. N. Bernal, M. Sanchez, *On Smooth Cauchy hypersurfaces and Geroch's splitting theorem*, Commun. Math. Phys. **243** (2003) 461–470, gr-qc/0306108.
47. A. N. Bernal, M. Sánchez, *Smoothness of Time Functions and the Metric Splitting of Globally Hyperbolic Spacetimes*, Commun. Math. Phys. **257** (2005), no. 1 43–50.
48. A. Grigor'yan, *Heat Kernel and Analysis on Manifolds*, AMS/IP Studies in Advanced Mathematics 47, American Mathematical Society, 2009.
49. R. Geroch, *Domain of Dependence*, J. Math. Phys. **11** (1970), no. 2 437–449.
50. J. K. Beem, P. Ehrlich and K. Easley, *Global Lorentzian Geometry*, CRC Press, 1996

Spectrum Perturbations of Linear Operators in a Banach Space

Michael Gil'

Abstract This chapter is a survey of the recent results of the author on the spectrum perturbations of linear operators in a Banach space. It consists of three parts. In the first part, for an integer $p \geq 1$, we introduce the approximative quasi-normed ideal Γ_p of compact operators A with a quasi-norm $N_{\Gamma_p}(.)$ and the property $\sum_k |\lambda_k(A)|^p \leq a_p N^p_{\Gamma_p}(A)$, where $\lambda_k(A)$ $(k = 1, 2, \ldots)$ are the eigenvalues of A and a_p is a constant independent of A. Let I be the unit operator. Assuming that $A \in \Gamma_p$ and $I - A^p$ is boundedly invertible, we obtain invertibility conditions for perturbed operators. Applications of these conditions to the spectrum perturbations of absolutely p-summing and absolutely $(p, 2)$ summing operators are also discussed. As examples, in the first part of the chapter, we consider the Hille–Tamarkin integral operators and Hille–Tamarkin infinite matrices. The second part of the chapter deals with the ideal of nuclear operators A in a Banach space satisfying the condition $\sum_k x_k(A) < \infty$, where $x_k(A)$ $(k = 1, 2, \ldots)$ are the Weyl numbers of A. The inequality between the resolvent and determinant of A is derived. That inequality gives us new perturbation results. The third part of the chapter is devoted to non-compact operators in a Banach space having maximal chains of invariant subspaces and admitting the so-called triangular representation. The representation for the resolvents of such operators via multiplicative operator integrals is established. That representation can be considered as a generalization of the representation for the resolvent of a normal operator in a Hilbert space. In addition, a norm estimate for the resolvent of operators admitting triangular representation is derived. It enables us to obtain a perturbation bound for the spectral variations and to show that the considered operators are Kreiss-bounded. Applications to operators in L^p are also discussed. In particular, a new bound for the spectral radius of an integral operator is obtained. Some of the results presented in this chapter are new.

M. Gil' (✉)
Department of Mathematics, Ben Gurion University of the Negev, Beer-Sheva, Israel
e-mail: gilmi@bezeqint.net

I. N. Parasidis et al. (eds.), *Mathematical Analysis in Interdisciplinary Research*, Springer Optimization and Its Applications 179,
https://doi.org/10.1007/978-3-030-84721-0_16

AMS (MOS) Subject Classification 47A10, 47A55, 47B10, 47A75, 47G10, 47A11, 47A30

1 Introduction

Roughly speaking, the spectrum perturbation theory for linear operators consists of two approaches. In the framework of the first one, some structure on the error is imposed; for example, they may be analytic functions of a complex variable. The problem is then to determine how this structure affects the perturbed spectrum: for example, when are they analytic functions of the variable, what kind of paths do they follow in the complex plane? That approach is well developed. For various results of this kind, see, for instance, the book by Kato [26]. About the recent relevant results, see the very interesting book [37] and references given therein.

In the framework of the second approach, the errors are unstructured and perturbations are bounded in terms of some norm of the errors. That approach in the case of operators in a Banach space to the best of our knowledge is at an early stage of development. Below we suggest perturbation results for compact and non-compact operators in a Banach space, which are connected with the second approach.

Throughout this chapter, $\mathscr{X}$ is a Banach space with the unit operator $I = I_{\mathscr{X}}$, a norm $\|.\|$ and the approximation property, that is, any compact operator in $\mathscr{X}$ is a limit in the operator norm of a sequence of operators with finite ranks [23, 35]. By $\mathscr{B}(\mathscr{X})$ we denote the algebra of all bounded linear operators in $\mathscr{X}$. For an $A \in \mathscr{B}(\mathscr{X})$, $\|A\|$ is the operator norm, A^{-1} is the inverse operator, $\sigma(A)$ is the spectrum and $R_\lambda(A) = (A - \lambda I)^{-1}$ $(\lambda \notin \sigma(A))$ is the resolvent. For a compact operator A, $\lambda_k(A)$ $(k = 1, 2, \ldots)$ are the eigenvalues enumerated with their algebraic multiplicities in the non-increasing order of their absolute values. A point $\lambda \in \mathbf{C}$ is said to be Φ-regular for A if $I - \lambda A$ is boundedly invertible; $\sigma_\Phi(A)$ denotes the Fredholm spectrum (the complement of all Φ-regular points in the closed complex plane).

This chapter consists of three parts. In the first part (Sects. 2–6), for an integer $p \geq 1$, we introduce the approximative quasi-normed ideal Γ_p of compact operators A with a quasi-norm $N_{\Gamma_p}(.)$ and the property $\sum_k |\lambda_k(A)|^p \leq a_p N^p_{\Gamma_p}(A)$, where a_p is a constant independent of A. Assuming that $A \in \Gamma_p$ and $I - A^p$ is boundedly invertible, we obtain invertibility conditions for perturbed operators. Applications of these conditions to the spectrum perturbations of absolutely p-summing and absolutely $(p, 2)$ summing operators are also discussed. As examples, in the first part, we consider the Hille–Tamarkin integral operators and Hille–Tamarkin infinite matrices.

Furthermore, Carleman in 1930s has established an inequality between the resolvent of a Schatten-von Neumann operator and its regularized characteristic determinant, cf. [33, p. 69, Theorem 4.14] and [7, p. 1023, Theorem XI.6.15]. In [16], that inequality has been slightly improved (see also [17, Section 7.3]). In the case of the nuclear operators in a Hilbert space, the relevant inequality has been proved in [24, Section V.5]. In the second part of the chapter (Sects. 7 and 8), an inequality between resolvents and determinants of nuclear operators in $\mathscr{X}$ is derived. That inequality is a generalization of the above-mentioned inequality for operators in a Hilbert space from [24, Section V.5]. Applications of the obtained inequality to spectrum perturbations are also discussed.

The third part of the chapter (Sects. 9–16) is devoted to non-compact operators in a Banach space having maximal chains of invariant subspaces and admitting the triangular representations.

The deep theory of triangular representations of non-selfadjoint operators in a Hilbert space $\mathscr{H}$ via integrals along maximal chains has been developed in the works of M.S. Brodskii, I. C. Gohberg, M.G. Krein, L.A. Sakhnovich and other mathematicians, cf. [3, 4, 22, 25, 38] and references therein. We particularly extend some of the representations investigated in the mentioned works to operators in $\mathscr{X}$.

In Sect. 9, we introduce the notion of the maximal chain of projections in $\mathscr{X}$ and consider some properties of operators with invariant maximal chains.

Section 10 is devoted to projection functions whose values form continuous maximal chains and to operators commuting with these projection functions. In addition, the notion of the triangular representation is introduced for operators having invariant continuous projection functions.

In Sect. 11, norm estimates are derived for the resolvents of the considered operators.

An operator $T \in \mathscr{B}(\mathscr{X})$ is said to be *Kreiss bounded* if

$$\|(\lambda I - T)^{-1}\| \leq \frac{c_0}{|\lambda| - 1} \quad (|\lambda| > 1, c_0 = const > 0),$$

cf. [34, 40]. In particular, in these papers, it was shown that the operator of the indefinite integration is Kreiss-bounded. In Sect. 11, we show that the considered operators are Kreiss-bounded.

Furthermore, the quantity

$$\mathrm{sv}_A(\tilde{A}) := \sup_{\mu \in \sigma(\tilde{A})} \inf_{\lambda \in \sigma(A)} |\mu - \lambda|$$

is said to be *the spectral variation of an operator* $\tilde{A} \in \mathscr{B}(\mathscr{X})$ *with respect to an operator* $A \in \mathscr{B}(\mathscr{X})$. It should be noted that the spectral variations mainly investigated in the cases of finite rank operators and operators in a Hilbert space, cf. [1, 39] (see also [17] and references therein). In Sect. 12, we estimate the spectral variations of the operators in $\mathscr{X}$ admitting the triangular representation.

Sections 13–15 are devoted to applications of the results from Sect. 12 to operators in L^p. In particular, a new bound for the spectral radius of an integral operator is obtained.

In Sect. 16, the representation for the resolvent via multiplicative operator integral is established. That representation can be considered as a generalization of the representation for the resolvent of a normal operator in a Hilbert space.

2 The Quasi-normed Ideal Γ_p

For an integer $p \geq 1$, introduce the two-sided quasi-normed ideal Γ_p of compact operators in $\mathscr{B}(\mathscr{X})$ with a quasi-norm $N_{\Gamma_p}(.)$ and the property

$$\sum_{k=1}^{\infty} |\lambda_k(A)|^p \leq a_p N_{\Gamma_p}^p(A) \quad (A \in \Gamma_p), \tag{1}$$

where a_p is a constant independent of A, and Γ_p is assumed to be approximative. (i.e. the set of all finite rank operators is dense in in the norm of Γ_p). Below b_p denotes the quasi-triangle constant in Γ_p:

$$N_{\Gamma_p}(A + \tilde{A}) \leq b_p \, (N_{\Gamma_p}(A) + N_{\Gamma_p}(\tilde{A})) \quad (A, \tilde{A} \in \Gamma_p). \tag{2}$$

For the theory of the approximative normed and quasi-normed ideals, see [28, 35] and references given therein. In the sequel, constant a_p in (1) will be called *the eigenvalue constant.*

Put

$$\Delta_p(A, \tilde{A}) := N_{\Gamma_p}(A - \tilde{A}) \, \exp\left[a_p b_p^p \left(1 + \frac{1}{2}(N_{\Gamma_p}(A + \tilde{A}) + N_{\Gamma_p}(A - \tilde{A}))\right)^p\right]$$

and

$$\psi_p(A) = \inf_{k=1,2,\ldots} |1 - \lambda_k^p(A)|.$$

Theorem 1 *For an integer $p \geq 1$, let $A, \tilde{A} \in \Gamma_p$ and $I - A^p$ be boundedly invertible. If, in addition,*

$$\Delta_p(A, \tilde{A}) \exp\left[\frac{a_p N_{\Gamma_p}^p(A)}{\psi_p(A)}\right] < 1,$$

then $I - \tilde{A}^p$ is also boundedly invertible.

For the proof of Theorem 1, see [20, Theorem 1.1].

Replacing in Theorem 1 A and $\tilde{A}$ by λA and $\lambda \tilde{A}$, respectively, we get the following result.

Corollary 1 *Let $A, \tilde{A} \in \Gamma_p$ and $\lambda^p \notin \sigma_\Phi(A^p)$. If, in addition,*

$$\Delta_p(\lambda A, \lambda \tilde{A}) \exp\left[\frac{a_p N^p_{\Gamma_p}(\lambda A)}{\psi_p(\lambda A)}\right] < 1,$$

then λ^p is Φ-regular also for $\tilde{A}^p$.

From this corollary, it follows

Corollary 2 *Let $A, \tilde{A} \in \Gamma_p$ and $\mu^p \in \sigma_\Phi(\tilde{A}^p)$. Then either $\mu^p \in \sigma_\Phi(A^p)$ or*

$$\Delta_p(\mu A, \mu \tilde{A}) \exp\left[\frac{a_p N^p_{\Gamma_p}(\mu A)}{\psi_p(\mu A)}\right] \geq 1. \tag{3}$$

Note that (3) can be rewritten as

$$|\mu| N_{\Gamma_p}(A - \tilde{A}) \exp[\frac{a_p |\mu|^p N^p_{\Gamma_p}(A)}{\psi_p(\mu A)} + a_p b_p^p \left(1 + \frac{|\mu|}{2}(N_{\Gamma_p}(A + \tilde{A}) + N_{\Gamma_p}(A - \tilde{A}))\right)^p] \geq 1. \tag{4}$$

3 Particular Cases

3.1 Absolutely p-Summing Operators

An operator $A \in \mathscr{B}(\mathscr{X})$ is said to be absolutely p-summing ($1 \leq p < \infty$), if there is a constant ν, such that regardless of a natural number m and regardless of the choice $x_1, \ldots, x_m \in \mathscr{X}$ we have

$$\left[\sum_{k=1}^m \|Ax_k\|^p\right]^{1/p} \leq \nu \sup\left\{\left[\sum_{k=1}^m |\langle x^*, x_k\rangle|^p\right]^{1/p} : x^* \in \mathscr{X}^*, \|x^*\| = 1\right\},$$

Here $\langle .,. \rangle$ means the functional on $\mathscr{X}$, $\mathscr{X}^*$ means the space adjoint to $\mathscr{X}$ [12, 28, 35]. The least ν for which this inequality holds is a norm and is denoted by $\pi_p(A)$. The set of absolutely p-summing operators in $\mathscr{X}$ with the finite norm π_p is a normed ideal in the set of bounded linear operators, which is denoted by Π_p, cf. [35].

As is well known,

$$\sum_{k=1}^{\infty} |\lambda_k(A)|^p \le \pi_p^p(A) \ \ (A \in \Pi_p;\ 2 \le p < \infty), \tag{5}$$

cf. Theorem 17.4.3 from [12] (see also Theorem 3.7.2 from [35, p. 159]). Thus, Π_p ($p \ge 2$) *has the properties of ideal* Γ_p. Besides, $N_{\Gamma_p}(A) = \pi_p(A)$, $b_p = 1$ and $a_p = 1$.

3.2 Ideal $\mathscr{E}_p$ and Absolutely $(p, 2)$-Summing Operators

Recall [35, p. 79] that $s_n(T)$ ($n = 1.2, \dots$) is called the n-th s-number (n-th singular number) of $T \in \mathscr{B}(\mathscr{X})$ if the following conditions are satisfied:

(S_1) $\|T\| = s_1(T) \ge s_2(T) \ge \dots \ge 0$;

(S_2) $s_{n+m-1}(S+T) \le s_m(T)+s_n(S)$ $(S \in \mathscr{B}(\mathscr{X}))$;

(S_3) $s_n(A_1TA_2) \le \|A_1\|s_n(T)\|A_2\|$ $(A_1, A_2 \in \mathscr{B}(\mathscr{X}))$;

(S_4) If rank $(T) < n$, then $s_n(T) = 0$;

(S_5) $s_n(I_{l_n^2}) = 1$.

Here $I_{l_n^2}$ is the unit operator in the n-dimensional Hilbert space l_n^2 with the traditional scalar product.

Let $L(l^2, \mathscr{X})$ denote the space of linear operators acting from the Hilbert space l^2 with the traditional scalar product into $\mathscr{X}$. The n-th Weyl number of $T \in \mathscr{B}(\mathscr{X}))$ is defined by

$$x_n(T) := \sup\{a_n(TZ) : Z \in L(l^2, \mathscr{X}), \|Z\| = 1\},$$

where $a_n(T)$ is the n-th approximation number defined by

$$a_n(T) := \inf\{\|T - T_n\| : T_n \in \mathscr{B}(\mathscr{X}), \text{rank } T_n < n\}.$$

$x_n(T)$ is an s-number with the sub-multiplicative property

$$(S_6)\ \ x_{n+m-1}(TS) \le x_n(T)x_m(S)\ \ (S, T \in \mathscr{B}(\mathscr{X})),$$

cf. [35, Theorem 2.4.14] and [35, Proposition 2.4.17]. For an integer $p \ge 1$, let $\mathscr{E}_p$ be the set of compact operators A acting in $\mathscr{X}$ and satisfying

$$N_{\mathscr{E}_p}(A) := (\sum_{k=1}^{\infty} x_k^p(A))^{1/p} < \infty.$$

Since $x_k(A) \le x_{k-1}(A)$ and $x_{2k-1}(A+\tilde{A}) \le x_k(A) + x_k(\tilde{A})$, we have

$$\sum_{k=1}^{\infty} x_k^p(A+\tilde{A}) = \sum_{j=1}^{\infty} x_{2j-1}^p(A+\tilde{A}) + x_{2j}^p(A+\tilde{A}) \le 2\sum_{j=1}^{\infty} x_{2j-1}^p(A+\tilde{A})$$

$$\le 2\sum_{j=1}^{\infty} (x_j(A) + x_j(\tilde{A}))^p.$$

By the Minkovsky inequalit

$$\left(\sum_{j=1}^{\infty} (x_j(A) + x_j(\tilde{A}))^p\right)^{1/p} \le \left(\sum_{j=1}^{\infty} x_j^p(A)\right)^{1/p} + \left(\sum_{j=1}^{\infty} x_j^p(\tilde{A})\right)^{1/p}.$$

Then

$$N_{\mathscr{E}_p}(A+\tilde{A}) \le 2^{1/p}(N_{\mathscr{E}_p}(A) + N_{\mathscr{E}_p}(\tilde{A})).$$

So $\mathscr{E}_p$ is a quasinormed ideal with the quasi-triangular constant $b_p = 2^{1/p}$. It is approximative, cf. [28, 35]. We need the following Weyl type inequality:

$$\sum_{k=1}^{\infty} |\lambda_k(A)|^p \le c_p^p \sum_{k=1}^{\infty} x_k^p(A) = c_p^p N_{\mathscr{E}_p}^p(A)$$

with

$$c_p = 2^{1/p}\sqrt{2e}.$$

cf. [28, Theorem 2.a.6, p. 85].

So $\mathscr{E}_p$ *is an example of ideal* Γ_p with $N_{\Gamma_p}(A) = N_{\mathscr{E}_p}(A)$, $a_p = c_p^p$ and $b_p = 2^{1/p}$.

Let us point an estimate for $N_{\mathscr{E}_p}(A)$. To this end, recall that an $A \in \mathscr{B}(\mathscr{X})$ is said to be absolutely (p,q)-summing $(p \ge q)$ if there is a constant ν such that regardless a natural number m and regardless of the choice $x_1, \ldots, x_m \in \mathscr{X}$ we have

$$\left[\sum_{k=1}^{m} \|Ax_k\|^p\right]^{1/p} \le \nu \sup\left\{\left[\sum_{k=1}^{m} |\langle x^*, x_k\rangle|^q\right]^{1/q} : x^* \in \mathscr{X}^*, \|x^*\| = 1\right\}$$

cf. [6, 12, 35]. The least ν for which this inequality holds is denoted by $\pi_{p,q}(A)$. The set of absolutely (p,q)-summing operators is denoted by $\Pi_{p,q}$.

Due to [12, Theorem 16.3.1], $\pi_{p,q}$ is a norm and $\Pi_{p,q}$ with that norm is a Banach space. If $A \in \Pi_{p,q}$, then $\|A\| \le \pi_{p,q}(A)$ since

$$\|Ax\| = [\|Ax\|^p]^{1/p} \le \pi_{p,q}(A)\ \sup\{[|\langle x^*, x\rangle|^q]^{1/q} : x^* \in \mathscr{X}^*, \|x^*\| = 1\}$$
$$\le \pi_{p,q}(A)\|x\|$$

for any $x \in \mathscr{X}$. If, in addition, R and S are bounded operators acting in $\mathscr{X}$, then $\pi_{p,q}(SAR) \le \|R\|_{\mathscr{X}}\|S\|_{\mathscr{X}}\pi_{p,q}(A)$.

We need Corollary 2.a.3 from [28, p. 81] (see also Corollary 17.2.2 from [12, p. 293]), which asserts the following: if $A \in \Pi_{p_0,2}$ $(2 \le p_0 < \infty)$, then

$$x_n(A) \le \frac{\pi_{p_0,2}(A)}{n^{1/p_0}} \quad (n = 1, 2, \dots).$$

Hence, for any $p > p_0$, we have

$$N_{\mathscr{E}_p}(A) = \left(\sum_{k=1}^{\infty} x_n^p(A)\right)^{1/p} \le \pi_{p_0,2}(A)\left(\sum_{k=1}^{\infty} \frac{1}{k^{p/p_0}}\right)^{1/p}$$
$$= \zeta^{1/p}(p/p_0)\pi_{p_0,2}(A) \quad (A \in \Pi_{p_0,2}),$$

where

$$\zeta(z) = \sum_{k=1}^{\infty} \frac{1}{k^z} \quad (\Re z > 1)$$

is the Riemann zeta-function.

4 Additional Upper Bounds for Determinants

Lemma 1 *For an integer $p \ge 1$ and $A \in \Gamma_p$, one has*

$$|\det(I - A^p)| \le \psi_p(A)\exp\,[a_p N_{\Gamma_p}^p(A)].$$

Proof Evidently,

$$|\det(I - A^p)| = |1 - \lambda_m^p(A)| \prod_{k=1, k\ne m}^{\infty} |1 - \lambda_k^p(A)|$$
$$\le |1 - \lambda_m^p(A)|\exp\left[\sum_{k=1}^{\infty} |\lambda_k(A)|^p\right]$$

for any $m \geq 1$. Taking into account (1) and choosing m in such a way that $|1 - \lambda_m^p(A)| = \psi_p(A)$, we prove the lemma. □

Furthermore, let $E_p(z)$ be the Weierstrass primary factor:

$$E_1(z) = (1 - z); \;\; E_p(z) = (1 - z) \exp\left[\sum_{m=1}^{p-1} \frac{z^m}{m}\right] \quad (p = 2, 3, \ldots; \; z \in \mathbf{C}).$$

Put

$$\gamma_p := \frac{p-1}{p} \quad (p \neq 1; \, p \neq 3) \text{ and } \gamma_1 = \gamma_3 = 1.$$

According to Theorem 1.5.3 [17],

$$|E_p(z)| \leq \exp[\gamma_p |z|^p] (z \in \mathbf{C}). \tag{6}$$

For an $A \in \Gamma_p$, $p \geq 2$, introduce the p-regularized determinant by

$$\det_p(I - A) := \prod_{k=1}^{\infty} E_p(\lambda_k(A)).$$

Due to (1) and (6)

$$|\det_p(I - A)| \leq \exp\left[\gamma_p \sum_{k=1}^{\infty} |\lambda_k(A)|^p\right] \leq \exp\left[a_0 \gamma_p N_{\Gamma_p}^p(A)\right] \quad (p \geq 2),$$

and therefore the product converges.

Lemma 2 *For an integer $p \geq 2$ and any $A \in \Gamma_p$, one has*

$$|\det_p(I - A)| \leq \psi_1(A) \exp\left[\sum_{k=1}^{p-1} \frac{r_s^k(A)}{k}\right] \exp\left[a_p \gamma_p N_{\Gamma_p}^p(A)\right],$$

where $r_s(A)$ is the spectral radius of A.

Proof By (6) and (1),

$$|\det_p(I - A)| = |E(\lambda_m)| \prod_{k=1, k \neq m}^{\infty} |E(\lambda_k)|$$

$$\leq |E(\lambda_m(A))| \exp \left[\gamma_p \sum_{k=1, k\neq m}^{\infty} |\lambda_k(A)|^p \right]$$

$$\leq |E(\lambda_m(A))| \exp \left[a_p \gamma_p N_{\Gamma_p}^p(A) \right]$$

for any $m \geq 1$. But

$$|E_p(\lambda_m(A))| = |1 - \lambda_m(A)|| \exp \left[\sum_{k=1}^{p-1} \frac{\lambda_m(A)^k}{k} \right] |$$

$$\leq |1 - \lambda_m(A)| \exp \left[\sum_{k=1}^{p-1} \frac{r_s^k(A)}{k} \right].$$

So

$$|\det_p(I - A)| \leq |1 - \lambda_m(A)| \exp \left[\sum_{k=1}^{p-1} \frac{r_s^k(A)}{k} \right] \exp \left[a_p \gamma_p N_{\Gamma_p}^p(A) \right].$$

Hence, choosing m in such a way that $|1 - \lambda_m(A)| = \psi_1(A)$, we prove the lemma. □

5 Hille–Tamarkin Integral Operators "Close" to Volterra Ones

In this section and in the next one, we consider some concrete integral and matrix operators. We need the following result.

Corollary 3 *Let $W \in \Gamma_p$ be a quasi-nilpotent operator (i.e. its spectrum is $\{0\}$). Then for an arbitrary $\tilde{A} \in \Gamma_p$, one has*

$$|\det(I - \tilde{A}^p) - 1| \leq \Delta_p(W, \tilde{A}).$$

Indeed, this result is due to Lemma 1, and the equality $\det(I - W^p) = 1$.

Let $L^p = L^p(0, 1)$ $(2 \leq p < \infty)$ be the space of scalar functions f defined on $[0, 1]$ and endowed the norm

$$\|f\| = [\int_0^1 |f(t)|^p dt]^{1/p}.$$

Let $K : L^p \to L^p$ be the operator defined by

$$(Kf)(t) = \int_0^1 k(t, s)f(s)ds \ \ (f \in L^p, 0 \le t \le 1),$$

whose kernel k defined on $[0, 1]^2$ satisfies the condition

$$\hat{k}_p(K) := \left[\int_0^1 (\int_0^1 |k(t, s)|^{p'} ds)^{p/p'} dt\right]^{1/p} < \infty,$$

where $1/p + 1/p' = 1$. Then K is called a (p, p')-Hille–Tamarkin integral operator.

As is well known, [6, p. 43], any (p, p')-Hille–Tamarkin operator K is an absolutely p-summing operator with $\pi_p(K) \le \hat{k}_p(K)$. Let the operator V be defined by

$$(Vf)(t) = \int_0^t k(t, s)f(s)ds \ \ (f \in L^p).$$

This operator is quasi-nilpotent. With $\Gamma_p = \Pi_p$, we have

$$\Delta_p(K, V) = \pi_p(K - V) \ \exp\left[\left(1 + \frac{1}{2}(\pi_p(K + V) + \pi_p(K - V))\right)^p\right]$$
$$\le \hat{\Delta}_p(K, V),$$

where

$$\hat{\Delta}_p(K, V) := \hat{k}_p(K - V) \ \exp\left[\left(1 + \frac{1}{2}(\hat{k}_p(K + V) + \hat{k}_p(K - V))\right)^p\right].$$

Note that

$$((K - V)f)(t) = \int_x^1 k(t, s)f(s)ds.$$

The previous corollary implies

Corollary 4 *Let K be a (p, p')-Hille–Tamarkin integral operator in $L^p(0, 1)$ for an integer $p \ge 2$ and $1/p + 1/p' = 1$. If $\hat{\Delta}_p(K, V) < 1$, then*

$$|det\ (I - K^p) - 1| \le \hat{\Delta}_p(K, V).$$

and therefore $\sqrt[p]{1} \notin \sigma_\Phi(K)$, provided $\hat{\Delta}_p(K, V) < 1$.

6 Hille–Tamarkin Infinite Matrices "Close" to Triangular Ones

Let us consider the linear operator T in l^p $(2 \le p < \infty)$ generated by an infinite matrix $(t_{jk})_{j,k=1}^{\infty}$, satisfying the condition

$$\tau_p(T) := \left[\sum_{j=1}^{\infty} \left(\sum_{k=1}^{\infty} |t_{jk}|^{p'} \right)^{p/p'} \right]^{1/p} < \infty,$$

where $1/p + 1/p' = 1$.

Then T is called a (p, p')-Hille–Tamarkin matrix. As is well known, any (p, p')-Hille–Tamarkin matrix T is an absolutely p-summing operator with $\pi_p(T) \le \tau_p(T)$, cf. [6, p. 43] and [35, Sections 5.3.2 and 5.3.3, p. 230]). So according to (5),

$$\sum_{k=1}^{\infty} |\lambda_k(T)|^p \le \tau_p^p(T) \;\; (2 \le p < \infty).$$

Let $T_+ = (\tau_{jk})_{j,k=1}^{\infty}$ be the upper-triangular part of T: $\tau_{jk} = t_{jk}$ for $1 \le j \le k \le \infty$ and $\tau_{jk} = 0$ otherwise. Since $p > p'$, we obtain

$$\left(\sum_{k=1}^{\infty} |t_{jk}|^{p'} \right)^{p/p'} \ge \sum_{k=1}^{\infty} |t_{jk}|^p$$

and thus

$$\sum_{j=1}^{\infty} \sum_{k=1}^{\infty} |t_{jk}|^p < \infty.$$

Since T_+ is triangular, its eigenvalues are the diagonal entries and

$$\det\left(I - T_+^p\right) = d_{+,p} := \prod_{k=1}^{\infty} (1 - t_{kk}^p).$$

Under consideration,

$$\begin{aligned} \Delta_p(T, T_+) &= \pi_p(T - T_+) \, \exp \left[\left(1 + \frac{1}{2}(\pi_p(T + T_+) + \pi_p(T - T_+)) \right)^p \right] \\ &\le \hat{\Delta}_p(T, T_+), \end{aligned}$$

where

$$\hat{\Delta}_p(T, T_+) := \tau_p(T - T_+) \exp \left[\left(1 + \frac{1}{2}(\tau_p(T + T_+) + \tau_p(T - T_+))\right)^p \right].$$

Note that $T - T_+$ is the strictly lower part of T.

Making use of Theorem 1, we arrive at

Corollary 5 *Let T be a (p, p')-Hille–Tamarkin matrix for an integer $p \geq 2$ and $1/p + 1/p' = 1$. Then $|\det(I - T^p) - d_{+,p}| \leq \hat{\Delta}_p(T, T_+)$, and therefore $\sqrt[p]{1} \notin \sigma_\Phi(T)$, provided $|d_{+,p}| > \hat{\Delta}_p(T, T_+)$.*

7 An Inequality Between Resolvents and Determinants for Nuclear Operators in a Banach Space

Recall that the s-numbers $s_n(T)$ and Weyl numbers $x_n(T)$ $(n = 1, 2, \ldots)$ are defined in Sect. 3.2. Recall also that the Weyl numbers are s-numbers with the submultiplicative property

$$(S_6)\ \ x_{n+m-1}(TS) \leq x_n(T)x_m(S) \ \ (T, S \in \mathscr{B}(\mathscr{X})),$$

cf. [35, Theorem 2.4.14] and [35, Proposition 2.4.17].

Our main object in this section is the set of compact operators $A \in \mathscr{B}(\mathscr{X})$ with the property

$$N_{\mathscr{W}}(A) := \sum_{k=1}^{\infty} x_k(A) < \infty,$$

which is denoted by $\mathscr{W}$. For $A, B \in \mathscr{W}$ due to (S_2), we have

$$N_{\mathscr{W}}(A + B) = \sum_{k=1}^{\infty} x_k(A + B) \leq 2\sum_{k=1}^{\infty} x_{2k-1}(A + B) \leq 2\sum_{k=1}^{\infty}(x_k(A) + x_k(B))$$

$$= 2N_{\mathscr{W}}(A) + 2N_{\mathscr{W}}(B)$$

and thus $N_{\mathscr{W}}(.)$ is a quasi-norm, and $\mathscr{W}$ is a quasi-normed ideal.

Following [23, Section II.1], we define the determinant det $(I_{\mathscr{X}} - A)$ $(A \in \mathscr{W})$ as the continuous extension of the determinants det $(I_{\mathscr{X}} - A_n)$, where A_n $(n = 1, 2, \ldots)$ are finite rank operators converging to A in the quasi-norm $N_{\mathscr{W}}(A)$. Besides, det $(I_{\mathscr{X}} - A) := \lim_{n\to\infty}$ det $(I_{\mathscr{X}} - A_n)$. Below we show that det $(I_{\mathscr{X}} - A)$ exists for any $A \in \mathscr{W}$.

Theorem 2 *If $A \in \mathscr{W}$ and $\lambda_k(A) \neq 1$ $(k = 1, 2, \dots)$, then*

$$\|(I_{\mathscr{X}} - A)^{-1} \det (I_{\mathscr{X}} - A)\| \leq c(1 + cx_2(A))^3 \prod_{k=3}^{\infty}(1 + cx_k(A))^2, \tag{7}$$

where $c = \sqrt{2e}$.

This theorem is proved in [21, Theorem 1.1].

Taking into account that $1 + y \leq e^y$ $(y \geq 0)$ and $x_2(A) \leq x_1(A)$, from 7 we get

$$\|(I_{\mathscr{X}} - A)^{-1} \det(I_{\mathscr{X}} - A)\| \leq c \exp\left[c(x_1(A) + 2x_2(A) + 2\sum_{k=3}^{n} x_k(A))\right]$$
$$\leq c \exp\,[2cN_{\mathscr{W}}(A)].$$

Hence, replacing A by $A\lambda^{-1}$ $(\lambda \in \mathbf{C})$, we arrive at

Corollary 6 *If $A \in \mathscr{W}$, then for any regular $\lambda \neq 0$ of A we have*

$$\|(\lambda I_{\mathscr{X}} - A)^{-1} \det (I_{\mathscr{X}} - \lambda^{-1}A)\| \leq \frac{c}{|\lambda|} \exp\,[\frac{2cN_{\mathscr{W}}(A)}{|\lambda|}].$$

8 Perturbations of Nuclear Operators

Let $A, \tilde{A} \in \mathscr{B}(\mathscr{X})$ and $q = \|A - \tilde{A}\|$. If $\lambda \notin \sigma(A)$ and $q\|(\lambda I - A)^{-1}\| < 1$, then it is simple to show that $\lambda \notin \sigma(\tilde{A})$ and

$$\|(\lambda I - \tilde{A})^{-1}\| \leq \frac{\|(\lambda I - A)^{-1}\|}{1 - q\|(\lambda I - A)^{-1}\|}.$$

Hence by Corollary 6, we obtain

Corollary 7 *If $A \in \mathscr{W}$, $\lambda \notin \sigma(A) \cup 0$, $\tilde{A} \in \mathscr{B}(\mathscr{X})$ and*

$$q\frac{c}{|\lambda|}\frac{\exp\,[\frac{2cN_{\mathscr{W}}(A)}{|\lambda|}]}{|\det (I - \lambda^{-1}A)|} < 1.$$

Then λ is regular for $\tilde{A}$, and

$$\|(\lambda I - \tilde{A})^{-1}\| \leq \frac{c \exp\left[\frac{2cN_{\mathscr{W}}(A)}{|\lambda|}\right]}{|\lambda \det (I - \lambda^{-1}A)| - cq \exp\left[\frac{2cN_{\mathscr{W}}(A)}{|\lambda|}\right]}.$$

This corollary supplements the recent investigations of resolvents [2, 5, 11, 29, 32].

To illustrate this corollary, assume that A is a quasinilpotent operator. Then the spectral radius $r_s(A)$ of A is equal to zero and $\det(I - \lambda^{-1}A) = 1$.

By Corollary 7, if

$$\frac{qc}{|\lambda|} \exp\left[\frac{2cN_{\mathscr{W}}(A)}{|\lambda|}\right] < 1,$$

then λ is regular for $\tilde{A}$, and therefore, the spectral radius $r_s(\tilde{A})$ of $\tilde{A}$ satisfies the inequality $r_s(\tilde{A}) \le y$, where y is a unique positive root of the equation

$$\frac{qc}{y} \exp\left[\frac{2cN_{\mathscr{W}}(A)}{y}\right] = 1.$$

If

$$qc \exp\left[2cN_{\mathscr{W}}(A)\right] \le 1,$$

then $qc < 1$, $y \ge 1$ and therefore

$$qc \exp\left[\frac{2cN_{\mathscr{W}}(A)}{y}\right] \ge 1.$$

Thus under consideration, we have

$$r_s(\tilde{A}) \le y \le \frac{2cN_{\mathscr{W}}(A)}{\ln(1/(qc))}.$$

Here we can take $\mathscr{X} = L^p(0,1)$ $(1 \le p < \infty)$ with the traditional norm,

$$(Af)(x) = \int_0^x K(x,s)f(s)ds \text{ and } (\tilde{A}f)(x) = \int_0^1 K(x,s)f(s)ds$$

$(f \in L^p(0,1),\ 0 \le x \le 1)$ with a sufficiently smooth kernel $K(.,.)$.

Note also that the following perturbation result for determinants is proved in section 5 of [21].

Corollary 8 *Let* $A, \tilde{A} \in \mathscr{W}$. *Then*

$$|\det(I-\tilde{A}) - \det(I-A)| \le N_{\mathscr{W}}(\tilde{A}-A)\ \exp\left[2 + cN_{\mathscr{W}}(\tilde{A}-A) + cN_{\mathscr{W}}(\tilde{A}+A)\right].$$

9 Maximal Chains of Projections

For two projections P_1, P_2 in $\mathscr{X}$ ($P_1 \neq P_2$), we write $P_1 < P_2$ if $P_1P_2 = P_2P_1 = P_1$ (and thus $P_1\mathscr{X} \subset P_2\mathscr{X}$). A set $\mathscr{P}$ of projections in $\mathscr{X}$ containing at least two projections is called *a chain (of projections)*, if from $P_1, P_2 \in \mathscr{P}$ with $P_1 \neq P_2$ it follows that either $P_1 < P_2$ or $P_1 > P_2$.

Let $P^-, P^+ \in \mathscr{P}$, and $P^- < P^+$. If for every $P \in \mathscr{P}$ we have either $P < P^-$ or $P > P^+$, then the pair (P^+, P^-) is called a gap of $\mathscr{P}$. Besides, dim $(P_+ - P_-)\mathscr{X}$ is the dimension of the gap. A chain which does not have gaps is called *a continuous chain*.

A projection P in $\mathscr{X}$ is called a limit projection of a chain $\mathscr{P}$ if exists a sequence $P_k \in \mathscr{P}$ $(k = 1, 2, \ldots)$ which strongly converges to P. A chain is said to be closed if it contains all its limit projections.

Definition 1 A chain $\mathscr{P}$ is said to be maximal if it is closed, contains 0 and I, all its gaps (if they exist) are one dimensional and

$$\sup_{P \in \mathscr{P}} \|P\| < \infty.$$

We will say that *a maximal chain $\mathscr{P}$ is invariant for $A \in \mathscr{B}(\mathscr{X})$, or A has a maximal invariant chain $\mathscr{P}$*, if $PAP = AP$ for any $P \in \mathscr{P}$.

Lemma 3 *Let P_1, P_2 be two invariant projections of A and $P_1 < P_2$. Then, $P_2 - P_1$ is also an invariant projection of A.*

Proof Since $P_2P_1 = P_1P_2 = P_1$, we have

$$(P_2 - P_1)A(P_2 - P_1) = P_2AP_2 - P_2AP_1 - P_1AP_2 + P_1AP_1 = AP_2 - P_2P_1AP_1$$
$$-P_1P_2AP_1 + AP_1 = AP_2 - P_1AP_1 - P_1AP_1 + AP_1 = A(P_2 - P_1).$$

As claimed. □

Let us prove the following result.

Lemma 4 *Let a compact operator $V \in \mathscr{B}(\mathscr{X})$ have a maximal invariant chain $\mathscr{P}$. If, in addition,*

$$(P^+ - P^-)V(P^+ - P^-) = 0 \tag{8}$$

for every gap (P^+, P^-) of $\mathscr{P}$ (if it exists), then V is a quasi-nilpotent operator, i.e. $\sigma(V) = \{0\}$.

Proof Indeed, since $(P^+ - P^-)V(P^+ - P^-)$ is one dimensional, by the previous lemma we have $(P^+ - P^-)V(P^+ - P^-)h = \mu(P^+ - P^-)h = V(P^+ - P^-)h$ $(\mu \in \mathbf{C}, h \in \mathscr{X})$. So μ is an eigenvalue of V and $(P^+ - P^-)h$ is the corresponding eigenvector. By (8) $\mu = 0$. Moreover, at the points of the continuity of $\mathscr{P}$, operator V does not have eigenvectors. So V does not have non-zero eigenvalues; but V is compact. So it is quasi-nilpotent. □

In particular, if a compact operator has a continuous invariant chain, then it is quasi-nilpotent.

We need also the following lemma.

Lemma 5 *Let V be a compact quasi-nilpotent operator having a maximal invariant chain $\mathscr{P}$. Then equality (8) holds for every gap (P^+, P^-) of $\mathscr{P}$ (if it exists).*

Proof As it is shown in the proof of the previous lemma, $(P^+ - P^-)V(P^+ - P^-)h = \mu(P_2 - P_1)h$, where μ is an eigenvalue of V. But V is quasi-nilpotent. So $\mu = 0$. This proves the lemma. □

In the sequel, the expression $(P^+ - P^-)T(P^+ - P^-)$ for a $T \in \mathscr{B}(\mathscr{X})$ will be called *the block of the gap (P^+, P^-) of $\mathscr{P}$ on T.*

Lemma 6 *Let V_1 and V_2 be compact quasi-nilpotent operators having a joint maximal invariant chain $\mathscr{P}$. Then $V_1 + V_2$ is a quasi-nilpotent operator having the same maximal invariant chain.*

Proof Since the blocks of the gaps of $\mathscr{P}$ on both V_1 and V_2, if they exist, are zero (due to Lemma 5), the blocks of the gaps of $\mathscr{P}$ on $V_1 + V_2$ are also zero. Now the required result is due to Lemma 4. □

Lemma 7 *Let V and B be bounded linear operators in $\mathscr{X}$ having a joint maximal invariant chain $\mathscr{P}$. In addition, let V be a compact quasi-nilpotent operator. Then VB and BV are quasi-nilpotent, and $\mathscr{P}$ is their maximal invariant chain.*

Proof It is obvious that

$$PVBP = VPBP = VBP \quad (P \in \mathscr{P}).$$

Now let $Q = P^+ - P^-$ for a gap (P^+, P^-). Then according to Lemma 5, equality (8) holds. Further, we have $QVP^- = QBP^- = 0$,

$$\begin{aligned} QVBQ &= QVB(P^+ - P^-) = QV(P^+BP^+ - P^-BP^-) \\ &= QV[(P^- + Q)B(P^- + Q) - P^-BP^-] = QVQBQ = 0. \end{aligned}$$

Due to Lemma 4, this relation implies that VB is a quasi-nilpotent operator. Similarly, we can prove that BV is quasi-nilpotent. □

Lemma 8 *Let V and B be bounded linear operators in $\mathscr{X}$ having a joint maximal invariant chain $\mathscr{P}$. In addition, let V be a compact quasi-nilpotent operator and the regular set of B be simply connected. Then $\sigma(B + V) = \sigma(B)$.*

Proof We have

$$PR_\lambda(B)P = -\sum_{k=0}^{\infty} P\frac{B^k}{\lambda^{k+1}}P = R_\lambda(B)P \quad (|\lambda| > \|B\|, P \in \mathscr{P}).$$

Since the set of regular points of B is simply connected, by the resolvent identity one can extend the equality $PR_\lambda(B)P = R_\lambda(B)P$ to all regular λ of B (see also [36, pp. 32–33]).

Put $T = B + V$. For any $\lambda \notin \sigma(B)$, operator $VR_\lambda(B)$ is quasi-nilpotent due to Lemma 7. So $I + VR_\lambda(B)$ is boundedly invertible and therefore,

$$R_\lambda(T) = (B + V - \lambda I)^{-1} = R_\lambda(B)(I + VR_\lambda(B))^{-1} \quad (\lambda \notin \sigma(B)).$$

Hence, it follows that λ is a regular point for T. Consequently,

$$\sigma(T) \subseteq \sigma(B). \tag{9}$$

So the regular set of T is also simply connected.

Now let $\lambda \notin \sigma(T)$. Since $\mathcal{P}$ is invariant for T, as above we can show that $\mathcal{P}$ is invariant for $R_\lambda(T)$. Then operator $VR_\lambda(T)$ is quasi-nilpotent due to Lemma 7. So $I - VR_\lambda(T)$ is boundedly invertible. Furthermore, according to the equality $B = T - V$, we get

$$R_\lambda(B) = (T - V - \lambda I)^{-1} = R_\lambda(T)(I - VR_\lambda(T))^{-1}.$$

Hence, it follows that λ is a regular point also for B, and therefore, $\sigma(B) \subseteq \sigma(T)$. Now (9) proves the result. □

10 Operators Having Continuous Maximal Chains

Definition 2 Let P_t $(t \in [a, b])$ be a function defined on a finite segment $[a, b]$, whose values form a maximal continuous chain $\mathcal{P}$ of projections, such that $P_{t_2}P_{t_1} = P_s$ $(s = \min\{t_2, t_1\})$, $P_a = 0$ and $P_b = I$. Then we will call P_t a continuous maximal projection function (CMPF).

So P_t is a particular case of a resolution of the identity.

It is assumed that there is a constant m_P dependent on P_t only, such that

$$\|\sum_{k=1}^{n} a_k \Delta P_k\| \le m_P \max_j |a_j| \tag{10}$$

$$(n < \infty;\ \Delta P_k = P_{t_k} - P_{k-1};\ a = t_0 < t_1 < \ldots < t_n = b)$$

for arbitrary numbers a_k and an arbitrary partitioning of $[a, b]$.

Let $\psi(t)$ be a bounded scalar function defined on $[a, b]$ and there exists a limit S of the operator sums

$$S_n = \sum_{k=1}^{n} \psi(t_k)\Delta P_k$$

in the operator norm. Then we write

$$S = \int_a^b \psi(t)dP_t, \tag{11}$$

$\psi(t)$ will be called a *P_t-integrable function*, and S will be called *a P_t-scalar operator*. We write $S = \psi(T_0)$, where

$$T_0 = \int_a^b t dP_t$$

is a scalar type spectral operator [8]. So a P_t-scalar operator is a function of a scalar type spectral operator.

Due to (10), $\|S_n\| \le m_p \sup_t |\psi(t)|$; by the Banach-Steinhaus theorem, S is bounded and $\|S\| \le \sup_n \|S_n\| \le m_P \sup_t |\psi(t)|$.

For example, let as usually $L^p = L^p(0,1)$ $(1 \le p < \infty)$ be the space of scalar-valued functions h defined on [0, 1] and equipped with the norm

$$|h|_{L^p} = [\int_0^1 |h(x)|^p dx]^{1/p}.$$

Let $\hat{P}_t$ $(0 \le t \le 1)$ be the truncation projection function, defined by the relations $\hat{P}_0 = 0$, $\hat{P}_1 = I$ and

$$(\hat{P}_t f)(x) = \begin{cases} f(x) & \text{if } 0 \le x < t, \\ 0 & \text{if } t < x \le 1 \end{cases} \quad (t \in (0,1);\ f \in L^p). \tag{12}$$

It is simple to check that the values of $\hat{P}_t$ form a continuous maximal chain and the operator $\hat{S}$ defined by

$$(\hat{S}f)(x) = \hat{\psi}(x)f(x) \quad (0 \le x \le 1,\ f \in L^p(0,1))$$

with an integrable function $\hat{\psi}$ can be written in the form (11).

Furthermore, if $\inf_{a \le t \le b} |\psi(t)| > 0$, then $1/\psi(t)$ is also P_t-integrable and according to (11),

$$S^{-1} = \int_a^b \frac{1}{\psi(t)} dP_t. \tag{13}$$

Indeed, put

$$B_n = \sum_{k=1}^{n} \frac{1}{\psi(t_k)} \Delta P_k.$$

Then

$$B_n S_n = S_n B_n = \sum_{k=1}^{n} \Delta P_k = I.$$

So $B_n = S_n^{-1}$ and by (10) $\|S_n^{-1}\| \le \frac{m_P}{\inf_t |\psi(t)|} < \infty$. Since

$$S_m^{-1} - S_n^{-1} = -S_m^{-1}(S_m - S_n)S_n^{-1} \to 0 \ \ (m, n \to \infty)$$

in the operator norm, we have $S_n^{-1} \to S^{-1}$. So (13) is valid and $\|S^{-1}\| \le \frac{m_P}{\inf_t |\psi(t)|} < \infty$.

It is simple to check that

$$\sigma(S) = \{z \in \mathbf{C} : z = \psi(t), t \in [a, b]\}.$$

Let $\lambda \ne \psi(t), t \in [a, b]$. Then according to (11) and (10),

$$(S - \lambda I)^{-1} = \int_a^b \frac{1}{\psi(t) - \lambda} dP_t \tag{14}$$

and

$$\|(S - \lambda I)^{-1}\| \le \frac{m_P}{\rho(S, \lambda)} \ \ (\lambda \notin \sigma(S)). \tag{15}$$

Here and below,

$$\rho(A, \lambda) = \inf_{t \in \sigma(A)} |A - \lambda|.$$

Definition 3 Let $A \in \mathscr{B}(\mathscr{X})$, P_t be a CMPF. If $P_t A P_t = A P_t$ $(a \le t \le b)$, then P_t is said to be an invariant CMPF of A, or A has a CMPF P_t.

Definition 4 Let $A \in \mathscr{B}(\mathscr{X})$ have a CMPF P_t defined on $[a, b]$ and there be a bounded P_t-integrable function ϕ, such that

$$A = D + V, \tag{16}$$

where

$$D = \int_a^b \phi(t) dP_t, \tag{17}$$

and V is a compact quasi-nilpotent operator in $\mathcal{X}$. In addition, let the regular set of A be simply connected. Then we will say that A is a P_t-triangular operator, equality (16) is its triangular representation, D and V are the diagonal and nilpotent parts of A, respectively, and $\phi(.)$ is a P_t- *diagonal function of* A.

Note that $P_t V P_t = P_t(A - D)P_t = V P_t$ $(a \leq t \leq b)$.

According to (17), we have

$$(D - \lambda I)^{-1} = \int_a^b \frac{1}{\phi(t) - \lambda} dP_t \quad (\lambda \notin \sigma(D)). \tag{18}$$

Corollary 9 *Let A be P_t-triangular, D and V be its diagonal part and nilpotent one, respectively. Then for any regular point λ of D, the operators $VR_\lambda(D)$ and $R_\lambda(D)V$ are quasi-nilpotent ones. Besides, P_t is invariant for $VR_\lambda(D)$ and $R_\lambda(D)V$.*

Indeed, due to (18), P_t is invariant for $R_\lambda(D)$. Now Lemma 7 ensures the required result.

From Lemma 8, it follows

Corollary 10 *Let A be P_t-triangular. Then $\sigma(A) = \sigma(D)$, where D is the diagonal part of A.*

Moreover, from (10), we have

$$R_\lambda(A) = (D + V - \lambda I)^{-1} = R_\lambda(D)(I + VR_\lambda(D))^{-1} \quad (\lambda \notin \sigma(A)). \tag{19}$$

Similarly, one can check that

$$R_\lambda(A) = (I + R_\lambda(D)V)^{-1} R_\lambda(D) \quad (\lambda \notin \sigma(A)).$$

Note that

$$\rho(A, \lambda) = \inf_{s \in \sigma(A)} |s - \lambda| = \rho(D, \lambda) = \inf_{s \in \sigma(D)} |s - \lambda| = \inf_t |\phi(t) - \lambda|.$$

11 Norm Estimates for Resolvents

Definition 5 Let P_t be a CMPF in $\mathcal{X}$ and E be a linear subspace of the set of compact operators in $\mathcal{X}$ endowed with a norm $N_E(.)$ having the following property: for arbitrary P_t-scalar operators $S, S_1 \in \mathcal{B}(\mathcal{X})$, the inequality

$$N_E(SBS_1) \leq \|S_1\| \|S\| N_E(B) \quad (B \in E) \tag{20}$$

is valid. Then E will be called a P_t-subset of compact operators.

For example, let E be the set of operators B in $L^p = L^p(0,1)$ $(1 \le p < \infty)$ defined by

$$(Bh)(x) = \int_0^1 k(x,s)h(s)ds \ (h \in L^p, x \in [0,1]),$$

where $k(x,s)$ is a scalar kernel defined on $[0,1]^2$ and satisfying the condition

$$M_p(B) := [\int_0^1 [\int_0^1 |k(x,s)|^{p'} ds]^{p/p'} dx]^{1/p}$$
$$< \infty \ (1 < p < \infty, 1/p + 1/p' = 1) \tag{21}$$

or

$$M_1(B) := ess \ \sup_x \int_0^1 |k(x,s)|dx < \infty.$$

Operators satisfying condition (21) are called (p, p')-Hille–Tamarkin operators [35, p.245].

It is not hard to check that $N_E(.) = M_p(.)$ is a norm. Take $P_t = \hat{P}_t$ as in (12). Then arbitrary P_t-scalar type operators S, S_1 are the operators of the multiplication by some scalar bounded measurable functions ψ and ψ_1, respectively. In this case, we have

$$M_p(SBS_1) = \left[\int_0^1 \left[\int_0^1 |\psi(x)k(x,s)\psi_1(s)|^{p'} ds\right]^{p/p'} dx\right]^{1/p}$$
$$\le \sup_x |\psi(x)| \sup_x |\psi_1(x)| \left[\int_0^1 \left[\int_0^1 |k(x,s)|^{p'} ds\right]^{p/p'} dx\right]^{1/p}$$
$$= \|S_1\| \|S\| M_p(B).$$

So condition (20) is satisfied.

Furthermore, let us suppose that for any quasi-nilpotent operator $W \in E$, there are positive numbers θ_k $(k = 1, 2, \ldots)$ independent of W (but dependent on E), such that

$$\|W^k\| \le \theta_k N_E^k(W) \ \ (k = 1, 2, \ldots)$$

and

$$\lim_{k\to\infty} \sqrt[k]{\theta_k} = 0. \tag{22}$$

Then

$$\|(I - W)^{-1}\| = \|\sum_{k=0}^{\infty} W^k\| \le \sum_{k=0}^{\infty} \theta_k N_E^k(W) < \infty. \tag{23}$$

Now we are in a position to formulate an prove the main result of this section.

Theorem 3 *Let A be a P_t-triangular operator, whose nilpotent part V belongs to a P_t-subset of compact operators E, such that the conditions*

$$\|V^k\| \le \theta_k N_E^k(V) \ \ (k = 1, 2, \ldots),$$

(20), (21) and (22) hold. Then

$$\|R_\lambda(A)\| \le \sum_{k=0}^{\infty} \frac{m_P^{k+1}\theta_k N_E^k(V)}{\rho^{k+1}(A,\lambda)} \ \ (\lambda \notin \sigma(A)). \tag{24}$$

Proof Let D be the diagonal part of A. Due to (15),

$$\|(D - \lambda I)^{-1}\| \le \frac{m_P}{\rho(D,\lambda)} \ \ (\lambda \notin \sigma(D)). \tag{25}$$

By Corollary 9, $VR_\lambda(D)$ $(\lambda \notin \sigma(D))$ is quasi-nilpotent, and according to (20), (21) and (22),

$$\|(VR_\lambda(D))^k\| \le \theta_k N_E^k(VR_\lambda(D))) \le \theta_k \|R_\lambda(D)\|^k N_E^k(V).$$

Now (25) implies

$$\|(VR_\lambda(D))^k\| \le \frac{m^k(D)\theta_k N_E^k(V)}{\rho^k(D,\lambda)}$$

and therefore by (23),

$$\|(I + VR_\lambda(D))^{-1}\| = \|\sum_{k=0}^{\infty}(-VR_\lambda(D))^k\| \le \sum_{k=0}^{\infty} \frac{m^k(D)\theta_k N_E^k(V)}{\rho^k(D,\lambda)}.$$

Hence (19) yields

$$\|R_\lambda(A)\| \le \sum_{k=0}^{\infty} \frac{m^{k+1}(D)\theta_k N_E^k(V)}{\rho^{k+1}(D,\lambda)} \ \ (\lambda \notin \sigma(D)).$$

Taking into account that by Corollary 10 $\rho(D,\lambda) = \rho(A,\lambda)$, we arrive at the required result. □

Observe that (24) implies

$$\|R_\lambda(A)\| \leq \sum_{k=0}^{\infty} \frac{m^{k+1}(D)\theta_k N_E^k(V)}{(|\lambda| - r_s(A))^{k+1}} \quad (|\lambda| > r_s(A)),$$

where $r_s(A)$ is the (upper) spectral radius. Assume that $r_s(A) < 1$. Then

$$\|R_\lambda(A)\| \leq \frac{c_A}{|\lambda| - 1} \quad (|\lambda| > 1)$$

with

$$c_A = \sum_{k=0}^{\infty} \frac{m^{k+1}(D)\theta_k N_E^k(V)}{(1 - r_s(A))^k}.$$

We thus arrive at

Corollary 11 *Under the hypothesis of Theorem 3, let $r_s(A) < 1$. Then A is Kreiss-bounded.*

12 Perturbations of Triangularizable Operators

Let $A, \tilde{A} \in \mathscr{B}(\mathscr{X})$ and $q := \|A - \tilde{A}\|$. Recall that the spectral variation of $\tilde{A}$ with respect to A is defined in Sect. 1.

Due to the Hilbert identity $R_\lambda(\tilde{A}) - R_\lambda(A) = R_\lambda(A)(A - \tilde{A})R_\lambda(\tilde{A})$, we have

$$\|R_\lambda(\tilde{A})\| \leq \|R_\lambda(A)\| + q\|R_\lambda(A)\|\|R_\lambda(\tilde{A})\|.$$

So if a $\lambda \in \mathbf{C}$ is regular for A and

$$q\|R_\lambda(A)\| < 1, \tag{26}$$

then λ is also regular for $\tilde{A}$. Moreover,

$$\|R_\lambda(\tilde{A})\| \leq \frac{\|R_\lambda(A)\|}{1 - q\|R_\lambda(A)\|}.$$

Assume that

$$\|R_\lambda(A)\| \leq F\left(\frac{1}{\rho(A, \lambda)}\right) \quad (\lambda \notin \sigma(A)), \tag{27}$$

where $F(t)$ is a monotonically increasing non-negative continuous function of a non-negative variable, such that $F(0) = 0$ and $F(\infty) = \infty$. We need the following technical lemma.

Lemma 9 *Let $A, \tilde{A} \in \mathscr{B}(\mathscr{X})$ and condition (27) hold. Then $\mathrm{sv}_A(\tilde{A}) \le z(F, q)$, where $z(F, q)$ is the unique positive root of the equation*

$$qF(1/z) = 1.$$

For the proof, see [18, Lemma 1.10]. Now Theorem 3 implies

Corollary 12 *Let $A \in \mathscr{B}(\mathscr{X})$ satisfy the hypothesis of Theorem 3. Then for any $\tilde{A} \in \mathscr{B}(\mathscr{X})$, we have $\mathrm{sv}_A(\tilde{A}) \le z_E(A, q)$, where $z_E(A, q)$ is the unique positive root of the equation*

$$q \sum_{k=0}^{\infty} \frac{m^{k+1}(D)\theta_k N_E^k(V)}{z^{k+1}} = 1.$$

13 Powers of Volterra Operators in L^p

The results of this section originally have been particularly published in [15, Chapters 16 and 17]. Throughout this section, W is a Volterra operator in $L^p \equiv L^p(0, 1)$ $(1 \le p \le \infty)$ defined by

$$(Wh)(x) = \int_0^x K(x, s)h(s)ds \; (h \in L^p, x \in [0, 1]), \tag{28}$$

where $K(x, s)$ is a scalar kernel defined on $0 \le s \le x \le 1$ and satisfying the inequalities pointed below.

13.1 Hille–Tamarkin Volterra Operators

Let $1 < p < \infty$ and

$$M_p(W) := \left[\int_0^1 \left[\int_0^x |K(x, s)|^{p'} ds \right]^{p/p'} dx \right]^{1/p} < \infty \;\; (1/p + 1/p' = 1). \tag{29}$$

That is, W is a (p, p')-Hille–Tamarkin Volterra operator.

Lemma 10 *Under condition (29), the operator W defined by (28) satisfies the inequality*

$$|W^k|_{L^p} \le \frac{M_p^k(W)}{\sqrt[p]{k!}} \quad (k = 1, 2, \ldots).$$

For the proof, see Lemma 6.1 from [19].

13.2 Volterra Operators in L^1 and L^∞

Lemma 11 *Assume that*

$$M_1(W) := ess - \sup_{s\in[0,1]} \int_s^1 |K(x,s)|dx < \infty.$$

Then the operator defined by (26) satisfies the inequality

$$|W^k|_{L^1} \le \frac{M_1^k(W)}{k!} \quad (k = 1, 2, \ldots).$$

For the proof, see Lemma 6.2 from [19].

Repeating the arguments of the of Lemma 6.2 from [19] with $p = \infty$ instead of $p = 1$, we arrive at

Lemma 12 *Assume that*

$$M_\infty(W) := ess \sup_{x\in[0,1]} \int_0^x |K(x,s)|ds < \infty.$$

Then the operator defined by (26) satisfies the inequality

$$|W|_{L^\infty} \le \frac{M_\infty^k(W)}{k!} \quad (k = 1, 2, \ldots).$$

Various aspects of powers of Volterra operators have been considered in papers [9, 10, 27, 30, 31, 34, 41], but mainly the convolution operators, in particular the operators of the indefinite integration, have been considered.

14 Triangularizable Operators in L^p

Consider in $L^p[0,1]$ $(1 \le p < \infty)$ the operator A defined by

$$(Ah)(x) = \phi(x)h(x) + \int_x^1 k(x,s)h(s)ds \ (h \in L^p, x \in [0,1]), \tag{30}$$

where $k(x,s)$ is a scalar kernel defined on $0 \le x \le s \le 1$ and having the properties pointed below, and $\phi(x)$ is a scalar bounded Riemann-integrable function, whose values lie on an unclosed Jordan curve. The Volterra operator in (30) is assumed to be compact.

Let $\hat{P}_t$ $(0 \le t \le 1)$ be the truncation projection function, defined by (12). It is simple to check that $\hat{P}_t A \hat{P}_t = A\hat{P}_t$. Define the operators $\hat{D}$ and $\hat{V}$ by

$$(\hat{D}h)(x) = \phi(x)h(x) \text{ and } (\hat{V}h)(x) = \int_x^1 k(x,s)h(s)ds \ (h \in L^p, x \in [0,1]).$$

Then $\hat{P}_t \hat{V} \hat{P}_t = \hat{V} \hat{P}_t$ and

$$\hat{D} = \int_0^1 \phi(s) d\hat{P}_s.$$

Omitting the obvious calculations, we arrive at

Lemma 13 *Let A be defined by (30). Then it is a $\hat{P}_s$-triangular operator, its diagonal part is $\hat{D}$ and its nilpotent part is $\hat{V}$.*

Assume that either

$$M_p(\hat{V}) := \left[\int_0^1 \left[\int_x^1 |k(x,s)|^{p'} ds \right]^{p/p'} dx \right]^{1/p}$$
$$< \infty \ (1 < p < \infty, 1/p + 1/p' = 1), \tag{31}$$

or

$$M_1(\hat{V}) := \int_0^1 ess - \sup_{s \in [x,1]} |k(x,s)| dx < \infty \tag{32}$$

Recall that absolutely p-summing ($1 \le p < \infty$) operators are defined in Sect. 3.1. As it was mentioned, the set of p-summing operators in $\mathscr{X}$ with the finite norm π_p is a two-sided normed ideal in the set of bounded linear operators, which is denoted by Π_p. In addition, any (p, p')-Hille–Tamarkin operator K is a p-summing operator with $\pi_p(K) \le M_p(K)$, cf. [35, Proposition 7.2.7], [6, p. 43]

Theorem 3 and Lemmas 10 and 11 imply

$$|R_\lambda(A)|_{L^p} \le \sum_{k=0}^{\infty} \frac{M_p^k(\hat{V})}{\sqrt[p]{k!}\rho^{k+1}(A,\lambda)} \quad (1 \le p < \infty, \lambda \notin \sigma(A)), \tag{33}$$

if condition (31) or (32) holds. Besides $\rho(A,\lambda) = \rho(\hat{D},\lambda) = \inf_{0 \le x \le 1} |\phi(x) - \lambda|$.

So

$$|R_\lambda(A)|_{L^1} \le \frac{1}{\rho(A,\lambda)} \exp[\frac{M_1(\tilde{V})}{\rho(A,\lambda)}]$$

if condition (32) holds.

Let $p > 1$. Then by the Hölder inequality for any $c > 1$, we get

$$\sum_{k=0}^{\infty} \frac{c^k M_p^k(\tilde{V})x^k}{c^k \sqrt[p]{k!}} \leq \left(\sum_{k=0}^{\infty} \frac{1}{c^{kp'}}\right)^{1/p'} \left(\sum_{k=0}^{\infty} \frac{c^{kp} M_p^{pk}(\tilde{V})x^{kp}}{k!}\right)^{1/p}$$
$$= \frac{c}{(c^{p'}-1)^{1/p'}} \exp[c^p M_p^p(\tilde{V})x^p/p] \ \ (x > 0).$$

By virtue of (33), we can write

$$|R_\lambda(A)|_{L^p} \leq \frac{1}{(1-c^{-p'})^{1/p'}\rho(A,\lambda)}$$
$$\times \exp\left[\frac{c^p M_p^p(\hat{V})}{\rho^p(A,\lambda)p}\right] \quad (1/p+1/p', c > 1, \lambda \notin \sigma(A)).$$

Take $c = p^{1/p}$. Then we obtain

$$|R_\lambda(A)|_{L^p} \leq \frac{b_p}{\rho(A,\lambda)} \exp\left[\frac{M_p^p(\hat{V})}{\rho^p(A,\lambda)}\right] \quad (1 < p < \infty, \lambda \notin \sigma(A)),$$

where

$$b_p := \frac{1}{(1-p^{-p'/p})^{1/p'}}.$$

Let $q_p = |A - \tilde{A}|_{L^p}$ and $z_p(\hat{V}, q_p)$ be the unique positive root of the equation

$$q_p F_p(\hat{V}, 1/z) = 1, \tag{34}$$

where

$$F_p(\hat{V}, x) = b_p x \exp[M_p^p(\hat{V})x^p] \quad (1 < p < \infty)$$

and

$$F_1(\hat{V}, x) = x \exp[x M_1(\hat{V})] \ \ (x \geq 0).$$

Note that one can take $b_1 = 1$. Now Corollary 12 implies

Lemma 14 *Let A be defined by (30) and satisfy one of the conditions (31) or (32). Then for any* $\tilde{A} \in \mathscr{B}(\mathscr{X})$ *we have* $\mathrm{sv}_A(\tilde{A}) \leq z_p(\hat{V}, q_p)$ $(1 \leq p < \infty)$.

To estimate $z_p(\hat{V}, q_p)$, we can apply the following lemma.

Lemma 15 *The unique positive root* z_a *of the equation*

$$y\, e^y = a \quad (a = const > 0) \tag{35}$$

satisfies the inequality $z_a \geq \delta_p(a)$, *where*

$$\delta_p(a) := \begin{cases} ae^{-1} & \text{if } ca \leq e, \\ \frac{1}{2}\ln(ae) & \text{if } a \geq e. \end{cases}$$

Proof Let $a \geq e$. Then $z_a \geq 1$. By the usual calculations, the function $f(y) = \frac{e^{y-1}}{y}$ has for $y \geq 1$ a unique extremum-minimum at $y = 1$ and $f(y) \geq 1$ for $y \geq 1$. We obtain $1 \leq z_a \leq e^{z_a-1}$, and

$$a = z_a e^{z_a} \leq e^{2z_a-1} \text{ and therefore } z_a \geq \frac{1}{2}\ln(ea).$$

Now let $a \leq e$. Then $z_a \leq 1$. Thus $e^{z_a} \leq e$ and therefore, $a = z_a e^{z_a} \leq e z_a$, as claimed. □

Rewrite Eq. (34) with $p = 1$ and $z = 1/x$ as

$$q_1 M_1(\hat{V}) x \exp[x M_1(\hat{V})] = M_1(\hat{V}).$$

Then we obtain Eq. (35) with $y = x M_1(\hat{V})$ and $a = M_1(\hat{V})/q_1$. So

$$z_1(\hat{V}, q_1) = \frac{M_1(\hat{V})}{z_a}.$$

Now Lemma 15 implies $z_1(A, q) \leq \delta_1(\hat{V}, q_1)$, where

$$\delta_1(\hat{V}, q_1) := \begin{cases} q_1 e & \text{if } M_1(\hat{V}) \leq q_1 e, \\ \dfrac{2M_1(\hat{V})}{\ln\left(\frac{M_1(\hat{V})e}{q_1}\right)} & \text{if } M_1(\hat{V}) > q_1 e. \end{cases}$$

Now let $1 < p < \infty$. Then Eq. (34) with $z = 1/x$ takes the form

$$q_p b_p x \exp[x^p M_p^p(\hat{V})] = 1 \text{ or } b_p^p q_p^p x^p \exp[p x^p M_p^p(\hat{V})] = 1.$$

Therefore

$$(q_p b_p)^p p M_p^p(\hat{V}) x^p \exp[p x^p M_p^p(\hat{V})] = p M_p^p(\hat{V}).$$

Hence we obtain Eq. (35) with $y = px^p M_p^p(\hat{V})$ and

$$a = \frac{pM_p^p(\hat{V})}{(q_p b_p)^p}.$$

So

$$z_p(\hat{V}, q_p) = \frac{p^{1/p} M_p(\hat{V})}{z_a^{1/p}}.$$

From Lemma 15 it follows that $z_p(\hat{V}, q_p) \le \delta_p(\hat{V}, q_p)$, where

$$\delta_p(\hat{V}, q_p) := \begin{cases} q_p b_p e^{1/p} & \text{if } M_p(\hat{V}) \le b_p q_p (e/p)^{1/p}, \\ \dfrac{2^{1/p} M_p(\hat{V})}{\ln^{1/p}\left(\frac{(pe)^{1/p} M_p(\hat{V}) e}{q_p b_p}\right)} & \text{if } M_p(\hat{V}) > b_p q_p (e/p)^{1/p}. \end{cases}$$

Now Lemma 14 implies

Corollary 13 *Let A be defined by (30) and satisfy one of the conditions (31) or (32). Then for any $\tilde{A} \in \mathscr{B}(\mathscr{X})$, we have* $\mathrm{sv}_A(\tilde{A}) \le \delta_p(\tilde{V}, q_p)$ $(1 \le p < \infty)$.

15 Integral Operators in L^p

Throughout this section, $\tilde{A}$ is a linear operator in $L^p = L^p(0, 1)$ $(1 \le p < \infty)$ defined by

$$(\tilde{A}h)(x) = \phi(x)h(x) + \int_0^1 k(x, s)h(s)ds \ (h \in L^p, x \in [0, 1]), \tag{36}$$

where ϕ is the same as in the previous section and $k(x, s)$ is a scalar kernel defined on $[0, 1]^2$ and having the properties pointed below.

15.1 The Case $1 < p < \infty$

Let

$$\left[\int_0^1 \left[\int_0^1 |k(x, s)|^{p'} ds\right]^{p/p'} dx\right]^{1/p} < \infty \ \ (1 < p < \infty; \ 1/p + 1/p' = 1). \tag{37}$$

So k is a Hille–Tamarkin kernel. Take A as in (30): $A = \hat{D} + \hat{V}$, where $\hat{D}$ and $\hat{V}$ are the same as in the previous section. By (37)

$$\tau_p := [\int_0^1 [\int_0^x |k(x,s)|^{p'} ds]^{p/p'} dx]^{1/p} < \infty.$$

So $q_p = |A - \tilde{A}|_{L^p} \le \tau_p$. Now Lemma 14 implies

Theorem 4 *Let $\tilde{A}$ be defined by (36) and condition (37) hold. Then*

$$\sigma(\tilde{A}) \subseteq \{z \in \mathbf{C} : |\phi(x) - z| \le z_p(A, \tau_p) \le \delta_p(\hat{V}, \tau_p), x \in [0, 1]\},$$

where $z(A, \tau_p)$ is the unique positive root of the equation

$$\frac{b_p \tau_p}{z} \exp[\frac{M_p^p(\hat{V})}{z^p}] = 1$$

and

$$\delta_p(\hat{V}, \tau_p) := \begin{cases} \tau_p b_p e^{1/p} & \text{if } M_p(\hat{V}) \le b_p \tau_p (e/p)^{1/p}, \\ \dfrac{2^{1/p} M_p(\hat{V})}{\ln^{1/p}\left(\frac{(pe)^{1/p} M_p(\hat{V}) e}{\tau_p b_p}\right)} & \text{if } M_p(\hat{V}) > b_p \tau_p (e/p)^{1/p}. \end{cases}$$

This result is sharp: if $\tau_p = 0$, then we have $\sigma(\tilde{A}) = \{z \in \mathbf{C} : z = \phi(x), x \in [0, 1]\}$. From Theorem 4, it follows that

Corollary 14 *Under condition (37), the (upper) spectral radius $r_s(\tilde{A})$ of the operator $\tilde{A}$ defined by ((36) satisfies the inequalities*

$$r_s(\tilde{A}) \le \sup_x |\phi(x)| + z_p(A, \tau_p) \le \sup_x |\phi(x)| + \delta_p(\hat{V}, \tau_p).$$

If, in addition,

$$\inf_x |\phi(x)| > z_p(A, \tau_p),$$

then the lower spectral radius $r_{low}(\tilde{A}) := \inf |\sigma(\tilde{A})|$ satisfies the inequality

$$r_{low}(\tilde{A}) \ge \inf_x |\phi(x)| - z_p(A, \tau_p).$$

Moreover, if

$$\inf_x |\phi(x)| > \delta_p(\hat{V}, \tau_p),$$

then $r_{low}(\tilde{A}) \ge \inf_x |\phi(x)| - \delta_p(\hat{V}, \tau_p)$.

15.2 The Case $p = 1$

Now suppose that

$$\int_0^1 \sup_{s\in[0,1]} |k(x,s)|dx < \infty. \tag{38}$$

Then according to (30)

$$|A - \tilde{A}|_{L^1} \le \tau_1 := \int_0^1 \sup_{s\in[0,x]} |k(x,s)|dx.$$

Now Lemma 14 implies

Theorem 5 *Let $\tilde{A}$ be defined by (36) and condition (38) hold. Then*

$$\sigma(\tilde{A}) \subseteq \{z \in \mathbf{C} : |\phi(x) - z| \le z_1(A, \tau_1) \le \delta_1(\hat{V}, \tau_1), x \in [0, 1]\},$$

where $z(A, \tau_1)$ is the unique positive root of the equation

$$\frac{\tau_1}{z} \exp\left[\frac{M_1(\hat{V})}{z}\right] = 1$$

and

$$\delta_1(\hat{V}, \tau_1) := \begin{cases} \tau_1 e & \text{if } M_1(\hat{V}) \le \tau_1 e, \\ \frac{2M_1(\hat{V})}{\ln\left(\frac{M_1(\hat{V})e}{\tau_1}\right)} & \text{if } M_1(\hat{V}) > \tau_1 e. \end{cases}$$

From this theorem, we obtain the following result.

Corollary 15 *Let condition (38) hold. Then*

$$r_s(\tilde{A}) \le \sup_x |\phi(x)| + z_1(A, \tau_1) \le \sup_x |\phi(x)| + \delta_1(\hat{V}, \tau_1).$$

If, in addition,

$$\inf_x |\phi(x)| > z_1(A, \tau_1),$$

then $r_{low}(\tilde{A}) \ge \inf_x |\phi(x)| - z_1(A, \tau_1)$. Moreover, if

$$\inf_x |\phi(x)| > \delta_1(\hat{V}, \tau_1),$$

then $r_{low}(\tilde{A}) \ge \inf_x |\phi(x)| - \delta_1(A, \tau_1)$.

16 Multiplicative Representations for Resolvents of Operators in a Banach Space

In this section, we suggest a representation for the resolvent of a P_t-triangular operator. We begin with the following lemma.

Lemma 16 *Let a sequence of compact quasi-nilpotent operators* $V_n \in \mathscr{B}(\mathscr{X})$ $(n = 1, 2, \ldots)$ *converge in the operator norm to an operator* V. *Then* V *is compact and quasi-nilpotent.*

Proof From the approximation property $\mathscr{X}$, it follows that the uniform limit of compact operators is compact. So V is compact. Assume that V has an eigenvalue $\lambda_0 \neq 0$. Since V is compact, λ_0 is an isolate point of $\sigma(V)$. Consequently, there is a circle L which contains λ_0 and does not contain zero and other points of $\sigma(V)$. For $z \in L$ we have

$$\|R_z(V_n)\| - \|R_z(V)\| \leq \|R_z(V_n) - R_z(V)\| \leq \|V - V_n\| \|R_z(V_n)\| \|R_z(V)\|.$$

Hence, for sufficiently large n,

$$\|R_z(V_n)\| \leq \frac{\|R_z(V)\|}{1 - \|V - V_n\| \|R_z(V_n)\| \|R_z(V)\|}.$$

Therefore, $\|R_z(V_n)\|$ are uniformly bounded on L. Since V_n $(n = 1, 2, \ldots)$ are quasi-nilpotent operators, we have

$$\int_L R_z(V_n)dz = 0$$

and

$$\int_L R_z(V)dz = \int_L (R_z(V) - R_z(V_n))dz = \int_L R_z(V)(V - V_n)R_z(V_n)dz \to 0.$$

So $\int_L R_z(V)dz = 0$, but this is impossible, since that integral represents the eigenprojection corresponding to λ_0. This contradiction proves the lemma. □

This lemma is well known for operators in a Hilbert space [4, Lemma 17.1]).

Let $\psi(.)$ be a scalar function defined and bounded on a finite real segment $[a, b]$, Q_t be a resolution of the identity defined on $[a, b]$, $B \in \mathscr{B}(\mathscr{X})$ and

$$M_n = \prod_{1 \leq k \leq n}^{\rightarrow} (I + \psi(t_k)B\Delta Q_k)$$
$$:= (I + \psi(t_1)B\Delta Q_1)(I + \psi(t_2)B\Delta Q_2) \cdots (I + \psi(t_n)B\Delta Q_n)$$
$$(\Delta Q_k = Q_{t_k} - Q_{t_{k-1}}, a = t_0 < t_1 < \ldots < t_n = b).$$

If the sequence of operators M_n converges in the operator norm to some $M \in \mathscr{B}(\mathscr{X})$, then M is called *the right multiplicative integral*. We write

$$M = \int_{[a,b]}^{\rightarrow} (I + \psi(t) B d Q_t).$$

Lemma 17 *Let V be a compact quasi-nilpotent operator in $\mathscr{X}$ having an invariant CMPF P_t $(a \le t \le b)$, and*

$$\sum_{k=1}^{n} \Delta P_k V \Delta P_k \to 0 \; as \; n \to \infty \tag{39}$$

$$(\Delta P_k = P_{t_k} - P_{t_{k-1}}, a = t_0 < t_1 < \ldots < t_n = b)$$

in the operator norm. Then

$$(I - V)^{-1} = \int_{[a,b]}^{\rightarrow} (I + V d P_t).$$

Proof Put

$$V_n = \sum_{k=1}^{n} P_{t_{k-1}} V \Delta P_k.$$

Since

$$V = \sum_{j=1}^{n} \Delta P_j V \sum_{k=1}^{n} \Delta P_k = \sum_{k=1}^{n} \sum_{j=1}^{k} \Delta P_j V \Delta P_k,$$

we have

$$V - V_n = \sum_{k=1}^{n} \Delta P_k V \Delta P_k.$$

Due to (39), the sequence of the operators V_n tends to V in the operator norm since V is compact. Besides, V_n is nilpotent, since with the notation $P_k = P_{t_k}$, we have

$$\begin{aligned} V_n^n &= V_n^n P_n = V_n^{n-1} P_{n-1} V_n = V_n^{n-2} P_{n-2} V_n P_{n-1} V_n = \ldots \\ &= V_n P_1 \cdots V_n P_{n-1} V_n = 0. \end{aligned}$$

Due to [17, Lemma 3.14],

$$(I - V_n)^{-1} = \prod_{2\le k\le n}^{\rightarrow} (I + V_n \Delta P_k).$$

That lemma is proved in a Hilbert space, but in $\mathscr{X}$ the proof is similar. In addition, $(I - V_n)^{-1} \to (I - V)^{-1}$ in the operator norm, cf. [8, p. 585, Lemma VII.6.3]. Hence the required result follows. □

Theorem 6 *Let A be a P_t-triangular operator. Then*

$$(A - \lambda I)^{-1} = \int_{[a,b]} \frac{dP_\tau}{\phi(\tau) - \lambda} \int_{[a,b]}^{\rightarrow} \left(I + \frac{V dP_t}{\phi(t) - \lambda}\right) \quad (\lambda \notin \sigma(A)),$$

where V is the nilpotent part of A and $\phi(.)$ is its P_t-diagonal function.

Proof Due to Corollary 9, $V(D - \lambda I)^{-1}$ is quasi-nilpotent. By the previous lemma

$$(I + V(D - \lambda I)^{-1})^{-1} = \int_{[a,b]}^{\rightarrow} (I + V(D - \lambda I)^{-1} dP_t).$$

According to (19), we have

$$(A - \lambda I)^{-1} = (D - \lambda I)^{-1} \int_{[a,b]}^{\rightarrow} (I + V(D - \lambda I)^{-1} dP_t) \quad (\lambda \notin \sigma(A)).$$

But

$$(D - \lambda I)^{-1} = \int_{[a,b]} \frac{dP_\tau}{\phi(\tau) - \lambda} \text{ and therefore } (D - \lambda I)^{-1} dP_t = \frac{1}{\phi(t) - \lambda} dP_t.$$

This yields the required result. □

Note that in Sect. 10 of the paper [19], in the case of a Hilbert space, the multiplicative representation for the resolvents of non-selfadjoint operators having maximal chains and Schatten-von Neumann Hermitian components has been derived. Besides, we do not assume that the chain is continuous. That representation generalizes the corresponding result from [18, Section 9.9].

Note that in paper [13], the representation of the resolvent of such operators via the spectral measure has been suggested without the proof. In [15, Chapter 10] and [17, Section 9.9], short proofs of the main result from [13] are given. In Sect. 10, we considerably refine the just mentioned results from [15] and [17].

References

1. Bhatia, R.: Perturbation Bounds for Matrix Eigenvalues, Classics in Applied Mathematics, Vol. 53, SIAM, Philadelphia (2007).
2. Boiti, M., Pempinelli, F. and Pogrebkov, A.K.: On the extended resolvent of the nonstationary Schrödinger operator for a Darboux transformed potential. J. Phys. A, Math. Gen. **39**, no. 8, 1877–1898 (2006).
3. Branges, L. de : Some Hilbert spaces of analytic functions II, J. Math. Analysis and Appl., **11** 44–72 (1965).
4. Brodskii, M.S.: Triangular and Jordan Representations of Linear Operators, Transl. Math. Monogr., Vol. 32, Amer. Math. Soc., Providence, R. I. (1971).
5. Chabi, G. Bio, S., Durand, G. and Goudjo, C.: Singularities of the resolvent at the thresholds of a stratified operator: a general method. Math. Methods Appl. Sci. **27**, no. 10, 1221–1239, (2004).
6. Diestel, J., Jarchow, H. and Tonge, A.: *Absolutely Summing Operators*, Cambridge University Press, Cambridge, (1995).
7. Dunford, N. and Schwartz J.T.: Linear Operators, part II. Spectral Theory Interscience, New York, (1963.
8. N. Dunford, and J.T. Schwartz :, Linear Operators, part III, Spectral Operators, Wiley-Interscience Publishers, Inc., New York, (1971.
9. S.P. Eveson, : Norms of iterates of Volterra operators on L^2. J. Operator Theory **50**, no. 2, 369–386 (2003).
10. S.P. Eveson, Asymptotic behaviour of iterates of Volterra operators on $L^p(0, 1)$. Integr. Equ. Oper. Theory **53**, 331–341 (2005).
11. H. Falomir, M.A. Muschietti and Pisani, P.A.G.: On the resolvent and spectral functions of a second order differential operator with a regular singularity. J. Math. Phys. **45**, no. 12, 4560–4577 (2004).
12. Garling D.J. :, Inequalities. A Journey into Linear Analysis, Cambridge: Cambridge University Press (2007), .
13. Gil', M.I.: On the representation of the resolvent of a nonselfadjoint operator by the integral with respect to a spectral function, Soviet Math. Dokl., **14**, 1214–1217 (1973).
14. Gil', M.I.: On an estimate for resolvents of nonselfadjoint operators which are "near" to selfadjoint and to unitary ones, Mathematical Notes, **33**, 81–84 (1983).
15. Gil', M.I.: Operator Functions and Localization of Spectra, Lecture Notes In Mathematics vol. 1830, Springer-Verlag, Berlin, (2003).
16. Gil', M.I.: Inequalities of the Carleman type for Neumann-Schatten operators, Asian-European J. of Math., **1**, no. 2, 203–212 (2008).
17. Gil, M.I.: Bounds for Determinants of Linear Operators and Their Applications, CRC Press, Taylor & Francis Group, London (2017).
18. Gil, M.I.: Operator Functions and Operator Equations, World Scientific, New Jersey, (2018).
19. Gil, M.I.: Norm estimates for resolvents of linear operators in a Banach space and spectral variations, Adv. Oper. Theory 4, no. 1, 113–139 (2019).
20. Gil, M.I.: Spectrum perturbations of compact operators in a Banach space Open Math. 17:1025–1034 (2019).
21. Gil, M.I.: An inequality between resolvents and determinants for operators in a Banach space, Annals of Functional Analysis, online from O1 January (2020).
22. Gohberg, I.C., Goldberg S. and Kaashoek M.A.: Classes of Linear Operators, Vol. 2, Birkhäuser Verlag, Basel (1993).
23. Gohberg, I.C., Goldberg S. and Krupnik N.: Traces and Determinants of Linear Operators, Birkhäuser Verlag, Basel (2000).
24. I.C. Gohberg, and M.G. Krein :, Introduction to the Theory of Linear Nonselfadjoint Operators, Trans. Mathem. Monographs, Vol. 18, Amer. Math. Soc., R. I. (1969).

25. I.C. Gohberg and M.G. Krein: Theory and Applications of Volterra Operators in a Hilbert Space, Trans. Mathem. Monographs, v. 24, Amer. Math. Soc., R. I, (1970).
26. Kato, T. : Perturbation Theory for Linear Operators, Berlin: Springer-Verlag, (1980).
27. D. Kershaw : Operator norms of powers of the Volterra operator, J. Integral Equations Appl. **11** , 351–362 (1999).
28. König, H.: Eigenvalue Distribution of Compact Operators, Operator Theory: Advances and Applications, Birkhäuser, Basel, (1986).
29. S. Kupin, and S. Treil : Linear resolvent growth of a weak contraction does not imply its similarity to a normal operator. Ill. J. Math. **45**, no.1, 229–242 (2001).
30. Lao, N. and Whitley, R.: Norms of powers of the Volterra operator, Integr. Equ. Oper. Theory **27** 419– 425 (1997).
31. Little, G. and Reade J.B.: Estimates for the norm of the n-th indefinite integral, Bull. London Math. Soc. **30**, 539–542 (1998).
32. Lizama C. and Poblete V.: On multiplicative perturbations of integral resolvent families, Math. Anal. Appl, **327**, 1335–1359, (2007).
33. Locker J.:, Spectral Theory of Nonselfadjoint Two Point Differential Operators. Amer. Math. Soc, Mathematical Surveys and Monographs, Volume 73, R.I. (1999).
34. Montes-Rodriguez, A. Sanchez-Alvarez, J. and Zemanek J. :, Uniform Abel–Kreiss boundedness and the extremal behaviour of the Volterra operator, Proc. London Math. Soc. (3) **91** 761–788 (2005).
35. Pietsch, A. : Eigenvalues and s-Numbers. Cambridge Univesity Press, Cambridge, (1987).
36. Radjavi, H. and Rosenthal, P.: Invariant Subspaces, Springer-Verlag, Berlin, (1973).
37. Rassias, Th.M. and Zagrebnov, V.A. (eds.), Analysis and Operator Theory. Dedicated in Memory of Tosio Kato's 100th Birthday. Foreword by Barry Simon, Springer, 2019.
38. Sakhnovich, L.: $(S + N)$-triangular operators: spectral properties and important examples, Math. Nachr. **289**, no. 13, 1680–1691 (2016).
39. Stewart, G.W. and Ji-guang Sun: Matrix Perturbation Theory, Academic Press, New York (1990).
40. Strikwerda J.C. and Wade B. A.:, A survey of the Kreiss matrix theorem for power bounded families of matrices and its extensions, Linear Operators (ed. J. Janas, F. H. Szafraniec and J. Zemanek), Banach Center Publications 38 (Institute of Mathematics, Polish Academy of Science, Warsaw, 339–360, (1997).
41. Thorpe B.:, The norm of powers of the indefinite integral operator on (0, 1), Bull. London Math. Soc. **30** 543–548 (1998).

Perturbations of Operator Functions: A Survey

Michael Gil'

Abstract The chapter is a survey of the recent results of the author on the perturbations of operator-valued functions. A part of the results presented in this chapter is new. Let A and $\tilde{A}$ be bounded linear operators in a Banach space $\mathscr{X}$ and $f(.)$ be a function analytic on neighborhoods of spectra of A and $\tilde{A}$. The chapter is devoted to norm estimates for $\Delta A = f(A) - f(\tilde{A})$ under various assumptions on functions and operators. In particular, we consider perturbations of entire operator-valued functions and Taylor series whose arguments are bounded operators in a Banach space. In the case of the separable Hilbert space, we derive a sharp perturbation bound for the Hilbert–Schmidt norm of Δf, provided $A - \tilde{A}$ is a Hilbert–Schmidt operator and the function is regular on the convex hull of the spectra A and $\tilde{A}$. In addition, operator functions in a Hilbert lattice are explored. Besides, two-sided estimates for $f(A)$ are established. These estimates enable us to obtain positivity conditions for functions of a given operator and of the perturbed one. As examples of concrete functions, we consider the operator fractional powers and operator logarithm. Moreover, applications of our results to infinite matrices and integral operators are discussed.

AMS (MOS) Subject Classification 47A56, 47A55, 47A60, 46B42, 47G10

1 Introduction

This chapter is a survey of the recent results of the author on the perturbations of operator-valued functions. Operator functions arise in numerous theoretical and practical applications, in particular, in the theory of differential and difference equations [1, 8, 17]. Besides, the solutions of autonomous linear ordinary differential equations can be represented by operator exponentials, and the solutions of

M. Gil' (✉)
Department of Mathematics, Ben Gurion University of the Negev, Beer-Sheva, Israel
e-mail: gilmi@bezeqint.net

I. N. Parasidis et al. (eds.), *Mathematical Analysis in Interdisciplinary Research*, Springer Optimization and Its Applications 179,
https://doi.org/10.1007/978-3-030-84721-0_17

autonomous linear difference equations can be represented by the operator powers. Estimating of an operator function is not always an easy task. In many cases, it is easier to obtain the norm of a function of a nearby operator and then to obtain the information about the function of the original operator.

Throughout this chapter, $\mathscr{X}$ is a Banach space with a norm $\|.\|$ and the unit operator $I = I_{\mathscr{X}}$. By $\mathscr{B}(\mathscr{X})$ we denote the algebra of all bounded linear operators in $\mathscr{X}$. For an $A \in \mathscr{B}(\mathscr{X})$, $\|A\|$ is the operator norm, A^{-1} is the inverse operator, A^* is the adjoint one, $\sigma(A)$ is the spectrum, and $R_\lambda(A) = (A - \lambda I)^{-1}$ $(\lambda \notin \sigma(A))$ is the resolvent.

Recall the definition of a function of $A \in \mathscr{B}(\mathscr{X})$ analytic on a neighborhood of the spectrum. To this end, denote by $\mathscr{F}(A)$ the family of all functions that are analytic on some neighborhood of $\sigma(A)$. (The neighborhood need not be connected.)

Definition 1 Let $f \in \mathscr{F}(A)$, and let U be an open set whose boundary L consists of a finite number of rectifiable Jordan curves, oriented in the positive sense customary in the theory of complex variables. Suppose that $\sigma(A) \subset U$ and that $U \cup L$ is contained in the domain of analyticity of f. Then, the operator $f(A)$ is defined by the equation

$$f(A) = -\frac{1}{2\pi i} \int_L f(z) R_z(A) dz.$$

The following result is well known, cf. Theorem VII.3.10 from [11].

Theorem 1 *If f and f_1 are in $\mathscr{F}(A)$, and α and β are complex numbers, then*

(a) $\alpha f + \beta f_1 \in \mathscr{F}(A)$ *and* $\alpha f(A) + \beta f_1(A) = (\alpha f + \beta f_1)(A)$.

(b) $f \cdot f_1 \in \mathscr{F}(A)$ *and* $f(A) f_1(A) = (f \cdot f_1)(A)$.

(c) *If f has the power series expansion*

$$f(z) = \sum_{k=0}^{\infty} c_k z^k,$$

valid in a neighborhood of $\sigma(A)$, then

$$f(A) = \sum_{k=0}^{\infty} c_k A^k.$$

The perturbation theory of operator functions in a Hilbert space has been developed in the works of M. Birman and M. Solomyak [3], K, Boyadzhiev [4], V. Matsaev [36], V. Peller [40], and other mathematicians. In particular, the remarkable results of Birman and Solomyak on double operator integrals reflected in [3] allow us to establish bounds for the norm of $f(A) - f(\tilde{A})$ in the case when A and $\tilde{A}$ are selfadjoint and $A - \tilde{A}$ belongs to some "nice" ideal. Besides, A and $\tilde{A}$ may be unbounded. The paper [45] should be mentioned; it deals with a trace class perturbation of a normal operator with the spectrum on a smooth curve. The results of that paper can be applied to perturbation theory, scattering theory, functional models, and others. The interesting inequality for $f(A) - f(\tilde{A})$ was derived in [4] under the assumption that f is a holomorphic function admitting certain integral representation. Some works are devoted to perturbations of concrete functions, such as the exponential function, sine and cosine operator functions [39].

Certainly, we could not survey here the whole subject and refer the reader to the above listed publications and references given therein. It should be noted that in the mentioned publications mainly it is assumed that A and $\tilde{A}$ are selfadjoint or normal operators in a Hilbert space. At the same time, below, we do not suppose that A and $\tilde{A}$ are selfadjoint or normal. For the simplicity, we have restricted ourselves by bounded operators, although in the appropriate situations our results can be directly extended to unbounded operators.

The chapter consists of 15 sections.

In Sect. 2, we have collected norm estimates for resolvents of various operators in a Hilbert space. Recall that Carleman in the 1930s obtained an estimate for the norm of the resolvent of operators belonging to the Neumann–Schatten ideal, cf. [10, p. 1038]. That estimate has been refined and extended to some classes of noncompact operators in [16, 27]. In this chapter, the mentioned estimates for resolvents are systematically applied to perturbation problems.

In Sect. 3, we present norm estimates for operator functions in a Hilbert space regular on the convex hull of the spectrum. Recall that in the book [12], I.M. Gel'fand and G.E. Shilov have established an estimate for the norm of a matrix-valued function in connection with their investigations of partial differential equations. However, that estimate is not sharp, it is not attained for any matrix. The problem of obtaining a precise estimate for the norm of a matrix function has been repeatedly discussed in the literature, cf. [8]. In the paper [13], the author has derived a sharp estimate for matrix-valued functions regular on the convex hull of the spectrum. It is attained for normal matrices. The results of the paper [13] were generalized to various classes of operators, cf. [16, 27].

Obviously, functions having singular points can be nonregular on the convex hull of the spectrum. But such functions, in particular, the logarithm, fractional powers, and meromorphic functions of operators, arise in many applications, cf. [5, 30, 42]. In Sect. 4, we extend some results from Sect. 3 to functions nonregular on the convex hull of the spectrum.

Sections 5 and 6 are devoted to perturbations of Taylor series whose arguments are operators in a Banach space.

In Sect. 7, we investigate entire operator-valued functions in a Banach space.

In Sect. 8, we consider perturbations of functions regular on the convex hull of the spectrum of a non-selfadjoint operator and derive a sharp perturbation bound for the Hilbert–Schmidt norm.

Section 9 is devoted to perturbations of analytic functions of infinite matrices.

Sections 10, 11, and 12 deal with operator functions in a Hilbert lattice. Besides, two-sided estimates are derived for a class of operator functions. These estimates enable us to obtain positivity conditions for operator functions.

Section 13 deals with perturbations of operators in a Hilbert lattice considered in Sect. 10.

As examples of concrete functions, in Sects. 14 and 15, we consider the fractional powers and logarithm, respectively.

The operator logarithm arises in numerous applications; in particular, its importance can be ascribed to it being the inverse function of the operator exponential. Moreover, if we consider a vector differential equation with a T-periodic operator, then according to the Floquet theory, its Cauchy operator $U(t)$ is equal to $V(t)e^{\Gamma t}$, where $V(t)$ is a T-periodic operator and $\Gamma = \frac{1}{T}\ln\ U(T)$, cf. [8]. The problems connected with the operator logarithms continue to attract the attention of many specialists, cf. the interesting recent papers [6, 43] and the references given therein. In particular, the paper [43] investigates the conditions under which the considered logarithm exists, is unique, and belongs to a particular class of operators. Moreover, the real Schur decomposition is used to compute the logarithm.

2 Norm Estimates for Resolvents of Operators in a Hilbert Space

Let $\mathscr{H}$ be a separable Hilbert space with a scalar product $\langle .,. \rangle$ and the norm $\|.\| = \sqrt{\langle .,. \rangle}$, and $\mathscr{B}(\mathscr{H})$ is the algebra of all bounded linear operators in $\mathscr{H}$. In this section, we have collected norm estimates for the resolvents of some classes of operators in $\mathscr{H}$, which will be used below.

For a compact operator $A \in \mathscr{B}(\mathscr{H})$, $\lambda_k(A)$ $(k = 1, 2, \ldots)$ are the eigenvalues of A taken with their multiplicities and ordered in the non-increasing way of their absolute values. $s_k(A)$ $(k = 1, 2, \ldots)$ are the singular numbers (i.e., the eigenvalues of $(A^*A)^{1/2}$), taken with their multiplicities and ordered in the decreasing way.

2.1 Properties of Singular Numbers

Throughout the rest of this section, *A and B are compact operators in* $\mathscr{H}$. The following results are well known, cf. [28, Section IV.4] and [29, Section II.4.2].

Lemma 1 *If C and D are bounded linear operators in* $\mathscr{H}$, *then*

$$s_k(CAD) \le \|C\|\|D\|s_k(A) \quad (k \ge 1).$$

Moreover,

$$\sum_{k=1}^{j} s_k(A+B) \leq \sum_{k=1}^{j} (s_k(A) + s_k(B)),$$

$$\sum_{k=1}^{j} s_k^p(AB) \leq \sum_{k=1}^{j} s_k^p(A) s_k^p(B) \ \ (p \geq 1)$$

and

$$\prod_{k=1}^{j} s_k(AB) \leq \prod_{k=1}^{j} s_k(A) s_k(B) \ \ (j = 1, 2, \ldots).$$

Recall that A is said to be normal if $AA^* = A^*A$.

Lemma 2 (Weyl's Inequalities) *The inequalities*

$$\prod_{j=1}^{k} |\lambda_j(A)| \leq \prod_{j=1}^{k} s_j(A)$$

and

$$\sum_{j=1}^{k} |\lambda_j(A)| \leq \sum_{j=1}^{k} s_j(A) \quad (k = 1, 2, \ldots)$$

are true. They become equalities if and only if A is normal.

For the proof, see Theorem IV.3.1 and Corollary IV.3.4 from [28] or Section II.3.1 from [29].

The set of compact operators $A \in \mathscr{B}(\mathscr{H})$ satisfying the condition

$$N_p(A) = \left[\sum_{k=1}^{\infty} s_k^p(A) \right]^{1/p} < \infty$$

for some $p \in [1, \infty)$ is called *the Schatten–von Neumann ideal and is denoted by* SN_p. $N_p(.)$ is called the Schatten–von Neumann p-norm. Besides, SN_1 is the ideal of *nuclear operators (the Trace class)*, SN_2 is the ideal of *Hilbert–Schmidt operators*, and $N_2(A)$ is the Hilbert–Schmidt norm.

From Lemma 1, we have

$$N_p(DAC) \leq N_p(A)\|D\|\|C\| \quad (A \in \mathrm{SN}_p; C, D \in \mathscr{B}(\mathscr{H})).$$

The following propositions are true (the proofs can be found, for instance, in the books [29, Section III.7] and [10]).

Lemma 3 *If* $A \in \mathrm{SN}_p$ *and* $B \in \mathrm{SN}_q$ $(1 < p, q < \infty)$, *then* $AB \in \mathrm{SN}_s$ *with* $1/s = 1/p + 1/q$. *Moreover,* $N_s(AB) \leq N_p(A)N_q(B)$.

Let $\{e_k\}$ be an orthogonal normal basis in $\mathscr{H}$, and the series

$$\sum_{k=1}^{\infty} \langle Ae_k, e_k \rangle$$

converges. Then, the sum of this series is called *the trace of* A. It is well known that

$$N_p(A) = \sqrt[p]{\text{trace } (AA^*)^{p/2}}.$$

The Schatten–von Neumann p-norms are non-increasing in p. In other words, for $1 \leq p \leq s$, we have $N_p(A) \geq N_s(A)$, provided $A \in \mathrm{SN}_p$.

The Schatten–von Neumann norm is unitarily invariant. This means that $N_p(UAU_1) = N_p(A)$ for any choice of linear unitary operators U and U_1.

For all $p, q \in (1, \infty)$, satisfying the equation $\frac{1}{p} + \frac{1}{q} = 1$, we have

$$N_p(A) = \sup \{\text{trace}\,(AB) : B \in \mathrm{SN}_q, N_q(B) \leq 1\}$$

and

$$|\text{trace}\,(AB)| \leq N_p(A)N_q(B).$$

From the Weyl inequalities, it directly follows

Corollary 1 *Let* $A \in \mathrm{SN}_p$, $1 \leq p < \infty$. *Then,*

$$\sum_{j=1}^{\infty} |\lambda_j(A)|^p \leq N_p^p(A).$$

2.2 *The Resolvent of a Hilbert–Schmidt Operator*

Let A be a Hilbert–Schmidt operator, i.e., $A \in \mathrm{SN}_2$. The following quantity plays an essential role in the sequel:

$$g(A) = \left[N_2^2(A) - \sum_{k=1}^{\infty} |\lambda_k(A)|^2 \right]^{1/2}.$$

Since

$$\sum_{k=1}^{\infty} |\lambda_k(A)|^2 \geq |\sum_{k=1}^{\infty} \lambda_k^2(A)| = |\text{trace } A^2|,$$

one can write

$$g^2(A) \leq N_2^2(A) - |\text{trace } A^2|. \tag{1}$$

If A is a normal Hilbert–Schmidt operator, then $g(A) = 0$, since

$$N_2^2(A) = \sum_{k=1}^{\infty} |\lambda_k(A)|^2$$

in this case.

Let $A_I = (A - A^*)/2i$. Due to Corollary 7.2 from [27],

$$g^2(A) = 2N_2^2(A_I) - 2\sum_{k=1}^{\infty} (\text{Im } \lambda_k(A))^2 \leq 2N_2^2(A_I) \tag{2}$$

for any $A \in \text{SN}_2$. Let $\rho(A, \lambda)$ be the distance between $\sigma(A)$ and a point $\lambda \in \mathbf{C}$:

$$\rho(A, \lambda) := \inf_{t \in \sigma(A)} |\lambda - t|.$$

Theorem 2 ([27, Theorem 7.1]) *Let A be a Hilbert–Schmidt operator. Then, the inequalities*

$$\|R_\lambda(A)\| \leq \sum_{k=0}^{\infty} \frac{g^k(A)}{\rho^{k+1}(A, \lambda)\sqrt{k!}} \tag{3}$$

and

$$\|R_\lambda(A)\| \leq \frac{1}{\rho(A, \lambda)} \exp\left[\frac{1}{2} + \frac{g^2(A)}{2\rho^2(A, \lambda)}\right] \quad (\lambda \notin \sigma(A)) \tag{4}$$

are true.

This theorem is sharp: if A is a normal operator, then $\|R_\lambda(A)\| = \frac{1}{\rho(A,\lambda)}$ and $g(A) = 0$. So, (3) is attained if we take $0^0 = 1$.

2.3 The Resolvent of a Schatten–von Neumann Operator

Theorem 3 *For some integer $p \geq 2$, let*

$$A \in \mathrm{SN}_{2p}. \tag{5}$$

Then,

$$\|R_\lambda(A)\| \leq \sum_{m=0}^{p-1} \sum_{k=0}^{\infty} \frac{(2N_{2p}(A))^{pk+m}}{\rho^{pk+m+1}(A,\lambda)\sqrt{k!}} \ (\lambda \notin \sigma(A)). \tag{6}$$

In addition,

$$\|R_\lambda(A)\| \leq \sqrt{e} \sum_{m=0}^{p-1} \frac{(2N_{2p}(A))^m}{\rho^{m+1}(A,\lambda)} \exp\left[\frac{(2N_{2p}(A))^{2p}}{2\rho^{2p}(A,\lambda)}\right] \quad (\lambda \notin \sigma(A)). \tag{7}$$

The proof of this theorem can be found in [27, Theorems 7.2 and 7.3].

Note that if (5) holds, then

$$A^p \text{ is a Hilbert–Schmidt operator.}$$

Use the identity

$$A^p - I\lambda^p = (A - I\lambda) \sum_{k=0}^{p-1} A^k \lambda^{p-k-1} = (A - I\lambda) T_{\lambda,p} \ \ (\lambda^p \notin \sigma(A^p)),$$

where

$$T_{\lambda,p} = \sum_{k=0}^{p-1} A^k \lambda^{p-k-1}.$$

Hence,

$$(A - I\lambda)^{-1} = T_{\lambda,p}(A^p - I\lambda^p)^{-1}.$$

Thus,

$$\|(A - I\lambda)^{-1}\| \leq \|T_{\lambda,p}\| \ \|(A^p - I\lambda^p)^{-1}\|.$$

Applying inequality (3) to the expression $(A^p - I\lambda^p)^{-1} = R_{\lambda^p}(A^p)$, we obtain

$$\|R_{\lambda^p}(A^p)\| \leq \sum_{k=0}^{\infty} \frac{g^k(A^p)}{\rho^{k+1}(A^p, \lambda^p)\sqrt{k!}} \quad (\lambda^p \notin \sigma(A^p)),$$

where

$$\rho(A^p, \lambda^p) = \inf_{t \in \sigma(A)} |t^p - \lambda^p|.$$

This implies

$$\|R_\lambda(A)\| \leq \|T_{\lambda,p}\| \sum_{k=0}^{\infty} \frac{g^k(A^p)}{\rho^{k+1}(A^p, \lambda^p)\sqrt{k!}} \quad (\lambda^p \notin \sigma(A^p)).$$

Similarly, making use of inequality (4), under condition (5), we obtain

$$\|R_\lambda(A)\| \leq \frac{\sqrt{e}\|T_{\lambda,p}\|}{\rho(A^p, \lambda^p)} \exp\left[\frac{g^2(A^p)}{2\rho^2(A^p, \lambda^p)}\right] \quad (\lambda^p \notin \sigma(A^p)).$$

2.4 The Resolvent of an Operator with a Hilbert–Schmidt Component

In this subsection, we suggest a norm estimate for the resolvent under the conditions

$$A \in \mathscr{B}(\mathscr{H}) \text{ and } A_I := (A - A^*)/(2i) \in \mathrm{SN}_2. \tag{8}$$

To this end, introduce the quantity

$$g_I(A) := \sqrt{2}\left[N_2^2(A_I) - \sum_{k=1}^{\infty} (\mathrm{Im}\hat{\lambda}_k(A))^2\right]^{1/2},$$

where $\hat{\lambda}_k(A)$ are the nonreal eigenvalues of A taken with their multiplicities and ordered in the following way: $|\mathrm{Im}\,\hat{\lambda}_{k+1}(A)| \leq |\mathrm{Im}\,\hat{\lambda}_k(A)|$ $(k = 1, 2, \ldots)$.

Obviously, $g_I(A) \leq \sqrt{2}N_2(A_I)$.

Theorem 4 ([27, Theorem 9.1]) *Let the conditions (8) hold. Then,*

$$\|R_\lambda(A)\| \leq \sum_{k=0}^{\infty} \frac{g_I^k(A)}{\rho^{k+1}(A, \lambda)\sqrt{k!}} \tag{9}$$

and

$$\|R_\lambda(A)\| \leq \frac{\sqrt{e}}{\rho(A,\lambda)} \exp\,[\,\frac{g_I^2(A)}{2\rho^2(A,\lambda)}]\quad (\lambda \notin \sigma(A)). \tag{10}$$

2.5 *The Resolvent of an Operator with a Schatten–von Neumann Component*

Now, assume that

$$A_I = (A - A^*)/2i \in \mathrm{SN}_{2p} \text{ for an integer } p \geq 2. \tag{11}$$

Put

$$\tau_p(A) = (1 + b_{2p})(N_{2p}(A_I) + N_{2p}(D_I)),$$

where b_{2p} is a constant defined in [27, Section 9.4] and dependent on p, only. As is shown in [27, Section 9.5, formulas (27) and (28)], in the general case, we have

$$\tau_p(A) \leq 2(1 + 2p)N_{2p}(A_I).$$

If A has a real spectrum, then

$$\tau_p(A) \leq (1 + 2p)N_{2p}(A_I).$$

Theorem 5 ([27, Theorem 9.5]) *Let $A \in \mathscr{B}(\mathscr{H})$ satisfy condition (11). Then,*

$$\|R_\lambda(A)\| \leq \sum_{m=0}^{p-1}\sum_{k=0}^{\infty} \frac{\tau_p^{pk+m}(A)}{\rho^{pk+m+1}(A,\lambda)\sqrt{k!}} \tag{12}$$

and

$$\|R_\lambda(A)\| \leq \sqrt{e}\sum_{m=0}^{p-1} \frac{\tau_p^m(A)}{\rho^{m+1}(A,\lambda)} \exp\left[\frac{\tau_p^{2p}(A)}{2\rho^{2p}(A,\lambda)}\right]\quad (\lambda \notin \sigma(A)). \tag{13}$$

For the norm estimates for the resolvent of compactly perturbed unitary and unbounded operators, see [27, Section 9.7] and [27, Chapter 11], respectively.

Recall the Hilbert identity

$$R_z(A) - R_z(\tilde{A}) = R_z(A)(\tilde{A} - A)R_z(\tilde{A})\quad (A, \tilde{A} \in \mathscr{B}(\mathscr{X}), z \notin \sigma(A) \cup \sigma(\tilde{A})),$$

and put $q = \|A - \tilde{A}\|$. If

$$q\|R_z(A)\| < 1,$$

then we have

$$\|R_z(\tilde{A})\| \leq \frac{\|R_z(A)\|}{1 - q\|R_z(A)\|}. \tag{14}$$

Definition 1 and (14) enable us to investigate perturbations of functions of operators via resolvents; namely, the following result is valid.

Lemma 4 *Let L be a boundary of an open set U, $\sigma(A) \cup \sigma(\tilde{A}) \subset U$, and $U \cup L$ is contained in the domain of analyticity of f. If, in addition,*

$$q \sup_{z \in L} \|R_z(A)\| < 1,$$

then

$$\|f(A) - f(\tilde{A})\| \leq \frac{q}{2\pi} \frac{\sup_{z \in L} \|R_z(A)\|^2}{(1 - q \sup_{z \in L} \|R_z(A)\|)} \int_L |f(z)||dz|.$$

Proof By the Hilbert identity, we have

$$\|f(A) - f(\tilde{A})\| \leq \frac{1}{2\pi} \int_L |f(z)|\|R_z(A)(A - \tilde{A})R_z(\tilde{A})\| \, |dz|$$

$$\leq q \sup_{z \in L} \|R_z(A)\| \sup_{z \in L} \|R_z(\tilde{A}))\| \frac{1}{2\pi} \int_L |f(z)||dz|.$$

Now, (14) yields

$$\sup_{z \in U} \|R_z(\tilde{A})\| \leq \frac{\sup_{z \in L} \|R_z(A)\|}{1 - q \sup_{z \in L} \|R_z(A)\|}.$$

This proves the lemma. □

2.6 *Resolvents of Finite-Dimensional and Nuclear Operators*

Finite-Dimensional Operators

Let $\mathbf{C}^n$ be the n-dimensional complex Euclidean space with a scalar product $(.,.)$ and the norm $\|.\| = \sqrt{(.,.)}$, $\mathbf{C}^{n \times n}$ is the set of $n \times n$-matrices, $\|A\|$ denotes the spectral norm of $A \in \mathbf{C}^{n \times n}$, i.e., the norm operator with respect to the Euclidean vector norm, and $N_2(A) = (\text{trace}\,(AA^*)^{1/2})$ is the Hilbert–Schmidt (Frobenius) norm. Recall that $s_1(A), \ldots, s_n(A)$ are the singular numbers of A taken with their multiplicities and enumerated in the non-increasing order.

Lemma 5 *Let $V \in \mathbf{C}^{n\times n}$ be a nilpotent matrix. Then,*

$$\|(I-V)^{-1}\| \le \prod_{k=1}^{n-1}(1+s_k(V)).$$

Proof Put $M=(I-V)^*(I-V)$. Obviously, $s_j(M)=s_j^2(I-V)$. Take into account that

$$\|M^{-1}\| = \frac{1}{s_n(M)} = \frac{s_1(M)\cdots s_{n-1}(M)}{s_1(M)s_2(M)\cdots s_n(M)} = \frac{s_1(M)\cdots s_{n-1}(M)}{\det(M)}.$$

Clearly, $\det(M)=\det(I-V)^*(I-V)=\det(I-V)^*\det(I-V)=1$. Consequently, $\|M^{-1}\|\le s_1(M)\cdots s_{n-1}(M)$, but $M^{-1}=(I-V)^{-1}((I-V)^*)^{-1}$, and therefore

$$\|((I-V)^*)^{-1}(I-V)^{-1}\| = \|(I-V)^{-1}\|^2 \le s_1(M)\cdots s_{n-1}(M),$$

but

$$s_1(M)\cdots s_{n-1}(M) = s_1^2(I-V)\cdots s_{n-1}^2(I-V).$$

Thus,

$$\|(I-V)^{-1}\| \le s_1(I-V)\cdots s_{n-1}(I-V) \le \prod_{k=1}^{n-1}(1+s_k(V)),$$

as claimed. □

According to the classical Schur theorem (see, for instance, [27, p. 44]), the triangular representation $A = D+V \quad (\sigma(A)=\sigma(D))$ is valid with a normal (diagonal) operator D and a nilpotent operator V. In addition, D and V have the joint invariant subspaces. Besides, D and V are called the diagonal part and nilpotent part of A, respectively.

Lemma 6 *Let V be the nilpotent part of $A \in \mathbf{C}^{n\times n}$. Then,*

$$\|R_\lambda(A)\| \le \frac{1}{\rho(A,\lambda)}\prod_{k=1}^{n-1}\left(1+\frac{s_k(V)}{\rho(A,\lambda)}\right) \quad (\lambda \notin \sigma(A)),$$

where $\rho(A,\lambda)=\min_k|\lambda-\lambda_k(A)|$.

Proof Due to the triangular representation $A=D+V$, we can write

$$A - \lambda I = D + V - \lambda I = (D - I\lambda)(I + (D - I\lambda)^{-1}V) \ \ (\lambda \notin \sigma(A)).$$

Hence, $R_\lambda(A) = (I + R_\lambda(D)V)^{-1}R_\lambda(D)$.

Since in the triangular representation of A, D is a diagonal matrix and V is a strictly upper (lower) triangular one, and $R_\lambda(D)V$ is a nilpotent operator. By virtue of the previous lemma,

$$\|R_\lambda(A)\| \leq \|R_\lambda(D)\|\|(I - R_\lambda(D)V)^{-1}\| \leq \|R_\lambda(D)\| \prod_{k=1}^{n-1}(1 + s_k(R_\lambda(D)V)).$$

Since D is a normal operator, we have $\|R_\lambda(D)\| = \rho^{-1}(D, \lambda)$, and therefore

$$s_k(R_\lambda(D)V) \leq \|R_\lambda(D)\|s_k(V) = \rho^{-1}(D, \lambda)s_k(V).$$

Hence,

$$\|R_\lambda(A)\| \leq \prod_{k=1}^{n-1}(1 + \rho^{-1}(D, \lambda)s_k(V))\rho^{-1}(D, \lambda).$$

Taking into account that $\sigma(A) = \sigma(D)$, we get the required result. □

Making use of the inequality between the arithmetical and geometrical means from the latter lemma, we obtain

$$\|R_\lambda(A)\| \leq \frac{1}{\rho(A, \lambda)}(1 + \frac{1}{(n-1)\rho(A, \lambda)}\sum_{k=1}^{n-1} s_k(V))^{1/(n-1)}.$$

Hence,

$$\|R_\lambda(A)\| \leq \frac{1}{\rho(A, \lambda)}(1 + \frac{N_1(V)}{(n-1)\rho(A, \lambda)})^{1/(n-1)}. \tag{15}$$

For an $n \times n$ matrix A introduce the quantity (the departure from normality)

$$g(A) = (N_2^2(A) - \sum_{k=1}^{n} |\lambda_k(A)|^2)^{1/2},$$

where $\lambda_k(A)$ are the eigenvalues of A taken with their multiplicities. For various properties of $g(A)$, see [27, Sec. 3.1]. In particular, $g(A) = N_2(V)$ and $g^2(A) \leq N_2^2(A) - |\text{trace } (A^2)|$.

Furthermore, due to the Schwarz inequality,

$$\sum_{k=1}^{n-1} s_k(V) \le ((n-1)\sum_{k=1}^{n-1} s_k^2(V))^{1/2} \le (n-1)^{1/2} N_2(V) = (n-1)^{1/2} g(A).$$

Now, (15) implies

Theorem 6 *For any $A \in \mathbf{C}^{n\times n}$, one has*

$$\|R_\lambda(A)\| \le \frac{1}{\rho(A,\lambda)}\left(1 + \frac{g(A)}{\sqrt{n-1}\,\rho(A,\lambda)}\right)^{1/(n-1)} \quad (\lambda \notin \sigma(A)).$$

This theorem is sharp: it is attained when A is normal, since $g(A) = 0$ in this case.

By the abovementioned Weyl inequalities, $N_1(D) \le N_1(A)$. Consequently,

$$N_1(V) = N_1(A - D) \le N_1(A) + N_1(D) \le 2N_1(A).$$

Now, Lemma 6 yields

$$\|R_\lambda(A)\| \le \frac{1}{\rho(A,\lambda)}(1 + \frac{2N_1(A)}{(n-1)\rho(A,\lambda)})^{1/(n-1)}. \tag{16}$$

Furthermore, taking into account that

$$\sum_{k=1}^{n} s_k(A) \le \sum_{k=1}^{n} \|Ad_k\| \quad (A \in \mathbf{C}^{n\times n}) \tag{17}$$

for an arbitrary orthonormal basis $\{d_k\}$, cf. [9, Theorem 4.7], we arrive at the following corollary.

Corollary 2 *For any $A \in \mathbf{C}^{n\times n}$ and an arbitrary orthonormal basis $\{d_k\}$ in $\mathbf{C}^n$, one has*

$$\|R_\lambda(A)\| \le \frac{1}{\rho(A,\lambda)}(1 + \frac{2}{(n-1)\rho(A,\lambda)}\sum_{k=1}^{n} \|Ad_k\|)^{1/(n-1)}.$$

Nuclear Operators

Inequality (16) enables us to prove the following result.

Theorem 7 *Let $A \in \mathrm{SN}_1$. Then,*

$$\|R_\lambda(A)\| \le \frac{1}{\rho(A,\lambda)} \exp\left[\frac{2N_1(A)}{\rho(A,\lambda)}\right] \quad (\lambda \notin \sigma(A)).$$

Proof Let $\{A_n\}$ be the sequence of n-dimensional operators converging to A in the norm $N_1(.)$. In view of the upper continuity of spectra [31], $\rho(A_n, \lambda) \geq \rho(A, \lambda)$ $(\lambda \notin \sigma(A))$ for sufficiently large n. Due to (16), we have

$$\|R_\lambda(A_n)\| \leq \frac{1}{\rho(A_n, \lambda)} \exp\left[\frac{2N_1(A_n)}{\rho(A_n, \lambda)}\right].$$

Hence, letting in $n \to \infty$, we arrive at the required result. □

Moreover, from (17) and Theorem 7 we get the following corollary.

Corollary 3 *For an $A \in \mathscr{B}(\mathscr{H})$ and an orthonormal basis $\hat{d} := \{d_k\}_{k=1}^\infty$ in $\mathscr{H}$, let*

$$\gamma_{\hat{d}}(A) := \sum_{k=1}^{\infty} \|Ad_k\| < \infty.$$

Then,

$$\|R_\lambda(A)\| \leq \frac{1}{\rho(A, \lambda)} \exp\left[\frac{2\gamma_{\hat{d}}(A)}{\rho(A, \lambda)}\right].$$

3 Norm Estimates for Operator Functions Regular on the Convex Hull of Spectra

Denote by $\mathrm{co}(A)$ the closed convex hull of the spectrum of A.

Theorem 8 ([27, Theorem 7.4]) *Let A be a Hilbert–Schmidt operator, and let f be a function holomorphic on a neighborhood of $\mathrm{co}(A)$. Then,*

$$\|f(A)\| \leq \sup_{\lambda \in \sigma(A)} |f(\lambda)| + \sum_{k=1}^{\infty} \sup_{\lambda \in \mathrm{co}(A)} |f^{(k)}(\lambda)| \frac{g^k(A)}{(k!)^{3/2}}.$$

This theorem is sharp: it is attained if A is normal because $g(A) = 0$ in this case. Note that, if A is normal, then it is required only that f is defined on $\sigma(A)$.

Now, assume that

$$A \in \mathscr{B}(\mathscr{H}) \text{ and } A_I = (A - A^*)/2i \in \mathrm{SN}_2. \tag{18}$$

Recall that

$$g_I(A) = \sqrt{2}\Big[N_2^2(A_I) - \sum_{k=1}^{\infty} (\mathrm{Im}\ \lambda_k(A))^2\Big]^{1/2} \leq \sqrt{2} N_2(A_I).$$

Theorem 9 ([27, Theorem 10.1]) *Let condition (18) hold and $f(z)$ be regular on a neighborhood of* $\operatorname{co}(A)$. *Then,*

$$\|f(A)\| \le \sup_{\lambda\in\sigma(A)} |f(\lambda)| + \sum_{k=1}^{\infty} \sup_{\lambda\in\operatorname{co}(A)} |f^{(k)}(\lambda)| \frac{g_I^k(A)}{(k!)^{3/2}}.$$

This theorem is sharp, it is attained if A is normal, since $\|f(A)\| = \sup_{\lambda\in\sigma(A)} |f(\lambda)|$ in this case, while from Theorem 9 we have $\|f(A)\| \le \sup_{\lambda\in\sigma(A)} |f(\lambda)|$.

For the norm estimates for functions of a compactly perturbed unitary and unbounded operators, see [27, Section 10.5] and [27, Section 11.5], respectively.

Corollary 4 *Let condition (18) hold. Then,*

$$\|e^{At}\| \le e^{\alpha(A)t} \sum_{k=0}^{\infty} \frac{t^k g_I^k(A)}{(k!)^{3/2}} \quad (t \ge 0),$$

where $\alpha(A) = \sup \operatorname{Re}\ \sigma(A)$. *In addition,*

$$\|A^m\| \le \sum_{k=0}^{m} \frac{m! r_s^{m-k}(A) g_I^k(A)}{(m-k)!(k!)^{3/2}} \quad (m = 1, 2, \ldots).$$

Assuming that $0 \notin \sigma(A)$ and following [8, Section V.1, formula (1.6)], define $\ln(A)$ by

$$\ln(A) = -\frac{1}{2\pi i} \int_C \ln(z) R_z(A) dz, \tag{19}$$

where the principal branch of the scalar logarithm is used, and the Jordan contour C surrounds $\sigma(A)$ and does not surround the origin.

Lemma 7 *Let the condition (18) hold and*

$$\hat{\beta}(A) := \min\{|z| : z \in \operatorname{co}(A)\} > 0.$$

Then,

$$\|\ln(A)\| \le \max_k |\ln(\lambda_k(A))| + \sum_{k=1}^{\infty} \frac{g^k(A)}{\hat{\beta}^k(A) k (k!)^{1/2}}.$$

Proof Since $\hat{\beta}(A) > 0$, $\ln(z)$ is regular on $\operatorname{co}(A)$. Moreover,

$$\left|\frac{d^k}{dz^k} \ln(z)\right| = (k-1)!|z|^{-k} \le \frac{(k-1)!}{\hat{\beta}^k(A)} \quad (z \in \operatorname{co}(A);\ k = 1, 2, \ldots).$$

Thus, due to Theorem 9, we get the required result. □

4 Functions Nonregular on the Convex Hull of the Spectrum

In this section, it is assumed that the spectrum of $A \in \mathscr{B}(\mathscr{H})$ is the union of two sets σ_1 and σ_2 separated by means of open disjoint simply connected sets M_1 and M_2:

$$\sigma(A) = \sigma_1 \cup \sigma_2, \sigma_j \subset M_j \ (j = 1, 2) \text{ and } M_1 \cap M_2 = \emptyset. \tag{20}$$

Note that our arguments below can be easily extended to the case

$$\sigma(A) = \cup_{j=1}^{m} \sigma_j \quad (2 \le m < \infty)$$

with $\sigma_j \cap \sigma_k = \emptyset$ $(j \neq k)$. Let $f(z)$ be a scalar function regular on $M = M_1 \cup M_2$. Then,

$$f(A) = -\frac{1}{2\pi i} \sum_{j=1}^{2} \int_{L_j} f(\lambda) R_\lambda(A) d\lambda, \tag{21}$$

where $L_j \subset M_j$ are closed Jordan contours surrounding σ_j and the integration is performed in the positive direction. It is also assumed that

$$A_I = (A - A^*)/2i \in SN_2. \tag{22}$$

Put

$$\delta := \text{distance}(\sigma_1, \sigma_2), \ \ p_t := \sum_{k=0}^{t} \frac{t!}{((t-k)!k!)^{3/2}} \quad (t = 1, 2, \ldots)$$

and

$$\xi(A) := \left(1 + \sum_{k=0}^{\infty} \frac{p_k (\sqrt{2} N_2(A_I))^{k+1}}{\delta^{k+1}}\right)^2.$$

Observe that

$$\frac{t!}{(t-k)!k!} \le 2^t,$$

and consequently,

$$p_t = \frac{1}{(t!)^{1/2}} \sum_{k=0}^{t} \frac{(t!)^{3/2}}{((t-k)!k!)^{3/2}} \le \frac{2^{t/2}}{(t!)^{1/2}} \sum_{k=0}^{t} \frac{t!}{(t-k)!k!} = \frac{2^{3t/2}}{(t!)^{1/2}} \ (t = 1, 2, \ldots).$$

So,

$$\xi(A) \le \left(1 + \sum_{k=0}^{\infty} \frac{2^{2k+1/2} N_2^{k+1}(A_I)}{(k!)^{1/2} \delta^{k+1}}\right)^2,$$

and therefore, the series in the definition of $\xi(A)$ converges. Moreover, by the Schwarz inequality,

$$\left(\sum_{k=0}^{\infty} \frac{2^{2k} N_2^k(A_I)}{(k!)^{1/2} \delta^{k+1}}\right)^2 = \left(\sum_{k=0}^{\infty} \frac{2^{3k} N_2^k(A_I)}{2^k (k!)^{1/2} \delta^k}\right)^2$$

$$\le \sum_{k=0}^{\infty} \frac{2^{6k} N_2^{2k}(A_I)}{k! \delta^{2k}} \sum_{j=0}^{\infty} \frac{1}{2^{2j}} = \exp\left[\frac{64 N_2^2(A_I)}{\delta^2}\right] \frac{4}{3}.$$

Thus,

$$\xi(A) \le \left(1 + \frac{2\sqrt{2} N_2(A_I)}{\sqrt{3}\delta} \exp\left[\frac{32 N_2^2(A_I)}{\delta^2}\right]\right)^2.$$

Let $\mathrm{co}(\sigma_j)$ be the closed convex hull of σ_j $(j = 1, 2)$.

Theorem 10 *Let conditions (20) and (22) hold. Let $f(z)$ be regular on a neighborhood of* $\mathrm{co}(\sigma_1) \cup \mathrm{co}(\sigma_2)$. *Then,*

$$\|f(A)\| \le \xi(A) \max_{j=1,2} \left(\sup_{s \in \sigma_j} |f(s)| + \sum_{k=1}^{\infty} \sup_{s \in \mathrm{co}(\sigma_j)} |f^{(k)}(s)| \frac{(\sqrt{2} N_2(A_I))^k}{(k!)^{3/2}}\right).$$

The proof of this theorem can be found in [26, Theorem 1.1]. In the finite-dimensional case, it has been proved in [21].

The series in Theorem 10 converges. Indeed, by the Cauchy formula

$$f^{(k)}(z) = \frac{k!}{2\pi i} \int_L \frac{f(s)ds}{(s-z)^{k+1}} \ (z \in \mathrm{co}(\sigma_j)),$$

where L is a closed Jordan contour surrounding $\mathrm{co}(\sigma_j)$ for a fixed $j = 1, 2$, we have

$$|f^{(k)}(z)| \le \frac{k! m_0}{v_0^{k+1}} \ (z \in \mathrm{co}(\sigma_j)), \ \text{where } m_0 = \frac{1}{2\pi} \int_L |f(s)||ds|$$

and $v_0 = \inf_{s \in L, z \in \mathrm{co}(\sigma_j)} |s - z|$. Since

$$\sum_{k=1}^{\infty} \frac{(\sqrt{2}N_2(A_I))^k}{v_0^{k+1}(k!)^{1/2}} < \infty,$$

the series in Theorem 10 really converges.

Theorem 10 is sharp: if A is selfadjoint, then $\xi(A) = 1$, and $\|f\| = \sup_{s \in \sigma(A)} |f(s)|$.

Example 1 Let

$$\sigma(A) = \sigma_1 \cup \sigma_2, \text{ with } \sigma_1 \subseteq [-b, -a], \sigma_2 \subseteq [a, b] \tag{23}$$

$(0 < a < b)$, and

$$\ln(A) = -\frac{1}{2\pi i} \sum_{j=1}^{2} \int_{L_j} \ln z \; R_z(A) dz,$$

where the principal branch of $\ln z$ is used, L_j is a closed Jordan contour surrounding σ_j, does not surrounding $z = 0$ and $L_1 \cap L_2 = \emptyset$.

Clearly, $\ln z$ is regular on $\mathrm{co}(\sigma_1) \cup \mathrm{co}(\sigma_2)$, but nonregular on $\mathrm{co}(A)$. We have $\delta = dist(\sigma_1, \sigma_2) > 2a$,

$$\xi(A) \leq \xi_1(A) := \left(1 + \sum_{k=0}^{\infty} \frac{p_k(\sqrt{2}N_2(A_I))^{k+1}}{(2a)^{k+1}}\right)^2. \tag{24}$$

In addition,

$$\sup_{s \in \sigma_j} |\ln\ s| \leq [\ln^2 b + \pi^2]^{1/2} \text{ and } \sup_{s \in \sigma_j} |(\ln\ s)^{(k)}|$$
$$\leq (k-1)!(2a)^{-k} \; (j = 1, 2; k = 1, 2, \ldots).$$

Now, Theorem 10 implies

$$\|\ln\ A\| \leq \xi_1(A) \left([\ln^2 b + \pi^2]^{1/2} + \sum_{k=1}^{\infty} \frac{(\sqrt{2}N_2(A_I))^k}{k(k!)^{1/2}(2a)^k}\right).$$

Example 2 Under condition (23), let

$$A^{\alpha} := -\frac{1}{2\pi i} \sum_{j=1}^{2} \int_{L_j} z^{\alpha} R_z(A) dz \quad (0 < \alpha < 1),$$

where the contours L_j are the same as in the previous example and the principal branch of z^{α} is used. Clearly, z^{α} is regular on $\text{co}(\sigma_1) \cup \text{co}(\sigma_2)$. As above, $\delta = \text{distance}(\sigma_1, \sigma_2) > 2a$. We have

$$\sup_{s \in \sigma_j} |s^{\alpha}| \leq b^{\alpha} = e^{\alpha \ln b} \text{ and } \sup_{s \in \sigma_j} |(s^{\alpha})^{(k)}| \leq \alpha(1-\alpha)\dots(k-\alpha+1)(2a)^{\alpha-k}$$

$(j = 1, 2, \dots)$. Now, Theorem 10 implies

$$\|A^{\alpha}\| \leq \xi_1(A) \left(b^{\alpha} + \sum_{k=1}^{\infty} \frac{(\sqrt{2} N_2(A_I))^k}{(k!)^{3/2}} \alpha(1-\alpha)\dots(k-\alpha+1)(2a)^{\alpha-k} \right).$$

5 Representations of Commutators

For $A, B, \tilde{A} \in \mathscr{B}(\mathscr{X})$, $[A, B] := AB - BA$ is the commutator, $[A, B, \tilde{A}] := AB - B\tilde{A}$ is the generalized commutator; $[f(A), B] := f(A)B - Bf(A)$ and $[f(A), B, f(\tilde{A})] := f(A)B - Bf(\tilde{A})$ will be called the function commutator and generalized function commutator, respectively.

In the present section, we discuss some representations of the generalized function commutators which will be used in the next section. We begin with the following lemma.

Lemma 8 *Let $A, \tilde{A}, B \in \mathscr{B}(\mathscr{X})$. Then, for any $z \notin \sigma(A) \cup \sigma(\tilde{A})$, we have*

$$(zI - A)^{-1} B - B(zI - \tilde{A})^{-1} = (Iz - A)^{-1} K (Iz - \tilde{A})^{-1}, \tag{25}$$

where

$$K := AB - B\tilde{A} = [A, B, \tilde{A}].$$

Proof Multiplying the both sides of (25) by $zI - A$ from the left and by $zI - \tilde{A}$ from the right, we have

$$B(zI - \tilde{A}) - (zI - A)B = K.$$

This proves the lemma. □

Lemma 9 *Let $A, \tilde{A}, B \in \mathscr{B}(\mathscr{X})$. Let $f(z)$ be regular on an open set U with a smooth boundary L, and let U contain $\sigma(A) \cup \sigma(\tilde{A})$. Then,*

$$f(A)B - Bf(\tilde{A}) = \frac{1}{2\pi i}\int_L f(z)R_z(A)KR_z(\tilde{A})dz. \tag{26}$$

Proof Lemma 8 implies

$$f(A)B - Bf(\tilde{A}) = -\frac{1}{2\pi i}\int_L f(z)(R_z(A)B - BR_z(\tilde{A}))dz$$
$$= \frac{1}{2\pi i}\int_L f(z)R_z(A)KR_z(\tilde{A})dz,$$

as claimed. □

For a positive $r < \infty$, put $\Omega(r) = \{z \in \mathbf{C} : |z| \le r\}$ and $\partial\Omega(r) = \{z \in \mathbf{C} : |z| = r\}$. In the rest of this section, it is assumed that $f(z)$ is regular on $\Omega(r)$ with

$$r > r_s(A, \tilde{A}) := \max\{r_s(A), r_s(\tilde{A})\},$$

where $r_s(A)$ is the spectral radius of A. Take into account that

$$R_\lambda(A) = -\sum_{k=0}^{\infty}\frac{A^k}{\lambda^{k+1}} \quad (|\lambda| > r_s(A)).$$

Then, by the previous lemma,

$$f(A)B - Bf(\tilde{A}) = \frac{1}{2\pi i}\int_{\partial\Omega(r)} f(z)R_z(A)KR_z(\tilde{A})dz$$
$$= \sum_{j,k=0}^{\infty}\frac{1}{2\pi i}\int_{\partial\Omega(r)}\frac{f(z)dz}{z^{k+j+2}}A^jK\tilde{A}^k.$$

Or

$$f(A)B - Bf(\tilde{A}) = \sum_{j,k=0}^{\infty} f_{j+k+1}A^jK\tilde{A}^k, \tag{27}$$

where f_j are the Taylor coefficients of f at zero. If, in particular, $f(z) = z^m$ for an integer $m \ge 1$, then we arrive at the following corollary.

Corollary 5 *Let $A, \tilde{A}, B \in \mathscr{B}(\mathscr{X})$. Then,*

$$A^mB - B\tilde{A}^m = \sum_{j=0}^{m-1} A^jK\tilde{A}^{m-j-1} \quad (m = 2, 3, \ldots). \tag{28}$$

Take $f(z) = e^{zt}, t \ge 0$. Then, the following result is true.

Lemma 10 *Let $A, B \in \mathscr{B}(\mathscr{X})$ and $K = [A, B]$. Then,*

$$[e^{At}, B] = \int_0^t e^{As} K e^{A(t-s)} ds \quad (t \geq 0).$$

Proof We have

$$\frac{d}{dt}([e^{At}, B]e^{-tA}) = \frac{d}{dt}(e^{At} B e^{-At} - B) = e^{At} K e^{-At}.$$

Integrating this equality, we get

$$[e^{At}, B]e^{-tA} = \int_0^t e^{At_1} K e^{-At_1} dt_1,$$

as claimed. □

6 Perturbations of Taylor Series with Operator Arguments

Let $A, \tilde{A} \in \mathscr{B}(\mathscr{X})$,

$$F(A) = \sum_{k=0}^{\infty} b_k A^k, \; F(\tilde{A}) = \sum_{k=0}^{\infty} b_k \tilde{A}^k, \tag{29}$$

where $b_k \in \mathbf{C}$ $(k = 0, 1, 2, \ldots)$ and each series in (29) converges in the operator norm. Due to Corollary 5 with $B = I$, we have

$$F(\tilde{A}) - F(A) = \sum_{k=0}^{\infty} b_k (\tilde{A}^k - A^k) = \sum_{k=1}^{\infty} b_k \sum_{j=0}^{k-1} \tilde{A}^j (\tilde{A} - A) A^{k-j-1}.$$

We thus arrive at the following lemma.

Lemma 11 *Let $F(\tilde{A})$ and $F(A)$ be defined by (29). Then,*

$$\|F(\tilde{A}) - F(A)\| \leq \|\tilde{A} - A\| \sum_{k=1}^{\infty} |b_k| \sum_{j=0}^{k-1} \|\tilde{A}^j\| \|A^{k-j-1}\|,$$

provided the series converges.

7 Entire Operator-Valued Functions

Let $C, \tilde{C} \in \mathscr{B}(\mathscr{X})$ and

$$F(C) = \sum_{k=0}^{\infty} b_k C^k, \tag{30}$$

where $b_k \in \mathbf{C}$ $(k = 0, 1, 2, \ldots)$ satisfy the condition

$$\sqrt[k]{|b_k|} \to 0 \ \ (k \to \infty). \tag{31}$$

Then,

$$F(C + \lambda\tilde{C}) = \sum_{k=0}^{\infty} b_k (C + \lambda\tilde{C})^k \ (\lambda \in \mathbf{C})$$

and

$$\sqrt[k]{|b_k|(\|(C + \lambda\tilde{C})^k\|} \leq \|C + \lambda\tilde{C}\| \sqrt[k]{|b_k|} \to 0 \ \ (k \to \infty).$$

So, $F(C + \lambda\tilde{C})$ is entire in λ.

Theorem 11 *Let F be defined by (30) and condition (31) hold. Let there be a monotone non-decreasing function $G : [0, \infty) \to [0, \infty)$, such that*

$$\|F(B)\| \leq G(\|B\|) \, \textit{for any } B \in \mathscr{B}(\mathscr{X}). \tag{32}$$

Then,

$$\|F(C) - F(\tilde{C})\| \leq \|C - \tilde{C}\| \ G\Big(1 + \frac{1}{2}\|C + \tilde{C}\| + \frac{1}{2}\|C - \tilde{C}\|\Big).$$

Proof Put

$$Z_1(\lambda) = F(\frac{1}{2}(C + \tilde{C}) + \lambda(C - \tilde{C})).$$

Then,

$$F(C) - F(\tilde{C}) = Z_1\Big(\frac{1}{2}\Big) - Z_1\Big(-\frac{1}{2}\Big).$$

Thanks to the Cauchy integral formula,

$$Z_1(1/2) - Z_1(-1/2) = \frac{1}{2\pi i} \int_{|z|=1/2+r} Z_1(z) \left(\frac{1}{z - 1/2} - \frac{1}{z + 1/2} \right) dz$$
$$= \frac{1}{2\pi i} \int_{|z|=1/2+r} Z_1(z) \frac{dz}{(z - 1/2)(z + 1/2)} \quad (r > 0).$$

Hence,

$$\|Z_1(1/2) - Z_1(-1/2)\| \le (1/2 + r) \sup_{|z|=1/2+r} \frac{\|Z_1(z)\|}{|z^2 - 1/4|} \le \frac{1}{r} \sup_{|z|=1/2+r} |Z_1(z)|. \tag{33}$$

In addition, by (32),

$$\|Z_1(z)\| = \left\| F\left(\frac{1}{2}(C + \tilde{C}) + z(C - \tilde{C}) \right) \right\|$$
$$\le G\left(\|\frac{1}{2}(C + \tilde{C}) + z(C - \tilde{C})\| \right)$$
$$\le G\left(\frac{1}{2}\|C + \tilde{C}\| + \left(\frac{1}{2} + r \right) \|C - \tilde{C}\| \right) \quad (|z| = 1/2 + r).$$

Therefore, according to (33),

$$\|F(C) - F(\tilde{C})\| = \|Z_1(1/2) - Z_1(-1/2)\|$$
$$\le \frac{1}{r} G\left(\frac{1}{2}\|C + \tilde{C}\| + (\frac{1}{2} + r)\|C - \tilde{C}\| \right).$$

Taking

$$r = \frac{1}{\|C - \tilde{C}\|},$$

we get the required result. □

For example, due to (30), one can take

$$\|G(x)\| = \sum_{k=0}^{\infty} |b_k| x^k \ (x \ge 0),$$

and thus (32) holds with

$$G(\|C\|) = \sum_{k=0}^{\infty} |b_k| \|C\|^k.$$

8 Perturbations of Operator Functions Regular on the Convex Hull of the Spectrum

Let $A, \tilde{A} \in \mathscr{B}(\mathscr{H})$. Denote by $\text{co}(A, \tilde{A})$ the closed convex hull of $\sigma(A) \cup \sigma(\tilde{A})$. In this section, we derive a bound for $\|f(A) - f(\tilde{A})\|$ for functions regular on a neighborhood of $\text{co}(A, \tilde{A})$ and the operators having the Hilbert–Schmidt Hermitian components. We begin with the finite-dimensional case.

8.1 *The Finite-Dimensional Operators*

In this subsection, A and $\tilde{A}$ are operators in an n-dimensional Euclidean space $\mathbf{C}^n$ ($n < \infty$) with a scalar product $(.,.)$ and the Euclidean norm $\|.\| = \sqrt{(.,.)}$. As in the case of the Hilbert–Schmidt operators, put

$$g(A) = (N_2^2(A) - \sum_{k=1}^{n} \lambda_k(A)|^2)^{1/2},$$

where $\lambda_k(A)$ are the eigenvalues of A enumerated with their multiplicities taken into account. For the properties of $g(A)$ in the finite-dimensional setting, see [27, Section 3.1].

Theorem 12 ([22, Theorem 1.1]) *Let A, $\tilde{A}$, and B be operators in $\mathbf{C}^n$ and $f(\lambda)$ be holomorphic on a neighborhood of* $\text{co}(A, \tilde{A})$. *Then, with the notations*

$$\eta_{j,k} := \sup_{z \in co\,(A,\tilde{A})} \frac{|f^{(k+j+1)}(z)|}{\sqrt{k!j!}(k+j+1)!} \quad (j, k = 0, 1, 2, \ldots),$$

we have the inequality

$$N_2(f(A)B - Bf(\tilde{A})) \leq N_2(K) \sum_{j,k=0}^{n-1} \eta_{j,k} g^j(A) g^k(\tilde{A}) \quad (K = AB - B\tilde{A}).$$

According to the abovementioned Schur theorem, the triangular representation

$$A = D + V \quad (\sigma(A) = \sigma(D)) \tag{34}$$

is valid, where D is a normal operator and V is a nilpotent one having the joint invariant subspaces. Recall that V and D are called the nilpotent and diagonal parts of A, respectively. Similarly,

$$\tilde{A} = \tilde{D} + \tilde{V} \quad (\sigma(\tilde{A}) = \sigma(\tilde{D})),$$

where $\tilde{D}$ is a normal operator and $\tilde{V}$ is a nilpotent one having joint invariant subspaces. The proof of Theorem 12 is based on the following lemma.

Lemma 12 *Under the hypothesis of Theorem 12, one has*

$$N_2(f(A)B - Bf(\tilde{A})) \leq N_2(K) \sum_{j,k=0}^{n-1} \eta_{j,k} N_2^j(V) N_2^k(\tilde{V}),$$

where V and $\tilde{V}$ are the nilpotent parts of A and $\tilde{A}$, respectively.

Proof By (34),

$$R_\lambda(A) = (D + V - I\lambda)^{-1} = (I + R_\lambda(D)V)^{-1}\ R_\lambda(D).$$

Note that $R_\lambda(D)V$ is a nilpotent matrix, and therefore $(R_\lambda(D)V)^n = 0$. Consequently,

$$R_\lambda(A) = \sum_{k=0}^{n-1}(-1)^k (R_\lambda(D)V)^k R_\lambda(D).$$

Similarly,

$$R_\lambda(\tilde{A}) = \sum_{k=0}^{n-1}(-1)^k (R_\lambda(\tilde{D})\tilde{V})^k R_\lambda(\tilde{D}).$$

So, by (26), we have

$$f(A)B - Bf(\tilde{A}) = \sum_{m,k=0}^{n-1} C_{mk}, \tag{35}$$

where

$$C_{mk} = (-1)^{k+m} \frac{1}{2\pi i} \int_L f(\lambda)(R_\lambda(D)V)^m R_\lambda(D) K (R_\lambda(\tilde{D})\tilde{V})^k R_\lambda(\tilde{D}) d\lambda.$$

Since D is a diagonal matrix in the orthonormal basis of the triangular representations of A (the Schur basis) $\{e_k\}$, and $\tilde{D}$ is a diagonal matrix in the Schur basis $\{\tilde{e}_k\}$ of $\tilde{A}$, we can write out

$$R_\lambda(D) = \sum_{j=1}^{n} \frac{Q_j}{\lambda_j - \lambda},\ R_\lambda(\tilde{D}) = \sum_{j=1}^{n} \frac{\tilde{Q}_j}{\tilde{\lambda}_j - \lambda},$$

where $\lambda_j = \lambda_j(A)$, $\tilde{\lambda}_j = \lambda_j(\tilde{A})$, $Q_k = (., e_k)e_k$, and $\tilde{Q}_k = (., \tilde{e}_k)\tilde{e}_k$. Besides,

$$Q_j V Q_k = \tilde{Q}_j \tilde{V} \tilde{Q}_k = 0 \quad (j \geq k).$$

We can write

$$C_{mk} = \sum_{i_1=1}^{n} Q_{i_1} V \sum_{i_2=1}^{n} Q_{i_2} V \dots V \sum_{i_{m+1}=1}^{n} Q_{i_{m+1}} K \sum_{j_1=1}^{n} \tilde{Q}_{j_1} \tilde{V} \sum_{j_2=1}^{n} \tilde{Q}_{j_2} \tilde{V} \dots$$

$$\tilde{V} \sum_{j_{k+1}=1}^{n} \tilde{Q}_{j_{k+1}} J_{i_1,i_2,\dots,i_{m+1},j_1 j_2 \dots j_{k+1}}. \tag{36}$$

Here,

$$J_{i_1,i_2,\dots,i_{m+1},j_1 j_2 \dots j_{k+1}} =$$

$$\frac{(-1)^{k+m}}{2\pi i} \int_L \frac{f(\lambda)d\lambda}{(\lambda_{i_1} - \lambda) \dots (\lambda_{i_{m+1}} - \lambda)(\tilde{\lambda}_{j_1} - \lambda) \dots (\tilde{\lambda}_{j_{k+1}} - \lambda)}.$$

Below, the symbol $|V|_e$ means the operator whose entries are absolute values of V in the basis $\{e_k\}$ and $|\tilde{V}|_{\tilde{e}}$ means the operator whose entries are absolute values of $\tilde{V}$ in the basis $\{\tilde{e}_k\}$. Furthermore, denote $K_{kj} = (K\tilde{e}_j, e_k)$ and $c_{kj}^{(ml)} = (C_{ml}\tilde{e}_j, e_k)$. Then,

$$K = \sum_{j,k=1}^{n} K_{kj}(., \tilde{e}_j)e_k \text{ and } C_{ml} = \sum_{j,k=1}^{n} c_{kj}^{(ml)}(., \tilde{e}_j)e_k.$$

Put

$$|K|_{e\tilde{e}} = \sum_{j,k=1}^{n} |K_{kj}|(., \tilde{e}_j)e_k \text{ and } |C_{ml}|_{e\tilde{e}} = \sum_{j,k=1}^{n} |c_{kj}^{(ml)}|(., \tilde{e}_j)e_k.$$

By Gil [27, Lemma 3.8],

$$|J_{i_1,i_2,\dots,i_{m+1},j_1 j_2 \dots j_{k+1}}| \leq \tilde{\eta}_{m,k} := \sup_{z \in co\,(A,\tilde{A})} \frac{|f^{(k+m+1)}(z)|}{(m+k+1)!}.$$

Now, (36) and the equality

$$\sum_{k=1}^{n} Q_k = I$$

imply

$$|C_{mk}|_{e\tilde{e}} \le \tilde{\eta}_{m,k} \sum_{i_1=1}^{n} Q_{i_1}|V|_e \sum_{i_2=1}^{n} Q_{i_2}|V|_e \cdots |V|_e \sum_{i_{m+1}=1}^{n} Q_{j_{m+1}}|K|_{e\tilde{e}} \sum_{j_1=1}^{n} \tilde{Q}_{j_2}|\tilde{V}|_{\tilde{e}} \cdots$$

$$\cdots |\tilde{V}|_{\tilde{e}} \sum_{j_{k+1}=1}^{n} \tilde{Q}_{j_{k+1}} = \tilde{\eta}_{m,k}|V|_e^m |K|_{e\tilde{e}}|\tilde{V}|_{\tilde{e}}^k. \tag{37}$$

The inequalities are understood in the entry-wise sense. Note that

$$N_2^2(|K|_{e\tilde{e}}) = \sum_{k=1}^{n} \||K|_{e\tilde{e}}\tilde{e}_k\|^2 = \sum_{k=1}^{n}\sum_{j=1}^{n} |K_{jk}|^2 = N_2^2(K).$$

Hence, (37) yields the inequality

$$N_2(C_{mk}) \le \tilde{\eta}_{m,k}\||V|_e^m\| N_2(K)\||\tilde{V}|_{\tilde{e}}^k\|.$$

By Gil [27, Lemma 3.4],

$$\|\,|V|_e^m\| \le \frac{N_2^m(|V|)}{\sqrt{m!}} = \frac{N_2^m(V)}{\sqrt{m!}}.$$

So,

$$N_2(C_{mk}) \le \tilde{\eta}_{m,k} N_2(K) \frac{N_2^m(V) N_2^k(\tilde{V})}{\sqrt{m!k!}}.$$

Now, (35) implies the required result. □

Proof of Theorem 12 By Lemma 3.2 from [27], $N_2(V) = g(A)$. Now, the required result is due to the preceding lemma. □

If A and $\tilde{A}$ are normal, then $g(A) = g(\tilde{A}) = 0$, and with $0^0 = 1$, Theorem 12 yields

$$N_2(f(A)B - Bf(\tilde{A})) \le N_2(K) \sup_{z \in co\,(A,\tilde{A})} |f'(z)|. \tag{38}$$

Taking in Theorem 12 $B = I$—the unit operator, we get the following corollary.

Corollary 6 *Let A and $\tilde{A}$ be n-dimensional and $f(\lambda)$ be holomorphic on a neighborhood of* $\mathrm{co}(A, \tilde{A})$. *Then,*

$$N_2(f(A) - f(\tilde{A})) \leq N_2(A - \tilde{A}) \sum_{j,k=0}^{n-1} \eta_{j,k} g^j(A) g^k(\tilde{A}).$$

If A and $\tilde{A}$ are normal, then according to (38),

$$N_2(f(A) - f(\tilde{A})) \leq N_2(A - \tilde{A}) \sup_{z \in co\,(A,\tilde{A})} |f'(z)|.$$

8.2 *Operators with Hilbert–Schmidt Components*

In the present subsection, we consider the generalized function commutator with $A, \tilde{A}, B \in \mathscr{B}(\mathscr{H})$ satisfying the conditions

$$A_I := (A - A^*)/2i \in \mathrm{SN}_2, \tilde{A}_I := (\tilde{A} - \tilde{A}^*)/2i \in \mathrm{SN}_2, \tag{39}$$

and

$$K = AB - B\tilde{A} \in \mathrm{SN}_2. \tag{40}$$

Recall that $g_I(A)$ is defined in Sect. 2.4 and $g_I(A) \leq \sqrt{2} N_2(A_I)$. Again, put

$$\eta_{j,k} := \sup_{z \in co\,(A,\tilde{A})} \frac{|f^{(k+j+1)}(z)|}{\sqrt{k!j!}(k+j+1)!} \quad (j, k = 0, 1, 2, \ldots).$$

Theorem 13 *Let conditions (39) and (40) hold. Let $f(\lambda)$ be holomorphic on a neighborhood of* $\mathrm{co}(A, \tilde{A})$. *Then,*

$$N_2(f(A)B - Bf(\tilde{A})) \leq N_2(K) \sum_{j,k=0}^{\infty} \eta_{j,k} g_I^j(A) g_I^k(\tilde{A}).$$

Proof By Corollary 10.1 from [27] under conditions (39), there are sequences A_n and $\tilde{A}_n$ $(n = 1, 2, \ldots)$ of n-dimensional operators strongly converging to A and $\tilde{A}$, respectively, such that $f(A_n) \to f(A)$ and $f(\tilde{A}_n) \to f(\tilde{A})$ in the strong topology. Note that

$$\begin{aligned}
&\|(f(A)B - Bf(\tilde{A}) - (f(A_n)B - Bf(\tilde{A}_n)))x\| \\
&\quad \leq \|(f(A) - f(A_n))Bx\| + \|B(f(\tilde{A}) - f(\tilde{A}_n))x\| \to 0 \ \ (x \in \mathscr{H}, n \to \infty).
\end{aligned}$$

So,

$$f(A_n)B - Bf(\tilde{A}_n) \to [f(A), B, f(\tilde{A})]$$

in the strong operator topology. Theorem 12 yields

$$N_2(f(A_n)B - Bf(\tilde{A}_n)) \le N_2(A_nB - B\tilde{A}_n) \sum_{j,k=0}^{n-1} \eta_{j,k} g^j(A_n) g^k(\tilde{A}_n).$$

Moreover, making using of [27, Theorem 3.1] and Corollary 10.2 from [27], we obtain $g(A_n) = g_I(A_n) \to g_I(A)$. Thus,

$$N_2(f(A_n)B - Bf(\tilde{A}_n)) \le N_2(A_nB - B\tilde{A}_n) \sum_{j,k=0}^{n-1} \eta_{j,k} g_I^j(A_n) g_I^k(\tilde{A}_n). \tag{41}$$

Letting $n \to \infty$ and taking into account (40), we get the required result. □

If A and $\tilde{A}$ are normal operators, then Theorem 13 implies the inequality

$$N_2(f(A)B - Bf(\tilde{A})) \le N_2(K) \sup_{z \in co\,(A,\tilde{A})} |f'(z)|.$$

Taking in the previous theorem $B = I$ we get the following corollary.

Corollary 7 *Let $f(\lambda)$ be holomorphic on a neighborhood of* $\mathrm{co}(A, \tilde{A})$ *and the conditions (39) and*

$$A - \tilde{A} \in \mathrm{SN}_2 \tag{42}$$

hold. Then,

$$N_2(f(A) - f(\tilde{A})) \le N_2(A - \tilde{A}) \sum_{j,k=0}^{\infty} \eta_{j,k} g_I^j(A) g_I^k(\tilde{A}).$$

Example 3 Let $f(A) = e^{At}$, $t \ge 0$. Then,

$$\sup_{z \in co\,(A,\tilde{A})} \left| \frac{d^{k+j+1} e^{zt}}{dz^{k+j+1}} \right| = e^{\alpha t} t^{k+j+1} \quad (j, k = 0, 1, 2, \ldots;\ t \ge 0),$$

where

$$\alpha := \max\{\alpha(A), \alpha(\tilde{A})\} \quad (\alpha(A) = \sup \mathrm{Re}\, \sigma(A)).$$

Thus,

$$\eta_{j,k} = \frac{e^{\alpha t} t^{k+j+1}}{\sqrt{k!j!}(k+j+1)!} \quad (j,k = 0,1,2,\ldots).$$

Under conditions (39) and (40), due to Theorem 13, we can write

$$N_2(e^{At}B - Be^{\tilde{A}t}) \le e^{\alpha t} N_2(K) \sum_{j,k=0}^{\infty} \frac{t^{k+j+1} g_I^k(A) g_I^j(\tilde{A})}{\sqrt{k!j!}(k+j+1)!} \quad (t \ge 0).$$

Note that in the appropriate situations, the norm estimates for functions of perturbed operators can be obtained via *the spectral variation* $\mathrm{sv}_A(\tilde{A})$ *of* $\tilde{A}$ *with respect to* A defined by

$$\mathrm{sv}_A(\tilde{A}) = \sup_{s \in \sigma(\tilde{A})} \inf_{t \in \sigma(A)} |t - s|.$$

For example, since

$$\alpha(\tilde{A}) \le \alpha(A) + \mathrm{sv}_A(\tilde{A}),$$

due to Corollary 4, we obtain

$$\|e^{\tilde{A}t}\| \le e^{\alpha(A)t + \mathrm{sv}_A(\tilde{A})t} \sum_{k=0}^{\infty} \frac{t^k g_I^k(\tilde{A})}{(k!)^{3/2}} \quad (t \ge 0),$$

provided $\tilde{A}^* - \tilde{A} \in SN_2$. The classical results on the spectrum perturbations can be found in [31]. For the recent results, see the books [27, 41] and the references given therein.

9 Perturbations of Functions of Infinite Matrices

9.1 Statement of the Result

Let $\{d_k\}$ be an orthogonal normal basis in a separable Hilbert space $\mathscr{H}$ and $A \in \mathscr{B}(\mathscr{H})$ be represented in $\{d_k\}$ by a matrix $(a_{jk})_{j,k=1}^{\infty}$. Denote that matrix also by A. We can write $A = D + V$, where $D = \mathrm{diag}\,[a_{11}, a_{22}, \ldots]$ and $V := A - D$ is the off-diagonal part of A. That is, the entries v_{jk} of V are $v_{jk} = a_{jk}$ $(j \ne k)$ and $v_{jj} = 0$ $(j,k = 1,2,\ldots)$.

Recall that in Sect. 8 D and V denote the diagonal and nilpotent parts (in the Schur basis) of a finite matrix A, respectively.

Furthermore, simultaneously, we consider another matrix

$$\tilde{A} = (\tilde{a}_{jk})_{j,k=1}^{\infty} = \tilde{D} + \tilde{V},$$

where $\tilde{D} = \text{diag}\,[\tilde{a}_{11}, \tilde{a}_{22}, \dots]$, $\tilde{V} = \tilde{A} - \tilde{D}$.

Put $|A| = (|a_{jk}|)_{j,l=1}^{\infty}$, i.e., $|A|$ is the matrix whose entries are the absolute values of A in basis $\{d_k\}$. Let B and C be matrices representing bounded operators in $\mathcal{H}$. In this section, we write $C \geq 0$ if all the entries of C are non-negative and $C \geq B$ if $C - B \geq 0$. Note that

$$r_s(D) = \sup_k |a_{kk}| \text{ and } r_s(\tilde{D}) = \sup_k |\tilde{a}_{kk}|,$$

and denote

$$a_0 = \max\{r_s(D), r_s(\tilde{D})\} \text{ and } v_0 = \max\{\||V|\|, \||\tilde{V}|\|\}.$$

Recall that $\Omega(r) := \{z \in \mathbf{C} : |z| \leq r\}$ for an $r > 0$, and assume that $f(\lambda)$ is a function holomorphic on a neighborhood of $\Omega(a_0 + v_0)$. Furthermore, let $\text{co}(D, \tilde{D})$ be the closed convex hull of the diagonal entries $a_{11}, a_{22}, \dots$ and $\tilde{a}_{11}, \tilde{a}_{22}, \dots$. Obviously,

$$\text{co}(D, \tilde{D}) \subseteq \Omega(a_0) \subseteq \Omega(a_0 + v_0).$$

Denote

$$\eta_{j,k} := \sup_{z \in \text{co}(D,\tilde{D})} \frac{|f^{(k+j+1)}(z)|}{(k+j+1)!} \quad (j, k = 0, 1, 2, \dots).$$

Theorem 14 *Let $f(\lambda)$ be holomorphic on a neighborhood of $\Omega(a_0 + v_0)$. Then,*

$$|f(A) - f(\tilde{A})| \leq \sum_{j,k=0}^{\infty} \eta_{j,k} |V|^j \, |A - \tilde{A}| |\tilde{V}|^k,$$

and the series converges in the operator norm.

Theorem 14 is proved in the next subsection. The relevant results can be found in [14, 15, 18]. Theorem 14 supplements the recent investigations of infinite matrices, cf. [38, 47] and the references therein.

If A and $\tilde{A}$ are diagonal: $V = \tilde{V} = 0$, then Theorem 14 implies

$$|f(A) - f(\tilde{A})| \leq \max_k |a_{kk} - \tilde{a}_{kk}| \sup_{z \in co\,(D,\tilde{D})} |f'(z)|.$$

Under the hypothesis of Theorem 14, we have

$$\|f(A)-f(\tilde{A})\| \le |||f(A)-f(\tilde{A})||| \le |||A-\tilde{A}||| \sum_{j,k=0}^{\infty} \eta_{j,k} |||V|||^j |||\tilde{V}|||^k,$$

and the series converges. Since $|V| \le |A|$, from Theorem 14, it follows

$$\|f(A)-f(\tilde{A})\| \le |||A-\tilde{A}||| \sum_{j,k=0}^{\infty} \eta_{j,k} |||A|||^j \,||||\tilde{A}|||^k,$$

provided the series converges.

9.2 Proof of Theorem 14

First, assume that A and $\tilde{A}$ are n-dimensional, $n < \infty$. With $r = a_0 + v_0 + \epsilon$, $\epsilon > 0$, let

$$C_{mk} = (-1)^{k+m} \frac{1}{2\pi i} \int_{|\lambda|=r} f(\lambda)(R_\lambda(D)V)^m R_\lambda(D) E (R_\lambda(\tilde{D})\tilde{V})^k R_\lambda(\tilde{D}) d\lambda,$$

where $E = \tilde{A} - A$. Obviously,

$$f(A)-f(\tilde{A}) = -\frac{1}{2\pi i} \int_{|\lambda|=r} f(\lambda)(R_\lambda(A) - R_\lambda(\tilde{A})) d\lambda$$

$$= \frac{1}{2\pi i} \int_{|\lambda|=r} f(\lambda) R_\lambda(\tilde{A}) E R_\lambda(A) d\lambda. \tag{43}$$

But

$$R_\lambda(A) = (D + V - I\lambda)^{-1} = (I + R_\lambda(D)V) R_\lambda(D).$$

Since $|\lambda| = r = a_0 + v_0 + \epsilon$, we can write

$$\|R_\lambda(D)V\| \le \frac{\|V\|}{\min_j |\lambda - a_{jj}|} \le \frac{\|V\|}{v_0 + \epsilon} < 1.$$

Consequently, the series

$$R_\lambda(A) = \sum_{k=0}^{\infty} (-1)^k (R_\lambda(D)V)^k R_\lambda(D)$$

converges in the operator norm. Similarly,

$$R_\lambda(\tilde{A}) = \sum_{k=0}^{\infty}(-1)^k (R_\lambda(\tilde{D})\tilde{V})^k R_\lambda(\tilde{D}).$$

So, by (43), we have

$$f(A) - f(\tilde{A}) = \sum_{m,k=0}^{\infty} C_{mk}. \tag{44}$$

Since D and $\tilde{D}$ are diagonal matrices, we can write

$$R_\lambda(D) = \sum_{j=1}^{n} \frac{Q_j}{a_{jj} - \lambda}, \; R_\lambda(\tilde{D}) = \sum_{j=1}^{n} \frac{Q_j}{\tilde{a}_{jj} - \lambda},$$

where $Q_k = (., d_k)d_k$. Recall that $(.,.)$ means the scalar product. Therefore,

$$C_{mk} = \sum_{i_1=1}^{n} Q_{i_1} V \sum_{i_2=1}^{n} Q_{i_2} V \dots V \sum_{i_{m+1}=1}^{n} Q_{i_{m+1}} E \sum_{j_1=1}^{n} Q_{j_1} \tilde{V} \sum_{j_2=1}^{n} Q_{j_2} \tilde{V} \dots$$

$$\tilde{V} \sum_{j_{k+1}=1}^{n} Q_{j_{k+1}} I_{i_1,i_2,\dots,i_{m+1},j_1 j_2 \dots j_{k+1}}.$$

Here,

$$I_{i_1,i_2,\dots,i_{m+1},j_1 j_2 \dots j_{k+1}} =$$

$$\frac{(-1)^{k+m}}{2\pi i} \int_{|\lambda|=\hat{r}} \frac{f(\lambda)d\lambda}{(a_{i_1 i_1} - \lambda)\dots(a_{i_{m+1} i_{m+1}} - \lambda)(\tilde{a}_{j_1 j_1} - \lambda)\dots(\tilde{a}_{j_{k+1} j_{k+1}} - \lambda)}.$$

By Lemma 1.5.1 [16],

$$|I_{i_1,i_2,\dots,i_{m+1},j_1 j_2 \dots j_{k+1}}| \le \eta_{m,k}.$$

Hence,

$$|C_{mk}| \le \eta_{mk} \sum_{j_1=1}^{n} Q_{j_1} |V| \sum_{j_2=1}^{n} Q_{j_2} |V| \dots |V| \sum_{j_k=1}^{n} Q_{j_m} |E|$$
$$\sum_{l_1=1}^{n} Q_{l_l} |\tilde{V}| \dots |\tilde{V}| \sum_{l_k=1}^{n} Q_{l_m}.$$

Thus,

$$|C_{mk}| \le \eta_{mk} |V|^m |E| |\tilde{V}|^k. \tag{45}$$

By the Cauchy inequality with $|z| \leq a_0$, we have

$$|f^{(m)}(z)| \leq \frac{1}{2\pi} \int_{|\lambda|=\hat{r}} \frac{|f(\lambda)|}{|\lambda - z|^{m+1}} |d\lambda| \leq \frac{c_0}{(r - a_0)^{m+1}} = \frac{c_0}{(v_0 + \epsilon)^{m+1}},$$

where

$$c_0 = \frac{1}{2\pi} \int_{|\lambda|=r} |f(\lambda)||d\lambda|.$$

Thus,

$$\eta_{mk} \leq \frac{c_0}{(v_0 + \epsilon)^{m+k+1}}.$$

Due to (45),

$$\|C_{mk}\| \leq \eta_{mk} |V|^m |E||\tilde{V}|^k \frac{c_0}{(v_0 + \epsilon)^{m+k+1}} \||V|^m\|\||E|\|\||\tilde{V}|^k\| \leq \text{const } t_0^{m+k+1},$$

where

$$t_0 = \frac{1}{v_0 + \epsilon} \||\tilde{V}|\|\||\tilde{V}|\| < 1.$$

Thus, the series

$$\sum_{m,k=1}^{\infty} C_{mk}$$

converges. Now, (44) and (45) imply the required result in the finite-dimensional setting. Letting $n \to \infty$, we get the required result due to the Banach–Schteihaus theorem. □

10 Positivity Conditions for Operator Functions in a Hilbert Lattice

Let H be a Hilbert lattice [37, p. 128] with a norm $\|.\|$ and the unit operator I, and let

$$D = \int_a^b s dP_s$$

be a bounded selfadjoint operator in H. Here, P_s is the orthogonal resolution of the identity defined on a finite segment $[a, b]$. In this section, we investigate functions of a bounded operator of the form

$$A = D + T, \tag{46}$$

where T is a positive operator in H. It is assumed that P_t *is non-negative*; namely, the operators $P(t_2) - P(t_1)$ $(a \le t_1 < t_2 \le b)$ are non-negative. Put $c_0 = \max\{|a|, |b|\}$.

Theorem 15 ([23, Theorem 1.1]) *Let A be defined by (46) and P_t be non-negative, and assume that a function $f(\lambda)$ is real on $[a, b]$ and holomorphic on a neighborhood of $\Omega(r_0) = \{z \in \mathbf{C} : |z| \le r_0\}$ with*

$$r_0 = c_0 + \|T\| + \epsilon \text{ for some } \epsilon > 0.$$

Then, with the notations,

$$\alpha_k := \min_{a \le s \le b} \frac{f^{(k)}(s)}{k!} \text{ and } \beta_k := \max_{a \le s \le b} \frac{f^{(k)}(s)}{k!} \quad (k = 0, 1, 2, \ldots),$$

the inequalities

$$\sum_{k=0}^{\infty} \alpha_k T^k \le f(A) \le \sum_{k=0}^{\infty} \beta_k T^k \ (T^0 = I) \tag{47}$$

hold, and both the series converge in the operator norm.

This theorem is proved in the next section. In the papers [24, 25], the similar results have been derived for operators in an ordered Banach space.

Corollary 8 *Under the conditions of Theorem 15, let $\alpha_k \ge 0$ $(k = 0, 1, 2, \ldots)$. Then, $f(A) \ge 0$.*

For the classical results on positive operators, see [32].

11 Proof of Theorem 15

By (46), we get

$$(A - I\lambda)^{-1} = (D + T - \lambda I)^{-1} = (I + R_\lambda(D)T)^{-1} R_\lambda(D). \tag{48}$$

If $r > \|T\| + \|D\|$, then

$$\frac{\|T\|}{\inf_{a\le t\le b}|\lambda-t|}\le\frac{\|T\|}{|\lambda|-\|D\|}<1$$

for $|\lambda|=r$. Hence,

$$\|R_\lambda(D)T\|\le\frac{\|T\|}{\inf_{t\in\sigma(D)}|\lambda-t|}<1,$$

and thus (48) implies

$$R_\lambda(A)=\sum_{k=0}^{\infty}(-R_\lambda(D)T)^k\ R_\lambda(D)\ \ (|\lambda|=r).$$

We can write

$$f(A)=-\frac{1}{2\pi i}\int_{|\lambda|=r}f(\lambda)R_\lambda(A)d\lambda.$$

Hence,

$$f(A)=-\frac{1}{2\pi i}\int_{|\lambda|=r}f(\lambda)R_\lambda(A)d\lambda=\sum_{k=0}^{\infty}B_k, \tag{49}$$

where

$$B_k=(-1)^{k+1}\frac{1}{2\pi i}\int_{|\lambda|=r}f(\lambda)(R_\lambda(D)T)^kR_\lambda(D)d\lambda.$$

But

$$R_\lambda(D)=\int_a^b\frac{dP(t)}{t-\lambda}$$

and

$$(R_\lambda(D)T)^kR_\lambda(D)=\int_a^b dP(t_1)T\int_a^b dP(t_2)T\dots$$

$$\dots T\int_a^b dP(t_k)T\int_a^b dP(t_{k+1})\frac{1}{(t_1-\lambda)\dots(t_{k+1}-\lambda)}.$$

We thus have

$$B_k = \int_a^b dP(t_1)T \int_a^b dP(t_2)T \dots T \int_a^b dP(t_k)T \int_a^b dP(t_{k+1})J(t_1, t_2, \dots, t_{k+1}). \tag{50}$$

Here,

$$J(t_1, t_2, \dots, t_{k+1}) = \frac{(-1)^{k+1}}{2\pi i} \int_{|\lambda|=r} \frac{f(\lambda)d\lambda}{(t_1 - \lambda)\dots(t_{k+1} - \lambda)}.$$

Lemma 1.5.2 from [16] gives us the inequalities

$$\alpha_k \leq J(t_1, t_2, \dots, t_{k+1}) \leq \beta_k. \tag{51}$$

Hence, by (50) and the positivity of the resolution $P(.)$,

$$B_k \geq \alpha_k \int_a^b dP(t_1)T \int_a^b dP(t_2)T \dots T \int_a^b dP(t_k)T \int_a^b dP(t_{k+1}) = \alpha_k T^k. \tag{52}$$

Similarly,

$$B_k \leq \beta_k T^k. \tag{53}$$

Due to (49), we get (47).

Let us check that both the series in (47) converge. Indeed, $f(z)$ is regular on $\Omega(r_0)$. So, the series

$$f(z) = \sum_{k=0}^{\infty} \frac{f^{(k)}(t)}{k!}(z - t)^k$$

converges, provided $t = c_0$, $|z| \leq r_0$. Take $z = \|T\| + c_0$. Then, $z - t = \|T\|$. So, the series

$$f(z) = \sum_{k=0}^{\infty} \frac{f^{(k)}(t)}{k!} \|T\|^k$$

converges for any $t \in \sigma(D)$. This proves the theorem. □

12 Examples to Theorem 15

Consider some operators having the form (46).

Example 4 Let $A = (a_{jk})$ be a real infinite matrix representing a bounded operator in a real space l^2 and $a_{jk} \geq 0$, $j \neq k$.

Take $D = \text{diag}\,(a_{jj})$, $T = (t_{jk})$ with $t_{jj} = 0$ and $t_{jk} = a_{jk}$, $j \neq k$. Then, (46) holds with $a = \inf_j a_{jj}$ and $b = \sup_j a_{jj}$. Now, we can directly apply Theorem 15. For the recent results on infinite matrices and their applications, see the papers [38, 47].

Example 5 Take $H = L^2[0,1]$, where the space $L^2[0,1]$ is real. Consider the integral operator A defined by

$$(Au)(x) = h(x)u(x) + \int_0^1 K(x,s)u(s)ds \quad (u \in L^2[0,1];\ x \in [0,1]), \tag{54}$$

where $h(.)$ is a real bounded measurable scalar-valued function and K is a non-negative kernel, providing the boundedness of the operator T defined by

$$(Tu)(x) = \int_0^1 K(x,s)u(s)ds.$$

Then, (46) holds with $(Du)(x) = h(x)u(x)$.

Furthermore, for finite real numbers c and d, $c < d$, let $\Lambda = \{c \leq x \leq d, 0 \leq y \leq 1\}$, and $L^2(\Lambda)$ is the Hilbert space of real functions defined on Λ with the scalar product

$$(f,g) = \int_c^d \int_0^1 f(x,y)g(x,y)dy\,dx.$$

Example 6 Our next object is the operator $\hat{A}$ defined by

$$(\hat{A}f)(x,y) = w(x,y)f(x,y) + \int_0^1 K(x,y,s)f(x,s)ds$$

$$(c \leq x \leq d;\ 0 \leq y \leq 1;\ f \in L^2(\Lambda)), \tag{55}$$

where $w(x,y)$ is a real bounded function and $K(x,y,s)$ is a positive function providing the boundedness in $L^2(\Lambda)$ of the operator T defined by

$$(Tf)(x,y) = \int_0^1 K(x,y,s)f(s)ds.$$

Taking $(Df)(x,y) = w(x,y)f(x,y)$, we obtain (46) with $A = \hat{A}$.

The operator $\hat{A}$ is called a partial integral operator, inasmuch as the integration is carried out only with respect to some arguments, while the other arguments of the integrand are "frozen." Such operators play an essential role in numerous applications, cf. [2]. The spectrum and norm estimates for functions of the partial integral operator were considered in [19, 20].

13 Perturbations of Operator Functions in a Hilbert Lattice

Let $\tilde{A}$ be a bounded linear operator on a Hilbert lattice H defined by

$$\tilde{A} = D + \tilde{T}, \tag{56}$$

where $\tilde{T}$ is a positive operator in H and D is the same as in Sect. 10. Let A and c_0 be the same as in Sect. 10 and $f(\lambda)$ be holomorphic on a neighborhood of

$$\Omega_0 := \{z \in \mathbf{C} : |z| \leq c_0 + \max\{\|T\|, \|\tilde{T}\|\}\}.$$

Then, by (49),

$$f(\tilde{A}) = \sum_{k=0}^{\infty} \tilde{B}_k$$

where $\tilde{B}_k$ are defined by (50) with $\tilde{T}$ instead of T. Due to (50) with $E = \tilde{T} - T \geq 0$, we have

$$\tilde{B}_k = \int_a^b dP(t_1)(T+E) \int_a^b dP(t_2)(T+E) \ldots$$

$$\ldots (T+E) \int_a^b dP(t_k)(T+E) \int_a^b dP(t_{k+1}) J(t_1, t_2, \ldots, t_{k+1}) \geq B_k.$$

Making use of Theorem 15, we arrive at the following result.

Theorem 16 *Let A and $\tilde{A}$ be defined by (46) and (56), respectively. Let $P(.)$ be non-negative, $T \leq \tilde{T}$, and $f(\lambda)$ be holomorphic on a neighborhood of Ω_0 and real on $[a, b]$. Then, $f(\tilde{A}) \geq f(A)$.*

14 Perturbations of Operator Fractional Powers

Let $A \in \mathscr{B}(\mathscr{X})$, $\beta(A) = \inf \operatorname{Re} \sigma(A) > 0$ and

$$\int_0^\infty t^{-\nu} \|(A + It)^{-1}\| dt < \infty \quad (0 < \nu < 1), \tag{57}$$

then the fractional power of A can be defined by the formula

$$A^{-\nu} = \frac{\sin(\pi\nu)}{\pi} \int_0^\infty t^{-\nu}(A + It)^{-1} dt, \tag{58}$$

cf. [33, Section I.5.2, formula (5.8)]. Let $\tilde{A} \in \mathscr{B}(\mathscr{X})$ and

$$q \sup_{t \geq 0} \|(A + It)^{-1}\| < 1 \quad (q = \|A - \tilde{A}\|). \tag{59}$$

Then, by (14),

$$\|(\tilde{A} + It)^{-1}\| \leq \frac{\|(A + It)^{-1}\|}{1 - q\|(A + It)^{-1}\|} \quad (t \geq 0).$$

We have $\beta(\tilde{A}) > 0$, and (57) and (58) are valid with $\tilde{A}$ instead of A. Consequently,

$$\tilde{A}^{-\nu} - A^{-\nu} = -\frac{\sin(\pi\nu)}{\pi} \int_0^\infty t^{-\nu}(\tilde{A} + It)^{-1}(\tilde{A} - A)(A + It)^{-1} dt.$$

Hence,

$$\|A^{-\nu} - \tilde{A}^{-\nu}\| \leq \frac{q \sin(\pi\nu)}{\pi} \int_0^\infty t^{-\nu} \|(\tilde{A} + It)^{-1}\| \|(A + It)^{-1}\| dt.$$

We thus arrive at the following lemma.

Lemma 13 *Let the conditions $\beta(A) > 0$, (57), and (59) hold. Then,*

$$\|A^{-\nu} - \tilde{A}^{-\nu}\| \leq \frac{q \sin(\pi\nu)|}{\pi} \int_0^\infty t^{-\nu} \frac{\|(A + It)^{-1}\|^2 dt}{(1 - q\|(A + It)^{-1}\|)} \quad (0 < \nu < 1).$$

Assume that $\mathscr{X} = \mathscr{H}$-separable Hilbert space,

$$\beta(A) > 0 \text{ and } A_I = (A - A^*)/2i \in \mathrm{SN}_2. \tag{60}$$

By Theorem 4,

$$\|R_\lambda(A)\| \leq \frac{\sqrt{e}}{\rho(A, \lambda)} \exp\left[\frac{g_I^2(A)}{2\rho^2(A, \lambda)}\right] \quad (\lambda \notin \sigma(A)).$$

But $\rho(A, -t) \geq t + \beta(A)$ $(t \geq 0)$, and therefore

$$\|(A + tI)^{-1}\| \leq \phi(t) \quad (t \geq 0),$$

where

$$\phi(t) := \frac{\sqrt{e}}{(t + \beta(A))} \exp\left[\frac{g_I^2(A)}{2(t + \beta(A))^2}\right].$$

Due to (58),

$$\|A^{-\nu}\| \leq \frac{\sin(\pi\nu)}{\pi} \int_0^\infty t^{-\nu}\phi(t)dt.$$

Moreover, Lemma 13 implies

Corollary 9 *Let the conditions (60) and* $q \sup_t \phi(t) < 1$ *hold. Then,*

$$\|A^{-\nu} - \tilde{A}^{-\nu}\| \leq \frac{q\sin(\pi\nu)}{\pi} \int_0^\infty t^{-\nu}\frac{\phi^2(t)}{1 - q\phi(t)}dt.$$

For the recent results on fractional powers of linear operators, see the book [35], papers [34, 44, 46], and the references therein.

15 Perturbations of the Operator Logarithm

Throughout this section, $A, \tilde{A} \in \mathscr{B}(\mathscr{X})$ and $q = \|A - \tilde{A}\|$. The results presented in this section are particularly based on the paper [21].

15.1 Definition via Contour Integral

Assume that $0 \notin \sigma(A)$, and define $\ln(A)$ by

$$\ln(A) = -\frac{1}{2\pi i} \int_C \ln(z) R_z(A)dz, \tag{61}$$

where the principal branch of the scalar logarithm is used, and the Jordan contour C surrounds $\sigma(A)$ and does not surround the origin. Hence, it follows that

$$\|\ln(A)\| \leq \frac{1}{2\pi} \int_C |\ln(z)| \|R_z(A)\| |dz| \leq m_C \int_C |\ln(z)||dz|,$$

where

$$m_C := \sup_{z \in C} \|R_z(A)\|.$$

Lemma 14 *Let the conditions* $0 \notin \sigma(A)$ *and* $qm_C < 1$ *hold. Then,*

$$\|\ln(A) - \ln(\tilde{A})\| \leq \frac{qm_C^2}{2\pi(1 - qm_C)} \int_C |\ln(z)||dz|.$$

Proof Due to (14),

$$\sup_{z\in C} \|R_z(\tilde{A})\| \le \frac{m_C}{1-qm_C} \text{ and } \sup_{z\in C} \|R_z(\tilde{A}) - R_z(A)\| \le \frac{qm_C^2}{1-qm_C}.$$

Thus, the integral in (61) with $\tilde{A}$ instead of A is finite. Hence, from (61), we get the required result. □

15.2 *Definition via Improper Integrals*

Assume that

$$\sigma(A) \cap (-\infty, 0] = \emptyset \tag{62}$$

and

$$\int_0^\infty \|(tI + A)^{-1}\| \frac{dt}{1+t} < \infty. \tag{63}$$

We need the following formula proved in [7, Theorem 10.1.3]:

$$\ln(A) = (A - I) \int_0^\infty (tI + A)^{-1} \frac{dt}{1+t}. \tag{64}$$

Thus,

$$\| \ln(A)\| \le \|(A - I)\| \int_0^\infty \|(tI + A)^{-1}\| \frac{dt}{1+t}.$$

Lemma 15 *Let the conditions (62), (63), and*

$$\eta(A) := \sup_{t\ge 0} \|(tI + A)^{-1}\| < \frac{1}{q} \tag{65}$$

hold. Then,

$$\| \ln(\tilde{A}) - \ln(A)\| \le \frac{q}{1 - q\eta(A)} (1 + \|A - I\|) \int_0^\infty \|(tI + A)^{-1}\| \frac{dt}{1+t}.$$

Proof Due to (65) and (14),

$$\|(\tilde{A} + t)^{-1}\| \le \frac{\|(A + It)^{-1}\|}{1 - q\eta(A)}$$

and

$$\|(\tilde{A}+t)^{-1}-(A+t)^{-1}\| \leq \frac{q\|(A+It)^{-1}\|}{1-q\eta(A)}, t \geq 0.$$

So, conditions (63) and (62) hold with $\tilde{A}$ instead of A. Due to (64),

$$\ln(\tilde{A}) = (\tilde{A}-I)\int_0^\infty (tI+\tilde{A})^{-1}\frac{dt}{1+t}$$

and

$$\|\ln(\tilde{A})-\ln(A)\| \leq \int_0^\infty \|(\tilde{A}-I)(tI+\tilde{A})^{-1}-(A-I)(tI+A)^{-1}\|\frac{dt}{1+t}.$$

$$\leq \int_0^\infty (\|\tilde{A}-A\|\|(tI+\tilde{A})^{-1}\|+\|A-I\|\|(tI+\tilde{A})^{-1}-(tI+A)^{-1}\frac{dt}{1+t}\|$$

$$\leq \int_0^\infty \left(q\frac{\|(A+t)^{-1}\|}{1-q\eta(A)}+(\frac{q}{1-q\eta(A)}\|A-I\|\|(tI+A)^{-1}\|\right)\frac{dt}{1+t},$$

as claimed. □

Finally, note that if $r_s(A-I)<1$, then one can use the obvious representation

$$\ln(A) = \sum_{k=1}^{\infty}\frac{1}{k}\,(I-A)^k$$

and Lemma 11 on perturbations of Taylor series.

References

1. Andrica, D. and Rassias, Th. M. (eds.): Differential and Integral Inequalities, Springer Optimization and Its Applications 151, Springer Nature Switzerland, (2019).
2. Appel, J., Kalitvin, A. and Zabreiko, P.: Partial Integral Operators and Integrodifferential Equations, Marcel Dekker, New York (2000).
3. Birman, M. and Solomyak, M.: Double operator integrals in a Hilbert space. Integral Equations Operator Theory **47**, no. 2, 131–168 (2003).
4. Boyadzhiev, K.N.: Some inequalities for generalized commutators, Publ. RIMS, Kyoto University, **26**, no 3, (1990), 521–527
5. Boyadzhiev, K.N.: Logarithms and imaginary powers of operators on Hilbert spaces, Collect. Math., 1994, 45, 287–300.
6. Cardoso, J.R. and Silva Leite, F.: Theoretical and numerical considerations about Padé approximants for the matrix logarithm. Linear Algebra Appl. **330**, no. 1–3, 31–42 (2001).

7. Carracedo, C.M. and Alix, M.S.: The Theory of Fractional Powers of Operators, Elsevier, Amsterdam, (2001).
8. Daleckii, Yu L. and Krein, M. G.: Stability of Solutions of Differential Equations in Banach Space, Amer. Math. Soc., Providence, R.I. (1974).
9. Diestel J., Jarchow, H. and Tonge, A.: Absolutely Summing Operators, Cambridge University Press, Cambridge (1995).
10. Dunford N. and Schwartz, J.T.: Linear Operators, part II , Wiley-Interscience, New York (1963).
11. Dunford N. and Schwartz, J.T.: Linear Operators, part I , Wiley-Interscience, New York, 1966.
12. Gel'fand, I.M. and Shilov, G.E.: Some Questions of Theory of Differential Equations. Nauka, Moscow (1958). (In Russian).
13. Gil', M.I.: Estimates for norm of matrix-valued functions, Linear and Multilinear Algebra, **35**, (1993) 65–73.
14. Gil', M.I.: Spectrum localization of infinite matrices. Math. Phys. Anal. Geom. **4**, no. 4, 379–394 (2001).
15. Gil', M.I.: Invertibility and spectrum of Hille-Tamarkin matrices, Mathematische Nachrichten, **244**, 1–11 (2002).
16. Gil', M.I.: Operator Functions and Localization of Spectra, Lectures Notes In Mathematics vol. 1830, Springer-Verlag, Berlin (2003).
17. Gil', M.I.: Difference Equations in Normed Spaces. Stability and Oscillations, North-Holland, Mathematics Studies **206**, Elsevier, Amsterdam, (2007).
18. Gil', M.I.: Estimates for entries of matrix valued functions of infinite matrices, Mathematical Physics, Analysis and Geometry, **11**, no. 2, 175–186 (2008).
19. Gil', M.I.: Spectrum and resolvent of a partial integral operator, Applicable Analysis, **87**, no. 5, 555–566 (2008).
20. Gil', M.I.: Spectrum and functions of operators on direct families of Banach spaces. Methods Appl. Anal. **16**, no. 4, 521–534 (2009).
21. Gil', M.I.: Matrix functions nonregular on the convex hull of the spectrum, Linear Multilinear Algebra, **60**, no. 4, 465–473 (2012).
22. Gil', M.I.: Perturbations of operator functions in a Hilbert Space, Communications in Mathematical Analysis, **13**, 108115 (2012).
23. Gil', M.I.: Bounds and positivity conditions for operator valued functions in a Hilbert space, Positivity, **17**, no. 3, 407–414 (2013).
24. Gil', M.I.: A Norm estimate for holomorphic operator functions in an ordered Banach space, Acta Sci. Math. (Szeged) **80**, no. 1–2 , 141–148 (2014).
25. Gil', M.I.: Two-sided estimates and positivity conditions for solutions of linear operator equations in a Banach lattice. Math. Nachr. **289**, no. 8–9, 974–981 (2016).
26. Gil', M.I.: Norm estimates for functions of non-selfadjoint operators nonregular on the convex hull of the spectrum. Demonstr. Math. **50**, no. 1, 267–277 (2017).
27. M.I. Gil: Operator Functions and Operator Equations, World Scientific, New Jersey (2018).
28. Gohberg, I.C., Goldberg, S. and Krupnik, N.: Traces and Determinants of Linear Operators Birkhäuser Verlag, Basel (2000).
29. Gohberg, I. C. and Krein, M. G.: Introduction to the Theory of Linear Nonselfadjoint Operators, Trans. Mathem. Monographs, v. 18, Amer. Math. Soc., Providence, R. I., (1969).
30. Haase, M.: Spectral properties of operator logarithms, Math. Z. **245**, 761–779 (2003).
31. Kato, T.: Perturbation Theory for Linear Operators, Springer-Verlag, Berlin (1980).
32. Krasnosel'skii, M.A., Lifshits, J. and Sobolev A.: Positive Linear Systems. The Method of Positive Operators, Heldermann Verlag, Berlin (1989).
33. Krein, S.G.: Linear Differential Equations in a Banach Space, Transl. Mathem. Monogr, vol 29, Amer. Math. Soc., (1971).
34. Kufner, A., Kuliev, K., Oguntuase, J.A. and Persson, L.-E.: Generalized weighted inequality with negative powers. J. Math. Inequal. **1**, no. 2, 269–280 (2007).
35. Martínez C. and Sanz M.: The Theory of Fractional Powers of Operators. North-Holland Mathematics Studies, 187. Elsevier, Amsterdam, (2001).

36. Matsaev, V.: Volterra operators obtained from self-adjoint operators by perturbation. Dokl. Akad. Nauk SSSR, **139**, 810–813 (1961) (Russian).
37. Meyer-Nieberg, P.: Banach Lattices, Springer - Verlag, Berlin (1991).
38. Mittal, M.L., Rhoades, B.E., Mishra, V.N. and Singh, U.: Using infinite matrices to approximate functions of class Lip(α, p) using trigonometric polynomials. J. Math. Anal. Appl. **326**, no. 1, 667–676 (2007).
39. Palencia, C. and Piskarev, S.: On multiplicative perturbations of C_0-groups and C_0-cosine operator functions. Semigroup Forum **63**, no.2, 127–152 (2001).
40. Peller, V.: Hankel operators in perturbation theory of unbounded self-adjoint operators, in the book Analysis and Partial Differential equations, Lecture Notes in Pure and Appl. Math., 122, Dekker, New York, 529–544 (1990).
41. Rassias, Th. M. and Zagrebnov, V.A. (eds.): Analysis and Operator Theory. Dedicated in Memory of Tosio Kato's 100th Birthday. Foreword by Barry Simon, Springer Optimization and Its Applications 146, Springer Nature Switzerland, (2019).
42. Schmoeger, C.: On logarithms of linear operators on Hilbert spaces, Demonstratio Math. **35**, no. 2, 375–384 (2002).
43. Sherif, N. and Morsy, E.: Computing real logarithm of a real matrix. Int. J. Algebra **2**, no. 1–4, 131–142 (2008).
44. Silvestre, L.: Regularity of the obstacle problem for a fractional power of the Laplace operator, Commun. Pure Appl. Math. **60**, no. 1, 67–112 (2007).
45. Tikhonov, A.: Boundary values of operator-valued functions and trace class perturbations. Rev. Roum. Math. Pures Appl. **47**, no. 5–6, 761–767 (2002).
46. Yang, Changsen and Zuo, Hongliang: A monotone operator function via Furuta-type inequality with negative powers, Math. Inequal. Appl. **6**, no. 2, 303–308 (2003).
47. Zhao, Xiqiang and Wang, Tianming: The algebraic properties of a type of infinite lower triangular matrices related to derivatives. J. Math. Res. Expo. **22**, no. 4, 549–554 (2002).

Representation Variety for the Rank One Affine Group

Ángel González-Prieto, Marina Logares, and Vicente Muñoz

Abstract The aim of this chapter is to study the virtual classes of representation varieties of surface groups onto the rank one affine group. We perform this calculation by three different approaches: the geometric method, based on stratifying the representation variety into simpler pieces; the arithmetic method, focused on counting their number of points over finite fields; and the quantum method, which performs the computation by means of a Topological Quantum Field Theory. We also discuss the corresponding moduli spaces of representations and character varieties, which turn out to be non-equivalent due to the non-reductiveness of the underlying group.

1 Introduction

Let Γ be a finitely presented group and G a complex algebraic group. A representation of Γ into G is a group homomorphism $\rho : \Gamma \longrightarrow G$. We shall denote the set of representations by

$$\mathfrak{X}_G(\Gamma) = \mathrm{Hom}\,(\Gamma, G),$$

Á. González-Prieto
Facultad de Ciencias Matemáticas, Universidad Complutense de Madrid, Madrid, Spain
e-mail: angelgonzalezprieto@ucm.es

M. Logares
Facultad de Ciencias Matemáticas, Universidad Complutense de Madrid, Madrid, Spain
e-mail: mlogares@ucm.es

V. Muñoz (✉)
Departamento Álgebra, Geometría y Topología, Facultad Ciencias, Universidad de Málaga, Málaga, Spain
e-mail: vicente.munoz@uma.es

I. N. Parasidis et al. (eds.), *Mathematical Analysis in Interdisciplinary Research*, Springer Optimization and Its Applications 179,
https://doi.org/10.1007/978-3-030-84721-0_18

which is a complex algebraic variety. Let X be a connected CW-complex with $\pi_1(X) = \Gamma$. Then, $\mathfrak{X}_G(\Gamma)$ parametrizes *local systems* over X, that is, G-principal bundles $P \to X$, which admit trivializations $P|_{U_\alpha} \simeq U_\alpha \times G$, for a covering $X = \bigcup U_\alpha$, such that the changes of charts are (locally) constant functions $g_{\alpha\beta} : U_\alpha \cap U_\beta \to G$. A local system can also be understood as a covering space with fiber G (with the discrete topology). From another perspective, we can take a principal G-bundle $P \to X$ and fix a base point $x_0 \in X$. Then, a local system is equivalent to a *flat connection* on P. Certainly, a flat connection ∇ on P determines the *monodromy* representation $\rho_\nabla : \pi_1(X, x_0) \to \mathrm{Aut}\,(P_{x_0}) \cong G$, given by associating with a path $[\gamma] \in \pi_1(X, x_0)$ the holonomy of ∇ along γ. Finally, if G admits a faithful representation $\kappa : G \hookrightarrow \mathrm{GL}_r(\mathbb{C})$, this can also be done with the vector bundle $E = P \times_\kappa \mathbb{C}^r \to X$ with G structure.

If we forget the trivialization at the base point, then we have the *coset space*

$$\widehat{\mathcal{M}}_G(\Gamma) = \mathfrak{X}_G(\Gamma)/G, \tag{1}$$

which is a topological space with the quotient topology. The action of G changes the isomorphism $\mathrm{Aut}\,(P_{x_0}) \cong G$, which corresponds to the action of G on P as principal bundle. This induces the adjoint action on the monodromy representation. The space (1) parametrizes isomorphism classes of local systems. In this case, we can forget the base point, due to the isomorphisms $\pi_1(X, x_0) \cong \pi_1(X, x_1)$, for two points $x_0, x_1 \in X$. In general, the coset space is badly behaved. It is not an algebraic variety, and it may be non-Hausdorff. From the algebro-geometric point of view, it is more natural to focus on the *moduli space* of representations $\mathcal{M}_G(\Gamma)$. This is defined as an algebraic variety with a "quotient map" $q : \mathfrak{X}_G(\Gamma) \to \mathcal{M}_G(\Gamma)$ such that (a) q is constant along orbits, that is, q is G-invariant and (b) it is an initial object for this property, that is, any other map $f : \mathfrak{X}_G(\Gamma) \to Y$, which is G-invariant factors through $\mathcal{M}_G(\Gamma)$. It turns out that the moduli space is defined by the *GIT quotient*

$$\mathcal{M}_G(\Gamma) = \mathrm{Spec}\ \mathcal{O}(\mathfrak{X}_G(\Gamma))^G,$$

that is, its ring of functions is given by the G-invariant functions on the representation variety.

In the case where G is a complex reductive group (e.g., $G = \mathrm{SL}_r(\mathbb{C})$ or $\mathrm{GL}_r(\mathbb{C})$), the GIT quotient has nice properties. Take a faithful representation $\kappa : G \hookrightarrow \mathrm{GL}_r(\mathbb{C})$. The natural map

$$\widehat{\mathcal{M}}_G(\Gamma) \to \mathcal{M}_G(\Gamma) \tag{2}$$

is a homeomorphism over the locus of irreducible representations (those that have no G-invariant proper subspaces $W \subset \mathbb{C}^r$). If $\rho : \Gamma \to G \subset \mathrm{GL}_r(\mathbb{C})$ is reducible, then it has a (maximal) filtration $W_0 = 0 \subsetneq W_1 \subsetneq \ldots \subsetneq W_m = \mathbb{C}^r$, such that the induced representations ρ_k on W_k/W_{k-1}, $k = 1, \ldots, m$, are irreducible. We call $\mathrm{Gr}(\rho) = \rho_1 \oplus \ldots \oplus \rho_m$ the semi-simplification of ρ, and we say that ρ and ρ' are

S-equivalent if they have the same semi-simplification. With all this said, the fibers of (2) are the S-equivalence classes [25, Theorem 1.28].

On the other hand, fixed an element $\gamma \in \Gamma$, we define the associated *character* as the map $\chi_\gamma : \mathfrak{X}_G(\Gamma) \longrightarrow \mathbb{C}$ given by $\chi_\gamma(\rho) = \operatorname{tr} \rho(\gamma)$. This defines a G-invariant function. The *character variety* is the algebraic space defined by these functions,

$$\chi_G(\Gamma) = \operatorname{Spec} \mathbb{C}[\chi_\gamma \,|\, \gamma \in \Gamma].$$

By the results of [23] and [25, Chapter 1], for $G =_n (\mathbb{C})$, $\mathrm{Sp}_{\not\models\ltimes}(\mathbb{C})$ or $\mathrm{SO}_{2n+1}(\mathbb{C})$ this is isomorphic to $\widehat{\mathcal{M}}_G(\Gamma)$.

The main focus of this chapter is the representation varieties for *surface groups*. Let Σ_g be a compact orientable surface of genus g. Its fundamental group is

$$\Gamma = \pi_1(\Sigma_g) = \left\langle a_1, b_1, \ldots, a_g, b_g \,\middle|\, \prod_{j=1}^{g} [a_j, b_j] = 1 \right\rangle. \tag{3}$$

The representation variety over the surface group $\pi_1(\Sigma_g)$, denoted by $\mathfrak{X}_G(\Sigma_g)$, parametrizes local systems over Σ_g. For $G = \mathrm{GL}_r(\mathbb{C})$, the variety $\mathfrak{X}_G(\Sigma_g)/\!\!/G$ is also known as the Betti moduli space in the context of non-abelian Hodge theory. Let $K = \mathrm{U}(r)$ be the maximal compact subgroup of $G = \mathrm{GL}_r(\mathbb{C})$. The celebrated theorem by Narasimhan and Seshadri in [33] establishes that if we give Σ_g a complex structure, then $\mathfrak{X}^{ss}_{\mathrm{U}(r)}(\Sigma_g)/\mathrm{U}(r)$ is isomorphic to the moduli space of (polystable) holomorphic bundles of degree 0 on Σ_g, where $\mathfrak{X}^{ss}_{\mathrm{U}(r)}(\Sigma_g)$ are the semi-simple representations. The Narasimhan–Seshadri correspondence can be considered an extension to higher ranks of the classical Hodge theorem. A representation $\rho : \pi_1(\Sigma_g) \to \mathrm{U}(1)$ can be regarded as a cohomology class $[\rho] \in H^1(\Sigma_g, \mathbb{C})$. Indeed, the $\mathfrak{X}_{\mathrm{U}(1)}(\Sigma_1)$ is isomorphic to

$$\mathrm{Hom}(\pi_1(\Sigma_g)/[\pi_1(\Sigma_g), \pi_1(\Sigma_g)], \mathrm{U}(1)) \cong \mathrm{Hom}(H_1(\Sigma_g), \mathbb{C}) \cong H^1(\Sigma_g, \mathbb{C}),$$

because U(1) is abelian. The classical Hodge theorem then says that there is a decomposition $\rho = \eta \oplus \omega$, where $\eta \in H^{0,1}(\Sigma_g)$ and $\omega \in H^{1,0}(\Sigma_g)$. Therefore, η provides us with a holomorphic line bundle, that is, a holomorphic object reflecting the algebraic structure of Σ_g.

In general, for a complex reductive group G, $\mathcal{M}_G(\Sigma_g) = \mathfrak{X}_G(\Sigma_g)/\!\!/G$ is a hyperkähler manifold, that is, a manifold, modelled on the quaternions, with three complex structures I, J, and K, where I is the complex structure inherited from the complex structure of the group G, in the same fashion as shown in Sect. 2.1, J is the complex structure provided by the complex structure of Σ_g as explained above, and K is the product JI. Therefore, by focusing on only one of the complex structures, three moduli spaces are obtained: the moduli space $\mathcal{M}_G(\Sigma_g)$ of representations of the fundamental group of Σ_g into G for complex structure I, also known as Betti moduli space; the moduli space of polystable G-Higgs bundles of degree 0 on Σ_g for complex structure J, called the Dolbeault moduli space; and the moduli space of polystable flat bundles on Σ_g with vanishing first Chern class, known as the de Rham moduli space. Moreover, the work of Corlette, Donaldson, Hitchin, and Simpson (see [4, 10, 21, 36–38]) proves that there are diffeomorphisms between the three moduli spaces: Betti, Dolbeault, and de Rham. These diffeomorphisms expand the Riemann–Hilbert correspondence and Narasimhan–Seshadri theorem into what is known as the *non-abelian Hodge correspondence*.

The diffeomorphism between $\mathcal{M}_G(\Sigma_g)$ and the Dolbeault moduli space has been largely exploited to obtain information on the topology of the character variety since Hitchin's work in [21]. Moreover, the rich interaction between string theory and the moduli space of G-Higgs bundles has driven the most recent research on character varieties. There exists a map, known as the Hitchin map, that shows the moduli space of Higgs bundles as a fibration over a vector space. This fibration was proved by Hausel and Thaddeus in [20] to be the first non-trivial example of mirror symmetry, following Strominger, Yau, and Zaslow's definition in [39]. That is, for Langlands dual groups G and ${}^LG$, the Hitchin map fibers over the same vector space in such a way that the fibers for the G-Higgs bundles moduli space are the dual Calabi–Yau manifolds to the fibers of the Hitchin map for ${}^LG$-Higgs bundles moduli space. In order to prove so, Hausel and Thaddeus studied the Hodge numbers for these moduli spaces. Since our non-abelian Hodge correspondence is not an algebraic isomorphism, it leads to one of the many motivations to study the Hodge numbers for character varieties. We introduce the Hodge numbers in Sect. 2.3.

This discussion is at the heart of much recent research that justifies the study of the geometry of character varieties of surface groups, in particular the Hodge numbers and E-polynomials (defined in Sect. 2.3), since they are algebro-geometric invariants associated with the complex structure. The first technique for this was the *arithmetic method* inspired in the Weil conjectures. Hausel and Rodríguez-Villegas started the computation of the E-polynomials of G-character varieties of surface groups for $G = \mathrm{GL}_n(\mathbb{C})$, $\mathrm{SL}_n(\mathbb{C})$ and $\mathrm{PGL}_n(\mathbb{C})$, using arithmetic methods. In [19], they obtained the E-polynomials of the Betti moduli spaces for $G = \mathrm{GL}_n(\mathbb{C})$ in terms of a simple generating function. Following these methods, Mereb [30] studied this case for $\mathrm{SL}_n(\mathbb{C})$, giving an explicit formula for the E-polynomial in the case $G = \mathrm{SL}_2(\mathbb{C})$. Recently, using this technique, explicit expressions of the E-polynomials have been computed [2] for orientable surfaces with $G = \mathrm{GL}_3(\mathbb{C})$, $\mathrm{SL}_3(\mathbb{C})$ and for non-orientable surfaces with $G = \mathrm{GL}_2(\mathbb{C})$, $\mathrm{SL}_2(\mathbb{C})$.

A *geometric method* to compute E-polynomials of character varieties of surfaces groups was initiated by Logares, Muñoz, and Newstead in [24]. In this method, the representation variety is chopped into simpler strata for which the E-polynomial can be computed. Following this idea, in the case $G = \mathrm{SL}_2(\mathbb{C})$, the E-polynomials were computed in a series of papers [24, 28, 29] and for $G = \mathrm{PGL}_2(\mathbb{C})$ in [27]. This method yields all the polynomials explicitly and not in terms of generating functions. Moreover, it allows to keep track of interesting properties, like the Hodge–Tate condition (cf. Remark 2) of these spaces.

In the papers [24, 29], the authors show that a recursive pattern underlies the computations. The E-polynomial of the $\mathrm{SL}_2(\mathbb{C})$-representation variety of Σ_g can be obtained from some data of the representation variety on the genus $g-1$ surface.

The recursive nature of character varieties is widely present in the literature as in [9, 18]. It suggests that some type of recursion formalism, in the spirit of a Topological Quantum Field Theory (TQFT for short), must hold. This leads to the third computational method, the *quantum method*, introduced in [13], which formalizes this setup and provides a powerful machinery to compute E-polynomials of character varieties. Moreover, this technique allows us to keep track of the classes in the Grothendieck ring of varieties (also known as virtual classes, as defined in section 2.4) of the representation varieties and had been successfully used in [15, 16] in the parabolic context, in which we deal with punctured surfaces with prescribed monodromy around the punctures.

This chapter applies the geometric, arithmetic, and quantum methods to the group of affine transformation of the line, $G = \mathrm{AGL}_1(\mathbb{C})$. The representations of this group parametrize (flat) rank one affine bundles $L \to \Sigma_g$, so it is a relevant space per se. Moreover, despite its simplicity, for $G = \mathrm{AGL}_1(\mathbb{C})$ the coincidence between the Bettin moduli space and the character variety is not granted by [5]. Nonetheless, we will directly prove in Sect. 3.2 that this isomorphism still holds. We shall see how the three methods apply, performing explicit computations of their virtual classes. In this way, our main result is the following.

Theorem 1 *Let $G = \mathrm{AGL}_1(\mathbb{C})$ and $g \geq 1$. The virtual class for the representation variety $\mathfrak{X}_{\mathrm{AGL}_1(\mathbb{C})}(\Sigma_g)$ is*

$$[\mathfrak{X}_{\mathrm{AGL}_1(\mathbb{C})}(\Sigma_g)] = q^{2g-1}(q-1)^{2g} + q^{2g} - q^{2g-1}.$$

2 General Background

2.1 Character Varieties

Let Γ be a finitely generated group and G an algebraic group over a ground field $\mathbb{K}$. A representation of Γ into G is a group homomorphism

$$\rho : \Gamma \longrightarrow G.$$

We shall denote the set of representations Hom (Γ, G), by $\mathfrak{X}_G(\Gamma)$. Since G is algebraic and Γ finitely presented, $\mathfrak{X}_G(\Gamma)$ inherits the structure of an algebraic variety. Indeed, if we consider a presentation $\Gamma = \langle \gamma_1, \ldots, \gamma_N \mid R_j(\gamma_1, \ldots, \gamma_N)\rangle$, then the homomorphism

$$\varphi : \mathfrak{X}_G(\Gamma) \longrightarrow G^N, \qquad \rho \mapsto (\rho(\gamma_1), \ldots, \rho(\gamma_N)),$$

describes an injection such that

$$\varphi(\mathfrak{X}_G(\Gamma)) = \left\{(g_1, \ldots, g_N) \in G^N \;\middle|\; R_j(g_1, \ldots, g_N)\right\},$$

so that $\varphi(\mathfrak{X}_G(\Gamma))$ is an affine algebraic variety.

The group G itself acts on $\mathfrak{X}_G(\Gamma)$ by conjugation, that is, $g \cdot \rho(\gamma) = g\rho(\gamma)g^{-1}$ for any $g \in G$, $\rho \in \mathfrak{X}_G(\Gamma)$ and $\gamma \in \Gamma$. We are interested on the orbits by this action since two representations are isomorphic if and only if they lie in the same orbit. But parametrizing these orbits requires the use of a subtler technique known as Geometric Invariant Theory (GIT). Let us explain this in some detail.

Example 1 Consider the simplest case where $\Gamma = \mathbb{Z}$, and let $G = \mathrm{SL}_2(\mathbb{C})$. Then, $\mathfrak{X}_{\mathrm{SL}_2(\mathbb{C})}(\mathbb{Z}) = \mathrm{SL}_2(\mathbb{C})$. The quotient $\mathrm{SL}_2(\mathbb{C})/\mathrm{SL}_2(\mathbb{C})$ contains the following orbits: if $g \in \mathrm{SL}_2(\mathbb{C})$ has two different eigenvalues λ and λ^{-1}, then the orbit of g is a closed one-dimensional space, namely the collection of matrices of trace $\lambda + \lambda^{-1}$. But in the case $\lambda = \lambda^{-1} = \pm 1$, we get a non-closed one-dimensional orbit and an orbit that consist of a point, which are, respectively,

$$\left[\begin{pmatrix} \pm 1 & 1 \\ 0 & \pm 1 \end{pmatrix}\right], \quad \left\{\begin{pmatrix} \pm 1 & 0 \\ 0 & \pm 1 \end{pmatrix}\right\}.$$

Moreover, for all $t \neq 0$, we have that the matrices

$$\begin{pmatrix} \pm 1 & t \\ 0 & \pm 1 \end{pmatrix} \in \left[\begin{pmatrix} \pm 1 & 1 \\ 0 & \pm 1 \end{pmatrix}\right]$$

but become the point orbit for $t = 0$. Therefore, $\mathrm{SL}_2(\mathbb{C})/\mathrm{SL}_2(\mathbb{C})$ is not an algebraic variety since its topology does not satisfy the T_1 separation axiom. The GIT quotient $\mathrm{SL}_2(\mathbb{C}) /\!\!/ \mathrm{SL}_2(\mathbb{C})$ solves this problem by collapsing the two 1-dimensional open orbits with the two orbits consisting on just a point. In this way, $\mathrm{SL}_2(\mathbb{C}) /\!\!/ \mathrm{SL}_2(\mathbb{C}) = \mathbb{C}$.

In general, for any algebraic group G acting on an affine variety X over $\mathbb{K}$, the action induces an action on the algebra of regular functions on X, $\mathcal{O}(X)$. In this case, the affine GIT quotient is defined as the morphism

$$\varphi : X \longrightarrow X /\!\!/ G := \operatorname{Spec} \mathcal{O}(X)^G$$

of affine schemes associated with the inclusion $\varphi^* : \mathcal{O}(X)^G \hookrightarrow \mathcal{O}(X)$, where $\mathcal{O}(X)^G$ is the subalgebra of G-invariant functions.

Remark 1 In general, the GIT quotient $X /\!\!/ G$ is only an affine scheme since $\mathcal{O}(X)^G$ might not be finitely generated (for an example of this phenomenon, see [31]). However, a theorem of Nagata [32] shows that, if G is a reductive group (cf. [34, Chapter 3]), then $\mathcal{O}(X)^G \subseteq \mathcal{O}(X)$ is finitely generated subalgebra, and, thus, $X /\!\!/ G$ is an affine variety. Many typical algebraic groups are reductive like $\mathrm{GL}_r(\mathbb{C})$, $\mathrm{SL}_r(\mathbb{C})$, or $\mathbb{C}^*$ with multiplication. However, an easy example of a non-reductive group is $\mathbb{C}$ with the sum.

The key point of the GIT quotient is that it is a quotient from a categorical point of view. A *categorical quotient* for X is a G-invariant regular map of algebraic varieties $\varphi : X \to Y$ such that for any G-invariant regular map of varieties $f : X \to Z$, there exists a unique $\tilde{f} : Y \to Z$ such that the following diagram commutes:

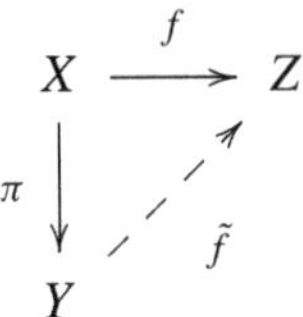

Using this universal property, it can be shown that if a categorical quotient exists, it is unique up to regular isomorphism. In this sense, it is straightforward (cf. [34, Corollary 3.5.1]) to check that the GIT quotient (if it is a variety, see Remark 1) is a categorical quotient. Thus, it is uniquely determined by this universal property.

Example 2 In Example 1, we have that the trace $\mathrm{tr} : \mathrm{SL}_2(\mathbb{C}) \longrightarrow \mathbb{C}$ is the only non-trivial $\mathrm{SL}_2(\mathbb{C})$-invariant function on $\mathrm{SL}_2(\mathbb{C})$. Therefore, $\mathrm{SL}_2(\mathbb{C}) /\!\!/ \mathrm{SL}_2(\mathbb{C}) = \mathrm{Spec}\, \mathbb{C}[\mathrm{tr}] = \mathbb{C}$. In general rank $r > 1$, we have that $\mathrm{SL}_r(\mathbb{C}) /\!\!/ \mathrm{SL}_r(\mathbb{C}) = \mathbb{C}^{r-1}$ with quotient map given by the coefficients of the characteristic polynomial.

Coming back to our case of study, we have an action of G on $\mathfrak{X}_G(\Gamma)$ by conjugation. The GIT quotient is called the *moduli space of representations*, and it is denoted as

$$\mathcal{M}_G(\Gamma) = \mathfrak{X}_G(\Gamma) /\!\!/ G.$$

By construction, there is a natural continuous map from the coset space $\widehat{\mathcal{M}}_G(\Gamma)$, which parametrizes the isomorphisms classes of representations of Γ into G, to this space $\widehat{\mathcal{M}}_G(\Gamma) \to \mathcal{M}_G(\Gamma)$.

However, if the ground ring is $\mathbb{K} = \mathbb{C}$ (or, in general, algebraically closed), we may consider another natural way of parametrize isomorphism classes of representations. Suppose that G is a linear algebraic group, so that $G < \mathrm{GL}_r(\mathbb{C})$. Given a representation $\rho : \Gamma \to G$, we define its character as the map

$$\chi_\rho : \Gamma \longrightarrow \mathbb{C}, \quad \gamma \mapsto \chi_\rho(\gamma) = \mathrm{tr}\, \rho(\gamma).$$

Note that two isomorphic representations ρ and ρ' have the same character, whereas the converse is also true if ρ and ρ' are *irreducible* (see [5, Proposition 1.5.2]). A representation is irreducible if it has no proper G-invariant subspaces of $\mathbb{C}^r$; otherwise, it is called *reducible*.

If ρ is reducible, let $\mathbb{C}^k \subset \mathbb{C}^r$ be a proper G-invariant subspace. Define $\rho_1 := \rho|_{\mathbb{C}^k}$, which is a representation on $\mathbb{C}^k$. There is an induced representation ρ_2 in the quotient $\mathbb{C}^{r-k} = \mathbb{C}^r / \mathbb{C}^k$. Then, we can write

$$\rho = \begin{pmatrix} \rho_1 & M \\ 0 & \rho_2 \end{pmatrix}.$$

Acting by conjugation by matrices $\begin{pmatrix} t\,\mathrm{Id} & 0 \\ 0 & \mathrm{Id} \end{pmatrix}$, we see that ρ is equivalent to $\rho_t = \begin{pmatrix} \rho_1 & tM \\ 0 & \rho_2 \end{pmatrix}$. When taking $t \to 0$, we have that ρ is in the same GIT orbit as $\begin{pmatrix} \rho_1 & 0 \\ 0 & \rho_2 \end{pmatrix} = \rho_1 \oplus \rho_2$. This is the same situation of Example 1. Repeating the argument with ρ_2, we have that any ρ is equivalent to some $\rho_1 \oplus \ldots \oplus \rho_l$, where ρ_j are irreducible. This is called a *semi-simple representation*. We say that they are S-equivalent and denote $\rho \sim \rho_1 \oplus \ldots \oplus \rho_l$. In this way, any point of the GIT quotient is determined by a unique class of semi-simple representation.

There is a character map

$$\chi : \mathfrak{X}_G(\Gamma) \longrightarrow \mathbb{C}^{\Gamma}, \quad \rho \mapsto \chi_\rho$$

whose image $\chi_G(\Gamma) = \chi(\mathfrak{X}_G(\Gamma))$ is called the *G-character variety* of Γ. Moreover, by the results in [5], there exists a collection $\gamma_1, \ldots, \gamma_a$ of elements of Γ such that χ_ρ is determined by $(\chi_\rho(\gamma_1), \ldots, \chi_\rho(\gamma_a))$, for any ρ. Such collection gives a map

$$\phi : \mathfrak{X}_G(\Gamma) \longrightarrow \mathbb{C}^a, \qquad \phi(\rho) = (\chi_\rho(\gamma_1), \ldots \chi_\rho(\gamma_a)),$$

and we have a bijection $\chi_G(\Gamma) \cong \phi(\mathfrak{X}_G(\Gamma))$, which endows $\chi_G(\Gamma)$ with the structure of an algebraic variety independent from the collection $\gamma_1, \ldots, \gamma_a$ chosen.

The character map $\chi : \mathfrak{X}_G(\Gamma) \to \chi_G(\Gamma)$ is a regular G-invariant map, so, since the GIT quotient is a categorical quotient, it induces a map

$$\tilde{\chi} : \mathcal{M}_G(\Gamma) \to \chi_G(\Gamma).$$

It is well-known that, when the group $G =_n (\mathbb{C})$, this map is an isomorphism [5]. This is the reason for the fact that sometimes the space $\widehat{\mathcal{M}}_G(\Gamma)$ is called the character variety. For different groups this isomorphism may still hold, as in this paper for $G = \mathrm{AGL}_1(\mathbb{K})$, or may not hold as in [11, Appendix A] for $G = \mathrm{SO}_2$. For a general discussion about the relation of $\widehat{\mathcal{M}}_G(\Gamma)$ and $\chi_G(\Gamma)$, see [23].

2.2 Representation Varieties of Orientable Surfaces

In this section, we shall focus on an important class of representation varieties, namely, those obtained by considering representations of the fundamental group of a closed surface, the so-called surface group. Let Σ_g be a compact orientable surface of genus g. We take $\Gamma = \pi_1(\Sigma_g)$, and we will focus on the representation variety $\mathfrak{X}_G(\pi_1(\Sigma_g))$, which we will shorten as $\mathfrak{X}_G(\Sigma_g)$. Using the presentation (3) of $\pi_1(\Sigma_g)$, we get that

$$\mathfrak{X}_G(\Sigma_g) = \left\{(A_1, B_1, \ldots, A_g, B_g) \in G^{2g} \;\middle|\; \prod_{j=1}^{g} [A_j, B_j]\right\} \subset G^{2g}.$$

The associated moduli space of representations, $\mathcal{M}_G(\Sigma_g) = \mathfrak{X}_G(\Sigma_g)/\!\!/G$, plays a fundamental role in the so-called non-abelian Hodge correspondence in the case $G = \mathrm{GL}_r(\mathbb{C})$ (respectively, $G = \mathrm{SL}_r(\mathbb{C})$). To be precise, consider a complex vector bundle

$$\pi : E \to \Sigma_g$$

of rank r and degree 0 (respectively, and trivial determinant line bundle) with a flat connection ∇ on E. By flatness, there is no local holonomy for ∇, so the holonomy map does not depend on the homotopy class of the loop, and hence it descends to a map, called the *monodromy*

$$\rho_\nabla : \pi_1(\Sigma_g) \to G.$$

This is a representation in $\mathfrak{X}_G(\Sigma_g)$. The isomorphism class of the pair (E, ∇) is given by changing the basis of the fiber $E_{x_0} = \mathbb{C}^r$ over the base point $x_0 \in \Sigma_g$. This produces the action by conjugation of G on $\mathfrak{X}_G(\Sigma_g)$.

In this way, the moduli of representations $\mathcal{M}_G(\Sigma_g) = \mathfrak{X}_G(\Sigma_g)/\!\!/G$ parametrizes the moduli space of classes of pairs (E, ∇) of flat connections on a vector bundle (modulo S-equivalence). In this context, the former space is usually referred to as the Betti moduli space (it captures topological information of Σ_g) and the later space is called the de Rham moduli space (it captures differentiable information of Σ_g).

2.3 *Mixed Hodge Structures*

In order to understand the geometry of representation varieties of surface groups, we will focus on an algebro-geometric invariant that is naturally present in the cohomology of complex varieties, the so-called *Hodge structure*. For this reason, in this section, we will consider that the ground ring is $\mathbb{C}$, and we will sketch briefly some remarkable properties of Hodge theory. For a more detailed introduction to Hodge theory, see [35].

A pure Hodge structure of weight k consists of a finite-dimensional rational vector space H whose complexification $H_\mathbb{C} = H \otimes_\mathbb{Q} \mathbb{C}$ is equipped with a decomposition

$$H_{\mathbb{C}} = \bigoplus_{k=p+q} H^{p,q},$$

such that $H^{q,p} = \overline{H^{p,q}}$, the bar meaning complex conjugation on H. A Hodge structure of weight k gives rise to the so-called Hodge filtration, which is a descending filtration $F^p = \bigoplus_{s \geq p} H^{s,k-s}$. From this filtration, we can recover the pieces via the graded complex $\mathrm{Gr}_F^p(H) := F^p/F^{p+1} = H^{p,k-p}$.

A mixed Hodge structure consists of a finite-dimensional rational vector space H, an ascending (weight) filtration $0 \subset \ldots \subset W_{k-1} \subset W_k \subset \ldots \subset H$ and a descending (Hodge) filtration $H_{\mathbb{C}} \supset \ldots \supset F^{p-1} \supset F^p \supset \ldots \supset 0$ such that F induces a pure Hodge structure of weight k on each $\mathrm{Gr}_k^W(H) = W_k/W_{k-1}$. We define the associated Hodge pieces as

$$H^{p,q} := \mathrm{Gr}_F^p \mathrm{Gr}_{p+q}^W(H)_{\mathbb{C}}$$

and write $h^{p,q}$ for the *Hodge number* $h^{p,q} := \dim_{\mathbb{C}} H^{p,q}$.

The importance of these mixed Hodge structures rises from the fact that the cohomology of complex algebraic varieties is naturally endowed with such structures, as proved by Deligne.

Theorem 2 (Deligne [6–8]) *Let X be any quasi-projective complex algebraic variety (maybe non-smooth or non-compact). The rational cohomology groups $H^k(X)$ and the cohomology groups with compact support $H_c^k(X)$ are endowed with mixed Hodge structures.*

In this way, for any complex algebraic variety X, we define the *Hodge numbers* of X by

$$h^{k,p,q}(X) = h^{p,q}(H^k(Z)) = \dim \mathrm{Gr}_F^p \mathrm{Gr}_{p+q}^W H^k(X)_{\mathbb{C}},$$

$$h_c^{k,p,q}(X) = h^{p,q}(H_c^k(Z)) = \dim \mathrm{Gr}_F^p \mathrm{Gr}_{p+q}^W H_c^k(X)_{\mathbb{C}}.$$

The E-polynomial (also called the Deligne–Hodge polynomial) is defined as

$$e(X) = e(X)(u,v) := \sum_{p,q,k} (-1)^k h_c^{k,p,q}(X) u^p v^q.$$

The key property of E-polynomials that permits their calculation is that they are additive for stratifications of X. If X is a complex algebraic variety and $X = \bigsqcup_{i=1}^{n} X_i$, where all X_i are locally closed in X, then $e(X) = \sum_{i=1}^{n} e(X_i)$. Moreover, if $X = F \times B$, the Küneth isomorphism implies that $e(X) = e(F)e(B)$.

An easy consequence of these two properties is that, indeed, for an algebraic bundle (that is, locally trivial in the Zariski topology)

$$F \longrightarrow X \stackrel{\pi}{\longrightarrow} B,$$

we have $e(X) = e(F)e(B)$. For this, just take a Zariski open subset $U \subset B$ so that $X|_U = \pi^{-1}(U) \cong U \times B$. Then, $B_1 = B - U$ is closed, and we can repeat the argument for $F \to X|_{B_1} \to B_1$. By the noethereanity, we get a finite chain

$$B_{n+1} = \emptyset \subsetneq B_n \subsetneq \ldots \subsetneq B_1 \subsetneq B = B_0,$$

where $U_k = B_{k-1} - B_k$ is Zariski open in B_{k-1} and $X|_{U_k} \cong U_k \times B$. Then,

$$e(X) = \sum_k e(X|_{U_k}) = \sum_k e(F)e(U_k) = e(F) \sum_k e(U_k) = e(F)e(B). \tag{4}$$

Example 3 Recall that the cohomology of the complex projective space, $H^\bullet(\mathbb{P}^n)$, is generated by the Fubini–Study Form, which is of type $(1, 1)$, so we get $h_c^{2p,p,p}(\mathbb{P}^n) = 1$ for $0 \leq p \leq n$, and 0 otherwise. Hence, its E-polynomial is $e(\mathbb{P}^n) = 1 + uv + u^2v^2 + \ldots + u^nv^n$. In particular, since $\mathbb{P}^1 = \mathbb{C} \sqcup \{\infty\}$, we get that $e(\mathbb{C}) = e(\mathbb{P}^1) - 1 = uv$. In this way, we get that $e(\mathbb{C}^n) = u^nv^n$, which is compatible with the usual decomposition $\mathbb{P}^n = \star \sqcup \mathbb{C} \sqcup \mathbb{C}^2 \sqcup \ldots \sqcup \mathbb{C}^n$.

Remark 2 When $h_c^{k,p,q}(X) = 0$ for $p \neq q$, the polynomial $e(X)$ depends only on the product uv. This will happen in all the cases that we shall investigate here. In this situation, it is conventional to use the variable $q = uv$. If this happens, we say that the variety is of Hodge–Tate type (also known as balanced type). For instance, $e(\mathbb{C}^n) = q^n$ is Hodge–Tate.

2.4 Grothendieck Ring of Algebraic Varieties

It is well known that from a (skeletally small) abelian category $\mathcal{A}$, it is possible to construct an abelian group, known as the Grothendieck group of $\mathcal{A}$. It is the abelian group $\mathrm{K}\mathcal{A}$ generated by the isomorphism classes $[A]$ of objects $A \in \mathcal{A}$, subject to the relations that whenever there exists a short exact sequence $0 \to B \to A \to C \to 0$, we declare $[A] = [B] + [C]$. Furthermore, if our abelian category is provided with a tensor product, i.e., $\mathcal{A}$ is monoidal, and the functors $- \otimes A : \mathcal{A} \to \mathcal{A}$ and $A \otimes - : \mathcal{A} \to \mathcal{A}$ are exact, then $\mathrm{K}\mathcal{A}$ inherits a ring structure by $[A] \cdot [B] = [A \otimes B]$ (see [40]), under which it is called the *Grothendieck ring* of $\mathcal{A}$. The elements $[A] \in \mathrm{K}\mathcal{A}$ are usually referred to as *virtual classes*.

In our case, we are interested on the category of algebraic varieties with regular morphisms $\mathbf{Var}_{\mathbb{K}}$ over a base field $\mathbb{K}$, which is not an abelian category. Nevertheless, we can still construct its Grothendieck group, $\mathrm{K}\mathbf{Var}_{\mathbb{K}}$, in an analogous manner, that is, as the abelian group generated by isomorphism classes of algebraic varieties with the relation that $[X] = [Y] + [U]$ if $X = Y \sqcup U$, with $Y \subset X$ a closed subvariety. Furthermore, the Cartesian product of varieties also provides $\mathrm{K}\mathbf{Var}_{\mathbb{K}}$ with a ring structure. A very important element is the class of the affine line, $q = [\mathbb{K}] \in \mathbf{KVar}_{\mathbb{K}}$, the so-called *Lefschetz motive*.

Remark 3 Despite the simplicity of its definition, the ring structure of $\mathbf{KVar}_{\mathbb{K}}$ is widely unknown. In particular, for almost 50 years, it was an open problem whether it is an integral domain. Indeed, the answer is no and, more strikingly, the Lefschetz motive q is a zero divisor [3].

Observe that, due to its additivity and multiplicativity properties, the E-polynomial defines a ring homomorphism

$$e : \mathbf{KVar}_{\mathbb{C}} \to \mathbb{Z}[u^{\pm 1}, v^{\pm 1}].$$

This homomorphism factorizes through mixed Hodge structures. To be precise, Deligne proved in [6] that the category of mixed Hodge structures **MHS** is an abelian category. Therefore, we may as well consider its Grothendieck group, **KMHS**, which again inherits a ring structure. The long exact sequence in cohomology with compact support and the Künneth isomorphism shows that there exist ring homomorphisms $\mathbf{KVar}_{\mathbb{C}} \to \mathbf{KMHS}$ given by $[X] \mapsto [H_c^\bullet(X)]$, as well as $\mathbf{KMHS} \to \mathbb{Z}[u^{\pm 1}, v^{\pm 1}]$ given by $[H] \mapsto \sum h^{p,q}(H) u^p v^q$ such that the following diagram commutes:

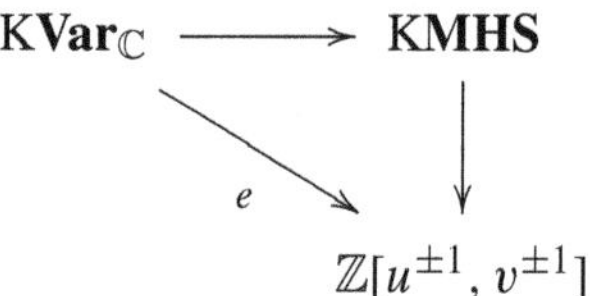

Remark 4 From the previous diagram, we get that the E-polynomial of the affine line is $q = e([\mathbb{C}])$, which justifies denoting by $q = [\mathbb{C}] \in \mathbf{KVar}_{\mathbb{C}}$ the Lefschetz motive. This implies that if the virtual class of a variety lies in the subring of $\mathbf{KVar}_{\mathbb{K}}$ generated by the affine line, then the E-polynomial of the variety coincides with the virtual class, seeing q as a variable. This will have deep implications, as we will explore in the arithmetic method in Sect. 4.

Example 4 As for E-polynomials, proceeding as in (4), we can show that if $F \to E \to B$ is an algebraic bundle, then $[E] = [F]\cdot[B]$ in $\mathbf{KVar}_{\mathbb{K}}$. This enables multiple computations. For instance, consider the fibration $\mathbb{C} \to \mathrm{SL}_2(\mathbb{C}) \to \mathbb{C}^2 - \{(0,0)\}$, $f \mapsto f(1,0)$. It is locally trivial in the Zariski topology, and therefore $[\mathrm{SL}_2(\mathbb{C})] = [\mathbb{C}] \cdot [\mathbb{C}^2 - \{(0,0)\}] = q(q^2 - 1) = q^3 - q$.

It is of interest to notice that one can compute $e(\mathrm{PGL}_2(\mathbb{C})) = e(\mathrm{SL}_2(\mathbb{C}))$, which is of no surprise since these groups are Langlands dual.

3 Geometric Method

Using the previous machinery, let us show in a simple situation how to compute the virtual classes of representation varieties for surface groups. We will do this computation by three different approaches, the so-called geometric, arithmetic, and quantum methods. The first geometric method, which we will follow in this section, is based on giving an explicit expression of the representation variety and chopping it into simpler pieces to ensemble the total virtual class. This is the method used in [24, 28, 29] to compute the $\mathrm{SL}_2(\mathbb{C})$-character varieties of surface groups. In Sect. 4, we shall use the arithmetic methods of [19], based on counting the number of points of the representation variety over finite fields. Finally, in Sect. 5, we shall use the machinery of the Topological Quantum Field Theories developed in [13] to offer an alternative approach.

Let Σ_g be the closed oriented surface of genus $g \geq 1$ as before. As target group we fix $G = \mathrm{AGL}_1(\mathbb{K})$, the group of $\mathbb{K}$-linear affine transformations of the affine line. Its elements are the matrices of the form $\begin{pmatrix} a & b \\ 0 & 1 \end{pmatrix}$, with $a \in \mathbb{K}^* = \mathbb{K} - \{0\}$ and $b \in \mathbb{K}$. The group operation is given by matrix multiplication. In this way, $\mathrm{AGL}_1(\mathbb{K})$ is isomorphic to the semidirect product $\mathbb{K}^* \ltimes_\varphi \mathbb{K}$ with the action $\varphi : \mathbb{K}^* \times \mathbb{K} \to \mathbb{K}$, $\varphi(a, b) = ab$.

The representation variety is given by

$$\mathfrak{X}_{\mathrm{AGL}_1(\mathbb{K})}(\Sigma_g) = \left\{(A_1, A_2, \ldots, A_{2g-1}, A_{2g}) \in \mathrm{AGL}_1(\mathbb{K})^{2g} \,\middle|\, \prod_{i=1}^{g} [A_{2i-1}, A_{2i}] = \mathrm{I}\right\}.$$

Therefore, if we write

$$A_i = \begin{pmatrix} a_i & b_i \\ 0 & 1 \end{pmatrix},$$

then the product of commutators is given by

$$\prod_{i=1}^{g} \left[\begin{pmatrix} a_{2i-1} & b_{2i-1} \\ 0 & 1 \end{pmatrix}, \begin{pmatrix} a_{2i} & b_{2i} \\ 0 & 1 \end{pmatrix} \right] = \begin{pmatrix} 1 & \sum_{i=1}^{g} (a_{2i-1} - 1)b_{2i} - (a_{2i} - 1)b_{2i-1} \\ 0 & 1 \end{pmatrix}. \tag{5}$$

We can identify this variety with a more familiar space. Consider the auxiliary variety

$$X_s = \left\{(\alpha_1, \ldots, \alpha_s, \beta_1, \ldots, \beta_s) \in (\mathbb{K} - \{-1\})^s \times \mathbb{K}^s \,\middle|\, \sum_{i=1}^{s} \alpha_i \beta_i = 0\right\}, \tag{6}$$

so that

$$\mathfrak{X}_{\mathrm{AGL}_1(\mathbb{K})}(\Sigma_g) \cong X_{2g}$$

via the morphism $(a_{2i-1}, b_{2i-1}, a_{2i}, b_{2i}) \mapsto (a_{2i-1} - 1, a_{2i} - 1, \ldots, b_{2i}, -b_{2i-1})$. Take $U = (\mathbb{K} - \{-1\})^s - \{(0, \ldots, 0)\}$ and $V = U \times \mathbb{K}^s$. We have that $X_s|_V$ is the pullback of the total space of the hyperplane bundle on $\mathbb{P}^{s-1}$, $\mathcal{O}_{\mathbb{P}^{s-1}}(1)$, via the natural quotient map $\pi : U \subset \mathbb{K}^s - \{0\} \to \mathbb{P}^{s-1}$; that is, we have a pullback

$$\begin{array}{ccc} X_s|_V = \pi^*\mathcal{O}_{\mathbb{P}^{s-1}}(1) & \longrightarrow & \mathcal{O}_{\mathbb{P}^{s-1}}(1) \\ \downarrow & & \downarrow \\ U & \xrightarrow[\pi]{} & \mathbb{P}^{s-1} \end{array}$$

On the special fiber, $X_s|_{\{(0,\ldots,0)\}\times\mathbb{K}^s} = \mathbb{K}^s$, which corresponds to the natural completion of the total space of the hyperplane bundle to the origin.

3.1 Stratification Analysis and Computation of Virtual Classes

Using this explicit description, we can compute the virtual class of the representation variety in a geometric way, by chopping the variety into simpler pieces, as shown in the following result.

Theorem 3 *The virtual class in the Grothendieck ring of algebraic varieties of the representation variety is*

$$[\mathfrak{X}_{\mathrm{AGL}_1(\mathbb{K})}(\Sigma_g)] = q^{2g-1}\left((q-1)^{2g} + q - 1\right).$$

Proof We stratify the varieties X_s in the following manner:

$$\begin{aligned} X_s &= \left\{\beta_s = \frac{1}{\alpha_s}\sum_{i=1}^{s-1}\alpha_i\beta_i, \alpha_s \neq 0\right\} \bigsqcup \left\{\sum_{i=1}^{s-1}\alpha_i\beta_i = 0, \alpha_s = 0\right\} \\ &= \left(((\mathbb{K} - \{-1\}) \times \mathbb{K})^{s-1} \times (\mathbb{K} - \{0, -1\})\right) \sqcup (X_{s-1} \times \mathbb{K}). \end{aligned}$$

This gives rise to the recursive formula for the virtual classes

$$[X_s] = (q-2)q^{s-1}(q-1)^{s-1} + q[X_{s-1}].$$

The base case is

$$X_1 = \{(\alpha, \beta)|\alpha\beta = 0\} = \left\{\beta = \frac{1}{\alpha}, \alpha \neq 0, -1\right\} \sqcup \{(0, \beta)\} = (\mathbb{K} - \{0, -1\}) \sqcup \mathbb{K},$$

which has $[X_1] = 2q - 2$. The induction gives

$$\begin{aligned}
[X_s] &= \sum_{t=1}^{s-1} (q-2)q^{s-t}(q-1)^{s-t}q^{t-1} + q^{s-1}(2q-2) \\
&= (q-2)q^{s-1}\frac{(q-1)^s - (q-1)}{(q-1)-1} + 2q^{s-1}(q-1) \\
&= q^{s-1}\big((q-1)^s - (q-1)\big) + 2q^{s-1}(q-1) \\
&= q^{s-1}(q-1)^s + q^s - q^{s-1}.
\end{aligned}$$

The representation variety is $\mathfrak{X}_{\mathrm{AGL}_1(\mathbb{K})}(\Sigma_g) \cong X_{2g}$, hence the result.

Remark 5 In the case that $\mathbb{K} = \mathbb{C}$, the same formula of Theorem 3 gives the E-polynomial of the representation variety by seeing q as a formal variable.

3.2 The Moduli Space of the Representations and the Character Variety

In this section, we will deal with the moduli space of representations, that is, the GIT quotient

$$\mathcal{M}_{\mathrm{AGL}_1(\mathbb{K})}(\Sigma_g) = \mathfrak{X}_{\mathrm{AGL}_1(\mathbb{K})}(\Sigma_1) /\!\!/ \mathrm{AGL}_1(\mathbb{K}).$$

For that purpose, let us write down the action explicitly. Consider elements

$$P = \begin{pmatrix} \lambda & \mu \\ 0 & 1 \end{pmatrix} \in \mathrm{AGL}_1(\mathbb{K}), \quad \rho = \left(\begin{pmatrix} a_1 & b_1 \\ 0 & 1 \end{pmatrix}, \ldots, \begin{pmatrix} a_{2g} & b_{2g} \\ 0 & 1 \end{pmatrix}\right) \in \mathfrak{X}_{\mathrm{AGL}_1(\mathbb{K})}(\Sigma_g),$$

then we have that

$$P\rho P^{-1} = \left(\begin{pmatrix} a_1 & \lambda b_1 + \mu(a_1 - 1) \\ 0 & 1 \end{pmatrix}, \ldots, \begin{pmatrix} a_{2g} & \lambda b_{2g} + \mu(a_{2g} - 1) \\ 0 & 1 \end{pmatrix}\right).$$

Remark 6 This action can also be understood in terms of X_{2g}. In these coordinates, the action of $(\lambda, \mu) \in \mathbb{K}^* \ltimes_\varphi \mathbb{K} = \mathrm{AGL}_1(\mathbb{K})$ is given by

$$\begin{aligned}
&(\lambda, \mu) \cdot (\alpha_1, \ldots, \alpha_{2g}, \beta_1, \ldots, \beta_{2g}) = \\
&\quad = (\alpha_1, \ldots, \alpha_{2g}, \lambda\beta_1 + \mu\alpha_2, \lambda\beta_2 - \mu\alpha_1, \ldots, \lambda\beta_{2g-1} + \mu\alpha_{2g}, \lambda\beta_{2g} - \mu\alpha_{2g-1}).
\end{aligned}$$

In particular, if we take $\mu = 0$, we have that the action is given by

$$P\rho P^{-1} = \left(\begin{pmatrix} a_1 & \lambda b_1 \\ 0 & 1 \end{pmatrix}, \ldots, \begin{pmatrix} a_{2g} & \lambda b_{2g} \\ 0 & 1 \end{pmatrix} \right) \stackrel{\lambda \to 0}{\longrightarrow} \left(\begin{pmatrix} a_1 & 0 \\ 0 & 1 \end{pmatrix}, \ldots, \begin{pmatrix} a_{2g} & 0 \\ 0 & 1 \end{pmatrix} \right).$$

Therefore, any representation is S-equivalent to a diagonal representation, which implies that

$$\mathcal{M}_{\mathrm{AGL}_1(\mathbb{K})}(\Sigma_g) = \mathfrak{X}_{\mathrm{AGL}_1(\mathbb{K})}(\Sigma_1) /\!\!/ \mathrm{AGL}_1(\mathbb{K}) = (\mathbb{K}^*)^{2g},$$

so we get that $[\mathcal{M}_{\mathrm{AGL}_1(\mathbb{K})}(\Sigma_g)] = (q-1)^{2g}$.

On the other hand, we also have the character variety $\chi_{\mathrm{AGL}_1(\mathbb{K})}(\Sigma_g)$ generated by the characters of the representations, as described in Sect. 2.1. Observe that given

$$\rho = (\rho(\gamma_1), \ldots, \rho(\gamma_{2g})) = \left(\begin{pmatrix} a_1 & b_1 \\ 0 & 1 \end{pmatrix}, \ldots, \begin{pmatrix} a_{2g} & b_{2g} \\ 0 & 1 \end{pmatrix} \right) \in \mathfrak{X}_{\mathrm{AGL}_1(\mathbb{K})}(\Sigma_g),$$

where $\gamma_1, \ldots, \gamma_{2g}$ are the standard generators of $\pi_1(\Sigma_g)$, its character is determined by the tuple

$$(\rho(\gamma_1), \ldots, \rho(\gamma_{2g})) = (a_1 + 1, \ldots, a_{2g} + 1) \in (\mathbb{K} - \{1\})^{2g}.$$

Reciprocally, any tuple of $(\mathbb{K} - \{1\})^{2g}$ is the character of an $\mathrm{AGL}_1(\mathbb{K})$-representation, namely, the diagonal one. Hence, we have that

$$\chi_{\mathrm{AGL}_1(\mathbb{K})}(\Sigma_g) = \phi(\mathfrak{X}_{\mathrm{AGL}_1(\mathbb{K})}) = (\mathbb{K} - \{1\})^{2g}.$$

In particular, this shows that $[\chi_{\mathrm{AGL}_1(\mathbb{K})}(\Sigma_g)] = (q-1)^{2g}$. Observe that we indeed have an isomorphism $\mathcal{M}_{\mathrm{AGL}_1(\mathbb{K})}(\Sigma_g) \cong \chi_{\mathrm{AGL}_1(\mathbb{K})}(\Sigma_g)$ given by $(a_1, \ldots, a_{2g}) \mapsto (a_1 + 1, \ldots, a_{2g} + 1)$. Notice that this isomorphism is not directly provided by [5], since the group $\mathrm{AGL}_1(\mathbb{K})$ is not directly provided by [5].

4 Arithmetic Method

In this section, we explore a different approach to the computation of E-polynomials with an arithmetic flavor. This approach was initiated with the works of Hausel and Rodríguez-Villegas [19]. The key idea is based on a theorem of Katz that, roughly speaking, states that if the number of points of a variety X over the finite field of q elements is a polynomial in q, $P(q) = |X(\mathbb{F}_q)|$, then the E-polynomial of $X(\mathbb{C})$ is also $P(q)$. Under this point of view, the computation of E-polynomials reduces to the arithmetic problem of counting points over finite fields.

4.1 Katz Theorem and E-Polynomials

Let us explain the result proved in [19, Appendix]. Start with a scheme $X/\mathbb{C}$ over $\mathbb{C}$. Let R be a subring of $\mathbb{C}$, which is finitely generated as a $\mathbb{Z}$-algebra, and let $\mathcal{X}$ be a separated R-scheme of finite type. We call $\mathcal{X}$ a *spreading out* of X if it yields X after extension of scalars from R to $\mathbb{C}$.

We say that $\mathcal{X}$ is *strongly polynomial count* if there exists a polynomial $P_{\mathcal{X}}(T) \in \mathbb{C}[T]$ such that for any finite field $\mathbb{F}_q$ and any ring homomorphism $\varphi : R \to \mathbb{F}_q$, the $\mathbb{F}_q$-scheme $\mathcal{X}^\varphi$ obtained from $\mathcal{X}$ by base change satisfies that for every finite extension $\mathbb{F}_{q^n}/\mathbb{F}_q$, we have

$$\#\mathcal{X}^\varphi(\mathbb{F}_{q^n}) = P_{\mathcal{X}}(q^n).$$

We say that a scheme $X/\mathbb{C}$ is *polynomial count* if it admits a spreading out $\mathcal{X}$ which is strongly polynomial count.

The following theorem is due to Katz [19, Appendix]. It computes the E-polynomial of X from the count of points of a spreading $\mathcal{X}$.

Theorem 4 *Assume that X is polynomial count with counting polynomial $P_{\mathcal{X}}(T) \in \mathbb{C}[T]$. Then,*

$$e(X) = P_{\mathcal{X}}(q),$$

where $q = uv$.

This is a powerful result that computes E-polynomials of varieties via arithmetic. For instance, it explains easily the equality $e(X) = e(U) + e(Y)$, when $Y \subset X$ is a closed subset and $U = X - Y$ is the (open) complement. Certainly, in this case,

$$\#\mathcal{X}^\varphi(\mathbb{F}_{q^n}) = \left(\#\mathcal{Y}^\varphi(\mathbb{F}_{q^n})\right) + \left(\#\mathcal{U}^\varphi(\mathbb{F}_{q^n})\right),$$

for spreadings $\mathcal{X}, \mathcal{Y}$, and $\mathcal{U}$ of X, Y, and Z, respectively. Therefore, $P_{\mathcal{X}}(T) = P_{\mathcal{Y}}(T) + P_{\mathcal{U}}(T)$, because they coincide on an infinity of values $T = q^n$. Note in particular that if $\mathcal{Y}$ and $\mathcal{U}$ are strongly polynomial count, then $\mathcal{X}$ is also strongly polynomial count. This also implies that the polynomial count only depends on the class in the Grothendieck ring.

The drawback of the arithmetic method is that it does not give information on the finer algebraic structure of the (mixed) Hodge polynomials or the classes in the Grothendieck ring of varieties. For instance, the E-polynomial of an elliptic curve X is $e(X) = 1 - u - v + uv$, which is not a polynomial in $q = uv$, and thus, X cannot be polynomial count.

Corollary 1 *Suppose that X has class in the Grothendieck ring $[X] = P(q)$, where P is a polynomial in the Lefschetz motive $q = [\mathbb{C}]$. Then, X is polynomial count with $e(X) = P(q)$, $q = uv$.*

Proof As the statement only depends on the class in the Grothendieck ring, it is enough to prove it for q^m, that is, $X = \mathbb{C}^m$, for $m \geq 0$, where $P(T) = T^m$. The spreading for X is given by $\mathcal{X} = \text{Spec}\, \mathbb{Z}[x_1, \dots, x_m]$ and $\mathcal{X}^\varphi = \text{Spec}\, \mathbb{F}_q[x_1, \dots, x_m] = \mathbb{F}_q^m$. Therefore, $\#\mathcal{X}^\varphi(\mathbb{F}_{q^n}) = \#\mathbb{F}_{q^n}^m = (q^n)^m = P(q^n)$. Hence, X is of polynomial count and its polynomial is $P(T) = T^m$. See also Remark 4.

In our situation, we start with an affine variety, which is of the form

$$X = \text{Spec}\, \frac{\mathbb{C}[x_1, \dots, x_N]}{I},$$

for some ideal $I = (p_1, \dots, p_M)$, defined by polynomials $p_1, \dots, p_M \in \mathbb{C}[x_1, \dots, x_N]$. Take the coefficients of the polynomials, which are complex numbers, and let $R \subset \mathbb{C}$ be the $\mathbb{Z}$-algebra generated by them. Then, $p_1, \dots, p_M \in R[x_1, \dots, x_N]$. A spreading of X is given by

$$\mathcal{X} = \text{Spec}\, \frac{R[x_1, \dots, x_N]}{(p_1, \dots, p_M)}.$$

A homomorphism $\varphi : R \to \mathbb{F}_q$ defines polynomials $\bar{p}_j = \varphi(p_j) \in \mathbb{F}_q[x_1, \dots, x_N]$, $j = 1, \dots, m$, and

$$\mathcal{X}^\varphi = \text{Spec}\, \frac{\mathbb{F}_q[x_1, \dots, x_N]}{(\bar{p}_1, \dots, \bar{p}_M)}.$$

This variety is

$$\mathcal{X}^\varphi = V(\bar{p}_1, \dots, \bar{p}_M) \subset \mathbb{F}_q^N,$$

and the $\mathbb{F}_{q^n}$-points of $\mathcal{X}^\varphi$ are the solutions over $\mathbb{F}_{q^n}$ to the equations:

$$\bar{p}_1(x_1, \dots, x_N) = 0, \dots, \bar{p}_M(x_1, \dots, x_N) = 0.$$

4.2 *Representation Variety for the Affine Group*

Let us take $G = \text{AGL}_1(\mathbb{C})$, the group of $\mathbb{C}$-linear affine transformations of the complex line. As mentioned before, the character variety is $\mathfrak{X}_{\text{AGL}_1(\mathbb{C})}(\Sigma_g) \cong X_{2g}$, where

$$X_s = \left\{ (\alpha_1, \dots, \alpha_s, \beta_1, \dots, \beta_s) \in (\mathbb{C} - \{-1\})^s \times \mathbb{C}^s \,\middle|\, \sum_{i=1}^{s} \alpha_i \beta_i = 0 \right\}.$$

The spreading of X_s is given by taking the base ring $R = \mathbb{Z}$ and the $\mathbb{Z}$-variety defined by

$$\mathcal{X}_s = \operatorname{Spec} \frac{\mathbb{Z}[\alpha_1, (\alpha_1+1)^{-1}, \ldots, \alpha_s, (\alpha_s+1)^{-1}, \beta_1, \ldots, \beta_s]}{\left(\sum_i \alpha_i \beta_i\right)}.$$

Take a prime q and the quotient map $\varphi : \mathbb{Z} \to \mathbb{Z}_q = \mathbb{F}_q$. This is followed by the embedding (scalar extension) $\mathbb{F}_q \subset \mathbb{F}_{q^n}$. Hence,

$$\mathcal{X}_s^{\varphi}(\mathbb{F}_{q^n}) = \left\{(\alpha_1, \ldots, \alpha_s, \beta_1, \ldots, \beta_s) \in (\mathbb{F}_{q^n} - \{-1\})^s \times \mathbb{F}_{q^n}^s \;\middle|\; \sum_{i=1}^{s} \alpha_i \beta_i = 0\right\},$$

and we want to count the number of points.

Theorem 5 *The variety $\mathcal{X}_s$ is strongly polynomial count with polynomial $P_{\mathcal{X}_s}(T) = T^{s-1}(T-1)^s + T^s - T^{s-1}$. In particular, the E-polynomial of $\mathfrak{X}_{\mathrm{AGL}_1(\mathbb{C})}(\Sigma_g) \cong X_{2g}$ is*

$$e(\mathfrak{X}_{\mathrm{AGL}_1(\mathbb{C})}(\Sigma_g)) = q^{2g-1}(q-1)^{2g} + q^{2g} - q^{2g-1}.$$

Proof Let

$$L = \left\{(\alpha_1, \ldots, \alpha_s, \beta_1, \ldots, \beta_s) \in \mathbb{F}_{q^n}^{2s} \;\middle|\; \sum \alpha_i \beta_i = 0\right\}.$$

There is a map

$$\varpi : L \to \mathbb{F}_{q^n}^s, \quad \varpi(\alpha_1, \ldots, \alpha_s, \beta_1, \ldots, \beta_s) = (\alpha_1, \ldots, \alpha_s).$$

This is surjective, and $\varpi^{-1}(\alpha)$ is a hyperplane of $(\mathbb{F}_{q^n})^s$ for $\alpha \neq (0, \ldots, 0)$, and all the space for $\alpha_0 = (0, \ldots, 0)$. Hence,

$$\begin{aligned}
\#L &= (\#\varpi^{-1}(\alpha)) \cdot (\#(\mathbb{F}_{q^n})^s - 1) + \#(\mathbb{F}_{q^n})^s \\
&= (q^n)^{s-1}((q^n)^s - 1) + (q^n)^s \\
&= (q^n)^{2s-1} + (q^n)^s - (q^n)^{s-1}.
\end{aligned}$$

Now, define the hyperplanes for $i = 1, \ldots, s$

$$\hat{H}_i = \{(\alpha_1, \ldots, \alpha_s) \in \mathbb{F}_{q^n}^s \mid \alpha_i = -1\}, \quad H_i = \hat{H}_i \times \mathbb{F}_{q^n}^s.$$

We have to remove the contributions to L of these hyperplanes. Observe that $H_{i_1} \cap \ldots \cap H_{i_t} \cap L = \varpi^{-1}(\hat{H}_{i_1} \cap \ldots \cap \hat{H}_{i_t})$, for $t \geq 1$, and in this case all fibers of ϖ are hyperplanes. Thus,

$$\#(H_{i_1} \cap \ldots \cap H_{i_t} \cap L) = (q^n)^{2s-t-1}.$$

Hence, by the inclusion–exclusion argument,

$$\begin{aligned}\#\big((\mathbb{F}_{q^n})^{2s} - (H_1 \cup \ldots \cup H_s)\big) \cap L &= \sum_{t=0}^{s} (-1)^t \binom{s}{t} (q^n)^{2s-t-1} + (q^n)^s - (q^n)^{s-1} \\ &= (q^n)^{s-1}(q^n - 1)^s + (q^n)^s - (q^n)^{s-1}.\end{aligned}$$

This means that $\mathcal{X}_s$ is strongly polynomial count with polynomial

$$P_{\mathcal{X}_s}(T) = T^{s-1}(T-1)^s + T^s - T^{s-1}.$$

4.3 Exhaustive Polynomial Count

There is a more computational method for finding the E-polynomial. Suppose that we know that the variety X is polynomial count. This may happen if we know that X is of Hodge–Tate type (in the sense of Remark 2) or that its virtual class $[X] \in \mathbf{KVar}_{\mathbb{C}}$ lies in the subring generated by the Lefschetz motive. Let N be a bound for the dimension of X; in the case of the representation variety $\mathfrak{X}_\Gamma(G)$, we can take $N = s \dim G - 1$, where s is the number of generators of the group Γ. Then, $P_X(T)$ is a polynomial of deg $P_X \leq N$. We can count the number of solutions to the defining equations of the variety over $\mathbb{Z}_{q_i}$, for a collection of $N+1$ prime powers $q_1, \ldots, q_{N+1}$. This will determine uniquely polynomial $P_X(T)$.

Let us see how we can implement this idea for computing $e(\mathfrak{X}_{\mathrm{AGL}_1(\mathbb{C})}(\Sigma_g))$ for arbitrary genus g. For this, we use the quantum method explained in Sect. 5 to gain some qualitative information on the structure of the E-polynomial and the arithmetic method to actually compute the E-polynomial. This is a nice combination of two methods.

As shown in Sect. 5, the quantum method tells us that all the information of the E-polynomial is encoded in a finitely generated $\mathbb{Z}[q]$-module W given in (8) and an endomorphism $\mathcal{Z}(L)$ on W given in (9). In our case, $\dim W = 2$, so in a certain basis we can write

$$\mathcal{Z}(L) = \begin{pmatrix} A(q) & B(q) \\ C(q) & D(q) \end{pmatrix},$$

for some polynomials $A, B, C, D \in \mathbb{Z}[q]$. The formula in Remark 13 and Eq. (10) tell us that we can recover the E-polynomial as

$$e(\mathfrak{X}_{\mathrm{AGL}_1(\mathbb{C})}(\Sigma_g)) = \frac{1}{q^g(q-1)^g} \begin{pmatrix} 1 & 0 \end{pmatrix} \begin{pmatrix} A(q) & B(q) \\ C(q) & D(q) \end{pmatrix}^g \begin{pmatrix} 1 \\ 0 \end{pmatrix}. \tag{7}$$

Observe that the upper-left entry of $\mathcal{Z}(L)^g$, which computes $e(\mathfrak{X}_{\mathrm{AGL}_1(\mathbb{C})}(\Sigma_g))$, only depends on the product BC for all $g \geq 1$. Hence, without lost of generality, we can take $C(q) = 1$. Now, observe that the first powers of $\mathcal{Z}(L)$ are given by

$$\mathcal{Z}(L)^2 = \begin{pmatrix} A^2 + B & AB + BD \\ A + D & D^2 + B \end{pmatrix}, \quad \mathcal{Z}(L)^3 = \begin{pmatrix} A^3 + 2AB + BD & \star \\ \star & \star \end{pmatrix}.$$

This implies that A, B, and D are completely determined by the three E-polynomials $e(\mathfrak{X}_{\mathrm{AGL}_1(\mathbb{C})}(\Sigma_1))$, $e(\mathfrak{X}_{\mathrm{AGL}_1(\mathbb{C})}(\Sigma_2))$, and $e(\mathfrak{X}_{\mathrm{AGL}_1(\mathbb{C})}(\Sigma_3))$, namely

$$A(q) = q(q-1)e(\mathfrak{X}_{\mathrm{AGL}_1(\mathbb{C})}(\Sigma_1)), \quad B(q) = q^2(q-1)^2 e(\mathfrak{X}_{\mathrm{AGL}_1(\mathbb{C})}(\Sigma_2)) - A^2,$$

$$D(q) = \frac{q^3(q-1)^3 e(\mathfrak{X}_{\mathrm{AGL}_1(\mathbb{C})}(\Sigma_3)) - A^3}{B} - 2A.$$

Now, observe that $\mathfrak{X}_{\mathrm{AGL}_1(\mathbb{C})}(\Sigma_g)$ is an affine subvariety of $\mathrm{AGL}_1(\mathbb{C})^{4g}$, so it has dimension at most $4g - 1$. Hence, $e(\mathfrak{X}_{\mathrm{AGL}_1(\mathbb{C})}(\Sigma_g))$ is a polynomial of degree at most $4g - 1$, and, thus, it is completely determined by its value at $4g$ points. Since $\mathfrak{X}_{\mathrm{AGL}_1(\mathbb{C})}(\Sigma_g)$ is polynomial counting, we can compute the number of points of $\mathfrak{X}_{\mathrm{AGL}_1(\mathbb{F}_{q_i})}(\Sigma_g)$ for $4g$ different prime powers $q_1, \ldots, q_{4g}$. For that purpose, we run a small counting script [14], and we obtain the results shown in Table 1.

This implies that the corresponding E-polynomials are

$$\begin{aligned} e(\mathfrak{X}_{\mathrm{AGL}_1(\mathbb{C})}(\Sigma_1)) &= q^3 - q^2, \\ e(\mathfrak{X}_{\mathrm{AGL}_1(\mathbb{C})}(\Sigma_2)) &= q^7 - 4q^6 + 6q^5 - 3q^4, \\ e(\mathfrak{X}_{\mathrm{AGL}_1(\mathbb{C})}(\Sigma_3)) &= q^{11} - 6q^{10} + 15q^9 - 20q^8 + 15q^7 - 5q^6. \end{aligned}$$

Therefore, we finally obtain that

$$\mathcal{Z}(L) = \begin{pmatrix} (q-1)^2 q^3 & (q-1)^3(q-2)^2 q^6 \\ 1 & (q^2 - 3q + 3)(q-1)q^3 \end{pmatrix}.$$

Plugging this matrix into Eq. (7), we recover the result of Theorem 3.

Remark 7 The philosophy behind this method is that, with the qualitative information provided by the TQFT, the E-polynomial of the representation variety for arbitrary genus g is completely determined by the result at small genus. And, moreover, this later value is determined by its number of points at finitely many genus and prime powers.

Table 1 Count of points of $\mathfrak{X}_{\mathrm{AGL}_1(\mathbb{F}_{q_i})}(\Sigma_g)$ for small prime powers q_i and genus g

q_i	*2*	*3*	*4*	*5*	*7*	*8*	*9*	*11*
$g = 1$	4	18	48	100	–	–	–	–
$g = 2$	16	486	5376	32500	446586	1232896	2991816	13323310
$g = 3$	64	16038	749568	12812500	784248234	3855351808	15479813448	161052610510
q_i	*13*		*16*		*17*		*19*	
$g = 3$	1108679412828		11943951728640		23821270295824		84217678403958	

5 Quantum Method

The last approach we will show for the problem of computing virtual classes of representation varieties is the so-called quantum method. The key idea of this method is to construct a geometric–categorical device, known as a Topological Quantum Field Theory (TQFT), and to use it for providing a precise method of computation.

5.1 *Definition of Topological Quantum Field Theories*

The origin of TQFTs dates back to the works of Witten [41], in which he showed that the Jones polynomial (a knot invariant) can be obtained through the Chern–Simons theory, a well-known Quantum Field Theory. Aware of the importance of this discovery, Atiyah formulated in [1] a description of TQFTs as a monoidal symmetric functor. This purely categorical definition is the one that we will review in this section. For a more detailed introduction, see [12, 22].

We will focus on *symmetric monoidal categories* $(\mathcal{C}, \otimes, I)$, which we recall that, by definition, are a category $\mathcal{C}$ with a symmetric associative bifunctor $\otimes : \mathcal{C} \times \mathcal{C} \to \mathcal{C}$ and a distinguished object $I \in \mathcal{C}$ that acts as left and right units for $\otimes$ (for further information, see [40]). A very important instance of a monoidal category is the category of R-modules and R-modules homomorphisms, R-**Mod**, for a given (commutative, unitary) ring R. The usual tensor product over R, $\otimes_R$, together with the ground ring $R \in R$-**Mod** as a unit, defines a symmetric monoidal category $(R\text{-}\mathbf{Mod}, \otimes_R, R)$.

In the same vein, a functor $\mathcal{F} : (\mathcal{C}, \otimes_\mathcal{C}, I_\mathcal{C}) \to (\mathcal{D}, \otimes_\mathcal{D}, I_\mathcal{D})$ is said to be *symmetric monoidal* if it preserves the symmetric monoidal structure, i.e., $\mathcal{F}(I_\mathcal{C}) = I_\mathcal{D}$, and there is an isomorphism of functors

$$\Delta : \mathcal{F}(-) \otimes_\mathcal{D} \mathcal{F}(-) \stackrel{\cong}{\Longrightarrow} \mathcal{F}(- \otimes_\mathcal{C} -).$$

For our purposes, we will focus on the category of bordisms. Let $n \geq 1$. We define the *category of n-bordisms*, $\mathbf{Bd}_n$, as the symmetric monoidal category given by the following data:

- Objects: The objects of $\mathbf{Bd}_n$ are an $(n-1)$-dimensional closed manifold, including the empty set.
- Morphisms: Given objects X_1 and X_2 of $\mathbf{Bd}_n$, a morphism $X_1 \to X_2$ is an equivalence class of bordisms $W : X_1 \to X_2$, i.e., of compact n-dimensional manifolds with $\partial W = X_1 \sqcup X_2$. Two bordisms W and W' are equivalent if there exists a diffeomorphism $F : W \to W'$ fixing the boundaries X_1 and X_2.
  For the composition, given $W : X_1 \to X_2$ and $W' : X_2 \to X_3$, we define

$W' \circ W = W \cup_{X_2} W' : X_1 \to X_3$, where $W \cup_{X_2} W'$ is the gluing of bordisms along X_2.

We endow $\mathbf{Bd}_n$ with the bifunctor given by disjoint union $\sqcup$ of both objects and bordisms. This bifunctor, with the unit $\emptyset \in \mathbf{Bd}_n$, turns $\mathbf{Bd}_n$ into a symmetric monoidal category.

Definition 1 Let R be a commutative ring with unit. An n-dimensional Topological Quantum Field Theory (shortened a TQFT) is a symmetric monoidal functor

$$\mathcal{Z} : \mathbf{Bd}_n \to R\text{-}\mathbf{Mod}.$$

Remark 8 This definition slightly differs from others presented in the literature, especially in those oriented to physics, where the objects and bordisms of $\mathbf{Bd}_n$ are required to be equipped with an orientation (which plays an important role in many physical theories).

The main application of TQFTs to algebraic topology comes from the following observation. Suppose that we are interested in an algebraic invariant that assigns to any closed n-dimensional manifold W an element $\chi(W) \in R$, for a fixed ring G. In principle, χ might be very hard to compute and very handcrafted arguments are needed for performing explicit computations.

However, suppose that we are able to *quantize* χ. This means that we are able to construct a TQFT, $\mathcal{Z} : \mathbf{Bd}_n \to R-\mathbf{Mod}$ such that $\mathcal{Z}(W)(1) = \chi(W)$ for any closed n-dimensional manifold. Note that the later formula makes sense since, as W is a closed manifold, it can be seen as a bordism $W : \emptyset \to \emptyset$ and, since $\mathcal{Z}$ is monoidal, $\mathcal{Z}(W) : \mathcal{Z}(\emptyset) = R \to \mathcal{Z}(\emptyset) = R$ is an R-module homomorphism, and, thus, it is fully determined by the element $\mathcal{Z}(W)(1) \in R$.

Such quantization gives rise to a new procedure for computing χ by decomposing W into simpler pieces. To illustrate the method, suppose that $n = 2$ and $W = \Sigma_g$ is the closed oriented surface of genus $g \geq 0$. We can decompose $\Sigma_g : \emptyset \to \emptyset$ as $\Sigma_g = D^\dagger \circ L^g \circ D$, where $D : \emptyset \to S^1$ is the disc, $D^\dagger : S^1 \to \emptyset$ is the opposite disc and $L : S^1 \to S^1$ is a twice holed torus, as shown in Fig. 1.

In that case, applying $\mathcal{Z}$, we get that

$$\chi(\Sigma_g) = \mathcal{Z}(D^\dagger) \circ \mathcal{Z}(L)^g \circ \mathcal{Z}(D)(1).$$

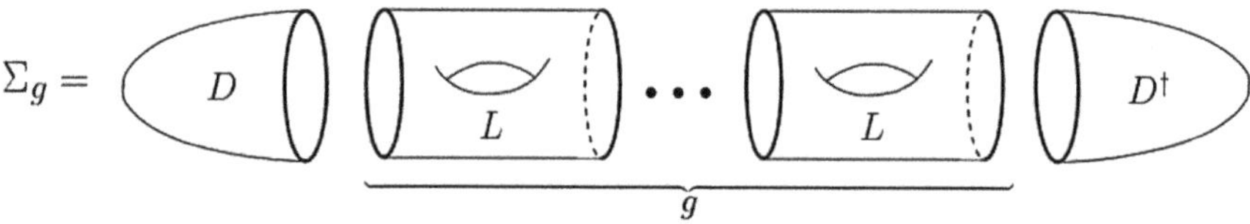

Fig. 1 Decomposition of Σ_g into simpler bordisms

That is, we can compute $\chi(\Sigma_g)$ for a surface of arbitrary genus just by computing three homomorphisms, $\mathcal{Z}(D) : R \to \mathcal{Z}(S^1)$ (which is determined by an element of $\mathcal{Z}(S^1)$), $\mathcal{Z}(D^\dagger) : \mathcal{Z}(S^1) \to R$ (which is essentially a projection), and an endomorphism $\mathcal{Z}(L) : \mathcal{Z}(S^1) \to \mathcal{Z}(S^1)$.

5.2 *Quantization of the Virtual Classes of Representation Varieties*

The aim of this section is to quantize the virtual classes of representation varieties. However, as we will see, our construction will not give a TQFT on the nose, but a kind of lax version.

The first ingredient we need to modify is the category of bordisms in order to include pairs of spaces. This might seem shocking at a first sight, but it is very natural if we think that we are dealing with fundamental groups of topological spaces, and the fundamental group is not a functor out of the category of topological spaces but out of the category of pointed topological spaces. The aim of this version for pairs is to track these base points.

Fix $n \geq 1$. We define the *category of n-bordisms of pairs*, $\mathbf{Bdp}_n$, as the symmetric monoidal category given by the following data:

- Objects: The objects of $\mathbf{Bdp}_n$ are pairs (X, A), where X is an $(n-1)$-dimensional closed manifold (maybe empty) together with a finite subset of points $A \subseteq X$ such that its intersection with each connected component of X is non-empty.
- Morphisms: Given objects (X_1, A_1) and (X_2, A_2) of $\mathbf{Bdp}_n$, a morphism $(X_1, A_1) \to (X_2, A_2)$ is an equivalence class of pairs (W, A), where $W : X_1 \to X_2$ is a bordism and $A \subseteq W$ is a finite set of points with $X_1 \cap A = A_1$ and $X_2 \cap A = A_2$. Two pairs (W, A) and (W', A') are equivalent if there exists a diffeomorphism of bordisms $F : W \to W'$ such that $F(A) = A'$. Finally, given $(W, A) : (X_1, A_1) \to (X_2, A_2)$ and $(W', A') : (X_2, A_2) \to (X_3, A_3)$, we define $(W', A') \circ (W, A) = (W \cup_{X_2} W', A \cup A') : (X_1, A_1) \to (X_3, A_3)$.

Remark 9 In this form, $\mathbf{Bdp}_n$ is not exactly a category since there is no unit morphism in $\mathrm{Hom}_{\mathbf{Bdp}_n}((X, A), (X, A))$. This can be solved by weakening slightly the notion of bordism, allowing that (X, A) itself could be seen as a bordism $(X, A) : (X, A) \to (X, A)$.

In order to construct the TQFT quantizing virtual classes of representation varieties, we need to introduce some notation. Fix a ground field $\mathbb{K}$ (not necessarily algebraically closed) and G an algebraic group over $\mathbb{K}$ (not necessarily reductive).

Given a topological space X and $A \subseteq X$, we denote by $\Pi(X, A)$ the fundamental groupoid of X with base points in A, that is, the groupoid of homotopy classes of paths in X between points in A. If X is compact and A is finite, we define the *G-representation variety* of the pair (X, A), $\mathfrak{X}_G(X, A)$, as the set of groupoids homomorphisms $\Pi(X, A) \to G$, i.e., $\mathfrak{X}_G(X, A) = \mathrm{Hom}(\Pi(X, A), G)$. Observe that, in particular, if A has a single point, then $\mathfrak{X}_G(X, A)$ is the usual G-representation variety.

As it happened for representation varieties with a single base point, $\mathfrak{X}_G(X, A)$ has a natural structure of algebraic variety given as follows. Let $X = \bigsqcup_{i=1}^{r} X_i$ be the decomposition of X into connected components, and let us order them so that $X_i \cap A \neq \emptyset$ for the first s components. Pick $x_i \in X_i \cap A$ and, for any i, choose a path α_i^x between x_i and any other $x \in X_i \cap A, x \neq x_i$. Then, a representation $\Pi(X, A) \to G$ is completely determined by the usual vertex representations $\pi_1(X, x_i) \to G$ for $1 \leq i \leq s$, together with an arbitrary element of G for any chosen path α_i^x. There are $|A| - s$ of such chosen paths, so we have a natural identification

$$\mathfrak{X}_G(X, A) = \prod_{i=1}^{s} \mathfrak{X}_G(X, x_i) \times G^{|A|-s}.$$

The right-hand side of this equality is naturally an algebraic variety, so $\mathfrak{X}_G(X, A)$ is endowed with the structure of an algebraic variety.

The second ingredient needed for quantizing representation varieties has a more algebraic nature. Given an algebraic variety S over $\mathbb{K}$, let us denote by $\mathbf{Var}/S$ the category of algebraic varieties over Z, that is, the category whose objects are regular morphisms $Z \to S$ and its morphisms are regular maps $Z \to Z'$ preserving the base projections. As in the usual category of algebraic varieties, together with the disjoint union $\sqcup$ of algebraic varieties, and the fibered product $\times_S$ over S, we may consider its associated Grothendieck ring $\mathbf{KVar}/S$. The element of $\mathbf{KVar}/S$ induced by a morphism $h : Z \to S$ will be denoted as $[(Z, h)]_S \in \mathbf{KVar}/S$, or just by $[Z]_S$ or $[Z]$ when the morphism h or the base variety is understood from the context. Recall that, in this notation, the unit of $\mathbf{KVar}/S$ is $\mathbb{1}_S = [S, \mathrm{Id}_S]_S$ and that, if $S = \star$ is the singleton variety, then $\mathbf{KVar}/\star = \mathbf{KVar}_{\mathbb{K}}$ is the usual Grothendieck ring of varieties.

This construction exhibits some important functoriality properties that will be useful for our construction. Suppose that $f : S_1 \to S_2$ is a regular morphism. It induces a ring homomorphism $f^*\mathbf{KVar}/S_2 \to \mathbf{KVar}/S_1$ given by $f^*[Z]_{S_2} = [Z \times_{S_2} S_1]_{S_1}$. In particular, taking the projection map $c : S \to \star$, we get a ring homomorphism $c^* : \mathbf{KVar}_{\mathbb{K}} \to \mathbf{KVar}/S$ that endows the rings $\mathbf{KVar}/S$ with a natural structure of $\mathbf{KVar}_{\mathbb{K}}$-module that corresponds to the Cartesian product. Finally, we also have the covariant version $f_! : \mathbf{KVar}/S_1 \to \mathbf{KVar}/S_2$ given by $f_![(Z, h)]_{S_1} = [(Z, f \circ h)]_{S_2}$. In general, $f_!$ is not a ring homomorphism, but the projection formula $f_!([Z_2] \times_{S_2} f^*[Z_1]) = f_![Z_2] \times_{S_1} [Z_1]$, for $[Z_1] \in \mathbf{KVar}/S_1$ and $[Z_2] \in \mathbf{KVar}/S_2$, implies that $f_!$ is a $\mathbf{KVar}_{\mathbb{K}}$-module homomorphism.

Remark 10 Some important properties that clarify the interplay between these two induced morphisms are listed below. They will be very useful for explicit computations in Sect. 5.3. Their proof is a straightforward computation using fibered products and it can be checked in [17].

- The induced morphisms are functorial, in the sense that $(g \circ f)^* = f^* \circ g^*$ and $(g \circ f)_! = g_! \circ f_!$. In particular, if $i : T \hookrightarrow S$ is an inclusion, then $i^* f^* = f|_T^*$.
- Suppose that we have a pullback of algebraic varieties (i.e., a fibered product diagram)

$$\begin{array}{ccc} S' = S_1 \times_S S_2 & \xrightarrow{g'} & S_1 \\ {\scriptstyle f'}\downarrow & & \downarrow{\scriptstyle f} \\ S_2 & \xrightarrow[g]{} & S \end{array}$$

  Then, it holds that $g^* \circ f_! = (f')_! \circ (g')^*$. This property is usually known as the base-change formula, or the Beck–Chevalley property, and it generalizes the projection formula.
- Suppose that we decompose $S = T \sqcup U$, where $i : T \hookrightarrow S$ is a closed embedding and $j : U \hookrightarrow S$ is an open subvariety. Then, we have that $i_! i^* + j_! j^* : \mathbf{KVar}/S \to \mathbf{KVar}/S$ is the identity map. This corresponds to the idea that virtual classes are compatible with chopping the space according to a stratification.

At this point, we are ready to define our TQFT. We take as ground ring $R = \mathbf{KVar}_{\mathbb{K}}$ the Grothendieck ring of algebraic varieties. We define a functor $\mathcal{Z} : \mathbf{Bdp}_n \to \mathbf{KVar}_{\mathbb{K}}-\mathbf{Mod}$ as follows:

- On an object $(X, A) \in \mathbf{Bdp}_n$, we set $\mathcal{Z}(X, A) = \mathbf{KVar}/\mathfrak{X}_G(X, A)$, the Grothendieck ring of algebraic varieties over $\mathfrak{X}_G(X, A)$.
- On a morphism $(W, A) : (X_1, A_1) \to (X_2, A_2)$, let us denote the natural restrictions $i : \mathfrak{X}_G(W, A) \to \mathfrak{X}_G(X_1, A_1)$ and $j : \mathfrak{X}_G(W, A) \to \mathfrak{X}_G(X_2, A_2)$. Then, we set

$$\mathcal{Z}(W, A) = j_! \circ i^* : \mathbf{KVar}/\mathfrak{X}_G(X_1, A_1) \to \mathbf{KVar}/\mathfrak{X}_G(W, A) \to \mathbf{KVar}/\mathfrak{X}_G(X_2, A_2).$$

Remark 11 Recall that, since in general $j_!$ is not a ring homomorphism, the induced map $\mathcal{Z}(W, A) : \mathbf{KVar}/\mathfrak{X}_G(X_1, A_1) \to \mathbf{KVar}/\mathfrak{X}_G(X_2, A_2)$ is only a $\mathbf{KVar}_{\mathbb{K}}$-module homomorphism.

It can be proven that, since the fundamental groupoid satisfies the Seifert–van Kampen theorem, $\mathcal{Z}$ is actually a functor (see [13, 15] for a detailed proof). However, it is not monoidal since, in general, for algebraic varieties S_1 and S_2, we have $\mathbf{KVar}/S_1 \otimes_{\mathbf{KVar}_{\mathbb{K}}} \mathbf{KVar}/S_2 \not\cong \mathbf{KVar}/S_1 \times S_2$. Nevertheless, we still have a map

$$\Delta_{S_1,S_2} : \mathbf{KVar}/S_1 \otimes_{\mathbf{KVar}_{\mathbb{K}}} \mathbf{KVar}/S_2 \to \mathbf{KVar}/S_1 \times S_2$$

given by "external product." That is, it is the map induced by

$$[Z_1] \otimes [Z_2] \in \mathbf{KVar}/S_1 \otimes_{\mathrm{KVar}_{\mathbb{K}}} \mathbf{KVar}/S_2 \mapsto \pi_1^*[Z_1] \times_{(S_1 \times S_2)} \pi_2^*[Z_2] \in \mathbf{KVar}/S_1 \times S_2,$$

where $\pi_i : S_1 \times S_2 \to S_i$ are the projections. In this situation, it is customary to say that $\mathcal{Z}$ is a *symmetric lax monoidal* functor.

Finally, in order to figure out what invariant is $\mathcal{Z}$ computing, first observe that for the empty set, we have $\mathfrak{X}_G(\emptyset) = \star$ is the singleton variety, and thus $\mathcal{Z}(\emptyset) = \mathbf{KVar}/\mathfrak{X}_G(\emptyset) = \mathbf{KVar}/\star = \mathbf{KVar}_{\mathbb{K}}$ is the usual Grothendieck ring of algebraic varieties. Now, let us take (W, A) a closed connected n-dimensional manifold. Seen as a morphism $(W, A) : \emptyset \to \emptyset$, it induces a $\mathbf{KVar}_{\mathbb{K}}$-module homomorphism $\mathcal{Z}(W, A) = c_! c^* : \mathbf{KVar}_{\mathbb{K}} \to \mathbf{KVar}_{\mathbb{K}}$, where $c : \mathfrak{X}_G(W, A) \to \star$ is projection onto a point. Therefore, we have that

$$\begin{aligned} \mathcal{Z}(W, A)(\mathbb{1}_\star) = c_! c^*(\mathbb{1}_\star) = c_! \mathbb{1}_{\mathfrak{X}_G(W,A)} = \\ = c_![\mathfrak{X}_G(W, A)]_{\mathfrak{X}_G(W,A)} = [\mathfrak{X}_G(W, A)]_\star = [\mathfrak{X}_G(W, A)], \end{aligned}$$

where the second equality follows from the fact that c^* is a ring homomorphism. Therefore, $\mathcal{Z}$ quantizes the virtual classes of representation varieties, so we have proven the following result.

Theorem 6 *Let $\mathbb{K}$ be a field, G an algebraic group over k, and $n \geq 1$. There exists a symmetric lax monoidal Topological Quantum Field Theory*

$$\mathcal{Z} : \mathbf{Bdp}_n \to \mathbf{KVar}_{\mathbb{K}}\text{--}\mathbf{Mod},$$

which quantizes the virtual classes of G-representation varieties.

Remark 12 To be precise, $\mathcal{Z}$ computes virtual classes of G-representation varieties of pairs. This implies that it computes virtual classes of classical G-representation varieties up to a known constant. For instance, let W be a compact connected n-dimensional manifold, and let $A \subseteq W$ be a finite set. Then, we have

$$\mathcal{Z}(W, A)(\mathbb{1}_\star) = [\mathfrak{X}_G(W, A)] = [\mathfrak{X}_G(W)] \times [G]^{|A|-1}.$$

Hence, $\mathcal{Z}(W, A)(\mathbb{1}_\star)$ computes $[\mathfrak{X}_G(W)]$ up to the factor $[G]^{|A|-1}$ (which is not a big problem since $[G]$ is known for most of the classical groups).

Unravelling the previous construction, we can describe precisely the morphisms induced by the TQFT. Let us focus on the case $n = 2$ and orientable surfaces. As we mentioned above, we need to understand the bordisms D, $D^\dagger$, and L, as depicted in Fig. 2. Observe that, in order to meet the requirements of $\mathbf{Bdp}_2$, we need to choose a base point on S^1, which we will loosely denote by $\star \in S^1$. In this way, $D : \emptyset \to (S^1, \star)$ and $D^\dagger : (S^1, \star) \to \emptyset$ have a marked base point, while $L : (S^1, \star) \to (S^1, \star)$ has two marked base points, one on each component of the boundary.

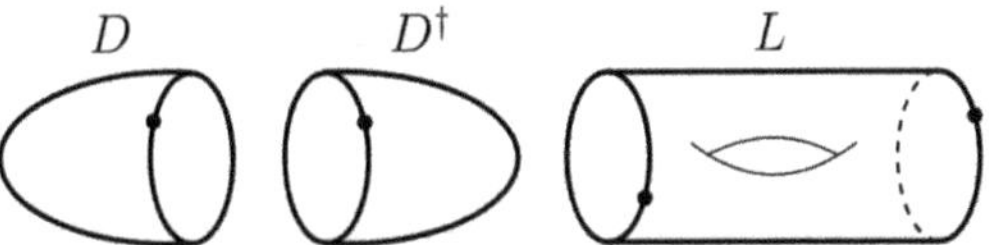

Fig. 2 The basic bordisms for orientable surfaces

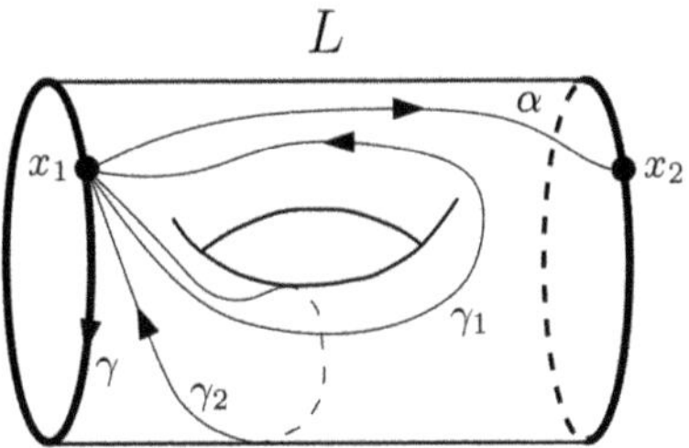

Fig. 3 Chosen paths for L

With respect to the object $(S^1, \star) \in \mathbf{Bdp}_2$, the associated representation variety is $\mathfrak{X}_G(S^1, \star) = \mathrm{Hom}\,(\mathbb{Z}, G) = G$. With respect to morphisms, the situation for D and $D^\dagger$ is very simple since they are simply connected. Therefore, the restriction maps at the level of fundamental groupoids are, respectively,

$$\star \longleftarrow \star \xrightarrow{\;i\;} G, \qquad G \xleftarrow{\;i\;} \star \longrightarrow \star,$$

where $i : \star \hookrightarrow G$ is the inclusion of the trivial representation. Hence, under $\mathcal{Z}$, we have that

$$\mathcal{Z}(D) = i_! : \mathbf{KVar}_{\mathbb{K}} \to \mathbf{KVar}/G, \qquad \mathcal{Z}(D^\dagger) = i^* : \mathbf{KVar}/G \to \mathbf{KVar}_{\mathbb{K}}.$$

For the holed torus $L : (S^1, \star) \to (S^1, \star)$, the situation is a bit more complicated. Let $L = (T, A)$, where $A = \{x_1, x_2\}$ is the set of marked points of L, with x_1 in the in-going boundary and x_2 in the out-going boundary. Recall that T is homotopically equivalent to a bouquet of three circles, so its fundamental group is the free group with three generators. Thus, we can take γ, γ_1, and γ_2 as the set of generators of $\pi_1(T, x_1)$ depicted in Fig. 3 and α the path between x_1 and x_2.

With this description, γ is a generator of $\pi_1(S^1, x_1)$ and $\alpha\gamma[\gamma_1, \gamma_2]\alpha^{-1}$ is a generator of $\pi_1(S^1, x_2)$, where $[\gamma_1, \gamma_2] = \gamma_1\gamma_2\gamma_1^{-1}\gamma_2^{-1}$ is the group commutator. Hence, since $\mathfrak{X}_G(L) = \mathrm{Hom}\,(\Pi(T, A), G) = G^4$, we have that restriction maps at the level of fundamental groupoids are

$$\begin{array}{ccccc} G & \xleftarrow{\;p\;} & G^4 & \xrightarrow{\;q\;} & G \\ g & \leftarrow & (g, g_1, g_2, h) & \mapsto & hg[g_1, g_2]h^{-1} \end{array}$$

where g, g_1, g_2, and h are the images of $\gamma, \gamma_1, \gamma_2$, and α, respectively. Hence, we obtain that

$$\mathcal{Z}(L) : \mathbf{KVar}/G \xrightarrow{p^*} \mathbf{KVar}/G^4 \xrightarrow{q_!} \mathbf{KVar}/G.$$

Remark 13 As we mentioned in Remark 12, the TQFT computes virtual classes of representation varieties of pairs. In particular, observe that if we decompose $\Sigma_g = D^\dagger \circ L^g \circ D$, we are forced to put on Σ_g a set of $g+1$ base points $A \subseteq \Sigma_g$. Hence, we have that

$$[\mathfrak{X}_G(\Sigma_g)] \times [G]^g = \mathcal{Z}(\Sigma_g, A)(\mathbb{1}_\star) = \mathcal{Z}(D^\dagger) \circ \mathcal{Z}(L)^g \circ \mathcal{Z}(D)(\mathbb{1}_\star).$$

Or equivalently, if we localize $\mathbf{KVar}_\mathbb{K}$ by $[G] \in \mathbf{KVar}_\mathbb{K}$, we have that

$$[\mathfrak{X}_G(\Sigma_g)] = \frac{1}{[G]^g} \mathcal{Z}(D^\dagger) \circ \mathcal{Z}(L)^g \circ \mathcal{Z}(D)(\mathbb{1}_\star).$$

5.3 Representation Varieties via the Quantum Method

In this section, as an application we will consider $G = \mathrm{AGL}_1(\mathbb{K})$, and we will focus on $\mathrm{AGL}_1(\mathbb{K})$-representation varieties. As in Sects. 3 and 4, we will compute the virtual classes of these representation varieties over any compact oriented surface, but, in this case, we will use the TQFT described above for performing the computation.

As mentioned in Remark 13, we only need to focus on the computation of the induced morphisms $\mathcal{Z}(D)$, $\mathcal{Z}(D^\dagger)$, and $\mathcal{Z}(L)$. For the disc $\mathcal{Z}(D) = i_! : \mathbf{KVar}_\mathbb{K} \to \mathbf{KVar}/\mathrm{AGL}_1(\mathbb{K})$, the situation is very simple since it is fully determined by the element $\mathcal{Z}(D)(\mathbb{1}_\star) = i_!\mathbb{1}_\star$. Along this section, we will denote the unit of $\mathbf{KVar}/S$ by $\mathbb{1}_S$, or just $\mathbb{1}$ is understood from the context. In particular, $\mathbb{1}_\star \in \mathbf{KVar}_\mathbb{K} = \mathbf{KVar}/\star$ is the unit of the ground ring.

In order to compute the morphism $\mathcal{Z}(L) : \mathbf{KVar}/\mathrm{AGL}_1(\mathbb{K}) \to \mathbf{KVar}/\mathrm{AGL}_1(\mathbb{K})$, recall that, with the notation of Sect. 5.2, $\mathcal{Z}(L) = q_!p^*$. We have a commutative diagram

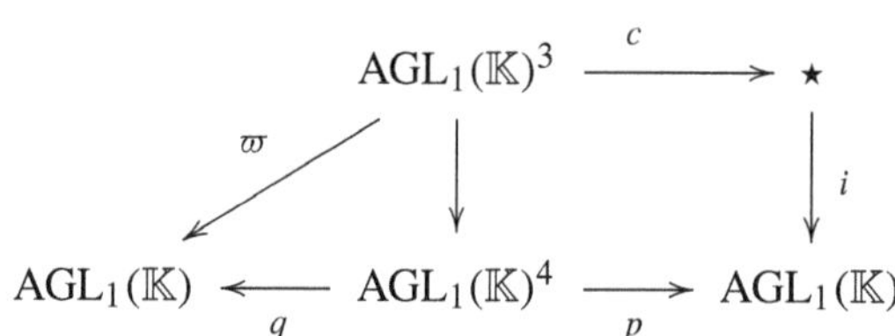

where c is the projection onto a point, the leftmost vertical arrow is given by $(A_1, A_2, B) \mapsto (\mathrm{I}, A_1, A_2, B)$ and $\varpi(A_1, A_2, B) = B[A_1, A_2]B^{-1}$, being $\mathrm{I} \in \mathrm{AGL}_1(\mathbb{K})$ the identity matrix. Moreover, the square is a pullback, so by Remark 10, we have

$$\mathcal{Z}(L) \circ \mathcal{Z}(D)(\mathbb{1}_\star) = q_! p^* i_! \mathbb{1}_\star = \varpi_! c^* \mathbb{1}_\star = \varpi_! \mathbb{1}_{\mathrm{AGL}_1(\mathbb{K})^3}.$$

In order to compute this later map, observe that, explicitly, the morphism ϖ is given by

$$\varpi\left(\begin{pmatrix} a_1 & b_1 \\ 0 & 1 \end{pmatrix}, \begin{pmatrix} a_2 & b_2 \\ 0 & 1 \end{pmatrix}, \begin{pmatrix} x & y \\ 0 & 1 \end{pmatrix}\right) = \begin{pmatrix} 1 & (a_1 - 1)b_2 x - (a_2 - 1)b_1 x \\ 0 & 1 \end{pmatrix}.$$

Therefore, ϖ is a projection onto $\mathrm{ASO}_1(\mathbb{K}) \subseteq \mathrm{AGL}_1(\mathbb{K})$, the subgroup of orthogonal orientation-preserving affine transformations. Outside $\mathrm{I} \in \mathrm{ASO}_1(\mathbb{K})$, ϖ is a locally trivial fibration in the Zariski topology with fiber, for $\alpha \neq 0$, given by

$$\begin{aligned} F &= \left\{(a_1, a_2, x, b_1, b_2, y) \in (\mathbb{K}^*)^3 \times \mathbb{K}^3 \mid (a_1 - 1)b_2 x - (a_2 - 1)b_1 x = \alpha\right\} \\ &= \left\{b_2 = \frac{\alpha + (a_2 - 1)b_1 x}{(a_1 - 1)x}, a_1 \neq 1\right\} \sqcup \left\{b_1 = -\frac{\alpha}{(a_2 - 1)x}, a_1 = 1\right\} \\ &\cong \left((\mathbb{K} - \{0, 1\}) \times (\mathbb{K}^*)^2 \times \mathbb{K}^2\right) \sqcup \left(\mathbb{K} - \{0, 1\} \times \mathbb{K}^* \times \mathbb{K}^2\right). \end{aligned}$$

Its virtual class is $[F] = (q-2)(q-1)^2q^2 + (q-2)(q-1)q^2 = q(q-1)(q^3 - 2q^2)$, where as always $q = [\mathbb{K}] \in \mathbf{KVar}_{\mathbb{K}}$.

On the other hand, on the identity matrix I, the special fiber is

$$\begin{aligned} \varpi^{-1}(\mathrm{I}) &= \left\{(a_1, a_2, x, b_1, b_2, y) \in (\mathbb{K}^*)^3 \times \mathbb{K}^3 \mid (a_1 - 1)b_2 = (a_2 - 1)b_1\right\} \\ &= \left\{b_2 = \frac{(a_2 - 1)b_1}{a_1 - 1}, a_1 \neq 1\right\} \sqcup \{a_1 = 1, a_2 = 1\} \sqcup \{a_1 = 1, a_2 \neq 1, b_1 = 0\} \\ &\cong \left((\mathbb{K} - \{0, 1\}) \times (\mathbb{K}^*)^2 \times \mathbb{K}^2\right) \sqcup \left(\mathbb{K}^* \times \mathbb{K}^3\right) \sqcup \left(\mathbb{K} - \{0, 1\} \times \mathbb{K}^* \times \mathbb{K}^2\right). \end{aligned}$$

Its virtual class is $[\varpi^{-1}(\mathrm{I})] = (q-2)(q-1)^2q^2 + (q-1)q^3 + (q-2)(q-1)q^2 = q(q-1)(q^3 - q^2)$.

Let us denote $\mathrm{ASO}_1(\mathbb{K})^* = \mathrm{ASO}_1(\mathbb{K}) - \{I\}$ with inclusion $j : \mathrm{ASO}_1(\mathbb{K})^* \hookrightarrow \mathrm{AGL}_1(\mathbb{K})$. Then, by Remark 10, we have that

$$\varpi_! \mathbb{1} = i_! i^* \varpi_! \mathbb{1} + j_! j^* \varpi_! \mathbb{1} = i_!(\varpi|_{\varpi^{-1}(\mathrm{I})})_! \mathbb{1} + j_!(\varpi|_{\varpi^{-1}(\mathrm{ASO}_1(\mathbb{K})^*)})_! \mathbb{1}.$$

For the first map, recall that ϖ is locally trivial in the Zariski topology over $\mathrm{ASO}_1(\mathbb{K})^*$. Thus, $(\varpi|_{\varpi^{-1}(\mathrm{ASO}_1(\mathbb{K})^*)})_! \mathbb{1}_{\mathrm{AGL}_1(\mathbb{K})^3} = [F]\, \mathbb{1}_{\mathrm{ASO}_1(\mathbb{K})^*}$. On the other hand, the map $\varpi|_{\varpi^{-1}(\mathrm{I})}$ is projection onto a point, so $(\varpi|_{\varpi^{-1}(\mathrm{I})})_! \mathbb{1}_{\mathrm{AGL}_1(\mathbb{K})^3} =$

$[\varpi^{-1}(\mathrm{I})]\mathbb{1}_\star$. Hence, putting all together, we obtain that

$$\begin{aligned}\mathcal{Z}(L) \circ \mathcal{Z}(D)(\mathbb{1}_\star) &= i_!(\varpi|_{\varpi^{-1}(\mathrm{I})})_!\mathbb{1} + j_!(\varpi|_{\varpi^{-1}(\mathrm{ASO}_1(\mathbb{K})^*)})_!\mathbb{1} \\ &= q(q-1)(q^3-q^2)\, i_!\mathbb{1}_\star + q(q-1)(q^3-2q^2)\, j_!\mathbb{1}_{\mathrm{ASO}_1(\mathbb{K})^*}.\end{aligned}$$

In this way, if we want to apply $\mathcal{Z}(L)$ twice, we need to compute the image $\mathcal{Z}(L)(j_!\mathbb{1}_{\mathrm{ASO}_1(\mathbb{K})^*})$. This computation is quite similar to the previous one. First, we again have a commutative diagram whose square is a pullback

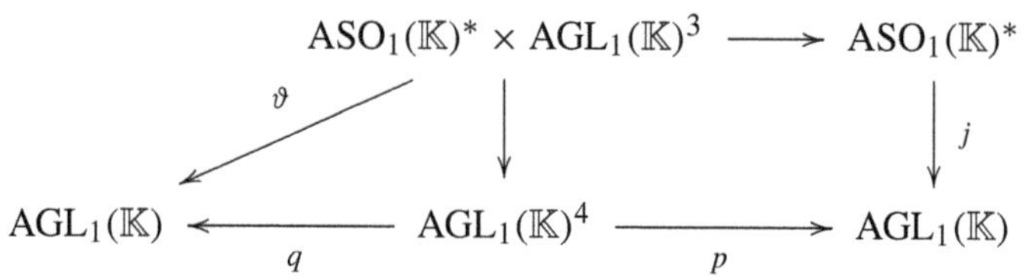

$$\begin{array}{ccccc} & & \mathrm{ASO}_1(\mathbb{K})^* \times \mathrm{AGL}_1(\mathbb{K})^3 & \longrightarrow & \mathrm{ASO}_1(\mathbb{K})^* \\ & \swarrow^{\vartheta} & \downarrow & & \downarrow j \\ \mathrm{AGL}_1(\mathbb{K}) & \xleftarrow[q]{} & \mathrm{AGL}_1(\mathbb{K})^4 & \xrightarrow[p]{} & \mathrm{AGL}_1(\mathbb{K}) \end{array}$$

The leftmost vertical arrow is the inclusion map and $\vartheta(A, A_1, A_2, B) = BA[A_1, A_2]B^{-1}$. Computing explicitly, we have that

$$\vartheta\left(\begin{pmatrix}1 & \beta\\ 0 & 1\end{pmatrix}, \begin{pmatrix}a_1 & b_1\\ 0 & 1\end{pmatrix}, \begin{pmatrix}a_2 & b_2\\ 0 & 1\end{pmatrix}, \begin{pmatrix}x & y\\ 0 & 1\end{pmatrix}\right) = \begin{pmatrix}1 & (a_1-1)b_2x - (a_2-1)b_1x + \beta x\\ 0 & 1\end{pmatrix}.$$

Hence, ϑ is again a morphism onto $\mathrm{ASO}_1(\mathbb{K}) \subseteq \mathrm{AGL}_1(\mathbb{K})$. Over $\mathrm{I} \in \mathrm{ASO}_1(\mathbb{K})$, the fiber is

$$\begin{aligned}\vartheta^{-1}(\mathrm{I}) &= \left\{(\beta, a_1, a_2, x, b_1, b_2, y) \in (\mathbb{K}^*)^4 \times \mathbb{K}^3 \mid (a_2-1)b_1x - (a_1-1)b_2x = \beta\right\} \\ &= \left((\mathbb{K}^*)^3 \times \mathbb{K}^3\right) - \{(a_1-1)b_2 - (a_2-1)b_1 = 0\} \\ &= \left((\mathbb{K}^*)^3 \times \mathbb{K}^3\right) - \varpi^{-1}(\mathrm{I}).\end{aligned}$$

Thus, $[\vartheta^{-1}(\mathrm{I})] = (q-1)^3q^3 - q(q-1)(q^3-q^2) = q(q-1)(q^4-3q^3+2q^2)$.

Analogously, on $\mathrm{ASO}_1(\mathbb{K})^*$, we have that ϑ is a locally trivial fibration in the Zariski topology with fiber over $\alpha \neq 0$ given by

$$\begin{aligned}F' &= \left\{(\beta, a_1, a_2, x, b_1, b_2, y) \in (\mathbb{K}^*)^4 \times \mathbb{K}^3 \mid (a_1-1)b_2x - (a_2-1)b_1x + \beta = \alpha\right\} \\ &= \left((\mathbb{K}^*)^3 \times \mathbb{K}^3\right) - \{(a_1-1)b_2 - (a_2-1)b_1 = \alpha\} = \left((\mathbb{K}^*)^3 \times \mathbb{K}^3\right) - F.\end{aligned}$$

Hence, the virtual class of the fiber is $[F'] = (q-1)^3q^3 - q(q-1)(q^3-2q^2) = q(q-1)(q^4-3q^3+3q^2)$. Putting together these computations, we obtain that

$$\begin{aligned}\mathcal{Z}(L)\left(j_!\mathbb{1}_{\mathrm{ASO}_1(\mathbb{K})^*}\right) &= \vartheta_!\mathbb{1} = i_!(\vartheta|_{\vartheta^{-1}(\mathrm{I})})_!\mathbb{1} + j_!(\vartheta|_{\vartheta^{-1}(\mathrm{ASO}_1(\mathbb{K})^*)})_!\mathbb{1} \\ &= q(q-1)(q^4-3q^3+2q^2)\, i_!\mathbb{1}_\star + q(q-1)(q^4-3q^3+3q^2)\, j_!\mathbb{1}_{\mathrm{ASO}_1(\mathbb{K})^*}.\end{aligned}$$

Let $W \subseteq \mathbf{KVar}/\mathrm{AGL}_1(\mathbb{K})$ be the submodule generated by the elements $i_!\mathbb{1}_\star$ and $j_!\mathbb{1}_{\mathrm{ASO}_1(\mathbb{K})^*}$. The previous computation shows that $\mathcal{Z}(L)(W) \subseteq W$. Furthermore, indeed we have

$$W = \langle \mathcal{Z}(L)^g(i_!\mathbb{1}_\star)\rangle_{g=0}^{\infty}. \tag{8}$$

On W, the morphism $\mathcal{Z}(D^\dagger) : W \to \mathbf{KMHS}$ is given by the projection $\mathcal{Z}(D^\dagger)(i_!\mathbb{1}_\star) = \mathbb{1}_\star$ and $\mathcal{Z}(D^\dagger)(j_!\mathbb{1}_{\mathrm{ASO}_1(\mathbb{K})^*}) = 0$. Hence, regarding the computation of virtual classes of representation varieties, we can restrict our attention to W.

If we want to compute explicitly these classes, observe that, by the previous calculations, on the set of generators $i_!\mathbb{1}_\star$, $j_!\mathbb{1}_{\mathrm{ASO}_1(\mathbb{K})^*}$ of W, the matrix of $\mathcal{Z}(L) : W \to W$ is

$$\mathcal{Z}(L) = q(q-1)\begin{pmatrix} q^3-q^2 & q^4-3q^3+2q^2 \\ q^3-2q^2 & q^4-3q^3+3q^2 \end{pmatrix}. \tag{9}$$

Since $[\mathrm{AGL}_1(\mathbb{K})] = [\mathbb{K}^* \times \mathbb{K}] = q(q-1)$, using the formula of Remark 13, we obtain that

$$\begin{aligned}
[\mathfrak{X}_{\mathrm{AGL}_1(\mathbb{K})}(\Sigma_g)] &= \begin{pmatrix} 1 & 0 \end{pmatrix}\begin{pmatrix} q^3-q^2 & q^4-3q^3+2q^2 \\ q^3-2q^2 & q^4-3q^3+3q^2 \end{pmatrix}^g \begin{pmatrix} 1 \\ 0 \end{pmatrix} \\
&= \begin{pmatrix} 1 & 0 \end{pmatrix}\begin{pmatrix} q-1 & q-1 \\ -1 & q-1 \end{pmatrix}\begin{pmatrix} q^{2g} & 0 \\ 0 & q^{2g}(q-1)^{2g} \end{pmatrix}\begin{pmatrix} q-1 & q-1 \\ -1 & q-1 \end{pmatrix}^{-1}\begin{pmatrix} 1 \\ 0 \end{pmatrix} \\
&= q^{2g-1}\left((q-1)^{2g} + q - 1\right).
\end{aligned} \tag{10}$$

Remark 14 Strictly speaking, this is not the virtual class of $\mathfrak{X}_{\mathrm{AGL}_1(\mathbb{K})}(\Sigma_g)$ on $\mathbf{KVar}_\mathbb{K}$ but on its localization by the multiplicative set S generated by q and $q-1$. This has some peculiarities since, as mentioned in Remark 3, $q = [\mathbb{C}]$ is a zero divisor of $\mathbf{KVar}_\mathbb{K}$. Hence, the morphism $\mathbf{KVar}_\mathbb{K} \to S^{-1}\mathbf{KVar}_\mathbb{K}$ is not injective, and indeed, its kernel is the annihilator of q or $q-1$. In this way, strictly we have computed the virtual class of the representation variety up to annihilators of q or $q-1$. This is a common feature of the quantum method, due to the requirement of Remark 13 of inverting $[G]$.

5.4 Concluding Remarks

The previous calculation agrees with the one of Sects. 3 and 4. It may seem that this quantum approach is lengthier than the other methods, but its strength lies in on the fact that it does not depend on finding good geometric descriptions. Therefore, it offers a systematic method that can be applied to more general contexts in which geometric or arithmetic methods fail. For instance, in [16], it is computed the virtual classes of $\mathrm{SL}_2(\mathbb{C})$-parabolic representation varieties in the general case by means of the quantum method. This result is unavailable using the geometric or the arithmetic approach due to very subtle interaction between the monodromies of the punctures that cannot be captured with the classical methods.

This calculation also shows a general feature of the quantum method. In principle, the $\mathbf{KVar}_{\mathbb{K}}$-module $\mathcal{Z}(S^1, \star) = \mathbf{KVar}/G$, in which we have to perform the computations, is infinitely generated. However, in all the known computations of $\mathcal{Z}$, it turns out that the computation can be restricted to a certain finitely generated submodule $W \subseteq \mathcal{Z}(S^1, \star)$ as it happened above.

This fact that $\mathcal{Z}(S^1, \star)$ is infinitely generated is in sharp contract with what happens for strict monoidal TQFTs. For $\mathcal{Z}$ a monoidal TQFT, a straightforward duality argument shows that $\mathcal{Z}(X)$ is forced to be a finitely generated module (see [22]). Indeed, this observation is the starting point of the later developments toward the classification of extended TQFTs [26], which show that the whole TQFT is determined by this "fully dualizable" object.

In this sense, the lax monoidal TQFT for representation varieties exhibits a mixed behavior, since it takes values in an infinitely generated module, but the calculations can be performed in a finitely submodule, mimicking a strict monoidal TQFT. On the other hand, when dealing with parabolic character varieties, the TQFT quantizing representation varieties is intrinsically infinitely generated. Definitely, further research is needed for shedding light on the interplay between lax monoidal and strict monoidal TQFTs.

Acknowledgments The authors want to thank Jesse Vogel for the very careful reading of this manuscript and for pointing out a mistake in the computation of Sect. 3.2, and to Sean Lawton for references. The first author acknowledges the hospitality of the ETSI de Sistemas Informáticos at Universidad Politécnica de Madrid, where this work was partially completed. The third author is partially supported by Project MINECO (Spain) PGC2018-095448-B-I00.

References

1. M. Atiyah, *Topological quantum field theories*, Inst. Hautes Études Sci. Publ. Math., **68** (1989), 175–186.
2. D. Baraglia and P. Hekmati, *Arithmetic of singular character varieties and their E-polynomials*, Proc. Lond. Math. Soc. (3), **114** (2017), 293–332.
3. L. Boriso, *Class of the affine line is a zero divisor in the Grothendieck ring*, J. Algebraic Geom., **27** (2018), 203–209.
4. K. Corlette, *Flat G-bundles with canonical metrics*, J. Diff. Geom., **28** (1988), 361–382.
5. M. Culler and P. B. Shalen, *Varieties of group representations and splittings of 3-manifolds*, Ann. of Math. (2), **117** (1983), 109–146.
6. P. Deligne, *Théorie de Hodge. I*, Actes du Congrès International des Mathématiciens (Nice, 1970), **1** (1971), 425–430.
7. P. Deligne, *Théorie de Hodge. II*, Inst. Hautes Études Sci. Publ. Math., **40** (1971), 5–58.
8. P. Deligne, *Théorie de Hodge. III*, Inst. Hautes Études Sci. Publ. Math., **44** (1974), 5–77.

9. D.-E. Diaconescu, *Local curves, wild character varieties, and degenerations*, Preprint arXiv:1705.05707, 2017.
10. S. K. Donaldson, *A new proof of a theorem of Narasimhan and Seshadri*, J. Diff. Geom., **18** (1983), 269–277.
11. C. Florentino and S. Lawton, *Singularities of free group character varieties*, Pacific J. Math., **260** (2012), 149–179.
12. D. S. Freed, M. J. Hopkins, J. Lurie and C. Teleman, *Topological quantum field theories from compact Lie groups*, In: CRM Proc. Lecture Notes, **50**, Amer. Math. Soc., 2010, 367–403.
13. Á. González-Prieto, M. Logares, and V. Muñoz, *A lax monoidal Topological Quantum Field Theory for representation varieties*, Bull. des Sci. Math., **161** (2020), 102871.
14. Á. González-Prieto, M. Logares, and V. Muñoz, *Arithmetic Method for* $\mathrm{AGL}_1(k)$, Available online: http://agt.cie.uma.es/ vicente.munoz/ArithmeticMethodAGL.ipynb (software) https://github.com/AngelGonzalezPrieto/ArithmeticMethodAGL.git (GitHub repository).
15. Á. González-Prieto, *Motivic theory of representation varieties via Topological Quantum Field Theories*, arxiv:1810.09714.
16. Á. González-Prieto, *Virtual classes of parabolic* $\mathrm{SL}_2(\mathbb{C})$*-character varieties*, Adv. Math., **368** (2020), 107–148.
17. R. Hartshorne, *Algebraic geometry*, Graduate Texts in Math., **52**, Springer-Verlag, 1977.
18. T. Hausel, E. Letellier and F. Rodríguez-Villegas, *Arithmetic harmonic analysis on character and quiver varieties II*, Adv. Math., **234** (2013), 85–128.
19. T. Hausel and F. Rodríguez-Villegas, *Mixed Hodge polynomials of character varieties. With an appendix by Nicholas M. Katz*, Invent. Math., **174** (2008), 555–624.
20. T. Hausel and M. Thaddeus, *Mirror symmetry, Langlands duality, and the Hitchin system*, Invent. Math., **153** (2003) 1:197–229.
21. N. J. Hitchin, *The self-duality equations on a Riemann surface*, Proc. London Math. Soc. (3), **55** (1987), 59–126.
22. J. Kock, *Frobenius algebras and 2D topological quantum field theories*, London Mathematical Society Student Texts, **59**, Cambridge University Press, 2004.
23. S. Lawton and A. S. Sikora, *Varieties of characters*, Algebr. Represent. Theory, **20** (2017), 1133–1141.
24. M. Logares, V. Muñoz, and P. E. Newstead, *Hodge polynomials of* $\mathrm{SL}(2,\mathbb{C})$*-character varieties for curves of small genus*, Rev. Mat. Complut., **26** (2013), 635–703.
25. A. Lubotzky and A. Magid, *Varieties of representations of finitely generated groups*, Mem. Amer. Math. Soc. **58** (1985).
26. J. Lurie, *On the classification of topological field theories*, In: Current Developments in Mathematics, Internat. Press, 2009, 129–280.
27. J. Martínez, *E-polynomials of* $\mathrm{PGL}(2,\mathbb{C})$*-character varieties of surface groups*, arxiv:1705.04649.
28. J. Martínez and V. Muñoz, *E-polynomials of* $\mathrm{SL}(2,\mathbb{C})$*-character varieties of complex curves of genus* 3, Osaka J. Math., **53** (2016), 645–681.
29. J. Martínez and V. Muñoz, *E-polynomials of the* $\mathrm{SL}(2,\mathbb{C})$*-character varieties of surface groups*, Int. Math. Res. Not., **2016** (2016), 926–961.
30. M. Mereb, *On the E-polynomials of a family of* SL_n*-character varieties*, Math. Ann., **363** (2015), 857–892.
31. M. Nagata, *On the fourteenth problem of Hilbert*. 1960 Proc. Internat. Congress Math. (1958) pp. 459–462 Cambridge Univ. Press, New York.
32. M. Nagata, *Invariants of a group in an affine ring*, J. Math. Kyoto Univ., **3** (1963/1964), 369–377.
33. M. S. Narasimhan and C. S. Seshadri, *Stable and unitary vector bundles on a compact Riemann surface*, Ann. of Math. (2), **82** (1965), 540–567.
34. P. E. Newstead, *Introduction to moduli problems and orbit spaces*, Tata Institute of Fundamental Research Lectures on Mathematics and Physics, **51** (1978), Tata Institute of Fundamental Research, Bombay; by the Narosa Publishing House, New Delhi.

35. C. A. M. Peters and J. H. M. Steenbrink. *Mixed Hodge structures*, Ergebnisse der Mathematik und ihrer Grenzgebiete. 3. Folge. A Series of Modern Surveys in Mathematics [Results in Mathematics and Related Areas. 3rd Series. A Series of Modern Surveys in Mathematics], **52** (2008), Springer-Verlag, Berlin.
36. C. T. Simpson, *Higgs bundles and local systems*, Inst. Hautes Études Sci. Publ. Math., **75** (1992), 5–95.
37. C. T. Simpson, *Moduli of representations of the fundamental group of a smooth projective variety. I*, Inst. Hautes Études Sci. Publ. Math., **79** (1994), 47–129.
38. C. T. Simpson, *Moduli of representations of the fundamental group of a smooth projective variety. II*, Inst. Hautes Études Sci. Publ. Math., **80** (1995), 5–79.
39. A. Strominger, S.-T. Yau, and E. Zaslow, *Mirror symmetry is T-duality*, Nuclear Phys. B, **479** (1996), 243–259.
40. C. A. Weibel, *The K-book. An introduction to algebraic K-theory*, Graduate Studies in Mathematics, **145**, Amer. Math. Soc., 2013.
41. E. Witten, *Topological quantum field theory*, Comm. Math. Phys., **102** (1988), 353–389.

A Regularized Stochastic Subgradient Projection Method for an Optimal Control Problem in a Stochastic Partial Differential Equation

Baasansuren Jadamba, Akhtar A. Khan, and Miguel Sama

Abstract This work studies an optimal control problem in a stochastic partial differential equation. We present a new regularized stochastic subgradient projection iterative method for a general stochastic optimization problem. By using the martingale theory, we provide a convergence analysis for the proposed method. We test the iterative scheme's feasibility on the considered optimal control problem. The numerical results are encouraging and demonstrate the utility of a stochastic approximation framework in control problems with data uncertainty.

2010 Mathematics Subject Classification 35R30, 49N45, 65J20, 65J22, 65M30

1 Introduction

Let $D \subset \mathbb{R}^n$ be a bounded domain and let ∂D be the sufficiently smooth boundary of Ω. Given a probability space $(\Omega, \mathscr{F}, \mathbb{P})$, and two random fields $a : \Omega \times D \mapsto R$ and $f : \Omega \times D \to \mathbb{R}$, the prototypical stochastic partial differential equation (SPDE) seeks a random field $y : \Omega \times D \to \mathbb{R}$ that almost surely satisfies:

$$-\nabla \cdot (a(\omega, x)\nabla y(\omega, x)) = f(\omega, x), \text{ in } D, \tag{1a}$$

$$y(\omega, x) = 0, \text{ on } \partial D. \tag{1b}$$

The above SPDE appears in important models and has been widely studied.

B. Jadamba · A. A. Khan (✉)
School of Mathematical Sciences, Rochester Institute of Technology, Rochester, NY, USA
e-mail: bxjsma@rit.edu; aaksma@rit.edu

M. Sama
Departamento de Matemática Aplicada, Universidad Nacional de Educación a Distancia, Madrid, Spain
e-mail: msama@ind.uned.es

I. N. Parasidis et al. (eds.), *Mathematical Analysis in Interdisciplinary Research*, Springer Optimization and Its Applications 179,
https://doi.org/10.1007/978-3-030-84721-0_19

Two inverse problems and a control problem are associated with (1). The first one is a linear inverse problem that estimates the source term f from a measurement of the solution y of (1). The second is a nonlinear inverse problem that identifies the parameter a from a measurement of the solution y of (1). The source identification becomes the optimal control problem when f is related to the control variable.

This work's primary motivation stems from the two recent papers: Geiersbach and Pflug [10] and Martin et al. [28]. The main novelty of [10, 28] is the use of a stochastic approximation framework for finding a deterministic optimal control in (1). In contrast, the optimal control problem was mostly explored using computationally demanding stochastic Galerkin/Collocation and related methodologies in the past. Inspired by this new research initiative, we study the same optimal control problem by means of a new iterative regularization method. For some of the recent developments in stochastic control problems, we refer the reader to [1, 5–7, 13, 21, 23, 26, 27, 29, 35, 36].

We pose the control problem as a stochastic optimization problem of the form:

$$\min_{a \in \mathbb{K}} \mathbb{J}(a) := \mathbb{E}\left[J(a, \omega)\right]. \tag{2}$$

Here $\mathbb{K}$ is subset of a real Hilbert space H, $J(a, \omega)$ is a misfit function, and $\mathbb{E}$ is the expectation with respect to the probability space $(\Omega, \mathscr{F}, \mathbb{P})$.

For (2), we develop a new regularized stochastic subgradient projection method in a Hilbert space setting and apply it for solving the optimal control problem described above. The proposed scheme falls under the umbrella of stochastic approximation. We recall that the dynamic field of stochastic approximation began by Robbins and Monro [31] and it has been used for a wide variety of research domains. In recent years, stochastic approximation has become very popular for machine learning algorithms, stochastic variational inequalities, and related developments. As a small sample, we cite [2–4, 8, 9, 11, 14, 15, 24, 25, 30, 33, 34, 38–41]. We note that in [22], an iterative regularization for stochastic variational inequalities posed in finite-dimensional setting was studied.

We organize the contents of this paper into five sections. Section 2 presents the regularized stochastic subgradient projection method for a general stochastic optimization problem. We provide a complete convergence analysis for the proposed method. In Sect. 3, we study the control problem formulated as a convex stochastic optimization problem. In Sect. 4, we present the numerical results, demonstrating the feasibility and the efficacy of the developed framework. The paper concludes with some remarks and open problems.

2 Regularized Stochastic Subgradient Projection Method

Let H be a real Hilbert space with inner product $\langle \cdot, \cdot \rangle$ and norm $\|\cdot\|$. Given $f : H \to \mathbb{R}$ and a closed, and convex set $K \subset H$, we consider the minimization problem:

$$\min_{u \in K} f(u). \tag{3}$$

Given $\{\varepsilon_n\}$ such that $\varepsilon_{n+1} < \varepsilon_n \leq 1$ and $\varepsilon_n \to 0$ as $n \to \infty$, we also consider the following regularized optimization problem of finding $y_{\varepsilon_n} \in K$ by solving

$$\min_{z \in K} f_{\varepsilon_n}(z) := f(z) + \frac{\varepsilon_n}{2}\|z\|^2, \quad \varepsilon_n > 0. \tag{4}$$

Let $(\Omega, \mathscr{F}, \mathbb{P})$ be a probability space, and $\{\omega_n\}$ be a H-valued sequence of random variables on $(\Omega, \mathscr{F}, \mathbb{P})$. Let $\{\alpha_n\}$ be a sequence of positive real step-lengths.

For a numerical solution of the optimization problem (3), we consider the iterative scheme: Given $\varepsilon_1 > 0, \alpha_1 > 0$, and $u_1 \in K$, at step n, compute $u_{n+1} \in K$ by

$$u_{n+1} = P_K\left[u_n - \alpha_n\left(g_n + \varepsilon_n u_n + \omega_n\right)\right], \tag{5}$$

where P_K is the projection onto K, $g_n \in \partial f(u_n)$, and ∂f is the subdifferential of f.

We recall that, given the probability space $(\Omega, \mathscr{F}, \mathbb{P})$, a filtration $\{\mathscr{F}_n\} \subset \mathscr{F}$ is an increasing sequence of σ-algebras. A sequence of random variables $\{\omega_n\}$ is said to be adapted to a filtration $\mathscr{F}_n$, if and only if, $\omega_n \in \mathscr{F}_n$ for all $n \in \mathbb{N}$, that is, ω_n is $\mathscr{F}_n$-measurable. Moreover, the natural filtration is the one generated by the sequence $\{\omega_n\}$ and is given by $\mathscr{F}_n = \sigma(\omega_n : m \leq n)$.

The following result by Robbins and Siegmund [32] will be used shortly:

Lemma 1 *Let $\mathscr{F}_n$ be an increasing sequence of σ-algebras, and V_n, a_n, b_n, and c_n be nonnegative random variables adapted to $\mathscr{F}_n$. Assume that $\sum_{n=1}^{\infty} a_n < \infty$ and $\sum_{n=1}^{\infty} b_n < \infty$, almost surely, and*

$$\mathbb{E}[V_{n+1}|\mathscr{F}_n] \leq (1 + a_n)V_n - c_n + b_n.$$

Then $\{V_n\}$ is almost surely convergent and $\sum_{n=1}^{\infty} c_n < \infty$, almost surely.

The following result provides the convergence of iteration scheme (5):

Theorem 1 *Let H be a Hilbert space, K be a nonempty, closed, and convex subset of H, and $f : H \mapsto \mathbb{R}$ be a convex and lower semicontinuous functional. Let the solution set $\mathscr{S}(f, K)$ of* (3) *be nonempty. Let $\{u_n\}$ be the sequence generated by iterative scheme* (5). *Let $\mathscr{F}_n$ be a filtration on $(\Omega, \mathscr{F}, \mathbb{P})$ such that $\{u_n\}$ and $\{g_n\}$ are $\mathscr{F}_n$-measurable. Assume that the following conditions hold:*

(A_1) *There exists $c > 0$ such that $\|g\| \leq c(1 + \|u\|)$, for every $u \in K$, $g \in \partial f(u)$.*
(A_2) *There exists $c_1 > 0$ such that for all $g_n \in \partial f(u_n)$, we have*

$$\mathbb{E}\left[\omega_n|\mathscr{F}_n\right] = 0, \tag{6}$$

$$\mathbb{E}\left[\|\omega_n\|^2|\mathscr{F}_n\right] \le c_1\left(1+\frac{1}{\delta_n}\|g_n\|^2\right), \quad \delta_n > 0. \tag{7}$$

(A_3) The sequences $\{\varepsilon_n\}$, $\{\alpha_n\}$, *and* $\{\delta_n\}$ *satisfy:*

$$\sum_{n\in\mathbb{N}}\varepsilon_n\alpha_n = \infty, \quad \sum_{n\in\mathbb{N}}\alpha_n^2 < \infty, \quad \sum_{n\in\mathbb{N}}\frac{\alpha_n^2}{\delta_n} < \infty, \quad \sum\alpha_n\delta_n < \infty,$$

$$\sum_{n\in\mathbb{N}}\left(\frac{1+\alpha_n\varepsilon_n}{\alpha_n\varepsilon_n}\right)\left|\frac{\varepsilon_{n-1}-\varepsilon_n}{\varepsilon_n}\right|^2 < \infty.$$

Then, $\|u_{n+1} - y_{\varepsilon_n}\| \to 0$, *almost sure, where* $\{y_{\varepsilon_n}\}$ *solves* (4).

Proof Since f_{ϵ_n} in (4) is strongly convex, the regularized optimization problem is uniquely solvable. Let $y_n := y_{\varepsilon_n} \in K$ be the unique solution of (4).

Then $y_n \in K$ satisfies the following variational inequality as a necessary and sufficient optimality condition:

$$\langle h_n + \varepsilon_n y_n, z - y_n\rangle \ge 0, \quad \text{for every } z \in K, \tag{8}$$

where $h_n \in \partial f(y_n)$ is arbitrary. Furthermore, using the fact that $\mathscr{S}(f, K) \neq \emptyset$, it can be shown that $\{y_n\}$ is uniformly bounded.

Analogously, for an arbitrary $h_{n+1} \in \partial f(y_{n+1})$, we also have

$$\langle h_{n+1} + \varepsilon_{n+1}y_{n+1}, z - y_{n+1}\rangle \ge 0, \quad \text{for every } z \in K. \tag{9}$$

We set $z = y_{n+1}$ in (8), $z = y_n$ in (9), and combine the resulting inequalities to get

$$\varepsilon_{n+1}\langle y_{n+1}, y_n - y_{n+1}\rangle + \varepsilon_n\langle y_n, y_{n+1} - y_n\rangle \ge \langle h_{n+1} - h_n, y_{n+1} - y_n\rangle \ge 0,$$

which implies

$$(\varepsilon_n - \varepsilon_{n+1})\langle y_n, y_{n+1} - y_n\rangle \ge \varepsilon_n\|y_n - y_{n+1}\|^2$$

confirming that there is a constant $C > 0$ such that

$$\|y_{n+1} - y_n\| \le C\frac{|\varepsilon_n - \varepsilon_{n+1}|}{\varepsilon_n}. \tag{10}$$

The variational characterization of the projection map and (8) implies that

$$y_n = P_K[y_n - \alpha_n(h_n + \varepsilon_n y_n)].$$

By using the iterative scheme (5) and the above identity, we obtain

$$\begin{aligned}
\|u_{n+1} - y_n\|^2 &= \| P_K(u_n - \alpha_n(g_n + \varepsilon_k u_n + \omega_n)) - P_K(y_n - \alpha_n(h_n + \varepsilon_n y_n))\|^2 \\
&\le \| [u_n - \alpha_n(g_n + \varepsilon_n u_n + \omega_n)] - [y_n - \alpha_n(h_n + \varepsilon_n y_n)] \|^2 \\
&= \|u_n - y_n - \alpha_n(g_n - h_n) - \alpha_n\varepsilon_n(u_n - y_n) - \alpha_n\omega_n\|^2 \\
&= \|u_n - y_n\|^2 + \alpha_n^2\|g_n - h_n\|^2 + \alpha_n^2\varepsilon_n^2\|u_n - y_n\|^2 + \alpha_n^2\|\omega_n\|^2 \\
&\quad - 2\alpha_n\varepsilon_n\|u_n - y_n\|^2 - 2\alpha_n\langle g_n - h_n, u_n - y_n\rangle \\
&\quad - 2\alpha_n\langle u_n - y_n, \omega_n\rangle + 2\alpha_n^2\langle g_n - h_n, \omega_n\rangle \\
&\quad + 2\alpha_n^2\varepsilon_n\langle u_n - y_n, \omega_n\rangle + 2\alpha_n^2\varepsilon_n\langle g_n - h_n, u_n - y_n\rangle,
\end{aligned}$$

and hence by taking the expectation past $\mathscr{F}_n$, we obtain

$$\begin{aligned}
\mathbb{E}\left[\| u_{n+1} - y_n\|^2|\mathscr{F}_n\right] &\le (1 - 2\alpha_n\varepsilon_n + \alpha_n^2\varepsilon_n^2)\|u_n - y_n\|^2 + \alpha_n^2\|g_n - h_n\|^2 \\
&\quad + \alpha_n^2\mathbb{E}\left[\|\omega_n\|^2|\mathscr{F}_n\right] + 2\alpha_n^2\varepsilon_n\langle u_n - y_n, g_n - h_n\rangle \\
&\le (1 - 2\alpha_n\varepsilon_n + 2\alpha_n^2)\|u_n - y_n\|^2 + 2\alpha_n^2\|g_n - h_n\|^2 \\
&\quad + \alpha_n^2\mathbb{E}\left[\|\omega_n\|^2|\mathscr{F}_n\right].
\end{aligned} \tag{11}$$

To find bounds on the terms in (11), we note that the sequence $\{y_n\}$ is bounded, and hence there exists a constant $c_0 > 0$ such that $\|y_n\| \le c_0$, for every $n \in \mathbb{N}$. Therefore,

$$\begin{aligned}
\|g_n - h_n\| &\le \|g_n\| + \|h_n\| \\
&\le c(1 + \|u_n\|) + c(1 + \|y_n\|) \\
&\le 2c(1 + \|y_n\|) + c\|u_n - y_n\| \\
&\le 2c(1 + c_0)(1 + \|u_n - y_n\|) \\
&\le k_1(1 + \|u_n - y_n\|),
\end{aligned} \tag{12}$$

where $k_1 := 2c(1 + c_0)$, and hence with $k_2 := 8c^2(1 + c_0)^2$, we obtain

$$\|g_n - h_n\|^2 \le k_2(1 + \|u_n - y_n\|^2). \tag{13}$$

Furthermore,

$$\alpha_n^2\mathbb{E}\left[\|\omega_n\|^2|\mathscr{F}_n\right] \le \alpha_n^2 c_1\left(1 + \frac{\|g_n\|^2}{\delta_n}\right)$$

$$\leq c_1\alpha_n^2\left(1+\frac{k_1^2(1+\|u_n-y_n\|^2)}{2\delta_n}\right)$$

$$\leq c_1\alpha_n^2+\frac{c_1k_1^2\alpha_n^2}{2\delta_n}(1+\|u_n-y_n\|^2),$$

and hence for a constant $k_3 := c_1 \max(1, k_1^2/2)$, we obtain

$$\alpha_n^2\mathbb{E}\left[\|\omega_n\|^2|\mathscr{F}_n\right]\leq k_3\alpha_n^2+\frac{k_3\alpha_n^2}{\delta_n}+\frac{k_3\alpha_n^2}{\delta_n}\|u_n-y_n\|^2. \tag{14}$$

Summarizing, due to (11), (13), and (14), there is a constant $k > 0$ with

$$\mathbb{E}\left[\| u_{n+1}-y_n\|^2|\mathscr{F}_n\right]\leq\left(1-2\alpha_n\varepsilon_n+k\alpha_n^2+k\frac{\alpha_n^2}{\delta_n}\right)\|u_n-y_n\|^2+k\alpha_n^2+k\frac{\alpha_n^2}{\delta_n}.$$

The above inequality, due to the following inequality, which holds for all $a, b \in \mathbb{R}$,

$$(a+b)^2\leq(1+\alpha_n\varepsilon_n)a^2+\left(1+\frac{1}{\alpha_n\varepsilon_n}\right)b^2,$$

yields

$$\begin{aligned}
\mathbb{E}\left[\| u_{n+1}-y_n\|^2|\mathscr{F}_n\right]&\leq\left(1-2\alpha_n\varepsilon_n+k\alpha_n^2+k\frac{\alpha_n^2}{\delta_n}\right)(1+\alpha_n\varepsilon_n)\|u_n-y_{n-1}\|^2\\
&\quad+k\alpha_n^2+k\frac{\alpha_n^2}{\delta_n}+\left(1-2\alpha_n\varepsilon_n+k\alpha_n^2+k\frac{\alpha_n^2}{\delta_n}\right)\left(1+\frac{1}{\alpha_n\varepsilon_n}\right)\|y_n-y_{n-1}\|^2\\
&\leq\left(1-2\alpha_n\varepsilon_n+k\alpha_n^2+k\frac{\alpha_n^2}{\delta_n}\right)(1+\alpha_n\varepsilon_n)\|u_n-y_{n-1}\|^2\\
&\quad+k\alpha_n^2+k\frac{\alpha_n^2}{\delta_n}+\left(1-2\alpha_n\varepsilon_n+k\alpha_n^2+k\frac{\alpha_n^2}{\delta_n}\right)\left(1+\frac{1}{\alpha_n\varepsilon_n}\right)\left|\frac{\varepsilon_{n-1}-\varepsilon_n}{\varepsilon_n}\right|^2\\
&\leq\left(1-\alpha_n\varepsilon_n-2\alpha_n^2\varepsilon_n^2+(1+\alpha_n\varepsilon_n)\left(k\alpha_n^2+k\frac{\alpha_n^2}{\delta_n}\right)\right)\|u_n-y_{n-1}\|^2\\
&\quad+k\alpha_n^2+k\frac{\alpha_n^2}{\delta_n}\\
&\quad+\left(1-2\alpha_n\varepsilon_n+k\alpha_n^2+k\frac{\alpha_n^2}{\delta_n}\right)\left(\frac{1+\alpha_n\varepsilon_n}{\alpha_n\varepsilon_n}\right)\left|\frac{\varepsilon_{n-1}-\varepsilon_n}{\varepsilon_n}\right|^2\\
&\leq\left(1-\alpha_n\varepsilon_n+s\alpha_n^2+s\frac{\alpha_n^2}{\delta_n}\right)\|u_n-y_{n-1}\|^2+k\alpha_n^2+k\frac{\alpha_n^2}{\delta_n}
\end{aligned}$$

$$+s\left(\frac{1+\alpha_n\varepsilon_n}{\alpha_n\varepsilon_n}\right)\left|\frac{\varepsilon_{n-1}-\varepsilon_n}{\varepsilon_n}\right|^2,$$

where s is constant such that

$$s=\sup_{n\in\mathbb{N}}\left(1+k\alpha_n^2+k\frac{\alpha_n^2}{\delta_n}\right).$$

Due to the summability condition on all the sequences involved, the terms in the above must converge to zero, and hence they must remain bounded.

Consequently, we have

$$\mathbb{E}\Big[\|u_{n+1}-y_n\|^2|\mathscr{F}_n\Big]\leq(1+t_n)\|u_n-y_{n-1}\|^2-\gamma_n+\kappa_n,$$

where

$$t_n:=s\alpha_n^2+s\frac{\alpha_n^2}{\delta_n},$$

$$\kappa_n:=k\alpha_n^2+k\frac{\alpha_n^2}{\delta_n}+s\left(\frac{1+\alpha_n\varepsilon_n}{\alpha_n\varepsilon_n}\right)\left|\frac{\varepsilon_{n-1}-\varepsilon_n}{\varepsilon_n}\right|^2,$$

$$\gamma_n:=\alpha_n\varepsilon_n\|u_n-y_{n-1}\|.$$

Since the sequence $\{t_n\}$ and $\{\kappa_n\}$ generate summable series, as a consequence of Lemma 1, $\|u_n-y_{n-1}\|$ converges almost surely and

$$\sum_{n\in\mathbb{N}}\alpha_n\varepsilon_n\|u_n-y_{n-1}\|^2<+\infty,$$

which due to the divergence of the series $\sum_{n\in\mathbb{N}}\alpha_n\varepsilon_n$ implies that $\|u_n-y_{n-1}\|\to 0$, almost surely. □

3 Optimal Control for Stochastic PDEs

We will apply the regularized stochastic subgradient projection method for an optimal control problem in a stochastic PDE. We first recall some function spaces. Given a bounded domain $D\subset\mathbb{R}^n$ with sufficiently smooth boundary ∂D, for $1\leq p<\infty$, by $L^p(D)$, we represent the space of pth Lebesgue integrable functions, that is,

$$L^p(D) = \left\{ y : D \to \mathbb{R} \text{ is measurable with } \int_D |y|^p \, dx < +\infty \right\}.$$

The space $L^\infty(D)$ contains the measurable functions that are bounded almost everywhere (a.e.) on D. We also recall that the Sobolev spaces are given by

$$H^1(D) = \left\{ y \in L^2(D), \ \ \partial_{x_i} y \in L^2(D), \ i = 1, \dots, n \right\},$$

$$H_0^1(D) = \left\{ y \in H^1(D), \ \ y|_{\partial D} = 0 \right\},$$

and $H^{-1}(D) = (H_0^1(D))^*$ is the dual of $H_0^1(D)$.

We aim to study the following stochastic unregularized PDE-constrained optimization problem:

$$\min_{u \in K} \mathbb{J}(u) := \mathbb{E}\left[J(u, \omega)\right] := \mathbb{E}\left[\frac{1}{2}\|y - z\|^2_{L^2(D)}\right], \tag{15}$$

subject to the stochastic PDE:

$$-\nabla \cdot (a(x, \omega) \nabla y(x, \omega)) = u(x), \quad (x, \omega) \in D \times \Omega, \tag{16a}$$

$$y(x, \omega) = 0, \quad (x, \omega) \in \partial D \times \Omega, \tag{16b}$$

where K is a closed, convex, and bounded set of feasible controls given by

$$K := \left\{ u \in L^2(D) | \ 0 < \alpha \le u(x) \le \beta, \quad \text{almost everywhere } x \in D \right\}.$$

In the following, we will assume that there are constants k_0 and k_1 such that

$$0 < k_0 \le a(\omega, x) \le k_1 < \infty, \text{ almost everywhere in } \Omega \times D. \tag{17}$$

In particular, $a \in L^\infty(\Omega \times D)$.

We will study (16) in the variational formulation that for a fixed $\omega \in \Omega$, seeks $y(\cdot, \omega) \in H_0^1(D)$ such that

$$\int_D a(x, \omega) \nabla y(x, \omega) \cdot \nabla v \, dx = \int_D uv \, dx, \text{ for all } v \in H_0^1(\Omega). \tag{18}$$

Problem (15) corresponds to unregularized optimal control problem which leads to non-unique bang-bang solutions. The following result summarizes the information necessary for (15). A proof of the above result is very similar to the well-known deterministic optimal control problems, see [10, 28, 37].:

Lemma 2 *Let* $u, z \in L^2(\Omega)$ *and let* a *satisfy* (17)*. Then, there exists a unique solution* $y(\cdot, \omega) \in H_0^1(D)$ *of* (18)*. Moreover, there is a constant* $C_1 > 0$ *such that*

$$\|y(\cdot\omega)\|_{L^2(D)} \leq C_1\|u\|_{L^2(D)}.$$

Furthermore, the optimization problem (15) *has a nonempty solution set. Finally, for* $\omega \in \Omega$*, the stochastic gradient* $\nabla_u J(u, \omega)$ *is given by*

$$\nabla_u J(u, \omega) = -p(\cdot, \omega),$$

where $p(\cdot, \omega) \in H_0^1(\Omega)$ *is the unique solution of the adjoint problem:*

$$\int_D a(x, \omega)\nabla p(x, \omega) \cdot \nabla v \, dx = \int_D (z(x) - y(x, \omega))v \, dx, \quad \text{for all } v \in H_0^1(D). \tag{19}$$

We consider the following algorithm:

Algorithm 1 Regularized stochastic projected subgradient scheme

---

1: At $n = 1$, start with a random initial point $u_1 \in L_2(D)$.
2: For $n = 1, 2, \cdots$ do
3: Generate random $a(\cdot, \omega_n)$, independent from previous observations, and $\alpha_n, \varepsilon_n > 0$.
4: Solve (18) with $a(\cdot, \omega) = a(\cdot, \omega_n)$.
5: Solve (19) with $y = y_n$ and $a(\cdot, \omega) = a(\cdot, \omega_n)$.
6: Set $G(u_n, \omega_n) := -p_n$.
7: Compute $u_{n+1} \in K$ by

$$u_{n+1} = P_K\left[u_n - \alpha_n\left(G(u_n, \omega_n) + \varepsilon_n u_n\right)\right]. \tag{20}$$

---

4 A Numerical Example

In this section, we test iterative scheme given in Algorithm 1 for stochastic optimal control problem (15). Our example is a slight modification of the unregularized example given in [10]. We set $D = [0, 1] \times [0, 1]$ and define the constraint set by

$$K = \{u \in L^2(D) : -1 \leq u(x) \leq 1, \text{ for every } x \in D\}.$$

We consider the parameter

$$a(x, \omega) = a(\omega) = 1 + Y(\omega),$$

where $Y(\omega) \sim U[0, 1]$ is uniformly distributed on $[0, 1]$. As in [10] (see also [37, Section 2]), we consider a slight modification of the state equation by introducing an extra term in the right-hand side, where the optimal control and state $(\bar{u}, \bar{y})$ satisfy the following equation

$$-a(\omega)\Delta\bar{y}(x,\omega) = \bar{u}(x) + u_D(x), \quad (x,\omega) \in D \times \Omega,$$
$$\bar{y}(x,\omega) = 0, \quad (x,\omega) \in \partial D \times \Omega.$$

Following [10], a deterministic adjoint state is defined by $\bar{p}(x) = -\sin(2\pi x)\sin(2\pi y)$. The solution set is characterized by the following expression

$$\bar{u}(x) = \text{sign}(\bar{p}(x)),$$

by taking into account $\text{sign}(0) := [-1, 1]$, $\bar{u}$ can be any value in $[-1, 1]$ on the set $\{x \in D : \bar{p}(x) = 0\}$. Since the adjoint state verifies

$$-a(\omega)\Delta\bar{p}(x,\omega) = z(x) - \bar{y}(x,\omega), \quad (x,\omega) \in D \times \Omega,$$
$$\bar{p}(x,\omega) = 0, \quad (x,\omega) \in \partial D \times \Omega,$$

the maps are defined by

$$\begin{aligned} u_D(x) &= \mathbb{E}[-a(\omega)\Delta\bar{p}(x,\omega) - \bar{u}(x)] \\ &= 3\pi^2 \sin(\pi x_1)\sin(\pi x_2) - \bar{u}(x), \\ z(x) &= \mathbb{E}[-a(\omega)\Delta\bar{p}(x,\omega) + \bar{y}(x,\omega)] \\ &= \frac{3}{2}\sin(2\pi x_1)\sin(2\pi x_2) + \sin(\pi x_1)\sin(\pi x_2). \end{aligned}$$

We discretize the problem by using a standard finite-element discretization on a uniform triangulation of 3600 nodes. The iterative scheme is implemented by using finite-element library FreeFem++ [12]. In numerical computations, we consider $\text{sign}(0) = 0$, and take $u_1 = 0$. After running 150 iterations, we get an acceptable reconstruction of a bang-bang solution, see Figs. 1, 2, 3, and 4.

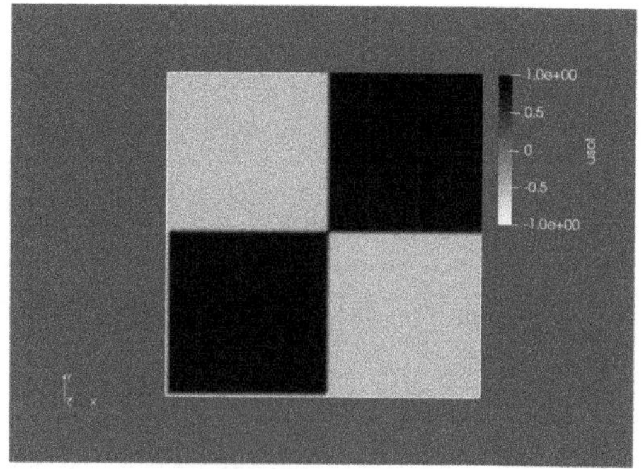

Fig. 1 A solution $\bar{u}$

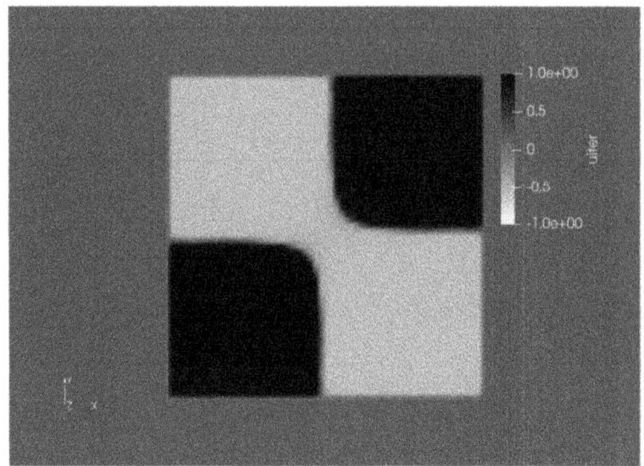

Fig. 2 Iterated u_{150}

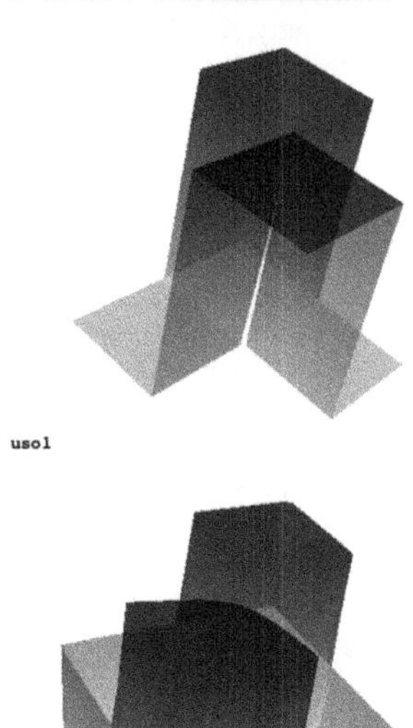

Fig. 3 A solution $\bar{u}$

Fig. 4 Iterated u_{150}

5 Concluding Remarks

We presented a regularized stochastic subgradient projection method for a general optimization problem and gave preliminary numerical results on an optimal control problem. The given application shows the utility of a stochastic approximation framework for control problems with uncertainty. It would be of interest to extend these results for state-constrained PDE-constrained optimization problems, see [16–20].

Acknowledgments B. Jadamba, and A. A. Khan are supported by the National Science Foundation under Award No. 1720067. This research is also supported by the Ministerio de Ciencia e Innovación, Agencia Estatal de Investigación (Spain) under project PID2020-112491GB-I00 /AEI/10.13039/501100011033. Miguel Sama is also supported by grant 2021-MAT11 (ETSI Industriales, UNED).

References

1. A. Alexanderian, N. Petra, G. Stadler, O. Ghattas, Mean-variance risk-averse optimal control of systems governed by PDEs with random parameter fields using quadratic approximations, SIAM/ASA J. Uncertain. Quantif. 5 (1) (2017) 1166–1192.
2. K. Barty, J.-S. Roy, C. Strugarek, Hilbert-valued perturbed subgradient algorithms, Math. Oper. Res. 32 (3) (2007) 551–562.
3. D. P. Bertsekas, J. N. Tsitsiklis, Gradient convergence in gradient methods with errors, SIAM J. Optim. 10 (3) (2000) 627–642.
4. R. I. Boţ, A. Böhm, Variable smoothing for convex optimization problems using stochastic gradients, J. Sci. Comput. 85 (2) (2020) Paper No. 33, 29.
5. A. Borzi, Multigrid and sparse-grid schemes for elliptic control problems with random coefficients, Comput. Vis. Sci. 13 (4) (2010) 153–160.
6. P. Chen, A. Quarteroni, G. Rozza, Multilevel and weighted reduced basis method for stochastic optimal control problems constrained by Stokes equations, Numer. Math. 133 (1) (2016) 67–102.
7. P. Chen, A. Quarteroni, G. Rozza, Reduced basis methods for uncertainty quantification, SIAM/ASA J. Uncertain. Quantif. 5 (1) (2017) 813–869.
8. J.-C. Culioli, G. Cohen, Decomposition/coordination algorithms in stochastic optimization, SIAM J. Control Optim. 28 (6) (1990) 1372–1403.
9. A. Dieuleveut, A. Durmus, F. Bach, Bridging the gap between constant step size stochastic gradient descent and Markov chains, Ann. Statist. 48 (3) (2020) 1348–1382.
10. C. Geiersbach, G. C. Pflug, Projected stochastic gradients for convex constrained problems in Hilbert spaces, SIAM J. Optim. 29 (3) (2019) 2079–2099.
11. L. Goldstein, Minimizing noisy functionals in Hilbert space: an extension of the Kiefer-Wolfowitz procedure, J. Theoret. Probab. 1 (2) (1988) 189–204.
12. F. Hecht, New development in freefem++, Journal of numerical mathematics 20 (3-4) (2012) 1–14.
13. M. Heinkenschloss, B. Kramer, T. Takhtaganov, Adaptive reduced-order model construction for conditional value-at-risk estimation, SIAM/ASA J. Uncertain. Quantif. 8 (2) (2020) 668–692.
14. A. N. Iusem, A. Jofré, R. I. Oliveira, P. Thompson, Extragradient method with variance reduction for stochastic variational inequalities, SIAM J. Optim. 27 (2) (2017) 686–724.
15. A. N. Iusem, A. Jofré, R. I. Oliveira, P. Thompson, Variance-based extragradient methods with line search for stochastic variational inequalities, SIAM J. Optim. 29 (1) (2019) 175–206.
16. B. Jadamba, A. Khan, M. Sama, Error estimates for integral constraint regularization of state-constrained elliptic control problems, Comput. Optim. Appl. 67 (1) (2017) 39–71.
17. B. Jadamba, A. A. Khan, M. Sama, Regularization for state constrained optimal control problems by half spaces based decoupling, Systems Control Lett. 61 (6) (2012) 707–713.
18. B. Jadamba, A. A. Khan, M. Sama, Stable conical regularization by constructible dilating cones with an application to L^p-constrained optimization problems, Taiwanese J. Math. 23 (4) (2019) 1001–1023.
19. B. Jadamba, A. A. Khan, M. Sama, C. Tammer, Regularization methods for scalar and vector control problems, in: Variational analysis and set optimization, CRC Press, Boca Raton, FL, 2019, pp. 296–321.

20. A. A. Khan, M. Sama, A new conical regularization for some optimization and optimal control problems: Convergence analysis and finite element discretization, Numerical Functional Analysis and Optimization 34 (8) (2013) 861–895.
21. P. Kolvenbach, O. Lass, S. Ulbrich, An approach for robust PDE-constrained optimization with application to shape optimization of electrical engines and of dynamic elastic structures under uncertainty, Optim. Eng. 19 (3) (2018) 697–731.
22. J. Koshal, A. Nedić, U. V. Shanbhag, Regularized iterative stochastic approximation methods for stochastic variational inequality problems, IEEE Trans. Automat. Control 58 (3) (2013) 594–609.
23. D. P. Kouri, M. Heinkenschloss, D. Ridzal, B. G. van Bloemen Waanders, A trust-region algorithm with adaptive stochastic collocation for PDE optimization under uncertainty, SIAM J. Sci. Comput. 35 (4) (2013) A1847–A1879.
24. H. J. Kushner, A. Shwartz, Stochastic approximation in Hilbert space: identification and optimization of linear continuous parameter systems, SIAM J. Control Optim. 23 (5) (1985) 774–793.
25. T. L. Lai, Stochastic approximation, Ann. Statist. 31 (2) (2003) 391–406.
26. H.-C. Lee, M. D. Gunzburger, Comparison of approaches for random PDE optimization problems based on different matching functionals, Comput. Math. Appl. 73 (8) (2017) 1657–1672.
27. C. Li, G. Stadler, Sparse solutions in optimal control of PDEs with uncertain parameters: the linear case, SIAM J. Control Optim. 57 (1) (2019) 633–658.
28. M. Martin, S. Krumschield, F. Nobile, Analysis of stochastic gradient methods for PDE-constrained optimal control problems with uncertain parameters, Preprint (2018) 1–39.
29. J. Martínez-Frutos, F. Periago E., Optimal control of PDEs under uncertainty, SpringerBriefs in Mathematics, Springer, 2018.
30. P. Mertikopoulos, M. Staudigl, Stochastic mirror descent dynamics and their convergence in monotone variational inequalities, J. Optim. Theory Appl. 179 (3) (2018) 838–867.
31. H. Robbins, S. Monro, A stochastic approximation method, Ann. Math. Statistics 22 (1951) 400–407.
32. H. Robbins, D. Siegmund, A convergence theorem for non negative almost supermartingales and some applications (1971) 233–257.
33. K. Scaman, F. Bach, S. Bubeck, Y. T. Lee, L. Massoulié, Optimal convergence rates for convex distributed optimization in networks, J. Mach. Learn. Res. 20 (2019) Paper No. 159, 31.
34. D. Scieur, A. d'Aspremont, F. Bach, Regularized nonlinear acceleration, Math. Program. 179 (1-2, Ser. A) (2020) 47–83.
35. R. E. Tanase, Parameter estimation for partial differential equations using stochastic methods, 2016, Thesis (Ph.D.)–University of Pittsburgh.
36. H. Tiesler, R. M. Kirby, D. Xiu, T. Preusser, Stochastic collocation for optimal control problems with stochastic PDE constraints, SIAM J. Control Optim. 50 (5) (2012) 2659–2682.
37. F. Tröltzsch, Optimal control of partial differential equations: theory, methods, and applications, vol. 112, American Mathematical Soc., 2010.
38. F. Q. Xia, Q. H. Ansari, J. C. Yao, A new incremental constraint projection method for solving monotone variational inequalities, Optim. Methods Softw. 32 (3) (2017) 470–502.
39. G. Yin, Y. M. Zhu, On H-valued Robbins-Monro processes, J. Multivariate Anal. 34 (1) (1990) 116–140.
40. X.-J. Zhang, X.-W. Du, Z.-P. Yang, G.-H. Lin, An infeasible stochastic approximation and projection algorithm for stochastic variational inequalities, J. Optim. Theory Appl. 183 (3) (2019) 1053–1076.
41. J. Zhu, J. C. Spall, Stochastic approximation with nondecaying gain: error bound and data-driven gain-tuning, Internat. J. Robust Nonlinear Control 30 (15) (2020) 5820–5870.

A Survey on Interpolative and Hybrid Contractions

Erdal Karapınar

Abstract In this chapter, we consider the distinct hybrid type contractions in various abstract spaces. In this work, hybrid contraction refers to combination of not only linear and nonlinear contractions, but also interpolative contractions. The main goal of the chapter is to clarify the metric fixed point theory literature by using the hybrid type contractions that unify several well-known results.

1 Introduction

Metric fixed point theory was initiated by the outstanding result of Banach [32] in the context of the complete normed spaces. In this pioneering result, Banach abstracted the method of successive approximations that was used to solve the concrete differential equations [94, 102]. The result of Banach was reconsidered in the setting of metric spaces by Caccioppoli [40] and this version is now known as the well-known Banach contraction mapping principle: "Every contraction in a complete metric space has a unique fixed point." As a second historical remark, we note that this result was known also Banach–Picard fixed point theorem in the early years of this century.

As we mention above, the fixed point theorem of Banach is an abstraction of the method of successive approximations. It is natural to ask the question: "Why such an abstraction is important/interesting?" The answer to this question may vary from researcher to researcher. First of all, we can say: It reduces a solution of differential equations to one point functional analysis problem. Furthermore, it is engaged in topology in solving this problem. Therefore, applied mathematics, topology, and functional analysis bring employees together in a common point. This can also

E. Karapınar (✉)
Department of Medical Research, China Medical University Hospital, China Medical University, Taichung, Taiwan

Department of Mathematics, Çankaya University, Etimesgut, Ankara, Turkey
e-mail: karapinar@mail.cmuh.org.tw; erdal.karapinar@cankaya.edu.tr

I. N. Parasidis et al. (eds.), *Mathematical Analysis in Interdisciplinary Research*, Springer Optimization and Its Applications 179,
https://doi.org/10.1007/978-3-030-84721-0_20

think the other way around. The fact that a relatively theoretical fixed point theorem can find a place in applied mathematics also helps it contribute to its development. Beside all these, the statement and proof of Banach fixed point theorem is amazing. In a simple way, not only the existence of the fixed point is guaranteed, but also it is shown "how to derive the desired point." Notice that most of the real-world problems can be considered as a fixed point problem, that is, $T(x) = x$ is equivalent to $F(x) = 0$ where $F(x) = x$, for the well-defined concrete mappings T, S. Hence, the desired fixed point is equivalent to the solution of the real-world problem.

Regarding the attraction of the topic and the wide application potential, a number of the authors published several fixed point results that improve, extend, and generalize the outstanding results of Banach in various direction, see, e.g., [1, 2, 7, 9–11, 27–31, 33, 34, 36–39, 41, 44–47, 50–52, 55, 58–60, 63, 67, 75–81, 84, 86–88, 92, 93, 96, 97, 100, 101, 104–107, 109, 112, 113, 115–118, 120, 123].

In this work, we shall deal with one of the recent and interesting fixed point result via interpolative contraction. The notion of the interpolative contraction was suggested in [66] to revisit Kannan type contraction in the context of the standard structure, complete metric spaces. Although interpolation theory is a significant tool in functional analysis, it was not involved to metric fixed point theory up to the publication [66]. The notion of the interpolative contraction not only opens a new frame but enriches the metric fixed point theory. After this initial result [66], this approach was generalized, extended, and improved in various direction in the setting of different structures, see e.g. [3, 22, 23, 49, 64, 65]. On the other side, Mitrovic et al. [99] proposed a new notion, hybrid contraction to combine the well-known linear contractions with interpolative contractions. The main result of this paper [99] yields several well-known linear contractions (Banach contraction, Kannan type contraction, Reich type contraction) and provides some new nonlinear contractions.

In this survey, we first indicate the improvement of interpolative contractions in different abstract spaces with some concrete examples. Then, we consider the advances on the hybrid contractions in various abstract contraction together with some immediate consequences.

2 Preliminaries

In this section we recollect and recall some basic notion, notations, and fundamental results that will play crucial roles on the upcoming sections and proofs. We presume that all considered sets and subsets are nonempty throughout the survey. We underline that the letters $\mathbb{N}$ and $\mathbb{R}$ reserved for the set of all positive integers and real numbers. Further, we shall use $\mathbb{N}_{\not{\kappa}} = \mathbb{N} \cup \{0\}$, $\mathbb{R}^+ = (0, \infty)$, and $\mathbb{R}_0^+ = [0, \infty)$.

2.1 Simulation Functions

First, we recall the notion of *simulation function* that was introduced in [89].

Definition 1 (See [89]) A mapping $\zeta : [0,\infty)\times[0,\infty) \to \mathbb{R}$ is called a *simulation function* if it fulfills

(ζ_1) $\zeta(0,0)=0$;
(ζ_2) $\zeta(\tau,s) < s-\tau$ for all $\tau, s>0$;
(ζ_3) if $\{\tau_n\}$, $\{s_n\}$ are sequences in $(0,\infty)$ such that $\lim_{n\to\infty} \tau_n = \lim_{n\to\infty} s_n > 0$, then

$$\limsup_{n\to\infty} \zeta(\tau_n, s_n) < 0. \tag{1}$$

It was understood that the first axiom (ζ_1) is superfluous by Argoubi et al. [15]. Note that (ζ_2) yields (ζ_1). Therefore, throughout the survey, we presume that any simulation function ζ satisfies only (ζ_2) and (ζ_3). Furthermore, we shall use the letter $\mathcal{Z}$ to indicate the class of all simulation functions $\zeta : [0,\infty)\times[0,\infty) \to \mathbb{R}$ that fulfill (ζ_2) and (ζ_3). In addition, we underline the following simple observation: The axiom (ζ_2) implies that

$$\zeta(r,r) < 0 \text{ for all } r > 0. \tag{2}$$

Example 1 (See e.g. [5, 89, 108]) We suppose that each mapping $\phi_i : [0,\infty) \to [0,\infty)$, for $i = 1,2,3,4,5,6$, is not only continuous but also satisfies

$$\phi_i(t) = 0 \text{ if, and only if, } t = 0.$$

For $i = 1,2,3,4,5,6$, we define the mappings $\zeta_i : [0,\infty)\times[0,\infty) \to \mathbb{R}$, as follows

(i) $\zeta_1(t,s) = \phi_1(s) - \phi_2(t)$ for all $t, s \in [0,\infty)$, where $\phi_1(t) < t \leq \phi_2(t)$ for all $t > 0$.
(ii) $\zeta_2(t,s) = s - \frac{f(t,s)}{g(t,s)} t$ for all $t, s \in [0,\infty)$, where $f, g : [0,\infty)^2 \to (0,\infty)$ are two continuous functions with respect to each variable such that $f(t,s) > g(t,s)$ for all $t, s > 0$.
(iii) $\zeta_3(t,s) = s - \phi_3(s) - t$ for all $t, s \in [0,\infty)$.
(iv) If $\varphi : [0,\infty) \to [0,1)$ is a function such that $\limsup_{t\to r^+} \varphi(t) < 1$ for all $r > 0$, and we define

$$\zeta_4(t,s) = s\varphi(s) - t \qquad \text{for all } s, t \in [0,\infty).$$

(v) If $\eta : [0,\infty) \to [0,\infty)$ is an upper semi-continuous mapping such that $\eta(t) < t$ for all $t > 0$ and $\eta(0) = 0$, and we define

$$\zeta_5(t,s) = \eta(s) - t \qquad \text{for all } s, t \in [0,\infty).$$

(vi) If $\phi : [0, \infty) \to [0, \infty)$ is a function such that $\int_0^{\varepsilon} \phi(u)du$ exists and $\int_0^{\varepsilon} \phi(u)du > \varepsilon$, for each $\varepsilon > 0$, and we define

$$\zeta_6(t, s) = s - \int_0^t \phi(u)du \qquad \text{for all } s, t \in [0, \infty).$$

It is straightforward to see that each ζ_i forms a simulation function for $(i = 1, 2, 3, 4, 5, 6)$.

For the detailed discussion and more samples on this notion, we refer to [5, 6, 8, 20, 21, 72–74, 89, 108].

2.2 Comparison and c-Comparison Functions

In this section, we shall recollect some basic properties of two interesting auxiliary function types: comparison and c-comparison functions [113].

The notion of comparison function is a crucial tool for obtained more general contractions.

Definition 2 ([113]) A mapping $\phi : [0, \infty) \to [0, \infty)$ is called a comparison if it is non-decreasing and

$$\phi^n(t) \to 0 \text{as} n \to \infty \text{ for every } t \in [0, \infty),$$

where ϕ^n is the n-th iterate of ϕ.

We reserve the letter Φ to denote the class of all comparison functions.

In the sequel, we recall fundamental properties of comparison functions:

Lemma 1 ([113]) *Suppose that* $\phi : [0, \infty) \to [0, \infty)$ *is a comparison function. Then, we have*

1. *for all* $k \geq 1$, ϕ^k *is also a comparison function (k-th iteration of* ϕ *);*
2. ϕ *is continuous at* 0*;*
3. $\phi(t) < t$ *for all* $t > 0$.

Let Ψ be the class of functions $\psi : [0, \infty) \to [0, \infty)$ which are non-decreasing and satisfying

(Ψ_2) $\sum_{n=1}^{+\infty} \psi^n(t) < \infty$ for all $t > 0$, where ψ^n is the nth iterate of ψ.

Here, each element (function) $\psi \in \Psi$ will be called (c)-comparison functions [113]. It is worth mentioning that $\Psi \subset \Phi$.

Lemma 2 (See e.g. [113]) *If* $\psi \in \Psi$, *then the following hold:*

(i) $(\psi^n(t))_{n\in\mathbb{N}}$ *converges to* 0 *as* $n \to \infty$ *for all* $t \in \mathbb{R}^+$*;*
(ii) $\psi(t) < t$*, for any* $t \in \mathbb{R}^+$;
(iii) ψ *is continuous at* 0*;*
(iv) *the series* $\sum_{k=1}^{\infty} \psi^k(t)$ *converges for any* $t \in \mathbb{R}^+$.

Now, we shall derive the definition of the b-comparison function:

Definition 3 ([35]) We say that a monotone increasing self-mapping $\varphi : [0, \infty) \to [0, \infty)$ is called a b-comparison if it satisfies the condition:

$$\text{there exist } k_0 \in \mathbb{N}, a \in (0, 1) \text{ and a convergent series of nonnegative terms } \sum_{k=1}^{\infty} v_k$$

$$\text{such that } s^{k+1}\varphi^{k+1}(t) \leq a s^k \varphi^k(t) + v_k, \text{ for } k \geq k_0 \text{ and any } t \in [0, \infty),$$

where $s \in [1, \infty)$.

Example 2 Let $s \geq 1$ and $\chi : [0, \infty) \to [0, \infty)$ be a self-mapping such that

$$\chi(\tau) = \lambda\tau \text{ for each } \tau \in [0, \infty) \text{ where } \lambda \in (0, \frac{1}{s}).$$

Thus, χ is a comparison function.

The upcoming technical result is useful in testing whether a given sequence is Cauchy.

Lemma 3 ([35]) *Suppose that a self-mapping* $\varphi : [0, \infty) \to [0, \infty)$ *is* b*-comparison. Then, we conclude that*

(1) $\sum_{k=0}^{\infty} s^k \varphi^k(t)$ *converges for any* $t \in [0, \infty)$*;*
(2) $S_b : [0, \infty) \to [0, \infty)$ *defined by* $S_b(t) = \sum_{k=0}^{\infty} s^k \varphi^k(t)$, $t \in [0, \infty)$*, is increasing and continuous at* 0*.*

It is a trivial observation that any b-comparison function forms a comparison function.

2.3 *Admissible Mappings*

Next, we consider the notion of α-admissible mappings and its extension see e.g. [114], [83], and [103] :

Definition 4 ([103]) Let $\alpha : S \times S \to [0, \infty)$ be a mapping. A self-mapping $T : S \to S$ is said to be an $\alpha-$orbital admissible if for all $s \in S$, we have

$$\alpha(s, Ts) \geq 1 \Rightarrow \alpha(Ts, T^2 s) \geq 1. \tag{3}$$

Furthermore, an $\alpha-$orbital admissible mapping T is called triangular α-orbital admissible if it holds the following condition:

(TO) $\alpha(s,t) \geq 1$ and $\alpha(s,Tt) \geq 1$ implies that $\alpha(s,Tt) \geq 1$, for all $s,t \in X$.

We underline that all $\alpha-$admissible mappings form α-orbital admissible mappings. On the other hand the converse is not true, in general, see, e.g., [103]. and also [4, 6, 12–19, 42, 54, 56, 57, 70, 71, 83] with the related references therein.

Lemma 4 *Let $T : X \to X$ be an $\alpha-$orbital admissible function. If there exists $x_0 \in X$ such that $\alpha(x_0, Tx_0) \geq 1$ and $\alpha(Tx_0, x_0) \geq 1$, then the sequence $(x_n)_{n\in\mathbb{N}}$, defined by $x_n = Tx_{n-1}, n \in \mathbb{N}$ satisfies the following relations:*

$$\alpha(x_n, x_{n+1}) \geq 1 \text{ and } \alpha(x_{n+1}, x_n) \geq 1, \text{ for all } n \in \mathbb{N}_0.$$

2.4 Branciari Distance Space

Definition 5 Let $d : X \times X \to [0,\infty)$ be a function such that for all $x, y \in X$ and all distinct points $u, v \in X$, each distinct from x and y:

$(d1)d(x,y) = 0$ if and only if $x = y$ (identification);
$(d2)d(x,y) = d(y,x)$ (symmetry);
$(d3)d(x,y) \leq d(x,u) + d(u,v) + d(v,y)$ (quadrilateral inequality).

Then d is called a Branciari distance and the pair (X,d) is called a Branciari distance space.

In some sources, the authors used “a rectangular metric” or “a generalized metric” to call Branciari distance. However, it was declared and proved in [119] that the topology of Branciari distance and standard metric are not comparable.

Definition 6 Let $\sigma := \{x_n\}$ be a sequence in a Branciari distance space (X,d).

(i) σ is convergent $x \in X$, that is, $sigma := \{x_n\} \to x$ if $\lim_{n\to\infty} d(x_n, x) = 0$.
(ii) σ is Cauchy if for every $\varepsilon > 0$, there exists a positive integer $N = N(\varepsilon)$ such that $d(x_n, x_m) < \varepsilon$ for all $n, m > N$.
(iii) Branciari distance space (X,d) is complete whenever each Cauchy sequence in X is convergent.

Although the notion the Branciari distance seems very close the concept metric, at the first glance, they are very different (for more details, see, e.g., [115, 119–122]). In particular, in the context of Branciari distance space:

1. A Branciari distance function d is not necessarily continuous.
2. A convergent sequence is not necessarily Cauchy sequence.
3. The limit of a sequence is not needed to be unique.
4. The topologies of a Branciari distance space and a metric space are incompatible.

Lemma 5 *A self-mapping T, defined on a Branciari distance space (X, d), is continuous at $u \in X$, if we have $Tx_n \to Tu$, for any sequence $\{x_n\}$ in X converges to $u \in X$. That is,*

$$x_n \to u \Rightarrow \lim_{n\to\infty} d(Tx_n, Tu) = 0.$$

In what follows we state a technical result that is crucial for the uniqueness of a limit in the setting of Branciari distance spaces.

Proposition 1 ([90]) *Let (X, d) be Branciari distance space. We presume that a Cauchy sequence $\{x_n\}$ satisfies*

$$\lim_{n\to\infty} d(x_n, u) = \lim_{n\to\infty} d(x_n, z) = 0,$$

where $u, z \in X$. Then $u = z$.

2.5 Partial Metric Spaces

Definition 7 (See [95]) . We say that a mapping $p : X \times X \to [0, \infty)$ is *partial metric* if, for each $x, y, w \in X$, we have

$$\begin{aligned} &(\text{P1})\ x = y \Leftrightarrow p(x, x) = p(y, y) = p(x, y);\\ &(\text{P2})\ p(x, x) \leq p(x, y);\\ &(\text{P3})\ p(x, y) = p(y, x);\\ &(\text{P4})\ p(x, y) \leq p(x, w) + p(w, y) - p(w, w). \end{aligned} \tag{4}$$

Here, the pair (X, p) is called partial metric space.

Note that there is a close relation between partial metric and standard metric. Indeed, a mapping $d_p : X \times X \to [0, \infty)$ such that

$$d_p(x, y) = 2p(x, y) - p(x, x) - p(y, y) \tag{5}$$

forms a standard metric on X. The topological concepts, induced by partial metric, are natural modifications and extensions of the corresponding notions in the standard metric topology. For more detailed discussion on these notions, we refer to [24–26] and related references therein.

Definition 8 Let $\sigma : \{x_n\}$ be a given sequence in a partial metric space (X, p). We say that

(i) a σ *converges to the limit* x (i.e., $\sigma : \{x_n\} \to x$) if $p(x, x) = \lim_{n\to\infty} p(x, x_n)$;

(ii) a σ is *fundamental* (or Cauchy) if $\lim_{n,m\to\infty} p(x_n, x_m)$ exists and is finite;

(iii) a (X, p) is *complete* if each fundamental (Cauchy) sequence $\{x_n\}$ converges to a point $x \in X$, that is

$$p(x, x) = \lim_{n,m\to\infty} p(x_n, x_m);$$

(iv) a mapping $F : X \to X$ is *continuous* at a point $x_0 \in X$ if, for each $\epsilon > 0$, there exists $\delta > 0$ such that $F(B_p(x_0, \delta)) \subseteq B_P(Fx_0, \epsilon)$.

Next, we state the easily derived technical result (see [95]).

Lemma 6 *Let σ : $\{x_n\}$ be a given sequence in a partial metric space (X, p). Suppose that d_p is a standard metric induced by the partial metric p.*

(a) A σ is fundamental (Cauchy) in a partial metric (X, p) if and only if it is a fundamental (Cauchy) sequence in the standard metric space (X, d_p).
(b) (X, p) is complete if and only if (X, d_p) is complete. Moreover,

$$\lim_{n\to\infty} d_p(x, x_n) = 0 \Leftrightarrow p(x, x) = \lim_{n\to\infty} p(x, x_n) = \lim_{n,m\to\infty} p(x_n, x_m). \quad (6)$$

(c) If $x_n \to w$ as $n \to \infty$ in a partial metric space (X, p) with $p(w, w) = 0$, then we have

$$\lim_{n\to\infty} p(x_n, y) = p(w, y) \; for\ every\ y \in X.$$

2.6 b-Metric Spaces

Definition 9 Let X be a nonempty set and let $b \geq 1$ be a given real number. A function $d : X \times X \to [0, \infty)$ is said to be a b-metric if and only if for all $x, y, z \in X$, the following conditions are satisfied:

(1) $d(x, y) = 0$ if and only if $x = y$;
(2) $d(x, y) = d(y, x)$;
(3) $d(x, z) \leq b[d(x, y) + d(y, z)]$.

A triplet (X, d, b) is called a b-metric space.

The following are the basic interesting examples of b-metric space that show how such spaces are fruitful.

Example 3 ([43]) Let $s \geq 1$ be arbitrary and A be any set that has more than three elements. Assume that A_1, A_2 are the subsets of A such that $A_1 \cap A_2 = \emptyset$ and $A = A_1 \cup A_2$. Define a functional $d : X \times X \to [0, \infty)$ such that

$$d(a,b) := \begin{cases} 0, & a = b, \\ 2s, & a, b \in A_1, \\ 1, & \text{otherwise.} \end{cases}$$

Thus, (X, d, s) is a b-metric space.

Example 4 ([43]) Let $X = \mathbb{R}$. The function $d : X \times X \to [0, \infty)$, defined as

$$d(x, y) = |x - y|^2, \tag{7}$$

is a b-metric on $\mathbb{R}$ with $s = 2$. Note that the first two axioms are fulfilled in a straightway. For the last axiom,

$$\begin{aligned} |x - y|^2 = |x - z + z - y|^2 &= |x - z|^2 + 2|x - z||z - y| + |z - y|^2 \\ &\leq 2[|x - z|^2 + |z - y|^2], \end{aligned}$$

since

$$2|x - z||z - y| \leq |x - z|^2 + |z - y|^2.$$

Thus, $(X, d, 2)$ is a b-metric space.

Example 5 ([43]) Let $X = \{u, v, w\}$ and $d : X \times X \to \mathbb{R}_0^+$ such that

$$\begin{aligned} &d\,(u, b) = d\,(v, u) = d\,(u, w) = d\,(w, u) = 1, \\ &d\,(b, c) = d\,(w, v) = \alpha \geq 2, \\ &d\,(u, u) = d\,(v, v) = d\,(w, w) = 0. \end{aligned}$$

Then,

$$d\,(x, y) \leq \frac{\alpha}{2}\,[d\,(x, z) + d\,(z, y)]\,, \text{ for } u, v, w \in X.$$

Thus $(X, d, \frac{\alpha}{2})$ forms a b-metric space.

Definition 10 Let (X, d, b) be a b-metric space, $\{x_n\}$ be a sequence in X, and $x \in X$.

(a) The sequence $\{x_n\}$ is said to be convergent in (X, d, b) to x, if for every $\varepsilon > 0$ there exists $n_0 \in \mathbb{N}$ such that $d(x_n, x) < \varepsilon$ for all $n > n_0$. This fact is represented by $\lim_{n\to\infty} x_n = x$ or $x_n \to x$ as $n \to \infty$.
(b) The sequence $\{x_n\}$ is said to be Cauchy in (X, d, b) if for every $\varepsilon > 0$ there exists $n_0 \in \mathbb{N}$ such that $d(x_n, x_{n+p}) < \varepsilon$ for all $n > n_0$, $p > 0$.
(c) (X, d, b) is said to be complete if every Cauchy sequence in X converges to some $x \in X$.

The following technical lemma was noted in Miculescu and Mihail [98] (Lemma 2.2) and Suzuki [119] (Lemma 6).

Lemma 7 *Let (X, d, b) be a b-metric space and let $\{x_n\}$ be a sequence in X. Assume that there exists $\gamma \in [0, 1)$ satisfying $d(x_{n+1}, x_n) \leq \gamma d(x_n, x_{n-1})$ for any $n \in \mathbb{N}$. Then $\{x_n\}$ is Cauchy.*

3 Quasi-Metric Spaces

A distance function $q : X \times X \to [0, \infty)$ is called a quasi-metric on X if

(q_1) $q(u, v) = 0 \Leftrightarrow u = v$;
(q_2) $q(u, w) \leq q(u, v) + q(v, w)$, for all $u, v, w \in X$.

In addition, the pair (X, q) is called a quasi-metric space.

Let q be a quasi-metric on X. Then, the function $q_* : X \times X \to [0, \infty)$ defined by $q_*(u, v) = q(v, u)$ forms a quasi-metric, too. This new quasi-metric is also called the dual (conjugate) of q. The functions $d_1, d_2 : X \times X \to [0, \infty)$, where

$$\begin{aligned} d_1(v, u) &= q(u, v) + q_*(u, v), \\ d_2(v, u) &= \max\{q(u, v), q_*(u, v)\} \end{aligned}$$

form standard metrics on X. Let $\{u_n\}$ be a sequence in X, and $u \in X$, where (X, q) a quasi-metric space. We say that:

1. $\{u_n\}$ converges to u if and only if

$$\lim_{n\to\infty} q(u_n, u) = \lim_{n\to\infty} q(u, u_n) = 0. \tag{8}$$

2. $\{u_n\}$ is left-Cauchy if and only if for every $\epsilon > 0$ there exists a positive integer $k = k(\epsilon)$ such that $q(u_n, u_m) < \epsilon$ for all $n \geq m > k$.
3. $\{u_n\}$ is right-Cauchy if and only if for every $\epsilon > 0$ there exists a positive integer $k = k(\epsilon)$ such that $q(u_n, u_m) < \epsilon$ for all $m \geq n > k$.
4. $\{u_n\}$ is Cauchy if and only if it is left-Cauchy and right-Cauchy.

We underline that in a quasi-metric space (X, q), the limit for a convergent sequence is unique. Indeed, if $u_n \to u$, for all $v \in X$, we have

$$\lim_{n\to\infty} q(u_n, v) = q(u, v) \text{ and } \lim_{n\to\infty} q(v, u_n) = q(v, u).$$

A quasi-metric space (X, q) is called complete (respectively, left-complete or right-complete) if and only if each Cauchy sequence (respectively, left-Cauchy sequence or right-Cauchy sequence) in X is convergent. Notice, in this context, that "right completeness" is equivalent to "Smyth completeness" [110]. See also [111].

A mapping $T : X \to X$ is continuously provided that for any sequence $\{u_n\}$ in X such that $u_n \to u \in X$, the sequence $\{Tu_n\}$ converges to Tu, that is,

$$\lim_{n\to\infty} q(Tu_n, Tu) = \lim_{n\to\infty} q(Tu, Tu_n) = 0 \tag{9}$$

3.1 Significant Contractions in Metric Fixed Point Theory

In this section we recall some crucial contractions that were published in the early and mid-twentieth century. The first metric fixed point theorem was given by Banach [32] in the context of normed spaces. The following characterization was given by Caccioppoli [40] but, in the literature, it was known as Banach's contraction mapping principle.

Theorem 1 ([40]) *Let (X, d) be a complete metric spaces and $T : X \to X$ be a contraction mapping, that is,*

$$d\,(Tx, Ty) \le \lambda d(x, y),$$

for all $x, y \in X$, where $\lambda \in [0, 1)$. Then T has a unique fixed point.

One of the first early outstanding extension of Banach's contraction mapping principle was given by Kannan [61, 62]. Notice that in Banach's theorem, the given mapping is necessarily continuous. On the other hand, in the consideration of Kannan's theorems, it is not necessary.

Theorem 2 ([61, 62]) *Let (X, d) be a complete metric spaces and $T : X \to X$ be a Kannan contraction mapping, that is,*

$$d\,(Tx, Ty) \le \lambda\,[d(x, Tx) + d(y, Ty)]\,,$$

for all $x, y \in X$, where $\lambda \in \left[0, \frac{1}{2}\right)$. Then T has a unique fixed point.

The following renowned results was proved independently by Rus, Reich, and Ćirić see, for example, [106, 112, 113, 115, 120].

Theorem 3 *Let (X, d) be a complete metric spaces and $T : X \to X$ be a Rus-Reich-Ćirić contraction mapping, that is,*

$$d\,(Tx, Ty) \le \lambda\,[d(x, y) + d(x, Tx) + d(y, Ty)]\,, \tag{10}$$

for all $x, y \in X$, where $\lambda \in \left[0, \frac{1}{3}\right)$. Then T has a unique fixed point.

Notice that several variation of Rus-Reich-Ćirić contraction (10) can be stated also as

$$d(Tx, Ty) \leq ad(x, y) + bd(x, Tx) + cd(y, Ty),$$

where a, b, c are nonnegative real numbers such that $0 \leq a + b + c < 1$.

In what follows, we state the well-known Hardy-Rogers [53]

Theorem 4 *Let (X, d) be a complete metric space. Let $T : X \to X$ be a given mapping such that*

$$d(T\theta, T\vartheta) \leq \alpha d(\theta, y) + \beta d(\theta, T\theta) + \gamma d(y, T\vartheta) + \delta[\frac{1}{2}(d(\theta, T\vartheta) + d(\vartheta, T\theta))],$$

for all $\theta, \vartheta \in X$, where $\alpha, \beta, \gamma, \delta$ are non-negative reals such that $\alpha+\beta+\gamma+\delta < 1$. Then T has a unique fixed point in X.

Let T be a self-mapping on a metric space (X, d) and $\zeta \in \mathcal{Z}$. We say that T is a $\mathcal{Z}$*-contraction* with respect to ζ [89], if

$$\zeta(d(Tx, Ty), d(x, y)) \geq 0 \qquad \text{for all } x, y \in X. \tag{11}$$

Theorem 5 *Every $\mathcal{Z}$-contraction on a complete metric space has a unique fixed point.*

4 Interpolative Contraction

4.1 Motivation

We shall begin the section by explaining how the concept of interpolative contraction emerged. For this purpose, we recall the notion of interpolation triple [91].

Suppose that Banach spaces A and B are algebraically and topologically imbedded in a separated topological linear space. In this case, the pair of A and B is called Banach couple and it is denoted by (A, B). If there is a Banach space E for the Banach couple (A, B) such that the imbedding $A \cap B \subset E \subset A + B$ holds, then E is called and intermediate space of (A, B). Let (C, D) be another Banach couple. A linear mapping T acting from the space $A + B$ to $C + D$ is called a bounded operator from (A, B) to (C, D) if the restrictions of T to the spaces A and B are bounded operators from A to C and B to D, respectively. We denote by $L(AB, CD)$ the linear space of all bounded operators from the couple (A, B) to the couple (C, D). This is a Banach space in the norm

$$\|T\|_{L(AB,CD)} = \max\{\|T\|_{A\to B}, \|T\|_{C\to D}\}.$$

Definition 11 ([91]) Let (A, B) and (C, D) be two Banach couples, and E (respectively F) be intermediate for the spaces of the Banach couple (A, B) (respectively

(C, D)). The triple (A, B, E) is called an interpolation triple, relative to (C, D, F), if every bounded operator from (A, B) to (C, D) maps E to F.

A triple (A, B, E) is said to be an interpolation triple of type γ $(0 \leq \alpha \leq 1)$ relative to (C, D, F) if it is an interpolation triple and the following inequality holds:

$$\|T\|_{E \to F} \leq c\|T\|_{A \to B}^{\gamma} \cdot \|T\|_{C \to D}^{1-\gamma},$$

for some constant c.

For more details on interpolation theory, we refer to [91].

4.2 A Pioneering Notion: An Interpolative Kannan Type Contraction

Inspired by Definition 11, the interpolation contraction was introduced in [66] by revisiting the well-known Kannan type contraction.

Definition 12 ([66]) Let (X, d) be a metric space. We say that the self-mapping $T : X \to X$ is an interpolative Kannan type contraction, if there exist a constant $\lambda \in [0, 1)$ and $\alpha \in (0, 1)$ such that

$$d(Tx, Ty) \leq \lambda [d(x, Tx)]^{\alpha} \cdot [d(y, Ty)]^{1-\alpha}. \tag{12}$$

for all $x, y \in X$ with $x \neq Tx$.

Theorem 6 ([66]) *Let (X, d) be a complete metric space and T be an interpolative Kannan type contraction. Then T has a unique fixed point in X.*

Proof For an arbitrary initial point $x_0 \in X$, we construct an iterative sequence $\{x_n\}$ by $x_{n+1} = T^n x_0$ for all positive integer n. In case there exist a nonnegative integer n_0 such that $x_{n_0} = x_{n_0+1} = Tx_{n_0}$, then x_{n_0} forms a fixed point that completes the proof. Consequently, throughout the proof, we presume that $x_n \neq x_{n+1}$ and hence $d(x_n, Tx_n) = d(x_n, x_{n+1}) > 0$ for each nonnegative integer n. By letting $x = x_n$ and $y = x_{n-1}$ in (12), we find that

$$\begin{aligned} d(x_{n+1}, x_n) = d(Tx_n, Tx_{n-1}) &\leq \lambda [d(x_n, Tx_n)]^{\alpha} \cdot [d(x_{n-1}, Tx_{n-1})]^{1-\alpha} \\ &= \lambda [d(x_{n-1}, x_n)]^{1-\alpha} \cdot [d(x_n, x_{n+1})]^{\alpha}, \end{aligned} \tag{13}$$

which turns into

$$[d(x_n, x_{n+1})]^{1-\alpha} \leq \lambda [d(x_{n-1}, x_n)]^{1-\alpha}. \tag{14}$$

So, we conclude that the sequence $\{d(x_{n-1}, x_n)\}$ is non-increasing and non-negative. So, there is a nonnegative constant L such that $\lim_{n\to\infty} d(x_{n-1}, x_n) = L$. On account of (22), we find that

$$d(x_n, x_{n+1}) \leq \lambda d(x_{n-1}, x_n) \leq \lambda^n d(x_0, x_1). \tag{15}$$

Letting $n \to \infty$ in the inequality above, we observe that $L = 0$.

By using the triangle inequality recursively, we deduce that sequence $\{x_n\}$ is Cauchy in a standard way. Since (X, d) is a complete metric space, there exists $x \in X$ such that $\lim_{n\to\infty} d(x_n, x) = 0$.

To finalize the proof, we substitute $x = x_n$ and $y = x$ in (12) which yields

$$d(Tx_n, Tx) \leq \lambda [d(x_n, Tx_n)]^{\alpha} \cdot [d(x, Tx)]^{1-\alpha}. \tag{16}$$

Taking $n \to \infty$ in the inequality above, we derive that $d(x, Tx) = 0$, that is, $Tx = x$.

Example 6 Let $S = \{a, b, c, e\}$ and $\mathbb{R}_0^+ := [0, \infty)$ be a set endowed with a metric d such that

$$\begin{aligned}
&d(a,a) = d(b,b) = d(c,c) = d(e,e) = 0,\\
&d(b,a) = d(a,b) = 3,\\
&d(c,a) = d(a,c) = 4,\\
&d(b,c) = d(c,b) = d(e,c) = d(c,e) = \tfrac{3}{2}\\
&d(e,a) = d(a,e) = \tfrac{5}{2}\\
&d(e,b) = d(b,e) = 2\\
&d(x,a) = d(a,x) = 0 \text{ whenever } x \in \mathbb{R}_0^+ \text{ and } a \in S,\\
&d(x,y) = |x - y| \text{ otherwise.}
\end{aligned}$$

We define a self-mapping T on X by $T : \begin{pmatrix} a & b & c & e \\ a & e & a & b \end{pmatrix}$ over S and $Tx = 0$, otherwise. It is clear that T is not Kannan contraction. Indeed, there is no $\lambda \in [0, \frac{1}{2})$ such that the following inequality is fulfilled:

$$d(Te, Tc) = d(b, a) = 3 \leq \lambda(d(Te, e) + d(c, Tc)) = 6\lambda.$$

On the other hand, for $\alpha = \frac{1}{8}$ and $\lambda = \frac{9}{10}$, the self-mapping T forms an interpolative Kannan type contraction and a and 0 are the desired unique fixed points of T. Notice that in the setting of interpolative Kannan type contraction, the constant lies between 0 and 1 although in the classical version it is restricted with 1/2.

4.3 An Interpolative Rus–Reich–Ćirić Type Contraction [3]

Definition 13 ([3]) Let T be a self-mapping defined on a metric space (X, d). If there exist $\zeta \in \mathcal{Z}$, $\psi \in \Psi$, $\gamma, \beta \in (0, 1)$ with $\gamma + \beta < 1$ and $\alpha : X \times X \to [0, \infty)$ such that

$$\zeta(\alpha(x, y)d(Tx, Ty), \psi(R(x, y))) \geq 0 \qquad \text{for all } x, y \in X \text{ with } x \neq Tx, \tag{17}$$

where

$$R(x, y) := [d(x, y)]^{\beta} \cdot [d(x, Tx)]^{\gamma} \cdot [d(y, Ty)]^{1-\gamma-\beta}, \tag{18}$$

then we say that T is an α-admissible interpolative Rus–Reich–Ćirić type $\mathcal{Z}$-*contraction* with respect to ζ.

Theorem 7 ([3]) *Let (X, d) be a complete metric space, $\zeta \in \mathcal{Z}$. If a self-mapping $T : X \to X$ forms an α-admissible interpolative Rus–Reich–Ćirić type $\mathcal{Z}$-* contraction *with respect to ζ and satisfying*

- (i) *T is triangular $\alpha-$orbital admissible;*
- (ii) *there exists $x_0 \in X$ such that $\alpha(x_0, Tx_0) \geq 1$;*
- (iii) *T is continuous.*

Then there exists $u \in X$ such that $Tu = u$.

Proof By (ii), there exists $x_0 \in X$ such that $\alpha(x_0, Tx_0) \geq 1$. By a standard way, we construct an iterative sequence $\{x_n\}$ by $x_{n+1} = Tx_n$ for all non-negative integers n. Regarding the corresponding discussion in the proof of Theorem 6, we suppose that

$$d(x_n, x_{n+1}) > 0, \text{ for all } n = 0, 1, \ldots. \tag{19}$$

Since T is $\alpha-$orbital admissible, assumption (ii) yields that

$$\alpha(x_n, x_{n+1}) \geq 1, \text{ for all } n = 0, 1, \ldots. \tag{20}$$

By letting $x = x_n$ and $y = x$ in (17), and keeping the inequalities (19) and (20) in mind, we find that

$$d(x_n, x_{n+1}) \leq \alpha(x_n, x_{n-1})d(x_n, x_{n+1}) < \psi(R(x_n, x_{n-1})) < R(x_n, x_{n-1}), \tag{21}$$

for all $n = 1, 2, \ldots$, where $R(x_n, x_{n-1}) = [d(x_n, x_{n+1})]^{\gamma} \cdot [d(x_{n-1}, x_n)]^{1-\gamma}$.

By a simple elimination, the inequality (21) turns into

$$[d(x_n, x_{n+1})]^{1-\gamma} \leq \lambda [d(x_{n-1}, x_n)]^{1-\gamma}. \tag{22}$$

which yields that the sequence $\{d(x_n, x_{n-1})\}$ is non-decreasing and bounded from below by zero. In addition the monotonicity of the sequence $\{d(x_n, x_{n-1})\}$ implied that $R(x_n, x_{n-1}) \leq d(x_n, x_{n-1})$. Accordingly, there exists $L \geq 0$ such that $\lim_{n\to\infty} d(x_n, x_{n-1}) = L \geq 0$. We shall prove that $L = 0$. Suppose, on the contrary, that $L > 0$. Note that from the inequality (13), we derive that

$$\lim_{n\to\infty} \alpha(x_n, x_{n-1})d(x_n, x_{n+1}) = L, \tag{23}$$

and

$$\lim_{n\to\infty} R(x_n, x_{n+1}) = L. \tag{24}$$

Letting $s_n = \alpha(x_n, x_{n-1})d(x_n, x_{n+1})$ and $t_n = R(x_n, x_{n-1})$ and taking (ζ_3) into account, we get that

$$0 \leq \limsup_{n\to\infty} \zeta(\alpha(x_n, x_{n-1})d(x_{n+1}, x_n), R(x_n, x_{n-1})) < 0 \tag{25}$$

which is a contradiction. Thus, we have $L = 0$.

Now, we shall prove that the iterative sequence $\{x_n\}$ is Cauchy. Again we use the method of *Reductio ad absurdum*. Suppose, on the contrary that $\{x_n\}$ is not a Cauchy sequence. Thus, there exists $\varepsilon > 0$, for all $N \in \mathbb{N}$, there exist $n, m \in \mathbb{N}$ with $n > m > N$ and $d(x_m, x_n) > \varepsilon$. On the other hand, from (16), there exists $n_0 \in \mathbb{N}$ such that

$$d(x_n, x_{n+1}) < \varepsilon \text{ for all } n > n_0. \tag{26}$$

Consider two partial subsequences x_{n_k} and x_{m_k} of x_n such that

$$n_0 \leq n_k < m_k < m_{k+1} \text{ and } d(x_{m_k}, x_{n_k}) > \varepsilon \text{ for all } k. \tag{27}$$

Notice that

$$d(x_{m_{k-1}}, x_{n_k}) \leq \varepsilon \text{ for all } k, \tag{28}$$

where m_k is chosen as a least number $m \in \{n_k, n_{k+1}, n_{k+2}, \ldots\}$ such that (27) is satisfied. We also mention that $n_k + 1 \leq m_k$ for all k. In fact, the case $n_k + 1 \leq m_k$ is impossible due to (26), (27). Thus, $n_k + 2 \leq m_k$ for all k. It yields that

$$n_k + 1 < m_k < m_k + 1 \text{ for all } k.$$

On account of (27) and (28) and the triangle inequality, we derive that

$$\begin{aligned} \varepsilon < d(x_{m_k}, x_{n_k}) &\leq d(x_{m_k}, x_{m_k-1}) + d(x_{m_k-1}, x_{n_k}) \\ &\leq d(x_{m_k}, x_{m_k-1}) + \varepsilon \text{ for all } k. \end{aligned} \tag{29}$$

Due to (16), we deduce that

$$\lim_{k\to\infty} d(x_{m_k}, x_{n_k}) = \varepsilon. \tag{30}$$

Again by the triangle inequality, together with (29), we derive that

$$d(x_{m_k}, x_{n_k}) \leq d(x_{m_k}, x_{m_k+1}) + d(x_{m_k+1}, x_{n_k+1}) + d(x_{n_k+1}, x_{n_k}) \text{ for all } k.$$

Analogously, we have

$$d(x_{m_k+1}, x_{n_k+1}) \leq d(x_{m_k+1}, x_{m_k}) + d(x_{m_k}, x_{n_k}) + d(x_{n_k}, x_{n_k+1}) \text{ for all } k.$$

Combining two inequalities above together with (16), we find that

$$\lim_{k\to\infty} d(x_{m_k+1}, x_{n_k+1}) = \varepsilon. \tag{31}$$

Particularly, there exists $n_1 \in \mathbb{N}$ such that for all $k \geq n_1$ we have

$$d(x_{m_k}, x_{n_k}) > \frac{\varepsilon}{2} > 0 \text{ and } d(x_{m_k+1}, x_{n_k+1}) > \frac{\varepsilon}{2} > 0. \tag{32}$$

Moreover, since T is triangular α-orbital admissible, we have

$$\alpha(x_{m_k}, x_{n_k}) \geq 1. \tag{33}$$

Regarding the fact T is an α-admissible $\mathcal{Z}$*-contraction* with respect to ζ, together with (32) and (33) we get that

$$\begin{aligned} 0 &\leq \zeta(\alpha(x_{m_k}, x_{n_k})d(Tx_{m_k}, Tx_{n_k}), \psi(R(x_{m_k}, x_{n_k}))) \\ &< \psi(R(x_{m_k}, x_{n_k})) - \alpha(x_{m_k}, x_{n_k})d(x_{m_k+1}, x_{n_k+1}), \end{aligned} \tag{34}$$

for all $k \geq n_1$, where

$$R(x_{m_k}, x_{n_k}) = \left[d\left(x_{m_k}, x_{n_k}\right)\right]^{\beta} \cdot \left[d\left(x_{m_k}, x_{m_k+1}\right)\right]^{\gamma} \cdot \left[d\left(x_{n_k}, x_{n_k+1}\right)\right]^{1-\gamma-\beta}. \tag{35}$$

Consequently, we have

$$\begin{aligned} 0 &< d(x_{m_k+1}, x_{n_k+1}) < \alpha(x_{m_k}, x_{n_k})d(x_{m_k+1}, x_{n_k+1}) \\ &< \psi(R(x_{m_k}, x_{n_k})) < R(x_{m_k}, x_{n_k}), \end{aligned}$$

for all $k \geq n_1$. Letting $n, m \to \infty$ in the inequality above, and keeping the observations in (16), (36), (31), (34), and (35), we find that

$$\lim_{k\to\infty} d(x_{m_k+1}, x_{n_k+1}) = 0, \tag{36}$$

which is a contradiction. Hence, $\{x_n\}$ is a Cauchy sequence. Owing to the fact that (X, d) is a complete metric space, there exists $u \in X$ such that

$$\lim_{n\to\infty} d(x_n, u) = 0. \tag{37}$$

Since T is continuous, we derive (37) that

$$\lim_{n\to\infty} d(x_{n+1}, Tu) = \lim_{n\to\infty} d(Tx_n, Tu) = 0. \tag{38}$$

From (37) and (38) and the uniqueness of the limit, we conclude that u is a fixed point of T, that is, $Tu = u$.

Theorem 8 ([3]) *Let (X, d) be a complete metric space and let $T : X \to X$ be an α-admissible $\mathcal{Z}$-*contraction *with respect to ζ. Suppose that*

(i) T is triangular α-orbital admissible;
(ii) there exists $x_0 \in X$ such that $\alpha(x_0, Tx_0) \geq 1$;
(iii) if $\{x_n\}$ is a sequence in X such that $\alpha(x_n, x_{n+1}) \geq 1$ for all n and $x_n \to x \in X$ as $n \to \infty$, then there exists a subsequence $\{x_{n(k)}\}$ of $\{x_n\}$ such that $\alpha(x_{n(k)}, x) \geq 1$ for all k.

Then there exists $u \in X$ such that $Tu = u$.

Proof Following the proof of Theorem 7, we know that the sequence $\{x_n\}$ defined by $x_{n+1} = Tx_n$ for all $n \geq 0$, converges for some $u \in X$. From (20) and condition (iii), there exists a subsequence $\{x_{n(k)}\}$ of $\{x_n\}$ such that $\alpha(x_{n(k)}, u) \geq 1$ for all k. Applying (17), for all k, we get that

$$\begin{aligned} 0 &\leq \zeta(\alpha(x_{n(k)}, u)d(Tx_{n(k)}, Tu), \psi(R(x_{n(k)}, u))) \\ &= \zeta(\alpha(x_{n(k)}, u)d(x_{n(k)+1}, Tu), \psi(R(x_{n(k)}, u))) \\ &< \psi(R(x_{n(k)}, u)) - \alpha(x_{n(k)}, u)d(x_{n(k)+1}, Tu), \end{aligned} \tag{39}$$

which is equivalent to

$$d(x_{n(k)+1}, Tu) = d(Tx_{n(k)}, Tu) \leq \alpha(x_{n(k)}, u)d(Tx_{n(k)}, Tu) \leq \psi(R(x_{n(k)}, u)). \tag{40}$$

Letting $k \to \infty$ in the above equality, we have $d(u, Tu) = 0$, that is, $u = Tu$.

Example 7 ([3]) Let $X = \{1, 3, 4, 7\}$ be a set endowed with a standard metric $d(x, y) = |x - y|$.

$d(x,y)$	1	3	4	7
1	0	2	3	6
3	2	0	1	4
4	3	1	0	3
7	6	4	3	0

We define a self-mapping T on X by $T : \begin{pmatrix} 1\,3\,4\,7 \\ 4\,7\,4\,3 \end{pmatrix}$. It is clear that T is not Rus–Reich–Ćirić contraction. Indeed, there is no $\lambda \in [0, \frac{1}{3})$ such that the following inequality is fulfilled:

$$\begin{aligned} d(T1, T3) = d(4, 7) = 3 &\leq \lambda(d(1,3) + d(T1, 1) + d(3, T3)) \\ &= \lambda(d(1,3) + d(4,1) + d(3,7)) \\ &= \lambda(2 + 3 + 4) = 9\lambda. \end{aligned}$$

On the other hand, for $\gamma = \beta = \frac{1}{16}$ and $\lambda = \frac{4}{5}$, the self-mapping T forms an interpolative Rus–Reich–Ćirić type contraction and 4 is the desired unique fixed point of T.

Consequences

In this section, we shall illustrate that several existing fixed point results in the literature can be derived from our main results by regarding Example 1.

If $\psi \in \Psi$ and we define

$$\zeta_E(t,s) = \psi(s) - t \qquad \text{for all } s, t \in [0, \infty),$$

then ζ_{BW} is a simulation function (cf. Example 1 (v)).

Corollary 1 ([3, 23]) *Let (X, d) be a complete metric space, $\zeta \in \mathcal{Z}$. If a self-mapping $T : X \to X$ satisfies*

$$\alpha(x,y)d(Tx,Ty) \leq \psi(R(x,y)), \textit{ for all } x, y \in X \setminus Fix(T).$$

Suppose also that

(i) T is triangular $\alpha-$orbital admissible;
(ii) there exists $x_0 \in X$ such that $\alpha(x_0, Tx_0) \geq 1$;
(iii) T is continuous.

Then there exists $u \in X$ such that $Tu = u$.

Proof Taking $\zeta_E(t,s) = \psi(s) - t$ for all $s, t \in [0, \infty)$ in Theorem 7, we get that

$$\alpha(x, y)d(Tx, Ty) \le \psi(R(x, y)), \text{ for all}$$

We skip the details.

Corollary 2 ([3, 23]) *Let (X, d) be a complete metric space, $\zeta \in \mathcal{Z}$. If a self-mapping $T : X \to X$ satisfies*

$$\alpha(x, y)d(Tx, Ty) \le \psi(R(x, y)), \text{ for all } x, y \in X \setminus Fix(T).$$

Suppose also that

- (*i*) *T is triangular $\alpha-$orbital admissible;*
- (*ii*) *there exists $x_0 \in X$ such that $\alpha(x_0, Tx_0) \ge 1$;*
- (*iii*) *if $\{x_n\}$ is a sequence in X such that $\alpha(x_n, x_{n+1}) \ge 1$ for all n and $x_n \to x \in X$ as $n \to \infty$, then there exists a subsequence $\{x_{n(k)}\}$ of $\{x_n\}$ such that $\alpha(x_{n(k)}, x) \ge 1$ for all k.*

Then there exists $u \in X$ such that $Tu = u$.

Proof Taking $\zeta_E(t, s) = \psi(s) - t$ for all $s, t \in [0, \infty)$ in Theorem 8, we get that

$$\alpha(x, y)d(Tx, Ty) \le \psi(R(x, y)), \text{ for all}$$

We skip the details.

By considering $\alpha(x, y) = 1$ in Corollary 1, we state the following.

Corollary 3 *Let T be a self-mapping on a complete metric space (X, d) such that:*

$$d(Tx, Ty) \le \psi\Big([d(x, y)]^{\beta} \cdot [d(x, Tx)]^{\gamma} \cdot [d(y, Ty)]^{1-\gamma-\beta} \Big), \tag{41}$$

for all $x, y \in X \setminus \mathrm{Fix}(T)$, where $\gamma, \beta > 0$ are positive reals satisfying $\gamma + \beta < 1$. Then, T admits a fixed point.

Corollary 4 *Let T be a self-mapping on a complete metric space (X, d) such that:*

$$d(Tx, Ty) \le \psi\Big([d(x, Tx)]^{\beta} \cdot [d(y, Ty)]^{1-\beta} \Big), \tag{42}$$

for all $x, y \in X \setminus \mathrm{Fix}(T)$, where $0 < \beta < 1$. Then, T admits a fixed point in X.

Taking $\psi(t) = \lambda t$ (where $\lambda \in [0, 1)$) in Corollary 3, we state:

Corollary 5 *Let T be a self-mapping on a complete metric space (X, d) such that:*

$$d(Tx, Ty) \le \lambda [d(x, y)]^{\beta} \cdot [d(x, Tx)]^{\gamma} \cdot [d(y, Ty)]^{1-\gamma-\beta}, \tag{43}$$

for all $x, y \in X \setminus \mathrm{Fix}(T)$, where γ, β are positive reals verifying $\gamma + \beta < 1$ and $\lambda \in [0, 1)$. Then, T has a fixed point in X.

Taking $\psi(t) = \lambda t$ (where $\lambda \in [0, 1)$) in Corollary 4, we state:

Corollary 6 *Let T be a self-mapping on a complete metric space (X, d) such that:*

$$d(Tx, Ty) \leq \lambda \cdot [d(x, Tx)]^{\beta} \cdot [d(y, Ty)]^{1-\beta}, \tag{44}$$

for all $x, y \in X \setminus \mathrm{Fix}(T)$, where $0 < \beta < 1$ and $\lambda \in [0, 1)$. Then, there exists a fixed point of T.

Remark 1 Corollary 5 corresponds to Corollary 2.1 in [85].

Let $(X, d, \preceq)$ be a partially ordered metric space. Let us consider the following condition.

(G) If $\{x_n\}$ is a sequence in X such that $x_n \preceq x_{n+1}$ for each n and $x_n \to x \in X$ as $n \to \infty$, then there exists a subsequence $\{x_{n(k)}\}$ of $\{x_n\}$ such that $x_{n(k)} \preceq x$ for each k.

Following [82], we may state the following consequences of Corollary 1.

Corollary 7 *Let $(X, d, \preceq)$ be a complete partially ordered metric space. Let $T : X \to X$ be the mapping such that:*

$$\alpha(x, y) d(Tx, Ty) \leq \psi\Big([d(x, y)]^{\beta} \cdot [d(x, Tx)]^{\gamma} \cdot [d(y, Ty)]^{1-\gamma-\beta}\Big),$$

for all $x, y \in X \setminus \mathrm{Fix}(T)$ with $x \preceq y$, where $\psi \in \Psi$ and $\gamma, \beta > 0$ are positive reals such that $\gamma + \beta < 1$. Assume that:

(i) T is non-decreasing with respect to $\preceq$;
(ii) there exists $x_0 \in X$ such that $x_0 \preceq Tx_0$;
(iii) either T is continuous on (X, d) or (G) holds.

Then, T has a fixed point in X.

Proof It suffices to take,

$$\alpha(x, y) = \begin{cases} 1 & \text{if } (x \preceq y) \text{ or } (y \preceq x), \\ 0 & \text{otherwise,} \end{cases}$$

in Corollaries 1 and 2.

The following is an immediate consequence of Corollary 7.

Corollary 8 *Let $(X, d, \preceq)$ be a complete partially ordered metric space and $T : X \to X$ be a given mapping satisfying:*

$$\alpha(x, y) d(Tx, Ty) \leq \psi\Big([d(x, Tx)]^{\beta} \cdot [d(y, Ty)]^{1-\beta}\Big),$$

for all $x, y \in X \setminus \mathrm{Fix}(T)$ *with* $x \preceq y$, *where* $\psi \in \Psi$ *and* $0 < \beta < 1$. *Assume that:*

(i) *T is non-decreasing with respect to $\preceq$;*
(ii) *there exists $x_0 \in X$ such that $x_0 \preceq Tx_0$;*
(iii) *either T is continuous on (X, d) or (G) holds.*

Then, T has a fixed point in X.

Corollary 9 *Suppose that the subsets A_1 and A_2 of a complete metric space (X, d) are closed. Suppose also that $T : A_1 \cup A_2 \to A_1 \cup A_2$ satisfies:*

$$\alpha(x, y) d(Tx, Ty) \le \psi\Big([d(x, y)]^{\beta} \cdot [d(x, Tx)]^{\gamma} \cdot [d(y, Ty)]^{1-\gamma-\beta}\Big)$$

for all $x \in A_1$ and $y \in A_2$, such that $x, y \notin Fix(T)$, where $\psi \in \Psi$ and $\gamma, \beta > 0$ are positive reals such that $\gamma + \beta < 1$. If $T(A_1) \subseteq A_2$ and $T(A_2) \subseteq A_1$, then there exists a fixed point of T in $A_1 \cap A_2$.

Proof It suffices to take,

$$\alpha(x, y) = \begin{cases} 1 & \text{if } (A_1 \times A_2) \cup (A_2 \times A_1), \\ 0 & \text{otherwise,} \end{cases}$$

in Corollary 1.

Corollary 10 *Let A_1 and A_2 be two nonempty closed subsets of a complete metric space (X, d). Suppose that $T : A_1 \cup A_2 \to A_1 \cup A_2$ satisfies:*

$$\alpha(x, y) d(Tx, Ty) \le \psi\Big([d(x, Tx)]^{\beta} \cdot [d(y, Ty)]^{1-\gamma-\beta}\Big)$$

for all $x \in A_1$ and $y \in A_2$ such that $x, y \notin Fix(T)$, where $\psi \in \Psi$ and $0 < \beta < 1$. If $T(A_1) \subseteq A_2$ and $T(A_2) \subseteq A_1$, then there exists a fixed point of T in $A_1 \cap A_2$.

Proof It suffices to take, in Theorem 10,

$$\alpha(x, y) = \begin{cases} 1, & \text{if } (A_1 \times A_2) \cup (A_2 \times A_1), \\ 0, & \text{otherwise.} \end{cases}$$

Corollary 1 is supported by the following.

Example 8 Let us consider the set $X = [0, 2]$ endowed with $d(x, y) = |x - y|$. Let T be a self-mapping on X defined by:

$$Tx = \begin{cases} \frac{3}{2}, & \text{if } x \in [1, 2] \\ \frac{1}{3}, & \text{if } x \in [0, 1). \end{cases}$$

Take:

$$\alpha(x, y) = \begin{cases} 1, & \text{if } x, y \in [1, 2] \\ 0, & \text{otherwise.} \end{cases}$$

Let $x, y \in X$ be such that $x \neq Tx$, $y \neq Ty$ and $\alpha(x, y) \geq 1$. Then, $x, y \in [1, 2]$ and $x, y \notin \{\frac{3}{2}\}$. We have $Tx = Ty = \dfrac{3}{2}$. Hence, (18) holds. For $x_0 = 2$, we have:

$$\alpha(2, T2) = \alpha\left(2, \frac{3}{2}\right) = 1.$$

Now, let $x, y \in X$ be such that $\alpha(x, y) \geq 1$. It yields that $x, y \in [1, 2]$, so $Tx = Ty \in [1, 2]$. Hence, $\alpha(Tx, Ty) \geq 1$, that is T is ω-orbital admissible. Notice that T is not continuous. We shall show that (H) holds. Let $\{x_n\}$ be a sequence in X such that $\alpha(x_n, x_{n+1}) \geq 1$ for each $n \in \mathbb{N}$. Then, $\{x_n\} \subset [1, 2]$. If $\{x_n\} \to u$ as $n \to \infty$, we have $|x_n - u| \to 0$ as $n \to \infty$. Hence, $u \in [1, 2]$, and so, $\alpha(x_n, u) = 1$. All conditions of Corollary 1 hold. Note that $\frac{1}{3}$ and $\frac{3}{2}$ are two fixed points of T.

Interpolative Rus–Reich–Ćirić Type Contractions on Branciari Distance Spaces

Theorem 9 ([22]) *Let $T : X \to X$ be an interpolative Rus–Reich–Ćirić type contraction on a complete Branciari distance space (X, p), then T has a fixed point in X.*

We skip the proof since it is similar to the proof of Theorem 6.

Theorem 10 ([22]) *Let $T : X \to X$ be an interpolative Kannan type contraction on a complete Branciari distance space (X, p), then T has a fixed point in X.*

We skip the proof since it is a slight modification of the proof of Theorem 6.

The following example illustrates Theorem 9.

Example 9 ([22]) Let $X = \{0, 1, 2, 3\}$ be a set endowed with the Branciari distance ρ given as

$\rho(x, y)$	0	1	2	3
0	0	0.1	0.8	0.9
1	0.1	0	1	0.7
2	0.8	1	0	0.2
3	0.9	0.7	0.2	0

Consider the self-mapping T on X as $T : \begin{pmatrix} 0 & 1 & 2 & 3 \\ 0 & 0 & 1 & 3 \end{pmatrix}$. We have $\rho(1,2) > \rho(1,3) + \rho(3,2)$, so ρ is not a metric. Let $x, y \in X \backslash Fix(T)$. Then $(x, y) \in \{(1,1), (2,2), (1,2), (2,1)\}$. By choosing $\lambda \in [0.4, 1)$, $\alpha = 0.6$ and $\beta = 0.3$, it is obvious that the self-mapping T is an interpolative Reich–Rus–Ćirić type contraction. Here, T has two fixed points, which are 0 and 3.

On the other hand, the inequality (10) does not hold for $x = 0$ and $y = 3$ (by taking the classical metric $d(x, y) =| x - y |$). That is, Theorem 3 is not applicable.

Interpolative Rus–Reich–Ćirić Type Contractions on Partial Metric Spaces

Theorem 11 ([65]) *In the framework of a partial metric space (X, p), if $T : X \to X$ is an interpolative Reich–Rus–Ćirić type contraction, then T has a fixed point in X.*

The following examples illustrate Theorem 11.

Example 10 ([65]) Let $X = \{1, 3, 4, 7\}$ be a set endowed with the classical partial metric $d(x, y) = \max\{x, y\}$, that is,

d(x, y)	1	3	4	7
1	1	3	4	7
3	3	3	4	7
4	4	4	4	7
7	7	7	7	7

We define a self-mapping T on X by $T : \begin{pmatrix} 1 & 3 & 4 & 7 \\ 1 & 3 & 1 & 3 \end{pmatrix}$. It is clear that T is not a Reich–Rus–Ćirić contraction. Indeed, there is no $\lambda \in [0, \frac{1}{3})$ such that the following inequality is fulfilled:

$$\begin{aligned} d(T1, T3) = d(1, 3) = 3 &\leq \lambda(d(1, 3) + d(T1, 1) + d(3, T3)) \\ &= \lambda(d(1, 3) + d(1, 1) + d(3, 3)) \\ &= 7\lambda. \end{aligned}$$

On the other hand, choose $\alpha = \frac{1}{2}$, $\beta = \frac{2}{5}$ and $\lambda = \frac{7}{10}$. Let $x, y \in X \backslash Fix(T)$; then, $(x, y) \in \{(4, 7), (7, 4), (4, 4), (7, 7)\}$. Without loss of generality, we have

Case 1: $x = y = 4$. Here,

$$d\,(Tx, Ty) = 1 \leq 4\lambda = \lambda\,[d\,(x, y)]^{\beta} \cdot [d\,(x, Tx)]^{\alpha} \cdot [d\,(y, Ty)]^{1-\alpha-\beta}.$$

Case 2: $x = y = 7$. we have

$$d(Tx,Ty)=3\leq 7\lambda=\lambda\,[d(x,y)]^{\beta}\cdot[d(x,Tx)]^{\alpha}\cdot[d(y,Ty)]^{1-\alpha-\beta}.$$

Case 3: $x=4$ and $y=7$. Here,

$$\begin{aligned}d(Tx,Ty)=&3\leq\lambda 7^{1-\alpha}4^{\alpha}\\ =&\lambda\,[d(x,y)]^{\beta}\cdot[d(x,Tx)]^{\alpha}\cdot[d(y,Ty)]^{1-\alpha-\beta}.\end{aligned}$$

Thus, the self-mapping T is an interpolative Reich–Rus–Ćirić type contraction and $1,3$ are the desired fixed points. Note that in the setting of interpolative Reich–Rus–Ćirić type contractions, the constant lies between 0 and 1, although in the classical version it is restricted by 1/3.

4.4 An Interpolative Hardy-Rogers Type Contraction

We start by introducing the notion of *interpolative Hardy-Rogers type contractions*.

Definition 14 ([64]) Let (X,d) be a metric space. We say that the self-mapping $T:X\to X$ is an *interpolative Hardy-Rogers type contraction* if there exists $\lambda\in[0,1)$ and $\alpha,\beta,\gamma\in(0,1)$ with $\alpha+\beta+\gamma<1$, such that

$$\begin{aligned}d(Tx,Ty)\leq\lambda\,[d(x,y)]^{\beta}\cdot[d(x,Tx)]^{\alpha}\cdot[d(y,Ty)]^{\gamma}\\ \cdot\left[\frac{1}{2}(d(x,Ty)+d(y,Tx))\right]^{1-\alpha-\beta-\gamma}\end{aligned}\tag{45}$$

for all $x,y\in X\backslash Fix(T)$.

Theorem 12 ([64]) *Let (X,d) be a complete metric space and T be an interpolative Hardy-Rogers type contraction. Then, T has a fixed point in X.*

The proof is a slight modification of the proof of Theorem 6, and hence we skipped it.

Example 11 Consider $X=\{0,1,2,3,5\}$ endowed with $d(x,y)=|x-y|$. Choose $\lambda=\frac{\sqrt{2}}{2}$, $\alpha=\frac{1}{3}$, $\beta=\frac{1}{2}$ and $\gamma=\frac{1}{7}$. It is obvious that

$$\begin{aligned}d(Tx,Ty)\leq\lambda\,[d(x,y)]^{\beta}\cdot[d(x,Tx)]^{\alpha}\\ \cdot[d(y,Ty)]^{\gamma}\cdot\left[\frac{1}{2}(d(x,Ty)+d(y,Tx))\right]^{1-\alpha-\beta-\gamma},\end{aligned}$$

for all $x,y\in X\backslash Fix(T)$; that is, (45) holds. All the hypotheses of Theorem 12 are satisfied, and so T has a fixed point. Here, we have two fixed points, which are 0 and 1.

On the other hand, for $x = 0$ and $y = 1$, we have

$$d(Tx, Ty) > \lambda \left[d(x, y) + d(x, Tx) + d(y, Ty) + \frac{1}{2}(d(x, Ty) + d(y, Tx)) \right],$$

for any $\lambda \in [0, \frac{1}{4})$, so Theorem 4 (for $\lambda = \alpha = \beta = \gamma = \delta$) is not applicable.

Example 12 ([64]) Let $X = [0, \infty)$ be endowed with the metric

$$d(x, y) = \begin{cases} 0 & \text{if} \quad x = y \\ 1 & \text{if} \quad x \neq y. \end{cases}$$

Define the self-mapping on X as

$$Tx = \begin{cases} 0 & \text{if} \quad x \in [0, 1) \\ x & \text{if} \quad x \geq 1. \end{cases}$$

Let $x, y \in X \backslash Fix(T)$. Then $x, y \notin (0, 1)$, and so $d(Tx, Ty) = 0$; that is, (45) holds. Thus, all the hypotheses of Theorem 12 hold, and so T has a fixed point. Here, we have an infinite number of fixed points.

On the other hand, Theorem 4 is not applicable (it suffices to take $x = 0$ and $y = 1$).

The following theorem is a characterization of Theorem 12, in the context of partial metric spaces.

Theorem 13 *Let (X, p) be a completed partial metric space. Let $T : X \to X$ be a given mapping. Suppose there exists $\lambda \in [0, 1)$ and $\alpha, \beta, \gamma \in (0, 1)$ with $\alpha + \beta + \gamma < 1$, such that*

$$p(Tx, Ty) \leq \lambda [p(x, y)]^{\beta} \cdot [p(x, Tx)]^{\alpha} \cdot [p(y, Ty)]^{\gamma} \cdot \left[\frac{1}{2}(p(x, Ty) + p(y, Tx)) \right]^{1-\alpha-\beta-\gamma}, \tag{46}$$

for all $x, y \in X \backslash Fix(T)$. Then, T has a fixed point in X.

The proof is a slight modification of the proof of Theorem 6.

5 Hybrid Contractions

The first hybrid contraction was given in [99] where the authors combined both linear, nonlinear and interpolative contraction in a successful way.

Definition 15 ([99]) A self-mapping T on b-metric space (X, d, b) is called an (r, a)-weight type (hybrid) contraction, if there exists $\lambda \in [0, 1)$ such that

$$d(Tx, Ty) \leq \lambda M^r(T, x, y, a), \tag{47}$$

where $r \geq 0$, $a = (a_1, a_2, a_3)$, $a_i \geq 0$, $i = 1, 2, 3$ such that $a_1 + a_2 + a_3 = 1$ and

$$M^r(T, x, y, a) = \begin{cases} [a_1(d(x, y))^r + a_2(d(x, Tx))^r + a_3(d(y, Ty))^r]^{1/r}, & r > 0 \\ (d(x, y))^{a_1}(d(x, Tx))^{a_2}(d(y, Ty))^{a_3}, & r = 0, \end{cases} \tag{48}$$

for all $x, y \in X \backslash Fix(T)$, where $Fix(T) = \{u \in X, Tu = u\}$.

Example 13 In all following cases, the $x, y \in X \setminus Fix(T)$.

(i) If $r = 1$, $a = (\frac{1}{3}, \frac{1}{3}, \frac{1}{3})$, we obtain Reich–Rus–Ćirić type contraction,

$$d(Tx, Ty) \leq \frac{\lambda}{3}[d(x, y) + d(x, Tx) + d(y, Ty)],$$

where $\lambda \in [0, 1)$, see [106, 112, 113, 115, 120].

(ii) If $r = 2$, $a = (\frac{1}{3}, \frac{1}{3}, \frac{1}{3})$, we obtain the following condition,

$$d(Tx, Ty) \leq \frac{\lambda}{\sqrt{3}}[d^2(x, y) + d^2(x, Tx) + d^2(y, Ty)]^{1/2}.$$

where $\lambda \in [0, 1)$.

(iii) If $r = 1$ and $a = (a_1, a_2, a_3)$, we have a Reich type contraction,

$$d(Tx, Ty) \leq \alpha d(x, y) + \beta d(x, Tx) + \gamma d(y, Ty)],$$

where $\alpha = \lambda a_1, \beta = \lambda a_2, \gamma = \lambda a_3, \alpha, \beta, \gamma, \lambda \in [0, 1)$, and $\alpha + \beta + \gamma < 1$, see [115].

(iv) If $r = 1$ and $a = (0, \frac{1}{2}, \frac{1}{2})$, we have a Kannan type contraction,

$$d(Tx, Ty) \leq \frac{\lambda}{2}[d(x, Tx) + d(y, Ty)],$$

see [61, 62].

(v) If $r = 2$ and $a = (0, \frac{1}{2}, \frac{1}{2})$, we have

$$d(Tx, Ty) \leq \frac{\lambda}{\sqrt{2}}[d^2(x, Tx) + d^2(y, Ty)]^{1/2}.$$

(vi) If $r = 0$ and $a = (0, \alpha, 1 - \alpha)$ with $\alpha \in (0, 1)$, we obtain an interpolative Kannan type contraction,

$$d(Tx, Ty) \leq \lambda(d(x, Tx))^{\alpha}(d(y, Ty))^{1-\alpha},$$

see [66].

(vii) If $r = 0$ and $a = (\beta, \alpha, 1-\alpha-\beta)$ with $\alpha, \beta \in (0, 1)$, we have an interpolative Reich-Rus-Ćirić type contraction,

$$d(Tx, Ty) \leq \lambda(d(x, y))^{\beta}(d(x, Tx))^{\alpha}(d(y, Ty))^{1-\alpha-\beta},$$

see [65].

Lemma 8 ([99]) *If $r \leq s$, then we have the following weighted inequality:*

$$M^r(T, x, y, a) \leq M^s(T, x, y, a). \tag{49}$$

Theorem 14 ([99]) *Let (X, d, b) be a complete b-metric space and $T : X \to X$ be an (r, a)-weight type contraction mapping. Then T has a fixed point $x^* \in X$ and for any $x_0 \in X$ the sequence $\{T^n x_0\}$ converges to x^* if one of the following conditions holds:*

(i) T is continuous at such point x^*;
(ii) $b^r a_2 < 1$;
(iii) $b^r a_3 < 1$.

Proof We built an iterative sequence $\{x_n\}$ by starting with an arbitrary point $x_0 \in X$ as follows:

$$x_{n+1} = Tx_n \text{ for all non}-\text{negative integers } n.$$

Since the existing of an integer n_0 with $x_{n_0} = x_{n_0+1}$ yields a fixed point (and hence complete the proof), we assume that successive terms are different, that is, $x_n \neq x_{n+1}$ for all non-negative integer n.

We shall examine two cases. Case 1. Suppose that $r > 0$. On (47), we find

$$d(x_{n+1}, x_n) \leq \lambda[a_1(d(x_n, x_{n-1}))^r + a_2(d(x_n, x_{n+1}))^r + a_3(d(x_{n-1}, x_n))^r]^{1/r}. \tag{50}$$

After a simple calculation, we get

$$d(x_{n+1}, x_n) \leq \left[\frac{\lambda^r(a_1 + a_3)}{1 - \lambda^r a_2}\right]^{1/r} d(x_n, x_{n-1}). \tag{51}$$

Set $\gamma = \left[\frac{\lambda^r(a_1+a_3)}{1-\lambda^r a_2}\right]^{1/r}$. We have that $\gamma \in [0, 1)$. It follows from Lemma 7 that $\{x_n\}$ is a Cauchy sequence in X. By completeness of (X, d, b), there exists $x^* \in X$ such that

$$\lim_{n\to\infty} x_n = x^*. \tag{52}$$

The rest is observed in a standard way.

Case 2. We presume that $r = 0$. By combining (47) and (48), we find

$$d(Tx, Ty) \leq \lambda(d(x, y))^{a_1}(d(x, Tx))^{a_2}(d(y, Ty))^{1-a_1-a_2}, \tag{53}$$

for all $x, y \in X \backslash Fix(T)$, where $\lambda \in [0, 1)$ and $a_1, a_2 \in (0, 1)$. By following [65] (Theorem 2.1 with its metric case), the map T has a fixed point in X.

Remark 2 It is clear that regarding Example 13, we can derive several corollaries from Theorem 14.

The following can be considered as a sample for this observation.

Corollary 11 *Let (X, d, b) be a complete b-metric space and $T : X \to X$ be a mapping such that*

$$d(Tx, Ty) \leq \lambda d^{a_1}(x, y) \cdot d^{a_2}(x, Tx) \cdot d^{a_3}(y, Ty), \tag{54}$$

for all $x, y \in X \backslash Fix(T)$, where $\lambda \in [0, 1)$, $a_1, a_2, a_3 \geq 0$, and $a_1 + a_2 + a_3 = 1$. Then T has a fixed point x^* and for any $x_0 \in X$ the sequence $\{T^n x_0\}$ converges to x^*.

Proof It is sufficient to take $r = 0$ and $a = (a_1, a_2, a_3)$ in Theorem 14.

5.1 Hybrid Contractions in the Context of Quasi-metric Spaces

Definition 16 [48] Let (X, q) be a quasi-metric space. We say that the mapping $T : X \to X$ is a *hybrid almost contraction of type* $\mathbb{I}$, if there exist $\zeta \in \mathcal{Z}$, $\psi \in \Psi$, $p \geq 0$, $L \geq 0$, and $a_1, a_2, a_3 \in [0, 1]$ with $a_1 + a_2 > 0$, $a_1 + a_2 + a_3 = 1$, such that for all distinct $u, v \in X$, we have

$$\begin{aligned} &\tfrac{1}{2}\min\{q(u, Tu), q(v, Tv)q(Tv, v)\} \leq q(u, v) \text{ implies} \\ &\zeta(\alpha(u, v)q(Tu, Tv), \psi(I_p(u, v) + LN(u, v))) \geq 0, \end{aligned} \tag{55}$$

where

$$I_p(u, v) = \begin{cases} [a_1(q(u, v))^p + a_2(q(u, Tu))^p + a_3(q(v, Tv))^p]^{1/p}, & \text{for } p > 0, \\ (q(u, v))^{a_1} \cdot (q(u, Tu))^{a_2} \cdot (q(v, Tv))^{a_3} & \text{for } p = 0 \end{cases}$$

and

$$N(u, v) = \min\{q(u, Tv), q(v, Tu)\}.$$

Theorem 15 ([48]) *Let (X, q) be a complete quasi-metric space and $\alpha : X \times X \to [0, \infty)$ be a mapping such that:*

(i) $u = Tu$ implies $\alpha(u, v) > 0$ for every $v \in X$;
(ii) $v = Tv$ implies $\alpha(u, v) > 0$ for every $u \in X$.

Suppose that $T : X \to X$ is an hybrid almost contraction of type $\mathbb{I}$ and

(C_1) T is $\alpha-$orbital admissible;
(C_2) there exists $u_0 \in X$ such that $\alpha(u_0, Tu_0) \geq 1$ and $\alpha(Tu_0, u_0) \geq 1$;
(C_3) T is continuous.

Then, T has a fixed point.

Corollary 12 ([48]) *Let (X, q) be a complete quasi-metric space, a function $\alpha : X \times X \to [0, \infty)$ and a mapping $T : X \to X$ such that there exist $\zeta \in \mathcal{Z}$ and $\psi \in \Psi$ such that for $p \geq 0$, $L \geq 0$ and $a_1, a_2, a_3 \in [0, 1)$ with $a_1 + a_2 > 0$ and $a_1 + a_2 + a_3 = 1$we have*

$$\zeta(\alpha(u, v)q(Tu, Tv), \psi(I_p(u, v) + LN(u, v))) \geq 0, \text{ for all distinct } u, v \in X. \tag{56}$$

Suppose also that the following assumptions hold:

(i) $u = Tu$ implies $\alpha(u, v) > 0$ for every $v \in X$;
(ii) $v = Tv$ implies $\alpha(u, v) > 0$ for every $u \in X$;
(iii) T is $\alpha-$orbital admissible;
(iv) there exists $u_0 \in X$ such that $\alpha(u_0, Tu_0) \geq 1$ and $\alpha(Tu_0, u_0) \geq 1$;
(v) T is continuous.

Then T has a fixed point.

In particular, letting $L = 0$ in the above Corollary we find Theorem 2.1 in [3].

Corollary 13 *[48] Let (X, q) be a complete quasi-metric space and a mapping $T : X \to X$ such that there exist $\zeta \in \mathcal{Z}$ and $\psi \in \Psi$ such that for $p \geq 0$, $L \geq 0$ and $a_1, a_2, a_3 \in [0, 1)$ with $a_1 + a_2 > 0$ and $a_1 + a_2 + a_3 = 1$ we have*

$$\zeta(q(Tu, Tv), \psi(I_p(u, v) + LN(u, v))) \geq 0, \text{ for all distinct } u, v \in X. \tag{57}$$

Then T has a fixed point.

Proof Let $\alpha(u, v) = 1$ in Corollary 12.

Corollary 14 ([48]) *Let (X, q) be a complete quasi-metric space, a function $\alpha : X \times X \to [0, \infty)$ and a continuous mapping $T : X \to X$ such that there exist $\psi \in \Psi$ such that for $p \geq 0$ and $a_1, a_2, a_3 \in [0, 1)$ with $a_1 + a_2 > 0$ and $a_1 + a_2 + a_3 = 1$we have*

$$\alpha(u, v)q(Tu, Tv) \leq \psi(I_p(u, v)), \text{ for all distinct } u, v \in X. \tag{58}$$

Suppose that there exists $u_0 \in X$ such that $\alpha(u_0, Tu_0) \geq 1$ and $\alpha(Tu_0, u_0) \geq 1$. Then T has a fixed point.

Proof Let $\zeta(t, s) = \psi(s) - t$ in Corollary 12.

Moreover, it easy to see that Theorem 7 is a generalization of Theorem 2.1 in [23] in the context of quasi-metric space. Indeed, if we take $L = 0$ and $p = 0$ in Corollary 14, we find:

Corollary 15 *Let (X, q) be a complete quasi-metric space, a function $\alpha : X\times X \to [0, \infty)$ and a continuous mapping $T : X \to X$ such that there exist $\psi \in \Psi$ such that for $a_1, a_2, a_3 \in [0, 1)$ with $a_1 + a_2 > 0$ and $a_1 + a_2 + a_3 = 1$ we have*

$$\alpha(u, v)q(Tu, Tv) \leq \psi((q(u, v))^{a_1} \cdot (q(u, Tu))^{a_2} \cdot (q(v, Tv))^{a_3}), \text{ for all distinct } u, v \in X. \tag{59}$$

Suppose that there exists $u_0 \in X$ such that $\alpha(u_0, Tu_0) \geq 1$ and $\alpha(Tu_0, u_0) \geq 1$. Then T has a fixed point.

In particular, for the case $p = 0$ the continuity condition of T can be replaced with the regularity condition of the space X.

Theorem 16 ([48]) *Let (X, q) be a complete quasi-metric space, a function $\alpha : X \times X \to [0, \infty)$ and a mapping $T : X \to X$ such that there exist $\zeta \in \mathcal{Z}$, $\psi \in \Psi$, $L \geq 0$ and $a_1, a_2, a_3 \in [0, 1]$ with $a_1 + a_2 + a_3 = 1$, such that for all distinct $u, v \in X$, we have*

$$\frac{1}{2}\min\{q(u, Tu), q(v, Tv), q(Tv, v)\} \leq q(u, v) \text{ implies}$$
$$\zeta(\alpha(u, v)q(Tu, Tv), \psi((q(u, v))^{a_1} \cdot (q(u, Tu))^{a_2} \cdot (q(v, Tv))^{a_3} + LN(u, v))) \geq 0, \tag{60}$$

Suppose also that

(i) $u = Tu$ implies $\alpha(u, v) > 0$ for every $v \in X$;
(ii) $v = Tv$ implies $\alpha(u, v) > 0$ for every $u \in X$;
(C_1) T is $\alpha-$orbital admissible;
(C_2) there exists $u_0 \in X$ such that $\alpha(u_0, Tu_0) \geq 1$ and $\alpha(Tu_0, u_0) \geq 1$;
(C_3) X is regular with respect to the mapping α.

Then, T has a fixed point.

Corollary 16 ([48]) *Let (X, q) be a complete quasi-metric space and $T : X \to X$ be a given mapping. Assume that there exist $L \geq 0$, $\zeta \in \mathcal{Z}$ and $\psi \in \Psi$ such that for all distinct $u, v \in X$, we have*

$$\tfrac{1}{2}\min\{q(u,Tu),q(v,Tv)q(Tv,v)\}\leq q(u,v) \text{ implies}$$
$$\zeta(q(Tu,Tv),\psi(I_p(u,v)+LN(u,v)))\geq 0,$$

for all distinct $u, v \in X$. Then T has a fixed point.

Proof It is sufficient to take $\alpha(u,v)=1$ for $u, v \in X$ in Theorem 15.

Corollary 17 ([48]) *Let (X, q) be a complete quasi-metric space and $T : X \to X$ be a given mapping. Assume that there exist $L \geq 0$, $\zeta \in \mathcal{Z}$ and $\psi \in \Psi$ such that for all distinct $u, v \in X$, we have*

$$\tfrac{1}{2}\min\{q(u,Tu),q(v,Tv)q(Tv,v)\}\leq q(u,v) \text{ implies} q(Tu,Tv)\leq kI_p(u,v)$$

for all distinct $u, v \in X$. Then T has a fixed point.

Proof It is sufficient to take $L=0$, $\zeta(t,s)=k_1s-t$, $\psi(u)=k_2u$ with $k_1,k_2\in(0,1)$ and $k=k_1k_2$ in Corollary 16.

Corollary 18 ([48]) *Let (X, q) be a complete quasi-metric space and $T : X \to X$ a continuous mapping such that*

$$\begin{aligned}&\tfrac{1}{2}\min\{q(u,Tu),q(v,Tv)q(Tv,v)\}\leq q(u,v) \text{ implies}\\ &q(Tu,Tv)\leq \tfrac{k}{\sqrt{3}}\cdot\sqrt{(q(u,v))^2+(q(u,Tu))^2+(q(v,Tv))^2}\end{aligned} \tag{61}$$

for all distinct $u, v \in X$ and some $k \in (0, 1)$. Then T has a fixed point in X.

Proof Let $p=2$ and $a_1=a_2=a_3=\frac{1}{3}$ in Corollary 17.

In the next theorem, we involve a Jaggi type expression to the hybrid contractions.

Definition 17 ([48]) Let (X, q) be a quasi-metric space. A mapping $T : X \to X$ is called a *hybrid almost contraction of type* $\mathbb{J}$, if there exist $\zeta \in \mathcal{Z}$ and $\psi \in \Psi$ such that for $p \geq 0$, $L \geq 0$ and $a_1, a_2 > 0$ with $a_1 + a_2 < 1$ we have

$$\begin{aligned}&\tfrac{1}{2}\min\{q(u,Tu),q(v,Tv)q(Tv,v)\}\leq q(u,v) \text{ implies}\\ &\zeta(\alpha(u,v)q(Tu,Tv),\psi(J_p(u,v)+LN(u,v)))\geq 0,\end{aligned} \tag{62}$$

for all distinct $u, v \in X$, where

$$J_p(u,v)=\begin{cases}[a_1(q(u,v))^p+a_2(\frac{q(u,Tu))\cdot(q(v,Tv)}{q(u,v)})^p]^{1/p}, & \text{for } p>0\\ (q(u,v))^{a_1}\cdot(q(u,Tu))^{a_1}\cdot(q(v,Tv))^{1-a_1-a_2}, & \text{for } p=0\end{cases}$$

and

$$N(u, v) = \min \{q(u, Tv), q(v, Tu)\}.$$

Theorem 17 ([48]) *Let (X, q) be a complete quasi-metric space and $\alpha : X \times X \to [0, \infty)$ be a mapping such that:*

(i) $u = Tu$ implies $\alpha(u, v) > 0$ for every $v \in X$;
(ii) $v = Tv$ implies $\alpha(u, v) > 0$ for every $u \in X$.

Suppose that $T : X \to X$ is a hybrid almost contraction of type $\mathbb{J}$ such that the following assumptions hold:

(i) T is $\alpha-$orbital admissible;
(ii) there exists $u_0 \in X$ such that $\alpha(u_0, Tu_0) \geq 1$ and $\alpha(Tu_0, u_0) \geq 1$;
(iii) there exists $\Delta > 0$ such that $(a_1 + a_2\Delta^{2p})^{1/p} \leq 1$ (where $p > 0$) and

$$\frac{1}{\Delta}q(u, v) \leq q(v, u) \leq \Delta q(u, v), \text{ for all } u, v \in X;$$

(iv) T is continuous.

Then T has a fixed point.

The following is a special case for $p = 0$.

Corollary 19 ([48]) *Let (X, q) be a complete quasi-metric space, a function $\alpha : X \times X \to [0, \infty)$ and a mapping $T : X \to X$ such that there exist $\zeta \in \mathcal{Z}$ and $\psi \in \Psi$ such that for $p \geq 0$, $L \geq 0$ and $a_1, a_2, \in [0, 1)$ with $a_1 + a_2 < 1$ we have*

$$\zeta(\alpha(u, v)q(Tu, Tv), \psi(J_p(u, v) + LN(u, v))) \geq 0, \text{ for all distinct } u, v \in X. \tag{63}$$

Suppose also that the following assumptions hold:

(i) $u = Tu$ implies $\alpha(u, v) > 0$ for every $v \in X$;
(ii) $v = Tv$ implies $\alpha(u, v) > 0$ for every $u \in X$;
(iii) T is $\alpha-$orbital admissible;
(iv) there exists $u_0 \in X$ such that $\alpha(u_0, Tu_0) \geq 1$ and $\alpha(Tu_0, u_0) \geq 1$;
(v) there exists $\Delta > 0$ such that $(a_1 + a_2\Delta^{2p})^{1/p} \leq 1$ (where $p > 0$) and

$$\frac{1}{\Delta}q(u, v) \leq q(v, u) \leq \Delta q(u, v), \text{ for all } u, v \in X;$$

1. (vi)] T is continuous.

Then T has a fixed point.

Corollary 20 ([48]) *Let (X, q) be a complete quasi-metric space and T be a continuous self-mapping on X. Suppose that there exist $\zeta \in \mathcal{Z}$, $\psi \in \Psi$ such that*

$$\zeta(q(Tu, Tv), \psi(J_p(u, v))) \geq 0,$$

for each distinct $u, v \in X$. If there exists $\Delta > 0$ such that $(a_1 + a_2 \cdot \Delta^{2p})^{1/p} \leq 1$ for $p > 0$, and $\frac{1}{\Delta}q(u, v) \leq q(v, u) \leq \Delta q(u, v)$ for all $u, v \in X$ then T has a fixed point.

Proof It is sufficient to take $L = 0$ and $\alpha(u, v) = 1$ for $u, v \in X$ in Corollary 19.

Corollary 21 ([48]) *Let (X, q) be a complete quasi-metric space and T be a self-mapping on X. Suppose that there exists $\Delta > 0$ such that $(a_1 + a_2 \cdot \Delta^{2p})^{1/p} \leq 1$ for $p > 0$, and $\frac{1}{\Delta}q(u, v) \leq q(v, u) \leq \Delta q(u, v)$ for all $u, v \in X$. The mapping T has a fixed point provided that*

$$q(Tu, Tv) \leq c \cdot J_p(u, v)$$

for each distinct $u, v \in X$ and some $c \in (0, 1)$.

Proof We set $\zeta(t, s) = c_1 s - t$, $\psi(u) = c_2 u$ with $c_1, c_2 \in [0, 1)$ and $c = c_1 + c_2$ in Corollary 20.

Letting $p = 0$ in Corollary 21 we find Theorem 2.2. in [66].

Corollary 22 ([48]) *Let (X, q) be a complete quasi-metric space and $T : X \to X$ a continuous mapping. Then T has a fixed point provided that*

$$q(Tu, Tv) \leq k_1 \cdot q(u, v) + k_2 \cdot \frac{q(u, Tu)q(v, Tv)}{q(u, v)} \tag{64}$$

for each $u, v \in X$ and $k_1, k_2 \in (0, 1)$ with $k_1 + k_2 < 1$

Proof Let $p = 1$ and $k_i = c \cdot a_i$, for $i \in \{1, 2\}$ in Corollary 21

Example 14 ([48]) Let (X, q) be the quasi-metric space where $X = [1, \infty)$ and

$$q(u, v) = \begin{cases} u - v, & \text{for } u \geq v \\ 2(v - u), & \text{for } u < v \end{cases}$$

Let

$$Tu = \begin{cases} u^3 - 8u^2 + 19u - 9, & \text{for } u \in [1, 5] \\ ln(u^2 - 24) + u + 6, & \text{for } u \in (5, \infty). \end{cases}$$

Consider the function ζ be arbitrary in $\mathcal{Z}$, $\psi \in \Psi$ with $\psi(t) = \frac{t}{\sqrt{3}}$ and $\alpha : X \times X \to [0, \infty)$ such that

$$\alpha(u, v) = \begin{cases} u^2 + 1, & \text{for } (u, v) \in \{(3, 3), (3, 4), (4, 3), (3, 1), (1, 3)\} \\ 1, & \text{for } (u, v) = (2, 1) \\ 0, & \text{otherwise.} \end{cases}$$

It is easily verified that T is $\alpha-$orbital admissible. Whereas $T1 = T3 = T4 = 3$, taking into account the definition of function α we have that the inequality (63) holds for every pair $(u, v) \in X^2 \setminus \{(2, 1)\}$. For the case $u = 2$ and $v = 1$, choosing $a_1 = \frac{1}{2}, a_2 = \frac{1}{48}$ and $p = 2$ we find that axiom (iii) holds. On the other hand,

$$J_p(2, 1) = \sqrt{\tfrac{25}{2}},$$

and

$$\alpha(2, 1)q(T2, T1) = q(5, 3) = 2 < \sqrt{\frac{25}{6}} = \psi(J_p(2, 1)).$$

Consequently, by Theorem 19 we have that the mapping T has a fixed point in X.

On the other hand we can observe that for $u = 1$ and $v = 5$,

$$q(T1, T(4.5)) = q(2, 5.625) = 7.25, \; q(1, T1) = q(1, 2) = 2,$$
$$q(4.5, T(4.5)) = q(4.5, 5.625) = 1.125,$$

so that since

$$q(T1, T(4.5)) > \lambda(q(1, T1))^{\alpha}(q(4.5, T(4.5)))^{1-\alpha}$$

for any $\lambda \in [0, 1)$ and $\alpha \in (0, 1)$, the Theorem 2.2 in [66] cannot be applied.

5.2 *Jaggi Type Hybrid Contraction*

Definition 18 ([69]) A self-mapping T on a metric space(X, d) is called a Jaggi type hybrid contraction if there is $\psi \in \Psi$ so that

$$d(Tx, Ty) \leq \psi\left(J_T^s(x, y)\right), \tag{65}$$

for all distinct $x, y \in X$ where $s \geq 0$ and $\sigma_i \geq 0, i = 1, 2$, such that $\sigma_1 + \sigma_2 = 1$ and

$$J_T^s(x, y) = \begin{cases} [\sigma_1\left(\frac{d(x,Tx)\cdot d(y,Ty)}{d(x,y)}\right)^s + \sigma_2(d(x, y))^s]^{1/s}, \\ \qquad \text{for } s > 0, \quad x, y \in X, x \neq y \\ \\ (d(x, Tx))^{\sigma_1}(d(y, Ty))^{\sigma_2}, \quad \text{for } s = 0, \quad x, y \in X\backslash F_T(X), \end{cases} \tag{66}$$

where $F_T(X) = \{z \in X : Tz = z\}$.

Theorem 18 ([69]) *A continuous self-mapping T on a complete metric space (X, d) possesses a fixed point x provided that T is a Jaggi type hybrid contraction. Moreover, for any $x_0 \in X$, the sequence $\{T^n x_0\}$ converges to x.*

Theorem 19 ([69]) *Let (X, d) be a complete metric space and $T : X \to X$ be a Jaggi type hybrid contraction. In the case where for some integer $p > 1$, T^p is continuous then T has a unique fixed point.*

The following are the immediate consequences of Theorems 20 and 19.
Indeed, letting $\psi(z) = \lambda z$, $z \geq 0$ in Theorem 20, for $p > 0$ we have:

Corollary 23 ([69]) *A continuous self-mapping T on (X, d) has a fixed point x^* if for any $x, y \in X$, $x \neq y$*

$$d(Tx, Ty) \leq \lambda \left[\sigma_1 \left(\frac{d(x, Tx)d(y, Ty)}{d(x, y)} \right)^s + \sigma_2 \left(d(x, y)\right)^s \right]^{1/s}, \tag{67}$$

where $\sigma_1, \sigma_2 \geq 0$ with $\sigma_1 + \sigma_2 = 1$, $s > 0$ and $\lambda \in (0, 1)$.

Corollary 24 ([69]) *A continuous self-mapping T on (X, d) has a fixed point if for any $x, y \in X$, $x \neq y$*

$$d(Tx, Ty) \leq \frac{\lambda}{\sqrt{2}} \left[\left(\frac{d(x, Tx)d(y, Ty)}{d(x, y)} \right)^2 + (d(x, y))^2 \right]^{1/2}, \tag{68}$$

where $\lambda \in (0, 1)$.

Proof Put in Corollary 23 $\sigma_1 = \sigma_2 = \frac{1}{2}$ and $s = 2$.

Corollary 25 ([69]) *A self-mapping T on (X, d) has a fixed point x^* if for $x, y \in X \backslash F_T(X)$,*

$$d(Tx, Ty) \leq \lambda [d(x, Tx)]^{\alpha} [d(y, Ty)]^{1-\alpha}, \tag{69}$$

where $\alpha, \lambda \in (0, 1)$.

Proof It follows from Theorem 20, letting $\psi(z) = \lambda z$ for any $z \geq 0$, $p = 0$, respectively, $\sigma_1 = \alpha$, $\sigma_2 = 1 - \alpha$.

Corollary 26 ([69]) *A self-mapping T on (X, d) has a fixed point if it satisfies the inequality*

$$d(Tx, Ty) \leq \lambda \sqrt{d(x, Tx)d(y, Ty)}, \tag{70}$$

for $x, y \in X \backslash F_T(X)$, where $\lambda \in (0, 1)$.

Proof Put in Corollary 23 $\alpha = 1/2$.

If in Corollary 23, we take $p = 1$, $\alpha = \lambda\sigma_1$, $\beta = \lambda\sigma_2$, where $\lambda \in (0, 1)$, we obtain the following:

Corollary 27 ([69]) *Let the space (X, d) and $T : X \to X$ be a continuous mapping such that for $x, y \in X$, $x \neq y$,*

$$d(Tx, Ty) \leq \alpha \frac{d(x, Tx)d(y, Ty)}{d(x, y)} + \beta d(x, y), \tag{71}$$

where $\alpha, \beta \in (0, 1)$ with $\alpha + \beta < 1$. Therefore, T possesses a fixed point.

6 Hybrid Contractions in b-Metric Spaces

Definition 19 A self-mapping T on a complete (b)-metric space (X, b, s) is said to be a (b)-hybrid contraction, if there is $\psi \in \Psi$ so that

$$b(Tx, Ty) \leq \psi\left(\mathcal{P}_T^r(x, y)\right), \tag{72}$$

where $r \geq 0$ and $\kappa_i \geq 0$, $i = 1, 2, 3, 4$, such that $\sum_{i=1}^{4} \kappa_i = 1$ and

$$\mathcal{P}_T^r(x, y) = \begin{cases} [\kappa_1(b(x, y))^r + \kappa_2(b(x, Tx))^r + \kappa_3(b(y, Ty))^r + \kappa_4\left(\frac{b(y,Tx)+b(x,Ty)}{2s}\right)^r]^{1/r}, \\ \qquad \text{for } r > 0, \quad x, y \in X \\ (b(x, y))^{\kappa_1}(b(x, Tx))^{\kappa_2}(b(y, Ty))^{\kappa_3}\left(\frac{(b(x,Ty)+b(y,Tx)}{2s}\right)^{\kappa_4}, \\ \qquad \text{for } r = 0, \quad x, y \in X \backslash F_T(X), \end{cases} \tag{73}$$

where $F_T(X) = \{\omega \in X : T\omega = \omega\}$.

Theorem 20 *Suppose that (X, b, s) is a complete (b)-metric space and $T : X \to X$ is a (b)-hybrid contraction. Then T has a fixed point ϱ and for any $x_0 \in X$, the sequence $\{T^n x_0\}$ converges to ϱ if either*

(a_1) T is continuous at ϱ;
(a_2) or T^2 is continuous at ϱ;
(a_3) or $\kappa_1 + \kappa_3 > 0$, (or $\kappa_2 + \kappa_3 > 0$).

Taking $s = 1$ in the above theorem we find the following corollary.

Corollary 28 *Let T be a self-mapping on a complete metric space (X, b) such that*

$$b(Tx, Ty) \leq \psi\left(\mathcal{P}_T^r(x, y)\right), \tag{74}$$

where $\psi \in \Psi$, $r \geq 0$ and $\kappa_i \geq 0$, $i = 1, 2, 3, 4$, such that $\sum_{i=1}^{4} \kappa_i = 1$ and

$$\mathcal{P}_T^r(x,y) = \begin{cases} [\kappa_1(b(x,y))^r + \kappa_2(b(x,Tx))^r + \kappa_3(b(y,Ty))^r + \kappa_4\left(\frac{b(y,Tx)+b(x,Ty)}{2}\right)^r]^{1/r}, \\ \qquad \text{for } r>0, \quad x,y \in X \\ (b(x,y))^{\kappa_1}(b(x,Tx))^{\kappa_2}(b(y,Ty))^{\kappa_3}\left(\frac{(b(x,Ty)+b(y,Tx)}{2}\right)^{\kappa_4}, \\ \qquad \text{for } r=0, \quad x,y \in X \backslash F_T(X), \end{cases} \tag{75}$$

where $F_T(X) = \{\omega \in X : T\omega = \omega\}$. *Then* T *has a fixed point* $\varrho \in X$, *if any of the following statements hold:*

(a_1) *T is continuous at ϱ;*
(a_2) *or T^2 is continuous at ϱ;*
(a_3) *or $\kappa_1+\kappa_3>0$, (or $\kappa_2+\kappa_3>0$).*

7 Admissible Hybrid $\mathcal{Z}$-Contractions in b-Metric Spaces

Definition 20 ([43]) Let (X,d) be a b-metric space with constant $s \geq 1$. A self-mapping f is called an admissible hybrid contraction, if there exist $\varphi : [0,\infty) \to [0,\infty)$ a b-comparison function and $\alpha : X \times X \to [0,\infty)$ such that

$$\alpha(x,y)d(fx,fy) \leq \varphi\left(\mathcal{R}_f^q(x,y)\right), \tag{76}$$

where $q \geq 0$ and $\lambda_i \geq 0$, $i = 1,2,3,4,5$ such that $\sum_{i=1}^5 \lambda_i = 1$ and

$$\mathcal{R}_f^q d(x,y) = \begin{cases} [N(x,y)]^{1/q}, & \text{for } q>0, x,y \in X, \\ P(x,y), & \text{for } q=0, x,y \in X. \end{cases} \tag{77}$$

where

$$N(x,y) := \lambda_1 d^q(x,y) + \lambda_2 d^q(x,fx) + \lambda_3 d^q(y,fy) + \lambda_4\left(\frac{d(y,fy)(1+d(x,fx))}{1+d(x,y)}\right)^q + \lambda_5\left(\frac{d(y,fx)(1+d(x,fy))}{1+d(x,y)}\right)^q,$$

and

$$P(x,y) := d^{\lambda_1}(x,y)\cdot d^{\lambda_2}(x,fx)\cdot d^{\lambda_3}(y,fy) \cdot\left(\frac{d(y,fy)(1+d(x,fx))}{1+d(x,y)}\right)^{\lambda_4}\cdot\left(\frac{d(x,fy)+d(y,fx)}{2s}\right)^{\lambda_5}.$$

Definition 21 ([68]) Let (X, d) be a b-metric space with constant $s \geq 1$. A mapping $f : X \to X$ is called admissible hybrid $\mathcal{Z}$-contraction mapping if there is $\varphi : [0, \infty) \to [0, \infty)$ a b-comparison function, $\alpha : X \times X \to [0, \infty)$ and $\zeta \in \mathcal{Z}$ such that

$$\zeta\left(\alpha(x, y)d(fx, fy), \varphi\left(\mathcal{R}_f^q(x, y)\right)\right) \geq 0, \text{ for all } x, y \in X, \tag{78}$$

where $\mathcal{R}_f^q(x, y)$ is as above.

Theorem 21 ([43]) *Let (X, d) be a complete b-metric space with constant $s \geq 1$ and let $f : X \to X$ be an admissible hybrid $\mathcal{Z}$-contraction. Suppose also that:*

(i) f is triangular α-orbital admissible;
(ii) there exists $x_0 \in X$ such that $\alpha(x_0, f(x_0)) \geq 1$;
(iii) either, f is continuous or
(iv) f^2 is continuous and $\alpha(fx, x) \geq 1$ for any $x \in Fix_{f^2}(X)$.

Then, f has a fixed point.

The idea of the proof can be deduced easily from the earlier proof of this chapter. So, we skipped it.

Theorem 22 *In the hypothesis of Theorem 21, if we assume supplementary that*

$$\alpha(x^*, y^*) \geq 1,$$

for any $x^*, y^* \in Fix_f(X)$, then the fixed point of f is unique.

Theorem 23 ([43]) *Let (X, d) be a complete b-metric space with constant $s \geq 1$, $f : X \to X$ and $\alpha : X \times X \to [0, \infty)$. Suppose that there exist two functions $\phi_1, \phi_2 \in \Phi$, with $\phi_1(t) < t \leq \phi_2(t)$, for all $t > 0$, such that*

$$\phi_2(\alpha(x, y)d(fx, fy)) \leq \phi_1\left(\mathcal{R}_f^q(x, y)\right). \tag{79}$$

Furthermore, we suppose that:

(i) f is triangular α-orbital admissible;
(ii) there exists $x_0 \in X$ such that $\alpha(x_0, f(x_0)) \geq 1$;
(iii) either, f is continuous or
(iv) f^2 is continuous and $\alpha(fx, x) \geq 1$ for any $x \in Fix_{f^2}(X)$.
(v) if $x^*, y^* \in Fix_f(X)$, then $\alpha(x^*, y^*) \geq 1$.

Then, f has a unique fixed point.

Proof Let $\zeta(t, s) = \phi_1(s) - \phi_2(t)$. According to Example 1, if $\phi_1, \phi_2 \in \Phi$ have the property $\phi_1(t) < t \leq \phi_2(t)$ for all $t > 0$, then $\zeta \in \mathcal{Z}$. Thus, the desired results follow from Theorems 21 and 22.

Remark 3 It is clear that we deduce more results for setting ζ properly, as in Example 1 and choosing admissible mapping α properly, as in Corollaries 7 and 9.

7.1 Hybrid Contractions in Branciari Distance Spaces

Definition 22 A self-mapping T on (X, d) is said to be a (p, c)-weight type ψ-contraction, if, there exists $\psi \in \Psi$ so that the following inequality holds for any s, $t \in X$ which are not fixed points of T

$$d(Ts, Tt) \leq \psi(\mathcal{W}_T^{p,c}(s, t)), \tag{80}$$

where $p \geq 0$, $c = (c_1, c_2, c_3)$, and c_1, c_2, and c_3 are positive numbers such that $c_1 + c_2 + c_3 = 1$, and

$$\mathcal{W}_T^{p,c}(s, t) = \begin{cases} (c_1 d^p(s, t) + c_2 d^p(s, Ts) + c_3 d^p(t, Tt))^{\frac{1}{p}}, & \text{if } p > 0 \\ d^{c_1}(s, t) d^{c_2}(s, Ts) d^{c_3}(t, Tt), & \text{if } p = 0. \end{cases}$$

Theorem 24 *Let (X, d) be a complete Branciari distance spaces and $T: X \to X$ be an (p, c)-weight type ψ-contraction mapping. Then the mapping T possesses a fixed point $\varkappa^*$.*

References

1. K. Abodayeh, E. Karapınar, A. Pitea, W. Shatanawi, *Hybrid Contractions on Branciari Type Distance Spaces*, Mathematics 2019, 7, 994.
2. J. Achari, *On Ćirić's non-unique fixed points*, Mat. Vesnik, **13** (28)no. 3, 255–257 (1976).
3. R. P. Agarwal and E. Karapınar, *Interpolative Rus-Reich-Ciric Type Contractions Via Simulation Functions*, An. St. Univ. Ovidius Constanta, Ser. Mat., Volume XXVII (2019) fascicola 3 Vol. **27**(3), 2019, 137–152.
4. U. Aksoy, E. Karapınar, İ. M. Erhan, *Fixed points of generalized α-admissible contractions on b-metric spaces with an application to boundary value problems*, J. Nonlinear and Convex A., **17** (2016). No: 6, 1095–1108
5. H.H. Alsulami, E. Karapınar, F. Khojasteh, A.F. Roldán-López-de-Hierro, *A proposal to the study of contractions in quasi-metric spaces*, Discrete Dynamics in Nature and Society 2014, Article ID 269286, 10 pages.
6. A.S. Alharbi, H.H. Alsulami, E. Karapınar, *On the Power of Simulation and Admissible Functions in Metric Fixed Point Theory*, J. Funct. Spaces, Volume **2017** (2017), Article ID 2068163, 7 pages.
7. N. Alharbi, H. Aydi, A. Felhi, C. Ozel, S. Sahmim, *α-contractive mappings on rectangular b-metric spaces and an application to integral equations*, J. Math. Anal., **2018**, 9, 47–60.
8. B. Alqahtani, A. Fulga, E. Karapınar, *Fixed Point Results On Δ-Symmetric Quasi-Metric Space Via Simulation Function With An Application To Ulam Stability*, Mathematics 2018, 6(10), 208.

9. M.U. Ali, T. Kamram, E. Karapınar, *An approach to existence of fixed points of generalized contractive multivalued mappings of integral type via admissible mapping*, Abstr. Appl. Anal. 2014, (2014) Article ID 141489.
10. M.U. Ali, T. Kamran, E. Karapınar, *On (α, ψ, η)-contractive multivalued mappings*, Fixed Point Theory Appl. (2014), 2014:7.
11. S. Almezel, C.M. Chen, E. Karapınar, V. Rakocev?? *Fixed point results for various α-admissible contractive mappings on metric-like spaces*, Abstr. Appl. Anal. 2014 (2014), Article ID 379358.
12. H. Alsulami, S. Gulyaz, E. Karapınar, I.M. Erhan, *Fixed point theorems for a class of α-admissible contractions and applications to boundary value problem*, Abstr. Appl. Anal. 2014 (2014) Article ID 187031.
13. B. Alqahtani, H. Aydi, E. Karapınar, V. Rakocevic, *A Solution for Volterra Fractional Integral Equations by Hybrid Contractions*, Mathematics 2019, 7, 694.
14. O. Alqahtani, E. Karapınar, *A Bilateral Contraction via Simulation Function*, Filomat 33:15 (2019), 4837.4843
15. H. Argoubi, B. Samet, C. Vetro, Nonlinear contractions involving simulation functions in a metric space with a partial order, J. Nonlinear Sci. Appl. **8** (2015), 1082–1094.
16. M. Arshad, E. Ameer, E. Karapınar, *Generalized contractions with triangular α-orbital admissible mapping on Branciari metric spaces*, J. Inequal. Appl. 2016, 2016:63
17. H. Aydi, E. Karapınar, H. Yazidi, *Modified F-Contractions via α-Admissible Mappings and Application to Integral Equations*, Filomat, 31 (5)(2017), 1141- 148.
18. H. Aydi, E. Karapınar, D. Zhang, *A note on generalized admissible-Meir-Keeler-contractions in the context of generalized metric spaces*, Results in Mathematics, **71** (2017) No. 1, 73–92.
19. H. Aydi, M. Jellali, E. Karapınar, *On fixed point results for α-implicit contractions in quasi-metric spaces and consequences*, Nonlinear Anal. Model. Control. 21 (1) (2016), 40–56.
20. H. Aydi, A. Felhi, E. Karapınar, F.A. Alojail, *Fixed points on quasi-metric spaces via simulation functions and consequences*, Journal of Mathematical Analysis, Volume **9** Issue 2 (2018), Pages 10–24.
21. H. Aydi, E. Karapınar and V. Rakočević, *Nonunique Fixed Point Theorems on b-Metric Spaces via Simulation Functions*, Jordan Journal of Mathematics and statistics, (in press).
22. H. Aydi, C.-M. Chen, E. Karapınar, *Interpolative Ciric-Reich-Rus type contractions via the Branciari distance*, Mathematics 2019 7(1), 84; 10.3390/math7010084
23. H. Aydi, E. Karapınar, A.F. Roldan Lopez de Hierro, *ω-Interpolative Reich-Rus-Ćirić-Type Contractions*, Mathematics 2019, 7, 57.
24. H. Aydi, S.H.; Amor, E. Karapınar, *Berinde Type generalized contractions on partial metric spaces*, Abstr. Appl. Anal., **2013**, 2013, doi:10.1155/2013/312479.
25. H. Aydi, E. Karapınar, W. Shatanawi, *Coupled fixed point results for (ϕ)-weakly contractive condition in ordered partial metric spaces*, Comput. Math. Appl., 2011, **62**, 4449–4460.
26. H. Aydi, E. Karapınar, *A Meir-Keeler common type fixed point theorem on partial metric spaces*, Fixed Point Theory Appl., **2012**, 2012, doi:10.1186/1687-1812-2012-26.
27. H. Aydi, E. Karapınar, A. Francisco Roldan Lopez de Hierro, *ω-interpolative Ciric-Reich-Rus type contractions*, Mathematics 2019, 7, 57.
28. H. Aydi, E. Karapınar, B. Samet, *Fixed points for generalized $(\alpha, \psi)-$contractions on generalized metric spaces*, J. Inequal. Appl., **2014**, *2014*, 229.
29. H. Aydi, E. Karapınar, W. Shatanawi, *Tripled fixed point results in generalized metric spaces*, J. Appl. Math., 2012, 314279.
30. H. Aydi, E. Karapınar, D. Zhang, *On common fixed points in the context of Brianciari metric spaces*, Results Math., 2019, *71*, 73–92.
31. A. Azam, M. Arshad, *Kannan fixed point theorem on generalized metric spaces*, J. Nonlinear Sci. Appl., 2008, **1**, 45–48.
32. S. Banach, *Sur les opérations dans les ensembles abstraits et leur application aux équations intégrales*, Fundamenta Mathematicae, 3 (1922), 133–181.
33. D.F. Bailey, *Some Theorems on Contractive Mappings*, Journal of the London Mathematical Society, 41 (1966) No:1, 101–106.

34. C. Bessaga, *On the converse of the Banach "fixed-point principle"*, Colloq. Math. 7, (1959), 41–43.
35. V. Berinde, *Generalized contractions in quasimetric spaces*, Semin. Fixed Point Theory, 1993, **3**, 3–9.
36. D.W. Boyd and J.S.W. Wong, *On nonlinear contractions*, Proc. Amer. Math. Soc. **20**(1969), 458–464.
37. A. Branciari, *A fixed point theorem of Banach-Caccioppoli type on a class of generalized metric spaces*, Publ. Math. Debrecen, 2000, **57**, 31–37.
38. F.E. Browder, *On the convergence of successive approximations for nonlinear functional equations*, Nederl. Akad. Wetensch. Ser. A71=Indag. Math. **30**(1968), 27–35.
39. P. S. Bullen, D. S. Mitrinović, *P.M. Vasić, Means and Their Inequalities*, D. Reidel Publ. Company, Dordrecht/Boston/Lancaster/Tokyo, 1988.
40. R. Caccioppoli, *Una teorema generale sull'esistenza di elementi uniti in una transformazione funzionale*, Ren. Accad. Naz Lincei **11**(1930), 794–799.
41. S.K. Chatterjea, *Fixed-point theorems*, Comptes Rendus de l'Acadmie Bulgare des Sciences, vol. **25**, pp. 727730, 1972
42. C.M. Chen, A. Abkar, S. Ghods, E. Karapınar, *Fixed Point Theory for the α-Admissible Meir-Keeler Type Set Contractions Having KKM* Property on Almost Convex Sets*, Appl. Math. Inf. Sci. 11 (1) (2017), 171–176.
43. I.C. Chifu, E. Karapınar, *Admissible Hybrid Z-Contractions in b-Metric Spaces*, Axioms, 2020, 9, 2.
44. S.C. Chu and J.B. Diaz, *Remarks on a Generalization of Banachs Principle of Contraction Mappings*, Journal Of Mathematical Analysis And Applications 11(1965) 440–446.
45. L.B.Ćirić, *Fixed point theory: Contraction mapping principle*, C-print, Beograd, 2003.
46. M. Edelstein, *An extension of Banach's contraction principle*, Proc. American Math. Soc, 12 (1961), 7–10.
47. M. Edelstein, *On Fixed and Periodic Points Under Contractive Mappings*, J. Lond. Math. Soc. 1962, 37: 74–79.
48. A. Fulga, E. Karapınar, G. Petru sel, *On Hybrid Contractions in the Context of Quasi-Metric Spaces*, Mathematics 2020, **8**, 675.
49. Y.U. Gaba, E. Karapınar, *A New Approach to the Interpolative Contractions*, Axioms 2019, 8, 110.
50. M. Geraghty, *On contractive mapping*s, Proc. Amer. Math. Soc. **40**(1973), 604–608.
51. K.M. Ghosh, *A generalization of contraction principle*, Int. J. Math. Math. Sci. 4(1), 201–207 (1981).
52. S. Gulyaz, E. Karapınar, I.M. Erhan, *Generalized α-Meir-Keeler Contraction Mappings on Branciari b-metric Spaces*, Filomat, 2017, **31**, 5445–5456.
53. G.E. Hardy, T.D. Rogers, *A generalization of a fixed point theorem of Reich*, Can. Math. Bull., 1973, **16**, 201–206.
54. K. Hammache, E. Karapınar, A. Ould-Hammouda, *On Admissible weak contractions in b-metric-like space*, J. Math. Anal. 8 (3) 2017), 167–180.
55. L. Janoş, *A converse of Banachs contraction theorem*, Proc. Amer. Math. Soc., 18, (1967), 287–289.
56. M. Jleli, E. Karapınar, B. Samet, *Best proximity points for generalized $\alpha - \psi$-proximal contractive type mappings*, J. Appl. Math. 2013 (2013) Article ID 534127, .
57. M. Jleli, E. Karapınar, B. Samet, *Fixed point results for $\alpha - \psi_\lambda$ contractions on gauge spaces and applications*, Abstr. Appl. Anal. 2013 (2013) Article ID 730825.
58. M. Jleli, E. Karapınar, B Samet, *Best proximity points for generalized $\alpha - \psi$-proximal contractive type mappings*, J. Appl. Math., **2013** (2013) Article ID 534127.
59. Z. Kadelburg, S. Radenović, *Pata-type common fixed point results in b-metric and b-rectangular metric spaces*, J. Nonlinear Sci. Appl. 2015, **8**, 944–954.
60. R. Kannan, *Some remarks on fixed points*, Bull. Calcutta Math. Soc. 60 (1960), 71–76.
61. R. Kannan, *Some results on fixed points*, Bull. Calcutta Math. Soc. 60, 71–76 (1968).
62. R. Kannan, *Some results on fixed points. II*, Am. Math. Mon. 76, 405–408 (1969).

63. E. Karapınar, *Discussion on $(\alpha, \psi)-$contractions on generalized metric spaces*, Abstr. Appl. Anal., **2014**, 962784.
64. E. Karapınar, O. Alqahtani, H. Aydi, *On Interpolative Hardy-Rogers Type Contractions*, Symmetry 2019, 11(1), 8; 10.3390/sym11010008
65. E. Karapınar, R. Agarwal, H. Aydi, *Interpolative Reich-Rus-Ćirić Type Contractions on Partial Metric Spaces*, Mathematics 2018, 6, 256. 10.3390/math6110256
66. E. Karapınar, *Revisiting the Kannan type contractions via interpolation*, Advances in the Theory of Nonlinear Analysis and its Applications, 2 (2) (2018), 85–87.
67. E. Karapınar and A. Fulga, *An admissible Hybrid contraction with an Ulam type stability*, Demonstr. Math. (2019); 52:428–436
68. E. Karapınar, A. Fulga, *New Hybrid Contractions on b-Metric Spaces*, Mathematics 2019, 7, 578.
69. E. Karapınar, A. Fulga, *A Hybrid Contraction that Involves Jaggi Type*, Symmetry 2019, 11, 715.
70. E. Karapınar, S. Czerwik, H. Aydi, (α, ψ)*-Meir-Keeler contraction mappings in generalized b-metric spaces*, J. Funct. Spaces, Volume **2018** (2018), Article ID 3264620, 4 pages.
71. E. Karapınar, B. Samet, *Generalized $(\alpha - \psi)$-contractive type mappings and related fixed point theorems with applications*, Abstr. Appl. Anal. 2012 (2012) Article iD 793486.
72. E. Karapınar, A. Roldan, D. Oregan, *Coincidence point theorems on quasi-metric spaces via simulation functions and applications to G-metric spaces*, Journal of Fixed Point Theory and Applications. 10.1007/s11784-018-0582-x
73. E. Karapınar, F. Khojasteh, *An approach to best proximity points results via simulation functions*, Journal of Fixed Point Theory and Applications, 19(3), 1983–1995, 2017
74. E. Karapınar, *Fixed points results via simulation functions*, Filomat, Volume 30, Number 8, 2016, 2343–2350
75. E. Karapınar, *Revisiting the Kannan type contractions via interpolation*, Advances in the Theory of Nonlinear Analysis and its Applications, 2 (2) (2018), 85–87.
76. E. Karapınar, R.P. Agarwal, H. Aydi, *Interpolative Reich-Rus-Ćirić Type Contractions on Partial Metric Spaces*, Mathematics 2018, 6, 256. 10.3390/math6110256
77. E. Karapınar, O. Alqahtani, H. Aydi, *On Interpolative Hardy-Rogers Type Contractions*, Symmetry 2019, 11(1), 8; 10.3390/sym11010008
78. E. Karapınar, H.H. Alsulami and M. Noorwali, *Some extensions for Geragthy type contractive mappings*, Journal of Inequalities and Applications 2015, 2015:303 (26 September 2015)
79. E. Karapınar, Discussion on (α, ψ) contractions on generalized metric spaces, Abstr. Appl. Anal., **2014** (2014) Article ID 962784.
80. E. Karapınar, *Fixed points results for α-admissible mapping of integral type on generalized metric spaces*, Abstr. Appl. Anal., **2014** (2014), Article Id: 141409
81. E. Karapınar, *On (α, ψ) contractions of integral type on generalized metric spaces*, in Proceedings of the 9th ISAAC Congress, V. Mityushevand, M. Ruzhansky, Eds., Springer, Krakow, Poland, 2013.
82. E. Karapınar, B. Samet, *Generalized α-ψ-contractive type mappings and related fixed point theorems with applications*, Abstr. Appl. Anal. **2012** (2012) Article ID 793486.
83. E. Karapınar, P. Kumam, P. Salimi, *On $\alpha - \psi$-Meir-Keeler contractive mappings*, Fixed Point Theory Appl. (2013), 2013:94 .
84. E. Karapınar, *Revisiting the Kannan Type Contractions via Interpolation*, Adv. Theory Nonlinear Anal. Appl., 2018, **2**, 85–87.
85. E. Karapınar, R.P. Agarwal, H. Aydi, *Interpolative Reich-Rus-Ćirić type contractions on partial metric spaces*, Mathematics **2018**, *6*, 256.
86. E. Karapınar, O. Alqahtani, H. Aydi, On interpolative Hardy-Rogers type contractions, Symmetry, **2018**, *11*, 8.
87. E. Karapınar, A. Pitea, *On alpha-psi-Geraghty contraction type mappings on quasi-Branciari*, metric spaces. J. Nonlinear Convex Anal., 2016, **17**, 1291–1301.
88. E. Karapınar, *A Short Survey on Dislocated Metric Spaces via Fixed-Point Theory*, In *Advances in Nonlinear Analysis via the Concept of Measure of Noncompactness*; Banas, J.,

Jleli, M., Mursaleen, M., Samet, B., Vetro, C., Eds.; Springer Nature Singapore Pte Ltd.: Singapore, 2017; Chapter 13, pp. 457–483, doi:10.1007/978-981-10-3722-1.

89. F. Khojasteh, S. Shukla, S. Radenović, *A new approach to the study of fixed point theorems via simulation functions*, Filomat 29:6 (2015), 1189–194.
90. Kirk, W.A.; Shahzad, N. *Generalized metrics and Caristi's theorem*, Fixed Point Theory Appl., **2013**, *2013*, 129.
91. Krein, S.G.; Petunin, J.I.; Semenov, E.M. *Interpolation Of Linear Operators*; American Mathematical Society: Providence, RI, USA, 1978.
92. A.N. Kolmogorov, and S.V. Fomin, *Elements of the Theory of Functions and Functional Analysis*, Volume I, Metric and Normed Spaces, Graylock Press, Rochester, New York, 1957.
93. H. Lakzian, B. Samet, *Fixed point for (ψ, φ)-weakly contractive mappings in generalized metric spaces*, *Appl. Math. Lett.*, 2012, **25**, 902–906.
94. J. Liouville, *Second mémoire sur le développement des fonctions ou parties de fonctions en séries dont divers termes sont assujettis á satisfaire a une m eme équation différentielle du second ordre contenant un paramétre variable*, J. Math. Pure et Appi., **2** (1837), 16–35.
95. S.G. Matthews, *Partial metric topology*, Ann. N. Y. Acad. Sci., 1994, **728**, 183–197.
96. N. Mlaiki, K. Abodayeh, H. Aydi, T. Abdeljawad, M. Abuloha, *Rectangular Metric-Like Type Spaces Related Fixed Points*, J. Math., **2018**, *2018*, 3581768, .
97. P. R. Meyers, *A converse to Banachs contraction theorem*, J. Res. Nat. Bur. Standards Sect. B 71B, (1967), 73–76.
98. R. Miculescu, A. Mihail, *New fixed point theorems for set-valued contractions in b-metric spaces*, J. Fixed Point Theory Appl. **19** (2017), 2153–2163.
99. Z.D. Mitrovic, H. Aydi, M.S. Noorani, H. Qawaqneh, *The weight inequalities on Reich type theorem in b-metric spaces*, J. Math. Computer Sci., **19** (2019), 51–57
100. V.V. Nemytskii, *The fixed point method in analysis*, Usp. Mat. Nauk 1 (1936) 141–174 (in Russian).
101. B.G. Pachpatte, *On Ćirić type maps with a nonunique fixed point, Indian J. Pure Appl. Math.*, **10**(8), 1039–1043 (1979).
102. E. Picard, *Memoire sur la theorie des equations aux derivees partielles et la methode des approximations successives*, J. Math. Pures et Appl., **6** (1890), 145–210.
103. O. Popescu, *Some new fixed point theorems for α-Geraghty contractive type maps in metric spaces*, Fixed Point Theory Appl. 2014, 2014:190
104. E. Rakotch, *A note on contractive mappings*, Proc. Amer. Math. Soc., **13**, (1962) 459–465.
105. S. Reich, *Some remarks concerning contraction mappings*, Can. math. Bull. **14** (1971), 121–124 .
106. S. Reich, *Fixed point of contractive functions*, Boll. Un. mat. Ital. **4** (5) (1972), 26–42 .
107. S. Reich, *Kannan's fixed point theorem*, Boll. Un. mat. Ital. 4 (4) (1971), 1–11 .
108. A.F. Roldán-López-de-Hierro, E. Karapınar, C. Roldán-López-de-Hierro, J. Martínez-Moreno, *Coincidence point theorems on metric spaces via simulation functions*, J. Comput. Appl. Math. 275 (2015) 345–355.
109. J.R. Roshan, N. Hussain, V. Parvaneh, Z. Kadelburg, *New fixed point results in rectangular b-metric spaces*, Nonlinear Anal., 2016, **21**, 614–634.
110. Romaguera, S.; Tirado, P. *A characterization of Smyth complete quasi-metric spaces via Caristi's fixed point theorem*, Fixed Point Theory Appl., **2015**, 2015:183.
111. Romaguera, S.; Tirado, P. *The Meir-Keeler fixed point theorems for quasi-metric spaces and some consequences*, Symmetry, 2019, **11(6)**, 741.
112. I.A. Rus, *Principles and Applications of the Fixed Point Theory* (in Romanian), Editura Dacia, Clui-Napoca, 1979.
113. I.A. Rus, *Generalized Contractions and Applications*, Cluj University Press, Cluj-Napoca, Romania, 2001.
114. B. Samet, C. Vetro, P. Vetro, *Fixed point theorem for $\alpha - \psi$ contractive type mappings*, Nonlinear Anal. 75 (2012) 2154–2165.
115. I.R. Sarma, J.M. Rao, S.S. Rao, *Contractions Over Generalized Metric Spaces*, J. Nonlinear Sci. Appl., 2009 **2**, 180–182.

116. W. Shatanawi, A. Al-Rawashdeh, H. Aydi, H.K. Nashine, *On a fixed point for generalized contractions in generalized metric spaces*, Abstr. Appl. Anal., 2012, 246085.
117. N. Shioji, T. Suzuki, W. Takahashi, *Contractive mappings, Kannan mappings and metric completeness*, Proc. Am. Math. Soc. 126, 3117–3124 (1998).
118. P.V. Subrahmanyam, *Completeness and fixed points*, Monatsh. Math. 80 (1975), 325–330.
119. T. Suzuki, *Generalized metric space do not have the compatible topology*, Abstr. Appl. Anal., **2014**, 458098.
120. T. Suzuki, B. Alamri, M. Kikkawa, *Only 3-generalized metric spaces have a compatible symmetric topology*, Open Math., 2015, **13**, 510–517.
121. T. Suzuki, *Completeness of 3-generalized metric spaces*, Filomat, 2016, **30**, 3575–3585.
122. T. Suzuki, *Some metrization problem on ν-generalized metric spaces*, Rev. R. Acad. Cienc. Exactas Fis. Nat. Ser. A Mater, 2019, **113**, 1267–1278.
123. T. Zamfirescu, *Fixed point theorems in metric spaces*, Arch. Math. **23**, 292–298 (1972).

Identifying the Computational Problem in Applied Statistics

Christos P. Kitsos and C. S. A. Nisiotis

1 Introduction

The problem of computation is it has two different lines of thought. The one is how to perform calculations, perhaps creating a computer program, or through a stat/math package. The second line of thought is to create a mathematical approach to the problem and then when calculations are needed, most of the times a stochastic model is created.

For the ladder two examples are:

The 13th problem of Hilbert can be stated as "can every continuous function of 3-variables to be expressed as a composition of finitely many continuous functions of 2-variables." The Russian Vladimir Arnold (1957–2010) when only 19 years old solved the problem. For this famous problem Rassias and Simsa [41] offered an extensively research for it coming across to nice results.

Moreover, part of work of S.L. Sobolev (1908–1989) and Kitsos and Tavoularis [32] was considered confidential, [4] work as theoretical framework was related to aeronautics. Moreover Kitsos and Tavoularis [33] working on logarithm Sobolev Inequalities (LSI) generalizing entropy type Fisher's information came across the Generalized Normal Distribution emerged from an LSI.

As far as the former line of thought concerns a simple but nice problem is the following

C. P. Kitsos (✉)
University of West Attica, Department of Informatics, Egaleo, Greece
e-mail: xkitsos@uniwa.gr

C. S. A. Nisiotis
University of West Attica, Department of Public and Community Health, Egaleo, Greece
e-mail: csnisiotis@uniwa.gr

I. N. Parasidis et al. (eds.), *Mathematical Analysis in Interdisciplinary Research*,
Springer Optimization and Its Applications 179,
https://doi.org/10.1007/978-3-030-84721-0_21

Hack's law is an empirical relationship between the length l of streams and the area S of their basins as:

$$l = cS^k$$

with c a constant and the exponent k around 0.6. The exponent k varies from region to region.

Moreover Rigon et al. [42] working on Hack's law were adopting probability theory to investigate the problem, while Arnold (2011) presented a number of calculations on Hack's exponent for different rivers.

2 Computational Difficulties: GLM

Consider the classical General Linear Model (GLM), Graybill [16], Seber [45], among others, of the form:

$$Y = X\beta + e \qquad \text{(GLM)}$$

where:

$Y \in \mathbb{R}^{n\times 1}$	the response vector
$X \in \mathbb{R}^{n\times(p+1)}$	the data matrix known as design matrix where $x_{ij} = 0 \text{or} 1, i = 1, 2, \ldots, n, i = 1, 2, \ldots, p$.
$\beta \in \mathbb{R}^{(p+1)\times 1}$	the vector of the involved linear parameters, and
$e \in \mathbb{R}^{n\times 1}$	the stochastic error usually assumed with i.i.d. (independent identically distributed) observations, with $E(e) = 0$, $E(ee') = I\sigma^2$ and $I = \text{diag}(1, 1, \ldots, 1) \in \mathbb{R}^{n\times n}$, $\sigma^2 > 0$ unknown.

We assume throughout this chapter that $\text{rank}(X) = p + 1 < n$.

It is well known that the Least Square Estimate (LSE) of β and σ^2 are:

$$\hat{\beta} = (X^T X)^{-1} X^T Y$$

$$\hat{\sigma}^2 = \frac{1}{n-(p+1)} Y^T (I - X(X^T X)^{-1} X^T) Y := s^2.$$

The well-known Gauss–Markov theorem states that within the class of linear unbiased estimators of β, the LSE has minimum variance.

Moreover when inference is needed the normal assumption is imposed, namely:

$$e \sim N(0, I\sigma^2)$$

with $N(\alpha, \beta)$ being the Normal Distribution with mean α and variance β. In such a case the LSE coincides with the MLE (= Maximum Likelihood Estimator) and they are complete, sufficient unbiased statistics for β and σ^2, respectively, with

$$\hat{\beta} \sim N(\beta, (X^T X)^{-1}\sigma^2) \quad \hat{\text{Cov}}(\hat{\beta}) = (X^T X)^{-1}s^2$$

and also $\hat{\beta}$ and s^2 are independent Grabill [16].

There are a number of techniques and criteria suggesting the way that the "best" Regression Equation is obtained, based on the early work of Hocking [25], while an early computer-oriented technique was due to LaMotte [34]. Moreover the difficulty to overpass the computation problem in Regression Analysis in 1980s is reflected to Graybill [16] who devotes a number of paragraphs on the subject.

We would like to clarify two important "hidden" calculations as far as a variable is getting "in the model" or "out of the model." Recently, in statistical packages are appeared only the results. Here is what is the theory behind.

At any step the i-th variable from the given set of all variables involving the GLM is eliminated from the p-term linear model if:

$$F_i = \underset{i}{min} \frac{RSS_{p-i} - RSS_p}{\hat{\sigma}_p^2} < F_{out}$$

where RSS_κ is the Residual sum of squares for the κ-term model. Notice that the value:

$$F_{out} = F_{1,n-p}(\alpha)$$

and Kennedy and Bancraft [28] at their early work recommended as the "best" significant level α, $\alpha = 0.10$.

Let $d(\beta, \hat{\beta})$ be the Euclidean distance of the estimate $\hat{\beta}$ from the true β, defined as:

$$d = d(\beta, \hat{\beta}) = (\hat{\beta} - \beta)^T(\hat{\beta} - \beta).$$

If we consider the expected value of d, $E(d)$ as:

$$L^2 = E\left[(\hat{\beta} - \beta)^T(\hat{\beta} - \beta)\right]$$

then it holds:

Proposition 1 *The expected value of the distance d is minimized when $\sigma^2\text{tr}(X^T X)^{-1}$ is minimum:*

$$\min L^2 = \min\left[\sigma^2\text{tr}(X^T X)^{-1}\right]$$

Indeed:

$$\begin{aligned}
L^2 &= E\left[\left((X^TX)^{-1}X^TY-\beta\right)^T\left((X^TX)^{-1}X^TY-\beta\right)\right]\\
&= E\left[e^TX\left((X^TX)^{-1}\right)^T\left((X^TX)^{-1}X^Te\right)\right]\\
&= \mathrm{tr}\left[X(X^TX)^{-1}(X^TX)^{-1}X^TI\sigma^2\right]\\
&= +\,E(e^T)X\left((X^TX)^{-1}\right)^T(X^TX)^{-1}XE(e)\\
&= \sigma^2\mathrm{tr}\left((X^TX)(X^TX)^{-1}(X^TX)^{-1}\right)+0\\
&= \sigma^2\mathrm{tr}(X^TX)^T.
\end{aligned}$$

Corollary 1 *If λ_i are the eigenvalues of $\left(X^TX\right)$, $i=1,2,\ldots,k$, then:*

$$\min L^2 = \sigma^2 \min \sum_{i=1}^{k} \lambda_i^{-1}$$

Corollary 2 *If there exists $\lambda_i \approx 0$, then L^2 dents to be large as $d(\beta,\hat{\beta})$ is getting large.*

Due to Corollary 2 the corresponding to λ_i variable X_i "does not offer" that much to the model, so it is not included and a re-calculation of the regression is attempted. Today computations it is not a problem, due to not only statistical packages, see Appendix 2, but the problem of getting easy computations can be faced though iteration techniques. We propose a method in the next paragraph.

2.1 Iterative $\bar{R}_p^2$ Calculation

One crucial parameter in Regression Analysis is the coefficient of determination R_p^2, for the p-term model, Seber [45], Draper and Smith [11], Helland [24], Lawrance [35], and Nelson [39] working on time series. In principle R_p^2 is "optimistic" in the sense that might have values, near to 1, but still the model fitted to be a problem. For n given observations the adjusted coefficient of determination, tries to be more "realistic," that is, to reflect the fit of the proposed regression linear model in a more "accurate manner." It is defined as:

$$\bar{R}_p^2 = 1 - \frac{n}{n-p}\left(1-R_p^2\right) \tag{1}$$

If we let ρ to be the population multiple correlation coefficient, that is, the correlation between y and $\sum_{i=1}^{p} \beta_i X_i$ that is ρ^2 is eventually equals to:

$$\rho^2 = \frac{\text{Var}\left(\sum_{i=1}^{p} \beta_i X_i\right)}{\text{Var}(y)} \tag{2}$$

then $\bar{R}^2$ is unbiased to R^2 when $\rho^2 = 0$. Notice that $\sum_{i=1}^{p} \beta_i X_i$ is the linear combination of p input variables from $X_1, \ldots, X_p, \ldots X_k$ that has the maximal correlation with the response Y.

Mallow's C_p-statistic for the p-term model, Mallows [37] has been a very popular technique in the computation aspect of the Regression Analysis to evaluate the number of variables in the model, due to its simplicity and to clear picture it provides thanks to a simple graph. It is:

$$C_p = \frac{RSS_p}{\hat{\sigma}^2} - (n - 2p) \tag{3}$$

with RSS_p being the residual sum of squares for the corresponding p-term model and $\hat{\sigma}^2$ a suitable estimator of the error, namely:

$$\hat{\sigma}^2 = \frac{RSS_{k+1}}{n - (k+1)} \tag{4}$$

when the "full model" contains k variables plus the constant term.

We state and prove a linear relation between $\bar{R}_p^2$ and $\bar{R}_{p-1}^2$ through the C_p-statistic.

We suppose that there exist k independent variables and we create a p-term linear model

$$Y = \beta_0 + \beta_1 X_1 + \cdots + \beta_p X_p + e$$

with $1 \leq p \leq k+1$. The β_i, $i = 1, \ldots, p$ line parameters are estimated and $\bar{R}_p^2$ is evaluated. Then one variable is deleted and for the $p-1$ term model holds:

Proposition 2 *There is a linear relation between* $\bar{R}_p^2$ *and* $\bar{R}_{p-1}^2$ *of the form:*

$$\bar{R}_{p-1}^2 = \alpha + \beta \bar{R}_p^2 \qquad 1 \leq p \leq k-1 \tag{5}$$

Indeed, when the "full model" is fitted the $\bar{R}_{k+1}^2$ can be evaluated. Then for the p-term model the following relation holds:

$$C_p = \frac{1 - \bar{R}_p^2}{1 - \bar{R}_{k+1}^2}(n - p) + p - 1 \tag{6}$$

see Seber [45, pg 368].

The C_p-statistic is related to $C - p - 1$ linearly, Gorman and Toman [17] as:

$$C_{p-1} = \frac{F_i RSS_p}{\hat{\sigma}^2(n-p)} + C_p - 2 \tag{7}$$

with F_i is the corresponding of the F-statistic when the i-th variable is deleted from the p-term model.

Considering C_{p-1} and C_p as in (6) and substituting into (7) we obtain (5) that:

$$\alpha = 1 - \frac{F_i RSS_p \left(1 - \bar{R}^2_{k+1}\right)}{\hat{\sigma}^2(n-p)(n-p+1)} + \frac{1}{n-p+1}$$

$$\beta = \frac{n-p}{n-p+1} \tag{8}$$

Corollary. For values of n much larger than p, $n >> p$, it holds:

$$\bar{R}^2_{p-1} = \alpha^* + \bar{R}^2_p, 1 < p \leq k \tag{9}$$

Indeed, from Seber [45, pg. 369] it holds

$$C_p \simeq p + 1 - \bar{R}^2_p$$

Adopting the same procedure as in the proposition above, it can be proved that:

$$\alpha^* = 1 - \frac{F_i RSS_p}{\hat{\sigma}^2(n-p)}, \qquad \beta = 1.$$

Example 1 Consider the linear model

$$Y = \beta_0 + \beta_1 X_1 + \beta_2 X_2 + \beta_3 X_3 + \beta_4 X_4 + e$$

with X_1 : amount of $3CaOAl_2O_3$
X_2 : amount of $3CaOSO_2$
X_3 : amount of $4CaO.Al_2O.Fe_2O_3$
X_3 : amount of $2CaOSO_2$
Y : temperature in cal/gr of ciment

See Draper and Smith [11]. The full model is
$Y = 62.4 + 7.46X_1 + 48.15X_2 + 11.7X_3 + 29.9X_4$. Table 1 summarizes the calculations.

Notice that $\bar{R}^2_3 = 0.97644747$ when variables X_1, X_2, X_3 participate in the model with corresponding $RSS_3 = 47.972589$ and $F = 5.0258974$, while $s^2 = 5.982$ so we can evaluate from (7)

Table 1 Results from Example 1

Variables in model	$100R_p^2$	$100\bar{R}_p^2$
X_1	53.394	49.157
X_2	66.626	63.529
X_3	28.587	22.2209
X_4	67.454	64.495
X_1 X_2	97.867	97.441
X_1 X_3	54.816	45.780
X_1 X_4	97.247	96.695
X_2 X_3	84.702	81.643
X_2 X_4	68.006	65.097
X_3 X_4	93.528	92.234
X_2 X_3 X_4	98.228	97.637
X_1 X_2 X_4	98.233	97.644
X_1 X_3 X_4	98.128	95.504
X_2 X_3 X_4	97.281	96.738
X_1 X_2 X_3 X_4	98.237	97.356

$$\alpha = 0.081628 \quad \beta = 0.9$$

so from (4) when variables X_1, and X_4 participants to the model

$$\bar{R}_2^2 = 0.9669534$$

which is as in Table 1, so the procedure has been verified correctly.

The "best" linear model can be the one with the maximum $\bar{R}_p^2$. This is equivalent to the corresponding p-term subset of estimates $\hat{\beta}$, and provides the best linear model. This gives rise to the following algorithm:

A1: Check if $n >> p$ so choose either (4) or (8)
A2: Fit the linear model with all the associated input variables
A3: Subtract one variable—better choose the one that F-test provides evidence to be subtract.
A4: Do A3 up to one variable model
A5: Choose that model with p-variables which corresponds to

$$\max \left\{ \bar{R}_p^2, p = 1, 2, \ldots, k \right\}.$$

The simplicity of the above algorithm and the fact that provides a save in calculations provide some evidence that can be easily adopted in real-life problems there is a need of the appropriate package to perform the calculations. Both SPSS and Minitab are extensively used for the Regression Analysis calculations, while Minitab provides easy calculations for Experimental Design Theory, Kitsos [29], among others. For the D-optimal Design for a Copolymer Reactivity ratio

Estimation, Burke et al. [8] adopted the symbolic algebra package MAPLE. That is, for real-life problems there is a need of appropriate package to perform the involved statistical calculations.

3 Difficulties: Adopting HF Calculation

In the sequence we define the Hypergeometric functions and discuss their difficulties in Applied Statistics.

3.1 *Hypergeometric Functions (HF) in Statistics*

The Hypergeometric Functions (HF) play an important role in applications, since the time that Gauss in 1812 presented his pioneering paper on "Disquisitiones Generales Circa seriem infinitam $1 + \frac{\alpha \cdot \beta}{\gamma \cdot 1} + \cdots$," in Statistics too, Kitsos [31]. It is true that J.K. Plaff was the first who referred to hypergeometric series, and we shall adopt the term HF. Let us consider the function

$$h_{1/2}(\alpha; z) = \sum_{r=0}^{\infty} \frac{\Gamma\left(\frac{\alpha+r}{2}\right)}{\Gamma\left(\frac{\alpha}{2}\right)} \frac{z^r}{r!}, \quad z \in \mathbb{R}, \quad \alpha > 0 \tag{10}$$

Function (10) defines the non-central t distribution with n degrees of freedom (df) and non-centrality parameter $\tau \in \mathbb{R}$, Graybill [16] among others.

From (10) we can create:

$$h_1(\alpha; z) = \sum_{r=0}^{\infty} \frac{\Gamma(\alpha+r)}{\Gamma(\alpha)} \frac{z^r}{r!}, \quad z \geq 0, \quad \alpha > 0 \tag{11}$$

which is part of the non-central $X_n^2(\delta)$, $\delta > 0$, the non-centrality parameter.

The confluent Hypergeometric function is "extending" (11) and is defined as:

$$H_{1,1}(\alpha; \beta; z) = \sum_{r=0}^{\infty} \frac{\Gamma(\alpha+r)}{\Gamma(\alpha)} \frac{\Gamma(\beta)}{\Gamma(\beta+r)} \frac{z^r}{r!}, \quad z \geq 0, \quad \alpha, \beta > 0 \tag{12}$$

Based on $H_{1,1}(\cdot, \cdot)$ the non-central $F_{m,n}(\phi)$ is defined, with ϕ the non-centrality parameter. Eventually the Hypergeometric Function (HF) is defined as:

$$H_{2,1}(\alpha; \beta; \gamma; z) = \sum_{r=0}^{\infty} \frac{\Gamma(\alpha+r)}{\alpha} \frac{\Gamma(\beta+r)}{\Gamma(\beta)} \frac{\Gamma(\gamma)}{\Gamma(\gamma+r)} \frac{z^r}{r!}, \quad |z| \leq 1 \tag{13}$$

and can be applied to various sequential statistical tests: t, F, X^2, T^2, Kitsos [29], when invariant sequential Probability Ratio Test (SPRT) are considered, Ghosh [15].

The HF can be applied to Linear Optimal Design theory, Kitsos [31].

The following examples clarify the above discussion.

Example 2 Let X_i, $i = 1, 2, \ldots, n$ be independent random variables from the Normal distribution, that is, $X_i \sim N(\mu_i, \sigma_i^2)$, $i = 1, 2, \ldots, n$. We define $z_i = \frac{x_i - \mu_i}{\sigma_i} \sim N(0, 1)$. Let us define:

$$X_*^2 := \sum_{i=1}^{n} \left(\frac{X_i}{\sigma_i} \right)^2, \quad t_n^* := \frac{\bar{x}_n}{s_n / \sqrt{n-1}}$$

with:

$$\bar{x}_n = n^{-1} \sum_{i=1}^{n} x_i, \qquad s_n^2 = (n-1)^{-1} \sum_{i=1}^{n} (x_i - \bar{x}_n)^2$$

Then X_*^2 follows a non-central chi-square distribution with n df and non-centrality parameter δ, t_ν^* follows a non-central t distribution with $n-1$ df and non-centrality parameter τ, that is,

$$X_*^2 \sim X_N^2(\delta), \qquad \delta^2 = \sum_{i=1}^{n} \left(\frac{\mu_i}{\sigma_i} \right)^2 \geq 0$$

$$t_n^* \sim t_{n-1}(\tau), \qquad \tau^2 = n \frac{\mu^2}{\sigma^2} \in \mathbb{R}.$$

The corresponding probability density function (pdf) follow the scheme:
pdf of non-central = pdf of central * exp $\left\{ \frac{1}{2} \sqrt{\text{non} - \text{centralparameter}} \right\}$ * HF.

Namely they are:

$$\exp \left\{ -\tfrac{1}{2} \text{sqofnon} - \text{centralparameter} \right\}$$

$$f_{X_*^2}(w) = f_{X^2}(w) \cdot \exp\left(-\tfrac{\delta^2}{2}\right) h_1\left(\tfrac{n}{2}; \tfrac{\delta^2 w}{4}\right) \tag{14}$$

$$f_{t_n^*}(w) = f_t(w) \cdot \exp\left(-\tfrac{\tau^2}{2}\right) h_{1/2}\left(n+1; \tfrac{\tau w \sqrt{2}}{\sqrt{n^2 + w^2}}\right) \tag{15}$$

Example 3 Consider the ratio of a non-central $X_n^2(\delta)$ and a central X_m^2, namely

$$F_{n,m}^* := \frac{X_n^2(\delta^2)/n}{X_m^2/m} \sim F_{n,m}(\delta^2)$$

that is, $F^*_{n,m}$ follows a non-central F with pdf:

$$f_{F^*_{n,m}}(w) = f_F(w)\exp\left\{-\frac{\delta^2}{2}\right\} H_{1,1}\left(\frac{m+n}{2}, \frac{m}{2}; \frac{\delta^2 mw}{2(m+n)}\right), \ \delta^2 \geq 0, \ \delta \in \mathbb{R} \tag{16}$$

Notice that $f_{X^2}(w)$, $f_t(w)$, $f_F(w)$ are the pdf of central X^2, t, F distributions.

Now, let us consider another optical angle of the computational problem applying non-central distributions. As non-central distributions are based on HF the computational effort is clear.

Suppose we have paired observations $z = (x_i, y_i)$, $i = 1, 2, \ldots, n$ coming from the Bivariate Normal distribution (BND).

Then for $n > 2$ the pdf $f_n(r; \rho)$ of the estimate of $r = r_n$, at stage n of the sequential procedure for evaluation of the correlation coefficient $\rho \in (-1, 1)$ of the (BND), is a function of the HF $H_{2,1}$, Kitsos [31], Anderson [2]:

$$f_n(r;\rho) = \gamma_n \alpha_{rr}^p \alpha_{\rho\rho}^q \alpha_{\rho r}^s H_{2,1}\left(\frac{1}{2}, \frac{1}{2}, n - \frac{1}{2}; \frac{1+r\rho}{2}\right) \tag{17}$$

with:

$$\gamma_n = \frac{(n-2)}{\sqrt{2n}} \frac{\Gamma(n-1)}{\Gamma\left(n-\frac{1}{2}\right)}, \quad \alpha_{\kappa\lambda} = 1 - \kappa\lambda, \quad p = \frac{n-4}{2}, q = \frac{n-1}{2}, s = -n + \frac{3}{2}$$

See also in Appendix 2, how HF can be evaluated.

Notice that $-1 < r, \rho < 1$ so $H_{2,1}(\cdot;\cdot)$ exists. Moreover the SPRT, Ψ_n say, for testing $H_0 : \rho = \rho_0$ vs $H_1 : \rho = \rho_1$, $-1 < \rho_0 < \rho_1 < 1$ is also a function of HF, Ghosh [15], Kitsos [31].

$$\Psi_n = \frac{n-1}{2} \ln \frac{\alpha_{\rho_1\rho_1}}{\alpha_{\rho_0\rho_0}} - \left(n - \frac{3}{2}\right) \ln \frac{\alpha_{\rho_1 r_n}}{\alpha_{\rho_0 r_n}} + \ln \frac{A(\rho_1, r_n)}{A(\rho_0, r_n)}, \tag{18}$$

with

$$A(\rho_i, r_n) = H_{2,1}\left(\frac{1}{2}, \frac{1}{2}, n - \frac{1}{2}; \frac{1+\rho_i r_n}{2}\right), \quad i = 0, 1, \quad n > 2$$

and $\alpha_{\kappa\lambda}$ as above.

The above discussion points out the difficulty in computations based on HF. In Kitsos [31] a number of calculations were proposed. The approximation in a reasonable accepted number of terms can be a solution, depending on the problem under investigation.

3.2 Transformations Can Reduce Calculations

The non-central F distribution appears a computational difficulty therefore an introduced transformation is reducing the computational effort.

The non-central $F_{n,m}(q)$ distribution can be approximated by the typical F distribution, Anderson [2], Patnaik [40], when Multiple—Multivariate—Sequential T^2—comparisons arise, Kitsos [29].

Indeed: Let $X_1, X_2, \ldots, X_n$ be independent identical distributed (i.i.d) observations from the k-variate Normal distribution $N_k(\mu, \Sigma)$ with both the mean μ, and covariance Σ unknown and $\det(\Sigma) > 0$. An invariant SPRT for testing

$$H_0 : \mu \Sigma \mu^T \leq \lambda_0 \text{vs} H_1 : \mu \Sigma \mu^T \geq \lambda_1$$

can be constructed from the statistic:

$$V_n = \bar{x}(n) S^{-1}(n) \bar{x}^T(n) \tag{19}$$

with $\bar{X}_k(n)$ the sample mean at stage n and $S^{-1}(n)$ the sample variance at stage n. Then V_n follows a non-central F, Kitsos [29] as:

$$F^* = \frac{n-k}{k} V_n \sim F_{k,n-k}(n\xi), \quad \xi = \mu \Sigma^{-1} \mu^T \tag{20}$$

with non-centrality parameter $\phi = n\xi$. But due to Patnaik [40], it is easier for the computational burden to work, with F distribution as:

$$\tilde{F} = \frac{n-k}{k+n\xi} V_n \sim F_{m,n-k}, \qquad m = \frac{(k+n\xi)^2}{k+2n\xi} \tag{21}$$

That is the transformed F is now a central one but the transformation influence the df of the distribution F as in (21), see the value m. Based on this result Kitsos (1994) proceeds and defines the distribution of the sequential likelihood function based on $H_{1,1}$ function.

Consider the General Linear Model (GLM).

As far as the application on non-central chi-square recall (2) for the GLM in Sect. 2 that

$$\rho^2 = \frac{\text{Var}(\sum_{i=1}^{p} \beta_i x_i)}{\text{Var}(y)}$$

Then, Helland [24] considered how useful the non-central chi-square distribution is when estimating the ratio $R^2/(1-R^2)$, defined as OR^2 in this chapter:

$$OR^2 := \frac{R^2}{1-R^2} = \frac{SSR}{SSE} = \frac{X_p^2(\delta)}{X_{n-p-1}^2} \tag{22}$$

with the non-centrality parameter δ:

$$\delta = \frac{\rho^2}{1-\rho^2} X_{n-1}^2$$

Eventually it can be proven that:

$$OR^2 = \frac{(n-1)k+p}{n-p-1} F_{\nu,n-p-1} \tag{23}$$

with ν appropriately and complicatedly defined, Helland [24]. See that the transformation again influences the df as above with m in (21). Moreover in (23) the approximation has been proved as an accurate one from the computational point of view, which started from a non-central X^2 in (22). That is the evaluation in (23) avoids eventually the non-central distribution through a theoretical inside, Helland [24]. This certainly is a real improvement, due to the computations needed to evaluate a non-central F-distribution, as to evaluate a non-central distribution it is not an easy task, see Appendix 1, where we elaborate the example.

4 Discussion

To proceed with the HF, Ledenev [36] in statistical problem is really very difficult. Most of the researchers need the final result and not the theoretical insight. The computational difficulties are not only computational. It can be mathematically tedious to proceed with Hyperbolic Functions (HF) which are involved in non-central distributions, Patnaik [40], as discussed already. That is, the calculations in non-central distributions is rather complicated and Appendix II might be useful.

Discussing the Industrial Statistics, Baines [6] devoted two sections on statistical computations and statistical packages. Now it is more easier to proceed on computations, not only in Statistics. The computation is more important when particular problems are discussed, Lawrence [35], Nelson [39], Mallows [37, 38].

It is true that “the ancient Greeks mathematicians have believed there is little—if anything—as unequivocal as a proved theorem”, Chaitin [10].

The Archimedean and Euclidean line of thought, we can say that, has been transformed to Gödel and Turing line of thought. But still you need calculations! You need numerical results despite an elegant Mathematical proof, Chaitin [9].

In this chapter we tried to discuss that the calculation problem, Sect. 2, is equally important as the evaluation problem, through a complicated mathematical form, Sect. 3.

The appropriate package support provides an intrinsic effort to the procedure we adopt to evaluate and the final solution of the problem under consideration.

Appendix 1: On Non-central Chi-Square

Lemma *Let the random variable* $z' = (z_1, \ldots, z_k) \sim N(\mu, \mathrm{Cov}(z))$ *where the form of* $\Sigma = Cov(z)$ *is:*

$$\Sigma = \mathrm{Cov}(z) = \begin{pmatrix} 1-\delta_1^2 & -\delta_1\delta_2 & \cdots & -\delta_1\delta_k \\ -\delta_1\delta_2 & 1-\delta_2^2 & & \\ \vdots & & \ddots & \\ -\delta_1\delta_k & & & 1-\delta_k^2 \end{pmatrix} = (\delta_{ij}) \tag{24}$$

$$\mu = (\mu_1, \ldots, \mu_k)', \qquad \sum_{j=1}^{k=1} \delta_j^2$$

with $\delta_{ij} = -\delta_i\delta_j$, $i \neq j = 1, 2, \ldots, n$, $\delta_{ii} = 1 - \delta_{ii}^2$, $i = 1, 2, \ldots, n$.

We want to prove that if:

$$\rho = \sum_{j=1}^{k} \mu_j^2 - \sum_{j=1}^{k} \delta_j, \qquad Q = z_1^2 + \ldots + z_k^2,$$

Then: $Q \sim X_{k-1}^2(\rho)$

Proof We notice that:

$$\Sigma = \mathrm{Cov}(z) = \begin{pmatrix} 1 & & 0 \\ & \ddots & \\ 0 & & 1 \end{pmatrix} - \begin{pmatrix} \delta_1^2 & \delta_1\delta_2 & \cdots & \delta_1\delta_k \\ \delta_1\delta_2 & \delta_2^2 & & \\ \vdots & & \ddots & \\ \delta_1\delta_k & & & \delta_k^2 \end{pmatrix} = I - \Delta \tag{25}$$

We are looking for $rank(\Sigma)$. Usually matrices of the form are independent. Let us check that:

$$(I - \Delta)^2 = (I - \Delta)(I - \Delta) = I - 2\Delta + \Delta^2$$

if: $\delta' = (\delta_1, \ldots, \delta_k)$ then $\Delta = \delta\delta' \Rightarrow \Delta^2 = (\delta\delta')(\delta\delta') = \delta \underbrace{\delta'\delta}_{\substack{Linear \\ Product}} \delta' =$

$\delta \left(\sum \delta_i^2\right) \delta' = \delta\delta' = \Delta$. Hence from (25)

$$(I - \Delta)^2 = I - 2\Delta + \Delta^2 = I - \Delta \qquad \text{i.e., independent}$$

Then: $rank(\Sigma) = tr(\Sigma) = tr(I - \Delta) = trI - tr\Delta = k - \sum_{i=1}^{k} \delta_i^2 = k - 1$ that is, the matrix Σ is not full of rank. □

We define:

$$Y_0 = \sum_{i=1}^{k} \delta_i y_i \qquad \text{i.e.} \qquad Y_0 = \delta' Y$$

Then, as $\delta\delta' = \Delta$ and $\delta'\delta = 1$, thus

$$Y_0^2 = (\delta' Y)'(\delta' Y) = Y'\delta\delta' Y = Y'\Delta Y$$

then if we let $U = Y'AY \sim X_p^2(\rho)$, $\rho = \mu' A \mu$.

Now due to ([16] Th. 4.43) $A\Sigma$ is independent and $rank(A\Sigma) = rank(I - \Delta) = k - 1$. So $U = Y'(I - \Delta)Y \sim X_{k-1}^2(\rho)$ with:

$$\rho = (\mu_1, \dots, \mu_k)(I - \Delta)\begin{pmatrix}\mu_1 \\ \vdots \\ \mu_k\end{pmatrix} = (\mu_1, \dots, \mu_k)I\begin{pmatrix}\mu_1 \\ \vdots \\ \mu_k\end{pmatrix} - (\mu_1, \dots, \mu_k)\Delta\begin{pmatrix}\mu_1 \\ \vdots \\ \mu_k\end{pmatrix}$$

$$= \sum_{i=1}^{k} \mu_i^2 - \left(\sum_{i=1}^{k} \delta_i \mu_i\right)^2$$

We have to prove that U and Y_0 are independent.

Recall, ([16] Th. 4.52). If $Y \sim N(\underset{\sim}{\mu}, \Sigma)$with$rank(\Sigma) = n$, then:

If $B\Sigma A = 0$ the quadratic form, $U = Y'AY$ is independent of the linear form BY

In this case: $Y_0 = \sum_{i=1}^{k} \delta_i y_i = \delta' Y$, that is, $B = \delta'$ and $\Sigma = I - \Delta$, $A = I - \Delta$

Hence:

$$B\Sigma A = \delta'(I - \Delta)(I - \Delta) = \delta'(I - \Delta) = \delta' - \delta'\Delta = \delta' - \delta' \underbrace{\delta\delta'}_{\sum \delta_i^2 = 1} = \delta' - \delta' = 0$$

So Y_0 is independent of $U = Y_1^2 + \dots + Y_k^2 - Y_0^2$ and as $Y_i \sim N(\mu_i, 1)$ and it holds:

$$Y_0 = \sum_{j=1}^{k} \delta_j Y_j \sim N\left(\sum_{j=1}^{k} \delta_j \mu_j, \sum_{j=1}^{k} \delta_j^2\right) = N(\mu_0, 1) \tag{26}$$

Let us denote:

$$\underset{\sim}{Y}^{(i)} = (Y_i, Y_0) \sim N\left((\mu_i, \mu_0), \Sigma^{(i)}\right)$$

where:

$$\Sigma^{(i)} = \begin{pmatrix} \operatorname{Var} Y_i & \operatorname{Cov}(Y_i, Y_0) \\ \operatorname{Cov}(Y_i, Y_0) & \operatorname{Var} Y_0 \end{pmatrix} = \begin{pmatrix} 1 & \operatorname{Cov}(Y_i, Y_0) \\ \operatorname{Cov}(Y_i, Y_0) & 1 \end{pmatrix}$$

We evaluate:

$$\operatorname{Cov}(Y^{(i)}) = \operatorname{Cov}(Y_i, Y_0) = \operatorname{Cov}\left(Y_i, \sum_{j=1}^{k} \operatorname{Cov}(Y_i, \delta_j Y_j)\right) = \sum_{j=1}^{k} \delta_j \operatorname{Cov}(Y_i, Y_j)$$
$$= 0 + 0 + \cdots + 0 + \delta_i \operatorname{Cov}(Y_i Y_j) + 0 + \cdots + 0 = \delta_i$$

Hence:

$$\Sigma^{(i)} = \begin{pmatrix} 1 & \delta_i \\ \delta_i & \sum \delta_i^2 \end{pmatrix} = \begin{pmatrix} 1 & \delta_i \\ \delta_i & 1 \end{pmatrix}$$

We evaluate:

$$E(Y_i|Y_0) = \int_{-\alpha}^{\alpha} y_i f(y_i|y_0) dy_i = \int_{-\alpha}^{\alpha} y_i \frac{f(y_i, y_0)}{f_{Y_0}(y_0)} dy_i$$

As Y_i, Y_0 are normal it holds:

$$f(y_i, y_0) = \frac{1}{2\pi\sqrt{1-\delta_i^2}} \exp\left\{-\frac{1}{2(1-\delta_i^2)}\left[(y_i - \mu_i)^2 \right.\right.$$
$$\left.\left. -2\delta_i (y_i - \mu_i)(y_0 - \mu_0) + (y_0 - \mu_0)^2\right]\right\}$$

and:

$$f_{Y_0}(y_0) = \frac{1}{\sqrt{2\pi}} \exp\left\{-(y_0 - \mu_0)^2\right\}$$

Taking the condition on $Y_0 = \mu_0$ we calculate:

$$\frac{f(y_i, y_0)}{f_{Y_0}(y_0)} = \frac{\frac{1}{2\pi\sqrt{1-\delta_i^2}} e^{\frac{1}{2(1-\delta_i^2)}\left[(y_i-\mu_i)^2 - 2\delta_i(y_i-\mu_i)(y_0-\mu_0)+(y_0-\mu_0)^2\right]}}{\frac{1}{\sqrt{2\pi}} e^{-(y_0-\mu_0)^2}}$$

$$= \frac{1}{2\pi\sqrt{1-\delta_i^2}} \exp\left\{ \frac{1}{2(1-\delta_i^2)} \left[(y_i - \mu_i)^2 - 2\delta_i(y_i - \mu_i)(y_0 - \mu_0) + (y_0 - \mu_0)^2 \right] \right\}$$

Hence:

$$E(Y_i|Y_0 = \mu_0) = \int_{-\infty}^{\infty} y_i \frac{1}{\sqrt{2\pi}\sqrt{1-\delta_i^2}} \exp\left\{ -\frac{1}{2(1-\delta_i^2)} (y_i - \mu_i)^2 \right\} dy_i$$

$$\Rightarrow E(Y_i) = \mu_i = E(Z_i)$$

For evaluating the $\mathrm{Cov}(Y_i, Y_j|Y_0)$ we evaluate the distribution of $(Y_i, Y_j|Y_0)$.

Notice that $(Y_i, Y_j, Y_0)'$ is multivariate normal, with mean vector $(\mu_i, \mu_j, \mu_0)'$ and covariance matrix:

$$\Sigma_{(i,j)} = \begin{pmatrix} \mathrm{Var}(Y_i) & \mathrm{Cov}(Y_i, Y_j) & \mathrm{Cov}(Y_i, Y_0) \\ \mathrm{Cov}(Y_j, Y_i) & \mathrm{Var}(Y_j) & \mathrm{Cov}(Y_j, Y_0) \\ \mathrm{Cov}(Y_i, Y_0) & \mathrm{Cov}(Y_j, Y_0) & \mathrm{Var}(Y_0) \end{pmatrix} = \begin{pmatrix} 1 & 0 & \delta_i \\ 0 & 1 & \delta_j \\ 0 & 0 & 1 \end{pmatrix}$$

Thus:

$$\Sigma_{(i,j)}^{-1} = \frac{1}{1-\delta_i^2-\delta_j^2} \begin{pmatrix} 1-\delta_j^2 & -\delta_i\delta_j & \delta_i \\ -\delta_i\delta_j & 1-\delta_i^2 & \delta_j \\ \delta_i & -\delta_j & 1 \end{pmatrix}$$

And hence:

$$f(y_i, y_j, y_0) = \frac{1}{(2\pi)^{3/2}(1-\delta_i^2-\delta_j^2)^{1/2}} \times \exp\left\{ \underbrace{-\frac{1}{2}(y_i - \mu_i, y_j - \mu_j, y_0 - \mu_0)\Sigma_{(i,j)}(y_i - \mu_i, y_j - \mu_j, y_0 - \mu_0)}_{T} \right\}$$

Conditioning again on $y_0 - \mu_0$ we evaluate

$$T = \frac{1}{1-\delta_i^2-\delta_j^2}(y_i-\mu_i, y_j-\mu_j, 0)\begin{pmatrix}1-\delta_i^2 & -\delta_i\delta_j & \delta_i\\ -\delta_i\delta_j & 1-\delta_i^2 & -\delta_j\\ \delta_i & -\delta_j & 1\end{pmatrix}\begin{pmatrix}y_i-\mu_i\\ y_j-\mu_j\\ 0\end{pmatrix}$$

$$= \frac{1}{1-\delta_i^2-\delta_j^2}(y_i-\mu_i, y_j-\mu_j)\begin{pmatrix}1-\delta_i^2 & -\delta_i\delta_j\\ -\delta_i\delta_j & 1-\delta_i^2\end{pmatrix}\begin{pmatrix}y_i-\mu_i\\ y_j-\mu_j\end{pmatrix}$$

$$= (y_i-\mu_i, y_j-\mu_j)\Lambda\begin{pmatrix}y_i-\mu_i\\ y_j-\mu_j\end{pmatrix}, \text{withthedefinitionof}\Lambda\text{obvious.}$$

Thus $f_Y(y_0=\mu_0) = \frac{1}{\sqrt{2\pi}}e^{-(y_0-\mu_0)^2} = \frac{1}{\sqrt{2\pi}}$. Hence:

$$\frac{f(y_i, y_j, y_0=\mu_0)}{f_y(y_0=\mu_0)} = \frac{\frac{1}{(2\pi)^{1/2}(1-\delta_i^2-\delta_j^2)^{1/2}}\exp\left\{-\frac{1}{2}(y_i-\mu_i, y_j-\mu_j)\Lambda(y_i-\mu_i, y_j-\mu_j)\right\}}{\sqrt{2\pi}}$$

$$= \frac{1}{2\pi(1-\delta_i^2-\delta_j^2)}\exp\left\{-\frac{1}{2}(y_i-\mu_i, y_j-\mu_j)\Lambda(y_i-\mu_i, y_j-\mu_j)\right\} \tag{27}$$

If we find matrix Σ^* such the $\Sigma^* = \Lambda^{-1}$ it is known from the multivariate statistics that the (27) will represent multivariate normal.

Indeed:

$$\Sigma^* = \Lambda^{-1} = \begin{pmatrix}1-\delta_j^2 & -\delta_i\delta_j\\ -\delta_i\delta_j & 1-\delta_i^2\end{pmatrix}$$

Hence:

$$(Y_i, Y_j|Y_0=\mu_0) \sim N((u_i, \mu_j), \Sigma^*)$$

That is:

$$\text{Cov}(Y_i, Y_j|Y_0) = \text{Cov}(z_i, z_j)$$

Now consider a normal random vector $(Z_1, \ldots, Z_k)$ with respective expectations $(\mu_1, \ldots, \mu_k)$ and the following covariances:

$$\text{Var}(Z_j) = 1-\tau_j^2, \quad j=1,\ldots,k, \quad \text{Cov}(Z_j, Z_g) = -\tau_j\tau_g, \quad 1 \le j \ne g \le k \tag{28}$$

where $\tau_1, \ldots, \tau_k$ are nonnegative numbers such that

$$\sum_{j=1}^{k} \tau_j^2 = 1 \tag{29}$$

Theorem *Under the above assumptions the statistic:*

$$Q = Z_1^2 + \cdots + Z_k^2 \tag{30}$$

Has a non-central X^2-distribution with $k - 1$ degrees of freedom and non-centrality parameter

$$\delta = \sum_{j=1}^{k} \mu_j^2 - \left(\sum_{j=1}^{k} \mu_j \tau_j \right)^2 \tag{31}$$

Proof Let $Y_1, \ldots, Y_k$ be independent normal random variables with respective expectations $\mu_1, \ldots, \mu_k$ and variances 1. Put:

$$Y_0 = \sum_{j=1}^{k} \tau_j Y_j \tag{32}$$

Then in view of (29),

$$Y_1^2 + \cdots + Y_k^2 - Y_0^2 \tag{33}$$

has a non-central X^2-distribution with $k - 1$ degrees of freedom and non-centrality parameter (31), and is independent of Y_0. Consequently, (33) has the same distribution conditionally for $Y_0 = 0$. However for $Y_0 = 0$ the conditional expectations and covariances of $Y_1, \ldots, Y_k$ coincide with those assumed concerning $Z_1, \ldots, Z_k$. □

Appendix 2: Numerical Evaluation of Hypergeometric Function

Statistical and mathematical software available for computers are nowadays the most convenient way to evaluate the Gauss Hypergeometric Function. There are more than one software, that is, a user can explore (HF) among which Mathematica (command `Hypergeometric2F1[a, b, c, z]`; for both inside and outside of the unit circle), Maple (command `hypergeom([n1, n2, ...], [d1, d2, ...], z)`; computes the generalized hypergeometric function), Maxima (command `hypergeometric([a1, ..., ap], [b1, ..., bq], x)`; the function supports evaluation outside the unit circle), Sage (command `hypergeometric([], [], x)`; the function implements manipulation of infinite hypergeometric series), and also R.

In the latter, there are three numerical implementations for (HF) included in packages gsl [21] (command `hyperg_2F1(a, b, c, x, give=FALSE, strict=TRUE)`; does not cover complex values), appell [7] (command `hyp2f1(a, b, c, z, algorithm = c("michel.stoitsov", "forrey"))`; fast computation with all the parameters complex) and hypergeo [23]. The third package, hypergeo (command `hypergeo(A, B, C, z, tol = 0, maxiter=2000)`) is offered as an R-centric suite of functionality with emphasis on multiple evaluation methodologies, and transparent coding with nomenclature and structure that of Abramowitz and Stegun [1]. The package implements a generalization of the method of Forrey [13] to the complex case. It utilizes the observation that the ratio of successive terms approaches z, and thus the strategy adopted is to seek a transformation which reduces the modulus of z to a minimum [22].

Recall relation (17). Suppose we are at the fourth iteration to evaluate sequentially $\rho = 0.475$. In this case $n = 5$ and we let $r_4 = r = 0.41$ (after loading the appropriate backage "hypergeo"). Then:

```
> install.packages("hypergeo", repos = "https://cloud.r-project.org/")
> library (hypergeo)
> A0 <– 1/2; B0 <– 1/2
> C0 <– 5 – 1/2; z0 <– (1 + 0.41 * 0.475) / 2
> hypergeo(A = A0, B = B0, C = C0, z = z0)
[1] 1.038289+0i
```

Moving $n + 1 = 6$ six observations we proceed with $n = 6$, $r = 0.412$, and $\rho = 0.475$:

```
> C0 <– 6 – 1/2; z0 <– (1 + 0.412 * 0.475) / 2
> hypergeo(A = A0, B = B0, C = C0, z = z0)
[1] 1.030591+0i
```

Moving on to $n = 7$, $r = 0.48$, and $p = 0.475$:

```
> C0 <– 7 – 1/2; z0 <– (1 + 0.48 * 0.475) / 2
> hypergeo(A = A0, B = B0, C = C0, z = z0)
[1] 1.026204+0i
```

So the evaluate HF through (17) functions are computed in [1], as above.

It should be clarified, however, that although more than enough options are available, the evaluation of the HF is hard, as evidenced by the extensive literature concerning its numerical evaluation [22].

References

1. Abramowitz M. and Stegun, A. I. (1965). Handbook of Mathematical Functions. Dover, New York.
2. Anderson, J.W. (2003) An introduction to Multivariate Statistical Analysis. Wiley, New York.

3. Arkfen, G. and Weber, H. (2000) Mathematical Methods for Physicists. Academic Press.
4. Arnold, V. I. (2014). Mathematical understanding of nature: essays on amazing physical phenomena and their understanding by mathematicians (Vol. 85). American Mathematical Soc.
5. Baharev, A., Schichl, H., & Rv, E. (2017). Computing the noncentral-F distribution and the power of the F-test with guaranteed accuracy. Computational Statistics, 32(2), 763–779.
6. Baines, A. (1984). Present Position and Potential Developments: Some personal views. Industrial Statistics and Operational Research. J.R. Statist. Soc. A., 147, 171–177.
7. Bove, S. D. et al. (2013). appell: Compute Appells F1 Hypergeometric Function, 2013. URL https://CRAN.Rproject.org/package=appell. R package version 0.0-4
8. Burke, A.L., Duever, T.A., Penlidis, A. (1993). Revisiting the Design of Experiments for Copolymer Reactivity Ratio Estimation. J. of Polymer Science, Part A: Polymer Chemistry, 31, 3065–3072.
9. Chaitin, J.G. (1975). Randomness and Mathematical Proof. Scientific American, 232(5), 47–52.
10. Chaitin, J.G. (1988). Randomness in Arithmetic. Scientific American, 259(1), 80–85.
11. Draper, N. and Smith, H. (1981). Applied Regression Analysis, New York.
12. Edwards, J.B. (1969). The relation between F-Test and $\bar{R}^2$. The American Statistician, 23, 28.
13. Forrey, R. C. (1997). Computing the hypergeometric function. Journal of Computational Physics, 137:79100.
14. Fraser, D. A. S., Wu, J., & Wong, A. C. M. (1998). An approximation for the noncentral chi-squared distribution. Communications in Statistics-Simulation and Computation, 27(2), 275–287.
15. Ghosh, B.K. (1970). Sequential Tests of Statistical Hypotheses. Addison - Wesley Pub.Co.
16. Graybill, A.F. (1976). Theory and Application of the Linear Model. Duxbury Press. Massachusetts.
17. Gorman, J.W. and Toman, R.J. (1966). Selection of variables for Fitting Equations to Data. Technometric, 8, 27–51.
18. Goutschi, W. (2002). Gauss quadrature approximates to hypergeometric and coefficient hypergeometric functions. J. of Comp. and Appl. Math., 139, 173–187.
19. Hack, J. (1957). Studies on longitudinal stream profiles in Virginia and Maryland. Vol. 294. US Government Printing Office.
20. Haitovsky, Y. (1969). A Note on the Maximization of $\bar{R}^2$. The American Statistician, 23, 20–21.
21. Hankin, R. K. (2006) Special functions in R: Introducing the gsl package. R News.
22. Hankin, R. K. (2015). Numerical Evaluation of the Gauss Hypergeometric Function with the hypergeo Package. R J., 7(2), 81.
23. Hankin, R. K. (2016). hypergeo: The Gauss Hypergeometric Function. R package version 1.2–13. https://CRAN.R-project.org/package=hypergeo
24. Helland, I.S. (1987). On the Interpretation and Use of R^2 in Regression Analysis. Biometrics, 43, 61–69.
25. Hocking, R.R. (1976). The Analysis and Selection of Variables in Linear Regression. Biometrics, 43, 61–69.
26. Johnson, N. L., & Welch, B. L. (1940). Applications of the non-central t-distribution. Biometrika, 31(3/4), 362–389.
27. Kennard, R.W. (1971). A Note on the C_p Statistic. Technometrics, Nov 1;13(4):899–900.
28. Kennedy, W.J. and Bancroft, T.A. (1971). Model building for prediction in regression based upon repeated significance tests. Ann. Math. Stat, 42, 1273–1284.
29. Kitsos, C.P. (1994a). Multiple - Multivariate - Sequential T^2 - comparisons. In: T. Calinski and R. Kala (eds) Proceedings of the International Conference on Linear Statistical Inference LINSTAT' 93, 47–51. Kluwer Ac-Pub.
30. Kitsos, C.P. (1994b) Statistical Analysis of Experiment Designs. New Technologies Pub. Co. (in Greek).
31. Kitsos, C.P. (2010). Adopting Hypergeometric Functions for Sequential Statistical Methods. Bulletin of the Greek Math. Soc., 57, 251 - 264.

32. Kitsos, C.P., Tavoularis, K.N. (2007). Logarithm Sobolev Inequalities and the Information Theory. Technical Report, 104-07, now Univ. of West Attica.
33. Kitsos, C.P., Tavoularis, K.N. (2009). Logarithm Sobolev Inequalities for Information Measures. IEEE Transactions on Information Theory, 55(6), 2554–2561.
34. La Motte, L.R. (1972). The SELECT routines: a program for identifying best subset regression. Appl. Stat, 21, 92–93.
35. Lawrence, A.J. (1979). Partial and Multiple Correlation for Time Series. The American Statistician, 33, 127–130.
36. Ledenev, N.N. (1956). Special functions and their Applications. Prentice-Hall.
37. Mallows, C.L. (1964). Choosing variables in a linear regression: A graphical aid. Presented at the Central Regional Meeting of the Institute of Mathematical Statistics, Manhatta, Kanas, May 7–9.
38. Mallows, C. L. (2000). Some comments on Cp. Technometrics, 42(1), 87–94.
39. Nelson, C.R. (1976). The Interpretation of R^2 in Autoregressive-Moving Average Time Series Models. The American Statistician, 30, 175–180.
40. Patnaik, P.B. (1949) The non-central X^2 and F-distributions and their application. Biometrika, 37, 78 - 87.
41. Rassias, Th. M., and Šimša, J. (1995). Finite sums decompositions in mathematical analysis (Vol. 25). John Wiley & Sons Inc.
42. Rigon, R., RodriguezIturbe, I., Maritan, A., Giacometti, A., Tarboton, D. G., & Rinaldo, A. (1996). On Hack's law. Water Resources Research, 32(11), 3367–3374.
43. Ronchetti, E., & Staudte, R. G. (1994). A robust version of Mallows's Cp. Journal of the American Statistical Association, 89, 550–559.
44. Schervish, J.M. (1995). Theory of Statistics. Springer.
45. Seber, G.A.F. (1977). Linear Regression Analysis. Wiley, New York.

Fractional Integral Operators in Linear Spaces

Jichang Kuang

Abstract In this chapter, we introduce some new fractional integral operators and fractional area balance operators in n-dimensional linear spaces. The corresponding integral operator inequalities are established. They are significant improvement and generalizations of many known and new classes of fractional integral operators.

Mathematics Subject Classification 26A33, 26D10, 26A51

1 Introduction

It is well-known that fractional integral operator is one of the important operators in harmonic analysis with background of partial differential equations. In fact, the solution of the Laplace equation $\Delta g = f$ for good functions on $\mathbb{R}^n$ can be represented by using the fractional integral operators acting on f. Recently, different versions of fractional integral operators have been developed which are useful in the study of different classes of differential and integral equations. These fractional integral operators act as ready tools to study the classes of differential and integral equations. Hence, fractional integral inequalities are very important in the theory and applications of differential equations. Such inequalities are also of great importance in the mathematical modeling of the fractional boundary value problems. First, we recall the following definitions and some related results.

Definition 1 (cf. [1, 2, 6, 7]) Let $f \in L[a, b]$, then Riemann–Liouville fractional integrals of f of order $\alpha > 0$ with $a \geq 0$ are defined by

$$T_1(f, x) = \frac{1}{\Gamma(\alpha)} \int_a^x (x - t)^{\alpha-1} f(t)dt \ x > a, \tag{1}$$

J. Kuang (✉)
Department of Mathematics, Hunan Normal University, Changsha, Hunan, P.R. China

I. N. Parasidis et al. (eds.), *Mathematical Analysis in Interdisciplinary Research*, Springer Optimization and Its Applications 179,
https://doi.org/10.1007/978-3-030-84721-0_22

and

$$T_2(f,x) = \frac{1}{\Gamma(\alpha)} \int_x^b (t-x)^{\alpha-1} f(t)dt, \ \ x < b, \tag{2}$$

respectively, where

$$\Gamma(\alpha) = \int_0^\infty t^{\alpha-1} e^{-t} dt \tag{3}$$

is the Gamma function and when $\alpha = 0$, $T_1(f,x) = T_2(f,x) = f(x)$.

Definition 2 (cf. [3]) Let $f \in L[a,b]$, then Riemann–Liouville k-fractional integrals of f of order $\alpha > 0$ with $a \geq 0$ are defined by

$$T_3(f,x) = \frac{1}{k\Gamma_k(\alpha)} \int_a^x (x-t)^{(\alpha/k)-1} f(t)dt, , \ \ x > a, \tag{4}$$

and

$$T_4(f,x) = \frac{1}{k\Gamma_k(\alpha)} \int_x^b (t-x)^{(\alpha/k)-1} f(t)dt, \ \ x < b, \tag{5}$$

respectively, where

$$\Gamma_k(\alpha) = \int_0^\infty t^{\alpha-1} e^{-(t^k/k)} dt, \ \ \alpha > 0, \tag{6}$$

is the k-Gamma function. Also, $\Gamma(x) = \lim_{k\to 1} \Gamma_k(x)$, $\Gamma_k(\alpha) = k^{(\alpha/k)-1}\Gamma(\alpha/k)$ and $\Gamma_k(\alpha + k) = \alpha\Gamma_k(\alpha)$.

It is well known that the Mellin transform of the exponential function $\exp -t^k/k$ is the k-Gamma function.

Definition 3 (cf. [4, 5]) Let $f \in L^{1,r}[a,b]$, $a \geq 0$, then the generalized Riemann–Liouville fractional integral of f of order (α, r) is defined by

$$T_5(f,x) = \frac{(r+1)^{1-\alpha}}{\Gamma(\alpha)} \int_a^x (x^{r+1} - t^{r+1})^{\alpha-1} t^r f(t)dt, \ \ x > a, \tag{7}$$

$$T_6(f,x) = \frac{(r+1)^{1-\alpha}}{\Gamma(\alpha)} \int_x^b (t^{r+1} - x^{r+1})^{\alpha-1} t^r f(t)dt, \ \ x < b, \tag{8}$$

and

$$T_7(f,x) = \frac{(r+1)^{1-(\alpha/k)}}{k\Gamma_k(\alpha)} \int_a^x (x^{r+1} - t^{r+1})^{(\alpha/k)-1} t^r f(t)dt, \ \ x > a, \tag{9}$$

$$T_8(f,x) = \frac{(r+1)^{1-(\alpha/k)}}{k\Gamma_k(\alpha)} \int_x^b (t^{r+1} - x^{r+1})^{(\alpha/k)-1} t^r f(t)dt, \ x < b, \tag{10}$$

respectively, where $k, \alpha > 0, r \geq 0, x \in [a, b]$.

In particular, if $r = 0$, then definition 3 reduces to definitions 1 and 2.

Definition 4 (cf. [11, 12]) Let f be a conformable integrable function on $[a, b] \subset [0, \infty)$.The right-sided and left-sided generalized conformable fractional integrals T_9 and T_{10} of f of order $\alpha > 0$ are defined by

$$T_9(f,x) = \frac{1}{\Gamma(\alpha)} \int_a^x \left(\frac{x^{r+s} - t^{r+s}}{r+s}\right)^{\alpha-1} t^{r+s-1} f(t)dt, \ x > a, \tag{11}$$

and

$$T_{10}(f,x) = \frac{1}{\Gamma(\alpha)} \int_x^b \left(\frac{t^{r+s} - x^{r+s}}{r+s}\right)^{\alpha-1} t^{r+s-1} f(t)dt \ x < b, \tag{12}$$

respectively, where $r, s \geq 0, r + s \neq 0$.

In particular, if $s = 1$, then T_9, T_{10} reduce to T_5, T_6 , respectively.

Definition 5 (cf. [8, 9]) Let $f \in L[a, b]$, $g : [a, b] \to (0, \infty)$ be an increasing function, and $g' \in C[a, b], \alpha > 0$.Then g-Riemann–Liouville fractional integrals of f with respect to the function g on $[a, b]$ are defined by

$$T_{11}(f,x) = \frac{1}{\Gamma(\alpha)} \int_a^x g'(t)[g(x) - g(t)]^{\alpha-1} f(t)dt, \ x > a, \tag{13}$$

and

$$T_{12}(f,x) = \frac{1}{\Gamma(\alpha)} \int_x^b g'(t)[g(t) - g(x)]^{\alpha-1} f(t)dt, \ x < b, \tag{14}$$

respectively .

In 2018, S.S. Dragomir [10] introduced the new notion of the area balance function:

Definition 6 ([10]) Let $f \in L[a, b]$, then the area balance function of f is defined by

$$T_{13}(f,x) = \frac{1}{2}\left\{\int_x^b f(t)dt - \int_a^x f(t)dt\right\}. \tag{15}$$

In 2020, Kuang [16] introduced the new notion of the generalized fractional integral operators and fractional area balance operators :

Definition 7 ([16]) Let $f \in L[a,b]$, $g : [a,b] \to (0,\infty)$ be an increasing function, and $g \in AC[a,b]$, $k, c, \alpha > 0$, $a \geq 0$. Then the generalized fractional integral operator T_{14} with respect to the function g on $[a,b]$ is defined by

$$T_{14}(f,x) = \frac{c}{k\Gamma_k(\alpha)} \int_a^b g'(t)|g(x) - g(t)|^{(\alpha/k)-1} f(t)dt, \tag{16}$$

where $\Gamma_k(\alpha)$ is defined by (6).

Let

$$T_{15}(f,x) = \frac{c}{k\Gamma_k(\alpha)} \int_a^x g'(t)[g(x) - g(t)]^{(\alpha/k)-1} f(t)dt, \ x > a, \tag{17}$$

and

$$T_{16}(f,x) = \frac{c}{k\Gamma_k(\alpha)} \int_x^b g'(t)[g(t) - g(x)]^{(\alpha/k)-1} f(t)dt, \ x < b. \tag{18}$$

Then

$$T_{14}(f,x) = T_{15}(f,x) + T_{16}(f,x). \tag{19}$$

In particular, if $c = k = 1$ in (19), then (19) reduces to

$$T_{14}(f,x) = T_{11}(f,x) + T_{12}(f,x). \tag{20}$$

If $c = (r+s)^{-\alpha}$, $g(t) = t^{r+s}$, $r, s \geq 0, r+s \neq 0, k = 1$ in (19), then (19) reduces to

$$T_{14}(f,x) = T_9(f,x) + T_{10}(f,x). \tag{21}$$

If $c = (r+1)^{-(\alpha/k)}$, $g(t) = t^{r+1}$, $r \geq 0$, in (19), then (19) reduces to

$$T_{14}(f,x) = T_7(f,x) + T_8(f,x). \tag{22}$$

If $s = 1$ in (21), then (21) reduces to

$$T_{14}(f,x) = T_5(f,x) + T_6(f,x). \tag{23}$$

If $r = 0$ in (22), then (22) reduces to

$$T_{14}(f,x) = T_3(f,x) + T_4(f,x). \tag{24}$$

If $k = 1$ in (24), then (24) reduces to

$$T_{14}(f, x) = T_1(f, x) + T_2(f, x). \tag{25}$$

We can also rewrite T_9 and T_{10} as

$$T_9(f, x) = \frac{(r+s)^{1-\alpha}}{\Gamma(\alpha)} \int_a^x (x^{r+s} - t^{r+s})^{\alpha-1} t^{r+s-1} f(t)dt, \ x > a,$$

and

$$T_{10}(f, x) = \frac{(r+s)^{1-\alpha}}{\Gamma(\alpha)} \int_x^b (t^{r+s} - x^{r+s})^{\alpha-1} t^{r+s-1} f(t)dt, \ x < b,$$

and then generalize them to

$$T_{17}(f, x) = \frac{(r+s)^{1-(\alpha/k)}}{k\Gamma_k(\alpha)} \int_a^x (x^{r+s} - t^{r+s})^{(\alpha/k)-1} t^{r+s-1} f(t)dt, \ x > a, \tag{26}$$

and

$$T_{18}(f, x) = \frac{(r+s)^{1-(\alpha/k)}}{k\Gamma_k(\alpha)} \int_x^b (t^{r+s} - x^{r+s})^{(\alpha/k)-1} t^{r+s-1} f(t)dt, \ x < b. \tag{27}$$

If $c = (r+s)^{-(\alpha/k)}$, $g(t) = t^{r+s}$, $r, s \geq 0, r+s \neq 0$ in (19), then (19) reduces to

$$T_{14}(f, x) = T_{17}(f, x) + T_{18}(f, x). \tag{28}$$

Definition 8 ([13, 14]) Let $f \in L[a, b], a \geq 0$. The left-sided and right-sided Hadamard fractional integrals T_{19} and T_{20} of f of order $\alpha > 0$ are defined by

$$T_{19}(f, x) = \frac{1}{\Gamma(\alpha)} \int_a^x (\log x - \log t)^{\alpha-1} t^{-1} f(t)dt, \ x > a,$$

and

$$T_{20}(f, x) = \frac{1}{\Gamma(\alpha)} \int_x^b (\log t - \log x)^{\alpha-1} t^{-1} f(t)dt, \ x < b,$$

respectively.

We can generalize them to

$$T_{21}(f, x) = \frac{1}{k\Gamma_k(\alpha)} \int_a^x (\log x - \log t)^{(\alpha/k)-1} t^{-1} f(t)dt, \ x > a,$$

and

$$T_{22}(f,x) = \frac{1}{k\Gamma_k(\alpha)} \int_x^b (\log t - \log x)^{(\alpha/k)-1} t^{-1} f(t)dt, \ \ x < b.$$

If $c = 1, g(t) = \log t$ in (19), then (19) reduces to

$$T_{14}(f,x) = T_{21}(f,x) + T_{22}(f,x).$$

In particular, if $k = 1$, then

$$T_{14}(f,x) = T_{19}(f,x) + T_{20}(f,x).$$

Definition 9 Under the assumptions of Definition 7, the fractional area balance operators T_{24} with respect to the function g on $[a,b]$ is defined by

$$T_{24}(f,x) = \frac{c}{k\Gamma_k(\alpha)} \{ \int_x^b g^{'}(t)[g(t) - g(x)]^{(\alpha/k)-1} f(t)dt$$
$$- \int_a^x g^{'}(t)[g(x) - g(t)]^{(\alpha/k)-1} f(t)dt\}, \tag{29}$$

where $\Gamma_k(\alpha)$ is defined by (6).

Using (17) and (18), we have

$$T_{24}(f,x) = T_{16}(f,x) - T_{15}(f,x). \tag{30}$$

In particular, if $c = (r+s)^{-(\alpha/k)}, g(t) = t^{r+s}, r, s \geq 0, r+s \neq 0$ in (30), then (30) reduces to

$$T_{24}(f,x) = T_{18}(f,x) - T_{17}(f,x). \tag{31}$$

If $g(t) = t, \alpha = k = 1, c = 1/2$ in (30), then T_{24} reduces to T_{13} . If $c = k = 1$ in (30), then (30) reduces to

$$T_{24}(f,x) = T_{12}(f,x) - T_{11}(f,x). \tag{32}$$

If $c = (r+s)^{-\alpha}, g(t) = t^{r+s}, r, s \geq 0, r+s \neq 0, k = 1$ in (30), then (30) reduces to

$$T_{24}(f,x) = T_{10}(f,x) - T_9(f,x). \tag{33}$$

If $c = (r+1)^{-(\alpha/k)}, g(t) = t^{r+1}, r \geq 0$ in (30), then (30) reduces to

$$T_{24}(f,x) = T_8(f,x) - T_7(f,x). \tag{34}$$

If $k = 1$ in (34), then (34) reduces to

$$T_{24}(f, x) = T_6(f, x) - T_5(f, x). \tag{35}$$

If $r = 0$ in (34), then (34) reduces to

$$T_{24}(f, x) = T_4(f, x) - T_3(f, x). \tag{36}$$

If $k = 1$ in (36), then (36) reduces to

$$T_{24}(f, x) = T_2(f, x) - T_1(f, x). \tag{37}$$

If $c = 1$, $g(t) = \log t$ in (30), then (30) reduces to

$$T_{24}(f, x) = T_{22}(f, x) - T_{21}(f, x).$$

In particular, if $k = 1$, then

$$T_{24}(f, x) = T_{20}(f, x) - T_{19}(f, x).$$

Hence, Definitions 7 and 9 unified and generalized many known and new classes of fractional integral operators. In 2020, by using Definition 10 and Lemma 1, Kuang [16] proves some inequalities for operators T_{14} and T_{24} .

Definition 10 ([1]) Let $[a, b] \subset [0, \infty)$, $h : [a, b] \to (0, \infty)$ be the given function. A function $f : [a, b] \to [0, \infty)$ is called exponentially (β, s, s_1, s_2, h)-strongly convex if

$$f(tx_1 + (1-t)x_2) \le \left\{ t^{ss_1} \left(\frac{f(x_1)}{e^{r_0 x_1}} \right)^{\beta} + (1 - t^{s_2})^s \left(\frac{f(x_2)}{e^{r_0 x_2}} \right)^{\beta} \right\}^{1/\beta} - t(1-t)h(|x_1 - x_2|), \tag{38}$$

where $x_1, x_2 \in [a, b]$, $t, s, s_1, s_2 \in [0, 1]$, $r_0, \beta \in \mathbb{R}$, $\beta \neq 0$.

Lemma 1 ([16]) *Let $[a, b] \subset [0, \infty)$, $f \in L[a, b]$, $g : [a, b] \to [0, \infty)$ be an increasing function, and $g \in AC[a, b]$, $k, \alpha, c > 0$, then*

$$T_{14}(f, x) = [T_{16}(1, x) + T_{15}(1, x)]f(x) + \int_x^b G_{16}(1, t)f'(t)dt - \int_a^x G_{15}(1, t)f'(t)dt, \tag{39}$$

and

$$T_{24}(f,x) = [T_{16}(1,x) - T_{15}(1,x)]f(x) + \int_x^b G_{16}(1,t)f^{'}(t)dt + \int_a^x G_{15}(1,t)f^{'}(t)dt \quad (40)$$

where $G_{15}(1,t)$ and $G_{16}(1,t)$ are defined by

$$G_{15}(1,t) = \frac{c}{k\Gamma_k(\alpha)} \int_a^t g^{'}(u)[g(x)-g(u)]^{(\alpha/k)-1}du, \quad (41)$$

and

$$G_{16}(1,t) = \frac{c}{k\Gamma_k(\alpha)} \int_t^b g^{'}(u)[g(u)-g(x)]^{(\alpha/k)-1}du, \quad (42)$$

and $T_{14}, T_{15}, T_{16}, T_{24}$ and $\Gamma_k(\alpha)$ are defined by (16), (17), (18), (29), and (6), respectively.

Theorem 1 ([16]) *Under the assumptions of Lemma 1, let $f^{'} \in L^p[a,b], a \geq 0$, $1 \leq p < \infty$, $\frac{1}{p}+\frac{1}{q}=1$, and for $p=1$, define $q=\infty$, $\frac{1}{\infty}=0$. If $1<p<\infty$, then*

$$|T_{14}(f,x) - [T_{15}(1,x)+T_{16}(1,x)]f(x)| \leq \{(\int_a^x |G_{15}(1,t)|^q dt)^{1/q} + (\int_x^b |G_{16}(1,t)|^q dt)^{1/q}\} \|f^{'}\|_p. \quad (43)$$

If $p=1$, then

$$|T_{14}(f,x) - [T_{15}(1,x)+T_{16}(1,x)]f(x)| \leq \|G\|_\infty \|f^{'}\|_1, \quad (44)$$

where

$$G(t) = G_{16}(1,t)\varphi_{D_2}(t) - G_{15}(1,t)\varphi_{D_1}(t), \quad (45)$$

$D_1=[a,x]$, $D_2=[x,b]$, and φ_D is the characteristic function of the set D, that is,

$$\varphi_D(t) = \begin{cases} 1, & t \in D \\ 0 & x \in D^c. \end{cases}$$

Theorem 2 ([16]) *Under the assumptions of Lemma 1, if $|f^{'}|^p$ is exponentially (β,s,s_1,s_2,h)-strongly convex on $[a,b]$, and $\frac{s}{\beta}+1>0$. If $1<p<\infty$, $\frac{1}{p}+\frac{1}{q}=1$, then*

$$|T_{24}(f,x) - [T_{16}(1,x) - T_{15}(1,x)]f(x)|$$

$$
\begin{aligned}
&\leq \left(\int_x^b |G_{16}(1,t)|^q dt\right)^{1/q} (b-x)^{1/p} \\
&\times \left\{ C_\beta [\frac{\beta}{ss_1+\beta}\frac{|f^{'}(b)|^p}{e^{r_0 b}} \right. \\
&+\frac{1}{s_2}B(\frac{s}{\beta}+1,\frac{1}{s_2})\frac{|f^{'}(x)|^p}{e^{r_0 x}}] - \frac{1}{6}h(b-x)\}^{1/p} \\
&+\left(\int_a^x |G_{15}(1,t)|^q dt\right)^{1/q} (x-a)^{1/p} \\
&\times \left\{ C_\beta [\frac{\beta}{ss_1+\beta}\frac{|f^{'}(x)|^p}{e^{r_0 x}} \right. \\
&\left. +\frac{1}{s_2}B(\frac{s}{\beta}+1,\frac{1}{s_2})\frac{|f^{'}(a)|^p}{e^{r_0 a}}] - \frac{1}{6}h(x-a) \right\}^{1/p}.
\end{aligned}
\tag{46}
$$

If $p = 1$, then

$$
\begin{aligned}
&|T_{24}(f,x) - [T_{16}(1,x) - T_{15}(1,x)]f(x)| \\
&\leq \|G_{16}\|_\infty (b-x) \left\{ C_\beta \left[\frac{\beta}{ss_1+\beta}\frac{|f^{'}(b)|}{e^{r_0 b}} \right.\right. \\
&\left.\left. +\frac{1}{s_2}B\left(\frac{s}{\beta}+1,\frac{1}{s_2}\right)\frac{|f^{'}(x)|}{e^{r_0 x}} \right] - \frac{1}{6}h(b-x) \right\} \\
&+\|G_{15}\|_\infty (x-a) \left\{ C_\beta \left[\frac{\beta}{ss_1+\beta}\frac{|f^{'}(x)|}{e^{r_0 x}} \right.\right. \\
&\left.\left. +\frac{1}{s_2}B\left(\frac{s}{\beta}+1,\frac{1}{s_2}\right)\frac{|f^{'}(a)|}{e^{r_0 a}} \right] - \frac{1}{6}h(x-a) \right\}
\end{aligned}
$$

where

$$
C_\beta = \begin{cases} 1 & \beta \geq 1, \\ 2^{(1/\beta)-1}, & 0 < \beta < 1. \end{cases}
\tag{47}
$$

and

$$
B(\alpha,\beta) = \int_0^1 t^{\alpha-1}(1-t)^{\beta-1}dt
$$

is the Beta function.

It is noted that all the fractional integral operators above are established for functions of one variable .We naturally ask, how do we generalize these results to functions of several variable? In 2014, Sarikaya [17] gives the definitions Riemann–Liouville fractional integrals of two variable functions:

Definition 11 ([17, 18]) Let $Q = [a, b] \times [c, d] = \bigcup_{k=1}^{4} Q_k$, where $Q_1 = [a, x] \times [c, y]$, $Q_2 = [a, x] \times [y, d]$, $Q_3 = [x, b] \times [c, y]$, $Q_4 = [x, b] \times [y, d]$. The Riemann–Liouville fractional integrals $I_k (1 \leq k \leq 4)$ are defined by

$$I_1(f; x, y) = \frac{1}{\Gamma(\alpha)\Gamma(\beta)} \int_{Q_1} (x-t)^{\alpha-1}(y-s)^{\beta-1} f(t, s)dsdt,$$

$$I_2(f; x, y) = \frac{1}{\Gamma(\alpha)\Gamma(\beta)} \int_{Q_2} (x-t)^{\alpha-1}(s-y)^{\beta-1} f(t, s)dsdt,$$

$$I_3(f; x, y) = \frac{1}{\Gamma(\alpha)\Gamma(\beta)} \int_{Q_3} (t-x)^{\alpha-1}(y-s)^{\beta-1} f(t, s)dsdt,$$

and

$$I_4(f; x, y) = \frac{1}{\Gamma(\alpha)\Gamma(\beta)} \int_{Q_4} (t-x)^{\alpha-1}(s-y)^{\beta-1} f(t, s)dsdt.$$

Obviously, the above definition does not apply to general functions on $\mathbb{R}_+^n$. The aim of this chapter is to introduce some new generalized fractional integral operators and fractional area balance operators on n-dimensional linear spaces E_n which includes $\mathbb{R}_+^n$ as the special case. In Sect. 2, we define generalized fractional integral operators and fractional area balance operators on E_n. In Sect. 3, some Lemmas are derived .The corresponding integral operator inequalities are established in Sects. 4 and 5. They are significant improvement and generalizations of many known and new classes of fractional integral operators.

2 Generalized Fractional Integral Operators and Fractional Area Balance Operators

Throughout this chapter, we write

$$E_n = \{x = (x_1, x_2, \cdots, x_n) : x_k \geq 0, 1 \leq k \leq n, \|x\| = \left(\sum_{k=1}^{n} |x_k|^r\right)^{1/r}, r > 0\}.$$

E_n is an n-dimensional linear space, when $1 \leq r < \infty$, E_n is a normed vector space. In particular, when $r = 2$, E_n is an n-dimensional Euclidean space $\mathbb{R}_+^n$. When $r = 1$, $\|x\| = \sum_{k=1}^{n} |x_k|$ is a Cartesian norm. Let

$$D = \{x = (x_1, x_2, \cdots, x_n) : x_k \geq 0, 1 \leq k \leq n, 0 \leq a < \|x\| < b\},$$

$$D_1 = \{y = (y_1, y_2, \cdots, y_n) : y_k \geq 0, 1 \leq k \leq n, 0 \leq a < \|y\| < \|x\|, x \in E_n\},$$

$$D_2 = \{y = (y_1, y_2, \cdots, y_n) : y_k \geq 0, 1 \leq k \leq n, \|x\| < \|y\| < b, x \in E_n\}.$$

Definition 12 Let $f \in L[a, b]$, $g : [a, b] \to (0, \infty)$ be an increasing function, and $g \in AC[a, b]$, $k, c, \alpha > 0, a \geq 0$.Then the generalized fractional integrals operator T_{25} with respect to the function g on D is defined by

$$T_{25}(f, x) = \frac{c}{k\Gamma_k(\alpha)} \int_D g'(\|y\|)|g(\|x\|) - g(\|y\|)|^{(\alpha/k)-1} f(\|y\|)dy, \tag{48}$$

where $\Gamma_k(\alpha)$ is defined by (6)

Let

$$T_{26}(f, x) = \frac{c}{k\Gamma_k(\alpha)} \int_{D_1} g'(\|y\|)[g(\|x\|) - g(\|y\|)]^{(\alpha/k)-1} f(\|y\|)dy, \tag{49}$$

and

$$T_{27}(f, x) = \frac{c}{k\Gamma_k(\alpha)} \int_{D_2} g'(\|y\|)[g(\|y\|) - g(\|x\|)]^{(\alpha/k)-1} f(\|y\|)dy. \tag{50}$$

Thus,

$$T_{25}(f, x) = T_{26}(f, x) + T_{27}(f, x). \tag{51}$$

The fractional area balance operator with respect to the function g on D is defined by

$$T_{28}(f, x) = T_{27}(f, x) - T_{26}(f, x). \tag{52}$$

In particular, if $n = r = 1$ in Definition 12, then Definition 12 reduces to Definitions 7 and 9.

3 Some Lemmas

We require the following Lemmas to prove our main results.

Lemma 2 ([15]) *If $a_k, b_k, p_k > 0, 1 \leq k \leq n$, f be a measurable function on $(0, \infty)$, then*

$$\int_{B(r_1,r_2)} f\left(\sum_{k=1}^{n}\left(\frac{x_k}{a_k}\right)^{b_k}\right) x_1^{p_1-1}\cdots x_n^{p_n-1}dx_1\cdots dx_n$$
$$= \frac{\prod_{k=1}^{n} a_k^{p_k}}{\prod_{k=1}^{n} b_k}\cdot\frac{\prod_{k=1}^{n}\Gamma\left(\frac{p_k}{b_k}\right)}{\Gamma\left(\sum_{k=1}^{n}\frac{p_k}{b_k}\right)}\int_{r_1}^{r_2} f(t)t^{\left(\sum_{k=1}^{n}\frac{p_k}{b_k}-1\right)}dt.$$

where $B(r_1, r_2) = \{x \in E_n : 0 \le r_1 < \|x\| < r_2\}$.

We get the following Lemma 3 by taking $a_k = 1, b_k = r > 0, p_k = 1, 1 \le k \le n, r_1 = a, r_2 = b$ in Lemma 2.

Lemma 3 *Let* f *be a measurable function on* $(0, \infty)$*, then*

$$\int_D f(\|x\|)dx = \frac{(\Gamma(1/r))^n}{r^n\Gamma(n/r)}\int_a^b f(t^{1/r})t^{(n/r)-1}dt. \tag{53}$$

Lemma 4 *Let* $[a, b] \subset [0, \infty)$*,* $f \in L[a, b]$*,* $g : [a, b] \to [0, \infty)$ *be an increasing function, and* $g \in AC[a, b]$*,* $k, \alpha, c > 0$*, then*

$$\begin{aligned} T_{25}(f, x) = {}& [T_{26}(1, x) + T_{27}(1, x)]f(\|x\|^{1/r}) \\ &+r\left[\int_{\|x\|}^{b} G_{27}(1, t)t^{1-(1/r)}f'(t^{1/r})dt\right. \\ &\left.-\int_a^{\|x\|} G_{26}(1, t)t^{1-(1/r)}f'(t^{1/r})dt\right], \end{aligned} \tag{54}$$

and

$$\begin{aligned} T_{28}(f, x) = {}& [T_{27}(1, x) - T_{26}(1, x)]f(\|x\|^{1/r}) \\ &+r\left[\int_{\|x\|}^{b} G_{27}(1, x)t^{1-(1/r)}f'(t^{1/r})dt\right. \\ &\left.+\int_a^{\|x\|} G_{26}(1, t)t^{1-(1/r)}f'(t^{1/r})dt\right], \end{aligned} \tag{55}$$

where $G_{26}(1, t)$ *and* $G_{27}(1, t)$ *are defined by*

$$\begin{aligned} G_{26}(1, t) = {}& \frac{c}{k\Gamma_k(\alpha)}\times\frac{\Gamma^n(1/r)}{r^n\Gamma(n/r)} \\ &\times\int_a^t g'(u^{1/r})\left[g(\|x\|) - g(u^{1/r})\right]^{(\alpha/k)-1}u^{(n/r)-1}du, \end{aligned} \tag{56}$$

and

$$G_{27}(1,t) = \frac{c}{k\Gamma_k(\alpha)} \times \frac{\Gamma^n(1/r)}{r^n\Gamma(n/r)} \times \int_t^b g'(u^{1/r})\left[g(u^{1/r}) - g(\|x\|)\right]^{(\alpha/k)-1} u^{(n/r)-1}du, \quad (57)$$

and T_{25}, T_{26}, T_{27}, and T_{28} are defined by (48), (49), (50), and (52), respectively.

Proof From (49), (50) and Lemma 3, we have

$$\begin{aligned} T_{26}(f,x) &= \frac{c}{k\Gamma_k(\alpha)} \int_{D_1} g'(\|y\|)[g(\|x\|) - g(\|y\|)]^{(\alpha/k)-1} f(\|y\|)dy \\ &= \frac{c}{k\Gamma_k(\alpha)} \times \frac{\Gamma^n(1/r)}{r^n\Gamma(n/r)} \int_a^{\|x\|} g'(t^{1/r})\left[g(\|x\|) - g(t^{1/r})\right]^{(\alpha/r)-1} \\ &\quad \times f(t^{1/r})t^{(n/r)-1}dt, \end{aligned} \quad (58)$$

and

$$\begin{aligned} T_{27}(f,x) &= \frac{c}{k\Gamma_k(\alpha)} \int_{D_2} g'(\|y\|)[g(\|y\|) - g(\|x\|)]^{(\alpha/k)-1} f(\|y\|)dy \\ &= \frac{c}{k\Gamma_k(\alpha)} \times \frac{\Gamma^n(1/r)}{r^n\Gamma(n/r)} \int_{\|x\|}^{b} g'(t^{1/r})\left[g(t^{1/r}) - g(\|x\|)\right]^{(\alpha/r)-1} \\ &\quad \times f(t^{1/r})t^{(n/r)-1}dt. \end{aligned} \quad (59)$$

Thus,

$$G_{26}(1,\|x\|) = T_{26}(1,x), G_{27}(1,\|x\|) = T_{27}(1,x). \quad (60)$$

Then making use of integration by parts, we get

$$\begin{aligned} &\int_a^{\|x\|} G_{26}(1,t)(rt^{1-(1/r)})f'(t^{1/r})dt \\ &= \int_a^{\|x\|} G_{26}df(t^{1/r}) = G_{26}(1,t)f(t^{1/r})|_a^{\|x\|} \\ &\quad - \int_a^{\|x\|} G_{26}'(1,t)f(t^{1/r})dt = G_{26}(1,\|x\|)f(\|x\|^{1/r}) \\ &\quad - \frac{c}{k\Gamma_k(\alpha)} \times \frac{\Gamma^n(1/r)}{r^n\Gamma(n/r)} \int_a^{\|x\|} g'\left(t^{1/r}\right)\left[g(\|x\|) - g(t^{1/r})\right]^{(\alpha/k)-1} t^{(n/r)-1} \\ &\quad \times f(t^{1/r})dt \\ &= T_{26}(1,x)f(\|x\|^{1/r}) - T_{26}(f,x), \end{aligned}$$

which leads to

$$T_{26}(f,x) = T_{26}(1,x)f(\|x\|^{1/r}) - r\int_a^{\|x\|} G_{26}(1,t)f^{'}(t^{1/r})t^{1-(1/r)}dt. \tag{61}$$

Similarly, we have

$$T_{27}(f,x) = T_{27}(1,x)f(\|x\|^{1/r}) + r\int_{\|x\|}^{b} G_{27}(1,t)f^{'}(t^{1/r})t^{1-(1/r)}dt. \tag{62}$$

Hence, (54) follows from (51), (61) and (62), as well as (55) follows from (52), (61), and (62). The proof is completed.

4 Some Inequalities for Operator T_{25}

Theorem 3 *Under the assumptions of Lemma 4, let* $f^{'} \in L^p(D)$, $1 \le p < \infty$, $\frac{1}{p} + \frac{1}{q} = 1$, *and for* $p = 1$, *define* $q = \infty$, $\frac{1}{\infty} = 0$. *If* $1 < p < \infty$, *then*

$$\begin{aligned}
&|T_{25}(f,x) - [T_{26}(1,x) + T_{27}(1,x)]f(\|x\|^{1/r})| \\
&\le r^{1+(1/p)}\left\{\left(\int_a^{\|x\|}|G_{26}(1,t)|^q dt\right)^{1/q} + \left(\int_{\|x\|}^{b}|G_{27}(1,t)|^q dt\right)^{1/q}\right\} \\
&\times\left(\int_{a^{1/r}}^{b^{1/r}}|f^{'}(u)|^p u^{(r-1)(p+1)}\right)^{1/p}.
\end{aligned} \tag{63}$$

If $p = 1$, *then*

$$|T_{25}(f,x) - [T_{26}(1,x) + T_{27}(1,x)]f(\|x\|^{1/r})| \le \|G\|_\infty\|f^{'}\|_1, \tag{64}$$

where

$$G(t) = G_{27}(1,t)\varphi_{D_2}(t) - G_{26}(1,t)\varphi_{D_1}(t), \tag{65}$$

$D_1 = [a, \|x\|]$, $D_2 = [\|x\|, b]$, *and* φ_D *is the characteristic function of the set* D, *that is,*

$$\varphi_D(t) = \begin{cases} 1, & t \in D, \\ 0, & t \in D^c, \end{cases}$$

Proof For $1 < p < \infty$, by using Lemma 4, we obtain

$$|T_{25}(f, x) - [T_{26}(1, x) + T_{27}(1, x)]f(\|x\|^{1/r})|$$
$$= r|\int_{\|x\|}^{b} G_{27}(1, t)t^{1-(1/r)} f'(t^{1/r})dt - \int_{a}^{\|x\|} G_{26}(1, t)t^{1-(1/r)} f'(t^{1/r})dt|$$
$$= r|\int_{a}^{b} G(t)t^{1-(1/r)} f'(t^{1/r})dt|. \tag{66}$$

Using the Hölder inequality, from (66), we obtain

$$|T_{25}(f, x) - [T_{26}(1, x) + T_{27}(1, x)]f(\|x\|^{1/r})|$$
$$\leq \left(\int_{a}^{b} |G(t)|^{q} dt\right)^{1/q} \left(\int_{a}^{b} |f'(t^{1/r})|^{p} t^{p(1-(1/r))} dt\right)^{1/p}$$
$$= r^{1+(1/p)} (\int_{a}^{b} |G(t)|^{q} dt)^{1/q} \left(\int_{a^{1/r}}^{b^{1/r}} |f'(u)|^{p} u^{(p+1)(r-1)} du\right)^{1/p}$$

and for $p = 1$, we have

$$|T_{25}(f, x) - [T_{26}(1, x) + T_{27}(1, x)]|f(\|x\|^{1/r})| \leq \|G\|_{\infty} \|f'\|_{1}$$

The proof is completed .

Taking $r = 1$ in Theorem 3, we get

Corollary 1 *Under the assumptions of Theorem 3, let $r = 1$. If $1 < p < \infty$, then*

$$|T_{25}(f, x) - [T_{26}(1, x) + T_{27}(1, x)]f(\|x\|)|$$
$$\leq \{\left(\int_{a}^{\|x\|} |G_{26}(1, t)|^{q} dt\right)^{1/q} + \left(\int_{\|x\|}^{b} |G_{27}(1, t)|^{q} dt\right)^{1/q}\} \|f'\|_{p};$$

If $p = 1$, then

$$|T_{25}(f, x) - [T_{26}(1, x) + T_{27}(1, x)]f(\|x\|)| \leq \|G\|_{\infty} \|f'\|_{1}$$

If $n = 1$ in Theorem 3, then Theorem 3 reduces to Theorem 1.

5 Some Inequalities for Operator T_{28}

Theorem 4 *Under the assumptions of Lemma 4, if $|f'|^{p}$ is exponentially (β, s, s_1, s_2, h)-strongly convex on $[a, b]$, and $\frac{s}{\beta} + 1 > 0$. If $1 < p < \infty$, $\frac{1}{p} + \frac{1}{q} = 1$, then*

$$
\begin{aligned}
&|T_{28}(f,x)-[T_{27}(1,x)-T_{26}(1,x)]f(\|x\|^{1/r})| \\
&\leq r^{1+(1/p)}\left(\int_{\|x\|}^{b}|G_{26}(1,t)|^{q}dt\right)^{1/q}\left(b^{1/r}-\|x\|^{1/r}\right)^{1/p} \\
&\times\left\{C_\beta\left[I_3\frac{|f'(b^{1/r})|^p}{e^{r_0 b^{1/r}}}\right.\right. \\
&\left.\left.+I_4\frac{|f'(\|x\|^{1/r})|^p}{e^{r_0\|x\|^{1/r}}}\right]-I_5 h\left(b^{1/r}-\|x\|^{1/r}\right)\right\}^{1/p} \\
&+r^{1+(1/p)}(\int_{a}^{\|x\|}|G_{25}(1,t)|^{q}dt)^{1/q}\left(\|x\|^{1/r}-a^{1/r}\right)^{1/p} \\
&\times\left\{C_\beta\left[I_3\frac{|f'(\|x\|^{1/r})|^p}{e^{r_0\|x\|^{1/r}}}\right.\right. \\
&\left.\left.+I_4\frac{|f'(a^{1/r})|^p}{e^{r_0 a^{1/r}}}\right]-I_5 h\left(\|x\|^{1/r}-a^{1/r}\right)\right\}^{1/p}. \qquad (67)
\end{aligned}
$$

If $p=1$, *then*

$$
\begin{aligned}
&|T_{28}(f,x)-[T_{26}(1,x)-T_{25}(1,x)]f(\|x\|^{1/r})| \\
&\leq \|G_{26}\|_\infty\left(b^{1/r}-\|x\|^{1/r}\right)\left\{C_\beta\left[I_3\frac{|f'(b^{1/r})|}{e^{r_0 b^{1/r}}}\right.\right. \\
&\left.\left.+I_4\frac{|f'(\|x\|^{1/r})|}{e^{r_0\|x\|^{1/r}}}\right]-I_5 h(b^{1/r}-\|x\|^{1/r})\right\} \\
&+\|G_{25}\|_\infty\left(\|x\|^{1/r}-a^{1/r}\right)\left\{C_\beta\left[I_3\frac{|f'(\|x\|^{1/r})|}{e^{r_0\|x\|^{1/r}}}\right.\right. \\
&\left.\left.+I_4\frac{f'(a^{1/r})}{e^{r_0 a^{1/r}}}\right]-I_5 h\left(\|x\|^{1/r}-a^{1/r}\right)\right\}, \qquad (68)
\end{aligned}
$$

where C_β *is defined by (47), and*

$$
I_3=\int_0^1 t^{(ss_1)/\beta}\left[a^{1/r}+(\|x\|^{1/r}-a^{1/r})t\right]^{(p+1)(r-1)}dt;
$$

$$
I_4=\int_0^1 (1-t^{s_2})^{s/\beta}\left[a^{1/r}+(\|x\|^{1/r}-a^{1/r})t\right]^{(p+1)(r-1)}dt
$$

$$I_5 = \int_0^1 t(1-t)\left[a^{1/r} + (\|x\|^{1/r} - a^{1/r})t\right]^{(p+1)(r-1)} dt.$$

In particular, if $r = 1$, then

$$I_3 = \int_0^1 t^{(ss_1)/\beta} dt = \frac{\beta}{ss_1 + \beta};$$
$$I_4 = \int_0^1 (1 - t^{s_2})^{s/\beta} dt = \frac{1}{s_2} B\left(\frac{s}{\beta} + 1, \frac{1}{s_2}\right);$$
$$I_5 = \int_0^1 t(1-t) dt = \frac{1}{6}.$$

Thus, we get

Corollary 2 *Under the assumptions of Theorem 4, let* $r = 1$. *If* $1 < p < \infty$, *then*

$$\begin{aligned}
&|T_{28}(f,x) - [T_{27}(1,x) - T_{26}(1,x)]f(\|x\|)| \\
&\leq \left(\int_{\|x\|}^b |G_{26}(1,t)|^q dt\right)^{1/q} (b - \|x\|)^{1/p} \\
&\quad \times \left\{ C_\beta \left[\frac{\beta}{ss_1 + \beta} \frac{|f'(b)|^p}{e^{r_0 b}} \right.\right. \\
&\quad \left.\left. + \frac{1}{s_2} B\left(\frac{s}{\beta} + 1, \frac{1}{s_2}\right) \frac{|f'(\|x\|)^p}{e^{r_0\|x\|}} \right] - \frac{1}{6} h(b - \|x\|) \right\}^{1/p} \\
&\quad + \left(\int_a^{\|x\|} |G_{25}(1,t)|^q dt\right)^{1/q} (\|x\| - a)^{1/p} \\
&\quad \times \left\{ C_\beta \left[\frac{\beta}{ss_1 + \beta} \frac{|f'(\|x\|)|^p}{e^{r_0\|x\|}} \right.\right. \\
&\quad \left.\left. + \frac{1}{s_2} B(\frac{s}{\beta} + 1, \frac{1}{s_2}) \frac{|f'(a)|^p}{e^{r_0 a}} \right] - \frac{1}{6} h(\|x\| - a) \right\}^{1/p}.
\end{aligned} \tag{69}$$

If $p = 1$, *then*

$$\begin{aligned}
&|T_{28}(f,x) - [T_{27}(1,x) - T_{26}(1,x)]f(\|x\|)| \\
&\leq \|G_{27}\|_\infty (b - \|x\|) \left\{ C_\beta \left[\frac{\beta}{ss_1 + \beta} \frac{|f'(b)|}{e^{r_0 b}} \right.\right.
\end{aligned}$$

$$+\frac{1}{s_2}B\left(\frac{s}{\beta}+1,\frac{1}{s_2}\right)\frac{|f^{'}(\|x\|)|}{e^{r_0\|x\|}}\Bigg]-\frac{1}{6}h(b-\|x\|\Bigg\}$$

$$+\|G_{26}\|_\infty(\|x\|-a)\Bigg\{C_\beta\Bigg[\frac{\beta}{ss_1+\beta}\frac{|f^{'}(\|x\|)|}{e^{r_0\|x\|}}$$

$$+\frac{1}{s_2}B\left(\frac{s}{\beta}+1,\frac{1}{s_2}\right)\frac{|f^{'}(a)|}{e^{r_0 a}}\Bigg]-\frac{1}{6}h(\|x\|-a)\Bigg\}^{'} \tag{70}$$

Proof of Theorem 4 For $1 < p < \infty$, by using Lemma 4, and the Hölder inequality, we obtain

$$|T_{28}(f,x)-[T_{27}(1,x)-T_{26}(1,x)]f(\|x\|^{1/r})|$$
$$\leq r\left\{\int_{\|x\|}^{b}|G_{27}(1,t)|\times|f^{'}(t^{1/r})|t^{1-(1/r)}dt\right.$$
$$\left.+\int_{a}^{\|x\|}|G_{26}(1,t)|\times|f^{'}(t^{1/r})|t^{1-(1/r)}dt\right\}$$
$$\leq r\left(\int_{\|x\|}^{b}|G_{27}(1,t)|^q dt\right)^{1/q}\left(\int_{\|x\|}^{b}|f^{'}(\xi^{1/r})|^p\xi^{p(1-(1/r))}d\xi\right)^{1/p}$$
$$+r\left(\int_{a}^{\|x\|}|G_{26}(1,t)|^q dt\right)^{1/q}\left(\int_{a}^{\|x\|}|f^{'}(\xi^{1/r})|^p\xi^{p(1-(1/r))}d\xi\right)^{1/p}. \tag{71}$$

Setting $\xi = u^r$, $u = \|x\|^{1/r} + (b^{1/r} - \|x\|^{1/r})t$ and using the exponentially (β, s, s_1, s_2, h)-strongly convexity of $|f^{'}|^p$ on $[a, b]$, we have

$$\int_{\|x\|}^{b}|f^{'}(\xi^{1/r})|^p\xi^{p(1-(1/r))}d\xi = r\int_{\|x\|^{1/r}}^{b^{1/r}}|f^{'}(u)|^p u^{(p+1)(r-1)}du$$
$$= r(b^{1/r}-\|x\|^{1/r})$$
$$\times\int_0^1|f^{'}(tb^{1/r}+(1-t)\|x\|^{1/r})|^p\{\|x\|^{1/r}+(b^{1/r}$$
$$-\|\|x\|^{1/r})t\}^{(p+1)(r-1)}dt$$
$$\leq r\left(b^{1/r}-\|x\|^{1/r}\right)\int_0^1\Bigg\{\Bigg[t^{ss_1}(\frac{|f^{'}(b^{1/r}|^p}{e^{r_0 b^{1/r}}})^\beta$$
$$+(1-t^{s_2})^s(\frac{|f^{'}(\|x\|^{1/r})|^p}{e^{r_0\|x\|^{1/r}}}))^\beta\Bigg]^{1/\beta}$$

$$
- t(1-t)h(b^{1/r} - \|x\|^{1/r})\Big\} \{\|x\|^{1/r} + (b^{1/r} - \|x\|^{1/r})t\}^{(p+1)(r-1)}dt
$$

$$
\leq r\left(b^{1/r} - \|x\|^{1/r}\right)\left\{C_\beta \int_0^1 \left[t^{(ss_1)/\beta}\left(\frac{|f'(b^{1/r})|^p}{e^{r_0 b^{1/r}}}\right)\right.\right.
$$

$$
\left. +(1-t^{s_2})^{s/\beta}\left(\frac{|f'(\|x\|^{1/r})|^p}{e^{r_0\|x\|^{1/r}}}\right)\right]
$$

$$
\left. \times [\|x\|^{1/r} + (b^{1/r} - \|x\|^{1/r})t]^{(p+1)(r-1)}dt\right\}
$$

$$
-h\left(b^{1/r} - \|x\|^{1/r}\right)\int_0^1 t(1-t)\{\|x\|^{1/r} + \left(b^{1/r} - \|x\|^{1/r}\right)t\}^{(p+1)(r-1)}dt
$$

$$
\leq r\left(b^{1/r} - \|x\|^{1/r}\right) \times \left\{C_\beta\left[I_3 \frac{|f'(b^{1/r})|^p}{e^{r_0 b^{1/r}}}\right.\right.
$$

$$
\left.\left. +I_4 \frac{|f'(\|x\|^{1/r})|^p}{e^{r_0\|x\|^{1/r}}}\right] - I_5 h(b^{1/r} - \|x\|^{1/r})\right\}. \tag{72}
$$

By letting $\xi = u^r$, $u = a^{1/r} + (\|x\|^{1/r} - a^{1/r})t$ and similar arguments, we get

$$
\int_a^{\|x\|} |f'(\xi^{1/r})|^p \xi^{p(1-(1/r))} d\xi = r\int_{a^{1/r}}^{\|x\|^{1/r}} |f'(u)|^p u^{(p+1)(r-1)}du
$$

$$
\leq r(\|x\|^{1/r} - a^{1/r}) \times \left\{C_\beta\left[I_3 \frac{|f'(\|x\|^{1/r})|^p}{e^{r_0\|x\|^{1/r}}}\right.\right.
$$

$$
\left.\left. +I_4 \frac{|f'(a^{1/r})|^p}{e^{r_0 a^{1/r}}}\right] - I_5 h(\|x\|^{1/r} - a^{1/r})\right\}. \tag{73}
$$

Hence, (67) follows from (71), (72) and (73). The case $p = 1$ can be treated analogously. The proof is completed.

If $n = 1$ in Theorem 4, then Theorem 4 reduces to Theorem 2. By giving particular values to the parameters in Theorem 4, we get the corresponding integral inequalities for different fractional integral operators. Such as, taking $r = c = \beta = 1$ and $g(t) = t$ in Theorem 4, then T_{28} reduces to $T_{30} - T_{29}$, where

$$
T_{29}(f, x) = \frac{1}{k(n-1)!\Gamma_k(\alpha)} \int_a^{\|x\|} [\|x\| - u]^{(\alpha/k)-1} f(u)u^{n-1}du,
$$

and

$$T_{30}(f,x) = \frac{1}{k(n-1)!\Gamma_k(\alpha)} \int_{\|x\|}^{b} [u - \|x\|]^{(\alpha/k)-1} f(u) u^{n-1} du.$$

Let

$$G_{29}(1,t) = \frac{1}{k(n-1)!\Gamma_k(\alpha)} \int_{a}^{t} [\|x\| - u]^{(\alpha/k)-1} u^{n-1} du,$$

and

$$G_{30}(1,t) = \frac{1}{k(n-1)!\Gamma_k(\alpha)} \int_{t}^{b} [u - \|x\|]^{(\alpha/k)-1} u^{n-1} du.$$

Thus, we have

$$G_{29}(1,\|x\|) = T_{29}(1,x);\ G_{30}(1,\|x\|) = T_{30}(1,x).$$

Hence, we get

Corollary 3 *Under the assumptions of Theorem 4, let* $r = c = \beta = 1, s > -1$, *and* $g(t) = t$. *If* $1 < p < \infty$, $\frac{1}{p} + \frac{1}{q} = 1$, *then*

$$\begin{aligned}
&|T_{28}(f,x) - [T_{30}(1,x) - T_{29}(1,x)] f(\|x\|)| \\
&\le \left(\int_{\|x\|}^{b} |G_{30}(1,t)|^q dt \right)^{1/q} (b - \|x\|)^{1/p} \times \left\{ \frac{1}{ss_1 + 1} \frac{|f'(b)|^p}{e^{r_0 b}} \right. \\
&\left. + \frac{1}{s_2} B(s+1, \frac{1}{s_2}) \frac{|f'(\|x\|)|^p}{e^{r_0\|x\|}} - \frac{1}{6} h(b - \|x\|) \right\}^{1/p} \\
&+ \left(\int_{a}^{\|x\|} |G_{29}(1,t)|^q dt \right)^{1/q} (\|x\| - a)^{1/p} \times \left\{ \frac{1}{ss_1 + 1} \frac{|f'(\|x\|)|^p}{e^{r_0\|x\|}} \right. \\
&\left. + \frac{1}{s_2} B(s+1, \frac{1}{s_2}) \frac{|f'(a)|^p}{e^{r_0 a}}] - \frac{1}{6} h(\|x\| - a) \right\}^{1/p}.
\end{aligned} \tag{74}$$

If $p = 1$, *then*

$$\begin{aligned}
&|T_{28}(f,x) - [T_{30}(1,x) - T_{29}(1,x)] f(\|x\|)| \\
&\le \|G_{30}\|_\infty (b - \|x\|) \left\{ \left[\frac{1}{ss_1 + 1} \frac{|f'(b)|}{e^{r_0 b}} \right. \right.
\end{aligned}$$

$$+\frac{1}{s_2}B(s+1,\frac{1}{s_2})\frac{|f^{'}(\|x\|)|}{e^{r_0}\|x\|}\Bigg] - \frac{1}{6}h(b-\|x\|)\Bigg\}$$

$$+\|G_{29}\|_{\infty}(\|x\|-a)\Bigg\{\Bigg[\frac{1}{ss_1+1}\frac{|f^{'}(\|x\|)|}{e^{r_0\|x\|}}$$

$$+\frac{1}{s_2}B(s+1,\frac{1}{s_2})\frac{|f^{'}(a)|}{e^{r_0 a}}\Bigg] - \frac{1}{6}h(\|x\|-a)\Bigg\}. \tag{75}$$

If $n = 1$ in Corollary 3, then T_{28} reduces to $T_4 - T_3$, where T_3 and T_4 are defined by (4)and (5), respectively. Thus we get

Corollary 4 *Under the assumptions of Corollary 3, let* $n = 1$*, and* $s > -1$ *. If* $1 < p < \infty, \frac{1}{p} + \frac{1}{q} = 1$*, then*

$$|T_4(f,x) - T_3(f,x) - \frac{1}{\alpha\Gamma_k(\alpha)}\left[(b-x)^{(\alpha/k)} - (x-a)^{\alpha/k}\right]f(x)|$$

$$\leq \frac{1}{\alpha\Gamma_k(\alpha)}(\int_x^b |(b-x)^{\alpha/k} - (t-x)^{\alpha/k}|^q dt)^{1/q}(b-x)^{1/p}$$

$$\times\Bigg\{\frac{1}{ss_1+1}\frac{|f^{'}(b)|^p}{e^{r_0 b}} + \frac{1}{s_2}B(s+1,\frac{1}{s_2})\frac{|f^{'}(x)|^p}{e^{r_0 x}}$$

$$-\frac{1}{6}h(b-x)\Bigg\}^{1/p}$$

$$+\frac{1}{\alpha\Gamma_k(\alpha)}\left(\int_a^x |(x-a)^{\alpha/k} - (x-t)^{\alpha/k}|^q dt\right)^{1/q}(x-a)^{1/p}$$

$$\times\Bigg\{\frac{1}{ss_1+1}\frac{|f^{'}(x)|^p}{e^{r_0 x}} + \frac{1}{s_2}B\left(s+1,\frac{1}{s_2}\right)\frac{|f^{'}(a)|^p}{e^{r_0 a}}$$

$$-\frac{1}{6}h(x-a)\Bigg\}^{1/p}. \tag{76}$$

If $p = 1$*, then*

$$|T_4(f,x) - T_3(f,x) - \frac{1}{\alpha\Gamma_k(\alpha)}\left[(b-x)^{(\alpha/k)} - (x-a)^{(\alpha/k)}\right]f(x)|$$

$$\leq \frac{1}{\alpha\Gamma_k(\alpha)}(b-x)^{1+(\alpha/k)}\Bigg\{\Bigg[\frac{1}{ss_1+1}\frac{|f^{'}(b)|}{e^{r_0 b}} + \frac{1}{s_2}B(s+1,\frac{1}{s_2})\frac{|f^{'}(x)|}{e^{r_0 x}}\Bigg]$$

$$-\frac{1}{6}h(b-x)\Bigg\}$$

$$+\frac{1}{\alpha\Gamma_k(\alpha)}(x-a)^{1+(\alpha/k)}\left\{\frac{1}{ss_1+1}\frac{|f^{'}(x)|}{e^{r_0x}}+\frac{1}{s_2}B\left(s+1,\frac{1}{s_2}\right)\frac{|f^{'}(a)|}{e^{r_0a}}\right.$$

$$\left.-\frac{1}{6}h(x-a)\right\}.$$

If $k=1$ in Corollary 4, then T_{28} reduce to T_2-T_1 . Thus, we get

Corollary 5 *Under the assumptions of Corollary 4, let* $k=1$ *. If* $1<p<\infty$*, then*

$$|T_2(f,x)-T_1(f,x)-\frac{1}{\alpha\Gamma(\alpha)}\left[(b-x)^{\alpha}-(x-a)^{\alpha}\right]f(x)|$$

$$\leq\frac{1}{\alpha\Gamma(\alpha)}(\int_x^b|(b-x)^{\alpha}-(t-x)^{\alpha}|^q dt)^{1/q}(b-x)^{1/p}$$

$$\times\left\{\frac{1}{ss_1+1}\frac{|f^{'}(b)|^p}{e^{r_0b}}+\frac{1}{s_2}B(s+1,\frac{1}{s_2})\frac{|f^{'}(x)|^p}{e^{r_0x}}-\frac{1}{6}h(b-x)\right\}^{1/p}$$

$$+\frac{1}{\alpha\Gamma(\alpha)}\left(\int_a^x|(x-a)^{\alpha}-(x-t)^{\alpha}|^q dt\right)^{1/q}(x-a)^{1/p}$$

$$\times\left\{\frac{1}{ss_1+1}\frac{|f^{'}(x)|^p}{e^{r_0x}}+\frac{1}{s_2}B(s+1,\frac{1}{s_2})\frac{|f^{'}(a)|^p}{e^{r_0a}}-\frac{1}{6}h(x-a)\right\}^{1/p}.$$

If $p=1$*, then*

$$|T_2(f,x)-T_1(f,x)-\frac{1}{\alpha\Gamma(\alpha)}\left[(b-x)^{\alpha}-(x-a)^{\alpha}\right]f(x)|$$

$$\leq\frac{1}{\alpha\Gamma(\alpha)}(b-x)^{1+\alpha}\left\{\frac{1}{ss_1+1}\frac{|f^{'}(b)|}{e^{r_0b}}+\frac{1}{s_2}B\left(s+1,\frac{1}{s_2}\right)\frac{|f^{'}(x)|}{e^{r_0x}}\right.$$

$$\left.-\frac{1}{6}h(b-x)\right\}$$

$$+\frac{1}{\alpha\Gamma(\alpha)}(x-a)^{1+\alpha}\left\{\frac{1}{ss_1+1}\frac{|f^{'}(x)|}{e^{r_0x}}+\frac{1}{s_2}B\left(s+1,\frac{1}{s_2}\right)\frac{|f^{'}(a)|}{e^{r_0a}}\right.$$

$$\left.-\frac{1}{6}h(x-a)\right\}.$$

References

1. Kuang J.C., Applied Inequalities, 5th.edu.Shangdong Science and Technology Press, Jinan(2021), (in Chinese).
2. Kilbas, A.A., Srivgstava, H.M., Trujillo, J.J., Theory and Applications of Fractional Differential Equations, North-Holland Mathematics Studies, Vol.204, Elsevier, New York, 2006.
3. Mubeen, S., Habibullah, G.M., $k-$ fractional integrals and applications, Int.J.Contemp.Math.Sci., 2012, 7:89–94.
4. Mubeen, S., Iqbal, S., Grüss type integral inequalities for generalized Riemann-Liouville $k-$ fractional integrals, J.Inequal.Appl., 2016:109.
5. Sarikaya, M.Z., Dahmani, Z., Kiris, M.E., Ahmad, F., $(k, s)-$Riemann-Liouville fractional integral and applications, Hacet.J.Math.Stat., 2016, 45(1):77–89.
6. Set.E., Tomar, M., Sarkaya, M.Z., On generalized Grüss type inequalities for $k-$ fractional integrals, Appl.Math.Comp., 2005, 8(269):29–34.
7. Abbas, G., et al., Generalizations of some fractional integral inequalities via generalized Mittag-Leffer function, J.Inequal.Appl., 2017:121.
8. da Sousa, J.V., de Oliveira, E.C., On the $\psi-$ Hilfer fractional derivative, Commun.Nonlinear Sci.Numer.Simul., 2018, 60:72–91.
9. Zhao Y., et al., Hermite-Hadamard type inequalities involving $\psi-$ Riemann-Liouville fractional integrals via $s-$ convex functions, J.Inequal.Appl., 2020:128.
10. Dragomir, S.S., Inequalities for the area balance of absolutely continuous functions, Stud.Univ.Babes-Bolyai Math., 2018, 63(1):37–57.
11. Khan, J.U., Khan, M.A., Generalized conformable fractional integral operators, J.Comput.Appl.Math., 2019, 346:378–389.
12. Rashid, S., et al., New generalized reverse Minkowski and related integral inequalities involving generalized fractional conformable integrals, J.Inequal.Appl., 2020:177.
13. Jarad, F., et al., On a new class of fractional operators, Adv.Differ.Equ., 2017, 247.
14. Iscan Imdat, Jensen-Mercer inequality for $GA-$ convex functions and some related inequalities, J.Inequal.Appl., 2020:212.
15. Kuang J.C., Norm inequalities for generalized Laplace transforms // Raigorodskii, A., Rassias, M.Th., Editors, Trigonometric Sums and Their Applications, Springer, 2020.
16. Kuang J.C., Some new inequalities for fractional integral operators, // Rassias, Th.M., Approximation and Computation in Science and Engineering, Springer, 2021.
17. Sarikaya, M, Z., On the Hermite-Hadarmard - type inequalities for co-ordinated convex function via fractional integrals . Integral Transforms Spec.Funct.2014, 25(2), 134–147.
18. Erden, S., Budak, H., Sarikaya, M.Z., Iftikhar, S., and Kumam, P., Fractional Ostrowski type inequalities for bounded functions, J.Inequal.Appl., 2020:123.

Anisotropic Elasticity and Harmonic Functions in Cartesian Geometry

D. Labropoulou, P. Vafeas, and G. Dassios

Abstract Linear elasticity in an isotropic space is a well-developed area of continuum mechanics. However, the situation is exactly opposite if the fundamental space exhibits anisotropic behavior. In fact, the area of linear anisotropic elasticity is not well developed at the quantitative level, where actual closed-form solutions are needed to be calculated. The present work aims to provide a little progress in this interesting branch of continuum mechanics. We provide a short review of isotropic elasticity in order to demonstrate in the sequel how the anisotropy modifies the final equations, via Hooke's and Newton's laws. The eight standard anisotropic structures are also reviewed for completeness. A simple technique is introduced that generates homogeneous polynomial solutions of the anisotropic equations in Cartesian form. In order to demonstrate how this technique is applied, we work out the case of cubic anisotropy, which is the simplest anisotropic structure, having three independent elasticities. This choice is dictated by the restricted number of calculations it requires, but it carries all the basic steps of the method. Isotropic elasticity accepts the differential representation of Papkovich, which expresses the displacement field in terms of a vector and a scalar harmonic function. Unfortunately, though, no such representation is known for the anisotropic elasticity, which can represent the anisotropic displacement field in terms of solutions of the anisotropic Laplacian, as also discussed in this work.

MSC 74E10; 74B05; 35Q74; 35J05; 42B37

D. Labropoulou · P. Vafeas (✉) · G. Dassios
Department of Chemical Engineering, University of Patras, Patras, Greece
e-mail: dlabropoulou@chemeng.upatras.gr; vafeas@chemeng.upatras.gr; gdassios@chemeng.upatras.gr

I. N. Parasidis et al. (eds.), *Mathematical Analysis in Interdisciplinary Research*, Springer Optimization and Its Applications 179,
https://doi.org/10.1007/978-3-030-84721-0_23

1 Introduction

Modern theoretical mechanics and engineering technology are often involved with the anisotropic behavior of elastic materials and structures [1–3]. The mathematical description of the physical quantities, associated with anisotropic media, constitutes an indivisible component in linear anisotropic elasticity [4, 5]. Classical linear elasticity, being a first approximation of the more general nonlinear theory of elasticity [6] and a branch of continuum mechanics, assumes small deformations and infinitesimal internal strains and stresses on solid bodies or objects, when they are subject to prescribed loading conditions. That leads to linear relationships between the components of strain and stress dyadic tensors. On the other hand, anisotropy is the property of a medium, which allows it to change or assume different properties in different directions as opposed to isotropy [7, 8]. Both these aspects are evidently used extensively in structural analysis and engineering design, often with the aid of standard numerical methods [9], offering a complete and comprehensive survey of the analysis with respect to anisotropic material theory. Once either the isotropic or the anisotropic character of linear or even nonlinear media is identified, the next step is the necessity of studying the elastic or inelastic wave propagation in such domains [10, 11] and of determining the corresponding scattered fields that are produced. As it is obvious, not only in the anisotropic theory but also for the isotropic case [12], the presence of external body forces renders the analysis more elaborate. Along this concept, it is apparent the continuous need of establishing novel theories of physical and mathematical interest, concerning the general ideas of elasticity.

The anisotropy of a medium has consequences on its mathematical depiction. Indeed, the effect of the dependence upon the direction is implied via an increase of the number of parameters to be used for the description of the phenomenon. However, the possible geometrical symmetries must be taken into account, because they significantly simplify the correlated relations. The basic vector function, which is associated with either isotropic or anisotropic elastic behavior of a medium, is the displacement field. Upon the introduction of this field, the strain is a linear and symmetric tensor that is given in terms of the displacement field. Once the strain is calculated, the stress tensor follows from the well-known Hooke's law, which connects the strain and the stress via the stiffness, being a tetratic that embodies the isotropic character of the medium. All the aforementioned fields are incorporated into the non-homogeneous (when external forces are present) and time-dependent fundamental equation of elasticity, i.e. Newton's law [4], in order to obtain a general equation of elasticity. Here, we present this analysis for reasons of clarity and completeness, emphasizing to the particular case where the body forces are absent, leading to the homogeneity of Newton's law. Following a handy notation for the stiffness tensor, we present its isotropic form and the eight systems of anisotropic elasticity, in order to provide a consistent background.

Historically, the already ample literature is full of references on different physical topics, which are interrelated with linear elasticity and its general features, most of them being invoked into the pre-mentioned list [1–12]. On one side, the linear isotropic behavior of elastic media has an inherent mathematical interest and has attracted many scientists due to the fact that even though the related theory is much simplified, many applications can accept the isotropic character without loss of robustness. For instance, a class of universal relations in isotropic linear elasticity theory in the absence of body forces has been studied in [13, 14], while a series of papers [15–17] demonstrate the efficiency of analytical methods in isotropic elasticity by dealing with elastic wave scattering at low frequencies around ellipsoidal solid bodies or cavities. Therein, under the assumption of no external body forces, the general solution Papkovich provided a closed-form solution for the displacement field in terms of differential operators, acting on harmonic functions. On the other side, linear anisotropic elasticity describes perfectly real-life problems in applied mechanics and engineering science. Indeed, among interesting references, we distinguish the research work in [18], where it is shown that the cubic symmetry of the stiffness tensor is the only situation in linear anisotropic elasticity for which a strain energy density extremum can exist for all stress states, rendering this class of anisotropy very useful. Reference [19] gathers the basic aspects of anisotropy in elasticity and their applications to composite materials, providing general relations for the strain and stress. An interesting paper [20] refers to the incompressible limit of anisotropic elasticity with applications to surface waves and elastostatics, meaning time-independent elasticity. Moreover, a new modified couple stress theory for anisotropic elasticity and microscale laminated Kirchhoff plate model has been developed in [21], while a very recent study [22] introduces the dynamic stiffness of three-dimensional anisotropic multi-layered media based on the continued-fraction method. It is important to mention that the theoretical analysis of wave propagation in anisotropic media [23] comprises a difficult task and sometimes numerical methods must be employed [24] in order to obtain solutions of three-dimensional formulations. Closed-form solutions of elastostatics in spherical and ellipsoidal geometry can be found in reference [25]. Finally, despite the applicability of linearity in classical elasticity, it is obvious that several materials have not linear properties [26] and, consequently, careful attention should be given.

This work includes in a brief manner the main mathematical aspects of the linear anisotropic elastic theory, whereas our aim is twofold. Primarily, after writing down Hooke's law, relating the stress components to the strain components via the stiffness tensor, we focus on Newton's law in its generalized formula for both the isotropic and anisotropic cases, the latter being analyzed by virtue of the most commonly utilized cubic-type anisotropy system. Henceforth, special attention is given in elastostatics, neglecting the temporal derivatives as for steady state and considering no external forces. Under these circumstances, closed-type solutions for the displacement field are displayed. Even though a general solution in terms of harmonic functions is valid for the isotropic case, namely the Papkovich differential representation, this is not the issue with the anisotropic situation, wherein a first attempt is shown by generating a solution in the form of a homogeneous second degree polynomial. However, in view of the ambitious goal to obtain such a representation in the anisotropic regime, we devoted a separate section to our second involvement, wherein we present a methodology, according to which we insert the anisotropic character of any medium into Laplace's operator and develop an analytic code in order to obtain the corresponding anisotropic harmonic functions. Our technique works perfectly for the first four eigenfunctions, but it can be generalized for any degree greater than four. Thus, we transfer the difficulty in proposing a general differential representation for the three-dimensional anisotropy into the implicated harmonic eigenfunctions, associated with this solution. Finally, comparison with the isotropic case validates the efficiency of the proposed formulation.

Elaborating with the fundamental mathematical formulation of anisotropic elasticity in an analytic or even semi-analytic manner leads to solid solutions that have important advantages compared to the pure numerical methods, since their validity can be technically verified. On the other hand, bearing in mind that important physical laws, such as Newton's law, can be derived from analytic methodologies, we can understand the necessity of a stable and secure mathematical basis for starting a brute computational procedure. Besides, it is to this end that analytic and numerical methods are considered as complementary. Hence, in this work we offer the minimum of the necessary mathematical tools, which coexist with pure numerical codes and can be found in bibliography, solving boundary value problems with physical applications, associated with anisotropic linear elasticity.

The chapter is planned as follows. In Sect. 2, the theoretical basis of linear anisotropic elasticity via an analytic mathematical formulation is sketched, while in Sect. 3, the case of complete isotropy is discussed as a demonstration of the general theory, recovering the fundamental equation of the displacement field in linear isotropic elasticity. Section 4 is devoted to the presentation of the special and commonly appeared type of anisotropic elasticity, which corresponds to the cubic system, and the relative governing equation is obtained, wherein a solution in a polynomial form is obtained. Aiming in developing a solid mathematical technique to produce harmonic functions in anisotropic elasticity, we invoke Sect. 5 in which we build up eigensolutions of certain degree that belong to the kernel space of Laplace's operator. Finally, an outline of our work and future steps follow in Sect. 6.

2 Basic Theory in Anisotropic Elasticity

In what follows, we shall refer to smooth, either bounded or unbounded, three-dimensional domains $V\left(\mathbb{R}^3\right)$, where every field or property will be generally written in terms of the position vector $\mathbf{r} = x_1\hat{\mathbf{x}}_1 + x_2\hat{\mathbf{x}}_2 + x_3\hat{\mathbf{x}}_3$, expressed via the Cartesian basis $\hat{\mathbf{x}}_p$, $p = 1, 2, 3$, in coordinates (x_1, x_2, x_3) and the time variable t, while the convenient Einstein convention regarding vectors and tensors is frequently used. On that account, the well-known gradient and Laplacian differential operators assume the forms

$$\nabla = \sum_{i=1}^{3} \hat{\mathbf{x}}_i \frac{\partial}{\partial x_i} \quad \text{and} \quad \Delta = \sum_{i=1}^{3} \frac{\partial^2}{\partial x_i^2}, \tag{1}$$

respectively.

The fundamental field under consideration in solid mechanics is the displacement field, defined by

$$\mathbf{u}(\mathbf{r},t)=\mathbf{u}\equiv\sum_{i=1}^{3}u_i\hat{\mathbf{x}}_i \quad \text{for every} \quad \mathbf{r}\in V\left(\mathbb{R}^3\right), \qquad t>0, \tag{2}$$

and it is a measure of deformation of the material. For small displacement gradients, the linearized theory of elasticity considers the strain dyadic tensor $\tilde{\boldsymbol{\varepsilon}}$ to be the linear and symmetric part of the displacement gradient, that is,

$$\tilde{\boldsymbol{\varepsilon}}=\frac{1}{2}\left[\nabla\otimes\mathbf{u}+(\nabla\otimes\mathbf{u})^{\top}\right]\equiv\sum_{i,j=1}^{3}\varepsilon_{ij}\hat{\mathbf{x}}_i\otimes\hat{\mathbf{x}}_j \quad \text{with} \quad \tilde{\boldsymbol{\varepsilon}}=\tilde{\boldsymbol{\varepsilon}}^{\top}, \tag{3}$$

wherein "$\top$" denotes transposition and "$\otimes$" stands for the classical tensor product. Once $\tilde{\boldsymbol{\varepsilon}}$ is determined via (3), the stress dyadic tensor $\tilde{\boldsymbol{\tau}}$ is expressed via Hooke's law as

$$\tilde{\boldsymbol{\tau}}=\tilde{\tilde{\mathbf{c}}}:\tilde{\boldsymbol{\varepsilon}}\equiv\sum_{i,j=1}^{3}\tau_{ij}\hat{\mathbf{x}}_i\otimes\hat{\mathbf{x}}_j \quad \text{with} \quad \tilde{\tilde{\mathbf{c}}}\equiv\sum_{i,j,k,l=1}^{3}c_{ijkl}\hat{\mathbf{x}}_i\otimes\hat{\mathbf{x}}_j\otimes\hat{\mathbf{x}}_k\otimes\hat{\mathbf{x}}_l, \tag{4}$$

where ":" refers to the double inner product, while $\tilde{\tilde{\mathbf{c}}}$ is the stiffness tetratic tensor. If an external force $\mathbf{f}$ is applied on the material or the structure, then Newton's law yields

$$\nabla\cdot\tilde{\boldsymbol{\tau}}+\mathbf{f}=\rho\frac{\partial^2\mathbf{u}}{\partial t^2}, \tag{5}$$

where $\nabla\cdot\tilde{\boldsymbol{\tau}}$ are the forces due to deformation and ρ is the constant mass density, while $\dfrac{\partial^2}{\partial t^2}\equiv\partial_{tt}$ denotes double derivation with respect to the time variable. Replacing Eq. (4) with (3) into Newton's law (5), we can also write

$$\nabla\cdot\left(\tilde{\tilde{\mathbf{c}}}:\tilde{\boldsymbol{\varepsilon}}\right)+\mathbf{f}=\rho\frac{\partial^2\mathbf{u}}{\partial t^2} \quad \Rightarrow \quad \frac{1}{2}\nabla\cdot\left\{\tilde{\tilde{\mathbf{c}}}:\left[\nabla\otimes\mathbf{u}+(\nabla\otimes\mathbf{u})^{\top}\right]\right\}+\mathbf{f}=\rho\frac{\partial^2\mathbf{u}}{\partial t^2} \tag{6}$$

or

$$\frac{1}{2}\nabla\cdot\left\{\tilde{\tilde{\mathbf{c}}}:\left[\nabla\otimes\mathbf{u}+(\nabla\otimes\mathbf{u})^{\top}\right]\right\}=\rho\frac{\partial^2\mathbf{u}}{\partial t^2}, \tag{7}$$

when $\mathbf{f}=\mathbf{0}$. Hence, in the absence of external forces, the second-order spatial derivatives are proportional to the second-order temporal derivatives, which is exactly the meaning of linearity.

The tetratic $\tilde{\tilde{\mathbf{c}}}$ has in total $3^4 = 81$ components, though due to the specific spatial symmetries

$$c_{ijkl} = c_{jikl} = c_{ijlk} = c_{klij} \quad \text{with} \quad i, j, k, l = 1, 2, 3, \tag{8}$$

it holds that the symmetry $ij \leftrightarrow ji$ restricts the values of ij from $3 \times 3 = 9$ to 6, since they correspond to a symmetric 3×3 matrix. For the same reason, the symmetry $kl \leftrightarrow lk$ restricts the values of kl to 6, as well. Hence, ij and kl are equal to 11, 22, 33, 23, 31, 12 and that makes the number of different components of $\tilde{\tilde{\mathbf{c}}}$ to be $6 \times 6 = 36$. Finally, the symmetry $ij \leftrightarrow kl$ in (8) reduces the independent components of $\tilde{\tilde{\mathbf{c}}}$ to 21. The components c_{ijkl} with the symmetries (8) are called elasticities. The linear anisotropic elastic medium is then characterized in the invariant Cartesian coordinate system by the 6×6 symmetric matrix

$$\tilde{\tilde{\mathbf{c}}}_a = \begin{bmatrix} c_{1111} & c_{1122} & c_{1133} & c_{1123} & c_{1131} & c_{1112} \\ c_{1122} & c_{2222} & c_{2233} & c_{2223} & c_{2231} & c_{2212} \\ c_{1133} & c_{2233} & c_{3333} & c_{3323} & c_{3331} & c_{3312} \\ c_{1123} & c_{2223} & c_{3323} & c_{2323} & c_{2331} & c_{2312} \\ c_{1131} & c_{2231} & c_{3331} & c_{2331} & c_{3131} & c_{3112} \\ c_{1112} & c_{2212} & c_{3312} & c_{2312} & c_{3112} & c_{1212} \end{bmatrix}, \quad \text{where} \quad \tilde{\tilde{\mathbf{c}}}_a^{\top} = \tilde{\tilde{\mathbf{c}}}_a, \tag{9}$$

which, introducing for clarity the notation $\underset{1}{\overset{11}{\downarrow}} \ \underset{2}{\overset{22}{\downarrow}} \ \underset{3}{\overset{33}{\downarrow}} \ \underset{4}{\overset{23}{\downarrow}} \ \underset{5}{\overset{31}{\downarrow}} \ \underset{6}{\overset{12}{\downarrow}}$, then the general anisotropy is represented by

$$\tilde{\tilde{\mathbf{c}}}_a = \begin{bmatrix} c_{11} & c_{12} & c_{13} & c_{14} & c_{15} & c_{16} \\ c_{12} & c_{22} & c_{23} & c_{24} & c_{25} & c_{26} \\ c_{13} & c_{23} & c_{33} & c_{34} & c_{35} & c_{36} \\ c_{14} & c_{24} & c_{34} & c_{44} & c_{45} & c_{46} \\ c_{15} & c_{25} & c_{35} & c_{45} & c_{55} & c_{56} \\ c_{16} & c_{26} & c_{36} & c_{46} & c_{56} & c_{66} \end{bmatrix}, \quad \text{where} \quad \tilde{\tilde{\mathbf{c}}}_a^{\top} = \tilde{\tilde{\mathbf{c}}}_a, \tag{10}$$

providing the initial mathematical tool, in order to construct the basic theory of any type of linear elasticity.

Under this aim, we introduce the 8 special anisotropies, which are characterized by 8 stiffness matrices. These are the monoclinic system (with 13 elasticities)

$$\tilde{\tilde{\mathbf{c}}}_1 = \begin{bmatrix} c_{11} & c_{12} & c_{13} & 0 & 0 & c_{16} \\ c_{12} & c_{22} & c_{23} & 0 & 0 & c_{26} \\ c_{13} & c_{23} & c_{33} & 0 & 0 & c_{36} \\ 0 & 0 & 0 & c_{44} & c_{45} & 0 \\ 0 & 0 & 0 & c_{45} & c_{55} & 0 \\ c_{16} & c_{26} & c_{36} & 0 & 0 & c_{66} \end{bmatrix}, \quad \text{where} \quad \tilde{\tilde{\mathbf{c}}}_1^{\top} = \tilde{\tilde{\mathbf{c}}}_1, \tag{11}$$

the rhombic system (with 9 elasticities)

$$\tilde{\tilde{\mathbf{c}}}_2 = \begin{bmatrix} c_{11} & c_{12} & c_{13} & 0 & 0 & 0 \\ c_{12} & c_{22} & c_{23} & 0 & 0 & 0 \\ c_{13} & c_{23} & c_{33} & 0 & 0 & 0 \\ 0 & 0 & 0 & c_{44} & 0 & 0 \\ 0 & 0 & 0 & 0 & c_{55} & 0 \\ 0 & 0 & 0 & 0 & 0 & c_{66} \end{bmatrix}, \quad \text{where} \quad \tilde{\tilde{\mathbf{c}}}_2^{\top} = \tilde{\tilde{\mathbf{c}}}_2, \tag{12}$$

the tetragonal system A (with 7 elasticities)

$$\tilde{\tilde{\mathbf{c}}}_3 = \begin{bmatrix} c_{11} & c_{12} & c_{13} & 0 & 0 & c_{16} \\ c_{12} & c_{11} & c_{13} & 0 & 0 & -c_{16} \\ c_{13} & c_{12} & c_{33} & 0 & 0 & 0 \\ 0 & 0 & 0 & c_{44} & 0 & 0 \\ 0 & 0 & 0 & 0 & c_{44} & 0 \\ c_{16} & -c_{16} & 0 & 0 & 0 & c_{66} \end{bmatrix}, \quad \text{where} \quad \tilde{\tilde{\mathbf{c}}}_3^{\top} = \tilde{\tilde{\mathbf{c}}}_3, \tag{13}$$

the tetragonal system B (with 6 elasticities)

$$\tilde{\tilde{\mathbf{c}}}_4 = \begin{bmatrix} c_{11} & c_{12} & c_{13} & 0 & 0 & 0 \\ c_{12} & c_{11} & c_{13} & 0 & 0 & 0 \\ c_{13} & c_{13} & c_{33} & 0 & 0 & 0 \\ 0 & 0 & 0 & c_{44} & 0 & 0 \\ 0 & 0 & 0 & 0 & c_{44} & 0 \\ 0 & 0 & 0 & 0 & 0 & c_{66} \end{bmatrix}, \quad \text{where} \quad \tilde{\tilde{\mathbf{c}}}_4^{\top} = \tilde{\tilde{\mathbf{c}}}_4, \tag{14}$$

the cubic system (with 3 elasticities)

$$\tilde{\tilde{\mathbf{c}}}_5 = \begin{bmatrix} c_{11} & c_{12} & c_{12} & 0 & 0 & 0 \\ c_{12} & c_{11} & c_{12} & 0 & 0 & 0 \\ c_{12} & c_{12} & c_{11} & 0 & 0 & 0 \\ 0 & 0 & 0 & c_{44} & 0 & 0 \\ 0 & 0 & 0 & 0 & c_{44} & 0 \\ 0 & 0 & 0 & 0 & 0 & c_{44} \end{bmatrix}, \quad \text{where} \quad \tilde{\tilde{\mathbf{c}}}_5^{\top} = \tilde{\tilde{\mathbf{c}}}_5, \tag{15}$$

the hexagonal system A (with 7 elasticities)

$$\tilde{\tilde{\mathbf{c}}}_6 = \begin{bmatrix} c_{11} & c_{12} & c_{13} & c_{14} & c_{15} & 0 \\ c_{12} & c_{11} & c_{13} & -c_{14} & -c_{15} & 0 \\ c_{13} & c_{13} & c_{33} & 0 & 0 & 0 \\ c_{14} & -c_{14} & 0 & c_{44} & 0 & -c_{15} \\ c_{15} & -c_{15} & 0 & 0 & c_{44} & c_{14} \\ 0 & 0 & 0 & -c_{15} & c_{14} & \frac{1}{2}(c_{11}-c_{12}) \end{bmatrix}, \quad \text{where} \quad \tilde{\tilde{\mathbf{c}}}_6^{\top} = \tilde{\tilde{\mathbf{c}}}_6, \tag{16}$$

the hexagonal system B (with 6 elasticities)

$$\tilde{\tilde{\mathbf{c}}}_7 = \begin{bmatrix} c_{11} & c_{12} & c_{13} & c_{14} & 0 & 0 \\ c_{12} & c_{11} & c_{13} & -c_{14} & 0 & 0 \\ c_{13} & c_{13} & c_{33} & 0 & 0 & 0 \\ c_{14} & -c_{14} & 0 & c_{44} & 0 & 0 \\ 0 & 0 & 0 & 0 & c_{44} & c_{14} \\ 0 & 0 & 0 & 0 & c_{14} & \frac{1}{2}(c_{11}-c_{12}) \end{bmatrix}, \quad \text{where} \quad \tilde{\tilde{\mathbf{c}}}_7^{\top} = \tilde{\tilde{\mathbf{c}}}_7, \tag{17}$$

and the hexagonal system C (with 5 elasticities)

$$\tilde{\tilde{\mathbf{c}}}_8 = \begin{bmatrix} c_{11} & c_{12} & c_{13} & 0 & 0 & 0 \\ c_{12} & c_{11} & c_{13} & 0 & 0 & 0 \\ c_{13} & c_{13} & c_{33} & 0 & 0 & 0 \\ 0 & 0 & 0 & c_{44} & 0 & 0 \\ 0 & 0 & 0 & 0 & c_{44} & 0 \\ 0 & 0 & 0 & 0 & 0 & \frac{1}{2}(c_{11}-c_{12}) \end{bmatrix}, \quad \text{where} \quad \tilde{\tilde{\mathbf{c}}}_8^{\top} = \tilde{\tilde{\mathbf{c}}}_8. \tag{18}$$

The aforementioned matrices (11)–(18) correspond to simplified cases of the generalized form (10).

3 Complete Isotropy in Elasticity and Fundamental Equation

Herein, we mention the simplest case of isotropy described by the matrix

$$\tilde{\tilde{\mathbf{c}}}_i = \begin{bmatrix} \lambda+2\mu & \lambda & \lambda & 0 & 0 & 0 \\ \lambda & \lambda+2\mu & \lambda & 0 & 0 & 0 \\ \lambda & \lambda & \lambda+2\mu & 0 & 0 & 0 \\ 0 & 0 & 0 & 2\mu & 0 & 0 \\ 0 & 0 & 0 & 0 & 2\mu & 0 \\ 0 & 0 & 0 & 0 & 0 & 2\mu \end{bmatrix}, \quad \text{where} \quad \tilde{\tilde{\mathbf{c}}}_i^{\top} = \tilde{\tilde{\mathbf{c}}}_i, \tag{19}$$

in which $\lambda, \mu \in \mathbb{R}$ are the standard elastic parameters of the isotropic theory. In order to obtain the constitutive relation for example of (19), we work as follows. We know from definitions (3) and (4) that

$$\tilde{\boldsymbol{\tau}} = \tilde{\tilde{\mathbf{c}}} : \tilde{\boldsymbol{\varepsilon}} = \frac{1}{2}\tilde{\tilde{\mathbf{c}}} : \left[\nabla \otimes \mathbf{u} + (\nabla \otimes \mathbf{u})^{\top}\right], \tag{20}$$

whereas, for notational simplicity, we incorporate each one of the derivatives of $\mathbf{u}$ with respect to the argument within the indexes, such as in terms of its components u_p for $p = 1, 2, 3$ (see also (2)), we may write $\dfrac{\partial u_p}{\partial x_q} \equiv u_{p,q}$ for $q = 1, 2, 3$, keeping consistency with the forthcoming steps. So, we multiply (19) by the vector

$$\mathbf{x} = \begin{bmatrix} u_{1,1} + u_{1,1} \\ u_{2,2} + u_{2,2} \\ u_{3,3} + u_{3,3} \\ u_{2,3} + u_{3,2} \\ u_{3,1} + u_{1,3} \\ u_{1,2} + u_{2,1} \end{bmatrix}, \tag{21}$$

we perform some trivial calculations based on the relation

$$\nabla \cdot \mathbf{u} = \sum_{i=1}^{3} \frac{\partial u_i}{\partial x_i} \equiv \sum_{i=1}^{3} u_{i,i}, \tag{22}$$

and we obtain the vector

$$\mathbf{y} = \begin{bmatrix} 2\lambda\nabla \cdot \mathbf{u} + 4\mu u_{1,1} \\ 2\lambda\nabla \cdot \mathbf{u} + 4\mu u_{2,2} \\ 2\lambda\nabla \cdot \mathbf{u} + 4\mu u_{3,3} \\ 2\mu\left(u_{2,3} + u_{3,2}\right) \\ 2\mu\left(u_{3,1} + u_{1,3}\right) \\ 2\mu\left(u_{1,2} + u_{2,1}\right) \end{bmatrix} \quad \text{or} \quad \mathbf{y}' = \begin{bmatrix} \alpha_{11}\ \alpha_{12}\ \alpha_{31} \\ \alpha_{12}\ \alpha_{22}\ \alpha_{23} \\ \alpha_{31}\ \alpha_{23}\ \alpha_{33} \end{bmatrix}, \tag{23}$$

considering the fact that the vector $\mathbf{y}$ represents the matrix $\mathbf{y}'$, where α_{11}, α_{22}, and α_{33} are the 1st, 2nd, and 3rd components of $\mathbf{y}$ and α_{23}, α_{31}, and α_{12} are the 4th, 5th, and 6th components of $\mathbf{y}$, as given in (23). Consequently, bearing in mind the above reasoning and in view of Hooke's law (20), we obtain

$$\begin{aligned} 2\tilde{\boldsymbol{\tau}} &= \left(2\lambda\nabla \cdot \mathbf{u} + 4\mu u_{1,1}\right)\hat{\mathbf{x}}_1 \otimes \hat{\mathbf{x}}_1 + \left(2\lambda\nabla \cdot \mathbf{u} + 4\mu u_{2,2}\right)\hat{\mathbf{x}}_2 \otimes \hat{\mathbf{x}}_2 \\ &\quad + \left(2\lambda\nabla \cdot \mathbf{u} + 4\mu u_{3,3}\right)\hat{\mathbf{x}}_3 \otimes \hat{\mathbf{x}}_3 \\ &\quad + 2\mu\left(u_{1,2} + u_{2,1}\right)\left(\hat{\mathbf{x}}_1 \otimes \hat{\mathbf{x}}_2 + \hat{\mathbf{x}}_2 \otimes \hat{\mathbf{x}}_1\right) \end{aligned}$$

$$+ 2\mu\left(u_{2,3}+u_{3,2}\right)\left(\hat{\mathbf{x}}_2\otimes\hat{\mathbf{x}}_3+\hat{\mathbf{x}}_3\otimes\hat{\mathbf{x}}_2\right)$$
$$+ 2\mu\left(u_{3,1}+u_{1,3}\right)\left(\hat{\mathbf{x}}_3\otimes\hat{\mathbf{x}}_1+\hat{\mathbf{x}}_1\otimes\hat{\mathbf{x}}_3\right) \tag{24}$$

or in terms of the unit dyadic $\tilde{\mathbf{I}}=\sum_{i=1}^{3}\hat{\mathbf{x}}_i\otimes\hat{\mathbf{x}}_i$ and after reorganizing the terms of the tensor (24), it is

$$\begin{aligned}
2\tilde{\boldsymbol{\tau}} &= 2\lambda\left(\nabla\cdot\mathbf{u}\right)\tilde{\mathbf{I}}\\
&+ 2\mu\left(u_{1,1}\hat{\mathbf{x}}_1\otimes\hat{\mathbf{x}}_1+u_{1,2}\hat{\mathbf{x}}_1\otimes\hat{\mathbf{x}}_2+u_{1,3}\hat{\mathbf{x}}_1\otimes\hat{\mathbf{x}}_3\right.\\
&\quad+u_{2,1}\hat{\mathbf{x}}_2\otimes\hat{\mathbf{x}}_1+u_{2,2}\hat{\mathbf{x}}_2\otimes\hat{\mathbf{x}}_2+u_{2,3}\hat{\mathbf{x}}_2\otimes\hat{\mathbf{x}}_3\\
&\quad\left.+u_{3,1}\hat{\mathbf{x}}_3\otimes\hat{\mathbf{x}}_1+u_{3,2}\hat{\mathbf{x}}_3\otimes\hat{\mathbf{x}}_2+u_{3,3}\hat{\mathbf{x}}_3\otimes\hat{\mathbf{x}}_3\right)\\
&+ 2\mu\left(u_{1,1}\hat{\mathbf{x}}_1\otimes\hat{\mathbf{x}}_1+u_{2,1}\hat{\mathbf{x}}_1\otimes\hat{\mathbf{x}}_2+u_{3,1}\hat{\mathbf{x}}_1\otimes\hat{\mathbf{x}}_3\right.\\
&\quad+u_{1,2}\hat{\mathbf{x}}_2\otimes\hat{\mathbf{x}}_1+u_{2,2}\hat{\mathbf{x}}_2\otimes\hat{\mathbf{x}}_2+u_{3,2}\hat{\mathbf{x}}_2\otimes\hat{\mathbf{x}}_3\\
&\quad\left.+u_{1,3}\hat{\mathbf{x}}_3\otimes\hat{\mathbf{x}}_1+u_{2,3}\hat{\mathbf{x}}_3\otimes\hat{\mathbf{x}}_2+u_{3,3}\hat{\mathbf{x}}_3\otimes\hat{\mathbf{x}}_3\right)\\
&= 2\lambda\left(\nabla\cdot\mathbf{u}\right)\tilde{\mathbf{I}}+2\mu\left[\nabla\otimes\mathbf{u}+\left(\nabla\otimes\mathbf{u}\right)^{\top}\right],
\end{aligned} \tag{25}$$

since $\nabla\otimes\mathbf{u}=\sum_{i,j=1}^{3}u_{j,i}\hat{\mathbf{x}}_i\otimes\hat{\mathbf{x}}_j$ and $\left(\nabla\otimes\mathbf{u}\right)^{\top}=\sum_{i,j=1}^{3}u_{i,j}\hat{\mathbf{x}}_i\otimes\hat{\mathbf{x}}_j$. Therefore, relationship (25) ends up to

$$\tilde{\boldsymbol{\tau}}=\lambda\left(\nabla\cdot\mathbf{u}\right)\tilde{\mathbf{I}}+\mu\left[\nabla\otimes\mathbf{u}+\left(\nabla\otimes\mathbf{u}\right)^{\top}\right]. \tag{26}$$

In order to facilitate our final task to recover Newton's law (6) for this special case, we proceed by writing the latter in components, i.e.

$$\tau_{pq}=\lambda\left(\sum_{i=1}^{3}u_{i,i}\right)\delta_{pq}+\mu\left(u_{p,q}+u_{q,p}\right)\quad\text{for any}\quad p,q=1,\ 2,\ 3, \tag{27}$$

in view of Kronecker's delta δ or

$$\tau_{pq}=\lambda\left(\sum_{i=1}^{3}\varepsilon_{i,i}\right)\delta_{pq}+2\mu\varepsilon_{pq}\quad\text{for any}\quad p,q=1,\ 2,\ 3, \tag{28}$$

where the components of (3) imply

$$\varepsilon_{pq}=\frac{1}{2}\left(u_{p,q}+u_{q,p}\right)\quad\text{with}\quad p,q=1,\ 2,\ 3. \tag{29}$$

Similarly, the stiffness tetratic for the isotropic case (see matrix (19) with definition (4) for instance) can be rewritten through its components as

$$c_{ijkl} = \lambda\delta_{ij}\delta_{kl} + \mu\left(\delta_{ik}\delta_{jl} + \delta_{il}\delta_{jk}\right) \quad \text{for every} \quad i, j, k, l = 1,\ 2,\ 3. \tag{30}$$

On the other hand, keeping the same notation and since $\mathbf{f} = \sum_{i=1}^{3} f_i\hat{\mathbf{x}}_i$, the general relationship for Newton's law (6) reads

$$\frac{\partial}{\partial x_i}\left(\sum_{i,j,k,l=1}^{3} c_{ijkl}u_{k,l}\right) + f_i = \rho\frac{\partial^2 u_i}{\partial t^2} \quad \text{for} \quad i = 1,\ 2,\ 3, \tag{31}$$

where, combining all together for the isotropic case, the partial differential equation (31) reduces to

$$\mu\sum_{j=1}^{3} u_{i,jj} + (\lambda + \mu)\sum_{j=1}^{3} u_{j,ij} + f_i = \rho\frac{\partial^2 u_i}{\partial t^2} \quad \text{for} \quad i = 1,\ 2,\ 3. \tag{32}$$

By means of the trivial relations $\sum_{j=1}^{3} u_{i,jj} \equiv \Delta\mathbf{u}$ and $\sum_{j=1}^{3} u_{j,ij} \equiv \nabla(\nabla\cdot\mathbf{u})$ for $i = 1, 2, 3$, Eq. (32) is equivalently given in its vector form

$$\mu\Delta\mathbf{u} + (\lambda + \mu)\nabla(\nabla\cdot\mathbf{u}) + \mathbf{f} = \rho\frac{\partial^2\mathbf{u}}{\partial t^2} \quad \text{with} \quad \lambda, \mu \in \mathbb{R}, \tag{33}$$

which is the well-known Navier equation in isotropic elasticity. The analysis described earlier in this paragraph to obtain (33) and demonstrate the presented general theory for anisotropic elastic media was based on calculations among invariant vectors and matrices, beginning with the isotropic form of the stiffness matrix (19). Nevertheless, we could also reach (33) using classical dyadic analysis by virtue of definitions (2)–(6), where our starting point should be mapping matrix (19) to the corresponding stiffness dyadic tensor in (4), by retaining the necessary elasticities c_{ijkl} with $i, j, k, l = 1, 2, 3$, as the tetratic $\tilde{\tilde{\mathbf{c}}}_i$ requires. Definitely, this could lead to a much more elaborate procedure due to the complexity of Newton's law.

Nonetheless, many applications concern the case of no body forces ($\mathbf{f} = \mathbf{0}$) and harmonic time dependence implied for the displacement field, that is, $\mathbf{u}(\mathbf{r}, t) = \mathbf{U}(\mathbf{r})e^{-i\omega t}$ for $\mathbf{r} \in V(\mathbb{R}^3)$ and $t > 0$, in terms of the angular frequency ω, where $i = \sqrt{-1}$ is the imaginary unit, reducing (33) to

$$\mu\Delta\mathbf{U} + (\lambda + \mu)\nabla(\nabla\cdot\mathbf{U}) + \rho\omega^2\mathbf{U} = \mathbf{0} \quad \text{with} \quad \lambda, \mu \in \mathbb{R}, \tag{34}$$

which is the non-homogeneous and time-independent linearized equation of dynamic elasticity. Moreover, if $\omega = 0$, then we recover the homogeneous analogous of (34), being

$$\mu\Delta\mathbf{U} + (\lambda + \mu)\nabla(\nabla\cdot\mathbf{U}) = \mathbf{0} \quad \text{with} \quad \lambda, \mu \in \mathbb{R}, \tag{35}$$

whose general solution is given via the Papkovich representation

$$\mathrm{U} = \mathbf{A} - \frac{\lambda + \mu}{2(\lambda + 2\mu)} \nabla (\mathbf{r} \cdot \mathbf{A} + \mathrm{B}), \quad \text{where} \quad \Delta \mathbf{A} = \mathbf{0} \quad \text{and} \quad \Delta \mathrm{B} = 0, \tag{36}$$

functions $\mathbf{A}$ and B are vector and scalar harmonic functions, respectively. It is not hard to prove that (36) satisfies (35), bearing in mind the interchange $\Delta\nabla = \nabla\Delta$ and the vector identity $\Delta(\mathbf{r} \cdot \mathbf{A}) = \Delta\mathbf{r} \cdot \mathbf{A} + \mathbf{r} \cdot \Delta\mathbf{A} + 2(\nabla \otimes \mathbf{r})^\top : (\nabla \otimes \mathbf{A}) = 2\tilde{\mathbf{I}}^\top : (\nabla \otimes \mathbf{A}) = 2\tilde{\mathbf{I}} : (\nabla \otimes \mathbf{A}) = 2\nabla \cdot \mathbf{A}$, since it readily holds $\Delta\mathbf{r} = \Delta\mathbf{A} = \mathbf{0}$. The general differential representation (36) provides a powerful analytical tool for solving the homogeneous and time-independent linearized equation of classical dynamic elasticity. However, no one of the anisotropic cases (11)–(18), as they were presented earlier, can accept a Papkovich-type representation, since the harmonic fields $\mathbf{A}$ and B are associated with the special Laplacian operator Δ, which is solely characterized of isotropy.

4 Anisotropy in Elasticity and Fundamental Equation: The Cubic System

Based on the step-by-step procedure of the previous section, our aim is to initially analyze certain systems among (11)–(18) that carry the particular anisotropy. In the sequel, we reconstruct the corresponding constitutive equation from Newton's law and finally develop a novel mechanism to produce anisotropic harmonics, which could provide us with the proper mathematical tool for analytic solutions.

The simpler and commonly used form of anisotropic elasticity corresponds to the stiffness tetratic tensor of cubic type (15), wherein 3 elasticities survive. For convenience, we rewrite them as

$$c_{11} = \alpha, \quad c_{12} = \beta \quad \text{and} \quad c_{44} = \gamma, \tag{37}$$

and hence, the relative matrix reads

$$\tilde{\tilde{\mathbf{c}}}_5 = \begin{bmatrix} \alpha & \beta & \beta & 0 & 0 & 0 \\ \beta & \alpha & \beta & 0 & 0 & 0 \\ \beta & \beta & \alpha & 0 & 0 & 0 \\ 0 & 0 & 0 & \gamma & 0 & 0 \\ 0 & 0 & 0 & 0 & \gamma & 0 \\ 0 & 0 & 0 & 0 & 0 & \gamma \end{bmatrix}, \quad \text{where} \quad \tilde{\tilde{\mathbf{c}}}_5^\top = \tilde{\tilde{\mathbf{c}}}_5. \tag{38}$$

Following similar steps like those in our previous detailed analysis and multiplying (38) by (21) from the right, we obtain

$$\mathbf{z}=\begin{bmatrix}2\beta\nabla\cdot\mathbf{u}+2(\alpha-\beta)u_{1,1}\\2\beta\nabla\cdot\mathbf{u}+2(\alpha-\beta)u_{2,2}\\2\beta\nabla\cdot\mathbf{u}+2(\alpha-\beta)u_{3,3}\\\gamma\left(u_{2,3}+u_{3,2}\right)\\\gamma\left(u_{3,1}+u_{1,3}\right)\\\gamma\left(u_{1,2}+u_{2,1}\right)\end{bmatrix},\quad\text{where}\quad\nabla\cdot\mathbf{u}=\sum_{i=1}^{3}\frac{\partial u_i}{\partial x_i}\equiv\sum_{i=1}^{3}u_{i,i},\tag{39}$$

and using the same argument as in the case of the isotropy that led to Eq. (24) by virtue of (20), we conclude to

$$\begin{aligned}\tilde{\boldsymbol{\tau}}&=\left(\beta\,\nabla\cdot\mathbf{u}+(\alpha-\beta)\,u_{1,1}\right)\hat{\mathbf{x}}_1\otimes\hat{\mathbf{x}}_1+\left(\beta\,\nabla\cdot\mathbf{u}+(\alpha-\beta)\,u_{2,2}\right)\hat{\mathbf{x}}_2\otimes\hat{\mathbf{x}}_2\\&+\left(\beta\,\nabla\cdot\mathbf{u}+(\alpha-\beta)\,u_{3,3}\right)\hat{\mathbf{x}}_3\otimes\hat{\mathbf{x}}_3\\&+\frac{\gamma}{2}\left(u_{1,2}+u_{2,1}\right)\left(\hat{\mathbf{x}}_1\otimes\hat{\mathbf{x}}_2+\hat{\mathbf{x}}_2\otimes\hat{\mathbf{x}}_1\right)\\&+\frac{\gamma}{2}\left(u_{2,3}+u_{3,2}\right)\left(\hat{\mathbf{x}}_2\otimes\hat{\mathbf{x}}_3+\hat{\mathbf{x}}_3\otimes\hat{\mathbf{x}}_2\right)\\&+\frac{\gamma}{2}\left(u_{3,1}+u_{1,3}\right)\left(\hat{\mathbf{x}}_3\otimes\hat{\mathbf{x}}_1+\hat{\mathbf{x}}_1\otimes\hat{\mathbf{x}}_3\right).\end{aligned}\tag{40}$$

Expression (40) can be reorganized to

$$\begin{aligned}\tilde{\boldsymbol{\tau}}&=\beta\,(\nabla\cdot\mathbf{u})\,\tilde{\mathbf{I}}+\frac{\gamma}{2}\left(\nabla\otimes\mathbf{u}+(\nabla\otimes\mathbf{u})^{\top}\right)\\&+(\alpha-\beta-\gamma)\left(u_{1,1}\hat{\mathbf{x}}_1\otimes\hat{\mathbf{x}}_1+u_{2,2}\hat{\mathbf{x}}_2\otimes\hat{\mathbf{x}}_2+u_{3,3}\hat{\mathbf{x}}_3\otimes\hat{\mathbf{x}}_3\right),\end{aligned}\tag{41}$$

since $\tilde{\mathbf{I}}=\hat{\mathbf{x}}_1\otimes\hat{\mathbf{x}}_1+\hat{\mathbf{x}}_2\otimes\hat{\mathbf{x}}_2+\hat{\mathbf{x}}_3\otimes\hat{\mathbf{x}}_3$ and $\nabla\otimes\mathbf{u}=\sum_{i,j=1}^{3}u_{j,i}\hat{\mathbf{x}}_i\otimes\hat{\mathbf{x}}_j$. Next, we calculate the equation satisfied by $\mathbf{u}$ corresponding to the stress tensor (41) via Newton's law (6). Eventually, this is

$$\begin{aligned}&\frac{\gamma}{2}\Delta\mathbf{u}+\left(\beta+\frac{\gamma}{2}\right)\nabla(\nabla\cdot\mathbf{u})+(\alpha-\beta-\gamma)\left[\hat{\mathbf{x}}_1\frac{\partial^2u_1}{\partial x_1^2}+\hat{\mathbf{x}}_2\frac{\partial^2u_2}{\partial x_2^2}+\hat{\mathbf{x}}_3\frac{\partial^2u_3}{\partial x_3^2}\right]+\mathbf{f}\\&\quad=\rho\frac{\partial^2\mathbf{u}}{\partial t^2}\end{aligned}\tag{42}$$

or in component form

$$\frac{\gamma}{2}\Delta u_p+\left(\beta+\frac{\gamma}{2}\right)\left(u_{1,1p}+u_{2,2p}+u_{3,3p}\right)+(\alpha-\beta-\gamma)\,u_{p,pp}+f_p=\rho\frac{\partial^2u_p}{\partial t^2}\quad\text{for}\quad p=1,2,3,\tag{43}$$

wherein we recall that $\mathbf{f} = \sum_{i=1}^{3} f_i \hat{\mathbf{x}}_i$ and $\mathbf{u} = \sum_{i=1}^{3} u_i \hat{\mathbf{x}}_i$, while (42) stands for the constitutive fundamental equation in anisotropic elasticity for the cubic system.

Here, we remark that the stiffness matrix (19) of the isotropic case is identified with the stiffness matrix (38) of the cubic case by choosing

$$\alpha = \lambda + 2\mu, \quad \beta = \lambda \quad \text{and} \quad \gamma = 2\mu, \tag{44}$$

where in this case

$$\alpha - \beta - \gamma = 0 \tag{45}$$

and Eq. (26) is identified with Eq. (41). Then, the constitutive equation (33) follows straightforwardly from (42), corresponding to isotropic elasticity.

In order to simplify our calculations and focus on the spatial structure of the solution, we consider the case of elastostatics where $\dfrac{\partial^2 \mathbf{u}}{\partial t^2} = \mathbf{0}$ and no external forces applied on the material, i.e. $\mathbf{f} = \mathbf{0}$, and thus (42) is rewritten as

$$\frac{\gamma}{2}\Delta\mathbf{u} + \left(\beta + \frac{\gamma}{2}\right)\nabla(\nabla\cdot\mathbf{u}) + (\alpha - \beta - \gamma)\left[\hat{\mathbf{x}}_1\frac{\partial^2 u_1}{\partial x_1^2} + \hat{\mathbf{x}}_2\frac{\partial^2 u_2}{\partial x_2^2} + \hat{\mathbf{x}}_3\frac{\partial^2 u_3}{\partial x_3^2}\right] = \mathbf{0}, \tag{46}$$

where α, β, and γ are the three elasticities of the cubic system. Unfortunately, it is difficult to construct a representation-type solution for (46) such as the Papkovich general solution, defined by relationship (36). However, in what follows, we will try to generate a solution of partial differential equation (46) in the form of a homogeneous second degree polynomial. So, let this solution be

$$\mathbf{u} = \mathbf{a}x_1^2 + \mathbf{b}x_2^2 + \mathbf{c}x_3^2 + \mathbf{d}x_1x_2 + \mathbf{e}x_2x_3 + \mathbf{f}x_3x_1, \tag{47}$$

in which

$$\begin{aligned}
\mathbf{a} &\equiv (a_1, a_2, a_3) = a_1\hat{\mathbf{x}}_1 + a_2\hat{\mathbf{x}}_2 + a_3\hat{\mathbf{x}}_3,\\
\mathbf{b} &\equiv (b_1, b_2, b_3) = b_1\hat{\mathbf{x}}_1 + b_2\hat{\mathbf{x}}_2 + b_3\hat{\mathbf{x}}_3,\\
\mathbf{c} &\equiv (c_1, c_2, c_3) = c_1\hat{\mathbf{x}}_1 + c_2\hat{\mathbf{x}}_2 + c_3\hat{\mathbf{x}}_3,\\
\mathbf{d} &\equiv (d_1, d_2, d_3) = d_1\hat{\mathbf{x}}_1 + d_2\hat{\mathbf{x}}_2 + d_3\hat{\mathbf{x}}_3,\\
\mathbf{e} &\equiv (e_1, e_2, e_3) = e_1\hat{\mathbf{x}}_1 + e_2\hat{\mathbf{x}}_2 + e_3\hat{\mathbf{x}}_3,\\
\mathbf{f} &\equiv (f_1, f_2, f_3) = f_1\hat{\mathbf{x}}_1 + f_2\hat{\mathbf{x}}_2 + f_3\hat{\mathbf{x}}_3
\end{aligned} \tag{48}$$

are arbitrary constant coefficients. With this solution, we obtain

$$\Delta\mathbf{u} = 2\mathbf{a} + 2\mathbf{b} + 2\mathbf{c} \tag{49}$$

$$\nabla\cdot\mathbf{u} = 2a_1x_1 + d_1x_2 + f_1x_3 + 2b_2x_2 + d_2x_1 + e_2x_3 + 2c_3x_3 + e_3x_2 + f_3x_1, \tag{50}$$

and hence

$$\nabla\left(\nabla\cdot\mathbf{u}\right)=\left(2a_1+d_2+f_3\right)\hat{\mathbf{x}}_1+\left(d_1+2b_2+e_3\right)\hat{\mathbf{x}}_2+\left(f_1+e_2+2c_3\right)\hat{\mathbf{x}}_3 \tag{51}$$

and

$$\hat{\mathbf{x}}_1\frac{\partial^2 u_1}{\partial x_1^2}+\hat{\mathbf{x}}_2\frac{\partial^2 u_2}{\partial x_2^2}+\hat{\mathbf{x}}_3\frac{\partial^2 u_3}{\partial x_3^2}=2a_1\hat{\mathbf{x}}_1+2b_2\hat{\mathbf{x}}_2+2c_3\hat{\mathbf{x}}_3. \tag{52}$$

Then, in view of (49)–(52), Eq. (46) provides the following algebraic linear system of equations

$$\gamma\left(a_1+b_1+c_1\right)+\left(\beta+\frac{\gamma}{2}\right)\left(2a_1+d_2+f_3\right)+2\left(\alpha-\beta-\gamma\right)a_1=0, \tag{53}$$

$$\gamma\left(a_2+b_2+c_2\right)+\left(\beta+\frac{\gamma}{2}\right)\left(d_1+2b_2+e_3\right)+2\left(\alpha-\beta-\gamma\right)b_2=0, \tag{54}$$

$$\gamma\left(a_3+b_3+c_3\right)+\left(\beta+\frac{\gamma}{2}\right)\left(f_1+e_2+2c_3\right)+2\left(\alpha-\beta-\gamma\right)c_3=0, \tag{55}$$

which is also written as

$$2\alpha a_1+\gamma\left(b_1+c_1\right)+\left(\beta+\frac{\gamma}{2}\right)\left(d_2+f_3\right)=0, \tag{56}$$

$$2\alpha b_2+\gamma\left(a_2+c_2\right)+\left(\beta+\frac{\gamma}{2}\right)\left(d_1+e_3\right)=0, \tag{57}$$

$$2\alpha c_3+\gamma\left(a_3+b_3\right)+\left(\beta+\frac{\gamma}{2}\right)\left(f_1+e_2\right)=0. \tag{58}$$

At this point, let us discuss on the system (56)–(58). The solution (47) involves the 6 unknown vectors $\mathbf{a},\ \mathbf{b},\ \mathbf{c},\ \mathbf{d},\ \mathbf{e}$, and $\mathbf{f}$, given in (48). Thus, we have to calculate 18 scalars. However, the scalars $e_1,\ f_2$, and d_3 do not appear in the system (56)–(58). On the other hand, each one of Eqs. (56)–(58) involves 5 unknown constants. Hence, each equation leaves 4 constants undetermined. Therefore, we have finally $3{\cdot}4+3=15$ undetermined constants, that, is 5 constants for each component of the vector solution $\mathbf{u}$ given in (47). This is in accordance to the fact that there are $2n+1$ independent solutions in the form of homogeneous polynomials of degree n. In our case, $n=2$, which gives $2{\cdot}2+1=5$ independent solutions. These 5 solutions come from the arbitrary values we can give to the undetermined constants. In fact, we can use (56) to determine a_1 in terms of $b_1,\ c_1,\ d_2$, and f_3 and the dummy constant e_1, we can use (57) to determine b_2 in terms of $a_2,\ c_2,\ d_1$, and e_3 and the dummy constant f_2, and we can use (58) to determine c_3 in terms of $a_3,\ b_3,\ f_1$, and e_2 and the dummy constant d_3. Note also that each one of the Eqs. (56)–(58) is independent of the other two, since no unknown constant appears in more than one equation.

Obviously, we can combine any solution of the x_1-direction with any solution in the x_2-direction and with any solution in the x_3-direction. Once the system (56)–(58) is solved as described, the displacement field $\mathbf{u}$, satisfying (46), is given by relationship (47).

In order to demonstrate our analytical approach, we provide the following example. Let us assume that

$$b_1 = a_2 = a_3 = 1 \quad \text{and} \quad a_1 b_2 c_3 \neq 0, \tag{59}$$

while all other constants are equal to zero. Then, the system (56)–(58) becomes

$$2\alpha a_1 + \gamma b_1 = 0, \tag{60}$$

$$2\alpha b_2 + \gamma a_2 = 0, \tag{61}$$

$$2\alpha c_3 + \gamma a_3 = 0 \tag{62}$$

or

$$a_1 = b_2 = c_3 = -\frac{\gamma}{2\alpha}, \tag{63}$$

where, in this case, we obtain from (47) with (48) the solution

$$\mathbf{u} = \left(-\frac{\gamma}{2\alpha}x_1^2 + x_2^2\right)\hat{\mathbf{x}}_1 + \left(x_1^2 - \frac{\gamma}{2\alpha}x_2^2\right)\hat{\mathbf{x}}_2 + \left(x_1^2 - \frac{\gamma}{2\alpha}x_3^2\right)\hat{\mathbf{x}}_3. \tag{64}$$

In order to verify this solution, we use (64) to calculate in sequence

$$\Delta\mathbf{u} = \left(-\frac{\gamma}{\alpha} + 2\right)\hat{\mathbf{x}}_1 + \left(2 - \frac{\gamma}{\alpha}\right)\hat{\mathbf{x}}_2 + \left(2 - \frac{\gamma}{\alpha}\right)\hat{\mathbf{x}}_3 = \left(2 - \frac{\gamma}{\alpha}\right)\left(\hat{\mathbf{x}}_1 + \hat{\mathbf{x}}_2 + \hat{\mathbf{x}}_3\right), \tag{65}$$

$$\nabla\cdot\mathbf{u} = \left(-\frac{\gamma}{\alpha}x_1\right) + \left(-\frac{\gamma}{\alpha}x_2\right) + \left(-\frac{\gamma}{\alpha}x_3\right) = -\frac{\gamma}{\alpha}\left(x_1 + x_2 + x_3\right), \tag{66}$$

$$\nabla\left(\nabla\cdot\mathbf{u}\right) = \left(-\frac{\gamma}{\alpha}\hat{\mathbf{x}}_1\right) + \left(-\frac{\gamma}{\alpha}\hat{\mathbf{x}}_2\right) + \left(-\frac{\gamma}{\alpha}\hat{\mathbf{x}}_3\right) = -\frac{\gamma}{\alpha}\left(\hat{\mathbf{x}}_1 + \hat{\mathbf{x}}_2 + \hat{\mathbf{x}}_3\right), \tag{67}$$

and

$$\hat{\mathbf{x}}_1\frac{\partial^2 u_1}{\partial x_1^2} + \hat{\mathbf{x}}_2\frac{\partial^2 u_2}{\partial x_2^2} + \hat{\mathbf{x}}_3\frac{\partial^2 u_3}{\partial x_3^2} = -\frac{\gamma}{\alpha}\hat{\mathbf{x}}_1 - \frac{\gamma}{\alpha}\hat{\mathbf{x}}_2 - \frac{\gamma}{\alpha}\hat{\mathbf{x}}_3 = -\frac{\gamma}{\alpha}\left(\hat{\mathbf{x}}_1 + \hat{\mathbf{x}}_2 + \hat{\mathbf{x}}_3\right), \tag{68}$$

and consequently, we insert (65)–(68) into the following expression, and we obtain

$$\begin{aligned}
&\frac{\gamma}{2}\Delta\mathbf{u}+\left(\beta+\frac{\gamma}{2}\right)\nabla\left(\nabla\cdot\mathbf{u}\right)+(\alpha-\beta-\gamma)\left[\hat{\mathbf{x}}_1\frac{\partial^2 u_1}{\partial x_1^2}+\hat{\mathbf{x}}_2\frac{\partial^2 u_2}{\partial x_2^2}+\hat{\mathbf{x}}_3\frac{\partial^2 u_3}{\partial x_3^2}\right]\\
&\qquad=\frac{\gamma}{2}\left(2-\frac{\gamma}{\alpha}\right)\left(\hat{\mathbf{x}}_1+\hat{\mathbf{x}}_2+\hat{\mathbf{x}}_3\right)\\
&\qquad\quad+\left(\beta+\frac{\gamma}{2}\right)\left(-\frac{\gamma}{\alpha}\right)\left(\hat{\mathbf{x}}_1+\hat{\mathbf{x}}_2+\hat{\mathbf{x}}_3\right)\\
&\qquad\quad+(\alpha-\beta-\gamma)\left(-\frac{\gamma}{\alpha}\right)\left(\hat{\mathbf{x}}_1+\hat{\mathbf{x}}_2+\hat{\mathbf{x}}_3\right)\\
&\qquad=\left(\hat{\mathbf{x}}_1+\hat{\mathbf{x}}_2+\hat{\mathbf{x}}_3\right)\left[\gamma-\frac{\gamma^2}{2\alpha}-\frac{\beta\gamma}{\alpha}-\frac{\gamma^2}{2\alpha}-\gamma+\frac{\beta\gamma}{\alpha}+\frac{\gamma^2}{\alpha}\right]\\
&\qquad=\left(\hat{\mathbf{x}}_1+\hat{\mathbf{x}}_2+\hat{\mathbf{x}}_3\right)\cdot 0=\mathbf{0},
\end{aligned}\tag{69}$$

a result that readily secures (46). Evidently, the presented analytical methodology is valid and provides us with a first attempt to obtain closed-form solutions in linear time-independent cubic-type anisotropic elasticity in the absence of body forces.

5 Anisotropic Harmonic Eigenfunctions

The theory of harmonic functions concerning anisotropic media is associated with the anisotropy tensor $\tilde{\boldsymbol{\sigma}}$, which admits

$$\tilde{\boldsymbol{\sigma}}=\sum_{i,j=1}^{3}\sigma_{ij}\hat{\mathbf{x}}_i\otimes\hat{\mathbf{x}}_j \tag{70}$$

in Cartesian coordinates. Given the constant dyadic (70) and aiming to find the so-called anisotropic harmonic functions $u=u(\mathbf{r})$ for $\mathbf{r}\in V\left(\mathbb{R}^3\right)$, we want to construct solutions of equation

$$\nabla\cdot\left(\tilde{\boldsymbol{\sigma}}\cdot\nabla u\right)=0\quad\text{with}\quad\nabla=\sum_{i=1}^{3}\hat{\mathbf{x}}_i\frac{\partial}{\partial x_i}, \tag{71}$$

which is the equivalent of Laplacian in anisotropic elasticity. Indeed, if $\tilde{\boldsymbol{\sigma}}=\tilde{\mathbf{I}}$, then (71) reduces to $\nabla\cdot\left(\tilde{\mathbf{I}}\cdot\nabla u\right)=0\ \Rightarrow\ \nabla\cdot\nabla u=0$ or $\Delta u=0$. In terms of (70), the differential operator within relationship (71) yields

$$\nabla\cdot\left(\tilde{\boldsymbol{\sigma}}\cdot\nabla\right)=\sum_{k=1}^{3}\hat{\boldsymbol{x}}_k\frac{\partial}{\partial x_k}\cdot\left[\sum_{i,j=1}^{3}\sigma_{ij}\hat{\boldsymbol{x}}_i\otimes\hat{\boldsymbol{x}}_j\cdot\sum_{l=1}^{3}\hat{\boldsymbol{x}}_l\frac{\partial}{\partial x_l}\right]=\sum_{i,j=1}^{3}\sigma_{ij}\frac{\partial}{\partial x_i}\frac{\partial}{\partial x_j} \tag{72}$$

or

$$\nabla \cdot (\tilde{\boldsymbol{\sigma}} \cdot \nabla) = \sigma_{11} \frac{\partial^2}{\partial x_1^2} + \sigma_{22} \frac{\partial^2}{\partial x_2^2} + \sigma_{33} \frac{\partial^2}{\partial x_3^2} + (\sigma_{12} + \sigma_{21}) \frac{\partial}{\partial x_1} \frac{\partial}{\partial x_2} + (\sigma_{23} + \sigma_{32}) \frac{\partial}{\partial x_2} \frac{\partial}{\partial x_3} + (\sigma_{31} + \sigma_{13}) \frac{\partial}{\partial x_3} \frac{\partial}{\partial x_1} \tag{73}$$

since $\tilde{\boldsymbol{\sigma}}$ is constant and $\hat{\boldsymbol{x}}_k \cdot \hat{\boldsymbol{x}}_i \otimes \hat{\boldsymbol{x}}_j \cdot \hat{\boldsymbol{x}}_l = 1$ when $k = i$, $l = j$ for any $i, j, k, l = 1, 2, 3$ and otherwise zero. In the sequel, we will try to develop a handy mathematical technique in order to build anisotropic harmonic eigenfunctions, which are solutions of (71) with (73). To this end, if u belongs to the kernel space of operator (73), then we assume the expansion

$$u(\mathbf{r}) \equiv \sum_{n=0}^{+\infty} u_n(\mathbf{r}) = \sum_{n=0}^{+\infty} \sum_{m=1}^{2n+1} c_n^m H_n^m(\mathbf{r}) \quad \text{for every} \quad \mathbf{r} \in V\left(\mathbb{R}^3\right), \tag{74}$$

written in terms of the $2n+1$ linearly independent eigenfunctions H_n^m that must be evaluated for each degree $n \geq 0$ and order $m = 1, 2, \ldots, 2n+1$, while $c_n^m \in \mathbb{R}$ are the arbitrary constant coefficients of the linear combination. The developed methodology is established by representing the function u as a homogeneous polynomial of nth degree via

$$u_n(\mathbf{r}) = \sum_{n_1+n_2+n_3=n} C_{n_1,n_2,n_3} x_1^{n_1} x_2^{n_2} x_3^{n_3} \text{ with } n \geq 0 \text{ and } \mathbf{r} \in V\left(\mathbb{R}^3\right), \tag{75}$$

where the constants C_{n_1,n_2,n_3} for $n_p \in \mathbb{N}$, $p = 1, 2, 3$, are unknown and need to be calculated after imposing (75) into (71), accompanied by (73). Then, matching (74) with (75), we are led to eigenfunctions H_n^m for $m = 1, 2, \ldots, 2n+1$ with $n \geq 0$ and an expansion of the form (74) is eventually feasible.

Our starting point is the first two trivial cases with respect to the monomial bases $u_0 \in \{1\}$ and $u_1 \in \{x_1, x_2, x_3\}$ that correspond to zeroth and first degree polynomials of the form (75), respectively, which immediately satisfy relationship (71) and operator (73). Hence, for $n = 0$, we have $2 \cdot 0 + 1 = 1$ ($m = 1$) eigenfunction, i.e.

$$H_0^1(\mathbf{r}) = 1 \tag{76}$$

for every $\mathbf{r} \in V\left(\mathbb{R}^3\right)$, while for $n = 1$, we readily obtain the $2 \cdot 1 + 1 = 3$ ($m = 1, 2, 3$) eigenfunctions

$$H_1^1(\mathbf{r}) = x_1, \tag{77}$$

$$H_1^2(\mathbf{r}) = x_2, \tag{78}$$

and

$$H_1^3(\mathbf{r}) = x_3 \tag{79}$$

for every $\mathbf{r} \in V(\mathbb{R}^3)$. In order to demonstrate a general methodology, the next three non-trivial cases for $n = 2, 3, 4$ follow.

Consequently, for $n = 2$, we assume the homogeneous second degree polynomial

$$u_2 = A_1x_1^2 + A_2x_2^2 + A_3x_3^2 + A_4x_1x_2 + A_5x_2x_3 + A_6x_3x_1 \tag{80}$$

from (75) and then by virtue of (73)

$$\begin{aligned}\nabla \cdot (\tilde{\boldsymbol{\sigma}} \cdot \nabla u_2) = {} & 2\sigma_{11}\mathrm{A}_1 + 2\sigma_{22}\mathrm{A}_2 + 2\sigma_{33}\mathrm{A}_3 + (\sigma_{12} + \sigma_{21})\,\mathrm{A}_4 \\ & + (\sigma_{23} + \sigma_{32})\,\mathrm{A}_5 + (\sigma_{31} + \sigma_{13})\,\mathrm{A}_6.\end{aligned} \tag{81}$$

Hence, u_2 is an anisotropic harmonic function if the right-hand side of (81) vanishes (see also (71)) and the 6 coefficients A_p, $p = 1, 2, \ldots, 6$, are connected by the condition

$$\begin{aligned}& 2\sigma_{11}\mathrm{A}_1 + 2\sigma_{22}\mathrm{A}_2 + 2\sigma_{33}\mathrm{A}_3 + (\sigma_{12} + \sigma_{21})\,\mathrm{A}_4 + (\sigma_{23} + \sigma_{32})\,\mathrm{A}_5 \\ & \quad + (\sigma_{31} + \sigma_{13})\,\mathrm{A}_6 = 0.\end{aligned} \tag{82}$$

Consequently, only 5 coefficients are independent and they define 5 anisotropic harmonics of degree $n = 2$. One possible way to construct these 5 harmonics is given via

$$\begin{array}{cccccc} A_1 & A_2 & A_3 & A_4 & A_5 & A_6 \\ \sigma_{31+13} & 0 & 0 & 0 & 0 & -2\sigma_{11} \\ \sigma_{23+32} & 0 & 0 & 0 & -2\sigma_{11} & 0 \\ \sigma_{12+21} & 0 & 0 & -2\sigma_{11} & 0 & 0 \\ \sigma_{33} & 0 & -\sigma_{11} & 0 & 0 & 0 \\ \sigma_{22} & -\sigma_{11} & 0 & 0 & 0 & 0 \end{array}, \tag{83}$$

which is a useful layout that leads to the following $2 \cdot 2 + 1 = 5$ ($m = 1, 2, \ldots, 5$) anisotropic harmonics of the second degree, those being

$$H_2^1(\mathbf{r}) = (\sigma_{31} + \sigma_{13})\,x_1^2 - 2\sigma_{11}x_3x_1, \tag{84}$$

$$H_2^2(\mathbf{r}) = (\sigma_{23} + \sigma_{32})\,x_1^2 - 2\sigma_{11}x_2x_3, \tag{85}$$

$$H_2^3(\mathbf{r}) = (\sigma_{12} + \sigma_{21})\,x_1^2 - 2\sigma_{11}x_1x_2, \tag{86}$$

$$H_2^4(\mathbf{r}) = \sigma_{33}x_1^2 - \sigma_{11}x_3^2, \tag{87}$$

and

$$H_2^5(\mathbf{r}) = \sigma_{22}x_1^2 - \sigma_{11}x_2^2. \tag{88}$$

Any other anisotropic harmonic of degree $n = 2$ has to be expressed as a linear combination of H_2^p, $p = 1, 2, \ldots, 5$.

In the sequel, for $n = 3$, let us assume the third degree homogeneous polynomial, which is rendered by

$$\begin{aligned} u_3 = B_1x_1^3 + B_2x_2^3 + B_3x_3^3 + B_4x_1^2x_2 + B_5x_1x_2^2 \\ + B_6x_2^2x_3 + B_7x_2x_3^2 \\ + B_8x_3^2x_1 + B_9x_3x_1^2 + B_{10}x_1x_2x_3 \end{aligned} \tag{89}$$

from expression (75), which, in view of (71) with (73), implies

$$\begin{aligned} &\nabla \cdot (\tilde{\boldsymbol{\sigma}} \cdot \nabla u_3) \\ &= [6\sigma_{11}B_1 + 2\sigma_{22}B_5 + 2\sigma_{33}B_8 + 2(\sigma_{12} + \sigma_{21})B_4 \\ &\quad + (\sigma_{23} + \sigma_{32})B_{10} + 2(\sigma_{31} + \sigma_{13})B_9]\,x_1 \\ &\quad + [2\sigma_{11}B_4 + 6\sigma_{22}B_2 + 2\sigma_{33}B_7 + 2(\sigma_{12} + \sigma_{21})B_5 \\ &\quad + 2(\sigma_{23} + \sigma_{32})B_6 + (\sigma_{31} + \sigma_{13})B_{10}]\,x_2 \\ &\quad + [2\sigma_{11}B_9 + 2\sigma_{22}B_6 + 6\sigma_{33}B_3 + (\sigma_{12} + \sigma_{21})B_{10} \\ &\quad + 2(\sigma_{23} + \sigma_{32})B_7 + 2(\sigma_{31} + \sigma_{13})B_8]\,x_3 = 0. \end{aligned} \tag{90}$$

Hence, according to (90), the function u_3 is anisotropic harmonic if the following three constrains hold true, i.e.

$$\begin{aligned} &6\sigma_{11}B_1 + 2\sigma_{22}B_5 + 2\sigma_{33}B_8 + 2(\sigma_{12} + \sigma_{21})B_4 \\ &\quad + (\sigma_{23} + \sigma_{32})B_{10} + 2(\sigma_{31} + \sigma_{13})B_9 = 0, \end{aligned} \tag{91}$$

$$\begin{aligned} &2\sigma_{11}B_4 + 6\sigma_{22}B_2 + 2\sigma_{33}B_7 + 2(\sigma_{12} + \sigma_{21})B_5 \\ &\quad +2(\sigma_{23} + \sigma_{32})B_6 + (\sigma_{31} + \sigma_{13})B_{10} = 0, \end{aligned} \tag{92}$$

$$\begin{aligned} &2\sigma_{11}B_9 + 2\sigma_{22}B_6 + 6\sigma_{33}B_3 + (\sigma_{12} + \sigma_{21})B_{10} \\ &\quad +2(\sigma_{23} + \sigma_{32})B_7 + 2(\sigma_{31} + \sigma_{13})B_8 = 0. \end{aligned} \tag{93}$$

Relations (91)–(92) reduce the independent coefficients B_p, $p = 1, 2, \ldots, 10$, to 7. In fact, they provide the following values of B_1, B_2, and B_3, respectively.

$$-6\sigma_{11}B_1 = 2\sigma_{22}B_5 + 2\sigma_{33}B_8 + 2(\sigma_{12}+\sigma_{21})B_4 + (\sigma_{23}+\sigma_{32})B_{10} + 2(\sigma_{31}+\sigma_{13})B_9, \quad (94)$$

$$-6\sigma_{22}B_2 = 2\sigma_{11}B_4 + 2\sigma_{33}B_7 + 2(\sigma_{12}+\sigma_{21})B_5 + 2(\sigma_{23}+\sigma_{32})B_6 + (\sigma_{31}+\sigma_{13})B_{10}, \quad (95)$$

$$-6\sigma_{33}B_3 = 2\sigma_{11}B_9 + 2\sigma_{22}B_6 + (\sigma_{12}+\sigma_{21})B_{10} + 2(\sigma_{23}+\sigma_{32})B_7 + 2(\sigma_{31}+\sigma_{13})B_8. \quad (96)$$

Then, we are left with the free to vary independently coefficients B_4, B_5, B_6, B_7, B_8, B_9, and B_{10}. Choosing the seven cases, where only one of the B_q for $q = 4, 5, 6, 7, 8, 9, 10$ is nonzero and all the others are equal to zero, we obtain, recalling (94)–(96),

- 1st Case: $B_4 \neq 0$ and $B_q = 0$ for $q = 5, 6, 7, 8, 9, 10$, then

$$B_1 = -\frac{\sigma_{12}+\sigma_{21}}{3\sigma_{11}}B_4, \quad (97)$$

$$B_2 = -\frac{\sigma_{11}}{3\sigma_{22}}B_4. \quad (98)$$

- 2nd Case: $B_5 \neq 0$ and $B_q = 0$ for $q = 4, 6, 7, 8, 9, 10$, then

$$B_1 = -\frac{\sigma_{22}}{3\sigma_{11}}B_5, \quad (99)$$

$$B_2 = -\frac{\sigma_{12}+\sigma_{21}}{3\sigma_{22}}B_5. \quad (100)$$

- 3rd Case: $B_6 \neq 0$ and $B_q = 0$ for $q = 4, 5, 7, 8, 9, 10$, then

$$B_2 = -\frac{\sigma_{23}+\sigma_{32}}{3\sigma_{22}}B_6, \quad (101)$$

$$B_3 = -\frac{\sigma_{22}}{3\sigma_{33}}B_6. \quad (102)$$

- 4th Case: $B_7 \neq 0$ and $B_q = 0$ for $q = 4, 5, 6, 8, 9, 10$, then

$$B_2 = -\frac{\sigma_{33}}{3\sigma_{22}}B_7, \quad (103)$$

$$B_3 = -\frac{\sigma_{23}+\sigma_{32}}{3\sigma_{33}}B_7. \quad (104)$$

– 5th Case: $B_8 \neq 0$ and $B_q = 0$ for $q = 4, 5, 6, 7, 9, 10$, then

$$B_1 = -\frac{\sigma_{33}}{3\sigma_{11}} B_8, \tag{105}$$

$$B_3 = -\frac{\sigma_{31} + \sigma_{13}}{3\sigma_{33}} B_8. \tag{106}$$

– 6th Case: $B_9 \neq 0$ and $B_q = 0$ for $q = 4, 5, 6, 7, 8, 10$, then

$$B_1 = -\frac{\sigma_{31} + \sigma_{13}}{3\sigma_{11}} B_9, \tag{107}$$

$$B_3 = -\frac{\sigma_{11}}{3\sigma_{33}} B_9. \tag{108}$$

– 7th Case: $B_{10} \neq 0$ and $B_q = 0$ for $q = 4, 5, 6, 7, 8, 9$, then

$$B_1 = -\frac{\sigma_{23} + \sigma_{32}}{6\sigma_{11}} B_{10}, \tag{109}$$

$$B_2 = -\frac{\sigma_{31} + \sigma_{13}}{6\sigma_{22}} B_{10}, \tag{110}$$

$$B_3 = -\frac{\sigma_{12} + \sigma_{21}}{6\sigma_{33}} B_{10}. \tag{111}$$

We handle now (97)–(111) so as to eliminate the dominators for notational convenience and without loss of generality as follows. Taking

$$B_4 = -3\sigma_{11}\sigma_{22} \tag{112}$$

in the 1st case, we obtain from (89) the anisotropic harmonic

$$\mathrm{H}_3^1(\mathbf{r}) = (\sigma_{12} + \sigma_{21})\,\sigma_{22}x_1^3 + \sigma_{11}^2 x_2^3 - 3\sigma_{11}\sigma_{22}x_1^2 x_2. \tag{113}$$

Taking again

$$B_5 = -3\sigma_{11}\sigma_{22} \tag{114}$$

in the 2nd case, we have from (89) that

$$\mathrm{H}_3^2(\mathbf{r}) = (\sigma_{12} + \sigma_{21})\,\sigma_{11}x_2^3 + \sigma_{22}^2 x_1^3 - 3\sigma_{11}\sigma_{22}x_1 x_2^2. \tag{115}$$

Taking

$$B_6 = -3\sigma_{22}\sigma_{33} \tag{116}$$

in the 3rd case, we obtain from (89) the anisotropic harmonic

$$\mathrm{H}_3^3(\mathbf{r}) = (\sigma_{23} + \sigma_{32})\,\sigma_{33}x_2^3 + \sigma_{22}^2 x_3^3 - 3\sigma_{22}\sigma_{33}x_2^2 x_3. \tag{117}$$

Taking again

$$B_7 = -3\sigma_{22}\sigma_{33} \tag{118}$$

in the 4th case, we have from (89) that

$$\mathrm{H}_3^4(\mathbf{r}) = (\sigma_{23} + \sigma_{32})\,\sigma_{22}x_3^3 + \sigma_{33}^2 x_2^3 - 3\sigma_{22}\sigma_{33}x_2 x_3^2. \tag{119}$$

Taking

$$B_8 = -3\sigma_{11}\sigma_{33} \tag{120}$$

in the 5th case, we obtain from (89) the anisotropic harmonic

$$\mathrm{H}_3^5(\mathbf{r}) = (\sigma_{31} + \sigma_{13})\,\sigma_{11}x_3^3 + \sigma_{33}^2 x_1^3 - 3\sigma_{11}\sigma_{33}x_3^2 x_1. \tag{121}$$

Taking again

$$B_9 = -3\sigma_{11}\sigma_{33} \tag{122}$$

in the 6th case, we have from (89) that

$$\mathrm{H}_3^6(\mathbf{r}) = (\sigma_{31} + \sigma_{13})\,\sigma_{33}x_1^3 + \sigma_{11}^2 x_3^3 - 3\sigma_{11}\sigma_{33}x_3 x_1^2. \tag{123}$$

Finally, taking

$$B_{10} = -6\sigma_{11}\sigma_{22}\sigma_{33} \tag{124}$$

in the 7th case, we recover from (89) the seventh anisotropic harmonic eigenfunction

$$\begin{aligned}\mathrm{H}_3^7(\mathbf{r}) = {} & (\sigma_{23} + \sigma_{32})\,\sigma_{22}\sigma_{33}x_1^3 \\ & + (\sigma_{31} + \sigma_{13})\,\sigma_{11}\sigma_{33}x_2^3 \\ & + (\sigma_{12} + \sigma_{21})\,\sigma_{11}\sigma_{22}x_3^3 - 6\sigma_{11}\sigma_{22}\sigma_{33}x_1 x_2 x_3,\end{aligned} \tag{125}$$

ending this task of computing the $2 \cdot 3 + 1 = 7$ ($m = 1, 2, \ldots, 7$) anisotropic harmonics of third degree. Any other harmonic function of degree $n = 3$ has to be expressed as a linear combination of H_3^p, $p = 1, 2, \ldots, 7$.

Proceeding to the next degree for $n = 4$, we consider the fourth degree homogeneous polynomial

$$
\begin{aligned}
u_4 = {} & \Gamma_1 x_1^4 + \Gamma_2 x_2^4 + \Gamma_3 x_3^4 + \Gamma_4 x_1^3 x_2 + \Gamma_5 x_1^3 x_3 + \Gamma_6 x_2^3 x_3 \\
& + \Gamma_7 x_1 x_2^3 + \Gamma_8 x_3^3 x_1 + \Gamma_9 x_2 x_3^3 \\
& + \Gamma_{10} x_1^2 x_2^2 + \Gamma_{11} x_3^2 x_1^2 + \Gamma_{12} x_2^2 x_3^2 + \Gamma_{13} x_1^2 x_2 x_3 \\
& + \Gamma_{14} x_1 x_2^2 x_3 + \Gamma_{15} x_1 x_2 x_3^2,
\end{aligned}
\tag{126}
$$

regarding expression (75), which, by virtue of (71) with (73), yields

$$
\begin{aligned}
& \nabla \cdot (\tilde{\boldsymbol{\sigma}} \cdot \nabla u_4) \\
& = [12\sigma_{11}\Gamma_1 + 2\sigma_{22}\Gamma_{10} + 2\sigma_{33}\Gamma_{11} + 3\,(\sigma_{12} + \sigma_{21})\,\Gamma_4 \\
& \quad + (\sigma_{23} + \sigma_{32})\,\Gamma_{13} + 3\,(\sigma_{31} + \sigma_{13})\,\Gamma_5]\, x_1^2 \\
& \quad + [12\sigma_{22}\Gamma_2 + 2\sigma_{11}\Gamma_{10} + 2\sigma_{33}\Gamma_{12} + 3\,(\sigma_{12} + \sigma_{21})\,\Gamma_7 \\
& \quad + 3\,(\sigma_{23} + \sigma_{32})\,\Gamma_6 + (\sigma_{31} + \sigma_{13})\,\Gamma_{14}]\, x_2^2 \\
& \quad + [12\sigma_{33}\Gamma_3 + 2\sigma_{11}\Gamma_{11} + 2\sigma_{22}\Gamma_{12} + (\sigma_{12} + \sigma_{21})\,\Gamma_{15} \\
& \quad + 3\,(\sigma_{23} + \sigma_{32})\,\Gamma_9 + 3\,(\sigma_{31} + \sigma_{13})\,\Gamma_8]\, x_3^2 \\
& \quad + [6\sigma_{11}\Gamma_4 + 6\sigma_{22}\Gamma_7 + 2\sigma_{33}\Gamma_{15} + 4\,(\sigma_{12} + \sigma_{21})\,\Gamma_{10} \\
& \quad + 2\,(\sigma_{23} + \sigma_{32})\,\Gamma_{14} + 2\,(\sigma_{31} + \sigma_{13})\,\Gamma_{13}]\, x_1 x_2 \\
& \quad + [6\sigma_{22}\Gamma_6 + 2\sigma_{11}\Gamma_{13} + 6\sigma_{33}\Gamma_9 + 2\,(\sigma_{12} + \sigma_{21})\,\Gamma_{14} \\
& \quad + 4\,(\sigma_{23} + \sigma_{32})\,\Gamma_{12} + 2\,(\sigma_{31} + \sigma_{13})\,\Gamma_{15}]\, x_2 x_3 \\
& \quad + [6\sigma_{11}\Gamma_5 + 2\sigma_{22}\Gamma_{14} + 6\sigma_{33}\Gamma_8 + 2\,(\sigma_{12} + \sigma_{21})\,\Gamma_{13} \\
& \quad + 2\,(\sigma_{23} + \sigma_{32})\,\Gamma_{15} + 4\,(\sigma_{31} + \sigma_{13})\,\Gamma_{11}]\, x_3 x_1 = 0.
\end{aligned}
\tag{127}
$$

Hence, similarly to the previous case, the function u_4 is anisotropic harmonic if the following six constraints hold true, that is,

$$
\begin{aligned}
& 12\sigma_{11}\Gamma_1 + 2\sigma_{22}\Gamma_{10} + 2\sigma_{33}\Gamma_{11} + 3\,(\sigma_{12} + \sigma_{21})\,\Gamma_4 \\
& \quad + (\sigma_{23} + \sigma_{32})\,\Gamma_{13} + 3\,(\sigma_{31} + \sigma_{13})\,\Gamma_5 = 0,
\end{aligned}
\tag{128}
$$

$$
\begin{aligned}
& 12\sigma_{22}\Gamma_2 + 2\sigma_{11}\Gamma_{10} + 2\sigma_{33}\Gamma_{12} + 3\,(\sigma_{12} + \sigma_{21})\,\Gamma_7 \\
& \quad + 3\,(\sigma_{23} + \sigma_{32})\,\Gamma_6 + (\sigma_{31} + \sigma_{13})\,\Gamma_{14} = 0,
\end{aligned}
\tag{129}
$$

$$
\begin{aligned}
& 12\sigma_{33}\Gamma_3 + 2\sigma_{11}\Gamma_{11} + 2\sigma_{22}\Gamma_{12} + (\sigma_{12} + \sigma_{21})\,\Gamma_{15} \\
& \quad + 3\,(\sigma_{23} + \sigma_{32})\,\Gamma_9 + 3\,(\sigma_{31} + \sigma_{13})\,\Gamma_8 = 0,
\end{aligned}
\tag{130}
$$

$$6\sigma_{11}\Gamma_4 + 6\sigma_{22}\Gamma_7 + 2\sigma_{33}\Gamma_{15} + 4(\sigma_{12} + \sigma_{21})\Gamma_{10} + 2(\sigma_{23} + \sigma_{32})\Gamma_{14} + 2(\sigma_{31} + \sigma_{13})\Gamma_{13} = 0, \tag{131}$$

$$6\sigma_{22}\Gamma_6 + 2\sigma_{11}\Gamma_{13} + 6\sigma_{33}\Gamma_9 + 2(\sigma_{12} + \sigma_{21})\Gamma_{14} + 4(\sigma_{23} + \sigma_{32})\Gamma_{12} + 2(\sigma_{31} + \sigma_{13})\Gamma_{15} = 0, \tag{132}$$

$$6\sigma_{11}\Gamma_5 + 2\sigma_{22}\Gamma_{14} + 6\sigma_{33}\Gamma_8 + 2(\sigma_{12} + \sigma_{21})\Gamma_{13} + 2(\sigma_{23} + \sigma_{32})\Gamma_{15} + 4(\sigma_{31} + \sigma_{13})\Gamma_{11} = 0. \tag{133}$$

Relations (128)–(133) reduce the independent coefficients Γ_p, $p = 1, 2, \ldots, 15$, to 9. As previously, we suppose that the 6 constants Γ_1, Γ_2, Γ_3, Γ_4, Γ_6, and Γ_8 within (128)–(133) are calculated in terms of the rest of the arbitrary chosen constants Γ_5, Γ_7, Γ_9, Γ_{10}, Γ_{11}, Γ_{12}, Γ_{13}, Γ_{14}, and Γ_{15} (9 constants in total). Bearing in mind the very same procedure described for the evaluation of the constants for $n = 3$, we consider the forthcoming nine cases, where only one of the Γ_p, $p = 5, 7, 9, 10, 11, 12, 13, 14, 15$, is not set to nil and all the others vanish. Those are

– 1st Case: $\Gamma_5 \neq 0$ and $\Gamma_q = 0$ for $q = 7, 9, 10, 11, 12, 13, 14, 15$, then

$$\Gamma_1 = -\frac{\sigma_{13} + \sigma_{31}}{4\sigma_{11}}\Gamma_5, \tag{134}$$

$$\Gamma_3 = \frac{\sigma_{11}(\sigma_{13} + \sigma_{31})}{4\sigma_{33}^2}\Gamma_5, \tag{135}$$

$$\Gamma_8 = -\frac{\sigma_{11}}{\sigma_{33}}\Gamma_5. \tag{136}$$

– 2nd Case: $\Gamma_7 \neq 0$ and $\Gamma_q = 0$ for $q = 5, 9, 10, 11, 12, 13, 14, 15$, then

$$\Gamma_1 = \frac{\sigma_{22}(\sigma_{12} + \sigma_{21})}{4\sigma_{11}^2}\Gamma_7, \tag{137}$$

$$\Gamma_2 = -\frac{\sigma_{12} + \sigma_{21}}{4\sigma_{22}}\Gamma_7, \tag{138}$$

$$\Gamma_4 = -\frac{\sigma_{22}}{\sigma_{11}}\Gamma_7. \tag{139}$$

– 3rd Case: $\Gamma_9 \neq 0$ and $\Gamma_q = 0$ for $q = 5, 7, 10, 11, 12, 13, 14, 15$, then

$$\Gamma_2 = \frac{\sigma_{33}(\sigma_{23} + \sigma_{32})}{4\sigma_{22}^2}\Gamma_9, \tag{140}$$

$$\Gamma_3 = -\frac{\sigma_{23} + \sigma_{32}}{4\sigma_{33}} \Gamma_9, \tag{141}$$

$$\Gamma_6 = -\frac{\sigma_{33}}{\sigma_{22}} \Gamma_9. \tag{142}$$

– 4th Case: $\Gamma_{10} \neq 0$ and $\Gamma_q = 0$ for $q = 5, 7, 9, 11, 12, 13, 14, 15$, then

$$\Gamma_1 = -\left[\frac{\sigma_{11}\sigma_{22} - (\sigma_{12} + \sigma_{21})^2}{6\sigma_{11}^2}\right] \Gamma_{10}, \tag{143}$$

$$\Gamma_2 = -\frac{\sigma_{11}}{6\sigma_{22}} \Gamma_{10}, \tag{144}$$

$$\Gamma_4 = -\frac{2\,(\sigma_{21} + \sigma_{12})}{3\sigma_{11}} \Gamma_{10}. \tag{145}$$

– 5th Case: $\Gamma_{11} \neq 0$ and $\Gamma_q = 0$ for $q = 5, 7, 9, 10, 12, 13, 14, 15$, then

$$\Gamma_1 = -\frac{\sigma_{33}}{6\sigma_{11}} \Gamma_{11}, \tag{146}$$

$$\Gamma_3 = -\left[\frac{\sigma_{11}\sigma_{33} - (\sigma_{13} + \sigma_{31})^2}{6\sigma_{33}^2}\right] \Gamma_{11}, \tag{147}$$

$$\Gamma_8 = -\frac{2\,(\sigma_{13} + \sigma_{31})}{3\sigma_{33}} \Gamma_{11}. \tag{148}$$

– 6th Case: $\Gamma_{12} \neq 0$ and $\Gamma_q = 0$ for $q = 5, 7, 9, 10, 11, 13, 14, 15$, then

$$\Gamma_2 = -\left[\frac{\sigma_{22}\sigma_{33} - (\sigma_{23} + \sigma_{32})^2}{6\sigma_{22}^2}\right] \Gamma_{12}, \tag{149}$$

$$\Gamma_3 = -\frac{\sigma_{22}}{\sigma_{33}} \Gamma_{12}, \tag{150}$$

$$\Gamma_6 = -\frac{2\,(\sigma_{23} + \sigma_{32})}{3\sigma_{22}} \Gamma_{12}. \tag{151}$$

– 7th Case: $\Gamma_{13} \neq 0$ and $\Gamma_q = 0$ for $q = 5, 7, 9, 10, 11, 12, 14, 15$, then

$$\Gamma_1 = -\left[\frac{\sigma_{11}\,(\sigma_{23} + \sigma_{32}) - (\sigma_{12} + \sigma_{21})\,(\sigma_{13} + \sigma_{31})}{12\sigma_{11}^2}\right] \Gamma_{13}, \tag{152}$$

$$\Gamma_2 = \frac{\sigma_{11}\,(\sigma_{23} + \sigma_{32})}{12\sigma_{22}^2} \Gamma_{13}, \tag{153}$$

$$\Gamma_3 = \frac{(\sigma_{13} + \sigma_{31})(\sigma_{12} + \sigma_{21})}{12\sigma_{33}^2} \Gamma_{13}, \tag{154}$$

$$\Gamma_4 = -\frac{\sigma_{13} + \sigma_{31}}{3\sigma_{11}} \Gamma_{13}, \tag{155}$$

$$\Gamma_6 = -\frac{\sigma_{11}}{3\sigma_{22}} \Gamma_{13}, \tag{156}$$

$$\Gamma_8 = -\frac{\sigma_{12} + \sigma_{21}}{3\sigma_{33}} \Gamma_{13}. \tag{157}$$

– 8th Case: $\Gamma_{14} \neq 0$ and $\Gamma_q = 0$ for $q = 5, 7, 9, 10, 11, 12, 13, 15$, then

$$\Gamma_1 = \frac{(\sigma_{12} + \sigma_{21})(\sigma_{23} + \sigma_{32})}{12\sigma_{11}^2} \Gamma_{14}, \tag{158}$$

$$\Gamma_2 = -\left[\frac{\sigma_{22}(\sigma_{13} + \sigma_{31}) - (\sigma_{23} + \sigma_{32})(\sigma_{21} + \sigma_{12})}{12\sigma_{22}^2}\right] \Gamma_{14}, \tag{159}$$

$$\Gamma_3 = \frac{\sigma_{22}(\sigma_{13} + \sigma_{31})}{12\sigma_{33}^2} \Gamma_{14}, \tag{160}$$

$$\Gamma_4 = -\frac{\sigma_{23} + \sigma_{32}}{3\sigma_{11}} \Gamma_{14}, \tag{161}$$

$$\Gamma_6 = -\frac{\sigma_{21} + \sigma_{12}}{3\sigma_{22}} \Gamma_{14}, \tag{162}$$

$$\Gamma_8 = -\frac{\sigma_{22}}{3\sigma_{33}} \Gamma_{14}. \tag{163}$$

– 9th Case: $\Gamma_{15} \neq 0$ and $\Gamma_q = 0$ for $q = 5, 7, 9, 10, 11, 12, 13, 14$, then

$$\Gamma_1 = \frac{\sigma_{33}(\sigma_{12} + \sigma_{21})}{12\sigma_{11}^2} \Gamma_{15}, \tag{164}$$

$$\Gamma_2 = \frac{(\sigma_{23} + \sigma_{32})(\sigma_{13} + \sigma_{31})}{12\sigma_{22}^2} \Gamma_{15}, \tag{165}$$

$$\Gamma_3 = -\left[\frac{\sigma_{33}(\sigma_{21} + \sigma_{12}) - (\sigma_{13} + \sigma_{31})(\sigma_{23} + \sigma_{32})}{12\sigma_{33}^2}\right] \Gamma_{15}, \tag{166}$$

$$\Gamma_4 = -\frac{\sigma_{33}}{3\sigma_{11}} \Gamma_{15}, \tag{167}$$

$$\Gamma_6 = -\frac{\sigma_{13} + \sigma_{31}}{3\sigma_{22}} \Gamma_{15}, \tag{168}$$

$$\Gamma_8 = -\frac{\sigma_{23} + \sigma_{32}}{3\sigma_{33}} \Gamma_{15}. \tag{169}$$

Similarly, we suggest values for Γ_p, $p = 5, 7, 9, 10, 11, 12, 13, 14, 15$ within (134)–(169) so as to eliminate the dominators and obtain handy relations, and thus we come up with the $2 \cdot 4 + 1 = 9$ ($m = 1, 2, \ldots, 9$) anisotropic harmonics of fourth degree, given by

$$\begin{aligned} H_4^1(\mathbf{r}) &= \sigma_{33}^2(\sigma_{13}+\sigma_{31})\,x_1^4 - \sigma_{11}^2(\sigma_{13}+\sigma_{31})\,x_3^4 \\ &\quad -4\sigma_{11}\sigma_{33}^2 x_3 x_1^3 + 4\sigma_{11}^2\sigma_{33}x_3^3 x_1, \end{aligned} \tag{170}$$

$$\begin{aligned} H_4^2(\mathbf{r}) &= -\sigma_{22}^2(\sigma_{12}+\sigma_{21})\,x_1^4 + \sigma_{11}^2(\sigma_{12}+\sigma_{21})\,x_2^4 \\ &\quad +4\sigma_{11}\sigma_{22}^2 x_1^3 x_2 - 4\sigma_{11}^2\sigma_{22}x_1 x_2^3, \end{aligned} \tag{171}$$

$$\begin{aligned} H_4^3(\mathbf{r}) &= -\sigma_{33}^2(\sigma_{23}+\sigma_{32})\,x_2^4 + \sigma_{22}^2(\sigma_{23}+\sigma_{32})\,x_3^4 \\ &\quad +4\sigma_{22}\sigma_{33}^2 x_2^3 x_3 - 4\sigma_{22}^2\sigma_{33}x_2 x_3^3, \end{aligned} \tag{172}$$

$$\begin{aligned} H_4^4(\mathbf{r}) &= \left[\sigma_{11}\sigma_{22}^2 - \sigma_{22}(\sigma_{12}+\sigma_{21})^2\right]x_1^4 + \sigma_{11}^3 x_2^4 + 4\sigma_{11}\sigma_{22}(\sigma_{21}+\sigma_{12})\,x_1^3 x_2 \\ &\quad -6\sigma_{11}^2\sigma_{22}x_1^2 x_2^2, \end{aligned} \tag{173}$$

$$\begin{aligned} H_4^5(\mathbf{r}) &= \sigma_{33}^3 x_1^4 + \left[\sigma_{11}^2\sigma_{33} - \sigma_{11}(\sigma_{13}+\sigma_{31})^2\right]x_3^4 + 4\sigma_{11}\sigma_{33}(\sigma_{13}+\sigma_{31})\,x_3^3 x_1 \\ &\quad -6\sigma_{11}\sigma_{33}^2 x_3^2 x_1^2, \end{aligned} \tag{174}$$

$$\begin{aligned} H_4^6(\mathbf{r}) &= \left[\sigma_{22}\sigma_{33}^2 - \sigma_{33}(\sigma_{23}+\sigma_{32})^2\right]x_2^4 + \sigma_{22}^3 x_3^4 + 4\sigma_{22}\sigma_{33}(\sigma_{23}+\sigma_{32})\,x_2^3 x_3 \\ &\quad -6\sigma_{22}^2\sigma_{33}x_2^2 x_3^2, \end{aligned} \tag{175}$$

$$\begin{aligned} H_4^7(\mathbf{r}) &= \left[\sigma_{11}\sigma_{22}^2\sigma_{33}^2(\sigma_{23}+\sigma_{32}) - \sigma_{22}^2\sigma_{33}^2(\sigma_{12}+\sigma_{21})(\sigma_{13}+\sigma_{31})\right]x_1^4 \\ &\quad -\sigma_{11}^3\sigma_{33}^2(\sigma_{23}+\sigma_{32})\,x_2^4 \\ &\quad -\sigma_{11}^2\sigma_{22}^2(\sigma_{13}+\sigma_{31})(\sigma_{12}+\sigma_{21})\,x_3^4 \\ &\quad +4\sigma_{11}\sigma_{22}^2\sigma_{33}^2(\sigma_{13}+\sigma_{31})\,x_1^3 x_2 + 4\sigma_{11}^3\sigma_{22}\sigma_{33}^2 x_2^3 x_3 \\ &\quad +4\sigma_{11}^2\sigma_{22}^2\sigma_{33}(\sigma_{12}+\sigma_{21})\,x_3^3 x_1 - 12\sigma_{11}^2\sigma_{22}^2\sigma_{33}^2 x_1^2 x_2 x_3, \end{aligned} \tag{176}$$

$$H_4^8(\mathbf{r}) = -\sigma_{22}^2\sigma_{33}^2(\sigma_{12}+\sigma_{21})(\sigma_{23}+\sigma_{32})x_1^4$$
$$+\left[(\sigma_{11}^2\sigma_{22}\sigma_{33}^2(\sigma_{13}+\sigma_{31})-\sigma_{11}^2\sigma_{33}^2(\sigma_{23}+\sigma_{32})(\sigma_{12}+\sigma_{21})\right]x_2^4$$
$$-\sigma_{11}^2\sigma_{22}^3(\sigma_{13}+\sigma_{31})x_3^4+4\sigma_{11}\sigma_{22}^2\sigma_{33}^2(\sigma_{23}+\sigma_{32})x_1^3x_2$$
$$+4\sigma_{11}^2\sigma_{22}\sigma_{33}^2(\sigma_{12}+\sigma_{21})x_2^3x_3$$
$$+4\sigma_{11}^2\sigma_{22}^3\sigma_{33}x_3^3x_1-12\sigma_{11}^2\sigma_{22}^2\sigma_{33}^2x_1x_2^2x_3, \quad (177)$$

and

$$H_4^9(\mathbf{r}) = -\sigma_{22}^2\sigma_{33}^3(\sigma_{12}+\sigma_{21})x_1^4-\sigma_{11}^2\sigma_{33}^2(\sigma_{23}+\sigma_{32})(\sigma_{13}+\sigma_{31})x_2^4$$
$$+\left[\sigma_{11}^2\sigma_{22}^2\sigma_{33}(\sigma_{12}+\sigma_{21})-\sigma_{11}^2\sigma_{22}^2(\sigma_{13}+\sigma_{31})(\sigma_{23}+\sigma_{32})\right]x_3^4$$
$$+4\sigma_{11}\sigma_{22}^2\sigma_{33}^3x_1^3x_2+4\sigma_{11}^2\sigma_{22}\sigma_{33}^2(\sigma_{13}+\sigma_{31})x_2^3x_3$$
$$+4\sigma_{11}^2\sigma_{22}^2\sigma_{33}(\sigma_{23}+\sigma_{32})x_3^3x_1-12\sigma_{11}^2\sigma_{22}^2\sigma_{33}^2x_1x_2x_3^2. \quad (178)$$

Obviously, any other harmonic function of degree $n = 4$ has to be written as a linear combination of H_4^p, $p = 1, 2, \ldots, 9$.

Notwithstanding the common sense that we developed mathematically in our previous steps that could be followed for every degree $n \geq 5$ and order $m = 1, 2, \ldots, 2n + 1$, it does not look possible to find a general methodology, incorporating the polynomial (75). However, a standard technique has been established, which could be efficiently used for higher degrees, no matter how complicated the calculations might be. Nevertheless, most physical applications require the first terms of series (74) in order for the solution to converge, and hence the above analysis is more than sufficient when dealing with such problems in anisotropic elasticity. Completing our analysis, note that while in the isotropic case the harmonic polynomials have only numerical values, in the anisotropic case the relative harmonic polynomials include the anisotropic characteristics of the space. Hence, the anisotropic harmonics are functions of the anisotropy tensor $\tilde{\boldsymbol{\sigma}}$, provided by (70).

6 Conclusions and Discussion

In this work, we developed a theoretical study of the basic mathematical components used in linear anisotropic elastostatics in the absence of body forces. Our survey was primarily oriented to the investigation of the fundamental equations of elasticity, i.e. Hooke's and Newton's law, in which the displacement field was interconnected with the strain, the stress, and the stiffness tensors, in order to derive the linearized equation of dynamic anisotropic elasticity. Henceforth, we considered the special anisotropies, which were characterized by the corresponding stiffness matrices. For the purpose of becoming familiar with the analytic manipulation of the anisotropic case, we begun elaborating Newton's law by virtue of the isotropic stiffness tensor, and we proceeded to the generation of the corresponding form of Newton's law, when the cubic-type system of anisotropy was implied. Closed-form solutions were presented for both the isotropic and anisotropic equations of linear time-independent elasticity.

Despite the fact that a differential-type representation exists for the complete spatial isotropy, providing a closed-form solution via harmonic functions, it was not possible at this stage to find such kind of solutions for the anisotropic case. Hence, we restricted ourselves in presenting a polynomial-type solution, referring to the cubic stiffness matrix, whereas we achieved to obtain a handy principal form. Though, since our intention is to explore the chance to prove the existence of more practical solution representations in linear anisotropic elastostatics, we demonstrated a mathematical technique, based on the definition of the anisotropic Laplace's operator. This led to the generation of anisotropic harmonic eigenfunctions of certain degree, which can be generalized for any degree with significantly longer calculational effort. Work under progress involves research directed toward the construction of general differential representations of solutions for describing the displacement field in anisotropic elastic media, written in terms of the anisotropic harmonic functions.

References

1. C. Truesdell, S. Flügge, S. Nemat-Nasser, W. Olmstead, *Mechanics of Solids II* (in *Encyclopedia of Physics*, vol. VIa/2), Springer-Verlag, New York (1972)
2. T.C.T. Ting, *Anisotropic elasticity. Theory and Applications*, Oxford University Press, New York (1996)
3. P. Vannucci, *Anisotropic Elasticity* (in *Lecture Notes in Applied and Computational Mechanics*, vol. 85), Springer, Singapore (2018)
4. I.S. Sokolnikoff and R.D. Specht, *Mathematical Theory of Elasticity*, McGraw-Hill, New York (1946)
5. O. Rand and V. Rovenski, *Analytical Methods in Anisotropic Elasticity*, Springer Science & Business Media, New York (2007)
6. P.M. Naghdi, A.J.M. Spencer and A.H. England, *Non-linear Elasticity and Theoretical Mechanics*, Oxford University Press, Oxford (1994)
7. A.E.H. Love, *A Treatise on the Mathematical Theory of Elasticity*, Cambridge University Press, Cambridge (2013)
8. A.E. Green and W. Zerna, *Theoretical Elasticity*, Oxford University Press, New York (1968); republished (1992) and reissued (2012) by Dover Publications unaltered; first published at the Clarendon Press, Oxford (1954)
9. C. Hwu, *Anisotropic Elasticity with Matlab* (in *Solid Mechanics and its Applications*, vol. 267), Springer Nature, Switzerland (2021)
10. J.M. Carcione, *Wave Fields in Real Media*, vol. 38, Elsevier, Amsterdam, (2015)
11. R.G. Payton, *Elastic Wave Propagation in Transversely Isotropic Media*, Kluwer Academic Publishers, New York (1983)
12. D. Danson, *Linear Isotropic Elasticity with Body Forces* (in *Progress in Boundary Element Methods*, chap. 4), Springer, New York (1983)

13. M.F. Beatty, "A class of universal relations in isotropic elasticity theory", *Journal of Elasticity*, **17**(2), 113–121 (1987)
14. A. Yavari, C. Goodbrake and A. Goriely, "Universal displacements in linear elasticity", *Journal of the Mechanics and Physics of Solids*, **135**, no. 103782 (2020)
15. G. Dassios and K. Kiriaki, "The low-frequency theory of elastic wave scattering", *Quarterly of Applied Mathematics*, **42**(2), 225–248 (1984)
16. G. Dassios and K.Kiriaki, "The rigid ellipsoid in the presence of a low frequency elastic wave", *Quarterly of Applied Mathematics*, **43**(4), 435–456 (1986)
17. G. Dassios and K. Kiriaki, "The ellipsoidal cavity in the presence of a low-frequency elastic wave", *Quarterly of Applied Mathematics*, **44**(4), 709–735 (1987)
18. S.C. Cowin, "Optimization of the strain energy density in linear anisotropic elasticity", *Journal of Elasticity*, **34**(1), 45–68 (1994)
19. T.C.T. Ting, "Recent developments in anisotropic elasticity", *International Journal of Solids and Structures*, **37**(1-2), 401–409 (2000)
20. M. Destrade, P.A. Martin and T.C.T. Ting, "The incompressible limit in linear anisotropic elasticity, with applications to surface waves and elastostatics", *Journal of the Mechanics and Physics of Solids*, **50**(7), 1453–1468 (2002)
21. W. Chen and X. Li, "A new modified couple stress theory for anisotropic elasticity and microscale laminated Kirchhoff plate model", *Archive of Applied Mechanics*, **84**(3), 323–341 (2014)
22. Z. Han, L. Yang, H. Fang and J. Zhang, "Dynamic stiffness of three-dimensional anisotropic multi-layered media based on the continued-fraction method", *Applied Mathematical Modelling*, **93**, 53–74 (2021)
23. G.A. Nariboli, "Wave propagation in anisotropic elasticity", *Journal of Mathematical Analysis and Applications* **16**(1), 108–122 (1966)
24. V. Lisitsa and D. Vishnevskiy, "Lebedev scheme for the numerical simulation of wave propagation in 3D anisotropic elasticity", *Geophysical Prospecting*, **58**(4), 619–635 (2010)
25. G. Dassios, *Low Frequency Scattering*, Oxford University Press, Oxford (2000)
26. A.A. Markin and M.Y. Sokolova, "Non-linear relations of anisotropic elasticity and a particular postulate of isotropy", *Journal of Applied Mathematics and Mechanics*, **71**(4), 536–542 (2007)

Hyers–Ulam Stability of Symmetric Biderivations on Banach Algebras

Jung Rye Lee, Choonkil Park, and Themistocles M. Rassias

Abstract In C. Park (Indian J Pure Appl Math 50:413–426, 2019), Park introduced the following bi-additive s-functional inequality:

$$\|f(x+y,z-w)+f(x-y,z+w)-2f(x,z)+2f(y,w)\| \leq \left\| s\left(2f\left(\frac{x+y}{2},z-w\right)+2f\left(\frac{x-y}{2},z+w\right)-2f(x,z)+2f(y,w)\right)\right\|, \tag{1}$$

where s is a fixed nonzero complex number with $|s|<1$. Using the fixed point method and the direct method, we prove the Hyers–Ulam stability of symmetric biderivations and a skew-symmetric biderivation on Banach algebras and unital C^*-algebras, associated with the bi-additive s-functional inequality (1).

1 Introduction and Preliminaries

The stability problem of functional equations originated from a question of Ulam [25] concerning the stability of group homomorphisms. Hyers [9] gave a first affirmative partial answer to the question of Ulam for Banach spaces. Hyers' theorem was generalized by Aoki [1] for additive mappings and by Rassias [24] for

J. R. Lee
Department of Mathematics, Daejin University, Pocheon, Korea
e-mail: jrlee@daejin.ac.kr

C. Park (✉)
Department of Mathematics, Hanyang University, Seoul, Korea
e-mail: baak@hanyang.ac.kr

Th. M. Rassias
Department of Mathematics, Zografou Campus, National Technical University of Athens, Athens, Greece
e-mail: trassias@math.ntua.gr

I. N. Parasidis et al. (eds.), *Mathematical Analysis in Interdisciplinary Research*, Springer Optimization and Its Applications 179,
https://doi.org/10.1007/978-3-030-84721-0_24

linear mappings by considering an unbounded Cauchy difference. A generalization of the Rassias theorem was obtained by Găvruta [8] by replacing the unbounded Cauchy difference by a general control function in the spirit of Rassias' approach. Park [17, 18] defined additive ρ-functional inequalities and proved the Hyers–Ulam stability of the additive ρ-functional inequalities in Banach spaces and non-Archimedean Banach spaces. The stability problems of various functional equations and functional inequalities have been extensively investigated by a number of authors (see [11, 20, 21]).

We recall a fundamental result in fixed point theory.

Theorem 1 ([3, 6]) *Let (X, d) be a complete generalized metric space and $J : X \to X$ be a strictly contractive mapping with Lipschitz constant $\alpha < 1$. Then, for each given element $x \in X$, either*

$$d(J^n x, J^{n+1} x) = \infty$$

for all nonnegative integers n or there exists a positive integer n_0 such that

(1) $d(J^n x, J^{n+1} x) < \infty, \qquad \forall n \geq n_0$.
(2) *The sequence $\{J^n x\}$ converges to a fixed point y^* of J.*
(3) *y^* is the unique fixed point of J in the set $Y = \{y \in X \mid d(J^{n_0} x, y) < \infty\}$.*
(4) *$d(y, y^*) \leq \frac{1}{1-\alpha} d(y, Jy)$ for all $y \in Y$.*

In 1996, Isac and Rassias [10] were the first to provide applications of stability theory of functional equations for the proof of new fixed point theorems with applications. By using fixed point methods, the stability problems of several functional equations have been extensively investigated by a number of authors (see [4, 5, 7, 19, 23]).

Maksa [13, 14] introduced and investigated biderivations and symmetric biderivations on rings. Öztürk and Sapanci [16], Vukman [26] and Yazarli [27] investigated some properties of symmetric biderivations on rings.

Definition 1 ([13, 14]) Let A be a ring. A bi-additive mapping $D : A \times A \to A$ is called a *symmetric biderivation* on A if D satisfies

$$D(xy, z) = D(x, z)y + xD(y, z),$$
$$D(x, y) = D(y, x)$$

for all $x, y, z \in A$.

In this chapter, we introduce a *symmetric biderivation* on a Banach algebra and a *skew-symmetric biderivation* on a Banach $*$-algebra.

Definition 2 Let A be a complex Banach algebra. A $\mathbf{C}$-bilinear mapping $D : A \times A \to A$ is called a *symmetric biderivation* on A if D satisfies

$$D(xy, z) = D(x, z)y + xD(y, z),$$

$$D(x, y) = D(y, x)$$

for all $x, y, z \in A$.

It is easy to show that if D is a symmetric biderivation, then

$$D(x, zw) = D(zw, x) = D(z, x)w + zD(w, x) = D(x, z)w + zD(x, w)$$

for all $x, z, w \in A$. So,

$$\begin{aligned} D(xy, zw) &= D(x, zw)y + xD(y, zw) \\ &= D(x, z)wy + zD(x, w)y + xD(y, z)w + xzD(y, w) \end{aligned}$$

for all $x, y, z, w \in A$.

Definition 3 Let A be a complex Banach $*$-algebra. A bi-additive mapping $D : A \times A \to A$ is called a *skew-symmetric biderivation* on A if D is $\mathbf{C}$-linear in the first variable and satisfies

$$\begin{aligned} D(xy, z) &= D(x, z)y + xD(y, z), \\ D(x, y) &= D(y, x)^* \end{aligned}$$

for all $x, y, z \in A$.

It is easy to show that if D is a skew-symmetric biderivation, then D is conjugate $\mathbf{C}$-linear in the second variable and

$$\begin{aligned} D(x, zw) &= D(zw, x)^* = (D(z, x)w + zD(w, x))^* = w^*D(z, x)^* + D(w, x)^*z^* \\ &= w^*D(x, z) + D(x, w)z^* \end{aligned}$$

for all $x, z, w \in A$. So,

$$\begin{aligned} D(xy, zw) &= D(x, zw)y + xD(y, zw) \\ &= w^*D(x, z)y + D(x, w)z^*y + xw^*D(y, z) + xD(y, w)z^* \end{aligned}$$

for all $x, y, z, w \in A$.

This chapter is organized as follows: in Sects. 2 and 3, we investigate symmetric biderivations on Banach algebras and unital C^*-algebras associated with the bi-additive s-functional inequality (1) by using the direct method. In Sects. 4 and 5, we investigate skew-symmetric biderivations on Banach $*$-algebras and unital C^*-algebras associated with the bi-additive s-functional inequality (1) by using the fixed point method.

Throughout this chapter, let X be a complex normed space and Y be a complex Banach space. Let A be a complex Banach algebra. Assume that s is a fixed nonzero complex number with $|s| < 1$.

2 Hyers–Ulam Stability of Symmetric Biderivations and Skew-Symmetric Derivations on Banach Algebras: Direct Method

In [22], Park solved the bi-additive s-functional inequality (1) in complex normed spaces.

Lemma 1 ([22, Lemma 2.1]) *If a mapping $f : X^2 \to Y$ satisfies $f(0, z) = f(x, 0) = 0$ and*

$$\begin{aligned} &\|f(x+y, z-w) + f(x-y, z+w) - 2f(x, z) + 2f(y, w)\| \\ &\quad \leq \left\| s\left(2f\left(\frac{x+y}{2}, z-w\right) + 2f\left(\frac{x-y}{2}, z+w\right) - 2f(x, z) + 2f(y, w)\right)\right\| \end{aligned} \tag{2}$$

for all $x, y, z, w \in X$, then $f : X^2 \to Y$ is bi-additive.

Using the direct method, we prove the Hyers–Ulam stability of the bi-additive s-functional inequality (2) in complex Banach spaces.

Theorem 2 ([22, Theprem 2.2]) *Let $\varphi : X^2 \to [0, \infty)$ be a function satisfying*

$$\Psi(x, y) := \sum_{j=1}^{\infty} 2^j \varphi\left(\frac{x}{2^j}, \frac{y}{2^j}\right) < \infty$$

for all $x, y \in X$. Let $f : X^2 \to Y$ be a mapping satisfying $f(x, 0) = f(0, z) = 0$ and

$$\begin{aligned} &\|f(x+y, z-w) + f(x-y, z+w) - 2f(x, z) + 2f(y, w)\| \\ &\quad \leq \left\| s\left(2f\left(\frac{x+y}{2}, z-w\right) + 2f\left(\frac{x-y}{2}, z+w\right) - 2f(x, z) + 2f(y, w)\right)\right\| \\ &\quad +\varphi(x, y)\varphi(z, w) \end{aligned} \tag{3}$$

for all $x, y, z, w \in X$. Then, there exists a unique bi-additive mapping $P : X^2 \to Y$ such that

$$\|f(x, z) - P(x, z)\| \leq \frac{1}{2}\Psi(x, x)\varphi(z, 0)$$

for all $x, z \in X$.

Using the direct method, we prove the Hyers–Ulam stability of symmetric biderivations on complex Banach algebras and unital C^*-algebras associated with the bi-additive s-functional inequality (1).

Lemma 2 ([2, Lemma 2.1]) *Let $f : X^2 \to Y$ be a bi-additive mapping such that $f(\lambda x, \mu z) = \lambda\mu f(x, z)$ for all $x, z \in X$ and $\lambda, \mu \in S^1 := \{\nu \in \mathbf{C} : |\nu| = 1\}$. Then, f is $\mathbf{C}$-bilinear.*

Theorem 3 *Let $\varphi : A^2 \to [0, \infty)$ be a function satisfying*

$$\sum_{j=1}^{\infty} 4^j \varphi\left(\frac{x}{2^j}, \frac{y}{2^j}\right) < \infty \tag{4}$$

for all $x, y \in A$ and $f : A^2 \to A$ be a mapping satisfying $f(x, 0) = f(0, z) = 0$ and

$$\begin{aligned} &\|f(\lambda(x+y), \mu(z-w)) + f(\lambda(x-y), \mu(z+w)) - 2\lambda\mu f(x,z) + 2\lambda\mu f(y.w)\| \\ &\leq \left\| s\left(2f\left(\frac{x+y}{2}, z-w\right) + 2f\left(\frac{x-y}{2}, z+w\right) - 2f(x,z) + 2f(y,w)\right)\right\| \\ &\quad + \varphi(x,y)\varphi(z,w) \end{aligned} \tag{5}$$

for all $\lambda, \mu \in S^1$ and all $x, y, z, w \in A$. Then, there exists a unique $\mathbf{C}$-bilinear mapping $D : A^2 \to A$ such that

$$\|f(x, z) - D(x, z)\| \leq \frac{1}{2}\Psi(x, x)\varphi(z, 0) \tag{6}$$

for all $x, z \in A$, where

$$\Psi(x, y) := \sum_{j=1}^{\infty} 2^j \varphi\left(\frac{x}{2^j}, \frac{y}{2^j}\right)$$

for all $x, y \in A$.

If, in addition, the mapping $f : A^2 \to A$ satisfies $f(2x, z) = 2f(x, z)$ and

$$\|f(xy, z) - f(x, z)y - xf(y, z)\| \leq \varphi(x, y)\varphi(z, 0), \tag{7}$$

$$\|f(x, z) - f(z, x)\| \leq \varphi(x, z) \tag{8}$$

for all $x, y, z \in A$, then the mapping $f : A^2 \to A$ is a symmetric biderivation.

Proof Let $\lambda = \mu = 1$ in (5). By Theorem 2, there is a unique bi-additive mapping $D : A^2 \to A$ satisfying (6) defined by

$$D(x, z) := \lim_{n\to\infty} 2^n f\left(\frac{x}{2^n}, z\right)$$

for all $x, z \in A$.

Letting $y = w = 0$ in (5), we get $f(\lambda x, \mu z) = \lambda\mu f(x, z)$ for all $x, z \in A$ and all $\lambda, \mu \in S^1$. By Lemma 2, the bi-additive mapping $D : A^2 \to A$ is **C**-bilinear.

If $f(2x, z) = 2f(x, z)$ for all $x, z \in A$, then we can easily show that $D(x, z) = f(x, z)$ for all $x, z \in A$.

It follows from (7) that

$$\begin{aligned} &\|D(xy, z) - D(x, z)y - xD(y, z)\| \\ &= \lim_{n\to\infty} 4^n \left\| f\left(\frac{xy}{2^n\cdot 2^n}, z\right) - f\left(\frac{x}{2^n}, z\right)\frac{y}{2^n} - \frac{x}{2^n} f\left(\frac{y}{2^n}, z\right)\right\| \\ &\le \lim_{n\to\infty} 4^n \varphi\left(\frac{x}{2^n}, \frac{y}{2^n}\right)\varphi(z, 0) = 0 \end{aligned}$$

for all $x, y, z \in A$. Thus,

$$D(xy, z) = D(x, z)y + xD(y, z)$$

for all $x, y, z \in A$.

It follows from (8) that

$$\begin{aligned} \|D(x, z) - D(z, x)\| &= \lim_{n\to\infty} 4^n \left\| D\left(\frac{x}{2^n}, \frac{z}{2^n}\right) - D\left(\frac{z}{2^n}, \frac{x}{2^n}\right)\right\| \\ &= \lim_{n\to\infty} 4^n \left\| f\left(\frac{x}{2^n}, \frac{z}{2^n}\right) - f\left(\frac{z}{2^n}, \frac{x}{2^n}\right)\right\| \\ &\le \lim_{n\to\infty} 4^n \varphi\left(\frac{x}{2^n}, \frac{z}{2^n}\right) = 0 \end{aligned}$$

for all $x, z \in A$. Thus,

$$D(x, z) = D(z, x)$$

for all $x, z \in A$. Hence, the mapping $f : A^2 \to A$ is a symmetric biderivation.

Corollary 1 *Let $r > 2$ and θ be nonnegative real numbers, and let $f : A^2 \to A$ be a mapping satisfying $f(x, 0) = f(0, z) = 0$ and*

$$\begin{aligned} &\|f(\lambda(x + y), \mu(z - w)) + f(\lambda(x - y), \mu(z + w)) - 2\lambda\mu f(x, z) + 2\lambda\mu f(y, w)\| \\ &\le \left\| s\left(2f\left(\frac{x + y}{2}, z - w\right) + 2f\left(\frac{x - y}{2}, z + w\right) - 2f(x, z) + 2f(y, w)\right)\right\| \\ &\quad + \theta(\|x\|^r + \|z\|^r)(\|y\|^r + \|w\|^r) \end{aligned} \tag{9}$$

for all $\lambda, \mu \in S^1$ *and all* $x, y, z, w \in A$*. Then, there exists a unique* $\mathbf{C}$*-bilinear mapping* $D : A^2 \to A$ *such that*

$$\|f(x,z) - D(x,z)\| \leq \frac{2\theta}{2^r - 2}\|x\|^r\|z\|^r \tag{10}$$

for all $x, z \in A$*.*

If, in addition, the mapping $f : A^2 \to A$ *satisfies* $f(2x, z) = 2f(x, z)$ *and*

$$\|f(xy,z) - f(x,z)y - xf(y,z)\| \leq \theta(\|x\|^r + \|y\|^r)\|z\|^r, \tag{11}$$

$$\|f(x,z) - f(z,x)\| \leq \theta(\|x\|^r + \|z\|^r) \tag{12}$$

for all $x, y, z \in A$*, then the mapping* $f : A^2 \to A$ *is a symmetric biderivation.*

Proof The proof follows from Theorem 3 by taking $\varphi(x, y) = \sqrt{\theta}(\|x\|^r + \|y\|^r)$ for all $x, y \in A$.

Theorem 4 *Let* $\varphi : A^2 \to [0, \infty)$ *be a function satisfying*

$$\Psi(x,y) := \sum_{j=0}^{\infty} \frac{1}{2^j}\varphi\left(2^j x, 2^j y\right) < \infty \tag{13}$$

for all $x, y \in A$*, and let* $f : A^2 \to A$ *be a mapping satisfying* $f(x, 0) = f(0, z) = 0$ *and (5). Then, there exists a unique* $\mathbf{C}$*-bilinear mapping* $D : A^2 \to A$ *such that*

$$\|f(x,z) - D(x,z)\| \leq \frac{1}{2}\Psi(x,x)\varphi(z,0) \tag{14}$$

for all $x, z \in A$*.*

If, in addition, the mapping $f : A^2 \to A$ *satisfies* $f(2x, z) = 2f(x, z)$*, (7) and (8), then the mapping* $f : A^2 \to A$ *is a symmetric biderivation.*

Proof The proof is similar to the proof of Theorem 3.

Corollary 2 *Let* $r < 1$ *and* θ *be nonnegative real numbers, and let* $f : A^2 \to A$ *be a mapping satisfying* (9) *and* $f(x, 0) = f(0, z) = 0$ *for all* $x, z \in A$*. Then, there exists a unique* $\mathbf{C}$*-bilinear mapping* $D : A^2 \to A$ *such that*

$$\|f(x,z) - D(x,z)\| \leq \frac{2\theta}{2 - 2^r}\|x\|^r\|z\|^r \tag{15}$$

for all $x, z \in A$*.*

If, in addition, the mapping $f : A^2 \to A$ *satisfies* (11), (12) *and* $f(2x, z) = 2f(x, z)$ *for all* $x, z \in A$*, then the mapping* $f : A^2 \to A$ *is a symmetric biderivation.*

Proof The proof follows from Theorem 4 by taking $\varphi(x, y) = \sqrt{\theta}(\|x\|^r + \|y\|^r)$ for all $x, y \in A$.

From now on, assume that A is a unital C^*-algebra with unit e and unitary group $U(A)$.

Theorem 5 *Let $\varphi : A^2 \to [0, \infty)$ be a function satisfying* (4) *and $f : A^2 \to A$ be a mapping satisfying* (5) *and $f(x, 0) = f(0, z) = 0$ for all $x, z \in A$. Then, there exists a unique* **C***-bilinear mapping $D : A^2 \to A$ satisfying* (6).

If, in addition, the mapping $f : A^2 \to A$ satisfies (8), *$f(2x, z) = 2f(x, z)$ and*

$$\|f(uy, z) - f(u, z)y - uf(y, z)\| \leq \theta(1 + \|y\|^r)\|z\|^r \tag{16}$$

for all $u, v \in U(A)$ and all $x, y, z \in A$, then the mapping $f : A^2 \to A$ is a symmetric biderivation.

Proof By the same reasoning as in the proof of Theorem 3, there is a unique **C**-bilinear mapping $D : A^2 \to A$ satisfying (6) defined by

$$D(x, z) := \lim_{n\to\infty} 2^n f\left(\frac{x}{2^n}, z\right)$$

for all $x, z \in A$.

If $f(2x, z) = 2f(x, z)$ for all $x, z \in A$, then we can easily show that $D(x, z) = f(x, z)$ for all $x, z \in A$.

By the same reasoning as in the proof of Theorem 3, $D(uy, z) = D(u, z)y + uD(y, z)$ for all $u, v \in U(A)$ and all $y, z \in A$.

Since D is **C**-linear in the first variable and each $x \in A$ is a finite linear combination of unitary elements (see [12]), i.e., $x = \sum_{j=1}^{m} \lambda_j u_j$ ($\lambda_j \in \mathbf{C}$, $u_j \in U(A)$),

$$\begin{aligned} D(xy, z) = D\left(\sum_{j=1}^{m} \lambda_j u_j y, z\right) &= \sum_{j=1}^{m} \lambda_j D(u_j y, z) = \sum_{j=1}^{m} \lambda_j (D(u_j, z)y + u_j D(y, z)) \\ &= \left(\sum_{j=1}^{m} \lambda_j\right) D(u_j, z)y + \left(\sum_{j=1}^{m} \lambda_j u_j\right) D(y, z) = D(x, z)y + xD(y, z) \end{aligned}$$

for all $x, y, z \in A$. So, by the same reasoning as in the proof of Theorem 3, $D : A^2 \to A$ is a symmetric biderivation. Thus, $f : A^2 \to A$ is a symmetric biderivation.

Corollary 3 *Let $r > 2$ and θ be nonnegative real numbers, and let $f : A^2 \to A$ be a mapping satisfying* (9) *and $f(x, 0) = f(0, z) = 0$ for all $x, z \in A$. Then, there exists a unique* **C***-bilinear mapping $D : A^2 \to A$ satisfying* (10).

If, in addition, the mapping $f : A^2 \to A$ satisfies (11), $f(2x, z) = 2f(x, z)$ *and*

$$\|f(uy, z) - f(u, z)y - uf(y, z)\| \leq \theta(1 + \|y\|^r)\|z\|^r \tag{17}$$

for all $u, v \in U(A)$ and all $x, y, z \in A$, then the mapping $f : A^2 \to A$ is a symmetric biderivation.

Theorem 6 *Let $\varphi : A^2 \to [0, \infty)$ be a function satisfying* (13) *and $f : A^2 \to A$ be a mapping satisfying $f(x, 0) = f(0, z) = 0$ for all $x, z \in A$ and* (5)*. Then, there exists a unique* **C**-*bilinear mapping $D : A^2 \to A$ satisfying* (14).

If, in addition, the mapping $f : A \to A$ satisfies (8), (16) *and $f(2x, z) = 2f(x, z)$ for all $x, z \in A$, then the mapping $f : A^2 \to A$ is a symmetric biderivation*

Proof The proof is similar to the proof of Theorem 5.

Corollary 4 *Let $r < 1$ and θ be nonnegative real numbers, and let $f : A^2 \to A$ be a mapping satisfying* (9) *and $f(x, 0) = f(0, z) = 0$ for all $x, z \in A$. Then, there exists a unique* **C**-*bilinear mapping $D : A^2 \to A$ satisfying* (15).

If, in addition, the mapping $f : A \to A$ satisfies (11), (17) *and $f(2x, z) = 2f(x, z)$ for all $x, z \in A$, then the mapping $f : A^2 \to A$ is a symmetric biderivation.*

3 Hyers–Ulam Stability of Skew-Symmetric Biderivations on Banach $*$-Algebras: Direct Method

In this section, using the direct method, we prove the Hyers–Ulam stability of skew-symmetric biderivations on complex Banach $*$-algebras and unital C^*-algebras associated with the bi-additive s-functional inequality (1).

Theorem 7 *Let $\varphi : A^2 \to [0, \infty)$ be a function satisfying* (4) *and $f : A^2 \to A$ be a mapping satisfying $f(x, 0) = f(0, z) = 0$ for all $x, z \in A$ and* (5)*. Then, there exists a unique* **C**-*bilinear mapping $D : A^2 \to A$ satisfying* (6).

If, in addition, the mapping $f : A^2 \to A$ satisfies (7), $f(2x, z) = 2f(x, z)$ *and*

$$\|f(x, z) - f(z, x)^*\| \leq \varphi(x, z) \tag{18}$$

for all $x, z \in A$, then the mapping $f : A^2 \to A$ is a skew-symmetric biderivation.

Proof By Theorem 3, there is a unique **C**-bilinear mapping $D : A^2 \to A$ satisfying (6) defined by

$$D(x, z) := \lim_{n\to\infty} 2^n f\left(\frac{x}{2^n}, z\right)$$

for all $x, z \in A$.

If $f(2x, z) = 2f(x, z)$ for all $x, z \in A$, then we can easily show that $D(x, z) = f(x, z)$ for all $x, z \in A$.

It follows from (18) that

$$\begin{aligned} \|D(x, z) - D(z, x)^*\| &= \lim_{n\to\infty} 4^n \left\| D\left(\frac{x}{2^n}, \frac{z}{2^n}\right) - D\left(\frac{z}{2^n}, \frac{x}{2^n}\right)^* \right\| \\ &= \lim_{n\to\infty} 4^n \left\| f\left(\frac{x}{2^n}, \frac{z}{2^n}\right) - f\left(\frac{z}{2^n}, \frac{x}{2^n}\right)^* \right\| \\ &\leq \lim_{n\to\infty} 4^n \varphi\left(\frac{x}{2^n}, \frac{z}{2^n}\right) = 0 \end{aligned}$$

for all $x, z \in A$. Thus,

$$D(x, z) = D(z, x)^*$$

for all $x, z \in A$.

The rest of the proof is similar to the proof of Theorem 3 and so the mapping $f : A^2 \to A$ is a skew-symmetric biderivation.

Corollary 5 *Let $r > 2$ and θ be nonnegative real numbers, and let $f : A^2 \to A$ be a mapping satisfying* (9) *and $f(x, 0) = f(0, z) = 0$ for all $x, z \in A$. Then, there exists a unique* **C***-bilinear mapping $D : A^2 \to A$ satisfying* (10).

If, in addition, the mapping $f : A^2 \to A$ satisfies (10), *$f(2x, z) = 2f(x, z)$ and*

$$\|f(x, z) - f(z, x)^*\| \leq \theta \|x\|^r \|z\|^r \tag{19}$$

for all $x, y, z \in A$, then the mapping $f : A^2 \to A$ is a skew-symmetric biderivation.

Theorem 8 *Let $\varphi : A^2 \to [0, \infty)$ be a function satisfying* (13) *and $f : A^2 \to A$ be a mapping satisfying $f(x, 0) = f(0, z) = 0$ and* (5). *Then, there exists a unique* **C***-bilinear mapping $D : A^2 \to A$ satisfying* (14).

If, in addition, the mapping $f : A^2 \to A$ satisfies $f(2x, z) = 2f(x, z)$ for all $x, z \in A$, (7) *and* (18), *then the mapping $f : A^2 \to A$ is a skew-symmetric biderivation.*

Proof The proof is similar to the proof of Theorem 7.

Corollary 6 *Let $r < 1$ and θ be nonnegative real numbers, and let $f : A^2 \to A$ be a mapping satisfying* (9) *and $f(x, 0) = f(0, z) = 0$ for all $x, z \in A$. Then, there exists a unique* **C***-bilinear mapping $D : A^2 \to A$ satisfying* (15).

If, in addition, the mapping $f : A^2 \to A$ satisfies (10), (19) *and $f(2x, z) = 2f(x, z)$ for all $x, z \in A$, then the mapping $f : A^2 \to A$ is a skew-symmetric biderivation.*

4 Hyers–Ulam Stability of Symmetric Biderivations and Skew-Symmetric Derivations on Banach Algebras: Fixed Point Method

Using the fixed point method, we prove the Hyers–Ulam stability of the bi-additive s-functional inequality (1) in complex Banach spaces.

Theorem 9 *Let $\varphi : X^2 \to [0, \infty)$ be a function such that there exists an $L < 1$ with*

$$\varphi\left(\frac{x}{2}, \frac{y}{2}\right) \leq \frac{L}{4}\varphi(x, y) \leq \frac{L}{2}\varphi(x, y) \tag{20}$$

for all $x, y \in X$. Let $f : X^2 \to Y$ be a mapping satisfying (3) *and $f(x, 0) = f(0, z) = 0$ for all $x, z \in X$. Then, there exists a unique bi-additive mapping $P : X^2 \to Y$ such that*

$$\|f(x, z) - P(x, z)\| \leq \frac{L}{2(1-L)}\varphi(x, x)\varphi(z, 0) \tag{21}$$

for all $x, z \in X$.

Proof Letting $w = 0$ and $y = x$ in (3), we get

$$\|f(2x, z) - 2f(x, z)\| \leq \varphi(x, x)\varphi(z, 0) \tag{22}$$

for all $x, z \in X$.

Consider the set

$$S := \{h : X^2 \to Y, \quad h(x, 0) = h(0, z) = 0 \ \ \forall x, z \in X\},$$

and introduce the generalized metric on S:

$$d(g, h) = \inf\{\mu \in \mathbf{R}_+ : \|g(x, z) - h(x, z)\| \leq \mu\varphi(x, x)\varphi(z, 0), \ \ \forall x, z \in X\},$$

where, as usual, $\inf \phi = +\infty$. It is easy to show that (S, d) is complete (see [15]).

Now, we consider the linear mapping $J : S \to S$ such that

$$Jg(x, z) := 2g\left(\frac{x}{2}, z\right)$$

for all $x, z \in X$.

Let $g, h \in S$ be given such that $d(g, h) = \varepsilon$. Then,

$$\|g(x, z) - h(x, z)\| \leq \varepsilon\varphi(x, x)\varphi(z, 0)$$

for all $x, z \in X$. Since

$$\|Jg(x,z) - Jh(x,z)\| = \left\|2g\left(\frac{x}{2}, z\right) - 2h\left(\frac{x}{2}, z\right)\right\| \leq 2\varepsilon\varphi\left(\frac{x}{2}, \frac{x}{2}\right)\varphi(z,0)$$
$$\leq 2\varepsilon\frac{L}{2}\varphi(x,x)\varphi(z,0) = L\varepsilon\varphi(x,x)\varphi(z,0)$$

for all $x, z \in X$, $d(Jg, Jh) \leq L\varepsilon$. This means that

$$d(Jg, Jh) \leq Ld(g,h)$$

for all $g, h \in S$.

It follows from (22) that

$$\left\|f(x,z) - 2f\left(\frac{x}{2}, z\right)\right\| \leq \varphi\left(\frac{x}{2}, \frac{x}{2}\right)\varphi(z,0) \leq \frac{L}{2}\varphi(x,x)\varphi(z,0)$$

for all $x, z \in X$. So, $d(f, Jf) \leq \frac{L}{2}$.

By Theorem 1, there exists a mapping $P : X^2 \to Y$ satisfying the following:

(1) P is a fixed point of J, i.e.,

$$P(x,z) = 2P\left(\frac{x}{2}, z\right) \tag{23}$$

for all $x, z \in X$. The mapping P is a unique fixed point of J. This implies that P is a unique mapping satisfying (23) such that there exists a $\mu \in (0, \infty)$ satisfying

$$\|f(x,z) - P(x,z)\| \leq \mu\varphi(x,x)\varphi(z,0)$$

for all $x, z \in X$;

(2) $d(J^l f, P) \to 0$ as $l \to \infty$. This implies the equality

$$\lim_{l\to\infty} 2^l f\left(\frac{x}{2^l}, z\right) = P(x,z)$$

for all $x, z \in X$;

(3) $d(f, P) \leq \frac{1}{1-L}d(f, Jf)$, which implies

$$\|f(x,z) - P(x,z)\| \leq \frac{L}{2(1-L)}\varphi(x,x)\varphi(z,0)$$

for all $x, z \in X$. So, we obtain (21).

It follows from (3) and (20) that

$$\|P(x+y, z-w) + P(x-y, z+w) - 2P(x,z) + 2P(y,w)\|$$

$$= \lim_{n\to\infty} \left\| 2^n \left(f\left(\frac{x+y}{2^n}, z-w\right) + f\left(\frac{x-y}{2^n}, z+w\right) \right.\right.$$
$$\left.\left. -2f\left(\frac{x}{2^n}, z\right) + 2f\left(\frac{y}{2^n}, w\right)\right)\right\|$$
$$\leq \lim_{n\to\infty} \left\| 2^n s \left(2f\left(\frac{x+y}{2^{n+1}}, z-w\right) + 2f\left(\frac{x-y}{2^{n+1}}, z+w\right) \right.\right.$$
$$\left.\left. -2f\left(\frac{x}{2^n}, z\right) + 2f\left(\frac{y}{2^n}, w\right)\right)\right\| + \lim_{n\to\infty} 2^n \varphi\left(\frac{x}{2^n}, \frac{x}{2^n}\right)\varphi(z,0)$$
$$\leq \left\| s \left(2P\left(\frac{x+y}{2}, z-w\right) + 2P\left(\frac{x-y}{2}, z+w\right) \right.\right.$$
$$\left.\left. -2P(x,z) + 2P(y,w)\right)\right\|$$

for all $x, y, z, w \in X$, since $2^n \varphi\left(\frac{x}{2^n}, \frac{x}{2^n}\right)\varphi(z,0) \leq \frac{2^n L^n}{2^n}\varphi(x,x)\varphi(z,0)$ tends to zero as $n \to \infty$. So,

$$\|P(x+y, z-w) + P(x-y, z+w) - 2P(x,z) + 2P(y,w)\|$$
$$\leq \left\| s \left(2P\left(\frac{x+y}{2}, z-w\right) + 2P\left(\frac{x-y}{2}, z+w\right) \right.\right.$$
$$\left.\left. -2P(x,z) + 2P(y,w)\right)\right\|$$

for all $x, y, z, w \in X$. By Lemma 1, the mapping $P : X^2 \to Y$ is bi-additive.

Using the fixed point method, we prove the Hyers–Ulam stability of symmetric biderivations on complex Banach algebras and unital C^*-algebras associated with the bi-additive s-functional inequality (1).

Theorem 10 *Let $\varphi : A^2 \to [0, \infty)$ be a function satisfying* (20) *with $A = X$ and $f : A^2 \to A$ be a mapping satisfying* (5) *and $f(x, 0) = f(0, z) = 0$ for all $x, z \in A$. Then, there exists a unique* **C***-bilinear mapping $D : A^2 \to A$ satisfying* (21) *with $X = A$.*

If, in addition, the mapping $f : A^2 \to A$ satisfies (7), (8) *and $f(2x, z) = 2f(x, z)$ for all $x, z \in A$, then the mapping $f : A^2 \to A$ is a symmetric biderivation.*

Proof Let $\lambda = \mu = 1$ in (5). By Theorem 9, there is a unique bi-additive mapping $D : A^2 \to A$ satisfying (21) defined by

$$D(x, z) := \lim_{n\to\infty} 2^n f\left(\frac{x}{2^n}, z\right)$$

for all $x, z \in A$.

Letting $y = w = 0$ in (5), we get $f(\lambda x, \mu z) = \lambda\mu f(x, z)$ for all $x, z \in A$ and all $\lambda, \mu \in S^1$. By Lemma 2, the bi-additive mapping $D : A^2 \to A$ is **C**-bilinear.

The rest of the proof is similar to the proof of Theorem 3.

Corollary 7 *Let $r > 2$ and θ be nonnegative real numbers, and let $f : A^2 \to A$ be a mapping satisfying* (9) *and $f(x, 0) = f(0, z) = 0$ for all $x, z \in A$. Then, there exists a unique* **C**-*bilinear mapping $D : A^2 \to A$ satisfying* (10).

If, in addition, the mapping $f : A^2 \to A$ satisfies (11), (12) *and $f(2x, z) = 2f(x, z)$ for all $x, z \in A$, then the mapping $f : A^2 \to A$ is a symmetric biderivation.*

Proof The proof follows from Theorem 10 by taking $L = 2^{1-r}$ and $\varphi(x, y) = \sqrt{\theta}(\|x\|^r + \|y\|^r)$ for all $x, y \in A$.

Theorem 11 *Let $\varphi : X^2 \to [0, \infty)$ be a function such that there exists an $L < 1$ with*

$$\varphi(x, y) \leq 2L\varphi\left(\frac{x}{2}, \frac{y}{2}\right) \tag{24}$$

for all $x, y \in X$. Let $f : X^2 \to Y$ be a mapping satisfying (3) *and $f(x, 0) = f(0, z) = 0$ for all $x, z \in X$. Then, there exists a unique bi-additive mapping $P : X^2 \to Y$ such that*

$$\|f(x, z) - P(x, z)\| \leq \frac{1}{2(1 - L)}\varphi(x, x)\varphi(z, 0)$$

for all $x, z \in X$.

Proof Let (S, d) be the generalized metric space defined in the proof of Theorem 9.

Now, we consider the linear mapping $J : S \to S$ such that

$$Jg(x, z) := \frac{1}{2}g(2x, z)$$

for all $x \in X$.

It follows from (22) that

$$\left\| f(x, z) - \frac{1}{2}f(2x, z) \right\| \leq \frac{1}{2}\varphi(x, x)\varphi(z, 0)$$

for all $x, z \in X$.

The rest of the proof is similar to the proof of Theorem 9.

Theorem 12 *Let $\varphi : A^2 \to [0, \infty)$ be a function satisfying* (24) *with $X = A$ and $f : A^2 \to A$ be a mapping satisfying $f(x, 0) = f(0, z) = 0$ and* (5). *Then, there exists a unique* **C**-*bilinear mapping $D : A^2 \to A$ satisfying* (6).

If, in addition, the mapping $f : A^2 \to A$ satisfies $f(2x, z) = 2f(x, z)$, (7) *and* (8), *then the mapping $f : A^2 \to A$ is a symmetric biderivation.*

Proof The proof is similar to the proof of Theorem 10.

Corollary 8 *Let $r < 1$ and θ be nonnegative real numbers, and let $f : A^2 \to A$ be a mapping satisfying* (9) *and $f(x, 0) = f(0, z) = 0$ for all $x, z \in A$. Then, there exists a unique* $\mathbf{C}$*-bilinear mapping $D : A^2 \to A$ satisfying* (15).

If, in addition, the mapping $f : A^2 \to A$ satisfies (11), (12) *and $f(2x, z) = 2f(x, z)$ for all $x, z \in A$, then the mapping $f : A^2 \to A$ is a symmetric biderivation.*

Proof The proof follows from Theorem 12 by taking $L = 2^{r-1}$ and $\varphi(x, y) = \sqrt{\theta}(\|x\|^r + \|y\|^r)$ for all $x, y \in A$.

From now on, assume that A is a unital C^*-algebra with unit e and unitary group $U(A)$.

Theorem 13 *Let $\varphi : A^2 \to [0, \infty)$ be a function satisfying* (20) *with $X = A$ and $f : A^2 \to A$ be a mapping satisfying* (5) *and $f(x, 0) = f(0, z) = 0$ for all $x, z \in A$. Then, there exists a unique* $\mathbf{C}$*-bilinear mapping $D : A^2 \to A$ satisfying* (21) *with $X = A$.*

If, in addition, the mapping $f : A^2 \to A$ satisfies (8), (16) *and $f(2x, z) = 2f(x, z)$ for all $x, z \in A$, then the mapping $f : A^2 \to A$ is a symmetric biderivation.*

Proof By the same reasoning as in the proof of Theorem 9, there is a unique $\mathbf{C}$-bilinear mapping $D : A^2 \to A$ satisfying (21) defined by

$$D(x, z) := \lim_{n\to\infty} 2^n f\left(\frac{x}{2^n}, z\right)$$

for all $x, z \in A$.

The rest of the proof is similar to the proofs of Theorems 3 and 5.

Corollary 9 *Let $r > 2$ and θ be nonnegative real numbers, and let $f : A^2 \to A$ be a mapping satisfying* (9) *and $f(x, 0) = f(0, z) = 0$ for all $x, z \in A$. Then, there exists a unique* $\mathbf{C}$*-bilinear mapping $D : A^2 \to A$ satisfying* (10).

If, in addition, the mapping $f : A^2 \to A$ satisfies (11), (17) *and $f(2x, z) = 2f(x, z)$ for all $x, z \in A$, then the mapping $f : A^2 \to A$ is a symmetric biderivation.*

Theorem 14 *Let $\varphi : A^2 \to [0, \infty)$ be a function satisfying* (24) *with $X = A$ and $f : A^2 \to A$ be a mapping satisfying $f(x, 0) = f(0, z) = 0$ for all $x, z \in A$ and* (5). *Then, there exists a unique* $\mathbf{C}$*-bilinear mapping $D : A^2 \to A$ satisfying* (6) *with $X = A$.*

If, in addition, the mapping $f : A \to A$ satisfies (8), (16) *and $f(2x, z) = 2f(x, z)$ for all $x, z \in A$, then the mapping $f : A^2 \to A$ is a symmetric biderivation.*

Proof The proof is similar to the proof of Theorem 13.

Corollary 10 *Let $r < 1$ and θ be nonnegative real numbers, and let $f : A^2 \to A$ be a mapping satisfying* (9) *and $f(x, 0) = f(0, z) = 0$ for all $x, z \in A$. Then, there exists a unique* **C**-*bilinear mapping $D : A^2 \to A$ satisfying* (15).

If, in addition, the mapping $f : A \to A$ satisfies (11), (17) *and $f(2x, z) = 2f(x, z)$ for all $x, z \in A$, then the mapping $f : A^2 \to A$ is a symmetric biderivation.*

5 Hyers–Ulam Stability of Skew-Symmetric Biderivations on Banach $*$-Algebras: Fixed Point Method

Using the fixed point method, we prove the Hyers–Ulam stability of skew-symmetric biderivations on complex Banach $*$-algebras and unital C^*-algebras associated with the bi-additive s-functional inequality (1).

Theorem 15 *Let $\varphi : A^2 \to [0, \infty)$ be a function satisfying* (20) *with $X = A$ and $f : A^2 \to A$ be a mapping satisfying $f(x, 0) = f(0, z) = 0$ for all $x, z \in A$ and* (5). *Then, there exists a unique* **C**-*bilinear mapping $D : A^2 \to A$ satisfying* (21) *with $X = A$.*

If, in addition, the mapping $f : A^2 \to A$ satisfies (7), (18) *and $f(2x, z) = 2f(x, z)$ for all $x, z \in A$, then the mapping $f : A^2 \to A$ is a skew-symmetric biderivation.*

Proof By Theorem 10, there is a unique **C**-bilinear mapping $D : A^2 \to A$ satisfying (21) defined by

$$D(x, z) := \lim_{n\to\infty} 2^n f\left(\frac{x}{2^n}, z\right)$$

for all $x, z \in A$.

The rest of the proof is similar to the proof of Theorem 7.

Corollary 11 *Let $r > 2$ and θ be nonnegative real numbers, and let $f : A^2 \to A$ be a mapping satisfying* (9) *and $f(x, 0) = f(0, z) = 0$ for all $x, z \in A$. Then, there exists a unique* **C**-*bilinear mapping $D : A^2 \to A$ satisfying* (10).

If, in addition, the mapping $f : A^2 \to A$ satisfies (10), (19) *and $f(2x, z) = 2f(x, z)$ for all $x, z \in A$, then the mapping $f : A^2 \to A$ is a skew-symmetric biderivation.*

Theorem 16 *Let $\varphi : A^2 \to [0, \infty)$ be a function satisfying* (24) *with $X = A$ and $f : A^2 \to A$ be a mapping satisfying $f(x, 0) = f(0, z) = 0$ and* (5). *Then, there exists a unique* **C**-*bilinear mapping $D : A^2 \to A$ satisfying* (6) *with $X = A$.*

If, in addition, the mapping $f : A^2 \to A$ satisfies $f(2x, z) = 2f(x, z)$ for all $x, z \in A$, (7) *and* (18), *then the mapping $f : A^2 \to A$ is a skew-symmetric biderivation.*

Proof The proof is similar to the proof of Theorem 15.

Corollary 12 *Let* $r < 1$ *and* θ *be nonnegative real numbers, and let* $f : A^2 \to A$ *be a mapping satisfying* (9) *and* $f(x, 0) = f(0, z) = 0$ *for all* $x, z \in A$. *Then, there exists a unique* $\mathbf{C}$*-bilinear mapping* $D : A^2 \to A$ *satisfying* (15).

If, in addition, the mapping $f : A^2 \to A$ *satisfies* (10), (19) *and* $f(2x, z) = 2f(x, z)$ *for all* $x, z \in A$, *then the mapping* $f : A^2 \to A$ *is a skew-symmetric biderivation.*

6 Conclusions

In this chapter, using the fixed point method and the direct method, we have proved the Hyers–Ulam stability of symmetric biderivations and a skew-symmetric biderivation on Banach algebras and unital C^*-algebras, associated with the bi-additive s-functional inequality (1).

References

1. T. Aoki, *On the stability of the linear transformation in Banach spaces*, J. Math. Soc. Japan **2** (1950), 64–66.
2. J. Bae and W. Park, *Approximate bi-homomorphisms and bi-derivations in* C^**-ternary algebras*, Bull. Korean Math. Soc. **47** (2010), 195–209.
3. L. Cădariu, V. Radu, *Fixed points and the stability of Jensen's functional equation*, J. Inequal. Pure Appl. Math. **4**, no. 1, Art. ID 4 (2003).
4. L. Cădariu, V. Radu, *On the stability of the Cauchy functional equation: a fixed point approach*, Grazer Math. Ber. **346** (2004), 43–52.
5. L. Cădariu, V. Radu, *Fixed point methods for the generalized stability of functional equations in a single variable*, Fixed Point Theory Appl. **2008**, Art. ID 749392 (2008).
6. J. Diaz, B. Margolis, *A fixed point theorem of the alternative for contractions on a generalized complete metric space*, Bull. Am. Math. Soc. **74** (1968), 305–309.
7. I. EL-Fassi, *Generalized hyperstability of a Drygas functional equation on a restricted domain using Brzdek's fixed point theorem,* J. Fixed Point Theory Appl. **19** (2017), 2529–2540.
8. P. Găvruta, *A generalization of the Hyers-Ulam-Rassias stability of approximately additive mappings*, J. Math. Anal. Appl. **184** (1994), 431–436.
9. D.H. Hyers, *On the stability of the linear functional equation*, Proc. Nat. Acad. Sci. U.S.A. **27** (1941), 222–224.
10. G. Isac, Th. M. Rassias, *Stability of* ψ*-additive mappings: Applications to nonlinear analysis*, Int.. J. Math. Math. Sci. **19** (1996), 219–228.
11. S. Jung, *On the quadratic functional equation modulo a subgroup*, Indian J. Pure Appl. Math. **36** (2005), 441–450.
12. R.V. Kadison and J.R. Ringrose, *Fundamentals of the Theory of Operator Algebras: Elementary Theory*, Academic Press, New York, 1983.
13. G. Maksa, *A remark on symmetric biadditive function having nonnegative diagonalization*, Glas. Mat. Ser. *III* **46** (1980), 279–282.
14. G. Maksa, *On the trace of symmetric bi-derivations*, C. R. Math. Rep. Acad. Sci. Canada **9** (1987), 303–307.

15. D. Mihet and V. Radu, *On the stability of the additive Cauchy functional equation in random normed spaces*, J. Math. Anal. Appl. **343** (2008), 567–572.
16. M.A. Öztürk and M. Sapanci, *Orthogonal symmetric bi-derivation on semi-prime gamma rings*, Hacet. Bull. Nat. Sci. Eng. Ser. B **26** (1997), 31–46.
17. C. Park, *Additive ρ-functional inequalities and equations*, J. Math. Inequal. **9** (2015), 17–26.
18. C. Park, *Additive ρ-functional inequalities in non-Archimedean normed spaces*, J. Math. Inequal. **9** (2015), 397–407.
19. C. Park, *Fixed point method for set-valued functional equations*, J. Fixed Point Theory Appl. **19** (2017), 2297–2308.
20. C. Park, *C^*-ternary biderivations and C^*-ternary bihomomorphisms*, Math. **6** (2018), Paper No. 30.
21. C. Park, *Bi-additive s-functional inequalities and quasi-$*$-multipliers on Banach algebras*, Math. **6** (2018), Paper No. 31.
22. C. Park, *Bi-additive s-functional inequalities and quasi-multipliers on Banach algebras*, Rocky Mountain J. Math. **49** (2019), 593–607.
23. V. Radu, *The fixed point alternative and the stability of functional equations*, Fixed Point Theory **4** (2003), 91–96.
24. Th. M. Rassias, *On the stability of the linear mapping in Banach spaces*, Proc. Am. Math. Soc. **72** (1978), 297–300.
25. S. M. Ulam, *A Collection of the Mathematical Problems*, Interscience Publ. New York, 1960.
26. J. Vukman, *Symmetric bi-derivations on prime and semi-prime rings*, Aequationes Math. **38** (1989), 245–254.
27. H. Yazarli, *Permuting triderivations of prime and semiprime rings*, Miskolc Math. Notes **18** (2017), 489–497.

Some New Classes of Higher Order Strongly Generalized Preinvex Functions

Muhammad Aslam Noor and Khalida Inayat Noor

Abstract In this chapter, we define and introduce some new concepts of the higher order strongly generalized preinvex functions and higher order strongly monotone operators with respect to two auxiliary bifunctions. Some new relationships among various concepts of higher order strongly generalized preinvex functions have been established. As special cases, one can obtain various new and known results from our results. Results obtained in this chapter can be viewed as refinement and improvement of previously known results.

1 Introduction

In recent years, several extensions and generalizations have been considered for classical convexity. Hanson [10] introduced the concept of invex function for the differentiable functions, which played significant part in the mathematical programming. Ben-Israel and Mond [5] introduced the concept of invex set and preinvex functions. It is known that the differentiable preinvex functions are invex functions. The converse also holds under certain conditions, see [13]. Noor [18] proved that the minimum of the differentiable preinvex functions on the invex set can be characterized by a class of variational inequalities, which is known as the variational-like inequality. For the recent developments, see [16, 18–26, 29–33], and the references therein. Noor [22–24] proved that a function f is preinvex function, if and only if it satisfies the Hermite–Hadamard-type integral inequality. This result has inspired a great deal of subsequent work which has expanded the role and applications of the invexity in nonlinear optimization and engineering sciences.

Strongly convex functions were introduced and studied by Polyak [32], which play an important part in the optimization theory and related areas, see, for example, [2–4, 11, 14, 19, 24, 27, 34], and the references therein. Noor et al. [25–29]

M. A. Noor (✉) · K. I. Noor
COMSATS University Islamabad, Islamabad, Pakistan

I. N. Parasidis et al. (eds.), *Mathematical Analysis in Interdisciplinary Research*, Springer Optimization and Its Applications 179,
https://doi.org/10.1007/978-3-030-84721-0_25

investigated the properties of the strongly preinvex functions and their variant forms. Adamek [1] introduced another class of convex function with respect to an arbitrary non-negative function, called relative strongly convex functions. With appropriate choice of non-negative function, one can obtain various known classes of convex functions. For the properties of the relative strongly convex functions, see [1, 2, 28].

Lin and Fukushima [12] introduced the concept of higher order strongly convex functions and used it in the study of mathematical program with equilibrium constraints. These mathematical programs with equilibrium constraints are defined by a parametric variational inequality or complementarity system and play an important role in many fields such as engineering design, economic equilibrium, and multilevel game. It is worth mentioning that the characterizations of the higher order strongly convex functions discussed in Lin and Fukushima [12] are not correct. To overcome this drawback, Noor and Noor [29] and Mohsen et al. [14] introduced and studied some classes of higher order strongly convex functions. These higher order strongly convex functions can be used to characterize the Banach spaces by the parallelogram laws. The parallelogram laws were obtained by Bynum [6] and Chen et al. [7–9] in the L^p-spaces and discussed their applications in the prediction theory. Xu [35] obtained these parallelogram laws using the function $\|.\|^p, \quad p > 1$, to characterize the Banach spaces.

Inspired by the research work going in this field, we introduce some new classes of higher order strongly preinvex functions involving two arbitrary bifunctions. For suitable and appropriate choice of these arbitrary bifunctions, one can obtain several new and known classes of strongly convex and strongly preinvex functions. Several new concepts of monotonicity are introduced. We establish the relationship between these classes and derive some new results under some mild conditions. As special cases, one can obtain various new and refined versions of known results. It is expected that the ideas and techniques of this chapter may stimulate further research in this field.

2 Preliminary Results

Let K be a nonempty closed set in a real Hilbert space H. We denote by $\langle \cdot, \cdot \rangle$ and $\| \cdot \|$ the inner product and norm, respectively. Let $F : K_\eta \to R$ be a continuous function, and let $\eta(.,.) : K_\eta \times K_\eta \to R$ be an arbitrary continuous bifunction. Let $h : [0, \infty) \to R$ be a non-negative function.

Definition 1 ([3]) The set K_η in H is said to be invex set with respect to an arbitrary bifunction $\eta(\cdot, \cdot)$, if

$$u + t\eta(v, u) \in K_\eta, \qquad \forall u, v \in K_\eta, t \in [0, 1].$$

The invex set K_η is also called η-connected set. Note that the invex set with $\eta(v, u) = v - u$ is a convex set K, but the converse is not true. For example, the set $K_\eta = R - (-\frac{1}{2}, \frac{1}{2})$ is an invex set with respect to η, where

$$\eta(v, u) = \begin{cases} v - u, \text{ for } & v > 0, u > 0 \quad \text{or} \quad v < 0, u < 0 \\ u - v, \text{ for } & v < 0, u > 0 \quad \text{or} \quad v < 0, u < 0. \end{cases}$$

It is clear that K is not a convex set.

From now onward, K_η is a nonempty closed invex set in H with respect to the bifunction $\eta(\cdot, \cdot)$, unless otherwise specified.

Definition 2 The function F on the invex set K_η is said to be higher order strongly generalized preinvex with respect to the bifunctions $\eta(\cdot, \cdot)$ and $G(., .)$, if there exists a constant $\mu > 0$, such that

$$\begin{aligned} F(u + t\eta(v, u)) &\leq (1 - t)F(u) + tF(v) \\ &-\mu\{t^p(1 - t) + t(1 - t)^p\}\|G(v, u)\|^p, \forall u, v \in K_\eta, t \in [0, 1], p \geq 1. \quad (1) \end{aligned}$$

The function F is said to be higher order strongly generalized preconcave, if and only if $-F$ is higher order strongly generalized preinvex. Note that every higher order strongly convex function is a higher order strongly preinvex function, but the converse is not true.

We now discuss some special cases.

I. If $G(v, u) = \eta(v, u)$, then the higher order strongly generalized preinvex function becomes higher order strongly preinvex functions, that is,

$$F(u + t\eta(v, u)) \leq (1 - t)F(u) + tF(v) - \mu\{t^p(1 - t) + t(1 - t)^p\}\|\eta(v, u)\|^p,$$
$$\forall u, v \in K_\eta, t \in [0, 1].$$

For the properties of the higher order strongly preinvex functions in variational inequalities and equilibrium problems, see Noor [13, 14, 16, 23, 24].

II. If $\eta(v, u) = v - u$. $G(v, u) = \xi(v - u)$, then the invex set becomes a convex set, and the preinvex function reduces to the convex function. In this case, Definition 2 becomes:

Definition 3 The function F on the convex set K is said to be higher order strongly convex function with respect to the arbitrary bifunction $\xi(. - .)$, if there exists a constant $\mu > 0$, such that

$$F((1 - t)u + tv) \leq (1 - t)F(u) + tF(v) - \mu\{t^p(1 - t) + t(1 - t)^p\}\|\xi(v - u)\|^p,$$
$$\forall u, v \in K, t \in [0, 1],$$

which was introduced and studied by Noor and Noor [29].

For the properties and other aspects of the higher order strongly functions, see Noor [17].

III. If $\eta(v,u) = v - u, \quad G(v,u) = v - u$, then the invex set becomes a convex set, and the preinvex function reduces to the convex function. In this case, Definition 2 becomes:

Definition 4 The function F on the convex set K is said to be higher order strongly convex function, if there exists a constant $\mu > 0$, such that

$$F((1-t)u+tv) \leq (1-t)F(u)+tF(v) - \mu\{t^p(1-t)+t(1-t)^p\}\|v-u\|^p,$$
$$\forall u, v \in K, t \in [0,1],$$

which was introduced and studied by Noor and Noor[].

IV. If $p = 2$, then Definition 2 becomes

$$F(u+t\eta(v,u)) \leq (1-t)F(u)+tF(v) - \mu\frac{1}{2^p}\|G(v,u)\|^p, \tag{2}$$
$$\forall u, v \in K_\eta, t \in [0,1], p \geq 1. \tag{3}$$

The function F is known as the higher order Jensen preinvex function.

Definition 5 The function F on the invex set K_η is said to be higher order strongly generalized affine preinvex with respect to the bifunctions $\eta(\cdot,\cdot)$ and $G(.,.)$, if there exists a constant $\mu > 0$, such that

$$F(u+t\eta(v,u)) = (1-t)F(u)+tF(v)$$
$$-\mu\{t^p(1-t)+t(1-t)^p\}\|G(v,u)\|^p, \forall u, v \in K_\eta, t \in [0,1], p \geq 1. \tag{4}$$

Definition 6 The function F on the invex set K is said to be higher order strongly generalized quasi-preinvex with respect to the bifunctions $\eta(\cdot,\cdot)$ and $G(.,.)$, if there exists a constant $\mu > 0$, such that

$$F(u+t\eta(v,u)) \leq \max\{F(u), F(v)\} - \mu\{t^p(1-t)+t(1-t)^p\}\|G(v,u)\|^p,$$
$$\forall u, v \in K_\eta, t \in [0,1].p \geq 1.$$

Definition 7 The function F on the invex set K is said to be higher order strongly generalized log-preinvex with respect to the bifunctions $\eta(\cdot,\cdot)$ and $G(.,.)$, if there exists a constant $\mu > 0$, such that

$$F(u+t\eta(v,u)) \leq (F(u))^{1-t}(F(v))^t - \mu\{t^p(1-t)+t(1-t)^p\}\|G(v,u)\|^p,$$
$$\forall u, v \in K_\eta, t \in [0,1],$$

where $F(\cdot) > 0$.

From the above definitions, we have

$$\begin{aligned}F(u+t\eta(v,u)) &\leq (F(u))^{1-t}(F(v))^{t}-\mu\{t^{p}(1-t)+t(1-t)^{p}\}\|G(v,u)\|^{p}\\ &\leq (1-t)F(u)+tF(v)-\mu\{t^{p}(1-t)+t(1-t)^{p}\}\|G(v,u)\|^{p}\\ &\leq \max\{F(u),F(v)\}-\mu\{t^{p}(1-t)+t(1-t)^{p}\}\|G(v,u)\|^{p}.\end{aligned}$$

This shows that every higher order strongly generalized log-preinvex function is a higher order strongly generalized preinvex function and every higher order strongly generalized preinvex function is a higher order strongly generalized quasi-preinvex function. However, the converse is not true.

For $t=1$, Definitions 2 and 7 reduce to the following condition, which is mainly due to Noor and Noor [5].

Condition A

$$F(u+\eta(v,u))\leq F(v),\quad \forall v\in K_{\eta}.$$

For the applications of Condition A, see [2, 4, 7, 8].

Definition 8 The differentiable function F on the invex set K_{η} is said to be higher order strongly invex function with respect to the bifunctions $\eta(\cdot,\cdot)$ and $G(.,.)$, if there exists a constant $\mu>0$, such that

$$F(v)-F(u)\geq\langle F'(u),\eta(v,u)\rangle+\mu\|G(v,u)||^{p},\quad \forall u,v\in K_{\eta},$$

where $F'(u))$ is the differential of F at u.

It is noted that, if $\mu=0$, then Definition 8 reduces to the definition of the invex function as introduced by Hanson [4]. It is well known that the concepts of preinvex and invex functions play a significant role in the mathematical programming and optimization theory, see [1–9], and the references therein.

Remark 1 Note that, if $\mu=0$, then Definitions 2–7 reduce to the ones in [3, 5].

Definition 9 An operator $T:K_{\eta}\to H$ is said to be

(i). higher order strongly generalized η-monotone, iff there exists a constant $\alpha>0$ such that

$$\langle Tu,\eta(v,u)\rangle+\langle Tv,\eta(u,v)\rangle\leq-\alpha\{G(v,u)+G(u,v)\},\quad u,v\in K_{\eta}.$$

(ii). η-monotone, iff

$$\langle Tu,\eta(v,u)\rangle+\langle Tv,\eta(u,v)\rangle\leq 0,\quad u,v\in K_{\eta}.$$

(iii). higher order strongly generalized η-pseudomonotone, iff there exists a constant $\nu>0$ such that

$$\langle Tu, \eta(v, u)\rangle + \nu G(v, u) \geq 0 \Rightarrow -\langle Tv, \eta(u, v)\rangle \geq 0, \quad u, v \in K_\eta.$$

(iv). higher order strongly generalized relaxed η-pseudomonotone, iff there exists a constant $\mu > 0$ such that

$$\langle Tu, \eta(v, u)\rangle \geq 0 \Rightarrow -\langle Tv, \eta(u, v)\rangle + \mu G(u, v) \geq 0, \quad u, v \in K_\eta.$$

(v). strictly η-monotone, iff

$$\langle Tu, \eta(v, u)\rangle + \langle Tv, \eta(u, v)\rangle < 0, \quad u, v \in K_\eta.$$

(vi). η-pseudomonotone, iff

$$\langle Tu, \eta(v, u)\rangle \geq 0 \Rightarrow \langle Tv, \eta(u, v)\rangle \leq 0, \quad u, v \in K_\eta.$$

(vii). quasi-η-monotone, iff

$$\langle Tu, \eta(v, u)\rangle > 0 \Rightarrow \langle Tv, \eta(u, v)\rangle \leq 0, \quad u, v \in K_\eta.$$

(viii). strictly η-pseudomonotone, iff

$$\langle Tu, \eta(v, u)\rangle \geq 0 \Rightarrow \langle Tv, \eta(u, v)\rangle < 0, \quad u, v \in K_\eta.$$

Note that, if $\eta)v, u) = v - u$, then the invex set K_η is a convex set K. This clearly shows that Definition 9 is more general than and includes the ones in [4–8] as special cases.

Definition 10 A differentiable function F on the invex set K_η is said to be higher order strongly generalized η-pseudo-invex function, iff, if there exists a constant $\mu > 0$ such that

$$\langle F'(u), \eta(v, u)\rangle + \mu G(u, v) \geq 0 \Rightarrow F(v) - F(u) \geq 0, \qquad \forall u, v \in K_\eta.$$

Definition 11 A differentiable function F on K_η is said to be higher order strongly generalized quasi-invex function, iff, if there exists a constant $\mu > 0$ such that

$$F(v) \leq F(u) \Rightarrow \langle F'(u), \eta(v, u)\rangle + \mu G(u, v) \leq 0, \qquad \forall u, v \in K_\eta.$$

Definition 12 The function F on the set K_η is said to be pseudo-invex, if

$$\langle F'(u), \eta(v, u)\rangle \geq 0 \Rightarrow F(v) \geq F(u), \qquad \forall u, v \in K_\eta.$$

Definition 13 The differentiable function F on the K_η is said to be quasi-invex function, if

$$F(v) \leq F(u) \Rightarrow \langle F'(u), \eta(v,u)\rangle \leq 0, \qquad \forall u, v \in K_\eta.$$

If $\eta(v,u) = -\eta(v,u), \forall u, v \in K_\eta$, that is, the function $\eta(\cdot,\cdot)$ is skew-symmetric, then Definitions 9–13 reduce to the ones in [6–8]. This shows that the concepts introduced in this chapter represent an improvement of the previously known ones. All these new concepts may play an important and fundamental part in the mathematical programming and optimization.

We also need the following assumption regarding the bifunction $\eta(\cdot,\cdot)$, which is due to Mohan and Neogy [8].

Condition C Let $\eta(\cdot,\cdot) : K_\eta \times K_\eta \to H$ satisfy assumptions

$$\eta(u, u + t\eta(v,u)) = -t\eta(v,u)$$
$$\eta(v, u + t\eta(v,u)) = (1-t)\eta(v,u), \quad \forall u, v \in K_\eta, t \in [0,1].$$

Clearly, for $t = 0$, we have $\eta(u,v) = 0$, if and only if $u = v, \forall u, v \in K_\eta$. One can easily show that $\eta(u + t\eta(v,u), u) = t\eta(v,u), \forall u, v \in K_\eta$.

3 Main Results

In this section, we consider some basic properties of higher order strongly generalized preinvex functions on the invex set K_η.

Theorem 1 *Let F be a differentiable function on the invex set K_η in H, and let the condition C hold. Then, the function F is a higher order strongly generalized preinvex function, if and only if F is a higher order strongly generalized invex function.*

Proof Let F be a higher order strongly generalized preinvex function on the invex set K_η. Then,

$$F(u + t\eta(v,u)) \leq (1-t)F(u) + tF(v) - \mu\{t^p(1-t) + t(1-t)^p\}\|G(v,u)\|^p, \quad \forall u, v \in K_\eta,$$

which can be written as

$$F(v) - F(u) \geq \left\{\frac{F(u + t\eta(v,u)) - F(u)}{t}\right\} + \mu\{t^{p-1}(1-t) + (1-t)^p\}\|G(v,u)\|^p.$$

Taking the limit in the above inequality as $t \to 0$, we have

$$F(v) - F(u) \geq \langle F'(u), \eta(v,u))\rangle + \mu\|G(v,u)\|^p.$$

This shows that F is a higher order strongly generalized invex function.

Conversely, let F be a higher order strongly generalized invex function on the invex set K_η. Then, $\forall u, v \in K_\eta, t \in [0, 1],\ v_t = u + t\eta(v, u) \in K_\eta$, and using the condition C, we have

$$\begin{aligned} &F(v) - F(u + t\eta(v, u)) \\ &\geq \langle F'(u + t\eta(v, u)), \eta(v, u + t\eta(v, u))\rangle + \mu\|G(v, u + t\eta(v, u))\|^p \\ &= (1 - t)F'(u + t\eta(v, u)), \eta(v, u)\rangle + \mu(1 - t)^p\|G(v, u)\|^p. \end{aligned} \tag{5}$$

In a similar way, we have

$$\begin{aligned} &F(u) - F(u + t\eta(v, u)) \\ &\geq \langle F'(u + t\eta(v, u)), \eta(u, u + t\eta(v, u))\rangle + \mu\|G(u, u + t\eta(v, u))\|^p \\ &= -tF'(u + t\eta(v, u)), \eta(v, u)\rangle + \mu t^p\|G(v, u)\|^p. \end{aligned} \tag{6}$$

Multiplying (5) by t and (6) by $(1 - t)$ and adding the resultant, we have

$$F(u + t\eta(v, u)) \leq (1 - t)F(u) + tF(v) - \{t^p(1 - t) + t(1 - t)^p\}\|G(v, u)\|^p,$$

showing that F is a higher order strongly generalized preinvex function. □

Theorem 2 *Let F be a differentiable higher order strongly generalized preinvex function on the invex set K_η. If F is a higher order strongly generalized invex function, then*

$$\langle F'(u), \eta(v, u))\rangle + \langle F'(v), \eta(u, v)\rangle \leq -\mu\{\|G(v, u) + G(u, v)\|^p\}, \forall u, v \in K_\eta. \tag{7}$$

Proof Let F be a higher order strongly generalized invex function on the invex set K_η. Then,

$$F(v) - F(u) \geq \langle F'(u), \eta(v, u))\rangle + \mu\|G(v, u)\|^p \quad \forall u, v \in K_\eta. \tag{8}$$

Changing the role of u and v in (8), we have

$$F(u) - F(v) \geq \langle F'(v), \eta(u, v)\rangle + \mu\|G(u, v)\|^p \quad \forall u, v \in K_\eta. \tag{9}$$

Adding (8) and (9), we have

$$\langle F'(u), \eta(v, u))\rangle + \langle F'(v), \eta(u, v)\rangle \leq -\mu\{\|G(v, u)) + G(u, v)\|^p\}, \forall u, v \in K_\eta, \tag{10}$$

which shows that $F'(.)$ is a higher order strongly η-monotone operator. □

We note that the converse of Theorem 2 is true only for $p = 2$. However, we have the following theorem:

Theorem 3 *If the differential $F'(.)$ is a higher order strongly η-monotone, then*

$$F(v) - F(u) \geq \langle F'(u), \eta(v, u)\rangle + \frac{2}{p}\mu\|G(v, u\|^p).$$

Proof Let $F'(.)$ be higher order strongly η-monotone. From (10), we have

$$\langle F'(v), \eta(u, v)\rangle \geq \langle F'(u), \eta(v, u))\rangle - \mu\{\|G(v, u))\|^p + \|G(u, v)\|^p)\}, \quad (11)$$

Since K is an invex set, $\forall u, v \in K_\eta$, $t \in [0, 1]$ $v_t = u + t\eta(v, u) \in K_\eta$. Taking $v = v_t$ in (11) and using Condition C, we have

$$\begin{aligned}\langle F'(v_t), \eta(u, u + t\eta(v, u))\rangle &\leq \langle F'(u), \eta(u + t\eta(v, u), u))\rangle - \mu\{\|G(u + t\eta(v, u), u)\|^p \\ &\quad + \|G(u, u + t\eta(v, u)\|^p\} \\ &= -t\langle F'(u), \eta(v, u)\rangle - 2t^p\mu\|G(v, u)\|^p,\end{aligned}$$

which implies that

$$\langle F'(v_t), \eta(v, u)\rangle \geq \langle F'(u), \eta(v, u) + 2\mu t^{p-1}\|G(v, u)\|^p. \quad (12)$$

Let $\xi(t) = F(u + t\eta(v, u))$. Then, from (12), we have

$$\begin{aligned}\xi'(t) &= \langle F'(u + t\eta(v, u)), \eta(v, u)\rangle \\ &\geq \langle F'(u), \eta(v, u) + 2\mu t^{p-1}\|G(v, u)\|^p. \quad (13)\end{aligned}$$

Integrating (13) between 0 and 1, we have

$$\xi(1) - \xi(0) \geq \langle F'(u), \eta(v, u) + \frac{2}{p}\mu\|G(v, u)\|^p;$$

that is,

$$F(u + t\eta(v, u)) - F(u) \geq \langle F'(u), \eta(v, u) + \frac{2}{p}\mu\|G(v, u\|^p).$$

By using Condition A, we have

$$F(v) - F(u) \geq \langle F'(u), \eta(v, u) + \frac{2}{p}\mu\|G(v, u)\|^p,$$

the required result. □

We now give a necessary condition for strongly η-pseudo-invex function.

Theorem 4 *Let F' be a higher order strongly generalized relaxed η- pseudomonotone operator, and Conditions A and C hold. Then, F is a higher order strongly generalized η-pseudo-invex function.*

Proof Let F' be higher order strongly generalized relaxed η-pseudomonotone. Then, $\forall u, v \in K_\eta$,

$$\langle F'(u), \eta(v, u)\rangle \geq 0$$

implies that

$$-\langle F'(v), \eta(u, v)\rangle \geq \alpha\|G(u, v)\|^p. \tag{14}$$

Since K is an invex set, $\forall u, v \in K_\eta,\ t \in [0, 1],\ v_t = u + t\eta(v, u) \in K_\eta$. Taking $v = v_t$ in (14) and using Condition C, we have

$$-\langle F'(u + t\eta(v, u)), \eta(u, v)\rangle \geq t\alpha\|G(v, u)\|^p. \tag{15}$$

Let

$$\xi(t) = F(u + t\eta(v, u)), \quad \forall u, v \in K_\eta, t \in [0, 1].$$

Then, using (15), we have

$$\xi'(t) = \langle F'(u + t\eta(v, u)), \eta(u, v)\rangle \geq t\alpha\|G(v, u)\|^p.$$

Integrating the above relation between 0 and 1, we have

$$\xi(1) - \xi(0) \geq \frac{\alpha}{2}\|G(v, u)\|^p,$$

that is,

$$F(u + t\eta(v, u)) - F(u) \geq \frac{\alpha}{2}\|G(v, u)\|^p,$$

which implies, using Condition A,

$$F(v) - F(u) \geq \frac{\alpha}{2}\|G(v, u)\|^p,$$

showing that F is a higher order strongly generalized η-pseudo-invex function. □

As special cases of Theorem 4, we have the following:

Theorem 5 *Let the differentiable $F'(u)$ of a function $F(u)$ on the invex set K_η be a higher order strongly η-pseudomonotone operator. If Conditions A and C hold, then F is a higher order strongly generalized η-pseudo-invex function.*

Theorem 6 *Let the differential $F'(u)$ of a function $F(u)$ on the invex set K_η be higher order strongly generalized η-pseudomonotone. If Conditions A and C hold, then F is a relative strongly η-pseudo-invex function.*

Theorem 7 *Let the differential $F'(u)$ of a function $F(u)$ on the invex set K_η be higher order strongly generalized η-pseudomonotone. If Conditions A and C hold, then F is a higher order strongly generalized η-pseudo-invex function.*

Theorem 8 *Let the differential $F'(u)$ of a function $F(u)$ on the invex set K_η be η-pseudomonotone. If Conditions A and C hold, then F is a higher order strongly generalized pseudo-invex function.*

Theorem 9 *Let the differential $F'(u)$ of a differentiable preinvex function $F(u)$ be Lipschitz continuous on the invex set K_η with a constant $\beta > 0$. Then,*

$$F(u + \eta(v, u)) - F(u) \leq \langle F'(u), \eta(v, u)\rangle + \frac{\beta}{2}\|\eta(v, u)\|^2, \quad u, v \in K_\eta.$$

Proof Its proof follows from Noor and Noor [21]. □

Definition 14 The function F is said to be sharply higher order strongly generalized pseudo-preinvex, if there exists a constant $\mu > 0$, such that

$$\langle F'(u), \eta(v, u)\rangle \geq 0$$

$$\Rightarrow$$

$$F(v) \geq F(v + t\eta(v, u)) + \mu t(1 - t)\|G\eta(v, u)\|^p, \forall u, v \in K_\eta, t \in [0, 1].$$

Theorem 10 *Let F be a sharply higher order strongly generalized pseudo-preinvex function on K_η with a constant $\mu > 0$. Then,*

$$-\langle F'(v), \eta(v, u)\rangle \geq \mu\|G(v, u)\|^p. \quad \forall u, v \in K_\eta.$$

Proof Let F be a sharply higher strongly generalized pseudo-preinvex function on K_η. Then,

$$F(v) \geq F(v + t\eta(v, u)) + \mu t(1 - t)\|G(v, u)\|^p, \quad \forall u, v \in K_\eta, t \in [0, 1],$$

from which we have

$$\frac{F(v + t\eta(v, u)) - F(v)}{t} + \mu t(1 - t)\|G(v, u\|^p) \leq 0.$$

Taking limit in the above inequality, as $t \to 0$, we have

$$-\langle F'(v), \eta(v, u)\rangle \geq \mu\|G(v, u)\|^p,$$

the required result. □

Definition 15 A function F is said to be a higher order strongly generalized pseudo-preinvex function, if there exists a strictly positive bifunction $B(.,.)$, such that

$$\begin{aligned} F(v) &< F(u) \\ &\Rightarrow \\ F(u + t\eta(v,u)) &< F(u) + t(t-1)B(v,u), \forall u, v \in K_\eta, t \in [0,1]. \end{aligned}$$

Theorem 11 *If the function F is a higher order strongly generalized preinvex function such that $F(v) < F(u)$, then the function F is higher order strongly generalized pseudo-preinvex.*

Proof Since $F(v) < F(u)$ and F is a higher order strongly generalized preinvex function, then $\forall u, v \in K_\eta, \quad t \in [0,1]$, we have

$$\begin{aligned} F(u + \lambda\eta(v,u)) &\leq F(u) + t(F(v) - F(u)) - \mu\{t^p(1-t) + t(1-t)^p\})\|G(v,u)\|^p \\ &< F(u) + t(1-t)(F(v) - F(u)) - \mu\{t^p(1-t) + t(1-t)^p\}\|G(v,u)\|^p \\ &= F(u) + t(t-1)F(u) - F(v)) - \mu\{t^p(1-t) + t(1-t)^p\}\|G(v,u)\|^p \\ &< F(u) + t(t-1)B(u,v) - \mu\{t^p(1-t) + t(1-t)^p\}\|G(v,u)\|^p, \end{aligned}$$

where $B(u,v) = e^{F(u)} - e^{F(v)} > 0$. This shows that F is a higher order strongly exponentially generalized preinvex function

□

We now discuss the optimality condition for the differentiable higher order strongly generalized preinvex functions, which is the main motivation of our next result.

Theorem 12 *Let F be a differentiable higher order strongly preinvex function with modulus $\mu > 0$. If $u \in K_\eta$ is the minimum of the function F, then*

$$F(v) - F(u) \geq \mu\|G(v,u)|^p, \quad \forall u, v \in K_\eta. \tag{16}$$

Proof *Let $u \in K_\eta$ be a minimum of the function F. Then,*

$$F(u) \leq F(v), \forall v \in K_\eta. \tag{17}$$

Since K_η is an invex set, so, $\forall u, v \in K_\eta, \quad t \in [0,1], \ v_t = u + t\eta(v,u) \in K_\eta$. Taking $v = v_t$ in (17), we have

$$0 \leq \lim_{t \to 0} \left\{ \frac{F(u + t\eta(v,u)) - F(u)}{t} \right\} = \langle F'(u), \eta(v,u) \rangle. \tag{18}$$

Since F is a differentiable higher order strongly preinvex function, so

$$F(u + t\eta(v, u)) \leq F(u) + t(F(v) - F(u)) \\ -\mu\{t^p(1-t) + t(1-t)^p\}\|G(v, u\|^p, \quad u, v \in K_\eta, t \in [0, 1],$$

from which, using (18), we have

$$F(v) - F(u) \geq \lim_{t\to 0}\left\{\frac{F(u + t\eta(v, u)) - F(u)}{t}\right\} + \mu\{t^{p-1}(1-t) + (1-t)^p\}\|G(v, u)\|^p. \\ = \langle F'(u), v - u\rangle + \mu\|G(v, u)\|^p) \\ \geq \mu\|G(v, u)\|^p,$$

the required result (5). □

Remark 2 If

$$\langle e^{F(u)}F'(u), \eta(v, u)\rangle + \mu\|\eta(v, u)\|^p \geq 0, \quad \forall u, v \in K_\eta, \tag{19}$$

then $u \in K_\eta$ is the minimum of the function F.

We would like to emphasize that inequality (19) is called the higher order strongly variational-like inequality and appears to be a new one. It is an interesting problem to study the existence of a unique solution of the variational-like inequality (19) and its applications [15, 36–38].

4 Applications

In this section, we discuss the relationship between the parallelogram law and higher order strongly generalized preinvex functions. From Definition 5 with $F(u) = \|u\|^p$, and $G(v, u) = v - u$,we have

$$\|u + t\eta(v, u)\|^p = (1-t)\|u\|^p + t\|v\|^p - \mu\{t^p(1-t) + t(1-t)^p\}\|v - u\|^p, \\ \forall u, v \in K_\eta, t \in [0, 1], p \geq 1. \tag{20}$$

At $t = \frac{1}{2}$, we obtain

$$\left\|\frac{2u + \eta(v, u)}{2}\right\|^p = \frac{1}{2}\{\|u\|^p + \|v\|^p\} - \mu\left\{\frac{1}{2^p}\right\}\|v - u\|^p, \forall u, v \in K_\eta, p \geq 1, \tag{21}$$

from which, it follows that

$$\|2u + \eta(v, u)\|^p + +\mu\|v - u\|^p = 2^{p-1}\{\|u\|^p + \|v\|^p\}, \forall u, v \in K_\eta, p \geq 1. \tag{22}$$

which is called the parallelogram-like law involving the preinvex functions and appears to the new one.

In particular, for $\eta(v, u) = v - u$, we obtain

$$\|u + v\|^p + +\mu\|v - u\|^p = 2^{p-1}\{\|u\|^p + \|v\|^p\}, \forall u, v \in K_\eta, p \geq 1, \quad (23)$$

which is well known as parallelogram law for Banach spaces. Bynum [7] and Chen et al. [8–10] have studied the properties and applications of the parallelogram laws (23) for the Banach spaces. Xu [31] discussed the characteristics of p-uniform convexity and q-uniform smoothness of a Banach space via the functionals $\|.\|^p$ and $\|.\|^q$, respectively. These results can be obtained from the concepts of higher order strongly affine convex(concave) functions, which can be viewed as novel application.

5 Conclusion

In this chapter, we have introduced and studied a new class of preinvex functions with respect to two arbitrary bifunctions. It is shown that several new classes of strongly preinvex and convex functions can be obtained as special cases of these relative strongly preinvex functions. We have studied the basic properties of these functions. We have shown that the optimality conditions of the higher order strongly generalized preinvex functions can be characterized by variational-like inequalities. This result motivated us to introduce higher order strongly variational inequalities. It is an interesting problem to investigate the analytical and numerical aspects of these variational-like inequalities. As novel applications of the higher order strongly generalized preinvex functions, one obtain the parallelogram-like laws for the L^p-spaces, which is itself a significant contribution. It is expected that the ideas and techniques of this chapter may motivate further research.

Acknowledgments The authors would like to thank the Rector, COMSATS Institute of Information Technology, Pakistan, for providing excellent research and academic environments.

References

1. M. Adamek, (2016), On a problem connected with strongly convex functions, Math. Inequ. Appl., 19(4)(2016), 1287–1293.
2. H. Angulo, J. Gimenez, A. M. Moeos and K. Nikodem, (2011), On strongly h-convex functions, Ann. Funct. Anal. 2(2)(2011), 85–91.
3. M. U. Awan, M. A. Noor, V. N. Mishra and K. I. Noor, Some characterizations of general preinvex functions, International J. Anal. Appl., 15(1)(2017), 46–56.
4. A. Azcar, J. Gimnez, K. Nikodem and J. L. Snchez, On strongly midconvex functions, Opuscula Math., 31(1)(2011), 15–26.

5. A. Ben-Isreal and B. Mond, What is invexity? J. Austral. Math. Soc., Ser. B, 28(1)(1986), 1–9.
6. W.L. Bynum, Weak parallelogram laws for Banach spaces. Can. Math. Bull. 19(1976), 269275.
7. R. Cheng, C.B. Harris, Duality of the weak parallelogram laws on Banach spaces. J. Math. Anal. Appl. 404(2013), 6470.
8. R. Cheng and W. T. Ross, Weak parallelogram laws on Banach spaces and applications to prediction, Period. Math. Hung. 71(2015), 45–58.
9. R. Cheng, J. Mashreghi and W. T. Ross, Optimal weak parallelogram constants for L_p space, Math. Inequal. Appl. 21(4)(2018), 10471058.
10. M. A. Hanson, On sufficiency of the Kuhn-Tucker conditions, J. Math. Anal. Appl., 80(1981), 545–550.
11. S. Karamardian, The nonlinear complementarity problems with applications, Part 2, J. Optim. Theory Appl. 4(3)(1969), 167–181
12. G. H. Lin AND M. Fukushima, Some exact penalty results for nonlinear programs and mathematical programs with equilibrium constraints, J. Optim. Theory Appl. 118(1)(2003), 6780.
13. S. R. Mohan and S. K. Neogy, On invex sets and preinvex functions, J. Math. Anal. Appl. 189(1995), 901–908.
14. B. B. Mohsen, M. A. Noor, K. I. Noor and M. Postolache, Strongly convex functions of higher order involving bifunction, Mathematics, 7(11)(2019):208.
15. C. P. Niculescu and L. E. Persson, Convex Functions and Their Applications, Springer-Verlag, New York, (2018).
16. K. Nikodem and Z. S. Pales, Characterizations of inner product spaces by strongly convex functions, Banach J. Math. Anal., 1(2011), 83–87.
17. M. A. Noor, Advanced Convex Analysis, Lecture Notes, COMSATS University Islamabad, Islamabad, Pakistan, (2008–2019).
18. M. A. Noor, Variational-like inequalities, Optimization, 30(1994), 323–330.
19. M. A. Noor, Invex Equilibrium problems, J. Math. Anal. Appl., 302(2005), 463–475.
20. M. A. Noor, On generalized preinvex functions and monotonicities, J. Inequal. Pure Appl. Math., 5(4)(2004), Article 110.
21. M. A. Noor, Fundamentals of equilibrium problems. Math. Inequal. Appl., 9(3)(2006), 529–566.
22. M. A. Noor, Hermite-Hadamard integral inequalities for log-preinvex functions, J. Math. Anal. Approx. Theory, 2(2007), 126–131.
23. M. A. Noor, On Hadamard type inequalities involving two log-preinvex functions, J. Inequal. Pure Appl. Math. 8(3)(2007), 1–14.
24. M. A. Noor, Hadamard integral inequalities for product of two preinvex functions, Nonl. Anal. Fourm, 14(2009), 167–173.
25. M. A. Noor and K. I. Noor, On strongly generalized preinvex functions, J. Inequal. Pure Appl. Math., 6(4)(2005), Article 102.
26. M. A. Noor and K. I. Noor, Some characterization of strongly preinvex functions. J. Math. Anal. Appl., 316(2)(2006), 697–706.
27. M. A. Noor and K. I. Noor, Generalized preinvex functions and their properties, J. Appl. Math. Stoch. Anal.,2006(2006), pp.1–13, doi:10.1155/JAMSA/2006/12736
28. M. A. Noor and K. I. Noor, On generalized strongly convex functions involving bifunction, Appl. Math. Inform. Sci. 13(3)(2019), 411–416.
29. M. A. Noor and K. I. Noor, Higher order strongly generalized convex functions, Appl. Math. Inform. Sci. 14(2)(2020).
30. M. A. Noor, K. I. Noor and M. U. Awan, Some quantum integral inequalities via preinvex functions, Appl. Math. Comput., 269(2015), 242–251.
31. M. A. Noor, K. I. Noor, S. Iftikhar and F. Safdar, Some properties of generalized strongly harmonic convex functions, Inter. J. Anal. Appl. 17(2018),
32. B. T. Polyak, Existence theorems and convergence of minimizing sequences in extremum problems with restrictions, Soviet Math. Dokl., 7(1966), 2–75.

33. G. Ruiz-Garzion, R. Osuna-Gomez and A. Rufian-Lizan, Generalized invex monotonicity, European J. Oper. Research, 144(2003), 501–512.
34. T. Weir and B. Mond, Preinvex functions in multiobjective optimization, J. Math. Anal. Appl., 136(1988), 29–38.
35. H-K, Xu, Inequalities in Banach spaces with applications, Nonl. Anal. Theory, Meth. Appl. 16(12)(1991), 1127–1138.
36. X. M. Yang, Q. Yang and K. L. Teo, Criteria for generalized invex monotonicities, European J. Oper. Research, 164(1)(2005), 115–119.
37. X. M. Yang, Q. Yang and K. L. Teo, Generalized invexity and generalized invariant monotonicity, J. Optim. Theory Appl., 117(2003), 607–625.
38. D. L. Zu and P. Marcotte, Co-coercivity and its role in the convergence of iterative schemes for solving variational inequalities. SIAM Journal on Optimization, 6(3)(1996), 714–726.

Existence of Global Solutions and Stability Results for a Nonlinear Wave Problem in Unbounded Domains

P. Papadopoulos, N. L. Matiadou, S. Fatouros, and G. Xerogiannakis

Abstract We investigate the asymptotic behavior of solutions for the nonlocal quasilinear hyperbolic problem of Kirchhoff type

$$u_{tt} - \phi(x)\|\nabla u(t)\|^2 \Delta u + \delta u_t = |u|^3 u, \quad x \in R^N, \quad t \geq 0,$$

with initial conditions $u(x, 0) = u_0(x)$ and $u_t(x, 0) = u_1(x)$, in the case where $N \geq 3$, $\delta > 0$, and $(\phi(x))^{-1} = g(x)$ is a positive function lying in $L^{N/2}(R^N) \cap L^\infty(R^N)$. It is proved that when the initial energy $E(u_0, u_1)$, which corresponds to the problem, is nonnegative and small, there exists a unique global solution in time in the space $X_0 =: D(A) \times D^{1,2}(R^N)$. When the initial energy $E(u_0, u_1)$ is negative, the solution blows up in finite time. For the proofs, a combination of the modified potential well method and the concavity method is used. Also, the existence of an absorbing set in the space $X_1 =: D^{1,2}(R^N) \times L^2_g(R^N)$ is proved and that the dynamical system generated by the problem possess an invariant compact set A in the same space.

Finally, for the generalized Kirchhoff's string problem with no dissipation

$$u_{tt} = -\|\mathrm{A}^{1/2}u\|_H^2 \mathrm{A}u + f(u), \quad x \in R^N, \quad t \geq 0,$$

with the same hypotheses as above, we study the stability of the trivial solution $u \equiv 0$. It is proved that if $f'(0) > 0$, then the solution is unstable for the initial Kirchhoff's system, while if $f'(0) < 0$, the solution is asymptotically stable.

P. Papadopoulos (✉) · N. L. Matiadou · G. Xerogiannakis
Department of Electrical and Electronics Engineering, University of West Attica, Athens, Greece
e-mail: ppapadop@uniwa.gr; lmatiadou@uniwa.gr; georgiosx@uniwa.gr

S. Fatouros
Department of Informatics and Computer Engineering, University of West Attica, Athens, Greece
e-mail: fatouros@uniwa.gr

I. N. Parasidis et al. (eds.), *Mathematical Analysis in Interdisciplinary Research*,
Springer Optimization and Its Applications 179,
https://doi.org/10.1007/978-3-030-84721-0_26

1 Introduction: Preliminaries

We study the following quasilinear hyperbolic initial value problem:

$$u_{tt} - \phi(x)\|\nabla u(t)\|^2 \Delta u + \delta u_t - |u|^3 u = 0, \tag{1.1}$$

$$u(x,0) = u_0(x), \quad u_t(x,0) = u_1(x), \quad x \in R^N, \quad t \geq 0, \tag{1.2}$$

with initial conditions u_0, u_1 in appropriate function spaces, $N \geq 3$ and $\delta \geq 0$. Throughout the chapter, we assume that the function ϕ and $g : R^N \to R$ satisfy the following condition:

$$\textbf{(G)}\phi(x) > 0, \text{ for all } \; x \in R^N, \text{and } \; (\phi(x))^{-1} = g(x) \in L^{N/2}(R^N) \cap L^\infty(R^N).$$

This class will include functions of the form:

$$\phi(x) \sim c_0 + \epsilon|x|^\alpha, \quad \epsilon > 0 \; \text{ and } \; \alpha > 0,$$

resembling phenomena of slowly varying wave speed around the constant speed c_0. G. Kirchhoff in 1883 proposed the so-called Kirchhoff's string model in the study of oscillations of stretched strings and plates

$$ph\frac{\partial^2 u}{\partial t^2} + \delta\frac{\partial u}{\partial t} = \left\{p_0 + \frac{Eh}{2L}\int_0^L \left(\frac{\partial u}{\partial x}\right)^2\right\}\frac{\partial^2 u}{\partial x^2} + f \; \text{ for } \; 0 < x < L, \quad t \geq 0,$$

where $u = u(x,t)$ is the lateral displacement at the space coordinate x and the time t, E the Young modulus, p the mass density, h the cross-section area, L the length, p_0 the initial axial tension, δ the resistance modulus, and f the external force (see [5]). When $p_0 = 0$, the equation is considered to be of degenerate type; otherwise, it is of nondegenerate type.

In the case of bounded domain, T. Kobayashi [6] constructed a unique weak solution by a Faedo-Galerkin method for a quasilinear wave equation with strong dissipation (see also [1], [8]). K. Nishihara [9] has derived a decay estimate from below of the potential of solutions. Also R. Ikehata [4] has shown that for sufficiently small initial data global existence can be obtained, even when the influence of the source terms is stronger than that of the damping terms. Finally K. Ono [10] for $\delta \geq 0$ has proved global existence and blow up results for a degenerate nonlinear wave equation of Kirchhoff type with strong dissipation.

In the case of unbounded domain, P. DAncona and S. Spagnolo [2] have shown the global existence of a unique solution for the nondegenerate type with small C^∞ data. T. Mizumachi (see [7]) studied the asymptotic behavior of solutions to the Kirchhoff equation with a viscous damping term with no external force. In our previous work (see [11]), we prove global existence and blow-up results of an equation of Kirchhoff type in all of R^N. Also, in [12] we prove the existence of compact invariant sets for the same equation. Finally, in [13] we study the stability of the trivial solution $u = 0$ for the generalized Kirchhoff's string equation, using the central manifold theory.

As we will see, the space setting for the initial conditions and the solutions of our problem is the product space $X_0 =: D(A) \times D^{1,2}(R^N)$.

By $D^{1,2}(R^N)$, we denote the closure of the $C_0^\infty(R^N)$ functions with respect to the energy norm:

$$\|u\|_{D^{1,2}} =: \int_{R^N} |\nabla u|^2 dx.$$

It is known that

$$D^{1,2}(R^N) = \left\{ u \in L^{\frac{2N}{N-2}}(R^N) : \nabla u \in (L^2(R^N))^N \right\}.$$

The weighted Lebesgue space $L_g^2(R^N)$ is the closure $C_0^\infty(R^N)$ function with respect to the inner product

$$(u, v)_{L_g^2(R^N)} =: \int_{R^N} guvdx \quad \text{(cf.[3])}.$$

We also have that the operator $\mathrm{A} = -\phi\Delta$ is self-adjoint and therefore graph closed. Its domain $D(\mathrm{A})$ is a Hilbert space with respect to the norm:

$$\|\mathrm{A}u\|_{L_g^2} =: \left\{ \int_{R^N} \phi|\Delta u|^2 dx \right\}^{1/2}.$$

Thus, we construct the following evolution quartet, with compact and dense embeddings:

$$D(\mathrm{A}) \subset D^{1,2}(R^N) \subset L_g^2(R^N) \subset D^{-1,2}(R^N).$$

For the positive self-adjoint operator $\mathrm{A} = -\phi\Delta$, we may define the fractional powers in the following way. For every $s > 0$, A^s is an unbounded self-adjoint operator in $L_g^2(R^N)$ with its domain $D(\mathrm{A}^s)$ to be a dense subset in $L_g^2(R^N)$. The operator A^s is strictly positive and injective. Also $D(\mathrm{A}^s)$, endowed with the scalar product

$$(u, v)_{D(\mathrm{A}^s)} = (u, v)_{L_g^2} + (\mathrm{A}^s u, \mathrm{A}^s v)_{L_g^2},$$

becomes a Hilbert space. We write as usual $V_{2s} = D(\mathrm{A}^s)$ and we have the following identifications:

$$D(\mathrm{A}^{-1/2}) = D^{-1,2}(R^N), \quad D(\mathrm{A}^0) = L_g^2, \quad D(\mathrm{A}^{1/2}) = D^{1,2}(R^N).$$

Moreover the mapping $A^{s/2} : V_x \to V_{x-s}$ is an isomorphism. Furthermore, we have that the injection $D(A^{s_1}) \subset D(A^{s_2})$ is compact and dense, for every $s_1, s_2 \in R, \ s_1 > s_2$. In order to clarify the kind of solutions we are going to obtain for our problem, we give the definition of the weak solution for the problem.

Definition 1.1 A weak solution of the problem (1.1) and (1.2) is a function u such that

$$u \in L^2[0,T; D(A)], \quad u_t \in L^2[0,T; D^{1,2}], \qquad \textbf{(i)}$$

$$u_{tt} \in L^2[0,T; L^2_g],$$

for all $v \in C_0^\infty\left([0,T] \times (R^N)\right)$, **(ii)**

satisfies the generalized formula:

$$\int_0^T (u_{tt}(\tau), v(\tau))_{L^2_g} d\tau + \int_0^T \left(\|\nabla u(\tau)\|^2 \int_{R^N} \nabla u(\tau)\nabla v(\tau) dx \right) d\tau$$
$$+ \delta \int_0^T (u_t(\tau), v(\tau))_{L^2_g} d\tau - \int_0^T f(u(\tau), v(\tau))_{L^2_g} d\tau = 0, \tag{1.3}$$

where $f(s) = |s|^3 \ s$, and (iii) satisfies the initial conditions:

$$u(x,0) = u_0(x) \in D(A), u_t(x,0) = u_1(x) \in D^{1,2}(R^N).$$

In the following section we briefly discuss the results concerning the asymptotic behavior of solutions for the problems (1.1) and (1.2). Among the global existence and blow-up results, we also prove the existence of a compact functional invariant set. We would like to mention that up to our knowledge, this is the first result concerning the existence of functional invariant sets for mathematical models of Kirchoff's string type.

2 Global Existence, Blow-Up Results, and Invariant Sets

In this section we provide global existence and blow-up results for the problems (1.1) and (1.2) in the space X_0. We also prove the existence of an attractor-like set. For the proofs, we refer on [12] and [13]. In order to obtain a local existence result for the problems (1.1) and (1.2), we need information concerning the solvability of the corresponding nonhomogeneous linearized problem around the function v, where $(v, v_t) \in C(0,T; D(A) \times D^{1,2})$ is given restricted in the sphere B_R,

$$
\begin{aligned}
u_{tt} - \phi(x)\|\nabla v\|^2 \Delta u + \delta u_t &= |v|^3 v, (x,t) \in B_R \times (0,T), u(x,0) \\
&= u_0(x), u_t(x,0) = u_1(x), \\
x \in B_R, u(x,t) &= 0, (x,t) \in \partial B_R \times (0,T), \\
& v \in C(0,T; D(\mathrm{A})) \text{ and } v_t \in C(0,T; D^{1,2}).
\end{aligned} \tag{2.1}
$$

Proposition 2.1 *Assume that $u_0 \in D(\mathrm{A})$, $u_1 \in D^{1,2}(R^N)$, and $0 \leq N \leq 10/3$, then the linear wave equation (2.1) has a unique solution such that $u \in C(0,T; D(\mathrm{A}))$ and $u_t \in C(0,T; D^{1,2})$.*

Proof The proof follows the spirit of [11]. The Galerkin method is used, based on the information taken from the eigenvalue problem.

Next, we have the following theorem (for the proof, see also [13]). □

Theorem 2.2 *If $(u_0, u_1) \in C(0,T; D^{1,2})$ and satisfy the nondegenerate condition $\|\nabla u_0\|^2 > 0$, then there exists $T > 0$ such that the problems (1.1) and (1.2) admit a unique local weak solution u satisfying*

$u \in C(0,T; D(\mathrm{A}))$, $u_t \in C(0,T; D^{1,2})$. Moreover, at least one of the following statements holds true, either

$$
\begin{aligned}
&T = +\infty, \quad or && \textbf{(i)}\\
&e(u(t)) =: \|u_t\|^2_{D^{1,2}} + \|u\|^2_{D(\mathrm{A})} \to \infty, \quad as\ t \to T_-. && \textbf{(ii)}
\end{aligned}
$$

The next theorem deals with the global existence, blow-up results, and the energy decay property of the problem.

First we define as the energy of the problems (1.1) and (1.2) the quantity

$$
E(t) =: E(u(t), u_t(t)) =: \|u(t)\|^2_{L^2_g} + \frac{1}{2}\|u(t)\|^4_{D^{1,2}} - \frac{2}{5}\|u(t)\|^5_{L^5_g}. \tag{2.2}
$$

Also, we introduce the potential of the problems (1.1) and (1.2), as

$$
J(u) =: \frac{1}{2}\|u(t)\|^4_{D^{1,2}} - \frac{2}{5}\|u(t)\|^5_{L^5_g}. \tag{2.3}
$$

Thus we derive the following relation:

$$
E(t) =: \|u(t)\|^2_{L^2_g} + J(u). \tag{2.4}
$$

Finally, we introduce a modified version of the modified potential well used in [11] (see also [12]), by

$$W =: \left\{ u \in D(\mathrm{A});\, K(u) = \|u\|^4_{D^{1,2}} - \|u\|^5_{L^5_g} > 0 \right\} \cup \{0\}. \tag{2.5}$$

Theorem 2.3 *Assume that* $N = 3$, $u_0 \in W(\subset D(\mathrm{A}))$, *and* $u_1 \in D^{1,2}$. *Also suppose that the following inequality holds:*

$$E(u_0, u_1) \le \left(\frac{1}{C_0 \mu_0^{p_1}} \right)^{1/p_2} \quad \text{and} \quad p_2 > 0. \tag{2.6}$$

Then (**a**) *for* $p_1 =: \frac{1}{2}$ *and* $p_2 =: \frac{1}{8}$, *there exists a unique global solution* $u \in W$ *of the problems (1.1) and (1.2) satisfying* $u \in C([0, +\infty); D(\mathrm{A}))$ *and* $u_t \in C([0, +\infty); D^{1,2})$.

(**b**) *Moreover, this solution obeys the following energy estimates:*

$$\|u(t)\|^2_{L^2_g} + d_*^{-1} \|\nabla u\|^4 \le E(u, u_t) \le \{E(u_0, u_1)^{-1/2} + d_0^{-1}[t-1]^+\}^{-2}, \tag{2.7}$$

where $d_* = 10$ *and* $d_0 \ge 1$; *that is,*

$$\|\nabla u\|^4 \le C_* (1+t)^{-1}, \tag{2.8}$$

where C_* *is some constant depending on* $\|u_0\|^4_{D^{1,2}}$ *and* $\|u_1\|_{L^2_g}$.

(**c**) *Suppose that* $N \ge 3$ *and the initial energy* $E(u_0, u_1)$ *is negative. Then there exists a time* T, *where*

$$\begin{aligned} 0 < T \le 3^{-2}(-E(u_0, U_1))^{-1} \Bigg[\Bigg\{ \Big(2\delta \|u_0\|^2_{L^2_g} - 3(u_0, u_1)_{L^2_g} \Big)^2 \\ +9(-E(u_0, u_1)) \|u_0\|^2_{L^2_g} \Bigg\}^{\frac{1}{2}} \\ +2\delta \|u_0\|^2_{L^2_g} - 3(u_0, u_1)_{L^2_g} \Bigg], \end{aligned} \tag{2.9}$$

such that the (unique) solution of the problems (1.1) and (1.2) blows up at T, *i.e.,*

$$\lim_{t \to T_-} \|u(t)\|^2_{L^2_g} = +\infty. \tag{2.10}$$

The existence of an absorbing set in X_0 *is given below. See also [12].*

Lemma 2.4 *Assume that* $p_1 > 4 \cdot 3^{-1/2} R^2 c_3^2$, $N > 3$, *and* $\|\nabla u_0\| > 0$. *Then the unique local solution defined by Theorem 2.2 exists globally in time.*

Remark 2.5 (Global Solutions) From the last Lemma 2.4, we may observe that solutions of the problems (1.1) and (1.2) (given by Theorem 2.2) belong to the space $C_b(R_+, X_0)$, i.e., we have achieved global solutions for the given problem. Let us remark that, in Theorem 2.3, using a modified potential well technique, we have proved global existence results under the condition $N = 3$ and the initial energy $E(0)$ been nonnegative and small. On the other hand, in Lemma 2.4, we could achieve global results for different type of nonlinearities, i.e., for any $N \geq 3$ and independent of the sign of the initial energy $E(0)$.

Lemma 2.4 has an immediate consequence:

Remark 2.6 A nonlinear semigroup $S(t) : X_0 \to X_0, t \geq 0$, may be associated to the problems (1.1) and (1.2) such that for

$$\psi = \{u_0, u_1\} \in X_0, \quad S(t)\psi = \{u(t), u_t(t)\}$$

is the weak solution of the problems (1.1) and (1.2). Moreover the ball $B_0 =: B_{X_0}(0, \overline{R_*})$ for any $\overline{R_*} > R_*$, where R_* is defined by Lemma 2.4 and is **an absorbing set** for the semigroup $S(t)$ in the energy space $X_0 \subset X_1$, compactly.

In the rest of the chapter we show that the ω-limit set of the absorbing set B_0 is a compact invariant set. To this end, we need to decompose the semigroup $S(t)$, in the form $S(t) = S_1(t) + S_2(t)$, where for a suitable bounded set $B \subset X_0$, the semigroups $S_1(t)$, $S_2(t)$ satisfy the following properties:

(S1) $S_1(t)$ is uniformly compact for t large, i.e., $\cup_{t \geq t_0} S_1(t)B$ is relatively compact in X_1.
(S2) $sup_{k \in B} \|S_2(t)k\|_{X_1} \to 0, \quad \text{as } t \to \infty.$

As a consequence of the above properties, we have the following result:

Theorem 2.7 *Let ϕ satisfy hypothesis (G). Then the semigroup $S(t)$ associated with the problems (1.1) and (1.2) possesses a functional invariant set $A = \omega(B_0)$, which is compact in the weak topology of X_1.*

Remark 2.8 We have that X_0 is compactly embedded in X_1, so the set $\cup_{t \geq t_0} S_1(t)B$ is compact with respect to the strong topology in X_1. For the functional invariant compact set $A = \omega(B_0)$, we observe that

$$(u_0, u_1) \in A, \quad \text{if } \|\nabla u_0\| > 0.$$

Then, A is **an attractor-like set**.

Remark 2.9 The above set $A = \omega(B_0)$ is a positively invariant set in the space X_0, because we have that $S(t)A \subset A$, from the definition of the absorbing set. This set is not invariant in the space X_0 because the semigroup $S(t)$ is weakly continuous in X_0 (see the following lemma), but it is not continuous in X_0. At the end, we prove the following lemma.

Lemma 2.10 *For every $t \in R$, the mapping $S(t)$ is weakly continuous from X_0 into X_0.*

Proof Let $\{u^n\}$ be a weakly convergent sequence in X_0 and u its (weak) limit. We fix $t \in R$; we have that the sequence $\{S(t)u^n\}$ is bounded in X_0. We extract a subsequence $\{S(t)u^{n'}\}$ that converges weakly to $v \in X_0$. On the other hand, the compactness of the injection of X_0 into X_1 ensures that $\{u^n\}$ converges strongly to u in X_1. Hence, $\{S(t)u^n\}$ converges strongly to $S(t)u$ in X_1 and then $v = S(t)u$. Therefore, the whole sequence $\{S(t)u^n\}$ weakly converges to $S(t)u$ in X_0 and the lemma is proved. □

Finally, in the following section we study the stability of the initial solution $u = 0$ for the generalized Kirchhoff equation with no dissipation.

3 Stability Results

We consider the generalized quasilinear Kirchhoff's string problem with no dissipation

$$u_{tt} = -\|\mathrm{A}^{1/2}u\|_H^2 \mathrm{A}u + f(u), \quad x \in R^N, t \geq 0,$$

under the same initial conditions as above and H is a Hilbert space. First, we prove the existence of solution for our problem, under small initial data (for the proof, we follow the lines of [12]).

Theorem 3.1 (Local Existence) *Let $f(u)$ be a C^1-function such that*

$$|f(u)| \leq k_1|u|^{\alpha+1}, \quad |f'(u)| \leq k_2|u|^{\alpha}, \quad 0 \leq \alpha \leq 4/(N-2), \quad N \geq 3.$$

Consider that $(u_0, u_1) \in D(\mathrm{A}) \times V$ and satisfy the nondegenerate condition

$$\|\mathrm{A}^{1/2}u_0\| > 0. \tag{3.1}$$

Then there exists $T_0 > 0$ such that our problem admits a unique local weak solution u satisfying $u \in C(0, T; V)$ and $u_t \in C(0, T; H)$.

Proof The proof follows the spirit of [11, Theorem 3.2]. In this case, because of the compact embedding $X_0 \subset X_1 =: V \times H$, we obtain for the associated norms that $e_1(u(t)) \leq e(u(t))$, where

$$e_1(u(t)) = \|u\|_V^2 + \|u'\|_H^2 \text{ and } e(u(t)) = \|u\|_{D(\mathrm{A})}^2 \times \|u'\|_V^2.$$

Following the same steps as in [11, Theorem 3.2], we take the inequality:

$$e_1(u(t)) \leq e(u(t)) \leq R^2,$$

where R is a positive parameter. Therefore, u is a solution such that

$$u \in L^\infty(0, T; V), \qquad u' \in L^\infty(0, T; H).$$

The continuity properties are also proved with the methods indicated in [15, sections II.3 and II.4]. Finally, the uniqueness of the solution can also be taken from [15, Proposition 4.1, p. 215].

Now, we have that the linearized equation of the system around the solution $u = 0$ is

$$\overline{u_t} + A^*\overline{u} = 0, \tag{3.2}$$

where

$$\overline{u_t} = (w, v)^T \text{ and } A^* = \begin{bmatrix} 0 & -f'(0) \\ -1 & 0 \end{bmatrix}. \tag{3.3}$$

Hence, in order to study the stability of the solution, we investigate the spectrum of the operator A^*. The characteristic polynomial of A^* is

$$\begin{bmatrix} \mu_j & f'(0) \\ 1 & \mu_j \end{bmatrix} = 0 \text{ or equivalently } \mu_j^2 - f'(0) = 0.$$

Then according to the sign of $f'(0)$, we have the following cases (see also [14], Theorem 5.1.1 and Theorem 5.1.3):

(I) Let $f'(0) > 0$, then we have that 0 is **unstable** for the initial Kirchhoff's system, because we have two real eigenvalues of different sign $\mu_j = \pm(f'(0))^{1/2}$ and we can easily see that the continuous spectrum of the operator A^* is empty.

(II) Let $f'(0) < 0$. This implies that the operator A^* admits two complex eigenvalues. Thus we obtain that the solution $u = 0$ is **asymptotically stable** for the initial Kirchhoff's system.

(III) Let $f'(0) = 0$. In this case we have that the initial solution is **stable** using the fact that the continuous spectrum of the operator A^* is equal to zero.

The above linearized equation in (3.2) and (3.3) can also be studied in an alternative way. Since $u_t = (w, v)^T$ and $\overline{u} = (wt, vt)^T = \overline{u}_t t$, the linearized equation (3.2) can be factorized as

$$\overline{u_t} + A^*\overline{u} = (I + tA^*)u_t = 0. \tag{3.4}$$

The eigenvalues λ_j of the matrix pencil $(I + tA^*)$ are equal to

$$\lambda_j = -\mu_j^{-1}. \tag{3.5}$$

A distribution of the eigenvalues λ_j for several real values of $f'(0)$ is shown in the following graphs. □

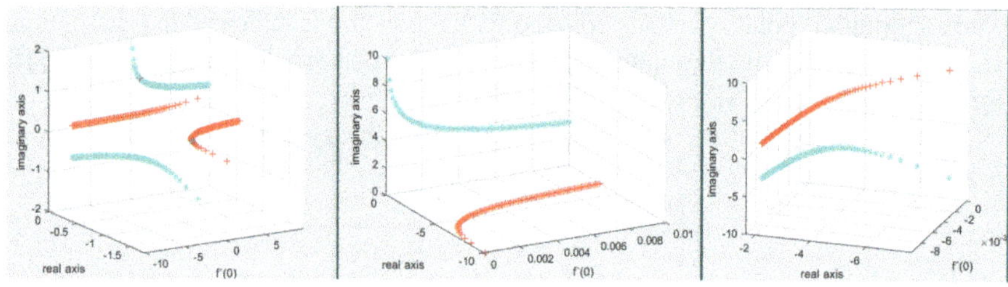

References

1. P. DAncona and Y. Shibata, *On global solvability for the degenerate Kirchhoff equation in the analytic category*, Math. Methods Appl. Sci., **17**, (1994), 477–489
2. P. DAncona and S. Spagnolo, *Nonlinear perturbations of the Kirchhoff equation*, Comm. Pure Appl. Math. **47**, (1994), 1005–1029.
3. K. J. Brown and N. M. Stavrakakis, *Global bifurcation results for a semilinear elliptic equation on all of R^N*, Duke Math. J., **85**, (1996), 77–94.
4. R. Ikehata, *Some remarks on the wave equations with nonlinear damping and source terms*, Nonlinear Analysis TMA, Vol 27,10, (1996), 1165–1175.
5. G. Kirchhoff, *Vorlesungen ber mechanik*, Leipzig: B.G. Teubner, (1883).
6. T. Kobayashi, H. Pecher and Y. Shibata, *On a global in time existence theorem of smooth solutions to a nonlinear wave equations with viscosity* , Math. Ann. 296, (1993), 215–234.
7. T. Mizumachi, *The asymptotic behavior of solutions to the Kirchhoff equation with a viscous damping term, Journal of Dynamics and Differential Equations*, 9, (1997), 211–247.
8. M. Nakao, *Energy decay for the quasilinear wave equation with viscosity*, Math. Z., 219, (1995), 289–299.
9. K. Nishihara, *Decay properties of solutions of some quasilinear hyperbolic equations with strong damping*, Nonlinear Analysis TMA, 21, (1993), 17–21.
10. K. Ono, *On global existence, asymptotic stability and blowing up of solutions for some degenerate non-linear wave equations of Kirchhoff type with a strong dissipation*, Math. Meth. Appl. Sci., 20, (1997), 151–177.
11. P. Papadopoulos, N. Stavrakakis *Global existence and blow-up results for an equation of Kirchhoff type on $\mathbb{R}^N$*, Topol. Methods Nonlin Analysis, **17** (2001), 91–109.
12. P. Papadopoulos, N. Stavrakakis *Compact invariant sets for some quasilinear nonlocal Kirchhoff strings on $\mathbb{R}^N$*, Applicable Analysis, **87** (2008), 133–148.
13. P. Papadopoulos, N. Stavrakakis *Central manifold theory for the generalized equation of Kirchhoff strings on $\mathbb{R}^N$*, Nonlinear Analysis TMA, **61** (2005), 1343–1362.
14. R. L. Pego, *Phase transitions in one-dimensional nonlinear viscoelasticity: admissibility and stability*, Arch. Rat. Anal. 97, (1987), 353–394.
15. R. Temam, *Infinite-Dimensional Dynamical Systems in Mechanics and Physics. Applications in Mathematical Science*, 2nd Edn, Vol. **68** (New York: Springer-Verlag).

Congestion Control and Optimal Maintenance of Communication Networks with Stochastic Cost Functions: A Variational Formulation

Mauro Passacantando and Fabio Raciti

Abstract We consider a game theory model of congestion control in communication networks, where each player is a user who wishes to maximize his/her flow over a path in the network. We allow for stochastic fluctuations of the cost function of each player, which consists of two parts: a pricing and a utility term. The solution concept we look for is the mean value of the (unique) variational Nash equilibrium of the game. Furthermore, we assume that it is possible to invest a certain amount of money to improve the network by enhancing the capacity of its links and, because of limited financial resources, an optimal choice of the links to improve has to be made. We model the investment problem as a nonlinear knapsack problem with generalized Nash equilibrium constraints in probabilistic Lebesgue spaces and solve it numerically for some examples.

1 Introduction

In this paper we first model a congestion control problem in communication networks within a game theory approach which permits to treat stochastic costs functions and then consider the problem of improving the overall network performance in an optimal way, by investing a given amount of money. The cost function of each player is the difference of a pricing term, which promotes congestion control, and a utility term which describes the user's profit.

Game theoretical models for network equilibrium problems are very popular and, in the case of communication networks, an interesting approach has been developed in the papers [1, 2, 19, 22]. Our starting point is the model introduced

M. Passacantando
Department of Computer Science, University of Pisa, Pisa, Italy
e-mail: mauro.passacantando@unipi.it

F. Raciti (✉)
Department of Mathematics and Computer Science, University of Catania, Catania, Italy
e-mail: fabio.raciti@unict.it

I. N. Parasidis et al. (eds.), *Mathematical Analysis in Interdisciplinary Research*,
Springer Optimization and Its Applications 179,
https://doi.org/10.1007/978-3-030-84721-0_27

in [1], where the players are network users who compete to send their flow from a given origin to a certain destination node along a route that has been computed previously by a routing algorithm. The fact that some links in the network are used by more than one player implies that the strategy space of each player also depends on the variables of other players. As a result, the game under consideration falls in the class of Generalized Nash Equilibrium Problems (GNEPs) with shared constrained, introduced by Rosen a long time ago [21], and further developed recently by using the theory of variational inequalities (see, e.g., [3, 4, 16, 18]). Indeed, it is well known (see, e.g., [6]) that GNEPs are equivalent to quasi-variational inequalities which are considered very difficult problems (for an L^2 approach to quasi-variational inequalities see, for instance, [20]). In our approach we allow for stochastic fluctuations of the players' cost function and apply the theory of variational inequalities in probabilistic Lebesgue spaces (see, e.g., [5, 7–12]), to find the unique variational equilibrium of the game, which is considered the most desirable from an economic point of view among the multiple equilibria [3]. Let us also mention that the theory of variational inequalities in probabilistic Lebesgue spaces has been recently applied to study the efficiency of road traffic networks [13, 17] and that a different approach to stochastic variational inequalities has been applied to communication networks in [15].

In our model the bandwidth is the most important characteristic of the network and, in this respect, a system manager may wish to improve the network by investing financial resources to enhance the capacity of the links. In a typical real situation the investment cannot cover all the links and a choice has to be made to decide which links to improve. The system manager makes his/her decision with the help of a network cost function associated with each set of improvements, which has the role of maximizing the aggregate utility, while minimizing the total delay at the links. Since this system function depends on the stochastic price and utility functions, the quantity of interest is its mean value. Thus, once the sets of variational equilibria, for all feasible improvements, is computed, we have to solve a knapsack-type problem which, for instances of reasonable dimensions can be solved by direct inspection, that is by ordering all the solutions according to their corresponding relative variation of the above mentioned system function.

The paper is organized in four sections and an appendix. In Sect. 2, we introduce some notations and the congestion control model proposed in [1], which we modify to include stochastic fluctuations of the cost functions; we also introduce the variational inequality whose (unique) solution gives the desired Nash equilibrium of the game, and define a system function which describes a global property of the network. Section 3 is devoted to describe the optimal investment strategy. In Sect. 4, we apply our model to some small-size problems which are solved numerically. The appendix has the role of providing the frame of stochastic variational inequalities in probabilistic Lebesgue space, but for the details of the numerical approximation scheme the interested reader can refer to the references mentioned in this introduction.

2 The Congestion Control Model and Its Stochastic Variational Inequality Formulation

Throughout the paper, vectors of $\mathbb{R}^n$ are thought of as rows, but in matrix operations they will be considered as columns and the superscript $^\top$ will denote transposition. The scalar product between two Euclidean vectors a and b will be denoted by $a^\top b$, while the scalar product between two square integrable functions f and g will be denoted in compact form by $\langle f, g\rangle_{L^2}$. The notation $E_P[f]$ will be used to denote the mean value of a random function f with respect to the probability measure P. The network topology consists of a set of links $\mathscr{L} = \{1, \ldots, L\}$ connecting the nodes in the set $\mathscr{N} = \{n_1, \ldots, n_N\}$. The users of the network belong to the set $\mathscr{G} = \{g_1, \ldots, g_M\}$. A route R in the network is a set of consecutive links and each user g_i wishes to send a flow x_i between a given pair $O_i - D_i$ of origin-destination nodes; $x \in \mathbb{R}^M$ is the (route) flow of the network; the notation $x = (x_i, x_{-i})$, common in game theory, will be used in the sequel when we need to distinguish the flow component of player g_i from all the others. We assume that the routing problem has already been solved and that there is only one route R_i assigned to user g_i. Each link l has a fixed capacity $C_l > 0$, so that user i cannot send a flow greater than the capacity of every link of his/her route, and we group these capacities into a vector $C \in \mathbb{R}^L$. To describe the link structure of each route, it is useful to introduce the link-route matrix whose entries are given by

$$A_{li} = \begin{cases} 1, & \text{if link } l \text{ belongs to route } R_i, \\ 0, & \text{otherwise.} \end{cases} \tag{1}$$

Using the link-route matrix the set of feasible flows can be written in compact form as

$$X := \left\{x \in \mathbb{R}^M : x \geq 0, \ \ Ax \leq C\right\}. \tag{2}$$

In order to better specify the feasible set of each player, we write by components the conservation of flow in X as

$$\sum_{i=1}^{M} A_{li} x_i \leq C_l, \qquad \forall\, l \in \mathscr{L}.$$

Therefore, because users share some links, the possible amount of flow x_i depends on the flows sent by the other users and is bounded from above by the quantity

$$m_i(x_{-i}) = \min_{l \in R_i} \left\{ C_l - \sum_{j=1,\ j \neq i}^{M} A_{lj} x_j \right\} \geq 0.$$

Now, let $(\Omega, \mathscr{A}, P)$ be a probability space, and define the cost function $J_i : \Omega \times \mathbb{R}^M \to \mathbb{R}$ of player g_i as

$$J_i(\omega, x) = P_i(\omega, x) - U_i(\omega, x_i), \tag{3}$$

where U_i represents the utility function of player g_i, which only depends on the flow that he/she sends through the network, while P_i is a pricing term which represents some kind of toll that g_i pays to exploit the network resources and depends on the flows of the players with common links to g_i. Stochastic fluctuations of both P_i and U_i are described by the random parameter $\omega \in \Omega$. Players compete in a non-cooperative manner, as it is assumed that they do not communicate, and act selfishly to increase their flow. Because the conservation law in X implies that users share the constraints, the solution concept adopted is the equilibrium introduced by Rosen in his seminal paper [21], which in the modern literature is known as generalized Nash equilibrium (with coupled constraints). Due to the presence of $\omega \in \Omega$, the Nash equilibrium is a random vector, according to the following definition:

$$\begin{gathered} x^* = (x_i^*(\omega), x_{-i}^*(\omega)) : \Omega \to \mathbb{R}^M \text{ is a generalized Nash equilibrium if} \\ \text{for each } i \in \{1, \dots, M\} \text{ and } P\text{–a.s. :} \\ J_i(\omega, x_i^*(\omega), x_{-i}^*(\omega)) = \min_{x_i \in X_i(x_{-i}^*(\omega))} J_i(\omega, x_i, x_{-i}^*(\omega)), \end{gathered} \tag{4}$$

where

$$X_i(x_{-i}^*(\omega)) := \left\{x_i \in \mathbb{R} : \ (x_i, x_{-i}^*(\omega)) \in X\right\} = \{x_i \in \mathbb{R} : \ 0 \le x_i \le m_i(x_{-i}^*(\omega))\}$$

and

$$m_i(x_{-i}^*(\omega)) = \min_{l \in R_i} \left\{ C_l - \sum_{j=1,\ j \neq i}^{M} A_{lj} x_j^*(\omega) \right\}.$$

For each fixed ω, it is well known (see, e.g., [6]) that, under standard differentiability and convexity assumptions, the above problem is equivalent to a quasi-variational inequality and that a particular subset of solutions (called variational equilibria) can be found by solving the variational inequality $VI(F, X)$, where X is the feasible set defined in (2) and F is the so-called *pseudogradient* of the game, defined by

$$F(\omega, x) = \left(\frac{\partial J_1(\omega, x)}{\partial x_1}, \dots, \frac{\partial J_M(\omega, x)}{\partial x_M} \right). \tag{5}$$

More precisely, the variational inequality under consideration is the problem of finding, for each $\omega \in \Omega$, a vector $x^*(\omega) \in X$ such that:

$$F(\omega, x^*(\omega))^\top (x - x^*(\omega)) \geq 0, \quad \forall x \in X, P - \text{ a. s.} \tag{6}$$

In (4) and (6) the solution is a random vector, i.e., a vector function which is merely measurable with respect to the probability measure P on Ω. Since we wish to compute statistical quantities associated with the solution, it is natural to require that x^* has finite first- and second-order moments. Following [7], we provide an L^2 formulation of both (4) and (6).

We will posit the following assumptions on the cost functions J_i, for each $i \in \{1, \dots, M\}$:

(A) $J_i(\cdot, x)$ is a random variable for each $x \in \mathbb{R}^M$, and $J_i(\omega, \cdot) \in C^1(\mathbb{R}^M)$, P-a.s.;
(B) $J_i(\omega, 0) \in L^1(\Omega, P)$;
(C) $J_i(\omega, \cdot, x_{-i})$ is convex P-a.s. and $\forall\, x_{-i} \in \mathbb{R}^{M-1}$;
(D) $|\nabla_x J_i(\omega, x)| \leq c(1 + |x|), \ \forall x \in \mathbb{R}^M, \ P -$ a.s..

Let us now introduce, for each $i \in \{1, \dots, M\}$, the mapping $T_i : L^2(\Omega, P, \mathbb{R}^M) \to \mathbb{R}$ defined by

$$T_i(u_i, u_{-i}) = \int_\Omega J_i(\omega, u_i(\omega), u_{-i}(\omega))\, dP_\omega. \tag{7}$$

The following Lemma specifies some fundamental properties of T_i and can be proved along the same lines as in [6].

Lemma 1 *Let us assume that, for each $i \in \{1, \dots, M\}$, J_i satisfies assumptions $(A) - (D)$. Then, for each $i \in \{1, \dots, M\}$, T_i is well defined in $L^2(\Omega, P, \mathbb{R}^M)$, $T_i(\cdot, u_{-i})$ is convex and Gateaux-differentiable in $L^2(\Omega, P)$, for each u_{-i}, and its derivative is given by*

$$D_i T_i(u_i, u_{-i})(v_i) = \int_\Omega \frac{\partial}{\partial x_i}[J_i(\omega, u_i(\omega), u_{-i}(\omega)]\; v_i(\omega)\, dP_\omega, \quad \forall\, v_i \in L^2(\Omega, P). \tag{8}$$

In order to provide the L^2-formulation of (4) and (6), we need to introduce the following sets:

$$K := \left\{ u \in L^2(\Omega, P, \mathbb{R}^M) : \ u(\omega) \geq 0, \ Au(\omega) \leq C, \ P - \text{ a.s.} \right\}$$

and

$$K_i(u_{-i}) := \left\{ u_i \in L^2(\Omega, P) : \ (u_i(\omega), u_{-i}(\omega)) \in K, \ P - \text{ a. s.} \right\}.$$

Thus, a vector $u^* = (u_i^*, u_{-i}^*) \in L^2(\Omega, P, \mathbb{R}^M)$ is a generalized Nash equilibrium iff, for each $i \in \{1, \dots, M\}$:

$$\int_{\Omega} J_i(\omega, u_i^*(\omega), u_{-i}^*(\omega))\, dP_\omega = \min_{u_i \in K_i(u_{-i}^*)} \int_{\Omega} J_i(\omega, u_i(\omega), u_{-i}^*(\omega))\, dP_\omega. \tag{9}$$

The variational solutions of (9) can be obtained by solving the following variational inequality $VI(\Gamma, K)$: find $u^* \in K$ such that

$$\int_{\Omega} \sum_{i=1}^{M} \left[\frac{\partial}{\partial x_i} J_i(\omega, u^*(\omega)) \right] (v_i(\omega) - u_i^*(\omega))\, dP_\omega \geq 0, \qquad \forall\, v \in K, \tag{10}$$

where $\Gamma : L^2(\Omega, P, \mathbb{R}^M) \to L^2(\Omega, P, \mathbb{R}^M)$ is given by

$$\Gamma(u) = (\Gamma_1(u), \ldots, \Gamma_M(u)) = \left(\frac{\partial}{\partial x_1} J_1(\omega, u(\omega)), \ldots, \frac{\partial}{\partial x_M} J_M(\omega, u(\omega)) \right). \tag{11}$$

It can be useful to write (10) in compact form by using the following notation:

$$\langle \Gamma(u), v - u \rangle_{L^2} \geq 0, \qquad \forall\, v \in L^2(\Omega, P, \mathbb{R}^M).$$

Problem (10) is a random (or stochastic) variational inequality in L^2 and the interested reader can refer to the articles mentioned in the introduction for a comprehensive treatment of this relatively new methodology as well as for several applications. In order to be self-consistent, we give in the appendix a short outline of the topic, in the general L^p setting ($p \geq 2$).

In what follows, we consider the specific functional form of P_i and U_i treated in [1], with a slight modification, and allowing for stochastic fluctuations. Furthermore, we show the existence of a unique variational equilibrium of the game. Specifically, the utility function U_i of player g_i is given by

$$U_i(\omega, x_i) = a_i(\omega) \log(x_i + 1), \tag{12}$$

where $a_i \in L^\infty(\Omega, P)$ and is bounded away from zero from below for each $i \in \{1, \ldots, m\}$. The route price function P_i of player g_i is the sum of the price functions of the links associated with route R_i:

$$P_i(\omega, x) = \sum_{l \in R_i} P_l \left(\omega, \sum_{j=1}^{M} A_{lj} x_j \right). \tag{13}$$

Let us notice that P_l is modeled so as to only depend on the variables of players who share the link l, namely:

$$P_l \left(\omega, \sum_{j=1}^{M} A_{lj} x_j \right) = \frac{k(\omega)}{C_l - \sum_{j=1}^{M} A_{lj} x_j + e}, \tag{14}$$

where $k \in L^\infty(\Omega, P)$ is a network function, bounded away from zero from below, and e is a small positive number which we introduce to allow capacity saturation, while obtaining a well behaved function. The price function of g_i is thus given by

$$P_i(\omega, x) = \sum_{l \in R_i} \frac{k(\omega)}{C_l - \sum_{j=1}^M A_{lj} x_j + e}, \tag{15}$$

and the resulting expression of the cost for g_i is

$$J_i(\omega, x) = \sum_{l \in R_i} \frac{k(\omega)}{C_l - \sum_{j=1}^M A_{lj} x_j + e} - a_i(\omega) \log(x_i + 1). \tag{16}$$

The following properties of the above functions are easy to check, for each fixed ω: (i) $U_i(\omega, \cdot)$ is twice continuously differentiable, non-decreasing, and strongly concave on any compact interval $[0, b]$ (the last condition means that there exists $\tau > 0$ such that $\partial^2 U_i(\omega, x_i)/\partial x_i^2 \leq -\tau$ for any $x_i \in [0, b]$); (ii) $P_i(\omega, \cdot)$ is twice continuously differentiable, convex and $P_i(\omega, \cdot, x_{-i})$ is non-decreasing. These properties of U_i and P_i entail an important monotonicity property of the pseudogradient F defined in (5), as the following theorem shows.

Theorem 1 *Let U_i and P_i be given as in* (12) *and* (15)*, then F is strongly monotone on X, uniformly with respect to $\omega \in \Omega$, i.e., there exists $\alpha > 0$ such that*

$$(F(\omega, x) - F(\omega, y))^\top (x - y) \geq \alpha \|x - y\|^2, \qquad \forall\, x, y \in X, \forall\, \omega \in \Omega.$$

Proof Similarly to [1], it can be shown that the Jacobian matrix of F is positive definite on X, uniformly with respect to x. Moreover, since the random parameters k and a_i are bounded, the Jacobian is positive definite, uniformly with respect to ω. Thus, F is strongly monotone on X, uniformly with respect to ω. □

The unique solvability of $VI(\Gamma, K)$ is based on standard arguments, as the following theorem shows.

Theorem 2 *There exists a unique variational equilibrium of the GNEP* (9).

Proof The variational equilibria of (9) are the solutions of (10), i.e., of $VI(\Gamma, K)$. Under assumptions (A)–(D), the operator Γ generated by F maps L^2 in L^2 and is norm-continuous, being P a probability measure. Moreover, the uniform strong monotonicity of F implies the uniform strong monotonicity of Γ. At last, the set K is a closed and convex subset of $L^2(\Omega, P, \mathbb{R}^M)$ and is norm-bounded, hence weakly compact. Then, applying monotone operator theory we get that (10) admits a unique solution (see, e.g., [14]), which is the unique variational equilibrium of (9). □

We now introduce a function f which describes a global property of the game:

$$f(\omega, x) = \sum_{l \in \mathscr{L}} P_l \left(\omega, \sum_{j=1}^{M} A_{lj} x_j \right) - \sum_{i=1}^{M} U_i(\omega, x_i), \tag{17}$$

which represents the aggregate delay at the links minus the sum of the utilities of all players.

The Carathéodory function f generates a functional $\Pi : L^2(\Omega, P, \mathbb{R}^M) \to \mathbb{R}$ through the position:

$$\Pi(u(\omega)) := E_P[f] = \int_{\Omega} f(\omega, u(\omega)) dP_{\omega}, \quad \forall u \in K \subset L^2(\Omega, P, \mathbb{R}^M). \tag{18}$$

The theorem which follows shows that Π plays the role of a potential for the game described by (9).

Theorem 3 *The unique variational equilibrium of the GNEP* (9) *coincides with the optimal solution of the system problem* $\min_{u \in K} \Pi(u)$.

Proof Since both Π and K are convex, $\bar{u}$ is a minimizer of Π on K if and only if

$$\langle D\Pi(\bar{u}), v - \bar{u} \rangle_{L^2} \geq 0, \quad \forall v \in K,$$

where $D\Pi(\bar{u})$ stands for the Gateaux derivative of Π in $\bar{u}$. Since $D\Pi = \Gamma$, the expression above is nothing else that the variational inequality $VI(\Gamma, K)$, whose solution gives the variational equilibrium of (9). □

To study our model from a numerical point of view, we need to pass from the abstract probability space $(\Omega, \mathscr{A}, P)$ to the probability space generated by the random variables under consideration: $(k, a_1, \ldots, a_M)$. The new probability space is then $(\mathbb{R}^{M+1}, \mathscr{B}, \mathbb{P})$, where $\mathscr{B}$ represents the Borel σ-algebra on $\mathbb{R}^{M+1}$ and $\mathbb{P} = P_k \otimes P_{a_1} \otimes, \ldots, \otimes P_{a_M} = P_k \otimes P_a$, where we assumed independence of all the random variables involved. In what follows, with a slight abuse of notation, we will continue to denote with K and $K_i(u_{-i})$ the sets previously defined but expressed now with new variables (k, a). The cost functions are thus expressed as $J_i(k, a, x)$, and problem (9) now reads as

$$\begin{aligned} &\int_{\mathbb{R}^{M+1}} J_i(k, a, u_i^*(k, a), u_{-i}^*(k, a))\, dP_k\, dP_a \\ &= \min_{u_i \in K_i(u_{-i}^*)} \int_{\mathbb{R}^{M+1}} J_i(k, a, u_i(k, a), u_{-i}^*(k, a))\, dP_k\, dP_a, \end{aligned} \tag{19}$$

while the variational solutions of (19) can be obtained by solving the following variational inequality: find $u^* \in K$ such that

$$\int_{\mathbb{R}^{M+1}} \sum_{i=1}^{M} \left[\frac{\partial}{\partial x_i} J_i(k, a, u^*(\omega)) \right] (v_i(k, a) - u_i^*(k, a)) dP_k \, dP_a \geq 0, \quad \forall\, v \in K. \tag{20}$$

Analogously, the system function, as a function of the random variables, reads as

$$f(k, a, x) = \sum_{l \in \mathcal{L}} P_l \left(k, a, \sum_{j=1}^{M} A_{lj} x_j \right) - \sum_{i=1}^{M} U_i(k, a, x_i), \tag{21}$$

and its mean value is expressed by

$$E_{\mathbb{P}}[f] = \Pi(u(k, a)) := \int_{\mathbb{R}^{M+1}} f(k, a, u(k, a)) dP_k \, dP_a. \tag{22}$$

Thus, the quantities of interest in our model are $E_{\mathbb{P}}[u^*] = \int_{\mathbb{R}^{M+1}} u^*(k, a) dP_k \, dP_a$ and $\Pi(u^*(k, a))$.

3 The Optimal Network Improvement Model

We now suppose that the network system manager has a budget B available to improve the network performance. He/she can only increase the capacity of a subset $\widetilde{\mathcal{L}} \subseteq \mathcal{L}$ of links and knows that I_l is the investment required to enhance the capacity of link l by a given ratio γ_l. Since the available budget is generally not sufficient to enhance the capacities of all the links of $\widetilde{\mathcal{L}}$, he/she has to decide which subset of links to invest in, in order to improve as much as possible the system cost Π computed at the variational equilibrium of the game with new link capacities, while satisfying the budget constraint. This problem can be formulated as an integer nonlinear program.

To this end, we define a binary variable y_l, for any $l \in \widetilde{\mathcal{L}}$, which takes on the value 1 if the investment is actually carried out on link l, and 0 otherwise. A vector $y = (y_l)_{l \in \widetilde{\mathcal{L}}}$ is feasible if the budget constraint $\sum_{l \in \widetilde{\mathcal{L}}} I_l y_l \leq B$ is satisfied. Given a feasible vector y, the new capacity of each link $l \in \widetilde{\mathcal{L}}$ is equal to

$$C'_l(y) := \gamma_l C_l y_l + (1 - y_l) C_l,$$

i.e., $C'_l(y) = \gamma_l C_l$ if $y_l = 1$ and $C'_l(y) = C_l$ if $y_l = 0$. The network manager aims to maximize the percentage relative variation of the system cost defined as

$$\varphi(y) := 100 \cdot \frac{\Pi(u_0^*(k, a)) - \Pi(u_y^*(k, a))}{\Pi(u_0^*(k, a))},$$

where $u_0^*(k, a)$ is the variational equilibrium of the GNEP before the investment, while $u_y^*(k, a)$ is the variational equilibrium of the GNEP on the improved network according to y. Therefore, the proposed optimization model is

$$\begin{aligned} \max \quad & \varphi(y) \\ \text{subject to} \quad & \sum_{l \in \widetilde{\mathcal{L}}} I_l y_l \leq B, \\ & y_l \in \{0, 1\} \quad l \in \widetilde{\mathcal{L}}. \end{aligned} \tag{23}$$

The above model can be considered a generalized knapsack problem because the computation of the nonlinear function φ at a given y requires to find the variational equilibrium of the GNEP both for the original and the improved network. Notice that, since the variational equilibrium of the GNEP is the minimizer of Π (see Theorem 3), the optimization problem (23) can be reformulated as the following mixed integer nonlinear program:

$$\begin{aligned} \min \quad & \sum_{l \in \widetilde{\mathcal{L}}} \int_{\mathbb{R}^{M+1}} \frac{k}{\gamma_l C_l y_l + (1 - y_l) C_l - \sum_{i=1}^{M} A_{li} u_i(k, a) + e} dP_k \, dP_a \\ & + \sum_{l \in \mathcal{L} \setminus \widetilde{\mathcal{L}}} \int_{\mathbb{R}^{M+1}} \frac{k}{C_l - \sum_{i=1}^{M} A_{li} u_i(k, a) + e} dP_k \, dP_a \\ & - \sum_{i=1}^{M} \int_{\mathbb{R}^{M+1}} a_i \log(u_i(k, a) + 1) dP_k \, dP_a \\ \text{subject to} \quad & \sum_{i=1}^{M} A_{li} u_i(k, a) \leq \gamma_l C_l y_l + (1 - y_l) C_l \qquad l \in \widetilde{\mathcal{L}}, \\ & \sum_{i=1}^{M} A_{li} u_i(k, a) \leq C_l \qquad l \in \mathcal{L} \setminus \widetilde{\mathcal{L}}, \ \mathbb{P} - \text{a.s.} \\ & \sum_{l \in \widetilde{\mathcal{L}}} I_l y_l \leq B, \\ & u_i(k, a) \geq 0, \qquad i = 1, \ldots, M, \ \mathbb{P} - \text{a.s.} \\ & y_l \in \{0, 1\} \qquad l \in \widetilde{\mathcal{L}}. \end{aligned}$$

4 Numerical Experiments

In this section, we show some preliminary numerical experiments on two test networks for the stochastic formulation of the congestion control problem and the optimal network improvement problem. The numerical approximation of random variational equilibria was performed by implementing in Matlab 2020a the discretization procedure described in [9, 10] and exploiting the Matlab Optimization toolbox. The nonlinear knapsack problem (23) has been solved evaluating the objective function at all the feasible solutions.

Example 1 We consider the network shown in Fig. 1 (see also [1]) with nine nodes and nine links. The origin-destination pairs of the users and their routes are described in Table 1. We set $e = 0.01$ and $C_l = 10$ for any $l \in \mathscr{L}$. Moreover, we assume that the random parameter k is equal to $k = 100 + \delta_k$, where δ_k is a random variable which varies in the interval $[-90, 90]$ with either uniform distribution or truncated normal distribution with mean 0 and standard deviation 9. Moreover, for any $i \in \{1, \dots, m\}$, the random parameters a_i are equal to $a_i = 100 + \delta_a$, where δ_a is a random variable which varies in the interval $[-90, 90]$ with either uniform distribution or truncated normal distribution with mean 0 and standard deviation 9. Both intervals $[-90, 90]$ have been partitioned into N^d subintervals in the approximation procedure. Tables 2, 3, 4, 5 show the convergence of the approximated mean values of the variational equilibrium u^* for different values of N^d by using the four different combinations of probability densities.

We now consider the optimal network improvement problem. We set $e = 0.01$ and $C_l = 10$ for any $l \in \mathscr{L}$. The random parameters are of the form: $k = 100 + \delta_k$ and $a_i = 100 + \delta_a$, where δ_k and δ_a vary in the interval $[-90, 90]$ with uniform

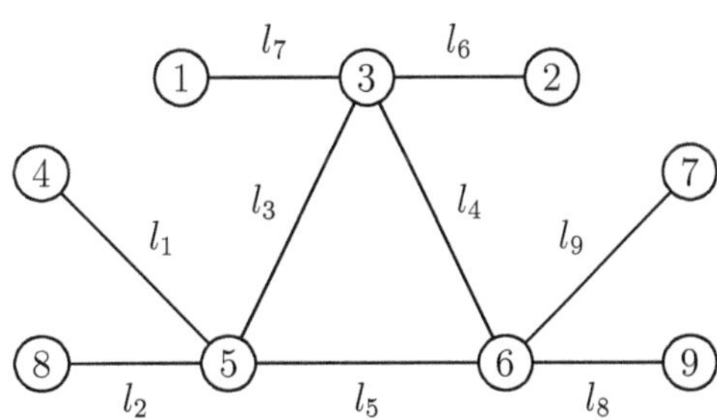

Fig. 1 Network topology of Example 1

Table 1 Origin-Destination pairs and routes (sequence of links) of the users in Example 1

User	Origin	Destination	Route
1	8	2	l_2, l_3, l_6
2	8	7	l_2, l_5, l_9
3	4	7	l_1, l_5, l_9
4	2	7	l_6, l_4, l_9
5	9	7	l_8, l_9

Table 2 Convergence of the approximated mean values of the variational equilibrium of Example 1; δ_k and δ_a vary in the interval $[-90, 90]$ with uniform distribution

$E_{\mathbb{P}}[u^*]$	$N^d = 10$	$N^d = 25$	$N^d = 50$	$N^d = 100$
$(E_{\mathbb{P}}[u^*])_1$	4.4478	4.4440	4.4434	4.4433
$(E_{\mathbb{P}}[u^*])_2$	1.6458	1.6433	1.6429	1.6428
$(E_{\mathbb{P}}[u^*])_3$	2.1001	2.0970	2.0965	2.0964
$(E_{\mathbb{P}}[u^*])_4$	1.7110	1.7081	1.7077	1.7076
$(E_{\mathbb{P}}[u^*])_5$	2.3924	2.3891	2.3886	2.3884

Table 3 Convergence of the approximated mean values of the variational equilibrium of Example 1; δ_k varies in the interval $[-90, 90]$ with uniform distribution, δ_a varies in the interval $[-90, 90]$ with truncated normal distribution with mean 0 and standard deviation 9

$E_{\mathbb{P}}[u^*]$	$N^d = 10$	$N^d = 25$	$N^d = 50$	$N^d = 100$
$(E_{\mathbb{P}}[u^*])_1$	4.7310	4.7326	4.7328	4.7328
$(E_{\mathbb{P}}[u^*])_2$	1.7183	1.7183	1.7183	1.7183
$(E_{\mathbb{P}}[u^*])_3$	2.2025	2.2026	2.2025	2.2025
$(E_{\mathbb{P}}[u^*])_4$	1.7811	1.7811	1.7811	1.7811
$(E_{\mathbb{P}}[u^*])_5$	2.4633	2.4634	2.4634	2.4634

Table 4 Convergence of the approximated mean values of the variational equilibrium of Example 1; δ_k varies in the interval $[-90, 90]$ with truncated normal distribution with mean 0 and standard deviation 9, δ_a varies in the interval $[-90, 90]$ with uniform distribution

$E_{\mathbb{P}}[u^*]$	$N^d = 10$	$N^d = 25$	$N^d = 50$	$N^d = 100$
$(E_{\mathbb{P}}[u^*])_1$	4.2698	4.2636	4.2627	4.2624
$(E_{\mathbb{P}}[u^*])_2$	1.6233	1.6203	1.6198	1.6196
$(E_{\mathbb{P}}[u^*])_3$	2.0642	2.0604	2.0598	2.0596
$(E_{\mathbb{P}}[u^*])_4$	1.6945	1.6912	1.6906	1.6905
$(E_{\mathbb{P}}[u^*])_5$	2.3775	2.3736	2.3730	2.3728

Table 5 Convergence of the approximated mean values of the variational equilibrium of Example 1; δ_k and δ_a vary in the interval $[-90, 90]$ with truncated normal distribution with mean 0 and standard deviation 9

$E_{\mathbb{P}}[u^*]$	$N^d = 10$	$N^d = 25$	$N^d = 50$	$N^d = 100$
$(E_{\mathbb{P}}[u^*])_1$	4.5685	4.5680	4.5679	4.5679
$(E_{\mathbb{P}}[u^*])_2$	1.6995	1.6993	1.6993	1.6993
$(E_{\mathbb{P}}[u^*])_3$	2.1726	2.1723	2.1722	2.1722
$(E_{\mathbb{P}}[u^*])_4$	1.7679	1.7677	1.7677	1.7677
$(E_{\mathbb{P}}[u^*])_5$	2.4504	2.4502	2.4502	2.4502

Table 6 Capacity enhancement factors and investments for links of Example 1

Links	l_1	l_2	l_3	l_4	l_5	l_6	l_7	l_8	l_9
γ_l	1.2	1.5	1.1	1.6	1.3	1.4	1.1	1.7	1.3
I_l (k€)	3	8	2	10	4	5	2	12	4

Table 7 The ten best feasible solutions for the optimal network improvement model in Example 1

Ranking	y	$\varphi(y)$	$I(y)$
1	(0,1,1,0,0,1,0,0,1)	28.0619	19
2	(1,1,0,0,0,1,0,0,1)	27.2448	20
3	(0,1,0,0,0,1,1,0,1)	26.8262	19
4	(0,1,0,0,0,1,0,0,1)	26.6207	17
5	(1,0,1,0,1,1,1,0,1)	23.2752	20
6	(0,1,1,0,1,0,1,0,1)	23.0862	20
7	(1,0,1,0,1,1,0,0,1)	23.0697	18
8	(0,1,1,0,1,0,0,0,1)	22.8807	18
9	(1,1,0,0,1,0,0,0,1)	22.8453	19
10	(0,0,1,0,1,1,1,0,1)	22.5908	17

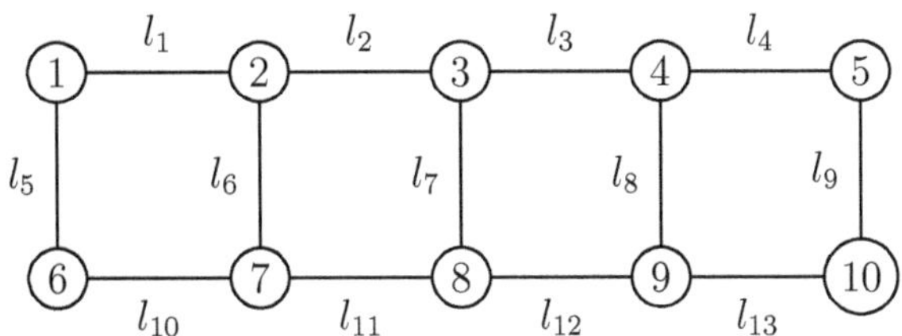

Fig. 2 Network topology of Example 2

distribution. Each interval $[-90, 90]$ has been partitioned into 25 subintervals in the approximation procedure. We assume that the available budget is $B = 20$ k€, the set of links to be improved is $\widetilde{\mathscr{L}} = \mathscr{L}$, while the values of γ_l and I_l are shown in Table 6.

Table 7 shows the ten best feasible solutions together with the percentage of total cost improvement $\varphi(y)$ and the corresponding investment $I(y) = \sum_{l \in \widetilde{\mathscr{L}}} I_l y_l$.

Example 2 We now consider the network shown in Fig. 2 with 10 nodes and 13 links. The O-D pairs of the ten users and their routes are described in Table 8. We report numerical experiments similar to Example 1. First, we show the convergence of the approximated mean values of the variational equilibrium with respect to different probability distributions of the random parameters k and a. Then, the solution of the optimal network improvement problem is reported.

We set $e = 0.01$ and $C_l = 10$ for any $l \in \mathscr{L}$. Moreover, we assume that the random parameter k is equal to $k = 10 + \delta_k$, where δ_k is a random variable which varies in the interval $[-9, 9]$ with either uniform distribution or truncated normal distribution with mean 0 and standard deviation 0.9. Moreover, for any $i \in \{1, \ldots, m\}$, the random parameters a_i are equal to $a_i = 10 + \delta_a$, where δ_a is a random variable which varies in the interval $[-9, 9]$ with either uniform distribution or truncated normal distribution with mean 0 and standard deviation 0.9. Both

Table 8 Origin-Destination pairs and routes (sequence of links) of the users in Example 2

User	Origin	Destination	Route	User	Origin	Destination	Route
1	1	5	l_1, l_2, l_3, l_4	6	5	1	l_4, l_3, l_2, l_1
2	6	10	$l_{10}, l_{11}, l_{12}, l_{13}$	7	10	6	$l_{13}, l_{12}, l_{11}, l_{10}$
3	2	10	$l_6, l_{11}, l_{12}, l_{13}$	8	5	8	l_9, l_{13}, l_{12}
4	8	5	l_7, l_3, l_4	9	4	6	$l_8, l_{12}, l_{11}, l_{10}$
5	6	5	l_5, l_1, l_2, l_3, l_4	10	8	1	l_7, l_2, l_1

Table 9 Convergence of the approximated mean values of the variational equilibrium of Example 2; δ_k and δ_a vary in the interval $[-9, 9]$ with uniform distribution

$E_{\mathbb{P}}[u^*]$	$N^d = 10$	$N^d = 25$	$N^d = 50$	$N^d = 100$
$(E_{\mathbb{P}}[u^*])_1$	1.3364	1.3342	1.3338	1.3337
$(E_{\mathbb{P}}[u^*])_2$	1.3470	1.3450	1.3447	1.3446
$(E_{\mathbb{P}}[u^*])_3$	1.4515	1.4491	1.4487	1.4486
$(E_{\mathbb{P}}[u^*])_4$	2.7526	2.7499	2.7495	2.7493
$(E_{\mathbb{P}}[u^*])_5$	1.2546	1.2523	1.2519	1.2518
$(E_{\mathbb{P}}[u^*])_6$	1.3364	1.3342	1.3338	1.3337
$(E_{\mathbb{P}}[u^*])_7$	1.3470	1.3450	1.3447	1.3446
$(E_{\mathbb{P}}[u^*])_8$	1.8810	1.8773	1.8768	1.8766
$(E_{\mathbb{P}}[u^*])_9$	1.6685	1.6652	1.6646	1.6645
$(E_{\mathbb{P}}[u^*])_{10}$	2.7526	2.7499	2.7495	2.7493

Table 10 Convergence of the approximated mean values of the variational equilibrium of Example 2; δ_k varies in the interval $[-9, 9]$ with uniform distribution, δ_a varies in the interval $[-9, 9]$ with truncated normal distribution with mean 0 and standard deviation 0.9

$E_{\mathbb{P}}[u^*]$	$N^d = 10$	$N^d = 25$	$N^d = 50$	$N^d = 100$
$(E_{\mathbb{P}}[u^*])_1$	1.4076	1.4076	1.4076	1.4076
$(E_{\mathbb{P}}[u^*])_2$	1.4204	1.4206	1.4206	1.4206
$(E_{\mathbb{P}}[u^*])_3$	1.5275	1.5276	1.5276	1.5276
$(E_{\mathbb{P}}[u^*])_4$	2.8993	2.9003	2.9004	2.9004
$(E_{\mathbb{P}}[u^*])_5$	1.3380	1.3380	1.3380	1.3380
$(E_{\mathbb{P}}[u^*])_6$	1.4076	1.4076	1.4076	1.4076
$(E_{\mathbb{P}}[u^*])_7$	1.4204	1.4206	1.4206	1.4206
$(E_{\mathbb{P}}[u^*])_8$	1.9536	1.9532	1.9532	1.9531
$(E_{\mathbb{P}}[u^*])_9$	1.7588	1.7585	1.7584	1.7584
$(E_{\mathbb{P}}[u^*])_{10}$	2.8993	2.9003	2.9004	2.9004

intervals $[-9, 9]$ have been partitioned into N^d subintervals in the approximation procedure. Tables 9, 10, 11, 12 show the convergence of the approximated mean values of the variational equilibrium for different values of N^d by using the four different combinations of probability densities.

Let us consider the optimal network improvement problem. We set $e = 0.01$ and $C_l = 10$ for any $l \in \mathscr{L}$. The random parameters are $k = 10 + \delta_k$ and $a_i = 10 + \delta_a$, where δ_k and δ_a vary in the interval $[-9, 9]$ with uniform distribution. Each interval $[-90, 90]$ has been partitioned into 25 subintervals in the approximation procedure. We assume that the available budget is $B = 20$ $k€$, the set of links to be improved is $\widetilde{\mathscr{L}} = \mathscr{L}$, while the values of γ_l and I_l are shown in Table 13.

Table 11 Convergence of the approximated mean values of the variational equilibrium of Example 2; δ_k varies in the interval $[-9, 9]$ with truncated normal distribution with mean 0 and standard deviation 0.9, δ_a varies in the interval $[-9, 9]$ with uniform distribution

$E_{\mathbb{P}}[u^*]$	$N^d = 10$	$N^d = 25$	$N^d = 50$	$N^d = 100$
$(E_{\mathbb{P}}[u^*])_1$	1.3135	1.3109	1.3104	1.3103
$(E_{\mathbb{P}}[u^*])_2$	1.3123	1.3098	1.3094	1.3093
$(E_{\mathbb{P}}[u^*])_3$	1.4258	1.4229	1.4224	1.4223
$(E_{\mathbb{P}}[u^*])_4$	2.6674	2.6632	2.6625	2.6624
$(E_{\mathbb{P}}[u^*])_5$	1.2263	1.2235	1.2231	1.2230
$(E_{\mathbb{P}}[u^*])_6$	1.3135	1.3109	1.3104	1.3103
$(E_{\mathbb{P}}[u^*])_7$	1.3123	1.3098	1.3094	1.3093
$(E_{\mathbb{P}}[u^*])_8$	1.8871	1.8831	1.8824	1.8823
$(E_{\mathbb{P}}[u^*])_9$	1.6590	1.6552	1.6546	1.6545
$(E_{\mathbb{P}}[u^*])_{10}$	2.6674	2.6632	2.6625	2.6624

Table 12 Convergence of the approximated mean values of the variational equilibrium of Example 2; δ_k and δ_a vary in the interval $[-9, 9]$ with truncated normal distribution with mean 0 and standard deviation 0.9

$E_{\mathbb{P}}[u^*]$	$N^d = 10$	$N^d = 25$	$N^d = 50$	$N^d = 100$
$(E_{\mathbb{P}}[u^*])_1$	1.3892	1.3890	1.3889	1.3889
$(E_{\mathbb{P}}[u^*])_2$	1.3888	1.3887	1.3886	1.3886
$(E_{\mathbb{P}}[u^*])_3$	1.5059	1.5057	1.5056	1.5056
$(E_{\mathbb{P}}[u^*])_4$	2.8207	2.8205	2.8204	2.8204
$(E_{\mathbb{P}}[u^*])_5$	1.3157	1.3155	1.3155	1.3155
$(E_{\mathbb{P}}[u^*])_6$	1.3892	1.3890	1.3889	1.3889
$(E_{\mathbb{P}}[u^*])_7$	1.3888	1.3887	1.3886	1.3886
$(E_{\mathbb{P}}[u^*])_8$	1.9665	1.9662	1.9661	1.9661
$(E_{\mathbb{P}}[u^*])_9$	1.7584	1.7580	1.7579	1.7579
$(E_{\mathbb{P}}[u^*])_{10}$	2.8207	2.8205	2.8204	2.8204

Table 13 Capacity enhancement factors and investments for links of Example 2

Links	l_1	l_2	l_3	l_4	l_5	l_6	l_7	l_8	l_9	l_{10}	l_{11}	l_{12}	l_{13}
γ_l	1.2	1.5	1.1	1.6	1.3	1.4	1.1	1.7	1.3	1.5	1.1	1.8	1.3
I_l (k€)	3	8	2	10	4	5	2	12	4	8	2	13	4

Table 14 The ten best feasible solutions for the optimal network improvement model in Example 2

Ranking	y	$\varphi(y)$	$I(y)$
1	(0,0,0,0,0,0,0,0,0,0,1,1,1)	19.3768	19
2	(1,0,0,0,0,0,0,0,0,0,0,1,1)	18.9122	20
3	(0,0,1,0,0,0,0,0,0,0,0,1,1)	18.1916	19
4	(0,0,0,0,0,0,1,0,0,0,0,1,1)	17.6307	19
5	(0,0,0,0,0,0,0,0,0,0,0,1,1)	17.0483	17
6	(1,0,1,0,0,0,0,0,0,0,1,1,0)	15.8179	20
7	(1,0,0,0,0,0,1,0,0,0,1,1,0)	15.2526	20
8	(1,0,0,0,0,0,0,0,0,0,1,1,0)	14.6397	18
9	(1,0,1,0,0,0,1,0,0,0,0,1,0)	14.5318	20
10	(0,0,1,0,0,0,1,0,0,0,1,1,0)	14.5210	19

Table 14 shows the ten best feasible solutions together with the value of φ and the corresponding investment.

Appendix

Let $(\Omega, \mathscr{A}, P)$ be a probability space, $A, B : \mathbb{R}^k \to \mathbb{R}^k$ two given mappings, and $b, c \in \mathbb{R}^k$ two given vectors in $\mathbb{R}^k$. Moreover, let R and S be two real-valued random variables defined on Ω, D a random vector in $\mathbb{R}^m$, and $G \in \mathbb{R}^{m\times k}$ a given matrix. For $\omega \in \Omega$, we define a random set

$$M(\omega) := \left\{x \in \mathbb{R}^k : \; Gx \leq D(\omega)\right\}.$$

Consider the following stochastic variational inequality: for almost every $\omega \in \Omega$, find $\hat{x} := \hat{x}(\omega) \in M(\omega)$ such that

$$(S(\omega)\, A(\hat{x}) + B(\hat{x}))^\top (z - \hat{x}) \geq (R(\omega)\, c + b)^\top (z - \hat{x}), \qquad \forall\, z \in M(\omega). \tag{24}$$

To facilitate the foregoing discussion, we set $T(\omega, x) := S(\omega)\, A(x) + B(x)$. We assume that A, B, and S are such that the map $T : \Omega \times \mathbb{R}^k \mapsto \mathbb{R}^k$ is a Carathéodory function. We also assume that $T(\omega, \cdot)$ is monotone for every $\omega \in \Omega$. Since we are only interested in solutions with finite first- and second-order moments, our approach is to consider an integral variational inequality instead of the parametric variational inequality (24).

Thus, for a fixed $p \geq 2$, consider the Banach space $L^p(\Omega, P, \mathbb{R}^k)$ of random vectors V from Ω to $\mathbb{R}^k$ such that the expectation (p-moment) is given by

$$E^P(\|V\|^p) = \int_\Omega \|V(\omega)\|^p dP(\omega) < \infty.$$

For subsequent developments, we need the following growth condition

$$\|T(\omega, z)\| \leq \alpha(\omega) + \beta(\omega)\|z\|^{p-1}, \qquad \forall\, z \in \mathbb{R}^k, \tag{25}$$

where $\alpha \in L^q(\Omega, P)$ and $\beta \in L^\infty(\Omega, P)$. Due to the above growth condition, the Nemytskii operator $\hat{T}$ associated with T acts from $L^p(\Omega, P, \mathbb{R}^k)$ to $L^q(\Omega, P, \mathbb{R}^k)$, where $p^{-1} + q^{-1} = 1$, and is defined by $\hat{T}(V)(\omega) := T(\omega, V(\omega))$, for any $\omega \in \Omega$. Assuming $D \in L^p_m(\Omega) := L^p(\Omega, P, \mathbb{R}^m)$, we introduce the following nonempty, closed, and convex subset of $L^p_k(\Omega)$:

$$M^P := \left\{V \in L^p_k(\Omega) : \; G\, V(\omega) \leq D(\omega), \;\; P - a.s.\right\}.$$

Let $S(\omega) \in L^\infty$, $0 < \underline{s} < S(\omega) < \overline{s}$, and $R(\omega) \in L^q$. Equipped with these notations, we consider the following L^p formulation of (24): find $\hat{U} \in M^P$ such that for every $V \in M^P$, we have

$$\int_{\Omega}(S(\omega)\,A[\hat{U}(\omega)]+B[\hat{U}(\omega))]^{\top}(V(\omega)-\hat{U}(\omega))\,dP(\omega)$$
$$\geq\int_{\Omega}(b+R(\omega)\,c)^{\top}(V(\omega)-\hat{U}(\omega))dP(\omega). \tag{26}$$

Classical theorems for the solvability of (26) can be found in [14].

To get rid of the abstract sample space Ω, we consider the joint distribution $\mathbb{P}$ of the random vector (R, S, D) and work with the special probability space $(\mathbb{R}^d, \mathscr{B}(\mathbb{R}^d), \mathbb{P})$, where $d := 2+m$ and $\mathscr{B}$ is the Borel σ-algebra on $\mathbb{R}^d$. For simplicity, we assume that R, S, and D are independent random vectors. We set

$$r=R(\omega),\quad s=S(\omega),\quad t=D(\omega),\quad y=(r,s,t).$$

For each $y \in \mathbb{R}^d$, we define the set

$$M(y):=\left\{x\in\mathbb{R}^k:\ Gx\leq t\right\}.$$

Consider the space $L^p(\mathbb{R}^d, \mathbb{P}, \mathbb{R}^k)$ and introduce the closed and convex set

$$M_{\mathbb{P}}:=\{v\in L^p(\mathbb{R}^d,\mathbb{P},\mathbb{R}^k):\ Gv(r,s,t)\leq t,\ \mathbb{P}-a.s.\}.$$

Without any loss of generality, we assume that $R \in L^q(\Omega, P)$ and $D \in L^p(\Omega, P, \mathbb{R}^m)$ are non-negative. Moreover, we assume that the support (i.e., the set of possible outcomes) of $S \in L^\infty(\Omega, P)$ is the interval $[\underline{s}, \overline{s}[\subset (0, \infty)$. With these ingredients, we consider the variational inequality problem of finding $\hat{u} \in M_{\mathbb{P}}$ such that for every $v \in M_{\mathbb{P}}$ we have

$$\int_0^\infty\int_{\underline{s}}^{\overline{s}}\int_{\mathbb{R}^m_+}(s\,A[\hat{u}(y)]+B[\hat{u}(y)])^{\top}(v(y)-\hat{u}(y))\,d\mathbb{P}(y)$$
$$\geq\int_0^\infty\int_{\underline{s}}^{\overline{s}}\int_{\mathbb{R}^m_+}(b+r\,c)^{\top}(v(y)-\hat{u}(y))\,d\mathbb{P}(y). \tag{27}$$

For the details on the numerical approximation of the solution $\hat{u}$ the interested reader can refer to the references in the introduction. Here, we only recall that the set $M_{\mathbb{P}}$ can be approximated by a sequence $\{M^n_{\mathbb{P}}\}$ of finite dimensional sets, and the functions r and s can be approximated by the sequences $\{\rho_n\}$ and $\{\sigma_n\}$ of step functions, with $\rho_n \to \rho$ in L^p and $\sigma_n \to \sigma$ in L^∞, respectively, where $\rho(r, s, t) = r$ and $\sigma(r, s, t) = s$. When the solution of (27) is unique, we can compute a sequence of step functions $\hat{u}_n$ which converges strongly to $\hat{u}$, under suitable hypotheses, by solving, for $n \in \mathbb{N}$, the following discretized variational inequality: find $\hat{u}_n := \hat{u}_n(y) \in M^n_{\mathbb{P}}$ such that, for every $v_n \in M^n_{\mathbb{P}}$, we have

$$\int_0^\infty \int_{\underline{s}}^{\overline{s}} \int_{\mathbb{R}_+^m} (\sigma_n(y)\, A[\hat{u}_n(y)] + B[\hat{u}_n(y)])^\top (v_n(y) - \hat{u}_n(y))\, d\mathbb{P}(y) \\ \geq \int_0^\infty \int_{\underline{s}}^{\overline{s}} \int_{\mathbb{R}_+^m} (b + \rho_n(y)\, c)^\top (v_n(y) - \hat{u}_n(y))\, d\mathbb{P}(y). \tag{28}$$

In absence of strict monotonicity, the solution of (26) and (27) can be not unique and the previous approximation procedure must be coupled with a regularization scheme as follows. We choose a sequence $\{\varepsilon_n\}$ of regularization parameters and choose the regularization map to be the duality map $J : L^p(\mathbb{R}^d, \mathbb{P}, \mathbb{R}^k) \to L^q(\mathbb{R}^d, \mathbb{P}, \mathbb{R}^k)$. We assume that $\varepsilon_n > 0$ for every $n \in \mathbb{N}$ and that $\varepsilon_n \downarrow 0$ as $n \to \infty$.

We can then consider the following regularized stochastic variational inequality: for $n \in \mathbb{N}$, find $w_n = w_n^{\varepsilon_n}(y) \in M_{\mathbb{P}}^n$ such that, for every $v_n \in M_{\mathbb{P}}^n$, we have

$$\int_0^\infty \int_{\underline{s}}^{\overline{s}} \int_{\mathbb{R}_+^m} \left(\sigma_n(y)\, A[w_n(y)] + B[w_n(y)] + \varepsilon_n J(w_n(y))\right)^\top (v_n(y) - w_n(y))\, d\mathbb{P}(y) \\ \geq \int_0^\infty \int_{\underline{s}}^{\overline{s}} \int_{\mathbb{R}_+^m} (b + \rho_n(y)\, c)^\top (v_n(y) - w_n(y))\, d\mathbb{P}(y). \tag{29}$$

As usual, the solution w_n will be referred to as the regularized solution. Weak and strong convergence of w_n to the minimal-norm solution of (27) can be proved under suitable hypotheses (see, e.g., [10]).

Acknowledgments The authors are members of the Gruppo Nazionale per l'Analisi Matematica, la Probabilità e le loro Applicazioni (GNAMPA - National Group for Mathematical Analysis, Probability and their Applications) of the Istituto Nazionale di Alta Matematica (INdAM - National Institute of Higher Mathematics).

References

1. Alpcan, T., Başar, T.: A game-theoretic framework for congestion control in general topology networks. Proceedings of the IEEE 41st Conference on Decision and Control Las Vegas, December 10–13, 2002
2. Altman, E. Başar, T., Jimenez, T., Shimkin, N.: Competitive routing in networks with polynomial costs. IEEE Trans. Autom. Control **47**, 92–96 (2002)
3. Facchinei, F., Fischer, A., Piccialli, V.: On generalized Nash games and variational inequalities. Oper. Res. Lett. **35**, 159–164 (2007)
4. Facchinei, F., Kanzov, C.: Generalized Nash equilibrium problems. Ann. Oper. Res. **175**, 177–211 (2010)
5. Faraci, F., Jadamba, B., Raciti, F.: On stochastic variational inequalities with mean value constraints. J. Optim. Theory Appl. **171**, 675–693 (2016)
6. Faraci, F., Raciti, F.: On generalized Nash equilibrium problems in infinite dimension: the Lagrange multipliers approach. Optimization **64**, 321–338 (2015)
7. Gwinner, J., Raciti, F.: On a class of random variational inequalities on random sets. Num. Funct. Anal. Optim. **27**, 619–636 (2006)

8. Gwinner, J., Raciti, F.: On Monotone Variational Inequalities with Random Data. J. Math. Inequal. **3**, 443–453 (2009)
9. Gwinner, J., Raciti, F.: Some equilibrium problems under uncertainty and random variational inequalities. Ann. Oper. Res. **200**, 299–319, (2012)
10. Jadamba, B., Khan, A.A., Raciti, F.: Regularization of Stochastic Variational Inequalities and a Comparison of an L_p and a Sample-Path Approach. Nonlinear Anal. Theory Methods Appl. **94**, 65–83 (2014)
11. Jadamba, B., Raciti,F.: Variational Inequality Approach to Stochastic Nash Equilibrium Problems with an Application to Cournot Oligopoly, J. Optim. Theory Appl. **165**, 1050–1070 (2015)
12. Jadamba, B., Raciti, F.: On the modelling of some environmental games with uncertain data. J. Optim. Theory Appl. **167**, 959–968 (2015)
13. Jadamba, B., Pappalardo, M., Raciti, F.: Efficiency and Vulnerability Analysis for Congested Networks with Random Data. J. Optim. Theory Appl. **177**, 563–583 (2018)
14. Kinderleher D., Stampacchia G.: An introduction to variational inequalities and their applications, Academic Press (1980)
15. Koshal, J., Nedić, A., Shanbhag, U.V.: Regularized Iterative Stochastic Approximation Methods for Stochastic Variational Inequality Problems. IEEE Trans. Autom. Control **58**, 594–609 (2013)
16. Mastroeni, G., Pappalardo M., Raciti, F.: Generalized Nash equilibrium problems and variational inequalities in Lebesgue spaces. Minimax Theory Appl. **5**, 47–64 (2020)
17. Passacantando, M., Raciti, F.: Optimal road maintenance investment in traffic networks with random demands. Optim. Lett. **15**, 1799–1819 (2021)
18. Nabetani, K., Tseng, P., Fukushima, M.: Parametrized variational inequality approaches to generalized Nash equilibrium problems with shared constraints. Comput. Optim. Appl. **48**, 423–452 (2011)
19. Orda, A., Rom, R., Shimkin, N.: Competitive routing in multiuser communication networks. IEEE/ACM Trans. Netw. **1**, 510–521 (1993)
20. Raciti, F., Scrimali L.: Time-dependent variational inequalities and applications to equilibrium problems. J. Global Optim. **28**, 387–400 (2004)
21. Rosen, J.B.: Existence and uniqueness of equilibrium points for concave n-person games. Econometrica **33**, 520–534 (1965)
22. Yin, H., Shanbhag, U.V., Metha, P.G.: Nash Equilibrium Problems with Congestion Costs and Shared Constraints. Proceedings of the 48h IEEE Conference on Decision and Control held jointly with 2009 28th Chinese Control Conference, Shanghai, China, December 16–18, (2009)

Nonlocal Problems for Hyperbolic Equations

L. S. Pulkina

Abstract In this chapter, we consider nonlocal problems for hyperbolic equations with integral conditions and discuss methods which can be applied to justify solvability. The chapter is based on some published papers and aims to introduce the reader to effective methods for investigating nonlocal problems. We also propose certain modifications in proving a unique solvability of nonlocal problems with integral conditions.

1 Introduction

Systematic studies of nonlocal problems with integral conditions originated with the papers by J.R. Cannon [6] and L.I. Kamynin [13] in 1963–1964. In these papers, both authors consider nonlocal problems for heat equation. Integral conditions often arise in practice when it is impossible to get boundary conditions by direct measurements. The most familiar examples are inverse problems. In this connection, we may observe that many papers deal with inverse problems for parabolic equations with over-determination of integral conditions (see for example [1, 7, 14] and elsewhere).

Here we focus our attention on nonlocal problems for hyperbolic equations with integral conditions. The study of nonlocal problems for hyperbolic equations started at the very end of the twentieth century. In the course of our discussion, we will mention many papers dealing with them.

L. S. Pulkina (✉)
Department of Mathematics, Samara National Research University, Samara, Russia
e-mail: louise@samdiff.ru

I. N. Parasidis et al. (eds.), *Mathematical Analysis in Interdisciplinary Research*, Springer Optimization and Its Applications 179,
https://doi.org/10.1007/978-3-030-84721-0_28

Nowadays, various nonlocal problems for partial differential equations have been actively studied. Investigations of nonlocal problems show that classical methods most widely used to prove solvability of initial-boundary problems break down as a rule when applied to nonlocal problems. The main reason of this is that the operators generated by nonlocal conditions are not self-adjoint. For this, several methods have been developed for overcoming the difficulties due to nonlocal conditions. We present here one of the most effective approaches and show how it works in various situations.

2 Nonlocal Problems for Hyperbolic Equations with Integral Conditions

In this section, we consider nonlocal problems for the hyperbolic equation

$$\mathcal{L}u \equiv u_{tt} - (a(x,t)u_x)_x + c(x,t)u = f(x,t). \tag{1}$$

The problem is to find a solution of (1) in $Q_T = (0,l) \times (0,T), \quad l, T < \infty$, satisfying the initial data

$$u(x,0) = 0, \quad u_t(x,0) = 0, \tag{2}$$

and the nonlocal conditions

$$\begin{aligned} l_1 u + \int_0^l k_1(x,t)u(x,t)dx = 0, \\ l_2 u + \int_0^l k_2(x,t)u(x,t)dx = 0, \end{aligned} \tag{3}$$

where $l_i u$ are boundary operators:

$$\begin{aligned} l_i u = {} & a_{i1}u_x(0,t) + a_{i2}u_x(l,t) + b_{i1}u(0,t) + b_{i2}u(l,t) \\ & + c_{i1}u_t(0,t) + c_{i2}u_t(l,t) + d_{i1}u_{tt}(0,t) + d_{i2}u_{tt}(l,t), \quad i = 1,2. \end{aligned}$$

Definition 1 If $\forall t \in [0,T]$, $l_i u = 0$, $i = 1,2$, then nonlocal conditions (3) are called *first-kind integral conditions*; otherwise, (3) are called *second-kind integral conditions*.

It appears that the choice of an effective method to show solvability of a problem with integral conditions depends on the kind of integral conditions and in particular on the form of $l_i u$.

2.1 A Problem with Second-Kind Integral Conditions

Let

$$l_1 u \equiv a_1 u_x(0,t) + b_1 u_x(l,t) + a_0 u(0,t) + b_0 u(l,t),$$
$$l_2 u \equiv c_1 u_x(0,t) + d_1 u_x(l,t) + c_0 u(0,t) + d_0 u(l,t).$$

Then the conditions (3) are of the second-kind. It is natural to ask why we begin with *second-kind* conditions. It will be clear soon.

Let in (3), $\forall t \in [0,T]$, $\Delta = a_1 d_1 - b_1 c_1 \neq 0$. Then we can solve (3) with respect to $u_x(0,t)$, $u_x(l,t)$ to get

$$\begin{array}{r} a(0,t)u_x(0,t) + \alpha_{11}(t)u(0,t) + \alpha_{12}(t)u(l,t) + \int_0^l K_1(x,t)u(x,t)dx = 0, \\ a(l,t)u_x(l,t) + \alpha_{21}(t)u(l,t) + \alpha_{22}(t)u(l,t) + \int_0^l K_2(x,t)u(x,t)dx = 0, \end{array} \tag{4}$$

where

$$\alpha_{11}(t) = \frac{a_0 d_1 - c_0 b_1}{\Delta} a(0,t), \quad \alpha_{12}(t) = \frac{b_0 d_1 - d_0 b_1}{\Delta} a(0,t),$$

$$\alpha_{21}(t) = \frac{a_0 c_1 - c_0 a_1}{\Delta} a(l,t), \quad \alpha_{22}(t) = \frac{b_0 c_1 - d_0 a_1}{\Delta} a(l,t),$$

$$K_1(x,t) = \frac{d_1 k_1(x,t) - b_1 k_2(x,t)}{\Delta} a(0,t), \ K_2(x,t) = \frac{c_1 k_1(x,t) - a_1 k_2(x,t)}{\Delta} a(l,t).$$

Problem 1 Find a solution $u(x,t)$ to Eq. (1) satisfying the initial data (2) and the nonlocal conditions (4).

Consider the Sobolev space $W_2^1(Q_T)$ and denote

$$\hat{W}_2^1(Q_T) = \left\{ v(x,t) : v \in W_2^1(Q_T), \ \ v(x,T) = 0 \right\}.$$

Let $u(x,t)$ be a solution to the Problem 1 and $v \in \hat{W}_2^1(Q_T)$. Using the standard method [18, p. 92] and taking into account (4) and $u_t(x,0) = 0$, we obtain the equality

$$\begin{aligned} &\int_0^T \int_0^l (-u_t v_t + a u_x v_x + c u v)dxdt - \int_0^T v(0,t)[\alpha_{11} u(0,t) + \alpha_{12} u(l,t)]dt \\ &+ \int_0^T v(l,t)[\alpha_{21} u(0,t) + \alpha_{22} u(l,t)]dt - \int_0^T v(0,t) \int_0^l K_1(x,t)u(x,t)dxdt \\ &+ \int_0^T v(l,t) \int_0^l K_2(x,t)u(x,t)dxdt = \int_0^T \int_0^l f v dxdt. \end{aligned} \tag{5}$$

Definition 2 A function $u \in W_2^1(Q_T)$ is said to be a weak solution to the Problem 1 if $u(x,0) = 0$, and for every $v \in \hat{W}(Q_T)$, the identity (5) holds.

Theorem 1 *Let*

(i) $a,\ a_t \in C(\bar{Q}_T),\ \ c \in C(\bar{Q}_T),\ \ a(x,t) > 0,\ \ \forall (x,t) \in \bar{Q}_T,$
(ii) $K_i \in C(\bar{Q}_T),\ \ f \in L_2(Q_T),$
(iii) $\alpha_{12} + \alpha_{21} = 0,\ \ \alpha_{ij} \in C^1[0,T].$

Then there exists a unique weak solution to the Problem 1.

Proof *Uniqueness.* Let $u_1(x,t)$ and $u_2(x,t)$ be two different solutions to the Problem 1. Then $u(x,t) = u_1(x,t) - u_2(x,t)$ satisfies the initial condition $u(x,0) = 0$ and the identity

$$\begin{aligned}
&\int_0^T \int_0^l (-u_t v_t + a u_x v_x + c u v) dx dt \\
&\quad - \int_0^T v(0,t)[\alpha_{11} u(0,t) + \alpha_{12} u(l,t)] dt + \int_0^T v(l,t)[\alpha_{21} u(0,t) + \alpha_{22} u(l,t)] dt \\
&\quad - \int_0^T v(0,t) \int_0^l K_1 u dx dt + \int_0^T v(l,t) \int_0^l K_2 u dx dt = 0. \qquad (6)
\end{aligned}$$

Let us substitute in (6)

$$v(x,t) = \begin{cases} \int\limits_\tau^t u(x,\eta) d\eta, & 0 \le t \le \tau, \\ 0, & \tau \le t \le T, \end{cases}$$

where $\tau \in [0,T]$ is arbitrary. We make some calculations.

$$\begin{aligned}
-\int_0^T \alpha_{11}(t) v(0,t) u(0,t) dt &= -\int_0^\tau \alpha_{11}(t) v(0,t) v_t(0,t) dt \\
&= \frac{1}{2} \int_0^\tau \alpha'_{11} v^2(0,t) dt + \frac{1}{2} \alpha_{11}(0) v^2(0,0).
\end{aligned}$$

In a similar way, we get

$$\begin{aligned}
-\int_0^T \alpha_{12} v(0,t) u(l,t) dt &= -\int_0^\tau \alpha_{12} v(0,t) v_t(l,t) dt, \\
\int_0^T \alpha_{21} v(l,t) u(0,t) dt &= -\int_0^\tau \alpha'_{21} v(l,t) v(0,t) dt - \int_0^\tau \alpha_{21} v_t(l,t) v(0,t) dt, \\
\int_0^T \alpha_{22}(t) v(l,t) u(l,t) dt &= -\frac{1}{2} \int_0^\tau \alpha'_{22} v^2(l,t) dt - \frac{1}{2} \alpha_{22}(0) v^2(l,0).
\end{aligned}$$

Taking into account that $\alpha_{12} + \alpha_{21} = 0$, we get

$$\begin{aligned}
&\int_0^l [u^2(x,\tau) + a(x,0)v_x^2(x,0)]dx \\
&= [\alpha_{11}(0)v^2(0,0) - 2\alpha_{21}v(0,0)v(l,0) - \alpha_{22}v^2(l,0)] \\
&\quad +2\int_0^\tau \int_0^l cuvdxdt - \int_0^\tau \int_0^l a_t v_x^2 dxdt - \\
&\quad + \int_0^\tau [\alpha'_{11}v^2(0,t) - 2\alpha'_{21}v(0,t)v(l,t) - \alpha'_{22}v^2(l,t)]dt \\
&\quad -2\int_0^\tau v(0,t)\int_0^l K_1(x,t)u(x,t)dxdt \\
&\quad +2\int_0^\tau v(l,t)\int_0^l K_2(x,t)u(x,t)dxdt.
\end{aligned}$$

It follows at once

$$\begin{aligned}
&\int_0^l [u^2(x,\tau) + a(x,0)v_x^2(x,0)]dx \\
&\le |\alpha_{11}(0)|v^2(0,0) + 2|\alpha_{21}v(0,0)v(l,0)| + |\alpha_{22}(0)|v^2(l,0) \\
&\quad + 2|\int_0^\tau \int_0^l cuvdxdt| + \int_0^\tau \int_0^l |a_t|v_x^2 dxdt \\
&\quad + \int_0^\tau [|\alpha'_{11}|v^2(0,t) + 2|\alpha'_{21}v(0,t)v(l,t)| + |\alpha'_{22}|v^2(l,t)]dt \\
&\quad + 2|\int_0^\tau v(0,t)\int_0^l K_1(x,t)u(x,t)dxdt| \\
&\quad + 2|\int_0^\tau v(l,t)\int_0^l K_2(x,t)u(x,t)dxdt|. \qquad (7)
\end{aligned}$$

Under the conditions of Theorem 1, the left side of (7) is nonnegative and there exist positive numbers a_0, a_1, a_2, c_0, k_0 such that

$$a(x,t) \ge a_0, \ \max_{\bar{Q}_T} |c(x,t)| \le c_0, \ \max_{\bar{Q}_T} |a_t(x,t)| \le a_1,$$

$$\max_{ij} \max_{[0,T]} |\alpha_{ij}, \alpha'_{ij}| \le a_2, \ \max_{i} \max_{[0,T]} \int_0^l K_{1i}^2 dx \le k_0.$$

Hence, it follows from (7) by means of Cauchy and Cauchy–Bunyakovskii-Schwartz inequalities:

$$2\left|\int_0^\tau\int_0^l cuvdxdt\right| \leq c_0\int_0^\tau\int_0^l (u^2+v^2)dxdt;$$

$$2\left|\int_0^\tau v(0,t)\int_0^l K_1(x,t)u(x,t)dxdt\right| + 2\left|\int_0^\tau v(l,t)\int_0^l K_2(x,t)u(x,t)dxdt\right|$$
$$\leq \int_0^\tau [v^2(0,t)+v^2(l,t)]dt + 2k_0\int_0^\tau\int_0^l u^2dxdt;$$

$$|\alpha_{11}(0)|v^2(0,0) + 2|\alpha_{21}v(0,0)v(l,0)| + |\alpha_{22}(0)|v^2(l,0)$$
$$\leq 2a_2[v^2(0,0)+v^2(l,0)];$$

$$\int_0^\tau [|\alpha'_{11}|v^2(0,t) + 2|\alpha'_{21}v(0,t)v(l,t)| + |\alpha'_{22}|v^2(l,t)]dt$$
$$\leq 2a_2\int_0^\tau [v^2(0,t)+v^2(l,t)]dt.$$

Thus

$$\int_0^l [u^2(x,\tau) + a_0v_x^2(x,0)]dx$$
$$\leq 2a_2[v^2(0,0)+v^2(l,0)] + (2a_2+1)\int_0^\tau [v^2(0,t)+v^2(l,t)]dt$$
$$+ c_0\int_0^\tau\int_0^l (u^2+v^2)dxdt + 2k_0\int_0^\tau\int_0^l u^2dxdt + a_1\int_0^\tau\int_0^l v_x^2dxdt.$$

To estimate the terms containing values of v on the boundary, we use the inequalities [21]:

$$v^2(z_i,t) \leq 2l\int_0^l v_x^2(x,t)dx + \frac{2}{l}\int_0^l v^2(x,t)dx, \quad z_1=0,\ z_2=l, \tag{8}$$

and in the special case for $n=1$ of the trace inequality [18] in the form

$$v^2(z_i,t) \leq \varepsilon\int_0^l v_x^2(x,t)dx + c(\varepsilon)\int_0^l v^2(x,t)dx, \quad z_1=0,\ z_2=l \tag{9}$$

and the obvious inequality

$$v^2(x,t) \leq \tau\int_0^\tau u^2dt, \tag{10}$$

which follows from the definition of $v(x,t)$.

From (7), we get

$$\int_0^l [u^2(x,\tau) + a_0 v_x^2(x,0)]dx \le 4a_2\varepsilon \int_0^l v_x^2(x,0)dx + M\int_0^\tau \int_0^l (u^2 + v_x^2)dxdt, \tag{11}$$

where M is a constant depending only on $a_0, a_1, a_2, c_0, k_0, l, T$. Choose ε such that $\nu = a_0 - 4a_2\varepsilon > 0$ and carry out $4a_2\varepsilon \int_0^l v_x^2(x,0)dx$ into the left side. Then

$$\int_0^l [(u^2(x,\tau)) + \nu(v_x^2(x,0))]dx \le M\int_0^\tau \int_0^l (u^2 + v_x^2)dxdt.$$

Introduce a function $w(x,t) = \int_0^t u_x(x,\eta)d\eta$. It is easy to see that

$$v_x(x,t) = w(x,t) - w(x,\tau), \quad v_x(x,0) = -w(x,\tau).$$

With the aid of these equalities, we obtain

$$\int_0^l [u^2(x,\tau)+\nu w^2(x,\tau)]dx \le 2M\int_0^\tau \int_0^l [u^2+w^2]dxdt + 2M\tau\int_0^l w^2(x,\tau)dxdt.$$

As τ is arbitrary, we choose it so that $\nu - 2M\tau > 0$. To be specific, let $\nu - 2M\tau \ge \frac{\nu}{2}$. Then for all $\tau \in [0, \frac{\nu}{4M}]$

$$m_0 \int_0^l [u^2(x,\tau) + w^2(x,\tau)]dx \le 2M\int_0^\tau \int_0^l (u^2 + w^2)dxdt,$$

where $m_0 = \min\{1, \nu/2\}$. From Gronwall's lemma, it follows immediately that $\int_0^l [u^2(x,\tau) + w^2(x,\tau)]dx = 0$. Hence $u(x,\tau) = 0 \; \forall \tau \in [0, \frac{\nu}{4M}]$. Considering now $t = \frac{\nu}{4M}$ as a line where initial data is given, we get $u(x,\tau) = 0$ for $[\frac{\nu}{4M}, \frac{\nu}{2M}]$. Continuing this process, we convince ourselves that $u(x,t) = 0$ in Q_T. This means that there cannot be more than one weak solution to the Problem 1.

Existence We prove the solution existence in several steps. First, we construct approximations of the weak solution by the Faedo-Galerkin method. Second, we obtain a priori estimates to guarantee weak convergence of approximations. Finally, we show that the limit of approximations is the required solution.

Step 1 Let $w_k(x) \in C^2[0,l]$ be a basis in $W_2^1(0,l)$. We define the approximations as follows

$$u^m(x,t) = \sum_{k=1}^m c_k(t)w_k(x), \tag{12}$$

and shall seek $c_k(t)$ from relations

$$\int_0^l (u_{tt}^m w_j + au_x^m w_j' + cu^m w_j)dx + w_j(l)\int_0^l K_1(x,t)u^m(x,t)dx$$
$$- w_j(0)\int_0^l K_1(x,t)u^m(x,t)dx + w_j(l)[\alpha_{21}u^m(0,t) + \alpha_{22}u^m(l,t)]$$
$$- w_j(0)[\alpha_{11}u^m(0,t) + \alpha_{12}u^m(l,t)] = \int_0^l f w_j dx. \tag{13}$$

For every m, (13) represents a system of second-order ODEs with respect to $c_k(t)$. Indeed, by substituting (12), we can rewrite (13) in the form

$$\sum_{k=1}^m A_{kj}c_k''(t) + B_{kj}c_k(t) = F_j(t), \tag{14}$$

where

$$A_{kj} = \int_0^l w_k w_j dx,$$
$$B_{kj}(t) = \int_0^l (aw_k' w_j' + cw_k w_j)dx + w_j(l)\int_0^l w_k K_1 dx - w_j(0)\int_0^l w_k K_1 dx$$
$$+ w_j(l)[\alpha_{21}w_k(0) + \alpha_{22}w_k(l)] - w_j(0)[\alpha_{11}w_k(0) + \alpha_{12}w_k(l)],$$
$$F_j(t) = \int_0^l f(x,t)w_j(x)dx.$$

Adding the initial data, $c_k(0) = c_k'(0) = 0$, we obtain the Cauchy problem. As $w_k(x)$ are linearly independent, the system (14) is solvable with respect to $c_k''(t)$. The conditions of Theorem 1 imply that the coefficients of (14) are bounded and $F_j \in L_1(0,T)$. These facts guarantee the solvability of the Cauchy problem. Moreover, $c_k'' \in L_1(0,T)$. Thus, the approximation $\{u^m(x,t)\}$ is constructed.

Step 2 To derive a priori estimate, we multiply (13) by $c_j'(t)$, sum with respect to $j = 1, \ldots, m$, integrate over $(0,\tau)$ and obtain

$$\int_0^\tau \int_0^l (u_{tt}^m u_t^m + au_x^m u_{xt}^m + cu^m u_t^m)dxdt + \int_0^\tau u_t^m(l,t)\int_0^l K_2(x,t)u^m(x,t)dxdt$$
$$- \int_0^\tau u_t^m(0,t)\int_0^l K_1(x,t)u^m(x,t)dxdt$$
$$+ \int_0^\tau u_t^m(l,t)[\alpha_{21}u^m(0,t) + \alpha_{22}u^m(l,t)]dt$$
$$- \int_0^\tau u_t^m(0,t)[\alpha_{11}u^m(0,t) + \alpha_{12}u^m(l,t)]dt = \int_0^\tau \int_0^l f u_t^m dxdt.$$

Integrating by parts and taking into account $u^m(x,0) = u_t^m(x,0) = 0$ and $\alpha_{12} + \alpha_{21} = 0$, we get

$$\frac{1}{2}\int_0^l [(u_t^m(x,\tau))^2 + a(u_x^m(x,\tau))^2]dx + \int_0^\tau \int_0^l c u^m u_t^m dxdt$$
$$- \frac{1}{2}\int_0^\tau \int_0^l a_x (u_x^m)^2 dxdt$$
$$- \int_0^\tau u^m(l,t)\int_0^l K_2(x,t)u_t^m(x,t)dxdt$$
$$- \int_0^\tau u^m(l,t)\int_0^l K_{2t}(x,t)u^m(x,t)dxdt$$
$$+ u^m(l,\tau)\int_0^l K_2(x,\tau)u^m(x,\tau)dx + \int_0^\tau u^m(0,t)\int_0^l K_1(x,t)u_t^m(x,t)dxdt$$
$$+ \int_0^\tau u^m(0,t)\int_0^l K_{1t}(x,t)u^m(x,t)dxdt - u^m(0,\tau)\int_0^l K_1(x,\tau)u^m(x,\tau)dx$$
$$- \frac{1}{2}\int_0^\tau [\alpha'_{22}(u^m(l,t))^2 + \alpha'_{11}(u^m(0,t))^2 + 2\alpha'_{12}u^m(0,t)u^m(l,t)]dt$$
$$+ \frac{1}{2}\alpha_{22}(\tau)(u^m(l,\tau))^2 - \alpha_{12}(\tau)u^m(0,\tau)u^m(l,\tau) - \frac{1}{2}\alpha_{11}(\tau)(u^m(0,\tau))^2$$
$$= \int_0^\tau \int_0^l f u_t^m dxdt.$$

It follows from this equality, the inequality

$$\int_0^l [(u_t^m(x,\tau))^2 + a(u_x^m(x,\tau))^2]dx \le \Big| \int_0^\tau \int_0^l a_x (u_x^m)^2 dxdt$$
$$- 2\int_0^\tau \int_0^l c u^m u_t^m dxdt$$
$$+ 2\int_0^\tau u^m(l,t)\int_0^l K_2 u_t^m dxdt + 2\int_0^\tau u^m(l,t)\int_0^l K_{2t}u^m dxdt$$
$$- 2u^m(l,\tau)\int_0^l K_2 u^m(x,\tau)dx$$
$$- 2\int_0^\tau u^m(0,t)\int_0^l K_1 u_t^m dxdt - 2\int_0^\tau u^m(0,t)\int_0^l K_{1t}u^m dxdt$$
$$+ 2u^m(0,\tau)\int_0^l K_1 u^m(x,\tau)dx$$

$$+\int_0^\tau \alpha'_{22}(u^m(l,t))^2dt - \int_0^\tau \alpha'_{11}(u^m(0,t))^2dt$$
$$-2\int_0^\tau \alpha'_{12}u^m(0,t)u^m(l,t)dt$$
$$+\alpha_{22}(\tau)(u^m(l,\tau))^2 - 2\alpha_{12}(\tau)u^m(0,\tau)u^m(l,\tau)$$
$$-\alpha_{11}(\tau)(u^m(0,\tau))^2$$
$$-2\int_0^\tau\int_0^l f(x,t)u_t^m(x,t)dxdt\Big|. \tag{15}$$

To estimate the right side of (15), we follow in essence the procedure demonstrated in the subsection *uniqueness*. Aside from basic inequalities used there, we also need the Cauchy inequality "with ε" :

$$2|u^m(0,\tau)\int_0^l K_1u^m(x,\tau)dx| + 2|u^m(l,\tau)\int_0^l K_2u^m(x,\tau)dx|$$
$$\leq \varepsilon[(u^m(0,\tau))^2 + (u^m(l,\tau))^2] + 2k_0c(\varepsilon)\int_0^l (u^m(x,\tau))^2dx.$$

The conditions of Theorem 1, inequalities (8), (9), and (10) and ε chosen with due care enable us to obtain the following inequality

$$\int_0^l [(u^m(x,\tau))^2 + (u_t^m(x,\tau))^2 + (u_x^m(x,\tau))^2]dx$$
$$\leq M\int_0^\tau\int_0^l [(u^m)^2 + (u_t^m)^2 + (u_x^m)^2]dxdt + N\int_0^l f^2(x,t)dxdt \tag{16}$$

where M and N do not depend on m. An application to this inequality of Gronwall's lemma leads to

$$||u^m||_{W_2^1(Q_T)} \leq P \tag{17}$$

where $P = e^{MT}T||f||^2_{L_2(Q_T)}$.

Step 3 As $W_2^1(Q_T)$ is Hilbert space, then the estimate (17) enables us to state that we can extract from approximations $\{u^m(x,t)\}$ a subsequence weakly convergent in $W_2^1(Q_T)$. It remains to show that the limit of this subsequence is the required weak solution to the Problem 1. To do this, multiply (13) by $d_j(t) \in W_2^1(0,T)$, $d_j(T) = 0$, sum with respect to $j = 1,\dots,m$ and integrate over $(0,T)$. After some manipulations, we get

$$\int_0^T \int_0^l (-u_t^n \eta_t + a u_x^n \eta_x + c u^n \eta) dx dt - \int_0^T \eta(0,t)[\alpha_{11} u^n(0,t) + \alpha_{12} u^n(l,t)] dt$$
$$+ \int_0^T \eta(l,t)[\alpha_{21} u^n(0,t) + \alpha_{22} u^n(l,t)] dt$$
$$- \int_0^T \eta(0,t) \int_0^l K_1(x,t) u^n(x,t) dx dt$$
$$+ \int_0^T \eta(l,t) \int_0^l K_2(x,t) u^n(x,t) dx dt = \int_0^T \int_0^l f \eta dx dt \quad (18)$$

where we denote $\eta(x,t) = \sum\limits_{j=1}^{m} d_j(t) w_j(x)$.

Passing to the limit in (18), we get (5) for $v(x,t) = \eta(x,t)$ and limit function $u(x,t)$. Since the union of all functions of the form $\sum\limits_{j=1}^{m} d_j(t) w_j(x)$ is dense in $\overset{\circ}{W}{}_2^1(Q_T)$, then the limit function $u(x,t)$ is the required weak solution to the Problem 1. This completes the proof. □

Remark 1 The special case of this problem with $\alpha_{ij} = 0$ in (4) is considered in [21].

2.2 A Problem with First-Kind Integral Conditions

Let now $\forall t \in [0,T]$, $a_i = b_i = c_i = d_i = 0$, $i = 0, 1$. Then (3) are first-kind integral conditions as both of them include only integral terms. We will assume that k_i depend only on x to simplify calculations.

Problem 2 Find a solution $u(x,t)$ to the Eq. (1) satisfying the initial data (2) and the nonlocal conditions

$$\int_0^l k_i(x) u(x,t) dx = 0, \quad i = 1, 2. \quad (19)$$

Such conditions like (19) cause a considerable difficulties when we try to show that the problem (1), (3), and (19) is solvable. In [21], a method has been developed for overcoming this difficulty. The essential idea of this technique is as follows. We transform the first-kind integral conditions to the second-kind ones. To do this, we suppose that $u(x,t)$ is a solution to (1), (2) and (19), multiply (1) by $k_i(x)$ and integrate over $(0,l)$ and get

$$k_i(0) a(0,t) u_x(0,t) - k_i(l) a(l,t) u_x(l,t)$$
$$- k_{ix}(0) a(0,t) u(0,t) + k_{ix}(l) a(l,t) u(l,t)$$

$$
-\int_0^l [(k_{ix}(x)a(x,t))_x - k_i(x)c(x,t)]u(x,t)dx
$$

$$
= \int_0^l k_i(x)f(x,t)dx. \tag{20}
$$

Denote

$$
\begin{aligned}
a_1(t) &= k_1(0)a(0,t),\ b_1(t) = -k_1(l)a(l,t),\\
a_0(t) &= -k_{1x}(0)a(0,t),\ b_0(t) = k_{1x}(l)a(l,t),\\
c_1(t) &= k_2(0)a(0,t),\ d_1(t) = -k_2(l)a(l,t),\\
c_0(t) &= -k_{2x}(0)a(0,t),\ d_0(t) = k_{2x}(l)a(l,t),\\
h_i(x,t) &= (k_{ix}(x)a(x,t))_x - k_i(x,t)c(x,t),\\
g_i(t) &= \int_0^l k_i(x)f(x,t)dx
\end{aligned}
$$

and write now (20) (omitting the arguments of coefficients) as follows

$$
\begin{aligned}
a_1u_x(0,t) + b_1u_x(l,t) + a_0u(0,t) + b_0u(l,t) - \int_0^l h_1udx = g_1(t),\\
c_1u_x(0,t) + d_1u_x(l,t) + c_0u(0,t) + d_0u(l,t) - \int_0^l h_2udx = g_2(t).
\end{aligned} \tag{21}
$$

Thus we arrive at second-kind integral conditions.

The reverse is also true. Indeed, let $u(x,t)$ be a solution of (1) and (21) holds. Then (20) holds too. Multiplying (1) by $k_i(x)$, integrating over $(0,l)$ and taking into account (20), we easily arrive to

$$
\frac{d^2}{dt^2}\int_0^l k_i(x)u(x,t)dx = 0.
$$

Since from (2) $\int_0^l k_i(x)u(x,0)dx = 0$ and $\frac{d}{dt}\left(\int_0^l k_i(x)u(x,t)dx\right)\big|_{t=0} = 0$, it follows that $\int_0^l k_i(x)u(x,t)dx = 0$ as a solution to Cauchy problem. Thus the nonlocal conditions (19) and (21) are equivalent.

The form (21) of nonlocal conditions enables us to introduce a notation of a weak solution as in subsection 2.1 and use all results of 2.1. Due to equivalence of (19) and (21), the solution of the problem with second-kind integral conditions (21) is the solution to the Problem 2.

Theorem 2 *Let*

(i) $a,\ a_t \in C(\bar{Q}_T),\ \ c \in C(\bar{Q}_T),\ \ a(x,t) > 0, \quad \forall (x,t) \in \bar{Q}_T,$
(ii) $k_i,\ k_{it},\ k_{itt} \in C(\bar{Q}_T),\ \ f \in L_2(Q_T),\ \ k_1(0,t)k_2(l,t) - k_1(l,t)k_2(0,t) \neq 0,$
(iii) $\alpha_{ij} \in C^1([0,T]),\ \ \alpha_{12} + \alpha_{21} = 0.$

Then there exists a unique weak solution to the problem (1), (2), and (19).

Note that under the conditions of Theorem 2, coefficients in (21) satisfy

$$\Delta = a_1 d_1 - b_1 c_1 \neq 0.$$

Indeed,

$$\begin{aligned}\Delta &= a_1 d_1 - b_1 c_1 \\ &= k_1(0)a(0,t)k_2(l)a(l,t) - k_1(l)a(l,t)k_2(0)a(0,t) \\ &= a(0,t)a(l,t)[k_1(0)k_2(l) - k_1(l)k_2(0)] \neq 0,\end{aligned}$$

and we can use all results of 2.1.

3 A Problem with Dynamical Nonlocal Conditions

Let now second-kind nonlocal conditions contain derivatives with respect to t. Such conditions are called dynamical ones. Dynamical boundary conditions arise in many applications [8, 17, 28].

Problem 3 Find a solution of Eq. (1), satisfying the initial conditions (2) and the following nonlocal conditions

$$\begin{aligned} a(0,t)u_x(0,t) &= \alpha_{11}u(0,t) + \alpha_{12}u(l,t) + \beta_{11}u_{tt}(0,t) + \beta_{12}u_{tt}(l,t) \\ &\quad + \int_0^l H_1(x,t)u(x,t)dx, \\ a(l,t)u_x(l,t) &= \alpha_{21}u(0,t) + \alpha_{22}u(l,t) + \beta_{21}u_{tt}(0,t) + \beta_{22}u_{tt}(l,t) \\ &\quad + \int_0^l H_2(x,t)u(x,t)dx. \end{aligned} \tag{22}$$

Denote

$$\begin{aligned} &\Gamma_0 = \{(x,t) : x = 0, t \in [0,T]\}, \\ &\Gamma_l = \{(x,t) :\ x = l, t \in [0,T]\},\ \Gamma = \Gamma_0 \cup \Gamma_l, \end{aligned}$$

$$W(Q_T) = \{u : u \in W_2^1(Q_T),\ \ u_t \in L_2(\Gamma)\},$$

$$\hat{W}(Q_T) = \{v(x,t) : v(x,t) \in W(Q_T),\ v(x,T) = 0\}.$$

Using the similar approach as in 2.1, we get the equality

$$\int_0^T\int_0^l (-u_t v_t + a u_x v_x + c u v)dxdt + \int_0^T v(0,t)[\alpha_{11}u(0,t) + \alpha_{12}u(l,t)]dt$$

$$+ \int_0^T v(0,t)\int_0^l H_1(x,t)u(x,t)dxdt - \int_0^T v_t(0,t)[\beta_{11}u_t(0,t) + \beta_{12}u_t(l,t)]dt$$

$$- \int_0^T v(l,t)[\alpha_{21}u(0,t) + \alpha_{22}u(l,t)]dt - \int_0^T v(l,t)\int_0^l H_2(x,t)u(x,t)dxdt$$

$$+ \int_0^T v_t(l,t)[\beta_{21}u_t(0,t) + \beta_{22}u_t(l,t)]dt = \int_0^T\int_0^l f(x,t)v(x,t)dxdt. \quad (23)$$

Definition 3 A function $u(x,t) \in W(Q_T)$ is said to be a weak solution to the Problem 3 if $u(x,0) = 0$ and for every $v \in \hat{W}(Q_T)$ (23) holds.

Theorem 3 *Suppose*

(i) $a \in C(\bar{Q}_T),\ a_t \in C(\bar{Q}_T),\ a(x,t) > 0,\ c \in C(\bar{Q}_T),\quad \forall (x,t) \in \bar{Q}_T,$
(ii) $H_i \in C(\bar{Q}_T),\quad S_i \in C[0,l],\quad f \in L_2(Q_T),\quad f_t \in L_2(Q_T),$
(iii) $\beta_{ij} \in C^1[0,T],\quad \beta_{11}(t) > 0,\quad \beta_{22}(t) < 0,\quad \beta_{11} - |\beta_{21}| > 0,\quad -\beta_{22} - |\beta_{21}| > 0,$
(iv) $\alpha_{ij} \in C^1[0,T],\quad \alpha_{12} + \alpha_{21} = 0,\quad \beta_{12} + \beta_{21} = 0.$

Then there exists a unique weak solution to Problem 3.

Proof *Uniqueness.* Suppose that u_1 and u_2 are two different solutions of Problem 3. Then $u = u_1 - u_2$ satisfies initial condition $u(x,0) = 0$ and the identity

$$\int_0^T\int_0^l (-u_t v_t + a u_x v_x + c u v)dxdt + \int_0^T v(0,t)[\alpha_{11}u(0,t) + \alpha_{12}u(l,t)]dt$$

$$- \int_0^T v_t(0,t)[\beta_{11}u_t(0,t) + \beta_{12}u_t(l,t)]dt + \int_0^T v(0,t)\int_0^l H_1 u dxdt$$

$$-\int_0^T v(l,t)[\alpha_{21}u(0,t)+\alpha_{22}u(l,t)]dt$$

$$+\int_0^T v_t(l,t)[\beta_{21}u_t(0,t)+\beta_{22}u_t(l,t)]dt-\int_0^T v(l,t)\int_0^l H_2udxdt=0. \tag{24}$$

Set in (24)

$$v(x,t)=\begin{cases}\int_\tau^t u(x,\eta)d\eta,\ 0\le t\le\tau,\\ 0,\ \tau\le t\le T,\end{cases}$$

$\tau\in[0,T]$ is arbitrary. After elementary manipulations and taking into account (iv) of Theorem 3, we get

$$\int_0^l [u^2(x,\tau)+a(x,0)v_x^2(x,0)]dx=-\int_0^\tau\int_0^l a_t v_x^2dxdt$$

$$+\int_0^\tau [\alpha'_{22}v^2(l,t)+2\alpha'_{21}v(0,t)v(l,t)-\alpha'_{11}v^2(0,t)]dt$$

$$+\alpha_{22}(0)v^2(l,0)+2\alpha_{21}v(0,0)v(l,0)-\alpha_{11}(0)v^2(0,0)$$

$$+\beta_{22}u^2(l,\tau)-2\beta_{21}u(0,\tau)u(l,\tau)-\beta_{11}u^2(0,\tau)$$

$$+\int_0^\tau v(0,t)\int_0^l H_1udxdt-\int_0^\tau v(l,t)\int_0^l H_2udxdt+\int_0^\tau\int_0^l cvv_tdxdt. \tag{25}$$

Under the assumptions of Theorem 3, it follows that there exist positive numbers c_0, h_0, a_1, a_2, b_1 such that

$$\max_{\bar{Q}_T}|c(x,t)|\le c_0,\ \max_{[0,T]}|\alpha_{ij},\alpha'_{ij}|\le a_2,\ \max_{[0,T]}|\beta_{ij},\beta'_{ij}|\le b_1, i,j=1,2,$$

$$\max_{\bar{Q}_T}|a_t(x,t)|\le a_1,\ \max_{[0,T]}\int_0^l H_i^2(x,t)dx\le h_0.$$

Using Cauchy, Cauchy–Bunyakovskii-Schwartz inequalities, trace inequalities (8), (9), and (10), we get

$$\int_0^l [u^2(x,\tau)+a(x,0)v_x^2(x,0)]dx \le A_1 \int_0^\tau \int_0^l v_x^2 dxdt + A_2 \int_0^\tau \int_0^l u^2 dxdt, \qquad (26)$$

where A_i depends only on c_0, h_0, a_1, a_2, b_1.

As in 2.1 we introduce a function $w(x,t) = \int_0^t u_x(x,\eta)d\eta$ and as above we obtain for such τ that $a_0 - 2A_1\tau > 0$

$$m_0 \int_0^l [u^2(x,\tau) + w^2(x,\tau)]dx \le A_3 \int_0^\tau \int_0^l [u^2(x,t) + w^2(x,t)]dxdt,$$

where $m_0 = \min\{1, \frac{a_0}{2}\}$, $A_3 = \max\{2A_1, A_2\}$. Now from Gronwall's lemma $u(x,t) = 0$, $t \in [0, \frac{a_0}{4A_1}]$. Proceeding as in 2.1, we can see that $u(x,t) = 0$ $\forall (x,t) \in Q_T$.

Existence We work in steps as in the proof of Theorem 1.

Step 1 Let $w_k(x) \in C^2[0,l]$ be the basis in $W_2^1(0,l)$. We seek approximations

$$u^m(x,t) = \sum_{k=1}^m c_k(t)w_k(x) \qquad (27)$$

from

$$\begin{aligned}
&\int_0^l (u_{tt}^m w_j + a u_x^m w_j' + c u^m w_j)dx \\
&+ w_k(0)[\alpha_{11}u^m(0,t) + \alpha_{12}u^m(l,t) \\
&+ \beta_{11}u_{tt}^m(0,t) + \beta_{12}u^m(l,t) + \int_0^l H_1(x,t)u(x,t)dx] \\
&- w_k(l)[\alpha_{21}u^m(0,t) + \alpha_{22}u^m(l,t) \\
&+ \beta_{21}u_{tt}^m(0,t) + \beta_{22}u^m(l,t) + \int_0^l H_2(x,t)u(x,t)dx] \\
&= \int_0^l f(x,t)w_j(x)dx. \qquad (28)
\end{aligned}$$

With initial conditions $c_k(0) = 0,\ c_k'(0) = 0$, (28) becomes the Cauchy problem with respect to $c_k(t)$. We can write (28) in the form

$$\sum_{k=1}^{m} A_{kj} c_k''(t) + \sum_{k=1}^{m} B_{kj}(t) c_k(t) = f_j(t),$$

where

$$A_{kj} = \int_0^l w_k(x) w_j(x) dx$$
$$+ \beta_{11} w_k(0) w_j(0) + \beta_{12} w_k(l) w_j(0) - \beta_{21} w_k(0) w_j(l) - \beta_{22} w_k(l) w_j(l),$$

$$B_{kj}(t) = \int_0^l (a(x,t) w_k'(x) w_j'(x) + c(x,t) w_k(x) w_j(x)) dx$$
$$+ \alpha_{11} w_k(0) w_j(0) + \alpha_{12} w_k(l) w_j(0) - \alpha_{21} w_k(0) w_j(l) - \alpha_{22} w_k(0) w_j(0),$$

$$f_j(t) = \int_0^l f(x,t) w_j(x) dx.$$

To show that the Cauchy problem has a solution, we consider a quadratic form $q = \sum_{k,l=1}^{m} A_{kl} \xi_k \xi_l$, where ξ_i are coefficients of $z = \sum_{i=1}^{m} \xi_i w_i(x)$. Substituting A_{kj}, we get

$$q = \int_0^l |z(x)|^2 dx + \beta_{11} |z(0)|^2 + 2\beta_{12} |z(0)||z(l)| - \beta_{22} |z(l)|^2.$$

It is easy to see that under the condition (iii) of Theorem 3 $q \geq 0$, moreover, $q = 0$ only if $z = 0$. In turn, $z = 0$ if $\xi_i = 0 \ \ \forall i = 1, \ldots, m$ only. Thus, q is positive definite. Therefore, the matrix (A_{kj}) is as well positive definite. Hence, (28) is solvable with respect to $c_k''(t)$. It is clear now that under the conditions of Theorem 3, the Cauchy problem for (28) has a solution and its approximation is constructed.

Step 2 To derive the estimate, we multiply (28) by $c_j'(t)$, sum over $j = 1, \ldots, m$ and integrate over $(0, \tau)$, where $\tau \in [0, T]$ is arbitrary:

$$\int_0^\tau \int_0^l (u_{tt}^m u_t^m + a u_x^m u_{xt}^m + c u^m u_t^m) dx dt$$

$$+\int_0^\tau u_t^m(0,t)[\alpha_{11}u^m(0,t)+\alpha_{12}u^m(l,t)+\beta_{11}u_{tt}^m(0,t)+\beta_{12}u_{tt}^m(l,t)]dt$$

$$+\int_0^\tau u_t^m(0,t)\int_0^l H_1(x,t)u^m(x,t)dxdt$$

$$-\int_0^\tau u_t^m(l,t)[\alpha_{21}u^m(0,t)+\alpha_{22}u^m(l,t)+\beta_{21}u_{tt}^m(0,t)+\beta_{22}u_{tt}^m(l,t)]dt$$

$$\int_0^\tau u_t^m(l,t)\int_0^l H_2(x,t)u^m(x,t)dxdt=\int_0^\tau\int_0^l f(x,t)u_t^m(x,t)dxdt. \quad (29)$$

Integration by parts and condition (iii) lead to

$$\int_0^l [(u_t^m(x,\tau))^2+a(x)(u_x^m(x,\tau))^2]dx+\beta_{11}(u_t^m(0,\tau))^2-\beta_{22}(u_t^m(l,\tau))^2$$
$$=\int_0^\tau\int_0^l a_t(u_x^m)^2dxdt-2\int_0^\tau\int_0^l cu^m u_t^m dxdt-2\beta_{21}u_t^m(0,\tau)u_t^m(l,\tau)$$
$$-[\alpha_{11}(u^m(0,\tau))^2+2\alpha_{21}u^m(0,\tau)u^m(l,\tau)-\alpha_{22}(u^m(l,\tau))^2]$$
$$+2\int_0^\tau u^m(0,t)\int_0^l H_1(x,t)u_t^m(x,t)dxdt$$
$$+2\int_0^\tau u^m(0,t)\int_0^l H_{1t}(x,t)u^m(x,t)dxdt$$
$$-2\int_0^\tau u^m(l,t)\int_0^l H_2(x,t)u_t^m(x,t)dxdt$$
$$-2\int_0^\tau u^m(l,t)\int_0^l H_{2t}(x,t)u_t^m(x,t)dxdt$$
$$-2u^m(0,\tau)\int_0^l H_1(x,\tau)u^m(x,\tau)+2u^m(l,\tau)\int_0^l H_2(x,\tau)u^m(x,\tau)dx$$
$$+\int_0^\tau[\alpha'_{11}(u^m(0,t))^2+2\alpha'_{21}u^m(0,t)u^m(l,t)-\alpha'_{22}(u^m(l,t))^2]dt$$
$$+\int_0^\tau[\beta'_{11}(u_t^m(0,t))^2+2\beta'_{21}u_t^m(0,t)u_t^m(l,t)-\beta'_{22}(u_t^m(l,t))^2]dt$$
$$+2\int_0^\tau\int_0^l fu_t^m dxdt. \quad (30)$$

As $\beta_{11} > 0$, $\beta_{22} < 0$ due to (iii) the left side of (30) is positive. To estimate the right-hand side of (30), we use in main the same technique as in subsection *uniqueness*. Since from (iii) $2|\beta_{21}u_t^m(0,\tau)u_t^m(l,\tau)| \leq |\beta_{21}((u_t^m(0,\tau))^2+(u_t^m(0,\tau))^2)$ and $\gamma_1 = \beta_{11} - |\beta_{21}| > 0$, $\gamma_2 = \beta_{22} - |\beta_{21}| > 0$ then

$$\int_0^l [(u_t^m(x,\tau))^2 + a(x)(u_x^m(x,\tau))^2]dx + \gamma_1(u_t^m(0,\tau))^2 + \gamma_2(u_t^m(l,\tau))^2$$

$$\leq \int_0^\tau \int_0^l |a_t|(u_x^m)^2 dxdt + 2|\int_0^\tau \int_0^l cu^m u_t^m dxdt|$$

$$+ |\alpha_{11}(u^m(0,\tau))^2 + 2\alpha_{21}u^m(0,\tau)u^m(l,\tau) - \alpha_{22}(u^m(l,\tau))^2|$$

$$+ 2|\int_0^\tau u^m(0,t)\int_0^l H_1(x,t)u_t^m(x,t)dxdt|$$

$$+ 2|\int_0^\tau u^m(0,t)\int_0^l H_{1t}(x,t)u^m(x,t)dxdt|$$

$$+ 2|\int_0^\tau u^m(l,t)\int_0^l H_2(x,t)u_t^m(x,t)dxdt|$$

$$+ 2|\int_0^\tau u^m(l,t)\int_0^l H_{2t}(x,t)u_t^m(x,t)dxdt|$$

$$2|u^m(0,\tau)\int_0^l H_1(x,\tau)u^m(x,\tau)| + 2|u^m(l,\tau)\int_0^l H_2(x,\tau)u^m(x,\tau)dx|$$

$$+ \int_0^\tau |\alpha'_{11}(u^m(0,t))^2 + 2\alpha'_{21}u^m(0,t)u^m(l,t) - \alpha'_{22}(u^m(l,t))^2|dt$$

$$+ \int_0^\tau |\beta'_{11}(u_t^m(0,t))^2 + 2\beta'_{21}u_t^m(0,t)u_t^m(l,t)$$

$$- \beta'_{22}(u_t^m(l,t))^2|dt + 2|\int_0^\tau \int_0^l fu_t^m dxdt|.$$

Firstly, we use trace inequalities (8) to estimate the terms containing boundary values of u^m under integrals. Then we consider

$$|\alpha_{11}(u^m(0,\tau))^2 + 2\alpha_{21}u^m(0,\tau)u^m(l,\tau) - \alpha_{22}(u^m(l,\tau))^2|$$

and use (9):

$$|\alpha_{11}(u^m(0,\tau))^2 + 2\alpha_{21}u^m(0,\tau)u^m(l,\tau) - \alpha_{22}(u^m(l,\tau))^2|$$
$$\leq 2a_1[(u^m(0,\tau))^2 + (u^m(l,\tau))^2]$$

$$\leq 8a_1\varepsilon \int_0^l (u_x^m(x,t))^2 dx + c(\epsilon) \int_0^l u^m(x,t)dx.$$

Choosing ε such that $\nu = a_0 - 8a_1\varepsilon > 0$, we can carry out $8a_1\varepsilon \int_0^l (u_x^m(x,t))^2 dx$ into the left side. Using Cauchy and Cauchy–Bunyakovskii-Schwartz inequalities and (10) to estimate the rest terms, we obtain

$$\begin{aligned}
&\int_0^l [(u^m(x,\tau))^2 + (u_t^m(x,\tau))^2 + \nu(u_x^m(x,\tau))^2]dx \\
&\quad + \gamma_1(u_t^m(0,\tau))^2 + \gamma_2(u_t^m(l,\tau))^2 \\
&\quad \leq M_1 \int_0^\tau \int_0^l ((u^m)^2 + (u_t^m)^2 + (u_x^m)^2)dxdt \\
&\quad\quad + M_2 \int_0^\tau (u_t^m(0,t))^2 dt + M_2 \int_0^\tau (u_t^m(l,t))^2 dt \\
&\quad\quad + N \int_0^\tau \int_0^l f^2 dxdt,
\end{aligned} \tag{31}$$

where γ_i, M_i, ν, N do not depend on m. Denote $m_0 = \min\{1, \nu\}$, $\gamma = \max\{\gamma 1, \gamma_2\}$, $M = \max\{M_1, M_2\}$. Now from Gronwall's lemma

$$\begin{aligned}
&\int_0^l [(u^m(x,\tau))^2 + (u_t^m(x,\tau))^2 + (u_x^m(x,\tau))^2]dx + \gamma[(u_t^m(0,\tau))^2 + (u_t^m(l,\tau))^2] \\
&\quad \leq Ne^{MT} \int_0^\tau \int_0^l f^2 dxdt.
\end{aligned}$$

We integrate this inequality over $(0, T)$ to obtain

$$||u^m||_{W_2^1(Q_T)} + ||u_t^m||_{L_2(\Gamma)} \leq P. \tag{32}$$

Thus, the approximation $\{u^m(x,t)\}$ is bounded in $W(Q_T)$ and there exists a subsequence $\{u^\mu(x,t)\}$ which converges weakly to a function $u \in W(Q_T)$.

Step 3 The result of Step 2 enables us to use the standard technique [18, pp. 214–215] to show that the limit of $\{u^\mu\}$ is the required weak solution to the Problem 3. The proof is completed. □

In concluding, only some of the results regarding the solvability of nonlocal problems with integral conditions have been included here. One can find more information in the papers (by no means not an exhaustive list) [2–5, 9–12, 15, 16, 19, 20, 22–27], and the references therein.

References

1. Ashyralyev, A., Sharifov, Y.A.: Optimal control problems for impulsive systems with integral boundary conditions. Electron. J. Differ. Equations **2013**, 1–11 (2013).
2. Assanova, A.T.: Nonlocal problem with integral conditions for the system of hyperbolic equations in the characteristic rectangle, Russ Math. **61**, 7–20 (2017).
3. Avalishvili, G., Avalishvili, M., Gordeziani, D.: On integral nonlocal boundary value problems for some partial differential equations. Bulletin of the Georgian National Academy of Sciences **5**, 31–37 (2011).
4. Beilin, S.A.: On a mixed nonlocal problem for a wave equation. Electron. J. Differ. Equations **2006**, 1–10 (2006).
5. Bouziani, A.: On the solvability of parabolic and hyperbolic problems with a boundary integral condition. Int. J. Math. Math. Sci. **31** 201–213 (2002).
6. Cannon, J.R.: The solution of the heat equation subject to the specification of energy. Quart. Appl. Math. **21**, 155–160 (1963).
7. Cannon, J.R., Lin, Y.: An Inverse Problem of Finding a Parameter in a Semi-linear Heat Equation. J. Math. Anal. Appl. **145**, 470–484 (1990).
8. Doronin, G.G., Larkin, N.A., Souza, A.J.: A hyperbolic problem with nonlinear second-order boundary damping. Electron. J. Differ. Equations **1998**, 1–10 (1998).
9. Gordeziani, D.G., Avalishvili, G.A.: Solutions of nonlocal problems for one-dimensional oscillations of the medium. Mat. Modelir. **12**, 94–103 (2000).
10. Ivanauskas, F.F., Novitski, Yu.A., Sapagovas, M.P.: On the stability of an explicit difference scheme for hyperbolic equations with nonlocal boundary conditions. Diff. Equations **49**, 849–856 (2013).
11. Ionkin, N.I.: A solution of certain boundary-value problem of heat conduction with nonclassical boundary condition. Diff. Equations **13**, 294–301 (1977). (in Russian)
12. Ionkin, N.I., Moiseev, E.I.: On a problem for heat equation with two-point boundary conditions. Diff. Equations **15**, 294–304 (1979). (in Russian)
13. Kamynin, L.I.: On certain boundary problem of heat conduction with nonclassical boundary conditions. Zh. Vychisl. Math. Math. Fiz.**4**, 1006–1024 (1964).
14. Kamynin, V.L.: Unique solvability of the inverse problem of determination of the leading coefficient in a parabolic equation. Diff. Equations **47**, 91–101 (2011).
15. Kozhanov, A.I., Pulkina, L.S.: On the solvability of boundary value problems with a nonlocal boundary condition of integral form for multidimentional hyperbolic equations. Diff. Equations **42**, 233–1246 (2006).
16. Korzyuk, V.I., Kozlovskaya, I.S., Naumavets, S.N.: Classical solution of a problem with integral conditions of the second kind for the one-dimensional wave equation. Diff. Equations **55**, 353–362 (2019).
17. Korpusov, O.M.: Blow-up in Nonclassical Wave Equations. Moscow, URSS (2010).
18. Ladyzhenskaya, O.A.: Boundary-value Problems of Mathematical Physics. Nauka, Moscow (1973).
19. Moiseev, E.I., Korzyuk, V.I., Kozlovskaya, I.S.: Classical solution of a problem with an integral condition for the one-dimensional wave equation. Diff. Equations **50**, 1364–1377 (2014).
20. Pulkina, L.S.: Initial-boundary value problem with a nonlocal boundary condition for a multidimensional hyperbolic equation. Diff. Equations **44**, 1119–1125 (2008).
21. Pulkina, L.S.: Boundary value problems for a hyperbolic equation with nonlocal conditions of the I and II kind. Russ Math. **56**, 62–69 (2012).
22. Pulkina, L.S.: Nonlocal problems for hyperbolic equations with degenerate integral conditions. Electron. J. Differ. Equations, **2016**, 1–12 (2016).
23. Pulkina, L.S., Beylin, A.B.: Nonlocal approach to problems on longitudinal vibration in a short bar. Electron. J. Differ. Equations **2019**, 1–9, (2019).
24. Pulkina, L.S., Kirichek, V.A.: Solvability of a nonlocal problem for a hyperbolic equation with degenerate integral conditions. Journal of Samara State Technical University, Ser. Physical and Mathematical Science **23**, 229–245 (2019).

25. Samarskii, A.A.: On certain problems of the modern theory of differential equations. Differ. Uravn. **16**, 1221–1228 (1980). (in Russian)
26. Steclov, V.A.: A promlem on cooling of a solid. Comm. of Kharkov Math. Society **5**, 136–181 (1896).
27. Tamarkin, J.D.: Some General Problems of the Theory of Linear Differential Equations and Expansions of an Arbitrary Function in Series. Petrograd (1917).
28. Zhang, Z.: Stabilization of the wave equation with variable coefficients and a dynamical boundary control. Electron. J. Differ. Equations **2016**, 1–10 (2016).

On the Solution of Boundary Value Problems for Loaded Ordinary Differential Equations

E. Providas and I. N. Parasidis

Abstract This chapter is devoted to the solution of the so-called *loaded ordinary differential equations* which arise in applications in sciences and engineering. We propose a direct operator method for examining existence and uniqueness and constructing the solution in closed form to a class of boundary value problems for loaded nth-order ordinary differential equations with multipoint and integral boundary conditions.

Mathematics Subject Classification: 34B10, 34L40

1 Introduction

The study of general differential boundary value problems has a long history and goes back to the early part of the twentieth century. A class of such problems are those involving the *differential-boundary equations* or *loaded differential equations*. The term *differential-boundary equations* is used by Krall; see, for example, his survey paper [13] where the development of this kind of problems is described from the beginning of the twentieth century until 1975. In 1971, Iskenderov published two articles [10, 11] referring to this type of problems by the name *loaded differential equations* (as it has been translated from Russian to English language). Specifically, Iskenderov states that a *loaded differential equation* is a differential equation which also includes the values of the desired function and its derivatives, taken at fixed points of the domain. Nakhushev in a series of papers investigates systematically boundary value problems for loaded functional, differential and integral equations, see [17, 19]. Moreover, he and his co-authors contemplate applications in engineering and sciences such as in heat transfer [6], ground fluid mechanics [18, 21], biology [20] and physics [22], where physical

E. Providas (✉) · I. N. Parasidis
Department of Environmental Sciences, University of Thessaly, Larissa, Greece
e-mail: providas@uth.gr; paras@uth.gr

I. N. Parasidis et al. (eds.), *Mathematical Analysis in Interdisciplinary Research*, Springer Optimization and Its Applications 179,
https://doi.org/10.1007/978-3-030-84721-0_29

phenomena and processes are modelled by loaded equations. Further applications are reported in [4, 8, 32] and [33]. For some very recent results on various aspects on the subject, one can look at, for example, [5, 12, 14, 29] and [30].

Loaded ordinary differential equations which model heat transfer phenomena and solved by the finite difference method are considered in [2]. Systems of loaded first-order ordinary differential equations and their solution by numerical methods are studied in [1, 3, 7], and the references therein. Exact explicit solutions to loaded ordinary differential equations are sporadically referred to the literature; see two examples of loaded first-order differential equations in [9] and [23] and a loaded second-order boundary value problem in [18]. Closed-form solutions to loaded ordinary linear and nonlinear difference equations are given in [26] and [27], respectively. Characteristic examples of general loaded even-order ordinary differential operators are presented in the works [15] and [16] where their spectral properties are investigated.

The aim of this study is to develop a method for establishing existence and uniqueness solvability criteria and obtaining the solution in closed form of boundary value problems for a loaded nth-order ordinary differential equation coupled with multipoint and integral boundary conditions. We consider the linear loaded nth-order ordinary differential equation of the most general type,

$$\begin{aligned} &a_0(x)u^{(n)}(x) + a_1(x)u^{(n-1)}(x) + \cdots + a_n(x)u(x) \\ &-\sum_{s=1}^{m_1}\sum_{k=0}^{n} g_{sk}(x)u^{(k)}(\check{x}_s) = f(x), \end{aligned} \tag{1}$$

for $x \in (a, b)$, where the coefficients $a_k(x)$, $k = 0, \ldots n$, the functions $g_{sk}(x)$, $s = 1, \ldots, m_1$, $k = 0, \ldots, n$, and the nonhomogeneous term $f(x)$ are continuous functions on $[a, b]$, and the leading coefficient $a_0(x)$ does not vanish at any point of that interval. The fixed points $a \leq \check{x}_1 < \check{x}_2 < \cdots < \check{x}_{m_1} \leq b$ designate the loading points. Equation (1) is subjected to the following nonlocal point and integral boundary conditions,

$$\sum_{j=1}^{l_1}\sum_{k=0}^{n-1} \mu_{ijk}u^{(k)}(\bar{x}_j) + \sum_{k=0}^{n-1}\int_a^b \gamma_{ik}(x)u^{(k)}(x)dx = \beta_i, \quad i = 1, \ldots, n, \tag{2}$$

where $\gamma_{ik}(x)$, $i = 1, \ldots, n$, $k = 0, \ldots, n-1$, are continuous functions on $[a, b]$, μ_{ijk}, $j = 1, \ldots, l_1$, $k = 0, \ldots, n-1$ and β_i, $i = 1, \ldots, n$, are real constants and $a \leq \bar{x}_1 < \bar{x}_2 < \cdots < \bar{x}_{l_1} \leq b$ are fixed points where boundary conditions are applied. The loading points $\check{x}_s$, $s = 1, \ldots, m_1$, may or may not be boundary points.

The method proposed is in analogy to the approach followed in [24] and [25] where an integro-differential operator is contemplated as a perturbation of a linear ordinary differential operator by an integral functional. Likewise, the nonlocal multipoint and integral boundary conditions are treated as perturbations of simpler conventional boundary conditions [28, 31]. Then the solution to the loaded boundary value problem can be constructed in closed form if the exact solution of the unperturbed boundary value problem is known.

The remainder of the chapter is organized as follows. In Sect. 2, the problem is formulated in an operator form in a Banach space and the notation used is explained. The main results including two key theorems are presented in Sect. 3. In Sect. 4, selected examples are solved to demonstrate the implementation and efficiency of the method suggested. Some conclusions are drawn in Sect. 5.

2 Formulation of the Problem

Let X, Y be complex Banach spaces and $A : X \to Y$ a linear ordinary differential operator of order n defined by

$$Au = a_0(x)u^{(n)}(x) + a_1(x)u^{(n-1)}(x) + \cdots + a_n(x)u(x). \tag{3}$$

Usually, $X = C[a, b]$, or $X = L_p(a, b)$, $p \geq 1$, and $X = Y$. The coefficients $a_k(x)$, $k = 0, \ldots, n$ are continuous functions on $[a, b]$ and the leading coefficient $a_0(x)$ does not vanish at any point of that interval. Let $D(A)$ and $R(A)$ indicate the domain and the range of A, respectively. We denote by $X_A^n = \left(D(A),\ \|\cdot\|_{X_A^n}\right)$ the Banach space of n times differentiable functions with the norm

$$\|u(x)\|_{X_A^n} = \sum_{i=0}^{n} \|u^{(i)}(x)\|_X. \tag{4}$$

Let $g_{sk}(x) \in Y$, $s = 1, \ldots, m_1$, $k = 0, \ldots, n$. We combine like terms and write the loading summands in (1) in vector form as

$$\mathbf{g}\Psi(u) = \sum_{i=1}^{m} g_i \Psi_i(u) = \sum_{s=1}^{m_1} \sum_{k=0}^{n} g_{sk}(x) u^{(k)}(\check{x}_s), \tag{5}$$

where $m \leq m_1 * (n + 1)$ and the m elements of the vector $\mathbf{g} = (g_1, \ldots, g_m) = (g_1(x), \ldots, g_m(x))$, $g_i \in Y$, are linearly independent; the m components of the vector $\Psi = \mathrm{col}(\Psi_1, \ldots, \Psi_m)$ are linear functionals, $\Psi_i \in [X_A^n]^*$ and each $\Psi_i(u)$ may contain the values of the function $u(x)$ and its up to nth-order derivatives at the m_1 fixed and ordered loading points $a \leq \check{x}_1 < \check{x}_2 < \cdots < \check{x}_{m_1} \leq b$. By means of (3) and (5), Eq. (1) is written as

$$Au - \mathbf{g}\Psi(u) = f, \tag{6}$$

where $f = f(x) \in Y$.

Likewise, the boundary conditions (2) may be recast meaningfully in terms of matrices as

$$\mathbf{M}\Phi(u) + \mathbf{N}\Theta(u) = \mathbf{b}, \tag{7}$$

where the n components of the vector $\Phi = \mathrm{col}(\Phi_1, \ldots, \Phi_n)$ and the l components of the vector $\Theta = \mathrm{col}(\Theta_1, \ldots, \Theta_l)$, $l \leq (l_1 + 1) * n$, are linear functionals, $\Phi_1, \ldots, \Phi_n, \Theta_1, \ldots, \Theta_l \in [X_A^{n-1}]^*$, $\mathbf{M}$ and $\mathbf{N}$ are respectively $n \times n$ and $n \times l$ constant matrices and the constant vector $\mathbf{b} = \mathrm{col}(\beta_1, \ldots, \beta_n)$. The functionals Φ_i, $i = 1, \ldots, n$, are chosen such that to formulate an initial or boundary value problem

$$Au = f, \quad \Phi(u) = \mathbf{0}, \tag{8}$$

which can be solved uniquely for any $f \in Y$. The matrix $\mathbf{M}$ may be the identity matrix, any other constant matrix, or even the zero matrix; see the example problems in Sect. 4.

As a consequence, we define the correct operator

$$\begin{gathered} \widehat{A}u = Au, \\ D(\widehat{A}) = \{u : u \in D(A),\ \Phi(u) = \mathbf{0}\} \end{gathered} \tag{9}$$

to be a restriction of A, viz. $\widehat{A} \subset A$. We recall here that an operator $\widehat{A} : X \to Y$ is called *correct* if $R(\widehat{A}) = Y$ and the inverse operator $\widehat{A}^{-1}$ exists and it is continuous on Y.

Finally, we define the operator $B : X \to Y$ by

$$\begin{aligned} Bu &= Au - \mathbf{g}\Psi(u), \\ D(B) &= \{u : u \in D(A),\ \mathbf{M}\Phi(u) + \mathbf{N}\Theta(u) = \mathbf{b}\}. \end{aligned} \tag{10}$$

Thus, the boundary value problem (1) and (2) may be written compactly in operator form as

$$Bu = Au - \mathbf{g}\Psi(u) = f, \quad f \in Y. \tag{11}$$

It is understood that throughout the chapter, bold face lower case Latin letters like $\mathbf{z} = (z_1, \ldots, z_n)$ and capital Greek letters as Ψ denote vectors, whereas bold face capital Latin letters such as $\mathbf{M}$ symbolize matrices. We also use the notations $\Psi(u)$ and $\Psi(\mathbf{z})$ to indicate the vector and matrix,

$$\Psi(u) = \begin{pmatrix} \Psi_1(u) \\ \vdots \\ \Psi_m(u) \end{pmatrix}, \quad \Psi(\mathbf{z}) = \begin{bmatrix} \Psi_1(z_1) & \cdots & \Psi_1(z_n) \\ \vdots & \ddots & \vdots \\ \Psi_m(z_1) & \cdots & \Psi_m(z_n) \end{bmatrix},$$

respectively. The zero vector is signified by $\mathbf{0}$, while the $n \times n$ zero and unit matrix by $\mathbf{0}_n$ and $\mathbf{I}_n$, respectively.

3 Main Results

We first consider the general boundary value problem (11) with homogeneous boundary conditions, namely

$$B_0 u = Au - \mathbf{g}\Psi(u) = f, \quad f \in Y, \tag{12}$$

where the operator $B_0 : X \to Y$ is defined by

$$\begin{aligned} B_0 u &= Au - \mathbf{g}\Psi(u), \\ D(B_0) &= \{u : u \in D(A),\ \mathbf{M}\Phi(u) + \mathbf{N}\Theta(u) = \mathbf{0}\}. \end{aligned} \tag{13}$$

We assume that a fundamental set of solutions $z_1, \dots, z_n$ of the homogeneous problem $Au = 0$ is known and that they are biorthogonal to functionals $\Phi_1, \dots, \Phi_n$, i.e. $\Phi_i(z_j) = \delta_{ij}$, where δ_{ij} is the Kronecker delta.

Theorem 1 *Let X, Y be complex Banach spaces, $A : X \to Y$ a linear operator defined by (3) and $\widehat{A}$ a correct restriction of A defined by (9). Further, let the components of the vector $\mathbf{z} = (z_1, \dots, z_n)$ constitute a basis of* $\ker A$ *and that $\Phi(\mathbf{z}) = \mathbf{I}_n$. Then:*

(i) The operator B_0 defined in (13) is injective if and only if

$$\det \mathbf{V} = \det \begin{bmatrix} \mathbf{I}_m - \Psi(\widehat{A}^{-1}\mathbf{g}) & -\Psi(\mathbf{z}) \\ \mathbf{N}\Theta(\widehat{A}^{-1}\mathbf{g}) & \mathbf{M} + \mathbf{N}\Theta(\mathbf{z}) \end{bmatrix} \neq 0. \tag{14}$$

(ii) If (i) is true, then B_0 is correct, and for all $f \in Y$, the unique solution to boundary value problem (12) is given by

$$\begin{aligned} u &= B_0^{-1} f \\ &= \widehat{A}^{-1} f + \left(\widehat{A}^{-1}\mathbf{g}\ \mathbf{z}\right) \mathbf{V}^{-1} \begin{pmatrix} \Psi(\widehat{A}^{-1} f) \\ -\mathbf{N}\Theta(\widehat{A}^{-1} f) \end{pmatrix}. \end{aligned} \tag{15}$$

Proof

(i) Suppose $\det \mathbf{V} \neq 0$ and we will show that the operator B_0 is injective, or equivalently $\ker B_0 = \{0\}$, which is to say the complete homogeneous boundary value problem

$$B_0 u = Au - \mathbf{g}\Psi(u) = 0, \quad \mathbf{M}\Phi(u) + \mathbf{N}\Theta(u) = \mathbf{0}, \tag{16}$$

possesses precisely one solution $u = 0$. Because $\mathbf{z} \in [\ker A]^n$ and $\Phi(\mathbf{z}) = \mathbf{I}_n$, and by noticing that $A(u - \mathbf{z}\Phi(u)) = Au - A\mathbf{z}\Phi(u) = Au$ and $\Phi(u - \mathbf{z}\Phi(u)) = \Phi(u) - \Phi(\mathbf{z})\Phi(u) = \mathbf{0}$, it follows that for every $u \in D(B_0)$, the element $u - \mathbf{z}\Phi(u) \in D(A)$ and moreover $u - \mathbf{z}\Phi(u) \in D(\widehat{A})$. Therefore, the first equation in (16) can be written as

$$\begin{aligned} B_0 u &= A\left(u - \mathbf{z}\Phi(u)\right) - \mathbf{g}\Psi(u) \\ &= \widehat{A}\left(u - \mathbf{z}\Phi(u)\right) - \mathbf{g}\Psi(u) = 0. \end{aligned} \tag{17}$$

Since $\widehat{A}$ is correct, there exists the inverse $\widehat{A}^{-1}$. Multiplying by $\widehat{A}^{-1}$ both sides of (17), we get

$$u - \mathbf{z}\Phi(u) - \widehat{A}^{-1}\mathbf{g}\Psi(u) = 0. \tag{18}$$

Acting by the vector Ψ on both sides of (18), we obtain successively

$$\begin{aligned} &\Psi(u) - \Psi(\mathbf{z})\Phi(u) - \Psi(\widehat{A}^{-1}\mathbf{g})\Psi(u) = \mathbf{0}, \\ &\left[\mathbf{I}_m - \Psi(\widehat{A}^{-1}\mathbf{g})\right]\Psi(u) - \Psi(\mathbf{z})\Phi(u) = \mathbf{0}. \end{aligned} \tag{19}$$

Solving (18) with respect to u and then substituting into the second equation in (16) and taking into account that the operator $\widehat{A}$ is correct, we get

$$\mathbf{M}\Phi(u) + \mathbf{N}\Theta(\mathbf{z})\Phi(u) + \mathbf{N}\Theta(\widehat{A}^{-1}\mathbf{g})\Psi(u) = \mathbf{0}. \tag{20}$$

Combining (19) and (20), we have the system

$$\begin{bmatrix} \mathbf{I}_m - \Psi(\widehat{A}^{-1}\mathbf{g}) & -\Psi(\mathbf{z}) \\ \mathbf{N}\Theta(\widehat{A}^{-1}\mathbf{g}) & \mathbf{M} + \mathbf{N}\Theta(\mathbf{z}) \end{bmatrix} \begin{pmatrix} \Psi(u) \\ \Phi(u) \end{pmatrix} = \mathbf{V} \begin{pmatrix} \Psi(u) \\ \Phi(u) \end{pmatrix} = \begin{pmatrix} \mathbf{0} \\ \mathbf{0} \end{pmatrix}. \tag{21}$$

Since $\det \mathbf{V} \neq 0$, it is implied that $\Psi(u) = \mathbf{0}$, $\Phi(u) = \mathbf{0}$ and so $u = 0$ by (18). Thus, $\ker B_0 = \{0\}$, and hence, the operator B_0 is injective.

Conversely, assume that B_0 is injective and we will prove that $\det \mathbf{V} \neq 0$, or equivalently let $\det \mathbf{V} = 0$ and we will show that B_0 is not injective. Since $\det \mathbf{V} = 0$, there exists a vector of constants $\mathbf{c} = \mathrm{col}(\mathbf{c}_1, \mathbf{c}_2)$, where $\mathbf{c}_1 = \mathrm{col}(c_{11}, \ldots, c_{1m})$ and $\mathbf{c}_2 = \mathrm{col}(c_{21}, \ldots, c_{2n})$, such that

$$\mathbf{V}\mathbf{c} = \begin{bmatrix} \mathbf{I}_m - \Psi(\widehat{A}^{-1}\mathbf{g}) & -\Psi(\mathbf{z}) \\ \mathbf{N}\Theta(\widehat{A}^{-1}\mathbf{g}) & \mathbf{M} + \mathbf{N}\Theta(\mathbf{z}) \end{bmatrix} \begin{pmatrix} \mathbf{c}_1 \\ \mathbf{c}_2 \end{pmatrix} = \begin{pmatrix} \mathbf{0} \\ \mathbf{0} \end{pmatrix}. \tag{22}$$

Consider the element $u_0 = \widehat{A}^{-1}\mathbf{g}\mathbf{c}_1 + \mathbf{z}\mathbf{c}_2 \in D(A)$ and observe that $u_0 \neq 0$ since the components of the vectors $\mathbf{g}$ and $\mathbf{z}$ are linearly independent. Furthermore, notice that

$$
\begin{aligned}
\mathbf{M}\Phi(u_0) + \mathbf{N}\Theta(u_0) &= \mathbf{M}\Phi(\widehat{A}^{-1}\mathbf{g}\mathbf{c}_1 + \mathbf{z}\mathbf{c}_2) + \mathbf{N}\Theta(\widehat{A}^{-1}\mathbf{g}\mathbf{c}_1 + \mathbf{z}\mathbf{c}_2) \\
&= \mathbf{N}\Theta(\widehat{A}^{-1}\mathbf{g})\mathbf{c}_1 + (\mathbf{M} + \mathbf{N}\Theta(\mathbf{z}))\,\mathbf{c}_2 \\
&= \mathbf{0},
\end{aligned}
\tag{23}
$$

by Eq. (22). Thus, $u_0 \in D(B_0)$, and therefore,

$$
\begin{aligned}
B_0 u_0 &= A u_0 - \mathbf{g}\Psi(u_0) \\
&= A(\widehat{A}^{-1}\mathbf{g}\mathbf{c}_1 + \mathbf{z}\mathbf{c}_2) - \mathbf{g}\Psi(\widehat{A}^{-1}\mathbf{g}\mathbf{c}_1 + \mathbf{z}\mathbf{c}_2) \\
&= \mathbf{g}\mathbf{c}_1 - \mathbf{g}\Psi(\widehat{A}^{-1}\mathbf{g}\mathbf{c}_1 + \mathbf{z}\mathbf{c}_2) \\
&= \mathbf{g}\left[\left(\mathbf{I}_m - \Psi(\widehat{A}^{-1}\mathbf{g})\right)\mathbf{c}_1 - \Psi(\mathbf{z})\mathbf{c}_2\right] \\
&= \mathbf{g}\mathbf{0} = 0,
\end{aligned}
\tag{24}
$$

by using Eq. (22). Hence, $u_0 \in \ker B_0$ and consequently $\ker B_0 \neq \{0\}$ which means that the operator B_0 is not injective.

(ii) Let $\det \mathbf{V} \neq 0$. Consider the problem $B_0 u = f$, or explicitly

$$
B_0 u = Au - \mathbf{g}\Psi(u) = f, \quad f \in Y, \quad \mathbf{M}\Phi(u) + \mathbf{N}\Theta(u) = \mathbf{0}. \tag{25}
$$

Following the same procedure as in (i), we have

$$
u - \mathbf{z}\Phi(u) - \widehat{A}^{-1}\mathbf{g}\Psi(u) = \widehat{A}^{-1} f, \tag{26}
$$

and then

$$
\left[\mathbf{I}_m - \Psi(\widehat{A}^{-1}\mathbf{g})\right]\Psi(u) - \Psi(\mathbf{z})\Phi(u) = \Psi(\widehat{A}^{-1} f). \tag{27}
$$

Also, substitution of (26) into the second equation of (25) yields

$$
\mathbf{N}\Theta(\widehat{A}^{-1}\mathbf{g})\Psi(u) + [\mathbf{M} + \mathbf{N}\Theta(\mathbf{z})]\,\Phi(u) = -\mathbf{N}\Theta(\widehat{A}^{-1} f). \tag{28}
$$

From (27) and (28), we obtain the system

$$
\mathbf{V}\begin{pmatrix} \Psi(u) \\ \Phi(u) \end{pmatrix} = \begin{pmatrix} \Psi(\widehat{A}^{-1} f) \\ -\mathbf{N}\Theta(\widehat{A}^{-1} f) \end{pmatrix}, \tag{29}
$$

where the matrix $\mathbf{V}$ is as in (21). Inverting (29) and substituting into (26), we get the solution formula (15).

Because f in (15) is arbitrary, we have $R(B_0) = Y$. Since the operator $\widehat{A}^{-1}$ and the functionals $\Psi_1, \ldots, \Psi_m$, $\Theta_1, \ldots, \Theta_l$ involved in (15) are bounded, it is implied that B_0^{-1} is bounded too. Hence, the operator B_0 is correct if and only if (14) holds, and in this case, the unique solution to the boundary value problem (12) is given explicitly by (15). The theorem is proved. □

Next, we look into the general boundary value problem (11) with nonhomogeneous boundary conditions, namely

$$
\begin{aligned}
Bu &= Au - \mathbf{g}\Psi(u) = f, \\
D(B) &= \{u : u \in D(A),\ \mathbf{M}\Phi(u) + \mathbf{N}\Theta(u) = \mathbf{b}\}.
\end{aligned}
\tag{30}
$$

We prove the following theorem.

Theorem 2 *Let X, Y be complex Banach spaces, $A : X \to Y$ a linear operator defined by (3) and $\widehat{A}$ a correct restriction of A defined by (9). Further, let the components of the vector $\mathbf{z} = (z_1, \ldots, z_n)$ constitute a basis of* $\ker A$ *and that $\Phi(\mathbf{z}) = \mathbf{I}_n$. Then:*

(i) The operator B defined in (30) is injective if and only if

$$
\det \mathbf{V} = \det \begin{bmatrix} \mathbf{I}_m - \Psi(\widehat{A}^{-1}\mathbf{g}) & -\Psi(\mathbf{z}) \\ \mathbf{N}\Theta(\widehat{A}^{-1}\mathbf{g}) & \mathbf{M} + \mathbf{N}\Theta(\mathbf{z}) \end{bmatrix} \neq 0. \tag{31}
$$

(ii) Under (i) the homogeneous in action but with nonhomogeneous conditions boundary value problem,

$$
Bu = 0, \tag{32}
$$

has a unique solution

$$
u = \left(\widehat{A}^{-1}\mathbf{g}\ \mathbf{z}\right)\mathbf{V}^{-1}\begin{pmatrix}\mathbf{0} \\ \mathbf{b}\end{pmatrix}. \tag{33}
$$

(iii) Under (i) the operator B is correct and the unique solution to the complete nonhomogeneous boundary value problem,

$$
Bu = f, \quad \forall f \in Y, \tag{34}
$$

is given by

$$
u = \widehat{A}^{-1}f + \left(\widehat{A}^{-1}\mathbf{g}\ \mathbf{z}\right)\mathbf{V}^{-1}\begin{pmatrix}\Psi(\widehat{A}^{-1}f) \\ \mathbf{b} - \mathbf{N}\Theta(\widehat{A}^{-1}f)\end{pmatrix}. \tag{35}
$$

Proof

(i) Let $\det \mathbf{V} \neq 0$ and we will prove that the operator B is injective. Suppose there exist $u_1, u_2 \in D(B)$ such as $Bu_1 = Bu_2$. Then

$$Au_1 - \mathbf{g}\Psi(u_1) = Au_2 - \mathbf{g}\Psi(u_2), \quad \mathbf{M}\Phi(u_1) + \mathbf{N}\Theta(u_1) = \mathbf{b},$$
$$\mathbf{M}\Phi(u_2) + \mathbf{N}\Theta(u_2) = \mathbf{b}. \tag{36}$$

By subtracting, we get

$$A(u_1 - u_2) - \mathbf{g}\Psi(u_1 - u_2) = 0, \quad \mathbf{M}\Phi(u_1 - u_2) + \mathbf{N}\Theta(u_1 - u_2) = \mathbf{0}. \tag{37}$$

Setting $v = u_1 - u_2$, we obtain

$$Av - \mathbf{g}\Psi(v) = 0, \quad \mathbf{M}\Phi(v) + \mathbf{N}\Theta(v) = \mathbf{0}, \tag{38}$$

and hence

$$B_0 v = 0, \tag{39}$$

by (13). Then from Theorem 1 follows that B_0 is injective and so $v = 0$. Subsequently, $u_1 = u_2$, and hence, the operator B is injective.

Conversely, we assume B is injective and we will show that $\det \mathbf{V} \neq 0$, or equivalently let $\det \mathbf{V} = 0$ and we will prove that B is not injective. Suppose there exist $u_1, u_2 \in D(B)$ with $Bu_1 = Bu_2$. Repeating the same sequence of operations as above, we obtain (39). Then from Theorem 1 follows that the operator B_0 is not injective and so the complete homogeneous equation $B_0 v = 0$ has a nonzero solution $v = u_1 - u_2 \neq 0$. That is, $u_1 \neq u_2$, and therefore, the operator B is not injective.

(ii) From (32), we have

$$Bu = Au - \mathbf{g}\Psi(u) = A\left(u - \widehat{A}^{-1}\mathbf{g}\Psi(u)\right) = 0, \quad \mathbf{M}\Phi(u) + \mathbf{N}\Theta(u) = \mathbf{b}. \tag{40}$$

We recall that the complementary solution to the homogeneous equation $Au = 0$ is given by $u = \mathbf{zc}$, where the components of $\mathbf{c} = \mathrm{col}(c_1, \ldots, c_n)$ are arbitrary constants, and therefore

$$u - \widehat{A}^{-1}\mathbf{g}\Psi(u) = \mathbf{zc}. \tag{41}$$

By applying the vector Ψ on the both sides of (41), we have

$$\left[\mathbf{I}_m - \Psi(\widehat{A}^{-1}\mathbf{g})\right]\Psi(u) - \Psi(\mathbf{z})\mathbf{c} = \mathbf{0}. \tag{42}$$

Moreover, solving (41) with respect to u and then substituting into the second equation of (40), we acquire

$$\mathbf{M}\Phi\left(\mathbf{zc} + \widehat{A}^{-1}\mathbf{g}\Psi(u)\right) + \mathbf{N}\Theta\left(\mathbf{zc} + \widehat{A}^{-1}\mathbf{g}\Psi(u)\right) = \mathbf{b}, \tag{43}$$

or, taking into account that $\Phi(\mathbf{z}) = \mathbf{I}_n$ and $\Phi(\widehat{A}^{-1}\mathbf{g}) = \mathbf{0}_n$,

$$\mathbf{N}\Theta(\widehat{A}^{-1}\mathbf{g})\Psi(u) + [\mathbf{M} + \mathbf{N}\Theta(\mathbf{z})]\,\mathbf{c} = \mathbf{b}. \tag{44}$$

From (42) and (44), we have the system made up of the $m + n$ equations

$$\begin{bmatrix} \mathbf{I}_m - \Psi(\widehat{A}^{-1}\mathbf{g}) & -\Psi(\mathbf{z}) \\ \mathbf{N}\Theta(\widehat{A}^{-1}\mathbf{g}) & \mathbf{M} + \mathbf{N}\Theta(\mathbf{z}) \end{bmatrix} \begin{pmatrix} \Psi(u) \\ \mathbf{c} \end{pmatrix} = \mathbf{V} \begin{pmatrix} \Psi(u) \\ \mathbf{c} \end{pmatrix} = \begin{pmatrix} \mathbf{0} \\ \mathbf{b} \end{pmatrix}. \tag{45}$$

Since $\det \mathbf{V} \neq 0$, Eq. (45) can be inverted to get

$$\begin{pmatrix} \Psi(u) \\ \mathbf{c} \end{pmatrix} = \mathbf{V}^{-1} \begin{pmatrix} \mathbf{0} \\ \mathbf{b} \end{pmatrix}. \tag{46}$$

Writing (41) in matrix form and substituting (46), we obtain the solution formula (33).

(iii) By the principle of superposition, the solution of the completely nonhomogeneous boundary value problem (34) can be constructed as the sum $u = v + w$, where v and w are solutions of the boundary value problems

$$Av - \mathbf{g}\Psi(v) = f, \quad \mathbf{M}\Phi(v) + \mathbf{N}\Theta(v) = \mathbf{0}, \quad \text{or} \quad B_0 v = f, \tag{47}$$

and

$$Aw - g\Psi(w) = 0, \quad \mathbf{M}\Phi(w) + \mathbf{N}\Theta(w) = \mathbf{b}, \quad \text{or} \quad Bw = 0, \tag{48}$$

respectively. Thus, using Theorem 1 and in particular (15) and (33), we obtain the solution formula (35).

Finally, the correctness of the operator B follows by the same arguments as for the operator B_0 in Theorem 1, i.e. because f in (35) is arbitrary, and hence $R(B_0) = Y$, and the operator $\widehat{A}^{-1}$ as well as the functionals $\Psi_1, \ldots, \Psi_m, \Theta_1, \ldots, \Theta_l$ involved are bounded, it is implied that B^{-1} is bounded too.

□

4 Applications

In this section, we implement the method presented in the previous section to solve some characteristic model problems which have been appeared in the literature.

4.1 Three-Point BVP with Nonhomogeneous Boundary Conditions

We begin by letting the following boundary value problem for a second-order integro-differential equation coupled with nonhomogeneous Dirichlet type boundary conditions,

$$\begin{aligned} &u''(x) + \lambda \int_0^1 u(t)dt = 0, \quad \lambda = \text{constant}, \quad 0 < x < 1, \\ &u(0) = u_0, \quad u(1) = u_1, \end{aligned} \tag{49}$$

proposed by Nakhushev [18]. If the integral term is replaced by

$$\int_0^1 u(t)dt \approx \frac{u(0) + u(1)}{6} + \frac{2}{3}u(\frac{1}{2}),$$

according to Simpson's integration rule, the problem (49) degenerates to the three-point loaded differential boundary value problem

$$\begin{aligned} &u''(x) + \frac{\lambda}{6}\left[u(0) + 4u(\frac{1}{2}) + u(1)\right] = 0, \quad \lambda = \text{constant}, \quad 0 < x < 1, \\ &u(0) = u_0, \\ &u(1) = u_1. \end{aligned} \tag{50}$$

To find the exact solution to problem (50), we apply Theorem 2. Thus, we take $X = Y = C[0, 1]$ and $X_A^2 = C^2[0, 1]$ and put the problem (50) in the form

$$Bu = Au - \mathbf{g}\Psi(u) = 0, \quad D(B) = \{u : u \in D(A),\ \mathbf{M}\Phi(u) + \mathbf{N}\Theta(u) = \mathbf{b}\}, \tag{51}$$

where

$$\begin{aligned} Au &= u''(x), \quad D(A) = \{u : u(x) \in X_A^2\}, \\ \mathbf{g} &= \left(g_1\right) = \left(-\tfrac{\lambda}{6}\right), \\ \Psi(u) &= \left(\Psi_1(u)\right) = \left(u(0) + 4u(\tfrac{1}{2}) + u(1)\right), \\ \widehat{A}u &= Au, \quad D(\widehat{A}) = \{u : u(x) \in D(A),\ \Phi(u) = \mathbf{0}\}, \\ \Phi(u) &= \begin{pmatrix} \Phi_1(u) \\ \Phi_2(u) \end{pmatrix} = \begin{pmatrix} u(0) \\ u'(0) \end{pmatrix}, \\ \Theta(u) &= \left(\Theta_1(u)\right) = \left(u(1)\right), \end{aligned}$$

$$\mathbf{M} = \begin{bmatrix} 1 & 0 \\ 0 & 0 \end{bmatrix},$$
$$\mathbf{N} = \begin{bmatrix} 0 \\ 1 \end{bmatrix},$$
$$\mathbf{b} = \begin{bmatrix} u_0 \\ u_1 \end{bmatrix}. \tag{52}$$

Since for obvious reasons $\lambda \neq 0$, the requirement for linear independence of g_1 is fulfilled. It is easy to verify that $\mathbf{z} = (1,\ x)$ is a fundamental set of solutions to homogeneous problem $Au = 0$, $\Psi_1 \in [X_A^2]^*$, $\Phi_1, \Phi_2, \Theta_1 \in [X_{\widehat{A}}^1]^*$ and $\Phi(\mathbf{z}) = \mathbf{I}_2$. Also, it is known that the unique solution to the correct problem $\widehat{A}u = f$ is given by

$$\widehat{A}^{-1} f = \int_0^x (x-t) f(t) dt, \quad \text{for any } f \in Y. \tag{53}$$

From (52) and (53), we construct the following matrices,

$$\begin{aligned}
\widehat{A}^{-1}\mathbf{g} &= \left(\widehat{A}^{-1} g_1\right) = \left(-\tfrac{\lambda}{6} \textstyle\int_0^x (x-t)dt\right) = \left(-\tfrac{\lambda}{12}x^2\right), \\
\Psi(\widehat{A}^{-1}\mathbf{g}) &= \left[\, \Psi_1(\widehat{A}^{-1} g_1) \,\right] = \left[-\tfrac{\lambda}{6}\right], \\
\Psi(\mathbf{z}) &= \left[\, \Psi_1(z_1)\ \Psi_1(z_2) \,\right] = \left[\, 6\ 3 \,\right], \\
\Theta(\widehat{A}^{-1}\mathbf{g}) &= \left[\, \Theta_1(\widehat{A}^{-1} g_1) \,\right] = \left[-\tfrac{\lambda}{12}\right], \\
\Theta(\mathbf{z}) &= \left[\, \Theta_1(z_1)\ \Theta_1(z_2) \,\right] = \left[\, 1\ 1 \,\right],
\end{aligned} \tag{54}$$

and then the 3×3 matrix

$$\mathbf{V} = \begin{bmatrix} \mathbf{I}_1 - \Psi(\widehat{A}^{-1}\mathbf{g}) & -\Psi(\mathbf{z}) \\ \mathbf{N}\Theta(\widehat{A}^{-1}\mathbf{g}) & \mathbf{M} + \mathbf{N}\Theta(\mathbf{z}) \end{bmatrix} = \begin{bmatrix} 1 + \frac{\lambda}{6} & -6 & -3 \\ 0 & 1 & 0 \\ -\frac{\lambda}{12} & 1 & 1 \end{bmatrix}. \tag{55}$$

Notice that $\det \mathbf{V} = \frac{12-\lambda}{12}$, and therefore, the operator B is injective if $\lambda \neq 12$.

If this is the case, then the unique solution to the boundary value problem (51) and hence to (50) follows from (33), viz.

$$u(x) = \left(\widehat{A}^{-1}\mathbf{g}\ \mathbf{z}\right) \mathbf{V}^{-1} \begin{pmatrix} \mathbf{0} \\ \mathbf{b} \end{pmatrix} \tag{56}$$

$$= \left(-\tfrac{\lambda x^2}{12}\ 1\ x\right) \mathbf{V}^{-1} \begin{pmatrix} 0 \\ u_0 \\ u_1 \end{pmatrix} \tag{57}$$

$$= \frac{3\lambda}{\lambda - 12}(u_1 + u_0)x^2 - \frac{1}{\lambda - 12}\left[2\lambda(u_1 + 2u_0) + 12(u_1 - u_0)\right]x + u_0.$$

4.2 Two-Point BVP with Point and Integral Boundary Conditions

Let us consider the loaded differential operator B_0 generated by the differential operation l and the boundary conditions,

$$\begin{aligned} lu(x) &= u''(x) + \alpha(x)u(0) + \beta(x)u(1), \quad x \in (0, 1), \\ u'(0) &= 0, \\ u'(1) &= \int_0^1 \gamma(x)u(x)dx, \end{aligned} \tag{58}$$

where $\alpha(x)$, $\beta(x)$ and $\gamma(x)$ are real continuous functions on $[0, 1]$, as it is has been presented by Lomov [15]. Here we examine the existence and uniqueness of the solution to the problem $B_0 u = f$, for any $f \in C[0, 1]$, and find it in closed form.

Since the boundary conditions are homogeneous, Theorem 1 is applicable. Accordingly, we take $X = Y = C[0, 1]$, $X_A^2 = C^2[0, 1]$ and

$$\begin{aligned} Au &= u''(x), \quad D(A) = \{u : u(x) \in X_A^2\}, \\ \mathbf{g} &= \begin{pmatrix} g_1 & g_2 \end{pmatrix} = \begin{pmatrix} -\alpha(x) & -\beta(x) \end{pmatrix}, \\ \Psi(u) &= \begin{pmatrix} \Psi_1(u) \\ \Psi_2(u) \end{pmatrix} = \begin{pmatrix} u(0) \\ u(1) \end{pmatrix}, \\ \widehat{A}u &= Au, \quad D(\widehat{A}) = \{u : u(x) \in D(A), \ \Phi(u) = \mathbf{0}\}, \\ \Phi(u) &= \begin{pmatrix} \Phi_1(u) \\ \Phi_2(u) \end{pmatrix} = \begin{pmatrix} u(0) \\ u'(0) \end{pmatrix}, \\ \Theta(u) &= \begin{pmatrix} \Theta_1(u) \\ \Theta_2(u) \end{pmatrix} = \begin{pmatrix} u'(1) \\ \int_0^1 \gamma(x)u(x)dx \end{pmatrix}, \\ \mathbf{M} &= \begin{bmatrix} 0 & 1 \\ 0 & 0 \end{bmatrix}, \\ \mathbf{N} &= \begin{bmatrix} 0 & 0 \\ 1 & -1 \end{bmatrix}. \end{aligned} \tag{59}$$

We assume that $g_1 = -\alpha(x)$ and $g_2 = -\beta(x)$ are linearly independent; otherwise, we have to combine them to one function and reformulate the problem. It is easy to verify that $\mathbf{z} = (1, \ x)$ consists a fundamental set of solutions to homogeneous problem $Au = 0$, $\Psi_1, \Psi_2 \in [X_A^2]^*$, $\Phi_1, \Phi_2, \Theta_1, \Theta_2 \in [X_A^1]^*$ and $\Phi(\mathbf{z}) = \mathbf{I}_2$. Lastly, notice that the unique solution to the correct problem $\widehat{A}u = f$ is given by

$$\widehat{A}^{-1}f = \int_0^x (x-t)f(t)dt, \quad \text{for any } f \in Y. \tag{60}$$

From (59) and (60), we construct the following matrices,

$$\begin{aligned}
\widehat{A}^{-1}\mathbf{g} &= \left(\widehat{A}^{-1}g_1 \; \widehat{A}^{-1}g_2\right) \\
&= \left(-\int_0^x (x-t)\alpha(t)dt \;\; -\int_0^x (x-t)\beta(t)dt\right), \\
\Psi(\widehat{A}^{-1}\mathbf{g}) &= \begin{bmatrix} \Psi_1(\widehat{A}^{-1}g_1) & \Psi_1(\widehat{A}^{-1}g_2) \\ \Psi_2(\widehat{A}^{-1}g_1) & \Psi_2(\widehat{A}^{-1}g_2) \end{bmatrix} \\
&= \begin{bmatrix} 0 & 0 \\ -\int_0^1 (1-t)\alpha(t)dt & -\int_0^1 (1-t)\beta(t)dt \end{bmatrix}, \\
\Psi(\mathbf{z}) &= \begin{bmatrix} \Psi_1(z_1) & \Psi_1(z_2) \\ \Psi_2(z_1) & \Psi_2(z_2) \end{bmatrix} = \begin{bmatrix} 1 & 0 \\ 1 & 1 \end{bmatrix}, \\
\Theta(\widehat{A}^{-1}\mathbf{g}) &= \begin{bmatrix} \Theta_1(\widehat{A}^{-1}g_1) & \Theta_1(\widehat{A}^{-1}g_2) \\ \Theta_2(\widehat{A}^{-1}g_1) & \Theta_2(\widehat{A}^{-1}g_2) \end{bmatrix} \\
&= \begin{bmatrix} -\int_0^1 \alpha(t)dt & -\int_0^1 \beta(t)dt \\ -\int_0^1 \gamma(x)\int_0^x (x-t)\alpha(t)dtdx & -\int_0^1 \gamma(x)\int_0^x (x-t)\beta(t)dtdx \end{bmatrix}, \\
\Theta(\mathbf{z}) &= \begin{bmatrix} \Theta_1(z_1) & \Theta_1(z_2) \\ \Theta_2(z_1) & \Theta_2(z_2) \end{bmatrix} \\
&= \begin{bmatrix} 0 & 1 \\ \int_0^1 \gamma(x)dx & \int_0^1 x\gamma(x)dx \end{bmatrix},
\end{aligned} \tag{61}$$

and eventually the 4×4 matrix

$$\mathbf{V} = \begin{bmatrix} \mathbf{I}_2 - \Psi(\widehat{A}^{-1}\mathbf{g}) & -\Psi(\mathbf{z}) \\ \mathbf{N}\Theta(\widehat{A}^{-1}\mathbf{g}) & \mathbf{M} + \mathbf{N}\Theta(\mathbf{z}) \end{bmatrix}. \tag{62}$$

If $\det \mathbf{V} \neq 0$, then the given boundary value problem $B_0 u = f$ has a unique solution.

In this case, by computing the inverse matrix $\mathbf{V}^{-1}$ and the vectors

$$\Psi(\widehat{A}^{-1}f) = \begin{pmatrix} \Psi_1(\widehat{A}^{-1}f) \\ \Psi_2(\widehat{A}^{-1}f) \end{pmatrix} = \begin{pmatrix} 0 \\ \int_0^1 (1-t)f(t)dt \end{pmatrix},$$

$$\Theta(\widehat{A}^{-1}f) = \begin{pmatrix} \Theta_1(\widehat{A}^{-1}f) \\ \Theta_2(\widehat{A}^{-1}f) \end{pmatrix} = \begin{pmatrix} \int_0^1 f(t)dt \\ \int_0^1 \gamma(x) \int_0^x (x-t) f(t)dtdx \end{pmatrix}, \tag{63}$$

and substituting, along with (60), the vector $\widehat{A}^{-1}\mathbf{g}$ in (61), the vector $\mathbf{z}$ and the matrix $\mathbf{N}$, into (15), we obtain the unique solution in closed form of the boundary value problem $B_0 u = f$ generated by (58) for an input function $f \in C[0, 1]$.

4.3 Four-Point BVP with Integral Boundary Conditions

From [15], consider the next loaded differential operator B_0 generated by the differential operation l and the homogeneous integral boundary conditions,

$$\begin{aligned} lu(x) &= u''(x) + \alpha(x)u(\frac{1}{4}) + \beta(x)u(\frac{1}{2}) + u(1), \quad x \in (0, 1), \\ u(0) &= \int_0^1 \gamma(x)u(x)dx, \\ u'(1) &= \int_0^1 \nu(x)u(x)dx, \end{aligned} \tag{64}$$

where $\alpha(x)$, $\beta(x)$, $\gamma(x)$ and $\nu(x)$ are real continuous functions on $[0, 1]$. Observe that the differential operation l encompasses loaded terms with values of the unknown function $u(x)$ at two interior points and one boundary point, specifically $\check{x}_1 = \frac{1}{4}$, $\check{x}_2 = \frac{1}{2}$ and $\check{x}_3 = 1$. We establish solvability conditions and find the unique solution to the problem $B_0 u = f$, for any $f \in C[0, 1]$.

Theorem 1 is applicable. Let $X = Y = C[0, 1]$, $X_A^2 = C^2[0, 1]$ and

$$\begin{aligned} Au &= u''(x), \quad D(A) = \{u : u(x) \in X_A^2\}, \\ \mathbf{g} &= \begin{pmatrix} g_1 & g_2 & g_3 \end{pmatrix} = \begin{pmatrix} -\alpha(x) & -\beta(x) & -1 \end{pmatrix}, \\ \Psi(u) &= \begin{pmatrix} \Psi_1(u) \\ \Psi_2(u) \\ \Psi_3(u) \end{pmatrix} = \begin{pmatrix} u(\frac{1}{4}) \\ u(\frac{1}{2}) \\ u(1) \end{pmatrix}, \\ \widehat{A}u &= Au, \quad D(\widehat{A}) = \{u : u(x) \in D(A), \ \Phi(u) = \mathbf{0}\}, \\ \Phi(u) &= \begin{pmatrix} \Phi_1(u) \\ \Phi_2(u) \end{pmatrix} = \begin{pmatrix} u(0) \\ u'(0) \end{pmatrix}, \\ \Theta(u) &= \begin{pmatrix} \Theta_1(u) \\ \Theta_2(u) \\ \Theta_3(u) \end{pmatrix} = \begin{pmatrix} \int_0^1 \gamma(x)u(x)dx \\ u'(1) \\ \int_0^1 \nu(x)u(x)dx \end{pmatrix}, \end{aligned}$$

$$\mathbf{M} = \begin{bmatrix} 1 & 0 \\ 0 & 0 \end{bmatrix},$$

$$\mathbf{N} = \begin{bmatrix} -1 & 0 & 0 \\ 0 & 1 & -1 \end{bmatrix}. \tag{65}$$

It is assumed that $g_1 = -\alpha(x),\ g_2 = -\beta(x)$ and $g_3 = -1$ are linearly independent; otherwise, we have to reduce them in number and reformulate the problem. The vector $\mathbf{z} = (1,\ x)$ is a fundamental set of solutions to homogeneous problem $Au = 0$, $\Psi_1,\ \Psi_2, \Psi_3 \in [X_A^2]^*,\ \Phi_1,\ \Phi_2,\ \Theta_1,\ \Theta_2,\ \Theta_3 \in [X_{\widehat{A}}^1]^*$ and $\Phi(\mathbf{z}) = \mathbf{I}_2$. Finally, recall that the unique solution to the correct problem $\widehat{A}u = f$ is given by

$$\widehat{A}^{-1}f = \int_0^x (x-t)f(t)dt, \quad \text{for any } f \in Y. \tag{66}$$

Using (65) and (66), we put up the following matrices,

$$\begin{aligned}
\widehat{A}^{-1}\mathbf{g} &= \left(\widehat{A}^{-1}g_1\ \widehat{A}^{-1}g_2\ \widehat{A}^{-1}g_3\right) \\
&= \left(-\int_0^x (x-t)\alpha(t)dt\ \ -\int_0^x (x-t)\beta(t)dt\ \ -\int_0^x (x-t)dt\right),
\end{aligned}$$

$$\Psi(\widehat{A}^{-1}\mathbf{g}) = \begin{bmatrix} \Psi_1(\widehat{A}^{-1}g_1) & \Psi_1(\widehat{A}^{-1}g_2) & \Psi_1(\widehat{A}^{-1}g_3) \\ \Psi_2(\widehat{A}^{-1}g_1) & \Psi_2(\widehat{A}^{-1}g_2) & \Psi_2(\widehat{A}^{-1}g_3) \\ \Psi_3(\widehat{A}^{-1}g_1) & \Psi_3(\widehat{A}^{-1}g_2) & \Psi_3(\widehat{A}^{-1}g_3) \end{bmatrix},$$

$$\Psi(\mathbf{z}) = \begin{bmatrix} \Psi_1(z_1) & \Psi_1(z_2) \\ \Psi_2(z_1) & \Psi_2(z_2) \\ \Psi_3(z_1) & \Psi_3(z_2) \end{bmatrix} = \begin{bmatrix} 1 & \frac{1}{4} \\ 1 & \frac{1}{2} \\ 1 & 1 \end{bmatrix},$$

$$\Theta(\widehat{A}^{-1}\mathbf{g}) = \begin{bmatrix} \Theta_1(\widehat{A}^{-1}g_1) & \Theta_1(\widehat{A}^{-1}g_2) & \Theta_1(\widehat{A}^{-1}g_3) \\ \Theta_2(\widehat{A}^{-1}g_1) & \Theta_2(\widehat{A}^{-1}g_2) & \Theta_2(\widehat{A}^{-1}g_3) \\ \Theta_3(\widehat{A}^{-1}g_1) & \Theta_3(\widehat{A}^{-1}g_2) & \Theta_3(\widehat{A}^{-1}g_3) \end{bmatrix},$$

$$\Theta(\mathbf{z}) = \begin{bmatrix} \Theta_1(z_1) & \Theta_1(z_2) \\ \Theta_2(z_1) & \Theta_2(z_2) \\ \Theta_3(z_1) & \Theta_3(z_2) \end{bmatrix} = \begin{bmatrix} \int_0^1 \gamma(x)dx & \int_0^1 x\gamma(x)dx \\ 0 & 1 \\ \int_0^1 \nu(x)dx & \int_0^1 x\nu(x)dx \end{bmatrix}, \tag{67}$$

and hence the 5×5 matrix

$$\mathbf{V} = \begin{bmatrix} \mathbf{I}_3 - \Psi(\widehat{A}^{-1}\mathbf{g}) & -\Psi(\mathbf{z}) \\ \mathbf{N}\Theta(\widehat{A}^{-1}\mathbf{g}) & \mathbf{M} + \mathbf{N}\Theta(\mathbf{z}) \end{bmatrix}. \tag{68}$$

If $\det \mathbf{V} \neq 0$, then the given boundary value problem $B_0 u = f$ has a unique solution.

To obtain the exact solution, we substitute (66), the vector $\widehat{A}^{-1}\mathbf{g}$ in (67), the vector $\mathbf{z}$, the inverse matrix $\mathbf{V}^{-1}$, the matrix $\mathbf{N}$ and the vectors

$$\Psi(\widehat{A}^{-1}f) = \begin{pmatrix} \Psi_1(\widehat{A}^{-1}f) \\ \Psi_2(\widehat{A}^{-1}f) \\ \Psi_3(\widehat{A}^{-1}f) \end{pmatrix} = \begin{pmatrix} \int_0^{1/4}(\frac{1}{4}-t)f(t)dt \\ \int_0^{1/2}(\frac{1}{2}-t)f(t)dt \\ \int_0^1(1-t)f(t)dt \end{pmatrix},$$

$$\Theta(\widehat{A}^{-1}f) = \begin{pmatrix} \Theta_1(\widehat{A}^{-1}f) \\ \Theta_2(\widehat{A}^{-1}f) \\ \Theta_3(\widehat{A}^{-1}f) \end{pmatrix} = \begin{pmatrix} \int_0^1 \gamma(x)\int_0^x(x-t)f(t)dtdx \\ \int_0^1 f(t)dt \\ \int_0^1 \nu(x)\int_0^x(x-t)f(t)dtdx \end{pmatrix}, \quad (69)$$

into formula (15).

5 Conclusions

Loaded differential equations appear in sciences, engineering, applied mathematics and economics. Recently, there has been a renewed interest in the study of their properties and solutions.

In this chapter, we have derived ready-to-use formulae for solving exactly general boundary value problems consisting of a *loaded* nth-order ordinary differential equation and nonlocal boundary conditions which may incorporate a finite number of interior points and integrals of the unknown function and its derivatives. The technique presented requires the knowledge of a set of fundamental solutions of the corresponding homogeneous differential equation and the exact solution of a simpler auxiliary correct problem such as a Cauchy problem.

The execution and the capability of the process have been shown by considering several boundary value problems and obtaining their solutions in closed form.

References

1. V.M. Abdullaev, K.R. Aida-Zade, *Numerical method of solution to loaded nonlocal boundary value problems for ordinary differential equations*, Comput. Math. Math. Phys., **54** (2014), 1096–1109.
2. A.A. Alikhanov, A.M. Berezgov, M.X. Shkhanukov-Lafishev, *Boundary value problems for certain classes of loaded differential equations and solving them by finite difference methods*, Comput. Math. and Math. Phys., **48** (2008), 1581–1590. https://doi.org/10.1134/S096554250809008X
3. A.T. Assanova, A.E. Imanchiyev, Z.M. Kadirbayeva, *Numerical solution of systems of loaded ordinary differential equations with multipoint conditions*, Comput. Math. and Math. Phys., **58** (2018), 508–516. https://doi.org/10.1134/S096554251804005X
4. V.R. Barseghyan, T.V. Barseghyan, *Control problem for a system of linear loaded differential equations*, J. Phys.: Conf. Ser., **991** (2018), 1–6. https://doi.org/10.1088/1742-6596/991/1/012010

5. M.K. Beshtokov, *Differential and difference boundary value problem for loaded third-order pseudo-parabolic differential equations and difference methods for their numerical solution*, Comput. Math. and Math. Phys., **57** (2017), 1973–1993. https://doi.org/10.1134/S0965542517120089
6. Kh.Zh. Dikinov, A.A. Kerefov, A.M. Nakhushev, *A certain boundary value problem for a loaded heat equation*, Differ. Uravn., **12** (1976), 177–179.
7. D. Dzhumabaev, *Computational methods for solving the boundary value problems for the loaded differential and Fredholm integro-differential equations*, Math. Meth. Appl. Sci., **41** (2018), 1439–1462. https://doi.org/10.1002/mma.4674
8. V.E. Fedorov, L.V. Borel, *Solvability of loaded linear evolution equations with a degenerate operator at the derivative*, St. Petersburg Math. J., **26** (2015), 487–497. https://doi.org/10.1090/S1061-0022-2015-01348-4
9. N. Imanbaev, B. Kalimbetov, Z. Khabibullayev, *To the eigenvalue problems of a special-loaded first-order differential operator*, Int. J. Math. Anal., **8** (2014), 2247–2254. https://doi.org/10.12988/ijma.2014.48263
10. A.D. Iskanderov, *On a mixed problem for loaded quasi-linear equations of hyperbolic type*, Dokl. Akad. Nauk SSSR, **199** (1971), 1237–1239.
11. A.D. Iskenderov, *The first boundary value problem for a charged system of quasi-linear parabolic equations*, Differ. Uravn., **7** (1971), 1911–1913.
12. B. Islomov, U.I. Baltaeva, *Boundary-value problems for a third-order loaded parabolic-hyperbolic equation with variable coefficients*, Electron. J. Diff. Equ., **2015** (2015), 1–10.
13. A.M. Krall, *The development of general differential and general differential-boundary systems*, Rocky Mountain J. Math., **5** (1975), 493–542. https://doi.org/10.1216/RMJ-1975-5-4-493
14. K.U. Khubiev, *Analogue of Tricomi problem for characteristically loaded hyperbolic-parabolic equation with variable coefficients*, Ufimsk. Mat. Zh., **9** (2017), 94–103. https://doi.org/10.13108/2017-9-2-92
15. I.S. Lomov, *Loaded differential operators: Convergence of spectral expansions*, Diff. Equat., **50** (2014), 1070–1079. https://doi.org/10.1134/S0012266114080060
16. I.S. Lomov, V.V. Chernov, *Study of spectral properties of a loaded second-order differential operator*, Diff. Equat., **51** (2015), 857–861. https://doi.org/10.1134/S0012266115070046
17. A.M. Nakhushev, *The Darboux problem for a certain degenerate second order loaded integrodifferential equation*, Differ. Uravn., **12** (1976), 103–108. [in Russian]
18. A.M. Nakhushev, *An approximate method for solving boundary value problems for differential equations and its application to the dynamics of ground moisture and ground water*, Differ. Uravn., **18** (1982), 72–81. [in Russian]
19. A.M. Nakhushev, *Loaded equations and their applications*, Differ. Uravn., **19** (1983), 86–94. [in Russian]
20. A.M. Nakhushev, *Equations of Mathematical Biology*, Vysshaya Shkola, 1995. [in Russian]
21. A.M. Nakhushev, V.N. Borisov, *Boundary value problems for loaded parabolic equations and their applications to the prediction of ground water level*, Differ. Uravn., **13** (1977), 105–110. [in Russian]
22. A.M. Nakhushev, V.A. Nakhusheva, *On some classes of loaded equations and their applications*, Caspian J. of Applied Mathematics, Economics and Ecology, **1**, (2013) 82–88.
23. Öztürk, İ., *On the nonlocal boundary value problem for one order loaded differential equation*, Indian J. Pure Appl. Math., **26** (1995), 309–314.
24. I.N. Parasidis, E. Providas, *Extension operator method for the exact solution of integro-differential equations*, In: Pardalos, P.M., Rassias, T.M., Eds, Contributions in Mathematics and Engineering, Berlin: Springer, 2016. https://doi.org/10.1007/978-3-319-31317-7
25. I.N. Parasidis, E. Providas, *Resolvent operators for some classes of integro-differential equations*, In: Rassias T.M., Gupta V., Eds., Mathematical Analysis, Approximation Theory and Their Applications, Cham: Springer, 2016. https://doi.org/10.1007/978-3-319-31281-1

26. I.N. Parasidis, E. Providas, *Closed-form solutions for some classes of loaded difference equations with initial and nonlocal multipoint conditions*, In: Daras N., Rassias T.M., Eds., Modern Discrete Mathematics and Analysis, Cham: Springer, 2018. https://doi.org/10.1007/978-3-319-74325-719
27. I.N. Parasidis, E. Providas, *An exact solution method for a class of nonlinear loaded difference equations with multipoint boundary conditions*, J. Difference Equ. Appl., **24** (2018), 1649–1663. https://doi.org/10.1080/10236198.2018.1515928
28. I.N. Parasidis E. Providas, *Exact Solutions to Problems with Perturbed Differential and Boundary Operators*, In: Rassias T., Zagrebnov V., Eds., Analysis and Operator Theory. Springer Optimization and Its Applications, vol. 146, Cham: Springer, 2019. https://doi.org/10.1007/978-3-030-12661-2
29. K.B. Sabitov, *The Dirichlet problem for higher-order partial differential equations*, Math. Notes, **97** (2015), 255–267. https://doi.org/10.1134/S0001434615010277
30. Y.K. Sabitova, *Dirichlet problem for Lavrent'ev–Bitsadze equation with loaded summands*, Russ. Math. **62** (2018), 35–51. https://doi.org/10.3103/S1066369X18090050
31. M.A. Sadybekov, N.S. Imanbaev, *A regular differential operator with perturbed boundary condition*, Math. Notes., **101** (2017), 878–887. https://doi.org/10.1134/S0001434617050133
32. J. Wiener, L. Debnath, *Partial differential equations with piecewise constant delay*, Internat. J. Math. and Math. Scz., **14** (1991), 485–496 .
33. E.N. Zhuravleva, E.A. Karabut, *Loaded complex equations in the jet collision problem*, Comput. Math. and Math. Phys., **51** (2011), 876–894. https://doi.org/10.1134/S0965542511050186

Set-Theoretic Properties of Generalized Topologically Open Sets in Relator Spaces

Themistocles M. Rassias, Muwafaq M. Salih, and Árpád Száz

Abstract A family $\mathscr{R}$ of binary relations on a set X is called a relator on X, and the ordered pair $X(\mathscr{R}) = (X, \mathscr{R})$ is called a relator space. Sometimes relators on X to Y are also considered.

By using an obvious definition of the generated open sets, each generalized topology $\mathscr{T}$ on X can be easily derived from the family $\mathscr{R}_{\mathscr{T}}$ of all Pervin's preorder relations $R_V = V^2 \cup (V^c \times X)$ with $V \in \mathscr{T}$, where $V^2 = V \times V$ and $V^c = X \setminus V$.

For a subset A of the relator space $X(\mathscr{R})$, we define

$$A^\circ = \operatorname{int}_{\mathscr{R}}(A) = \{x \in X : \quad \exists\, R \in \mathscr{R} : \quad R(x) \subseteq A\}$$

and $A^- = \operatorname{cl}_{\mathscr{R}}(A) = \operatorname{int}_{\mathscr{R}}(A^c)^c$. And, for instance, we also define

$$\mathscr{T}_{\mathscr{R}} = \{A \subseteq X : \quad A \subseteq A^\circ\} \qquad \text{and} \qquad \mathscr{F}_{\mathscr{R}} = \{A \subseteq X : \quad A^c \in \mathscr{T}_{\mathscr{R}}\}.$$

Moreover, motivated by some basic definitions in topological spaces, for a subset A of the relator space $X(\mathscr{R})$ we shall write

(1) $A \in \mathscr{T}_{\mathscr{R}}^{r}$ if $A = A^{-\circ}$;
(2) $A \in \mathscr{T}_{\mathscr{R}}^{p}$ if $A \subseteq A^{-\circ}$; (3) $A \in \mathscr{T}_{\mathscr{R}}^{s}$ if $A \subseteq A^{\circ-}$;
(4) $A \in \mathscr{T}_{\mathscr{R}}^{\alpha}$ if $A \subseteq A^{\circ-\circ}$; (5) $A \in \mathscr{T}_{\mathscr{R}}^{\beta}$ if $A \subseteq A^{-\circ-}$;
(6) $A \in \mathscr{T}_{\mathscr{R}}^{a}$ if $A \subseteq A^{-\circ} \cap A^{\circ-}$; (7) $A \in \mathscr{T}_{\mathscr{R}}^{b}$ if $A \subseteq A^{-\circ} \cup A^{\circ-}$;
(8) $A \in \mathscr{T}_{\mathscr{R}}^{q}$ if there exists $V \in \mathscr{T}_{\mathscr{R}}$ such that $V \subseteq A \subseteq V^-$;
(9) $A \in \mathscr{T}_{\mathscr{R}}^{ps}$ if there exists $V \in \mathscr{T}_{\mathscr{R}}$ such that $A \subseteq V \subseteq A^-$;

Th. M. Rassias
Department of Mathematics, Zografou Campus, National Technical University of Athens, Athens, Greece
e-mail: trassias@math.ntua.gr

M. M. Salih · Á. Száz (✉)
Department of Mathematics, University of Debrecen, Debrecen, Hungary
e-mail: muwafaq.salih@science.unideb.hu; szaz@science.unideb.hu

I. N. Parasidis et al. (eds.), *Mathematical Analysis in Interdisciplinary Research*, Springer Optimization and Its Applications 179,
https://doi.org/10.1007/978-3-030-84721-0_30

(10) $A \in \mathcal{T}_{\mathcal{R}}^{\gamma}$ if there exists $V \in \mathcal{T}_{\mathcal{R}}^{s}$ such that $A \subseteq V \subseteq A^-$;
(11) $A \in \mathcal{T}_{\mathcal{R}}^{\delta}$ if there exists $V \in \mathcal{T}_{\mathcal{R}}^{p}$ such that $V \subseteq A \subseteq V^-$.

And, the members of the above families will be called the topologically regular open, preopen, semi-open, α-open, β-open, a-open, b-open, quasi-open, pseudo-open, γ-open, and δ-open subsets of the relator space $X(\mathcal{R})$, respectively.

In a former paper, we have systematically investigated the various relationships among the families $\mathcal{T}_{\mathcal{R}}^{\kappa}$. Moreover, we have tried to establish several illuminating characterizations of the families $\mathcal{T}_{\mathcal{R}}^{\kappa}$.

Here, we shall mainly be interested in the most simple set-theoretic properties of the families $\mathcal{T}_{\mathcal{R}}^{\kappa}$. First of all, we shall briefly investigate their dual families $\mathcal{F}_{\mathcal{R}}^{\kappa} = \{A \subseteq X : \ A^c \in \mathcal{T}_{\mathcal{R}}^{\kappa}\}$.

Then, we shall establish some intrinsic characterizations of the families $\mathcal{T}_{\mathcal{R}}^{\kappa}$. Moreover, we shall give some necessary and sufficient conditions in order that $\emptyset$, $\{x\}$, with $x \in X$, and X could be contained in $\mathcal{T}_{\mathcal{R}}^{\kappa}$.

Finally, we shall show that, with the exception of $\mathcal{T}_{\mathcal{R}}^{r}$, the families $\mathcal{T}_{\mathcal{R}}^{\kappa}$ are closed under arbitrary unions. Moreover, for every $\mathcal{T}_{\mathcal{R}}^{\kappa}$, we shall try to determine those subsets A of X which satisfy $A \cap B \in \mathcal{T}_{\mathcal{R}}^{\kappa}$ for all $B \in \mathcal{T}_{\mathcal{R}}^{\kappa}$.

Furthermore, we shall indicate that, analogously to the family $\mathcal{T}_{\mathcal{R}}$ of all topologically open subsets of the relator spaces $X(\mathcal{R})$, the families $\mathcal{T}_{\mathcal{R}}^{\kappa}$ can also be used to introduce some interesting classifications of relators.

1 Introduction

If $\mathcal{T}$ is a family of subsets of a set X such that $\mathcal{T}$ is closed under finite intersections and arbitrary unions, then the family $\mathcal{T}$ is called a *topology* on X, and the ordered pair $X(\mathcal{T}) = (X, \mathcal{T})$ is called a *topological space*.

The members of $\mathcal{T}$ are called the *open subsets* of X. While, the members of $\mathcal{F} = \{A^c : \ A \in \mathcal{T}\}$, where $A^c = X \setminus A$, are called the *closed subsets* of X. And, the members of $\mathcal{T} \cap \mathcal{F}$ are called the *clopen subsets* of X.

Since, $\emptyset = \bigcup \emptyset$ and $X = \bigcap \emptyset$, we necessarily have $\{\emptyset, \ X\} \subseteq \mathcal{T} \cap \mathcal{F}$. Therefore, if, in particular, $\mathcal{T} = \{\emptyset, \ X\}$, then $\mathcal{T}$ is called *minimal* [73] instead of indiscrete. While, if $\mathcal{T} \cap \mathcal{F} = \{\emptyset, \ X\}$, then $\mathcal{T}$ is called *connected* [102, p. 31].

For a subset A of $X\ (\mathcal{T})$, the sets $A^\circ = \operatorname{int}(A) = \bigcup \big(\mathcal{T} \cap \mathcal{P}(A)\big)$,

$$A^- = \operatorname{cl}(A) = \operatorname{int}(A^c)^c \qquad \text{and} \qquad A^\dagger = \operatorname{res}(A) = \operatorname{cl}(A) \setminus A$$

are called the *interior, closure, and residue of* A, respectively.

Thus, $-$ is a *Kuratowski closure operation* on $\mathcal{P}(X)$. That is, $\emptyset^- = \emptyset$, and $-$ is *extensive, idempotent, and additive* in the sense that, for any $A, B \subseteq X$, we have $A \subseteq A^-$, $A^{--} = A^-$ and $(A \cup B)^- = A^- \cup B^-$.

In particular, the members of the families

$$\mathscr{D} = \{A \subseteq X : \ A^- = X\} \qquad \text{and} \qquad \mathscr{N} = \{A \subseteq X : \ A^{-\circ} = \emptyset\}$$

are called the *dense and rare (or nowhere dense) subsets* of $X(\mathscr{T})$, respectively.

In 1922, a subset A of a closure space $X(-)$ was called *regular open* by Kuratowski [50] if $A = A^{-\circ}$. While, in 1937, a subset A of a topological space $X(\mathscr{T})$ was called regular open by Stone [77] if $A = B^\circ$ for some $B \in \mathscr{F}$.

The importance of regular open subsets of $X(\mathscr{T})$ lies mainly in the fact that their family forms a complete *Boolean algebra* [38, p. 66] with respect to the operations defined by $A' = A^{-c}$, $A \wedge B = A \cap B$, and $A \vee B = (A \cup B)''$.

In 1982, a subset A of $X(\mathscr{T})$ was called *preopen* by Mashhour et al. [60] if $A \subseteq A^{-\circ}$. However, by Dontchev [28], preopen sets, under different names, were much earlier studied by several mathematicians.

For instance, in 1964, Corson and Michael [11] called a subset A of $X(\mathscr{T})$ *locally dense* if it is a dense subset of some $V \in \mathscr{T}$ in the sense that $A \subseteq V \subseteq A^-$. Moreover, they noted that this property is equivalent to the inclusion $A \subseteq A^{-\circ}$.

This equivalence was later also stated by Jun at al. [48]. Moreover, Ganster [35] proved that A is preopen if and only if there exist $V \in \mathscr{T}$ and $B \in \mathscr{D}$ such that $A = V \cap B$. (See also Dontchev [28].)

In 1963, a subset A of $X(\mathscr{T})$ was called *semi-open* by Levine [54] if there exists $V \in \mathscr{T}$ such that $V \subseteq A \subseteq V^-$. First of all, he showed that the set A is semi-open if and only if $A \subseteq A^{\circ -}$.

Moreover, he also proved that if A is a semi-open subset of $X(\mathscr{T})$, then there exist $V \in \mathscr{T}$ and $B \in \mathscr{N}$ such that $A = V \cup B$ and $V \cap B = \emptyset$. In addition, he also noted that the converse statement is false.

Levine's statement closely resembles to a famous stability theorem of Hyers [43] which says that an ε–approximately additive function of one Banach space to another is the sum of an additive function and an ε–small function.

Analogously to the paper of Hyers, Levine's paper has also attracted the interest of a surprisingly great number of mathematicians. For instance, by the Google Scholar, it has been cited by 3096 works.

Moreover, the above statement of Levine was improved by Dlaska et al. [27] who observed that a subset A of $X(\mathscr{T})$ is semi-open if and only if there exist $V \in \mathscr{T}$ and $B \subseteq V^\dagger$ such that $A = V \cup B$.

The latter observation was later reformulated, in a more convenient form, by Duszyńki and Noiri [29] who noted that a subset A of $X(\mathscr{T})$ is semi-open if and only if there exists $B \subseteq A^{\circ\,\dagger}$ such that $A = A^\circ \cup B$.

In particular, in 1965 and 1971, Njåstad [63] and Isomichi [44], being not aware of the paper of Levine, studied semi-open sets under the names *β-sets* and *subcondensed sets*, respectively.

Moreover, Njåstad called a subset A of $X(\mathscr{T})$ to be an *α–set* if $A \subseteq A^{\circ - \circ}$. And, he proved that the set A is an α–set if and only if there exist $V \in \mathscr{T}$ and $B \in \mathscr{N}$ such that $A = V \setminus B$.

He also proved that A is an α–set if and only if its intersection with every β–set is a β–set. Thus, the family of all α–sets is a topology. The fact that the semi-open sets form only a generalized topology was already noticed by Levine.

A further important property of α-sets was established by Noiri [65] and Reilly and Wamanamurthy [71], in 1984 and 1985, respectively, who proved that a set is α–open if and only if it is both preopen and semi-open.

In 1983, the subset A was called *β–open* by Abd El-Monsef et al. [1] if $A \subseteq A^{-\circ-}$. Moreover, in 1986 Andrijević [3] used the term *semi-preopen* instead of β–open without knowing of [1].

Actually, Andrijević called a subset A of $X(\mathscr{T})$ to be semi-preopen if there exists a preopen subset V of $X(\mathscr{T})$ such that $V \subseteq A \subseteq V^{-}$. And, he showed that this is equivalent to the inclusion $A \subseteq A^{-\circ-}$.

Moreover, in 1996, a subset A of $X(\mathscr{T})$ was called *b–open* by Andrijević [4] if $A \subseteq A^{\circ-} \cup A^{-\circ}$. He proved that A is b–open if and only if there exist a preopen subset B and a semi-open subset C of $X(\mathscr{T})$ such that $A = B \cup C$.

In a former paper [70], we have shown that the above definitions and several characterization theorems of generalized open sets can be naturally extended not only to generalized topological and closure spaces but also to relator spaces.

In the sequel, following a terminology introduced by the third author [78], a family $\mathscr{R}$ of binary relations on a set X will be called a *relator* on X, and the ordered pair $X(\mathscr{R}) = (X, \mathscr{R})$ will be called a *relator space*.

Thus, relator spaces are generalizations of not only ordered sets [25] and uniform spaces [34] but also topological, closure, and proximity spaces [62]. However, to include context spaces [36] a further generalization is needed [83, 84].

For instance, by Száz [87], each generalized topology $\mathscr{T}$ on X can be easily derived from the family $\mathscr{R}_{\mathscr{T}}$ of all Pervin's preorder relations $R_V = V^2 \cup (V^c \times X)$ with $V \in \mathscr{T}$. Thus, generalized topologies need not be studied separately.

Here, we shall mainly be interested in the most simple set-theoretic properties of the various families $\mathscr{T}_{\mathscr{R}}^{\kappa}$ of generalized topologically open subsets of $X(\mathscr{R})$. First of all, we shall briefly investigate their dual families $\mathscr{F}_{\mathscr{R}}^{\kappa} = \{A \subseteq X : \; A^c \in \mathscr{T}_{\mathscr{R}}^{\kappa}\}$.

Then, we shall establish some intrinsic characterizations of the families $\mathscr{T}_{\mathscr{R}}^{\kappa}$. Moreover, we shall give some necessary and sufficient conditions in order that $\emptyset$, $\{x\}$, with $x \in X$, and X could be contained in $\mathscr{T}_{\mathscr{R}}^{\kappa}$.

Finally, we shall show that, with the exception of $\mathscr{T}_{\mathscr{R}}^{r}$, the families $\mathscr{T}_{\mathscr{R}}^{\kappa}$ are closed under arbitrary unions. Moreover, for every $\mathscr{T}_{\mathscr{R}}^{\kappa}$, we shall try to determine those subsets A of X which satisfy $A \cap B \in \mathscr{T}_{\mathscr{R}}^{\kappa}$ for all $B \in \mathscr{T}_{\mathscr{R}}^{\kappa}$.

Furthermore, we shall indicate that, analogously to the family $\mathscr{T}_{\mathscr{R}}$ of all topologically open subsets of a relator spaces $X(\mathscr{R})$, the families $\mathscr{T}_{\mathscr{R}}^{\kappa}$ can also be used to introduce some interesting classifications of relators.

The necessary prerequisites on relations and relators, which are certainly unfamiliar to the reader, will be briefly laid out in the subsequent preparatory sections which will also contain several new observations.

These sections may also be useful for all those readers who are not very much interested in the various generalizations of open sets having been studied, as the extensive References show, by a surprisingly great number of topologists.

2 A Few Basic Facts on Relations

A subset F of a product set $X \times Y$ is called a *relation on* X *to* Y. In particular, a relation on X to itself is called a *relation on* X. And, $\Delta_X = \{(x, x) : x \in X\}$ is called the *identity relation of* X.

If F is a relation on X to Y, then by the above definitions we can also state that F is a relation on $X \cup Y$. However, for several purposes, the latter view of the relation F would be quite unnatural.

If F is a relation on X to Y, then for any $x \in X$ and $A \subseteq X$ the sets $F(x) = \{y \in Y : (x, y) \in F\}$ and $F[A] = \bigcup \{F(x) : x \in A\}$ are called the *images or neighborhoods* of x and A under F, respectively.

If $(x, y) \in F$, then instead of $y \in F(x)$, we may also write $x\,F\,y$. However, instead of $F[A]$, we cannot write $F(A)$. Namely, it may occur that, in addition to $A \subseteq X$, we also have $A \in X$.

Now, the sets $D_F = \{x \in X : F(x) \neq \emptyset\}$ and $R_F = F[X]$ may be called the *domain* and *range* of F, respectively. If, in particular, $D_F = X$, then we may say that F is a *relation of* X *to* Y, or that F is a *non-partial relation on* X *to* Y.

In particular, a relation f on X to Y is called a *function* if for each $x \in D_f$ there exists $y \in Y$ such that $f(x) = \{y\}$. In this case, by identifying singletons with their elements, we may simply write $f(x) = y$ instead of $f(x) = \{y\}$.

Moreover, a function $\star$ of X to itself is called a *unary operation on* X. While, a function $*$ of X^2 to X is called a *binary operation on* X. And, for any $x, y \in X$, we usually write $x^\star$ and $x * y$ instead of $\star(x)$ and $*((x, y))$, respectively.

If F is a relation on X to Y, then a function f of D_F to Y is called a *selection function* of F if $f(x) \in F(x)$ for all $x \in D_F$. By using the Axiom of Choice, it can be shown that every relation is the union of its selection functions.

For a relation F on X to Y, we may naturally define two *set-valued functions* φ_F of X to $\mathscr{P}(Y)$ and Φ_F of $\mathscr{P}(X)$ to $\mathscr{P}(Y)$ such that $\varphi_F(x) = F(x)$ for all $x \in X$ and $\Phi_F(A) = F[A]$ for all $A \subseteq X$.

Functions of X to $\mathscr{P}(Y)$ can be naturally identified with relations on X to Y. While, functions of $\mathscr{P}(X)$ to $\mathscr{P}(Y)$ are more general objects than relations on X to Y. In [91, 97, 98], they were briefly called *corelations* on X to Y.

However, a relation on $\mathscr{P}(X)$ to Y should be rather called a *super relation* on X to Y, and a relation on $\mathscr{P}(X)$ to $\mathscr{P}(Y)$ should be rather called a *hyper relation* on X to Y. Thus, $\mathrm{cl}_{\mathscr{R}}$ is a super relation and $\mathrm{Cl}_{\mathscr{R}}$ is a hyper relation on Y to X.

If F is a relation on X to Y, then one can easily see that $F = \bigcup_{x \in X} \big(\{x\} \times F(x)\big)$. Therefore, the images $F(x)$, where $x \in X$, uniquely determine F. Thus, a relation F on X to Y can also be naturally defined by specifying $F(x)$ for all $x \in X$.

For instance, the *complement* F^c and the *inverse* F^{-1} can be defined such that $F^c(x) = F(x)^c = Y \setminus F(x)$ for all $x \in X$ and $F^{-1}(y) = \{x \in X : y \in F(x)\}$ for all $y \in Y$. Thus, it can be easily seen that $F^c = \big(X \times Y\big) \setminus F$.

Moreover, if in addition G is a relation on Y to Z, then the *composition* $G \circ F$ can be defined such that $(G \circ F)(x) = G[F(x)]$ for all $x \in X$. Thus, it can be easily seen that $(G \circ F)[A] = G[F[A]] = \bigcup_{y \in F[A]} G(y)$ for all $A \subseteq X$.

While, if G is a relation on Z to W, then the *box product* $F \boxtimes G$ can be defined such that $(F \boxtimes G)(x, z) = F(x) \times G(z)$ for all $x \in X$ and $z \in Z$. Thus, it can be shown that $(F \boxtimes G)[A] = G \circ A \circ F^{-1}$ for all $A \subseteq X \times Z$ [89].

Hence, by taking $A = \{(x, z)\}$, and $A = \Delta_Y$ if $Y = Z$, one can at once see that the box and composition products are actually equivalent tools. However, the box product can be immediately defined for any family of relations.

Now, a relation R on X may be briefly defined to be *reflexive* on X if $\Delta_X \subseteq R$, and *transitive* if $R \circ R \subseteq R$. Moreover, R may be briefly defined to be *symmetric* if $R \subseteq R^{-1}$, and *antisymmetric* if $R \cap R^{-1} \subseteq \Delta_X$.

Thus, a reflexive and transitive (symmetric) relation may be called a *preorder (tolerance) relation*. And, a symmetric (antisymmetric) preorder relation may be called an *equivalence (partial order) relation*.

For any relation R on X, we may also naturally define $R^0 = \Delta_X$ and $R^n = R \circ R^{n-1}$ if $n \in \mathbb{N}$. Moreover, we may also naturally define $R^\infty = \bigcup_{n=0}^\infty R^n$. Thus, R^∞ is the smallest preorder relation on X containing R [39].

For $A \subseteq X$, the Pervin relation $R_A = A^2 \cup \left(A^c \times X\right)$ is an important preorder on X [69]. While, for a *pseudometric* d on X, the *Weil surrounding* $B_r = \{(x, y) \in X^2 : d(x, y) < r\}$, with $r > 0$, is an important tolerance on X [103].

Note that $S_A = R_A \cap R_A^{-1} = R_A \cap R_{A^c} = A^2 \cap \left(A^c\right)^2$ is already an equivalence relation on X. And, more generally if $\mathcal{A}$ is a *cover (partition)* of X, then $S_{\mathcal{A}} = \bigcup_{A \in \mathcal{A}} A^2$ is a tolerance (equivalence) relation on X.

As an important generalization of the Pervin relation R_A, for any $A \subseteq X$ and $B \subseteq Y$, we may also naturally consider the *Hunsaker-Lindgren relation* $R_{(A,B)} = (A \times B) \cap \left(A^c \times Y\right)$ [42]. Namely, thus we evidently have $R_A = R_{(A,A)}$.

The Pervin relations R_A and the Hunsaker-Lindgren relations $R_{(A,B)}$ were actually first used by Davis [26] and Császár [15, pp. 42 and 351] in some less explicit and convenient forms, respectively.

3 A Few Basic Facts on Relators

A family $\mathcal{R}$ of relations on one set X to another Y is called a *relator on X to Y*, and the ordered pair $(X, Y)(\mathcal{R}) = \left((X, Y), \mathcal{R}\right)$ is called a *relator space*. For the origins of this notion, see [78, 83], and the references in [78].

If, in particular, $\mathcal{R}$ is a relator on X to itself, then $\mathcal{R}$ is simply called a *relator on X*. Thus, by identifying singletons with their elements, we may naturally write $X(\mathcal{R})$ instead of $(X, X)(\mathcal{R})$. Namely, $(X, X) = \{\{X\}, \{X, X\}\} = \{\{X\}\}$.

Relator spaces of this simpler type are already substantial generalizations of the various *ordered sets* [25] and *uniform spaces* [34]. However, they are insufficient for some important purposes. (See, for instance, [36] and [83].)

A relator $\mathcal{R}$ on X to Y, or the relator space $(X, Y)(\mathcal{R})$, is called *simple* if $\mathcal{R} = \{R\}$ for some relation R on X to Y. Simple relator spaces of the forms $(X, Y)(R)$ and $X(R)$ were called *formal contexts* and *gosets* in [36] and [93], respectively.

Moreover, a relator $\mathscr{R}$ on X, or the relator space $X(\mathscr{R})$, may, for instance, be naturally called *reflexive* if each member of $\mathscr{R}$ is reflexive on X. Thus, we may also naturally speak of *preorder, tolerance, and equivalence relators.*

For instance, for a family $\mathscr{A}$ of subsets of X, the family $\mathscr{R}_{\mathscr{A}} = \{R_A : \ A \in \mathscr{A}\}$, where $R_A = A^2 \cup \left(A^c \times X\right)$, is an important preorder relator on X. Such relators were first used by Pervin [69] and Levine [57]. (See also [87, 8].)

While, for a family $\mathscr{D}$ of *pseudo-metrics* on X, the family $\mathscr{R}_{\mathscr{D}} = \{B_r^d : \ r > 0, \ d \in \mathscr{D}\}$, where $B_r^d = \{(x, y) : \ d(x, y) < r\}$, is an important tolerance relator on X. Such relators were first considered by Weil [103].

Moreover, if $\mathfrak{S}$ is a family of *partitions* of X, then the family $\mathscr{R}_{\mathfrak{S}} = \{S_{\mathscr{A}} : \ \mathscr{A} \in \mathfrak{S}\}$, where $S_{\mathscr{A}} = \bigcup_{A \in \mathscr{A}} A^2$, is an equivalence relator on X. Such practically important relators were first investigated by Levine [56].

If $\star$ is a unary operation for relations on X to Y, then for any relator $\mathscr{R}$ on X to Y we may naturally define $\mathscr{R}^\star = \left\{R^\star : \ R \in \mathscr{R}\right\}$. However, this plausible notation may cause some confusions whenever, for instance, $\star = c$.

In particular, for any relator $\mathscr{R}$ on X, we may naturally define $\mathscr{R}^\infty = \left\{R^\infty : \ R \in \mathscr{R}\right\}$. Moreover, we may also naturally define $\mathscr{R}^\partial = \left\{S \subseteq X^2 : \ S^\infty \in \mathscr{R}\right\}$. These operations were first introduced by Mala [58, 59] and Pataki [67, 68].

While, if $*$ is a binary operation for relations, then for any two relators $\mathscr{R}$ and $\mathscr{S}$ we may naturally define $\mathscr{R} * \mathscr{S} = \left\{R * S : \ R \in \mathscr{R}, \ S \in \mathscr{S}\right\}$. However, this plausible notation may again cause some confusions whenever, for instance, $* = \cap$.

Therefore, in general, we rather write $\mathscr{R} \wedge \mathscr{S} = \left\{R \cap S : \ R \in \mathscr{R}, \ S \in \mathscr{S}\right\}$. Moreover, for instance, we also write $\mathscr{R} \Delta \mathscr{R}^{-1} = \left\{R \cap R^{-1} : \ R \in \mathscr{R}\right\}$. Note that thus $\mathscr{R} \Delta \mathscr{R}^{-1}$ is a symmetric relator such that $\mathscr{R} \Delta \mathscr{R}^{-1} \subseteq \mathscr{R} \wedge \mathscr{R}^{-1}$.

A function $\Box$ of the family of all relators on X to Y is called a *direct (indirect) unary operation for relators* if, for every relator $\mathscr{R}$ on X to Y, the value $\mathscr{R}^\Box = \Box(\mathscr{R})$ is a relator on X to Y (on Y to X).

For instance, c and -1 are *involution operations* for relators. While, ∞ and ∂ are *projection operations* for relators. Moreover, the operation $\Box = c$, ∞ or ∂ is inversion compatible in the sense that $\mathscr{R}^{\Box -1} = \mathscr{R}^{-1 \Box}$.

More generally, a function $\mathfrak{F}$ of the family of all relators on X to Y is called a *structure for relators* if, for every relator $\mathscr{R}$ on X to Y, the value $\mathfrak{F}_{\mathscr{R}} = \mathfrak{F}(\mathscr{R})$ is in a power set depending only on X and Y.

For instance, if $\operatorname{cl}_{\mathscr{R}}(B) = \bigcap \{R^{-1}[B] : \ R \in \mathscr{R}\}$ for every relator $\mathscr{R}$ on X to Y and $B \subseteq Y$, then the function $\mathfrak{F}$, defined by $\mathfrak{F}(\mathscr{R}) = \operatorname{cl}_{\mathscr{R}}$, is a structure for relators such that $\mathfrak{F}(\mathscr{R}) \subseteq \mathscr{P}(Y) \times X$, and thus $\mathfrak{F}(\mathscr{R}) \in \mathscr{P}\left(\mathscr{P}(Y) \times X\right)$.

A structure $\mathfrak{F}$ for relators is called *increasing* if $\mathscr{R} \subseteq \mathscr{S}$ implies $\mathfrak{F}_{\mathscr{R}} \subseteq \mathfrak{F}_{\mathscr{S}}$ for any two relators $\mathscr{R}$ and S on X to Y. And, $\mathfrak{F}$ is called *quasi-increasing* if $R \in \mathscr{R}$ implies $\mathfrak{F}_R \subseteq \mathfrak{F}_{\mathscr{R}}$ for any relator $\mathscr{R}$ on X to Y. Note that here $\mathfrak{F}_R = \mathfrak{F}_{\{R\}}$.

Moreover, the structure $\mathfrak{F}$ is called *union-preserving* if $\mathfrak{F}_{\bigcup_{i \in I} \mathscr{R}_i} = \bigcup_{i \in I} \mathfrak{F}_{\mathscr{R}_i}$ for any family $(\mathscr{R}_i)_{i \in I}$ of relators on X to Y. It can be shown that $\mathfrak{F}$ is union-preserving if and only if $\mathfrak{F}_{\mathscr{R}} = \bigcup_{R \in \mathscr{R}} \mathfrak{F}_R$ for every relator $\mathscr{R}$ on X to Y [91].

In particular, an increasing operation $\Box$ for relators on X to Y is called a *projection or modification operation* for relators if it is idempotent in the sense that $\mathscr{R}^{\Box\Box} = \mathscr{R}^{\Box}$ holds for any relator $\mathscr{R}$ on X to Y.

Moreover, a projection operation $\Box$ for relators on X to Y is called a *closure or refinement operation* for relators if it is extensive in the sense that $\mathscr{R} \subseteq \mathscr{R}^{\Box}$ holds for any relator $\mathscr{R}$ on X to Y. (For the origins, see [82].)

By using Pataki connections [67, 99], several closure operations can be derived from union-preserving structures. However, more generally, one can find first the Galois adjoint $\mathfrak{G}$ of such a structure $\mathfrak{F}$, and then take $\Box_{\mathfrak{F}} = \mathfrak{G} \circ \mathfrak{F}$ [86].

Now, for an operation $\Box$ for relators, a relator $\mathscr{R}$ on X to Y may be naturally called $\Box$*–fine* if $\mathscr{R}^{\Box} = \mathscr{R}$. And, for some structure $\mathfrak{F}$ for relators, two relators $\mathscr{R}$ and $\mathscr{S}$ on X to Y may be naturally called $\mathfrak{F}$*–equivalent* if $\mathfrak{F}_{\mathscr{R}} = \mathfrak{F}_{\mathscr{S}}$.

Moreover, for a structure $\mathfrak{F}$ for relators, a relator $\mathscr{R}$ on X to Y may, for instance, be naturally called $\mathfrak{F}$*–simple* if $\mathfrak{F}_{\mathscr{R}} = \mathfrak{F}_R$ for some relation R on X to Y. Thus, in particular, singleton relators have to be actually called *properly simple*.

4 Structures Derived from Relators

Definition 1 If $\mathscr{R}$ is a relator on X to Y, then for any $A \subseteq X$, $B \subseteq Y$ and $x \in X$, $y \in Y$ we define :

(1) $A \in \operatorname{Int}_{\mathscr{R}}(B)$ if $R[A] \subseteq B$ for some $R \in \mathscr{R}$;
(2) $A \in \operatorname{Cl}_{\mathscr{R}}(B)$ if $R[A] \cap B \neq \emptyset$ for all $R \in \mathscr{R}$;
(3) $x \in \operatorname{int}_{\mathscr{R}}(B)$ if $\{x\} \in \operatorname{Int}_{\mathscr{R}}(B)$; (4) $x \in \sigma_{\mathscr{R}}(y)$ if $x \in \operatorname{int}_{\mathscr{R}}(\{y\})$;
(5) $x \in \operatorname{cl}_{\mathscr{R}}(B)$ if $\{x\} \in \operatorname{Cl}_{\mathscr{R}}(B)$; (6) $x \in \rho_{\mathscr{R}}(y)$ if $x \in \operatorname{cl}_{\mathscr{R}}(\{y\})$;
(7) $B \in \mathscr{E}_{\mathscr{R}}$ if $\operatorname{int}_{\mathscr{R}}(B) \neq \emptyset$; (8) $B \in \mathscr{D}_{\mathscr{R}}$ if $\operatorname{cl}_{\mathscr{R}}(B) = X$.

Remark 1 The relations $\operatorname{Int}_{\mathscr{R}}$, $\operatorname{int}_{\mathscr{R}}$ and $\sigma_{\mathscr{R}}$ are called *the proximal, topological, and infinitesimal interiors* generated by $\mathscr{R}$, respectively. While, the members of the families, $\mathscr{E}_{\mathscr{R}}$ and $\mathscr{D}_{\mathscr{R}}$ are called the *fat and dense subsets* of the relator space $(X, Y)(\mathscr{R})$, respectively.

The origins of the relations $\operatorname{Cl}_{\mathscr{R}}$ and $\operatorname{Int}_{\mathscr{R}}$ go back to Efremović's proximity δ [30] and Smirnov's strong inclusion $\Subset$ [76], respectively. While, the convenient notations $\operatorname{Cl}_{\mathscr{R}}$ and $\operatorname{Int}_{\mathscr{R}}$, and family $\mathscr{E}_{\mathscr{R}}$, together with its dual $\mathscr{D}_{\mathscr{R}}$, were first explicitly used by the third author [78, 80, 81, 85].

The following theorem shows that the big interior and closure are equivalent tools in a relator space.

Theorem 1 *If $\mathscr{R}$ is a relator on X to Y, then for any $B \subseteq Y$ we have*

(1) $\operatorname{Cl}_{\mathscr{R}}(B) = \operatorname{Int}_{\mathscr{R}}(B^c)^c$; *(2)* $\operatorname{Int}_{\mathscr{R}}(B) = \operatorname{Cl}_{\mathscr{R}}(B^c)^c$.

Remark 2 By using the notation $\mathscr{C}_Y(B) = B^c$, assertion (1) can be expressed in the more concise form that $\operatorname{Cl}_{\mathscr{R}} = \big(\operatorname{Int}_{\mathscr{R}} \circ \mathscr{C}_Y\big)^c$ or $\operatorname{Cl}_{\mathscr{R}} = \big(\operatorname{Int}_{\mathscr{R}}\big)^c \circ \mathscr{C}_Y$.

From Theorem 1, we can easily derive the following

Theorem 2 *If $\mathscr{R}$ is a relator on X to Y, then for any $B \subseteq Y$ we have*

(1) $\mathrm{cl}_{\mathscr{R}}(B) = \mathrm{int}_{\mathscr{R}}(B^c)^c$; *(2)* $\mathrm{int}_{\mathscr{R}}(B) = \mathrm{cl}_{\mathscr{R}}(B^c)^c$.

Remark 3 By using the convenient notations $B^- = \mathrm{cl}_{\mathscr{R}}(B)$ and $B^\circ = \mathrm{int}_{\mathscr{R}}(B)$, assertion (1) can be expressed in the more concise form that $- = c \ \circ \ c$, or equivalently $- \ c = c \ \circ$.

The small closures and interiors are, in general, much weaker tools than the big ones. Namely, we can only prove

Theorem 3 *If $\mathscr{R}$ is a relator on X to Y, then for any $A \subseteq X$ and $B \subseteq Y$*

(1) $A \in \mathrm{Int}_{\mathscr{R}}(B)$ *implies* $A \subseteq \mathrm{int}_{\mathscr{R}}(B)$;
(2) $A \cap \mathrm{cl}_{\mathscr{R}}(B) \neq \emptyset$ *implies* $A \in \mathrm{Cl}_{\mathscr{R}}(B)$.

Concerning closures and interiors, we can also prove the following two theorems which show that, despite their equivalences, closures are sometimes more convenient tools than interiors.

Theorem 4 *For any relator $\mathscr{R}$ on X to Y, we have*

(1) $\mathrm{Cl}_{\mathscr{R}^{-1}} = \mathrm{Cl}_{\mathscr{R}}^{-1}$; *(2)* $\mathrm{Int}_{\mathscr{R}^{-1}} = \mathscr{C}_Y \circ \mathrm{Int}_{\mathscr{R}}^{-1} \circ \mathscr{C}_X$.

Theorem 5 *If $\mathscr{R}$ is a relator on X to Y, then for any $B \subseteq Y$, we have*

(1) $\mathrm{cl}_{\mathscr{R}}(B) = \bigcap_{R \in \mathscr{R}} R^{-1}[B]$; *(2)* $\mathrm{int}_{\mathscr{R}}(B) = \bigcup_{R \in \mathscr{R}} R^{-1}[B^c]^c$.

From the $B = \{y\}$ particular case of this theorem, we can easily derive

Corollary 1 *For any relator $\mathscr{R}$ on X to Y, we have*

$$\rho_{\mathscr{R}} = \bigcap \mathscr{R}^{-1} = \left(\bigcap \mathscr{R}\right)^{-1}.$$

Moreover, by using the $\mathscr{R} = \{R\}$ particular case of Theorem 5, we can prove

Theorem 6 *If R is a relation on X to Y, then for any $A \subseteq X$ and $B \subseteq Y$*

$$A \subseteq \mathrm{int}_R(B) \iff \mathrm{cl}_{R^{-1}}(A) \subseteq B.$$

Remark 4 This shows that the mappings $A \mapsto \mathrm{cl}_{R^{-1}}(A)$ and $B \mapsto \mathrm{int}_R(B)$ establish a Galois connection between the posets $\mathscr{P}(X)$ and $\mathscr{P}(Y)$.

The above important closure-interior Galois connection, used first in [96], is not independent from the upper and lower bound Galois connection [88].

The following two closely related theorems show that the fat and dense sets are also equivalent tools in a relator space.

Theorem 7 *If $\mathscr{R}$ is a relator on X to Y, then for any $B \subseteq Y$ we have*

(1) $B \in \mathscr{D}_{\mathscr{R}} \iff B^c \notin \mathscr{E}_{\mathscr{R}}$; *(2)* $B \in \mathscr{E}_{\mathscr{R}} \iff B^c \notin \mathscr{D}_{\mathscr{R}}$.

Theorem 8 *If $\mathscr{R}$ is a relator on X to Y, then for any $B \subseteq Y$ we have*

(1) $B \in \mathscr{D}_{\mathscr{R}}$ *if and only if* $B \cap E \neq \emptyset$ *for all* $E \in \mathscr{E}_{\mathscr{R}}$;
(2) $B \in \mathscr{E}_{\mathscr{R}}$ *if and only if* $B \cap D \neq \emptyset$ *for all* $D \in \mathscr{D}_{\mathscr{R}}$.

Remark 5 By the corresponding definitions, we have $R(x) \in \mathscr{E}_{\mathscr{R}}$, and thus also $R(x)^c \notin \mathscr{D}_{\mathscr{R}}$, for all $x \in X$ and $R \in \mathscr{R}$.

While, by using the notation $\mathscr{U}_{\mathscr{R}}(x) = \operatorname{int}_{\mathscr{R}}^{-1}(x) = \{B \subseteq Y : \ x \in \operatorname{int}_{\mathscr{R}}(B)\}$, we can note that $\mathscr{E}_{\mathscr{R}} = \bigcup_{x \in X} \mathscr{U}_{\mathscr{R}}(x)$.

By using Definition 1, we may easily introduce several further important definitions. For instance, we may also naturally have the following

Definition 2 If $\mathscr{R}$ is a relator on X to Y, then for any $B \subseteq Y$, we define

(1) $\operatorname{bnd}_{\mathscr{R}}(B) = \operatorname{cl}_{\mathscr{R}}(B) \setminus \operatorname{int}_R(B)$.

Moreover, if, in particular, $\mathscr{R}$ is a relator on X, then for any $A \subseteq X$ we also define
(2) $\operatorname{res}_{\mathscr{R}}(A) = \operatorname{cl}_{\mathscr{R}}(A) \setminus A$; (3) $\operatorname{bor}_{\mathscr{R}}(A) = A \setminus \operatorname{int}_{\mathscr{R}}(A)$.

Remark 6 Somewhat differently, the *border, boundary, and residue of a set* in neighborhood and closure spaces were also introduced by Hausdorff and Kuratowski [50, pp. 4–5]. (See also Elez and Papaz [33] for a recent treatment.)

If, in particular, $\mathscr{R}$ is a reflexive relator on X, then by Definition 1, for any $A \subseteq X$, we have $A^\circ \subseteq A \subseteq A^-$. Therefore,

$$\operatorname{bnd}_{\mathscr{R}}(A) = \operatorname{res}_{\mathscr{R}}(A) \cup \operatorname{bor}_{\mathscr{R}}(A) = \operatorname{res}_{\mathscr{R}}(A) \cup \operatorname{res}_{\mathscr{R}}(A^c).$$

Namely, by using Definition 2 and Theorem 2, we can easily see that

$$\operatorname{res}_{\mathscr{R}}(A^c) = A^{c-} \setminus A^c = A^{c-} \cap A^{cc} = A^{\circ c} \cap A = A \setminus A^\circ = \operatorname{bor}_{\mathscr{R}}(A).$$

Note that if, in particular, $A \in \mathscr{T}_{\mathscr{R}}$ in the sense that $A \subseteq A^\circ$, then $\operatorname{bor}_{\mathscr{R}}(A) = \emptyset$. Therefore, in this particular case, by the above equality, we can simply state that $\operatorname{bnd}_{\mathscr{R}}(A) = \operatorname{res}_{\mathscr{R}}(A)$.

5 Further Structures Derived from Relators

By using Definition 1, we may also naturally introduce the following

Definition 3 If $\mathscr{R}$ is a relator on X, then for any $A \subseteq X$ we also define :

(1) $A \in \tau_{\mathscr{R}}$ if $A \in \operatorname{Int}_{\mathscr{R}}(A)$; (2) $A \in \tau\!\!\!-_{\mathscr{R}}$ if $A^c \notin \operatorname{Cl}_{\mathscr{R}}(A)$;
(3) $A \in \mathscr{T}_{\mathscr{R}}$ if $A \subseteq \operatorname{int}_{\mathscr{R}}(A)$; (4) $A \in \mathscr{F}_{\mathscr{R}}$ if $\operatorname{cl}_{\mathscr{R}}(A) \subseteq A$;
(5) $A \in \mathscr{N}_{\mathscr{R}}$ if $\operatorname{cl}_{\mathscr{R}}(A) \notin \mathscr{E}_{\mathscr{R}}$; (6) $A \in \mathscr{M}_{\mathscr{R}}$ if $\operatorname{int}_{\mathscr{R}}(A) \in \mathscr{D}_{\mathscr{R}}$.

Remark 7 The members of the families, $\tau_{\mathscr{R}}$ and $\mathscr{T}_{\mathscr{R}}$ and $\mathscr{N}_{\mathscr{R}}$, are called the *proximally open, topologically open, and rare (or nowhere dense) subsets* of the relator space $X(\mathscr{R})$, respectively.

The family $\tau_{\mathscr{R}}$ was first introduced by the third author in [80, 81]. While, the practical notation $\bar{\tau}_{\mathscr{R}}$ was suggested by János Kurdics who first noticed that "connectedness" is a particular case of "well-chainedness." (See [52, 53, 68, 73].)

By using the corresponding results of Section 4, we can easily establish the following theorems.

Theorem 9 *If $\mathscr{R}$ is a relator on X, then for any $A \subseteq X$, we have*

(1) $A \in \bar{\tau}_{\mathscr{R}} \iff A^c \in \tau_{\mathscr{R}}$; *(2)* $A \in \tau_{\mathscr{R}} \iff A^c \in \bar{\tau}_{\mathscr{R}}$.

Theorem 10 *For any relator $\mathscr{R}$ on X, we have*

(1) $\bar{\tau}_{\mathscr{R}} = \tau_{\mathscr{R}^{-1}}$; *(2)* $\tau_{\mathscr{R}} = \bar{\tau}_{\mathscr{R}^{-1}}$.

Theorem 11 *If $\mathscr{R}$ is a relator on X, then for any $A \subseteq X$, we have*

(1) $A \in \mathscr{F}_{\mathscr{R}} \iff A^c \in \mathscr{T}_{\mathscr{R}}$; *(2)* $A \in \mathscr{T}_{\mathscr{R}} \iff A^c \in \mathscr{F}_{\mathscr{R}}$.

Corollary 2 *If $\mathscr{R}$ is a relator on X and $A \subseteq X$ and $V \in \mathscr{T}_{\mathscr{R}}$ such that $A \cap V = \emptyset$, then $\mathrm{cl}_{\mathscr{R}}(A) \cap V = \emptyset$ also hold.*

Proof By Theorem 11, we have $V^c \in \mathscr{F}_{\mathscr{R}}$. Thus, by Definition 3, we also have $V^{c-} \subseteq V^c$. Hence, by using the increasingness of the operation $-$, we can see that $A \cap V = \emptyset \Rightarrow A \subseteq V^c \Rightarrow A^- \subseteq V^{c\,-} \Rightarrow A^- \subseteq V^c \Rightarrow A^- \cap V = \emptyset$.

Remark 8 Note that if $\mathscr{R}$ is a reflexive relator on X, then $A \subseteq A^-$ for any $A \subseteq X$. Therefore, $A^- \cap V = \emptyset$ trivially implies $A \cap V = \emptyset$ for any $A, V \subseteq X$.

Theorem 12 *For any relator $\mathscr{R}$ on X, we have*

(1) $\tau_{\mathscr{R}} \subseteq \mathscr{T}_{\mathscr{R}}$; *(2)* $\bar{\tau}_{\mathscr{R}} \subseteq \mathscr{F}_{\mathscr{R}}$.

Remark 9 In particular, for any relation R on X, we have

(1) $\tau_R = \mathscr{T}_R$; (2) $\bar{\tau}_R = \mathscr{F}_R$.

Theorem 13 *For any relator $\mathscr{R}$ on X, we have*

(1) $\mathscr{T}_{\mathscr{R}} \setminus \{\emptyset\} \subseteq \mathscr{E}_{\mathscr{R}}$; *(2)* $\mathscr{D}_{\mathscr{R}} \cap \mathscr{F}_{\mathscr{R}} \subseteq \{X\}$.

Remark 10 Hence, by using global complementations, we can easily infer that $\mathscr{F}_{\mathscr{R}} \subseteq \left(\mathscr{D}_{\mathscr{R}}\right)^c \cup \{X\}$ and $\mathscr{D}_{\mathscr{R}} \subseteq \left(\mathscr{F}_{\mathscr{R}}\right)^c \cup \{X\}$.

Theorem 14 *If $\mathscr{R}$ is a relator on X, then for any $A \subseteq X$ we have*

(1) $A \in \mathscr{E}_{\mathscr{R}}$ *if* $V \subseteq A$ *for some* $V \in \mathscr{T}_R \setminus \{\emptyset\}$;
(2) $A \in \mathscr{D}_{\mathscr{R}}$ *only if* $A \setminus W \neq \emptyset$ *for all* $W \in \mathscr{F}_{\mathscr{R}} \setminus \{X\}$.

Remark 11 The fat sets are frequently more convenient tools than the topologically open ones. For instance, if $\leq$ is a relation on X, then $\mathscr{T}_{\leq}$ and $\mathscr{E}_{\leq}$ are the families of all ascending and residual subsets of the goset $X(\leq)$, respectively.

Moreover, if, in particular, $X = \mathbb{R}$ and $R(x) = \{x-1\} \cup [\, x, +\infty\, [$ for all $x \in X$, then R is a reflexive relation on X such that $\mathscr{T}_R = \{\emptyset, X\}$, but $\mathscr{E}_R$ is quite a large family. Namely, the supersets of each $R(x)$ are also contained in $\mathscr{E}_R$.

However, the importance of fat and dense lies mainly in the following

Definition 4 If $\mathscr{R}$ is a relator on X to Y, and φ and ψ are functions of a relator space $\Gamma(\mathscr{U})$ to X and Y, respectively, then using the notation

$$(\varphi,\ \psi)(\gamma) = \big(\varphi(\gamma),\ \psi(\gamma)\big)$$

for all $\gamma \in \Gamma$, we may also naturally define

(1) $\varphi \in \operatorname{Lim}_{\mathscr{R}}(\psi)$ if $(\varphi,\ \psi)^{-1}[\, R\,] \in \mathscr{E}_{\mathscr{U}}$ for all $R \in \mathscr{R}$,
(2) $\varphi \in \operatorname{Adh}_{\mathscr{R}}(\psi)$ if $(\varphi,\ \psi)^{-1}[\, R\,] \in \mathscr{D}_{\mathscr{U}}$ for all $R \in \mathscr{R}$.

Moreover, for any $x \in X$, we may also naturally define:
(3) $x \in \lim_{\mathscr{R}}(\psi)$ if $x_\Gamma \in \operatorname{Lim}_{\mathscr{R}}(\psi)$, (4) $x \in \operatorname{adh}_{\mathscr{R}}(\psi)$ if $x_\Gamma \in \operatorname{Adh}_{\mathscr{R}}(\psi)$,

where x_Γ is a function of Γ to X such that $x_\Gamma(\gamma) = x$ for all $\gamma \in \Gamma$.

Remark 12 Fortunately, the small limit and adherence relations are equivalent to the small closure and interior ones.

However, the big limit and adherence relations, suggested by Efremović and Švarc [31], are usually stronger tools than the big closure and interior ones.

In this respect, it seems convenient to only mention here the following

Theorem 15 *If $\mathscr{R}$ is a relator on X to Y, then for any $A \subseteq X$ and $B \subseteq Y$ the following assertions are equivalent:*

(1) $A \in \operatorname{Cl}_{\mathscr{R}}(B)$;
(2) *there exist functions φ and ψ of the poset $\mathscr{R}(\supseteq)$ to A and B, respectively, such that $\varphi \in \operatorname{Lim}_{\mathscr{R}}(\psi)$;*
(3) *there exist functions φ and ψ of a relator space $\Gamma(\mathscr{U})$ to A and B, respectively, such that $\varphi \in \operatorname{Lim}_{\mathscr{R}}(\psi)$.*

Hint. If (1) holds, then for each $R \in \mathscr{R}$, we have $R\,[\, A\,] \cap B \neq \emptyset$. Therefore, there exist $\varphi(R) \in A$ and $\psi(R) \in B$ such that $\psi(R) \in R(\varphi(R))$. Hence, we can see that $(\varphi, \psi)(R) = (\varphi(R), \psi(R)) \in R$, and thus $R \in (\varphi \boxtimes \psi)^{-1}[\, R\,]$.

Therefore, if $R \in \mathscr{R}$, then for any $S \in \mathscr{R}$, with $R \supseteq S$, we have

$$S \in (\varphi,\ \psi)^{-1}[\, S\,] \subseteq (\varphi,\ \psi)^{-1}[\, R\,].$$

This shows that $(\varphi, \psi)^{-1}[\, R\,]$ is a fat subset of $\mathscr{R}(\supseteq)$, and thus $\varphi \in \operatorname{Lim}_{\mathscr{R}}(\psi)$.

Remark 13 Finally, we note that if $\mathscr{R}$ is a relator on X to Y, then according to [84], for any $A \subseteq X$ and $B \subseteq Y$, we may also naturally write $A \in \mathrm{Lb}_{\mathscr{R}}(B)$ and $B \in \mathrm{Ub}_{\mathscr{R}}(A)$ if there exists $R \in \mathscr{R}$ such that $A \times B \subseteq R$.

However, the algebraic structures $\mathrm{Lb}_{\mathscr{R}}$ and $\mathrm{Ub}_{\mathscr{R}}$, and the structures derivable from them, are not independent of the former topological ones. Namely, it can be easily shown that $\mathrm{Lb}_{\mathscr{R}} = \mathrm{Int}_{\mathscr{R}^c} \circ \mathscr{C}_Y$ and $\mathrm{Int}_{\mathscr{R}} = \mathrm{Lb}_{\mathscr{R}^c} \circ \mathscr{C}_Y$.

6 Reflexive, Non-partial, and Non-degenerated Relators

Definition 5 A relator $\mathscr{R}$ on X is called *reflexive* if each member R of $\mathscr{R}$ is a reflexive relation on X.

Remark 14 Thus, the following assertions are equivalent :

(1) $\mathscr{R}$ is reflexive ;
(2) $x \in R(x)$ for all $x \in X$ and $R \in \mathscr{R}$;
(3) $A \subseteq R[A]$ for all $A \subseteq X$ and $R \in \mathscr{R}$.

The importance of reflexive relators is also apparent from the following two obvious theorems.

Theorem 16 *For a relator $\mathscr{R}$ on X, the following assertions are equivalent:*

(1) $\rho_{\mathscr{R}}$ is reflexive; *(2) $\mathscr{R}$ is reflexive;*
(3) $A \subseteq \mathrm{cl}_{\mathscr{R}}(A)$ $\big(\mathrm{int}_{\mathscr{R}}(A) \subseteq A\big)$ for all $A \subseteq X$.

Proof To prove the equivalence of (1) and (2), recall that by Corollary 1 we have $\rho_{\mathscr{R}} = \left(\bigcap \mathscr{R}\right)^{-1}$.

Remark 15 Thus, the relator $\mathscr{R}$ is reflexive if and only if $A^\circ \subseteq A$ $(A \subseteq A^-)$ for all $A \subseteq X$.

Therefore, if $\mathscr{R}$ is a reflexive relator on X, then for any $A \subseteq X$ we have $A \in \mathscr{T}_{\mathscr{R}}$ $(A \in \mathscr{F}_{\mathscr{R}})$ if and only if $A^\circ = A$ $(A^- = A)$.

Theorem 17 *For a relator $\mathscr{R}$ on X, the following assertions are equivalent:*

(1) $\mathscr{R}$ is reflexive;
(2) $A \in \mathrm{Int}_{\mathscr{R}}(B)$ implies $A \subseteq B$ for all $A, B \subseteq X$;
(3) $A \cap B \neq \emptyset$ implies $A \in \mathrm{Cl}_{\mathscr{R}}(B)$ for all $A, B \subseteq X$.

Remark 16 In addition to the above two theorems, it is also worth mentioning that if $\mathscr{R}$ is a reflexive relator on X, then

(1) $\mathrm{Int}_{\mathscr{R}}$ is a transitive relation on $\mathscr{P}(X)$;
(2) $B \in \mathrm{Cl}_{\mathscr{R}}(A)$ implies $\mathscr{P}(X) = \mathrm{Cl}_{\mathscr{R}}(A)^c \cup \mathrm{Cl}_{\mathscr{R}}^{-1}(B)$;
(3) $\mathrm{int}_{\mathscr{R}}\big(\mathrm{bor}_{\mathscr{R}}(A)\big) = \emptyset$ and $\mathrm{int}_{\mathscr{R}}\big(\mathrm{res}_{\mathscr{R}}(A)\big) = \emptyset$ for all $A \subseteq X$.

Thus, for instance, for any $A \subseteq X$ we have $\mathrm{res}_{\mathscr{R}}(A) \in \mathscr{T}_{\mathscr{R}}$ if and only if $A \in \mathscr{F}_{\mathscr{R}}$.

Analogously to Definition 5, we may also naturally have the following

Definition 6 A relator $\mathscr{R}$ on X to Y is called *non-partial* if each member R of $\mathscr{R}$ is a non-partial relation on X to Y.

Remark 17 Thus, the following assertions are equivalent :

(1) $\mathscr{R}$ is non-partial ;
(2) $R^{-1}[\,Y\,]=X$ for all $R\in\mathscr{R}$;
(3) $R(x)\neq\emptyset$ for all $x\in X$ and $R\in\mathscr{R}$.

The importance of non-partial relators is apparent from the following

Theorem 18 *For a relator $\mathscr{R}$ on X to Y, the following assertions are equivalent:*

(1) $\mathscr{R}$ is non-partial;
(2) $\emptyset\notin\mathscr{E}_{\mathscr{R}}$; (3) $\mathscr{D}_{\mathscr{R}}\neq\emptyset$; (4) $Y\in\mathscr{D}_{\mathscr{R}}$; (5) $\mathscr{E}_{\mathscr{R}}\neq\mathscr{P}(Y)$.

In the sequel, we shall also need the following localized form of Definition 6.

Definition 7 A relator $\mathscr{R}$ on X will be called *locally non-partial* if for each $x\in X$ there exists $R\in\mathscr{R}$ such that for any $y\in R\,(x)$ and $S\in\mathscr{R}$ we have $S\,(y)\neq\emptyset$.

Remark 18 Thus, if either $X=\emptyset$ or $\mathscr{R}$ is nonvoid and non-partial, then $\mathscr{R}$ is locally non-partial.

Moreover, by using the corresponding definitions, we can also easily prove

Theorem 19 *For a relator $\mathscr{R}$ on X, the following assertions are equivalent:*

(1) $\mathscr{R}$ is locally non-partial; *(2) $X=\operatorname{int}_{\mathscr{R}}\big(\operatorname{cl}_{\mathscr{R}}(X)\big)$.*

Proof To prove the implication (1) $\Rightarrow$ (2), note that if (1) holds, then for each $x\in X$ there exists $R\in\mathscr{R}$ such that for any $y\in R\,(x)$ and for any $S\in\mathscr{R}$ we have $S\,(y)\cap X=S\,(y)\neq\emptyset$, and thus $y\in\operatorname{cl}_{\mathscr{R}}(X)$.

Therefore, for each $x\in X$ there exists $R\in\mathscr{R}$ such that $R\,(x)\subseteq\operatorname{cl}_{\mathscr{R}}(X)$, and thus $x\in\operatorname{int}_{\mathscr{R}}\big(\operatorname{cl}_{\mathscr{R}}(X)\big)$. Hence, we can already see that $X\subseteq\operatorname{int}_{\mathscr{R}}\big(\operatorname{cl}_{\mathscr{R}}(X)\big)$, and thus (2) also holds. (Therefore, by a former notation, $X\in\mathscr{T}^{\,r}_{\mathscr{R}}$.)

In addition to Definition 6, it is also worth introducing the following

Definition 8 A relator $\mathscr{R}$ on X to Y is called *non-degerated* if $X\neq\emptyset$ and $\mathscr{R}\neq\emptyset$.

Thus, analogously to Theorem 18, we can also easily establish the following

Theorem 20 *For a relator $\mathscr{R}$ on X to Y, the following assertions are equivalent:*

(1) $\mathscr{R}$ is non-degenerated;
(2) $\emptyset\notin\mathscr{D}_{\mathscr{R}}$; (3) $\mathscr{E}_{\mathscr{R}}\neq\emptyset$; (4) $Y\in\mathscr{E}_{\mathscr{R}}$; (5) $\mathscr{D}_{\mathscr{R}}\neq\mathscr{P}(Y)$.

Remark 19 In addition to Theorems 18 and 20, it is also worth mentioning that if a relator $\mathscr{R}$ on X to Y is paratopologically simple in the sense that $\mathscr{E}_{\mathscr{R}}=\mathscr{E}_{R}$ for some relation R on X to Y, then the stack $\mathscr{E}_{\mathscr{R}}$ has a base $\mathscr{B}$ with $\operatorname{card}(\mathscr{B})\leq\operatorname{card}(X)$. (See [66, Theorem 5.9] of Pataki.)

The existence of a non-paratopologically simple (actually finite equivalence) relator, proved first by Pataki [66, Example 5.11], shows that in our definitions of the relations $\mathrm{Lim}_{\mathscr{R}}$ and $\mathrm{Adh}_{\mathscr{R}}$ we cannot restrict ourselves to functions of gosets (generalized ordered sets) without some loss of generality.

7 Topological and Quasi-Topological Relators

The following improvement of [79, Definition 2.1] was first considered in [80].

Definition 9 A relator $\mathscr{R}$ on X is called :

(1) *quasi-topological* if $x \in \mathrm{int}_{\mathscr{R}}\big(\mathrm{int}_{\mathscr{R}}\big(R(x)\big)\big)$ for all $x \in X$ and $R \in \mathscr{R}$;
(2) *topological* if for any $x \in X$ and $R \in \mathscr{R}$ there exists $V \in \mathscr{T}_{\mathscr{R}}$ such that $x \in V \subseteq R(x)$.

The appropriateness of these definitions is already quite obvious from the following four theorems.

Theorem 21 *For a relator $\mathscr{R}$ on X, the following assertions are equivalent:*

(1) *$\mathscr{R}$ is quasi-topological;*
(2) $\mathrm{int}_{\mathscr{R}}\big(R(x)\big) \in \mathscr{T}_{\mathscr{R}}$ *for all $x \in X$ and $R \in \mathscr{R}$;*
(3) $\mathrm{cl}_{\mathscr{R}}(A) \in \mathscr{F}_{\mathscr{R}}$ $\big(\mathrm{int}_{\mathscr{R}}(A) \in \mathscr{T}_{\mathscr{R}}\big)$ *for all $A \subseteq X$.*

Remark 20 Hence, by Definition 3, we can see that the relator $\mathscr{R}$ is quasi-topological if and only if $A^{\circ} \subseteq A^{\circ\circ}$ $(A^{--} \subseteq A^{-})$ for all $A \subseteq X$.

Theorem 22 *For a relator $\mathscr{R}$ on X, the following assertions are equivalent:*

(1) $\mathscr{R}$ is topological; *(2) $\mathscr{R}$ is reflexive and quasi-topological.*

Remark 21 By Theorem 21, the relator $\mathscr{R}$ may be called *weakly (strongly) quasi-topological* if $\rho_{\mathscr{R}}(x) \in \mathscr{F}_{\mathscr{R}}$ $\big(R(x) \in \mathscr{T}_{\mathscr{R}}\big)$ for all $x \in X$ and $R \in \mathscr{R}$.

Moreover, by Theorem 22, the relator $\mathscr{R}$ may be called *weakly (strongly) topological* if it is reflexive and weakly (strongly) quasi-topological.

Theorem 23 *For a relator $\mathscr{R}$ on X, the following assertions are equivalent:*

(1) *$\mathscr{R}$ is topological;*
(2) $\mathrm{int}_{\mathscr{R}}(A) = \bigcup\big(\mathscr{T}_{\mathscr{R}} \cap \mathscr{P}(A)\big)$ *for all $A \subseteq X$;*
(3) $\mathrm{cl}_{\mathscr{R}}(A) = \bigcap\big(\mathscr{F}_{\mathscr{R}} \cap \mathscr{P}^{-1}(A)\big)$ *for all $A \subseteq X$.*

Now, as an immediate consequence of this theorem, we can also state

Corollary 3 *If $\mathscr{R}$ is topological relator on X, then for any $A \subset X$, we have*

(1) $A \in \mathscr{E}_{\mathscr{R}}$ *if and only if there exists $V \in \mathscr{T}_{\mathscr{R}} \setminus \{\emptyset\}$ such that $V \subseteq A$;*
(2) $A \in \mathscr{D}_{\mathscr{R}}$ *if and only if for all $W \in \mathscr{F}_{\mathscr{R}} \setminus \{X\}$ we have $A \setminus W \neq \emptyset$.*

However, it is now more important to note that we can also prove the following

Theorem 24 *For a relator $\mathscr{R}$ on X, the following assertions are equivalent:*

(1) *$\mathscr{R}$ is topological;*
(2) *$\mathscr{R}$ is topologically equivalent to a preorder relator on X.*

Proof To prove the implication (1) $\Rightarrow$ (2), note that if (1) holds, then by Definition 9, for any $x \in X$ and $R \in \mathscr{R}$, there exists $V \in \mathscr{T}_{\mathscr{R}}$ such that $x \in V \subseteq R(x)$. Thus, by using the Pervin preorder relator

$$\mathscr{S} = \mathscr{R}_{\mathscr{T}_{\mathscr{R}}} = \{R_V : \quad V \in \mathscr{T}_{\mathscr{R}}\}, \qquad \text{where} \qquad R_V = V^2 \cup (V^c \times X),$$

we can show that $\operatorname{int}_{\mathscr{R}}(A) = \operatorname{int}_{\mathscr{S}}(A)$ for all $A \subseteq X$, and thus (2) also holds.

For this, we have to note that

$$R_V(x) = V \quad \text{if} \quad x \in V \qquad \text{and} \qquad R_V(x) = X \quad \text{if} \quad x \in V^c.$$

Remark 22 In addition to Theorems 21 and 22, it is also worth proving that a relator $\mathscr{R}$ on X is quasi-topological if and only if its topological closure

$$\mathscr{R}^{\wedge} = \{S \subseteq X^2 : \quad \forall\ x \in X : \quad x \in \operatorname{int}_{\mathscr{R}}(S(x))\}$$

is topologically transitive in the sense that $\mathscr{R}^{\wedge} \subseteq (\mathscr{R}^{\wedge} \circ \mathscr{R}^{\wedge})^{\wedge}$.

This property can be reformulated in the simpler form that $\mathscr{R} \subseteq (\mathscr{R}^{\wedge} \circ \mathscr{R})^{\wedge}$. That is, for each $x \in X$ and $R \in \mathscr{R}$ there exist $S \in \mathscr{R}$ and $T \in \mathscr{R}^{\wedge}$ such that $T[S(x)] \subseteq R(x)$.

Remark 23 Analogously to Definition 9, the relator $\mathscr{R}$ may be called

(1) *quasi-proximal* if $A \in \operatorname{Int}_{\mathscr{R}}[\tau_{\mathscr{R}} \cap \operatorname{Int}_{\mathscr{R}}(R[A])]$ for all $A \subseteq X$ and $R \in \mathscr{R}$;
(2) *proximal* if for any $A \subseteq X$ and $R \in \mathscr{R}$ there exists $V \in \tau_{\mathscr{R}}$ such that $A \subseteq V \subseteq R[A]$.

Thus, in addition to the counterparts of our former theorems, we can prove that $\mathscr{R}$ is topological if and only if its topological refinement $\mathscr{R}^{\wedge}$ is proximal.

Therefore, several theorems on topological relators can be derived from those on the proximal ones by using that $\tau_{\mathscr{R}^{\wedge}} = \mathscr{T}_{\mathscr{R}}$ whenever $\mathscr{R} \neq \emptyset$.

8 A Few Basic Facts on Filtered Relators

Intersection properties of relators were also first investigated in [79, 80].

Definition 10 A relator $\mathscr{R}$ on X to Y is called

(1) *properly filtered* if for any $R,\ S \in \mathscr{R}$ we have $R \cap S \in \mathscr{R}$;

(2) *uniformly filtered* if for any $R,\ S \in \mathscr{R}$ there exists $T \in \mathscr{R}$ such that $T \subseteq R \cap S$;
(3) *proximally filtered* if for any $A \subseteq X$ and $R,\ S \in \mathscr{R}$ there exists $T \in \mathscr{R}$ such that $T\,[\,A\,] \subseteq R\,[\,A\,] \cap S\,[\,A\,]$;
(4) *topologically filtered* if for any $x \in X$ and $R,\ S \in \mathscr{R}$ there exists $T \in \mathscr{R}$ such that $T\,(x) \subseteq R\,(x) \cap S\,(x)$.

Remark 24 By using the binary operation $\wedge$ and the basic closure operations on relators, the above properties can be reformulated in some more concise forms.

For instance, we can see that $\mathscr{R}$ is topologically filtered if and only if any one of the properties $\mathscr{R} \wedge \mathscr{R} \subseteq \mathscr{R}^{\wedge}$, $(\mathscr{R} \wedge \mathscr{R})^{\wedge} = \mathscr{R}^{\wedge}$ and $\mathscr{R}^{\wedge} \wedge \mathscr{R}^{\wedge} = \mathscr{R}^{\wedge}$ holds.

However, in general, we only have $(R \cap S)[\,A\,] \subseteq R\,[\,A\,] \cap S\,[\,A\,]$. Therefore, the corresponding proximal filteredness properties are, unfortunately, not equivalent.

Despite this, we can easily prove the following theorem which shows the appropriateness of the above proximal filteredness property.

Theorem 25 *For a relator $\mathscr{R}$ on X to Y, the following assertions are equivalent:*

(1) *$\mathscr{R}$ is proximally filtered;*
(2) $\mathrm{Cl}_{\mathscr{R}}\,(A \cup B) = \mathrm{Cl}_{\mathscr{R}}\,(A) \cup \mathrm{Cl}_{\mathscr{R}}\,(B)$ *for all $A, B \subseteq Y$;*
(3) $\mathrm{Int}_{\mathscr{R}}\,(A \cap B) = \mathrm{Int}_{\mathscr{R}}\,(A) \cap \mathrm{Int}_{\mathscr{R}}\,(B)$ *for all $A, B \subseteq Y$.*

Proof To prove the implication (3) $\Rightarrow$ (1), note that if $A \subseteq X$ and $R,\ S \in \mathscr{R}$, then by the definition of $\mathrm{Int}_{\mathscr{R}}$ we trivially have $A \in \mathrm{Int}_{\mathscr{R}}\big(R\,[\,A\,]\big)$ and $A \in \mathrm{Int}_{\mathscr{R}}\big(S\,[\,A\,]\big)$. Therefore, if (3) holds, then we also have $A \in \mathrm{Int}_{\mathscr{R}}\big(R\,[\,A\,] \cap S\,[\,A\,]\big)$. Thus, by the definition of $\mathrm{Int}_{\mathscr{R}}$, there exists $T \in \mathscr{R}$ such that $T\,[\,A\,] \subseteq R\,[\,A\,] \cap S\,[\,A\,]$.

Now, as an immediate consequence of this theorem, we can also state

Corollary 4 *If $\mathscr{R}$ is a proximally filtered relator on X, then the families $\bar{\tau}_{\mathscr{R}}$ and $\tau_{\mathscr{R}}$ are closed under binary unions and intersections, respectively.*

Analogously to Theorem 25, we can also easily prove the following

Theorem 26 *For a relator $\mathscr{R}$ on X to Y, the following assertions are equivalent:*

(1) *$\mathscr{R}$ is topologically filtered;*
(2) $\mathrm{cl}_{\mathscr{R}}\,(A \cup B) = \mathrm{cl}_{\mathscr{R}}\,(A) \cup \mathrm{cl}_{\mathscr{R}}\,(B)$ *for all $A, B \subseteq Y$;*
(3) $\mathrm{int}_{\mathscr{R}}\,(A \cap B) = \mathrm{int}_{\mathscr{R}}\,(A) \cap \mathrm{int}_{\mathscr{R}}\,(B)$ *for all $A, B \subseteq Y$.*

Thus, in particular, we can also state the following

Corollary 5 *If $\mathscr{R}$ is a topologically filtered relator on X, then the families $\mathscr{F}_{\mathscr{R}}$ and $\mathscr{T}_{\mathscr{R}}$ are closed under binary unions and intersections, respectively.*

The following example shows that, for a non-topological relator $\mathscr{R}$, the converse of the above corollary need not be true.

Example 1 If $X = \{1, 2, 3\}$ and R_i is relation on X, for each $i = 1, 2$, such that

$$R_i(1) = \{1,\ i+1\} \qquad \text{and} \qquad R_i(2) = R_i(3) = \{2, 3\},$$

then $\mathscr{R} = \{R_1, R_2\}$ is reflexive relator on X such that $\mathscr{T}_{\mathscr{R}}$ is closed under arbitrary intersections, but $\mathscr{R}$ is still not topologically filtered.

By the corresponding definitions, it is clear that $\mathscr{T}_{\mathscr{R}} = \{\emptyset, \{2, 3\}, X\}$. Moreover, we can note that $R_i(1) \not\subseteq R_1(1) \cap R_2(1)$ for each $i = 1, 2$, and thus by Definition 10 the relator $\mathscr{R}$ is not topologically filtered.

9 A Few Basic Facts on Quasi-Filtered Relators

Since $R \subseteq R^{\infty}$ for every relation R on X, in addition to Definition 10, we may also naturally introduce the following

Definition 11 A relator $\mathscr{R}$ on X is called

(1) *quasi-uniformly filtered* if for any $R, S \in \mathscr{R}$ there exists $T \in \mathscr{R}$ such that $T \subseteq R^{\infty} \cap S^{\infty}$;
(2) *quasi-proximally filtered* if for any $A \subseteq X$ and $R, S \in \mathscr{R}$ there exists $T \in \mathscr{R}$ such that $T[A] \subseteq R^{\infty}[A] \cap S^{\infty}[A]$;
(3) *quasi-topologically filtered* if for any $x \in X$ and $R, S \in \mathscr{R}^{\wedge}$ there exists $T \in \mathscr{R}$ such that $T(x) \subseteq R^{\infty}(x) \cap S^{\infty}(x)$.

Remark 25 Analogously to Remark 24, the above quasi-filteredness properties can also be reformulated in some more concise forms.

For instance, we can see that $\mathscr{R}$ is quasi-topologically filtered if and only if $\mathscr{R}^{\wedge\infty} \wedge \mathscr{R}^{\wedge\infty} \subseteq \mathscr{R}^{\wedge}$, $(\mathscr{R}^{\wedge\infty} \wedge \mathscr{R}^{\wedge\infty})^{\wedge\infty} = \mathscr{R}^{\wedge\infty}$ or $\mathscr{R}^{\wedge\infty} \wedge \mathscr{R}^{\wedge\infty} = \mathscr{R}^{\wedge\infty}$.

However, it is now more important to note that, by using some former results, we can also prove the following two theorems which show the appropriateness of the above quasi-proximal and quasi-topological filteredness properties.

Theorem 27 *For any relator $\mathscr{R}$ on X, the following assertions are equivalent:*

(1) *$\mathscr{R}$ is a quasi-proximally filtered;*
(2) *$\tau\!\!\!-_{\mathscr{R}}$ is closed under binary unions;*
(3) *$\tau_{\mathscr{R}}$ is closed under binary intersections.*

Theorem 28 *For any relator $\mathscr{R}$ on X, the following assertions are equivalent:*

(1) *$\mathscr{R}$ is a quasi-topologically filtered;*
(2) *$\mathscr{F}_{\mathscr{R}}$ is closed under binary unions;*
(3) *$\mathscr{T}_{\mathscr{R}}$ is closed under binary intersections.*

Remark 26 In this respect it is also worth mentioning that if $\mathscr{R}$ is a relator on X to Y, then the family $\mathscr{E}_{\mathscr{R}}$ is closed under binary intersections if and only if $\mathscr{R}$ is *quasi-directed* in the sense that for any $x, y \in X$ and $R, S \in \mathscr{R}$ we have $R(x) \cap S(y) \in \mathscr{E}_{\mathscr{R}}$.

From the above two theorems, by using Corollaries 4 and 5, we can derive

Corollary 6 *If $\mathcal{R}$ is a proximally (topologically) filtered relator on X, then $\mathcal{R}$ is also quasi-proximally (quasi-topologically) filtered.*

Now, by using Theorem 27, we can also easily prove the following

Theorem 29 *If $\mathcal{R}$ is a quasi-proximally filtered, proximal relator on X, then $\mathcal{R}$ is proximally filtered.*

Proof Suppose that $A \subseteq X$ and $R,\ S \in \mathcal{R}$. Then, by Remark 23, there exist $U,\ V \in \tau_{\mathcal{R}}$ such that $A \subseteq U \subseteq R\,[\,A\,]$ and $A \subseteq V \subseteq S\,[\,A\,]$. Moreover, by Theorem 27, we can state that $U \cap V \in \tau_{\mathcal{R}}$. Therefore, by the definition of $\tau_{\mathcal{R}}$, there exists $T \in \mathcal{R}$ such that $T\,[\,U \cap V\,] \subseteq U \cap V$. Hence, we can already see that

$$T\,[\,A\,] \subseteq T\,[\,U \cap V\,] \subseteq U \cap V \subseteq R\,[\,A\,] \cap S\,[\,A\,].$$

Moreover, by using Theorem 28, we can quite similarly prove the following

Theorem 30 *If $\mathcal{R}$ is a quasi-topologically filtered, topological relator on X, then $\mathcal{R}$ is topologically filtered.*

Remark 27 Our former Example 1 shows that even a quasi-proximally filtered, reflexive relator need not be topologically filtered.

Namely, if X and $\mathcal{R}$ are as in Example 1, then by the corresponding definitions it is clear that $\tau_{\mathcal{R}} = \big\{\emptyset, \{2,\ 3\},\ X\,\big\}$, and thus by Theorem 27 the relator $\mathcal{R}$ is quasi-proximally filtered.

10 Some Further Theorems on Topologically Filtered Relators

In our former paper [70], by using the arguments of Kuratowski [51, pp. 39, 45], we have proved the following basic theorems whose slightly shortened proofs are included here for the reader's convenience.

Theorem 31 *If $\mathcal{R}$ is a topologically filtered relator on X to Y, then for any $A, B \subseteq Y$ we have*

$$\mathrm{cl}_{\mathcal{R}}(A) \setminus \mathrm{cl}_{\mathcal{R}}(B) = \mathrm{cl}_{\mathcal{R}}(A \setminus B) \setminus \mathrm{cl}_{\mathcal{R}}(B).$$

Proof By using Theorem 26, we can see that

$$A^{-} \cup B^{-} = (A \cup B)^{-} = \big((A \setminus B) \cup B\,\big)^{-} = (A \setminus B)^{-} \cup B^{-}.$$

Hence, because of the identity $(U \cup V\,) \setminus V = U \setminus V$, the required equality follows.

This theorem can be derived from its subsequent corollary which can be proved directly, without using Theorem 26.

Corollary 7 *If* $\mathscr{R}$ *is a topologically filtered relator on* X *to* Y, *then for any* $A, B \subseteq Y$ *we have* $\mathrm{cl}_{\mathscr{R}}(A) \setminus \mathrm{cl}_{\mathscr{R}}(B) \subseteq \mathrm{cl}_{\mathscr{R}}(A \setminus B)$.

This corollary already allows us to easily prove the following

Theorem 32 *If* $\mathscr{R}$ *is a topologically filtered relator on X, then for any* $A \subseteq X$ *and* $U \in \mathscr{T}_{\mathscr{R}}$ *we have*

$$\mathrm{cl}_{\mathscr{R}}(A) \cap U = \mathrm{cl}_{\mathscr{R}}(A \cap U) \cap U.$$

Proof By Definition 3 and Theorem 2, we have $U \subseteq U^{\circ} = U^{c-c}$. Hence, by using Corollary 7, we can infer that

$$A^{-} \cap U \subseteq A^{-} \cap U^{c-c} = A^{-} \setminus U^{c-} \subseteq (A \setminus U^{c})^{-} = (A \cap U)^{-}.$$

Therefore, $A^{-} \cap U = A^{-} \cap U \cap U \subseteq (A \cap U)^{-} \cap U$.

Moreover, by using the increasingness of $-$, we can see that $(A \cap U)^{-} \subseteq A^{-}$, and thus $(A \cap U)^{-} \cap U \subseteq A^{-} \cap U$ is always true. Therefore, we actually have $A^{-} \cap U = (A \cap U)^{-} \cap U$.

Hence, we can see that this theorem can also be derived from its

Corollary 8 *If* $\mathscr{R}$ *is a topologically filtered relator on X, then for any* $A \subseteq X$ *and* $U \in \mathscr{T}_{\mathscr{R}}$ *we have* $\mathrm{cl}_{\mathscr{R}}(A) \cap U \subseteq \mathrm{cl}_{\mathscr{R}}(A \cap U)$.

A direct proof. Assume that $x \in A^{-} \cap U$ and $R \in \mathscr{R}$. Then, since $x \in U \in \mathscr{T}_{\mathscr{R}}$, there exists $S \in \mathscr{R}$ such that $S(x) \subseteq U$. Moreover, since $\mathscr{R}$ is topologically filtered, there exists $T \in \mathscr{R}$ such that $T(x) \subseteq R(x) \cap S(x)$. Furthermore, since $x \in A^{-}$, there exists $y \in A$ such that $y \in T(x)$. Hence, we can already infer that

$$y \in A \cap T(x) \subseteq A \cap S(x) \subseteq A \cap U \qquad \text{and} \qquad y \in T(x) \subseteq R(x).$$

Therefore, $R(x) \cap (A \cap U) \neq \emptyset$, and thus $x \in (A \cap U)^{-}$ also holds.

Remark 28 The importance of the closure space counterpart of Corollary 8 was also recognized Császár [16–19, 22, 23] and Sivagami [75] who assumed it as an axiom for an increasing set-to-set function γ.

Moreover, it is also worth noticing that, by using Theorem 2, Corollary 8 can be reformulated in the dual form that if $\mathscr{R}$ is a topologically filtered relator on X, then for any $A \subseteq X$ and $V \in \mathscr{F}_{\mathscr{R}}$ we have $\mathrm{int}_{\mathscr{R}}(A \cup V) \subseteq \mathrm{int}_{\mathscr{R}}(A) \cup V$.

11 Some More Particular Theorems on Topologically Filtered Relators

By using Corollary 8, we can also easily prove the following

Theorem 33 *If $\mathcal{R}$ is a topologically filtered, topological relator on X, then for any $A \subseteq X$ and $U \in \mathcal{T}_{\mathcal{R}}$ we have*

$$\mathrm{cl}_{\mathcal{R}}(A \cap U) = \mathrm{cl}_{\mathcal{R}}\big(\mathrm{cl}_{\mathcal{R}}(A) \cap U\big).$$

Proof By Corollary 8 we have $A^- \cap U \subseteq (A \cap U)^-$. Hence, by using Theorems 22 and 21, we can infer that

$$(A^- \cap U)^- \subseteq (A \cap U)^{--} \subseteq (A \cap U)^-.$$

On the other hand, by Theorem 16, we have $A \subseteq A^-$, and thus also $A \cap B \subseteq A^- \cap B$. Hence, we can infer that $(A \cap B)^- \subseteq (A^- \cap B)^-$. Therefore, the corresponding equality is also true.

From this theorem, we can immediately derive

Corollary 9 *If $\mathcal{R}$ is a topologically filtered, topological relator on X, then for any $A \in \mathcal{D}_{\mathcal{R}}$ and $U \in \mathcal{T}_{\mathcal{R}}$ we have*

$$\mathrm{cl}_{\mathcal{R}}(U) = \mathrm{cl}_{\mathcal{R}}(A \cap U).$$

Proof By Definition 1 and Theorem 33, we evidently have

$$U^- = (X \cap U)^- = (A^- \cap U)^- = (A \cap U)^-.$$

Now, by modifying an argument of Levine [55], we can also prove

Theorem 34 *If $\mathcal{R}$ is a nonvoid, topological relator on X and $A \subseteq X$ such that $\mathrm{cl}_{\mathcal{R}}(U) = \mathrm{cl}_{\mathcal{R}}(A \cap U)$ for all $U \in \mathcal{T}_{\mathcal{R}}$, then $A \in \mathcal{D}_{\mathcal{R}}$.*

Proof Assume on the contrary that $A \notin \mathcal{D}_{\mathcal{R}}$. Then, by Definition 1, there exists $x \in X$ such that $x \notin A^-$. Thus, by Definition 1, there exists $R \in \mathcal{R}$ such that $A \cap R(x) = \emptyset$. Moreover, by Definition 9, there exists $U \in \mathcal{T}_{\mathcal{R}}$ such that $x \in U \subseteq R(x)$. Thus, in particular, we also have $A \cap U = \emptyset$.

Hence, by using the assumptions of the theorem, we can infer that

$$U^- = (A \cap U)^- = \emptyset^- = \emptyset.$$

Note that the latter equality already requires that $\mathcal{R} \neq \emptyset$.

On the other hand, from the inclusion $x \in U$, by using Theorems 22 and 16 and the increasingness of $-$, we can infer that $x \in \{x\}^- \subseteq U^-$, and thus $U^- \neq \emptyset$. This contradiction proves that $A \in \mathcal{D}_{\mathcal{R}}$.

Remark 29 If $\mathscr{R}$ is a nonvoid, reflexive relator on X and $A \subseteq X$ such that

$$\operatorname{cl}_{\mathscr{R}}\left(R(x)\right) = \operatorname{cl}_{\mathscr{R}}\left(A \cap R(x)\right)$$

for all $x \in X$ and $R \in \mathscr{R}$, then we can even more easily prove that $A \in \mathscr{D}_{\mathscr{R}}$.

In addition to Theorem 34, we can also prove the following

Theorem 35 *If $\mathscr{R}$ is a topologically filtered, topological relator on X, then for any $U \in \mathscr{T}_{\mathscr{R}}$ we have*

$$\operatorname{res}_{\mathscr{R}}(U) \in \mathscr{F}_{\mathscr{R}} \setminus \mathscr{E}_{\mathscr{R}}.$$

Proof By Theorem 11, we have $U^c \in \mathscr{F}_{\mathscr{R}}$. Moreover, by Theorems 22 and 21, we also have $U^- \in \mathscr{F}_{\mathscr{R}}$. Hence, by using the notation $U^\dagger = \operatorname{res}_{\mathscr{R}}(U)$ and Corollary 5, we can already infer that

$$U^\dagger = U^- \setminus U = U^- \cap U^c \in \mathscr{F}_{\mathscr{R}}.$$

Moreover, by using Theorems 26 and 16 and the increasingness of $-$, we can also see that

$$U^{\dagger\circ} = \left(U^- \setminus U\right)^\circ = \left(U^- \cap U^c\right)^\circ = U^{-\circ} \cap U^{c\circ} = U^{-\circ} \cap U^{-c} \subseteq U^- \cap U^{-c} = \emptyset,$$

and thus $U^{\dagger\circ} = \emptyset$. Therefore, $U^\dagger \notin \mathscr{E}_{\mathscr{R}}$, and thus $U^\dagger \in \mathscr{F}_{\mathscr{R}} \setminus \mathscr{E}_{\mathscr{R}}$.

Now, as an immediate consequence of this theorem, we can also state

Corollary 10 *If $\mathscr{R}$ is a topologically filtered, topological relator on X, then $\operatorname{res}_{\mathscr{R}}(U) \in \mathscr{N}_{\mathscr{R}}$ for all $U \in \mathscr{T}_{\mathscr{R}}$.*

Remark 30 Note that if $\mathscr{R}$ is a topological relator on X and $U \in \mathscr{T}_{\mathscr{R}}$, then by Theorems 22, 21, and 16 we have $U = U^\circ$. Therefore, under the notation $U^\ddagger = \operatorname{bnd}_{\mathscr{R}}(U)$, we have $U^\dagger = U^- \setminus U = U^- \setminus U^\circ = U^\ddagger$.

Moreover, in Theorem 35 and Corollary 10, it is also enough to assume only that $\mathscr{R}$ is a quasi-topologically filtered, topological relator on X. Namely, in this case, $\mathscr{R}$ is already topologically filtered by Theorem 30.

12 Some Generalized Topologically Open Sets

Notation 1 *In the sequel, we shall always assume that X is a set and $\mathscr{R}$ is a relator on X.*

Moreover, to shorten the subsequent proofs, we shall again use the notations

$$A^- = \operatorname{cl}_{\mathscr{R}}(A), \qquad A^\circ = \operatorname{int}_{\mathscr{R}}(A) \qquad \text{and} \qquad A^\dagger = \operatorname{res}_{\mathscr{R}}(A).$$

Motivated by the corresponding definitions in topological spaces, listed in the Introduction, we shall use the following

Definition 12 For a subset A of the relator space $X(\mathscr{R})$, we shall write

(1) $A \in \mathcal{T}_{\mathscr{R}}^{s}$ if $A \subseteq \operatorname{cl}_{\mathscr{R}}\left(\operatorname{int}_{\mathscr{R}}(A)\right)$;
(2) $A \in \mathcal{T}_{\mathscr{R}}^{p}$ if $A \subseteq \operatorname{int}_{\mathscr{R}}\left(\operatorname{cl}_{\mathscr{R}}(A)\right)$;
(3) $A \in \mathcal{T}_{\mathscr{R}}^{\alpha}$ if $A \subseteq \operatorname{int}_{\mathscr{R}}\left(\operatorname{cl}_{\mathscr{R}}\left(\operatorname{int}_{\mathscr{R}}(A)\right)\right)$;
(4) $A \in \mathcal{T}_{\mathscr{R}}^{\beta}$ if $A \subseteq \operatorname{cl}_{\mathscr{R}}\left(\operatorname{int}_{\mathscr{R}}\left(\operatorname{cl}_{\mathscr{R}}(A)\right)\right)$;
(5) $A \in \mathcal{T}_{\mathscr{R}}^{a}$ if $A \subseteq \operatorname{cl}_{\mathscr{R}}\left(\operatorname{int}_{\mathscr{R}}(A)\right) \cap \operatorname{int}_{\mathscr{R}}\left(\operatorname{cl}_{\mathscr{R}}(A)\right)$;
(6) $A \in \mathcal{T}_{\mathscr{R}}^{b}$ if $A \subseteq \operatorname{cl}_{\mathscr{R}}\left(\operatorname{int}_{\mathscr{R}}(A)\right) \cup \operatorname{int}_{\mathscr{R}}\left(\operatorname{cl}_{\mathscr{R}}(A)\right)$;
(7) $A \in \mathcal{T}_{\mathscr{R}}^{q}$ if there exists $V \in \mathcal{T}_{\mathscr{R}}$ such that $V \subseteq A \subseteq \operatorname{cl}_{\mathscr{R}}(V)$;
(8) $A \in \mathcal{T}_{\mathscr{R}}^{ps}$ if there exists $V \in \mathcal{T}_{\mathscr{R}}$ such that $A \subseteq V \subseteq \operatorname{cl}_{\mathscr{R}}(A)$;
(9) $A \in \mathcal{T}_{\mathscr{R}}^{\gamma}$ if there exists $V \in \mathcal{T}_{\mathscr{R}}^{s}$ such that $A \subseteq V \subseteq \operatorname{cl}_{\mathscr{R}}(A)$;
(10) $A \in \mathcal{T}_{\mathscr{R}}^{\delta}$ if there exists $V \in \mathcal{T}_{\mathscr{R}}^{p}$ such that $V \subseteq A \subseteq \operatorname{cl}_{\mathscr{R}}(V)$.

And, the members of the above families will be called the *topologically semi-open, preopen, α-open, β-open, a-open, b-open, quasi-open, pseudo-open, γ-open, and δ-open subsets* of the relator space $X(\mathscr{R})$, respectively.

Remark 31 The inclusions $A \subseteq A^{\circ -}$ and $A \subseteq A^{-\circ}$ mean only that the set A is open with respect to the composite operations $\circ -$ and $- \circ$, respectively.

While, the inclusions $V \subseteq A \subseteq V^{-}$ and $A \subseteq V \subseteq A^{-}$ mean that A is near to V from above and below or can be approximated by V from below and above.

Concerning the families $\mathcal{T}_{\mathscr{R}}^{\kappa}$, in our former paper [70], we have proved the following simple, but important theorems, with substantial references to the enormous literature on generalized open sets in topological and closure spaces and their straightforward generalizations.

Theorem 36 *We have*

(1) $\mathcal{T}_{\mathscr{R}}^{q} \subseteq \mathcal{T}_{\mathscr{R}}^{s}$; *(2)* $\mathcal{T}_{\mathscr{R}}^{ps} \subseteq \mathcal{T}_{\mathscr{R}}^{p}$;
(3) $\mathcal{T}_{\mathscr{R}}^{a} = \mathcal{T}_{\mathscr{R}}^{s} \cap \mathcal{T}_{\mathscr{R}}^{p}$; *(4)* $\mathcal{T}_{\mathscr{R}}^{s} \cup \mathcal{T}_{\mathscr{R}}^{p} \subseteq \mathcal{T}_{\mathscr{R}}^{b}$; *(5)* $\mathcal{T}_{\mathscr{R}}^{\gamma} \cup \mathcal{T}_{\mathscr{R}}^{\delta} \subseteq \mathcal{T}_{\mathscr{R}}^{\beta}$.

Theorem 37 *If $\mathscr{R}$ is a reflexive relator on X, then*

(1) $\mathcal{T}_{\mathscr{R}}^{\alpha} \subseteq \mathcal{T}_{\mathscr{R}}^{a}$; *(2)* $\mathcal{T}_{\mathscr{R}}^{b} \subseteq \mathcal{T}_{\mathscr{R}}^{\beta}$; *(3)* $\mathcal{T}_{\mathscr{R}}^{s} \cup \mathcal{T}_{\mathscr{R}}^{p} \subseteq \mathcal{T}_{\mathscr{R}}^{\beta}$;
(4) $\mathcal{T}_{\mathscr{R}}^{s} \cup \mathcal{T}_{\mathscr{R}}^{ps} \subseteq \mathcal{T}_{\mathscr{R}}^{\gamma}$; *(5)* $\mathcal{T}_{\mathscr{R}}^{p} \cup \mathcal{T}_{\mathscr{R}}^{q} \subseteq \mathcal{T}_{\mathscr{R}}^{\delta}$; *(6)* $\mathcal{T}_{\mathscr{R}}^{\alpha} \subseteq \mathcal{T}_{\mathscr{R}}^{\gamma} \cap \mathcal{T}_{\mathscr{R}}^{\delta}$.

Theorem 38 *If $\mathscr{R}$ is a reflexive relator on X, then $\mathcal{T}_{\mathscr{R}} \subseteq \mathcal{T}_{\mathscr{R}}^{\kappa}$ for all $\kappa = s, p, \alpha, \beta$, a, b, q, ps, γ, and δ.*

Remark 32 Note that, by Theorems 36 and 37, it is enough to prove the inclusion $\mathcal{T}_{\mathscr{R}} \subseteq \mathcal{T}_{\mathscr{R}}^{\kappa}$ only for $\kappa = q, ps$ and α.

Theorem 39 *If $\mathscr{R}$ is a quasi-topological relator on X and A, B ⊆ X such that*

$$A \subseteq B \subseteq \mathrm{cl}_{\mathscr{R}}(A),$$

then

(1) $A \in \mathscr{T}_{\mathscr{R}}^{s}$ *implies* $B \in \mathscr{T}_{\mathscr{R}}^{s}$; *(2)* $A \in \mathscr{T}_{\mathscr{R}}^{q}$ *implies* $B \in \mathscr{T}_{\mathscr{R}}^{q}$.

Theorem 40 *If $\mathscr{R}$ is a quasi-topological relator on X and A, B ⊆ X such that*

$$B \subseteq A \subseteq \mathrm{cl}_{\mathscr{R}}(B),$$

then

(1) $A \in \mathscr{T}_{\mathscr{R}}^{p}$ *implies* $B \in \mathscr{T}_{\mathscr{R}}^{p}$; *(2)* $A \in \mathscr{T}_{\mathscr{R}}^{ps}$ *implies* $B \in \mathscr{T}_{\mathscr{R}}^{ps}$.

Corollary 11 *If $\mathscr{R}$ is a topological relator on X and A ⊆ X, then*

(1) $A \in \mathscr{T}_{\mathscr{R}}^{s}$ *implies* $\mathrm{cl}_{\mathscr{R}}(A) \in \mathscr{T}_{\mathscr{R}}^{s}$; *(2)* $\mathrm{cl}_{\mathscr{R}}(A) \in \mathscr{T}_{\mathscr{R}}^{p}$ *implies* $A \in \mathscr{T}_{\mathscr{R}}^{p}$.

Theorem 41 *If $\mathscr{R}$ is a topological relator on X, then for any A ⊆ X, the following assertions are equivalent:*

(1) $A \in \mathscr{T}_{\mathscr{R}}^{s}$;
(2) $\mathrm{cl}_{\mathscr{R}}(A) \subseteq \mathrm{cl}_{\mathscr{R}}\big(\mathrm{int}_{\mathscr{R}}(A)\big)$; *(3)* $\mathrm{cl}_{\mathscr{R}}(A) = \mathrm{cl}_{\mathscr{R}}\big(\mathrm{int}_{\mathscr{R}}(A)\big)$;
(4) there exists $V \in \mathscr{T}_{\mathscr{R}}$ *such that* $V \subseteq A$ *and* $\mathrm{cl}_{\mathscr{R}}(A) = \mathrm{cl}_{\mathscr{R}}(V)$.

Theorem 42 *If $\mathscr{R}$ is a topological relator on X, then*

(1) $\mathscr{T}_{\mathscr{R}}^{q} = \mathscr{T}_{\mathscr{R}}^{s}$; *(2)* $\mathscr{T}_{\mathscr{R}}^{ps} = \mathscr{T}_{\mathscr{R}}^{p}$;
(3) $\mathscr{T}_{\mathscr{R}}^{\alpha} = \mathscr{T}_{\mathscr{R}}^{a}$; *(4)* $\mathscr{T}_{\mathscr{R}}^{\gamma} = \mathscr{T}_{\mathscr{R}}^{\beta}$.

Theorem 43 *If $\mathscr{R}$ is a topological relator on X, then*

(1) $\mathscr{T}_{\mathscr{R}} = \big\{ \mathrm{int}_{\mathscr{R}}(A) : \ A \in \mathscr{T}_{\mathscr{R}}^{s} \big\}$; *(2)* $\mathscr{T}_{\mathscr{R}} = \big\{ \mathrm{int}_{\mathscr{R}}(A) : \ A \in \mathscr{T}_{\mathscr{R}}^{p} \big\}$.

Theorem 44 *If $\mathscr{R}$ is a topologically filtered, topological relator on X, then* $\mathscr{T}_{\mathscr{R}}^{\delta} = \mathscr{T}_{\mathscr{R}}^{\beta}$.

13 The Duals of the Families $\mathscr{T}_{\mathscr{R}}^{\kappa}$ with $\kappa = q, ps, s$, and p

To introduce the corresponding generalized topologically closed sets, we shall use the following plausible notation.

Definition 13 For any $\kappa = s, p, \alpha, \beta, a, b, q, ps, \gamma$, and δ, we define

$$\mathscr{F}_{\mathscr{R}}^{\kappa} = \big\{ A \subseteq X : \quad A^{c} \in \mathscr{T}_{\mathscr{R}}^{\kappa} \big\}.$$

Thus, by using Theorem 2 and Remark 3, we can easily prove the following theorems

Theorem 45 *For any $A \subseteq X$ the following assertions are equivalent:*

(1) $A \in \mathscr{F}^{q}_{\mathscr{R}}$;
(2) *there exists* $W \in \mathscr{F}_{\mathscr{R}}$ *such that* $\operatorname{int}_{\mathscr{R}}(W) \subseteq A \subseteq W$.

Proof To prove that (1) ⇒ (2), note that if (1) holds, then $A^c \in \mathscr{T}^{q}_{\mathscr{R}}$. Thus, by Definition 12, there exists $V \in \mathscr{T}_{\mathscr{R}}$ such that

$$V \subseteq A^c \subseteq V^{-}.$$

Hence, by using that $c\, \circ = -\, c$, we can infer that

$$V^{c\circ} = V^{-c} \subseteq A \subseteq V^c.$$

Thus, by taking $W = V^c$, we can see that $W \in \mathscr{F}_{\mathscr{R}}$ such that

$$W^{\circ} \subseteq A \subseteq W,$$

and thus assertion (2) also holds.

Theorem 46 *For any $A \subseteq X$ the following assertions are equivalent:*

(1) $A \in \mathscr{F}^{ps}_{\mathscr{R}}$;
(2) *there exists* $W \in \mathscr{F}_{\mathscr{R}}$ *such that* $\operatorname{int}_{\mathscr{R}}(A) \subseteq W \subseteq A$.

Proof To prove that (2) ⇒ (1), note that if (2) holds, then there exists $W \in \mathscr{F}_{\mathscr{R}}$ such that

$$A^{\circ} \subseteq W \subseteq A.$$

Hence, by using that $\circ\, c = c\, -$, we can infer that

$$A^c \subseteq W^c \subseteq A^{\circ c} = A^{c-}.$$

Thus, by taking $V = W^c$, we can see that $V \in \mathscr{T}_{\mathscr{R}}$ such that

$$A^c \subseteq V \subseteq A^{c-}.$$

Therefore, by Definition 12, we have $A^c \in \mathscr{T}^{ps}_{\mathscr{R}}$, and thus (1) also holds.

Theorem 47 *For any $A \subseteq X$, the following assertions are equivalent:*

(1) $A \in \mathscr{F}^{s}_{\mathscr{R}}$; *(2)* $\operatorname{int}_{\mathscr{R}}\left(\operatorname{cl}_{\mathscr{R}}(A)\right) \subseteq A$.

Proof By the corresponding definitions, we have

$$(1) \iff A^c \in \mathscr{T}^s_{\mathscr{R}} \iff A^c \subseteq A^{c\,\circ\,-} \iff A^{c\,\circ\,-\,c} \subseteq A.$$

Moreover, by using the equalities $-\ c = c\ \circ$ and $c\ \circ c = -$, we can see that

$$A^{c\,\circ\,-\,c} = A^{c\,\circ\,c\,\circ} = A^{-\,\circ}.$$

Therefore, we actually have (1) $\iff A^{-\,\circ} \subseteq A \iff$ (2).

Theorem 48 *For any $A \subseteq X$, the following assertions are equivalent:*

(1) $A \in \mathscr{F}^p_{\mathscr{R}}$*;* *(2)* $\mathrm{cl}_{\mathscr{R}}\left(\mathrm{int}_{\mathscr{R}}(A)\right) \subseteq A$.

Proof By the corresponding definitions, we have

$$(1) \iff A^c \in \mathscr{T}^p_{\mathscr{R}} \iff A^c \subseteq A^{c-\circ} \iff A^{c-\circ\,c} \subseteq A.$$

Moreover, by using the equalities $c\ - = \circ\ c$ and $c\ \circ c = -$, we can see that

$$A^{c-\circ\,c} = A^{\circ\,c\,\circ\,c} = A^{\circ-}.$$

Therefore, we actually have (1) $\iff A^{\circ-} \subseteq A \iff$ (2).

Now, by using Theorem 41, we can also easily establish the following

Theorem 49 *If $\mathscr{R}$ is a topological relator on X, then for any $A \subseteq X$ the following assertions are equivalent:*

(1) $A \in \mathscr{F}^s_{\mathscr{R}}$*;*
(2) $\mathrm{int}_{\mathscr{R}}\left(\mathrm{cl}_{\mathscr{R}}(A)\right) \subseteq \mathrm{int}_{\mathscr{R}}(A)$*;* *(3)* $\mathrm{int}_{\mathscr{R}}\left(\mathrm{cl}_{\mathscr{R}}(A)\right) = \mathrm{int}_{\mathscr{R}}(A)$*;*
(4) there exists $W \in \mathscr{F}_{\mathscr{R}}$ *such that* $A \subseteq W$ *and* $\mathrm{int}_{\mathscr{R}}(A) = \mathrm{int}_{\mathscr{R}}(W)$.

14 The Duals of the Families $\mathscr{T}^{\kappa}_{\mathscr{R}}$ with $\kappa = \gamma, \delta, \alpha, \beta, a$, and b

Analogously to Theorems 46 and 45, we can also prove the following two theorems.

Theorem 50 *For any $A \subseteq X$ the following assertions are equivalent:*

(1) $A \in \mathscr{F}^{\gamma}_{\mathscr{R}}$;
(2) *there exists* $W \in \mathscr{F}^s_{\mathscr{R}}$ *such that* $\mathrm{int}_{\mathscr{R}}(A) \subseteq W \subseteq A$.

Theorem 51 *For any $A \subseteq X$ the following assertions are equivalent:*

(1) $A \in \mathscr{F}^{\delta}_{\mathscr{R}}$;
(2) *there exists* $W \in \mathscr{F}^p_{\mathscr{R}}$ *such that* $\mathrm{int}_{\mathscr{R}}(W) \subseteq A \subseteq W$.

Proof To prove that (1) $\Rightarrow$ (2), note that if (1) holds, then $A^c \in \mathscr{T}_{\mathscr{R}}^{\delta}$. Thus, by Definition 12, there exists $V \in \mathscr{T}_{\mathscr{R}}^{p}$ such that

$$V \subseteq A^c \subseteq V^{-}.$$

Hence, by using that $c \circ = - c$, we can infer that

$$V^{c\circ} = V^{-c} \subseteq A \subseteq V^c.$$

Thus, by taking $W = V^c$, we can see that $W \in \mathscr{F}_{\mathscr{R}}^{p}$ such that

$$W^{\circ} \subseteq A \subseteq W,$$

and thus (2) also holds.

Moreover, analogously to Theorems 47 and 48, we can also prove the following two theorems.

Theorem 52 *For any $A \subseteq X$, the following assertions are equivalent:*

(1) $A \in \mathscr{F}_{\mathscr{R}}^{\alpha}$; *(2) $\mathrm{cl}_{\mathscr{R}}\left(\mathrm{int}_{\mathscr{R}}\left(\mathrm{cl}_{\mathscr{R}}(A)\right)\right) \subseteq A$.*

Theorem 53 *For any $A \subseteq X$ the following assertions are equivalent:*

(1) $A \in \mathscr{F}_{\mathscr{R}}^{\beta}$; *(2) $\mathrm{int}_{\mathscr{R}}\left(\mathrm{cl}_{\mathscr{R}}\left(\mathrm{int}_{\mathscr{R}}(A)\right)\right) \subseteq A$.*

Proof By the corresponding definitions, we have

$$(1) \iff A^c \in \mathscr{T}_{\mathscr{R}}^{\beta} \iff A^c \subseteq A^{c-\circ-} \iff A^{c-\circ-c} \subseteq A.$$

Moreover, by using the equalities $c\, - = \circ\, c$, $-\, c = c\, \circ$, and $c\, \circ\, c = -$, we can see that

$$A^{c-\circ-c} = A^{\circ c \circ - c} = A^{\circ c \circ c \circ} = A^{\circ - \circ}.$$

Therefore, we actually have (1) $\iff A^{\circ - \circ} \subseteq A \iff$ (2).

No, by using Theorem 36, we can also easily prove the following

Theorem 54 *We have*

$$\mathscr{F}_{\mathscr{R}}^{a} = \mathscr{F}_{\mathscr{R}}^{s} \cap \mathscr{F}_{\mathscr{R}}^{p}.$$

Proof By the corresponding definitions and Theorem 36, for any $A \subseteq X$, we have

$$\begin{aligned} A \in \mathscr{F}_{\mathscr{R}}^{a} &\iff A^c \in \mathscr{T}_{\mathscr{R}}^{a} \iff A^c \in \mathscr{T}_{\mathscr{R}}^{s}, \quad A^c \in \mathscr{T}_{\mathscr{R}}^{p} \\ &\iff A \in \mathscr{F}_{\mathscr{R}}^{s}, \quad A \in \mathscr{F}_{\mathscr{R}}^{p} \iff A \in \mathscr{F}_{\mathscr{R}}^{s} \cap \mathscr{F}_{\mathscr{R}}^{p}. \end{aligned}$$

Therefore, the required equality is true.

Hence, by using Theorems 46 and 47, we can immediately derive

Corollary 12 *For any $A \subseteq X$, the following assertions are equivalent:*

(1) $A \in \mathscr{F}^{a}_{\mathscr{R}}$; *(2)* $\operatorname{int}_{\mathscr{R}}\big(\operatorname{cl}_{\mathscr{R}}(A)\big) \cup \operatorname{cl}_{\mathscr{R}}\big(\operatorname{int}_{\mathscr{R}}(A)\big) \subseteq A$.

The latter statement can also be easily proved directly, by using only the corresponding definitions.

Moreover, by using a direct argument, we can also easily prove the following counterpart of this corollary.

Theorem 55 *For any $A \subseteq X$, the following assertions are equivalent:*

(1) $A \in \mathscr{F}^{b}_{\mathscr{R}}$; *(2)* $\operatorname{int}_{\mathscr{R}}\big(\operatorname{cl}_{\mathscr{R}}(A)\big) \cap \operatorname{cl}_{\mathscr{R}}\big(\operatorname{int}_{\mathscr{R}}(A)\big) \subseteq A$.

Proof By the corresponding definitions, we have

$$(1) \iff A^{c} \in \mathscr{T}^{b}_{\mathscr{R}} \iff A^{c} \subseteq A^{c\circ-} \cup A^{c-\circ} \iff \big(A^{c\circ-} \cup A^{c-\circ}\big)^{c} \subseteq A.$$

Moreover, by using De Morgan's law and the equalities established in the proofs of Theorems 46 and 47, we can see that

$$\big(A^{c\circ-} \cup A^{c-\circ}\big)^{c} = A^{c\circ-c} \cap A^{c-\circ c} = A^{-\circ} \cap A^{\circ-}.$$

Therefore, we actually have $(1) \iff A^{-\circ} \cap A^{\circ-} \subseteq A \iff (2)$.

15 Topologically Regular Open Sets

Regular open sets were first introduced by Kuratowski [50] with reference to a paper of Henri Lebesgue. However, their importance became completely clear only after the considerations of Stone [77].

Following Kuratowski's definition, in our former paper [70], we have also introduced the following

Definition 14 A subset A of the relator space $X(\mathscr{R})$ will be called *topologically regular open* if

$$A = \operatorname{int}_{\mathscr{R}}\big(\operatorname{cl}_{\mathscr{R}}(A)\big).$$

And, the family of all such subsets of $X(\mathscr{R})$ will be denoted by $\mathscr{T}^{r}_{\mathscr{R}}$.

Thus, in contrast to the topological case, $\mathscr{T}^{r}_{\mathscr{R}}$ need not be a subfamily of $\mathscr{T}_{\mathscr{R}}$. To show this, we can use the following

Example 2 If $X = \{1, 2\}$ and R is a relation on X such that

$$R(1) = \{2\} \qquad \text{and} \qquad R(2) = \{1\},$$

then $\mathscr{R} = \{R\}$ is a symmetric relator on X such that

$$\mathscr{T}_{\mathscr{R}} = \{\emptyset,\ X\} \qquad \text{and} \qquad \mathscr{T}^{r}_{\mathscr{R}} = \mathscr{P}(X).$$

Of course, by Theorem 21, we evidently have the following

Theorem 56 *If $\mathscr{R}$ is a quasi-topological relator on X, then $\mathscr{T}^{r}_{\mathscr{R}} \subseteq \mathscr{T}_{\mathscr{R}}$.*

Moreover, by using Theorem 16, we can easily establish the following

Theorem 57 *If $\mathscr{R}$ is a reflexive relator on X, then $\mathscr{T}_{\mathscr{R}} \cap \mathscr{F}_{\mathscr{R}} \subseteq \mathscr{T}^{r}_{\mathscr{R}}$.*

Proof If $A \in \mathscr{T}_{\mathscr{R}} \cap \mathscr{F}_{\mathscr{R}}$, then $A \in \mathscr{T}_{\mathscr{R}}$ and $A \in \mathscr{F}_{\mathscr{R}}$. Thus, by Definition 3 and Theorem 16, we have $A^{\circ} = A$ and $A = A^{-}$. Therefore, $A^{\circ} = A^{\circ} = A$, and thus by Definition 14 we also have $A \in \mathscr{T}^{r}_{\mathscr{R}}$.

From the above two theorems, by using Theorem 22, we can derive

Corollary 13 *If $\mathscr{R}$ is a topological relator on X, then*

$$\mathscr{T}_{\mathscr{R}} \cap \mathscr{F}_{\mathscr{R}} \subseteq \mathscr{T}^{r}_{\mathscr{R}} \subseteq \mathscr{T}_{\mathscr{R}}.$$

The appropriateness of Definition 14 is also apparent from the following generalization of a statement of Dontchev [28, p. 4].

Theorem 58 *We have*

$$\mathscr{T}^{r}_{\mathscr{R}} = \mathscr{T}^{p}_{\mathscr{R}} \cap \mathscr{F}^{s}_{\mathscr{R}}.$$

Proof Namely, by the corresponding definitions and Theorem 47,

$$\begin{aligned} A \in \mathscr{T}^{r}_{\mathscr{R}} \iff A = A^{-\circ} &\iff A \subseteq A^{-\circ},\ \ A^{-\circ} \subseteq A \\ \iff A \in \mathscr{T}^{p}_{\mathscr{R}},\ A \in \mathscr{F}^{s}_{\mathscr{R}} &\iff A \in \mathscr{T}^{p}_{\mathscr{R}} \cap \mathscr{F}^{s}_{\mathscr{R}}. \end{aligned}$$

Remark 33 Now, if $\mathscr{R}$ is a reflexive relator on X, then by Theorems 38 and 58, we can also state that $\mathscr{T}_{\mathscr{R}} \cap \mathscr{F}^{s}_{\mathscr{R}} \subseteq \mathscr{T}^{r}_{\mathscr{R}}$.

Thus, by Definition 3 and Theorem 47, we can also state

Corollary 14 *If $\mathscr{R}$ is a reflexive relator on X and $A \subseteq X$ such that*

$$\operatorname{int}_{\mathscr{R}}\big(\operatorname{cl}_{\mathscr{R}}(A)\big) \subseteq A \subseteq \operatorname{int}_{\mathscr{R}}(A),$$

then $A \in \mathscr{T}^{r}_{\mathscr{R}}$.

Now, by using our former results, we can also easily prove

Theorem 59 *If $\mathscr{R}$ is a topological relator on X, then*

$$\mathscr{T}^{r}_{\mathscr{R}} = \mathscr{T}_{\mathscr{R}} \cap \mathscr{F}^{s}_{\mathscr{R}}.$$

Proof By Theorem 22, the relator $\mathscr{R}$ is reflexive and quasi-topological. Thus, by Theorems 38 and 56, we have

$$\mathscr{T}_{\mathscr{R}} \subseteq \mathscr{T}_{\mathscr{R}}^{p} \qquad \text{and} \qquad \mathscr{T}_{\mathscr{R}}^{r} \subseteq \mathscr{T}_{\mathscr{R}}.$$

Hence, by using Theorem 58, we can already infer that

$$\mathscr{T}_{\mathscr{R}}^{r} = \mathscr{T}_{\mathscr{R}} \cap \mathscr{T}_{\mathscr{R}}^{r} = \mathscr{T}_{\mathscr{R}} \cap \mathscr{T}_{\mathscr{R}}^{p} \cap \mathscr{F}_{\mathscr{R}}^{s} = \mathscr{T}_{\mathscr{R}} \cap \mathscr{F}_{\mathscr{R}}^{s}.$$

From this theorem, by using Definition 3 and Theorem 47, we can obtain

Corollary 15 *If $\mathscr{R}$ is a topological relator on X, then for any $A \subseteq X$ the following assertions are equivalent:*

(1) $A \in \mathscr{T}_{\mathscr{R}}^{r}$*;* *(2)* $\operatorname{int}_{\mathscr{R}}\big(\operatorname{cl}_{\mathscr{R}}(A)\big) \subseteq A \subseteq \operatorname{int}_{\mathscr{R}}(A)$.

From Theorems 59 and 58, by using Theorem 42, we can also derive

Theorem 60 *If $\mathscr{R}$ is a topological relator on X, then*

(1) $\mathscr{T}_{\mathscr{R}}^{r} = \mathscr{T}_{\mathscr{R}} \cap \mathscr{F}_{\mathscr{R}}^{q}$*;* *(2)* $\mathscr{T}_{\mathscr{R}}^{r} = \mathscr{T}_{\mathscr{R}}^{ps} \cap \mathscr{F}_{\mathscr{R}}^{q}$.

Proof Namely, by Theorem 42, we have not only $\mathscr{T}_{\mathscr{R}}^{p} = \mathscr{T}_{\mathscr{R}}^{ps}$ but also

$$A \in \mathscr{F}_{\mathscr{R}}^{s} \iff A^{c} \in \mathscr{T}_{\mathscr{R}}^{s} \iff A^{c} \in \mathscr{T}_{\mathscr{R}}^{q} \iff A \in \mathscr{F}_{\mathscr{R}}^{q},$$

and thus $\mathscr{F}_{\mathscr{R}}^{s} = \mathscr{F}_{\mathscr{R}}^{q}$.

Remark 34 Counterparts of Theorem 58 were also proved by Ekici [32, Theorem 8] and Jamunarani et al. [46, Theorem 2.2] by using the weak structures of Császár [24] and the generalized weak structures of Ávila and Molina [9]. Classes of regular open sets, in various generalized topological spaces introduced by Császár, have also been intensively investigated in [45] and [40].

16 Some Further Theorems on the Family $\mathscr{T}_{\mathscr{R}}^{r}$

By using Theorem 59, we can also prove the following generalization of a statement of Kuratowski [50].

Theorem 61 *If $\mathscr{R}$ is a topological relator on X, then for any $A \in \mathscr{T}_{\mathscr{R}}^{s}$ we have*

$$\operatorname{cl}_{\mathscr{R}}(A)^{c} \in \mathscr{T}_{\mathscr{R}}^{r}.$$

Proof By Theorems 22 and 21, we have $A^{-} \in \mathscr{F}_{\mathscr{R}}$. Hence, by Theorem 11, we infer that $A^{-c} \in \mathscr{T}_{\mathscr{R}}$.

Moreover, by Corollary 11, we have $A^{-} \in \mathcal{T}^{s}_{\mathcal{R}}$, and thus $A^{-\,c} \in \mathcal{F}^{s}_{\mathcal{R}}$. Hence, by Theorem 59, we can see that $A^{-\,c} \in \mathcal{T}^{r}_{\mathcal{R}}$.

From this theorem, by using that $\mathcal{T}_{\mathcal{R}} \subseteq \mathcal{T}^{s}_{\mathcal{R}}$ whenever $\mathcal{R}$ is reflexive, we can easily derive the following

Corollary 16 *If $\mathcal{R}$ is a topological relator on X and $\mathcal{A} = \mathcal{T}_{\mathcal{R}}$ or $\mathcal{T}^{s}_{\mathcal{R}}$, then*

$$\mathcal{T}^{r}_{\mathcal{R}} = \big\{ \operatorname{cl}_{\mathcal{R}}(A)^{c} : \quad A \in \mathcal{A} \big\}.$$

Proof Namely, if, for instance, $B \in \mathcal{T}^{r}_{\mathcal{R}}$, then by choosing $A = B^{c\,\circ}$, we can see that $A \in \mathcal{T}_{\mathcal{R}}$, and thus also $A \in \mathcal{T}^{s}_{\mathcal{R}}$, such that

$$A^{-\,c} = B^{c\,\circ\,-\,c} = B^{c\,\circ\,c\,\circ} = B^{-\,\circ} = B.$$

Remark 35 Following an observation of Halmos [38, p. 61], it is also worth noticing that, for a topological relator $\mathcal{R}$ on X, we have $\mathcal{T}_{\mathcal{R}} = \big\{ \operatorname{cl}_{\mathcal{R}}(A)^{c} : \quad A \subseteq X \big\}$.

Namely, if, for instance, $V \in \mathcal{T}_{\mathcal{R}}$, then by choosing $A = V^{c}$, we can see that $A \in \mathcal{F}_{\mathcal{R}}$, and thus $A = A^{-}$. Therefore, $V = A^{c} = A^{-c}$ even if $\mathcal{R}$ is assumed to be only reflexive.

From Theorem 61, by Theorem 2, we can see that $A^{-\circ} = A^{-c-c} \in \mathcal{F}^{r}_{\mathcal{R}}$ for all $A \in \mathcal{T}^{s}_{\mathcal{R}}$. However, this fact is of no importance for us. Namely, by using Theorem 59, we can prove a better statement.

Theorem 62 *If $\mathcal{R}$ is a topological relator on X, then for any $A \subseteq X$ we have*

$$\operatorname{int}_{\mathcal{R}}\big(\operatorname{cl}_{\mathcal{R}}(A)\big) \in \mathcal{T}^{r}_{\mathcal{R}}.$$

Proof By Theorems 22 and 21, we have $A^{-\,\circ} \in \mathcal{T}_{\mathcal{R}}$. Moreover, quite similarly, we also have $A^{c\circ} \in \mathcal{T}_{\mathcal{R}}$. Hence, by using Theorem 38 and Corollary 11, we can infer that $A^{c\circ-} \in \mathcal{T}^{s}_{\mathcal{R}}$, and thus $A^{c\circ-c} \in \mathcal{F}^{s}_{\mathcal{R}}$. However, by using the equalities $c\,\circ = -\,c$ and $c\,-\,c = \circ$, we can see that $A^{c\circ-c} = A^{-c-c} = A^{-\circ}$. Therefore, we actually have $A^{-\circ} \in \mathcal{F}^{s}_{\mathcal{R}}$. Hence, by Theorem 59, we can already see that $A^{-\circ} \in \mathcal{T}^{r}_{\mathcal{R}}$.

Remark 36 The topological counterparts of Theorems 61 and 62 are usually proved directly, by using only the corresponding properties of the operations $-$ and $\circ$.

Now, by using Theorem 62, we can also easily establish the following

Corollary 17 *If $\mathcal{R}$ is a topological relator on X, then*

$$\mathcal{T}^{r}_{\mathcal{R}} = \big\{ \operatorname{int}_{\mathcal{R}}(A) : \quad A \in \mathcal{F}_{\mathcal{R}} \big\} = \big\{ \operatorname{int}_{\mathcal{R}}\big(\operatorname{cl}_{\mathcal{R}}(A)\big) : \quad A \subseteq X \big\}.$$

Remark 37 Hence, it is clear that Stone's definition [77, p. 376] of a regular open set coincides with that of Kuratowski [50, p. 9].

However, it is now more important to note that, by using Theorem 59 we can also prove the following counterpart of [32, Theorem 7] of Ekici and [46, Theorem 2.1] of Jamunarani et al.

Theorem 63 *If $\mathscr{R}$ is a topological relator on X, then*

$$\mathscr{T}_{\mathscr{R}}^{r} = \mathscr{T}_{\mathscr{R}}^{\alpha} \cap \mathscr{F}_{\mathscr{R}}^{\beta}.$$

Proof By Theorems 38 and 37, we have $\mathscr{T}_{\mathscr{R}} \subseteq \mathscr{T}_{\mathscr{R}}^{\alpha}$ and $\mathscr{T}_{\mathscr{R}}^{s} \subseteq \mathscr{T}_{\mathscr{R}}^{\beta}$, and thus also $\mathscr{F}_{\mathscr{R}}^{s} \subseteq \mathscr{F}_{\mathscr{R}}^{\beta}$. Hence, by Theorem 59, we can see that

$$\mathscr{T}_{\mathscr{R}}^{r} = \mathscr{T}_{\mathscr{R}} \cap \mathscr{F}_{\mathscr{R}}^{s} \subseteq \mathscr{T}_{\mathscr{R}}^{\alpha} \cap \mathscr{F}_{\mathscr{R}}^{\beta}.$$

On the other hand, if $A \in \mathscr{T}_{\mathscr{R}}^{\alpha} \cap \mathscr{F}_{\mathscr{R}}^{\beta}$, and thus $A \in \mathscr{T}_{\mathscr{R}}^{\alpha}$ and $A \in \mathscr{F}_{\mathscr{R}}^{\beta}$, then by Definition 12 and Theorem 53 we have $A \subseteq A^{\circ - \circ}$ and $A^{\circ - \circ} \subseteq A$, and thus $A = A^{\circ - \circ}$. Hence, by using Theorem 62, we can infer that $A \in \mathscr{T}_{\mathscr{R}}^{r}$. Therefore, $\mathscr{T}_{\mathscr{R}}^{\alpha} \cap \mathscr{F}_{\mathscr{R}}^{\beta} \subseteq \mathscr{T}_{\mathscr{R}}^{r}$, and thus the required equality is also true.

Finally, we note that, analogously to Theorem 48, we can also prove

Theorem 64 *For any $A \subseteq X$, the following assertions are equivalent:*

(1) $A \in \mathscr{F}_{\mathscr{R}}^{r}$; *(2) $A = \mathrm{cl}_{\mathscr{R}}\big(\mathrm{int}_{\mathscr{R}}(A)\big)$.*

Remark 38 Several further properties of the family $\mathscr{F}_{\mathscr{R}}^{r}$ can be directly derived from those of the family $\mathscr{T}_{\mathscr{R}}^{r}$.

17 Characterizations of the Families $\mathscr{T}_{\mathscr{R}}^{\kappa}$ with $\kappa = s, p, \alpha, \beta$, and b

In our former paper [70], we have also proved the following theorems.

Theorem 65 *If $\mathscr{R}$ is a reflexive relator on X, then for any $A \subseteq X$ the following assertions are equivalent:*

(1) $A \in \mathscr{T}_{\mathscr{R}}^{s}$;
(2) *there exists $B \subseteq X$ such that*

$$A = \mathrm{int}_{\mathscr{R}}(A) \cup B \qquad \text{and} \qquad B \subseteq \mathrm{res}_{\mathscr{R}}\big(\mathrm{int}_{\mathscr{R}}(A)\big).$$

Theorem 66 *If $\mathscr{R}$ is a topological relator on X, then for any $A \subseteq X$ the following assertions are equivalent:*

(1) $A \in \mathscr{T}_{\mathscr{R}}^{s}$;
(2) *there exist $V \in \mathscr{T}_{\mathscr{R}}$ and $B \subseteq X$ such that*

$$A = V \cup B \qquad \text{and} \qquad B \subseteq \text{res}_{\mathcal{R}}(V).$$

Theorem 67 *If $\mathcal{R}$ is a topologically filtered, topological relator on X and $A \in \mathcal{T}_{\mathcal{R}}^{s}$, then there exist $V \in \mathcal{T}_{\mathcal{R}}$ and $B \in \mathcal{N}_{\mathcal{R}}$ such that*

$$A = V \cup B \qquad \text{and} \qquad V \cap B = \emptyset.$$

The above theorems are straightforward generalizations of [29, Lemma 1] of Duszyński and Noiri, an observation of Dlaska et al. [27, p. 1163] and [54, Theorem 7] of Levine, respectively.

While, the following theorem is a counterpart of [35, Proposition 1] of Ganster.

Theorem 68 *If $\mathcal{R}$ is a topologically filtered, topological relator on X, then for any $A \subseteq X$ the following assertions are equivalent:*

(1) $A \in \mathcal{T}_{\mathcal{R}}^{p}$;
(2) *there exist $V \in \mathcal{T}_{\mathcal{R}}$ and $B \in \mathcal{D}_{\mathcal{R}}$ such that $A = V \cap B$;*
(3) *there exists $V \in \mathcal{T}_{\mathcal{R}}$ such that $A \subseteq V$ and $\text{cl}_{\mathcal{R}}(A) = \text{cl}_{\mathcal{R}}(V)$.*

Remark 39 In this theorem, we may write $\mathcal{T}_{\mathcal{R}}^{r}$ instead of $\mathcal{T}_{\mathcal{R}}$.

The following theorem is an improvement of [63, Proposition 4] of Njåstad.

Theorem 69 *If $\mathcal{R}$ is a topologically filtered, topological relator on X, then for any $A \subseteq X$ the following assertions are equivalent:*

(1) $A \in \mathcal{T}_{\mathcal{R}}^{\alpha}$;
(2) *there exist $V \in \mathcal{T}_{\mathcal{R}}$ and $B \in \mathcal{N}_{\mathcal{R}}$ such that $A = V \setminus B$;*
(3) *there exist $V \in \mathcal{T}_{\mathcal{R}}$ and $B \subseteq \text{res}_{\mathcal{R}}\big(\text{int}_{\mathcal{R}}(A)\big)$ such that $A = V \setminus B$.*

The following two theorems are closely related to [32, Theorems 26 and 23] of Ekici and [46, Theorem 3.7 and 3.5] of Jamunarani et al.

Theorem 70 *If $\mathcal{R}$ is a topological relator on X, then for any $A \subseteq X$ the following assertions are equivalent:*

(1) $A \in \mathcal{T}_{\mathcal{R}}^{\beta}$; *(2)* $\text{cl}_{\mathcal{R}}(A) \in \mathcal{T}_{\mathcal{R}}^{s}$;
(3) there exists $V \in \mathcal{T}_{\mathcal{R}}$ such that $\text{cl}_{\mathcal{R}}(A) = \text{cl}_{\mathcal{R}}(V)$;
(4) there exist $V \in \mathcal{T}_{\mathcal{R}}$ and $B \subseteq X$ such that

$$\text{cl}_{\mathcal{R}}(A) = V \cup B \qquad \text{and} \qquad B \subseteq \text{res}_{\mathcal{R}}(V).$$

Remark 40 In assertion (2), instead of $\mathcal{T}_{\mathcal{R}}^{s}$, we may write not only $\mathcal{T}_{\mathcal{R}}^{q}$ but also $\mathcal{T}_{\mathcal{R}}^{r}$.

Theorem 71 *If $\mathcal{R}$ is a topological relator on X and $A \in \mathcal{T}_{\mathcal{R}}^{\beta}$, then there exist $V \in \mathcal{T}_{\mathcal{R}}^{s}$ and $B \in \mathcal{D}_{\mathcal{R}}$ such that $A = V \cap B$.*

Remark 41 In [70], it was also shown that, analogously to Theorem 67, the converse of Theorem 71 need not as well be true.

Moreover, in accordance, with [4, Remark 1] Andrijević, we have also proved

Theorem 72 *If $\mathscr{R}$ is a topologically filtered, topological relator on X, then for any $A \subseteq X$ the following assertions are equivalent:*

(1) $A \in \mathscr{T}_{\mathscr{R}}^{b}$;
(2) *there exist $B \in \mathscr{T}_{\mathscr{R}}^{s}$ and $C \in \mathscr{T}_{\mathscr{R}}^{p}$ such that $A = B \cup C$.*

Remark 42 Now, from the equalities

$$\mathscr{T}_{\mathscr{R}}^{a} = \mathscr{T}_{\mathscr{R}}^{s} \cap \mathscr{T}_{\mathscr{R}}^{p}, \quad \mathscr{T}_{\mathscr{R}}^{r} = \mathscr{T}_{\mathscr{R}}^{p} \cap \mathscr{F}_{\mathscr{R}}^{s} \text{ and } \mathscr{T}_{\mathscr{R}}^{a} = \mathscr{T}_{\mathscr{R}}^{s} \cap \mathscr{T}_{\mathscr{R}}^{p}, \quad \mathscr{T}_{\mathscr{R}}^{r} = \mathscr{T}_{\mathscr{R}}^{\alpha} \cap \mathscr{F}_{\mathscr{R}}^{\beta}$$

one can also derive some characterization theorems.

18 Intrinsic Characterizations of the Families $\mathscr{T}_{\mathscr{R}}$ and $\mathscr{T}_{\mathscr{R}}^{\kappa}$ with $\kappa = q, ps, s, p$

By the corresponding definitions, we evidently have the following

Theorem 73 *For any $A \subseteq X$, the following assertions are equivalent:*

(1) $A \in \mathscr{T}_{\mathscr{R}}$;
(2) *for each $x \in A$ there exists $R \in \mathscr{R}$ such that $R(x) \subseteq A$.*

By using the corresponding definitions, we can also easily prove

Theorem 74 *For any $A \subseteq X$, the following assertions are equivalent:*

(1) $A \in \mathscr{T}_{\mathscr{R}}^{q}$;
(2) *there exists $V \in \mathscr{T}_{\mathscr{R}}$ such that $V \subseteq A$ and, for any $x \in A$ and $R \in \mathscr{R}$, we have $R(x) \cap V \neq \emptyset$.*

Proof By the definition of $\mathscr{T}_{\mathscr{R}}^{q}$, assertion (1) means only that there exists $V \in \mathscr{T}_{\mathscr{R}}$ such that $V \subseteq A$ and $A \subseteq V^{-}$.

Moreover, by using the definition of $-$, we can see that the following assertions are equivalent :

(a) $A \subseteq V^{-}$;
(b) $\forall\, x \in A: \qquad x \in V^{-}$;
(c) $\forall\, x \in A: \qquad \forall\, R \in \mathscr{R}: \qquad R(x) \cap V \neq \emptyset.$

Therefore, assertions (1) and (2) are also equivalent.

Quite similarly, we can also prove the following

Theorem 75 *For any $A \subseteq X$, the following assertions are equivalent:*

(1) $A \in \mathscr{T}_{\mathscr{R}}^{ps}$;

(2) *there exists* $V \in \mathscr{T}_{\mathscr{R}}$ *such that* $A \subseteq V$ *and, for any* $x \in V$ *and* $R \in \mathscr{R}$, *we have* $R(x) \cap A \neq \emptyset$.

Remark 43 Note that if $\mathscr{R}$ is a reflexive relator on X and $A \in \mathscr{T}_{\mathscr{R}}$, then by taking $V = A$ we can see that $V \in \mathscr{T}_{\mathscr{R}}$ such that $V \subseteq A$ and, for any $x \in A$ and $R \in \mathscr{R}$, we have $x \in R(x) \cap V$, and thus $R(x) \cap V \neq \emptyset$. Therefore, by Theorem 74, we also have $A \in \mathscr{T}_{\mathscr{R}}^{q}$.

A quite similar application of Theorem 75 shows that now $A \in \mathscr{T}_{\mathscr{R}}^{ps}$ also holds.

By using the corresponding definitions, we can also easily prove the following two theorems.

Theorem 76 *For any* $A \subseteq X$, *the following assertions are equivalent:*

(1) $A \in \mathscr{T}_{\mathscr{R}}^{s}$;
(2) *for each* $x \in A$ *and* $R \in \mathscr{R}$, *there exist* $y \in R(x)$ *and* $S \in \mathscr{R}$ *such that* $S(y) \subseteq A$.

Proof By using the corresponding definitions, assertion (1) can be reformulated in the following equivalent forms :

(a) $A \subseteq A^{\circ -}$;
(b) $\forall\, x \in A: \quad x \in A^{\circ -}$;
(c) $\forall\, x \in A: \quad \forall\, R \in \mathscr{R}: \quad R(x) \cap A^{\circ} \neq \emptyset$;
(d) $\forall\, x \in A: \quad \forall\, R \in \mathscr{R}: \quad \exists\, y \in R(x): \quad y \in A^{\circ}$;
(e) $\forall\, x \in A: \quad \forall\, R \in \mathscr{R}: \quad \exists\, y \in R(x): \quad \exists\, S \in \mathscr{R}: \quad S(y) \subseteq A$.

Therefore, assertions (1) and (2) are also equivalent.

Theorem 77 *For any* $A \subseteq X$, *the following assertions are equivalent:*

(1) $A \in \mathscr{T}_{\mathscr{R}}^{p}$;
(2) *for each* $x \in A$, *there exists* $R \in \mathscr{R}$ *such that, for any* $y \in R(x)$ *and* $S \in \mathscr{R}$, *we have* $S(y) \cap A \neq \emptyset$.

Proof By using the corresponding definitions, assertion (1) can be reformulated in the following equivalent forms :

(a) $A \subseteq A^{-\circ}$;
(b) $\forall\, x \in A: \quad x \in A^{-\circ}$;
(c) $\forall\, x \in A: \quad \exists\, R \in \mathscr{R}: \quad R(x) \subseteq A^{-}$;
(d) $\forall\, x \in A: \quad \exists\, R \in \mathscr{R}: \quad \forall\, y \in R(x): \quad y \in A^{-}$;
(e) $\forall\, x \in A: \quad \exists\, R \in \mathscr{R}: \quad \forall\, y \in R(x): \quad \forall\, S \in \mathscr{R}: \quad S(y) \cap A \neq \emptyset$.

Therefore, assertions (1) and (2) are also equivalent.

Remark 44 Note that if $A \in \mathscr{T}_{\mathscr{R}}^{q}$, then by Theorem 74 there exists $V \in \mathscr{T}_{\mathscr{R}}$ such that $V \subseteq A$ and, for any $x \in A$ and $R \in \mathscr{R}$, we have $R(x) \cap V \neq \emptyset$. Therefore, there exists $y \in R(x)$ such that $y \in V$. Now, by Theorem 73, we can also state that there exists $S \in \mathscr{R}$ such that $S(y) \subseteq V$. Thus, by Theorem 76, we also have $A \in \mathscr{T}_{\mathscr{R}}^{s}$.

A quite similar application of Theorems 75, 73, and 77 shows that $A \in \mathscr{T}_{\mathscr{R}}^{ps}$ also implies $A \in \mathscr{T}_{p}^{p}$.

19 Intrinsic Characterizations of the Families $\mathscr{T}_{\mathscr{R}}^{\kappa}$ with $\kappa = \gamma, \delta, \alpha, \beta, a, b$

Analogously, to Theorems 75 and 74, we can also easily establish the following two theorems.

Theorem 78 *For any $A \subseteq X$, the following assertions are equivalent:*

(1) $A \in \mathscr{T}_{\mathscr{R}}^{\gamma}$;
(2) *there exists $V \in \mathscr{T}_{\mathscr{R}}^{s}$ such that $A \subseteq V$ and, for any $x \in V$ and $R \in \mathscr{R}$, we have $R(x) \cap A \neq \emptyset$.*

Theorem 79 *For any $A \subseteq X$, the following assertions are equivalent:*

(1) $A \in \mathscr{T}_{\mathscr{R}}^{\delta}$;
(2) *there exists $V \in \mathscr{T}_{\mathscr{R}}^{p}$ such that $V \subseteq A$ and, for any $x \in A$ and $R \in \mathscr{R}$, we have $R(x) \cap V \neq \emptyset$.*

Moreover, by using the corresponding definitions and the proofs of Theorems 76 and 77, we can also easily prove the following two theorems.

Theorem 80 *For any $A \subseteq X$, the following assertions are equivalent:*

(1) $A \in \mathscr{T}_{\mathscr{R}}^{\alpha}$;
(2) *for every $x \in A$, there exists $R \in \mathscr{R}$ such that, for any $y \in R(x)$ and $S \in \mathscr{R}$, there exist $z \in S(y)$ and $T \in \mathscr{R}$ such that $T(z) \subseteq A$.*

Proof By using the corresponding definitions, assertion (1) can be reformulated in the following equivalent forms :

(a) $A \subseteq A^{\circ - \circ}$;
(b) $\forall\, x \in A: \quad x \in A^{\circ - \circ}$;
(c) $\forall\, x \in A: \quad \exists\, R \in \mathscr{R}: \quad R(x) \subseteq A^{\circ -}$;
(d) $\forall\, x \in A: \quad \exists\, R \in \mathscr{R}: \quad \forall\, y \in R(x): \quad y \in A^{\circ -}$.

Moreover, from the proof of Theorem 76, we can see that

$$y \in A^{\circ -} \iff \forall\, S \in \mathscr{R}: \quad \exists\, z \in S(y): \quad \exists\, T \in \mathscr{R}: \quad T(z) \subseteq A.$$

Therefore, assertions (1) and (2) are also equivalent.

Theorem 81 *For any $A \subseteq X$, the following assertions are equivalent:*

(1) $A \in \mathscr{T}_{\mathscr{R}}^{\beta}$;
(2) *for each $x \in A$ and $R \in \mathscr{R}$, there exist $y \in R(x)$ and $S \in \mathscr{R}$ such that, for any $z \in S(y)$ and $T \in \mathscr{R}$ we have $T(z) \cap A \neq \emptyset$.*

Proof By using the corresponding definitions, assertion (1) can be reformulated in the following equivalent forms :

(a) $A \subseteq A^{- \circ -}$;

(b) $\forall\, x \in A: \qquad x \in A^{-\circ-}$;
(c) $\forall\, x \in A: \qquad \forall\, R \in \mathscr{R}: \qquad R(x) \cap A^{-\circ} \neq \emptyset$;
(d) $\forall\, x \in A: \qquad \forall\, R \in \mathscr{R}: \qquad \exists\, y \in R(x): \qquad y \in A^{-\circ}$.

Moreover, from the proof of Theorem 77, we can see that

$$y \in A^{-\circ} \iff \exists\, S \in \mathscr{R}: \quad \forall\, z \in S(y): \quad \forall\, T \in \mathscr{R}: \quad T(z) \cap A \neq \emptyset.$$

Therefore, assertions (1) and (2) are also equivalent.

Now, as an immediate consequence of Theorem 36, 76 and 77, we can also state the following

Theorem 82 *For any $A \subseteq X$, we have $A \in \mathscr{T}^{a}_{\mathscr{R}}$ if and only if the following assertions hold:*

(1) *for each $x \in A$ and $R \in \mathscr{R}$, there exist $y \in R(x)$ and $S \in \mathscr{R}$ such that $S(y) \subseteq A$;*
(2) *for each $x \in A$, there exists $R \in \mathscr{R}$ such that, for any $y \in R(x)$ and $S \in \mathscr{R}$, we have $S(y) \cap A \neq \emptyset$.*

Moreover, by using the corresponding definitions and the proofs of Theorems 76 and 77, we can also easily prove the following

Theorem 83 *For any $A \subseteq X$, we have $A \in \mathscr{T}^{b}_{\mathscr{R}}$ if and only if for each $x \in A$, any one of the following assertions holds:*

(1) *for each $R \in \mathscr{R}$, there exist $y \in R(x)$ and $S \in \mathscr{R}$ such that $S(y) \subseteq A$;*
(2) *there exists $R \in \mathscr{R}$ such that, for any $y \in R(x)$ and $S \in \mathscr{R}$, we have $S(y) \cap A \neq \emptyset$.*

Proof By using the corresponding definitions, the assertion $A \in \mathscr{T}^{b}_{\mathscr{R}}$ can be reformulated in the following equivalent forms :

(a) $A \subseteq A^{\circ-} \cup A^{-\circ}$;
(b) $\forall\, x \in A: \qquad x \in A^{\circ-} \cup A^{-\circ}$;
(c) $\forall\, x \in A: \qquad x \in A^{\circ-}$ or $x \in A^{-\circ}$.

Moreover, from the proofs of Theorems 76 and 77, we can see that

$$x \in A^{\circ-} \iff \forall\, R \in \mathscr{R}: \quad \exists\, y \in R(x): \quad \exists\, S \in \mathscr{R}: \quad S(y) \subseteq A,$$

and

$$x \in A^{-\circ} \iff \exists\, R \in \mathscr{R}: \quad \forall\, y \in R(x): \quad \forall\, S \in \mathscr{R}: \quad S(y) \cap A \neq \emptyset.$$

Therefore, the assertion of the theorem is also true.

20 Intrinsic Characterizations of the Family $\mathcal{T}^{r}_{\mathcal{R}}$

From Theorems 58, 59, 60, and 63, by using the corresponding results of Sects. 18 and 19, we can immediately derive several intrinsic characterizations of the family $\mathcal{T}^{r}_{\mathcal{R}}$.

Theorem 84 *For any $A \subseteq X$, we have $A \in \mathcal{T}^{r}_{\mathcal{R}}$ if and only if*

(1) *for each $x \in A^{c}$ and $R \in \mathcal{R}$, there exist $y \in R(x)$ and $S \in \mathcal{R}$ such that $S(y) \cap A = \emptyset$;*
(2) *for each $x \in A$, there exists $R \in \mathcal{R}$ such that, for any $y \in R(x)$ and $S \in \mathcal{R}$, we have $S(y) \cap A \neq \emptyset$.*

Proof By Theorem 58, we have $A \in \mathcal{T}^{r}_{\mathcal{R}}$ if and only if $A \in \mathcal{T}^{p}_{\mathcal{R}}$ and $A \in \mathcal{F}^{s}_{\mathcal{R}}$, i. e., $A^{c} \in \mathcal{T}^{s}_{\mathcal{R}}$. Hence, by using Theorems 77 and 76, we can see that the $A \in \mathcal{T}^{r}_{\mathcal{R}}$ if and only if both (2) and (1) hold.

Theorem 85 *If $\mathcal{R}$ is a topological relator on X, then for any $A \subseteq X$ we have $A \in \mathcal{T}^{r}_{\mathcal{R}}$ if and only if*

(1) *for each $x \in A$ there exists $R \in \mathcal{R}$ such that $R(x) \subseteq A$;*
(2) *for each $x \in A^{c}$ and $R \in \mathcal{R}$, there exist $y \in R(x)$ and $S \in \mathcal{R}$ such that $S(y) \cap A = \emptyset$;*

Proof By Theorem 59, we have $A \in \mathcal{T}^{r}_{\mathcal{R}}$ if and only if $A \in \mathcal{T}_{\mathcal{R}}$ and $A^{c} \in \mathcal{T}^{s}_{\mathcal{R}}$. Hence, by using Theorems 73 and 76, we can see that $A \in \mathcal{T}^{r}_{\mathcal{R}}$ if and only if both (1) and (2) hold.

Theorem 86 *If $\mathcal{R}$ is a topological relator on X, then for any $A \subseteq X$ we have $A \in \mathcal{T}^{r}_{\mathcal{R}}$ if and only if*

(1) *for each $x \in A$ there exists $R \in \mathcal{R}$ such that $R(x) \subseteq A$;*
(2) *there exists $W \in \mathcal{F}_{\mathcal{R}}$ such that $A \subseteq W$ and for any $x \in A^{c}$ and $R \in \mathcal{R}$ we have $R(x) \setminus W \neq \emptyset$.*

Proof By Theorem 60, we have $A \in \mathcal{T}^{r}_{\mathcal{R}}$ if and only if $A \in \mathcal{T}_{\mathcal{R}}$ and $A^{c} \in \mathcal{T}^{q}_{\mathcal{R}}$. Hence, by using Theorems 73 and 74, we can see that $A \in \mathcal{T}^{r}_{\mathcal{R}}$ if and only if (1) holds and

(a) $\exists\ V \in \mathcal{T}_{\mathcal{R}} : \quad V \subseteq A^{c} \quad \text{and} \quad \forall\ x \in A^{c} : \quad \forall\ R \in \mathcal{R} : \quad R(x) \cap V \neq \emptyset.$

Moreover, by noticing that

$$\big(V \in \mathcal{T}_{\mathcal{R}} \iff V^{c} \in \mathcal{F}_{\mathcal{R}}\big) \qquad \text{and} \qquad \big(V \subseteq A^{c} \iff A \subseteq V^{c}\big),$$

and $R(x) \cap V = R(x) \cap V^{cc} = R(x) \setminus V^{c}$, we can see that assertion (a) can be reformulated in the form that

(b) $\exists\ W \in \mathcal{F}_{\mathcal{R}} : \quad A \subseteq W \quad \text{and} \quad \forall\ x \in A^{c} : \quad \forall\ R \in \mathcal{R} : \quad R(x) \setminus W \neq \emptyset.$

Therefore, $A \in \mathcal{T}^{r}_{\mathcal{R}}$ if and only if both (1) and (2) hold.

Theorem 87 *If $\mathscr{R}$ is a topological relator on X, then for any $A \subseteq X$ we have $A \in \mathscr{T}_{\mathscr{R}}^{r}$ if and only if*

(1) *there exists $V \in \mathscr{T}_{\mathscr{R}}$ such that $A \subseteq V$ and, for any $x \in V$ and $R \in \mathscr{R}$, we have $R(x) \cap A \neq \emptyset$;*
(2) *there exists $W \in \mathscr{F}_{\mathscr{R}}$ such that $A \subseteq W$ and for any $x \in A^{c}$ and $R \in \mathscr{R}$ we have $R(x) \setminus W \neq \emptyset$.*

Proof By Theorem 60, we have $A \in \mathscr{T}_{\mathscr{R}}^{r}$ if and only if $A \in \mathscr{T}_{\mathscr{R}}^{ps}$ and $A^{c} \in \mathscr{T}_{\mathscr{R}}^{q}$. Hence, by using Theorem 75 and the proof of Theorem 86, we can see that $A \in \mathscr{T}_{\mathscr{R}}^{r}$ if and only if both (1) and (2) hold.

Theorem 88 *If $\mathscr{R}$ is a topological relator on X, then for any $A \subseteq X$ we have $A \in \mathscr{T}_{\mathscr{R}}^{r}$ if and only if*

(1) *for each $x \in A^{c}$ and $R \in \mathscr{R}$, there exist $y \in R(x)$ and $S \in \mathscr{R}$ such that, for any $z \in S(y)$ and $T \in \mathscr{R}$, we have $T(z) \setminus A \neq \emptyset$;*
(2) *for every $x \in A$, there exists $R \in \mathscr{R}$ such that, for any $y \in R(x)$ and $S \in \mathscr{R}$, there exist $z \in S(y)$ and $T \in \mathscr{R}$ such that $T(z) \subseteq A$.*

Proof By Theorem 63, we have $A \in \mathscr{T}_{\mathscr{R}}^{r}$ if and only if $A \in \mathscr{T}_{\mathscr{R}}^{\alpha}$ and $A^{c} \in \mathscr{T}_{\mathscr{R}}^{\beta}$. Hence, by using Theorem 80 and 81, we can see that $A \in \mathscr{T}_{\mathscr{R}}^{r}$ if and only if both (2) and (1) hold.

Remark 45 In principle each of the above characterizations can be used to prove several properties of the family $\mathscr{T}_{\mathscr{R}}^{r}$.

However, for instance, these characterizations cannot certainly be used to easily prove our former Theorems 61 and 62.

21 Conditions in Order That $\emptyset$ Could Be in $\mathscr{T}_{\mathscr{R}}^{\kappa}$

By using the corresponding definitions, we can easily establish the following

Theorem 89 *We have*

(1) $\emptyset \in \mathscr{T}_{\mathscr{R}}$; *(2)* $\emptyset \in \mathscr{T}_{\mathscr{R}}^{s} \cap \mathscr{T}_{\mathscr{R}}^{p}$;
(3) $\emptyset \in \mathscr{T}_{\mathscr{R}}^{a} \cap \mathscr{T}_{\mathscr{R}}^{b}$; *(4)* $\emptyset \in \mathscr{T}_{\mathscr{R}}^{\alpha} \cap \mathscr{T}_{\mathscr{R}}^{\beta}$.

Proof Clearly, $\emptyset \subseteq A$ for all $A \subseteq X$. Therefore, by the corresponding definitions, the required assertions are true.

For instance, from the inclusions $\emptyset \subseteq \emptyset^{\circ}$ and $\emptyset \subseteq \emptyset^{\circ - \circ}$, we can at once see that $\emptyset \in \mathscr{T}_{\mathscr{R}}$ and $\emptyset \in \mathscr{T}_{\mathscr{R}}^{\alpha}$ are true.

Now, by using this theorem, we can also easily prove the following

Theorem 90 *We have*

(1) $\emptyset \in \mathscr{T}_{\mathscr{R}}^{q} \cap \mathscr{T}_{\mathscr{R}}^{ps}$; *(2)* $\emptyset \in \mathscr{T}_{\mathscr{R}}^{\gamma} \cap \mathscr{T}_{\mathscr{R}}^{\delta}$.

Proof By taking $V=\emptyset$, we have $V \in \mathscr{T}_{\mathscr{R}}$, $V \in \mathscr{T}_{\mathscr{R}}^{p}$ and $V \in \mathscr{T}_{\mathscr{R}}^{s}$. Moreover, we also have

$$V \subseteq \emptyset \subseteq V^{-} \qquad \text{and} \qquad \emptyset \subseteq V \subseteq \emptyset^{-}.$$

Therefore, by the corresponding definitions, the required assertions are also true. For instance, from $V \in \mathscr{T}_{\mathscr{R}}^{s}$ and $\emptyset \subseteq V \subseteq \emptyset^{-}$, we can see that $\emptyset \in \mathscr{T}_{\mathscr{R}}^{\gamma}$.

The next simple example shows that, in contrast to the above theorems, $\emptyset$ need not be contained in $\mathscr{T}_{\mathscr{R}}^{r}$

Example 3 If $X=\{1, 2\}$ and R is a relation on X such that

$$R\,(1) = \emptyset \qquad \text{and} \qquad R\,(2) = \{2\},$$

then $\emptyset \notin \mathscr{T}_{R}^{r}$.

Namely, by the corresponding definitions, we have $\emptyset^{-}=\emptyset$ and $\emptyset^{\circ}=\{1\}$. Therefore, $\emptyset^{-\circ}=\emptyset^{\circ}=\{1\}\neq\emptyset$, and thus $\emptyset \notin \mathscr{T}_{R}^{r}$.

Remark 46 Note that if $\mathscr{R}$ is a nonvoid, non-partial relator on X, then by the corresponding definitions we have $\emptyset^{-}=\emptyset$ and $\emptyset^{\circ}=\emptyset$. Therefore, $\emptyset^{-\circ}=\emptyset^{\circ}=\emptyset$, and thus $\emptyset \in \mathscr{T}_{\mathscr{R}}^{r}$.

However, by using Theorem 73, we can now prove a much better statement.

Theorem 91 *The following assertions are equivalent:*

(1) $\emptyset \in \mathscr{T}_{\mathscr{R}}^{r}$ *;* *(2)* $\mathscr{R}$ *is non-partial.*

Proof By using the corresponding definitions, assertion (2) can be reformulated in the form :

(a) $\forall\ x \in X: \quad \forall\ R \in \mathscr{R}: \quad \exists\ y \in R\,(x).$

Moreover, from Theorem 73, we can see that assertion (1) is equivalent to the statement :

(b) $\forall\ x \in X: \quad \forall\ R \in \mathscr{R}: \quad \exists\ y \in R\,(x): \quad \exists\ S \in \mathscr{R}: \quad S\,(y) \cap \emptyset = \emptyset.$

Clearly, (b) implies (a), and thus (1) implies (2). On the other hand, if (a) holds and $\mathscr{R} \neq \emptyset$, then by taking any $S \in \mathscr{R}$, we can see that (b) also holds. While, if $\mathscr{R}=\emptyset$, then (b) trivially holds. Therefore, (2) also implies (1).

From this theorem, by using Theorem 18, we can immediately derive

Corollary 18 *The following assertions are equivalent:*

(1) $\emptyset \in \mathscr{T}_{\mathscr{R}}^{r}$ *;* *(2)* $\emptyset \notin \mathscr{E}_{\mathscr{R}}$ *;* *(3)* $X \in \mathscr{D}_{\mathscr{R}}$.

Actually, by using the corresponding definitions and Theorems 2 and 7, we can also prove the following

Theorem 92 *The following assertions are equivalent:*

(1) $\emptyset \in \mathscr{T}_{\mathscr{R}}^{r}$; *(2)* $\emptyset \in \mathscr{N}_{\mathscr{R}}$;
(3) $\mathrm{cl}_{\mathscr{R}}(\emptyset) \notin \mathscr{E}_{\mathscr{R}}$; *(4)* $\mathrm{int}_{\mathscr{R}}(X) \in \mathscr{D}_{\mathscr{R}}$.

Proof By the corresponding definitions, it is clear that each of the assertions (1), (2), and (3) is equivalent to the equality $\emptyset^{-\circ} = \emptyset$. Therefore, assertions (1), (2), and (3) are equivalent.

Moreover, by using Theorem 7 and the equality $-\,c = c\ \circ$, we can see that

$$\emptyset^{-} \notin \mathscr{E}_{\mathscr{R}} \iff \emptyset^{-c} \in \mathscr{D}_{\mathscr{R}} \iff \emptyset^{c\,\circ} \in \mathscr{D}_{\mathscr{R}} \iff X^{\circ} \in \mathscr{D}_{\mathscr{R}}.$$

Therefore, assertions (3) and (4) are also equivalent.

Thus, in addition to Theorem 18, we can also state

Corollary 19 *The following assertions are equivalent:*

(1) $\mathscr{R}$ *is non-partial;* *(2)* $\mathrm{cl}_{\mathscr{R}}(\emptyset) \notin \mathscr{E}_{\mathscr{R}}$; *(3)* $\mathrm{int}_{\mathscr{R}}(X) \in \mathscr{D}_{\mathscr{R}}$.

22 Conditions in Order That a Singleton Could Be in $\mathscr{T}_{\mathscr{R}}^{\kappa}$

By specializing the results of Sects. 15, 16, and 17 to a singleton, we can easily establish the following theorems.

Theorem 93 *For any* $x \in X$, *the following assertions are equivalent:*

(1) $\{x\} \in \mathscr{T}_{\mathscr{R}}$; *(2) there exists* $R \in \mathscr{R}$ *such that* $R(x) \subseteq \{x\}$.

Theorem 94 *For any* $x \in X$, *the following assertions are equivalent:*

(1) $\{x\} \in \mathscr{T}_{\mathscr{R}}^{q}$;
(2) *there exists* $V \in \mathscr{T}_{\mathscr{R}}$ *such that* $V \subseteq \{x\}$ *and, for each* $R \in \mathscr{R}$, *we have* $R(x) \cap V \neq \emptyset$.

Theorem 95 *For any* $x \in X$, *the following assertions are equivalent:*

(1) $\{x\} \in \mathscr{T}_{\mathscr{R}}^{ps}$;
(2) *there exists* $V \in \mathscr{T}_{\mathscr{R}}$ *such that* $x \in V$ *and, for each* $y \in V$ *and* $R \in \mathscr{R}$ *we have* $x \in R(y)$.

Theorem 96 *For any* $x \in X$, *the following assertions are equivalent:*

(1) $\{x\} \in \mathscr{T}_{\mathscr{R}}^{s}$;
(2) *for every* $R \in \mathscr{R}$ *there exist* $y \in R(x)$ *and* $S \in \mathscr{R}$ *such that* $S(y) \subseteq \{x\}$.

Theorem 97 *For any* $x \in X$, *the following assertions are equivalent:*

(1) $\{x\} \in \mathscr{T}_{\mathscr{R}}^{p}$;
(2) *there exists* $R \in \mathscr{R}$ *such that for any* $y \in R(x)$ *and* $S \in \mathscr{R}$ *we have* $x \in S(y)$.

Theorem 98 *For any* $x \in X$, *we have* $\{x\} \in \mathcal{T}^{r}_{\mathcal{R}}$ *if and only if*

(1) *there exists* $R \in \mathcal{R}$ *such that for any* $y \in R(x)$ *and* $S \in \mathcal{R}$ *we have* $x \in S(y)$;
(2) *for each* $y \in \{x\}^{c}$ *and* $U \in \mathcal{R}$ *there exist* $z \in U(y)$ *and* $V \in \mathcal{R}$ *such that* $x \in V(z)$.

Theorem 99 *If* $\mathcal{R}$ *is a topological relator on X, then for any* $x \in X$ *we have* $\{x\} \in \mathcal{T}^{r}_{\mathcal{R}}$ *if and only if*

(1) *there exists* $V \in \mathcal{T}_{\mathcal{R}}$ *such that* $x \in V$ *and, for any* $y \in \{x\}^{c}$ *and* $R \in \mathcal{R}$, *we have* $R(y) \cap V \neq \emptyset$;
(2) *there exists* $W \in \mathcal{T}_{\mathcal{R}}$ *such that* $x \in W$ *and, for any* $y \in W$ *and* $R \in \mathcal{R}$, *we have* $x \in R(y)$.

Theorem 100 *For any* $x \in X$, *the following assertions are equivalent:*

(1) $\{x\} \in \mathcal{T}^{\gamma}_{\mathcal{R}}$;
(2) *there exists* $V \in \mathcal{T}^{s}_{\mathcal{R}}$ *such that* $x \in V$ *and, for any* $y \in V$ *and* $R \in \mathcal{R}$ *we have* $x \in R(y)$.

Theorem 101 *For any* $x \in X$, *the following assertions are equivalent:*

(1) $\{x\} \in \mathcal{T}^{\delta}_{\mathcal{R}}$;
(2) *there exists* $V \in \mathcal{T}^{p}_{\mathcal{R}}$ *such that* $V \subseteq \{x\}$ *and, for any* $R \in \mathcal{R}$, *we have* $R(x) \cap V \neq \emptyset$.

Theorem 102 *For any* $x \in X$, *the following assertions are equivalent:*

(1) $\{x\} \in \mathcal{T}^{\alpha}_{\mathcal{R}}$;
(2) *there exists* $R \in \mathcal{R}$ *such that for any* $y \in R(x)$ *and* $S \in \mathcal{R}$ *there exist* $z \in S(y)$ *and* $T \in \mathcal{R}$ *such that* $T(z) \subseteq \{x\}$.

Theorem 103 *For any* $x \in X$, *the following assertions are equivalent:*

(1) $\{x\} \in \mathcal{T}^{\beta}_{\mathcal{R}}$;
(2) *for each* $R \in \mathcal{R}$ *there exist* $y \in R(x)$ *and* $S \in \mathcal{R}$ *such that for any* $z \in S(y)$ *and* $T \in \mathcal{R}$ *we have* $x \in T(z)$.

Theorem 104 *For any* $x \in X$, *we have* $\{x\} \in \mathcal{T}^{a}_{\mathcal{R}}$ *if and only if*

(1) *for every* $R \in \mathcal{R}$ *there exist* $y \in R(x)$ *and* $S \in \mathcal{R}$ *such that* $S(y) \subseteq \{x\}$;
(2) *there exists* $R \in \mathcal{R}$ *such that for any* $y \in R(x)$ *and* $S \in \mathcal{R}$ *we have* $x \in S(y)$.

Corollary 20 *If* $x \in X$ *such that* $\{x\} \in \mathcal{T}^{a}_{\mathcal{R}}$, *then there exist* $R \in \mathcal{R}$, $y \in R(x)$ *and* $S \in \mathcal{R}$ *such that* $S(y) = \{x\}$.

Theorem 105 *For any* $x \in X$, *we have* $\{x\} \in \mathcal{T}^{b}_{\mathcal{R}}$ *if and only if any one of the following assertions holds:*

(1) *for every* $R \in \mathcal{R}$ *there exist* $y \in R(x)$ *and* $S \in \mathcal{R}$ *such that* $S(y) \subseteq \{x\}$;
(2) *there exists* $R \in \mathcal{R}$ *such that for any* $y \in R(x)$ *and* $S \in \mathcal{R}$ *we have* $x \in S(y)$.

23 Some Further Conditions in Order That a Singleton Could Be in $\mathscr{T}_{\mathscr{R}}^{\kappa}$

In addition to Theorem 93, we can also prove the following

Theorem 106 *If $\mathscr{R}$ is a reflexive relator on X, then for any $x \in X$ the following assertions are equivalent:*

(1) $\{x\} \in \mathscr{T}_{\mathscr{R}}$; *(2)* $\operatorname{int}_{\mathscr{R}}\left(\{x\}\right) \neq \emptyset$.

Proof If (1) holds, then by Definition 3, we have $\{x\} \subseteq \{x\}^{\circ}$. Hence, we can infer that $x \in \{x\}^{\circ}$, and thus $\{x\}^{\circ} \neq \emptyset$. Therefore, (2) also holds even if $\mathscr{R}$ is not assumed to be reflexive.

Conversely, if (2) holds, then there exists $y \in X$ such that $y \in \{x\}^{\circ}$. Thus, by Definition 1, there exists $R \in \mathscr{R}$ such that $R(y) \subseteq \{x\}$. Hence, by using that $y \in R(y)$, we can infer that $y \in \{x\}$, and thus $y = x$. Therefore, we have $x \in \{x\}^{\circ}$, and thus also $\{x\} \subseteq \{x\}^{\circ}$. Consequently, (1) also holds.

From Theorem 94, we can easily derive the following

Theorem 107 *If $\mathscr{R}$ is a nonvoid relator on X, then for any $x \in X$, the following assertions are equivalent:*

(1) $\{x\} \in \mathscr{T}_{\mathscr{R}}^{q}$; *(2)* $\{x\} \in \mathscr{T}_{\mathscr{R}}$ *and, for any $R \in \mathscr{R}$, we have $x \in R(x)$.*

Proof If (1) holds, then by Theorem 94, there exists $V \in \mathscr{T}_{\mathscr{R}}$ such that $V \subseteq \{x\}$ and, for each $R \in \mathscr{R}$, we have $R(x) \cap V \neq \emptyset$. Hence, by choosing an arbitrary $R \in \mathscr{R}$, we can infer that $V \neq \emptyset$. Therefore, we necessarily have $V = \{x\}$. Hence, we can already see that $\{x\} \in \mathscr{T}_{\mathscr{R}}$. Moreover, we can also note that, for any $R \in \mathscr{R}$, we have $R(x) \cap \{x\} = R(x) \cap V \neq \emptyset$, and thus $x \in R(x)$. Therefore, (2) also holds.

Conversely, if (2) holds, then by choosing $V = \{x\}$, we can see that $V \in \mathscr{T}_{\mathscr{R}}$ such that $V \subseteq \{x\}$, and for each $R \in \mathscr{R}$ we have $x \in R(x) \cap V$, and thus $R(x) \cap V \neq \emptyset$. Therefore, by Theorem 94, assertion (1) also holds even if $\mathscr{R}$ is not supposed to be nonvoid.

Now, as an improvement of [74, Proposition 3.1], of Sarsak, we can also prove

Theorem 108 *If $\mathscr{R}$ is a nonvoid, reflexive relator on X, then for any $x \in X$, the following assertions are equivalent:*

(1) $\{x\} \in \mathscr{T}_{\mathscr{R}}$; *(2)* $\{x\} \in \mathscr{T}_{\mathscr{R}}^{q}$; *(3)* $\{x\} \in \mathscr{T}_{\mathscr{R}}^{s}$.

Proof From Theorem 36, we can see that (2) always implies (3). While, from Theorem 38, we can see that (1) implies (2) even if $\mathscr{R}$ is assumed to be only reflexive. Therefore, we need to only prove that (3) also implies (1).

For this, note that if (3) holds, then by Theorem 96, for every $R \in \mathscr{R}$, there exist $y \in R(x)$ and $S \in \mathscr{R}$ such that $S(y) \subseteq \{x\}$. Hence, by using that $y \in S(y)$, we can infer that $y \in \{x\}$, and thus $y = x$. Therefore, $S(x) \subseteq \{x\}$. Thus, by Theorem 73, assertion (1) also holds.

By using an argument of Sarsak [74], we can also prove the following analogue of Theorem 106.

Theorem 109 *If $\mathscr{R}$ is a topological relator on X, then for any $x \in X$ the following assertions are equivalent:*

(1) $\{x\} \in \mathscr{T}_{\mathscr{R}}^{p}$; *(2)* $\operatorname{int}_{\mathscr{R}}\big(\operatorname{cl}_{\mathscr{R}}(\{x\})\big) \neq \emptyset$.

Proof If (1) holds, then by Definition 12, we have $\{x\} \subseteq \{x\}^{-\circ}$. Hence, it is clear that $x \in \{x\}^{-\circ}$, and thus (2) also holds even if $\mathscr{R}$ is not assumed to have any particular property. Therefore, we need actually prove the converse implication.

For this, note that if (1) does not hold, then $\{x\} \not\subseteq \{x\}^{-\circ}$. Hence, we can infer that $x \notin \{x\}^{-\circ}$, and thus

$$\{x\} \cap \{x\}^{-\circ} = \emptyset.$$

Moreover, from Theorems 22 and 21, we can see that $\{x\}^{-\circ} \in \mathscr{T}_{\mathscr{R}}$. Hence, by using Corollary 2, we can infer that

$$\{x\}^{-} \cap \{x\}^{-\circ} = \emptyset.$$

Moreover, from Theorems 22 and 16, we can see that $\{x\}^{-\circ} \subseteq \{x\}^{-}$. Therefore, we actually have

$$\{x\}^{-\circ} = \{x\}^{-} \cap \{x\}^{-\circ} = \emptyset,$$

and thus (2) does not also holds.

Now, as a counterpart of [74, Proposition 3.3], of Sarsak, we can also prove

Theorem 110 *If $\mathscr{R}$ is a nonvoid topological relator on X, then for any $x \in X$ the following assertions are equivalent:*

(1) $\{x\} \in \mathscr{T}_{\mathscr{R}}^{p}$; *(2)* $\{x\} \in \mathscr{T}_{\mathscr{R}}^{b}$; *(3)* $\{x\} \in \mathscr{T}_{\mathscr{R}}^{\beta}$.

Proof From Theorem 36, we can see that (1) always implies (2). Moreover, from Theorem 37, we can see that (2) implies (3) even if $\mathscr{R}$ is assumed to be reflexive. Therefore, we need to only prove that (3) also implies (1).

For this, note that if (1) does not hold, then by Theorem 109 we have $\{x\}^{-\circ} = \emptyset$. Hence, by using that $\mathscr{R} \neq \emptyset$, we can already infer that $\{x\}^{-\circ-} = \emptyset^{-} = \emptyset$. Therefore, $\{x\} \not\subseteq \{x\}^{-\circ-}$, and thus (3) does not also hold.

Moreover, by using Theorem 109, we can also easily establish the following two theorems.

Theorem 111 *If $\mathscr{R}$ is a topological relator on X, then for any $x \in X$ we have* $\{x\} \in \mathscr{N}_{\mathscr{R}}^{p} \cup \mathscr{T}_{\mathscr{R}}^{p}$.

Theorem 112 *If $\mathscr{R}$ is a topological relator on X, then for any $x \in X$ we have $\{x\} \in \mathscr{T}_{\mathscr{R}}^{p} \cup \mathscr{F}_{\mathscr{R}}^{\alpha}$.*

Proof If $\{x\} \notin \mathscr{T}_{\mathscr{R}}^{p}$, then by Theorem 109 we have $\{x\}^{-\circ-} = \emptyset \subseteq \{x\}$. Thus, by Theorem 52, we also have $\{x\} \in \mathscr{F}_{\mathscr{R}}^{\alpha}$.

Remark 47 If $\mathscr{R}$ is a reflexive relator on X, then by Theorems 36 and 37, we have $\mathscr{T}_{\mathscr{R}}^{\alpha} \subseteq \mathscr{T}_{\mathscr{R}}^{s} \cup \mathscr{T}_{\mathscr{R}}^{p}$, and thus also $\mathscr{F}_{\mathscr{R}}^{\alpha} \subseteq \mathscr{F}_{\mathscr{R}}^{s} \cup \mathscr{F}_{\mathscr{R}}^{p}$,

Therefore, as an immediate consequence of Theorem 112, we can also state

Corollary 21 *If $\mathscr{R}$ is a topological relator on X, then for any $x \in X$ we have $\{x\} \in \mathscr{T}_{\mathscr{R}}^{p} \cup \mathscr{F}_{\mathscr{R}}^{p}$.*

Remark 48 Note that Theorem 111 is a generalization of [47, Lemma 2] of Janković and Reilly.

While, Corollary 21 can be used to obtain a generalization of [72, Theorem 3] of Reilly and Vamanamurthy and [61, Theorem 2] of Mukharjee and Roy.

24 Conditions in Order That X Could Be in $\mathscr{T}_{\mathscr{R}}^{\kappa}$

By Theorem 73, we evidently have the following

Theorem 113 *The following assertions are equivalent:*

(1) $X \in \mathscr{T}_{\mathscr{R}}$; *(2) either $X = \emptyset$ or $\mathscr{R} \neq \emptyset$.*

In addition to Theorem 91, we can also prove the following

Theorem 114 *The following assertions are equivalent:*

(1) $X \in \mathscr{T}_{\mathscr{R}}^{q}$; *(2) $\mathscr{T}_{\mathscr{R}} \cap \mathscr{D}_{\mathscr{R}} \neq \emptyset$;*
(3) $X \in \mathscr{T}_{\mathscr{R}}^{s}$; *(4) $\mathscr{R}$ is non-partial;* *(5) $X \in \mathscr{T}_{\mathscr{R}}^{\gamma}$.*

Proof By using the corresponding definitions, assertion (1) can be reformulated in the following equivalent forms:

(a) $\exists\ V \in \mathscr{T}_{\mathscr{R}}: \quad V \subseteq X \subseteq V^{-}$;
(b) $\exists\ V \in \mathscr{T}_{\mathscr{R}}: \quad X = V^{-}$; (c) $\exists\ V \in \mathscr{T}_{\mathscr{R}}: \quad V \in \mathscr{D}_{\mathscr{R}}$.

Therefore, assertions (1) and (2) are equivalent.

On the other hand, from Theorem 36, it is clear that (1) implies (3). Moreover, by using Theorems 89, 58, and 91, we can see that

$$(3) \iff \emptyset^{c} \in \mathscr{T}_{\mathscr{R}}^{s} \iff \left(\emptyset \in \mathscr{T}_{\mathscr{R}}^{p},\ \emptyset^{c} \in \mathscr{T}_{\mathscr{R}}^{s}\right) \iff \emptyset \in \mathscr{T}_{\mathscr{R}}^{r} \iff (4).$$

Now, to obtain the equivalence of assertions (1)–(4), we need to only show that if (4) holds, then (1) also holds. That is, by Theorem 74,

(d) there exists $V \in \mathscr{T}_{\mathscr{R}}$ such that $V \subseteq X$ and, for any $x \in X$ and $R \in \mathscr{R}$, we have $R(x) \cap V \neq \emptyset$.

For this, note that if $\mathscr{R} = \emptyset$, then by taking $V = \emptyset$ condition (d) can be trivially satisfied. While, if $\mathscr{R} \neq \emptyset$, then by taking $V = X$ condition (d) can be trivially satisfied.

Finally, to complete the proof, we note that, by Theorem 78, assertion (5) holds if and only if

(e) there exists $V \in \mathscr{T}_{\mathscr{R}}^{s}$ such that $X \subseteq V$ and, for any $x \in V$ and $R \in \mathscr{R}$, we have $R(x) \cap X \neq \emptyset$.

That is, $X \in \mathscr{T}_{\mathscr{R}}^{s}$ and for any $x \in X$ and $R \in \mathscr{R}$ we have $R(x) \neq \emptyset$. Therefore, (5) holds if and only if both (3) and (4) hold. Hence, since (3) and (4) are equivalent, it is clear that assertions (4) and (5) are also equivalent.

Now, by using Theorem 19 and 80, we can also prove

Theorem 115 *The following assertions are equivalent:*

(1) $X \in \mathscr{T}_{\mathscr{R}}^{r}$; *(2)* $X \in \mathscr{T}_{\mathscr{R}}^{p}$;
(3) $X \in \mathscr{T}_{\mathscr{R}}^{\alpha}$; *(4)* $\mathscr{R}$ *is locally non-partial.*

Proof From Definition 12 and Theorem 19, we can see that

$$(1) \iff X = X^{-\circ} \iff (4).$$

Moreover, by the corresponding definitions, it is clear that

$$(1) \iff X = X^{-\circ} \iff X \subseteq X^{-\circ} \iff (2).$$

Therefore, we need to only prove that (3) and (4) are also equivalent. For this, note that, by Theorem 80, assertion (3) holds if and only if for every $x \in X$ there exists $R \in \mathscr{R}$ such that, for any $y \in R(x)$ and $S \in \mathscr{R}$, there exist $z \in S(y)$ and $T \in \mathscr{R}$ such that $T(z) \subseteq X$. Hence, since the latter inclusion gives no requirement, it is clear that assertions (3) and (4) are also equivalent.

Moreover, by using our former results, we can also prove the following theorems.

Theorem 116 *The following assertions are equivalent:*

(1) $X \in \mathscr{T}_{\mathscr{R}}^{a}$; *(2)* $\mathscr{R}$ *is non-partial and locally non-partial.*

Proof From Theorems 36, 114, and 115, we can see that

$$(1) \iff \left(X \in \mathscr{T}_{\mathscr{R}}^{s},\ X \in \mathscr{T}_{\mathscr{R}}^{p}\right) \iff (2).$$

Theorem 117 *The following assertions are equivalent:*

(1) $X \in \mathscr{T}_{\mathscr{R}}^{ps}$; *(2)* $\mathscr{R}$ *is non-partial and either* $X = \emptyset$ *or* $\mathscr{R} \neq \emptyset$.

Proof From Theorem 75, we can see that (1) holds if and only if there exists $V \in \mathscr{T}_{\mathscr{R}}$ such that $X \subseteq V$ and, for any $x \in V$ and $R \in \mathscr{R}$ we have $R(x) \cap X \neq \emptyset$.

However, this is equivalent to the requirement that $X \in \mathscr{T}_{\mathscr{R}}$ and, for any $x \in X$ and $R \in \mathscr{R}$, we have $R(x) \neq \emptyset$. Hence, by Theorem 113 and Definition 7, it is clear that assertions (1) and (2) are equivalent.

Theorem 118 *The following assertions are equivalent:*

(1) $X \in \mathscr{T}_{\mathscr{R}}^{\delta}$;
(2) *there exists* $V \in \mathscr{T}_{\mathscr{R}}^{p}$ *such that for any* $x \in X$ *and* $R \in \mathscr{R}$ *we have* $R(x) \cap V \neq \emptyset$.

Proof From Theorem 79, we can see that assertion (1) holds if and only if there exists $V \in \mathscr{T}_{\mathscr{R}}^{p}$ such that $V \subseteq X$ and, for any $x \in X$ and $R \in \mathscr{R}$, we have $R(x) \cap V \neq \emptyset$. Therefore, assertions (1) and (2) are also equivalent.

From the latter three theorems, by using Remark 18 and Theorem 115, we can derive the following

Corollary 22 *If* $\mathscr{R}$ *is a nonvoid relator on X, then the following assertions are equivalent:*

(1) $X \in \mathscr{T}_{\mathscr{R}}^{a}$; *(2)* $X \in \mathscr{T}_{\mathscr{R}}^{ps}$; *(3)* $X \in \mathscr{T}_{\mathscr{R}}^{\delta}$; *(4)* $\mathscr{R}$ *is non-partial.*

Proof To prove the equivalence of (3) and (4), note that if (3) holds, then by Theorem 118 and Definition 6 assertion (4) also holds.

While, if (4) holds, then Remark 18 the relator $\mathscr{R}$ is locally non-partial. Thus, by Theorem 115, we have $X \in \mathscr{T}_{\mathscr{R}}^{p}$. Therefore, by choosing $V = X$, we can state that $V \in \mathscr{T}_{\mathscr{R}}^{p}$ such that, for every $x \in X$ and $R \in \mathscr{R}$, we have $R(x) \cap V = R(x) \neq \emptyset$. Thus, by Theorem 118, assertion (3) also holds.

Moreover, in addition to Theorems 81 and 114, we can also prove

Theorem 119 *The following assertions are equivalent:*

(1) $X \in \mathscr{T}_{\mathscr{R}}^{\beta}$; *(2)* $\mathscr{R}$ *is non-partial.*

Proof From Theorem 81, we can see that assertion (1) holds if and only if for each $x \in X$ and $R \in \mathscr{R}$ there exist $y \in R(x)$ and $S \in \mathscr{R}$ such that for any $z \in S(y)$ and $T \in \mathscr{R}$ we have $T(z) \cap X \neq \emptyset$, i. e., $T(z) \neq \emptyset$. Therefore, assertions (1) and (2) are also equivalent.

25 Union Properties of the Families $\mathscr{T}_{\mathscr{R}}^{\kappa}$

By using the corresponding definitions and the increasingness of the operations $\circ$ and $-$, we can easily prove the following

Theorem 120 *The families* $\mathscr{T}_{\mathscr{R}}$ *and* $\mathscr{T}_{\mathscr{R}}^{\kappa}$, *with* $\kappa = s, p, \alpha$, *and* β, *are closed under arbitrary unions.*

Proof Note that if $\Diamond = \circ,\ \circ-,\ -\circ,\ \circ-\circ$ or $-\circ-$, then $\Diamond$ is an increasing unary operation on $\mathscr{P}(X)$.

Moreover, if $A_i \subseteq A_i^{\Diamond}$ for all $i \in I$, then we also have

$$\bigcup_{i\in I} A_i \subseteq \bigcup_{i\in I} A_i^{\Diamond} \subseteq \bigcup_{i\in I} \Big(\bigcup_{i\in I} A_i \Big)^{\Diamond} \subseteq \Big(\bigcup_{i\in I} A_i \Big)^{\Diamond}.$$

Hence, it is clear that the required assertions are true.

Theorem 121 *The families $\mathscr{T}_{\mathscr{R}}^{\kappa}$, with $\kappa = q$, ps, γ, and δ, are also closed under arbitrary unions.*

Proof For instance, if $A_i \in \mathscr{T}_{\mathscr{R}}^{q}$ for all $i \in I$, then by Definition 12, for each $i \in I$, there exists $V_i \in \mathscr{T}_{\mathscr{R}}$ such that

$$V_i \subseteq A_i \subseteq V_i^{-}.$$

Hence, by using the increasingness of the operation $-$, we can infer that

$$\bigcup_{i\in I} V_i \subseteq \bigcup_{i\in I} A_i \subseteq \bigcup_{i\in I} V_i^{-} \subseteq \bigcup_{i\in I} \Big(\bigcup_{i\in I} V_i \Big)^{-} \subseteq \Big(\bigcup_{i\in I} V_i \Big)^{-}.$$

Moreover, by Theorem 93, we also have $\bigcup_{i\in I} V_i \in \mathscr{T}_{\mathscr{R}}$. Therefore, by Definition 12, we also have $\bigcup_{i\in I} A_i \in \mathscr{T}_{\mathscr{R}}^{q}$.

While, if, for instance, $A_i \in \mathscr{T}_{\mathscr{R}}^{\gamma}$ for all $i \in I$, for each $i \in I$, there exists $V_i \in \mathscr{T}_{\mathscr{R}}^{s}$ such that

$$A_i \subseteq V_i \subseteq A_i^{-}.$$

Hence, by using the increasingness of the operation $-$, we can infer that

$$\bigcup_{i\in I} A_i \subseteq \bigcup_{i\in I} V_i \subseteq \bigcup_{i\in I} A_i^{-} \subseteq \bigcup_{i\in I} \Big(\bigcup_{i\in I} A_i \Big)^{-} \subseteq \Big(\bigcup_{i\in I} A_i \Big)^{-}.$$

Moreover, by Theorem 93, we also have $\bigcup_{i\in I} V_i \in \mathscr{T}_{\mathscr{R}}^{s}$. Therefore, by Definition 12, we also have $\bigcup_{i\in I} A_i \in \mathscr{T}_{\mathscr{R}}^{\gamma}$.

The next simple example shows that, in contrast to the above theorems, the family $\mathscr{T}_{\mathscr{R}}^{r}$ need not be closed even under pairwise unions.

Example 4 If $X = \mathbb{R}$ and

$$R_n = \big\{ (x,\ y) \in X^2 : \quad d\,(x,\ y) < n^{-1} \big\}$$

for all $n \in \mathbb{N}$, then $\mathscr{R} = \{R_n : \ n \in \mathbb{N}\}$ is a properly filtered, strongly topological, tolerance relator on X such that, for the sets

$$A = \,]0,\ 1[\qquad \text{and} \qquad B = \,]1,\ 2[,$$

we have $A,\ B \in \mathcal{T}_{\mathcal{R}}^{r}$ such that $A \cup B \notin \mathcal{T}_{\mathcal{R}}^{r}$.

To check the latter statement, note that

$$A^{-\circ} = \,]0,\ 1[^{\,-\circ} = [0\ 1]^{\circ} = \,]0,\ 1[\, = A,$$

and quite similarly $B^{-\circ} = B$. Therefore, $A,\ B \in \mathcal{T}_{\mathcal{R}}^{r}$.

However, for the set $C = A \cup B$, we have

$$C^{-} = A^{-} \cup B^{-} = [0,\ 1] \cup [1,\ 2] = [0, 2], \quad \text{and thus} \quad C^{-\circ} = [0, 2]^{\circ} = \,]0, 2[.$$

Therefore, $C^{-\circ} \not\subseteq C$, and thus $C \notin \mathcal{T}_{\mathcal{R}}^{r}$, despite that $C \in \mathcal{T}_{\mathcal{R}}$, and thus by Theorem 38 we have $C \in \mathcal{T}_{\mathcal{R}}^{\kappa}$ for all $\kappa \neq r$.

Now, in addition to Theorems 93 and 94, we can also easily prove

Theorem 122 *The families $\mathcal{T}_{\mathcal{R}}^{a}$ and $\mathcal{T}_{\mathcal{R}}^{b}$ are also closed under arbitrary unions.*

Proof By Theorem 36, we have $\mathcal{T}_{\mathcal{R}}^{a} = \mathcal{T}_{\mathcal{R}}^{s} \cap \mathcal{T}_{\mathcal{R}}^{p}$. Moreover, by Theorem 93, the families $\mathcal{T}_{\mathcal{R}}^{s}$ and $\mathcal{T}_{\mathcal{R}}^{p}$ are closed under unions. Hence, it can be easily seen that $\mathcal{T}_{\mathcal{R}}^{a}$ is also closed under unions.

While, if $A_i \in \mathcal{T}_{\mathcal{R}}^{b}$ for all $i \in I$, then by Definition 12 we have

$$A_i \subseteq A_i^{\circ-} \cup A_i^{-\circ}$$

for all $i \in I$. Hence, by using the increasingness of the operations $\circ\,-$ and $-\,\circ$, we can infer that

$$\bigcup_{i\in I} A_i \subseteq \bigcup_{i\in I} \left(A_i^{\circ-} \cup A_i^{-\circ} \right)$$

$$\subseteq \bigcup_{i\in I} \left(\Big(\bigcup_{i\in I} A_i \Big)^{\circ-} \cup \Big(\bigcup_{i\in I} A_i \Big)^{-\circ} \right) \subseteq \Big(\bigcup_{i\in I} A_i \Big)^{\circ-} \cup \Big(\bigcup_{i\in I} A_i \Big)^{-\circ}.$$

Therefore, by Definition 12, we also have $\bigcup_{i\in I} A_i \in \mathcal{T}_{\mathcal{R}}^{b}$.

Remark 49 From the above three theorems, by taking $I = \emptyset$, we can also infer that $\emptyset \in \mathcal{T}_{\mathcal{R}}$ and $\emptyset \in \mathcal{T}_{\mathcal{R}}^{\kappa}$ for all $\kappa = s, p, \alpha, \beta, q, ps, \gamma, \delta, a$, and b.

Therefore, $\mathcal{T}_{\mathcal{R}}$ and $\mathcal{T}_{\mathcal{R}}^{\kappa}$, with $\kappa = s, p, \alpha, \beta, q, ps, \gamma, \delta, a$, and b, are generalized topologies in the sense of a recent terminology of Császár [20, 21].

However, it is now more important to note that, as an immediate consequence of the results of the present and the previous sections, we can also state the following

Theorem 123 *If $\mathcal{R}$ is a nonvoid, non-partial relator on X, then $\mathcal{T}_{\mathcal{R}}$ and $\mathcal{T}_{\mathcal{R}}^{\kappa}$, with $\kappa = s, p, \alpha, \beta, q, ps, \gamma, \delta, a$, and b, are generalized topologies on X in the narrower sense that they are closed under arbitrary unions and contain X.*

Namely, from this theorem, by using [87, Theorem 3.9], we can immediately derive the following

Theorem 124 *If $\mathscr{R}$ is a nonvoid, non-partial relator on X and $\mathscr{A} = \mathscr{T}_{\mathscr{R}}$ or $\mathscr{T}_{\mathscr{R}}^{\kappa}$, with $\kappa = s, p, \alpha, \beta, q, ps, \gamma, \delta, a$, and b, then*

$$\mathscr{S} = \mathscr{R}_{\mathscr{A}} = \left\{ R_A : \quad A \in \mathscr{A} \right\},$$

with $R_A = A^2 \cup A^c \times X$, is a nonvoid preorder relator on X such that $\mathscr{T}_{\mathscr{S}} = \mathscr{A}$.

Remark 50 Therefore, the properties of the generalized topologically open sets with respect to a nonvoid, non-partial relator can, in principle, be derived from those of the topologically open sets with respect to a nonvoid, preorder relator.

26 An Illustrating Example to Theorem 124

For an instructive illustration of Theorem 124, we can now easily establish an improvement of [13, Example 1.1] of Crossley and Hildebrandt.

This example was later used by Hamlett [41] to show that [54, Theorem 10] of Levine, who used the same interior sign for different topologies, is false.

Example 5 If $X = \{1, 2, 3\}$, and R_1, R_2 , and R_3 are relations on X such that

$$\begin{array}{lll} R_1(1) = \{1\}, & R_1(2) = X, & R_1(3) = X; \\ R_2(1) = \{1, 2\}, & R_2(2) = \{1, 2\}, & R_2(3) = X; \\ R_3(1) = \{1, 3\}, & R_3(2) = X, & R_3(3) = \{1, 3\}; \end{array}$$

then

$$\mathscr{R} = \left\{ R_1, R_2 \right\} \qquad \text{and} \qquad \mathscr{S} = \left\{ R_1, R_2, R_3 \right\}$$

are topologically filtered preorder relators on X such that, under the notations

$$V_1 = \{1\}, \qquad V_2 = \{1, 2\}, \qquad V_3 = \{1, 3\},$$

we have

(1) $\mathscr{T}_{\mathscr{R}}^{r} = \{\emptyset, X\}$; (2) $\mathscr{T}_{\mathscr{S}}^{r} = \{\emptyset, X\}$;
(3) $\mathscr{T}_{\mathscr{R}} = \{\emptyset, V_1, V_2, X\}$; (4) $\mathscr{T}_{\mathscr{S}} = \{\emptyset, V_1, V_2, V_3, X\}$;
(5) $\mathscr{T}_{\mathscr{R}}^{s} = \mathscr{T}_{\mathscr{R}}^{q} = \mathscr{T}_{\mathscr{R}}^{p} = \mathscr{T}_{\mathscr{R}}^{ps} = \mathscr{T}_{\mathscr{R}}^{\alpha} = \mathscr{T}_{\mathscr{R}}^{\beta} = \mathscr{T}_{\mathscr{R}}^{\gamma} = \mathscr{T}_{\mathscr{R}}^{\delta} = \mathscr{T}_{\mathscr{R}}^{a} = \mathscr{T}_{\mathscr{R}}^{b} = \mathscr{T}_{\mathscr{S}}$;
(6) $\mathscr{T}_{\mathscr{S}}^{s} = \mathscr{T}_{\mathscr{S}}^{q} = \mathscr{T}_{\mathscr{S}}^{p} = \mathscr{T}_{\mathscr{S}}^{ps} = \mathscr{T}_{\mathscr{S}}^{\alpha} = \mathscr{T}_{\mathscr{S}}^{\beta} = \mathscr{T}_{\mathscr{S}}^{\gamma} = \mathscr{T}_{\mathscr{S}}^{\delta} = \mathscr{T}_{\mathscr{S}}^{a} = \mathscr{T}_{\mathscr{S}}^{b} = \mathscr{T}_{\mathscr{S}}$.

To prove assertions (1), (3), and (5), note that

$$R_i = R_{V_i} = V_i^2 \cup V_i^c \times X$$

for $i = 1, 2$, and

$$R_1(1) \subseteq R_1(1) \cap R_2(1), \qquad R_2(2) \subseteq R_1(2) \cap R_2(2), \qquad R_1(3) \subseteq R_1(3) \cap R_2(3).$$

Moreover, for any $x \in X$ and $A \subseteq X$, we have

$$x \in A^\circ \iff \exists\, R \in \mathscr{R}: \quad R(x) \subseteq A$$

and

$$x \in A^- \iff \forall\, R \in \mathscr{R}: \quad R(x) \cap A \neq \emptyset.$$

Therefore, concerning the operations $\circ$ and $-$ with respect to $\mathscr{R}$, we have :

A	A°	A^-	$A^{\circ -}$	$A^{-\circ}$	$A^{\circ - \circ}$	$A^{-\circ -}$
$\emptyset$	$\emptyset$	$\emptyset$	$\emptyset$	$\emptyset$	$\emptyset$	$\emptyset$
$\{1\}$	$\{1\}$	X	X	X	X	X
$\{2\}$	$\emptyset$	$\{2, 3\}$	$\emptyset$	$\emptyset$	$\emptyset$	$\emptyset$
$\{3\}$	$\emptyset$	$\{3\}$	$\emptyset$	$\emptyset$	$\emptyset$	$\emptyset$
$\{1, 2\}$	$\{1, 2\}$	X	X	X	X	X
$\{1, 3\}$	$\{1\}$	X	X	X	X	X
$\{2, 3\}$	$\emptyset$	$\{2, 3\}$	$\emptyset$	$\emptyset$	$\emptyset$	$\emptyset$
X	X	X	X	X	X	X

Remark 51 To determine topological interiors and closures, with respect to Pervin relators, instead of the corresponding definitions, we can also use Theorem 5 or 23.

For instance, if $\mathscr{A}$ is a family of subsets of X and

$$\mathscr{R} = \{R_A : \quad A \in \mathscr{A}\},$$

then by Theorem 5 and the corresponding results of [87], for any $B \subseteq X$, we have

$$\operatorname{cl}_{\mathscr{R}}(B) = \bigcap_{A \in \mathscr{A}} R_A^{-1}[B] = \bigcap_{A \in \mathscr{A}} R_{A^c}[B]$$

with

$$R_{A^c}[B] = A^c \text{ if } B \subseteq A^c, \; B \neq \emptyset \quad \text{and} \quad R_{A^c}[B] = X \text{ if } B \not\subseteq A^c, \; B \subseteq X.$$

Remark 52 [13, Example 1.1] of Crossley and Hildebrandt shows only that two different topologies on the same set may have the same collection of semi-open sets.

While, our Example 5 shows that two different topologically filtered preorder relators on the same set may have the same collections of various generalized topologically open sets.

27 Intersection Properties of the Families $\mathscr{T}_{\mathscr{R}}^{q}$ and $\mathscr{T}_{\mathscr{R}}^{s}$

Levine [54, Remark 5] already noticed that the intersection of two semi-open subsets of a topological space need not be semi-open. Thus, the family of all semi-open sets of a topological space does not, in general, form a topology.

While, Crossley and Hildebrand [13, Theorem 1.9] proved that, in a topological space, the intersection of an open and a semi-open set is semi-open. Later, this statement was also proved independently by Noiri [64, Lemma 1].

In the proofs the subsequent two generalizations of [13, Theorem 1.9] of Crosley and Hildebrand, we shall rather use the more simple argument of Noiri.

Theorem 125 *If $\mathscr{R}$ is a topologically filtered relator on X, and moreover $U \in \mathscr{T}_{\mathscr{R}}$ and $A \in \mathscr{T}_{\mathscr{R}}^{q}$, then $U \cap A \in \mathscr{T}_{\mathscr{R}}^{q}$ also holds.*

Proof By Definition 12, there exists $V \in \mathscr{T}_{\mathscr{R}}$ such that

$$V \subseteq A \subseteq V^{-}.$$

Hence, by using Corollaries 5 and 8, we can infer that $U \cap V \in \mathscr{T}_{\mathscr{R}}$ and

$$U \cap V \subseteq U \cap A \subseteq U \cap V^{-} \subseteq (U \cap V)^{-}.$$

Therefore, by Definition 12, the required assertion is also true.

Theorem 126 *If $\mathscr{R}$ is a topologically filtered relator on X, and moreover $U \in \mathscr{T}_{\mathscr{R}}$ and $A \in \mathscr{T}_{\mathscr{R}}^{s}$, then $U \cap A \in \mathscr{T}_{\mathscr{R}}^{s}$ also holds.*

Proof By the corresponding definitions, we have

$$U \subseteq U^{\circ} \qquad \text{and} \qquad A \subseteq A^{\circ -}.$$

Hence, by using Corollary 8, the increasingness of the operation $-$, and Theorem 26, we can see that

$$U \cap A \subseteq U \cap A^{\circ -} \subseteq \left(U \cap A^{\circ}\right)^{-} \subseteq \left(U^{\circ} \cap A^{\circ}\right)^{-} = (U \cap A)^{\circ -}.$$

Therefore, by Definition 12, the required assertion is true.

The fact that the families $\mathcal{T}_{\mathcal{R}}^{q}$ and $\mathcal{T}_{\mathcal{R}}^{s}$ are not, in general, closed under pairwise intersections can be easily demonstrated with the help of the following

Example 6 If X and $\mathcal{R}$ are as in Example 4 and

$$A = [\,0, 1\,] \qquad \text{and} \qquad B = [1,\ 2\,],$$

then $\mathcal{R}$ is a properly filtered, strongly topological, tolerance relator on X such that $A,\ B \in \mathcal{T}_{\mathcal{R}}^{s}$, but $A \cap B \notin \mathcal{T}_{\mathcal{R}}^{s}$.

Namely, we evidently have

$$A^{\circ -} = [\,0, 1]^{\circ} = \,]\,0, 1[^{-} = [\,0, 1] = A$$

and quite similarly $B^{\circ -} = B$. Thus, in particular, $A,\ B \in \mathcal{T}_{\mathcal{R}}^{s}$. However,

$$\big(A \cap B\,\big)^{\circ -} = \{1\}^{\circ -} = \emptyset^{-} = \emptyset,$$

and thus $A \cap B \notin \mathcal{T}_{\mathcal{R}}^{s}$.

Note that $\mathcal{R}$ is a topological relator on X. Therefore, by Theorem 43, the same assertions hold for the family $\mathcal{T}_{\mathcal{R}}^{q}$.

Curiously enough, [13, Theorem 1.9] of Crossley and Hildebrand is actually a very particular case of a part of a former [63, Proposition 1] of Njåstad who used the term "β–set" instead of "semi-open set."

He defined a subset A of a topological space X to be an α–set if $A \subseteq A^{\circ - \circ}$. And, he proved that A is an α–set if and only if $A \cap B$ is β–subset of X for all β–subset B of X.

Now, by using a more simple argument than that of Njåstad, we can also prove the following

Theorem 127 *If $\mathcal{R}$ is a topologically filtered, topological relator on X, $A \in \mathcal{T}_{\mathcal{R}}^{\alpha}$ and $B \in \mathcal{T}_{\mathcal{R}}^{s}$, then $A \cap B \in \mathcal{T}_{\mathcal{R}}^{s}$ also holds.*

Proof By the corresponding definitions, we have

$$A \subseteq A^{\circ - \circ} \qquad \text{and} \qquad B \subseteq B^{\circ -}.$$

Hence, by using Corollary 8 and Theorem 26 and some basic properties of topological relators, we can see that

$$\begin{aligned} A \cap B &\subseteq A^{\circ - \circ} \cap B^{\circ -} \subseteq \big(A^{\circ - \circ} \cap B^{\circ}\big)^{-} \subseteq \big(A^{\circ -} \cap B^{\circ}\big)^{-} \\ &\subseteq \big(A^{\circ} \cap B^{\circ}\big)^{--} = \big(A^{\circ} \cap B^{\circ}\big)^{-} = \big(A \cap B\big)^{\circ -}. \end{aligned}$$

Therefore, the corresponding definition, the required assertion is also true.

Remark 53 Note that, by Theorem 42, in Theorem 127 we may again write $\mathscr{T}_{\mathscr{R}}^{q}$ in place of $\mathscr{T}_{\mathscr{R}}^{s}$.

28 Intersection Properties of the Families $\mathscr{T}_{\mathscr{R}}^{ps}$ and $\mathscr{T}_{\mathscr{R}}^{p}$

Analogously, to Theorems 125 and 126, we can also prove the following two theorems.

Theorem 128 *If $\mathscr{R}$ is a topologically filtered relator on X, and moreover $U \in \mathscr{T}_{\mathscr{R}}$ and $A \in \mathscr{T}_{\mathscr{R}}^{ps}$, then $U \cap A \in \mathscr{T}_{\mathscr{R}}^{ps}$ also holds.*

Proof By Definition 12, there exists $V \in \mathscr{T}_{\mathscr{R}}$ such that

$$A \subseteq V \subseteq A^{-}.$$

Hence, by using Corollaries 5 and 8, we can infer that $U \cap V \in \mathscr{T}_{\mathscr{R}}$ and

$$U \cap A \subseteq U \cap V \subseteq U \cap A^{-} \subseteq (U \cap A)^{-}.$$

Therefore, by Definition 12, the required assertion is also true.

Theorem 129 *If $\mathscr{R}$ is a topologically filtered relator on X, $U \in \mathscr{T}_{\mathscr{R}}$ and $A \in \mathscr{T}_{\mathscr{R}}^{p}$, then $U \cap A \in \mathscr{T}_{\mathscr{R}}^{p}$ also holds.*

Proof By the corresponding definitions, we have

$$U \subseteq U^{\circ} \qquad \text{and} \qquad A \subseteq A^{-\circ}.$$

Hence, by using Theorem 26 and Corollary 8, and the increasingness of the operation $\circ$, we can see that

$$U \cap A \subseteq U^{\circ} \cap A^{-\circ} = \left(U \cap A^{-}\right)^{\circ} \subseteq \left(U \cap A\right)^{-\circ}.$$

Thus, by Definition 12, the required assertion is true.

The fact that the families $\mathscr{T}_{\mathscr{R}}^{ps}$ and $\mathscr{T}_{\mathscr{R}}^{p}$ are also not, in general, closed under binary intersections can now be easily demonstrated with the help of the simple observation that dense sets with respect to a nonvoid relator are, in particular, topologically preopen.

Example 7 If X and $\mathscr{R}$ are as in Example 4 and

$$A = \mathbb{Q} \qquad \text{and} \qquad B = \{1\} \cup \mathbb{Q}^{c},$$

then $\mathscr{R}$ is a properly filtered, strongly topological, tolerance relator on X such that $A,\ B \in \mathscr{T}_{\mathscr{R}}^{p}$, but $A \cap B \notin \mathscr{T}_{\mathscr{R}}^{p}$.

To check this, recall that families of all rational and irrational numbers are dense in $\mathbb{R}$. Therefore,

$$A^{-\circ} = \mathbb{R}^{\circ} = \mathbb{R} = X$$

and quite similarly $B^{-\circ} = X$. Thus, in particular, $A,\ B \in \mathscr{T}_{\mathscr{R}}^{p}$. However,

$$(A \cap B)^{-\circ} = \{1\}^{-\circ} = \{1\}^{\circ} = \emptyset,$$

and thus $A \cap B \notin \mathscr{T}_{\mathscr{R}}^{p}$.

Note that $\mathscr{R}$ is a topological relator on X. Therefore, by Theorem 42, the same assertions hold for the family $\mathscr{T}_{\mathscr{R}}^{ps}$.

Now, analogously to Theorem 127, we can also prove the following theorem whose topological counterpart was already stated by Dontchev [28, p. 3].

Theorem 130 *If $\mathscr{R}$ is a topologically filtered, topological relator on X, $A \in \mathscr{T}_{\mathscr{R}}^{\alpha}$ and $B \in \mathscr{T}_{\mathscr{R}}^{p}$, then $A \cap B \in \mathscr{T}_{\mathscr{R}}^{p}$ also holds.*

Proof By the corresponding definitions, we have

$$A \subseteq A^{\circ-\circ} \qquad \text{and} \qquad B \subseteq B^{-\circ}.$$

Hence, by using Theorem 26 and the equality $\circ\circ = \circ$, we can infer that

$$A \cap B \subseteq A^{\circ-\circ} \cap B^{-\circ} = \left(A^{\circ-} \cap B^{-\circ}\right)^{\circ}.$$

Moreover, by using Corollary 8 and some basic properties of topological relators, we can see that

$$\begin{aligned} A^{\circ-} \cap B^{-\circ} &\subseteq \left(A^{\circ} \cap B^{-\circ}\right)^{-} \subseteq \left(A^{\circ} \cap B^{-}\right)^{-} \\ &\subseteq \left(A^{\circ} \cap B\right)^{--} = \left(A^{\circ} \cap B\right)^{-} \subseteq \left(A \cap B\right)^{-}. \end{aligned}$$

Therefore,

$$A \cap B \subseteq \left(A^{\circ-\circ} \cap B^{-\circ}\right)^{\circ} \subseteq \left(A \cap B\right)^{-\circ}.$$

Thus, by the corresponding definition, the required assertion is also true.

Remark 54 Note that, by Theorem 42, in Theorem 130 we may again write $\mathscr{T}_{\mathscr{R}}^{ps}$ in place of $\mathscr{T}_{\mathscr{R}}^{p}$.

29 Intersection Properties of the Family $\mathscr{T}_{\mathscr{R}}^{r}$

In contrast to Examples 4, 6, and 7, we can prove a counterpart of the dual of the first part of [50, Theorem 10] of Kuratowski.

Theorem 131 *If $\mathscr{R}$ is a topologically filtered, topological relator on X and $A,\ B \in \mathscr{T}_{\mathscr{R}}^{r}$, then $A \cap B \in \mathscr{T}_{\mathscr{R}}^{r}$ also holds.*

Proof Define $C = A \cap B$. From Theorem 22, we know that $\mathscr{R}$ is reflexive and quasi-topological. Therefore, by Theorem 56, we have $\mathscr{T}_{\mathscr{R}}^{r} \subseteq \mathscr{T}_{\mathscr{R}}$, and thus $A,\ B \in \mathscr{T}_{\mathscr{R}}$. Hence, by using Corollary 5, we can infer that $C \in \mathscr{T}_{\mathscr{R}}$, and thus $C \subseteq C^{\circ}$. Therefore, to prove that $C \in \mathscr{T}_{\mathscr{R}}^{r}$, by Corollary 14 it is enough to show only that $C^{-\circ} \subseteq C$.

For this, define $D = C^{-\circ}$. Then, by Theorem 21, we have $D \in \mathscr{T}_{\mathscr{R}}$, and thus $D \subseteq D^{\circ}$. Moreover, by using Theorem 16 and the increasingness of $-$, we can see that

$$D = C^{-\circ} \subseteq C^{-} \subseteq A^{-}$$

and quite similarly $D \subseteq B^{-}$. Hence, by using the increasingness of $\circ$ and Definition 12, we can infer that

$$D \subseteq D^{\circ} \subseteq A^{-\circ} = A$$

and quite similarly $D \subseteq B$. Therefore, $D \subseteq A \cap B$, and thus $C^{-\circ} \subseteq C$.

Somewhat more generally, we can also prove the following

Theorem 132 *If $\mathscr{R}$ is a topological relator on X and $A_i \in \mathscr{T}_{\mathscr{R}}^{r}$ for all $i \in I$, then $\operatorname{int}_{\mathscr{R}}\left(\bigcap_{i \in I} A_i\right) \in \mathscr{T}_{\mathscr{R}}^{r}$ also holds.*

Proof Define $C = \left(\bigcap_{i \in I} A_i\right)^{\circ}$. From Theorem 22, we know that $\mathscr{R}$ is reflexive and quasi-topological. Therefore, by Theorem 21, we have $C \in \mathscr{T}_{\mathscr{R}}$, and thus $C \subseteq C^{\circ}$. Now, to prove that $C \in \mathscr{T}_{\mathscr{R}}^{r}$, by Corollary 14 it is enough to show only that $C^{-\circ} \subseteq C$.

For this, define $D = C^{-\circ}$. Then, by Theorem 21, we have $D \in \mathscr{T}_{\mathscr{R}}$, and thus $D \subseteq D^{\circ}$. Moreover, by using Theorem 16 and the incresingness of $-$, we can see that

$$D = C^{-\circ} \subseteq C^{-} = \Big(\bigcap_{i \in I} A_i\Big)^{\circ -} \subseteq \Big(\bigcap_{i \in I} A_i\Big)^{-} \subseteq A_i^{-}$$

for all $i \in I$. Hence, by using the increasingness of $\circ$ and Definition 12, we can infer that

$$D \subseteq D^{\circ} \subseteq A_i^{-\circ} = A_i$$

for all $i \in I$, and thus $D \subseteq \bigcap_{i \in I} A_i$. Hence, by using the increasingness of $\circ$, we can infer that

$$D \subseteq D^\circ \subseteq \left(\bigcap_{i \in I} A_i \right)^\circ,$$

and thus $C^{-\circ} \subseteq C$.

Remark 55 Note that if, in particular, $\mathscr{R}$ and A, B are as in Theorem 131, then by Theorems 22, 21, 16, 26 and Theorem 132 we have

$$A \cap B = A^\circ \cap B^\circ = (A \cap B)^\circ \in \mathscr{T}_{\mathscr{R}}^{r}.$$

Therefore, Theorem 131 can be derived from Theorem 132.

Finally, we note that, for the family $\mathscr{T}_{\mathscr{R}}^{r}$, an analogue of Theorems 126 and 129 need not be true.

Example 8 If X and $\mathscr{R}$ are as in Example 4 and

$$U = \{1\}^c \qquad \text{and} \qquad A = \,]0,\ 2[\,,$$

then $\mathscr{R}$ is a properly filtered, strongly topological, tolerance relator on X, such that $U \in \mathscr{T}_{\mathscr{R}}$ and $A \in \mathscr{T}_{\mathscr{R}}^{r}$, but $U \cap A \notin \mathscr{T}_{\mathscr{R}}^{r}$.

To check this, note that, by Theorem 73, we have $U,\ A \in \mathscr{T}_{\mathscr{R}}$. Thus, by Theorem 38, we also have $U,\ A \in \mathscr{T}_{\mathscr{R}}^{\kappa}$ for all $\kappa \neq r$. Moreover, as in Example 4, we can see that $A \in \mathscr{T}_{\mathscr{R}}^{r}$ also holds.

However, we evidently have

$$U \cap A = \{1\}^c \cap \,]0,\ 2[\, = \,]0,\ 1[\, \cup \,]1,\ 2[,$$

and thus by Example 4 we can also state that $U \cap A \notin \mathscr{T}_{\mathscr{R}}^{r}$, despite that $U \cap A \in \mathscr{T}_{\mathscr{R}}^{\kappa}$ for all $\kappa \neq r$.

30 Intersection Properties of the Families $\mathscr{T}_{\mathscr{R}}^{\alpha}$ and with $\mathscr{T}_{\mathscr{R}}^{\beta}$

Analogously to Theorems 126, 127, 129, and 130, we can also prove the following theorems.

Theorem 133 *If $\mathscr{R}$ is a topologically filtered relator on X, $U \in \mathscr{T}_{\mathscr{R}}$ and $A \in \mathscr{T}_{\mathscr{R}}^{\alpha}$, then $U \cap A \in \mathscr{T}_{\mathscr{R}}^{\alpha}$.*

Proof By Definitions 3 and 12, we have

$$U \subseteq U^\circ \qquad \text{and} \qquad A \subseteq A^{\circ - \circ}.$$

Hence, by using Theorem 26, an inclusion established in the proof of Theorem 126 and the increasingness of $\circ$, we can see that

$$U \cap A \subseteq U^{\circ} \cap A^{\circ-\circ} = \left(U \cap A^{\circ-}\right)^{\circ} \subseteq (U \cap A)^{\circ-\circ}.$$

Thus, by Definition 12, the required assertion is true.

Theorem 134 *If $\mathscr{R}$ is a topologically filtered, topological relator on X and $A, B \in \mathscr{T}_{\mathscr{R}}^{\alpha}$, then $A \cap B \in \mathscr{T}_{\mathscr{R}}^{\alpha}$.*

Proof By Definition 12, we have

$$A \subseteq A^{\circ-\circ} \qquad \text{and} \qquad B \subseteq B^{\circ-\circ}.$$

Hence, by using Theorem 26 and the equality $\circ\,\circ = \circ$, we can see that

$$A \cap B \subseteq A^{\circ-\circ} \cap B^{\circ-\circ} = \left(A^{\circ-\circ} \cap B^{\circ-\circ}\right)^{\circ}.$$

Moreover, by using Theorem 16 and an inclusion established in the proof of Theorem 127, we can see that

$$A^{\circ-\circ} \cap B^{\circ-\circ} \subseteq A^{\circ-\circ} \cap B^{\circ-} \subseteq (A \cap B)^{\circ-}.$$

Hence, by using the increasingness of the operation $\circ$, we can see that

$$A \cap B \subseteq \left(A^{\circ-\circ} \cap B^{\circ-\circ}\right)^{\circ} \subseteq (A \cap B)^{\circ-\circ}.$$

Thus, by Definition 12, the required assertion is true.

Now, as an immediate consequence of Theorems 123 and 134, we can also state the following counterpart of [63, Proposition 2] of Njåstad.

Corollary 23 *If $\mathscr{R}$ is a nonvoid, topologically filtered, topological relator on X, then $\mathscr{T}_{\mathscr{R}}^{\alpha}$ is a topology on X.*

Theorem 135 *If $\mathscr{R}$ is a topologically filtered relator on X, $U \in \mathscr{T}_{\mathscr{R}}$ and $A \in \mathscr{T}_{\mathscr{R}}^{\beta}$, then $U \cap A \in \mathscr{T}_{\mathscr{R}}^{\beta}$.*

Proof By Definition 12, we have $A \subseteq A^{-\circ-}$. Hence, by using an inclusion established in the proof of Theorem 126, and moreover Corollary 8, we can see that

$$U \cap A \subseteq U \cap A^{-\circ-} \subseteq \left(U \cap A^{-}\right)^{\circ-} \subseteq \left(U \cap A\right)^{-\circ-}.$$

Thus, by Definition 12, the required assertion is also true.

Theorem 136 *If $\mathscr{R}$ is a topologically filtered, topological relator on X, $A \in \mathscr{T}_{\mathscr{R}}^{\alpha}$ and $B \in \mathscr{T}_{\mathscr{R}}^{\beta}$, then $A \cap B \in \mathscr{T}_{\mathscr{R}}^{\beta}$.*

Proof By Definition 12, we have

$$A \subseteq A^{\circ-\circ} \qquad \text{and} \qquad B \subseteq B^{-\circ-}.$$

Hence, by using Corollary 8, Theorem 26 and the equality $\circ\circ=\circ$, an inclusion established in the proof of Theorem 130 and the increasingness of the operation $\circ\,-$, we can see that

$$A \cap B \subseteq A^{\circ-\circ} \cap B^{-\circ-}$$
$$\subseteq \left(A^{\circ-\circ} \cap B^{-\circ}\right)^{-} = \left(A^{\circ-\circ} \cap B^{-\circ}\right)^{\circ-} \subseteq (A \cap B)^{-\circ-}.$$

Therefore, by Definition 12, the required assertion is also true.

31 Intersection Properties of the Families $\mathcal{T}_{\mathcal{R}}^{\kappa}$ with $\kappa = \gamma, \delta, a$, and b

Theorem 137 *If $\mathcal{R}$ is a topologically filtered relator on X, $U \in \mathcal{T}_{\mathcal{R}}$ and $A \in \mathcal{T}_{\mathcal{R}}^{\gamma}$, then $U \cap A \in \mathcal{T}_{\mathcal{R}}^{\gamma}$.*

Proof By Definition 12, there exists $V \in \mathcal{T}_{\mathcal{R}}^{s}$ such that

$$A \subseteq V \subseteq A^{-}.$$

Hence, by using Theorem 126 and Corollary 8, we can see that $U \cap V \in \mathcal{T}_{\mathcal{R}}^{s}$ and

$$U \cap A \subseteq U \cap V \subseteq U \cap A^{-} \subseteq (U \cap A)^{-}.$$

Thus, by Definition 12, the required assertion is also true.

Theorem 138 *If $\mathcal{R}$ is a topologically filtered, topological relator on X, $A \in \mathcal{T}_{\mathcal{R}}^{\alpha}$ and $B \in \mathcal{T}_{\mathcal{R}}^{\gamma}$, then $A \cap B \in \mathcal{T}_{\mathcal{R}}^{\gamma}$.*

Proof By Theorem 42, we have $\mathcal{T}_{\mathcal{R}}^{\gamma} = \mathcal{T}_{\mathcal{R}}^{\beta}$. Therefore, Theorem 136 can be applied to obtain the required assertion.

Theorem 139 *If $\mathcal{R}$ is a topologically filtered relator on X, $U \in \mathcal{T}_{\mathcal{R}}$ and $A \in \mathcal{T}_{\mathcal{R}}^{\delta}$, then $U \cap A \in \mathcal{T}_{\mathcal{R}}^{\delta}$.*

Proof By Definition 12, there exists $V \in \mathcal{T}_{\mathcal{R}}^{p}$ such that

$$V \subseteq A \subseteq V^{-}.$$

Hence, by using Theorem 129 and Corollary 8, we can see that $U \cap V \in \mathcal{T}_{\mathcal{R}}^{p}$ and

$$U \cap V \subseteq U \cap A \subseteq U \cap V^{-} \subseteq (U \cap V)^{-}.$$

Thus, by Definition 12, the required assertion is also true.

Theorem 140 *If $\mathscr{R}$ is a topologically filtered, topological relator on X, $A \in \mathscr{T}_{\mathscr{R}}^{\alpha}$ and $B \in \mathscr{T}_{\mathscr{R}}^{\delta}$, then $A \cap B \in \mathscr{T}_{\mathscr{R}}^{\delta}$.*

Proof By Theorem 44, we have $\mathscr{T}_{\mathscr{R}}^{\delta} = \mathscr{T}_{\mathscr{R}}^{\beta}$. Therefore, Theorem 136 can be applied to obtain the required assertion.

Theorem 141 *If $\mathscr{R}$ is a topologically filtered relator on X, $U \in \mathscr{T}_{\mathscr{R}}$ and $A \in \mathscr{T}_{\mathscr{R}}^{a}$, then $U \cap A \in \mathscr{T}_{\mathscr{R}}^{a}$.*

Proof By Theorem 36, we have $A \in \mathscr{T}_{\mathscr{R}}^{s}$ and $A \in \mathscr{T}_{\mathscr{R}}^{p}$. Hence, by using Theorems 114 and 121, we can infer that $U \cap A \in \mathscr{T}_{\mathscr{R}}^{s}$ and $U \cap A \in \mathscr{T}_{\mathscr{R}}^{p}$. Thus, by Theorem 36, the required assertion is also true.

Theorem 142 *If $\mathscr{R}$ is a topologically filtered, topological relator on X, $A \in \mathscr{T}_{\mathscr{R}}^{\alpha}$ and $B \in \mathscr{T}_{\mathscr{R}}^{a}$, then $A \cap B \in \mathscr{T}_{\mathscr{R}}^{a}$.*

Proof By Theorem 36, we have $B \in \mathscr{T}_{\mathscr{R}}^{s}$ and $B \in \mathscr{T}_{\mathscr{R}}^{p}$. Hence, by using Theorems 127 and 130, we can infer that $A \cap B \in \mathscr{T}_{\mathscr{R}}^{s}$ $A \cap B \in \mathscr{T}_{\mathscr{R}}^{p}$. Therefore, by Theorem 36, the required assertion is also true.

Theorem 143 *If $\mathscr{R}$ is a topologically filtered relator on X, $U \in \mathscr{T}_{\mathscr{R}}$ and $A \in \mathscr{T}_{\mathscr{R}}^{b}$, then $U \cap A \in \mathscr{T}_{\mathscr{R}}^{b}$.*

Proof By Definition 3 and 12, we have

$$U \subseteq U^{\circ} \qquad \text{and} \qquad A \subseteq A^{\circ -} \cup A^{- \circ}.$$

Hence, by using the inclusions established in the proofs of Theorems 126 and 129, we can see that

$$U \cap A \subseteq U \cap \left(A^{\circ -} \cup A^{- \circ}\right) = \left(U \cap A^{\circ -}\right) \cup \left(U \cap A^{- \circ}\right) \subseteq \left(U \cap A\right)^{\circ -} \cup \left(U \cap A\right)^{- \circ}.$$

Thus, by Definition 12, the required assertion is also true.

Theorem 144 *If $\mathscr{R}$ is a topologically filtered, topological relator on X, $A \in \mathscr{T}_{\mathscr{R}}^{\alpha}$ and $B \in \mathscr{T}_{\mathscr{R}}^{b}$, then $A \cap B \in \mathscr{T}_{\mathscr{R}}^{b}$.*

Proof By Definition 12, we

$$A \subseteq A^{\circ - \circ} \qquad \text{and} \qquad B \subseteq B^{\circ -} \cup B^{- \circ}.$$

Hence, by using the corresponding inclusions established in the proofs of Theorems 127 and 130, we can see that

$$\begin{aligned} A \cap B &\subseteq A^{\circ - \circ} \cap \left(B^{\circ -} \cup B^{- \circ}\right) \\ &= \left(A^{\circ - \circ} \cap B^{\circ -}\right) \cup \left(A^{\circ - \circ} \cap B^{- \circ}\right) \subseteq (A \cap B)^{\circ -} \cup (A \cap B)^{- \circ}. \end{aligned}$$

Therefore, by Definition 12, the required assertion is also true.

32 A Further Intersection Property of the Families $\mathscr{T}_{\mathcal{R}}^{s}$ and $\mathscr{T}_{\mathcal{R}}^{\alpha}$

To prove the following converse of Theorem 127, we could not simplify the argument of Njåstad used in the second part of the proof of [63, Proposition 1].

Theorem 145 *If $\mathcal{R}$ is a nonvoid, topologically filtered, topological relator on X and $A \subseteq X$ such that $A \cap B \in \mathscr{T}_{\mathcal{R}}^{s}$ for all $B \in \mathscr{T}_{\mathcal{R}}^{s}$, then $A \in \mathscr{T}_{\mathcal{R}}^{\alpha}$.*

Proof Assume, on the contrary, that $A \notin \mathscr{T}_{\mathcal{R}}^{\alpha}$. Then, by Definition 12, we have $A \not\subseteq A^{\circ - \circ}$. Therefore, there exists $x \in X$ such that

$$x \in A \qquad \text{and} \qquad x \notin A^{\circ - \circ}.$$

Hence, by using that $\circ\, c = c\, -$, we can infer that

$$x \in A^{\circ - \circ\, c} = A^{\circ - c\, -}.$$

Thus, by defining

$$V = A^{\circ - c},$$

we can note that $x \in V^{-}$. Moreover, by Theorems 22 and 21, we can also state that $A^{\circ -} \in \mathscr{F}_{\mathcal{R}}$, and thus $V \in \mathscr{T}_{\mathcal{R}}$.

Now, by defining

$$B = V \cup \{x\},$$

we can note that $V \subseteq B$. Moreover, by Theorems 22, 16 and 26, we can also state that

$$B \subseteq B^{-} = \big(V \cup \{x\}\big)^{-} = V^{-} \cup \{x\}^{-}.$$

Now, since $\{x\} \subseteq V^{-} \in \mathscr{F}_{\mathcal{R}}$, and thus $\{x\}^{-} \subseteq V^{--} \subseteq V^{-}$, we can also note that

$$V^{-} \cup \{x\}^{-} \subseteq V^{-}.$$

Therefore, $B \subseteq V^{-}$ also holds. Thus, by Definition 12 and Theorem 42, we have $B \in \mathscr{T}_{\mathcal{R}}^{q} = \mathscr{T}_{\mathcal{R}}^{s}$. Hence, by using the assumptions of the theorem, we can infer that $A \cap B \in \mathscr{T}_{\mathcal{R}}^{s}$.

However, by the corresponding definitions, we also have

$$A \cap B = A \cap \big(V \cup \{x\}\big) = (A \cap V) \cup \big(A \cap \{x\}\big) = \big(A \cap A^{\circ - c}\big) \cup \{x\}.$$

Moreover, by Theorems 22 and 114, we can also state that $X \in \mathscr{T}_{\mathscr{R}}^{s}$. Hence, by using the assumption of the theorem, we can infer that $A = A \cap X \in \mathscr{T}_{\mathscr{R}}^{s}$. Therefore, by Definition 12, $A \subseteq A^{\circ -}$, and thus

$$A \cap A^{\circ - c} = \emptyset.$$

Thus, we actually have $A \cap B = \{x\}$, and thus $\{x\} \in \mathscr{T}_{\mathscr{R}}^{s}$.

Hence, by using Theorems 22 and 107, we can infer that $\{x\} \in \mathscr{T}_{\mathscr{R}}$, and thus $\{x\} \subseteq \{x\}^{\circ}$. On the other hand, from the former inclusions $\{x\} \subseteq A$ and $A \subseteq A^{\circ -}$ we can infer that $\{x\}^{\circ} \subseteq A^{\circ} \subseteq A^{\circ - \circ}$. Therefore, we also have $\{x\} \subseteq \{x\}^{\circ} \subseteq A^{\circ - \circ}$, and thus $x \in A^{\circ - \circ}$. This contradiction proves the required assertion.

Now, as an immediate consequence of Theorems 127 and 145, we can also state the following counterpart of [63, Proposition 1] of Njåstad.

Corollary 24 *If $\mathscr{R}$ is a nonvoid, topologically filtered, topological relator on X, then*

$$\mathscr{T}_{\mathscr{R}}^{\alpha} = \big\{ A \subseteq X : \quad \forall\ B \in \mathscr{T}_{\mathscr{R}}^{s} : \quad A \cap B \in \mathscr{T}_{\mathscr{R}}^{s} \big\}.$$

Remark 56 This corollary can also be used to prove Corollary 23 which says that if $\mathscr{R}$ is a nonvoid, topologically filtered, topological relator on X, then $\mathscr{T}_{\mathscr{R}}^{\alpha}$ is a topology on X.

Namely, if $\mathscr{B}$ is a generalized topology on X in the Császár sense that it is closed under arbitrary unions, and moreover $\mathscr{T} = \{ A \subseteq X : \ \forall\ B \in \mathscr{B} : \ A \cap B \in \mathscr{B} \}$, then $\mathscr{T}$ is already an ordinary topology on X.

However, in contrast to [63, Proposition 2] of Njåstad, this fact seems to be of no particular importance for us now. Since the induced topologically open sets cannot, in general, play an essential role in the theory of relator spaces.

Also, it seems now not to be an important question that which relators can be quasi-topologically equivalent to a relator derived from a metric. Namely, we can usually more easily work with the induced surroundings than with a metric.

33 Minimality Properties of the Families $\mathscr{T}_{\mathscr{R}}^{\kappa}$

The following definition has been mainly suggested by the papers of Reilly and Vamanamuthy [72], Mukharjee and Roy [61], and Salih and Száz [73].

Definition 15 For $\kappa = s$, p, α, β, a, b, q, ps, γ ,and δ, a relator $\mathscr{R}$ on X, will be called

(1) *κ–minimal* if $\mathscr{T}_{\mathscr{R}}^{\kappa} \subseteq \{\emptyset, X\}$;
(2) *relatively κ–minimal* if $\mathscr{T}_{\mathscr{R}}^{\kappa} \subseteq \mathscr{T}_{\mathscr{R}}$.

Thus, by the results of Sections 22 and 24 and Theorems 38 and 37, we can at once state the following three theorems.

Theorem 146 *If $\mathscr{R}$ is a nonvoid, κ–minimal relator on X, then $\mathscr{R}$ is also relatively κ–minimal.*

Theorem 147 *If $\mathscr{R}$ is a nonvoid, non-partial relator on X, then $\mathscr{R}$ is κ–minimal if and only if $\mathscr{T}_{\mathscr{R}}^{\kappa} = \{\emptyset,\ X\}$.*

Theorem 148 *If $\mathscr{R}$ is a reflexive relator on X, then $\mathscr{R}$ is relatively κ–minimal if and only if $\mathscr{T}_{\mathscr{R}} = \mathscr{T}_{\mathscr{R}}^{\kappa}$.*

Remark 57 In [73], following the terminologies of Bourbaki [10, p. 139] and Kelley [49, p. 76], a relator $\mathscr{R}$ on X has been called a *quasi-topologically*

(1) *submaximal relator* if $\mathscr{D}_{\mathscr{R}} \subseteq \mathscr{T}_{\mathscr{R}}$;
(2) *door relator* if $\mathscr{P}(X) = \mathscr{T}_{\mathscr{R}} \cup \mathscr{F}_{\mathscr{R}}$.

Hence, it is clear that $\mathscr{R}$ is a quasi-topologically door relator if and only if $\mathscr{P}(X) \setminus \mathscr{T}_{\mathscr{R}} \subseteq \mathscr{F}_{\mathscr{R}}$, or equivalently $\mathscr{P}(X) \setminus \mathscr{F}_{\mathscr{R}} \subseteq \mathscr{T}_{\mathscr{R}}$.

Moreover, it can be shown that $\mathscr{R}$ is quasi-topologically submaximal if and only if any one of the assertions $\mathscr{D}_{\mathscr{R}} \setminus \mathscr{T}_{\mathscr{R}} = \emptyset$, $\mathscr{P}(X) = \mathscr{F}_{\mathscr{R}} \cup \mathscr{E}_{\mathscr{R}}$, $\mathscr{P}(X) \setminus \mathscr{E}_{\mathscr{R}} \subseteq \mathscr{F}_{\mathscr{R}}$ and $\mathscr{P}(X) \setminus \mathscr{F}_{\mathscr{R}} \subseteq \mathscr{E}_{\mathscr{R}}$ holds.

Now, by using the above definitions, we can also prove the following counterparts of [72, Theorems 2, 3 and 4] of Reilly and Vamanamuthy and [61, Theorems 1, 3 and 5] of Mukharjee and Roy.

Theorem 149 *If $\mathscr{R}$ is a quasi-topologically door relator on X, then $\mathscr{R}$ is relatively p–minimal.*

Proof If $A \in \mathscr{T}_{\mathscr{R}}^{p}$, then $A \subseteq A^{-\circ}$. While, if $A \notin \mathscr{T}_{\mathscr{R}}$, then by the assumed door property of $\mathscr{R}$ we have $A \in \mathscr{F}_{\mathscr{R}}$, and thus $A^{-} \subseteq A$. Hence, by using the increasingness of $\circ$, we can infer that $A \subseteq A^{-\circ} \subseteq A^{\circ}$, and thus $A \in \mathscr{T}_{\mathscr{R}}$. This contradiction proves that $A \in \mathscr{T}_{\mathscr{R}}$. Therefore, we have $\mathscr{T}_{\mathscr{R}}^{p} \subseteq \mathscr{T}_{\mathscr{R}}$, and thus $\mathscr{R}$ is relatively p-minimal.

Theorem 150 *If $\mathscr{R}$ is a nonvoid, relatively p–minimal, reflexive relator on X, then for any $x \in X$ we have $\{x\} \in \mathscr{T}_{\mathscr{R}} \cup \mathscr{F}_{\mathscr{R}}$.*

Proof If this is not the case, then there exists $x \in X$ such that $\{x\} \notin \mathscr{T}_{\mathscr{R}}$ and $\{x\} \notin \mathscr{F}_{\mathscr{R}}$. Thus, in particular, by Theorem 106 we necessarily have $\{x\}^{\circ} = \emptyset$. Hence, by using that $\mathscr{R} \neq \emptyset$, we can infer that $\{x\}^{\circ -} = \emptyset^{-} = \emptyset \subseteq \{x\}$. Therefore, by Theorem 48, we have $\{x\} \in \mathscr{F}_{\mathscr{R}}^{p}$. Hence, by using that $\mathscr{T}_{\mathscr{R}}^{p} \subseteq \mathscr{T}_{\mathscr{R}}$, and thus also $\mathscr{F}_{\mathscr{R}}^{p} \subseteq \mathscr{F}_{\mathscr{R}}$, we can already infer that $\{x\} \in \mathscr{F}_{\mathscr{R}}$. This contradiction proves the theorem.

Theorem 151 *If $\mathscr{R}$ is a nonvoid, relatively p–minimal relator on X, then $\mathscr{R}$ is quasi-topologically submaximal.*

Proof If $A \in \mathscr{D}_{\mathscr{R}}$, then $A^{-\circ} = X^{\circ} = X$. Therefore, we trivially have $A \subseteq A^{-\circ}$, and thus $A \in \mathscr{T}_{\mathscr{R}}^{p}$. Hence, by the assumed p–minimality of $\mathscr{R}$, it follows that $A \in \mathscr{T}_{\mathscr{R}}$. Therefore, $\mathscr{D}_{\mathscr{R}} \subseteq \mathscr{T}_{\mathscr{R}}$, and thus $\mathscr{R}$ is quasi-topologically submaximal.

Theorem 152 *If $\mathscr{R}$ is a topologically filtered, quasi-topologically submaximal, topological relator on X, then $\mathscr{R}$ is relatively p–minimal.*

Proof If $A \in \mathscr{T}_{\mathscr{R}}^{p}$, then by Theorem 68 there exists $V \in \mathscr{T}_{\mathscr{R}}$ such that

$$A \subseteq V \qquad \text{and} \qquad A^{-} = V^{-}.$$

Now, by using Theorems 26 and 18, we can see that

$$\left(A \cup V^{c}\right)^{-} = A^{-} \cup V^{c-} = V^{-} \cup V^{c-} = \left(V \cup V^{c}\right)^{-} = X^{-} = X,$$

and thus $A \cup V^{c} \in \mathscr{D}_{\mathscr{R}}$. Hence, by using the assumed submaximality of $\mathscr{R}$, we can infer that $A \cup V^{c} \in \mathscr{T}_{\mathscr{R}}$. Now, by using Corollary 5, we can also see that

$$A = V \cap A = \left(V \cap A\right) \cup \left(V \cap V^{c}\right) = V \cap \left(A \cup V^{c}\right) \in \mathscr{T}_{\mathscr{R}}.$$

Therefore, $\mathscr{T}_{\mathscr{R}}^{p} \subseteq \mathscr{T}_{\mathscr{R}}$, and thus $\mathscr{R}$ is relatively p–minimal.

Remark 58 Hence, we see that a nonvoid, topologically filtered, topological relator $\mathscr{R}$ is relatively p–minimal if and only if it is quasi-topologically submaximal.

Now, by using Theorem 69, we can also prove the following improvement of [63, Corollary] of Njåstad.

Theorem 153 *For a nonvoid, topologically filtered, topological relator $\mathscr{R}$ on X, the following assertions are equivalent:*

(1) $\mathscr{R}$ is relatively α–minimal;
(2) $\mathscr{T}_{\mathscr{R}} = \mathscr{T}_{\mathscr{R}}^{\alpha}$; *(3) $\mathscr{N}_{\mathscr{R}} \subseteq \mathscr{F}_{\mathscr{R}}$;* *(4) $\mathscr{N}_{\mathscr{R}} = \mathscr{F}_{\mathscr{R}} \setminus \mathscr{E}_{\mathscr{R}}$.*

Proof From Theorem 147, we can see that (1) and (2) are equivalent even if $\mathscr{R}$ is only reflexive.

Moreover, if $A \in \mathscr{N}_{\mathscr{R}}$, then we can note that $A^{\circ} \subseteq A^{-\circ} = \emptyset$, and thus $A^{\circ} = \emptyset$. Therefore, $A \notin \mathscr{E}_{\mathscr{R}}$. Moreover, if (3) holds, then we also have $A \in \mathscr{F}_{\mathscr{R}}$. Therefore, $A \in \mathscr{F}_{\mathscr{R}} \setminus \mathscr{E}_{\mathscr{R}}$, and thus $\mathscr{N}_{\mathscr{R}} \subseteq \mathscr{F}_{\mathscr{R}} \setminus \mathscr{E}_{\mathscr{R}}$.

On the other hand, if $A \in \mathscr{F}_{\mathscr{R}} \setminus \mathscr{E}_{\mathscr{R}}$, then $A \in \mathscr{F}_{\mathscr{R}}$ and $A \notin \mathscr{E}_{\mathscr{R}}$. Therefore, $A^{-} \subseteq A$ and $A^{\circ} = \emptyset$. Hence, we can see that $A^{-\circ} \subseteq A^{\circ} = \emptyset$, and so $A^{-\circ} = \emptyset$. Therefore, $A \in \mathscr{N}_{\mathscr{R}}$, and thus $\mathscr{F}_{\mathscr{R}} \setminus \mathscr{E}_{\mathscr{R}} \subseteq \mathscr{N}_{\mathscr{R}}$ is always true. Hence, we can see that (3) implies (4) even if $\mathscr{R}$ is only reflexive. Now, since, (4) trivially implies (3), it is clear that (3) and (4) are also equivalent even if $\mathscr{R}$ is only reflexive.

Thus, to complete the proof, we need to only show that now (1) and (3) are also equivalent. For this, note that $X \in \mathscr{T}_{\mathscr{R}}$. Therefore, if $B \in \mathscr{N}_{\mathscr{R}}$, then by Theorem 69 we also have $B^{c} = X \setminus B \in \mathscr{T}_{\mathscr{R}}^{\alpha}$. Hence, if (1) holds, i. e., $\mathscr{T}_{\mathscr{R}}^{\alpha} \subseteq \mathscr{T}_{\mathscr{R}}$, we can infer that $B^{c} \in \mathscr{T}_{\mathscr{R}}$, and thus $B \in \mathscr{F}_{\mathscr{R}}$. Therefore, (3) also holds.

On the other hand, if $A \in \mathscr{T}_{\mathscr{R}}^{\alpha}$, then by Theorem 69 there exist $V \in \mathscr{T}_{\mathscr{R}}$ and $B \in \mathscr{N}_{\mathscr{R}}$ such that $A = V \setminus B$. Moreover, if (3) holds, we can also state that $B \in \mathscr{F}_{\mathscr{R}}$, and thus $B^{c} \in \mathscr{T}_{\mathscr{R}}$. Hence, by using Corollary 5, we can infer that $A = V \setminus B = V \cap B^{c} \in \mathscr{T}_{\mathscr{R}}$. Therefore, $\mathscr{T}_{\mathscr{R}}^{\alpha} \subseteq \mathscr{T}_{\mathscr{R}}$, and thus (1) also holds.

34 Maximality Property of the Families $\mathscr{T}^{\kappa}_{\mathscr{R}}$

Analogously to Definition 15, we may also naturally introduce the following

Definition 16 For $\kappa = s, p, \alpha, \beta, a, b, q, ps, \gamma$, and δ, a relator $\mathscr{R}$ on X, will be called

(1) *κ–maximal* if $\mathscr{T}^{\kappa}_{\mathscr{R}} = \mathscr{P}(X)$;
(2) *relatively κ–maximal* if $\mathscr{T}^{\beta}_{\mathscr{R}} \subseteq \mathscr{T}^{\kappa}_{\mathscr{R}}$.

Remark 59 Thus, every κ–maximal relator is evidently also relatively κ–maximal.

Moreover, by Theorems 36, 37, 42 and 44, we can at once state the following three theorems.

Theorem 154 *If $\mathscr{R}$ is a reflexive relator on X, then $\mathscr{R}$ is relatively κ–maximal if and only if $\mathscr{T}^{\kappa}_{\mathscr{R}} = \mathscr{T}^{\beta}_{\mathscr{R}}$.*

Theorem 155 *If $\mathscr{R}$ is a topological relator on X, then $\mathscr{R}$ is relatively γ–maximal.*

Theorem 156 *If $\mathscr{R}$ is a topologically filtered, topological relator on X, then $\mathscr{R}$ is relatively δ–maximal.*

Now, as a generalization of [72, Theorem 5] of Reilly and Vamanamurthy and [61, Theorem 5] of Mukharjaee and Roy, we can also prove the following

Theorem 157 *If $\mathscr{R}$ is a topological relator on X, then the following assertions are equivalent:*

(1) $\mathscr{R}$ is p–maximal;
(2) $\mathscr{T}_{\mathscr{R}} \subseteq \mathscr{F}_{\mathscr{R}}$; *(3) $\mathscr{F}_{\mathscr{R}} \subseteq \mathscr{T}_{\mathscr{R}}$;* *(4) $\mathscr{T}_{\mathscr{R}} = \mathscr{F}_{\mathscr{R}}$.*

Proof If $V \in \mathscr{T}_{\mathscr{R}}$, then $V^c \in \mathscr{F}_{\mathscr{R}}$. Moreover, if (1) holds, then $V^{c-} \in \mathscr{T}^{p}_{\mathscr{R}}$ also holds. Hence, by using Corollary 11, we can infer that $V^c \in \mathscr{T}^{p}_{\mathscr{R}}$. Now, by the corresponding definitions, it is clear that

$$V^c \subseteq V^{c-\circ} = V^{c\circ},$$

and thus $V^c \in \mathscr{T}_{\mathscr{R}}$. Therefore, $V \in \mathscr{F}_{\mathscr{R}}$, and thus (2) also holds.

On the other hand, if $A \subseteq X$, then $A^- \in \mathscr{F}_{\mathscr{R}}$, and thus $A^{-c} \in \mathscr{T}_{\mathscr{R}}$. Moreover, if (2) holds, then we also have $A^{-c} \in \mathscr{F}_{\mathscr{R}}$. Hence, by using the corresponding definitions and the equality $c\, - = \circ\, c$, we can see that

$$A^{-c} = A^{-c-} = A^{-\circ c}, \qquad \text{and thus} \qquad A^- = A^{-\circ}.$$

Therefore, in particular, we have $A \subseteq A^{-\circ}$, and thus $A \in \mathscr{T}^{p}_{\mathscr{R}}$. Consequently, $\mathscr{P}(X) \subseteq \mathscr{T}^{p}_{\mathscr{R}}$, and thus (1) also holds.

Hence, since assertions (3) and (4) are immediate formulations of (2), it is clear that the assertion of the theorem is true.

Remark 60 Note that if $\mathscr{R}$ is a p–maximal relator on X, then by Theorem 36 $\mathscr{R}$ is also b–maximal.

While, if $\mathscr{R}$ is a p-maximal reflexive relator on X, then by Theorem 37 $\mathscr{R}$ is both β-maximal and δ–maximal.

Moreover, if $\mathscr{R}$ is a topological relator on X, then by Theorem 42 $\mathscr{R}$ is p–maximal if and only if it is ps-maximal.

The following example shows that the latter assertion need not be true for a non-topological relator.

Example 9 If X and $\mathscr{R}$ are as in Example 2, then $\mathscr{R}$ is a symmetric, p–maximal relator on X such that $\mathscr{R}$ is ps–minimal.

Namely, by using the corresponding definitions, we can see that $A^{-\circ} = A$ for all $A \subseteq X$, and thus $\mathscr{T}_{\mathscr{R}}^{p} = \mathscr{P}(X)$. Moreover, $\mathscr{T}_{\mathscr{R}} = \{\emptyset, X\}$, and thus also $\mathscr{T}_{\mathscr{R}}^{ps} = \{\emptyset, X\}$.

Remark 61 Quite similarly, we can also show that $\mathscr{R}$ is, in addition, s–maximal and q–minimal.

Concerning the relative p–maximality of a relator $\mathscr{R}$, we can only prove

Theorem 158 *If $\mathscr{R}$ is a topological relator on X such that $\mathscr{T}_{\mathscr{R}}^{r} \subseteq \mathscr{F}_{\mathscr{R}}$, then $\mathscr{R}$ is relatively p–maximal.*

Proof If $A \in \mathscr{T}_{\mathscr{R}}^{\beta}$, then by Definition 12 we have $A \subseteq A^{-\circ-}$. Moreover, by Theorem 62, we have $A^{-\circ} \in \mathscr{T}_{\mathscr{R}}^{r}$. Hence, by using the assumption $\mathscr{T}_{\mathscr{R}}^{r} \subseteq \mathscr{F}_{\mathscr{R}}$, we can infer that $A^{-\circ} \in \mathscr{F}_{\mathscr{R}}$, and thus $A^{-\circ-} \subseteq A^{-\circ}$. Therefore, $A \subseteq A^{-\circ}$, and thus $A \in \mathscr{T}_{\mathscr{R}}^{p}$ also holds. Consequently, $\mathscr{T}_{\mathscr{R}}^{\beta} \subseteq \mathscr{T}_{\mathscr{R}}^{p}$, and thus by Definition 16 the relator $\mathscr{R}$ is relatively p–maximal.

Remark 62 In this respect, it is also worth noticing that if $\mathscr{R}$ is an arbitrary relator on X, then we already have $\mathscr{T}_{\mathscr{R}}^{p} \cap \mathscr{F}_{\mathscr{R}}^{\alpha} \subseteq \mathscr{F}_{\mathscr{R}}$, and thus also $\mathscr{T}_{\mathscr{R}}^{r} \cap \mathscr{F}_{\mathscr{R}}^{\alpha} \subseteq \mathscr{F}_{\mathscr{R}}$.

Namely, if $A \in \mathscr{T}_{\mathscr{R}}^{p} \cap \mathscr{F}_{\mathscr{R}}^{\alpha}$, then $A \in \mathscr{T}_{\mathscr{R}}^{p}$ and $A \in \mathscr{F}_{\mathscr{R}}^{\alpha}$. Hence, by using Definition 12 and Theorem 52, we can infer that $A \subseteq A^{-\circ}$ and $A^{-\circ-} \subseteq A$. Therefore, $A^{-} \subseteq A^{-\circ-} \subseteq A$, and thus $A \in \mathscr{F}_{\mathscr{R}}$ also holds.

Note 1 By [2, Theorem 2.1] of Aho and Nieminen, for some κ_1 and κ_2, the inclusion $\mathscr{T}_{\mathscr{R}}^{\kappa_1} \subseteq \mathscr{T}_{\mathscr{R}}^{\kappa_2}$ could also be investigated.

Moreover, by Andrijević [5, 6], Andrijević and Ganster [7], Crossley [12], Crossley and Hildebrandt [13, 14], and Njåstad [63], for any two relators $\mathscr{R}$ and $\mathscr{S}$ on X, the inclusion $\mathscr{T}_{\mathscr{R}}^{\kappa} \subseteq \mathscr{T}_{\mathscr{S}}^{\kappa}$ could also be investigated.

More generally, by using the families $\mathscr{T}_{\mathscr{R}}^{\kappa}$, we can introduce several interesting continuity properties of relations on one space to another.

These continuity properties cannot certainly be included in the general frameworks worked out in the former papers [83, 90, 92, 94, 95, 101].

Moreover, we note that, by using the ideas of Gargouri and Rezgui [37] and the third author [100], our present results could also be generalized.

References

1. M. E. Abd El-Monsef, S.N. El-Deeb, R. A. Mahmoud, β–open sets and β–continuous mappings. Bull. Fac. Sci. Assiut Univ. **12**, 77–90 (1983)
2. T. Aho, T. Nieminen, Spaces in which preopen subsets are semiopen. Richerche Mat. **43**, 45–49 (1994)
3. D. Andrijević, Semi-preopen sets. Mat. Vesnik **38**, 24–32 (1986)
4. D. Andrijević, On b–open sets. Mat. Vesnik **48**, 59–64 (1996)
5. D. Andrijević, On SPO–equivalent topologies. Suppl. Rend. Circ. Mat. Palermo **29**, 317–328 (1992)
6. D. Andrijević, A note on α–equivalent topologies. Mat. Vesnik **45**, 65–69 (1993)
7. D. Andrijević, M. Ganster, On PO–equivalent topologies. Suppl. Rend. Circ. Mat. Palermo **24**, 251–256 (1990)
8. H. Arianpoor, Preorder relators and generalized topologies. J. Lin. Top. Algebra **5**, 271–277 (2016)
9. J. Ávila, F. Molina, Generalized weak structures. Int. Math. Forum **7**, 2589–2595 (2012)
10. N. Bourbaki, *General Topology, Chapters 1–4* (Springer-Verlag, Berlin, 1989)
11. H. H. Corson and E. Michael, Metrizability of countable unions. Illinois J. Math. **8**, 351–360 (1964)
12. S. G. Crossley, A note on semitopological classes. Proc. Amer. Math. Soc. **43**, 416–420 (1974)
13. S. G. Crossley, S. K. Hildebrandt, Semi-closure. Texas J. Sci. **22**, 99–112 (1971)
14. S. G. Crossley, S. K. Hildebrandt, Semi-topological properties. Fund. Math. **74**, 233–254 (1972)
15. Á. Császár, *Foundations of General Topology* (Pergamon Press, London, 1963)
16. Á. Császár, Generalized open sets. Acta Math. Hungar. **75**, 65–87 (1997)
17. Á. Császár, On the γ–interior and γ–closure of a set. Acta Math. Hungar. **80**, 89–93 (1998)
18. Á. Császár, γ–quasi-open sets. Studia Sci. Math. Hungar. **38**, 171–176 (2001)
19. Á. Császár, Remarks on γ-quasi-open sets. Studia Sci. Math. Hungar. **39**, 137–141 (2002)
20. Á. Császár, Generalized topology, generalized continuity. Acta Math. Hungar. **96**, 351–357 (2002)
21. Á. Császár, Generalized open sets in generalized topologies. Acta Math. Hungar. **106**, 53–66 (2005)
22. Á. Császár, Further remarks on the formula for the γ–interior. Acta Math. Hungar. **113**, 325–332 (2006)
23. Á. Császár, Remarks on quasi-topologies. Acta Math. Hungar. **119**, 197–200 (2008)
24. Á. Császár, Weak structures. Acta Math. Hungar. **131**, 193–195 (2011)
25. B. A. Davey, H. A. Priestley, *Introduction to Lattices and Order* (Cambridge University Press, Cambridge, 2002)
26. A. S. Davis, Indexed systems of neighbordoods for general topological spaces. Amer. Math. Monthly **68**, 886–893 (1961)
27. K. Dlaska, N. Ergun, M. Ganster, On the topology generated by semi-regular sets. Indian J. Pure Appl. Math. **25**, 1163–1170 (1994)
28. J. Dontchev, Survey on preopen sets. Meetings on Topological Spaces, Theory and Applications, Yatsushiro College of Technology, Kumamoto, Japan, 18 pp. (1998)
29. Z. Duszyński, T. Noiri, Semi-open, semi-closed sets and semi-continuity of functions. Math. Pannon. **23**, 195–200 (2012)
30. V. A. Efremovič, The geometry of proximity. Mat. Sb. **31**, 189–200 (1952) (Russian)
31. V. A. Efremović, A. S. Švarc, A new definition of uniform spaces. Metrization of proximity spaces. Dokl. Acad. Nauk. SSSR **89**, 393–396 (1953) (Russian)
32. E. Ekici, On weak structures due to Császár. Acta Math. Hungar. **134**, 565–570 (2012)
33. N. Elez, O. Papaz, The new operators in topological spaces. Math. Moravica **17**, 63–68 (2013)
34. P. Fletcher, W. F. Lindgren, *Quasi-Uniform Spaces* (Marcel Dekker, New York, 1982)

35. M. Ganster, Preopen sets and resolvable spaces. Kyungpook J. **27**, 135–143 (1987)
36. B. Ganter, R. Wille, *Formal Concept Analysis* (Springer-Verlag, Berlin, 1999)
37. R. Gargouri, A. Rezgui, A unification of weakening of open and closed subsets in a topological space. Bull. Malays. Math. Sci. Soc. **40**, 1219–1230 (2017)
38. S. Givant, P. Halmos, *Introduction to Boolean Algebras* (Springer-Verlag, Berlin, 2009)
39. T. Glavosits, Generated preorders and equivalences. Acta Acad. Paed. Agrienses, Sect. Math. **29**, 95–103 (2002)
40. A. Gupta, R. D. Sarma, PS–regular sets in topology and generalized topology. Hindawi J. Math. **2014**, 6 pp. (2014)
41. T. R. Hamlett, Correction to the paper "Semi-open sets and semi-continuity in topological spaces" by Norman Levine. Proc. Amer. Math. Soc. **70**, 36–41 (1963)
42. W. Hunsaker, W. Lindgren, Construction of quasi-uniformities. Math. Ann. **188**, 39–42 (1970)
43. D. H. Hyers, On the stability of the linear functional equation. Proc. Nat. Acad. Sci. U.S.A **27**, 222–224 (1941)
44. Y. Isomichi, New concept in the theory of topological spaces–Supercondensed set, subcondensed set, and condensed set. Pacific J. Math. **38**, 657–668 (1971)
45. R. Jamunarani, P. Jeyanthi, Regular sets in generalized topological spaces. Acta Math. Hungar. **135**, 342–349 (2012)
46. R. Jamunarani, P. Jeyanthi, T. Noiri, On generalized weak structures. J. Algorithms Comput. **47**, 21–26 (2016)
47. D. S. Janković, I. L. Reilly, On semi separation properties. Indian J. Pure Apppl. Math. **16**, 957–964 (1985)
48. Y. B. Jun, S. W. Jeong, H. j. Lee, J. W. Lee, Applications of pre-open sets. Appl. Gen. Top. **9**, 213–228 (2008)
49. J. L. Kelley, *General Topology* (Van Nostrand Reinhold Company, New York, 1955)
50. K. Kuratowski, Sur l'opération $\overline{A}$ de l'analysis situs. Fund. Math. **3**, 182–199 (1922) (An English translation: On the operation $\overline{A}$ in analysis situs, prepared by M. Bowron in 2010, is available on the Internet.)
51. K. Kuratowski, *Topology I* (Academic Press, New York, 1966)
52. J. Kurdics, *A note on connection properties*, Acta Math. Acad. Paedagog. Nyházi. **12**, 57–59 (1990).
53. J. Kurdics, Á. Száz, *Well-chainedness characterizations of connected relators*, Math. Pannon. **4**, 37–45 (1993)
54. N. Levine, Semi-open sets and semi-continuity in topological spaces. Amer. Math. Monthly **70**, 36–41 (1963)
55. N. Levine, Some remarks on the closure operator in topological spaces. Amer. Math. Monthly **70**, p. 553 (1963)
56. N. Levine, On uniformities generated by equivalence relations. Rend. Circ. Mat. Palermo **18**, 62–70 (1969)
57. N. Levine, On Pervin's quasi uniformity. Math. J. Okayama Univ. **14**, 97–102 (1970)
58. J. Mala, Relators generating the same generalized topology. Acta Math. Hungar. **60**, 291–297 (1992)
59. J. Mala, Á. Száz, Modifications of relators. Acta Math. Hungar. **77**, 69–81 (1997)
60. A. S. Mashhour, M. E. Abd El-Monsef, S. N. El-Deeb, On precontinuous and weak precontinuous mappings. Proc. Math. Phys. Soc. Egypt **53**, 47–53 (1982)
61. A. Mukharjee, R. M. Roy, On generalized preopen sets. Mat. Stud. **51**, 195–199 (2019)
62. S. A. Naimpally, B. D. Warrack, *Proximity Spaces*. (Cambridge University Press, Cambridge, 1970)
63. O. Njåstad, On some classes of nearly open sets. Pacific J. Math. **15**, 195–213 (1965)
64. T. Noiri, On semi-continuous mappings. Lincei-Rend. Sci. Fis. Mat. Nat. **54**, 210–214 (1973)
65. T. Noiri, Hyperconnectedness and preopen sets. Rev. Roum. Math. Pures Appl. **29**, 329–334 (1984)

66. G. Pataki, Supplementary notes to the theory of simple relators. Radovi Mat. **9**, 101–118 (1999)
67. G. Pataki, On the extensions, refinements and modifications of relators. Math. Balk. **15**, 155–186 (2001)
68. G. Pataki, Á. Száz, A unified treatment of well-chainedness and connectedness properties. Acta Math. Acad. Paedagog. Nyházi. (N.S.) **19**, 101–165 (2003)
69. W. J. Pervin, Quasi-uniformization of topological spaces. Math. Ann. **147**, 316–317 (1962)
70. Th. M. Rassias, M. Salih, Á. Száz, Characterizations of generalized topologically open sets in relator spaces. In: G. V. Milovanovic, Thm. M. Rassias, Y. Simsek (Eds.), Recent Trends on Pure and Applied Mathematics, Special Issue of the Montes Taurus J. Pure Appl. Math., Dedicated to Professor Hari Mohan Srivastava on the occasion of his 80th Birthday, Montes Taurus J. Pure Appl. Math. **3**, 39–94 (2021)
71. I. L. Reilly, M. K. Vamanamurthy, On α-continuity in topological spaces. Acta Math. Hungar. **45**, 27–32 (1985)
72. I. L. Reilly, M. K. Vamanamurthy, On some questions concerning preopen sets. Kyungpook Math. J. **30**, 87–93 (1990)
73. M. Salih, Á. Száz, Generalizations of some ordinary and extreme connectedness properties of topological spaces to relator spaces. Elec. Res. Arch. **28**, 471–548 (2020)
74. M. S. Sarsak, On some properties of generalized open sets in generalized topological spaces. Demonstr. Math. **46**, 415–427 (2013)
75. P. Sivagami, Remarks on γ–interior. Acta Math. Hungar. **119**, 81–94 (2008)
76. Yu. M. Smirnov, On proximity spaces. Math. Sb. **31**, 543–574 (1952) (Russian.)
77. M. H. Stone, Application of the theory of Boolean rings to general topology. Trans. Amer. Math. Soc. **41**, 374–481 (1937)
78. Á. Száz, Basic tools and mild continuities in relator spaces. Acta Math. Hungar. **50**, 177–201 (1987)
79. Á. Száz, Directed, topological and transitive relators. Publ. Math. Debrecen **35**, 179–196 (1988)
80. Á. Száz, Relators, Nets and Integrals. Unfinished doctoral thesis, Debrecen, 126 pp. (1991)
81. Á. Száz, Structures derivable from relators. Singularité **3**, 14–30 (1992)
82. Á. Száz, Refinements of relators. Tech. Rep., Inst. Math., Univ. Debrecen **76**, 19 pp. (1993)
83. Á. Száz, Somewhat continuity in a unified framework for continuities of relations. Tatra Mt. Math. Publ. **24**, 41–56 (2002)
84. Á. Száz, Upper and lower bounds in relator spaces. Serdica Math. J. **29**, 239–270 (2003)
85. Á. Száz, Rare and meager sets in relator spaces. Tatra Mt. Math. Publ. **28**, 75–95 (2004)
86. Á. Száz, Galois-type connections on power sets and their applications to relators. Tech. Rep., Inst. Math., Univ. Debrecen **2005/2**, 38 pp.
87. Á. Száz, Minimal structures, generalized topologies, and ascending systems should not be studied without generalized uniformities. Filomat **21**, 87–97 (2007)
88. Á. Száz, Galois type connections and closure operations on preordered sets. Acta Math. Univ. Comenian. (N.S.) **78**, 1–21 (2009)
89. Á. Száz, Inclusions for compositions and box products of relations. J. Int. Math. Virt. Inst. **3**, 97–125 (2013)
90. Á. Száz, Lower semicontinuity properties of relations in relator spaces. Adv. Stud. Contemp. Math. (Kyungshang) **23**, 107–158 (2013)
91. Á. Száz, A particular Galois connection between relations and set functions. Acta Univ. Sapientiae, Math. **6**, 73–91 (2014)
92. Á. Száz, Generalizations of Galois and Pataki connections to relator spaces. J. Int. Math. Virtual Inst. **4**, 43–75 (2014)
93. Á. Száz, Basic tools, increasing functions, and closure operations in generalized ordered sets. In: P. M. Pardalos and Th. M. Rassias (Eds.), Contributions in Mathematics and Engineering: In Honor of Constantin Caratheodory, Springer, 551–616 (2016)

94. Á. Száz, Four general continuity properties, for pairs of functions, relations and relators, whose particular cases could be investigated by hundreds of mathematicians. Tech. Rep., Inst. Math., Univ. Debrecen **2017/1**, 17 pp.
95. Á. Száz, Contra continuity properties of relations in relator spaces. Tech. Rep., Inst. Math., Univ. Debrecen **2017/5**, 48 pp.
96. Á. Száz, The closure-interior Galois connection and its applications to relational equations and inclusions. J. Int. Math. Virt. Inst. **8**, 181–224 (2018)
97. Á. Száz, Corelations are more powerful tools than relations. In: Th. M. Rassias (Ed.), Applications of Nonlinear Analysis, Springer Optimization and Its Applications **134**, 711–779 (2018)
98. Á. Száz, Relationships between inclusions for relations and inequalities for corelations. Math. Pannon. **26**, 15–31 (2018)
99. Á. Száz, Galois and Pataki connections on generalized ordered sets. Earthline J. Math. Sci. **2**, 283–323 (2019)
100. Á. Száz, Birelator spaces are natural generalizations of not only bitopological spaces, but also ideal topological spaces. In: Th. M. Rassias and P. M. Pardalos (Eds.), Mathematical Analysis and Applications, Springer Optimization and Its Applications **154**, Springer Nature Switzerland AG, 543–586 (2019)
101. Á. Száz, A. Zakaria, Mild continuity properties of relations and relators in relator spaces. In: P. M. Pardalos and Th. M. Rassias (Eds.), Essays in Mathematics and its Applications: In Honor of Vladimir Arnold, Springer, 439–511 (2016)
102. W. J. Thron, *Topological Structures* (Holt, Rinehart and Winston, New York, 1966)
103. A. Weil, Sur les espaces á structure uniforme et sur la topologie générale. Actual. Sci. Ind. **551** (Herman and Cie, Paris 1937)

On Degenerate Boundary Conditions and Finiteness of the Spectrum of Boundary Value Problems

Victor A. Sadovnichii, Yaudat T. Sultanaev, and Azamat M. Akhtyamov

Abstract It is shown that for the asymmetric diffusion operator the case when the characteristic determinant is identically equal to zero is impossible and the only possible degenerate boundary conditions are the Cauchy conditions. In the case of a symmetric diffusion operator, the characteristic determinant is identically equal to zero if and only if the boundary conditions are false–periodic boundary conditions and is identically equal to a constant other than zero if and only if its boundary conditions are generalized Cauchy conditions. All degenerate boundary conditions for a spectral problem with a third-order differential equation $y'''(x) = \lambda\, y(x)$ are described. The general form of degenerate boundary conditions for the fourth-order differentiation operator D^4 is found. Twelve classes of boundary value eigenvalue problems are described for the operator D^4, the spectrum of which fills the entire complex plane. It is known that spectral problems whose spectrum fills the entire complex plane exist for differential equations of any even order. John Locker posed the following problem (eleventh problem): Are there similar problems for odd-order differential equations? A positive answer is given to this question. It is proved that spectral problems, the spectrum of which fills the entire complex plane, exist for differential equations of any odd order. Thus, the problem of John Locker is resolved. John Locker posed a problem (tenth problem): Can a spectral boundary value problem have a finite spectrum? Boundary value problems with a polynomial occurrence of a spectral parameter in a differential equation are considered. It is shown that the corresponding boundary value problem can have a predetermined finite spectrum in the case when the roots of the characteristic equation are multiple. If the roots of the characteristic equation are not multiple, then there can be no finite spectrum. Thus, John Locker's tenth problem is resolved.

V. A. Sadovnichii · Y. T. Sultanaev (✉)
Department of Mechanics and Mathematics, Lomonosov Moscow State University, Moscow, Russia
e-mail: rector@rector.msu.su

A. M. Akhtyamov
Mavlyutov Institute of Mechanics, Russian Academy of Science, Moscow, Russia

I. N. Parasidis et al. (eds.), *Mathematical Analysis in Interdisciplinary Research*, Springer Optimization and Its Applications 179,
https://doi.org/10.1007/978-3-030-84721-0_31

1 Introduction

The boundary conditions in a spectral problem are said to be degenerate if the characteristic determinant $\Delta(\lambda)$ of the problem is identically constant [1, p. 35]. Direct and inverse problems with nondegenerate boundary conditions are sufficiently well studied (e.g., see [2, 3]). The case of degenerate boundary conditions has been studied much less. Apparently, only the Stone's example of a second-order differential operator whose spectrum fills the entire complex plane is well known [4]. Namely, Stone showed that if the potential function $q(x)$ is symmetric (i.e., $q(x) = q(\pi - x)$) and $a = 1$, then every complex number belongs to the spectrum of the boundary value problem

$$- y'' + q(x)\, y = \lambda\, y, \quad y(0) \pm a\, y(\pi) = 0, \quad y'(0) \mp a\, y'(\pi) = 0; \tag{1}$$

i.e., its spectrum coincides with the entire complex plane. The first results about degenerate boundary conditions for differential operators of arbitrary even order were obtained by Sadovnichii and Kanguzhin [5] (see also the monograph [6, pp.273–275]) who showed that there exist differential operators of arbitrary even order n whose spectrum fills the entire complex plane. These boundary conditions have the form

$$U_j(y) = y^{(j-1)}(0) + (-1)^{j-1}\, y^{(j-1)}(1) = 0, \quad j = 1, 2, \ldots, n. \tag{2}$$

It was shown in [7] that the boundary conditions

$$U_j(y) = y^{(j-1)}(0) + d \cdot (-1)^{j-1}\, y^{(j-1)}(1) = 0, \quad j = 1, 2, \ldots, n$$

for differential equations of even order n are degenerate for $d \neq \pm 1$ as well, but in this case the spectrum of the corresponding boundary value problem is empty ($\Delta(\lambda) \equiv C = \text{const} \neq 0$). It was shown in [8] that there also exist degenerate boundary conditions for boundary value problems with a differential equation of arbitrary odd order. The question of describing all boundary value problems with degenerate boundary conditions is related to a description of all Volterra problems. The problem for operator L is called Volterra problem if the inverse operator L^{-1} is Volterra operator (see [9, p. 208]). In the case of nondegenerate boundary conditions for an arbitrary continuous function $q(x)$, the system of eigen-vectors of the operator L is complete in $L_2(0, \pi)$ (see [1, p. 29]). Therefore, Volterra problems are among problems with degenerate boundary conditions. In [10] it is shown that all Volterra problems for operator D^2 with common boundary conditions have the form

$$y(0) \mp a\, y(\pi) = 0, \quad y'(0) \pm a\, y'(\pi) = 0, \tag{3}$$

where $a \neq 1$. A similar result is obtained in [11] for Sturm–Liouville problems with the differential equation $-y'' + q(x)\, y = \lambda\, y$ and the symmetric potential ($q(x) =$

$q(\pi - x))$. All degenerate boundary conditions for the Sturm–Liouville problem were described in [12]. More precisely, it was shown in [12] that the following assertions are true. If $q(x) \neq q(\pi - x)$ for x in some subinterval of $[0,\pi]$, then the case of $\Delta(\lambda) \equiv 0$ is impossible and the only possible degenerate boundary conditions are the Cauchy conditions $y(0) = y'(0) = 0$ and $y(\pi) = y'(\pi) = 0$. If $q(x) = q(\pi - x)$ a.e. on $[0,\pi]$, then the case of $\Delta(\lambda) \equiv 0$ is realized if and only if the boundary conditions of problem (1) are pseudoperiodic boundary conditions with $a = 1$ and the case of $\Delta(\lambda) \equiv C \neq 0$ is realized if and only if the boundary conditions (1) are the generalized Cauchy conditions (1) with $a \neq 1$. *The second section* is devoted to the degenerate boundary conditions of the Sturm–Liouville problem. In [13] the diffusion operator is considered:

$$ly = y'' + \left(\lambda^2 - 2\lambda\, p(x) - q(x)\right) y = 0, \tag{4}$$

$$U_i(y) = a_{i1}\, y(0) + a_{i2}\, y'(0) + a_{i3}\, y(\pi) + a_{i4}\, y'(\pi) = 0, \qquad i = 1, 2, \tag{5}$$

where $p(x) \in W_2^1(0, \pi)$ and $q(x) \in L_2(0, \pi)$ are real functions and a_{ij}, $i = 1, 2$, $j = 1, 2, 3, 4$, are complex constants.

If $p(x) \neq p(\pi - x)$ and/or $q(x) \neq q(\pi - x)$ on some subinterval of the closed interval $[0,\pi]$, then the case of $\Delta(\lambda) \equiv 0$ is impossible and the only possible degenerate boundary conditions are the Cauchy conditions $y(0) = y'(0) = 0$ and $y(\pi) = y'(\pi) = 0$. If $p(x) = p(\pi - x)$ and $q(x) = q(\pi - x)$, then the case of $\Delta(\lambda) \equiv 0$ is realized if and only if the boundary conditions $a_{i1}\, y(0) + a_{i2}\, y'(0) + a_{i3}\, y(\pi) + a_{i4}\, y'(\pi) = 0$ are the falsely periodic boundary conditions $y(0) \mp y(\pi) = 0, \quad y'(0) \pm y'(\pi) = 0$ and the case of $\Delta(\lambda) \equiv C \neq 0$ is realized if and only if conditions $a_{i1}\, y(0) + a_{i2}\, y'(0) + a_{i3}\, y(\pi) + a_{i4}\, y'(\pi) = 0$ are the generalized Cauchy conditions $y(0) \mp a\, y(\pi) = 0, \quad y'(0) \pm a\, y'(\pi) = 0$ with $a \neq 0$. *The third section* is devoted to the degenerate boundary conditions of the diffusion operator.

Only the differential operator of any even order for which the spectrum fills the whole complex plane is rather well known [5, 6]. The question was posed by Jogn Locker [6]: Do there exist similar differential operators of odd order? We give an answer to this question. It is shown that, for any odd integer n, there exist differential operators of order n whose spectrum fills the whole complex plane. *The fourth section* is devoted to questions of degenerate boundary conditions for boundary value problems with an odd-order differential equation. *The fifth section* is devoted to questions of degenerate boundary conditions for boundary value problems with a three-order differential equation. It is well known, perhaps the only one, an example for differential operator of any even order for which the spectrum fills the entire complex plane [5] (see also [6]). In this example the boundary conditions have the following form:

$$U_j(y) = y^{(j-1)}(0) + (-1)^{j-1}\, y^{(j-1)}(1) = 0, \qquad j = 1, 2, 3, 4.$$

However, in connection with this, the question arises: Are there other examples of such operators? *In the sixth section*, for operator D^4 we find other examples of such operators and describe all boundary value problems for operator D^4 whose spectrum fills the entire complex plane. The form of degenerate boundary conditions is found too. *In the seventh section*, we study the boundary conditions of the Sturm–Liouville problem posed on a star-shaped geometric graph consisting of three edges with a common vertex. We show that the Sturm–Liouville problem has no degenerate boundary conditions in the case of pairwise distinct edge lengths. However, if the edge lengths coincide and all potentials are the same, then the characteristic determinant of the Sturm–Liouville problem cannot be a nonzero constant and the set of Sturm–Liouville problems whose characteristic determinant is identically zero and whose spectrum accordingly coincides with the entire plane is infinite (a continuum). It is shown that, for one special case of the boundary conditions, this set consists of eighteen classes, each having from two to four arbitrary constants, rather than of two problems as in the case of the Sturm–Liouville problem on an interval.

The case of finite spectrum of boundary eigenvalue problem has not been studied well enough. It was shown in [6] and [14, p. 556] that the differentiation operators D^2 and D^4 with the corresponding boundary conditions cannot have a finite spectrum. In 2008, Locker [6] posed the following question for the boundary problem:

$$y^{(n)} + a_1(x)\, y^{n-1} + \cdots + a_{n-1}(x)\, y' + a_n(x)\, y = \lambda\, y(x), \qquad x \in [0, 1] \tag{6}$$

$$U_j(y) = \sum_{k=0}^{n-1} b_{jk}\, y^{(k)}(0) + \sum_{k=0}^{n-1} b_{j\,k+n}\; y^{(k)}(1) = 0, \qquad j = 1, 2, \ldots, n, \tag{7}$$

where $\operatorname{rank} ||b_{jk}||_{n\times 2n} = n$, $b_{jk} \in \mathbb{C}$. Can the boundary value problem (6) and (7) have finite spectrum? In the same year, Kalmenov and Suragan [15] proved that the spectrum of regular partial differential boundary value problems, including problems (6) and (7) is either empty or infinite. *The eighth section* is devoted to the study of a finite spectrum of boundary value problems. We consider boundary value problems with spectral parameter polynomially occurring in the differential equation or the boundary conditions. It is shown that some of these problems have a prescribed finite spectrum. A wide class of boundary value problems which do not have finite spectrum is found.

2 On Degenerate Boundary Conditions in the Sturm–Liouville Problem

We describe all degenerate boundary conditions in the homogeneous Sturm–Liouville problem.

By L we denote the Sturm–Liouville problem

$$ly = -y'' + q(x)\,y = \lambda\,y = s^2\,y, \tag{8}$$

$$U_i(y) = a_{i1}\,y(0) + a_{i2}\,y'(0) + a_{i3}\,y(\pi) + a_{i4}\,y'(\pi) = 0, \qquad i = 1, 2, \tag{9}$$

where $q(x) \in L_1(0, \pi)$ is a real function and the $a_{ij},\ i = 1,\ 2,\ j = 1,\ 2,\ 3,\ 4$, are complex constants.

The boundary conditions in the problem L are said to be nondegenerate if the characteristic determinant of the problem L is not a constant [1, p. 35]. If the boundary conditions of the problem L are such that the characteristic determinant of the problem L is a constant, then such conditions are said to be degenerate. In the present paper, we find all degenerate boundary conditions in the problem L.

We denote the matrix consisting of the coefficients a_{lk} in the boundary conditions (9) by A and the minor consisting of the ith and jth columns of this matrix by M_{ij}:

$$A = \begin{Vmatrix} a_{11}\ a_{12}\ a_{13}\ a_{14} \\ a_{21}\ a_{22}\ a_{23}\ a_{24} \end{Vmatrix}, \quad M_{ij} = \begin{vmatrix} a_{1i}\ a_{1j} \\ a_{2i}\ a_{2j} \end{vmatrix}, \qquad i, j = 1, 2, 3, 4. \tag{10}$$

In what follows, we assume that the rank of the matrix A is equal to 2, rank$A$=2.

The eigenvalues of the problem L are the roots of the entire function [1, pp. 33–36], [16, p. 29]

$$\begin{aligned} \Delta(\lambda) = M_{12} + M_{34} + M_{32}\,y_1(\pi,\ \lambda) + M_{42}\,y_1'(\pi,\ \lambda) \\ + M_{13}\,y_2(\pi,\ \lambda) + M_{14}\,y_2'(\pi,\ \lambda), \end{aligned} \tag{11}$$

where $y_1(x,\ \lambda)$ and $y_2(x,\ \lambda)$ are the linearly independent solutions of Eq. (8) satisfying the conditions

$$y_1(0,\ \lambda) = 1, \quad y_1'(0,\ \lambda) = 0, \quad y_2(0,\ \lambda) = 0, \quad y_2'(0,\ \lambda) = 1.$$

The asymptotic formulas

$$\begin{aligned} y_1(x,\ \lambda) &= \cos sx + \tfrac{1}{s}\,u(x)\,\sin sx + \mathcal{O}\left(\tfrac{1}{s^2}\right), \\ y_2(x,\ \lambda) &= \tfrac{1}{s}\,\sin sx - \tfrac{1}{s^2}\,u(x)\,\cos sx + \mathcal{O}\left(\tfrac{1}{s^3}\right), \\ y_1'(x,\ \lambda) &= -s\,\sin sx + u(x)\,\cos sx + \mathcal{O}\left(\tfrac{1}{s}\right), \\ y_2'(x,\ \lambda) &= \cos sx + \tfrac{1}{s}\,u(x)\,\sin sx + \mathcal{O}\left(\tfrac{1}{s^2}\right), \end{aligned} \tag{12}$$

where $u(x) = \frac{1}{2}\int_0^x q(t)\,dt$ hold for $\lambda \in \mathbb{R}$ and for sufficiently large λ ([16, pp. 62–65]).

The identity $y_1(\pi, \lambda) \equiv y_2'(\pi, \lambda)$ holds if and only if $q(x) = q(x - \pi)$ almost everywhere on $[0,\pi]$ [17, Lemma 4].

If $q(x) \neq q(\pi - x)$ in some interval contained in $[0,\pi]$ and $\Delta(\lambda) \equiv C = \text{const}$, then it follows from relations (11) and (12) that

$$M_{12} + M_{34} = C, \quad M_{32} = 0, \quad M_{42} = 0, \quad M_{13} = 0, \quad M_{14} = 0. \tag{13}$$

To find the minors M_{12} and M_{34} we use the fact that the minors of a matrix cannot be arbitrary numbers. Given numbers M_{12}, M_{13}, M_{14}, M_{23}, M_{24}, and M_{34} are the minors of some matrix if and only if the following Plücker relations hold [18]:

$$M_{12}\,M_{34} - M_{13}\,M_{24} + M_{14}\,M_{23} = 0. \tag{14}$$

The minors M_{23}, M_{24} occurring in relations (14) differ from the minors M_{32} and M_{42} in relations (13) only in sign. From relations (13) and (14), we obtain two sets of minors,

$$\begin{aligned} &M_{12} = C \neq 0, \quad M_{34} = 0, \quad M_{32} = 0, \\ &M_{42} = 0, \quad M_{13} = 0, \quad M_{14} = 0; \end{aligned} \tag{15}$$

$$\begin{aligned} &M_{12} = 0, \quad M_{34} = C \neq 0, \quad M_{32} = 0, \\ &M_{42} = 0, \quad M_{13} = 0, \quad M_{14} = 0. \end{aligned} \tag{16}$$

The case in which $C = 0$ (and hence $\Delta(\lambda) \equiv 0$) cannot be realized, because the vanishing of all second-order determinants contradicts the condition rank $A = 2$. With the use of methods for the reconstruction of a matrix from its minors [18], the sets of minors (15) and (16) uniquely determine the boundary conditions (9) (i.e., the matrix A can be found up to a linear transformation of its rows). The set of minors (15) corresponds to the Cauchy conditions $y(0) = y'(0) = 0$, and the set of minors (16) corresponds to the Cauchy conditions $y(\pi) = y'(\pi) = 0$.

We have thereby proved the following assertion.

Theorem 2.1 *If $q(x) \neq q(\pi - x)$ on some interval, then the case in which $\Delta(\lambda) \equiv 0$ is impossible, and the Cauchy conditions $y(0) = y'(0) = 0$ and $y(\pi) = y'(\pi) = 0$ are the only possible degenerate boundary conditions.*

If $q(x) = q(\pi - x)$ almost everywhere and $\Delta(\lambda) \equiv C = \text{const}$, then it follows from relations (11) and (12) that

$$M_{12} + M_{34} = C, \quad M_{32} + M_{14} = 0, \quad M_{42} = 0, \quad M_{13} = 0. \tag{17}$$

From relations (14) and (17), we obtain two sets of minors

$$\begin{aligned} M_{12} = C_1, \quad M_{34} = C - C_1, \quad M_{32} = \mp\sqrt{C_1\,(C_1 - C)}, \\ M_{42} = 0, \quad M_{13} = 0, \quad M_{14} = \pm\sqrt{C_1\,(C_1 - C)}. \end{aligned} \tag{18}$$

If $C = 0$ [the case in which $\Delta(\lambda) \equiv 0$], then from (18) we obtain the relations

$$\begin{aligned} M_{12} = C_1, \quad M_{34} = -C_1, \quad M_{32} = \mp C_1, \\ M_{42} = 0, \quad M_{13} = 0, \quad M_{14} = \pm C_1. \end{aligned} \tag{19}$$

By using methods for the reconstruction of a matrix from its minors [18], for these sets of minors we uniquely find two forms of boundary conditions

$$y(0) \mp y(\pi) = 0, \quad y'(0) \pm y'(\pi) = 0. \tag{20}$$

Conditions (20) are said to be falsely periodic, because they are degenerate and differ from nondegenerate periodic or antiperiodic boundary conditions by the change of only one sign (plus is replaced by minus, or minus is replaced by plus).

If $C \neq 0$ [the case in which $\Delta(\lambda) \not\equiv 0$], then the form of boundary conditions depends on nonzero minors in (18). If $C - C_1 = 0$, then we obtain the Cauchy conditions $y(0) = y'(0) = 0$; if $C_1 = 0$, we have the Cauchy conditions $y(\pi) = y'(\pi) = 0$; and if $C - C_1 \neq 0$ and $C_1 \neq 0$, then we obtain the conditions

$$y(0) \mp a\, y(\pi) = 0, \quad y'(0) \pm a\, y'(\pi) = 0, \tag{21}$$

where $a = \sqrt{\frac{C_1 - C}{C_1}}$. We have thereby proved the following assertion.

Theorem 2.2 *If $q(x) = q(\pi - x)$ almost everywhere, then the condition $\Delta(\lambda) \equiv 0$ is realized in the only case where the boundary conditions (9) are the falsely periodic boundary conditions (20), and the case $\Delta(\lambda) \equiv C \neq 0$ is realized only in the case where conditions (9) are generalized Cauchy conditions, i.e., conditions of the form (21), where $0 \leq a < \infty$ and $a \neq 1$.*

3 Degenerate Boundary Conditions for the Diffusion Operator

We describe all degenerate two-point boundary conditions possible in a homogeneous spectral problem for the diffusion operator. We show that the case in which the characteristic determinant is identically zero is impossible for the nonsymmetric diffusion operator and that the only possible degenerate boundary conditions are the Cauchy conditions. For the symmetric diffusion operator, the characteristic determinant is zero if and only if the boundary conditions are falsely periodic boundary conditions; the characteristic determinant is identically a nonzero constant if and only if the boundary conditions are generalized Cauchy conditions.

Let L be the following problem for the diffusion operator:

$$ly = y'' + \left(\lambda^2 - 2\lambda\, p(x) - q(x)\right) y = 0, \tag{22}$$

$$U_i(y) = a_{i1}\, y(0) + a_{i2}\, y'(0) + a_{i3}\, y(\pi) + a_{i4}\, y'(\pi) = 0, \qquad i = 1, 2, \tag{23}$$

where $p(x) \in W_2^1(0, \pi)$ and $q(x) \in L_2(0, \pi)$ are real functions and a_{ij}, $i = 1, 2$, $j = 1, 2, 3, 4$, are complex constants.

We denote the matrix formed by the coefficients a_{lk} of the boundary conditions (23) by A and its minors formed by the ith and jth columns by M_{ij}:

$$A = \left\| \begin{matrix} a_{11}\ a_{12}\ a_{13}\ a_{14} \\ a_{21}\ a_{22}\ a_{23}\ a_{24} \end{matrix} \right\|, \quad M_{ij} = \left| \begin{matrix} a_{1i}\ a_{1j} \\ a_{2i}\ a_{2j} \end{matrix} \right|, \qquad i, j = 1, 2, 3, 4. \tag{24}$$

In what follows, we assume that the rank of A is equal to two, rank A=2.

The eigenvalues of the problem L are the zeros of the following entire function [1, pp. 33–36], [16, p. 29]:

$$\begin{aligned} \Delta(\lambda) = M_{12} + M_{34} + M_{32}\, y_1(\pi, \lambda) + M_{42}\, y_1'(\pi, \lambda) \\ + M_{13}\, y_2(\pi, \lambda) + M_{14}\, y_2'(\pi, \lambda), \end{aligned} \tag{25}$$

where $y_1(x, \lambda)$ and $y_2(x, \lambda)$ are the linearly independent solutions of Eq. (22) satisfying the conditions

$$y_1(0, \lambda) = 1, \quad y_1'(0, \lambda) = 0, \quad y_2(0, \lambda) = 0, \quad y_2'(0, \lambda) = 1.$$

The following asymptotic formulas hold for sufficiently large $\lambda \in \mathbb{R}$ ([19, 20]):

$$\begin{aligned} y_1(x, \lambda) &= \cos\pi\,(\lambda - a) - a_1\, \frac{\cos\pi\,(\lambda - a)}{\lambda} \\ &\quad + \pi\, c_1\, \frac{\sin\pi\,(\lambda - a)}{\lambda} + \frac{1}{\lambda} \int_{-\pi}^{\pi} \psi_1(t)\, e^{i\lambda t}\, dt, \\ y_2(x, \lambda) &= \frac{\sin\pi\,(\lambda - a)}{\lambda} + a_0\, \frac{\sin\pi\,(\lambda - a)}{\lambda^2} \\ &\quad - \pi\, c_1\, \frac{\cos\pi\,(\lambda - a)}{\lambda^2} + \frac{1}{\lambda^2} \int_{-\pi}^{\pi} \psi_2(t)\, e^{i\lambda t}\, dt, \\ y_1'(x, \lambda) &= -\lambda\, \sin\pi\,(\lambda - a) + a_0\, \sin\pi\,(\lambda - a) \\ &\quad + \pi\, c_1\, \cos\pi\,(\lambda - a) + \int_{-\pi}^{\pi} \psi_3(t)\, e^{i\lambda t}\, dt, \\ y_2'(x, \lambda) &= \cos\pi\,(\lambda - a) + a_1\, \frac{\cos\pi\,(\lambda - a)}{\lambda} \\ &\quad + \pi\, c_1\, \frac{\sin\pi\,(\lambda - a)}{\lambda} + \frac{1}{\lambda} \int_{-\pi}^{\pi} \psi_4(t)\, e^{i\lambda t}\, dt, \end{aligned}$$

where

$$a = \frac{1}{\pi} \int_0^{\pi} p(t)\, dt,$$

$$a_0 = \frac{1}{2} \big(p(0) + p(\pi)\big), \quad a_1 = \frac{1}{2} \big(p(0) - p(\pi)\big),$$

$$c_1 = \frac{1}{2\pi} \int_0^{\pi} \big(q(t) + p^2(t)\big)\, dt,$$

$$\psi_i(t) \in L_2[0, \pi], \quad i = 1, 2, 3, 4.$$

It follows from these relations that the functions $y_1(\pi, \lambda)$, $y_1'(\pi, \lambda)$, $y_2(\pi, \lambda)$, and 1, which occur in the representation (25) of the function $\Delta(\lambda)$, are linearly independent. If we supplement these functions with the function $y_2'(\pi, \lambda)$, then the resulting function system will be independent if and only if $p(x) \neq p(\pi - x)$ and/or $q(x) \neq q(\pi - x)$ on some subinterval of the closed interval $[0,\pi]$. This follows from the fact that the identity $y_1(\pi, \lambda) \equiv y_2'(\pi, \lambda)$ holds if and only if $p(x) = p(\pi - x)$ and $q(x) = q(\pi - x)$ [21, Lemma 3] (equalities of functions are understood as equalities in the function spaces where these functions are given).

If $p(x) \neq p(\pi - x)$ and/or $q(x) \neq q(\pi - x)$ and if $\Delta(\lambda) \equiv C = \text{const}$, then the representation (25) and the linear independence of the corresponding functions imply the relations

$$M_{12} + M_{34} = C, \quad M_{32} = 0, \quad M_{42} = 0, \quad M_{13} = 0, \quad M_{14} = 0. \tag{26}$$

To find the minors M_{12} and M_{34}, we use the fact that not every finite sequence of numbers can be represented as the sequence of minors of a matrix. A necessary and sufficient condition that numbers M_{12}, M_{13}, M_{14}, M_{23}, M_{24}, M_{34} be the minors of a 2×4 matrix is that the following Plücker relations be satisfied [18]:

$$M_{12}\, M_{34} - M_{13}\, M_{24} + M_{14}\, M_{23} = 0. \tag{27}$$

The minors M_{23} and M_{24} in (27) differ only in sign from the minors M_{32} and M_{42} in (26). Relations (26) and (27) imply that only the following two sequences of minors are possible:

$$M_{12} = C \neq 0, \quad M_{34} = 0, \quad M_{32} = 0, \quad M_{42} = 0, \quad M_{13} = 0, \quad M_{14} = 0; \tag{28}$$

$$M_{12} = 0, \quad M_{34} = C \neq 0, \quad M_{32} = 0, \quad M_{42} = 0, \quad M_{13} = 0, \quad M_{14} = 0. \tag{29}$$

The case of $C = 0$ [and hence the case of $\Delta(\lambda) \equiv 0$] cannot be realized, because all the second-order determinants being zero contradicts the condition that rank A=2. One can uniquely determine the boundary conditions (28) and (29) of minors by methods for the identification of a matrix from its minors [18]. The sequence

(23) of minors corresponds to the Cauchy conditions $y(0) = y'(0) = 0$, and the sequence (29) corresponds to the Cauchy conditions $y(\pi) = y'(\pi) = 0$.

Thus, the following theorem holds.

Theorem 3.1 *If $p(x) \neq p(\pi - x)$ and/or $q(x) \neq q(\pi - x)$ on some subinterval of the closed interval $[0,\pi]$, then the case $\Delta(\lambda) \equiv 0$ is impossible and the only possible degenerate boundary conditions are the Cauchy conditions $y(0) = y'(0) = 0$ and $y(\pi) = y'(\pi) = 0$.*

If $p(x) = p(\pi - x)$, $q(x) = q(\pi - x)$, and $\Delta(\lambda) \equiv C = \text{const}$, then the representation (25) and the linear independence of the corresponding functions imply the relations

$$M_{12} + M_{34} = C, \quad M_{32} + M_{14} = 0, \quad M_{42} = 0, \quad M_{13} = 0. \tag{30}$$

It follows from Eqs. (27) and (30) that only the following two sequences of minors are possible:

$$\begin{gathered} M_{12} = C_1, \quad M_{34} = C - C_1, \quad M_{32} = \mp\sqrt{C_1\,(C_1 - C)}, \\ M_{42} = 0, \quad M_{13} = 0, \quad M_{14} = \pm\sqrt{C_1\,(C_1 - C)}. \end{gathered} \tag{31}$$

(where one takes either the upper or the lower signs in both occurrences simultaneously).

If $C = 0$ (the case of $\Delta(\lambda) \equiv 0$), then relations (31) imply that

$$\begin{gathered} M_{12} = C_1, \quad M_{34} = -C_1, \quad M_{32} = \mp C_1, \\ M_{42} = 0, \quad M_{13} = 0, \quad M_{14} = \pm C_1. \end{gathered} \tag{32}$$

We uniquely determine the following two forms of boundary conditions from these sequences of minors by methods for the identification of a matrix from its minors [18]

$$y(0) \mp y(\pi) = 0, \quad y'(0) \pm y'(\pi) = 0. \tag{33}$$

Conditions (33) were named falsely periodic in the paper [12], because they are degenerate and differ from the nondegenerate periodic or antiperiodic boundary conditions by a change in exactly one sign (from plus to minus or from minus to plus).

If $C \neq 0$ (the case of $\Delta(\lambda) \not\equiv 0$), then the form of the boundary conditions depends on which of the minors (31) are nonzero. If $C - C_1 = 0$, then we obtain the Cauchy conditions $y(0) = y'(0) = 0$; if $C_1 = 0$, then we obtain the Cauchy conditions $y(\pi) = y'(\pi) = 0$; and if $C - C_1 \neq 0$ and $C_1 \neq 0$, then we obtain the conditions

$$y(0) \mp a\,y(\pi) = 0, \quad y'(0) \pm a\,y'(\pi) = 0, \tag{34}$$

where $a = \sqrt{\frac{C_1 - C}{C_1}} \neq 1$.

These conditions (conditions of the form (34) with $0 \leq a \leq \infty$ and $a \neq 1$) were dubbed generalized Cauchy conditions in the paper [12].

Consequently, the following theorem holds.

Theorem 3.2 *If $p(x) = p(\pi - x)$ and $q(x) = q(\pi - x)$, then the case of $\Delta(\lambda) \equiv 0$ is realized if and only if the boundary conditions are defined by (23), and the case $\Delta(\lambda) \equiv C \neq 0$ is realized if and only if conditions (33) are the generalized Cauchy conditions (34).*

4 The Degenerate Boundary Conditions for Boundary Value Problems with an Odd-Order Differential Equation

Operators generated by a differential expression on a finite closed interval are considered. It is shown that, for any odd integer n, there exist differential operators of order n whose spectrum fills the whole complex plane.

Consider the following spectral problem for the differential operator:

$$i^{-n}\, y^{(n)}(x) = \lambda\, y(x) = s^n\, y(x), \qquad x \in [0, 1] \tag{35}$$

of odd order n with the boundary conditions

$$\begin{gathered} y(0) + \alpha_0\, y(1) = 0, \quad y'(0) + \alpha_1\, y'(1) = 0, \quad \ldots, \\ y^{(n-1)}(0) + \alpha_{n-1}\, y^{(n-1)}(1) = 0. \end{gathered} \tag{36}$$

Theorem 4.1 *The spectrum of problem (35) and (36) fills the whole complex plane if*

$$\begin{gathered} \alpha_k = e^{\pi\, i} = -1, \quad \alpha_{k+1} = e^{\pi\, i + \frac{2\pi\, i}{n}}, \quad \alpha_{k+2} = e^{\pi\, i + \frac{4\pi\, i}{n}}, \quad \ldots, \\ \alpha_{n-1} = e^{\pi\, i + \frac{2\pi\, k\, i}{n}}, \end{gathered}$$

$$\begin{gathered} \alpha_0 = e^{\pi\, i + \frac{2\pi\, (k+1)\, i}{n}}, \quad \alpha_1 = e^{\pi\, i + \frac{2\pi\, (k+2)\, i}{n}}, \quad \ldots, \\ \alpha_{k-1} = e^{\pi\, i + \frac{2\pi\, (n-1)\, i}{n}} \quad \left(k = \frac{n-1}{2}\right). \end{gathered}$$

Proof Problem (35) and (36) has the following characteristic determinant $\Delta(\lambda)$ [16, p. 26]:

$$\begin{vmatrix} y_1(0)+\alpha_0\,y_1(1) & y_2(0)+\alpha_0\,y_2(1) & \dots & y_n(0)+\alpha_0\,y_n(1) \\ y_1'(0)+\alpha_1\,y_1'(1) & y_2'(0)+\alpha_1\,y_2'(1) & \dots & y_n'(0)+\alpha_1\,y_n'(1) \\ \vdots & \vdots & \ddots & \vdots \\ y_1^{(n-1)}(0) & y_2^{(n-1)}(0) & \dots & y_n^{(n-1)}(0) \\ +\alpha_{n-1}\,y_1^{(n-1)}(1) & +\alpha_{n-1}\,y_2^{(n-1)}(1) & & +\alpha_{n-1}\,y_n^{(n-1)}(1) \end{vmatrix}, \tag{37}$$

where

$$y_j(x,\lambda)=\begin{cases}\frac{x^{j-1}}{(j-1)!}, & \text{if } \lambda=0,\\ e^{\omega_j s x}, & \text{if } \lambda\neq 0,\end{cases} \qquad \omega_j=e^{\frac{\pi i}{2}+\frac{2\pi i(j-1)}{n}}, \qquad j=1,2,\dots,n.$$

Consider the case $\lambda \neq 0$. Expanding the determinant (37) in the sum [26, p. 41], we obtain

$$\Delta(\lambda)=\begin{vmatrix} y_1(0) & y_2(0) & \dots & y_n(0) \\ y_1'(0) & y_2'(0) & \dots & y_n'(0) \\ \vdots & \vdots & \ddots & \vdots \\ y_1^{(n-1)}(0) & y_2^{(n-1)}(0) & \dots & y_n^{(n-1)}(0) \end{vmatrix}$$

$$+\begin{vmatrix} y_1(0) & y_2(0) & \dots & \alpha_0\,y_n(1) \\ y_1'(0) & y_2'(0) & \dots & \alpha_1\,y_n'(1) \\ \vdots & \vdots & \ddots & \vdots \\ y_1^{(n-1)}(0) & y_2^{(n-1)}(0) & \dots & \alpha_{n-1}\,y_n^{(n-1)}(1) \end{vmatrix}+\dots$$

$$+\begin{vmatrix} \alpha_0\,y_1(1) & \alpha_0\,y_2(1) & \dots & \alpha_0\,y_n(1) \\ \alpha_1\,y_1'(1) & \alpha_1\,y_2'(1) & \dots & \alpha_1\,y_n'(1) \\ \vdots & \vdots & \ddots & \vdots \\ \alpha_{n-1}\,y_1^{(n-1)}(1) & \alpha_{n-1}\,y_2^{(n-1)}(1) & \dots & \alpha_{n-1}\,y_n^{(n-1)}(1) \end{vmatrix}. \tag{38}$$

Since

$$\begin{vmatrix} y_1(0) & y_2(0) & \dots & y_n(0) \\ y_1'(0) & y_2'(0) & \dots & y_n'(0) \\ \vdots & \vdots & \ddots & \vdots \\ y_1^{(n-1)}(0) & y_2^{(n-1)}(0) & \dots & y_n^{(n-1)}(0) \end{vmatrix}$$
$$=\begin{vmatrix} y_1(1) & y_2(1) & \dots & y_n(1) \\ y_1'(1) & y_2'(1) & \dots & y_n'(1) \\ \vdots & \vdots & \ddots & \vdots \\ y_1^{(n-1)}(1) & y_2^{(n-1)}(1) & \dots & y_n^{(n-1)}(1) \end{vmatrix}$$

and $\alpha_0 \cdot \alpha_1 \cdot \dots \cdot \alpha_{n-1} = -1$, it follows that the sum of the first and last summands in the last sum is zero. All the other determinant summands in (38) vanish, because there are proportional columns in these determinants. □

Let us prove that proportional columns exist. In each of the determinants summands in the sum (38), except the first and the last one, there is at least one column with values of the linearly independent solutions at the point $x = 0$ and at least one column with values of the solutions at the point $x = 1$. Moreover, among them, there are two adjacent columns, the left one containing values of the solution at the point $x = 1$ and the right one containing values of the solution at the point $x = 0$. Denote this left column vector (with values of the function y_p at the point $x = 1$) by $\mathbf{y}_p$ and the right column vector (with values of y_{p+1} at the point $x = 0$) by $\mathbf{y}_{p+1}$. If $\mathbf{y}_p$ is the right column of the corresponding determinant, then for the column $\mathbf{y}_{p+1}$ we take the first column of the corresponding determinant. Let us show that these columns are linearly independent. We have

$$\begin{aligned}\mathbf{y}_p &= (\alpha_0\, y_p(1), \alpha_1\, y_p'(1), \quad \dots, \quad \alpha_{n-1}\, y_p^{(n-1)}(1))^T \\ &= e^{\omega_p s}\,(\alpha_0, \alpha_1\, \omega_p, \quad \dots, \quad \alpha_{n-1}\, \omega_p^{n-1}(1))^T \\ \mathbf{y}_{p+1} &= (y_{p+1}(0),\ y_{p+1}'(0), \quad \dots, \quad y_{p+1}^{(n-1)}(0))^T \\ &= (1,\ \omega_{p+1}, \quad \dots, \quad \omega_{p+1}^{n-1}(1))^T.\end{aligned}$$

It is easy to see that

$$\begin{aligned}&(\alpha_0, \alpha_1\, \omega_p, \quad \dots, \quad \alpha_{n-1}\, \omega_p^{n-1}(1))^T \\ &= \alpha_0\,(1,\ \omega_{p+1}, \quad \dots, \quad \omega_{p+1}^{n-1}(1))^T.\end{aligned}$$

Indeed, let us verify this equality component-wise. For the first coordinate, the equality is obvious: $\alpha_0 = \alpha_0 \cdot 1$. For the second coordinate, we also have the valid equality:

$$\alpha_1\, \omega_p = e^{\pi\, i + \frac{2\pi\,(k+2)\,i}{n}} \cdot e^{\frac{\pi\, i}{2} + \frac{2\pi\, i\,(p-1)}{n}} = e^{\pi\, i + \frac{2\pi\,(k+1)\,i}{n}} \cdot e^{\frac{\pi\, i}{2} + \frac{2\pi\, i\, p}{n}} = \alpha_0\, \omega_{p+1}.$$

For the third component, we have

$$\alpha_2\, \omega_p^2 = e^{\pi\, i + \frac{2\pi\,(k+3)\,i}{n}} \cdot e^{\pi\, i + \frac{4\pi\, i\,(p-1)}{n}} = e^{\pi\, i + \frac{2\pi\,(k+1)\,i}{n}} \cdot e^{\pi\, i + \frac{4\pi\, i\, p}{n}} = \alpha_0\, \omega_{p+2}.$$

For the subsequent components, we will also have the valid equalities. Indeed, for an arbitrary $m \in \mathbb{N}$ we have

$$\begin{aligned}\alpha_m\, \omega_p^m &= e^{\pi\, i + \frac{2\pi\,(k+m+1)\,i}{n}} \cdot e^{\frac{m\pi\, i}{2} + \frac{2m\pi\, i\,(p-1)}{n}} \\ &= e^{\pi\, i + \frac{2\pi\,(k+1)\,i}{n}} \cdot e^{\frac{m\pi\, i}{2} + \frac{2m\pi\, i\, p}{n}} = \alpha_0\, \omega_{p+1}^m.\end{aligned}$$

Therefore,

$$\mathbf{y}_p = \alpha_0\, e^{\omega_p\, s}\, \mathbf{y}_{p+1}$$

and, in the case $\lambda \neq 0$ the theorem is proved.

Now let $\lambda = 0$. Then the functions

$$y_1(x) = 1, \quad y_2(x) = x, \quad y_3(x) = \tfrac{x^2}{2}, \quad \ldots, \quad y_k(x) = \tfrac{x^k}{k!}, \quad \ldots, \quad y_{n-1} = \tfrac{x^{n-1}}{(n-1)!},$$

will be the linearly independent solutions of Eq. (35), while the characteristic determinant $\Delta(\lambda)$ for problem (35) and (36) will be of upper-triangular form (there will be zeros below of the main diagonal). In addition, since $\alpha_k = -1$, we will have zero in the middle of the main diagonal (in the row and column numbered k + 1). Therefore, the characteristic determinant is zero in the case $\lambda = 0$ as well. Therefore, for problem (35) and (36), the characteristic determinant $\Delta(\lambda)$ is identically zero for all $\lambda \in \mathbb{C}$. The theorem is proved.

Example 4.1 For the third-order differential equation

$$-i\, y'''(x) = \lambda\, y(x), \qquad x \in [0, 1]$$

the characteristic determinant is identically zero for the problem with the boundary conditions

$$y(0) + \frac{1 - i\sqrt{3}}{2}\, y(1) = 0, \quad y'(0) - y'(1) = 0,$$
$$y''(0) + \frac{1 + i\sqrt{3}}{2}\, y''(1) = 0$$

and for the problem with the boundary conditions

$$y(0) + \frac{1 + i\sqrt{3}}{2}\, y(1) = 0, \quad y'(0) - y'(1) = 0,$$
$$y''(0) + \frac{1 - i\sqrt{3}}{2}\, y''(1) = 0.$$

5 The Degenerate Boundary Conditions for Boundary Value Problems with a Third-Order Differential Equation

We consider the spectral problem $y'''(x) = \lambda\, y(x)$ with general two-point boundary conditions that do not contain the spectral parameter λ. We prove that the boundary *conditions in this problem are degenerate if and only if their* 3×6 *coefficient matrix can be reduced by a linear row transformation to a matrix consisting of two diagonal* 3×3 *matrices, one of which is the identity matrix and the diagonal entries of the other are all cubic roots of some number. Further, the characteristic determinant of the problem is identically zero if and only if that number is* -1. *We also show that the problem in question cannot have finite spectrum.*

Consider the two-point boundary value problem

$$y'''(x) = \lambda\, y(x) = s^3\, y(x), \qquad x \in [0, 1], \tag{39}$$

$$U_j(y) = \sum_{k=1}^{3} a_{jk}\, y^{(k-1)}(0) + \sum_{k=1}^{3} a_{j\,k+n}\; y^{(k-1)}(1) = 0, \qquad j = 1, 2, 3. \tag{40}$$

The coefficient matrix of the boundary conditions (40) will be denoted by

$$A = \begin{Vmatrix} a_{11} & a_{12} & a_{13} & a_{14} & a_{15} & a_{16} \\ a_{21} & a_{22} & a_{23} & a_{24} & a_{25} & a_{26} \\ a_{31} & a_{32} & a_{33} & a_{34} & a_{35} & a_{36} \end{Vmatrix}, \tag{41}$$

where $a_{jm} \in \mathbb{C}, \quad j = 1, \ldots, 3, \quad m = 1, \ldots, 6$, and

$$\text{rank}\, A = 3. \tag{42}$$

In this section, we obtain the following results.

Theorem 5.1 *The boundary value problem (39) and (40) cannot have a spectrum consisting of finitely many eigenvalues.*

Theorem 5.2 *The boundary conditions (40) can be degenerate only for problems with matrices (41), that are reduced to the following two types using linear transformation*

$$A_1 = \begin{Vmatrix} 1 & 0 & 0 & a_1 & 0 & 0 \\ 0 & 1 & 0 & 0 & a_2 & 0 \\ 0 & 0 & 1 & 0 & 0 & a_3 \end{Vmatrix}, \tag{43}$$

and

$$A_2 = \begin{Vmatrix} a_1 & 0 & 0 & 1 & 0 & 0 \\ 0 & a_2 & 0 & 0 & 1 & 0 \\ 0 & 0 & a_3 & 0 & 0 & 1 \end{Vmatrix}. \tag{44}$$

Theorem 5.3 *The boundary conditions in problem (39) and (40) are degenerate if and only if the coefficient matrix (41) of the boundary conditions (40) are reduced to the matrix (43) or (44) using linear transformation, (43) or with the matrix (44),* *where $\{a_1, a_2, a_3\}$ is the set of cubic roots of some number, and, $\Delta(\lambda) \equiv 0$ if and only if the numbers a_1, a_2, and a_3 are three distinct cubic roots of -1.*

These theorems are proved by the next scheme.
Consider the characteristic determinant $\Delta(\lambda)$ of the spectral problem (39) and (40), i.e., the determinant

$$\Delta(\lambda) = \begin{vmatrix} U_1(y_1) & U_1(y_2) & U_1(y_3) \\ U_2(y_1) & U_2(y_2) & U_2(y_3) \\ U_3(y_1) & U_3(y_2) & U_3(y_3) \end{vmatrix}, \tag{45}$$

where $y_j = y_j(x, s)$, $j = 1, 2, 3$, are the linearly independent solutions of Eq. (39) with the conditions

$$y_j^{(k-1)}(0, \lambda) = \begin{cases} 1, & \text{if } k = j \\ 0, & \text{if } k \neq j \end{cases}$$

where $j, k = 1, 2, 3$. It is easily seen that

$$y_1 = \frac{1}{3}\exp(-s\,x) + \frac{2}{3}\exp\left(\frac{1}{2}s\,x\right)\cos\left(\frac{\sqrt{3}}{2}s\,x\right),$$

$$y_2 = -\frac{1}{3\,s}\exp(-s\,x) + \frac{\sqrt{3}}{3\,s}\exp\left(\frac{1}{2}s\,x\right)\sin\left(\frac{\sqrt{3}}{2}s\,x\right)$$
$$+\frac{1}{3\,s}\exp\left(\frac{1}{2}s\,x\right)\cos\left(\frac{\sqrt{3}}{2}s\,x\right),$$

$$y_3 = \frac{1}{3\,s^2}\exp(-s\,x) + \frac{\sqrt{3}}{3\,s^2}\exp\left(\frac{1}{2}s\,x\right)\sin\left(\frac{\sqrt{3}}{2}s\,x\right)$$
$$-\frac{1}{3\,s^2}\exp\left(\frac{1}{2}s\,x\right)\cos\left(\frac{\sqrt{3}}{2}s\,x\right).$$

Lemma 5.1 *The functions $y_{j-1}^{(k-1)}(1, \lambda)$ and $y_j^{(k)}(1, \lambda)$ coincide identically; i.e.,*

$$y_{j-1}^{(k-1)}(1, \lambda) \equiv y_j^{(k)}(1, \lambda), \qquad k = 1, 2, 3, \quad j = 2, 3. \tag{46}$$

Lemma 5.1 can be verified by straightforward computations.
Let B be the matrix

$$B = \begin{Vmatrix} y_1(0) & y_1'(0) & y_1''(0) & y_1(1) & y_1'(1) & y_1''(1) \\ y_2(0) & y_2'(0) & y_2''(0) & y_2(1) & y_2'(1) & y_2''(1) \\ y_3(0) & y_3'(0) & y_3''(0) & y_n(1) & y_3'(1) & y_3''(1) \end{Vmatrix}$$

$$= \left\| \begin{matrix} 1 & 0 & 0 & y_1(1) & y_1'(1) & y_1''(1) \\ 0 & 1 & 0 & y_2(1) & y_2'(1) & y_2''(1) \\ 0 & 0 & 1 & y_3(1) & y_3'(1) & y_3''(1) \end{matrix} \right\| = \|B_1, B_2\|.$$

Thus, B consists of two 3×3 block matrices B_1 and B_2, where B_1 is the identity matrix. The determinants of these matrices give the Wronskian $W(x)$ of the fundamental solution system $y_j(x, s)$, $k, j = 1, 2, 3$ of Eq. (39) calculated at the points $x = 0$ and $x = 1$, respectively; i.e., $\det(B_1) = W(0)$, $\det(B_2) = W(1)$. It follows from the Liouville formula for the Wronskian [22, p. 95–96] that

$$\det(B_1) = W(0) = 1, \quad \det(B_2) = W(1) = 1.$$

We use the matrices A and B to represent the determinant (45) as

$$\Delta(\lambda) \equiv \det(A \cdot B^T).$$

We expand this determinant with the use of the Binet–Cauchy formula and obtain [27, section 1.14, p.41–42]

$$\Delta(\lambda) = \sum_{1 \le i_1 < i_2 < i_3 \le 2n} A_{i_1,i_2,i_3} B_{i_1,i_2,i_3}(s) = 0. \tag{47}$$

Here A_{i_1,i_2,i_3} is the minor formed by the i_1th, i_2th, and i_3th columns of the matrix A and B_{i_1,i_2,i_3} is the minor formed by the i_1th, i_2th, and i_3th columns of the matrix B, or, which is the same, by the corresponding rows of the transpose matrix B^T.

Let $P(s)$ be the sum $A_{123} B_{123} + A_{456} B_{456}$ of the first and last terms in the expansion (47), and let $S(s)$ be the sum of all other terms; i.e.,

$$\Delta(\lambda) = S(s) + P(s).$$

The function $P(s)$ is identically constant,

$$P(s) \equiv A_{123} \cdot W(0) + A_{456} \cdot W(1) = A_{123} + A_{456} = \text{const}.$$

Each term in the sum $S(s)$ contains linear combinations of $e^{\omega_j s x}$ or $e^{\sum_j \omega_j s x}$. The exponents of these exponentials do not vanish, and the coefficients multiplying the exponentials do not coincide. In what follows, we show that the characteristic determinant (47) is identically constant only if $S(s) \equiv 0$, the constant being $A_{123} + A_{456}$.

It follows that

$$\begin{aligned} \Delta(\lambda) = {} & A_{123} + A_{456} + A_{124} B_{124} \\ & + A_{125} B_{125} + A_{126} B_{126} \end{aligned}$$

$$
\begin{aligned}
&+ A_{134} B_{134} + A_{135} B_{135} + A_{136} B_{136} \\
&+ A_{145} B_{145} + A_{146} B_{146} + A_{156} B_{156} \\
&+ A_{234} B_{234} + A_{235} B_{235} \\
&+ A_{236} B_{236} + A_{245} B_{245} + A_{246} B_{246} + A_{256} B_{256} \\
&+ A_{345} B_{345} + A_{346} B_{346} + A_{356} B_{356}.
\end{aligned}
$$

From Lemma 5.1, we obtain

$$
B_2 = \left\| \begin{matrix} y_1(1) & y_1'(1) & y_1''(1) \\ y_2(1) & y_1(1) & y_1'(1) \\ y_3(1) & y_2(1) & y_1(1) \end{matrix} \right\|.
$$

Moreover, we have $y_1'(1)\, y_2(1) = y_1''(1)\, y_3(1)$. Thus, the characteristic determinant becomes

$$
\begin{aligned}
\Delta(\lambda) = {} & (A_{123} + A_{456}) + A_{124}\, y_3(1) + (A_{125} - A_{134})\, y_2(1) \\
& + (A_{126} - A_{135} + A_{234})\, y_1(1) + (-A_{136} + A_{235})\, y_1'(1) \\
& + (A_{156} - A_{246} + A_{345})\, (y_1^2(1) - y_1'(1)\, y_2(1)) \\
& + A_{145}\, (y_2^2(1) - y_1(1)\, y_3(1)) + (A_{146} - A_{245})\, (y_1(1)\, y_2(1) - y_1'(1)\, y_3(1)) \\
& + A_{236}\, y_1''(1) + (-A_{256} + A_{346})\, (y_1(1)\, y_1'(1) - y_1''(1)\, y_2(1)) \\
& + A_{356}\, ((y_1'(1))^2 - y_1(1)\, y_1''(1)),
\end{aligned}
$$

where

$$
\begin{aligned}
y_1(1) &= \frac{1}{3} e^{-s} + \frac{2}{3} e^{\frac{s}{2}} \cos\left(\frac{\sqrt{3}}{2} s\right), \\
y_2(1) &= -\frac{1}{3s} e^{-s} + \frac{\sqrt{3}}{3s} e^{\frac{s}{2}} \sin\left(\frac{\sqrt{3}s}{2}\right) + \frac{\sqrt{3}}{3s} e^{\frac{s}{2}} \cos\left(\frac{\sqrt{3}s}{2}\right), \\
y_3(1) &= \frac{1}{3s^2} e^{-s} + \frac{\sqrt{3}}{3s^2} e^{\frac{s}{2}} \sin\left(\frac{\sqrt{3}s}{2}\right) - \frac{\sqrt{3}}{3s^2} e^{\frac{s}{2}} \cos\left(\frac{\sqrt{3}s}{2}\right), \\
y_1'(1) &= -\frac{s}{3} e^{-s} + \frac{s}{3} e^{\frac{s}{2}} \cos\left(\frac{\sqrt{3}}{2} s\right) - \frac{\sqrt{3}s}{3} e^{\frac{s}{2}} \sin\left(\frac{\sqrt{3}}{2} s\right), \\
y_1''(1) &= \frac{s^2}{3} e^{-s} - \frac{s^2}{3} e^{\frac{s}{2}} \cos\left(\frac{\sqrt{3}s}{2}\right) - \frac{\sqrt{3}s^2}{3} e^{\frac{s}{2}} \sin\left(\frac{\sqrt{3}}{2} s\right),
\end{aligned}
$$

$$y_1^2(1) - y_1'(1)\, y_2(1) = \frac{1}{3}\, e^s + \frac{2}{3}\, e^{-\frac{s}{2}} \cos\left(\frac{\sqrt{3}}{2} s\right),$$

$$\begin{aligned} y_1^2(1) - y_1(1)\, y_3(1) = -\frac{1}{9\, s^2} \Bigg(& -e^{-2\,s}\, s^2 - 4\, e^{-\frac{s}{2}}\, s^2 \cos\left(\frac{s\,\sqrt{3}}{2}\right) \\ & + e^{-2\,s} + e^{-\frac{s}{2}} \cos\left(\frac{s\,\sqrt{3}}{2}\right) + e^{-\frac{s}{2}} \sin\left(\frac{s\sqrt{3}}{2}\right) \sqrt{3} \\ & - 2\, e^s \cos^2\left(\frac{s\sqrt{3}}{2}\right) (2\, s^2 + 1) \\ & +2\, \sqrt{3}\, e^s \cos\left(\frac{s\sqrt{3}}{2}\right) \sin\left(\frac{s\sqrt{3}}{2}\right)\Bigg), \end{aligned}$$

$$y_1(1)\, y_2(1) - y_1'(1)\, y_3(1) =$$

$$\frac{1}{3s}\, e^{\frac{s}{2}} \left(e^{-s} \sin\left(\frac{s\,\sqrt{3}}{2}\right) \sqrt{3} - e^{-s} \cos\left(\frac{s\,\sqrt{3}}{2}\right) + e^{\frac{s}{2}} \right),$$

$$y_1(1)\, y_1'(1) - y_1''(1)\, y_2(1) =$$

$$-\frac{s}{3}\, e^{\frac{s}{2}} \left(e^{-s} \cos\left(\frac{s\,\sqrt{3}}{2}\right) + e^{-s} \sin\left(\frac{s\,\sqrt{3}}{2}\right) \sqrt{3} - e^{\frac{s}{2}} \right),$$

$$(y_1'(1))^2 - y_1''(1)\, y_1(1) =$$

$$\frac{s^2}{3}\, e^{\frac{s}{2}} \left(\sqrt{3}\, e^{-s} \sin\left(\frac{s\,\sqrt{3}}{2}\right) - e^{-s} \cos\left(\frac{s\,\sqrt{3}}{2}\right) + e^{\frac{s}{2}} \right).$$

It follows from the representation of the characteristic determinant $\Delta(\lambda)$ that it is an entire function of the class K [23, 24] and hence it has infinitely many roots, whose asymptotic representations can be found in [23, 24]. The proof of Theorem 5.1 is complete.

We need the following three definitions.

1. *Characteristic sum* r_0 of the determinant $B_{i_1,\,i_2,\,i_3}$ (and of the corresponding determinant $A_{i_1,\,i_2,\,i_3}$) is the sum of all of its indices; i.e., $r_0 = i_1 + i_2 + i_3$.
2. Let the number of indices i_k of the determinant $B_{i_1,\,i_2,\,i_3}$ that satisfy the inequality $1 \leq i_k \leq 3$ be r_1, and let the number of indices $i_k$of the determinant $B_{i_1,\,i_2,\,i_3}$ that satisfy the inequality $4 \leq i_k \leq 6$ be r_2. Then the ordered triple (r_0, r_1, r_2) is called the *characteristic index* of the determinant $B_{i_1,\,i_2,\,i_3}$ $(A_{i_1,\,i_2,\,i_3})$.
3. It follows from the expansion obtained above for the characteristic determinant and from Lemma 5.1 that the determinants $B_{i_1,\,i_2,\,i_3}$ with distinct characteristic indices are linearly independent and differ in the exponents s^k, while the

determinants B_{i_1,i_2,i_3} with the same characteristic indices coincide or differ only in the sign. The determinants B_{i_1,i_2,i_3} (A_{i_1,i_2,i_3})) with the same characteristic indices are said to be *similar*.

If $S(s) \equiv 0$, then one of the minors A_{123} and A_{456} is nonzero. Otherwise, all third-order minors of the matrix A would be zero, which contradicts the condition $\operatorname{rank} A = 3$. Indeed, the representation of the characteristic determinant and the linear independence of the corresponding functions imply the relations

$$\begin{gathered} A_{124} = A_{145} = A_{236} = A_{356} = 0, \\ A_{126} - A_{135} + A_{234} = A_{156} - A_{246} + A_{345} = 0, \\ A_{125} - A_{134} = A_{235} - A_{136} = A_{146} - A_{245} = A_{346} - A_{256} = 0. \end{gathered} \tag{48}$$

It is well known in algebraic geometry that given numbers $A_{i_1i_2i_3}$ are the minors of a matrix A if and only if the Pücker relations are satisfied (e.g., see [18]). For the 3×6 matrix A, these relations read

$$\begin{aligned} &A_{i_1i_4i_5}A_{i_1i_2i_3} - A_{i_1i_4i_3}A_{i_1i_2i_5} + A_{i_1i_5i_3}A_{i_1i_2i_4} = 0, \\ &A_{i_1i_4i_6}A_{i_1i_2i_3} - A_{i_1i_4i_3}A_{i_1i_2i_6} + A_{i_1i_6i_3}A_{i_1i_2i_4} = 0, \\ &A_{i_1i_5i_6}A_{i_1i_2i_3} - A_{i_1i_5i_3}A_{i_1i_2i_6} + A_{i_1i_6i_3}A_{i_1i_2i_5} = 0, \\ &A_{i_2i_4i_5}A_{i_1i_2i_3} - A_{i_2i_4i_3}A_{i_1i_2i_5} + A_{i_2i_5i_3}A_{i_1i_2i_4} = 0, \\ &A_{i_2i_4i_6}A_{i_1i_2i_3} - A_{i_2i_4i_3}A_{i_1i_2i_6} + A_{i_2i_6i_3}A_{i_1i_2i_4} = 0, \\ &A_{i_2i_5i_6}A_{i_1i_2i_3} - A_{i_2i_5i_3}A_{i_1i_2i_6} + A_{i_2i_6i_3}A_{i_1i_2i_5} = 0, \\ &A_{i_3i_4i_5}A_{i_1i_2i_3} - A_{i_2i_4i_3}A_{i_1i_3i_5} + A_{i_2i_5i_3}A_{i_1i_3i_4} = 0, \\ &A_{i_3i_4i_6}A_{i_1i_2i_3} - A_{i_2i_4i_3}A_{i_1i_3i_6} + A_{i_2i_6i_3}A_{i_1i_3i_4} = 0, \\ &A_{i_3i_5i_6}A_{i_1i_2i_3} - A_{i_2i_5i_3}A_{i_1i_3i_6} + A_{i_2i_6i_3}A_{i_1i_3i_5} = 0, \\ &A_{i_4i_5i_6}A_{i_1i_2i_3} - A_{i_1i_2i_4}A_{i_3i_5i_6} + A_{i_1i_2i_5}A_{i_3i_4i_6} - A_{i_1i_2i_6}A_{i_3i_4i_5} = 0, \end{aligned}$$

where $(i_1, i_2, i_3, i_4, i_5, i_6)$ is a permutation such that $A_{i_1,i_2,i_3} \neq 0$. If $A_{123} = A_{456} = 0$, then it follows from these relations and from (48) that all minors $A_{i_1i_2i_3}$ of the matrix A are zero, which contradicts the relation $\operatorname{rank} A = 3$. Therefore, if $S(s) \equiv 0$, then one of the minors A_{123} and A_{456} is nonzero.

Let $A_{123} \neq 0$. Then, up to linear row transformations, the matrix (41) has the form

$$A = \begin{Vmatrix} 1 & 0 & 0 & a_{14} & a_{15} & a_{16} \\ 0 & 1 & 0 & a_{24} & a_{25} & a_{26} \\ 0 & 0 & 1 & a_{34} & a_{35} & a_{36} \end{Vmatrix}.$$

Here the entries a_{ij} of the matrix A do not generally coincide with the entries a_{ij} of the matrix (41). We do not denote them by different symbols, because it is always clear from the context which matrix entries we mean. Let us show that the condition $S(s) \equiv 0$ implies that, up to linear row transformations, the matrix A has the form (43), i.e., that the submatrix composed of the last three columns is diagonal.

Note that the sum $S(s)$ of determinants B_{i_1,i_2,i_3} does not contain determinants similar to $B_{236} = y_1''(1)$, and hence B_{236} is linearly independent of any other term in the sum $S(s)$. Then the identity $S(s) \equiv 0$ implies that $A_{236} = 0$. Let us calculate this determinant. Using A, we obtain

$$A_{236} = \begin{vmatrix} 0 & 0 & a_{16} \\ 1 & 0 & a_{26} \\ 0 & 1 & a_{36} \end{vmatrix} = a_{16} = 0. \tag{49}$$

Now let us show that the entries a_{15} and a_{26} on the upper secondary diagonal of the matrix A are zero as well. In the sum $S(s)$ of the determinants $B_{i_1 i_2 i_3}$, there are no determinants similar to $B_{356} = (y_1'(1))^2 - y_1(1)\, y_1''(1)$, and hence B_{356} is linearly independent of any other term in the sum $S(s)$. Then the identity $S(s) \equiv 0$ implies that $A_{356} = 0$. Let us calculate this determinant. Using A, we obtain

$$A_{356} = \begin{vmatrix} 0 & a_{15} & 0 \\ 0 & a_{25} & a_{26} \\ 1 & a_{35} & a_{36} \end{vmatrix} = a_{15}\, a_{26} = 0;$$

i.e.,

$$a_{15}\, a_{26} = 0. \tag{50}$$

Further, the determinants $B_{235} = y_1'(1)$ and B_{136} are similar. No other determinants B_{i_1,i_2,i_3} in the sum $S(s)$ are similar to them. Therefore, we have

$$A_{235} + A_{136} = \begin{vmatrix} 0 & 0 & a_{15} \\ 1 & 0 & a_{25} \\ 0 & 1 & a_{35} \end{vmatrix} + \begin{vmatrix} 1 & 0 & 0 \\ 0 & 0 & a_{26} \\ 0 & 1 & a_{36} \end{vmatrix} = 0,$$

$$a_{15} + a_{26} = 0. \tag{51}$$

Obviously, it follows from (50) and (51) that

$$a_{15} = a_{26} = 0, \tag{52}$$

i.e., the diagonal consisting of the entries a_{15} and a_{26} is zero.

We can show in a similar way that any diagonal below the main diagonal consists of zeros. If $A_{456} \neq 0$, then it follows from the condition that $S(s) \equiv 0$ for all $\lambda = s^2 \neq 0$ that the matrix A has the form (44), i.e., that the submatrix composed of the first n columns is diagonal. This can be proved by analogy with the case of $A_{123} \neq 0$. Thus, the proof of Theorem 5.2 is complete.

By Theorem 5.2, in the set of all boundary value problems (39) and (40) only the boundary value problems with matrices A, up to linear row transformations, coincide with the matrices of one of the two forms (43) or (44). The spectral problem (39) and (40) with the matrix A_1 has the characteristic determinant

$$\Delta(\lambda) = \big(a_1\, a_2 + a_1\, a_3 + a_2\, a_3\big)\left(\big(y_3''(1)\big)^2 - y_3'(1)\, y_3'''(1)\right) + \big(a_1 + a_2 + a_3\big)\, y_3''(1) + a_1\, a_2\, a_3 + 1.$$

The identity

$$\Delta(\lambda) \equiv C = \text{const}$$

holds if and only if the numbers a_1, a_2, and a_3 are a solution of the system of equations

$$a_1 + a_2 + a_3 = 0, \quad a_1\, a_2 + a_1\, a_3 + a_2\, a_3 = 0, \quad a_1\, a_2\, a_3 + 1 = C = \text{const}. \tag{53}$$

By the Vieta theorem, the set $\{a_1, a_2, a_3\}$ coincides with the set of roots of the cubic equation $\lambda^3 - (C-1) = 0$, i.e., with the set of cubic roots of the number $C-1$. In particular, $\Delta(\lambda) \equiv 0$ for problem (39) and (40) with coefficient matrix (43) if and only if the numbers a_1, a_2, and a_3 are the cubic roots of -1. Since the endpoints are equivalent, we see that similar conclusions can be made for problem (39) and (40) with coefficient matrix (44). Thus, the degenerate boundary conditions (40 for Eq. 39) can only be the boundary conditions with coefficient matrix (5) or (6), where the set $\{a_1, a_2, a_3\}$ coincides with the set of cubic roots of some number, and $\Delta(\lambda) \equiv 0$ if and only if the numbers a_1, a_2, and a_3 are three distinct cubic roots of -1.

6 On Degenerate Boundary Conditions for Operator D^4

The common form for degenerate boundary conditions for the operator D^4 (D^n) is found. It is shown that the matrix for coefficients of degenerate boundary conditions has a two diagonal form and the elements for one of the diagonal are units. Operator D^4 whose spectrum fills the entire complex plane are studied too. Earlier, examples of eigenvalue problems for the differential operator of even order with common boundary conditions (not containing a spectral parameter) whose spectrum fills the entire complex plane were given. However, in connection with this, another question arises whether there are other examples of such operators. In this paper we show that such examples exist. Moreover, all eigenvalue boundary problems for the operator D^4 whose spectrum fills the entire complex plane are described. It is proved that the characteristic determinant is identically equal to zero if and only if the matrix of coefficients of boundary conditions has a two diagonal form. The elements of this matrix for one of the diagonal are units, and the elements of the other diagonal are 1, -1 and an arbitrary constant.

Consider the following problem for operator D^4:

$$y^{(4)}(x) = \lambda\, y(x) = s^4\, y(x), \qquad x \in [0, 1], \tag{54}$$

$$U_j(y) = \sum_{k=1}^{4} a_{jk}\, y^{(k-1)}(0) + \sum_{k=1}^{4} a_{j\,k+4}\, y^{(k-1)}(1) = 0, \quad j, k = 1, 2, 3, 4. \tag{55}$$

We denote the matrix consisting of the coefficients a_{lk} in the boundary conditions (55) by A and the minor consisting of the i_1th, i_2th, i_3th, and i_4th columns of this matrix A by A_{i_1,i_2,i_3,i_4},

$$A = \begin{Vmatrix} a_{11} & a_{12} & a_{13} & a_{14} & a_{15} & a_{16} & a_{17} & a_{18} \\ a_{21} & a_{22} & a_{23} & a_{24} & a_{25} & a_{26} & a_{27} & a_{28} \\ a_{31} & a_{32} & a_{33} & a_{34} & a_{35} & a_{36} & a_{37} & a_{38} \\ a_{41} & a_{42} & a_{43} & a_{44} & a_{45} & a_{46} & a_{47} & a_{48} \end{Vmatrix}, \tag{56}$$

$$A_{i_1,i_2,i_3,i_4} = \begin{vmatrix} a_{1,i_1} & a_{1,i_2} & a_{1,i_3} & a_{1,i_4} \\ a_{2,i_1} & a_{2,i_2} & a_{2,i_3} & a_{2,i_4} \\ a_{3,i_1} & a_{3,i_2} & a_{3,i_3} & a_{3,i_4} \\ a_{4,i_1} & a_{4,i_2} & a_{4,i_3} & a_{4,i_4} \end{vmatrix}. \tag{57}$$

In what follows, we assume that the rank of the matrix A is equal to 4,

$$\operatorname{rank} A = 4. \tag{58}$$

The aim of this paper is to prove the following theorems:

Theorem 6.1 (Matrix (56)) *for coefficients of degenerate boundary conditions (55) has the following form:*

$$A_1 = \begin{Vmatrix} 1 & 0 & 0 & 0 & a_1 & 0 & 0 & 0 \\ 0 & 1 & 0 & 0 & 0 & a_2 & 0 & 0 \\ 0 & 0 & 1 & 0 & 0 & 0 & a_3 & 0 \\ 0 & 0 & 0 & 1 & 0 & 0 & 0 & a_4 \end{Vmatrix} \tag{59}$$

or

$$A_2 = \begin{Vmatrix} a_1 & 0 & 0 & 0 & 1 & 0 & 0 & 0 \\ 0 & a_2 & 0 & 0 & 0 & 1 & 0 & 0 \\ 0 & 0 & a_3 & 0 & 0 & 0 & 1 & 0 \\ 0 & 0 & 0 & a_4 & 0 & 0 & 0 & 1 \end{Vmatrix}, \tag{60}$$

where a_i $(i = 1, 2, 3, 4)$ are some numbers.

Proof The eigenvalues of the problem (54) and (55) are the roots of the entire function [16, P. 26] $\Delta(\lambda)$:

$$\Delta(\lambda) = \begin{vmatrix} U_1(y_1) & U_1(y_2) & U_1(y_3) & U_1(y_4) \\ U_2(y_1) & U_2(y_2) & U_2(y_3) & U_2(y_4) \\ U_3(y_1) & U_3(y_2) & U_3(y_3) & U_3(y_4) \\ U_4(y_1) & U_4(y_2) & U_4(y_3) & U_4(y_4) \end{vmatrix}, \tag{61}$$

where

$$y_1 = \frac{1}{4}\exp(s\,x) + \frac{1}{4}\exp(-s\,x) + \frac{1}{2}\cos(s\,x),$$
$$y_2 = \frac{1}{4\,s}\exp(s\,x) - \frac{1}{4\,s}\exp(-s\,x) + \frac{1}{2\,s}\sin(s\,x),$$
$$y_3 = \frac{1}{4\,s^2}\exp(s\,x) + \frac{1}{4\,s^2}\exp(-s\,x) - \frac{1}{2\,s^2}\cos(s\,x),$$
$$y_4 = \frac{1}{4\,s^3}\exp(s\,x) - \frac{1}{4\,s^3}\exp(-s\,x) - \frac{1}{2\,s^3}\sin(s\,x),$$

are the linearly independent solutions of Eq. (54) satisfying the conditions

$$y_j^{(r-1)}(0,\lambda) = \begin{cases} 0 \text{ for } j \neq r, \\ 1 \text{ for } j = r, \end{cases} \qquad j,\, r = 1,\, 2,\, 3,\, 4. \tag{62}$$

By B, B_1 and B_2 denote the following matrixes:

$$B = \begin{Vmatrix} y_1(0) & y_1'(0) & y_1''(0) & y_1'''(0) & y_1(1) & y_1'(1) & y_1''(1) & y_1'''(1) \\ y_2(0) & y_2'(0) & y_2''(0) & y_2'''(0) & y_2(1) & y_2'(1) & y_2''(1) & y_2'''(1) \\ y_3(0) & y_3'(0) & y_3''(0) & y_3'''(0) & y_3(1) & y_3'(1) & y_3''(1) & y_3'''(1) \\ y_4(0) & y_4'(0) & y_4''(0) & y_4'''(0) & y_4(1) & y_4'(1) & y_4''(1) & y_4'''(1) \end{Vmatrix},$$

$$B_1 = \begin{Vmatrix} y_1(0) & y_1'(0) & y_1''(0) & y_1'''(0) \\ y_2(0) & y_2'(0) & y_2''(0) & y_2'''(0) \\ y_3(0) & y_3'(0) & y_3''(0) & y_3'''(0) \\ y_4(0) & y_4'(0) & y_4''(0) & y_4'''(0) \end{Vmatrix},$$

$$B_2 = \begin{Vmatrix} y_1(1) & y_1'(1) & y_1''(1) & y_1'''(1) \\ y_2(1) & y_2'(1) & y_2''(1) & y_2'''(1) \\ y_3(1) & y_3'(1) & y_3''(1) & y_3'''(1) \\ y_4(1) & y_4'(1) & y_4''(1) & y_4'''(1) \end{Vmatrix},$$

where

$$y_1(1) = \frac{1}{4}\left(e^s + e^{-s} + 2\cos(s)\right), \quad y_1'(1) = \frac{1}{4}s\left(e^s - e^{-s} - 2\sin(s)\right),$$

$$y_1''(1) = \frac{1}{4}s^2\left(e^s + e^{-s} - 2\cos(s)\right), \; y_1'''(1) = \frac{1}{4}s^3\left(e^s - e^{-s} + 2\sin(s)\right),$$

$$y_2(1) = \frac{1}{4s}\left(e^s - e^{-s} + 2\sin(s)\right), \quad y_2'(1) = \frac{1}{4}\left(e^s + e^{-s} + 2\cos(s)\right),$$

$$y_2''(1) = \frac{1}{4}s\left(e^s - e^{-s} - 2\sin(s)\right), \quad y_2'''(1) = \frac{1}{4}s^2\left(e^s - e^{-s} - 2\cos(s)\right),$$

$$y_3(1) = \frac{1}{4s^2}\left(e^s - e^{-s} - 2\cos(s)\right), \; y_3'(1) = \frac{1}{4s}\left(e^s - e^{-s} + 2\sin(s)\right),$$

$$y_3''(1) = \frac{1}{4}\left(e^s + e^{-s} + 2\cos(s)\right), \quad y_3'''(1) = \frac{1}{4}s\left(e^s - e^{-s} - 2\sin(s)\right),$$

$$y_4(1) = \frac{1}{4s^3}\left(e^s - e^{-s} - 2\sin(s)\right), \; y_4'(1) = \frac{1}{4s^2}\left(e^s + e^{-s} - 2\cos(s)\right),$$

$$y_4''(1) = \frac{1}{4s}\left(e^s - e^{-s} + 2\sin(s)\right), \quad y_4'''(1) = \frac{1}{4}\left(e^s + e^{-s} + 2\cos(s)\right).$$

Note that

$$y_{j-1}^{(k-1)}(1,\lambda) \equiv y_j^{(k)}(1,\lambda), \qquad j = 2,3,4, \quad k = 1,2,3,4. \tag{63}$$

From (62) and (63), it follows that

$$B = \|B_1, B_2\| = \begin{Vmatrix} 1\,0\,0\,0 & y_1(1) & y_1'(1) & y_1''(1) & y_1'''(1) \\ 0\,1\,0\,0 & y_2(1) & y_1(1) & y_1'(1) & y_1''(1) \\ 0\,0\,1\,0 & y_3(1) & y_2(1) & y_1(1) & y_1'(1) \\ 0\,0\,0\,1 & y_4(1) & y_3(1) & y_2(1) & y_1(1) \end{Vmatrix}. \tag{64}$$

□

Using A and B the determinant (61) represents in the form

$$\Delta(\lambda) \equiv \det(A \cdot B^T).$$

It follows from Cauchy–Binet formula [27, 1.14] that

$$\Delta(\lambda) = \sum_{1 \le i_1 < i_2 < i_3 < i_4 \le 8} A_{i_1,i_2,i_3,i_4} B_{i_1,i_2,i_3,i_4} = 0. \tag{65}$$

Here we denote by $B_{i_1,i_2,i_3,i_4} = B_{i_1,i_2,i_3,i_4}(\lambda)$ the minor consisting of the i_1th, i_2th, i_3th, and i_4th columns of the matrix B (lines of the matrix B^T).

By $P(s)$ denote $P(s) = A_{1234} B_{1234} + A_{5678} B_{5678}$. From the Liouville–Ostrogradsky formula for the Wronskian determinant it follows that ([22], 17.1)

$B_{1234} = \det(B_1) = W(0) = 1$, $B_{5678} = \det(B_2) = W(1) = 1$, and $P(s) = A_{1234} + B_{5678} = \text{const}$.

All other functions $B_{i_1,i_2,i_3,i_4} = B_{i_1,i_2,i_3,i_4}(s)$ (except B_{1234} and B_{5678}) are not constants. So if $\Delta(\lambda) \equiv C = \text{const}$, then $\Delta(\lambda) - P(s) \equiv 0$ and one of minors A_{1234} or A_{5678} are not equal to zero. Assume the converse. Then all minors A_{i_1,i_2,i_3,i_4} of the matrix are equal to zero. This contradicts the condition rank $A = 4$. Suppose that $A_{1234} \neq 0$. Then the matrix (56) have the following form:

$$A = \begin{Vmatrix} 1 & 0 & 0 & 0 & a_{15} & a_{16} & a_{17} & a_{18} \\ 0 & 1 & 0 & 0 & a_{25} & a_{26} & a_{27} & a_{28} \\ 0 & 0 & 1 & 0 & a_{35} & a_{36} & a_{37} & a_{38} \\ 0 & 0 & 0 & 1 & a_{45} & a_{46} & a_{47} & a_{48} \end{Vmatrix}.$$

(In order not to introduce new notations by a_{ij} we denote other coefficients a_{ij} than (56)). Let us remark that the determinant $B_{2348} = y_1'''(1)$ and any other determinant B_{i_1,i_2,i_3,i_4} are linear independent. Suppose $\Delta(\lambda) \equiv C = \text{const}$, then $\Delta(\lambda) - P(s) \equiv 0$ and $A_{2348} = 0$. From this it follows that

$$A_{2348} = \begin{vmatrix} 0 & 0 & 0 & a_{18} \\ 1 & 0 & 0 & a_{28} \\ 0 & 1 & 0 & a_{38} \\ 0 & 0 & 1 & a_{48} \end{vmatrix} = -a_{18} = 0. \tag{66}$$

Let us show that a_{17} and a_{28} are equal to zero too. Indeed, $B_{3478} = y_1'(1)\, y_1'''(1) - (y_1''(1))^2$ and any other determinant B_{i_1,i_2,i_3,i_4} are linear independent. Suppose $\Delta(\lambda) \equiv C = \text{const}$, then $\Delta(\lambda) - P(s) \equiv 0$ and $A_{3478} = 0$. From this it follows that

$$A_{3478} = \begin{vmatrix} 0 & 0 & a_{17} & 0 \\ 0 & 0 & a_{27} & a_{28} \\ 1 & 0 & a_{37} & a_{38} \\ 0 & 1 & a_{47} & a_{48} \end{vmatrix} = a_{17} \cdot a_{28} = 0. \tag{67}$$

In addition, $B_{2347} = -B_{1348} = -y_1''(1)$ and any other determinant B_{i_1,i_2,i_3,i_4} are linear independent. This implies that

$$A_{2347} - A_{1348} = -(a_{17} + a_{28}) = 0. \tag{68}$$

Combining (67) and (68), we get

$$a_{17} = a_{28} = 0.$$

Likewise,

$$a_{16} = a_{27} = a_{38} = 0.$$

Further, $B_{1235} = y_4(1)$ and any other determinant B_{i_1,i_2,i_3,i_4} are linear independent. So if $\Delta(\lambda) - P(s) \equiv 0$, then the minor $A_{1235} = a_{45} = 0$. As before, we have

$$a_{34} = a_{46} = a_{25} = a_{36} = a_{47} = 0.$$

Therefore if $A_{1234} \neq 0$, then the matrix A has the form A_1.

Arguing as above, we see that if $A_{5678} \neq 0$, then the matrix A has the form A_2. This completes the proof of Theorem 6.1.

Theorem 6.2 *The characteristic determinant of problem (54) and (55) is identically equal to zero if and only if the matrix (56) has the form (59) or (60), where $\{a_i\}$ $(i = 1, 2, 3, 4)$ are the sets:*

$$\begin{array}{l}
1.\ \ a_1 = C_1, \quad a_2 = -1, \quad a_3 = C_1^{-1}, \quad a_4 = 1, \\
2.\ \ a_1 = C_2, \quad a_2 = 1, \quad a_3 = C_2^{-1}, \quad a_4 = -1, \\
3.\ \ a_1 = C_3, \quad a_2 = -1, \quad a_3 = 1, \quad a_4 = -1, \\
4.\ \ a_1 = C_4, \quad a_2 = 1, \quad a_3 = -1, \quad a_4 = 1, \\
5.\ \ a_1 = -1, \quad a_2 = C_5, \quad a_3 = -1, \quad a_4 = 1, \\
6.\ \ a_1 = -1, \quad a_2 = C_6, \quad a_3 = 1, \quad a_4 = C_6^{-1}, \\
7.\ \ a_1 = 1, \quad a_2 = C_7, \quad a_3 = -1, \quad a_4 = C_7^{-1}, \\
8.\ \ a_1 = 1, \quad a_2 = C_8, \quad a_3 = 1, \quad a_4 = -1, \\
9.\ \ a_1 = -1, \quad a_2 = 1, \quad a_3 = C_9, \quad a_4 = 1, \\
10.\ \ a_1 = 1, \quad a_2 = -1, \quad a_3 = C_{10}, \quad a_4 = -1, \\
11.\ \ a_1 = -1, \quad a_2 = 1, \quad a_3 = -1 \quad a_4 = C_{11}, \\
12.\ \ a_1 = 1, \quad a_2 = -1, \quad a_3 = 1, \quad a_4 = C_{12},
\end{array} \tag{69}$$

where C_j $(j = 1, 2, \ldots, 12)$ are arbitrary constants.

Proof If $A_{1234} \neq 0$ and $\Delta(\lambda) \equiv 0$ it follows from Theorem 6.1 that

$$\begin{aligned}
0 \equiv \Delta(\lambda) = \det(A_1 \cdot B^T) &= 1 + \frac{1}{2}\,(a_1\,a_2 + a_1\,a_4 + a_2\,a_3 + a_3\,a_4) \\
&\quad + a_1\,a_2\,a_3\,a_4 \\
&\quad + \frac{1}{4}\,(a_1\,a_2 + a_1\,a_4 + a_2\,a_3 + a_3\,a_4 + 2\,a_1\,a_3 + 2\,a_2\,a_4)\,\left(e^s + e^{-s}\right)\,\cos s \\
&\quad + \frac{1}{4}\,(a_1 + a_2 + a_3 + a_4 + a_1\,a_2\,a_3 \\
&\quad + a_1\,a_2\,a_4 + a_1\,a_3\,a_4 + a_2\,a_3\,a_4)\,\left(e^s + e^{-s} + 2\,\cos s\right).
\end{aligned} \tag{70}$$

The functions 1, $\left(e^s + e^{-s}\right)\,\cos s$ and $\left(e^s + e^{-s} + 2\,\cos s\right)$ are linear independent. So characteristic determinant (70) is identically equal to zero if and only if the coefficients a_1, a_2, a_3, a_4 are the solutions of the following system of the equations:

$$2 + a_1 a_2 + a_1 a_4 + a_2 a_3 + a_3 a_4 + 2 a_1 a_2 a_3 a_4 = 0,$$

$$a_1 a_2 + a_1 a_4 + a_2 a_3 + a_3 a_4 + 2 a_1 a_3 + 2 a_2 a_4 = 0, \tag{71}$$

$$a_1 + a_2 + a_3 + a_4 + a_1 a_2 a_3 + a_1 a_2 a_4 + a_1 a_3 a_4 + a_2 a_3 a_4 = 0.$$

By direct calculation we find the solutions of the system of the Eqs. (71). This solutions are (69).

If $A_{5678} \neq 0$ and $\Delta(\lambda) \equiv 0$ it follows from Theorem 6.1 that

$$0 \equiv \Delta(\lambda) = \det(A_2 \cdot B^T). \tag{72}$$

From this we have the system of Eqs. (71), the solutions of whose are (69). This concludes the proof of Theorem 6.2. □

Remark 6.1 Theorem 6.1 may be generalized for any order $n \geq 2$. If n is an order of the differential equation, then the matrix A for coefficients of the boundary conditions has the following form:

$$A = \begin{Vmatrix} a_{11} & a_{12} & \dots & a_{1n} & a_{1\,n+1} & a_{1\,n+2} & \dots & a_{1\,2n} \\ a_{21} & a_{22} & \dots & a_{2n} & a_{2\,n+1} & a_{2\,n+2} & \dots & a_{2\,2n} \\ \vdots & \vdots & \ddots & \vdots & \vdots & \vdots & \ddots & \vdots \\ a_{n1} & a_{n2} & \dots & a_{nn} & a_{n\,n+1} & a_{n\,n+2} & \dots & a_{n\,2n} \end{Vmatrix}, \tag{73}$$

where rank $A = n$.

If the matrix A determines degenerate boundary conditions, then it has the forms:

$$A_1 = \begin{Vmatrix} 1 & 0 & \dots & 0 & a_1 & 0 & \dots & 0 \\ 0 & 1 & \dots & 0 & 0 & a_2 & \dots & 0 \\ \vdots & \vdots & \ddots & \vdots & \vdots & \vdots & \ddots & \vdots \\ 0 & 0 & \dots & 1 & 0 & 0 & \dots & a_n \end{Vmatrix} \tag{74}$$

or

$$A_2 = \begin{Vmatrix} a_1 & 0 & \dots & 0 & 1 & 0 & \dots & 0 \\ 0 & a_2 & \dots & 0 & 0 & 1 & \dots & 0 \\ \vdots & \vdots & \ddots & \vdots & \vdots & \vdots & \ddots & \vdots \\ 0 & 0 & \dots & a_n & 0 & 0 & \dots & 1 \end{Vmatrix}. \tag{75}$$

This statement can be proved similarly as in Theorem 6.1.

Remark 6.2 Equation (2) is associated only with the four solutions: 1, 3, 8, 10 in Theorem 6.2.

In this section we proved that the characteristic determinant is identically equal to zero if and only if the matrix of coefficients of boundary conditions has a two diagonal form. The elements of this matrix for one of the diagonal are units, and the elements of the other diagonal are numbers (69). Also is shown that the matrix for coefficients of degenerate boundary conditions has a two diagonal form and the elements for one of the diagonal are units. All eigenvalue boundary problems for the operator D^4 whose spectrum fills the entire complex plane are described.

Let us remark that if

$$2 + a_1 a_2 + a_1 a_4 + a_2 a_3 + a_3 a_4 + 2 a_1 a_2 a_3 a_4 = C \neq 0$$

in (71), then solving of the new system of equations reduces to solving a sixth-degree equation, and therefore is no longer analytical. Therefore, we cannot write specific expressions for the coefficients in Theorem 6.1. The system (71) can be solved analytically in view of the fact that the coefficients of odd powers vanish, and therefore the sixth-degree equation reduces to a three-degree equation. Such specific expressions for the coefficients are given in Theorem 6.2.

7 Degenerate Boundary Conditions for the Sturm–Liouville Problem on a Geometric Graph

We study the boundary conditions of the Sturm–Liouville problem posed on a star-shaped geometric graph consisting of three edges with a common vertex. We show that the Sturm–Liouville problem has no degenerate boundary conditions in the case of pairwise distinct edge lengths. However, if the edge lengths coincide and all potentials are the same, then the characteristic determinant of the Sturm–Liouville problem cannot be a nonzero constant and the set of Sturm–Liouville problems whose characteristic determinant is identically zero and whose spectrum accordingly coincides with the entire plane is infinite (a continuum). It is shown that, for one special case of the boundary conditions, this set consists of eighteen classes, each having from two to four arbitrary constants, rather than of two problems as in the case of the Sturm–Liouville problem on an interval.

By L we denote the following Sturm–Liouville problem on the graph Γ:

$$ly = -y_i'' + q_i(x)\, y_i = \lambda\, y_i = s^2\, y_i, \qquad i = 1, 2, 3, \tag{76}$$

$$y_1(0) = y_2(0) = y_3(0), \quad y_1'(0) + y_2'(0) + y_3'(0) = 0, \tag{77}$$

$$a_{i1}\, y_1(l_1) + a_{i2}\, y_2(l_2) + a_{i3}\, y_3(l_3) + a_{i4}\, y_1'(l_1) + a_{i5}\, y_2'(l_2) + a_{i6}\, y_3'(l_3) = 0, \tag{78}$$

where λ is a spectral parameter, the real function $q_i(\cdot)$ belongs to the space $L_1(0, \pi)$, and the a_{ij} ($i = 1, 2, 3$, $j = 1, \ldots, 6$) are complex constants.

Let A be the matrix of the coefficients a_{ij} in the boundary conditions (78) and let the A_{lmn} be its minor consisting of the lth, mth, and nth columns,

$$A = \begin{Vmatrix} a_{11} & a_{12} & a_{13} & a_{14} & a_{15} & a_{16} \\ a_{21} & a_{22} & a_{23} & a_{24} & a_{25} & a_{26} \\ a_{31} & a_{32} & a_{33} & a_{34} & a_{35} & a_{36} \end{Vmatrix}, \qquad A_{lmn} = \begin{vmatrix} a_{1l} & a_{1m} & a_{1n} \\ a_{2l} & a_{2m} & a_{2n} \\ a_{3l} & a_{3m} & a_{3n} \end{vmatrix}. \tag{79}$$

Throughout the paper, we assume that the following condition is satisfied:

$$\operatorname{rank} A = 3. \tag{80}$$

The problem is to find coefficients a_{ij} for which the boundary conditions (78) are degenerate.

By $z_{i1}(x, \lambda)$ and $z_{i2}(x, \lambda)$ we denote the linearly independent solutions of Eqs. (76) satisfying the conditions

$$z_{i1}(0, \lambda) = 1, \quad z_{i1}'(0, \lambda) = 0, \quad z_{i2}(0, \lambda) = 0, \quad z_{i2}'(0, \lambda) = 1.$$

Then the general solutions (76) can be written as

$$y_i = C_{i1} z_{i1} + C_{i2} z_{i2}, \quad i = 1, 2, 3. \tag{81}$$

From the transmission conditions (77), we obtain

$$C_{11} = C_{21} = C_{31} = C, \quad C_{32} = -C_{12} - C_{22}. \tag{82}$$

By substituting the representations (81) into the boundary conditions (78) into the boundary conditions (82), we obtain the linear algebraic system

$$\begin{aligned} &a_{i1}\,(C\, z_{11}(l_1) + C_{12}\, z_{12}(l_1)) + a_{i2}\,(C\, z_{21}(l_2) + C_{22}\, z_{22}(l_2)) \\ &\quad + a_{i3}\,(C\, z_{31}(l_3) - (C_{12} + C_{22})\, z_{32}(l_3)) \\ &\quad + a_{i4}\,(C\, z_{11}'(l_1) + C_{12}\, z_{12}'(l_1)) + a_{i5}\,(C\, z_{21}'(l_2) + C_{22}\, z_{22}'(l_2)) \\ &\quad + a_{i6}\,(C\, z_{31}'(l_3) - (C_{12} + C_{22})\, z_{32}'(l_3)) = 0, \qquad i = 1, 2, 3, \end{aligned} \tag{83}$$

with determinant

$$\Delta(\lambda) = \begin{vmatrix} d_{11} & d_{12} & d_{13} \\ d_{21} & d_{22} & d_{23} \\ d_{31} & d_{32} & d_{33} \end{vmatrix}, \tag{84}$$

where

$$\begin{aligned} d_{i1} &= a_{i1}\, z_{11}(l_1) + a_{i2}\, z_{21}(l_2) + a_{i3}\, z_{31}(l_3) + a_{i4}\, z_{11}'(l_1) \\ &\qquad + a_{i5}\, z_{21}'(l_2) + a_{i6}\, z_{31}'(l_3), \\ d_{i2} &= a_{i1}\, z_{12}(l_1) - a_{i3}\, z_{32}(l_3) + a_{i4}\, z_{12}'(l_1) - a_{i6}\, z_{32}'(l_3), \\ d_{i3} &= a_{i2}\, z_{22}(l_2) - a_{i3}\, z_{32}(l_3) + a_{i5}\, z_{22}'(l_2) - a_{i6}\, z_{32}'(l_3), \qquad i = 1, 2, 3. \end{aligned}$$

The Liouville formula ([25]) implies the relations

$$z_{i1}(l_i)\,z'_{i2}(l_i) - z'_{i1}(l_i)\,z_{i2}(l_i) = \begin{vmatrix} z_{i1}(l_i) & z_{i2}(l_i) \\ z'_{i1}(l_i) & z'_{i2}(l_i) \end{vmatrix} = \begin{vmatrix} z_{i1}(0) & z_{i2}(0) \\ z'_{i1}(0) & z'_{i2}(0) \end{vmatrix} = 1. \tag{85}$$

We calculate the determinant (84) with regard to notation (79) and relations (85) and obtain the following representation of the characteristic determinant of problem L

$$\Delta(\lambda) = \sum_{l<m<n} A_{lmn}\, Z_{lmn}, \tag{86}$$

where

$$\begin{aligned}
Z_{123} &= z_{11}(l_1)\,z_{22}(l_2)\,z_{32}(l_3) + z_{12}(l_1)\,z_{21}(l_2)\,z_{32}(l_3) \\
&\quad + z_{12}(l_1)\,z_{22}(l_2)\,z_{31}(l_3), \\
Z_{124} &= -z_{11}(l_1)\,z'_{12}(l_1)\,z_{22}(l_2) + z'_{11}(l_1)\,z_{12}(l_1)\,z_{22}(l_2) = -z_{22}(l_2), \\
Z_{125} &= -z_{21}(l_2)\,z_{12}(l_1)\,z'_{22}(l_2) + z_{12}(l_1)\,z'_{21}(l_2)\,z_{22}(l_2) = -z_{12}(l_1), \\
Z_{126} &= z_{11}(l_1)\,z_{22}(l_2)\,z'_{32}(l_3) + z_{12}(l_1)\,z_{21}(l_2)\,z'_{32}(l_3) \\
&\quad + z_{12}(l_1)\,z_{22}(l_2)\,z'_{31}(l_3), \\
Z_{134} &= z_{11}(l_1)\,z'_{12}(l_1)\,z_{32}(l_3) - z'_{11}(l_1)\,z_{12}(l_1)\,z_{32}(l_3) = z_{32}(l_3), \\
Z_{135} &= -z_{11}(l_1)\,z'_{22}(l_2)\,z_{32}(l_3) - z_{12}(l_1)\,z'_{22}(l_2)\,z_{31}(l_3) \\
&\quad - z_{12}(l_1)\,z'_{21}(l_2)\,z_{32}(l_3), \\
Z_{136} &= z_{12}(l_1)\,z_{31}(l_3)\,z'_{32}(l_3) - z_{12}(l_1)\,z'_{31}(l_3)\,z_{32}(l_3) = z_{12}(l_1), \\
Z_{145} &= z_{11}(l_1)\,z'_{12}(l_1)\,z'_{22}(l_2) - z'_{11}(l_1)\,z_{12}(l_1)\,z'_{22}(l_2) = z'_{22}(l_2), \\
Z_{146} &= -z_{11}(l_1)\,z'_{12}(l_1)\,z'_{32}(l_3) + z'_{11}(l_1)\,z_{12}(l_1)\,z'_{32}(l_3) = -z'_{32}(l_3), \\
Z_{156} &= z_{11}(l_1)\,z'_{22}(l_2)\,z'_{32}(l_3) + z_{12}(l_1)\,z'_{21}(l_2)\,z'_{32}(l_3) \\
&\quad + z_{12}(l_1)\,z'_{22}(l_2)\,z'_{31}(l_3), \\
Z_{234} &= z'_{12}(l_1)\,z_{21}(l_2)\,z_{32}(l_3) + z'_{12}(l_1)\,z_{22}(l_2)\,z_{31}(l_3) \\
&\quad + z'_{11}(l_1)\,z_{22}(l_2)\,z_{32}(l_3), \\
Z_{235} &= -z_{21}(l_2)\,z'_{22}(l_2)\,z_{32}(l_3) + z'_{21}(l_2)\,z_{22}(l_2)\,z_{32}(l_3) = -z_{32}(l_3), \\
Z_{236} &= -z_{22}(l_2)\,z_{31}(l_3)\,z'_{32}(l_3) + z_{22}(l_2)\,z'_{31}(l_3)\,z_{32}(l_3) = -z_{22}(l_2), \\
Z_{245} &= z'_{12}(l_1)\,z_{21}(l_2)\,z'_{22}(l_2) - z'_{12}(l_1)\,z'_{21}(l_2)\,z_{22}(l_2) = z'_{12}(l_1), \\
Z_{246} &= -z'_{12}(l_1)\,z_{21}(l_2)\,z'_{32}(l_3) - z'_{11}(l_1)\,z_{22}(l_2)\,z'_{32}(l_3) \\
&\quad - z'_{12}(l_1)\,z_{22}(l_2)\,z'_{31}(l_3), \\
Z_{256} &= z_{21}(l_2)\,z'_{22}(l_2)\,z'_{32}(l_3) - z'_{21}(l_2)\,z_{22}(l_2)\,z'_{32}(l_3) = z'_{32}(l_3), \\
Z_{345} &= z'_{12}(l_1)\,z'_{22}(l_2)\,z_{31}(l_3) + z'_{11}(l_1)\,z'_{22}(l_2)\,z_{32}(l_3) \\
&\quad + z'_{12}(l_1)\,z'_{21}(l_2)\,z_{32}(l_3), \\
Z_{346} &= -z'_{12}(l_1)\,z_{31}(l_3)\,z'_{32}(l_3) + z'_{12}(l_1)\,z'_{31}(l_3)\,z_{32}(l_3) = -z'_{12}(l_1), \\
Z_{356} &= z'_{22}(l_2)\,z_{31}(l_3)\,z'_{32}(l_3) - z'_{22}(l_2)\,z'_{31}(l_3)\,z_{32}(l_3) = z'_{22}(l_2), \\
Z_{456} &= z'_{12}(l_1)\,z'_{21}(l_2)\,z'_{32}(l_3) + z'_{12}(l_1)\,z'_{22}(l_2)\,z'_{31}(l_3) \\
&\quad + z'_{11}(l_1)\,z'_{22}(l_2)\,z'_{32}(l_3).
\end{aligned} \tag{87}$$

The asymptotic formulas

$$\begin{aligned} z_{i1}(l_i, \lambda) &= \cos s\, l_i + \tfrac{1}{s}\, u_i(l_i)\, \sin s + \mathscr{O}\left(\tfrac{1}{s^2}\right), \\ z_{i2}(l_i, \lambda) &= \tfrac{1}{s}\, \sin s\, l_i - \tfrac{1}{s^2}\, u_i(l_i)\, \cos s\, l_i + \mathscr{O}\left(\tfrac{1}{s^3}\right), \\ z'_{i1}(l_i, \lambda) &= -s\, \sin s\, l_i + u_i(l_i)\, \cos s\, l_i + \mathscr{O}\left(\tfrac{1}{s}\right), \\ z'_{i2}(l_i, \lambda) &= \cos s\, l_i + \tfrac{1}{s}\, u_i(l_i)\, \sin s\, l_i + \mathscr{O}\left(\tfrac{1}{s^2}\right), \end{aligned} \tag{88}$$

where $u_i(l_i) = \frac{1}{2}\int_0^{l_i} q(t)\, dt$, hold for sufficiently large $\lambda \in \mathbb{R}$ ([16, pp. 62–65]).

It follows that

$$\begin{aligned} Z_{124} &= Z_{236} = -z_{22}(l_2) \sim -\tfrac{\sin s\, l_2}{s}, \\ Z_{125} &= -Z_{136} = -z_{12}(l_1) \sim -\tfrac{\sin s\, l_1}{s}, \\ Z_{134} &= -Z_{235} = z_{32}(l_3) \sim \tfrac{\sin s\, l_3}{s}, \\ Z_{145} &= Z_{356} = z'_{22}(l_2) \sim \cos s\, l_2, \\ Z_{146} &= -Z_{256} = -z'_{32}(l_3) \sim -\cos s\, l_3, \\ Z_{245} &= -Z_{346} = z'_{12}(l_1) \sim \cos s\, l_1, \end{aligned} \tag{89}$$

$$\begin{aligned} Z_{126} &\sim \tfrac{1}{s}\,(\sin s\, l_1\, \cos s\, l_2\, \cos s\, l_3 + \sin s\, l_2\, \cos s\, l_1\, \cos s\, l_3 - a(s)), \\ Z_{135} &\sim \tfrac{1}{s}\,(-\sin s\, l_1\, \cos s\, l_2\, \cos s\, l_3 - \sin s\, l_3\, \cos s\, l_1\, \cos s\, l_2 + a(s)), \\ Z_{234} &\sim \tfrac{1}{s}\,(\sin s\, l_2\, \cos s\, l_1\, \cos s\, l_3 + \sin s\, l_3\, \cos s\, l_1\, \cos s\, l_2 - a(s)), \end{aligned} \tag{90}$$

where $a(s) = \sin s\, l_1\, \sin s\, l_2\, \sin s\, l_3$,

$$\begin{aligned} Z_{156} &\sim b(s) - \sin s\, l_1\, \sin s\, l_2\, \cos s\, l_3 - \sin s\, l_1\, \sin s\, l_3\, \cos s\, l_2, \\ Z_{246} &\sim -b(s) + \sin s\, l_1\, \sin s\, l_2\, \cos s\, l_3 + \sin s\, l_2\, \sin s\, l_3\, \cos s\, l_1, \\ Z_{345} &\sim b(s) - \sin s\, l_1\, \sin s\, l_3\, \cos s\, l_2 - \sin s\, l_2\, \sin s\, l_3\, \cos s\, l_1, \end{aligned} \tag{91}$$

where $b(s) = \cos s\, l_1\, \cos s\, l_2\, \cos s\, l_3$,

$$\begin{aligned} Z_{123} &\sim \tfrac{1}{s^2}\,(\sin s\, l_1\, \sin s\, l_2\, \cos s\, l_3 + \sin s\, l_1\, \sin s\, l_3\, \cos s\, l_2 \\ &\quad + \sin s\, l_2\, \sin s\, l_3\, \cos s\, l_1), \\ Z_{456} &\sim -s\,(\cos s\, l_1\, \cos s\, l_2\, \sin s\, l_3 + \cos s\, l_1\, \cos s\, l_3\, \sin s\, l_2, \\ &\quad + \cos s\, l_2\, \cos s\, l_3\, \sin s\, l_1). \end{aligned} \tag{92}$$

If the numbers l_i $(i = 1, 2, 3)$ are pairwise distinct, then these asymptotic relations imply the linear independence of the functions

$$\begin{aligned} &Z_{124} = Z_{236}, \quad Z_{125} = -Z_{136}, \quad Z_{134} = -Z_{235}, \\ &Z_{145} = Z_{356}, \quad Z_{146} = -Z_{256}, \quad Z_{245} = -Z_{346}, \\ &Z_{126}, \quad Z_{135}, \quad Z_{234}, \quad Z_{126}, \quad Z_{135}, \quad Z_{234}, \quad Z_{156}, \quad Z_{246}, \quad Z_{345}. \end{aligned} \tag{93}$$

It follows from the asymptotic relations (89)–(92) that the identity

$$\Delta(\lambda) \equiv \sum_{l<m<n} A_{lmn}\, Z_{lmn} \equiv C \neq 0 \tag{94}$$

is impossible, and the linear independence of the functions (93) implies that the identity

$$\Delta(\lambda) \equiv \sum_{l<m<n} A_{lmn}\, Z_{lmn} \equiv 0 \tag{95}$$

is possible if and only if

$$\begin{aligned}
&A_{124} + A_{236} = A_{125} - A_{136} = A_{134} - A_{235} = 0,\\
&A_{123} = 0, \quad A_{456} = 0,\\
&A_{145} + A_{356} = A_{146} - A_{256} = A_{245} - A_{346} = 0,\\
&A_{126} = A_{135} = A_{234} = A_{156} = A_{246} = A_{345} = 0.
\end{aligned} \tag{96}$$

It is well known that given numbers $A_{i_1 i_2 i_3}$ are the minors of a matrix A if and only if the Plücker relations are satisfied [18]. For the 3×6 -matrix A, these relations are

$$\begin{aligned}
&A_{i_1 i_4 i_5} A_{i_1 i_2 i_3} - A_{i_1 i_4 i_3} A_{i_1 i_2 i_5} + A_{i_1 i_5 i_3} A_{i_1 i_2 i_4} = 0,\\
&A_{i_1 i_4 i_6} A_{i_1 i_2 i_3} - A_{i_1 i_4 i_3} A_{i_1 i_2 i_6} + A_{i_1 i_6 i_3} A_{i_1 i_2 i_4} = 0,\\
&A_{i_1 i_5 i_6} A_{i_1 i_2 i_3} - A_{i_1 i_5 i_3} A_{i_1 i_2 i_6} + A_{i_1 i_6 i_3} A_{i_1 i_2 i_5} = 0,\\
&A_{i_2 i_4 i_5} A_{i_1 i_2 i_3} - A_{i_2 i_4 i_3} A_{i_1 i_2 i_5} + A_{i_2 i_5 i_3} A_{i_1 i_2 i_4} = 0,\\
&A_{i_2 i_4 i_6} A_{i_1 i_2 i_3} - A_{i_2 i_4 i_3} A_{i_1 i_2 i_6} + A_{i_2 i_6 i_3} A_{i_1 i_2 i_4} = 0,\\
&A_{i_2 i_5 i_6} A_{i_1 i_2 i_3} - A_{i_2 i_5 i_3} A_{i_1 i_2 i_6} + A_{i_2 i_6 i_3} A_{i_1 i_2 i_5} = 0,\\
&A_{i_3 i_4 i_5} A_{i_1 i_2 i_3} - A_{i_2 i_4 i_3} A_{i_1 i_3 i_5} + A_{i_2 i_5 i_3} A_{i_1 i_3 i_4} = 0,\\
&A_{i_3 i_4 i_6} A_{i_1 i_2 i_3} - A_{i_2 i_4 i_3} A_{i_1 i_3 i_6} + A_{i_2 i_6 i_3} A_{i_1 i_3 i_4} = 0,\\
&A_{i_3 i_5 i_6} A_{i_1 i_2 i_3} - A_{i_2 i_5 i_3} A_{i_1 i_3 i_6} + A_{i_2 i_6 i_3} A_{i_1 i_3 i_5} = 0,\\
&A_{i_4 i_5 i_6} A_{i_1 i_2 i_3} - A_{i_1 i_2 i_4} A_{i_3 i_5 i_6} + A_{i_1 i_2 i_5} A_{i_3 i_4 i_6} - A_{i_1 i_2 i_6} A_{i_3 i_4 i_5} = 0,
\end{aligned} \tag{97}$$

where$(i_1, i_2, i_3, i_4, i_5, i_6)$ is a permutation of the numbers 1, 2, 3, 4, 5, 6 such that $A_{i_1,i_2,i_3} \neq 0$.

Assume that $A_{124} = 1 \neq 0$. Then by the fifth relation in (97)

$$A_{236}\, A_{124} - A_{234}\, A_{126} + A_{264}\, A_{123} = 0$$

which implies that $A_{124} = 0$. This is a contradiction.

In a similar way, assuming that one of the minors A_{ijk} is nonzero, we see that relations (96) and (97) imply that $A_{ijk} = 0$. Therefore, if the l_i are pairwise distinct, then all third-order minors of the matrix A are zero, and this contradicts the condition (80). Therefore, the identity $\Delta(\lambda) \equiv 0$ is impossible, and we have the following assertion.

Theorem 7.1 *If the lengths l_i $(i = 1, 2, 3)$ are pairwise distinct, then the boundary value problem (76)–(78) has no degenerate boundary conditions.*

Now consider the case in which $l_i = l$ and $q_i(x) = q(x)$ $(i = 1, 2, 3)$. Then $z_{i1} = z_{11}$, $z_{i2} = z_{12}$,

$$\begin{aligned}
Z_{126} &= -Z_{135} = Z_{234} = 2\,z_{11}(l)\,z_{12}(l)\,z'_{12}(l) \\
&\quad -z'_{11}(l)\,z^2_{12}(l), \\
Z_{156} &= -Z_{246} = Z_{345} = 2\,z'_{11}(l)\,z_{12}(l)\,z'_{12}(l) \\
&\quad +z_{11}(l)\,(z'_{12}(l))^2, \\
Z_{124} &= Z_{236} = Z_{125} = Z_{235} = -Z_{136} = -Z_{134} = -z_{12}(l), \\
Z_{145} &= Z_{356} = Z_{256} = Z_{245} = -Z_{146} = -Z_{346} = z'_{12}(l).
\end{aligned}$$

It follows from the asymptotic relations that (94) is impossible, and identity (95) is possible if and only if

$$\begin{aligned}
&A_{123} = 0, \quad A_{456} = 0, \\
&A_{126} - A_{135} + A_{234} = 0, \quad A_{156} - A_{246} + A_{345} = 0, \\
&A_{124} + A_{236} + A_{125} + A_{235} - A_{136} - A_{134} = 0, \\
&A_{145} + A_{356} + A_{256} + A_{245} - A_{146} - A_{346} = 0,
\end{aligned} \tag{98}$$

and relations (97) are satisfied. Equations (97) and (98) form an algebraic system of equations that must be satisfied by the minors of the matrix A to ensure identity (95). Equations (97) depend on which of the minors A_{ijk} is nonzero. If $A_{124} = 1 \neq 0$, then we obtain precisely 18 solutions of system (97) and (98):

$$1. \quad A_{125} = 1, \quad A_{126} = 1, \quad A_{134} = 0,$$

$$A_{135} = 0, \quad A_{136} = 0, \quad A_{145} = C_2,$$

$$A_{146} = C_1, \quad A_{156} = C_1 - C_2, \quad A_{234} = -1, \quad A_{235} = -1,$$

$$A_{236} = -1, \quad A_{246} = C_1,$$

$$A_{245} = C_2, \quad A_{256} = C_1 - C_2, \quad A_{345} = C_2, \quad A_{346} = C_1,$$

$$A_{356} = C_1 - C_2.$$

$$2. \quad A_{125} = -C_1, \quad A_{126} = 1,$$

$$A_{134} = 0,$$

$$A_{135} = 0, \quad A_{136} = 0, \quad A_{145} = C_3,$$

$$A_{146} = 0, \quad A_{156} = -C_3, \quad A_{234} = -1, \quad A_{235} = C_1,$$

$$A_{236} = -1, \quad A_{245} = C_2,$$

$$A_{246} = 0,$$

$$A_{256} = -C_2, \quad A_{345} = C_3, \quad A_{346} = 0, \quad A_{356} = -C_3.$$

3. $A_{125} = -1 + C_1, \quad A_{126} = -C_1, \quad A_{134} = 0, \quad A_{135} = 0, \quad A_{136} = 0,$

$$A_{145} = -C_3/C_1, \quad A_{234} = C_1, \quad A_{146} = C_2, \quad A_{156} = -C_2 + C_1 \cdot C_2 - C_3,$$

$$A_{235} = -C_1 + C_1^2, \quad A_{236} = -C_1^2,$$

$$A_{245} = -(-1 + C_1) \cdot C_3/C_1, \quad A_{246} = -C_2 + C_1 \cdot C_2,$$

$$A_{256} = C_3 - C_3 \cdot C_1 + C_2 - 2 \cdot C_1 \cdot C_2 + C_1^2 \cdot C_2, \quad A_{345} = C_3,$$

$$A_{346} = -C_1 \cdot C_2, \quad A_{356} = C_1 \cdot C_2 - C_1^2 \cdot C_2 + C_3 \cdot C_1.$$

4. $A_{125} = -1, \quad A_{126} = 0,$

$$A_{134} = 0, \quad A_{135} = 0, \quad A_{136} = 0, \quad A_{145} = -C_1,$$

$$A_{146} = -C_2, \quad A_{156} = C_2,$$

$$A_{234} = 0, \quad A_{235} = 0, \quad A_{236} = 0, \quad A_{245} = C_1,$$

$$A_{246} = C_2, \quad A_{256} = -C_2,$$

$$A_{345} = 0, \quad A_{346} = 0, \quad A_{356} = 0.$$

5. $A_{125} = 1, \quad A_{126} = 1,$

$$A_{134} = C_1, \quad A_{135} = C_1, \quad A_{136} = C_1, \quad A_{145} = C_2,$$

$$A_{146} = C_3, \quad A_{156} = -C_2 + C_3,$$

$$A_{234} = -1 + C_1, \quad A_{235} = -1 + C_1, \quad A_{236} = -1 + C_1,$$

$$A_{245} = C_2, \quad A_{246} = C_3, \quad A_{256} = -C_2 + C_3,$$

$$A_{345} = C_2, \quad A_{346} = C_3, \quad A_{356} = -C_2 + C_3.$$

6. $$A_{125} = C_2/C_1, \quad A_{126} = -C_1 + C_2 + 1,$$

$$A_{134} = C_1, \quad A_{135} = C_2,$$

$$A_{136} = C_1 \cdot C_2 - C_1^2 + C_1, \quad A_{145} = C_3/C_1,$$

$$A_{146} = C_3, \quad A_{156} = C_3 \cdot (-1 + C_1)/C_1,$$

$$A_{234} = -1 + C_1, \quad A_{235} = C_2 \cdot (-1 + C_1)/C_1,$$

$$A_{236} = -C_2 + C_1 \cdot C_2 - 1 + 2 \cdot C_1 - C_1^2,$$

$$A_{245} = C_4/C_1, \quad A_{246} = C_4,$$

$$A_{256} = C_4 \cdot (-1 + C_1)/C_1,$$

$$A_{345} = -(-C_3 + C_3 \cdot C_1 - A246 \cdot C_1)/C_1,$$

$$A_{346} = C_3 - C_3 \cdot C_1 + C_4 \cdot C_1,$$

$$A_{356} = -(C_3 - 2 \cdot C_3 \cdot C_1 + C_3 \cdot C_1^2 + C_4 \cdot C_1 - C_4 \cdot C_1^2)/C_1.$$

7. $$A_{125} = (-2 + C_1)/(C_1 + 1), \quad A_{126} = -(-1 + 2 \cdot C_1)/(C_1 + 1),$$

$$A_{134} = C_1,$$

$$A_{135} = C_1 \cdot (-2 + C_1)/(C_1 + 1), \quad A_{136} = -C_1 \cdot (-1 + 2 \cdot C_1)/(C_1 + 1),$$

$$A_{145} = (C_1 + 1) \cdot C_3/(-2 + C_1), \qquad A_{146} = C_2,$$

$$A_{156} = (4 \cdot C_2 - C_3 + C_3 \cdot C_1 - 4 \cdot C_2 \cdot C_1 +$$

$$2 \cdot C_3 \cdot C_1^2 + C_2 \cdot C_1^2)/((C_1 + 1) \cdot (-2 + C_1)),$$

$$A_{234} = -1 + C_1, \quad A_{235} = (2 - 3 \cdot C_1 + C_1^2)/(C_1 + 1),$$

$$A_{236} = -(-1 + C_1) \cdot (-1 + 2 \cdot C_1)/(C_1 + 1),$$

$$A_{245} = C_3, \quad A_{246} = C_2 \cdot (-2 + C_1)/(C_1 + 1),$$

$$A_{256} = (4 \cdot C_2 - C_3 + C_3 \cdot C_1 - 4 \cdot C_2 \cdot C_1 +$$

$$2 \cdot C_3 \cdot C_1^2 + C_2 \cdot C_1^2)/(C_1 + 1)^2,$$

$$A_{345} = -C_3 \cdot (-1 + 2 \cdot C_1)/(-2 + C_1),$$

$$A_{346} = -C_2 \cdot (-1 + 2 \cdot C_1)/(C_2 + 1),$$

$$A_{356} =$$

$$\frac{-(-1 + 2 \cdot C_1) \cdot (4 \cdot C_2 - C_3 + C_3 \cdot C_1 - 4 \cdot C_2 \cdot C_1 + 2 \cdot C_3 \cdot C_1^2 + C_2 \cdot C_1^2)}{(C_1 + 1)^2 \cdot (-2 + C_1)}.$$

8. $A_{125} = -C_1, \quad A_{126} = C_1 - 1, \quad A_{134} = -1,$

$$A_{135} = C_1, \quad A_{136} = 1 - C_1,$$

$$A_{145} = -C_2, \quad A_{146} = C_2, \quad A_{156} = -C_2,$$

$$A_{234} = 1, \quad A_{235} = -C_1, \quad A_{236} = C_1 - 1,$$

$$A_{245} = -C_3, \quad A_{246} = C_3, \quad A_{256} = -C_3, \quad A_{345} = C_2 + C_3,$$

$$A_{346} = -C_3 - C_2, \quad A_{356} = C_2 + C_3.$$

9. $A_{125} = C_2/C_1, \quad A_{126} = -(C_1 + C_2)/C_1, \quad A_{134} = C_1, \quad A_{135} = C_2,$

$$A_{136} = -C_2 - C_1, \quad A_{145} = C_3 \cdot C_1/C_2,$$

$$A_{146} = C_4 \cdot C_1/C_2,$$

$$A_{156} = (C_4 \cdot C_2 + C_3 \cdot C_1 + C_3 \cdot C_2)/C_2,$$

$$A_{234} = (C_1 \cdot C_2 + C_1 + C_2)/C_1,$$

$$A_{235} = (C_1 \cdot C_2 + C_1 + C_2) \cdot C_2/C_1^2,$$

$$A_{236} = -(C_1 \cdot C_2 + C_1 + C_2) \cdot (C_2 + C_1)/C_1^2,$$

$$A_{245} = C_3, \quad A_{246} = C_4,$$

$$A_{256} = (C_4 \cdot C_2 + C_3 \cdot C_1 + C_3 \cdot C_2)/C_1,$$

$$A_{345} = -(C_1 + C_2) \cdot C_3/C_2,$$

$$A_{346} = -C_4 \cdot (C_1 + C_2)/C_2,$$

$$A_{356} = -(C_1 + C_2) \cdot (C_4 \cdot C_2 + C_3 \cdot C_1 + C_3 \cdot C_2)/(C_1 \cdot C_2).$$

10. $$A_{125} = (-1 + C_1)/C_1, \quad A_{126} = 0,$$

$$A_{134} = C_1, \quad A_{135} = -1 + C_1, \quad A_{136} = 0,$$

$$A_{145} = C_2/C_1, \quad A_{146} = C_2,$$

$$A_{156} = C_2 \cdot (-1 + C_1)/C_1, \quad A_{234} = -1 + C_1,$$

$$A_{235} = (1 - 2 \cdot C_1 + C_1^2)/C_1, \quad A_{236} = 0,$$

$$A_{245} = C_3/C_1, \quad A_{246} = C_3,$$

$$A_{256} = C_3 \cdot (-1 + C_1)/C_1,$$

$$A_{345} = -(-C_2 + C_1 \cdot C_2 - C_1 \cdot C_3)/C_1,$$

$$A_{346} = C_2 - C_2 \cdot C_1 + C_3 \cdot C_1,$$

$$A_{356} = -(C_2 - 2 \cdot C_1 \cdot C_2 + C_2 \cdot C_1^2 + C_3 \cdot C_1 - C_3 \cdot C_1^2)/C_1.$$

11. $$A_{125} = C_2/C_1, \quad A_{126} = -C_1 + C_2 + 1,$$

$$A_{134} = C_1, \quad A_{135} = C_2,$$

$$A_{136} = C_1 \cdot C_2 - C_1^2 + C_1, \quad A_{145} = C_3/C_1,$$

$$A_{146} = C_3, \quad A_{156} = C_3 \cdot (-1 + C_1)/C_1,$$

$$A_{234} = -1 + C_1, \quad A_{235} = C_2 \cdot (-1 + C_1)/C_1,$$

$$A_{236} = -C_2 + C_1 \cdot C_2 - 1 + 2 \cdot C_1 - C_1^2,$$

$$A_{245} = 0, \quad A_{246} = 0, \quad A_{256} = 0, \quad A_{345} = -C_3 \cdot (-1 + C_1)/C_1,$$

$$A_{346} = -C_3 \cdot (-1 + C_1), \quad A_{356} = -C_3 \cdot (-1 + C_1)^2/C_1.$$

12. $A_{125} = (-2 + C_1)/C_1, \quad A_{126} = -1,$

$$A_{134} = C_1, \quad A_{135} = -2 + C_1,$$

$$A_{136} = -C_1, \quad A_{145} = C_2/C_1, \quad A_{146} = C_2,$$

$$A_{156} = C_2 \cdot (-1 + C_1)/C_1,$$

$$A_{234} = -1 + C_1,$$

$$A_{235} = (2 - 3 \cdot C_1 + C_1^2)/C_1,$$

$$A_{236} = 1 - C_1, \quad A_{245} = 0,$$

$$A_{246} = 0, \quad A_{256} = 0,$$

$$A_{345} = -C_2 \cdot (-1 + C_1)/C_1,$$

$$A_{346} = -C_2 \cdot (-1 + C_1),$$

$$A_{356} = -C_2 \cdot (-1 + C_1)^2/C_1.$$

13. $A_{125} = 1, \quad A_{126} = 1, \quad A_{134} = C_1,$

$$A_{135} = C_1, \quad A_{136} = C_1, \quad A_{145} = C_2,$$

$$A_{146} = 0, \quad A_{156} = -C_2, \quad A_{234} = -1 + C_1,$$

$$A_{235} = -1 + C_1, \quad A_{236} = -1 + C_1,$$

$$A_{245} = C_2, \quad A_{246} = 0, \quad A_{256} = -C_2,$$

$$A_{345} = C_2, \quad A_{346} = 0, \quad A_{356} = -C_2.$$

$$14.\quad A_{125} = (-2 + C_1)/(C_1 + 1),$$

$$A_{126} = -(-1 + 2 \cdot C_1)/(C_1 + 1), \quad A_{134} = C_1,$$

$$A_{135} = C_1 \cdot (-2 + C_1)/(C_1 + 1),$$

$$A_{136} = -C_1 \cdot (-1 + 2 \cdot C_1)/(C_1 + 1),$$

$$A_{145} = -C_2 \cdot (C_1 + 1)/(-1 + 2 \cdot C_1),$$

$$A_{146} = 0, \quad A_{156} = -C_2, \quad A_{234} = -1 + C_1,$$

$$A_{235} = (2 - 3 \cdot C_1 + C_1^2)/(C_1 + 1),$$

$$A_{236} = -(-1 + C_1) \cdot (-1 + 2 \cdot C_1)/(C_1 + 1),$$

$$A_{245} = -C_2 \cdot (-2 + C_1)/(-1 + 2 \cdot C_1),$$

$$A_{246} = 0, \quad A_{256} = -C_2 \cdot (-2 + C_1)/(C_1 + 1),$$

$$A_{345} = C_2, \quad A_{346} = 0,$$

$$A_{356} = C_2 \cdot (-1 + 2 \cdot C_1)/(C_1 + 1).$$

$$15.\quad A_{125} = 0, \quad A_{126} = -1,$$

$$A_{134} = C_1, \quad A_{135} = 0,$$

$$A_{136} = -C_1, \quad A_{145} = -C_3,$$

$$A_{146} = C_2, \quad A_{156} = -C_3,$$

$$A_{234} = 1, \quad A_{235} = 0, \quad A_{236} = -1, \quad A_{245} = 0,$$

$$A_{246} = 0, \quad A_{256} = 0, \quad A_{345} = C_3,$$

$$A_{346} = -C_2, \quad A_{356} = C_3.$$

$$16.\quad A_{125} = 0, \quad A_{126} = -1, \quad A_{134} = -1,$$

$$A_{135} = 0, \quad A_{136} = 1, \quad A_{145} = C_1 - C_2,$$

$$A_{146} = -C_1 + C_2, \quad A_{156} = C_1 - C_2, \quad A_{234} = 1,$$

$$A_{235} = 0, \quad A_{236} = -1,$$

$$A_{245} = -C_1, \quad A_{246} = C_1, \quad A_{256} = -C_1,$$

$$A_{345} = C_2, \quad A_{346} = -C_2, \quad A_{356} = C_2.$$

17. $A_{125} = -2, \quad A_{126} = 1, \quad A_{134} = -1,$

$$A_{135} = 2, \quad A_{136} = -1, \quad A_{145} = C_1 - C_2,$$

$$A_{146} = -C_1 + C_2, \quad A_{156} = C_1 - C_2,$$

$$A_{234} = 1, \quad A_{235} = -2, \quad A_{236} = 1,$$

$$A_{245} = -C_1, \quad A_{246} = C_1,$$

$$A_{256} = -C_1,$$

$$A_{345} = C_2, \quad A_{346} = -C_2, \quad A_{356} = C_2.$$

18. $A_{125} = -2, \quad A_{126} = 1, \quad A_{134} = C_1,$

$$A_{135} = -2 \cdot C_1, \quad A_{136} = C_1, \quad A_{145} = C_3,$$

$$A_{146} = -(1/2) \cdot C_2, \quad A_{156} = C_2 - C_3,$$

$$A_{234} = -2 \cdot C_1 - 1, \quad A_{235} = 2 + 4 \cdot C_1,$$

$$A_{236} = -2 \cdot C_1 - 1, \quad A_{245} = -2 \cdot C_3,$$

$$A_{246} = C_2, \quad A_{256} = 2 \cdot C_3 - 2 \cdot C_2,$$

$$A_{345} = C_3, \quad A_{346} = -(1/2) \cdot C_2, \quad A_{356} = C_2 - C_3.$$

Here C_1, C_2, C_3, C_4 are arbitrary constants. If the solution components 1–18 are known, then one can readily calculate the coefficient matrix of the boundary conditions and the boundary conditions themselves (e.g., see [18]). We show this for the components of solution 2 (which contains three arbitrary constants). If

$\mathbf{x} = (x_1, x_2, x_3, x_4, x_5, x_6)$ is an arbitrary row of the desired matrix A, then it must satisfy condition (80); i.e.,

$$\operatorname{rank} \begin{Vmatrix} a_{11} & a_{12} & a_{13} & a_{14} & a_{15} & a_{16} \\ a_{21} & a_{22} & a_{23} & a_{24} & a_{25} & a_{26} \\ a_{31} & a_{32} & a_{33} & a_{34} & a_{35} & a_{36} \\ x_1 & x_2 & x_3 & x_4 & x_5 & x_6 \end{Vmatrix} = 3. \tag{99}$$

Since the minor A_{124} is nonzero, it follows that the fourth-order minors bordering it must be zero. This implies the equations

$$\begin{vmatrix} a_{11} & a_{12} & a_{13} & a_{14} \\ a_{21} & a_{22} & a_{23} & a_{24} \\ a_{31} & a_{32} & a_{33} & a_{34} \\ x_1 & x_2 & x_3 & x_4 \end{vmatrix} = -x_1 \cdot A_{234} + x_2 \cdot A_{134} - x_3 \cdot A_{124} + x_4 \cdot A_{123} = 0, \tag{100}$$

$$\begin{vmatrix} a_{11} & a_{12} & a_{14} & a_{15} \\ a_{21} & a_{22} & a_{24} & a_{25} \\ a_{31} & a_{32} & a_{34} & a_{35} \\ x_1 & x_2 & x_4 & x_5 \end{vmatrix} = -x_1 \cdot A_{245} + x_2 \cdot A_{145} - x_4 \cdot A_{125} + x_5 \cdot A_{124} = 0, \tag{101}$$

$$\begin{vmatrix} a_{11} & a_{12} & a_{14} & a_{16} \\ a_{21} & a_{22} & a_{24} & a_{26} \\ a_{31} & a_{32} & a_{34} & a_{36} \\ x_1 & x_2 & x_4 & x_6 \end{vmatrix} = -x_1 \cdot A_{246} + x_2 \cdot A_{146} - x_4 \cdot A_{126} + x_6 \cdot A_{124} = 0. \tag{102}$$

Since $A_{124} \neq 0$, we can assume that $A_{124} = 1$. Moreover, $A_{123} = 0$. By substituting these values of the minors and the solution 2 into Eqs. (100)–(102), we obtain the system of equations

$$\begin{aligned} x_1 - x_3 &= 0, \\ -x_1 C_2 + x_2 \cdot C_3 + x_4 \cdot C_1 + x_5 &= 0, \\ -x_4 + x_6 &= 0. \end{aligned} \tag{103}$$

Taking into account relations (103) and the fact that $A_{124} = 1$, for linearly independent rows we can take the rows with elements

$$\begin{aligned} &x_1 = 1, \quad x_2 = 0, \quad x_4 = 0, \quad x_3 = 1, \quad x_5 = C_2, \quad x_6 = 0, \\ &x_1 = 0, \quad x_2 = 1, \quad x_4 = 0, \quad x_3 = 0, \quad x_5 = -C_3, \quad x_6 = 0, \\ &x_1 = 0, \quad x_2 = 0, \quad x_4 = 1, \quad x_3 = 0, \quad x_5 = -C_1, \quad x_6 = 1. \end{aligned} \tag{104}$$

Therefore, the desired degenerate boundary conditions have the form

$$\begin{aligned} y_1(l) + y_3(l) + C_2\, y_2'(l) &= 0, \\ y_2(l) - C_3\, y_2'(l) &= 0, \\ y_1'(l) - C_1\, y_2'(l) + y_3'(l) &= 0. \end{aligned} \tag{105}$$

By substituting the other components of the solutions into Eqs. (100)–(102), we obtain 17 additional classes of degenerate boundary conditions,

$$\begin{aligned} y_1(l) + y_3(l) + C_2\, y_2'(l) + C_1\, y_3'(l) &= 0, \\ y_2(l) - C_2\, y_2'(l) - C_1\, y_3'(l) &= 0, \\ y_1'(l) + y_2'(l) + y_3'(l) &= 0. \end{aligned} \tag{106}$$

$$\begin{aligned} y_1(l) - C_1\, y_3(l) - (-1 + C_1)\frac{C_3}{C_1}\, y_2'(l) + (C_1\, C_2 - C_2)\, y_3'(l) &= 0, \\ y_2(l) - \frac{C_3}{C_1}\, y_2'(l) - C_2\, y_3'(l) &= 0, \\ y_1'(l) + (-1 + C_1)\, y_2'(l) - C_1\, y_3'(l) &= 0. \end{aligned} \tag{107}$$

$$\begin{aligned} y_1(l) + C_1\, y_2'(l) + C_2\, y_3'(l) &= 0, \\ y_2(l) + C_1\, y_2'(l) + C_2\, y_3'(l) &= 0, \\ y_1'(l) - y_2'(l) &= 0. \end{aligned} \tag{108}$$

$$\begin{aligned} y_1(l) + (C_1 - 1)\, y_3(l) + C_2\, y_2'(l) + C_3\, y_3'(l) &= 0, \\ y_2(l) + C_1\, y_3(l) - C_2\, y_2'(l) - C_3\, y_3'(l) &= 0, \\ y_1'(l) + y_2'(l) + y_3'(l) &= 0. \end{aligned} \tag{109}$$

$$\begin{aligned} y_1(l) + (1 - C_1)\, y_3(l) + \frac{C_4}{C_1}\, y_2'(l) + C_4\, y_3'(l) &= 0, \\ y_2(l) + C_1\, y_3(l) - \frac{C_3}{C_1}\, y_2'(l) - C_3\, y_3'(l) &= 0, \\ y_1'(l) + \frac{C_2}{C_1}\, y_2'(l) + (1 + C_2 - C_1)\, y_3'(l) &= 0. \end{aligned} \tag{110}$$

$$y_1(l) + (1 - C_1)\, y_3(l) + C_3\, y_2'(l) + \frac{(C_1 - 2)\, C_2}{C_1 + 1}\, y_3'(l) = 0,$$
$$y_2(l) + C_1\, y_3(l) - \frac{(C_1 + 1)\, C_3}{C_1 - 2}\, y_2'(l) - C_2\, y_3'(l) = 0, \quad (111)$$
$$y_1'(l) + \frac{C_1 - 2}{C_1 + 1}\, y_2'(l) + \frac{1 - C_1}{C_1 + 1}\, y_3'(l) = 0.$$

$$y_1(l) - y_3(l) - C_3\, y_2'(l) + C_3\, y_3'(l) = 0,$$
$$y_2(l) - y_3(l) + C_2\, y_2'(l) - C_2\, y_3'(l) = 0, \quad (112)$$
$$y_1'(l) - C_1\, y_2'(l) + (C_1 - 1)\, y_3'(l) = 0.$$

$$y_1(l) - (C_1\, C_2 + C_1 + C_2)\, y_3(l) + C_3\, y_2'(l) + C_4\, y_3'(l) = 0,$$
$$y_2(l) + C_1\, y_3(l) - \frac{C_1\, C_3}{C_2}\, y_2'(l) - \frac{C_1\, C_4}{C_2}\, y_3'(l) = 0, \quad (113)$$
$$y_1'(l) + \frac{C_2}{C_1}\, y_2'(l) - \frac{C_1 + C_2}{C_1}\, y_3'(l) = 0.$$

$$y_1(l) + (1 - C_1)\, y_3(l) + \frac{C_3}{C_1}\, y_2'(l) + C_3\, y_3'(l) = 0,$$
$$y_2(l) + C_1\, y_3(l) - \frac{C_2}{C_1}\, y_2'(l) - C_2\, y_3'(l) = 0, \quad (114)$$
$$y_1'(l) + \frac{C_1 - 1}{C_1}\, y_2'(l) = 0.$$

$$y_1(l) + (C_1 - 1)\, y_3(l) = 0,$$
$$y_2(l) + C_1\, y_3(l) - \frac{C_3}{C_1}\, y_2'(l) - C_3\, y_3'(l) = 0, \quad (115)$$
$$y_1'(l) + \frac{C_2}{C_1}\, y_2'(l) + (1 + C_2 - C_1)\, y_3'(l) = 0.$$

$$y_1(l) + (C_1 - 1)\, y_3(l) = 0,$$
$$y_2(l) - C_1\, y_3(l) - \frac{C_2}{C_1}\, y_2'(l) - C_2\, y_3'(l) = 0, \quad (116)$$
$$y_1'(l) + \frac{C_1 - 2}{C_1}\, y_2'(l) - y_3'(l) = 0.$$

$$
\begin{aligned}
y_1(l) + (1 - C_1)\, y_3(l) + C_2\, y_2'(l) &= 0,\\
y_2(l) + C_1\, y_3(l) - C_2\, y_2'(l) &= 0,\\
y_1'(l) + y_2'(l) + y_3'(l) &= 0.
\end{aligned}
\tag{117}
$$

$$
\begin{aligned}
y_1(l) + (C_1 - 1)\, y_3(l) - \frac{(C_1 - 2)\, C_2}{2\, C_1 - 1}\, y_2'(l) &= 0,\\
y_2(l) + C_1\, y_3(l) + \frac{(C_1 + 1)\, C_2}{2\, C_1 - 1}\, y_2'(l) &= 0,\\
y_1'(l) + \frac{C_1 - 2}{C_1 + 1}\, y_2'(l) + \frac{1 - 2\, C_1}{C_1 + 1}\, y_3'(l) &= 0.
\end{aligned}
\tag{118}
$$

$$
\begin{aligned}
y_1(l) - y_3(l) &= 0,\\
y_2(l) + C_1\, y_3(l) + C_3\, y_2'(l) - C_2\, y_3'(l) &= 0,\\
y_1'(l) - y_3'(l) &= 0.
\end{aligned}
\tag{119}
$$

$$
\begin{aligned}
y_1(l) - y_3(l) - C_1\, y_2'(l) + C_1\, y_3'(l) &= 0,\\
y_2(l) - y_3(l) + (C_2 - C_1)\, y_2'(l) + (C_1 - C_2)\, y_3'(l) &= 0,\\
y_1'(l) - y_3'(l) &= 0.
\end{aligned}
\tag{120}
$$

$$
\begin{aligned}
y_1(l) - y_3(l) - C_1\, y_2'(l) + C_1\, y_3'(l) &= 0,\\
y_2(l) - y_3(l) + (C_2 - C_1)\, y_2'(l) + (C_1 - C_2)\, y_3'(l) &= 0,\\
y_1'(l) - 2\, y_2'(l) + y_3'(l) &= 0.
\end{aligned}
\tag{121}
$$

$$
\begin{aligned}
y_1(l) + (2\, C_1 + 1)\, y_3(l) - 2\, C_3\, y_2'(l) + C_2\, y_3'(l) &= 0,\\
y_2(l) + C_1\, y_3(l) - C_3\, y_2'(l) + \frac{1}{2}\, C_2\, y_3'(l) &= 0,\\
y_1'(l) - 2\, y_2'(l) + y_3'(l) &= 0.
\end{aligned}
\tag{122}
$$

Thus, the following assertion holds.

Theorem 7.2 *If $l_i = l$ $(i = 1, 2, 3)$ and $q_i(x) = q(x)$, then identity (94) is impossible for the boundary value problem (76)–(78) and identity (95) is possible for infinitely many problems (76)–(78). In particular, for $A_{124} = 1 \neq 0$ the set of degenerate boundary conditions on the star-shaped graph consists not of two problems as in the case of the Sturm–Liouville problem posed on an interval but of the eighteen classes (105)–(122), each of which contains from two to four arbitrary constants.*

8 Finiteness of the Spectrum of Boundary Value Problems

We consider boundary value problems with spectral parameter polynomially occurring in the differential equation or the boundary conditions. It is shown that some of these problems have a prescribed finite spectrum. A wide class of boundary value problems which do not have finite spectrum is found.

For the differential equation

$$y^{(n)}+a_1(x,\lambda)\,y^{n-1}+\cdots+a_{n-1}(x,\lambda)\,y'+a_n(x,\lambda)\,y=0, \qquad x\in[0,1] \tag{123}$$

consider the boundary value problem

$$U_j(y)=\sum_{k=0}^{n-1} b_{jk}\,y^{(k)}(0)+\sum_{k=0}^{n-1} b_{j\,k+n}\;y^{(k)}(1)=0,$$
$$j=1,2,\ldots,n, \tag{124}$$

where rank $||b_{jk}||_{n\times 2n}=n$, the functions $a_q(x,\lambda)$ $(q=1,\ldots,n)$ are continuous in x on the interval [0,1] and polynomial in the parameter λ, $b_{jk}\in\mathbb{C}$, and the coefficients b_{jk} are complex. It is well known [16, p. 27] that the following two situations are only possible for the spectrum of problem (123) and (124): (1) there exists at most countably many eigenvalues, and they have no limit points in $\mathbb{C}$; (2) every $\lambda\in\mathbb{C}$ is an eigenvalue.

The direct and inverse problems for the case in which the spectrum consists of infinitely many eigenvalues have been studied sufficiently well [2, 3], but the case of finite spectrum of problem (123) and (124) has not been studied well enough.

It was shown in [14, p. 556] and [6] that the differentiation operators D^2 and D^4 with the corresponding boundary conditions (124) cannot have a finite spectrum. In 2008, Locker [6] posed the following question for the equations:

$$y^{(n)}+a_1(x)\,y^{n-1}+\cdots+a_{n-1}(x)\,y'+a_n(x)\,y=\lambda\,y(x), \qquad x\in[0,1]. \tag{125}$$

Can the boundary value problem (125) and (124) have finite spectrum? In the same year, Kalmenov and Suragan [15] proved that the spectrum of regular partial differential boundary value problems including problems (125) and (124) is either empty or infinite. The following assertion shows that this result also holds for one general class of problems (123) and (124).

Theorem 8.1 *If the functions $a_q(x,\lambda)$ have the form*

$$a_q(x,\lambda)=\lambda^q\sum\nolimits_{j=0}^{q}\lambda^{-j}\,a_{qj}(x),\quad a_{q0}(x)=a_{q0}\cdot r(x),$$
$$r(x)>0,\quad q=1,2,\ldots,n,$$

and the polynomial $\pi(\lambda) = \lambda^n + a_{10}\,\lambda^{n-1} + \cdots + a_{q0}$ *does not have multiple roots, then the spectrum of boundary value problem (123) and (124) is either empty or infinite.*

The proof of Theorem 8.1 follows from the results obtained by Lidskii and Sadovnichii [23, 24], who showed that the characteristic determinant $\Delta(\lambda)$ of problem (123) and (124) satisfying the assumptions of Theorem 8.1 is an entire function of class K and the number of roots (if any) of this function is infinite. (Their asymptotic representations are given in [23, 24] as well.). Now assume that the polynomial $\pi(\lambda)$ has multiple roots. The following question arises: Can the boundary value problem (123) and (124) have finite spectrum in this case? It is shown below that the answer is yes. Moreover, we prove that there exist boundary value problems (123) and (124) with any prescribed finite spectrum.

Theorem 8.2 *Let* $\lambda_1, \lambda_2, \ldots, \lambda_n$ *be given complex numbers. There exists a boundary value problem (123) and (124), whose spectrum consists precisely of the numbers* $\lambda_1, \lambda_2, \ldots, \lambda_n$.

Proof We denote the product $(\lambda - \lambda_1) \cdot \ldots \cdot (\lambda - \lambda_n)$ by $p(\lambda)$, $p(\lambda) - 1$ by d, and for the differential equation

$$y'' - 2\,d(\lambda)\,y' + d^2(\lambda)\,y = 0, \tag{126}$$

we consider the boundary value problem

$$U_1(y) = y(0) = 0, \quad U_2(y) = y'(1) = 0. \tag{127}$$

Equation (126) has the characteristic equation $(\omega - d)^2 = 0$. Therefore, if $d \neq 0$, then the following functions are linearly independent solutions of Eq. (126): $y_1 = \exp(d\,x)$, $y_2 = x\,\exp(d\,x)$. Then the characteristic determinant $\Delta(\lambda)$ of problem (126) and (127) is

$$\Delta(\lambda) = \begin{vmatrix} y_1(0) & y_2(0) \\ y_1'(1) & y_2'(1) \end{vmatrix} = \begin{vmatrix} 1 & 0 \\ d\,e^d & (1+d)\,e^d \end{vmatrix}$$
$$= (1+d)\,e^d = p(\lambda)\,e^{d(\lambda)}.$$

Thus, if $p(\lambda) - 1 \neq 0$, $\Delta(\lambda) = p(\lambda)\,e^{d(\lambda)}$. If $p(\lambda) - 1 = 0$, then, substituting the linearly independent solutions $y_1 = 1$, $y_2 = x$ of Eq. (126) into the characteristic determinant, we obtain $\Delta(\lambda) \neq 0$. Therefore, the roots of the characteristic determinant $\Delta(\lambda)$ are precisely the roots of the polynomial $p(\lambda)$. The proof of the theorem is complete. □

It follows from Theorem 8.2 that, varying the polynomial $p(\lambda)$ one can ensure that the corresponding boundary value problem has a prescribed finite spectrum. The following question arises: If the polynomial $p(\lambda)$ occurs in the boundary conditions (124) rather than in the differential equation, can one ensure that the boundary value problem has a prescribed finite spectrum by varying $p(\lambda)$? It was shown that if the characteristic determinant of the boundary value problem is not identically zero, then, varying one of the boundary conditions of the problem, one can ensure that the spectral problem has a prescribed (and finite) spectrum. But the boundary condition considered can contain a function that is not a polynomial in general. Our question is different: Can the boundary value problem (125) and (124), where the coefficients b_{jk} in the boundary conditions are polynomials of the spectral parameter λ, have a prescribed finite spectrum?

Theorem 8.3 *Let $\lambda_1, \lambda_2, \ldots, \lambda_n$ be given complex numbers. There exists a problem (125) and (124), that the coefficients b_{jk} in the boundary conditions are polynomials of the spectral parameter λ and the spectrum of the problem consists precisely of the numbers $\lambda_1, \lambda_2, \ldots, \lambda_n$.*

Proof By $z_j(x)$ we denote the linearly independent solutions of Eq. (125) satisfying the conditions $z_i^{(j-1)}(x) = \delta_{ij}$ $(i, j = 1, 2, \ldots, n)$, where δ_{ij} is the Kronecker delta. Assume that the boundary conditions (124) have the form

$$U_1(y) = p(\lambda)\, y(0) = 0, \quad U_k(y) = y^{(k-1)}(0) = 0, \qquad k = 2, 3, \ldots, n. \tag{128}$$

Then for the characteristic determinant of problem (125) and (128) we have

$$\Delta(\lambda) = \begin{vmatrix} p(\lambda) & 0 & \ldots & 0 \\ 0 & 1 & \ldots & 0 \\ \vdots & \vdots & \ddots & \vdots \\ 0 & 0 & \ldots & 1 \end{vmatrix} = p(\lambda).$$

The proof of the theorem is complete. □

References

1. V. A. Marchenko *Sturm-Liouville operators and their applications*, Kiev, Naukova Dumka, (1977), 332 p. (in Russian)
2. A. A. Shkalikov, *Boundary problems for ordinary differential equations with parameter in the boundary conditions*, Journal of Soviet Mathematics, **33 (6)** (1986), 1311–1342.
3. V. A. Sadovnichii, Ya.T. Sultanaev, A. M. Akhtyamov, *General Inverse Sturm-Liouville Problem with Symmetric Potential*, Azerbaijan Journal of Mathematics, **5 (2)** (2015), 96–108.
4. M. H. Stone *Irregular differential systems of order two and the related expansion problems*, Trans. Amer. Math. Soc. Vol. 29, no 1 (1927), 23–53.
5. V. A. Sadovnichy, B. E. Kanguzhin, *On the connection between the spectrum of a differential operator with symmetric coefficients and boundary conditions*, Dokl. Akad. Nauk SSSR. **267 (2)** (1982), 310–313 (in Russian).
6. *Locker J. Eigenvalues and completeness for regular and simply irregular two-point differential operators*, Providence, American Mathematical Society (2008), vii, 177 p. (Memoirs of the American Mathematical Society; Vol.195, N 911).

7. A. Makin *Two-point boundary-value problems with nonclassical asymptotics on the spectrum*, Electronic Journal of Differential Equations, No. 95 (2018), 1–7.
8. A. M. Akhtyamov, *On spectrum for differential operator of odd order*, Mathematical Notes, **101 (5)** (2017), 755–758.
9. N. Dunford, and J. T. Schwartz, Linear Operators, Part III: Spectral Operators, Wiley-Interscience, New York (1971).
10. A. A. Dezin, *Spectral characteristics of general boundary-value problems for operator D2*, Mathematical notes of the Academy of Sciences of the USSR, Vol. 37, 142–146 (1985)
11. B. N. Biyarov, S. A. Dzhumabaev, *A criterion for the Volterra property of boundary value problems for Sturm-Liouville equations*, Mathematical Notes, **56 (1)** (1994), 751–753.
12. A. M. Akhtyamov *On Degenerate Boundary Conditions in the Sturm-Liouville Problem* , Differential Equations, Vol.52, No. 8, pp. 1085–1087 (2016)
13. A. M. Akhtyamov, *Degenerate Boundary Conditions -for the Diffusion Operator*, Differential Equations, Vol. 53, No.11 (2017),1515–1518.
14. P. Lang, and J.'Locker, *Spectral theory of two-point differential operators determined by D^2. I. Spectral properties*, Journal of Mathematical Analysis and Applications, Vol. 141 J.:, 538–558.
15. T. Sh. Kal'menov and D. Suragan, *Determination of the Structure of the Spectrum of Regular Boundary Value Problems for Differential Equations by V.A. Il'in's Method of Priori Estimates*, Doklady Mathematics, Vol. 78, No. 3 92008), 913–915.
16. Naimark M.A.: *Linear Differential Operators*, Nauka, Moscow (1969) (in Russian)
17. V. A. Yurko, *The inverse problem for differential operators of second order with regular boundary conditions*, Math. Notes, 18:4 (1975), 928–932
18. *A. Akhtyamov, M. Amram, A. Mouftakhov*, *On reconstruction of a matrix by its minors*, International Journal of Mathematical Education in Science and Technology, Vol. 49, No. 2 (2018), 268–321.
19. G.'Sh. Guseinov, *Inverse spectral problems for a quadratic pencil of Sturm-Liouville operators on a finite interval*, Spektralaya teoriya operatorov i ee prilozheniya (Spectral Theory of Operators and Its Applications), Baku, 1986, no. 7, pp. 51–101.
20. G.'Sh. Guseinov, and I. M. Nabiev, *The inverse spectral problem for pencils of differential operators*, Sb. Math., 2007, vol. 198, no. 11, pp. 1579–1598.
21. I. M. Nabiev, A. Sh. Shukurov, *Solution of inverse problems for the diffusion operator in the symmetric case*, Izv. Sarat. un-that. New ser. Ser. Maths. Mechanics. Informatics, vol. 9:4 (1) (2009), 36–40 (in Russian)
22. E. Kamke, *Reference Book in Ordinary Differential Equations*, New York, Van Nostrand (19600
23. V. B. Lidskii, V.'A. Sadovnichii, *Regularized sums of roots of a class of entire functions*, Funct. Anal. Appl., vol. 1, no. 2 (1967), 133–139.
24. V. B. Lidskii, V.'A. Sadovnichii, *Asymptotic formulas for the roots of a class of entire functions*, Math. USSR Sb., vol. 4, no. 4 (1968), 519–527.
25. G. Sansone; *Sopra una famiglia di cubiche con infiniti punti razionali,* (Italian) Rendiconti Istituto Lombardo, **62** (1929), 354–360.
26. A. G. Kurosh, *A Course in Higher Algebra* Moscow, Nauka (1963), (in Russian).
27. P. Lancaster, *Theory of Matrices*, Academic Press, NY (1969)

Deceptive Systems of Differential Equations

Martin Schechter

Abstract We study nonlinear steady state Schrödinger systems of equations arising in the study of photonic lattices.

1 Introduction

We study a system of equations arising in optics describing the propagation of a light wave in induced photonic lattices. This leads to the following system of equations over a periodic domain $\Omega \subset \mathbb{R}^2$:

$$\Delta v = \frac{Pv}{1 + v^2 + w^2} + \lambda v, \tag{1}$$

$$\Delta w = \frac{Qw}{1 + v^2 + w^2} + \lambda w, \tag{2}$$

where P, Q, λ are parameters. The solutions v, w are to be periodic in Ω with the same periods. One wishes to obtain intervals of the parameter λ for which there are nontrivial solutions. This will provide continuous energy spectrum that allows the existence of steady state solutions. This system was studied in [2], where it was shown that

1. If P, Q, λ are all positive, then the only solution is trivial.
2. If $P < 0$ and $0 < \lambda < -P$, then the system (1) and (2) has a nontrivial solution.
3. If $P, Q > 0$, there is a constant $\delta > 0$ such that the system (1) and (2) has a nontrivial solution provided $0 < -\lambda < \delta$.
4. All of these statements are true if we replace P by Q.

M. Schechter (✉)
Department of Mathematics, University of California, Irvine, CA, USA
e-mail: mschecht@math.uci.edu

I. N. Parasidis et al. (eds.), *Mathematical Analysis in Interdisciplinary Research*, Springer Optimization and Its Applications 179,
https://doi.org/10.1007/978-3-030-84721-0_32

Wave propagation in nonlinear periodic lattices has been studied by many researchers (cf., e.g., [1–10, 18–21] and their bibliographies.)

The system (1) and (2) has some interesting properties. If one has solved this system, it can very well be that all that one has solved is the system

$$\Delta v = \frac{Pv}{1+v^2} + \lambda v, \tag{3}$$

$$\Delta w = \frac{Qw}{1+w^2} + \lambda w, \tag{4}$$

even if we are guaranteed that we have a nontrivial solution. As noted in [2], to prove the existence of a nontrivial solution of system (1) and (2), it suffices to obtain a nontrivial solution of either

$$\Delta v = \frac{Pv}{1+v^2} + \lambda v, \tag{5}$$

or

$$\Delta w = \frac{Qw}{1+w^2} + \lambda w. \tag{6}$$

This stems from the fact that $(v, 0)$ is a solution of (1),(2) if v is a solution of (5) and $(0, w)$ is a solution of (1),(2) if w is a solution of (6). As a consequence, when one has solved system (1) and (2), it is not clear whether one has solved system (1) and (2) or system (5) and (6). For this reason we have called the system (1) and (2) "deceptive." This raises the question whether there are values of P, Q, λ for which the system (1) and (2) has a solution (v, w) where $v \neq 0$, $w \neq 0$.

It is clear that any set of hypotheses that does not involve both P and Q can solve only (5) and (6). If it involves only P, then it can solve only (5) with a similar statement for Q. Even if the hypotheses involve both P and Q, will the method of solution fail unless both of them satisfy the hypotheses? Therefore, we shall only consider theorems in which the hypotheses involve all three parameters in such a way that the method of solution fails otherwise.

We consider the following situation. Let Ω be a bounded periodic domain in $\mathbb{R}^n$, $n \geq 1$. Consider the operator $-\Delta$ on functions in $L^2(\Omega)$ having the same periods as Ω. The spectrum of $-\Delta$ consists of isolated eigenvalues of finite multiplicity:

$$0 = \lambda_0 < \lambda_1 < \cdots < \lambda_\ell < \cdots,$$

with eigenfunctions in $L^\infty(\Omega)$. Let λ_ℓ, $\ell \geq 0$, be one of these eigenvalues, and define

$$N = \bigoplus_{\lambda \leq \lambda_\ell} E(\lambda), \quad M = N^\perp.$$

We shall prove

Theorem 1 *If* $0 < P \leq \sigma = -\lambda < \min[P + \lambda_1, Q]$, *then (1) and (2) has a nontrivial solution.*

Theorem 2 *If* $0 < Q \leq \sigma = -\lambda < \min[Q + \lambda_1, P]$, *then (1) and (2) has a nontrivial solution.*

Theorem 3 *If* $0 < P + \lambda_\ell \leq \sigma = -\lambda < \min[P + \lambda_{\ell+1}, Q]$, *then (1) and (2) has a nontrivial solution.*

Theorem 4 *If* $0 < Q + \lambda_\ell \leq \sigma = -\lambda < \min[Q + \lambda_{\ell+1}, P]$, *then (1) and (2) has a nontrivial solution.*

Proofs of these theorems will be given after a series of lemmas.

2 Some Lemmas

In proving our results we shall make use of the following lemmas (cf., e.g., [11, 14, 17]). For the definition of linking, cf. [11].

Lemma 1 *Let* M, N *be closed subspaces such that* $\dim N < \infty$ *and* $E = M \oplus N$. *Let* $w_0 \neq 0$ *be an element of* M, *and take*

$$A = \{v \in N : \|v\| \leq R\} \cup$$

$$\{sw_0 + v : v \in N, s \geq 0, \|sw_0 + v\| = R\},$$

$$B = \partial \mathscr{B}_\delta \cap M, \ 0 < \delta < R.$$

Then A *and* B *link each other.*

Lemma 2 *If* A *links* B, *and* $G(u) \in C^1(E, \mathbb{R})$ *satisfies*

$$a_0 = \sup_A G \leq b_0 = \inf_B G, \tag{7}$$

then there is a sequence $\{u_k\}$ *such that*

$$G(u_k) \to c \geq b_0, \quad (1 + \|u_k\|_E)\|G'(u_k)\| \to 0. \tag{8}$$

We let E be the subspace of $H^{1,2}(\Omega)$ consisting of those functions having the same periodicity as Ω with norm given by

$$\|w\|_E^2 = \|\nabla w\|^2 + \|w\|^2.$$

Assume $P \neq 0$, $Q \neq 0$, $\lambda \neq 0$. Let

$$a(u) = \frac{1}{P}[\ \|\nabla v\|^2 + \lambda\ \|v\|^2] + \frac{1}{Q}[\ \|\nabla w\|^2 + \lambda\ \|w\|^2], \quad v, w \in E \tag{9}$$

and

$$G(u) = a(u) + \int_\Omega \ln(1 + u^2)\, dx. \tag{10}$$

We have

Lemma 3 *If G(u) is given by (10), then every sequence satisfying (8) has a subsequence converging in E. Consequently, there is a $u \in E$ such that G(u)=c and $G'(u) = 0$.*

Proof The sequence satisfies

$$G(u_k) = \frac{1}{P}\|\nabla v_k\|^2 + \frac{\lambda}{P}\|v_k\|^2 + \frac{1}{Q}\|\nabla w_k\|^2 + \frac{\lambda}{Q}\|w_k\|^2 \tag{11}$$
$$+ \int_\Omega \ln\{1 + |u_k|^2\}\, dx \to c,$$

$$(G'(u_k), q)/2 = \frac{1}{P}(\nabla v_k, \nabla g) + \frac{\lambda}{P}(v_k, g) \tag{12}$$
$$+ \frac{1}{Q}(\nabla w_k, \nabla h) + \frac{\lambda}{Q}(w_k, h)$$
$$+ \int_\Omega \frac{u_k q}{1 + u_k^2}\, dx \to 0, \quad q = (g, h),$$

$$(G'(u_k), v_k)/2 = \frac{1}{P}(\nabla v_k, \nabla v_k) + \frac{\lambda}{P}(v_k, v_k) \tag{13}$$
$$+ \int_\Omega \frac{u_k v_k}{1 + u_k^2}\, dx \to 0.$$

and

$$(G'(u_k), w_k)/2 = \frac{1}{Q}(\nabla w_k, \nabla w_k) + \frac{\lambda}{Q}(w_k, w_k) \tag{14}$$
$$+ \int_\Omega \frac{u_k w_k}{1 + u_k^2}\, dx \to 0.$$

Thus,

$$\int_{\Omega} H(x, u_k)\, dx \to c, \tag{15}$$

where

$$H(x,t) = \ln(1+t^2) - \frac{t^2}{1+t^2}. \tag{16}$$

Let $\rho_k = \|u_k\|_H$, where

$$\begin{aligned} \|u\|_H^2 = & \frac{1}{|P|}[\|\nabla v\|^2 + |\lambda| \ \|v\|^2] \\ & + \frac{1}{|Q|}[\|\nabla w\|^2 + |\lambda| \ \|w\|^2], \quad u = (v, w) \in E. \end{aligned} \tag{17}$$

Assume first that $\rho_k \to \infty$. Let $\tilde{u}_k = u_k/\rho_k$. Then $\|\tilde{u}_k\|_H = 1$. Hence, there is a renamed subsequence such that $\tilde{u}_k \rightharpoonup \tilde{u}$ in E, and $\tilde{u}_k \to \tilde{u}$ in $L^2(\Omega)$ and a.e. Now

$$\|u_k\|_H^2 = \frac{1}{|P|}[\|\nabla v_k\|^2 + |\lambda| \ \|v_k\|^2] + \frac{1}{|Q|}[\|\nabla w_k\|^2 + |\lambda| \ \|w_k\|^2]. \tag{18}$$

By (13) and (14),

$$\begin{aligned} \|u_k\|_H^2 \le & |(G'(u_k), v_k)|/2 + |(G'(u_k), w_k)|/2 \\ & + \frac{|\lambda| - \lambda}{|P|}\|v_k\|^2 + \frac{|\lambda| - \lambda}{|Q|}\|w_k\|^2 \\ & + \int_{\Omega} \frac{u_k^2}{1+u_k^2}\, dx. \end{aligned}$$

Hence,

$$1 = \|\tilde{u}_k\|_H^2 \le [|(G'(u_k), v_k)|/2 + |(G'(u_k), w_k)|/2]/\rho_k^2 + C\|\tilde{u}_k\|^2. \tag{19}$$

In the limit we have,

$$1 \le C\|\tilde{u}\|^2.$$

This shows that $\tilde{u} \not\equiv 0$. Let Ω_0 be the subset of Ω where $\tilde{u}(x) \neq 0$. Then $|\Omega_0| \neq 0$. Thus

$$\int_{\Omega} H(x, u_k)\, dx = \int_{\Omega_0} H(x, u_k)\, dx + \int_{\Omega \setminus \Omega_0} H(x, u_k)\, dx$$

$$\geq \int_{\Omega_0} H(x, u_k)\, dx \to \infty.$$

This contradicts (15). Thus, the sequence satisfying (8) is bounded in E. Hence, there is a renamed subsequence such that $u_k \rightharpoonup u_0$ in E, and $u_k \to u_0$ in $L^2(\Omega)$ and a.e. Taking the limit in (13), we obtain

$$\begin{aligned}(G'(u_0), q)/2 = {} & \frac{1}{P}(\nabla v_0, \nabla g) + \frac{\lambda}{P}(v_0, g) \\ & + \frac{1}{Q}(\nabla w_0, \nabla h) + \frac{\lambda}{Q}(w_0, h) \\ & + \int_\Omega \frac{u_0 q}{1+u_0^2}\, dx = 0, \quad q = (g, h),\end{aligned} \tag{20}$$

Thus, u_0 satisfies $G'(u_0) = 0$. Since $u_0 \in E$, it satisfies

$$\begin{aligned}(G'(u_0), u_0)/2 = {} & \frac{1}{P}(\nabla v_0, \nabla v_0) + \frac{\lambda}{P}(v_0, v_0) \\ & + \frac{1}{Q}(\nabla w_0, \nabla w_0) + \frac{\lambda}{Q}(w_0, w_0) \\ & + \int_\Omega \frac{u_0^2}{1+u_0^2}\, dx = 0.\end{aligned} \tag{21}$$

Also, from the limit in (13), we have

$$\begin{aligned}\lim \frac{1}{P}\|\nabla v_k\|^2 & = \lim(G'(u_k), v_k)/2 \\ & \quad - \lim[\frac{\lambda}{P}\|v_k\|^2 + \int_\Omega \frac{v_k^2}{1+u_k^2}\, dx] \\ & = -[\frac{\lambda}{P}\|v\|^2 + \int_\Omega \frac{v^2}{1+u^2}\, dx] \\ & = \frac{1}{P}\|\nabla v\|^2,\end{aligned}$$

with a similar statement for $\|\nabla w\|^2$. Consequently, $\nabla u_k \to \nabla u$ in $L^2(\Omega)$. This shows that $G(u_k) \to G(u_0)$. Hence, $G(u_0) = c$.

Lemma 4 *If $G'(u) = 0$, then $u = (v, w)$ is a solution of (1) and (2).*

Proof It satisfies

$$
\begin{aligned}
(G'(u), q)/2 = {} & \frac{1}{P}(\nabla v, \nabla g) + \frac{\lambda}{P}(v, g) \\
& + \frac{1}{Q}(\nabla w, \nabla h) + \frac{\lambda}{Q}(w, h) \\
& + \int_{\Omega} \frac{uq}{1+u^2}\, dx,
\end{aligned}
$$

where $u = (v, w),\ q = (g, h) \in H$. If $G'(u) = 0$, this expression vanishes for all $q = (g, h)$. This implies that $u = (v, w)$ is a solution of (1) and (2).

Lemma 5

$$
\int_{\Omega} \ln(1+u^2)dx/\|u\|_H^2 \to 0, \quad \|u\|_H \to \infty. \tag{22}
$$

Proof Suppose $u_k \in H$ is a sequence such that $\rho_k = \|u_k\|_H \to \infty$. Let $\tilde{u}_k = u_k/\rho_k$. Then $\|\tilde{u}_k\|_H = 1$. Hence, there is a renamed subsequence such that $\tilde{u}_k \rightharpoonup \tilde{u}$ in H, and $\tilde{u}_k \to \tilde{u}$ in $L^2(\Omega)$ and a.e. Now

$$
\frac{\ln(1+u_k^2)}{\rho_k^2} = \frac{\ln(1+u_k^2)}{u_k^2}\tilde{u}_k^2 \to \ 0 \ a.e.
$$

and it is dominated a.e. by $\tilde{u}_k^2 \to \tilde{u}^2$ in $L^1(\Omega)$. Thus

$$
\int_{\Omega} \frac{\ln(1+u_k^2)}{\rho_k^2} dx \to \ 0.
$$

Since this is true for any sequence satisfying $\|u_k\|_H \to \infty$, we see that (22) holds.

Corollary 1 *If*

$$
I(u) = \|u\|_H^2 - \int_{\Omega} \ln(1+u^2)\, dx,
$$

then

$$
I(v) \to \infty \ as \ \|v\|_H \to \infty. \tag{23}
$$

Proof We have

$$
I(u)/\|u\|_H^2 = 1 - \int_{\Omega} \ln(1+u^2)dx/\|u\|_H^2 \to 1, \quad \|u\|_H \to \infty
$$

by Lemma 5. This gives (23).

Lemma 6

$$\int_\Omega [u^2 - \ln(1+u^2)]dx/\|u\|_H^2 \to 0, \quad \|u\|_H \to 0. \tag{24}$$

Proof Suppose $u_k \in H$ is a sequence such that $\rho_k = \|u_k\|_H \to 0$. In particular, there is a renamed subsequence such that $u_k \to 0$ a.e. Let $\tilde{u}_k = u_k/\rho_k$. Then $\|\tilde{u}_k\|_H = 1$. Hence, there is a renamed subsequence such that $\tilde{u}_k \rightharpoonup \tilde{u} \in H$, and $\tilde{u}_k \to \tilde{u}$ in $L^2(\Omega)$ and a.e. Now

$$\frac{u_k^2 - \ln(1+u_k^2)}{\rho_k^2} \leq \frac{u_k^2}{1+u_k^2}\tilde{u}_k^2 \to 0 \; a.e.$$

and it is dominated a.e. by $\tilde{u}_k^2 \to \tilde{u}^2$ in $L^1(\Omega)$. Thus

$$\int_\Omega \frac{u_k^2 - \ln(1+u_k^2)}{\rho_k^2} dx \to 0.$$

Since this is true for any sequence satisfying $\|u_k\|_H \to 0$, we see that (24) holds.

3 Proofs

Proof of Theorem 1 We let E be the subspace of $H^{1,2}(\Omega)$ consisting of those functions having the same periodicity as Ω with norm given by

$$\|w\|_E^2 = \|\nabla w\|^2 + \|w\|^2.$$

Let $u = (v, w)$, where $v, w \in E$ and $u^2 = v^2 + w^2$. If $q = (g, h)$, we write $uq = vg + wh$. Define

$$\|u\|_H^2 = \frac{1}{|P|}[\|\nabla v\|^2 + |\lambda| \; \|v\|^2] + \frac{1}{|Q|}[\|\nabla w\|^2 + |\lambda| \; \|w\|^2], \quad v, w \in E. \tag{25}$$

Assume that P, Q, λ do not vanish. Then $\|u\|_H^2$ is a norm on $H = E \times E$ having a scalar product $(u, h)_H$.

Let $a(u)$ and $G(u)$ be defined by (9) and (10), respectively. If $G'(u) = 0$, then $u = (v, w)$ satisfies (1) and (2) (Lemma 4). Let Y be the subspace of E consisting of the constants. We let $N = \{c, 0) \in H, c \in Y\}$. First, we note that

$$G(u) \leq 0, \quad u \in N,$$

if $\sigma \geq P$. To see this, let $u = (c, 0) \in N$. Then

$$a(c, 0) = -\frac{\sigma}{P}c^2|\Omega|$$

and

$$\int_\Omega \ln(1 + c^2)dx \leq c^2|\Omega|.$$

Thus,

$$G(u) \leq [1 - \frac{\sigma}{P}]c^2|\Omega|.$$

This means that

$$G(u) \leq 0, \quad u \in N, \tag{26}$$

provided $\sigma \geq P$.

Next, let $M = N^\perp$ and let $u = (v, w)$ be any function in M. Then $\|\nabla v\|^2 \geq \lambda_1 \|v\|^2$. Then

$$a(u) + \|u\|^2 \geq \frac{1}{P}[1 - \frac{\sigma - P}{\lambda_1}] \|v\|_H^2 + \frac{1}{Q}[\|\nabla w\|^2 + (Q - \sigma)\|w\|^2].$$

Thus, there is an $\varepsilon > 0$ such that

$$a(u) + \|u\|^2 \geq 2\varepsilon\|u\|_H^2, \quad u \in M, \tag{27}$$

when $0 < \sigma < \min[P + \lambda_1, Q]$.

Now

$$\int_\Omega [u^2 - \ln(1 + u^2)]dx/\|u\|_H^2 \to 0, \quad \|u\|_H \to 0 \tag{28}$$

by Lemma 6. If we combine (27) and (28), we see that there is an $\varepsilon > 0$ such that

$$G(u) \geq \varepsilon\|u\|_H^2, \quad u \in M, \tag{29}$$

when $\|u\|_H^2 \leq \rho$ is small and $0 < \sigma < \min[P + \lambda_1, Q]$.

Next, let $u = (c, d)$, where $c, d \in Y$. Then

$$a(c, d) = -\frac{\sigma}{P}c^2|\Omega| - \frac{\sigma}{Q}d^2|\Omega|.$$

Thus,

$$G(u) \le -\frac{\sigma}{P}c^2|\Omega| - \frac{\sigma}{Q}d^2|\Omega| + \int_\Omega \ln(1+c^2+d^2)dx.$$

By Lemma 5,

$$\int_\Omega \ln(1+c^2+d^2)dx/(c^2+d^2) \to 0$$

as $R^2 = c^2 + d^2 \to \infty$. Hence $\liminf G(c, d) < 0$ as $R^2 = c^2 + d^2 \to \infty$ when $\sigma > 0$.

Let $B = M \cap \partial\mathscr{B}_\rho$. Then for ρ sufficiently small there is an $\varepsilon > 0$ such that

$$\inf_B G \ge \varepsilon > 0.$$

Take

$$A = N \cap \mathscr{B}_R \oplus \{(c, d) \in H : c, d \in Y,\ R^2 = c^2 + d^2\}.$$

Then

$$\sup_A G \le 0.$$

By Lemma 1, A links B. Now we can apply Lemmas 2, 3, and 4 to reach the conclusion that (1) and (2) has a solution u such that $G(u) \ge \varepsilon > 0$. Since $G(0, 0) = 0$, we see that u is a nontrivial solution.

The proof of Theorem 2 is similar to that of Theorem 1 and is omitted.

Proof of Theorem 3 We let $N = \{v, 0) \in H, v \in \bigoplus_{\lambda\le\lambda_\ell} E(\lambda)\}$. First, we note that

$$G(u) \le 0, \quad u \in N,$$

if $\sigma \ge P + \lambda_\ell$. To see this, let $u = (v, 0) \in N$. Then

$$G(u) \le \frac{1}{P}\Big[1 - \frac{\sigma - P}{\lambda_\ell}\Big] \|v\|_H^2.$$

This means that

$$G(u) \le 0, \quad u \in N, \tag{30}$$

provided $\sigma \ge P + \lambda_\ell$.

Next, let $M = N^\perp$ and let $u = (v, w)$ be any function in M. Then $\|\nabla v\|^2 \ge \lambda_{\ell+1}\|v\|^2$ and

$$a(u) + \|u\|^2 \geq \frac{1}{P}[1 - \frac{\sigma - P}{\lambda_{\ell+1}}]\ \|v\|_H^2$$
$$+ \frac{1}{Q}[\|\nabla w\|^2 + (Q - \sigma)\|w\|^2].$$

Thus, there is an $\varepsilon > 0$ such that

$$a(u) + \|u\|^2 \geq 2\varepsilon\|u\|_H^2, \quad u \in M, \tag{31}$$

when $\sigma < \min[P + \lambda_{\ell+1}, Q]$.

Now

$$\int_\Omega [u^2 - \ln(1 + u^2)]dx/\|u\|_H^2 \to 0, \quad \|u\|_H \to 0 \tag{32}$$

by Lemma 6. If we combine (31) and (32), we see that there is an $\varepsilon > 0$ such that

$$G(u) \geq \varepsilon\|u\|_H^2, \quad u \in M, \tag{33}$$

when $\|u\|_H^2 \leq \rho$ is small and $\sigma < \min[P + \lambda_{\ell+1}, Q]$.

Next, let $u = (v, d)$, where $(v, 0) \in N$ and $d \in Y$. Then

$$a(v, d) = \frac{1}{P}[\ \|\nabla v\|^2 - \sigma\ \|v\|^2] - \frac{\sigma}{Q}d^2|\Omega|.$$

Thus,

$$G(u) \leq \frac{1}{P}\Big[1 - \frac{\sigma - P}{\lambda_\ell}\Big]\ \|v\|_H^2 - \frac{\sigma}{Q}d^2|\Omega| + \int_\Omega \ln(1 + v^2 + d^2)dx.$$

By Lemma 5,

$$\int_\Omega \ln(1 + v^2 + d^2)dx/(\|v\|_H^2 + d^2) \to 0$$

as $R^2 = \|v\|_H^2 + d^2 \to \infty$. Hence $\liminf G(v, d) < 0$ as $R^2 = \|v\|_H^2 + d^2 \to \infty$ when $\sigma > 0$.

Let $B = M \cap \partial\mathscr{B}_\rho$. Then for ρ sufficiently small there is an $\varepsilon > 0$ such that

$$\inf_B G \geq \varepsilon > 0.$$

Take

$$A = N \cap \mathscr{B}_R \oplus \{(v, d) \in H : (v, 0) \in N,\ d \in Y,\ R^2 = \|v\|_H^2 + d^2\}.$$

Then

$$\sup_A G \leq 0.$$

By Lemma 1, A links B. Now we can apply Lemmas 2, 3 and 4 to reach the conclusion that (1) and (2) has a solution u such that $G(u) \geq \varepsilon > 0$. Since $G(0, 0) = 0$, we see that u is a nontrivial solution.

The proof of Theorem 4 is similar to that of Theorem 3 and is omitted.

References

1. G. Bartal, O. Manela, O. Cohen, J.W. Fleischer, and M. Segev, Observation of second-band vortex solitons in 2D photonic lattices, Phys. Rev. Lett. 95(2005) 053904.
2. S. Chen and Y. Lei, Existence of steady-state solutions in a nonlinear photonic lattice model, J. Math. Phys. 52 (2011), no. 6, 063508.
3. W. Chen and D.L. Mills, Gap solitons and the nonlinear optical response of superlattices. Phys. Rev. Lett. 62 (1989) 1746–1749.
4. N.K. Efremidis, S. Sears and D.N. Christodoulides, Discrete solitons in photorefractive optically-induced photonic lattices. Phys.Rev.Lett. 85 (2000) 1863–1866.
5. J.W. Fleischer, G. Bartal, O. Cohen, O. Manela, M. Segev, J. Hudock, and D.N. Christodoulides, Observation of vortex-ring discrete solitons in photonic lattices, Phys. Rev. Lett. 92(2004), 123904.
6. W. J.W. Fleischer, M. Segev, N.K. Efremidis and D.N. Christodolides, Observation of two-dimensional discrete solitons in optically induced nonlinear photonic lattices. Nature,(2003) 147–149.
7. P. Kuchment, The mathematics of photonic crystals. Mathematical modeling in optical science, 207–272, Frontiers Appl. Math., 22, SIAM, Philadelphia, PA, 2001.
8. H. Martin, E.D. Eugenieva and Z. Chen, Discrete Solitons and Soliton-Induced Dislocations in Partially Coherent Photonic Lattices. Martin et al. Phys. Rev. Lett. 92 (2004) 123902.
9. D.N. Neshev, T.J. Alexander, E.A. Ostrovskaya, Y.S. Kivshar, H. Martin, I. Makasyuk, and Z. Chen, Observation of discrete vortex solitons in optically induced photonic lattices, Phys. Rev. Lett. 92(2004), 123903.
10. A. Pankov, Periodic nonlinear Schrodinger equation with application to photonic crystals. Milan J. Math. 73 (2005), 259–287.
11. M. Schechter, Linking Methods in Critical Point Theory, Birkhauser Boston, 1999.
12. M. Schechter, An Introduction to Nonlinear Analysis. Cambridge Studies in Advanced Mathematics, 95. Cambridge University Press, Cambridge, 2004.
13. M. Schechter, The use of Cerami sequences in critical point theory, Abstr. Appl. Anal. 2007 (2007), Art. ID 58948, 28 pp.
14. M. Schechter, Minimax Systems and Critical Point Theory, Birkhauser Boston, 2009.
15. M. Schechter, Steady state solutions for Schrodinger equations governing nonlinear optics., J. Math. Phys., 53 (2012), 043504, 8 pp.
16. M. Schechter, Photonic lattices., J. Math. Phys., 54 (2013) 061502, 7 pp.
17. M. Schechter, Critical Point Theory, Sandwich and Linking Systems, Birkhauser, 2020.
18. Y. Yang, Solition in Field Theory and Nonlinear Analysis, Springer-Verlag, New York, 2001.

19. J. Yang, A. Bezryadina, Z. Chen, and I. Makasyuk, Observation of two-dimensional lattice vector solitons. Opt. Lett. 29 (2004) 1656.
20. J. Yang, I. Makasyuk, A. Bezryadina and Z. Chen, Dipole and Quadrupole Solitons in Optically Induced Two-Dimensional Photonic Lattices: Theory and Experiment, Studies in Applied Mathmatics 113 (2004) 389–412.
21. Y. Yang and R. Zhang, Steady state solutions for nonlinear Schrödinger equation arising in optics, J. Math. Phys. 50 (2009) 053501–9.

Some Certain Classes of Combinatorial Numbers and Polynomials Attached to Dirichlet Characters: Their Construction by p-Adic Integration and Applications to Probability Distribution Functions

Yilmaz Simsek and Irem Kucukoglu

Abstract The aim of this chapter is to survey on old and new identities for some certain classes of combinatorial numbers and polynomials derived from the non-trivial Dirichlet characters and p-adic integrals. This chapter is especially motivated by the recent papers (Simsek, Turk J Math 42:557–577, 2018; Srivastava et al., J Number Theory 181:117–146, 2017; Kucukoglu et al. Turk J Math 43:2337–2353, 2019; Axioms 8(4):112, 2019) in which the aforementioned combinatorial numbers and polynomials were extensively investigated and studied in order to obtain new results. In this chapter, after recalling the origin of the aforementioned combinatorial numbers and polynomials, which goes back to the paper (Simsek, Turk J Math 42:557–577, 2018), a compilation has been made on what has been done from the paper (Simsek, Turk J Math 42:557–577, 2018) up to present days about the main properties and relations of these combinatorial numbers and polynomials. Moreover, with the aid of some known and new formulas, relations, and identities, which involve some kinds of special numbers and polynomials such as the Apostol-type, the Peters-type, the Boole-type numbers and polynomials the Bernoulli numbers and polynomials, the Euler numbers and polynomials, the Genocchi numbers and polynomials, the Stirling numbers, the Cauchy numbers (or the Bernoulli numbers of the second kind), the binomial coefficients, the falling factorial, etc., we give further new formulas and identities regarding these combinatorial numbers and polynomials. Besides, some derivative and integral formulas, involving not only these combinatorial numbers and polynomials, but also their generating functions, are presented in addition to those given for their positive and negative higher-

---

Y. Simsek (✉)
Department of Mathematics, Faculty of Science, University of Akdeniz, Antalya, Turkey
e-mail: ysimsek@akdeniz.edu.tr

I. Kucukoglu
Department of Engineering Fundamental Sciences, Faculty of Engineering, Alanya Alaaddin Keykubat University, Antalya, Turkey
e-mail: irem.kucukoglu@alanya.edu.tr

I. N. Parasidis et al. (eds.), *Mathematical Analysis in Interdisciplinary Research*, Springer Optimization and Its Applications 179,
https://doi.org/10.1007/978-3-030-84721-0_33

order extensions. By using Wolfram programming language in Mathematica, we present some plots for these combinatorial numbers and polynomials with their generating functions. Finally, in order to do mathematical analysis of the results in an interdisciplinary way, we present some observations on a few applications of the positive and negative higher-order extension of the generating functions for combinatorial numbers and polynomials to the probability theory for researchers to shed light on their future interdisciplinary studies.

1 Introduction and Preliminaries

When we look at the developments in the history of mathematics, we see that in almost every period, the mystery of numbers, problems involving numbers and games involving numbers, have always remained the leading role of the historical period. With what was discovered at every stage of development of mathematics, every branch of science was also naturally affected by this development. This interaction has led scientists to study interdisciplinary and has brought them together in order to develop new mathematical models, algorithms, and other mathematical techniques and methods. In the construction of mathematical modeling, it has been realized that it is possible to take advantage of some concepts such as polynomials, generating functions for numbers and polynomials, moment-generating functions, differential calculus and equations, and matrices, etc. In recent years, it has been seen that special numbers and polynomials are used frequently in almost all fields of mathematics, physics, engineering, medical sciences, economics, and social sciences. So, it is well known that the special numbers and polynomials have many vital applications in almost all branches of mathematics, physics, engineering, and other relevant areas. Due to which, it is pretty easy to perform mathematical calculations and operations by using polynomials and their generating functions with their functional equations are commonly used inside of the techniques of mathematics, physics, biology, and engineering in order to solve real-world problems. Therefore, studying on properties and relations regarding any family of polynomials is pretty important to provide a technical infrastructure in order to solve real-world problems. In this context, with this chapter, we give survey on some certain classes of combinatorial numbers and polynomials constructed with the aid of non-trivial Dirichlet character and the p-adic integral methods including the p-adic bosonic (Volkenborn) integral and the p-adic fermionic integral. These combinatorial-type numbers and polynomials and their generating functions are especially associated with some special numbers and polynomials such as the Apostol-type numbers and polynomials, the Bernstein basis functions, the Poisson–Charlier polynomials, the Peters polynomials, the Boole numbers and polynomials, the Changhee numbers and polynomials, the Daehee numbers and polynomials, the Stirling numbers, the Bell polynomials (i.e., exponential polynomials), the Bernoulli numbers, the Euler numbers, the Cauchy numbers (or the Bernoulli numbers of the second kind), the binomial coefficients, the falling factorial, etc.

Thus, here we present some evaluations on what has been studied from the past to the present and what can be found new for the aforementioned classes of combinatorial numbers and polynomials are given here.

After giving a brief explanation about the motivation of this chapter, we continue with presenting the notations and definitions needed to be more clearly understood in the results:

Let $\mathbb{N}$, $\mathbb{Z}$, $\mathbb{N}_0$, $\mathbb{Q}$, $\mathbb{R}$, and $\mathbb{C}$ denote as usual the set of natural numbers, the set of integers, the set of nonnegative integers, the set of rational numbers, the set of real numbers, and the set of complex numbers, respectively. Let $\log z$ denote the principal branch of the multi-valued function $\log z$ with the imaginary part $\mathrm{Im}(\log z)$ constrained by the interval $(-\pi, \pi]$. We also assume that

$$0^n = \begin{cases} 1 & \text{if } n = 0 \\ 0 & \text{if } n \in \mathbb{N}. \end{cases}$$

Besides,

$$\binom{z}{w} = \frac{(z)_w}{w!} = \frac{(-1)^w (-z)^{(w)}}{w!} \quad (w \in \mathbb{N}_0, z \in \mathbb{C}), \tag{1}$$

in which $(z)_w$ and $(z)^{(w)}$ denote, respectively, the falling factorial and the rising factorial defined, respectively, by

$$(z)_w = z(z-1)(z-2)\ldots(z-w+1)$$

and

$$(z)^{(w)} = z(z+1)(z+2)\ldots(z+w-1)$$

such that $(z)_0 = 1$ and $(z)^{(0)} = 1$ (*cf.* [5–113]).

Since the main motivation of this chapter is to survey on the results on some certain classes of combinatorial numbers and polynomials including the generalized Apostol-type polynomials, let us recall some members from the class of Apostol-type polynomials with their generating functions:

The Apostol–Bernoulli polynomials, $\mathcal{B}_n(x; \lambda)$, are defined by Apostol in [2] with the following generating function:

$$\frac{te^{tx}}{\lambda e^t - 1} = \sum_{n=0}^{\infty} \mathcal{B}_n(x; \lambda) \frac{t^n}{n!}, \tag{2}$$

where $|t| < 2\pi$ when $\lambda = 1$ and $|t| < |\log \lambda|$ when $\lambda \neq 1$ and $\lambda \in \mathbb{C}$. One can easily see that for $x = 0$, these polynomials are reduced to the Apostol–Bernoulli numbers $\mathcal{B}_n(\lambda)$, which are given by the following generating function:

$$\frac{t}{\lambda e^t - 1} = \sum_{n=0}^{\infty} \mathcal{B}_n(\lambda)\frac{t^n}{n!} \tag{3}$$

(*cf.* [2, 68, 108, 111, 112]), and for $\lambda = 1$, these numbers are reduced to the classical Bernoulli numbers (the Bernoulli numbers of the first kind):

$$B_n = \mathcal{B}_n(1),$$

which is defined by means of the following generating function:

$$\frac{t}{e^t - 1} = \sum_{n=0}^{\infty} B_n\frac{t^n}{n!}, \qquad (t < |2\pi|), \tag{4}$$

which arise in not only analytic number theory, but also other related areas (*cf.* [2–112], and the references cited therein).

By using the above generating functions for the Apostol–Bernoulli numbers and polynomials with the method of umbral calculus convention, few of these numbers and polynomials are computed as follows, respectively:

$$\mathcal{B}_0(\lambda) = 0, \quad \mathcal{B}_1(\lambda) = \frac{1}{\lambda - 1},$$
$$\mathcal{B}_2(\lambda) = \frac{-2\lambda}{(\lambda-1)^2}, \quad \mathcal{B}_3(\lambda) = \frac{3\lambda(\lambda+1)}{(\lambda-1)^3}, \ldots$$

and

$$\mathcal{B}_0(x;\lambda) = 0, \quad \mathcal{B}_1(x;\lambda) = \frac{1}{\lambda - 1},$$
$$\mathcal{B}_2(x;\lambda) = \frac{1}{\lambda-1}x - \frac{2\lambda}{(\lambda-1)^2},$$
$$\mathcal{B}_3(x;\lambda) = \frac{3}{\lambda-1}x^2 - \frac{6\lambda}{(\lambda-1)^2}x + \frac{3\lambda(\lambda+1)}{(\lambda-1)^3},$$

and so on (*cf.* [2–112]; and the references cited therein).

The Apostol–Bernoulli numbers of higher order, $\mathcal{B}_n^{(k)}(\lambda)$, are defined by the following generating function:

$$\left(\frac{t}{\lambda e^t - 1}\right)^k = \sum_{n=0}^{\infty} \mathcal{B}_n^{(k)}(\lambda)\frac{t^n}{n!}, \tag{5}$$

where $|t| < 2\pi$ when $\lambda = 1$ and $|t| < |\log(\lambda)|$ when $\lambda \neq 1$ so that in the special case when $k = 1$, (5) reduces (3) (*cf.* [2–112]; and the references cited therein).

The Apostol–Euler polynomials, $\mathcal{E}_n(x;\lambda)$, are given by the following generating function:

$$\frac{2e^{tx}}{\lambda e^t+1}=\sum_{n=0}^{\infty}\mathcal{E}_n(x;\lambda)\frac{t^n}{n!}, \tag{6}$$

$|t|<\pi$ when $\lambda=1$ and $|t|<|\log(-\lambda)|$ when $\lambda\neq 1$ and $\lambda\in\mathbb{C}$ (*cf.* [11, 39, 74, 108, 110, 111]; and the references cited therein). One can easily see that for $x=0$, these polynomials are reduced to the Apostol–Euler numbers $\mathcal{E}_n(\lambda)=\mathcal{E}_n(0;\lambda)$ which are given by the following generating function:

$$\frac{2}{\lambda e^t+1}=\sum_{n=0}^{\infty}\mathcal{E}_n(\lambda)\frac{t^n}{n!}, \tag{7}$$

and also for $\lambda=1$, the Apostol–Euler polynomials are reduced to the classical (the first kind) Euler polynomials:

$$E_n(x)=\mathcal{E}_n(x;1) \tag{8}$$

(*cf.* [12–112]; and the references cited therein), and it is clear that for $\lambda=1$, we have

$$E_n=\mathcal{E}_n(1), \tag{9}$$

where E_n denotes the Euler numbers of the first kind defined by means of the following generating function:

$$\frac{2}{e^t+1}=\sum_{n=0}^{\infty}E_n\frac{t^n}{n!}, \qquad (t<|\pi|) \tag{10}$$

(*cf.* [12–112]; and the references cited therein).

By using the above generating functions for the Apostol–Euler polynomials and numbers with the method of umbral calculus convention, few of these numbers are computed as follows, respectively:

$$\mathcal{E}_0(\lambda)=\frac{2}{\lambda+1}, \quad \mathcal{E}_1(\lambda)=-\frac{2\lambda}{(\lambda+1)^2},$$
$$\mathcal{E}_2(\lambda)=\frac{2\lambda(\lambda-1)}{(\lambda+1)^3}, \quad \mathcal{E}_3(\lambda)=-\frac{2\lambda\left(\lambda^2-4\lambda+1\right)}{(\lambda+1)^4},$$

and so on (*cf.* [12–112]; and the references cited therein).

The Apostol–Euler numbers of higher order, $\mathcal{E}_n^{(k)}(\lambda)$, are defined by the following generating function:

$$\left(\frac{2}{\lambda e^t+1}\right)^k=\sum_{n=0}^{\infty}\mathscr{E}_n^{(k)}(\lambda)\frac{t^n}{n!}, \tag{11}$$

where $|t|<|\log(-\lambda)|$ so that in the special case when $k=1$, (11) reduces (7) (*cf.* [2–112]; and the references cited therein).

The Apostol–Genocchi polynomials of higher order, $\mathscr{G}_n^{(k)}(x;\lambda)$, are defined by the following generating function:

$$\left(\frac{2t}{\lambda e^t+1}\right)^k e^{xt}=\sum_{n=0}^{\infty}\mathscr{G}_n^{(k)}(x;\lambda)\frac{t^n}{n!}, \tag{12}$$

where $\lambda\in\mathbb{C}$ and $|t|<|\log(-\lambda)|$, so that

$$\begin{aligned}\mathscr{G}_n^{(k)}(\lambda)&=\mathscr{G}_n^{(k)}(0;\lambda),\\ \mathscr{G}_n(x;\lambda)&=\mathscr{G}_n^{(1)}(x;\lambda),\\ \mathscr{G}_n(\lambda)&=\mathscr{G}_n^{(1)}(\lambda)\\ G_n&=\mathscr{G}_n(1)\end{aligned}$$

in which $\mathscr{G}_n^{(k)}(\lambda)$, $\mathscr{G}_n(x;\lambda)$, $\mathscr{G}_n(\lambda)$, and G_n denote, respectively, the Apostol–Genocchi numbers of higher order, the Apostol–Genocchi polynomials, the Apostol–Genocchi numbers, and the Genocchi numbers (*cf.* [8, 63, 65, 68, 73, 74, 83, 103, 112]; and see also the references cited therein).

For $n,k\in\mathbb{N}_0$ and $0\le k\le n$, the relation among higher-order versions of the Apostol–Bernoulli polynomials, the Apostol–Euler polynomials, and the Apostol–Genocchi polynomials is given as follows (*cf.* [68, Lemma 2–3, p. 5707]):

$$\mathscr{G}_n^{(k)}(x;\lambda)=(n)_k\,\mathscr{E}_{n-k}^{(k)}(x;\lambda)=(-2)^k\,\mathscr{B}_n^{(k)}(x;-\lambda), \tag{13}$$

which, for $x=0$, yields

$$\mathscr{G}_n^{(k)}(\lambda)=(n)_k\,\mathscr{E}_{n-k}^{(k)}(\lambda)=(-2)^k\,\mathscr{B}_n^{(k)}(-\lambda) \tag{14}$$

(*cf.* [68, 112]; and cited reference therein).

1.1 Basic Properties of Dirichlet Characters

One of the most frequently used concepts in this chapter is the Dirichlet characters. Thus, let us recall its definition below:

Let $d\in\mathbb{N}$ and $(\mathbb{Z}/d\mathbb{Z})^*$ denote the unit group of reduced residue class modulo d. Throughout of this chapter, χ is a Dirichlet character with modulo d, which is a group homomorphism, i.e.,

$$\chi : (\mathbb{Z}/d\mathbb{Z})^* \to \mathbb{C}\setminus\{0\}$$

(*cf.* [3]).

Let φ denote the Euler totient function. Note that *there are $\varphi(d)$ distinct Dirichlet characters with modulo d, each of which is completely multiplicative and periodic with period d* (*cf.* [3, Theorem 6.15]).

Some of the Dirichlet characters are given by Tables 1, 2, 3, 4, and 5 as follows (see, for detail, [3, Theorem 6.15]):

When $d = 1$ or $d = 2$, then $\varphi(d) = 1$ and the only Dirichlet character is the principal character χ_1. For $d \geq 3$, there are at least two Dirichlet characters since $\varphi(d) \geq 2$.

Tables 1, 2, and 3 display all the Dirichlet characters with conductors $d = 3, 4,$ and 5 (see, for detail, [3]):

Table 1 All the Dirichlet characters with conductor $d = 3$ since $\varphi(d) = 2$

n	0	1	2
$\chi_1(n)$	0	1	1
$\chi_2(n)$	0	1	-1

Table 2 All the Dirichlet characters with conductor $d = 4$ since $\varphi(d) = 2$

n	0	1	2	3
$\chi_1(n)$	0	1	0	1
$\chi_2(n)$	0	1	0	-1

Table 3 All the Dirichlet characters with conductor $d = 5$ since $\varphi(d) = 4$

n	0	1	2	3	4
$\chi_1(n)$	0	1	1	1	1
$\chi_2(n)$	0	1	-1	-1	1
$\chi_3(n)$	0	1	i	$-i$	-1
$\chi_4(n)$	0	1	$-i$	i	-1

Tables 4 and 5 display all the Dirichlet characters with conductors $d = 6$ and $d = 7$ (see, for detail, [3]):

Dirichlet characters are used in the construction of many special numbers and polynomial families. Let us briefly give some of special numbers and polynomials attached to the Dirichlet character as follows:

The generalized Apostol–Bernoulli numbers attached to the Dirichlet character, $\mathcal{B}_{n,\chi}(\lambda)$, are defined as follows:

Table 4 $d = 6, \varphi(d) = 2$

n	0	1	2	3	4	5
$\chi_1(n)$	0	1	0	0	0	1
$\chi_2(n)$	0	1	0	0	0	-1

Table 5 $d = 7, \varphi(d) = 6$, $\omega = \exp(\pi i/3)$

n	0	1	2	3	4	5	6
$\chi_1(n)$	0	1	1	1	1	1	1
$\chi_2(n)$	0	1	1	-1	1	-1	-1
$\chi_3(n)$	0	1	ω^2	ω	$-\omega$	$-\omega^2$	-1
$\chi_4(n)$	0	1	ω^2	$-\omega$	$-\omega$	ω^2	1
$\chi_5(n)$	0	1	$-\omega$	ω^2	ω^2	$-\omega$	1
$\chi_6(n)$	0	1	$-\omega$	$-\omega^2$	ω^2	ω	-1

$$\sum_{j=0}^{d-1} \frac{\lambda^j e^{tj} t\chi(j)}{\lambda^d e^{td} - 1} = \sum_{n=0}^{\infty} \mathcal{B}_{n,\chi}(\lambda) \frac{t^n}{n!} \tag{15}$$

(*cf.* [2, 32, 40, 41, 45, 111], and the references cited therein).

By combining (15) with (2), one can easily get

$$\mathcal{B}_{n,\chi}(\lambda) = d^{n-1} \sum_{j=0}^{d-1} \lambda^j \chi(j) \mathcal{B}_n\left(\frac{j}{d}; \lambda^d\right).$$

If χ is a trivial character in (15), then the numbers $\mathcal{B}_{n,\chi}(\lambda)$ reduce to the Apostol–Bernoulli numbers, that is

$$\mathcal{B}_n(\lambda) = \mathcal{B}_{n,1}(\lambda)$$

(*cf.* [2, 32, 40, 41, 61, 71, 111], and the references cited therein).

The generalized Apostol–Euler numbers attached to the Dirichlet character, $\mathcal{E}_{n,\chi}(\lambda)$, are defined as follows:

$$2\sum_{j=0}^{d-1} \frac{\lambda^j e^{tj} \chi(j)}{\lambda^d e^{td} + 1} = \sum_{n=0}^{\infty} \mathcal{E}_{n,\chi}(\lambda) \frac{t^n}{n!} \tag{16}$$

(*cf.* [40, 41, 111], and the references cited therein).

By combining (16) with (6), one easily see that

$$\mathscr{E}_{n,\chi}(\lambda) = d^n \sum_{j=0}^{d-1} \lambda^j \chi(j) \mathscr{E}_n\left(\frac{j}{d}; \lambda^d\right).$$

When $\chi \equiv 1$ in (16), one has

$$\mathscr{E}_n(\lambda) = \mathscr{E}_{n,1}(\lambda)$$

(*cf.* [40, 41, 111]).

1.2 Some Basic Properties of p-Adic Integration Method

One of the most frequently used methods in this chapter is the method of p-adic integrals including Volkenborn integral and fermionic integral. We now give a brief introduction about notations of the p-adic integrals. Thus, we next recall some definitions and notations associated with p-adic integrals as follows:

Let $\mathbb{Z}_p$, $\mathbb{Q}_p$, and $\mathbb{C}_p$ denote, respectively, the ring of p-adic integers, the set of p-adic rational numbers, and the set of the completion of algebraic closure of $\mathbb{Q}_p$. Let $\mathbb{K}$ be a field with a complete valuation and $C^1(\mathbb{Z}_p \to \mathbb{K})$ be a set of continuous differentiable functions. Namely, $C^1(\mathbb{Z}_p \to \mathbb{K})$ is contained in the following set:

$$\left\{ f : \mathbb{Z}_p \to \mathbb{K} : f(x) \text{ is differentiable and } \frac{d}{dx}\{f(x)\} \text{ is continuous} \right\}.$$

The p-adic q-integral of a function $f \in C^1(\mathbb{Z}_p \to \mathbb{K})$ is defined by Kim [35] as follows:

$$\int_{\mathbb{Z}_p} f(x) d\mu_q(x) = \lim_{N \to \infty} \frac{1}{[p^N]} \sum_{x=0}^{p^N-1} f(x) q^x, \tag{17}$$

where $q \in \mathbb{C}_p$ with $| 1 - q |_p < 1$, $\mu_q(x)$ denotes the q-Haar distribution defined by Kim [35] as follows:

$$\mu_q(x) = \mu_q(x + p^N \mathbb{Z}_p) = \frac{q^x}{[p^N]},$$

and

$$[x] = [x : q] = \begin{cases} \frac{1-q^x}{1-q} & \text{if } q \neq 1 \\ x & \text{if } q = 1, \end{cases}$$

such that

$$\lim_{q\to 1}[x:q]=x.$$

Taking limit $q\to 1$, (17) reduces to the Volkenborn integral (the p-adic bosonic integral) of the uniformly differentiable function f on $\mathbb{Z}_p$ as follows:

$$\int_{\mathbb{Z}_p} f(x)d\mu_1(x)=\lim_{N\to\infty}\frac{1}{p^N}\sum_{x=0}^{p^N-1} f(x), \tag{18}$$

where

$$\mu_1(x)=\mu_1(x+p^N\mathbb{Z}_p)=\frac{1}{p^N}$$

(*cf.* [81, Definition 55.1, p. 167]; see also [33, 35, 39, 41, 98]).

The Volkenborn integral in terms of the Mahler coefficients is given by the following formula:

$$\int_{\mathbb{Z}_p} f(x)\,d\mu_1(x)=\sum_{n=0}^{\infty}\frac{(-1)^n}{n+1}a_n,$$

where

$$f(x)=\sum_{n=0}^{\infty}a_n\binom{x}{j}\in C^1(\mathbb{Z}_p\to\mathbb{K})$$

(*cf.* [81, p. 168, Proposition 55.3]). Due to the above fact, in the case when

$$f(x)=\binom{x}{j}, \tag{19}$$

(18) yields

$$\int_{\mathbb{Z}_p}\binom{x}{j}d\mu_1(x)=\frac{(-1)^j}{j+1} \tag{20}$$

(*cf.* [81, p. 168, Proposition 55.3]).

Let $f:\mathbb{Z}_p\to\mathbb{K}$ be an analytic function and $f(x)=\sum\limits_{n=0}^{\infty}a_nx^n$ with $x\in\mathbb{Z}_p$. The Volkenborn integral of this analytic function is given by

$$\int_{\mathbb{Z}_p} \left(\sum_{n=0}^{\infty} a_n x^n \right) d\mu_1(x) = \sum_{n=0}^{\infty} a_n \int_{\mathbb{Z}_p} x^n d\mu_1(x)$$

(*cf.* [81, p. 168, Proposition 55.4]).

The following identity is very important to derive results for special numbers:

$$\int_{\mathbb{Z}_p} f(x+m) d\mu_1(x) = \int_{\mathbb{Z}_p} f(x) d\mu_1(x) + \sum_{j=0}^{m-1} f'(j), \tag{21}$$

where

$$f'(j) = \frac{d}{dx}\{f(x)\}\bigg|_{x=j}$$

(*cf.* [35, 39, 81]; see also the references cited therein).

It is well known that the p-adic bosonic integral enables to construct Bernoulli-type numbers. For example, in the case when $f(x) = x^n$, we have (*cf.* [35, 81])

$$B_n = \int_{\mathbb{Z}_p} x^n d\mu_1(x), \tag{22}$$

which is called Witt's formula for the Bernoulli numbers, and the Witt's formula for the Bernoulli polynomials is given as follows:

$$B_n(y) = \int_{\mathbb{Z}_p} (y+x)^n d\mu_1(x) \tag{23}$$

(*cf.* [35, 39, 81]; see also the references cited therein).

Kim [39] also defined the p-adic fermionic integral of the function f as follows:

$$\int_{\mathbb{Z}_p} f(x) d\mu_{-1}(x) = \lim_{N\to\infty} \sum_{x=0}^{p^N-1} (-1)^x f(x) \tag{24}$$

where $p \neq 2$ and

$$\mu_{-1}(x) = \mu_{-1}\left(x + p^N \mathbb{Z}_p\right) = (-1)^x$$

(*cf.* [33, 39]).

It is well known that the p-adic fermionic integral enables to construct Euler-type numbers. For example, in the case when $f(x) = x^n$, we have (*cf.* [35])

$$E_n = \int_{\mathbb{Z}_p} x^n d\mu_{-1}(x). \tag{25}$$

By setting

$$E^d\{f(x)\} = f(x+d),$$

Kim [41, Theorem 1] defined the following functional equation for the q-bosonic p-adic Volkenborn integral on $\mathbb{Z}_p$ as follows:

$$\begin{aligned} q^n \int_{\mathbb{Z}_p} E^n\{f(x)\} d\mu_q(x) - \int_{\mathbb{Z}_p} f(x)\, d\mu_q(x) \\ = \frac{q-1}{\log q}\left(\sum_{j=0}^{n-1} q^j f'(j) + \log q \sum_{j=0}^{n-1} q^j f(j)\right), \end{aligned} \tag{26}$$

where n is a positive integer.

Also, Kim gave the following integral equation for the q-fermionic p-adic integral on $\mathbb{Z}_p$ as follows [41, Theorem 3]:

$$q^d \int_{\mathbb{Z}_p} E^d\{f(x)\} d\mu_{-q}(x) - (-1)^d \int_{\mathbb{Z}_p} f(x)\, d\mu_{-q}(x) = [2]\sum_{j=0}^{d-1} (-1)^{d-l-1} q^j f(j), \tag{27}$$

where d is a positive integer. Substituting $d = 1$ into (27), one has

$$\int_{\mathbb{Z}_p} (qf(x+1) + f(x))\, d\mu_{-q}(x) = (q+1)\, f(0).$$

When $q \to 1$ in the above integral equation, one may easily see that

$$\int_{\mathbb{Z}_p} (f(x+1) + f(x))\, d\mu_{-1}(x) = 2f(0)$$

(*cf.* [41]). Substituting (19) into the above integral equation yields

$$\int_{\mathbb{Z}_p} \binom{x}{j} d\mu_{-1}(x) = \frac{(-1)^j}{2^j}, \tag{28}$$

which was given by Kim et al. [25].

In order to give integral of a function associated with the Dirichlet character with conductor d, the following notations are also needed:

Let p be a fixed prime. Let d be a fixed positive integer with $(p, d) = 1$; we have

$$\mathbb{X} = \mathbb{X}_d = \lim_{\overleftarrow{N}} \mathbb{Z}/dp^N\mathbb{Z},$$

$$\mathbb{X}_1 = \mathbb{Z}_p$$

$$\mathbb{X}^* = \bigcup_{\substack{0<a<dp \\ (a,p)=1}} a + dp\mathbb{Z}_p$$

and

$$a + dp^N\mathbb{Z}_p = \left\{x \in \mathbb{X} \mid x \equiv a \left(mod \left(dp^N\right)\right)\right\}$$

such that $a \in \mathbb{Z}$ satisfies the condition $0 \leq a < dp^N$. Thus, we have

$$\int_{\mathbb{Z}_p} f(x)d\mu_1(x) = \int_{\mathbb{X}} f(x)d\mu_1(x) \tag{29}$$

for the uniformly differentiable function $f : \mathbb{Z}_p \to \mathbb{C}_p$ (*cf.* [32, 33, 35, 37, 81]).

1.3 Other Needed Notations and Definitions

In order to give the results in this chapter, we next recall other needed notations and definitions regarding the well-known classical numbers and polynomials with their generating functions:

The Stirling numbers of the first kind, $S_1(n, k)$, are defined as follows:

$$(x)_n = \sum_{k=0}^{n} S_1(n, k)\, x^k \tag{30}$$

and

$$\frac{(\log(1+t))^k}{k!} = \sum_{n=k}^{\infty} S_1(n, k)\frac{t^n}{n!}; \quad (k \in \mathbb{N}_0), \tag{31}$$

and also these numbers satisfy the following recurrence relation:

$$S_1(n+1, k) = -nS_1(n, k) + S_1(n, k-1)$$

such that $S_1(0, 0) = 1$, $S_1(0, k) = 0$ if $k > 0$, $S_1(n, 0) = 0$ if $n > 0$, $S_1(n, k) = 0$ if $k > n$ (*cf.* [5, 7, 9, 10, 77, 79, 85, 109, 110, 112]; and the references cited therein).

The Stirling numbers of the first kind have the following well-known computation formula:

$$S_1(n,k) = \sum_{j=0}^{n-k} \sum_{m=0}^{j} (-1)^m \binom{2n-k}{n-k-j, n-k+j, k} \binom{j}{m} \frac{km^{j+n-k}}{(n+j)j!}, \tag{32}$$

where

$$\binom{2n-k}{n-k-j, n-k+j, k} = \frac{(2n-k)!}{(n-k-j)!\,(n-k+j)!k!}$$

(*cf.* [10, 95]; and the references cited therein).

The λ-Stirling numbers of the second kind, $S_2(n,k;\lambda)$, are defined with generating function given below (*cf.* [85, 112]):

$$F_{S_2}(t; v; \lambda) = \frac{\left(\lambda e^t - 1\right)^v}{v!} = \sum_{n=0}^{\infty} S_2(n, v; \lambda) \frac{t^n}{n!}, \quad (v \in \mathbb{N}_0), \tag{33}$$

which, for $\lambda = 1$, reduces to the Stirling numbers of the second kind, $S_2(n,k)$, given by

$$F_S(t,k) = \frac{\left(e^t - 1\right)^k}{k!} = \sum_{n=0}^{\infty} S_2(n,k) \frac{t^n}{n!}, \tag{34}$$

and for these numbers, the following explicit formula holds true:

$$S_2(n,k) = \frac{1}{k!} \sum_{j=0}^{k} (-1)^{k-j} \binom{k}{j} j^n. \tag{35}$$

A relation between the Stirling numbers of the second kind, $S_2(n,k)$, and the combinatorial numbers $y_6(n,k;\lambda,p)$, defined in [96], is given as follows:

$$S_2(n,k) = (-1)^k y_6(n,k;-1,1),$$

where

$$y_6(n,k;\lambda,p) = \frac{1}{k!} \sum_{j=0}^{k} \binom{k}{j}^p \lambda^j j^n. \tag{36}$$

Also these numbers satisfy the following recurrence relation:

$$S_2(n+1,k) = S_2(n,k-1) + kS_2(n,k)$$

such that $S_2(0,0) = 1$, $S_2(n,k) = 0$ if $k > n$, $S_2(n,0) = 0$ if $n > 0$ (*cf.* [5, 12, 79, 85, 112]; and the references cited therein).

The Bell polynomials (i.e., exponential polynomials), $Bl_n(x)$, is defined by

$$Bl_n(x) = \sum_{v=1}^{n} S_2(n,v)\, x^v \tag{37}$$

so that the generating function for the Bell polynomials is given by

$$F_{Bell}(t,x) = e^{(e^t-1)x} = \sum_{n=0}^{\infty} Bl_n(x) \frac{t^n}{n!} \tag{38}$$

(*cf.* [10, 79]).

The Daehee numbers D_n and the Daehee polynomials $D_n(x)$ are defined, respectively, by the following generating function:

$$F_D(t) = \frac{\log(1+t)}{t} = \sum_{n=0}^{\infty} D_n \frac{t^n}{n!} \tag{39}$$

and

$$F_D(x,t) = F_D(t)(1+t)^x = \sum_{n=0}^{\infty} D_n(x) \frac{t^n}{n!} \tag{40}$$

(*cf.* [24, 90, 91]).

By using (39), the explicit formula for the Daehee numbers is given by

$$D_n = D_n(0) = \frac{(-1)^n n!}{n+1} \tag{41}$$

(*cf.* [13, 24]).

The Peters polynomials $s_k(x;\lambda,\mu)$, which is a member of the family of the Sheffer polynomials, are given by the following generating functions:

$$\frac{1}{\left(1+(1+t)^{\lambda}\right)^{\mu}}(1+t)^x = \sum_{n=0}^{\infty} s_n(x;\lambda,\mu)\frac{t^n}{n!} \tag{42}$$

(*cf.* [4, p.128], [22, 26, 28, 29, 31, 36, 54, 79, 84, 92, 95, 99, 102, 105, 106]; and also see cited references therein).

In the special case of (42) when $\mu = 1$, the Peters polynomials are reduced to the Boole polynomials $\xi_n(x;\lambda) = s_n(x;\lambda,1)$ (*cf.* [22, 79, 99]).

Setting $x = 0$ into (42) yields the generating functions for the Peters numbers denoted by $s_n(\lambda,\mu) = s_n(0;\lambda,\mu)$ (*cf.* [22, 79, 99]).

Setting $\lambda = \mu = 1$ into (42) yields the numbers $r_n(x) = s_n(x; 1, 1)$ studied by Jordan [22].

In addition, for $\lambda = \mu = 1$, we also have the Changhee polynomials $Ch_n(x) = 2s_n(x; 1, 1)$ defined by

$$F_{Ch}(x,t) = F_{Ch}(t)(1+t)^x = \sum_{n=0}^{\infty} Ch_n(x)\frac{t^n}{n!} \tag{43}$$

(*cf.* [25, 46, 49]). Observe that substituting $x = 0$ into (43) yields

$$F_{Ch}(t) = \frac{2}{t+2} = \sum_{n=0}^{\infty} Ch_n \frac{t^n}{n!} \tag{44}$$

in which Ch_n denotes the Changhee numbers whose explicit formula is given by

$$Ch_n = \frac{(-1)^n n!}{2^n} \tag{45}$$

(*cf.* [25, 46]).

The Bernoulli numbers of the second kind $b_n(0)$ (also called the Cauchy numbers) are defined by means of the following generating function (*cf.* [79, p. 116]):

$$\frac{t}{\log(1+t)} = \sum_{n=0}^{\infty} b_n(0)\frac{t^n}{n!}, \tag{46}$$

and these numbers are also calculated by the definite integral of the falling factorial $(x)_n$, from 0 to 1, as follows:

$$b_n(0) = \int_0^1 (x)_n \, dx$$

(*cf.* [79, pp. 113–117]). By using the above formula, few of the Cauchy numbers are given as follows:

$$b_0(0) = 1, b_1(0) = \frac{1}{2}, b_2(0) = -\frac{1}{12}, b_3(0) = \frac{1}{24}, b_4(0) = -\frac{19}{720},$$

and so on (cf. [48, 51, 77, 79]; and the references cited therein).

The Humbert polynomials $\Pi_{n,m}^{(\lambda)}(x)$ are defined by Humbert in [19] with the following generating function:

$$\left(1 - mxt + t^m\right)^{-\lambda} = \sum_{n=0}^{\infty} \Pi_{n,m}^{(\lambda)}(x)\, t^n$$

(cf. [19], [109, p. 86, Eq-(26)], [72]), and the recurrence relation for these polynomials is given as follows:

$$(n+1)\,\Pi_{n+1,m}^{(\lambda)}(x) - mx\,(n+\lambda)\,\Pi_{n,m}^{(\lambda)}(x) - (n+m\lambda-m+1)\,\Pi_{n-m+1,m}^{(\lambda)}(x) = 0$$

(cf. [12, 70]; and the references cited therein).

The generalized Humbert polynomials $P_n(m,x,y,p,C)$ are defined by the following generating functions:

$$\left(C - mxt + yt^m\right)^p = \sum_{n=0}^{\infty} P_n(m,x,y,p,C)t^n,$$

and it is clear that

$$P_n(m,x,1,-\lambda,1) = \Pi_{n,m}^{(\lambda)}(x)$$

(*cf.* [12, 15, 70, 72]).

The Poisson–Charlier polynomials $C_n(x;a)$, which are members of the family of Sheffer-type sequences, are defined as below:

$$F_{pc}(t,x;a) = e^{-t}\left(\frac{t}{a}+1\right)^x = \sum_{n=0}^{\infty} C_n(x;a)\frac{t^n}{n!}, \tag{47}$$

where

$$C_n(x;a) = \sum_{j=0}^{n} (-1)^{n-j}\binom{n}{j}\frac{(x)_j}{a^j} \tag{48}$$

(*cf.* [79, p. 120], [97]).

Let $t \in \mathbb{C}$, $x \in [0,1]$, and $k \in \mathbb{N}_0$. Then, the generating function for the Bernstein Basis functions, $B_k^n(x)$, is given as follows:

$$\frac{(xt)^k e^{(1-x)t}}{k!} = \sum_{n=0}^{\infty} B_k^n(x)\frac{t^n}{n!}, \tag{49}$$

where

$$B_k^n(x) = \binom{n}{k}x^k(1-x)^{n-k}, \qquad (k=0,1,\ldots,n;\ \ n\in\mathbb{N}_0), \tag{50}$$

which have relationships with a large number of concepts including the Catalan numbers, the binomial distribution, the Poisson distribution, etc.; see, for details, [1, 60, 86–88, 104] and also cited references therein.

2 A Certain Class of Combinatorial Numbers $Y_{n,\chi}(\lambda, q)$ and Polynomials $Y_{n,\chi}(z; \lambda, q)$ Attached to Dirichlet Characters

In this section, we recall a certain class of combinatorial numbers and polynomials attached to Dirichlet characters as follows:

Let $x, \lambda \in \mathbb{Z}_p$ and χ be a non-trivial Dirichlet character with conductor d.

With the application of the p-adic q-integrals to the following continuous differentiable function on the ring of p-adic integers,

$$f(x, t; \lambda) = \lambda^x (1 + \lambda t)^x \chi(x), \tag{51}$$

Simsek [93] gave the construction of the generating functions for the generalized Apostol-type numbers and polynomials attached to the Dirichlet character, respectively, with odd and even conductors.

Substituting (51) into (27), we get

$$\int_{\mathbb{X}} \lambda^x (1 + \lambda t)^x \chi(x) d\mu_{-q}(x) = \frac{1+q}{(\lambda q)^d (1 + \lambda t)^d - (-1)^d} \times \sum_{j=0}^{d-1} (-1)^j \chi(j) (\lambda q)^j (1 + \lambda t)^j, \tag{52}$$

where $\lambda \in \mathbb{Z}_p$.

In [93], Simsek investigated (52) in two-folds:

In the first fold, by selecting the conductor d as odd number and using the integral formula arising from this selection, Simsek defined the generalized Apostol–Changhee numbers and polynomials, see, for details, [93].

Furthermore, in the second fold, by selecting the conductor d as even number, Simsek also defined the generalized Apostol-type numbers attached to the Dirichlet character with even conductor whose construction is given below in detail:

Let d be an even integer. If χ is the Dirichlet character with even conductor d, then Eq. (52) reduces to the following integral equation:

$$\int_{\mathbb{X}} \lambda^x (1+\lambda t)^x \chi(x) d\mu_{-q}(x) = \frac{1+q}{(\lambda q)^d (1 + \lambda t)^d - 1} \sum_{j=0}^{d-1} (-1)^j \chi(j) (\lambda q)^j (1+\lambda t)^j.$$

By the aid of the right-hand side of the above equation, Simsek constructed the generating functions for the family of the generalized Apostol-type numbers $Y_{n,\chi}(\lambda, q)$ and the generalized Apostol-type polynomials $Y_{n,\chi}(z; \lambda, q)$, respectively, as follows:

$$H(t;\lambda,q,\chi)=\frac{1+q}{(\lambda q)^{d}(1+\lambda t)^{d}-1}\sum_{j=0}^{d-1}(-1)^{j}\chi(j)(\lambda q)^{j}(1+\lambda t)^{j}$$
$$=\sum_{n=0}^{\infty}Y_{n,\chi}(\lambda,q)\frac{t^{n}}{n!}, \tag{53}$$

and

$$H(t,z;\lambda,q,\chi)=(1+\lambda t)^{z}H(t;\lambda,q,\chi)=\sum_{n=0}^{\infty}Y_{n,\chi}(z;\lambda,q)\frac{t^{n}}{n!}, \tag{54}$$

where d is an even positive integer and $\lambda\in\mathbb{Z}_p$ with $\lambda\neq 1$.

By (53) and (54), we have the relation between the numbers $Y_{n,\chi}(\lambda,q)$ and the polynomials $Y_{n,\chi}(z;\lambda,q)$ given by the following theorem:

Theorem 1 (*cf.* [93]) *Let $n\in\mathbb{N}_0$. Let d be an even positive integer. Then we have*

$$Y_{n,\chi}(z;\lambda,q)=\sum_{j=0}^{n}\binom{n}{j}\lambda^{n-j}(z)_{n-j}Y_{j,\chi}(\lambda,q). \tag{55}$$

Observe that

$$Y_{n,\chi}(\lambda,q)=Y_{n,\chi}(0;\lambda,q).$$

Remark 1 For further properties regarding the numbers $Y_{n,\chi}(\lambda,q)$ and the polynomials $Y_{n,\chi}(z;\lambda,q)$, the interested reader may refer to [93].

In the special case when $q\to 1$, (53) reduces to the following generating functions for the numbers $Y_{n,\chi}(\lambda)$:

$$\frac{2\sum_{j=0}^{d-1}(-1)^{j}\chi(j)\lambda^{j}(1+\lambda t)^{j}}{\lambda^{d}(1+\lambda t)^{d}-1}=\sum_{n=0}^{\infty}Y_{n,\chi}(\lambda)\frac{t^{n}}{n!}. \tag{56}$$

Some numerical applications of (56) are given as follows:

Let F_n be the Fibonacci numbers. It is known from [50, Lemma 5.1, p. 78] that the Fibonacci numbers satisfy

$$\left(\frac{1+\sqrt{5}}{2}\right)^{n}=\left(\frac{1+\sqrt{5}}{2}\right)F_n+F_{n-1} \tag{57}$$

and

$$\left(\frac{1-\sqrt{5}}{2}\right)^{n}=\left(\frac{1-\sqrt{5}}{2}\right)F_n+F_{n-1} \tag{58}$$

where $n \in \mathbb{N}$. Thus, by substituting

$$\lambda=\frac{1+\sqrt{5}}{2} \quad \text{and} \quad \lambda=\frac{1-\sqrt{5}}{2} \tag{59}$$

into (56), we, respectively, get

$$\frac{2\sum\limits_{j=0}^{d-1}(-1)^{j}\chi(j)\sum\limits_{l=0}^{j}\binom{j}{l}\left(\left(\frac{1+\sqrt{5}}{2}\right)F_{j+l}+F_{j+l-1}\right)t^{l}}{\sum\limits_{l=0}^{d}\binom{d}{l}\left(\left(\frac{1+\sqrt{5}}{2}\right)F_{d+l}+F_{d+l-1}\right)t^{l}-1}=\sum_{n=0}^{\infty}Y_{n,\chi}\left(\frac{1+\sqrt{5}}{2}\right)\frac{t^{n}}{n!},$$

and

$$\frac{2\sum\limits_{j=0}^{d-1}(-1)^{j}\chi(j)\sum\limits_{l=0}^{j}\binom{j}{l}\left(\left(\frac{1-\sqrt{5}}{2}\right)F_{j+l}+F_{j+l-1}\right)t^{l}}{\sum\limits_{l=0}^{d}\binom{d}{l}\left(\left(\frac{1-\sqrt{5}}{2}\right)F_{d+l}+F_{d+l-1}\right)t^{l}-1}=\sum_{n=0}^{\infty}Y_{n,\chi}\left(\frac{1-\sqrt{5}}{2}\right)\frac{t^{n}}{n!}.$$

Therefore, we obtain the generating functions for the numbers $S_{n,\chi}$:

$$S_{n,\chi}=Y_{n,\chi}\left(\frac{1+\sqrt{5}}{2}\right)+Y_{n,\chi}\left(\frac{1-\sqrt{5}}{2}\right), \tag{60}$$

as follows:

$$\sum_{n=0}^{\infty}S_{n,\chi}\frac{t^{n}}{n!}=\frac{2\sum\limits_{j=0}^{d-1}(-1)^{j}\chi(j)\sum\limits_{l=0}^{j}\binom{j}{l}\left(\left(\frac{1+\sqrt{5}}{2}\right)F_{j+l}+F_{j+l-1}\right)t^{l}}{\sum\limits_{l=0}^{d}\binom{d}{l}\left(\left(\frac{1+\sqrt{5}}{2}\right)F_{d+l}+F_{d+l-1}\right)t^{l}-1}$$
$$+\frac{2\sum\limits_{j=0}^{d-1}(-1)^{j}\chi(j)\sum\limits_{l=0}^{j}\binom{j}{l}\left(\left(\frac{1-\sqrt{5}}{2}\right)F_{j+l}+F_{j+l-1}\right)t^{l}}{\sum\limits_{l=0}^{d}\binom{d}{l}\left(\left(\frac{1-\sqrt{5}}{2}\right)F_{d+l}+F_{d+l-1}\right)t^{l}-1}.$$

Here, we note that investigation of the fundamental properties of the numbers $S_{n,\chi}$ and their relationships with other special numbers are not addressed in this chapter.

We now set

$$W(t,\chi)=\frac{2\sum\limits_{j=0}^{d-1}(-1)^{j}\chi(j)\sum\limits_{l=0}^{j}\binom{j}{l}\left(\left(\frac{1+\sqrt{5}}{2}\right)F_{j+l}+F_{j+l-1}\right)t^{l}}{\sum\limits_{l=0}^{d}\binom{d}{l}\left(\left(\frac{1+\sqrt{5}}{2}\right)F_{d+l}+F_{d+l-1}\right)t^{l}-1}+\frac{2\sum\limits_{j=0}^{d-1}(-1)^{j}\chi(j)\sum\limits_{l=0}^{j}\binom{j}{l}\left(\left(\frac{1-\sqrt{5}}{2}\right)F_{j+l}+F_{j+l-1}\right)t^{l}}{\sum\limits_{l=0}^{d}\binom{d}{l}\left(\left(\frac{1-\sqrt{5}}{2}\right)F_{d+l}+F_{d+l-1}\right)t^{l}-1},$$

and

$$V(t,x,\chi)=W(t,\chi)(1+t)^{x}. \tag{61}$$

Then, by using the above functions we define a new class of special polynomials $S_{n,\chi}(x)$, which are a linear combination of the numbers $S_{n,\chi}$, by the following generating functions:

$$V(t,x,\chi)=\sum_{n=0}^{\infty}S_{n,\chi}(x)\frac{t^{n}}{n!}. \tag{62}$$

Here, we also state that investigation of the fundamental properties of the polynomials $S_{n,\chi}(x)$ and their relationships with other special numbers and polynomials are not addressed in this chapter.

3 Illustrations of the Generating Functions for the Numbers $Y_{n,\chi}(\lambda)$ by Dirichlet Characters with Different Conductors d

In this section, by using Wolfram programming language in Mathematica [114], we also provide some two-dimensional and three-dimensional illustrations, involving surface plots and parametric plots, for the generating functions given by (56).

Let $\chi_{d,m}$ denote the m-th Dirichlet character with conductor d. In the case when $d=8$, due to the fact that $\varphi(8)=4$, there exist four distinct Dirichlet characters with conductor 8, namely $\chi_{8,1}$, $\chi_{8,2}$, $\chi_{8,3}$, and $\chi_{8,4}$, which are, respectively, given in each row of Table 6 by using the following Mathematica code:

Implementation 1 Mathematica code to create the rows of Table 6

```
Table[DirichletCharacter[8, j ,n ],{ j ,1, EulerPhi[8]},{n ,0,7}]
```

Table 6 All Dirichlet characters with conductor 8

n	0	1	2	3	4	5	6	7
$\chi_{8,1}(n)$	0	1	0	1	0	1	0	1
$\chi_{8,2}(n)$	0	1	0	−1	0	−1	0	1
$\chi_{8,3}(n)$	0	1	0	−1	0	1	0	−1
$\chi_{8,4}(n)$	0	1	0	1	0	−1	0	−1

Thus, in conjunction with the Dirichlet characters given by Table 6, we have exactly four different cases of the generating functions for the numbers $Y_{n,\chi}(\lambda)$.

Here, let us rewrite (56) as in the following equation:

$$H(t;\lambda,\chi_{d,m}):=\frac{2\sum_{j=0}^{d-1}(-1)^j\chi_{d,m}(j)\lambda^j(1+\lambda t)^j}{\lambda^d(1+\lambda t)^d-1}=\sum_{n=0}^{\infty}Y_{n,\chi_{d,m}}(\lambda)\frac{t^n}{n!}, \tag{63}$$

where $\chi_{d,m}$ denotes the m-th Dirichlet character with conductor d.

By application of Table 6 to Eq. (63), we get the following generating functions for the numbers $Y_{n,\chi_{d,m}}(\lambda)$ for $d=8$ and $m=1$, $m=2$, $m=3$, and $m=4$:

Implementation 2 Let the letter l denote the parameter λ. Then, the following Mathematica code including the procedure `GenFuncH` returns the generating functions $H(t;\lambda,\chi_{d,m})$

```
GenFuncH[l_,t_,d_,m_]:=(2/((l^d)*(1+l*t)^d−1))* Sum[((−1)^j)* DirichletCharacter[d,m,j]
    *(l^j)*(1+l*t)^j, {j,0,d−1}].
```

By using the Mathematica code given in Implementation 2, in the case when $d=8$, we have the following four different generating functions for the numbers $Y_{n,\chi_{8,m}}(\lambda)$ with $m=1$, $m=2$, $m=3$, and $m=4$ (or equivalently all generating functions for the numbers $Y_{n,\chi}(\lambda)$ attached to the Dirichlet characters with conductor 8 so that $\chi\equiv\{\chi_{8,1},\chi_{8,2},\chi_{8,3},\chi_{8,4}\}$):

$$H(t;\lambda,\chi_{8,1})=\frac{2\left(-\lambda(1+\lambda t)-\lambda^3(1+\lambda t)^3-\lambda^5(1+\lambda t)^5-\lambda^7(1+\lambda t)^7\right)}{\lambda^8(1+\lambda t)^8-1}$$

$$=\sum_{n=0}^{\infty}Y_{n,\chi_{8,1}}(\lambda)\frac{t^n}{n!},$$

$$H(t;\lambda,\chi_{8,2})=\frac{2\left(-\lambda(1+\lambda t)+\lambda^3(1+\lambda t)^3+\lambda^5(1+\lambda t)^5-\lambda^7(1+\lambda t)^7\right)}{\lambda^8(1+\lambda t)^8-1}$$

$$= \sum_{n=0}^{\infty} Y_{n,\chi_{8,2}}(\lambda)\frac{t^n}{n!},$$

$$H(t;\lambda,\chi_{8,3}) = \frac{2\left(-\lambda(1+\lambda t)+\lambda^3(1+\lambda t)^3-\lambda^5(1+\lambda t)^5+\lambda^7(1+\lambda t)^7\right)}{\lambda^8(1+\lambda t)^8-1}$$

$$= \sum_{n=0}^{\infty} Y_{n,\chi_{8,3}}(\lambda)\frac{t^n}{n!},$$

and

$$H(t;\lambda,\chi_{8,4}) = \frac{2\left(-\lambda(1+\lambda t)-\lambda^3(1+\lambda t)^3+\lambda^5(1+\lambda t)^5+\lambda^7(1+\lambda t)^7\right)}{\lambda^8(1+\lambda t)^8-1}$$

$$= \sum_{n=0}^{\infty} Y_{n,\chi_{8,4}}(\lambda)\frac{t^n}{n!}.$$

Observe that the number of the generating functions $H(t;\lambda,\chi)$ is based upon the number of the Dirichlet characters in conjunction with the image of the conductor d by the Euler totient function.

By implementing the above special generating functions in Mathematica (See Implementation 3), we present Fig. 1 that includes three-dimensional plots (Plot3D) of the functions $H(t;\lambda,\chi_{d,m})$ for the randomly selected special cases $d=8$, $t\in[-1,1]$ and $\lambda\in\left[-\frac{1}{2},\frac{1}{2}\right]$ for all possible cases of the Dirichlet characters with conductor 8.

Implementation 3 Let the letter l denote the parameter λ. Then, the following Mathematica code returns three-dimensional plots (Plot3D) of the generating functions $H(t;\lambda,\chi_{d,m})$ for the randomly selected special cases. For plots, see Fig. 1

```
Plot3D[GenFuncH[l,t,8,1], {l,-0.5,0.5}, {t,-1,1}, AxesLabel->{ToString[
    ToExpression["{HoldForm}[\\lambda]",TeXForm],TraditionalForm],"t",ToString[
    ToExpression["{HoldForm}[H\\left(\\lambda;t,\\chi_{_{8,1}}\\right)]", TeXForm],
    TraditionalForm]}, LabelStyle->Directive[Black, Bold], ColorFunction->"
    BlueGreenYellow"]

Plot3D[GenFuncH[l,t,8,2], {l,-0.5,0.5}, {t,-1,1}, AxesLabel->{ToString[
    ToExpression["{HoldForm}[\\lambda]",TeXForm],TraditionalForm],"t",ToString[
    ToExpression["{HoldForm}[H\\left(\\lambda;t,\\chi_{_{8,2}}\\right)]", TeXForm],
    TraditionalForm]}, LabelStyle->Directive[Black, Bold], ColorFunction->"
    BlueGreenYellow"]

Plot3D[GenFuncH[l,t,8,3], {l,-0.5,0.5}, {t,-1,1}, AxesLabel->{ToString[
    ToExpression["{HoldForm}[\\lambda]",TeXForm],TraditionalForm],"t",ToString[
    ToExpression["{HoldForm}[H\\left(\\lambda;t,\\chi_{_{8,3}}\\right)]", TeXForm],
```

```
TraditionalForm]}, LabelStyle->Directive[Black, Bold], ColorFunction->"
    BlueGreenYellow"]

Plot3D[GenFuncH[l,t,8,4], {l,-0.5,0.5}, {t,-1,1}, AxesLabel->{ToString[
    ToExpression["{HoldForm}[\\lambda]",TeXForm],TraditionalForm],"t",ToString[
    ToExpression["{HoldForm}[H\\left(\\lambda;t,\\chi_{_{8,4}}\\right)]", TeXForm],
    TraditionalForm]}, LabelStyle->Directive[Black, Bold], ColorFunction->"
    BlueGreenYellow"]
```

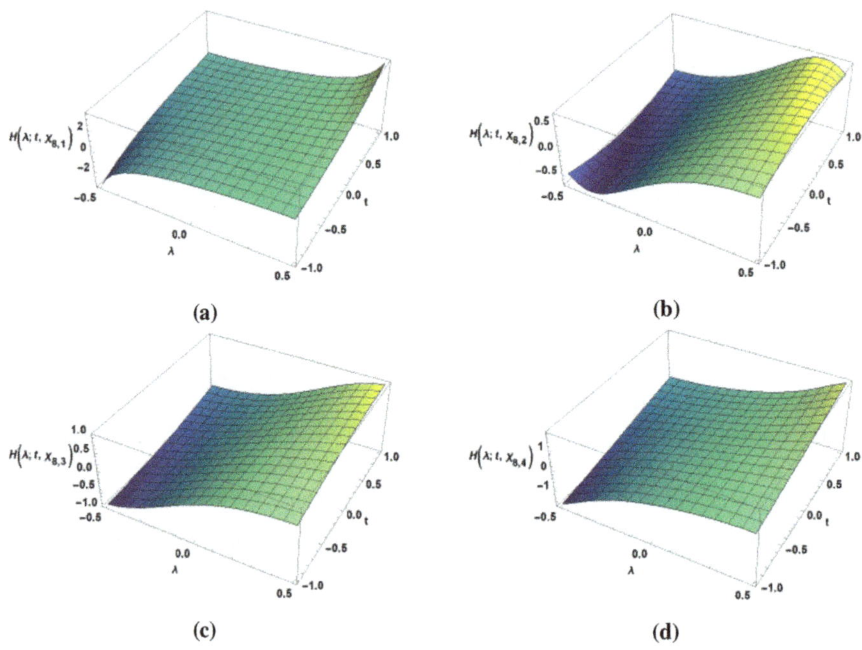

Fig. 1 Three-dimensional plots (Plot3D) of the functions $H(t; \lambda, \chi_{d,m})$ for the randomly selected special cases $d = 8$, $t \in [-1, 1]$ and $\lambda \in \left[-\frac{1}{2}, \frac{1}{2}\right]$ with (**a**) $m = 1$; (**b**) $m = 2$; (**c**) $m = 3$; (**d**) $m = 4$

Implementation 4 The following Mathematica code returns two-dimensional plots of the generating functions $H(t; \lambda, \chi_{d,m})$ for the randomly selected special cases. For plots, see Fig. 2

```
Plot[GenFuncH[1.5,t,8,1], {t, -1,1}, AxesLabel -> {Style["t", Bold, 10],Style[ToString
    [ToExpression["{HoldForm}[H\\left(\\frac{3}{2};t ,\\ chi_{_ {8,1}}\\ right ) ]",
    TeXForm], TraditionalForm], Bold, 10] }, LabelStyle -> Directive[Black, Bold],
    PlotStyle -> {Red, Thick}]

Plot[GenFuncH[1.5,t,8,2], {t, -1,1}, AxesLabel -> {Style["t", Bold, 10],Style[ToString
    [ToExpression["{HoldForm}[H\\left(\\frac{3}{2};t ,\\ chi_{_ {8,2}}\\ right ) ]",
```

```
    TeXForm], TraditionalForm], Bold, 10] }, LabelStyle -> Directive[Black, Bold],
    PlotStyle -> {Red, Thick}]

Plot[GenFuncH[1.5,t,8,3], {t, -1,1}, AxesLabel -> {Style["t", Bold, 10],Style[ToString
    [ToExpression["{HoldForm}[H\\left(\\frac{3}{2};t ,\\ chi_{_ {8,3}}\\ right)]",
    TeXForm], TraditionalForm], Bold, 10] }, LabelStyle -> Directive[Black, Bold],
    PlotStyle -> {Red, Thick}]

Plot[GenFuncH[1.5,t,8,4], {t, -1,1}, AxesLabel -> {Style["t", Bold, 10],Style[ToString
    [ToExpression["{HoldForm}[H\\left(\\frac{3}{2};t ,\\ chi_{_ {8,4}}\\ right)]",
    TeXForm], TraditionalForm], Bold, 10] }, LabelStyle -> Directive[Black, Bold],
    PlotStyle -> {Red, Thick}]
```

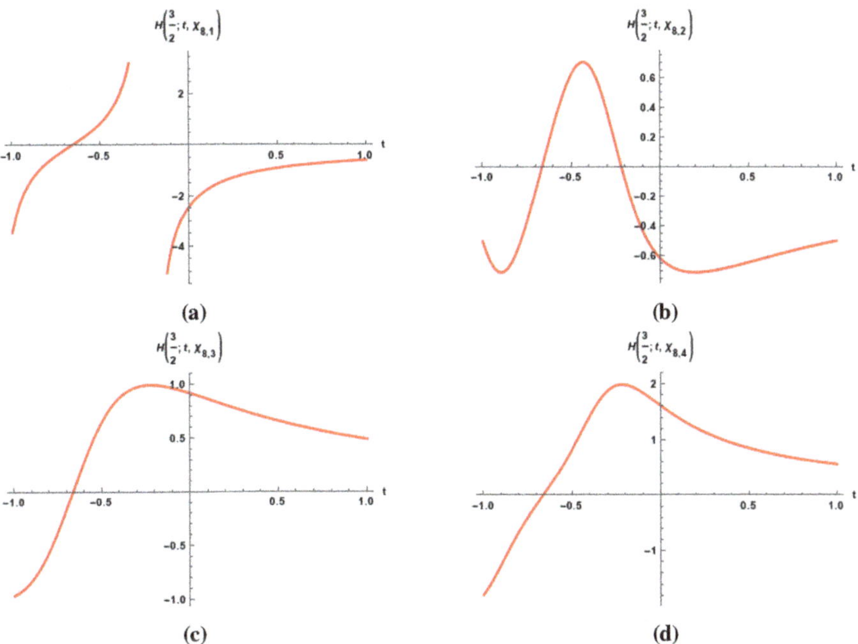

Fig. 2 Surface plots of the functions $H(t; \lambda, \chi_{d,m})$ for the randomly selected special cases $d = 8$, $\lambda = \frac{3}{2}$ and $t \in [-1, 1]$ with **(a)** $m = 1$; **(b)** $m = 2$; **(c)** $m = 3$; **(d)** $m = 4$

Figure 2 includes some two-dimensional plots of the functions $H(t; \lambda, \chi_{d,m})$ for the randomly selected special cases $d = 8$, $\lambda = \frac{3}{2}$ and $t \in [-1, 1]$ for all possible cases of the Dirichlet characters with conductor 8.

As another example, in the case when $d = 14$, due to the fact that $\varphi(14) = 6$, there exist six distinct Dirichlet characters with conductor 14, namely $\chi_{14,1}, \chi_{14,2}, \ldots, \chi_{14,6}$, which are, respectively, given in each row of Table 7 below:

Table 7 For $\omega = \exp(\pi i/3)$, all Dirichlet characters with conductor 14

n	1	3	5	9	11	13
$\chi_{14,1}(n)$	1	1	1	1	1	1
$\chi_{14,2}(n)$	1	ω	$-\omega^2$	ω^2	$-\omega$	-1
$\chi_{14,3}(n)$	1	ω^2	$-\omega$	$-\omega$	ω^2	1
$\chi_{14,4}(n)$	1	-1	-1	1	1	-1
$\chi_{14,5}(n)$	1	$-\omega$	ω^2	ω^2	$-\omega$	1
$\chi_{14,6}(n)$	1	$-\omega^2$	ω	$-\omega$	ω^2	-1

In conjunction with the Dirichlet characters given by Table 7 and using Eq. (63), we get the following generating functions for the numbers $Y_{n,\chi_{d,m}}(\lambda)$ for $d = 14$ and $m = 1, m = 2, m = 3, m = 4, m = 5$, and $m = 6$:

In the case when $d = 14$, we have the following six different generating functions for the numbers $Y_{n,\chi_{14,m}}(\lambda)$ with $m = 1$, $m = 2$, $m = 3$, $m = 4$, $m = 5$, and $m = 6$ (or equivalently all generating functions for the numbers $Y_{n,\chi}(\lambda)$ attached to the Dirichlet characters with conductor 14 so that $\chi \equiv \{\chi_{14,1}, \chi_{14,2}, \chi_{14,3}, \chi_{14,4}, \chi_{14,5}, \chi_{14,6}\}$):

Let $\omega = \exp(\pi i/3)$. Then we have

$$\begin{aligned} H(t;\lambda,\chi_{14,1}) &= \left(\frac{2}{\lambda^{14}(1+\lambda t)^{14}-1}\right) \\ &\quad \times \Big(-\lambda(1+\lambda t) - \lambda^3(1+\lambda t)^3 - \lambda^5(1+\lambda t)^5 - \lambda^9(1+\lambda t)^9 \\ &\quad -\lambda^{11}(1+\lambda t)^{11} - \lambda^{13}(1+\lambda t)^{13}\Big) \\ &= \sum_{n=0}^{\infty} Y_{n,\chi_{14,1}}(\lambda)\frac{t^n}{n!}, \end{aligned}$$

$$\begin{aligned} H(t;\lambda,\chi_{14,2}) &= \left(\frac{2}{\lambda^{14}(1+\lambda t)^{14}-1}\right) \\ &\quad \times \Big(-\lambda(1+\lambda t) - \omega\lambda^3(1+\lambda t)^3 + \omega^2\lambda^5(1+\lambda t)^5 - \omega^2\lambda^9(1+\lambda t)^9 \\ &\quad +\omega\lambda^{11}(1+\lambda t)^{11} + \lambda^{13}(1+\lambda t)^{13}\Big) \\ &= \sum_{n=0}^{\infty} Y_{n,\chi_{14,2}}(\lambda)\frac{t^n}{n!}, \end{aligned}$$

$$\begin{aligned} H(t;\lambda,\chi_{14,3}) &= \left(\frac{2}{\lambda^{14}(1+\lambda t)^{14}-1}\right) \\ &\quad \times \Big(-\lambda(1+\lambda t) - \omega^2\lambda^3(1+\lambda t)^3 + \omega\lambda^5(1+\lambda t)^5 + \omega\lambda^9(1+\lambda t)^9 \end{aligned}$$

$$-\omega^2\lambda^{11}(1+\lambda t)^{11}+\lambda^{13}(1+\lambda t)^{13}\Big)$$

$$=\sum_{n=0}^{\infty} Y_{n,\chi_{14,3}}(\lambda)\frac{t^n}{n!},$$

$$H(t;\lambda,\chi_{14,4})=\left(\frac{2}{\lambda^{14}(1+\lambda t)^{14}-1}\right)$$
$$\times\Big(-\lambda(1+\lambda t)+\lambda^3(1+\lambda t)^3+\lambda^5(1+\lambda t)^5-\lambda^9(1+\lambda t)^9$$
$$-\lambda^{11}(1+\lambda t)^{11}+\lambda^{13}(1+\lambda t)^{13}\Big)$$
$$=\sum_{n=0}^{\infty} Y_{n,\chi_{14,4}}(\lambda)\frac{t^n}{n!},$$

$$H(t;\lambda,\chi_{14,5})=\left(\frac{2}{\lambda^{14}(1+\lambda t)^{14}-1}\right)$$
$$\times\Big(-\lambda(1+\lambda t)+\omega\lambda^3(1+\lambda t)^3-\omega^2\lambda^5(1+\lambda t)^5-\omega^2\lambda^9(1+\lambda t)^9$$
$$\omega\lambda^{11}(1+\lambda t)^{11}-\lambda^{13}(1+\lambda t)^{13}\Big)$$
$$=\sum_{n=0}^{\infty} Y_{n,\chi_{14,5}}(\lambda)\frac{t^n}{n!},$$

and

$$H(t;\lambda,\chi_{14,6})=\left(\frac{2}{\lambda^{14}(1+\lambda t)^{14}-1}\right)$$
$$\times\Big(-\lambda(1+\lambda t)+\omega^2\lambda^3(1+\lambda t)^3-\omega\lambda^5(1+\lambda t)^5+\omega\lambda^9(1+\lambda t)^9$$
$$-\omega^2\lambda^{11}(1+\lambda t)^{11}+\lambda^{13}(1+\lambda t)^{13}\Big)$$
$$=\sum_{n=0}^{\infty} Y_{n,\chi_{14,6}}(\lambda)\frac{t^n}{n!}.$$

By implementing the above special generating functions in Mathematica (See Implementation 6), we present Fig. 3 that contains two parametric plots (ParametricPlot) and (ParametricPlot3D) derived from the real and imaginary parts of the functions $H(t;\lambda,\chi_{d,m})$ for the third Dirichlet character with conductor 14, $\chi_{14,3}$ in the cases when $\lambda=\frac{3}{2}$ and $t\in[-1,1]$. The reason why we select $\chi_{14,3}$ to plot is that $\chi_{14,3}$ is one of the Dirichlet characters with complex numbers.

Implementation 5 The following Mathematica code returns two-dimensional parametric plot and three-dimensional space curve derived from the real and imaginary parts of the generating functions $H(t; \lambda, \chi_{d,m})$ for the randomly selected case. For plots, see Fig. 3

```
ParametricPlot[{Re[GenFuncH[1.5,t,14,3]], Im[GenFuncH[1.5,t,14,3]]}, {t, -1,1},
    LabelStyle -> Directive[Black, Bold], PlotStyle -> {Red, Thick}]
```

```
ParametricPlot3D[{Re[GenFuncH[1.5,t,14,3]], Im[GenFuncH[1.5,t,14,3]], t}, {t, -1,1},
    LabelStyle -> Directive[Black, Bold],ColorFunction -> Function[{x, y, z, u}, Hue
    [u]]]
```

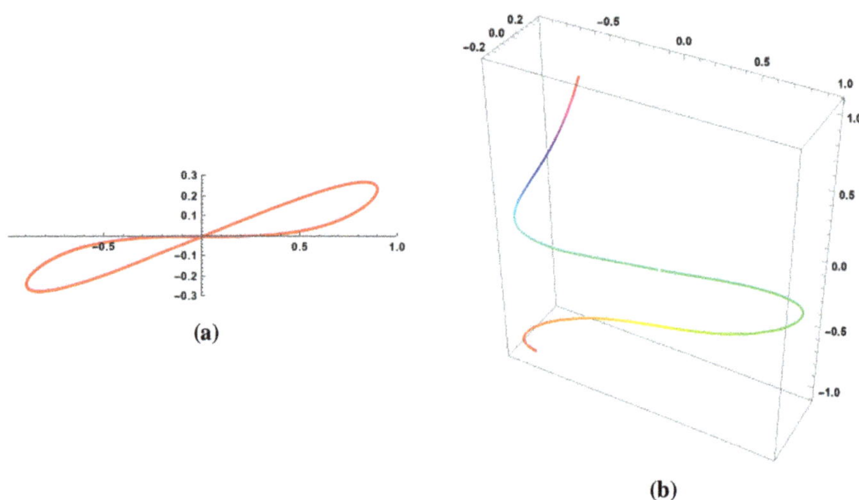

Fig. 3 For the case when $d = 14, \lambda = \frac{3}{2}$, and $t \in [-1, 1]$. (**a**) Two-dimensional parametric plot, as a function of t, derived from the real and imaginary parts of the generating functions $H(t; \lambda, \chi_{d,m})$. (**b**) Three-dimensional space curve parametrized by the variable t, which runs from the real part of the generating functions $H(t; \lambda, \chi_{d,m})$ to its imaginary part

Potential applications of the curves given especially in Fig. 3 may be investigated in the theory of the splines including the B-Spline, Euler spline, and other spline functions. Such an investigation to be done contributes some branches of computational geometry, such as computer-aided geometric design and other related areas, in the phase of method development for the mathematical description of some concepts.

Implementation 6 The following Mathematica code returns two-dimensional parametric plot and three-dimensional space curve derived from the real and imaginary parts of the generating functions $H(t; \lambda, \chi_{d,m})$ for the randomly selected case. For plots, see Fig. 4

```
Plot3D[Re[GenFuncH[x+ I y,1,14,3]], {x,−10, 10}, {y, −10, 10}, LabelStyle −> Directive[
    Black, Bold], ColorFunction −>"BlueGreenYellow"]
```

```
Plot3D[Im[GFYX[x+ I y,1,14,3]], {x,−10, 10}, {y, −10, 10}, LabelStyle −> Directive[
    Black, Bold], ColorFunction −>"BlueGreenYellow"]
```

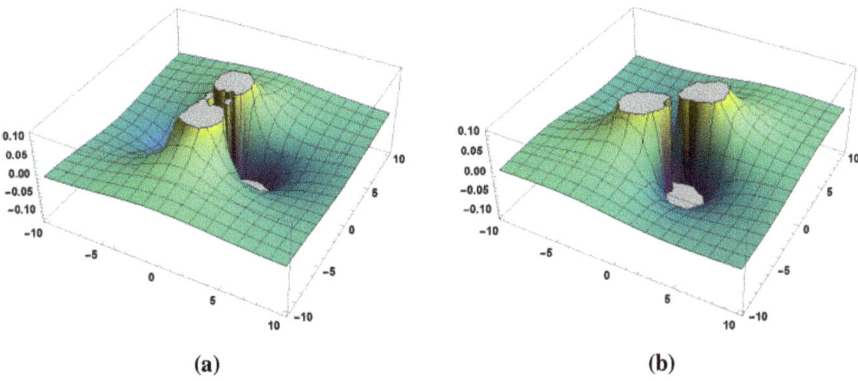

Fig. 4 For the case when $d = 14, m = 3, t = 3, \lambda = x+iy$ with $x \in [-10, 10]$ and $y \in [-10, 10]$ (**a**) Three-dimensional plot of the real part of the functions $H(t; \lambda, \chi_{d,m})$. (**b**) Three-dimensional plot of the imaginary part the functions $H(t; \lambda, \chi_{d,m})$

4 Derivative and Integral Formulas for the Polynomials $Y_{n,\chi}(z; \lambda, q)$

Here, derivative and integral formulas for the polynomials $Y_{n,\chi}(x; \lambda, q)$ are given. Moreover, by the aid of p-adic Volkenborn integral, some identities and combinatorial sums are derived.

Theorem 2 (*cf.* [93]) *Let n be a positive integer. Then we have*

$$\frac{\partial}{\partial z}\{Y_{n+1,\chi}(z; \lambda, q)\} = \sum_{j=0}^{n}(-1)^j\binom{n+1}{j+1}j!\lambda^{j+1}Y_{n-j,\chi}(z; \lambda, q).$$

Proof (cf. [93]) We shall give just a brief sketch of the proof as the details are similar to those in [93]: Differentiating (54) with respect to the parameter z yields the following partial differential equation:

$$\frac{\partial}{\partial z}\{H(t, z; \lambda, q, \chi)\} = H(t, z; \lambda, q, \chi)\log(1 + \lambda t).$$

The above partial differential equation gives us the following series equation:

$$\sum_{n=0}^{\infty}\frac{\partial}{\partial z}\{Y_{n,\chi}(z;\lambda,q)\}\frac{t^n}{n!}=\sum_{n=0}^{\infty}(-1)^n\frac{(\lambda t)^{n+1}}{n+1}\sum_{n=0}^{\infty}Y_{n,\chi}(z;\lambda,q)\frac{t^n}{n!}.$$

After some algebraic calculations with the aid of the Cauchy product rule for the related series and comparing the coefficients of $\frac{t^n}{n!}$ on both sides of the final equation, the desired result is obtained.

If we integrate both sides of (55), from 0 to 1, then we get a formula for the Riemann integral of the polynomials $Y_{n,\chi}(z;\lambda,q)$ by the following theorem:

Theorem 3 (*cf.* **[93]**)

$$\int_0^1 Y_{n,\chi}(z;\lambda,q)dz=\sum_{j=0}^{n}\binom{n}{j}\lambda^{n-j}b_{n-j}(0)Y_{j,\chi}(\lambda,q).$$

The p-adic bosonic integral representation for the polynomials $Y_{n,\chi}(z;\lambda,q)$ is given as follows:

$$\int_{\mathbb{X}} Y_{n,\chi}(z;\lambda,q)d\mu_1(z)=\sum_{j=0}^{n}\binom{n}{j}\lambda^{n-j}D_{n-j}Y_{j,\chi}(\lambda,q)$$

(*cf.* [93]). With the combination of (41), the following combinatorial sum is obtained:

$$\int_{\mathbb{X}} Y_{n,\chi}(z;\lambda,q)d\mu_1(z)=\sum_{j=0}^{n}(-1)^{n-j}\binom{n}{j}\frac{(n-j)!\lambda^{n-j}}{n+1-j}Y_{j,\chi}(\lambda,q)$$

(*cf.* [93]).

The p-adic fermionic integral representation for the polynomials $Y_{n,\chi}(z;\lambda,q)$ is given as follows:

$$\int_{\mathbb{X}} Y_{n,\chi}(z;\lambda,q)d\mu_{-1}(z)=\sum_{j=0}^{n}\binom{n}{j}\lambda^{n-j}Ch_{n-j}Y_{j,\chi}(\lambda,q).$$

With the combination of (45), we also get the following combinatorial sums:

$$\int_{\mathbb{X}} Y_{n,\chi}(z;\lambda,q)d\mu_{-1}(z)=\sum_{j=0}^{n}(-1)^{n-j}\binom{n}{j}\frac{(n-j)!\lambda^{n-j}}{2^{n-j}}Y_{j,\chi}(\lambda,q)$$

(*cf.* [93]).

5 Reduction to the Numbers $Y_n(\lambda)$ and the Polynomials $Y_n(x;\lambda)$

The occurrence and identification of the numbers $Y_n(\lambda)$ and the polynomials $Y_n(x;\lambda)$ are as follows:

When $q \rightarrow 1$ and $\chi \equiv 1$, the family of the numbers $Y_{n,\chi}(\lambda, q)$ and the polynomials $Y_{n,\chi}(z;\lambda, q)$ reduces another family of the numbers $Y_n(\lambda)$ and the polynomials $Y_n(x;\lambda)$ defined, respectively, by the following generating functions:

$$F(t,\lambda) = \frac{2}{\lambda(1+\lambda t)-1} = \sum_{n=0}^{\infty} Y_n(\lambda)\frac{t^n}{n!}, \tag{64}$$

and

$$F(t,x,\lambda) = \frac{2(1+\lambda t)^x}{\lambda(1+\lambda t)-1} = \sum_{n=0}^{\infty} Y_n(x;\lambda)\frac{t^n}{n!} \tag{65}$$

(*cf.* [93]).

By (64) and (65), we have the relation between the numbers $Y_n(\lambda)$ and the polynomials $Y_n(x;\lambda)$ given by

$$Y_n(x;\lambda) = \sum_{j=0}^{n} \binom{n}{j} \lambda^{n-j} (x)_{n-j} Y_j(\lambda) \tag{66}$$

(*cf.* [93]).

Observe that

$$Y_n(\lambda) = Y_n(0;\lambda).$$

As stated by Simsek in [93] that there exist some significant combinatorial identities essentially associated with the numbers $Y_n(\lambda)$, the polynomials $Y_n(x;\lambda)$ and some special numbers and polynomials such as the Apostol-type numbers and polynomials, the Stirling numbers and the Bernoulli numbers of the second kind. Recently, many applications of these numbers and polynomials have been studied and investigated by many researchers (*cf.* [23, 30, 93, 105, 106, 113]). Among others, in [23], Khan et al. called the polynomials $Y_n(x;\lambda)$ as "*Simsek polynomials*," and they constructed a 2-variable extension of the Simsek polynomials by the following generating functions (*cf.* [23]):

$$H(x,y,t;\lambda,\delta) = \frac{2(1+\lambda t)^x\left(1+\delta t^2\right)^y}{\lambda(1+\lambda t)-1}$$
$$= \sum_{n=0}^{\infty} Y_n(x,y;\lambda,\delta)\frac{t^n}{n!},$$

which, for $x = y = 0$, yields the Simsek numbers, i.e.:

$$Y_n(0, 0; \lambda, \delta) = Y_n(\lambda),$$

in which the parameter δ acts as a free variable.

Remark 2 In [23], Khan et al. gave not only quasimonomial properties of the 2-variable Simsek polynomials $Y_n(x, y; \lambda, \delta)$ on the Weyl group structure, but also differential equations satisfied by these polynomials. The interested reader may refer to [23] to see further details regarding these polynomials.

6 Some Properties of the Numbers $Y_n(\lambda)$ and the Polynomials $Y_n(x; \lambda)$ with Their Generating Functions

In this section, we give some properties of the numbers $Y_n(\lambda)$ and the polynomials $Y_n(x; \lambda)$.

With the application of the umbral calculus convention to (64), a recurrence relation for the numbers $Y_n(\lambda)$ is obtained as in the following theorem:

Theorem 4 (*cf.* [93]) *Let $n \in \mathbb{N}$. Then the numbers $Y_n(\lambda)$ are given by the following recurrence relation:*

$$Y_n(\lambda) = \frac{n\lambda^2}{1-\lambda} Y_{n-1}(\lambda), \tag{67}$$

with the initial condition:

$$Y_0(\lambda) = \frac{2}{\lambda - 1}.$$

In addition to the recurrence relation in (67), by using (64), an explicit formula for the number $Y_n(\lambda)$ is given as in the following theorem:

Theorem 5 (*cf.* [93]) *Let $n \in \mathbb{N}_0$. Then we have*

$$Y_n(\lambda) = 2(-1)^n \frac{n!}{\lambda - 1} \left(\frac{\lambda^2}{\lambda - 1}\right)^n. \tag{68}$$

Thus, by using not only (67), but also (68), first few values of the numbers $Y_n(\lambda)$ are computed as follows:

$$Y_0(\lambda) = \frac{2}{\lambda-1}, \quad Y_1(\lambda) = -\frac{2\lambda^2}{(\lambda-1)^2}, \quad Y_2(\lambda) = \frac{4\lambda^4}{(\lambda-1)^3},$$

$$Y_3(\lambda) = -\frac{12\lambda^6}{(\lambda-1)^4}, \quad Y_4(\lambda) = \frac{48\lambda^8}{(\lambda-1)^5},$$

and so on (*cf.* [93]).

Theorem 6 (**cf. [93]**) *Let $n \in \mathbb{N}_0$. Then the polynomials $Y_n(x;\lambda)$ are given by the following recurrence relation:*

$$2\,(x)_n\,\lambda^n = n\lambda^2 Y_{n-1}(x;\lambda) + (\lambda-1)\,Y_n(x;\lambda). \tag{69}$$

Proof (cf. [93]) Here, we shall give just a brief sketch of the proof as the details are similar to those in [93]. By making cross multiplication in (65), and then using the binomial theorem, we have

$$2\sum_{n=0}^{\infty}(x)_n\,\lambda^n\frac{t^n}{n!} = \lambda^2\sum_{n=0}^{\infty}Y_n(x;\lambda)\frac{t^{n+1}}{n!} + (\lambda-1)\sum_{n=0}^{\infty}Y_n(x;\lambda)\frac{t^n}{n!}.$$

Comparing the coefficients of $\frac{t^n}{n!}$ on both sides of the equation just above, the desired result is obtained.

Thus, by using not only (69), but also (66), first few values of the polynomials $Y_n\,(x;\lambda)$ are computed as follows:

$$Y_0(x;\lambda) = \frac{2}{\lambda-1},$$

$$Y_1(x;\lambda) = \frac{2\lambda}{\lambda-1}x - \frac{2\lambda^2}{(\lambda-1)^2},$$

$$Y_2(x;\lambda) = \frac{2\lambda^2}{\lambda-1}x^2 - \frac{6\lambda^3-2\lambda^2}{(\lambda-1)^2}x + \frac{4\lambda^4}{(\lambda-1)^3},$$

$$Y_3(x;\lambda) = \frac{2\lambda^3}{\lambda-1}x^3 - \frac{12\lambda^4-6\lambda^3}{(\lambda-1)^2}x^2 + \frac{22\lambda^5-14\lambda^4+4\lambda^3}{(\lambda-1)^3}x - \frac{12\lambda^6}{(\lambda-1)^4},$$

and so on (*cf.* [93]).

By (65), one has the following functional equation:

$$F(t,x+w;\lambda) = F(t,x;\lambda)(1+\lambda t)^w,$$

which yields the following result:

Theorem 7 (**cf. [93]**) *Let $n \in \mathbb{N}_0$. Then we have*

$$Y_n(x+w;\lambda) = \sum_{j=0}^{n}\binom{n}{j}\lambda^{n-j}(w)_{n-j}Y_j(x;\lambda). \tag{70}$$

7 Illustrations for the Numbers $Y_n(\lambda)$ and the Polynomials $Y_n(x; \lambda)$

Here, we give some illustrations for the numbers $Y_n(\lambda)$ and the polynomials $Y_n(x; \lambda)$ as follows:

Note that the numbers $Y_n(\lambda)$ are rational functions of real variable λ. Thus, by (68), we shall give some plots of the rational functions $Y_n(\lambda)$ in Fig. 5. Observe that there exists a vertical asymptote for all curves in Fig. 5, and its equation is $\lambda = 1$.

The implementation of Eq. (68) in Mathematica is given in Implementation 8:

Implementation 7 Let the letter l denote the parameter λ. Then, the following Mathematica code returns the rational functions $Y_n(\lambda)$

```
YNum[l_,n_]:=2*((−1)^n)* (Factorial[n]/(l−1))*(((l^2)/(l−1))^n)
```

Implementation 8 Let the letter l denote the parameter λ. Then, the following Mathematica code returns plots of the rational functions $Y_n(\lambda)$ for randomly selected special cases. For plots, see Fig. 5

```
Plot[Evaluate[Table[YNum[l,n],{n,0,4}] ], {l,−5,5},AxesLabel −> {Style[lparameter,
    Bold, 10],Style[expr2, Bold, 10] }, PlotLegends −> {ToString[ToExpression["{
    HoldForm}[{Y}_{0}\\left(\\lambda\\right)]",TeXForm],TraditionalForm], ToString[
    ToExpression["{HoldForm}[{Y}_{1}\\left(\\lambda\\right)]",TeXForm],
    TraditionalForm], ToString[ToExpression["{HoldForm}[{Y}_{2}\\left(\\lambda\\
    right)]",TeXForm], TraditionalForm], ToString[ToExpression["{HoldForm}[{Y}_
    {3}\\left(\\lambda\\right)]",TeXForm],TraditionalForm], ToString[ToExpression["{
    HoldForm}[{Y}_{4}\\left(\\lambda\\right)]",TeXForm], TraditionalForm]}, LabelStyle
    −> Directive[Black, Bold]]
```

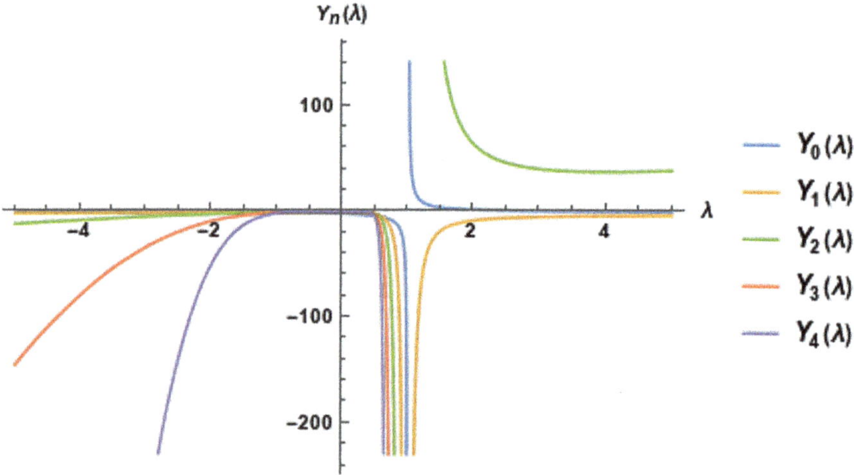

Fig. 5 Plots of the rational functions $Y_n(\lambda)$ for randomly selected special cases when $\lambda \in [-5, 5]$ and $n \in \{0, 1, 2, 3, 4\}$

By (68) and (66), we shall give some two-dimensional plots of the polynomials $Y_n(x; \lambda)$ in Figs. 6 and 7.

The implementation of Eq. (66) in Mathematica is given in Implementation 11:

Implementation 9 Let the letter l denote the parameter λ. Then, the following Mathematica code returns the polynomials $Y_n(x; \lambda)$

```
YPoly[x_,l_,n_]:=Sum[Binomial[n,j]*(l^(n-j))*FactorialPower[x, n-j, 1]*YNum[l,j], {j,0,
    n}]
```

Implementation 10 Let the letter l denote the parameter λ. Then, the following Mathematica code returns plots of the polynomials $Y_n(x; \lambda)$ for the randomly selected special cases. For plots, see Figs. 6 and 7

```
Plot[Evaluate[Table[YPoly[3,l,n],{n ,0,4}]], {l,-5,5}, AxesLabel -> {Style[lparameter,
    Bold, 10],Style[ToString[ToExpression["{HoldForm}[{Y}_{n}\\left(3;\\lambda\\right)
    ]",TeXForm], TraditionalForm], Bold, 10] }, PlotLegends -> {ToString[
    ToExpression["{HoldForm}[{Y}_{0}\\left(3;\\lambda\\right)]",TeXForm],
    TraditionalForm], ToString[ToExpression["{HoldForm}[{Y}_{1}\\left(3;\\lambda\\
    right)]",TeXForm], TraditionalForm], ToString[ToExpression["{HoldForm}[{Y}_
    {2}\\left(3;\\lambda\\right)]",TeXForm], TraditionalForm], ToString[ToExpression[
    "{HoldForm}[{Y}_{3}\\left(3;\\lambda\\right)]",TeXForm], TraditionalForm],ToString
    [ToExpression["{HoldForm}[{Y}_{4}\\left(3;\\lambda\\right)]",TeXForm],
    TraditionalForm]}, LabelStyle -> Directive[Black, Bold]]
```

```
Plot[Evaluate[Table[YPoly[x,0.5,n],{n ,0,4}]], {x,-5,5}, AxesLabel -> {Style["x", Bold,
    10],Style[ToString[ToExpression["{HoldForm}[{Y}_{n}\\left(x;\\frac {1}{2}\\ right)]
    ",TeXForm], TraditionalForm], Bold, 10] }, PlotLegends -> {ToString[
    ToExpression["{HoldForm}[{Y}_{0}\\left(x;\\frac{1}{2}\\right)]",TeXForm],
    TraditionalForm], ToString[ToExpression["{HoldForm}[{Y}_{1}\\left(x;\\frac{1}{2}\\
    right)]",TeXForm], TraditionalForm], ToString[ToExpression["{HoldForm}[{Y}_
    {2}\\left(x;\\frac{1}{2}\\right)]",TeXForm], TraditionalForm], ToString[
    ToExpression["{HoldForm}[{Y}_{3}\\left(x;\\frac{1}{2}\\right)]",TeXForm],
    TraditionalForm], ToString[ToExpression["{HoldForm}[{Y}_{4}\\left(x;\\frac{1}{2}\\
    right)]",TeXForm], TraditionalForm]}, LabelStyle -> Directive[Black, Bold]]
```

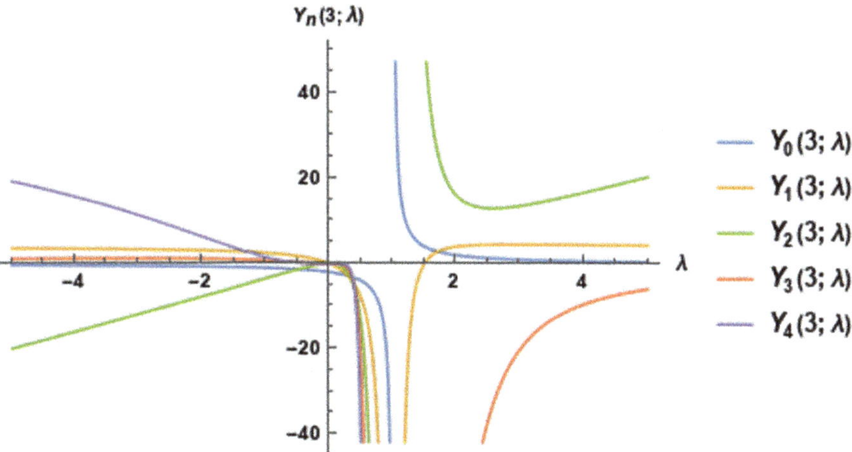

Fig. 6 Plots of the polynomials $Y_n(x; \lambda)$ for the case when $x = 3$ and $\lambda \in [-5, 5]$ with $n \in \{0, 1, 2, 3, 4\}$

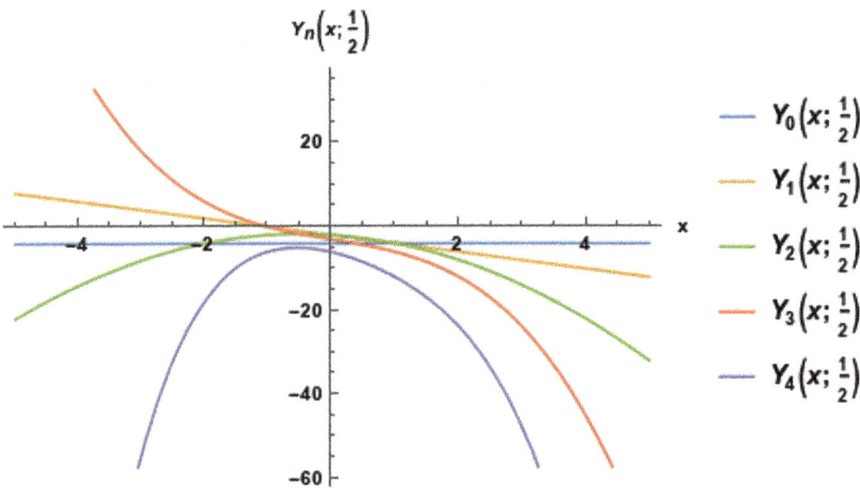

Fig. 7 Plots of the polynomials $Y_n(x;\lambda)$ for randomly selected case when $\lambda=\frac{1}{2}$ and $x\in[-5,5]$ with $n\in\{0,1,2,3,4\}$

In Fig. 8, we also give surface plots of the polynomials $Y_n(x;\lambda)$.

Implementation 11 Let the letter l denote the parameter λ. Then, the following Mathematica code returns surface plots of the polynomials $Y_n(x;\lambda)$ for the randomly selected special cases. For plots, see Fig. 8

```
Plot3D[YPoly[x,l,0],{x,-5,5},{l,-0.5,0.5}, AxesLabel ->{"x",ToString[ToExpression[
    "{HoldForm}[\\lambda]",TeXForm], TraditionalForm], ToString[ToExpression["{
    HoldForm}[{Y}_{0}\\left(x;\\lambda\\right)]",TeXForm], TraditionalForm]},
    LabelStyle -> Directive[Black, Bold], ColorFunction ->"BlueGreenYellow"]

Plot3D[YPoly[x,l,1],{x,-5,5},{l,-0.5,0.5}, AxesLabel ->{"x",ToString[ToExpression[
    "{HoldForm}[\\lambda]",TeXForm], TraditionalForm], ToString[ToExpression["{
    HoldForm}[{Y}_{1}\\left(x;\\lambda\\right)]",TeXForm], TraditionalForm]},
    LabelStyle -> Directive[Black, Bold], ColorFunction ->"BlueGreenYellow"]

Plot3D[YPoly[x,l,2],{x,-5,5},{l,-0.5,0.5}, AxesLabel ->{"x",ToString[ToExpression[
    "{HoldForm}[\\lambda]",TeXForm], TraditionalForm], ToString[ToExpression["{
    HoldForm}[{Y}_{2}\\left(x;\\lambda\\right)]",TeXForm], TraditionalForm]},
    LabelStyle -> Directive[Black, Bold], ColorFunction ->"BlueGreenYellow"]

Plot3D[YPoly[x,l,3],{x,-5,5},{l,-0.5,0.5}, AxesLabel ->{"x",ToString[ToExpression[
    "{HoldForm}[\\lambda]",TeXForm], TraditionalForm], ToString[ToExpression["{
    HoldForm}[{Y}_{3}\\left(x;\\lambda\\right)]",TeXForm], TraditionalForm]},
    LabelStyle -> Directive[Black, Bold], ColorFunction ->"BlueGreenYellow"]
```

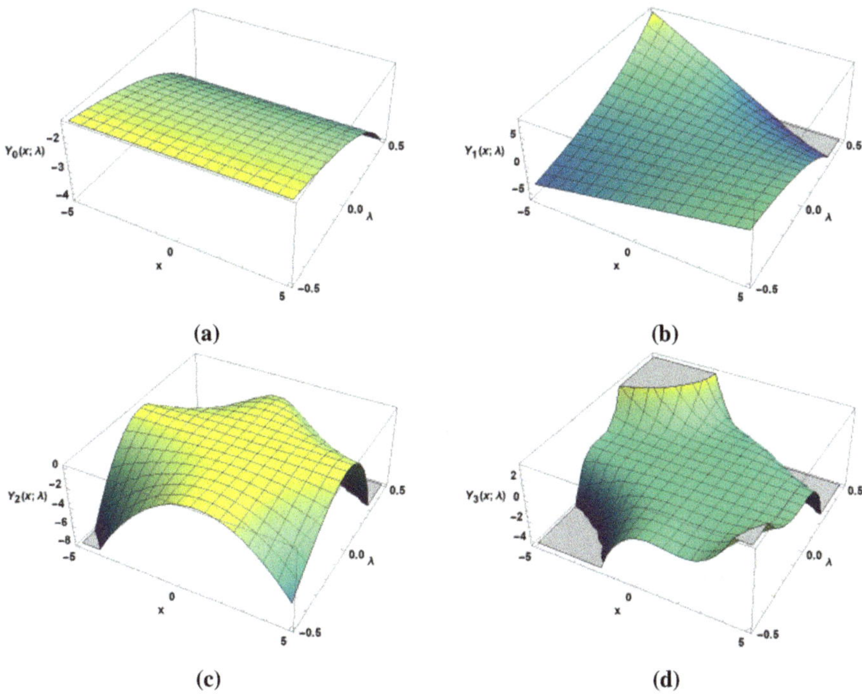

Fig. 8 Surface plots of the polynomials $Y_n(x; \lambda)$ for the randomly selected special cases when $\lambda \in \left[-\frac{1}{2}, \frac{1}{2}\right]$ and $x \in [-5, 5]$ **(a)** $n = 0$; **(b)** $n = 1$; **(c)** $n = 2$; **(d)** $n = 3$

8 Some Identities Derived from Derivative Formulas of the Generating Functions for the Numbers $Y_n(\lambda)$ and the Polynomials $Y_n(x; \lambda)$

Derivation of some partial derivative formulas involving the generating functions $F(t, x, \lambda)$ is given as follows.

Differentiating both sides of (65), with respect to the parameter t, yields the following partial differential equations:

$$\frac{\partial F(t, x, \lambda)}{\partial t} = F(t, x, \lambda)\left(\frac{\lambda x}{1+\lambda t} - \frac{\lambda^2}{2} F(t, \lambda)\right) \tag{71}$$

and

$$\frac{\partial F(t, x, \lambda)}{\partial t} = F(t, x, \lambda)\left(\lambda(\lambda t+1)^{-1} x - \lambda^2\left(\lambda^2 t + \lambda - 1\right)^{-1}\right) \tag{72}$$

(*cf.* [113]).

If the above second derivative operation is repeated v times, we obtain a higher-order partial differential equation as in the following theorem:

Theorem 8 (*cf.* [113]) *Let $v \in \mathbb{N}_0$. Then*

$$\frac{\partial^{(v)} F(t,x,\lambda)}{\partial t^v} = F(t,x,\lambda) \sum_{j=0}^{v} (-1)^j (v)_j (x)_{v-j} \lambda^{v+j} (1+\lambda t)^{j-v} \times \left(\lambda^2 t + \lambda - 1\right)^{-j}.$$

Theorem 9 (*cf.* [113])

$$Y_{n+1}(x;\lambda) = \frac{1}{2} \sum_{k=0}^{n} \binom{n}{k} Y_k(x;\lambda) \left(2(-1)^{n-k} \lambda^{n-k+1} x (n-k)! - \lambda^2 Y_{n-k}(\lambda)\right).$$

Proof (cf. [113]) We shall give just a brief sketch of the proof as the details are similar to those in [113]: Combining (65) and (64) with (71), such that $|\lambda t| < 1$, and using the Cauchy product in the final equation yield

$$\sum_{n=0}^{\infty} Y_{n+1}(x;\lambda) \frac{t^n}{n!} = \lambda x \sum_{n=0}^{\infty} \left(\sum_{k=0}^{n} \frac{(-1)^{n-k} \lambda^{n-k} Y_k(x;\lambda)}{k!} \right) t^n - \frac{\lambda^2}{2} \sum_{n=0}^{\infty} \left(\sum_{k=0}^{n} \binom{n}{k} Y_k(x;\lambda) Y_{n-k}(\lambda) \right) \frac{t^n}{n!}.$$

Comparing the coefficients of $\frac{t^n}{n!}$ on both sides of the equation just above, the desired result is obtained.

Corollary 1 (*cf.* [113])

$$Y_{n+1}(x;\lambda) = \sum_{k=0}^{n} \binom{n}{k} Y_k(x;\lambda) \left((-1)^{n-k} \lambda^{n-k+1} x (n-k)! - 2^{n-k} \left(\frac{\lambda^2}{\lambda-1}\right)^{n-k+1} Ch_{n-k}\right).$$

Proof (cf. [113]) We shall give just a brief sketch of the proof as the details are similar to those in [113]: By using (64), one has the following functional equation:

$$\frac{\lambda^2}{2} F(t,\lambda) = \frac{\lambda^2}{\lambda-1} F_{Ch}\left(\frac{2\lambda^2}{\lambda-1} t\right). \tag{73}$$

Combining (73) with Eq. (71) such that $|\lambda t| < 1$ gives us the following series equations:

$$\sum_{n=0}^{\infty} Y_{n+1}(x;\lambda)\frac{t^n}{n!} = \sum_{n=0}^{\infty} Y_n(x;\lambda)\frac{t^n}{n!}\left(\lambda x \sum_{n=0}^{\infty}(-1)^n \lambda^n t^n \right.$$
$$\left. -\frac{\lambda^2}{\lambda-1}\sum_{n=0}^{\infty}\left(\frac{2\lambda^2}{\lambda-1}\right)^n Ch_n \frac{t^n}{n!}\right).$$

After some elementary calculations with the Cauchy product and comparing the coefficients of $\frac{t^n}{n!}$ on both sides of the final equation, the desired result is obtained.

Theorem 10 (*cf.* [113]) *Let $n \in \mathbb{N}$. Then we have*

$$Y_{n-1}(x;\lambda) = \frac{1}{\lambda n}\sum_{k=0}^{n}\binom{n}{k}\frac{\partial}{\partial x}\{Y_k(x;\lambda)\}\lambda^{n-k} b_{n-k}(0).$$

Proof (cf. [113]) We shall give just a brief sketch of the proof as the details are similar to those in [113]: If we differentiate both sides of (65), with respect to the parameter x, we have the following partial differential equation:

$$\frac{\partial F(t,x,\lambda)}{\partial x} = F(t,x,\lambda)\log(1+\lambda t), \tag{74}$$

which, by using (46), yields

$$\sum_{n=0}^{\infty}\left(\sum_{k=0}^{n}\binom{n}{k}\frac{\partial}{\partial x}\{Y_k(x;\lambda)\}\lambda^{n-k} b_{n-k}(0)\right)\frac{t^n}{n!} = \sum_{n=0}^{\infty}\lambda n Y_{n-1}(x;\lambda)\frac{t^n}{n!}.$$

Comparing the coefficients of $\frac{t^n}{n!}$ on both sides of the equation just above, the desired result is obtained.

Differentiating both sides of (72), with respect to the parameter x, yields the following partial differential equation:

$$\frac{\partial^2}{\partial x \partial t}\{F(t,x,\lambda)\} = F(t,x,\lambda)\{\lambda(\lambda t+1)^{-1} x \log(\lambda t+1)$$
$$-\lambda^2\left(\lambda^2 t+\lambda-1\right)^{-1}\log(\lambda t+1)$$
$$+\lambda(\lambda t+1)^{-1}\}. \tag{75}$$

Moreover, differentiating both sides of (72), with respect to the parameter λ, yields the following another partial differential equation:

$$\frac{\partial^2}{\partial\lambda\partial t}\{F(t,x,\lambda)\} = F(t,x,\lambda)\{\lambda t(\lambda t+1)^{-2}(x-1)x + (\lambda t+1)^{-1}x$$
$$-\lambda^2 t(\lambda t+1)^{-1}\left(\lambda^2 t+\lambda-1\right)^{-1}x$$
$$-\lambda(\lambda t+1)^{-1}\left(\lambda^2 t+\lambda-1\right)^{-1}(2\lambda t+1)x - 2\lambda\left(\lambda^2 t+\lambda-1\right)^{-1}$$
$$+2\lambda^2\left(\lambda^2 t+\lambda-1\right)^{-2}(2\lambda t+1)\}. \quad (76)$$

On the other hand, differentiating both sides of Eq. (74), with respect to the parameter λ, yields the following another partial differential equation:

$$\frac{\partial^2 F(t,x,\lambda)}{\partial\lambda\partial x} = F(t,x,\lambda)\{tx(\lambda t+1)^{-1}\log(\lambda t+1)$$
$$-(2\lambda t+1)\left(\lambda^2 t+\lambda-1\right)^{-1}\log(\lambda t+1)$$
$$+t(\lambda t+1)^{-1}\}. \quad (77)$$

Remark 3 To see further formulas derived from the above partial differential equations involving the generating functions for the numbers $Y_n(\lambda)$ and the polynomials $Y_n(x;\lambda)$, the interested reader may refer to [113].

9 Other Relations of the Numbers $Y_n(\lambda)$ and the Polynomials $Y_n(x;\lambda)$ with Some Special Numbers and Polynomials

Here, we give other relations of some special numbers and polynomials with the numbers $Y_n(\lambda)$ and the polynomials $Y_n(x;\lambda)$.

Substituting $\lambda = -1$ into (65) yields the following relation between the polynomials $Y_n(x;\lambda)$ and the Changhee polynomials:

$$Y_n(x;-1) = (-1)^{n+1} Ch_n(x) \quad (78)$$

(*cf.* [113]).

Moreover, substituting $x = 0$ into (78), we have the following relation between the numbers $Y_n(\lambda)$ and the Changhee numbers:

$$Y_n(-1) = (-1)^{n+1} Ch_n \quad (79)$$

(*cf.* [113]).

Substituting $\lambda t = e^u - 1$ into (64) yields a computation formula for the Apostol–Bernoulli numbers: Let $m \in \mathbb{N}$. Then we have

$$\mathcal{B}_m(\lambda) = \frac{m}{\lambda - 1} \sum_{n=0}^{m-1} \sum_{k=0}^{n} (-1)^k \binom{n}{k} \left(\frac{\lambda}{\lambda - 1} \right)^n k^{m-1} \tag{80}$$

(*cf.* [93]).

Notice that the different proofs of the above formula were also given by Apostol [2, Eq (3.7)] and Boyadzhiev [6].

Combining (68) with (80), another relation between the numbers $Y_n(\lambda)$ and the Apostol–Bernoulli numbers is obtained as in the following theorem:

Theorem 11

$$\mathcal{B}_m\left(\lambda^2\right) = \frac{m}{2} \sum_{n=0}^{m-1} \sum_{k=0}^{n} (-1)^{k-n} \frac{k^{m-1}}{(n-k)!k!} \left(\frac{1}{\lambda + 1} \right)^{n+1} Y_n(\lambda). \tag{81}$$

The relation among the numbers $Y_n(\lambda)$, the Stirling numbers of the first kind, and the Apostol–Bernoulli numbers is given by (*cf.* [113])

$$Y_n(\lambda) = 2\lambda^n \sum_{k=0}^{n} \frac{S_1(n,k)\, \mathcal{B}_{k+1}(\lambda)}{k+1}; \quad (n \in \mathbb{N}_0). \tag{82}$$

By combining (32) with (82), we arrive at the following theorem:

Theorem 12 *Let $n \in \mathbb{N}_0$. Then we have*

$$Y_n(\lambda) = 2\lambda^n \sum_{k=0}^{n} \frac{\mathcal{B}_{k+1}(\lambda)}{k+1} \times \sum_{j=0}^{n-k} \sum_{m=0}^{j} (-1)^m \binom{2n-k}{n-k-j,\, n-k+j,\, k} \binom{j}{m} \frac{km^{j+n-k}}{(n+j)j!}. \tag{83}$$

The relation among the numbers $Y_n(\lambda)$, the Stirling numbers of the first kind, and the Apostol–Euler numbers is given by (*cf.* [113])

$$Y_n(-\lambda) = (-1)^{n+1} \lambda^n \sum_{k=0}^{n} \mathcal{E}_k(\lambda)\, S_1(n,k); \quad (n \in \mathbb{N}_0). \tag{84}$$

By combining (32) with (84), we arrive at the following theorem:

Theorem 13 *Let $n \in \mathbb{N}_0$. Then we have*

$$Y_n(-\lambda) = (-1)^{n+1}\lambda^n \sum_{k=0}^{n} \mathscr{E}_k(\lambda)$$
$$\times \sum_{j=0}^{n-k}\sum_{m=0}^{j}(-1)^m \binom{2n-k}{n-k-j, n-k+j, k}\binom{j}{m}\frac{km^{j+n-k}}{(n+j)j!}. \tag{85}$$

Remark 4 Setting $\lambda = 1$ in Eq. (85) and combining the final equation with (32), (79), and (9) give the following formula:

$$Ch_n = \sum_{k=0}^{n} E_k S_1(n,k),$$

which was proven by Kim et al. [25, Theorem 2.7].

In addition, by combining the case of (14) when $k = 1$ with not only (82), but also (84), we get a relation among the numbers $Y_n(\lambda)$, the Stirling numbers of the first kind, and the Apostol–Genocchi numbers as in the following theorem:

Theorem 14 *Let* $n \in \mathbb{N}_0$. *Then we have*

$$Y_n(\lambda) = -\lambda^n \sum_{k=0}^{n} \frac{S_1(n,k)\mathscr{G}_{k+1}(-\lambda)}{k+1}; \quad (n \in \mathbb{N}_0). \tag{86}$$

By combining (32) with (86), we arrive at by the following theorem:

Theorem 15 *Let* $n \in \mathbb{N}_0$. *Then we have*

$$Y_n(\lambda) = -\lambda^n \sum_{k=0}^{n} \frac{\mathscr{G}_{k+1}(-\lambda)}{k+1}$$
$$\times \sum_{j=0}^{n-k}\sum_{m=0}^{j}(-1)^m \binom{2n-k}{n-k-j, n-k+j, k}\binom{j}{m}\frac{km^{j+n-k}}{(n+j)j!}. \tag{87}$$

The relation among the polynomials $Y_n(x;\lambda)$, the Stirling numbers of the first kind, and the Apostol–Euler polynomials is given by (*cf.* [113])

$$Y_n(x;-\lambda) = (-1)^{n+1}\lambda^n \sum_{k=0}^{n} \mathscr{E}_k(x;\lambda) S_1(n,k); \quad (n \in \mathbb{N}_0). \tag{88}$$

By combining (32) with (88), we arrive at the following theorem:

Theorem 16 *Let $n \in \mathbb{N}_0$. Then we have*

$$Y_n(x; -\lambda) = (-1)^{n+1} \lambda^n \sum_{k=0}^{n} \mathscr{E}_k(x; \lambda)$$
$$\times \sum_{j=0}^{n-k} \sum_{m=0}^{j} (-1)^m \binom{2n-k}{n-k-j, n-k+j, k} \binom{j}{m} \frac{km^{j+n-k}}{(n+j)j!}. \tag{89}$$

Remark 5 Setting $\lambda = 1$ in Eq. (89) and combining the final equation with (32), (78), and (8) give the following formula:

$$Ch_m(x) = \sum_{n=0}^{m} E_n(x) S_1(m, n),$$

which was proven by Kim et al. [25, Theorem 2.5].

By combining the case of (13) when $k = 1$ with not only (88), we get a relation among the polynomials $Y_n(x; \lambda)$, the Stirling numbers of the first kind, and the Apostol–Genocchi polynomials as in the following theorem:

Theorem 17 *Let $n \in \mathbb{N}_0$. Then we have*

$$Y_n(x; -\lambda) = (-1)^{n+1} \lambda^n \sum_{k=0}^{n} \frac{S_1(n, k) \mathscr{G}_{k+1}(x; \lambda)}{k+1}. \tag{90}$$

By combining (32) with (90), we arrive at the following theorem:

Theorem 18 *Let $n \in \mathbb{N}_0$. Then we have*

$$Y_n(x; -\lambda) = (-1)^{n+1} \lambda^n \sum_{k=0}^{n} \frac{\mathscr{G}_{k+1}(x; \lambda)}{k+1}$$
$$\times \sum_{j=0}^{n-k} \sum_{m=0}^{j} (-1)^m \binom{2n-k}{n-k-j, n-k+j, k} \binom{j}{m} \frac{km^{j+n-k}}{(n+j)j!}. \tag{91}$$

10 Some Relations on Hypergeometric Functions Derived from the Integral of the Numbers $Y_n(\lambda)$ and the Polynomials $Y_n(x; \lambda)$

Here, we give some relations on hypergeometric functions derived from the integral of the numbers $Y_n(\lambda)$ and the polynomials $Y_n(x; \lambda)$.

Theorem 19 **(*cf.* [113])**

$$\int_0^u Y_n(\lambda)d\lambda = \frac{-2n!u^{2n+1}}{2n+1} {}_2F_1\left(-n-1, -2n-1; -2n-2; -u\right),$$

where ${}_2F_1$ *denotes the Gauss hypergeometric functions.*

Theorem 20 **(*cf.* [113])**

$$\int_0^u Y_n(x;\lambda)d\lambda = -2n!u^{2n+1} \sum_{k=0}^{n} \binom{x}{k} \frac{u^{-k}}{2n-k+1}$$
$$\times {}_2F_1\left(k-n-1, k-2n-1; k-2n-2; -u\right).$$

11 Some Infinite Series Containing the Numbers $Y_n(\lambda)$

Here, we present some infinite series containing the numbers $Y_n(\lambda)$, the Changhee numbers, the Daehee numbers, and the Lucas numbers. In addition, relations of these infinite series with the Humbert polynomials are given.

By (68), an infinite series representation for the reciprocal of the numbers $Y_n(\lambda)$ is obtained as in the following theorem:

Theorem 21 **(*cf.* [113])**

$$\sum_{n=0}^{\infty} \frac{1}{Y_n(\lambda)} = \frac{\lambda-1}{2} e^{\frac{1-\lambda}{\lambda^2}}.$$

By (68) and (41), an infinite series, including the ratio of the numbers $Y_n(\lambda)$ to the Daehee numbers, is obtained as in the following theorem:

Theorem 22 **(*cf.* [113])** *Let* $\left|\frac{\lambda^2}{\lambda-1}\right| < 1$. *Then*

$$\sum_{n=0}^{\infty} \frac{Y_n(\lambda)}{D_n} = \frac{2\lambda^2}{\left(1-\lambda+\lambda^2\right)^2} - \frac{2}{1-\lambda+\lambda^2}. \tag{92}$$

In addition, by (68) and (41), another infinite series, including the ratio of the Daehee numbers to the numbers $Y_n(\lambda)$, is obtained as in the following theorem:

Theorem 23 **(*cf.* [113])**

$$\sum_{n=0}^{\infty} \frac{D_n}{Y_n(\lambda)} = -\frac{\lambda^2}{2} \log\left(1+\frac{1-\lambda}{\lambda^2}\right). \tag{93}$$

We next give some examples for Eq. (93) as follows:

By substituting

$$\lambda = -\frac{1+\sqrt{5}}{2} \quad \text{and} \quad \lambda = -\frac{1-\sqrt{5}}{2}$$

into (93), we get, respectively, the following two infinite series:

$$\sum_{n=0}^{\infty} \frac{D_n}{Y_n\left(-\frac{1+\sqrt{5}}{2}\right)} = -\frac{3+\sqrt{5}}{4} \log 2$$

and

$$\sum_{n=0}^{\infty} \frac{D_n}{Y_n\left(-\frac{1-\sqrt{5}}{2}\right)} = \frac{\sqrt{5}-3}{4} \log 2.$$

Let L_n be the Lucas numbers. Then, it is well known that

$$\sum_{n=1}^{\infty} \frac{L_n}{n2^n} = 2 \log 2$$

(*cf.* [69, p. 7]). Combining the above identity with the previous two infinite series, we get

$$\sum_{n=0}^{\infty} \frac{D_n}{Y_n\left(-\frac{1+\sqrt{5}}{2}\right)} + \left(\frac{1+\sqrt{5}}{2}\right)^2 \sum_{n=1}^{\infty} \frac{L_n}{n2^{n+2}} = 0$$

and

$$\sum_{n=0}^{\infty} \frac{D_n}{Y_n\left(-\frac{1-\sqrt{5}}{2}\right)} + \left(\frac{1-\sqrt{5}}{2}\right)^2 \sum_{n=1}^{\infty} \frac{L_n}{n2^{n+2}} = 0.$$

Thus, we get

$$\sum_{n=1}^{\infty} \left(\frac{D_n}{Y_n\left(-\frac{1+\sqrt{5}}{2}\right)} + \left(\frac{1+\sqrt{5}}{2}\right)^2 \frac{L_n}{n2^{n+2}} \right) = \frac{3+\sqrt{5}}{4}$$

and

$$\sum_{n=1}^{\infty} \left(\frac{D_n}{Y_n\left(-\frac{1-\sqrt{5}}{2}\right)} + \left(\frac{1-\sqrt{5}}{2}\right)^2 \frac{L_n}{n2^{n+2}} \right) = \frac{3-\sqrt{5}}{4}.$$

By (68) and (45), an infinite series, including the ratio of the numbers $Y_n(\lambda)$ to the Changhee numbers, is obtained as in the following theorem:

Theorem 24 (*cf.* [113]) *Let* $\left|\frac{\lambda^2}{\lambda-1}\right| < \frac{1}{2}$. *Then*

$$\sum_{n=0}^{\infty} \frac{Y_n(\lambda)}{Ch_n} = \frac{2}{\lambda - 1 - 2\lambda^2}. \tag{94}$$

Likewise, by (68) and (45), an infinite series, including the ratio of the Changhee numbers to the numbers $Y_n(\lambda)$, is obtained as in the following theorem:

Theorem 25 (*cf.* [113]) *Let* $\left|\frac{\lambda-1}{2\lambda^2}\right| < 1$. *Then*

$$\sum_{n=0}^{\infty} \frac{Ch_n}{Y_n(\lambda)} = \frac{\lambda^2(\lambda - 1)}{2\lambda^2 - \lambda + 1}. \tag{95}$$

In addition to the above infinite series, by (41) and (45), an infinite series, including the ratio of the Changhee numbers to the Daehee numbers, is obtained as in the following theorem:

Theorem 26 (*cf.* [113])

$$\sum_{n=0}^{\infty} \frac{Ch_n}{D_n} = 4.$$

By rewriting the right-hand side of equation (92) in terms of the Humbert polynomials, one has

$$\sum_{n=0}^{\infty} \frac{Y_n(\lambda)}{D_n} = 2\lambda^2 \sum_{n=0}^{\infty} \Pi_{n,2}^{(2)}\left(\frac{1}{2}\right)\lambda^n - 2\sum_{n=0}^{\infty} \Pi_{n,2}^{(1)}\left(\frac{1}{2}\right)\lambda^n$$

(*cf.* [113]).

Similarly, by rewriting the right-hand side of equation (94) and (95) in terms of the generalized Humbert polynomials, one has

$$\sum_{n=0}^{\infty} \frac{Y_n(\lambda)}{Ch_n} = -2\sum_{n=0}^{\infty} P_n\left(2, \frac{1}{2}, 2, -1, 1\right)\lambda^n \tag{96}$$

and

$$\sum_{n=0}^{\infty} \frac{Ch_n}{Y_n(\lambda)} = \lambda^2(\lambda - 1)\sum_{n=0}^{\infty} P_n\left(2, \frac{1}{2}, 2, -1, 1\right)\lambda^n \tag{97}$$

(*cf.* [113]).

Remark 6 As seen above, the infinite series obtained from the ratios of the numbers $Y_n(\lambda)$ to other special numbers have become the ordinary generating functions for a case of the generalized Humbert polynomials. Undoubtedly, many more infinite series not covered here but involving the numbers $Y_n(\lambda)$ will lead to obtaining different applications and usage areas.

12 Positive Higher-Order Extension of the Numbers $Y_n(\lambda)$ and the Polynomials $Y_n(x;\lambda)$ with Their Generating Functions

Generating functions for positive higher-order extension of the numbers $Y_n(\lambda)$ and the polynomials $Y_n(x;\lambda)$ have been constructed in [56] as follows:

Let $k \in \mathbb{N}_0$ and $\lambda \in \mathbb{R}$ (or $\mathbb{C}$). Generating functions for the numbers $Y_n^{(k)}(\lambda)$ and the polynomials $Y_n^{(k)}(x;\lambda)$ are given by

$$\mathscr{F}(t,k;\lambda) = \left(\frac{2}{\lambda(1+\lambda t)-1}\right)^k = \sum_{n=0}^{\infty} Y_n^{(k)}(\lambda)\frac{t^n}{n!} \tag{98}$$

and

$$\mathscr{F}(t,x,k;\lambda) = \mathscr{F}(t,k;\lambda)(1+\lambda t)^x = \sum_{n=0}^{\infty} Y_n^{(k)}(x;\lambda)\frac{t^n}{n!} \tag{99}$$

(*cf.* [56]).

By (98) and (99), we have

$$Y_n^{(k)}(\lambda) = Y_n^{(k)}(0;\lambda). \tag{100}$$

Therefore,

$$Y_n(\lambda) = Y_n^{(1)}(\lambda)$$

and

$$Y_n(x;\lambda) = Y_n^{(1)}(x;\lambda).$$

By using (98), the computation formula for the numbers $Y_n^{(k)}(\lambda)$ is given as in the following theorem:

Theorem 27 (*cf.* [56])

$$Y_n^{(k)}(\lambda) = (-1)^n \binom{n+k-1}{n} \frac{2^k n! \lambda^{2n}}{(\lambda-1)^{k+n}}. \tag{101}$$

Notice that in the case when $k = 1$, (101) reduces to (68).

By making use of (101), few values of the numbers $Y_n^{(k)}(\lambda)$ are computed by

$$Y_0^{(2)}(\lambda) = \frac{4}{(\lambda-1)^2}, \quad Y_1^{(2)}(\lambda) = -\frac{8\lambda^2}{(\lambda-1)^3},$$
$$Y_2^{(2)}(\lambda) = \frac{24\lambda^4}{(\lambda-1)^4}, \quad Y_3^{(2)}(\lambda) = -\frac{96\lambda^6}{(\lambda-1)^5}, \ldots$$

$$Y_0^{(3)}(\lambda) = \frac{8}{(\lambda-1)^3}, \quad Y_1^{(3)}(\lambda) = -\frac{24\lambda^2}{(\lambda-1)^4},$$
$$Y_2^{(3)}(\lambda) = \frac{96\lambda^4}{(\lambda-1)^5}, \quad Y_3^{(3)}(\lambda) = -\frac{480\lambda^6}{(\lambda-1)^6},$$

and so on (*cf.* [56]).

By using (101), we also obtain a recurrence relation of the numbers $Y_n^{(k)}(\lambda)$ by the following theorem:

Theorem 28 (*cf.* [56]) *Let $n \in \mathbb{N}$. Then, the numbers $Y_n^{(k)}(\lambda)$ are given by the following recurrence relation:*

$$Y_n^{(k)}(\lambda) = \frac{\lambda^2}{1-\lambda}(n+k-1)\, Y_{n-1}^{(k)}(\lambda),$$

with the initial condition:

$$Y_0^{(k)}(\lambda) = \frac{2^k}{(\lambda-1)^k}.$$

Another recurrence relation for the numbers $Y_n^{(k)}(\lambda)$ is given as follows (*cf.* [56]):

$$\sum_{j=0}^{k} (-1)^{k-j} (n)_j \binom{k}{j} \lambda^{2j} (1-\lambda)^{k-j}\, Y_{n-j}^{(k)}(\lambda) = 0. \tag{102}$$

Remark 7 The interested reader may refer to [56] to see a computational algorithm that gives the values of the polynomials $Y_n^{(k)}(x;\lambda)$.

First few values of the polynomials $Y_n^{(k)}(x;\lambda)$ are given as follows:

$$Y_1^{(k)}(x;\lambda)=\lambda\left(\frac{2}{\lambda-1}\right)^k\left(x-\frac{k\lambda}{\lambda-1}\right),$$

$$Y_2^{(k)}(x;\lambda)=\lambda^2\left(\frac{2}{\lambda-1}\right)^k\left(x^2-\left(1+\frac{2k\lambda}{\lambda-1}\right)x+k(k+1)\left(\frac{\lambda}{\lambda-1}\right)^2\right),$$

$$\begin{aligned}Y_3^{(k)}(x;\lambda)=\lambda^3\left(\frac{2}{\lambda-1}\right)^k&\left(x^3-3\left(1+\frac{k\lambda}{\lambda-1}\right)x^2+\left(2+\frac{3k\lambda}{\lambda-1}\right.\right.\\&\left.+3k(k+1)\left(\frac{\lambda}{\lambda-1}\right)^2\right)x\\&\left.-k(k+1)(k+2)\left(\frac{\lambda}{\lambda-1}\right)^3\right),\end{aligned}$$

and so on (*cf.* [56]).

The relation among the numbers $Y_n^{(k)}(\lambda)$, the Stirling numbers of the first kind, and the Apostol–Bernoulli numbers of positive higher order is given by

$$Y_n^{(k)}(\lambda)=(-1)^{k+1}2^k\lambda^n\sum_{m=0}^{n}\frac{S_1(n,m)\mathscr{B}_{m+1}^{(k)}(\lambda)}{m+1};\quad(n\in\mathbb{N})\tag{103}$$

(*cf.* [56]).

By combining (32) with (103), we arrive at the following theorem:

Theorem 29 *Let $n\in\mathbb{N}$. Then we have*

$$\begin{aligned}Y_n^{(k)}(\lambda)&=(-1)^{k+1}2^k\lambda^n\sum_{m=0}^{n}\frac{\mathscr{B}_{m+1}^{(k)}(\lambda)}{m+1}\\&\times\sum_{j=0}^{n-m}\sum_{r=0}^{j}(-1)^r\binom{2n-m}{n-m-j,\,n-m+j,\,m}\binom{j}{r}\frac{mr^{j+n-m}}{(n+j)j!}.\end{aligned}\tag{104}$$

The relation among the numbers $Y_n^{(k)}(\lambda)$, the Stirling numbers of the first kind, and the Apostol–Euler numbers of positive higher order is given by

$$Y_n^{(k)}(-\lambda)=(-1)^{n+k}\lambda^n\sum_{m=0}^{n}\mathscr{E}_m^{(k)}(\lambda)S_1(n,m)\tag{105}$$

(*cf.* [56]).

By combining (32) with (105), we arrive at the following theorem:

Theorem 30 *Let $n \in \mathbb{N}$. Then we have*

$$Y_n^{(k)}(-\lambda) = (-1)^{n+k}\lambda^n \sum_{m=0}^{n} \mathscr{E}_m^{(k)}(\lambda) \times \sum_{j=0}^{n-m}\sum_{r=0}^{j}(-1)^r \binom{2n-m}{n-m-j,\, n-m+j,\, m}\binom{j}{r}\frac{mr^{j+n-m}}{(n+j)j!}. \tag{106}$$

Remark 8 Notice that in the special case when $k = 1$, (103) reduces to (82) and (105) reduces to (84). Also, in the special case when $k = 1$, (104) reduces to (83) and (106) reduces to (85).

In addition, by combining (14) with not only (103), but also (105), we get a relation among the numbers $Y_n^{(k)}(\lambda)$, the Stirling numbers of the first kind, and the Apostol–Genocchi numbers of positive higher order as in the following theorem:

Theorem 31 *Let $n \in \mathbb{N}$. Then we have*

$$Y_n^{(k)}(\lambda) = -\lambda^n \sum_{m=0}^{n} \frac{S_1(n,m)\,\mathscr{G}_{m+1}^{(k)}(-\lambda)}{m+1}; \quad (n \in \mathbb{N}). \tag{107}$$

Remark 9 Notice also that substituting $k = 1$ into (107) yields (86).

By using the partial derivatives of the functions $\mathscr{F}(t, k; \lambda)$, with respect to the parameters t and λ, some derivative formulas and identities for the numbers $Y_n^{(k)}(\lambda)$ were obtained in [56] as follows:

Differentiating the functions $\mathscr{F}(t, k; \lambda)$ with respect to the parameter t, we have

$$\frac{d}{dt}\{\mathscr{F}(t,k;\lambda)\} = -\frac{k}{2}\lambda^2 \mathscr{F}(t,k+1;\lambda),$$

which yields a formula given by the following theorem:

Theorem 32 (*cf.* [56]) *Let $n, k, v \in \mathbb{N}_0$. Then we have*

$$Y_{n+v}^{(k)}(\lambda) = \frac{(-1)^v (k)^{(v)} \lambda^{2v}}{2^v} Y_n^{(k+v)}(\lambda). \tag{108}$$

Differentiating the functions $\mathscr{F}(t, k; \lambda)$ with respect to the parameter λ, we obtain the following derivative formula:

$$\frac{d}{d\lambda}\{\mathscr{F}(t,k;\lambda)\} = -\frac{k}{2}(2\lambda t + 1)\mathscr{F}(t,k+1;\lambda),$$

which yields a formula given by the following theorem:

Theorem 33 (*cf.* [56])

$$\frac{d}{d\lambda}Y_n^{(k)}(\lambda) = -\frac{k}{2}\left(2\lambda n Y_{n-1}^{(k+1)}(\lambda) + Y_n^{(k+1)}(\lambda)\right).$$

Remark 10 The well-known Chu–Vandermonde identity is given as follows:

$$\binom{x+a}{k} = \sum_{j=0}^{k}\binom{x}{j}\binom{a}{k-j} \tag{109}$$

(*cf.* [15, 22, 88]). The interested reader may refer to [56] for some Chu–Vandermonde-type convolution formulas derived from the functional equations of the generating function for the numbers $Y_n^{(k)}(\lambda)$.

13 Negative Higher-Order Extension of the Numbers $Y_n(\lambda)$ and the Polynomials $Y_n(x;\lambda)$ with Their Generating Functions

Generating functions for negative higher-order extension of the numbers $Y_n(\lambda)$ and the polynomials $Y_n(x;\lambda)$ have been constructed in [58] as follows:

Let $k \in \mathbb{N}_0$ and $\lambda \in \mathbb{R}$ (or $\mathbb{C}$). Generating functions for the numbers $Y_n^{(-k)}(\lambda)$ and the polynomials $Q_n(x;\lambda,k)$ are given by

$$\mathcal{G}(t,k;\lambda) = 2^{-k}(\lambda(1+\lambda t)-1)^k = \sum_{n=0}^{\infty} Y_n^{(-k)}(\lambda)\frac{t^n}{n!} \tag{110}$$

and

$$\mathcal{G}(t,x,k;\lambda) = \mathcal{G}(t,k;\lambda)(1+\lambda t)^x = \sum_{n=0}^{\infty} Q_n(x;\lambda,k)\frac{t^n}{n!} \tag{111}$$

(*cf.* [58]).

By (110) and (111), we have

$$Q_n(x;\lambda,k) = \sum_{j=0}^{n}\binom{n}{j}\lambda^{n-j}Y_j^{(-k)}(\lambda)(x)_{n-j}, \tag{112}$$

and

$$Y_n^{(-k)}(\lambda) = \begin{cases} 2^{-k} n! \binom{k}{n} \lambda^{2n} (\lambda - 1)^{k-n} & \text{if} \quad n \le k \\ 0 & \text{if} \quad n > k \end{cases} \tag{113}$$

where $k, n \in \mathbb{N}_0$ (*cf.* [58]).

By (113), first few values of the numbers $Y_n^{(-k)}(\lambda)$ are given as follows:

$$Y_0^{(-k)}(\lambda) = 2^{-k} (\lambda - 1)^k,$$

$$Y_1^{(-k)}(\lambda) = 2^{-k} \binom{k}{1} \lambda^2 (\lambda - 1)^{k-1},$$

$$Y_2^{(-k)}(\lambda) = 2^{-k} 2! \binom{k}{2} \lambda^4 (\lambda - 1)^{k-2},$$

$$\vdots$$

$$Y_j^{(-k)}(\lambda) = 2^{-k} j! \binom{k}{j} \lambda^{2j} (\lambda - 1)^{k-j} \quad \text{for} \quad j \le k,$$

$$\vdots$$

$$Y_k^{(-k)}(\lambda) = 2^{-k} k! \lambda^{2k},$$

$$Y_j^{(-k)}(\lambda) = 0 \quad \text{for} \quad j > k$$

(*cf.* [58]).

By (112) and (113), we also have the following first few values of the polynomials $Q_n(x; \lambda, k)$:

$$Q_0(x; \lambda, k) = 2^{-k} (\lambda - 1)^k,$$

$$Q_1(x; \lambda, k) = 2^{-k} (\lambda - 1)^k \lambda x + 2^{-k} k \lambda^2 (\lambda - 1)^{k-1},$$

$$Q_2(x; \lambda, k) = 2^{-k} (\lambda - 1)^k \lambda^2 x^2 + \left(-2^{-k} (\lambda - 1)^k \lambda^2 + 2^{-k+1} k \lambda^3 (\lambda - 1)^{k-1} \right) x$$
$$+ 2^{-k} k (k-1) \lambda^4 (\lambda - 1)^{k-1},$$

and so on (*cf.* [58]).

Remark 11 It follows from (50) and (113) that there exists a relationship between the numbers $Y_n^{(-k)}(\lambda)$ and the Bernstein basis functions as follows:

$$Y_n^{(-k)}(\lambda) = \frac{(-1)^{k-n} n!}{2^k} \lambda^n B_n^k(\lambda), \tag{114}$$

where $n, k \in \mathbb{N}_0$ and $\lambda \in [0, 1]$ (*cf.* [58]). The interested reader may refer to [58] for further identities containing the numbers $Y_n^{(-k)}(\lambda)$, the Poisson–Charlier polynomials, the Bell polynomials (i.e., exponential polynomials), and other kinds of combinatorial numbers.

13.1 Derivative Formulas and Recurrence Relations Derived from Partial Derivatives of the Functions $\mathcal{G}(t, k; \lambda)$ and $\mathcal{G}(t, x, k; \lambda)$

By using partial derivatives of the generating functions $\mathcal{G}(t, k; \lambda)$ and $\mathcal{G}(t, x, k; \lambda)$, with respect to the parameters t, λ, and x, some derivative formulas and recurrence relations for the numbers $Y_n^{(-k)}(\lambda)$ and the polynomials $Q_n(x; \lambda, k)$ were obtained in [58] as follows:

Differentiating both sides of (110) with respect to the parameter λ, we have the following partial derivative equation:

$$\frac{\partial}{\partial\lambda}\{\mathcal{G}(t, k; \lambda)\} = \frac{k}{2}(2\lambda t + 1)\mathcal{G}(t, k-1; \lambda), \tag{115}$$

which, combining the right-hand side of (110), yields a derivative formula given by the following theorem:

Theorem 34 **(*cf.* [58])** *Let $n \in \mathbb{N}$. Then, we have*

$$\frac{d}{d\lambda}\{Y_n^{(-k)}(\lambda)\} = \frac{k}{2}\left(2n\lambda Y_{n-1}^{(-k+1)}(\lambda) + Y_n^{(-k+1)}(\lambda)\right). \tag{116}$$

Differentiating both sides of (110) with respect to the parameter t, we also have the following another partial derivative equation:

$$\frac{\partial}{\partial t}\{\mathcal{G}(t, k; \lambda)\} = \frac{k\lambda^2}{2}\mathcal{G}(t, k-1; \lambda), \tag{117}$$

which, combining the right-hand side of (110), yields a formula given by following theorem:

Theorem 35 **(*cf.* [58])** *Let $n \in \mathbb{N}_0$. Then, we have*

$$Y_{n+1}^{(-k)}(\lambda) = \frac{k\lambda^2}{2}Y_n^{(-k+1)}(\lambda). \tag{118}$$

Differentiating both sides of (111) with respect to the parameter λ, we also have the following another partial derivative equation:

$$\frac{\partial}{\partial\lambda}\{\mathcal{G}(t,x,k;\lambda)\} = \frac{k}{2}(2\lambda t+1)\mathcal{G}(t,x,k-1;\lambda) + xt\mathcal{G}(t,x-1,k;\lambda), \tag{119}$$

which, combining the right-hand side of (111), yields a derivative formula given by the following theorem:

Theorem 36 (*cf.* [58]) *Let $n \in \mathbb{N}$. Then, we have*

$$\frac{\partial}{\partial\lambda}\{Q_n(x;\lambda,k)\} = kn\lambda Q_{n-1}(x;\lambda,k-1) + \frac{k}{2}Q_n(x;\lambda,k-1) + xnQ_{n-1}(x-1;\lambda,k). \tag{120}$$

Differentiating both sides of (111) with respect to the parameter t, we also have the following another partial derivative equation:

$$\frac{\partial}{\partial t}\{\mathcal{G}(t,x,k;\lambda)\} = \frac{k\lambda^2}{2}\mathcal{G}(t,x,k-1;\lambda) + x\lambda\mathcal{G}(t,x-1,k;\lambda), \tag{121}$$

which, combining the right-hand side of (111), yields a formula given by the following theorem:

Theorem 37 (*cf.* [58]) *Let $n \in \mathbb{N}_0$. Then, we have*

$$Q_{n+1}(x;\lambda,k) = \frac{k\lambda^2}{2}Q_n(x;\lambda,k-1) + x\lambda Q_n(x-1;\lambda,k). \tag{122}$$

Moreover, when we differentiate both sides of (111) with respect to the parameter x, we also have the following another partial derivative equation:

$$\frac{\partial}{\partial x}\{\mathcal{G}(t,x,k;\lambda)\} = \log(1+\lambda t)\mathcal{G}(t,x,k;\lambda), \tag{123}$$

which, combining the right-hand side of (111), yields a derivative formula given by the following theorem:

Theorem 38 (*cf.* [58]) *Let $n \in \mathbb{N}$. Then, we have*

$$\frac{\partial}{\partial x}\{Q_n(x;\lambda,k)\} = n\sum_{j=0}^{n-1}(-1)^j\binom{n-1}{j}\frac{j!\lambda^{j+1}}{j+1}Q_{n-1-j}(x;\lambda,k). \tag{124}$$

Notice that setting (41) into (124) gives us the following formula:

$$\frac{\partial}{\partial x}\{Q_n(x;\lambda,k)\} = n\sum_{j=0}^{n-1}\binom{n-1}{j}\lambda^{j+1}D_jQ_{n-1-j}(x;\lambda,k) \tag{125}$$

(*cf.* [58]).

By combining (40) with (123), we also have

$$\frac{\partial}{\partial x}\{\mathcal{G}(t,x,k;\lambda)\} = \lambda t\mathcal{G}(t,k;\lambda)\,F_D(x,\lambda t)\,, \tag{126}$$

which, combining the right-hand side of (40) and (111), yields a derivative formula given by the following theorem:

Theorem 39 (*cf.* [58]) *Let $n \in \mathbb{N}$. Then, we have*

$$\frac{\partial}{\partial x}\{Q_n(x;\lambda,k)\} = n\sum_{j=0}^{n-1}\lambda^{j+1}\binom{n-1}{j}Y_{n-j}^{(-k)}(\lambda)\,D_j(x)\,.$$

14 Some Applications to the Probability Distribution Functions

Generating functions and moment-generating functions play a vital role in theory of probability and statistics. For this reason, special numbers and special polynomials have very important application areas in theory of probability and statistics.

The above observations show that combinatorial-type numbers and polynomials and their generating functions have a wide range of applications. Among others, some relations of combinatorial-type numbers and polynomials with probability distribution functions have been investigated by Kucukoglu et al. in their recent papers [56] and [58] by using positive and negative higher-order extension of combinatorial-type numbers and polynomials.

One of the mentioned applications is related to the probability functions for negative hypergeometric-type probability distribution, and we shall give just a brief sketch of its construction as the details are similar to those in [56]:

Multiplying the generating functions for the positive higher-order extension of combinatorial-type numbers given by (98), a Chu–Vandermonde- type convolution formula is obtained as in the following theorem:

Theorem 40 (*cf.* [56]) *Let $m \in \mathbb{N}$, $k_1, k_2, \ldots, k_m \in \mathbb{N}$ and $n \in \mathbb{N}_0$. Then we have*

$$\begin{aligned}&\binom{k_1+k_2+\cdots+k_m+n-1}{n}\\ &= \sum_{v_1+v_2+\cdots+v_{m-1}=n}\binom{k_m+v_{m-1}-1}{v_{m-1}}\binom{k_{m-1}+v_{m-2}-1}{v_{m-2}}\cdots\\ &\quad\times\binom{k_1+v_1-1}{v_1}\binom{k_2+n-v_1-v_2-\cdots-v_{m-1}-1}{n-v_1-v_2-\cdots-v_{m-1}}.\end{aligned}$$

By using the Chu–Vandermonde- type convolution formula given in Theorem 40, if we set

$$f(v_1, \ldots, v_n; k_1 + \ldots + k_m + n - 1, n, k_1, \ldots, k_n)$$
$$= \frac{\binom{k_m+v_{m-1}-1}{v_{m-1}}\binom{k_{m-1}+v_{m-2}-1}{v_{m-2}} \cdots \binom{k_1+v_1-1}{v_1}\binom{k_2+n-v_1-v_2-\cdots-v_{m-1}-1}{n-v_1-v_2-\cdots-v_{m-1}}}{\binom{k_1+k_2+\cdots+k_m+n-1}{n}}, \quad (127)$$

then due mainly to the fact that

$$\sum_{v_1+v_2+\cdots+v_{m-1}=n} f(v_1, \ldots, v_n; k_1 + \cdots + k_m + n - 1, n, k_1, \ldots, k_n) = 1,$$

the function $f(v_1, \ldots, v_n; k_1 + \cdots + k_m + n - 1, n, k_1, \ldots, k_n)$ is the probability functions for negative hypergeometric-type distribution with the parameters $k_1, k_2, \ldots, k_m$ and n with the random variable $(v_1, v_2, \ldots, v_n)$ (*cf.* [56]).

Remark 12 In the case when $m = 2$, (127) reduces to the well-known probability function for negative hypergeometric distribution given by

$$f(v_1, k_1 + k_2 + n - 1, n, k_1) = \frac{\binom{v_1+k_1-1}{v_1}\binom{k_2+n-v_1-1}{n-v_1}}{\binom{k_1+k_2+n-1}{n}}, \quad (128)$$

where $k_1 + k_2 + n - 1$ is the population size, n is the number of success states in the population, k_1 is the number of failures, v_1 is the number of observed successes for $0 \leq v_1 \leq n$; $0 \leq k_1 \leq n$ (*cf.* [56]). For further details about the negative hypergeometric distribution, the reader may refer to [21, 59].

The aforementioned negative hypergeometric-type distribution (with the parameters $k_1, k_2, \ldots, k_m$ and n with the random variable $(v_1, v_2, \ldots, v_n)$) has the following moment-generating function (*cf.* [56]):

$$M(t; k_1, \ldots, k_m, n) = \sum_{v_1+v_2+\cdots+v_{m-1}=n} e^{t(v_1+v_2+\cdots+v_{m-1})}$$
$$\times f(v_1, \ldots, v_n; k_1 + \cdots + k_m + n - 1, n, k_1, \ldots, k_n),$$

whose j-th moment (or the j-th derivative of $M(t; k_1, \ldots, k_m, n)$ computed at $t = 0$) is given as follows (*cf.* [56]):

$$\left. \frac{d^j}{dt^j} \{M(t; k_1, \ldots, k_m, n)\} \right|_{t=0} = \mu_j,$$

which, in the special case when $m = 2$, yields

$$\mu_1 = \frac{nk_1}{k_1 + k_2},$$

$$\mu_2 = \frac{nk_1 (n (1 + k_1) + k_2)}{(k_1 + k_2) (1 + k_1 + k_2)},$$

because of the reduction to the negative hypergeometric distribution given in (128), see, for details, [21, 59].

Second application mentioned above is especially related to an approach to the binomial (or Newton) distribution and the Poisson distribution by negative higher-order extension of combinatorial-type numbers and polynomials. Next, we shall give just a brief sketch of its construction as the details are similar to those in [58]:

With assumption of $0 < p \leq 1$ and $n = 0, 1, 2, \ldots, k$, if we set

$$f (p; k, n) = \frac{(-1)^{k-n} 2^k}{n!p^n} Y_n^{(-k)} (p), \tag{129}$$

then the above function is corresponding to the discrete probability distribution such that p is a probability of success, k is the number of trials, n is the number of successes in k trials, and $n = 0, 1, 2, \ldots, k$. Thus, the discrete probability distribution function $f (p; k, n)$ is binomially distributed with parameters (k, p), which leads us to conclude that the probability distribution function $f (p; k, n)$ is a binomial-type probability distribution function with parameters (k, p); see, for details, [58].

Some properties of the discrete probability distribution function $f (p; k, n)$, with a random variable with parameters k, n, and p, are given as follows:

- For all k, n, p with $0 \leq n \leq k$ and $0 < p \leq 1$, $0 \leq f (p; k, n) \leq 1$. Thus,

$$f (p; k, n) \geq 0.$$

- For the discrete probability distribution function $f (p; k, n)$, the following equality

$$\sum_{n=0}^{\infty} f (p; k, n) = 1$$

holds true.
- With the assumption that X denotes a binomial random variable with parameters (k, p). The distribution function of the random variable X is computed by

$$P(X \leq j) = \sum_{n=0}^{j} f (p; k, n); \quad (j = 0, 1, \ldots, k).$$

- The expected value and variance for random variable with parameters k and p are computed with the aid of

$$E\left[X^{v}\right]=\sum_{n=0}^{k}n^{v}f\left(p;k,n\right),$$

which, respectively, gives the expected value as

$$E\left[X\right]=kp$$

and the variance as

$$E\left[X^{2}\right]-\left(E\left[X\right]\right)^{2}=kp\left(1-p\right).$$

- In the case when $k \to \infty$, the discrete probability distribution function $f\left(p;k,n\right)$ goes to the Poisson distribution, according to which the Poisson–Charlier polynomials are orthogonal polynomials; see, for details, [79, 97].

Remark 13 For further applications of generating functions for Apostol-type, Peters-type, Boole-type combinatorial numbers and polynomials related to the generating functions for the numbers $Y_n\left(\lambda\right)$ and the polynomials $Y_n\left(x;\lambda\right)$ and their positive and negative higher-order extension, the interested reader may glance at the recent papers [93, 100, 105, 105, 106], which will shed light on the readers for their future studies.

15 Further Remarks and Observations

In this chapter, we deal with old and new results arising from the generating functions for some certain classes of combinatorial numbers and polynomials attached to Dirichlet characters that are one of the most important tools of the analytic number theory. In addition, we also present a mathematical analysis on the construction of these generating functions by p-adic integration, which is an elegant method allowing us to obtain results to be potentially used in mathematical physics. Besides, the applications of some results were thoroughly examined within the probability theory. As a result of these examinations, it has been seen that the aforementioned applications are based upon the probability distribution functions such as negative hypergeometric-type probability distribution, binomial-type probability, and Poisson distribution. Hence, the results given about the classes of combinatorial numbers and polynomials are the outputs of an interdisciplinary study conducted with the blending of many fields such as analytic number theory, mathematical physics, and probability theory. There is no doubt that the results covered here will lead future studies, potential applications, and new interdisciplinary usage areas. Overall, this chapter and its content will serve as a resource for scientists to study interdisciplinary and have brought them together in order to develop new mathematical models, algorithms, and other mathematical techniques and methods. In future studies, the further applications of the combinatorial numbers and polynomials in combinatoric analysis and discrete mathematics can be explored. As a result of this exploration very useful results can be presented.

References

1. M. Acikgoz, S. Araci, *On generating function of the Bernstein polynomials*, AIP Conf. Proc., **1281**(1) (2010), 1141; https://doi.org/10.1063/1.3497855.
2. T. M. Apostol, *On the Lerch zeta function*, Pacific J. Math., **1** (1951), 161–167.
3. T. M. Apostol, *Introduction to Analytic Number Theory* (Narosa Publishing, Springer-Verlag, New Delhi, Chennai, Mumbai, 1998).
4. R. P. Boas, R. C. Buck, *Polynomial Expansions of Analytic Functions*, (Academic Press, New York, 1964).
5. M. Bona, *Introduction to Enumerative Combinatorics* (The McGraw-Hill Companies Inc., New York, NY, USA, 2007).
6. K. N. Boyadzhiev, *Apostol–Bernoulli functions, derivative polynomials and Eulerian polynomials*, arXiv:0710.1124v1.
7. N. P. Cakic, G. V. Milovanovic, *On generalized Stirling numbers and polynomials*, Mathematica Balkanica **18** (2004), 241–248.
8. I. N. Cangul, H. Ozden, Y. Simsek, *A new approach to q-Genocchi numbers and their interpolation functions*, Nonlinear Analysis, **71** (2009), e793–e799.
9. C. A. Charalambides, *Enumerative Combinatorics* (Chapman and Hall/ CRC Press Company, London, UK; New York, NY, USA, 2002).
10. L. Comtet, *Advanced Combinatorics: The Art of Finite and Infinite Expansions* (D. Reidel Publishing Company, Dordrecht, The Netherlands; Boston, MA, USA, 1974).
11. R. Dere, Y. Simsek, H. M. Srivastava, *A unified presentation of three families of generalized Apostol type polynomials based upon the theory of the umbral calculus and the umbral algebra*, J. Number Theory **133** (2013), 3245–3263.
12. G. B. Djordjevic, G. V. Milovanović, *Special Classes of Polynomials* (Faculty of Technology, University of Nis, Leskovac, Serbia, 2014).
13. B. S. El-Desouky, A. Mustafa, *New results and matrix representation for Daehee and Bernoulli numbers and polynomials*, arXiv:1412.8259v1.
14. R. Golombek, *Aufgabe 1088*, El. Math. **49** (1994), 126–127.
15. H. W. Gould, *Inverse series relations and other expansions involving Humbert polynomials*, Duke Math. J. **32**(4) (1965), 697–712.
16. H. W. Gould, *Combinatorial Identities: Table I: Intermediate Techniques for Summing Finite Series*; http://math.wvu.edu/~hgould/Vol.4.PDF.
17. H. W. Gould, *Fundamentals of Series: Table III: Basic Algebraic Techniques*; http://math.wvu.edu/~hgould/Vol.3.PDF.
18. H. Haruki, Th. M. Rassias, New integral representations for Bernoulli and Euler polynomials, J. Math. Anal. Appl., **175** (1993), 81–90.
19. P. Humbert, *Some extensions of Pincherle's polynomials*, Proc. Edinburgh Math. Soc., **39**(1) (1921), 21–24.
20. L. C. Jang, H. K. Pak, *Non-archimedean integration associated with q-Bernoulli numbers*, Proc. Jangjeon Math. Soc., **5** (2002), 125–129.
21. N. L Johnson, A. W. Kemp, S. Kotz, *Univariate Discrete Distributions* (3rd ed.) (Wiley Series in Probability and Statistics, New Jersey, USA, 2005).

22. C. Jordan, *Calculus of Finite Differences* (2nd ed.) (Chelsea Publishing Company, New York, 1950).
23. S. Khan, T. Nahid, M. Riyasat, *Partial derivative formulas and identities involving 2-variable Simsek polynomials*, Bol. Soc. Mat. Mex., **26** (2020), 1–13.
24. D. S. Kim, T. Kim, *Daehee numbers and polynomials*, Appl. Math. Sci. (Ruse), **7**(120) (2013), 5969–5976.
25. D. S. Kim, T. Kim, J. Seo, *A note on Changhee numbers and polynomials*, Adv. Stud. Theor. Phys. **7** (2013), 993–1003.
26. D. S. Kim, T. Kim, *A note on Boole polynomials*, Integral Transforms Spec. Funct., **25**(8) (2014), 627–633.
27. D. S. Kim, T. Kim, *Some identities of degenerate special polynomials*, Open Math **13** (2015), 380–389.
28. D. S. Kim, T. Kim, J.-W. Park, J. J. Seo, *Differential equations associated with Peters polynomials*, Glob. J. Pure Appl. Math., **12**(4) (2016), 2915–2922.
29. D. S. Kim, T. Kim, T. Komatsu, H. I. Kwon, S.-H. Lee, *Barnes-type Peters polynomials associated with poly-Cauchy polynomials of the second kind*, J. Comput. Anal. Appl., **20**(1) (2016), 151–174.
30. D. S. Kim, T. Kim, *Differential equations associated with degenerate Changhee numbers of the second kind*, Rev. R. Acad. Cienc. Exactas Fís. Nat. Ser. A Mat. RACSAM, **113**(3) (2019), 1785–1793.
31. D. S. Kim, T. Kim, H. I. Kwon, T. Mansour, J. Seo, *Barnes-type Peters polynomial with umbral calculus view point*, J. Inequal. Appl., **2014**(324) (2014), 1–16.
32. M. S. Kim, J. W. Son, *Analytic properties of the q-Volkenborn integral on the ring of p-adic integers*, Bull. Korean Math. Soc., **44** (2007), 1–12.
33. M. S. Kim, *On Euler numbers, polynomials and related p-adic integrals*, J. Number Theory, **129** (2009), 2166–2179.
34. T. Kim, *On a q-analogue of the p-adic log gamma functions and related integrals*, J. Number Theory, **76** (1999), 320–329.
35. T. Kim, *q-Volkenborn integration*, Russ. J. Math. Phys., **19** (2002), 288–299.
36. T. Kim, *An invariant p-adic integral associated with Daehee numbers*, Integral Transforms Spec. Funct., **13** (2002), 65–69.
37. T. Kim, *Non-archimedean q-integrals associated with multiple Changhee q-Bernoulli polynomials*, Russ. J. Math. Phys. **10** (2003), 91–98.
38. T. Kim, *p-adic q-integrals associated with the Changhee-Barnes' q-Bernoulli polynomials*, Integral Transform Spec. Funct. **15** (2004), 415–420.
39. T. Kim, *q-Euler numbers and polynomials associated with p-adic q-integral and basic q-zeta function*, Trends Math. (Information Center for Mathematical Sciences), **9** (2006), 7–12.
40. T. Kim, *On the analogs of Euler numbers and polynomials associated with p-adic q-integral on $\mathbb{Z}_p$ at $q = 1$*, J. Math. Anal. Appl., **331** (2007), 779–792.
41. T. Kim, *An invariant p-adic q-integral on $\mathbb{Z}_p$*, Appl. Math. Letters, **21** (2008), 105–108.
42. T. Kim, *p-adic l-functions and sums of powers*, arXiv:math/0605703v1.
43. T. Kim, *On the q-extension of Euler and Genocchi numbers*, J. Math. Anal. Appl., **326** (2007), 1458–1465.
44. T. Kim, S. H. Rim, *Some q-Bernoulli numbers of higher order associated with the p-adic q-integrals*, Indian J. Pure Appl. Math., **32** (2001), 1565–1570.
45. T. Kim, S. H. Rim, Y. Simsek, D. Kim, *On the analogs of Bernoulli and Euler numbers, related identities and zeta and l-functions*, J. Korean Math. Soc., **45** (2008), 435–453.
46. T. Kim, D. S. Kim, T. Mansour, S. H. Rim, M. Schork, *Umbral calculus and Sheffer sequences of polynomials*, J. Math. Phys., **54** (2013), 083504; https://doi.org/10.1063/1.4817853.
47. T. Kim, T. Mansour, S. H. Rim, J. J. Soo, *A note on q-Changhee polynomials and numbers*, Adv. Studies Theor. Phys., **8** (2014), 35–41.
48. T. Kim, D. S. Kim, D. V. Dolgy, J.-J. Seo, *Bernoulli polynomials of the second kind and their identities arising from umbral calculus*, J. Nonlinear Sci. Appl., **9** (2016), 860–869.

49. T. Kim, D. V. Dolgy, D. S. Kim, J. J. Seo, *Differential equations for Changhee polynomials and their applications*, J. Nonlinear Sci. Appl., 9 (2016), 2857–2864.
50. T. Koshy, *Fibonacci and Lucas numbers with applications* (John Wiley & Sons, Inc., New York, 2001).
51. T. Komatsu, *Convolution identities for Cauchy numbers*, Acta Math. Hungar., **144** (2014), 76–91.
52. I. Kucukoglu, Y. Simsek, *Combinatorial identities associated with new families of the numbers and polynomials and their approximation value*, arXiv:1711.00850v1.
53. I. Kucukoglu, *A note on combinatorial numbers and polynomials*, Proceedings Book of the Mediterranean International Conference of Pure & Applied Mathematics and Related Areas 2018 (MICOPAM 2018), 103–106.
54. I. Kucukoglu, *Derivative formulas related to unification of generating functions for Sheffer type sequences*, AIP Conf. Proc., **2116** (2019), 100016; https://doi.org/10.1063/1.5114092
55. I. Kucukoglu, Y. Simsek, *Observations on identities and relations for interpolation functions and special numbers*, Adv. Stud. Contemp. Math., **28** (2018), 41–56.
56. I. Kucukoglu, B. Simsek, Y. Simsek, *An approach to negative hypergeometric distribution by generating function for special numbers and polynomials*, Turk. J. Math., **43** (2019), 2337–2353.
57. I. Kucukoglu, Y. Simsek, *On a family of special numbers and polynomials associated with Apsotol-type numbers and polynomials and combinatorial numbers*, Appl. Anal. Discrete Math., **13**(2) (2019), 478–494.
58. I. Kucukoglu, B. Simsek, Y. Simsek, *Generating Functions for New Families of Combinatorial Numbers and Polynomials: Approach to Poisson–Charlier Polynomials and Probability Distribution Function*, Axioms, **8**(4) (2019), 112; https://doi.org/10.3390/axioms8040112.
59. L. Lawrence, *Univariate Distribution Relationships-Negative hypergeometric distribution*; http://www.math.wm.edu/~leemis/chart/UDR/UDR.html.
60. G. G. Lorentz, *Bernstein Polynomials* (Chelsea Pub. Comp., New York, NY, USA, 1986).
61. D. Q. Lu, H. M. Srivastava, *Some series identities involving the generalized Apostol type and related polynomials*, Comput. Math. Appl., **62** (2011), 3591–3602.
62. Q.-M. Luo, *Apostol-Euler polynomials of higher order and Gaussian hypergeometric functions*, Taiwanese J. Math., **10** (2006), 917–925.
63. Q.-M. Luo, *Fourier expansions and integral representations for the Genocchi polynomials*, J. Integer Seq., **12** (2009), Article 09.1.4.
64. Q.-M. Luo, *The multiplication formulas for the Apostol–Bernoulli and Apostol–Euler polynomials of higher order*, Integral Transforms Spec. Funct., **20** (2009), 377–391.
65. Q.-M. Luo, *Extension for the Genocchi polynomials and its Fourier expansions and integral representations*, Osaka J. Math., **48**(2) (2011), 291–309.
66. Q.-M. Luo, H. M. Srivastava, *Some generalizations of the Apostol-Bernoulli and Apostol-Euler polynomials*, J. Math. Anal. Appl., **308** (2005), 290–302.
67. Q.-M. Luo, H. M. Srivastava, *Some relationships between the Apostol-Bernoulli and Apostol-Euler polynomials*, Comput. Math. Appl., **51** (2006), 631–642.
68. Q. M. Luo, H. M. Srivastava, *Some generalizations of the Apostol-Genocchi polynomials and the Stirling numbers of the second kind*, Appl. Math. Comput., **217** (2011), 5702–5728.
69. I. Mezo, *Several Generating Functions for Second-Order Recurrence Sequences*, J. Integer Sequences, **12** (2009), Article 09.3.7.
70. G. V. Milovanovic, G. P. Djordevic, *On some properties of Humbert's polynomials*, Fibonacci Quart., **25** (1987), 356–360.
71. M. A. Özarslan, *Unified Apostol-Bernoulli, Euler and Genocchi polynomials*, Comput. Math. Appl., **62** (2011), 2452–2462.
72. G. Ozdemir, Y. Simsek, G. V. Milovanovic, *Generating functions for special polynomials and numbers including Apostol-type and Humbert-type polynomials*, Mediterr. J. Math., **14** (2017), 117; https://doi.org/10.1007/s00009-017-0918-6.
73. H. Ozden, Y. Simsek, I. N. Cangul, V. Kurt, *On the higher-order w-q-Genocchi numbers*, Adv. Stud. Contemp. Math., **19**(1) (2009), 39–57.

74. H. Ozden, Y. Simsek, H. M. Srivastava, *A unified presentation of the generating functions of the generalized Bernoulli, Euler and Genocchi polynomials*, Comput. Math. Appl., **60** (2010), 2779–2787.
75. H. Ozden, Y. Simsek, *A new extension of q-Euler numbers and polynomials related to their interpolation functions*, Appl. Math. Letters, **21** (2008), 934–939.
76. H. Ozden, Y. Simsek, *Modification and unification of the Apostol-type numbers and polynomials and their applications*, Appl. Math. Compute. **235** (2014), 338–351.
77. F. Qi, *Explicit formulas for computing Bernoulli numbers of the second kind and Stirling numbers of the first kind*, Filomat, **28**(2) (2014), 319–327.
78. Th. M. Rassias, H. M. Srivastava, Some classes of infinite series associated with the Riemann zeta and polygamma functions and generalized harmonic numbers, Appl. Math. Comput. **131** (2–3) (2002), 593–605.
79. S. Roman, *The Umbral Calculus* (Dover Publ. Inc., New York, 2005).
80. S. M. Ross, *A First Course in Probability* (8th ed.) (Pearson Education, Inc., London, UK, 2010).
81. W. H. Schikhof, *Ultrametric Calculus: an introduction to p-adic analysis* (Cambridge University Press, Cambridge Studies in Advanced Mathematics 4, Cambridge, 1984).
82. Y. Simsek, *q-analogue of the twisted l-series and q-twisted Euler numbers*, J. Number Theory, **100** (2005), 267–278.
83. Y. Simsek, *q-Hardy-Berndt type sums associated with q-Genocchi type zeta and q-l-functions*, Nonlinear Analysis, **71** (2009), e377–e395.
84. Y. Simsek, *p-adic (h, q)-L-functions*, Comput. Math. Appl. **59**(6) (2010), 2097–2110.
85. Y. Simsek, *Generating functions for generalized Stirling type numbers, array type polynomials, Eulerian type polynomials and their applications*, Fixed Point Theory Appl., **87** (2013), 343–355.
86. Y. Simsek, *Functional equations from generating functions: A novel approach to deriving identities for the Bernstein basis functions*, Fixed Point Theory Appl., **2013** (2013), 1–13.
87. Y. Simsek, Generating functions for the Bernstein type polynomials: A new approach to deriving identities and applications for the polynomials, Hacet. J. Math. Stat., **43**(1) (2014), 1–14
88. Y. Simsek, *Analysis of the Bernstein basis functions: an approach to combinatorial sums involving binomial coefficients and Catalan numbers*, Math. Methods Appl. Sci., **38**(14) (2015), 3007–3021.
89. Y. Simsek, *Analysis of the p-adic q-Volkenborn integrals: An approach to generalized Apostol-type special numbers and polynomials and their applications*, Cogent Math. Stat., **3** (2016), 1269393; https://doi.org/10.1080/23311835.2016.1269393.
90. Y. Simsek, *Apostol type Daehee numbers and polynomials*, Adv. Studies Contemp. Math., **26**(3) (2016), 1–12.
91. Y. Simsek, *Identities on the Changhee numbers and Apostol-Daehee polynomials*, Adv. Stud. Contemp. Math., **27**(2) (2017), 199–212.
92. Y. Simsek, *Computation methods for combinatorial sums and Euler-type numbers related to new families of numbers*, Math. Methods Appl. Sci., **40**(7) (2017), 2347–2361.
93. Y. Simsek, *Construction of some new families of Apostol-type numbers and polynomials via Dirichlet character and p-adic q-integrals*, Turk. J. Math., **42** (2018), 557–577.
94. Y. Simsek, *New families of special numbers for computing negative order Euler numbers and related numbers and polynomials*, Appl. Anal. Discr. Math., **12** (2018), 1–35.
95. Y. Simsek, *Construction method for generating functions of special numbers and polynomials arising from analysis of new operators*, Math. Methods Appl. Sci., **41** (2018), 6934–6954.
96. Y. Simsek, *Generating functions for finite sums involving higher powers of binomial coefficients: Analysis of hypergeometric functions including new families of polynomials and numbers*, J. Math. Anal. Appl., **477** (2019), 1328–1352.
97. Y. Simsek, *Formulas for Poisson–Charlier, Hermite, Milne-Thomson and other type polynomials by their generating functions and p-adic integral approach*, Rev. R. Acad. Cienc. Exactas Fís. Nat. Ser. A Mat. RACSAM, **113** (2019), 931–948.

98. Y. Simsek, *Explicit formulas for p-adic integrals: Approach to p-adic distributions and some families of special numbers and polynomials*, Montes Taurus J. Pure Appl. Math., **1**(1) (2019), 1–76.
99. Y. Simsek, *Peters type polynomials and numbers and their generating functions: Approach with p-adic integral method*, Math. Meth. Appl. Sci., **42** (2019), 7030–7046.
100. Y. Simsek, *Remarks and some formulas associated with combinatorial numbers*, AIP Conf. Proc., **2116** (2019), 100002; https://doi.org/10.1063/1.5114078.
101. Y. Simsek, Analysis of Apostol-type numbers and polynomials with their approximations and asymptotic behavior, In: Rassias T.M. (eds) Approximation Theory and Analytic Inequalities. Springer, Cham, pp. 435–486, https://doi.org/10.1007/978-3-030-60622-0_23.
102. Y. Simsek, Peters type polynomials and numbers and their generating functions: Approach with p-adic integral method, Math Meth Appl Sci., **42** (2019), 7030–7046.
103. Y. Simsek, I. N. Cangul, V. Kurt, D. Kim, *q-Genocchi numbers and polynomials associated with q-Genocchi-type l-functions*, Adv. Difference Equ., **2008** (2008), Article ID 815750; https://doi.org/10.1155/2008/815750.
104. Y. Simsek, M. Acikgoz, *A new generating function of (q-) Bernstein-type polynomials and their interpolation function*, Abstr. Appl. Anal., **2010** (2010), 769095; https://doi.org/10.1155/2010/769095.
105. Y. Simsek, J. S. So, *Identities, inequalities for Boole-type polynomials: approach to generating functions and infinite series*, J. Inequal. Appl., **2019** (2019), 62; https://doi.org/10.1186/s13660-019-2006-x.
106. Y. Simsek, J. S. So, *On generating functions for Boole type polynomials and numbers of higher order and their applications*, Symmetry, **11**(3) (2019), 352; https://doi.org/10.3390/sym11030352.
107. H. M. Srivastava, *Some formulas for the Bernoulli and Euler polynomials at rational arguments*, Math. Proc. Cambridge Philos. Soc., **129** (2000), 77–84.
108. H. M. Srivastava, *Some generalizations and basic (or q-) extensions of the Bernoulli, Euler and Genocchi polynomials*, Appl. Math. Inf. Sci., **5** (2011), 390–444.
109. H. M. Srivastava, H. L. Manocha, *A Treatise on Generating Functions* (Ellis Horwood Limited Publisher, Chichester, 1984).
110. H. M. Srivastava, J. Choi, *Series Associated with the Zeta and Related Functions* (Kluwer Academic Publishers, Dordrecht, Boston, London, 2001).
111. H. M. Srivastava, T. Kim, Y. Simsek, *q-Bernoulli numbers and polynomials associated with multiple q-zeta functions and basic L-series*, Russ. J. Math. Phys., **12** (2005), 241–268.
112. H. M. Srivastava, J. Choi, *Zeta and q-Zeta Functions and Associated Series and Integrals* (Elsevier Science Publishers, Amsterdam, London and New York, 2012).
113. H. M Srivastava, I. Kucukoglu, Y. Simsek, *Partial differential equations for a new family of numbers and polynomials unifying the Apostol-type numbers and the Apostol-type polynomials*, J. Number Theory, **181** (2017), 117–146.
114. Wolfram Research Inc., Mathematica Online (Wolfram Cloud), Champaign, IL, 2020; https://www.wolframcloud.com

Pathwise Stability and Positivity of Semi-Discrete Approximations of the Solution of Nonlinear Stochastic Differential Equations

Ioannis S. Stamatiou

Abstract We use the main idea of the semi-discrete method, originally proposed in (N. Halidias, International Journal of Computer Mathematics, **89**(6) (2012), 780–794), to reproduce qualitative properties of a class of nonlinear stochastic differential equations with non-negative, non-globally Lipschitz coefficients and a unique equilibrium solution. The proposed fixed-time step method preserves the positivity of the solution and reproduces the almost sure asymptotic stability behavior of the equilibrium with no time-step restrictions. In particular, we are interested in the following class of scalar stochastic differential equations,

$$x_t = x_0 + \int_0^t x_s a(x_s)ds + \int_0^t x_s b(x_s)dW_s,$$

where $a(\cdot), b(\cdot)$ are non-negative functions with $b(u) \neq 0$ for $u \neq 0, x_0 \geq 0$ and $\{W_t\}_{t\geq 0}$ is a one-dimensional Wiener process adapted to the filtration $\{\mathcal{F}_t\}_{t\geq 0}$.

1 Introduction

We are interested in the following class of scalar stochastic differential equations (SDEs),

$$x_t = x_0 + \int_0^t x_s a(x_s)ds + \int_0^t x_s b(x_s)dW_s, \tag{1}$$

where $a(\cdot), b(\cdot)$ are non-negative functions with $b(u) \neq 0$ for $u \neq 0, x_0 \geq 0$ and $\{W_t\}_{t\geq 0}$ is a one-dimensional Wiener process adapted to the filtration $\{\mathcal{F}_t\}_{t\geq 0}$. We want to reproduce dynamical properties of (1). We use a fixed-time step explicit

I. S. Stamatiou (✉)
Department of Biomedical Sciences, University of West Attica, Athens, Greece
e-mail: istamatiou@uniwa.gr

I. N. Parasidis et al. (eds.), *Mathematical Analysis in Interdisciplinary Research*, Springer Optimization and Its Applications 179,
https://doi.org/10.1007/978-3-030-84721-0_34

numerical method, in particular the exponential truncated semi-discrete method (expTSD), see [15], which reads

$$y_{n+1}^{\Delta} = y_n^{\Delta} \exp\left\{\left(a(\pi_\Delta(y_n^\Delta)) - \frac{b^2(\pi_\Delta(y_n^\Delta))}{2}\right)\Delta + b(\pi_\Delta(y_n^\Delta))\Delta W_n\right\} \tag{2}$$

with $y_0 = x_0$ a.s., where $\Delta = t_{n+1} - t_n$ is the time step-size and $\Delta W_n := W_{t_{n+1}} - W_{t_n}$ are the increments of the Wiener process. The function π_Δ appearing in the argument of a and b stands for

$$\pi_\Delta(x) := \left(|x| \wedge \mu^{-1}(h(\Delta))\right)\frac{x}{|x|}, \tag{3}$$

where the strictly increasing function μ and the strictly decreasing function h are defined later on, in Sect. 2. For the derivation of (2) see Sect. 2.

The scopes of this article are two. Our main goal is to reproduce the almost sure (a.s.) stability of the unique equilibrium solution of (1), i.e., for the trivial solution $x_t \equiv 0$. The positivity of the drift pushes the solution to explosive situations and the diffusion stabilizes this effect in a way we want to mimic.

On the other hand, SDE (1) has unique positive solutions when $x_0 > 0$. The truncated semi-discrete scheme (2) preserves positivity; the assertion is obvious for the (expTSD) scheme by construction.

Explicit fixed-step Euler methods fail to strongly converge to solutions of (1) when the drift or diffusion coefficient grows superlinearly [5, Theorem 1]. A proposed fix to this problem is the so-called Tamed Euler methods, c.f. [6, (4)], [16, (3.1)], [11] and the references therein. However, they usually do not preserve positivity. On the other hand, adaptive time-stepping strategies applied to explicit Euler method are an alternative way to address the problem and there is an ongoing research on that approach, see [2, 8], and [9]. The fixed-step method we propose reproduces the almost sure asymptotic stability behavior of the equilibrium with no time-step restrictions, compare Theorem 2 with [9, Theorem 4.1], respectively.

Our proposed fixed-step method is explicit, strongly convergent, non-explosive, and positive. The semi-discrete method (SD) was originally proposed in [3] and recently in [12] and [13]; see also [14] for a review of the method. The key idea behind the SD method is freezing on each integration interval of size Δ, parts of the drift and diffusion coefficients of the solution at the endpoints of the subinterval, obtaining explicitly solved SDEs. Here, we freeze the nonlinear parts in (1), that is a and b, obtaining a linear SDE with explicit solution of exponential type, see (2).

Let us now assume some minimal additional conditions for the functions $a(\cdot)$ and $b(\cdot)$. In particular, we assume locally Lipschitz continuity of $a(\cdot)$ and $b(\cdot)$, which in turn implies the existence of a unique, continuous $\mathcal{F}_t$-measurable process x (cf. [10, Chapter 2]) satisfying (1) up to the explosion time $\tau_e^{x_0}$, i.e., on the interval $[0, \tau_e^{x_0})$, where

$$\tau_e^{x_0} := \inf\{t > 0 : |x_t^{x_0}| \notin [0, \infty)\}.$$

Denoting $\theta_e^{x_0}$ the first hitting time of zero, i.e.,

$$\theta_e^{x_0} := \inf\{t > 0 : |x_t^{x_0}| = 0\},$$

it was shown in [1, Section 3] that in the case

$$\sup_{u \neq 0} \frac{2a(u)}{b^2(u)} = \beta < 1, \tag{4}$$

then $\tau_e^{x_0} = \theta_e^{x_0} = \infty$, i.e., there exist unique positive solutions. The equilibrium zero solution of (1) is a.s. stable if (see again [1, Section 3])

$$\lim_{u \to 0} \frac{2a(u)}{b^2(u)} < 1, \tag{5}$$

i.e., for all $x_0 > 0$

$$\mathbb{P}(\{\omega : \lim_{t \to \infty} x_t(\omega) = 0\}) > 0.$$

Condition (5) shows how condition (4) is close to being sharp. Furthermore, the presence of a sufficiently intense stochastic perturbation (because of the positivity of the function $a(\cdot)$) is necessary for the existence of a unique global solution and stability of the zero equilibrium. The proposed fixed-time step method (expTSD) reproduces the almost sure asymptotic stability behavior of the equilibrium with no time-step restrictions, as stated in Theorem 2. Moreover, as already discussed, it preserves the positivity of the solution.

The outline of the article is the following. In Sect. 2 we present some preliminaries and our main results, that is Theorems 1 and 2, the proofs of which are deferred to Sects. 3 and 4, respectively. Theorem 1 concerns the convergence properties of the proposed scheme and Theorem 2 deals with its stability properties. Section 5 provides a numerical example.

2 Setting and Main Results

Let $T > 0$ and let $(\Omega, \mathcal{F}, \{\mathcal{F}_t\}_{0 \le t \le T}, \mathbb{P})$ be a complete probability space. Let also $W_{t,\omega} : [0, T] \times \Omega \to \mathbb{R}$ be a one-dimensional Wiener process adapted to the filtration $\{\mathcal{F}_t\}_{0 \le t \le T}$. The notation and the setting is standard and we refer the reader in [14] and the references therein.

We recall the semi-discrete scheme. The auxiliary functions $f(x, y), g(x, y) : \mathbb{R}^+ \times \mathbb{R}^+ \to \mathbb{R}^+$ are such that $f(x, x) = xa(x)$ and $g(x, x) = xb(x)$. In particular, we take

$$f(x,y) = a(x)y \quad \text{and} \quad g(x,y) = b(x)y. \tag{6}$$

We discretize $[0, T]$ in N equidistant intervals $[t_j, t_{j+1}]$ of size Δ and in each subinterval consider the following SDEs for $n = 0, 1, \ldots, N-1$

$$\begin{aligned} y_t &= y_{t_n} + \int_{t_n}^t f(y_{t_n}, y_s)ds + \int_{t_n}^t g(y_{t_n}, y_s)dW_s \\ &= y_{t_n} + a(y_{t_n})\int_{t_n}^t y_s ds + b(y_{t_n})\int_{t_n}^t y_s dW_s \\ &= y_{t_n} \exp\left\{\left(a(y_{t_n}) - \frac{b^2(y_{t_n})}{2}\right)(t - t_n) + b(y_{t_n})(W_t - W_{t_n})\right\}, \end{aligned} \tag{7}$$

where $t \in (t_n, t_{n+1}]$ with $y_0 = x_0$ a.s. The construction of the truncated semi-discrete scheme, see [15], uses the truncated functions f_Δ and g_Δ where $f_\Delta(x,y) := f(\pi_\Delta(x), y)$, $g_\Delta(x,y) := g(\pi_\Delta(x), y)$, for $x, y \in \mathbb{R}^+$ where we set $x/|x| = 0$ when $x = 0$ and π_Δ is defined in (3). The strictly increasing function $\mu : \mathbb{R}^+ \to \mathbb{R}^+$ is such that

$$\sup_{|x|\le u} (|f(x,y)| \vee |g(x,y)|) \le \mu(u)(1 + |y|), \qquad u \ge 1, \tag{8}$$

and the strictly decreasing function $h : (0, 1] \to [\mu(1), \infty)$ satisfies

$$\lim_{\Delta \to 0} h(\Delta) = \infty \quad \text{and} \quad \Delta^{1/4}h(\Delta) \le \hat{h} \quad \text{for every} \quad \Delta \in (0, 1], \tag{9}$$

for a constant $\hat{h} \ge 1 \vee \mu(1)$. We end up with the process defining the (expTSD) scheme (2), that is

$$y_t^\Delta = y_{t_n}^\Delta \exp\left\{\left(a(\pi_\Delta(y_{t_n}^\Delta)) - \frac{b^2(\pi_\Delta(y_{t_n}^\Delta))}{2}\right)(t - t_n) + b(\pi_\Delta(y_{t_n}^\Delta))(W_t - W_{t_n})\right\}, \tag{10}$$

where $t \in (t_n, t_{n+1}]$. We make the following hypotheses for the coefficients in (1).

Assumption 2.1 *The locally Lipschitz continuous functions $a(\cdot)$ and $b(\cdot)$ in (1) are non-negative with $b(u) \neq 0$ for $u \neq 0$. Moreover there exists $\beta < 1$ such that (4) holds, i.e., $\sup_{u\neq 0} \frac{2a(u)}{b^2(u)} = \beta < 1$.*

At this point we may tell more about the function μ satisfying (8). Note that

$$\sup_{|x|\le u} |f(x,y)| = \sup_{|x|\le u} a(x)y \le \frac{\beta}{2} \sup_{|x|\le u} b^2(x)y$$

and

$$\sup_{|x|\leq u} |g(x,y)| = \sup_{|x|\leq u} b(x)y \leq (1 + \sup_{|x|\leq u} b^2(x))y,$$

which suggests that μ should be such that $1 + \sup_{|x|\leq u} b^2(x) \leq \mu(u)$. Moreover, $\pi_\Delta(x)$ reads

$$\pi_\Delta(x) = |x| \wedge \mu^{-1}(h(\Delta)). \tag{11}$$

We present the compact form of the approximation process (10)

$$y_t^\Delta = y_0 + \int_0^t a(\pi_\Delta(y_{\hat{s}}^\Delta))y_s^\Delta ds + \int_0^t b(\pi_\Delta(y_{\hat{s}}^\Delta))y_s^\Delta dW_s, \tag{12}$$

where $\hat{s} = t_n$ when $s \in [t_n, t_{n+1})$ which is used in the statement of our first convergence result.

Theorem 1 (Positivity and Convergence) *Let $a(\cdot)$ and $b(\cdot)$ satisfy Assumption 2.1 and define for $\gamma > 0$*

$$\mu(u) = \overline{C}u^{1+\gamma}, \quad u \geq 1 \quad \text{and} \quad h(\Delta) = \overline{C} + \sqrt{\ln \Delta^{-\epsilon}}, \quad \Delta \in (0, 1], \tag{13}$$

with $\epsilon \in (0, 1/4)$, where $\Delta \leq 1$ and $\hat{h}$ are such that (9) holds. The truncated semi-discrete numerical scheme (12) converges to the true solution of (1) in the $\mathcal{L}^1$-sense with order arbitrarily close to $\frac{1-\beta}{4(1+\gamma)}$, i.e.,

$$\lim_{\Delta\to 0} \mathbb{E} \sup_{0\leq t\leq T} |y_t^\Delta - x_t| \leq C\Delta^{\frac{1-\beta}{4(1+\gamma)}(1-\epsilon)}. \tag{14}$$

We proceed with a Lemma related to a notion of stability of a stochastic difference equation.

Lemma 1 (c.f. [4], Lemma 5.1) *Given a sequence $(\Psi_n)_{n\in\mathbb{N}}$ of non-negative, independent, and identically distributed random variables define a sequence of random variables $(M_n)_{n\in\mathbb{N}}$ by*

$$M_n = \left(\prod_{i=0}^{n-1} \Psi_i\right) Z_0,$$

with $M_0 = \theta > 0$. Assume that $\ln(\Psi_i)$ are square-integrable for all i. Then

$$\lim_{n\to\infty} M_n = 0, \quad \textit{a.s. iff } \mathbb{E}\ln(\Psi_i) < 0 \quad \textit{for all } i.$$

The following result provides sufficient conditions for solutions of the truncated semi-discrete scheme (2) to demonstrate a.s. stability.

Theorem 2 (a.s. stability) *Let $a(\cdot)$ and $b(\cdot)$ satisfy Assumption 2.1. Let μ and h be as in (13) with $\epsilon \in (0, 1/4)$. Let also $\{y_n^\Delta\}_{n\in\mathbb{N}}$ be a solution of (2) with $y_0^\Delta = x_0 > 0$. Then for all $\Delta < 1$,*

$$\lim_{n\to\infty} y_n^\Delta = 0, \qquad a.s. \tag{15}$$

Note that there is no time-step restriction in the result of Theorem 2, that is (15) holds for all $\Delta < 1$.

3 Proof of Convergence

In this section we provide the proof of Theorem 1. Note that since a and b are locally Lipschitz the same applies to the auxiliary functions f and g, see (6). Denote the truncated version of the auxiliary functions with a subscript Δ, that is $f_\Delta(x, y) = a(\pi_\Delta(x))y$ and $g_\Delta(x, y) = b(\pi_\Delta(x))y$. Note that

$$|f_\Delta(x_1, y_1) - f_\Delta(x_2, y_2)| \leq h(\Delta)(|x_1 - x_2| + |y_1 - y_2|) \tag{16}$$

and

$$|f_\Delta(x, y)| \vee |g_\Delta(x, y)| \leq h(\Delta)(1 + |y|). \tag{17}$$

In the next subsection we prove the $\mathcal{L}^1$-convergence result for the expTSD scheme (y_n^Δ). To avoid heavy notation, we occasionally write y instead of y^Δ.

3.1 Convergence of (y_n^Δ)

We essentially follow the proof in [15] pointing out the main differences. Defining the stopping time

$$\rho_{\Delta,R} = \inf\{t \in [0, T] : |y_t^\Delta| > R \text{ or } |y_{\hat{t}}^\Delta| > R\}, \tag{18}$$

we provide the error bound for the truncated semi-discrete scheme, see Lemma 2 in [15].

Lemma 2 *Let $R > 1$, and $\rho_{\Delta,R}$ as in (18). Then the following estimate holds*

$$\mathbb{E}|y_{s\wedge\rho_{\Delta,R}} - y_{\widehat{s\wedge\rho_{\Delta,R}}}|^{\hat{p}} \leq C(\Delta^{1/2}h(\Delta)R)^{\hat{p}},$$

for any $\hat{p} > 0$, where C does not depend on Δ.

We next show moment bounds for the approximation process (y_t^{Δ}).

Lemma 3 *For any $R \leq h(\Delta)$ and any $p \leq 1 - \beta$*

$$\sup_{0 \leq \Delta \leq 1} \sup_{0 \leq t \leq T} \mathbb{E}|y_t^{\Delta}|^p \leq C, \tag{19}$$

for all $T > 0$.

Proof We fix a $\Delta \in (0, 1]$ and a $T > 0$. We take advantage of the analytic expression of the approximation process. We rewrite (2) as

$$\begin{aligned} |y_{n+1}^{\Delta}|^p &= |y_n^{\Delta}|^p \exp\left\{\frac{pb^2(\pi_\Delta(y_n^{\Delta}))}{2}\left(\frac{2a(\pi_\Delta(y_n^{\Delta}))}{b^2(\pi_\Delta(y_n^{\Delta}))} - 1 + p\right)\Delta\right\} \\ &\quad \times \exp\left\{-\frac{p^2b^2(\pi_\Delta(y_n^{\Delta}))}{2}\Delta + pb(\pi_\Delta(y_n^{\Delta}))\Delta W_n\right\} \\ &= |y_n^{\Delta}|^p \mathcal{E}(y_n^{\Delta})\xi_{n+1}, \end{aligned} \tag{20}$$

where we used the notation y_n^{Δ} for $y_{t_n}^{\Delta}$ and the exponential function $\mathcal{E}(\cdot)$ reads

$$\mathcal{E}(u) = \exp\left\{\frac{pb^2(u)}{2}\left(\frac{2a(u)}{b^2(u)} - 1 + p\right)\Delta\right\}$$

and for $t \in (t_n, t_{n+1}]$ we consider the SDE

$$d\xi_t = pb(\pi_\Delta(y_n^{\Delta}))dW_t,$$

with $\xi_n = 1$. Therefore $\mathbb{E}\xi_{n+1} = 1$, (cf. [10], [7]) and for $0 < p \leq 1 - \beta$, with the β appearing in Assumption 2.1 we get that $\mathcal{E}(u) \leq 1$ for any $\Delta > 0$ implying the boundness of the moments of $(y_n)_{n \in \mathbb{N}}$ since

$$\mathbb{E}|y_{n+1}^{\Delta}|^p \leq \mathbb{E}(|y_n^{\Delta}|^p \xi_{n+1}) \leq \mathbb{E}|y_n^{\Delta}|^p \leq \ldots \leq \mathbb{E}|y_0^{\Delta}|^p \leq C.$$

□

We proceed with the proof of Theorem 1.

Proof We denote the difference $\mathcal{E}_t^{\Delta} := y_t^{\Delta} - x_t$ and define the stopping times

$$\tau_R = \inf\{t \in [0, T] : |x_t| > R\}, \quad \theta_{\Delta,R} := \tau_R \wedge \rho_{\Delta,R}, \tag{21}$$

for some $R > 1$ big enough. Moreover the events Ω_R are

$$\Omega_R := \{\omega \in \Omega : \sup_{0 \leq t \leq T} |x_t| \leq R, \sup_{0 \leq t \leq T} |y_t^{\Delta}| \leq R\}.$$

We have that

$$
\begin{aligned}
\mathbb{E} \sup_{0\le t\le T} |\mathcal{E}_t| &= \mathbb{E} \sup_{0\le t\le T} |\mathcal{E}_t|\mathbb{I}_{\Omega_R} + \mathbb{E} \sup_{0\le t\le T} |\mathcal{E}_t|\Delta^{m/2}\Delta^{-m/2}\mathbb{I}_{(\Omega_R)^c} \\
&\le \mathbb{E} \sup_{0\le t\le T} |\mathcal{E}_{t\wedge\theta_{\Delta,R}}| + \frac{\Delta^m}{2}\mathbb{E} \sup_{0\le t\le T} |\mathcal{E}_t| + \frac{\Delta^{-m}}{2}\mathbb{E}\mathbb{I}_{(\Omega_R)^c} \\
&\le \mathbb{E} \sup_{0\le t\le T} |\mathcal{E}_{t\wedge\theta_{\Delta,R}}| + \frac{\Delta^{-m}}{2}\mathbb{P}(\Omega_R)^c, \qquad (22)
\end{aligned}
$$

where we applied Young's inequality with $m > 0$. It holds that

$$
\begin{aligned}
\mathbb{P}(\Omega_R)^c &\le \mathbb{P}(\sup_{0\le t\le T} |y_t| > R) + \mathbb{P}(\sup_{0\le t\le T} |x_t| > R) \\
&\le (\mathbb{E} \sup_{0\le t\le T} |y_t|^k)R^{-k} + (\mathbb{E} \sup_{0\le t\le T} |x_t|^k)R^{-k},
\end{aligned}
$$

for any $k > 0$ by the subadditivity of the measure $\mathbb{P}$ and the Markov inequality. Setting $k = (1-\beta)$ we get

$$
\mathbb{P}(\Omega_R)^c \le 2AR^{-(1-\beta)}. \qquad (23)
$$

Now, we turn to the estimation of the difference $|\mathcal{E}_{t\wedge\theta_{\Delta,R}}|$. Using the triangle inequality we get

$$
\begin{aligned}
|\mathcal{E}_{t\wedge\theta_{\Delta,R}}| &= \Bigg| \int_0^{t\wedge\theta_{\Delta,R}} (f_\Delta(y_{\hat{s}}, y_s) - f(x_s, x_s))\, ds \\
&\quad + \int_0^{t\wedge\theta_{\Delta,R}} (g_\Delta(y_{\hat{s}}, y_s) - g(x_s, x_s))\, dW_s \Bigg| \\
&\le \int_0^{t\wedge\theta_{\Delta,R}} |f_\Delta(y_{\hat{s}}, y_s) - f_\Delta(x_s, y_s) + f_\Delta(x_s, y_s) - f(x_s, x_s)|\, ds + |M_t|,
\end{aligned}
$$

where $M_t := \int_0^{t\wedge\theta_{\Delta,R}} (g_\Delta(y_{\hat{s}}, y_s) - g(x_s, x_s))\, dW_s$. By (16) and the fact that $f_\Delta(x_s, x_s) = f(x_s, x_s)$ when $R \le \mu^{-1}(h(\Delta))$ we arrive at

$$
|\mathcal{E}_{t\wedge\theta_{\Delta,R}}| \le h(\Delta) \int_0^{t\wedge\theta_{\Delta,R}} (|y_{\hat{s}} - y_s| + 2|\mathcal{E}_s|)\, ds + |M_t|. \qquad (24)
$$

It holds that

$$
\mathbb{E} \sup_{0\le t\le T} |M_t| \le \sqrt{32}\cdot\mathbb{E}\sqrt{\int_0^{T\wedge\theta_{\Delta,R}} (g_\Delta(y_{\hat{s}}, y_s) - g(x_s, x_s))^2\, ds}
$$

$$
\begin{aligned}
&\leq \mathbb{E}\sqrt{64h^2(\Delta)\int_0^{T\wedge\theta_{\Delta,R}}\left(|y_{\hat{s}}-y_s|^2+4|\mathcal{E}_s|^2\right)ds} \\
&\leq 8h(\Delta)\mathbb{E}\sqrt{\int_0^{T\wedge\theta_{\Delta,R}}|y_{\hat{s}}-y_s|^2 ds} \\
&\quad+\mathbb{E}\sqrt{\sup_{0\leq s\leq T}|\mathcal{E}_{s\wedge\theta_{\Delta,R}}|\cdot 256h^2(\Delta)\int_0^{T\wedge\theta_{\Delta,R}}|\mathcal{E}_s|ds} \\
&\leq 8h(\Delta)\mathbb{E}\sqrt{\int_0^{T\wedge\theta_{\Delta,R}}|y_{\hat{s}}-y_s|^2 ds}+\frac{1}{2}\mathbb{E}\sup_{0\leq s\leq T}|\mathcal{E}_{s\wedge\theta_{\Delta,R}}| \\
&\quad+256h^2(\Delta)\int_0^{T\wedge\theta_{\Delta,R}}\mathbb{E}|\mathcal{E}_s|ds,
\end{aligned}
$$

by the Young inequality. Taking the supremum and then expectations in (24) and rearranging the terms we reach

$$
\begin{aligned}
\mathbb{E}\sup_{0\leq t\leq T}|\mathcal{E}_{t\wedge\theta_{\Delta,R}}| &\leq 2h(\Delta)\mathbb{E}\int_0^{T\wedge\theta_{\Delta,R}}\mathbb{E}|y_{\hat{s}}-y_s|ds+4h(\Delta)\int_0^{T\wedge\theta_{\Delta,R}}\mathbb{E}\sup_{0\leq l\leq s}|\mathcal{E}_l|ds \\
&\quad+8h^2(\Delta)\Delta^{\lambda}+\frac{1}{2}\Delta^{-\lambda}\int_0^{T\wedge\theta_{\Delta,R}}\mathbb{E}|y_{\hat{s}}-y_s|^2 ds \\
&\quad+512h^2(\Delta)\int_0^{T\wedge\theta_{\Delta,R}}\mathbb{E}\sup_{0\leq l\leq s}|\mathcal{E}_l|ds,
\end{aligned}
\tag{25}
$$

for any $\lambda > 0$. Using two times Lemma 2 and collecting the terms we arrive at

$$
\begin{aligned}
\mathbb{E}\sup_{0\leq t\leq T}|\mathcal{E}_{t\wedge\theta_{\Delta,R}}| &\leq C\Delta^{1/2}h^2(\Delta)R+Ch(\Delta)\int_0^{T\wedge\theta_{\Delta,R}}\mathbb{E}\sup_{0\leq l\leq s}|\mathcal{E}_l|ds \\
&\quad+C\Delta^{\lambda}h^2(\Delta)+C\Delta^{1-\lambda}h^2(\Delta)R^2+Ch^2(\Delta)\int_0^{T\wedge\theta_{\Delta,R}}\mathbb{E}\sup_{0\leq l\leq s}|\mathcal{E}_l|ds \\
&\leq C\Delta^{1/2}h^3(\Delta)+C\Delta^{\lambda}h^2(\Delta)+C\Delta^{1-\lambda}h^4(\Delta) \\
&\quad+Ch^2(\Delta)\int_0^{T\wedge\theta_{\Delta,R}}\mathbb{E}\sup_{0\leq l\leq s}|\mathcal{E}_l|ds \\
&\leq C\left(\Delta^{1/2}h^3(\Delta)+\Delta^{\lambda}h^2(\Delta)+\Delta^{1-\lambda}h^4(\Delta)\right)e^{h^2(\Delta)},
\end{aligned}
\tag{26}
$$

where we used that $1 < R \leq h(\Delta)$ and in the last step applied the Gronwall inequality and C is a constant independent of Δ and R varying from line to line. Plugging estimates (26) and (23) in (22) we have

$$
\mathbb{E}\sup_{0\leq t\leq T}|\mathcal{E}_t| \leq C\left(\Delta^{1/2}h^3(\Delta)+\Delta^{\lambda}h^2(\Delta)+\Delta^{1-\lambda}h^4(\Delta)\right)e^{h^2(\Delta)}+C\Delta^{-m}R^{-(1-\beta)}
$$

$$\leq C\Delta^{1/2}h^4(\Delta)e^{h^2(\Delta)} + C\Delta^{-m}R^{-(1-\beta)}, \tag{27}$$

where we chose $\lambda = 1/2$. Bearing in mind the definitions of μ and h, see (13), we have

$$C\Delta^{1/2}h^4(\Delta)e^{h^2(\Delta)} \leq C\Delta^{1/2}(\ln \Delta^{-\epsilon})^2\Delta^{-\epsilon} \leq C\Delta^{\frac{1}{2}-2\epsilon},$$

by choosing $\epsilon < 1/4$, where we used the fact that $0 \leq z(\ln z)^2 \leq z^2$ for big enough z. Moreover, by

$$\hat{h} > \Delta^{1/4}h(\Delta) > \overline{C}\Delta^{1/4} > \Delta^{\frac{1+\gamma}{4q}},$$

whenever $(1+\gamma) < q$, which implies

$$h(\Delta) \geq \Delta^{\frac{1+\gamma}{4q}-\frac{1}{4}}.$$

By the monotone property of μ^{-1} we have

$$\mu^{-1}(h(\Delta)) \geq \overline{C}^{-\frac{1}{1+\gamma}}\Delta^{\frac{1}{4q}-\frac{1}{4(1+\gamma)}} = R.$$

Estimate (27) becomes

$$\mathbb{E}\sup_{0\leq t\leq T}|\mathcal{E}_t| \leq C\Delta^{\frac{1}{2}(1-4\epsilon)} + C\Delta^{\frac{1-\beta}{4(1+\gamma)}\frac{q-(1+\gamma)}{q}-m}. \tag{28}$$

Taking big $q > 1+\gamma$ and small m we reach inequality (14). □

4 Proof of Stability

In this section we provide the proof of Theorem 2. We examine the stability behavior of the equilibrium solution of the expTSD scheme (y_n^Δ).

4.1 *Stability of* (y_n^Δ)

Using representation (20) we may write

$$|y_{n+1}^\Delta|^p = |y_n^\Delta|^p\mathcal{E}(y_n^\Delta)\xi_{n+1} = |y_{n-1}^\Delta|^p\mathcal{E}(y_n^\Delta)\mathcal{E}(y_{n-1}^\Delta)\xi_{n+1}\xi_n$$
$$= \ldots = |y_0^\Delta|^p\prod_{i=0}^{n}\mathcal{E}(y_i^\Delta)\xi_{i+1},$$

or using the notation $M_n := |y_n^\Delta|^p$ and $\Psi_i := \mathcal{E}(y_i^\Delta)\xi_{i+1}$ we have

$$M_n = \left(\prod_{i=0}^{n-1} \Psi_i\right) M_0. \tag{29}$$

Obviously, Ψ_i are non-negative, independent random variables such that

$$\begin{aligned}
\ln \Psi_i &= \frac{pb^2(\pi_\Delta(y_n^\Delta))}{2}\left(\frac{2a(\pi_\Delta(y_n^\Delta))}{b^2(\pi_\Delta(y_n^\Delta))} - 1 + p\right)\Delta \\
&\quad - \frac{p^2 b^2(\pi_\Delta(y_n^\Delta))}{2}\Delta + pb(\pi_\Delta(y_n^\Delta))\Delta W_n \\
&\leq \frac{pb^2(\pi_\Delta(y_n^\Delta))}{2}(\beta - 1 + p)\Delta + pb(\pi_\Delta(y_n^\Delta))\Delta W_n \\
&\leq pb(\pi_\Delta(y_n^\Delta))\Delta W_n,
\end{aligned}$$

for any $0 < p \leq 1-\beta$, therefore taking expectations in the above inequality implies $\mathbb{E}\ln \Psi_i \leq 0$ which in turn by application of Lemma 1 leads to $\lim_{n\to\infty} M_n = 0$ a.s. and consequently (15).

5 Numerical Illustration

We will use the numerical example of [9, Section 5], that is we take $a(x) = x^2$ and $b(x) = \sigma x$ and $x_0 = 1$ in (1), i.e.,

$$x_t = 1 + \int_0^t (x_s)^3 ds + \sigma \int_0^t (x_s)^2 dW_s, \qquad t \geq 0. \tag{30}$$

Note that the value of σ determines the value of the ratio $2a(u)/b^2(u) = 2/\sigma^2$. Moreover, (13) holds with $\mu(u) = (1+\sigma^2)|u|^2$ since

$$\sup_{|x|\leq u}\left(|x^2| \vee \sigma|x|\right) \leq (1+\sigma^2)|u|^2, \qquad u \geq 1.$$

In the notation of Theorem 1, $\gamma = 1$ and $\overline{C} = 1+\sigma^2$. Finally, $h(\Delta) = 1+\sigma^2 + \sqrt{\ln \Delta^{-\epsilon}}$ for any $\Delta \in (0,1]$. Clearly $h(1) \geq \mu(1)$ and

$$\Delta^{1/6}h(\Delta) \leq (1+\sigma^2)\Delta^{1/6} + \sqrt{\Delta^{1/3}\ln \Delta^{-\epsilon_1}} \leq 2+\sigma^2,$$

for any $\Delta \in (0,1]$ and $0 < \epsilon \leq 1/4$. Therefore we take $\hat{h} = 2+\sigma^2$. The exponential truncated semi-discrete method (2) reads

$$y_{n+1}^{\Delta} = y_n^{\Delta} \exp\left\{ \left(1 - \frac{\sigma^2}{2}\right) \pi_{\Delta}^2(y_n^{\Delta})\Delta + \sigma \pi_{\Delta}(y_n^{\Delta})\Delta W_n \right\}, \tag{31}$$

with $y_0 = 1$ where

$$\pi_{\Delta}(x) = x \wedge \sqrt{\frac{h(\Delta)}{1+\sigma^2}}$$

and therefore

$$\pi_{\Delta}^2(x) = x^2 \wedge \frac{h(\Delta)}{1+\sigma^2}.$$

First, we examine numerically the order of convergence of the truncated semi-discrete method. The numerical results suggest that the expTSD scheme converges in the $\mathcal{L}^1$-sense with order at least $\frac{1-\beta}{4(1+\gamma)}$, see Fig. 1. Actually we see that it is close to $1/2$.

We also examine the stability behavior of the method. In this case, we also use the exponential semi-discrete method

$$y_{n+1} = y_n \exp\left\{ \left(1 - \frac{\sigma^2}{2}\right) (y_n)^2\Delta + \sigma y_n \Delta W_n \right\}. \tag{32}$$

We begin with the stable case, that is when $\beta < 1$ or $\sigma > \sqrt{2}$. Figure 2 displays trajectories of the expTSD method (31) for the cases $\sigma = 2$ and $\sigma = 3$ accordingly. We observe the asymptotic stability in each case as well as the positivity of the paths. There is no need for time-step restriction for scheme (31) as in [9, Fig. 2]. Note that as σ takes bigger values the paths go to zero faster.

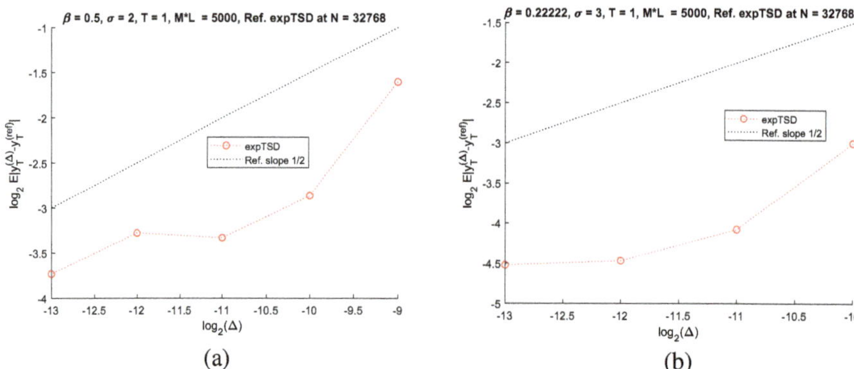

Fig. 1 Convergence of the truncated semi-discrete method (31) for the approximation of (30) for different σ. (**a**) $\sigma = 2$. (**b**) $\sigma = 3$

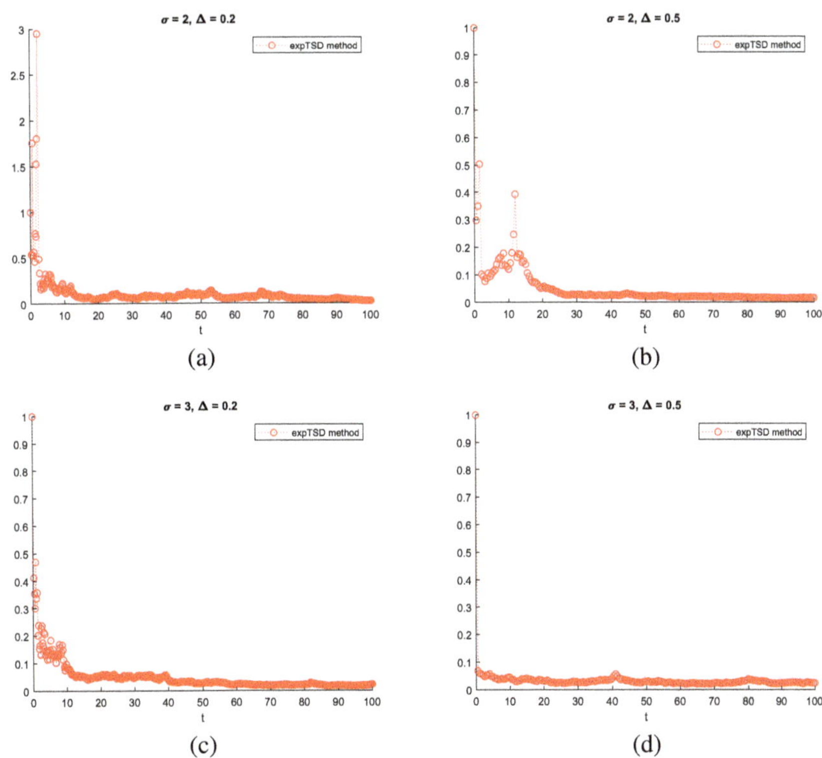

Fig. 2 Trajectories of (31) for different values of σ, Δ. (**a**) Trajectory of (31) with $\sigma = 2, \Delta = 0.2$. (**b**) Trajectory of (31) with $\sigma = 2, \Delta = 0.5$. (**c**) Trajectory of (31) with $\sigma = 3, \Delta = 0.2$. (**d**) Trajectory of (31) with $\sigma = 3, \Delta = 0.5$

Figures 3 and 4 display the case when $\sigma < \sqrt{2}$. We consider the cases $\sigma = 0$ and $\sigma = 1$ accordingly. Now, we observe instability and an apparent finite-time explosion. The apparent explosion time in the ordinary differential equation (case $\sigma = 0$) is very close to the computed one

$$\tau_e^1 := \int_1^\infty u^{-3} du = 0.5,$$

and becomes closer as we lower the step-size Δ. In fact, the exponential scheme (32) can better detect this behavior. In the case $\sigma = 1$ we observe again the apparent explosion time for the SDE which is now random.

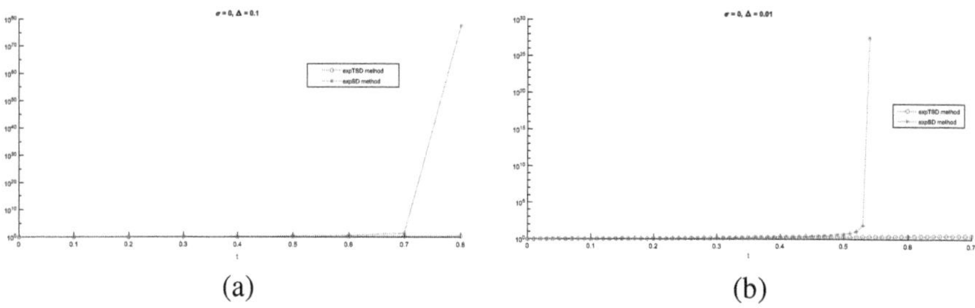

(a) (b)

Fig. 3 Trajectories of (32) and (31) for $\sigma = 0$ and different values of Δ. (**a**) Paths of (31) and (32) with $\sigma = 0$, $\Delta = 0.1$. (**b**) Paths of (31) and (32) with $\sigma = 0$, $\Delta = 0.01$

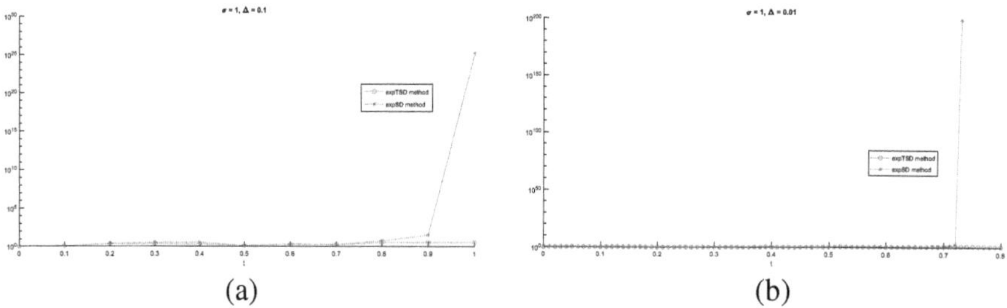

(a) (b)

Fig. 4 Trajectories of (31) and (32) for $\sigma = 1$ and different values of Δ. (**a**) Paths of (31) and (32) with $\sigma = 1$, $\Delta = 0.1$. (**b**) Paths of (31) and (32) with $\sigma = 1$, $\Delta = 0.01$

6 Discussion and Future Work

In this paper we studied a class of SDEs with non-globally Lipschitz coefficients, non-negative solutions and a unique equilibrium solution. We proposed a numerical scheme that preserves the domain of the solution process and reproduces the asymptotic stability behavior of the equilibrium without imposing restrictions on the time-step size. In particular we applied the truncated semi-discrete method producing an exponential scheme. We proved the $\mathcal{L}^1$-convergence of the scheme to the solution of the SDE and its asymptotic stability behavior. The non-truncated scheme works better if our aim is to detect instability. One may argue about the computational time consumed by application of the exponential scheme, mainly because of the exponential calculations. We aim to answer to that question in future work.

References

1. J.A.D. Appleby, X. Mao, A. Rodkina, *Stabilization and destabilization of nonlinear differential equations by noise*, IEEE Transactions on Automatic Control, **53**(3) (2008), 683–691.
2. W. Fang, M.B.Giles, *Adaptive Euler–Maruyama Method for SDEs with Non-globally Lipschitz Drift*, Monte Carlo and Quasi-Monte Carlo Methods, Springer International Publishing, (2018), 217–234.
3. N. Halidias, *Semi-discrete approximations for stochastic differential equations and applications*, International Journal of Computer Mathematics, **89**(6) (2012), 780–794.
4. D.J. Higham, *Mean-square and asymptotic stability of the stochastic theta method*, SIAM J. Numer. Anal, **38**(3) (2000), 753–769
5. M. Hutzenthaler, A. Jentzen, P.E. Kloeden, *Strong and weak divergence in finite time of Euler's method for stochastic differential equations with non-globally Lipschitz continuous coefficients*, Proceedings of the Royal Society of London A: Mathematical, Physical and Engineering Sciences, **467**(2130) (2011), 1563–1576.
6. M. Hutzenthaler, A. Jentzen, *Numerical approximations of stochastic differential equations with non-globally Lipschitz continuous coefficients*, Memoirs of the American Mathematical Society, **236**(1112) (2015), 1611–1641.
7. I. Karatzas, S.E. Shreve, *Brownian motion and stochastic calculus*, Springer-Verlag, New York, (1988).
8. C.Kelly, G.Lord, *Adaptive time-stepping strategies for nonlinear stochastic systems*, IMA Journal of Numerical Analysis, (2017)
9. C.Kelly, A. Rodkina, E.M. Rapoo, *Adaptive timestepping for pathwise stability and positivity of strongly discretised nonlinear stochastic differential equations*, Journal of Computational and Applied Mathematics, **334** (2018), 39–57.
10. X. Mao, *Stochastic differential equations and applications*, Horwood Publishing, 2nd edition (2007).
11. S. Sabanis, *Euler approximations with varying coefficients: the case of superlinearly growing diffusion coefficients*, Annals of Applied Probability, **26**(4) (2016).
12. I.S. Stamatiou, *A boundary preserving numerical scheme for the Wright–Fisher model*, Journal of Computational and Applied Mathematics, **328** (2018), 132–150.
13. I.S. Stamatiou, *An explicit positivity preserving numerical scheme for CIR/CEV type delay models with jump*, Journal of Computational and Applied Mathematics, **360** (2019), 78–98.
14. I.S. Stamatiou, *The Semi-Discrete Method for the approximation of the solution of stochastic differential equations*, Nonlinear Analysis, Differential Equations, and Applications, Springer, (2021), 625–638.
15. I.S. Stamatiou, N. Halidias, *Convergence rates of the semi-discrete method for stochastic differential equations*, Theory of Stochastic Processes, **24**(40), (2019), 89–100.
16. M.V. Tretyakov, Z. Zhang, *A fundamental mean-square convergence theorem for SDEs with locally Lipschitz coefficients and its applications*, SIAM Journal on Numerical Analysis, **51**(6) (2013), 3135–3162.

Solution of Polynomial Equations

N. Tsirivas

Abstract We present a method for the solution of polynomial equations. We do not intend to present one more method among several others, because today there are many excellent methods. Our main aim is educational. Here we attempt to present a method with elementary tools in order to be understood and useful by students and educators. For this reason, we provide a self-contained approach. Our method is a variation of the well-known method of resultant, which has its origin back to Euler. Our goal, in the present chapter, is in the spirit of calculus and secondary school mathematics.

MSC (2010) 65H04

1 Introduction

It is well known that many problems in Physics, Chemistry, and Science lead generally to a polynomial equation.

In pure mathematics also, there are classical problems that lead to a polynomial equation.

N. Tsirivas (✉)
Department of Mathematics, University of Thessaly, Volos, Greece

I. N. Parasidis et al. (eds.), *Mathematical Analysis in Interdisciplinary Research*,
Springer Optimization and Its Applications 179,
https://doi.org/10.1007/978-3-030-84721-0_35

Let us give two examples:

1. If we are to compute the integral $\int_{\alpha}^{\beta} \frac{p(x)}{q(x)} dx$, where $\alpha, \beta \in \mathbb{R}$, $\alpha < \beta$, $p(x), q(x)$ are two real polynomials of one variable, and $q(x)$ is a nonzero polynomial that does not have any root in the interval $[\alpha, \beta]$, then we are led to the problem of finding the real roots of $q(x)$.
2. Let $n \in \mathbb{N}$, $a_i \in \mathbb{R}$ for $i = 1, \ldots, n$, where $\mathbb{N}, \mathbb{R}$ are the sets of natural and real numbers, respectively.

   We can consider the differential equation

$$a_n y^{(n)} + a_{n-1} y^{(n-1)} + \cdots + a_1 y + a_0 = 0,$$

where y is the unknown function.

In order to solve this simple equation we have to find all the roots of the polynomial

$$p(x) = a_n x^n + a_{n-1} x^{n-1} + \cdots + a_1 x + a_0.$$

So, the utility to solve a polynomial equation, or in other words to find the roots of a polynomial, is undoubted. This problem is a very old classical problem in mathematics and Numerical Analysis, especially. For this reason, there exist many methods that solve it.

However, if a scientist wants to solve an equation for his work, it is sufficient to use programs as "mathematica" and "maple," nowadays. So, the utility of the problem has another direction, which is the finding of better algorithms and programs. This is the main line of research among the area experts, nowadays.

We are moving in another direction in this chapter.

Our main aim is educational.

In this chapter we present a method of solving a polynomial equation with full details for educational reasons, so that a student of positive sciences can improve the level of knowledge in the subject. First of all, let us state our problem. We denote $\mathbb{C}$ as the set of complex numbers. Let $n \in \mathbb{N}$ and $a_i \in \mathbb{C}$ for $i = 0, 1, \ldots, n$. We then consider the polynomial

$$p(z) = a_n z^n + a_{n-1} z^{n-1} + \cdots + a_1 z + a_0,$$

which is a polynomial of one complex variable z with complex coefficients. We suppose that $a_n \neq 0$. The natural number n is called the degree of $p(z)$ and it is denoted by $degp(z) = n$. The number $a \in \mathbb{C}$ is a root of $p(z)$ when it is applicable: $p(a) = 0$. Our problem is to find all the roots of $p(z)$, or in other words to solve the equation $p(z) = 0$. Polynomials are simple and specific functions that have the following fundamental property.

Fundamental Theorem of Algebra Every polynomial of one complex variable with complex coefficients and a degree greater than or equal to one has at least one root in $\mathbb{C}$.

This result is central. It is the basis of our method.

However, even if this theorem is fundamental, its proof is not trivial. Its simplest proof comes from complex analysis that many students do not learn in university. In the appendix we give one of the simplest proofs of the fundamental theorem.

Many of the best methods of our problem are iterative. They are based on the construction of specific sequences that approach to the roots of the supposed polynomial. Our method here uses algebra as much as possible, and when algebra cannot go further, analysis takes its role in solving the problem. Here we do not deal with the problem of speed of convergence. We use numerical analysis as little as we can. It is sufficient for us to use the simplest method in order to find a root in a specific real open interval, the bisection method.

Most of the books on numerical analysis describe the bisection method with details. For example, see [8, 11].

There are some formulas that provide bounds for the roots of a polynomial. Cauchy had given such a bound; see [8]. In the frame of our method we provide such a bound.

There are some results that give information about the number of positive or real roots, for example Descart's law of signs and Sturm's sequence [11]. A basic problem is to find disjoint real intervals, so that every one of them contains one root exactly. There are, also, many methods for this.

An extensive discussion of the theory of zeros of polynomials and extremal problems for polynomials the reader can find in the books [10] and [13].

Let us describe now, roughly, the stages of our method:

1. In the first stage we find all the real roots of a polynomial. For this reason, we are based on two results. First of all the bisection method and second by the following result:

   If we have a polynomial p of one real variable with real coefficients with a degree greater than or equal to one for which we know the roots of p', then using the bisection method we can find all the real roots of p.

   The first stage is simple. It uses only elementary knowledge and it is also convenient for students of secondary school!

   We think that it is very useful for students of secondary school to know a method that finds all the real roots of an arbitrary real polynomial with their knowledge base.
2. In the second stage, we provide a method that gives all the real roots of a system of the form:

$$\begin{cases} p(x, y) = 0 \\ q(x, y) = 0 \end{cases}, \tag{A}$$

where $p(x, y), q(x, y)$ are polynomials of two real variables x and y with real coefficients. Our method here is a variation of the well-known method of resultant (see [6, 13]), which has its origin in Euler. With this method, the solution of the above system A is reduced to the first stage. As in the first stage, the second stage is also convenient for students of secondary school (except for Theorem 4.17 in our prerequisites).

3. In the third stage we show that the solution of our problem is reduced to the second stage.

So, roughly speaking, our main aim in this chapter is to present a method that is in the frame of the usual lessons of calculus in secondary school or in university and present it with all the necessary details in order for it to be understood by students.

As for the notation. Let $p(x, y)$ be a polynomial of two real variables x and y, with real coefficients. We denote by $deg_x p(x, y)$ the greatest degree of $p(x, y)$ with respect to x and $deg_y p(x, y)$ the greatest degree of $p(x, y)$ with respect to y. If $deg_x p(x, y) \geq 1$ and $deg_y p(x, y) \geq 1$, we call the polynomial $p(x, y)$ a pure polynomial. If $p(z)$, $q(z)$ are two complex polynomials, we write $p(z) \equiv q(z)$, when they are equal by identity. We also write $p(z) \equiv 0$, when $p(z)$ is equal to zero polynomial by identity. We write $p(z) \not\equiv q(z)$ when polynomials $p(z), q(z)$ are not equal by identity and $p(z) \not\equiv 0$ when $p(z)$ is not the zero polynomial.

There are many methods and algorithms to the solution of polynomial equations. Some of them are very old like the methods of Horner, Graeffe, and Bernoulli, whereas today there are some others like the methods of Rutishauser, Lehmer, Lin, Bairstow, Bareiss, and many others. Another method, similar to Bernoulli method, is the QD method. A classical and popular method today is that of Muller. It is a general method, not only for polynomials.

The interested reader can find the details of some of the above methods in the books of our references; see [1, 3, 4, 7, 9–11], and [12, 14, 15]. As we said, there exist many algorithms and programs to our problem.

One of the best is the subroutine ZEROIN. One can find the details of this program in [4].

As we said formerly, the basis of our method is the resultant (or eliminent). With this method we can convert a system of polynomial equations in one equation with only one unknown!

Theoretically, we can succeed in that, but the complexity of calculations is enormous, so its value today is only for polynomial equations with a low degree, and is used as a theoretical tool. For details of the resultant, see [6, 12]. Apart from this, there are some cases where the resultant fails. This can happen, for example, when we have to solve a system of two equations with two unknowns and one of the two equations is a multiple of the other, and the system has a finite number of solutions. See, for example, the equation: $(x^2 - 1)^2 + (y^2 - 2)^2 = 0$, which has the set of solutions

$$L = \{(1, \sqrt{2}), (1, -\sqrt{2}), (-1, \sqrt{2}), (-1, -\sqrt{2})\}.$$

We describe with details how we handle these cases in our method here. An alternate method for our problem is to solve it with Gröbner bases. Gröbner bases is a method that was developed in 1960 for the division of polynomials with more than one variable. With Gröbner bases, we can also convert a system of equations in an equation with only one unknown, as the resultant does. This is the main application of Gröbner bases. This can be done in most cases.

However, there are some cases where Gröbner bases fail to succeed, like the above case.

For Gröbner bases, see [2]. Many books of secondary school contain the elementary theory of polynomials and Euclidean division that we refer to in our prerequisites.

The structure of our chapter is as follows:

In Sect. 2 we give a rough description of our method. In Sect. 3 we give the complete description of our method. In Sect. 4 we collect all the prerequisite tools of our method from Algebra and Analysis and present them with all the necessary proofs, especially for results that someone cannot find easily in books.

Finally, in Appendix we give one of the simplest proofs of the Fundamental Theorem of Algebra that one cannot find easily in books.

We, also, give a short description of the solution of binomial equation: $x^n = a$, where $n \in \mathbb{N}$, $n \geq 2$, and a is a positive number.

2 General Description of the Solution of Our Problem

For methodological reasons, we divide the solution of our problem into the following three stages.

2.1 *First Stage*

In this stage we find all the real roots of the polynomial equation

$$a_v x^v + a_{v-1} x^{v-1} + \cdots + a_1 x + a_0 = 0,$$

where $a_i \in \mathbb{R}$, for every $i = 0, 1, \ldots, v$, where $v \in \mathbb{N}$.

2.2 *Second Stage*

Let $p_1(x, y)$, $p_2(x, y)$ be two polynomials of two real variables x and y whose coefficients are in $\mathbb{R}$. We consider the system of equations:

$$\begin{cases} p_1(x, y) = 0 \ (1) \\ p_2(x, y) = 0 \ (2) \end{cases} \tag{A}$$

Let L_A be the set of solutions of the above system (A), in $\mathbb{R}^2$. That is, we consider the set

$$L_A := \{(x, y) \in \mathbb{R}^2 | p_1(x, y) = 0 \text{ and } p_2(x, y) = 0\},$$

of solutions of the above system (A), in $\mathbb{R}^2$. In the second stage we find the set L_A under the following supposition (S):

(S): Supposition We suppose that the set L_A is finite.

That is, we solve the above system (A), in $\mathbb{R}^2$, in case if supposition (S) holds. We note that we succeed the second stage using first stage.

2.3 Third Stage

In the third stage we completely solve our initial problem of finding all the roots of the polynomial equation $a_n z^n + a_{n-1} z^{n-1} + \cdots + a_1 z + a_0 = 0$, where $n \in \mathbb{N}$, $a_i \in \mathbb{C}$, for every $i = 0, 1, \ldots, n$, using the previous two stages.

The first stage is the analytical part, whereas the second and third stages are the algebraic parts of our method. The prerequisites of our method are few. Elementary calculus and the elementary linear algebra of secondary school are enough, except only for a specific case, where we use Theorem 4.17 from our prerequisites (a very well-known result from calculus of several variables).

In the following paragraph, we give the complete description of our method.

3 Complete Description of Our Method

3.1 First Stage

Let $a_i \in \mathbb{R}$, for $i = 0, 1, \ldots, v$, $v \in \mathbb{N}$ and a polynomial

$$p(x) = a_v x^v + a_{v-1} x^{v-1} + \cdots + a_1 x + a_0,$$

where $a_v \neq 0$, so $degp(x) = v$.

Here we find all the real roots of $p(x)$. If $v = 1$, or $v = 2$, we know how to find the real roots of $p(x)$ from secondary school. Let us suppose that $v \geq 3$. We find all the real roots of p' (if any) and then we find the roots of p by applying basic Lemma 4.8 or Corollaries 4.9 and 4.10.

More generally, we suppose that p has degree $\upsilon \in \mathbb{N}$, $\upsilon \geq 3$. We consider polynomials $p', p'', \ldots, p^{(\upsilon-3)}$ $p^{(\upsilon-2)}$. Polynomial p' has degree $\upsilon - 1$, p'' has degree $\upsilon - 2$, and polynomial $p^{(\upsilon-2)}$ has degree 2.

We find the roots of $p^{(\upsilon-2)}$ (if any). After using basic Lemma 4.8, or Corollaries 4.9 and 4.10, we find the roots of $p^{(\upsilon-3)}$ and going inductively, after a finite number of steps, we find the roots of p' and finally in the same way the roots of p, and thus we complete our first stage.

3.2 *Second Stage*

We will now consider the system of two polynomials $p_1(x, y)$, $p_2(x, y)$ of two real variables x and y with coefficients in $\mathbb{R}$. We solve the system (A), where

$$\begin{cases} p_1(x, y) = 0 \ (1) \\ p_2(x, y) = 0 \ (2) \end{cases} . \tag{A}$$

We solve system (A) with the following supposition.

Supposition We suppose that system (A) has a finite number of solutions; that is, the set

$$L_A = \{(x, y) \in \mathbb{R}^2 | p_1(x, y) = p_2(x, y) = 0\}$$

is nonvoid and finite.

First, we notice that one of the polynomials $p_1(x, y)$, $p_2(x, y)$, at least, is nonzero, or else if $p_1(x, y) \equiv p_2(x, y) \equiv 0$ for every $(x, y) \in \mathbb{R}$, then we have $L_A = \mathbb{R}^2$, which is false because the set L_A is finite. We will examine some cases.

First of all, we suppose that at least one of the polynomials is of one variable only. We can distinguish some cases here. Let $p_1(x, y) \equiv q_1(x)$, $p_2(x, y) \equiv q_2(x)$. Then, we solve the equations $q_1(x) = 0$ and $q_2(x) = 0$ with the method of the first stage and later we conclude that set L_A is the set of all (x, y), where x is one of the common solutions of equations $q_1(x) = 0$ and $q_2(x) = 0$ and $y \in \mathbb{R}$; that is, L_A is an infinite set, which is false by our supposition. So this case itself cannot occur. Similarly, we cannot have the case where $p_1(x, y) \equiv r_1(y)$ and $p_2(x, y) \equiv r_2(y)$. Now we consider the case where:

$$p_1(x, y) \equiv q_1(x) \text{ and } p_2(x, y) \equiv q_2(y).$$

Then, we can solve the equations $q_1(x) = 0$ and $q_2(y) = 0$ with the method of the first stage, and we find the finite sets $A_1 = \{\rho_1, \rho_2, \ldots, \rho_\upsilon\}$ and $B_1 =$

$\{\lambda_1, \lambda_2, \ldots, \lambda_m\}$, $A_1 \cup B_1 \subseteq \mathbb{R}$, where A_1 is the set of roots of q_1 and B_1 is the set of roots of q_2, $v, m \in \mathbb{N}$. Then, we have $L_A = \{(\rho_i, \lambda_j), i = 1, \ldots, v, j = , \ldots, m\}$. In a similar way, we can solve the system A, when $p_1(x, y) = r_1(y)$ and $p_2(x, y) = r_2(x)$, for some polynomials $r_1(y)$, $r_2(x)$.

Now, we consider the case where $p_1(x, y)$, $p_2(x, y)$ are two pure polynomials.

(i) The simplest case is when $deg_y p_1(x, y) = deg_y p_2(x, y) = 1$. Then we have

$$p_1(x, y) = \alpha_1(x)y + \alpha_2(x) \quad \text{and}$$

$$p_2(x, y) = \beta_1(x)y + \beta_2(x),$$

where $\alpha_1(x), \alpha_2(x), \beta_1(x), \beta_2(x)$ are some polynomials of real variable x only and $\alpha_1(x) \not\equiv 0$ an $\beta_1(x) \not\equiv 0$, because $p_1(x, y)$, $p_2(x, y)$ are pure polynomials.

So we have to solve the system:

$$\begin{cases} \alpha_1(x)y + \alpha_2(x) = 0 \ (3) \\ \beta_1(x)y + \beta_2(x) = 0 \ (4) \end{cases}.$$

We can distinguish some cases here. There exists a $(x_0, y_0) \in L_A$, so that:

1. $\alpha_1(x_0) = \beta_1(x_0) = 0$. Then with (3) and (4), we get $\alpha_2(x_0) = \beta_2(x_0) = 0$. We get $(x_0, y) \in L_A$ for every $y \in \mathbb{R}$, which is false because L_A is finite. So, this case cannot occur.
2. $\alpha_1(x_0) = 0$ and $\beta_1(x_0) \neq 0$. Then with (4) we take $y_0 = -\dfrac{\beta_2(x_0)}{\beta_1(x_0)}$ (5). With (3), we have $\alpha_2(x_0) = 0$.

   So, in this case we find the common roots of polynomials $\alpha_1(x)$ and $\alpha_2(x)$, and for every common root x_0 of $\alpha_1(x)$ and $\alpha_2(x)$, so that $\beta_1(x_0) \neq 0$, the couple $(x_0, y_0) \in L_A$, where y_0 is given from (5). Of course we find the real roots of polynomials $\alpha_1(x)$ and $\alpha_2(x)$ with the method of our first stage. In a similar way we find the roots $(x_0, y_0) \in L_A$, so that $\alpha_1(x_0) \neq 0$ and $\beta_1(x_0) = 0$.
3. $\alpha_1(x_0) \neq 0$ and $\beta_1(x_0) \neq 0$.

   Here, we have some cases:

   (i) $\alpha_2(x_0) = \beta_2(x_0) = 0$. Then with (3), (4), and our supposition, we get $y = 0$. So, in this case we find the common roots of polynomials $\alpha_2(x)$ and $\beta_2(x)$, so that they are not roots of polynomials $\alpha_1(x)$ and $\beta_1(x)$, with the method of the first stage. If x_0 is such a root, that is, $\alpha_2(x_0) = \beta_2(x_0) = 0$ and $\alpha_1(x_0) \neq 0$ and $\beta_2(x_0) \neq 0$, then $(x_0, 0) \in L_A$.

   (ii) $\alpha_2(x_0) = 0$ and $\beta_2(x_0) \neq 0$.

   Then, with (3), and the facts $\alpha_2(x_0) = 0$ and $\alpha_1(x_0) \neq 0$, we get $y = 0$. Then, because $y = 0$, by (4) we get $\beta_2(x_0) = 0$, which is a contradiction by our supposition. So, this case cannot occur.

(iii) $\alpha_2(x_0) \neq 0$ and $\beta_2(x_0) = 0$. As in the previous case (ii), this case cannot occur.

(iv) $\alpha_2(x_0) \neq 0$ and $\beta_2(x_0) \neq 0$.
Then, with (3) and (4) we get
$y_0 = -\dfrac{\alpha_2(x_0)}{\alpha_1(x_0)}$ (6) and $y_0 = -\dfrac{\beta_2(x_0)}{\beta_1(x_0)}$ (7).
With (6) and (7), we get
$-\dfrac{\alpha_2(x_0)}{\alpha_1(x_0)} = -\dfrac{\beta_2(x_0)}{\beta_1(x_0)} \Leftrightarrow \alpha_2(x_0)\beta_1(x_0) - \alpha_1(x_0)\beta_2(x_0) = 0.$ (8)

Now, we can consider two systems of relations (A_1) and (A_2) as follows:

$$\begin{cases} \alpha_1(x)y + \alpha_2(x) = 0, & (i) \\ \beta_1(x)y + \beta_2(x) = 0, & (ii) \\ \alpha_1(x) \neq 0, \alpha_2(x) \neq 0, \beta_1(x) \neq 0, \beta_2(x) \neq 0 \end{cases} \tag{A$_1$}$$

and

$$\begin{cases} \alpha_2(x)\beta_1(x) - \alpha_1(x)\beta_2(x) = 0, & (i) \\ y = -\dfrac{\alpha_2(x)}{\alpha_1(x)}, & (ii) \\ \alpha_1(x) \neq 0, \alpha_2(x) \neq 0, \beta_1(x) \neq 0, \beta_2(x) \neq 0 \end{cases} \tag{A$_2$}$$

Let L_{A_1}, L_{A_2} be the two set of solutions of systems A_1 and A_2, respectively. We prove that $L_{A_1} = L_{A_2}$.

By previous procedure and equalities (6) and (8), we get

$$L_{A_1} \subseteq L_{A_2}. \tag{9}$$

Now let $(x, y) \in L_{A_2}$. Then equality (ii) of A_2 gives equality (ii) of A_1. By equality (i) of (A_2), we get $\alpha_2(x)\beta_1(x) = \alpha_1(x)\beta_2(x)$ and by the fact that $\alpha_1(x) \neq 0$ and $\beta_1(x) \neq 0$, we get

$$-\frac{\alpha_2(x)}{\alpha_1(x)} = -\frac{\beta_2(x)}{\beta_1(x)}. \tag{10}$$

Through the equality (ii) of (A_2) and (10), we get

$$y = -\frac{\beta_2(x)}{\beta_1(x)}. \tag{11}$$

Equality (11) gives equality (ii) of (A_1). So we have $(x, y) \in L_{A_1}$; that is, $L_{A_2} \subseteq L_{A_1}$ (12).

By (9) and (12), we get $L_{A_1} = L_{A_2}$.

So, we proved that in order to solve system (A_1) it suffices to solve system (A_2). Thus, we solve equation (i) of (A_2) with the method of the first stage, and for every root x of polynomial $\alpha_2(x)\beta_1(x) - \alpha_1(x)\beta_2(x)$ so that $\alpha_1(x) \neq 0$, $\alpha_2(x) \neq 0$, $\beta_1(x) \neq 0$, $\beta_2(x) \neq 0$, we get the respective y from equality (ii) of (A_2).

So far we have completely solved the system (A), in the case of $deg_y p_1(x, y) = deg_y p_2(x, y) = 1$.

For the sequel, we solve the case ii) where $deg_y p_1(x, y) \leq 2$ and $deg_y p_2(x, y) \leq 2$ and $p_1(x, y)$, $p_2(x, y)$ are two pure polynomials. Of course, we have $deg_y p_1(x, y) \geq 1$ and $deg_y p_2(x, y) \geq 1$, because $p_1(x, y)$, $p_2(x, y)$ are pure polynomials.

We have already examined the case $deg_y p_1(x, y) = deg_y p_2(x, y) = 1$.

So we, here, examine the case where at least one of natural numbers $deg_y p(x, y)$, $deg_y p_2(x, y)$ is equal to 2.

We examine, firstly, the case where:

$$deg_y p_1(x, y) = 2 \text{ and } deg_y p_1(x, y) = 1.$$

Then, we can write the system (A) as follows:

$$\begin{cases} \alpha_2(x)y^2 + \alpha_1(x)y + \alpha_0(x) = 0 \ (13) \\ \beta_1(x)y + \beta_0(x) = 0 \ (14) \end{cases} . \tag{A}$$

If $\alpha_2(x) \equiv 0$, we have the previous system. So we suppose that $\alpha_2(x) \not\equiv 0$.

Now let some $(x_0, y_0) \in L_A$ as above. We distinguish some cases:

1. $\alpha_2(x_0) = 0$. Then, we solve the system $\begin{cases} \alpha_1(x)y + \alpha_0(x) = 0 \\ \beta_1(x)y + \beta_0(x) = 0 \end{cases}$ as previously, and we take only the solutions (x, y) of this system so that $\alpha_2(x) = 0$ holds, solving the equation $\alpha_2(x) = 0$ with the method of the first stage.
2. $\alpha_2(x_0) \neq 0$. We distinguish some cases:

   (i) $\alpha_1(x_0) = \beta_1(x_0) = 0$. Then we have to solve the system:

$$\begin{cases} \alpha_2(x)y^2 + \alpha_2(x) = 0 \ (15) \\ \beta_0(x) = 0 \ (16) \end{cases} . \tag{B$_1$}$$

   By (15), we take

$$y^2 = -\frac{\alpha_0(x)}{\alpha_2(x)}. \tag{17}$$

   So, in order to solve this system we do the following.

   First of all, we find all the common roots x of three polynomials $\alpha_1(x)$, $\beta_1(x)$, and $\beta_0(x)$ that are not roots of polynomial $\alpha_2(x)$.

If $x \in \mathbb{R}$ and $\alpha_1(x) = \beta_1(x) = \beta_0(x) = 0$ and $\alpha_2(x) \neq 0$, we consider the number $-\frac{\alpha_0(x)}{\alpha_2(x)}$. If $-\frac{\alpha_0(x)}{\alpha_2(x)} \geq 0$, then we set $\left(y_1 = \sqrt{-\frac{\alpha_0(x)}{\alpha_2(x)}} \text{ and } y_2 = -\sqrt{-\frac{\alpha_0(x)}{\alpha_2(x)}}, \text{ if } -\frac{\alpha_0(x)}{\alpha_2(x)} > 0\right)$ and $(y = 0$ if $\alpha_0(x) = 0)$, and then under the above conditions $(x, y) \in L_A$.

We find the roots of polynomials $\alpha_1(x)$, $\beta_1(x)$, $\beta_0(x)$ with the method of the first stage.

Of course, if we cannot find couples $(x, y) \in \mathbb{R}^2$ so that all the above conditions hold, this means that we do not have solutions to this case.

(ii) $\alpha_1(x_0) = 0$, $\beta_1(x_0) \neq 0$.

We consider the system:

$$\begin{cases} \alpha_2(x)y^2 + \alpha_0(x) = 0 \ (17) \\ \beta_1(x)y + \beta_0(x) = 0 \ (18) \end{cases} . \tag{B$_1$}$$

Through (17) and (18), we get

$$y^2 = -\frac{\alpha_2(x)}{\alpha_2(x)}, \tag{19}$$

$$y = -\frac{\beta_0(x)}{\beta_1(x)} \quad (20) \quad \Rightarrow \quad y^2 = \left(\frac{\beta_0(x)}{\beta_1(x)}\right)^2 \quad , (21)$$

Through (19) and (21), we get

$$-\frac{\alpha_0(x)}{\alpha_2(x)} = \left(\frac{\beta_0(x)}{\beta_1(x)}\right)^2 \Leftrightarrow \alpha_2(x)\beta_0(x)^2 + \alpha_0(x)\beta_1(x)^2 = 0. \tag{22}$$

From the above, in order to find a solution of system (B_2) we do the following.

We find all the common roots of two polynomials $\alpha_2(x)\beta_0(x)^2 + \alpha_0(x)\beta_1(x)^2$ and $\alpha_1(x)$, which are not roots of polynomials $\alpha_2(x)$ and $\beta_1(x)$ (if any). Let x be such a root. We set $y = -\frac{\beta_0(x)}{\beta_1(x)}$, and then (x, y) is a solution of (B_2) and we get all the other solutions of (B_2) in the same way.

(iii) $\alpha_1(x_0) \neq 0$, $\beta_1(x_0) = 0$.

Then, through (14) we get $\beta_0(x_0) = 0$. So, in order to solve system (A) in this case, we do the following.

We find all the common roots (if any) x of polynomials $\beta_1(x)$, $\beta_0(x)$, so that $\alpha_2(x) \neq 0$ and $\alpha_1(x) \neq 0$. Of course this is a finite set of numbers x.

For such a root x_0, we solve the equation $\alpha_2(x_0)y^2 + \alpha_1(x_0)y + \alpha_0(x) = 0$ and we find the respective number y (if any).

All these couples $(x, y) \in \mathbb{R}^2$ (if any) are the set of solutions of system (A) in this case.

(iv) $\alpha_1(x_0) \neq 0$, $\beta_1(x_0) \neq 0$.

We leave this case for later. In a similar way we examine the case where $deg_y p_1(x, y) = 1$ and $deg_y p_2(x, y) = 2$.

Now, we examine the case where:

$$deg_y p_1(x, y) = deg_y p_2(x, y) = 2.$$

We have the system:

$$\begin{cases} \alpha_2(x)y^2 + \alpha_1(x)y + \alpha_0(x) = 0 \ (23) \\ \beta_2(x)y^2 + \beta_1(x)y + \beta_0(x) = 0 \ (24) \end{cases}. \tag{B$_3$}$$

Here we examine some cases:

1. Let $(x_0, y_0) \in L_{B_3}$.
   If $\alpha_2(x_0) = 0$, or $\beta_2(x_0) = 0$, we have a system as in the previous case.
   So, we suppose that:

$$\alpha_2(x_0) \neq 0 \ \text{ and } \ \beta_2(x_0) \neq 0.$$

Now, we can distinguish some cases:

(i) $\alpha_1(x_0) = \beta_1(x_0) = 0$.

So, we are to solve the system:

$$\alpha_2(x_0)y^2 + \alpha_0(x_0) = 0 \ (25) \ \text{ and}$$
$$\beta_2(x_0)y^2 + \beta_0(x_0) = 0 \ (26).$$

Through (25), we have $y^2 = -\dfrac{\alpha_0(x_0)}{\alpha_2(x_0)}$ (27), and by (26), we get

$$y^2 = -\frac{\beta_0(x_0)}{\beta_2(x_0)}. \tag{28}$$

Through (27) and (28), we get

$$-\frac{\alpha_0(x_0)}{\alpha_2(x_0)} = -\frac{\beta_0(x_0)}{\beta_2(x_0)} \Leftrightarrow \alpha_0(x_0)\beta_2(x_0) - \beta_0(x_0)\alpha_2(x_0) = 0. \tag{29}$$

From the above, we have the following solution:

We find the common roots of polynomials $\alpha_1(x)$, $\beta_1(x)$, $\alpha_0(x)\beta_2(x) - \beta_0(x)\alpha_2(x)$, so that $\alpha_2(x) \neq 0$ and $\beta_2(x) \neq 0$.

We suppose that there exists such a root x_0. If $\alpha_0(x_0) = 0$, we get $y_0 = 0$. If $\frac{\alpha_0(x_0)}{\alpha_2(x_0)} < 0$, we consider

$$y_1 = \sqrt{-\frac{\alpha_0(x_0)}{\alpha_2(x_0)}}, \quad y_2 = -\sqrt{-\frac{\alpha_0(x_0)}{\alpha_2(x_0)}},$$

and $(x_0, y_1), (x_0, y_2) \in L_{B_3}$. We get all the other solutions of B_3 in the same way. Of course, if (x, y) does not exist with the above conditions, we do not have solutions of B_3 in this case.

(ii) $\alpha_1(x_0) = 0$ and $\beta_1(x_0) \neq 0$.
We will postpone this case for later.

(iii) $\alpha_1(x_0) \neq 0$ and $\beta_1(x_0) = 0$.
We will also postpone this case for later.

(iv) $\alpha_1(x_0) \neq 0$ and $\beta_1(x_0) \neq 0$.
This is the central case of system (B_3).

We consider the number:

$$D = \alpha_2(x_0)\beta_1(x_0) - \alpha_1(x_0)\beta_2(x_0) = \begin{vmatrix} \alpha_2(x_0) & \alpha_1(x_0) \\ \beta_2(x_0) & \beta_1(x_0) \end{vmatrix}$$

that we call it: the determinant of system (B_3).

We distinguish two cases:

(a) $D \neq 0$.

We consider the linear system:

$$\begin{cases} \alpha_2(x_0)z + \alpha_1(x_0)\omega = -\alpha_0(x_0) \ (30) \\ \beta_2(x_0)z + \beta_1(x_0)\omega = -\beta_0(x_0) \ (31) \end{cases} . \qquad (\text{B}_4)$$

This linear system has determinant $D \neq 0$, so it has exactly one solution.

We set

$$D_1 = \begin{vmatrix} -\alpha_0(x_0) & \alpha_1(x_0) \\ -\beta_0(x_0) & \beta_1(x_0) \end{vmatrix} = \alpha_1(x_0)\beta_0(x_0) - \alpha_0(x_0)\beta_1(x_0) \qquad (32)$$

and

$$D_2 = \begin{vmatrix} \alpha_2(x_0) & -\alpha_0(x_0) \\ \beta_2(x_0) & -\beta_0(x_0) \end{vmatrix} = \alpha_0(x_0)\beta_2(x_0) - \alpha_2(x_0)\beta_2(x_0) \tag{33}$$

Through Cramer's law of linear algebra, we get the unique solution (z, ω) of system B_4, that is:

$$z = \frac{D_1}{D} \text{ and } \omega = \frac{D_2}{D}.$$

From our supposition, the couple $(x_0, y_0) \in L_{B_3}$. This means that the numbers y_0^2 and y_0 satisfy equations (23) and (24) of (B_3), or differently, in other words the couple (y_0^2, y_0) is a solution of the linear system (B_4). But because of our supposition $D \neq 0$, the couple (z, ω), where $z = \frac{D_1}{D}$ (34) and $\omega = \frac{D_2}{D}$ (35), is the unique solution of system (B_4), as it is well known in linear algebra. So, we have $z = y_0^2$ and $\omega = y_0$, and by (34) and (35), we get

$$y_0^2 = \frac{D_1}{D} \quad (36) \quad \text{and} \quad y_0 = \frac{D_2}{D}. \tag{37}$$

Now, we exploit the inner relation that numbers y_0 and y_0^2 have, that is:

$$y_0^2 = y_0 \cdot y_0. \tag{38}$$

Replacing (36) and (37) in relation (38), we get

$$\frac{D_1}{D} = y_0 \cdot \frac{D_2}{D} \Rightarrow D_1 - y_0 D_2 = 0. \tag{39}$$

By (37), we have

$$D y_0 - D_2 = 0. \tag{40}$$

So, the couple $(x_0, y_0) \in L_{B_3}$ also satisfies the system:

$$\begin{cases} D_1 - y D_2 = 0 \quad (39) \\ D y - D_2 = 0 \quad (40) \end{cases}$$

From the above, we have the two systems:

$$\begin{cases} \alpha_2(x)y^2 + \alpha_1(x)y + \alpha_0(x) = 0 \\ \beta_2(x)y^2 + \beta_1(x)y + \beta_0(x) = 0 \\ \alpha_2(x) \neq 0, \beta_2(x) \neq 0, \ \alpha_1(x) \neq, \ \beta_1(x) \neq 0, \\ D = \begin{vmatrix} \alpha_2(x) & \alpha_1(x) \\ \beta_2((x) & \beta_1(x) \end{vmatrix} \neq 0 \end{cases} \quad (B_5)$$

and

$$\begin{cases} D_1 = yD_2 = 0 \\ Dy - D_2 = 0 \\ \alpha_2(x) \neq 0, \ \beta_2(x) \neq 0, \ \alpha_1(x) \neq 0, \ \beta_1(x) \neq 0, \\ D \neq 0 \end{cases} . \quad (B_6)$$

Let L_{B_5}, L_{B_6} be the set of solutions of systems (B_5) and (B_6). We now show that $L_{B_5} = L_{B_6}$. Of course we have $L_{B_5} \subseteq L_{B_6}$ from the previous procedure, because we obtained equalities (39) and (40) of system (B_6) from system B_5.

Reversely, let $(x_0, y_0) \in L_{B_6}$. From the first two equalities of (B_6), we get

$$y_0 = \frac{D_1}{D_2} \text{ and } y_0 = \frac{D_2}{D}. \quad (37)$$

We multiply these equalities and we take $y_0^2 = \frac{D_1}{D}$ (36).

Now, we consider the linear system (B_4). Because $D \neq 0$ (by our supposition), this system has a unique solution $(z, \omega) = \left(\frac{D_1}{D}, \frac{D_2}{D}\right)$, (41) as it is well known, in Cramer's law.

From (36), (37), and (41), we have $z = y_0^2$ (42) and $\omega = y_0$ (43).

Replacing (42) and (43) in (30) and (31) of (B_4), we get the first two equalities of (B_5); that is, $(x_0, y_0) \in L_{B_5}$. So, we have $L_{B_6} \subseteq L_{B_5}$.

From the above, we have $L_{B_5} = L_{B_6}$. So, we are led to solve system B_6, which we have examined previously, in the system: $\left.\begin{matrix} p_1(x, y) = 0 \\ p_2(x, y) = 0 \end{matrix}\right\}$, where

$deg_y p_1(x, y) \leq 1$ and

$deg_y p_2(x, y) \leq 1$.

(b) $D = 0$

The solution of these cases is taken as follows.

We take the roots of polynomial $D\,x_1$; that is, $\alpha_2(x_1)\beta_1(x_1)-\alpha_1(x_1)\beta_2(x_1)=0$, so that $\alpha_2(x_1)\neq 0$, $\beta_2(x_1)\neq 0$, $\alpha_1(x_1)\neq 0$, and $\beta_1(x_1)\neq 0$. We get y that satisfies one of the Eqs. (23), or (24) of (B_3); that is:

$$\alpha_2(x_1)y^2+\alpha_1(x_1)y+\alpha_0(x_1)=0.$$

This holds because the two Eqs. (23) and (24) are equivalent (as we have shown in prerequisites of linear algebra), and each of them is a multiple of the other.

Remark 3.1 We note that the three remaining cases we have left are similar to case (a) above where $D\neq 0$.

So far, we have examined system (A), where $deg_y p_1(x,y)\leq 2$ and $deg_y p_2(x,y)\leq 2$. We set

$$m_0:=\max\{deg_y p_1(x,y), deg_y p_2(x,y)\}.$$

We solve system (A) in the general case with induction above the number m_0. We have examined the cases where $m_0=1$ or $m_0=2$.

We suppose that for $k_0\in\mathbb{N}$, $k_0\geq 3$, we have solved system (A) for every system, so that $m_0\leq k_1-1$. We now solve system (A) when $m_0=k_0$.

We can write polynomials $p_1(x,y)$, $p_2(x,y)$ as follows:

$$\begin{aligned}\alpha_{m_0}(x)y^{m_0}+\alpha_{v_0}(x)y^{v_0}+q_1(x,y)=p_1(x,y)\ \text{ and }\ &\beta_{n_0}(x)y^{n_0}\\ &+\beta_{\mu_0}(x)y^{\mu_0}+q_2(x,y)\\ &=p_2(x,y),\end{aligned}$$

where $v_0<m_0$, $v_0,m_0\in\mathbb{N}$, $deg_y q_1(x,y)<v_0$ and $\mu_0<n_0$, $\mu_0,n_0\in\mathbb{N}$, $n_0\leq m_0$, $deg_y q_2(x,y)<\mu_0$.

So, the initial system can be written as follows:

$$\begin{cases}\alpha_{m_0}(x)y^{m_0}+\alpha_{v_0}(x)y^{v_0}+q_1(x,y)=0\ (1)\\ \beta_{n_0}(x)y^{n_0}+\beta_{\mu_0}(x)y^{\mu_0}+q_2(x,y)=0\ (2)\end{cases},\tag{A}$$

where $\alpha_{m_0}(x)$, $q_{v_0}(x)$, $\beta_{n_0}(x)$, and $\beta_{\mu_0}(x)$ are polynomials of the real variable x only and $q_1(x,y)$, $q_2(x,y)$ be polynomials of real variables x and y.

We, also, suppose that $a_{m_0}(x)\not\equiv 0$. We can distinguish some cases as previously:

1. Let $\alpha_{v_0}(x)\equiv q_1(x,y)\equiv\beta_{n_0}(x)\equiv\beta_{\mu_0}(x)\equiv q_2(x,y)\equiv 0$. Then we have the system: $\alpha_{m_0}(x)y^{m_0}=0$. If $m_0=0$ and $\alpha_{m_0}(x)=c\in\mathbb{R}$, then every couple $(x,0)\in L_A$ and the system has an infinite set of solutions, which is false. So, this case cannot occur.
2. $\beta_{n_0}(x)\equiv\beta_{\mu_0}(x)\equiv q_2(x,y)\equiv 0$.

We will examine this case later.

3. $\beta_{n_0}(x) \not\equiv 0, \alpha_{v_0}(x) \equiv q_1(x, y) \equiv \beta_{\mu_0}(x) \equiv q_2(x, y) \equiv 0.$
   We have the system:

$$\begin{cases} \alpha_{m_0}(x)y^{m_0} = 0 \\ \beta_{n_0}(x)y^{n_0} = 0 \end{cases}. \tag{A}$$

If $deg\alpha_{m_0}(x) = deg\beta_{n_0}(x) = 0$, then any couple $(x, 0) \in L_A$ and the set of solutions is infinite, which is false. So, this case cannot occur.

4. $\beta_{n_0}(x) \not\equiv 0, \alpha_{v_0}(x) \equiv \beta_{\mu_0}(x) \equiv 0 \equiv q_1(x, y)$ and $q_2(x, y) \equiv r(x) \not\equiv 0.$
   So, we have the system:

$$\begin{cases} \alpha_{m_0}(x)y^{m_0} = 0 \ (3) \\ \beta_{n_0}(x)y^{n_0} + r(x) = 0 \ (4) \end{cases}. \tag{A}$$

We can distinguish some cases here.

First, we suppose that (A) has a solution $(x_0, y_0) \in L_A$.

(i) $\alpha_{m_0}(x_0) = 0 = \beta_{n_0}(x_0)$. Then, of course $r(x_0) = 0$. So, if the polynomials $\alpha_{m_0}(x), \beta_{n_0}(x), r(x)$ have a common root x_0, then any couple $(x_0, y) \in L_A$, for every $y \in \mathbb{R}$, which is false of course.

So, this case cannot happen.

(ii) $\alpha_{m_0}(x_0) = 0, \beta_{n_0}(x_0) \neq 0.$

By (4), we take $y_0 = -\dfrac{r(x_0)}{\beta_{n_0}(x_0)}$.

Thus, in this case we solve the system as follows.

We find the roots x of $\alpha_{m_0}(x)$, so that $\beta_{n_0}(x) \neq 0$. For every such root, the couple $(x, y) = \left(x, -\dfrac{r(x)}{\beta_{n_0}(x)}\right) \in L_A$.

We get all the other solutions of this system in the same way.

(iii) $\alpha_{m_0}(x_0) \neq 0$ and $\beta_{n_0}(x_0) = 0$.

Through (3), we get $y = 0$. By (4), we take $r(x_0) = 0$.

So, in this case we can solve the system as follows: We find the roots of $r(x)$ such that $\alpha_{m_0}(x) \neq 0$ and $\beta_{n_0}(x) = 0$. For every such root x, the couple $(x, 0) \in L_A$.

(iv) $\alpha_{m_0}(x_0) \neq 0$ and $\beta_{n_0}(x_0) \neq 0$.

By (3), we get $y_0 = 0$, and by (4), for $y_0 = 0$ we get $r(x_0) = 0$.

So, in this case we can solve the system as follows: We find the roots x of $r(x)$, so that $\alpha_{m_0}(x) \neq 0$ and $\beta_{n_0} \neq 0$. Then the couple $(x, 0)$ is a solution of (A).

(v) In a similar way we can solve a system of the form:

$$\begin{cases} \alpha_{m_0}(x)y^{m_0} + r(x) = 0 \\ \beta_{n_0}(x)y^{n_0} = 0. \end{cases}$$

5. $\beta_{n_0}(x) \not\equiv 0$, $\alpha_{v_0}(x) \equiv \beta_{\mu_0}(x) \equiv 0$, $q_1(x, y) \equiv r_v(x) \not\equiv 0$, $q_2(x, y) \equiv r_2(x) \not\equiv 0$.
   So, we have the system:

$$\begin{cases} \alpha_{m_0}(x)y^{m_0} + r_1(x) = 0 \\ \beta_{n_0}(x)y^{n_0} + r_2(x) = 0 \end{cases}. \qquad (A)$$

Let $(x_0, y_0) \in L_A$.

If $r_1(x_0) = 0$ or $r_2(x_0) = 0$, then we have the system of the previous case 4. So, we suppose that $r_1(x_0) \neq 0$ and $r_2(x_0) \neq 0$. We can distinguish some cases:

(i) $\alpha_{m_0}(x_0) = \beta_{n_0}(x_0) = 0$.
   So, we have the system:

$$\begin{cases} r_1(x) = 0 \\ r_2(x) = 0 \end{cases}. \qquad (A)$$

Let x_0 be a common root of $r_1(x)$, $r_2(x)$. Then, we have $(x_0, y) \in L_A$, for every $y \in \mathbb{R}$, which is false of course, because the set L_A is finite.

So, this case cannot occur.

(ii) $\alpha_{m_0}(x_0) = 0$ and $\beta_{n_0}(x_0) \neq 0$. Then, we get $r_1(x_0) = 0$ and $y_0^{n_0} = -\dfrac{r_2(x_0)}{\beta_{n_0}(x_0)}$, and if n_0 is odd, we have $y_0 = \sqrt[n_0]{-\dfrac{r_2(x_0)}{\beta_{n_0}(x_0)}}$ if $\dfrac{r_2(x_0)}{\beta_{n_0}(x_0)} \leq 0$ and $y_0 = -\sqrt[n_0]{\dfrac{r_2(x_0)}{\beta_{n_0}(x_0)}}$
if $\dfrac{r_2(x_0)}{\beta_{n_0}(x_0)} > 0$.

If n_0 is even and $\dfrac{r_2(x_0)}{\beta_{n_0}(x_0)} \leq 0$, we have

$$y_1 = \sqrt[n_0]{-\frac{r_2(x_0)}{\beta_{n_0}(x_0)}} \text{ and } y_2 = -\sqrt[n_0]{-\frac{r_2(x_0)}{\beta_{n_0}(x_0)}}.$$

So, in this case we solve the system as follows.

We find the common roots of polynomials $\alpha_{m_0}(x)$ and $r_1(x_0)$, such that $\beta_{n_0}(x) \neq 0$. Let x_0 be such a root.

Then, if n_0 is odd and $\dfrac{r_2(x_0)}{\beta_{n_0}(x_0)} \leq 0$, then the couple $(x_0, y_0) \in L_A$, where

$$y_0 = \sqrt[n_0]{-\frac{r_2(x_0)}{\beta_{n_0}(x_0)}},$$

whereas if $\dfrac{r_2(x_0)}{\beta_{n_0}(x_0)} > 0$, then the couple $(x_0, y_0) \in L_A$, where

$$y_0 = -\sqrt[n_0]{\frac{r_2(x_0)}{\beta_{n_0}(x_0)}}.$$

If n_0 is even, we set

$$y_1 = \sqrt[n_0]{-\frac{r_2(x_0)}{\beta_{n_0}(x_0)}} \text{ and } y_2 = -\sqrt[n_0]{-\frac{r_2(x_0)}{\beta_{n_0}(x_0)}},$$

and the couples $(x_0, y_1), (x_0, y_2) \in L_A$, where $\dfrac{r_2(x_0)}{\beta_{n_0}(x_0)} \leq 0$. This case can happen if the above conditions hold, of course.

(iii) $\alpha_{m_0}(x_0) \neq 0, \beta_{n_0}(x_0) = 0$.

This case is similar to the previous case (ii).

(iv) $\alpha_{m_0}(x_0) \neq 0$ and $\beta_{n_0}(x_0) \neq 0$.

Through the equations of (A), we get

$$y_0^{m_0} = -\frac{r_1(x_0)}{\alpha_{m_0}(x_0)} \text{ and } y_0^{n_0} = -\frac{r_2(x_0)}{\beta_{n_0}(x_0)}.$$

Through these equations, we get

$$y_0^{n_0 m_0} = (-1)^{n_0}\left(\frac{r_1(x_0)}{\alpha_{m_0}(x_0)}\right)^{n_0} \text{ and } y_0^{n_0 m_0} = (-1)^{m_0}\left(\frac{r_2(x_0)}{\beta_{n_0}(x_0)}\right)^{m_0}$$

and by these equations, we get

$$\begin{aligned}(-1)^{n_0}\left(\frac{r_1(x_0)}{\alpha_{m_0}(x_0)}\right)^{n_0} &= (-1)^{m_0}\left(\frac{r_2(x_0)}{\beta_{n_0}(x_0)}\right)^{m_0}\\ &\Leftrightarrow (-1)^{m_0-n_0} r_2(x_0)^{m_0}\alpha_{m_0}(x_0)^{n_0}\\ &\quad - r_1(x_0)^{n_0}\beta_{n_0}(x_0)^{m_0} = 0.\end{aligned}$$

So, we solve this case as follows:

We find the real roots of polynomial $(-1)^{m_0-n_0}r_2(x)^{m_0}\alpha_{m_0}(x)^{n_0} - r_1(x)^{n_0}\beta_{n_0}(x)^{m_0}$, so that $\alpha_{m_0}(x) \neq 0$ and $\beta_{n_0}(x) \neq 0$.

For every x_0, we get y_0, so that $y_0^{m_0} = -\dfrac{r_1(x_0)}{\alpha_{m_0}(x_0)}$ as in the previous case (ii).

6. $\beta_{n_0}(x) \not\equiv 0$, $\alpha_{v_0}(x) \equiv \beta_{\mu_0}(x) \equiv 0$, $q_1(x, y)$, $q_2(x, y)$ are two pure polynomials. So, we have to solve the system:

$$\begin{cases} \alpha_{m_0}(x)y^{m_0} + q_1(x, y) = 0 \\ \beta_{n_0}(x)y^{n_0} + q_2(x, y) = 0 \end{cases}, \qquad (A)$$

where $deg_y q_1(x, y) \geq 1$, $deg_y q_2(x, y) \geq 1$, and $q_1(x, y)$, $q_2(x, y)$ are two monomials. So, in this case we have the system:

$$\begin{cases} \alpha_{m_0}(x)y^{m_0} + \alpha_0(x)y^{\lambda_1} = 0 \\ \beta_{n_0}(x)y^{n_0} + \beta_0(x)y^{\lambda_2} = 0 \end{cases},$$

where $\alpha_0(x)$, $\beta_0(x)$ are two polynomials, so that one of them (at least) is nonzero and $\lambda_1, \lambda_2 \in \mathbb{N}$, so that $\lambda_1 < m_0$ and $\lambda_2 < n_0$. We can write the system as follows:

$$\begin{cases} y^{\lambda_1}(\alpha_{m_0}(x)y^{m_0-\lambda_1} + \alpha_0(x)) = 0 \\ y^{\lambda_2}(\beta_{n_0}(x)^{n_0-\lambda_2} + \beta_0(x)) = 0 \end{cases}, \qquad (A)$$

so for every $x \in \mathbb{R}$, the couple $(x, 0) \in L_A$, which is false, because L_A is finite. Thus, this cannot occur.

7. We suppose that $\beta_{n_0}(x) \not\equiv 0$ $q_1(x, y) \equiv r_1(x)$, $\alpha_{v_0}(x) \not\equiv 0$ or $\beta_{\mu_0}(x) \not\equiv 0$, $q_2(x, y) \equiv r_2(x)$, where $r_1(x) \not\equiv 0$ or $r_2(x) \not\equiv 0$. So, we get the system:

$$\begin{cases} \alpha_{m_0}(x)y^{m_0} + \alpha_{v_0}(x)y^{v_0} + r_1(x) = 0 \\ \beta_{n_0}(x)y^{n_0} + \beta_{\mu_0}(x)y^{\mu_0} + r_2(x) = 0 \end{cases}. \qquad (A)$$

We distinguish some cases:

(i) $r_1(x) \equiv 0$ and $r_2(x) \not\equiv 0$. So, we get the system:

$$\begin{cases} \alpha_{m_0}(x)y^{m_0} + \alpha_{v_0}(x)y^{v_0} = 0 \\ \beta_{n_0}(x)y^{n_0} + \beta_{\mu_0}(x)y^{\mu_0} + r_2(x) = 0 \end{cases}. \qquad (A)$$

Let $(x_0, y_0) \in L_A$.

Through the first equation, we get

$$y^{v_0}(\alpha_{m_0}(x)y^{m_0-v_0} + \alpha_{v_0}(x)) = 0.$$

If $y = 0$, by the second equation we get $r_2(x_0) = 0$. So, if x_0 is a root of $r_2(x)$, then the couple $(x_0, 0) \in L_A$.

Let $y_0 \neq 0$. Then we get

$$\alpha_{m_0}(x_0)y_0^{m_0-v_0} + \alpha_{v_0}(x_0) = 0 \text{ and } \beta_{n_0}(x_0)y_0^{n_0} + \beta_{\mu_0}(x_0)y_0^{\mu_0} + r_2(x_0) = 0.$$

If $r_2(x_0) = 0$, we have some of the previous cases that we have already examined. So, we suppose that $r_2(x_0) \neq 0$. If $n_0 < m_0$, we have a system that we supposedly can solve through the induction step. Thus, we suppose that $n_0 = m_0$. So we have the system:

$$\begin{cases} \alpha_{m_0}(x_0)y_0^{m_0-v_0} + \alpha_{v_0}(x_0) & = 0 \\ \beta_{n_0}(x_0)y_0^{m_0} + \beta_{\mu_0}(x_0)y^{\mu_0} + r_2(x_0) = 0 \end{cases}.$$

We will postpone this case, because we will examine a more general case later, which covers this case.

(ii) $r_1(x) \not\equiv 0$ and $r_2(x) \equiv 0$.

This case is similar to the previous.

(iii) $r_1(x) \not\equiv 0$ and $r_2(x) \not\equiv 0$.

Let $(x_0, y_0) \in L_A$. If $r_1(x_0) = 0$, or $r_2(x_0) = 0$, we have some of the previous cases. So, we suppose that $r_1(x_0) \neq 0$ and $r_2(x_0) \neq 0$. We have some cases:

(i) $\alpha_{v_0}(x) \not\equiv 0$ and $\beta_{\mu_0}(x) \equiv 0$.

So, we have the system:

$$\begin{cases} \alpha_{m_0}(x)y^{m_0} + \alpha_{v_0}(x)y^{v_0} + r_1(x) = 0 \\ \beta_{n_0}(x)y^{m_0} \qquad\qquad + r_2(x) = 0 \end{cases}. \tag{A}$$

We consider some cases:

(a) Let $(x_0, y_0) \in L_A$, $\alpha_{m_0}(x_0) = \beta_{n_0}(x_0) = 0$. We have examined this case previously.

(b) $\alpha_{m_0}(x_0) = 0$ and $\beta_{n_0}(x_0) \neq 0$. We have examined this case previously.

(c) $\alpha_{m_0}(x_0) \neq 0$ and $\beta_{n_0}(x_0) = 0$.

If $\alpha_{v_0}(x_0) = 0$, we have examined this case previously. So, we suppose that $\alpha_{v_0}(x_0) \neq 0$. Then we get $r_2(x_0) = 0$ by the second equation of (A) because $\beta_{n_0}(x_0) = 0$, which is false by our supposition. So, this case cannot occur.

(d) $\alpha_{m_0}(x_0) \neq 0$ and $\beta_{n_0}(x_0) \neq 0$.

If $\alpha_{v_0}(x_0) = 0$, we have examined this case previously. So, we suppose that $\alpha_{v_0}(x_0) \neq 0$. We will examine this case later.

(ii) $\alpha_{v_0}(x) \equiv 0$ and $\beta_{\mu_0}(x) \not\equiv 0$.

This case is similar to the previous one.

(iii) $\alpha_{v_0}(x) \not\equiv 0$ and $\beta_{\mu_0}(x) \not\equiv 0$.

We have some cases here:

(a) We suppose that $\beta_{n_0}(x_0) = \beta_{\mu_0}(x_0) = 0$, and $\alpha_{m_0}(x_0) \neq 0$, $\alpha_{v_0}(x_0) \neq 0$. So, we have the system:

$$\begin{cases} \alpha_{m_0}(x_0)y^{m_0} + \alpha_{v_0}(x_0)y^{v_0} + r_1(x_0) = 0 \\ r_2(x_0) = 0 \end{cases}. \tag{A}$$

This cannot happen because $r_2(x_0) \neq 0$ by our supposition.

(b) We suppose that $\alpha_{m_0}(x_0) = \alpha_{v_0}(x_0) = 0$. Then we get that $r_1(x_0) = 0$, which is false by our supposition. Thus, this case cannot occur.

(c) If $\alpha_{m_0}(x_0) = 0$, or $\alpha_{v_0}(x_0) = 0$, or $\beta_{n_0}(x_0) = 0$, or $\beta_{\mu_0}(x_0) = 0$, then we get some of the previous cases.

(d) $\alpha_{m_0}(x_0) \neq 0$ and $\alpha_{v_0}(x_0) \neq 0$ and $\beta_{n_0}(x_0) \neq 0$ and $\beta_{\mu_0}(x_0) \neq 0$.

We have to solve the system:

$$\begin{cases} \alpha_{m_0}(x_0)y^{m_0} + \alpha_{v_0}(x_0)y^{v_0} + r_1(x_0) = 0 \\ \beta_{n_0}(x_0)y^{m_0} + \beta_{\mu_0}(x_0)y^{\mu_0} + r_2(x_0) = 0 \end{cases}. \tag{A}$$

We have here the basic case of this system.

We will examine this case later in a more general case.

Now, we will examine the system:

$$\begin{cases} \alpha_{m_0}(x)y^{m_0} + \alpha_{v_0}(x)y^{v_0} + q_1(x, y) = 0 \\ \beta_{n_0}(x)y^{n_0} + \beta_{\mu_0}(x)y^{\mu_0} + q_2(x, y) = 0 \end{cases},$$

where $\alpha_{m_0}(x) \not\equiv 0$, $m_0 \geq 3$, $v_0 < m_0$, $n_0 \leq m_0$ $n_0 > \mu_0$, $q_1(x, y)$, $q_2(x, y)$ are two pure polynomials, $\alpha_{v_0}(x) \not\equiv 0$, and $\beta_{\mu_0}(x) \not\equiv 0$.

We can distinguish the following cases:

(1) $\beta_{n_0}(x) \equiv 0$.

We get $(x_0, y_0) \in L_A$.

If $\alpha_{m_0}(x_0) = 0$, then we have a system from the induction step. So, we suppose $\alpha_{m_0}(x_0) \neq 0$.

We have some cases:

(i) $\alpha_{v_0}(x_0) = \beta_{\mu_0}(x_0) = 0$.

Then, we analyze polynomials $q_1(x, y), q_2(x, y)$ and we reach a system of the following form:

$$\begin{cases} \alpha_{m_0}(x_0)y^{m_0} + \alpha_{v_1}(x_0)y^{v_1} + q_3(x_0, y_0) = 0 \\ \beta_{m_1}(x_0)y^{\mu_1} + q_3(x_0, y_0) = 0 \end{cases}, \quad \text{(A)}$$

where $\alpha_{v_1}(x_0) \neq 0$, $\beta_{\mu_1}(x_0) \neq 0$, $v_1 < m_0$, $\mu_1 \in \mathbb{N}$, $deg_y q_3(x, y) < v_1$, $deg_y q_3(x, y) < \mu_1$.

We will see later how we can solve such a system.

(ii) $\alpha_{v_0}(x_0) \neq 0$ and $\beta_{\mu_0}(x_0) = 0$.

Then, we analyze polynomial $q_2(x, y)$ and we reach a system of the following form:

$$\begin{cases} \alpha_{v_0}(x_0)y_0^{m_0} + \alpha_{v_0}(x_0)y_0^{v_0} + q_1(x_0, y_0) = 0 \ (1) \\ \beta_{\mu_1}(x_0)y_0^{\mu_1} + r_2(x_0) = 0 \ (2) \end{cases}. \quad \text{(B)}$$

If $\beta_{\mu_1}(x_0) = 0$, we get from (2) that $r_2(x_0) = 0$.

In this case we solve system (B) as follows.

We take x_0 that is a common root of polynomials in the second equation of Eq. (B), so that $\alpha_{m_0}(x_0) \neq 0$ and $\alpha_{v_0}(x_0) \neq 0$, and we find y_0 from the first equation of Eq. (B).

(iii) $\alpha_{v_0}(x_0) \neq 0$ and $\beta_{\mu_0}(x_0) \neq 0$. We will see later how we solve this system.

After all the above cases, we reach now to the most important case.

We have the system:

$$\begin{cases} \alpha_{m_0}(x)y^{m_0} + \alpha_{v_0}(x)y^{v_0} + q_1(x, y) = 0 \\ \beta_{n_0}(x)y^{n_0} + \beta_{\mu_0}(x)y^{\mu_0} + q_2(x, y) = 0 \end{cases}, \quad \text{(A)}$$

where we have

$$m_0 > v_0 > deg_y q_1(x, y),$$

$$n_0 > \mu_0 > deg_y q_2(x, y),$$

$\alpha_{m_0}(x) \not\equiv 0$, $\alpha_{v_0}(x) \not\equiv 0$, $\beta_{n_0}(x) \not\equiv 0$, $\beta_{\mu_0}(x) \not\equiv 0$, and $q_1(x, y), q_2(x, y)$ be two pure polynomials.

We distinguish some cases:

(i) $m_0 = n_0$.

We also have some cases here:

(a) $v_0 = \mu_0$.

So, we have the system:

$$\begin{cases} \alpha_{m_0}(x)y^{m_0} + \alpha_{v_0}(x)y^{v_0} + q_1(x,y) = 0 \\ \beta_{m_0}(x)y^{m_0} + \beta_{v_0}(x)y^{v_0} + q_2(x,y) = 0 \end{cases}. \quad \text{(A)}$$

First, we examine the case where:

$$(x_0, y_0) \in L_A \text{ and } \alpha_{m_0}(x_0) \cdot \alpha_{v_0}(x_0) \cdot \beta_{m_0}(x_0) \cdot \beta_{m_0}(x_0) \cdot \beta_{v_0}(x_0) \neq 0.$$

Let

$$D = \begin{vmatrix} \alpha_{m_0}(x_0) & \alpha_{v_0}(x_0) \\ \beta_{m_0}(x_0) & \beta_{v_0}(x_0) \end{vmatrix} = \alpha_{m_0}(x_0)\beta_{v_0}(x_0) - \alpha_{v_0}(x_0)\beta_{m_0}(x_0).$$

We suppose that $D \neq 0$.

This exactly **is the first basic case**. We will study three basic cases overall. We consider the linear system:

$$\begin{cases} \alpha_{m_0}(x_0)z + \alpha_{v_0}(x_0)\omega = -q_1(x_0, y_0) \ (1) \\ \beta_{m_0}(x_0)z + \beta_{v_0}(x_0)\omega = -q_2(x_0, y_0) \ (2) \end{cases}. \quad \text{(B)}$$

We set

$$D_1 = \begin{vmatrix} -\alpha_1(x_0, y_0) & \alpha_{v_0}(x_0) \\ -q_2(x_0, y_0) & \beta_{v_0}(x_0) \end{vmatrix} \text{ and } D_2 = \begin{vmatrix} \alpha_{m_0}(x_0) & -q_1(x_0, y_0) \\ \beta_{m_0}(x_0) & -q_2(x_0, y_0) \end{vmatrix}.$$

That is, we have

$$D_1 = \alpha_{v_0}(x_0)q_2(x_0, y_0) - \beta_{v_0}(x_0)q_1(x_0, y_0) \text{ and}$$

$$D_2 = \beta_{m_0}(x_0)q_1(x_0, y_0) - \alpha_{m_0}(x_0)q_2(x_0, y_0).$$

Because $D \neq 0$, by our supposition, we take it that system (B) has only one solution (z_0, ω_0), where $z_0 = \dfrac{D_1}{D}$ (3) and $\omega_0 = \dfrac{D_2}{D}$ (4), as it is well known by linear algebra by Cramer's law.

Because $(x_0, y_0) \in L_A$ (by our supposition), this means that the couple $(y_0^{m_0}, y_0^{v_0})$ is a solution of system (B).

But, (z_0, ω_0) is the unique solution of system (B). So, we have $(z_0, \omega_0) = (y_0^{m_0}, y_0^{v_0}) \Leftrightarrow z_0 = y_0^{m_0}$ (5) and $\omega_0 = y_0^{v_0}$ (6). By (3), (4), (5), and (6), we get

$$y_0^{m_0} = \frac{D_1}{D} \ (7) \text{ and } y_0^{v_0} = \frac{D_2}{D} \ (8).$$

Now, we use the obvious relation of numbers $y_0^{m_0}$ and $y_0^{v_0}$; that is, $y_0^{m_0} = y_0^{m_0-v_0} \cdot y_0^{v_0}$ (9), where $v_0 < m_0$, by our supposition.

Replacing by (7) and (8) in (9), we get

$$\frac{D_1}{D} = y_0^{m_0-v_0} \cdot \frac{D_2}{D} \Leftrightarrow D_2 y_0^{m_0-v_0} - D_1 = 0. \tag{10}$$

From the above, we see that (x_0, y_0) satisfies the two equations:

$$\begin{cases} D y_0^{v_0} - D_2 = 0 \ (11) \\ D_2 y_0^{m_0-v_0} - D_1 = 0 \ (12) \end{cases} . \tag{C}$$

We notice that polynomials in (11) and (12) have degree with respect to y lower than m_0.

Let us consider now the following systems:

$$\begin{cases} \alpha_{m_0}(x) y^{m_0} + \alpha_{v_0}(x) y^{v_0} + q_1(x, y) = 0 \quad (13) \\ \beta_{m_0}(x) y^{m_0} + \beta_{v_0}(x) y^{v_0} + q_2(x, y) = 0 \quad (14) \ , \\ y \cdot \alpha_{m_0}(x) \cdot \alpha_{v_0}(x) \beta_{m_0}(x) \cdot \beta_{v_0}(x) D \neq 0 \end{cases} \tag{A}$$

$$\begin{cases} D y^{v_0} - D_2 = 0 & (15) \\ D_2 y^{m_0-v_0} - D_1 = 0 & (16) \ , \\ y \cdot \alpha_{m_0}(x) \cdot \alpha_{v_0}(x) \cdot \beta_{m_0}(x) \cdot \beta_{v_0}(x) \cdot D \neq 0 \end{cases} \tag{B}$$

where $D = \alpha_{m_0}(x)\beta_{v_0}(x) - \alpha_{v_0}(x)\beta_{m_0}(x)$.

$$D_1 = \alpha_{v_0}(x) q_2(x, y) - \beta_{v_0}(x) q_1(x, y),$$

$$D_2 = \beta_{m_0}(x) q_1(x, y) - \alpha_{m_0}(x) q_2(x, y).$$

It is obvious that $deg_y(D y^{v_0}) = v_0 < m_0$, because $D \neq 0$ and $deg_y(D_2 y^{m_0-v_0}) < m_0$, as $deg_y D_2 < v_0$, by our suppositions. We will prove now that $L_A = L_B$. It is obvious that $L_A \subseteq L_B$ (17) from the previous procedure, because we got Eqs. (11) and (12) of system (Γ) from equations of system (A).

Now, let $(x_0, y_0) \in L_B$.

Through Eqs. (15) and (16) of (B) and the fact that $y_0 \neq 0$, we get

$$y_0^{v_0} = \frac{D_2}{D} \ (18) \ \text{ and } \ y_0^{m_0-v_0} = \frac{D_1}{D_2} \ . \tag{19}$$

Through Eqs. (18) and (19), we get $y_0^{m_0} = \dfrac{D_1}{D}$ (20). Now, we consider system (B). Because $D \neq 0$, this system has a unique solution $(z_0, \omega_0) = \left(\dfrac{D_1}{D}, \dfrac{D_2}{D}\right)$ (21), from Cramer's law. Through (18), (20), and (21), we get $z_0 = y_0^{m_0}$ (22) and $\omega_0 = y_0^{v_0}$ (23).

Replacing (22) and (23) in equations of (B), we take it that $(x_0, y_0) \in L_A$, so $L_B \subseteq L_A$ (24). By (17) and (24), we have $L_A = L_B$ (25). The equality (25) means that: in order to solve system (A), it suffices to solve system (B), whose degree with respect to y is smaller than m_0, that is the degree of system (A) with respect to y. But with the induction step, we can solve a system whose degree with respect to y is smaller than m_0, and thus we complete this case.

The second basic case is the following $D \not\equiv 0$, but

$$D(x_0) = \alpha_{m_0}(x_0)\beta_{v_0}(x_0) - \alpha_{v_0}(x_0)\beta_{m_0}(x_0) = 0.$$

In this case the two equations of system (A) are equivalent to those of linear algebra, as we have shown in prerequisites.

So, we can solve this case as follows.

We find the roots of polynomial $D = \alpha_{m_0}(x)\beta_{v_0}(x) - \alpha_{v_0}(x)\beta_{m_0}(x)$, so that: $\alpha_{m_0}(x) \cdot \alpha_{v_0}(x) \cdot \beta_{n_0}(x) \cdot \beta_{\mu_0}(x)$.

For every such root x_0, we find y_0 from one of the equations of (A) that are equivalent. We can complete this case by finding the solutions of the form $(x, 0)$ (if any).

Third Basic Case (Singular Case)

We suppose that $D \equiv 0 \equiv \alpha_{m_0}(x)\beta_{v_0}(x) - \alpha_{v_0}(x)\beta_{m_0}(x)$. We call this case **the singular case**.

We consider the system:

$$\begin{cases} \alpha_{m_0}(x)y^{m_0} + \alpha_{v_0}(x)y^{v_0} + q_1(x, y) = 0 \ (1) \\ \beta_{m_0}(x)y^{m_0} + \beta_{v_0}(x)y^{v_0} + q_2(x, y) = 0 \ (2) \\ \alpha_{m_0}(x)\alpha_{v_0}(x)\beta_{m_0}(x)\beta_{v_0}(x) \neq 0 \end{cases} . \qquad \text{(A)}$$

We consider our general supposition. That is, we suppose $L_A \neq \emptyset$. Let $(x_0, y_0) \in L_A$. Then we get

$$\begin{cases} \alpha_{m_0}(x_0)y_0^{m_0} + \alpha_{v_0}(x_0)y_0^{v_0} = -q_1(x_0, y_0) \ (3) \\ \beta_{m_0}(x_0)y_0^{m_0} + \beta_{v_0}(x_0)y_0^{v_0} = -q_2(x_0, y_0) \ (4) \end{cases} . \qquad \text{(B)}$$

We get

$$D(x_0) = \alpha_{m_0}(x_0)\beta_{v_0}(x_0) - \alpha_{v_0}(x_0)\beta_{m_0}(x_0) = 0.$$

Let

$$D_1(x_0, y_0) = \alpha_{v_0}(x_0)q_2(x_0, y_0) - \beta_{v_0}(x_0)q_1(x_0.y_0),$$

$$D_2(x_0, y_0) = \beta_{m_0}(x_0)q_1(x_0, y_0) - \alpha_{m_0}(x_0)q_2(x_0, y_0).$$

We now consider the following system:

$$\begin{cases} \alpha_{m_0}(x_0)z + \alpha_{v_0}(x_0)\omega = -q_1(x_0, y_0) \ (5) \\ \beta_{m_0}(x_0)z + \beta_{v_0}(x_0)\omega = -q_2(x_0, y_0) \ (6) \end{cases}. \tag{Γ}$$

Through the previous system (B), we have that $(y_0^{m_0}, y_0^{v_0})$ is a solution of (Γ). That is, (Γ) is a linear system that has a solution and $D(x_0) = 0$. So, we have that $D_1(x_0, y_0) = 0$ through linear algebra.

We consider now the following two systems: (A) and the following:

$$\begin{cases} \alpha_{m_0}(x)y^{m_0} + \alpha_{v_0}(x)y^{v_0} + q_1(x, y) = 0 \ (7) \\ \alpha_{v_0}(x)q_2(x, y) - \beta_{v_0}(x)q_1(x, y) = 0 \quad (8) \\ y \cdot \alpha_{m_0}(x)\alpha_{v_0}(x)\beta_{m_0}(x)\beta_{v_0}(x) \neq 0 \end{cases}. \tag{Δ}$$

From the above, we have $L_A \subseteq L_\Delta$ (9).

We can now prove the reverse inclusion of (9).

Let $(x_0, y_0) \in L_\Delta$. Of course (x_0, y_0) satisfies Eq. (1) of (A). We distinguish two cases:

(i) $q_1(x_0, y_0) \neq 0$.

We can consider linear system (Γ). This system has $D = D_1 = 0$ by our supposition $D \equiv 0$ and $D_1 = 0$ because (8) holds for (x_0, y_0); that is, $D_1(x_0, y_0) = 0$. So, system (Γ) has an infinity of solutions and $D_2(x_0, y_0) = 0$, because $D = D_1(x_0, y_0) = 0$. Of course we have $\alpha_{m_0}(x_0)q_{v_0}(x_0)\beta_{m_0}(x_0)\beta_{v_0}(x_0) \neq 0$, by our supposition.

By relation $D(x_0) = 0$, we take

$$\alpha_{m_0}(x_0)\beta_{v_0}(x_0) - \alpha_{v_0}(x_0)\beta_{m_0}(x_0) = 0 \Leftrightarrow \frac{\alpha_{m_0}(x_0)}{\beta_{m_0}(x_0)} = \frac{\alpha_{v_0}(x_0)}{\beta_{v_0}(x_0)}. \tag{10}$$

By equation $D_1(x_0, y_0) = 0$, we take

$$\alpha_{v_0}(x_0)q_2(x_0, y_0) = \beta_{v_0}(x_0)q_1(x_0, y_0) \Rightarrow \frac{\alpha_{v_0}(x_0)}{\beta_{v_0}(x_0)} = \frac{q_1(x_0, y_0)}{q_2(x_0, y_0)}. \tag{11}$$

We have $q_2(x_0, y_0) \neq 0$ or else if $q_2(x_0, y_0) = 0 \Rightarrow q_1(x_0, y_0) = 0$, which is false by our supposition. So, (11) holds. By (10) and (11), we set

$$0 \neq \lambda = \frac{\alpha_{m_0}(x_0)}{\beta_{m_0}(x_0)} = \frac{\alpha_{v_0}(x_0)}{\beta_{v_0}(x_0)} = \frac{q_1(x_0, y_0)}{q_2(x_0, y_0)} \Rightarrow \beta_{m_0}(x_0) = \frac{1}{\lambda}\alpha_{m_0}(x_0), \tag{12}$$

$$\beta_{v_0}(x_0) = \frac{1}{\lambda}\alpha_{v_0}(x_0), \tag{13}$$

$$q_2(x_0, y_0) = \frac{1}{\lambda}q_1(x_0, y_0). \tag{14}$$

By (12), (13), and (14), we get

$$\begin{aligned}
&\beta_{m_0}(x_0)y_0^{m_0} + \beta_{v_0}(x_0)y_0^{v_0} + q_2(x_0, y_0)\\
&= \frac{1}{\lambda}\alpha_{m_0}(x_0)y_0^{m_0} + \frac{1}{\lambda}\alpha_{v_0}(x_0)y_0^{v_0} + \frac{1}{\lambda}q_1(x_0, y_0)\\
&= \frac{1}{\lambda}(\alpha_{m_0}(x_0)y_0^{m_0} + \alpha_{v_0}(x_0)y_0^{v_0} + q_1(x_0, y_0)\\
&= \frac{1}{\lambda}\cdot 0 = 0,
\end{aligned}$$

because $(x_0, y_0) \in L_\Delta$, which means that (x_0, y_0) satisfies equality (7). So, we proved that if $(x_0, y_0) \in L_\Delta$ and $q_1(x_0, y_0) \neq 0$, then $(x_0, y_0) \in L_A$.

(ii) $q_1(x_0, y_0) = 0$.

Then, because $(x_0, y_0) \in L_\Delta$, through equality (8) we get $q_2(x_0, y_0) = 0$, because $\alpha_{v_0}(x_0) \neq 0$, by our supposition.

As previously, because $D(x_0) = 0$ and $\beta_{m_0}(x_0)\beta_{v_0}(x_0) \neq 0$, we take it that (12) and (13) hold, so

$$\begin{aligned}
&\beta_{m_0}(x_0)y_0^{m_0} + \beta_{v_0}(x_0)y_0^{v_0} + q_2(x_0, y_0)\\
&= \frac{1}{\lambda}\alpha_{m_0}(x_0)y_0^{m_0} + \frac{1}{\lambda}\alpha_{v_0}(x_0)y_0^{v_0} + 0\\
&= \frac{1}{\lambda}(\alpha_{m_0}(x_0)y_0^{m_0} + \alpha_{v_0}(x_0)y_0^{v} + q_1(x_0, y_0) = 0,
\end{aligned}$$

by equality (7) of (Δ) because $(x_0, y_0) \in L_\Delta$ by our supposition.

So, equality (2) of (A) holds; that is, $(x_0, y_0) \in L_A$. So, we have $L_\Delta \subseteq L_A$ (15). Through (9) and (15), we get $L_A = L_\Delta$. So, in order to solve system (A), it suffices to solve system (Δ). What is the profit from system (Δ)? The profit is that polynomial in Eq. (8) of (Δ); that is, D_1 has $deg_y D_1(x, y) < v_0$ or $D_1(x, y) \equiv 0$. We examine now how we exploit these facts.

We leave the case $D_1(x, y) \equiv 0$ for the end.

We examine now the case where $D_1(x, y) \not\equiv 0$. We can write $D_1(x, y)$ in the following form:

$$D_1(x, y) = \alpha_{v_1}(x)y^{v_1} + \alpha_{v_2}(x)y^{v_2} + q_3(x, y),$$

where $v_0 > v_1 > v_2$, $deg_y q_3(x, y) < v_2$, or $q_3(x, y) \equiv 0$. This is the general case.

We suppose, also, that $\alpha_{v_1}(x) \not\equiv 0$, $\alpha_{v_2}(x) \not\equiv 0$, and $q_3(x, y)$ is a pure polynomial.

We get

$$y^{m_0 - v_1} D_1(x, y) = \alpha_{v_1}(x)y^{m_0} + \alpha_{v_2}(x)y^{m_0 - v_1 + v_2} + y^{m_0 - v_1} q_3(x, y),$$

where $deg_y(y^{m_0 - v_1} q_3(x, y)) < m_0 - v_1 + v_2$ because $deg_y q_3(x, y) < v_2$ by our supposition.

We consider the system:

$$\begin{cases} \alpha_{m_0}(x)y^{m_0} + \alpha_{v_0}(x)y^{v_0} + q_1(x, y) = 0 \\ \alpha_{v_1}(x)y^{m_0} + \alpha_{v_2}(x)y^{m_0 - v_1 + v_2} + y^{m_0 - v_1} q_3(x, y) = 0 \\ y\alpha_{m_0}(x)\alpha_{v_0}(x)\beta_{m_0}(x)\beta_{v_0}(x) \neq 0 \end{cases} . \tag{E}$$

If $\alpha_{v_1}(x) = 0$, or $\alpha_{v_2}(x) = 0$ for $x \in \mathbb{R}$, we examine whether system (E) has a root of $\alpha_{v_1}(x)$ or $\alpha_{v_2}(x)$ that satisfies system (E). So, we examine the case where $\alpha_{v_1}(x) \cdot \alpha_{v_2}(x) \neq 0$.

Let

$$D = \begin{vmatrix} \alpha_{m_0}(x) & \alpha_{v_0}(x) \\ \alpha_{v_1}(x) & \alpha_{v_2}(x) \end{vmatrix} = \alpha_{m_0}(x)\alpha_{v_2}(x) - \alpha_{v_0}(x)\alpha_{v_1}(x).$$

Then, system (E) is a system similar to system (A).

So, we examine the similar cases in the same way.

Here, we examine only the case where $D = \alpha_{m_0}(x)\alpha_{v_2}(x) - \alpha_{v_0}(x)\alpha_{v_1}(x) \equiv 0$. In this case we again reach a system similar to (Δ), so that the respective Eq. (8) of the new system has $D_1(x, y) \not\equiv 0$.

We handle this case as follows.

In system (A) of page 23, we can take any of the two equations in order to get an equivalent system as (Δ). So it helps us to take the equation in which the respective pure polynomial $q_1(x, y)$ or $q_2(x, y)$ has the smallest number of terms. For this reason in system (E) (that is similar to A), we take as a first equation (of system (Δ)) the second equation because this polynomial $y^{m_0 - v_1} q_3(x, y)$ has at most v_2 terms with respect to y (by its definition), where $v_2 < v_1 < v_0 \Rightarrow v_2 \leq v_0 - 2$.

In the new system (Δ) we take that the respective $D_1(x, y)$ polynomial of (Δ) has $deg_y D_1(x, y) < v_0$, so if we write this polynomial again in the form:

$$D_1(x, y) = \alpha_{v_1}(x)y^{v_1} + \alpha_{v_2}(x)y^{v_2} + q_4(x, y),$$

the new polynomial $q_4(x, y)$ has at most $v_2 \leq v_0 - 2$ terms.

So, the profit, is that the new polynomials $q_1(x, y)$, $q_2(x, y)$ of the new system (Δ) will have at most $\upsilon_0 - 2$ terms each one of them and the respective new polynomial $D_1(x, y)$ also. So, the profit is the following.

In system (Δ) polynomial $D_1(x, y)$ has at most υ_0 terms with respect to y, whereas in a new system like (Δ) in a following stage the respective polynomial $D_1(x, y)$ of the new system (Δ) will have at most $\upsilon_0 - 2$ terms with respect to y.

With the same procedure, we can see that the terms of the respective polynomials $D_1(x, y)$ are decreasing, so that after a finite number of steps we reach a polynomial $D_1(x, y) \equiv 0$ or $D_1(x, y) \equiv r(x)$ for polynomial $r(x) \not\equiv 0$. If $D_1(x, y) \equiv 0$, we solve this case in the final step, or else if $r(x) \not\equiv 0$, it suffices to find the roots of polynomial $r(x)$; otherwise, we have some of the previous cases that we have already examined.

Now we will examine the remaining case. In system (A), page 21, if $\upsilon_0 \neq \mu_0$ and $n_0 = m_0$, we have the first basic case where $D(x_0) \neq 0$.

Now, let $m_0 \neq n_0$; that is, $n_0 < m_0$. If $n_0 \geq \upsilon_0$, we have the first basic case. So, we can examine the case $\upsilon_0 > n_0$. In this case we have

$$y^{m_0-n_0}(\beta_{n_0}(x)y^{n_0} + \beta_{\mu_0}(x)y^{\mu_0} + q_2(x, y)) = 0$$

$$\Leftrightarrow \beta_{n_0}(x)y^{m_0} + \beta_{\mu_0}(x)y^{m_0-n_0+\mu_0} + y^{m_0-n_0}q_2(x, y) = 0$$

and instead of (A) we examine the system:

$$\begin{cases} \alpha_{m_0}(x)y^{m_0} + \alpha_{\upsilon_0}(x)y^{\upsilon_0} + q_1(x, y) = 0 \\ \beta_{n_0}(x)y^{m_0} + \beta_{\mu_0}(x)y^{m_0-n_0+\mu_0} + y^{m_0-n_0}q_2(x, y) = 0 \end{cases}.$$

This system is of the case where $m_0 = n_0$, which we have already examined. So, up to now, we have examined all the possible cases of the initial system except only one, which we will examine now.

In the third basic case we will examine now the case where $D_1(x, y) \equiv 0$. Then, as in pages 23, 24 we take it that $D_2(x, y) \equiv 0$, also that for every $(x, y) \in \mathbb{R}^2$ there exists $c \in \mathbb{R}$, such that

$$\alpha_{m_0}(x)y^{m_0} + \alpha_{\upsilon_0}(x)y^{\upsilon_0} + q_1(x, y) = c \cdot (\beta_{m_0}(x)y^{m_0} + \beta_{\upsilon_0}(x)y^{\upsilon_0} + q_2(x, y)) \quad (*)$$

and $c \neq 0$. The number c depends on the couple (x, y), so it is better to write $c(x, y)$, instead of c.

Now, we will consider system (A*)

$$\begin{cases} \alpha_{m_0}(x)y^{m_0} + \alpha_{\upsilon_0}(x)y^{\upsilon_0} + q_1(x, y) = 0 \\ \alpha_{m_0}(x)\alpha_{\upsilon_0}(x)\beta_{m_0}(x)\beta_{\upsilon_0}(x) \neq 0 \end{cases}. \quad (A^*)$$

Equality $(*)$ gives us that

$$L_A = L_{A^*}.$$

So, in order to solve system (A) it suffices to solve the "simpler" system (A*) that has only one equation.

Now, it is the time to exploit the unique supposition that we have not used up to now.

That is, the set $L_A = L_{A^*}$ is finite. As we have seen in the prerequisites, there are polynomials $p(x, y)$ of two real variables that have a finite set of roots only. For example, let:

$$p(x, y) = (x^2 - 4)^2 + (y^2 - 9)^2.$$

It is easy to see that

$$L_{p(x,y)} = \{(2, 3), (2, -3), (-2, 3), (-2, -3)\}.$$

We denote

$$R(x, y) = \alpha_{m_0}(x)y^{m_0} + \alpha_{v_0}(x)y^{v_0} + q_1(x, y),$$

for simplicity.

So, we solve the system:

$$\begin{cases} R(x, y) = 0 \\ \alpha_{m_0}(x)\alpha_{v_0}(x)\beta_{m_0}(x)\beta_{v_0}(x) \neq 0 \end{cases}. \tag{A*}$$

Of course, we get $R(x, y) \not\equiv 0$, because $\alpha_{m_0} \not\equiv 0$.

Now, it is the time to use the results of our prerequisites.

By the suppositions of the third case, we get $\alpha_{m_0}(x) \not\equiv 0$ and $m_0 > 1$, which gives that $R(x, y)$ is a pure polynomial that has a finite set of roots, nonempty.

We apply Corollary 4.16 by our prerequisites and we take it that 0 is the global maximum or minimum of $R(x, y)$.

Without loss of generality, we suppose that 0 is the global minimum of $R(x, y)$. This means that if we consider the function $F : U \to \mathbb{R}$ (where $U = \{(x, y) \in \mathbb{R}^2 \mid \alpha_{m_0}(x)\alpha_{v_0}(x)\beta_{m_0}(x)\beta_{v_0}(x) \neq 0\}$ is an open subset of $\mathbb{R}^2$) $F((x, y)) = R(x, y)$ for every $(x, y) \in U$, then it holds $F((x, y)) \geq 0$ for every $(x, y) \in U$, and there exists $(x_0, y_0) \in U$, so that $F((x_0, y_0)) = 0$.

Let $(x_0, y_0) \in \mathbb{R}^2$, so that $(x_0, y_0) \in L_{A^*}$. Then, we have $F((x_0, y_0)) = 0$ and function F has a global minimum in (x_0, y_0). Then, by Theorem 4.17, we get $\nabla F(x_0, y_0) = (0, 0)$. So we have $\dfrac{\partial F}{\partial y}((x_0, y_0)) = 0$.

We can now consider the system:

$$\begin{cases} F((x,y)) = 0 \\ \dfrac{\partial F}{\partial y}((x,y)) = 0 \\ \alpha_{m_0}(x)\alpha_{v_0}(x)\beta_{m_0}(x)\beta_{v_0}(x) \neq 0 \end{cases} . \tag{A$_1$}$$

Of course we get $L_{A_1} \subseteq L_{A^*} = L_A$ and by the above we also get $L_A^* \subseteq L_{A_1}$. So we get

$$L_{A_1} = L_A.$$

So, in order to solve system (A*) it suffices to solve system (A_1). We need to write a more analytic system (A_1). We get

$$\begin{cases} \alpha_{m_0}(x)y^{m_0} + \alpha_{v_0}(x)y^{v_0} + q_1(x,y) = 0 \\ m_0\alpha_{m_0}(x)y^{m_0-1} + v_0\alpha_{v_0}(x)y^{v_0-1} + \dfrac{\partial q_1}{\partial y}(x,y) = 0 \\ \alpha_{m_0}(x)\alpha_{v_0}(x)\beta_{m_0}(x)\beta_{v_0}(x) \neq 0 \end{cases} \cdot \tag{A$_1$}$$

Because $m_0 > v_0 \Rightarrow m_0 - 1 \geq v_0$. This shows that system (A_1) is the first basic case, and so we can transfer system (A_1) to a system that has smaller than m_0 degree with respect to y, which we can solve with the induction step. So, inductively we have managed to solve the initial system in any case. So, we have completed our second stage.

3.3 *Third Stage*

Let a polynomial

$$p(z) = \alpha_0 + \alpha_1 z + \cdots + \alpha_{v-1}z^{v-1} + \alpha_v z^v,$$

for $v \in \mathbb{N}$, $\alpha_1 \in \mathbb{C}$, for $i = 0, 1, \ldots, v$, $\alpha_v \neq 0$, of one complex variable.

We are now ready to solve completely the equation $p(z) = 0$, or in other words to find the roots of polynomial $p(z)$ with degree v.

We distinguish two cases:

(i) $\alpha_i \in \mathbb{R}$ for every $i = 0, 1, \ldots, v$, and (ii) $\alpha_i \in \mathbb{C}$, $i = 0, 1, \ldots, v$. First, we prove the following lemma.

Lemma 3.3.1 (A Well-Known Lemma) *Let $p(z)$, be a polynomial as above with degree $v = degp(z) \in \mathbb{N}$. Then, there exist two polynomials $p_1(x,y)$, $p_2(x,y)$ of two real variables with real coefficients, so that it holds*

$$p(x + yi) = p_1(x, y) + ip_2(x, y)$$

for every $(x, y) \in \mathbb{R}^2$.

Proof We can prove this lemma with induction above the degree v of $p(z)$. Let $p(z) = \alpha_0 + \alpha_1 z, \alpha_0, \alpha_1 \in \mathbb{R}, \alpha_1 \neq 0$. Let $(x, y) \in \mathbb{R}^2$. We get

$$p(x + yi) = \alpha_0 + \alpha_1(x + yi) = (\alpha_0 + \alpha_1 x) + \alpha_1 yi, \quad \text{for} \quad v = 1$$

so for $p_1(x, y) = \alpha_0 + \alpha_1 x$ and $p_2(x, y) = \alpha_1 y$, the result holds.

For $v = 2$.

Let $p(z) = \alpha_0 + \alpha_1 z + \alpha_2 z^2$, where $\alpha_0, \alpha_1, \alpha_2 \in \mathbb{R}, \alpha_2 \neq 0$.

Let $z = x + yi$, $(x, y) \in \mathbb{R}^2$. We get

$$\begin{aligned} p(z) = p(x + yi) &= \alpha_0 + \alpha_1(x + yi) + \alpha_2(x + yi)^2 \\ &= (\alpha_0 + \alpha_1 + \alpha_2 x^2 - \alpha_2 y^2) + (\alpha_1 y + 2\alpha_2 xy)i, \end{aligned}$$

so for $p_1(x, y) = \alpha_0 + \alpha_1 x + \alpha_2 x^2 - \alpha_2 y^2$ and $p_2(x, y) = \alpha_1 y + 2\alpha xy$, the result holds. We suppose now that the result holds for any $1 \leq i \leq k_0 \in \mathbb{N}$. We can prove that the result holds for $k_0 + 1$.

Let

$$p(z) = \alpha_0 + \alpha_1 z + \cdots + \alpha_{n_0} z^{k_0} + \alpha_{k_0+1} z^{k_0+1}$$

be a polynomial with $\alpha_{k_0+1} \neq 0$, $\alpha_i \in \mathbb{R}$, for every $i = 0, 1, \ldots, k_0 + 1$.

Let $(x, y) \in \mathbb{R}^2$. We have

$$p(z) = q(z) + \alpha_{k_0+1} z^{k_0+1},$$

and we distinguish two cases:

(a) $q(z) \not\equiv 0$. Then, through the induction step we can show that there exist two polynomials $p_1(x, y)$, $p_2(x, y)$ of two real variables x and y with real coefficients, so that:

$$q(x + y_i) = p_1(x, y) + p_2(x, y)i \ \text{ forevery } \ (x, y) \in \mathbb{R}^2. \tag{1}$$

We get

$$\begin{aligned} \alpha_{k_0+1} z^{k_0+1} &= \alpha_{k_0+1}(x + yi)^{k_0+1} = \alpha_{k_0+1} \sum_{j=0}^{k_0+1} \binom{k_0 + 1}{j} x^j \cdot (yi)^{k_0+1-j} \\ &= \alpha_{k_0+1} \sum_{j=0}^{k_0+1} \binom{k_0 + 1}{j} x^j y^{k_0+1-j} i^{k_0+1-j} \end{aligned}$$

$$= \sum_{\substack{k_0+1-j=2\rho \\ \rho\in\mathbb{N} \\ \alpha<j\le k_0+1}} \alpha_{k_0+1}\binom{k_0+1}{j} x^j y^{k_0+1-j}(-1)^{(n_0+1-j)/2}$$

$$+ \sum_{\substack{k_0+1-j=2\rho+1 \\ \rho\in\mathbb{N} \\ 0\le j\le k_0+1}} \alpha_{k_0+1} x^j y^{k_0+1-j} i^{k_0+1-j}$$

$$= q_1(x,y) + i q_2(x,y), \tag{2}$$

where

$$q_1(x,y) = \sum_{\substack{k_0+1-j=2\rho \\ \rho\in\mathbb{N} \\ 0\le j\le k_0+1}} \alpha_{k_0+1}\binom{k_0+1}{j} x^j y^{k_0+1-j}(-1)^{(n_0+1-j)/2} \text{ and}$$

$$i q_2(x,y) = \sum_{\substack{k_0+1-j=2\rho+1 \\ \rho\in\mathbb{N} \\ 0\le j\le k_0+1}} \alpha_{k_0+1} x^j y^{k_0+1-j} i^{k_0+1-j},$$

where $q_1(x, y)$, $q_2(x, y)$ are two polynomials of the two real variables with real coefficients because $i^{2v+1} = i$ or $-i$, $v \in \mathbb{N}$.

So, we get $\alpha_{k_0+1}z^{k_0+1} = q_1(x, y) + iq_2(x, y)$. So, we get by (1) and (2),

$$p(z) = q(z) + \alpha_{k_0+1}z^{k_0+1} = (p_1(x, y)) + p_2(x, y)i) + (q_1(x, y) + q_2(x, y)i)$$
$$= (p_1(x)y) + q_1(x, y)) + (p_2(x, y) + q_2(x, y))i$$

and the result also holds for every $(x, y) \in \mathbb{R}^2$.

(b) $q(z) \equiv 0$. Then, with the above equality (2) we get $p(z) = \alpha_{k_0+1}z^{k_0+1} = q_1(x, y) + iq_2(x, y)$ for every $(x, y) \in \mathbb{R}^2$ and the result also holds. So, by induction we see that the result holds in this case.

Now, we suppose that $\alpha_i \in \mathbb{C}$ for every $i = 0, 1, \ldots, v$.

Let $\alpha_j = \beta_j + \gamma_j i$ for every $j = 0, 1, \ldots, v$, where $\beta_j, \gamma_j \in \mathbb{R}$ for every $j = 0, 1, \ldots, v$. Let $z = x + yi \in \mathbb{C}$, $(x, y) \in \mathbb{R}^2$. We get

$$p(z) = p(x + yi) = \alpha_0 + \alpha_1 z + \cdots + \alpha_{v-1}z^{v-1} + \alpha_v z^v$$
$$= (\beta_0 + \gamma_0 i) + (\beta_1 + \gamma_1 i)z + \cdots + (\beta_{v-1} + \gamma_{v-1}i)z^{v-1} + (\beta_v + \gamma_v i)z^v$$
$$= (\beta_0 + \beta_1 z + \cdots + \beta_{v-1}z^{v-1} + \beta_v z^v)$$

$$+ (\gamma_0 + \gamma_1 z + \cdots + \gamma_{v-1} z^{v-1} + \gamma_v z^v)i. \tag{3}$$

In the previous case (i), we see that there exist polynomials $p_1(x, y)$, $p_2(x, y)$, $q_1(x, y)$, $q_2(x, y)$ of the two real variables x and y with real coefficients, so that:

$$\beta_0 + \beta_1 z + \cdots + \beta_{v-1} z^{v-1} + \beta_v z^v = p_1(x, y) + p_2(x, y)i \tag{4}$$

and

$$\gamma_0 + \gamma_1 z + \cdots + \gamma_{v-1} z^{v-1} + \gamma_v z^v = q_1(x, y) + q_2(x, y)i \tag{5}$$

for every $(x, y) \in \mathbb{R}^2$.

By (3), (4), and (5), we get

$$\begin{aligned} p(z) &= (p_1(x, y) + p_2(x, y)i) + (q_1(x, y) + q_2(x, y)i)i \\ &= (p_1(x, y) - q_2(x, y)) + (p_2(x, y) + q_1(x, y))i \end{aligned}$$

and the result also holds. ■

With the help of this lemma, we can now solve the equation $p(z) = 0$ as follows. We examine the general case where

$$p(z) = \alpha_0 + \alpha_1 z + \cdots + \alpha_{v-1} z^{v-1} + \alpha_v z^v, \quad v \in \mathbb{N}, \quad \alpha_v \neq 0, \quad \alpha_i \in \mathbb{C},$$

for every $i = 0, 1, \ldots, v$.

With the help of the above lemma, we write

$$p(x + yi) = q_1(x, y) + q_2(x, y)i, \tag{$*$}$$

for every $(x, y) \in \mathbb{R}^2$, where $q_1(x, y), q_2(x, y)$ are two polynomials of two real variables x and y with real coefficients.

Let A be the set of roots of $p(z)$. We consider the system:

$$\begin{cases} q_1(x, y) = 0 \\ q_2(x, y) = 0 \end{cases}. \tag{B}$$

It is obvious from the above equality $(*)$ that $A = L_B$. So, in order to find all the roots of A, it suffices to find all the real roots of system (B).

So, we solve system B with the method we have developed in the second stage, and thus we find all the roots of polynomial $p(z)$.

Our method has been completed now because our supposition (S) (that system (B) has a solution) is satisfied because the same holds for (A). So, in all the cases we can reduce our initial system to a system in which the two polynomials have a lower degree than that of the polynomials of the initial system. Thus, we apply the induction step and the system is solved inductively.

4 Prerequisites

(a) Prerequisites from Algebra.

We use some basic tools and results from the theory of polynomials.

We denote $\mathbb{C}[z]$ as the set of complex polynomials. We denote $\mathbb{R}[x]$ as the set of real polynomials, which is the set of polynomials of one real variable with coefficients in the set of real numbers $\mathbb{R}$.

We begin with the following basic result, which is a simple implication of the algorithm of Euclidean division.

Proposition 4.1 *Let $p(z) \in \mathbb{C}[z]$, $degp(z) \geq 1$. The number $r \in \mathbb{C}$ is a root of $p(z)$ if and only if there exists a unique polynomial $q(z) \in \mathbb{C}[z]$, so that:*

$$p(z) = (z - r)q(z). \quad \blacksquare$$

We need the definition of multiplicity of a root of a polynomial.

Definition 4.2 Let $p(z) \in \mathbb{C}[z]$. Let $\rho \in \mathbb{C}$ be a root of $p(z)$. The natural number m is a multiplicity of the root ρ of $p(z)$ if polynomial $(z - \rho)^m$ divides $p(z)$, whereas polynomial $(z - \rho)^{m+1}$ does not divide $p(z)$.

As consequence of Proposition 4.1, there is the following proposition.

Proposition 4.3 *Every root of a polynomial $p(z) \in \mathbb{C}[z]$ has a multiplicity, which is unique.* $\blacksquare$

We state now the fundamental theorem of algebra, whose proof is not simple and needs some tools from analysis.

Theorem 4.4 *Every complex polynomial $p(z)$, with $degp(z) \geq 1$, has at least one root.* $\blacksquare$

From Theorem 4.4 and Proposition 4.1, we get the following fundamental result.

Theorem 4.5 *Let $p(z) \in \mathbb{C}[z]$ be a complex polynomial with $degp(z) \geq 1$. Then $p(z)$ has a finite number of different roots.*

Let $\rho_1, \rho_2, \ldots, \rho_\upsilon$ be the different roots of $p(z)$ with respect to multiplicities $m_1, m_2, \ldots, m_\upsilon$. Then, the following formula holds:

$$p(z) = \alpha \cdot (z - \rho_1)^{m_1}(z - \rho_2)^{m_2} \cdots (z - \rho_\upsilon)^{m_\upsilon},$$

where $\alpha \neq 0$ and α is the coefficient of the monomial of greater grade $m_0 = degp(z)$, and $m_0 = m_1 + m_2 + \cdots + m_\upsilon$. $\blacksquare$

Now, we describe a simple algorithm in order to find the multiplicity of a root of a complex polynomial.

4.1 An Algorithm for the Multiplicity of a Root

Let $p(z) \in \mathbb{C}[z]$ be a complex polynomial of degree $degp(z) \geq 1$.

By Theorem 4.5, polynomial $p(z)$ has a finite number of roots. Let ρ be a root of $p(z)$. We describe with details a way in order to find the multiplicity of ρ.

By Proposition 4.1, there exists a unique polynomial $q(z)$, so that:

$$p(z) = (z - \rho)q(z). \tag{1}$$

We find the polynomial $q(z)$ through the algorithm of Euclidean division, for example, using Horner's scheme.

Afterwards, we compute the number $q(\rho)$, for example, with Horner's scheme. If $q(\rho) \neq 0$, then the root ρ has multiplicity 1. In order to prove this, we suppose that the root ρ does not have multiplicity 1. By Proposition 4.3, the root ρ has a unique multiplicity, $m \in \mathbb{N}$ (see Definition 4.2). Because of $m \neq 1$, we have that $m \geq 2$. By the definition of multiplicity, we have that polynomial $(z - \rho)^m$ divides $p(z)$. This means (by the definition of division) that there exists a polynomial $R(z) \in \mathbb{C}[z]$, such that:

$$p(z) = (z - \rho)^m R(z). \tag{2}$$

By relations (1) and (2), we get

$$(z - \rho)q(z) = (z - \rho)^m R(z) \Leftrightarrow (z - \rho)(q(z) - (z - \rho)^{m-1} R(z) = 0. \tag{3}$$

The expressions $z - \rho$ and $q(z) - (z - \rho)^{m-1} R(z)$ are polynomials in $\mathbb{C}[z]$ of course, because $m \geq 2$ (as we have seen). Because of $z - \rho \not\equiv 0$, we take it that

$$q(z) - (z - \rho)^{m-1} R(z) = 0, \tag{4}$$

because the Ring of polynomials $\mathbb{C}[z]$ is an integer neighborhood, as is well known from Algebra. Relation (4) gives $q(\rho) = 0$ (because $m \geq 2$), which is false because we have supposed that $q(\rho) \neq 0$. So, if $q(\rho) \neq 0$, then root ρ has multiplicity 1.

Whereas if $q(\rho) = 0$, then through Proposition 4.1, we take it that there exists a polynomial $q_1(z) \in \mathbb{C}[z]$, so that:

$$q(z) = (z - \rho)q_1(z). \tag{5}$$

By (1) and (5), we take that

$$p(z) = (z - \rho)^2 q_1(z). \tag{6}$$

Relation (6) tells us that polynomial $(z - \rho)^2$ divides $p(z)$. Afterwards, we find polynomial $q_1(z)$ by (5) with the Euclidean Algorithm, for example, from Horner's scheme, because we have found polynomial $q(z)$ previously. After that, we compute number $q_1(\rho)$, for example, with Horner's scheme. If $q_1(\rho) \neq 0$, then the multiplicity of ρ is 2, with a proof similar to what we had found previously. Or otherwise if $q_1(\rho) = 0$, then again through Proposition 4.1 there exists a polynomial $q_2(z) \in \mathbb{C}[z]$, so that:

$$q_1(z) = (z - \rho)q_2(z). \tag{7}$$

Through (6) and (7), we take it that

$$p(z) = (z - \rho)^3 q_2(z). \tag{8}$$

We inductively continue this procedure of finding a sequence of polynomials

$$q_j(z) \in \mathbb{C}[z],$$

for $j = 1, 2, \ldots$, where $q_j(z) = (z - \rho)q_{j+1}(z)$, for $j = 1, 2, \ldots$, and

$$p(z) = (z - \rho)^{j+1} q_j(z).$$

If $p(z) = (z - \rho)^{j+1} q_j(z)$ for $j \in \mathbb{N}$ (where $q_j(\rho) \neq 0$), then the multiplicity of ρ is $j + 1$, with a proof similar to what we have shown previously. This procedure stops if some natural number $j \in \mathbb{N}$, or if $degp(z) = v_0 \in \mathbb{N}$, then we take it that $p(z) = (z - \rho)^{v_0+1} q_{v_0}(z)$, where $q_{v_0}(z) \neq 0$ (or else $p(z) = 0$, which is false because $degp(r) \geq 1$, by supposition), so $deg((z - \rho)^{v_0+1} q_{v_0}(z)) \geq v_0 + 1$, which is false of course because $degp(z) = v_0$.

That is, we take it that $p(z) = (z - \rho)^{j+1} q_j(z)$ for some $j \in \mathbb{N}$, $j < v_0 - 1$, and $q_j(\rho) \neq 0$, which gives that the multiplicity of ρ is $j + 1 < v_0$; otherwise, we take it that

$$p(z) = (z - \rho)^{v_0} q_{v_0 - 1}(z). \tag{9}$$

Relation (9) gives that $q_{v_0-1}(z) \neq 0$ (or else $p(z) = 0$, which is false of course), and by relation (9), we take it also that $q_{v_0-1}(z)$ is a constant polynomial with value, say c_0. That is, $p(z) = (z - \rho)^{v_0} c_0$. Of course, polynomial $(z - \rho)^{v_0+1}$ cannot divide $p(z)$, because this polynomial has a degree $deg((z - \rho)^{v_0+1}) > v_0 = degp(z)$, which gives that multiplicity of ρ is v_0 (by the definition of multiplicity). So we have described a complete algorithm that gives us the multiplicity of a root of a complex polynomial. ■

Remark 4.7 We can combine Proposition 4.1 with Theorem 4.4 and the previous algorithm (and of course Proposition 4.3), in order to prove Theorem 4.5. We leave it as an easy exercise for the reader. So far we have developed all we need from polynomials of one complex variable. We also obtained some basic results from Linear Algebra. Here we will now consider the following linear system of two equations:

$$\begin{cases} \alpha_1 x + \beta_1 y = \gamma_1 \ (1) \\ \alpha_2 x + \beta_2 y = \gamma_2 \ (2) \end{cases}, \qquad \text{(A)}$$

where $\alpha_i, \beta_i, \gamma_i \in \mathbb{C}$ for $i = 1, 2$. We consider the determinants D, D_x, D_y where

$$D = \begin{vmatrix} \alpha_1 & \beta_1 \\ \alpha_2 & \beta_2 \end{vmatrix} = \alpha_1\beta_2 - \alpha_2\beta_1,$$

$$D_x = \begin{vmatrix} \gamma_1 & \beta_1 \\ \gamma_2 & \beta_2 \end{vmatrix} = \gamma_1\beta_2 - \beta_1\gamma_2, \; D_y = \begin{vmatrix} \alpha_1 & \gamma_1 \\ \alpha_2 & \gamma_2 \end{vmatrix} = \alpha_1\gamma_2 - \alpha_2\gamma_1.$$

When $D \neq 0$, then system (A) has only one solution (x_0, y_0), where $x_0 = \frac{D_x}{D}$, $y_0 = \frac{D_y}{D}$. When $D = 0$ and $D_x \neq 0$, or $D_y \neq 0$, then system (A) does not have any solution, whereas when $D = D_x = D_y = 0$, then system (A) has an infinite number of solutions except only in the case where $\alpha_1 = \alpha_2 = \beta_1 = \beta_2 = 0$ and only one of the numbers γ_1, γ_2 is nonzero. We need the case where $D \neq 0$ and the case where $D = D_x = D_y = 0$. We consider the case where $D = D_x = D_y = 0$. We suppose that system (A) is a pure system of two variables x and y; that is, we suppose that at least one of the numbers α_1, α_2 is also nonzero. That is, $\alpha_1 \neq 0$ or $\alpha_2 \neq 0$ and $\beta_1 \neq 0$ or $\beta_2 \neq 0$; otherwise, we do not have a system of equations of two different variables.

We have two cases:

(i) One from the six numbers $\alpha_i, \beta_i, \gamma_i, i = 1, 2$ is zero.

Let $\alpha_1 = 0$ (3). We have $D = 0$; that is, $\alpha_1\beta_2 - \alpha_2\beta_1 = 0 \Rightarrow \alpha_1\beta_2 = \alpha_2\beta_1$ (4). Through (3) and (4), we have $\alpha_2\beta_1 = 0$ (5). Because of $\alpha_1 = 0$ and our hypothesis, we have $\alpha \neq 0$ (6). By (5) and (6), we get $\beta_1 = 0$ (7).

Through (3) and the fact that $D_y = 0$, we get in a similar way that $\gamma_1 = 0$. That is, Eq. (1) is the equation $0 \cdot x + 0 \cdot y = 0$, with set of solutions in the set $\mathbb{R}^2$. This means that system (A) is equivalent to Eq. (2) of (A) only. If $\beta_1 = 0$, or $\gamma_1 = 0$, we get in a similar way that $\alpha_1 = \beta_1 = \gamma_1 = 0$ and we have similarly the same implication; that is, system (A) is equivalent to Eq. (2) of (A) only. If $\alpha_2 = 0$, or $\beta_2 = 0$, or $\gamma_2 = 0$, we take it that $\alpha_2 = \beta_2 = \gamma_2 = 0$ in an analogous way and finally system (A) is equivalent to Eq. (1) of (A) only.

Now, we suppose that $\alpha_1\beta_1\gamma_1\alpha_2\beta_2\gamma_2 \neq 0$; that is, none of the six numbers $\alpha_i, \beta_i, \gamma_i$, $i = 1, 2$, is zero. We have $D = 0 \Leftrightarrow \alpha_1\beta_2 - \alpha_2\beta_1 = 0 \Leftrightarrow \frac{\alpha_2}{\alpha_1} = \frac{\beta_2}{\beta_1}$ $(\alpha_1 \neq 0, \beta_1 \neq 0)$. We set $\lambda = \frac{\alpha_2}{\alpha_1} = \frac{\beta_2}{\beta_1}$ (8).

We have $D_x = 0 \Leftrightarrow \gamma_1\beta_2 - \beta_1\gamma_2 = 0 \Leftrightarrow \lambda = \frac{\beta_2}{\beta_1} = \frac{\gamma_2}{\gamma_1}$ (9). With (8) and (9), we get $\alpha_2 = \lambda\alpha_1$, $\beta_2 = \lambda\beta_1$, $\lambda_2 = \lambda\gamma_1$; that is, $\alpha_2 x + \beta_2 y = \gamma_2 \Leftrightarrow (\lambda\alpha_1)x + (\lambda\beta_1)y = (\lambda\gamma_1) \Leftrightarrow \lambda \cdot (\alpha_1 x + \beta_1 y) = \lambda\gamma_1 \overset{\lambda\neq 0}{\Leftrightarrow} \alpha_1 x + \beta_y = \gamma_1$; that is, Eqs. (1) and (2) of (A) are equivalent, which means that they have the same set of solutions, meaning that system (A) is equivalent to one only from Eqs. (1) and (2), whichever of the two.

So, we have proved that in the case of $D = D_x = D_y = 0$, system (A) has an infinite number of solutions and it is equivalent to only one from Eqs. (1) and (2). So we have stated our prerequisites from Algebra.

(b) Prerequisites from Analysis

As it is well known, by Galois theory, that there are no formulas that give the roots of an arbitrary polynomial as a function of its coefficients with radicals. So, for an arbitrary polynomial the only way to find its roots is to approximate them with a numerical method. Perhaps, the simplest numerical method for algebraic equations is the bisection method, which is presented in all classical books of Numerical Analysis.

It is a simple method, and here we have based in it in our problem. The bisection method has very weak suppositions, and it is convenient for secondary students also.

Let $\alpha, \beta \in \mathbb{R}$, $\alpha < \beta$, and $f : [\alpha, \beta] \rightarrow \mathbb{R}$ be a continuous function. We suppose that $f(\alpha) \cdot f(\beta) < 0$. Then, function f has a root, at least in the interval (α, β), and bisection method approximates a root of f in (α, β), as closely as we want with a specific minor error.

There are many different numerical methods that find the roots in a specific interval. We will not discuss this subject. This is a vast subject in Numerical Analysis. In this text, it is enough for us to find only one root in a specific interval and approximate it using bisection method.

The solution to all the real roots of a polynomial will be based on the following basic lemma.

Basic Lemma 4.8 *Let $v \in \mathbb{N}$, $v \geq 3$, $p(x) = \alpha_v x^v + \alpha_{v-1} x^{v-1} + \cdots + \alpha_1 x + \alpha_0$, be a polynomial $p(x) \in \mathbb{R}[x]$, with degree $degp(x) = v$.*

Let $\rho_1, \rho_2, \ldots, \rho_k$ be all the different real roots of polynomial $p'(x)$, $k \in \mathbb{N}$, $k \geq 2$, $\rho_i \neq \rho_j$, for all $i, j \in \{1, 2, \ldots, k\}$, $i \neq j$.

Then, we can find, with an algorithm, all the real roots of $p(x)$ with their multiplicities.

Proof Let $L = \{\rho_1, \rho_2, \ldots, \rho_k\}$ be the set of all real roots of $p'(x)$. We suppose, also, without loss of generality that $\rho_1 < \rho_2 < \cdots < \rho_k$.

Let $i_0 \in \{1, \ldots, k-1\}$. Then $p'(x) > 0$ for every $x \in (\rho_{i_0}, \rho_{i_0+1})$ or $p'(x) < 0$ for every $x \in (\rho_{i_0}, \rho_{i_0+1})$. This gives that p is a strictly decreasing or strictly increasing function on $[\rho_{i_0}, \rho_{i_0+1}]$. If $p(\rho_{i_0}) = 0$, then ρ_{i_0} is the unique root of p in $[\rho_{i_0}, \rho_{i_0+1}]$. The same holds if $p(\rho_{i_0+1}) = 0$; that is, ρ_{i_0+1} is the unique root of p in $[\rho_{i_0}, \rho_{i_0+1}]$, if $p(\rho_{i_0+1}) = 0$.

Of course polynomial p cannot have the numbers ρ_{i_0} and ρ_{i_0+1} as roots simultaneously, by its monotonicity. We suppose now that $p(\rho_{i_0}) \cdot p(\rho_{i_0+1}) \neq 0$. Then, if $p(\rho_{i_0}) \cdot p(\rho_{i_0+1}) > 0$, polynomial p does not have any root in $[\rho_{i_0}, \rho_{i_0+1}]$. If $p(\rho_{i_0}) \cdot p(\rho_{i_0+1}) < 0$, then p has one root exactly in the interval $[\rho_{i_0}, \rho_{i_0+1}]$, and more specifically this root belongs in $(\rho_{i_0}, \rho_{i_0+1})$.

Applying the bisection method, we find this root, because the suppositions of bisection method are satisfied now. We do the same in every interval $[\rho_i, \rho_{i+1}]$.

So we find all the roots of p in the interval $[\rho_1, \rho_k]$. We examine the roots in $[\rho_k, +\infty)$. Because $\alpha_v \neq 0$, we have two cases:

(i) If $\alpha_v > 0$, then $\lim_{x \to +\infty} p(x) = +\infty$.

Then p is a strictly increasing function in $[\rho_k, +\infty)$.

(a) $p(\rho_k) = 0$, then ρ_k is the unique root of p in $[\rho_k, +\infty)$.
(b) If $p(\rho_k) > 0$, then p does not have any root in $[\rho_k, +\infty)$.
(c) If $p(\rho_k) < 0$, then p has one root exactly (say ρ_{k+1}) in $[\rho_k, +\infty)$ and more specifically $\rho_{k+1} \in (\rho_k, +\infty)$.

Because $\lim_{x \to +\infty} p(x) = +\infty$, there exists some $x_0 \in \mathbb{R}$, $x_0 > \rho_k$, so that $p(x_0) > 0$. Then $p(\rho_k) \cdot p(x_0) < 0$ and thus $\rho_{k+1} \in (\rho_k, x_0)$.

Applying bisection method in $[\rho_{k+1}, x_0]$, we approximate the root ρ_{k+1}. Later, we will see how we compute a number like x_0, in order to apply bisection method.

(ii) If $\alpha_v < 0$, then $\lim_{x \to +\infty} p(x) = -\infty$. Polynomial p is a strictly decreasing function in $[\rho_k, +\infty)$.

(a) If $p(\rho_k) = 0$, then ρ_k is the unique root of p in $[\rho_k, +\infty)$.
(b) If $p(\rho_k) < 0$, then p does not have any root in $[\rho_k, +\infty)$.
(c) If $p(\rho_k) > 0$, then p has unique one root in $[\rho_k, +\infty)$ (say ρ_{k+1}) and more specifically $\rho_{k+1} \in (\rho_k, +\infty)$.

Because $\lim_{x \to +\infty} p(x) = -\infty$, there exists some $x_0 \in (\rho_k, +\infty)$, so that $p(x_0) < 0$.

Then, $p(\rho_k) \cdot p(x_0) < 0$, and $\rho_{k+1} \in (\rho_k, x_0)$, and applying bisection method, we approximate the unique root ρ_{k+1} in (ρ_k, x_0). Now we examine the roots in $(-\infty, \rho_1]$. Whether $p(\rho_1) = 0$, then ρ_1 is the unique root of p in $(-\infty, \rho_1]$.

Now we suppose that $p(\rho_1) \neq 0$. We examine two cases:

(i) $\lim_{x \to -\infty} p(x) = +\infty$.

This happens when v is even and $\alpha_v > 0$, or v is odd and $\alpha_v < 0$. Then p is a strictly decreasing function in $(-\infty, \rho_1]$.

(i), (1) If $p(\rho_1) > 0$, then p does not have any root in $(-\infty, \rho_1]$.

(i), (2) If $p(\rho_1) < 0$, then p has a unique root in $(-\infty, \rho_1]$ (say ρ_{k+2}) and more specifically $\rho_{k+2} \in (-\infty, \rho_1)$. Because $\lim_{x\to-\infty} p(x) = +\infty$, there exists some

$x_0 < \rho_1$, so that $p(x_0) > 0$. Then $\rho_{k+2} \in (x_0, \rho_1)$ and applying bisection method in $[x_0, \rho_1]$, we approximate root ρ_{k+2}.

(ii) $\lim_{x\to-\infty} p(x) = -\infty$. This is happened when v is even and $\alpha_v < 0$, or v is odd and $\alpha_v > 0$. Then p is a strictly increasing function in $(-\infty, \rho_1]$.

We have two cases:

(ii), (1)$p(\rho_1) < 0$. Then, p does not have any root in $(-\infty, \rho_1]$.

(ii), (2) $p(\rho_1) > 0$. Then p has unique root in $(-\infty, \rho_1]$ (say ρ_{k+2}) and more specifically $\rho_{k+2} \in (-\infty, \rho_1)$. Because $\lim_{x\to-\infty} p(x) = -\infty$, there exists some $x_0 < \rho_1$, such that $p(x_0) < 0$. Then $p(x_0) \cdot p(\rho_1) < 0$ and $\rho_{k+2} \in (x_0, \rho_1)$.

Applying bisection method in $[x_0, \rho_1]$, we approximate root ρ_{k+2}. All the implications of this lemma are easy to prove and are left as an easy exercise for the interested reader. The proofs are of secondary school.

Corollary 4.9 *Basic Lemma 4.8 holds again, in the case when polynomial p' has only one root.*

Proof The proof is similar to that of basic lemma for the intervals $(-\infty, \rho_1]$ and $[\rho_1, +\infty)$, where $p'(\rho_1) = 0$. ■

Corollary 4.10 *Let $v \in \mathbb{N}$, $v \geq 3$, $p(x) = \alpha_v x^v + \alpha_{v-1}x^{v-1} + \cdots + \alpha_1 x + \alpha_0$, be a polynomial $p(x) \in \mathbb{R}[x]$, with degree $degp(x) = v$.*

We suppose that p' does not have any root. Then p has unique real root and we can construct an algorithm in order to find it.

Proof Of course p' is a polynomial of even degree $degp' = v - 1$, so p is a polynomial of odd degree. Thus p has, at least, one real root. Because p' does not have any root, we have $p'(x) \neq 0$, for every $x \in \mathbb{R}$. Thus, $p'(x) > 0$ for every $x \in \mathbb{R}$, or $p'(x) < 0$ for every $x \in \mathbb{R}$, or else if there exist $\alpha, \beta \in \mathbb{R}$, so that $p'(\alpha) < 0$ and $p'(\beta) > 0$ (of course $\alpha \neq \beta$), then because p' is a continuous function (as a polynomial) and $p'(\alpha) \cdot p'(\beta) < 0$, we take it that there exists $\gamma \in (\alpha, \beta)$ (if $\alpha < \beta$) or $\gamma \in (\beta, \alpha)$ (if $\beta < \alpha$), so that $p'(\gamma) = 0$; that is, a contradiction because $p'(x) \neq 0$ for every $x \in \mathbb{R}$. Thus, p is a strictly increasing function in $\mathbb{R}$, if $p'(x) > 0$, for every $x \in \mathbb{R}$, or else p is a strictly decreasing function in $\mathbb{R}$ if $p'(x) < 0$ for every $x \in \mathbb{R}$. If p is a strictly increasing function, then $\lim_{x\to+\infty} p(x) = +\infty$ and $\lim_{x\to-\infty} p(x) = -\infty$, or else if p is a strictly decreasing function in $\mathbb{R}$, then $\lim_{x\to+\infty} p(x) = -\infty$ and $\lim_{x\to-\infty} p(x) = +\infty$.

Polynomial p is a strictly increasing function if $\alpha_v > 0$, or else if $\alpha_v < 0$, then p is a strictly decreasing function.

If $\alpha_v > 0$, then because $\lim\limits_{x \to +\infty} p(x) = +\infty$, there exists $y_0 \in \mathbb{R}$, so that $p(y_0) > 0$, and because $\lim\limits_{x \to -\infty} = -\infty$, there exists $x_0 \in \mathbb{R}$, $x_0 < y_0$, so that $p(x_0) < 0$. So $p(x_0) \cdot p(y_0) < 0$ and p has unique root in $\mathbb{R}$ (say ρ), so that $\rho \in (x_0, y_0)$.

If $\alpha_v < 0$, then because $\lim\limits_{x \to +\infty} p(x) = -\infty$, there exists $y_0 \in \mathbb{R}$, so that $p(y_0) < 0$. Because $\lim\limits_{x \to -\infty} p(x) = +\infty$, there exists $x_0 \in \mathbb{R}$, $x_0 < y_0$, so that $p(x_0) > 0$. Thus $p(x_0) \cdot p(y_0) < 0$, and p has unique root in $\mathbb{R}$ (say ρ), so that $\rho \in (x_0, y_0)$.

We will see later how we compute numbers x_0, y_0 as above.

In any of the cases above we apply the bisection method in the interval $[x_0, y_0]$, to find the unique real root of p. ■

Remark 4.11 The multiplicity of a root is found with the algebraic algorithm 3.6.

However, we can find the multiplicity of a root in an analytic way.

More specifically,

Let $p(x) \in \mathbb{C}[x]$ be a polynomial and ρ be a root of p, where $degp(x) = v \in \mathbb{N}$.

Then, there exists a unique natural number $k \in \mathbb{N} \cup \{0\}$ $k \leq v - 1$, so that: $p(\rho) = 0$, $p'(\rho) = 0, \ldots, p^{(k)}(\rho) = 0$ and $p^{(k+1)}(\rho) \neq 0$, that is $p^{(i)}(\rho) = 0$, for all $i = 0, 1, \ldots, k$ and $p^{(k+1)}(\rho) = 0$, where $p^{(0)}(\rho) = p(\rho)$.

The natural number $k + 1$ is the multiplicity of root ρ of p. (Of course we have always $p^{(v)}(\rho) \neq 0$.)

This is a classical result in calculus, which is proven easily.

Now, we cover the gap from basic Lemma 4.8, computing a number like x_0 in this lemma.

Remark 4.12 Let $p(x) \in \mathbb{R}[x]$ be a real polynomial:

$$p(x) = \alpha_0 + \alpha_1 x + \cdots + \alpha_{v-1} x^{v-1} + \alpha_v x^v, \quad v = degp(x), \quad v \geq 3.$$

We suppose that $\alpha_v > 0$ and that $p'(x)$ has real roots.

Proof Let ρ be the greater real root of $p'(x)$. We suppose that $p(\rho) < 0$. We consider an arbitrary real number x_0, so that $x_0 > \rho$, $x_0 > 1$ and $x_0 > \dfrac{|\alpha_0| + |\alpha_1| + \cdots + |\alpha_{v-1}|}{\alpha_v}$. We prove that $p(x_0) > 0$.

We have of course $-|y| \leq y$ for every $y \in \mathbb{R}$ (1). We apply (1) for

$$y = \frac{\alpha_0}{x^v} + \frac{\alpha_1}{x^{v-1}} + \cdots + \frac{\alpha_{v-1}}{x} \tag{1}$$

for some $x \in \mathbb{R} - \{0\}$ and we have

$$-\left|\frac{\alpha_0}{x^v} + \frac{\alpha_1}{x^{v-1}} + \cdots + \frac{\alpha_{v-1}}{x}\right| \leq \frac{\alpha_0}{x^v} + \frac{\alpha_1}{x^{v-1}} + \cdots + \frac{\alpha_{v-1}}{x}. \tag{2}$$

Adding the number α_v in two members of (2), we get

$$\alpha_v - \left|\frac{\alpha_0}{x^v} + \frac{\alpha_1}{x^{v-1}} + \cdots + \frac{\alpha_{v-1}}{x}\right| \leq \alpha_v + \frac{\alpha_0}{x^v} + \frac{\alpha_1}{x^{v-1}} + \cdots + \frac{a_{v-1}}{x}. \tag{3}$$

By the triangle inequality, we take for $x > 0$

$$\left|\frac{\alpha_0}{x^v} + \frac{\alpha_1}{x^{v-1}} + \cdots + \frac{\alpha_{v-1}}{x}\right| \leq \left|\frac{\alpha_0}{x^v}\right| + \left|\frac{\alpha_1}{x^{v-1}}\right| + \cdots + \left|\frac{\alpha_{v-1}}{x}\right|$$
$$\Leftrightarrow -\left(\frac{|\alpha_0|}{x^v} + \frac{|\alpha_v|}{x^{v-1}} + \cdots + \frac{|\alpha_{v-1}|}{x}\right)$$
$$\leq -\left|\frac{\alpha_0}{x^v} + \frac{\alpha_1}{x^{v-1}} + \cdots + \frac{\alpha_{v-1}}{x}\right|. \tag{4}$$

Adding the number α_v in two members of (4), we get

$$\alpha_v - \left(\frac{|\alpha_0|}{x^v} + \frac{|\alpha_1|}{x^{v-1}} + \cdots + \frac{|\alpha_{v-1}|}{x}\right) \leq \alpha_v - \left|\frac{\alpha_0}{x^v} + \frac{\alpha_1}{x^{v-1}} + \cdots + \frac{\alpha_{v-1}}{x}\right|,$$
$$\text{for } x > 0. \tag{5}$$

Let some $x > 1$. Then we have

$$x \geq x, x^2 \geq x, \ldots, x^v \geq x \Rightarrow \frac{1}{x} \leq \frac{1}{x}, \frac{1}{x^2} < \frac{1}{x}, \ldots, \frac{1}{x^v} < \frac{1}{x}$$
$$\Rightarrow \frac{|\alpha_{v-1}|}{x} \leq \frac{|\alpha_{v-1}|}{x}, \frac{|\alpha_{v-2}|}{x^2} \leq, \frac{|\alpha_{v-2}|}{x}, \ldots, \frac{|\alpha_1|}{x^{v-1}} \leq \frac{|\alpha_1|}{x}, \frac{|\alpha_0|}{x^v} \leq \frac{|\alpha_0|}{x}.$$

Adding the above inequalities in pairs, we get

$$\frac{|\alpha_{v-1}|}{x} + \cdots + \frac{|\alpha_1|}{x^{v-1}} + \frac{|\alpha_0|}{x^v} \leq \frac{|\alpha_0| + |\alpha_1| + \cdots + |\alpha_{v-1}|}{x} \Rightarrow$$
$$-\frac{|\alpha_0| + |\alpha_1| + \cdots + |\alpha_{v-1}|}{x} \leq -\left(\frac{|\alpha_{v-1}|}{x} + \cdots + \frac{|\alpha_1|}{x^{v-1}} + \frac{|\alpha_0|}{x^v}\right) \Rightarrow$$
$$\alpha_v - \frac{|\alpha_0| + |\alpha_1| + \cdots + |\alpha_{v-1}|}{x} \leq \alpha_v - \left(\frac{|\alpha_{v-1}|}{x} + \cdots + \frac{|\alpha_1|}{x^{v-1}} + \frac{|\alpha_0|}{x^v}\right). \tag{6}$$

Through inequalities (3), (5), and (6), we get

$$\alpha_v - \frac{|\alpha_0| + |\alpha_1| + \cdots + |\alpha_{v-1}|}{x} \leq \alpha_v + \frac{\alpha_0}{x^v} + \frac{\alpha_1}{x^{v-1}} + \cdots + \frac{\alpha_{v-1}}{x} \text{ for } x > 1. \tag{7}$$

Now, for every $x > 1$, $x > \dfrac{|\alpha_0| + |\alpha_1| + \cdots + |\alpha_{v-1}|}{\alpha_v}$, we take

$$\alpha_v > \frac{|\alpha_0| + |\alpha_1| + \cdots + |\alpha_{v-1}|}{x} \Rightarrow \alpha_v - \frac{|\alpha_0| + |\alpha_1| + \cdots + |\alpha_{v-1}|}{x} > 0. \quad (8)$$

Through (7) and (8), we take it that for every $x > 1, x > \dfrac{|\alpha_0| + |\alpha_1| + \cdots + |\alpha_{v-1}|}{\alpha_v}$ we get

$$\alpha_v + \frac{\alpha_{v-1}}{x} + \cdots + \frac{\alpha_1}{x^{v-1}} + \frac{\alpha_0}{x^v} > 0. \quad (*)$$

This gives

$$x^v \cdot \left(\alpha_v + \frac{\alpha_v - 1}{x} + \cdots + \frac{\alpha_1}{x^{v-1}} + \frac{\alpha_0}{x^v}\right) > 0$$

$$\Leftrightarrow p(x) > 0 \text{ (by the definition of } p(x)). \quad (9)$$

We apply (9) for the number $x_0 \in \mathbb{R}$, so that $x_0 > \rho$, $x_0 > 1$ and $x_0 > \dfrac{|\alpha_0| + |\alpha_1| + \cdots + |\alpha_{v-1}|}{\alpha_v}$ and we take it that $p(x_0) > 0$.

Thus, we have $p(\rho) \cdot p(x_0) < 0$. This means that the unique real root of $p(x)$ in $[\rho, +\infty)$ belongs in (ρ, c_0). Applying the bisection method in the interval $[\rho, x_0]$, we compute the unique real root x_0^* of $p(x)$ in $[\rho, +\infty)$; that is, $x_0^* \in (\rho, x_0)$. Of course if $p(\rho) > 0$, polynomial p does not have any real root in $[\rho, +\infty)$ as we have seen in basic Lemma 4.8, and if $p(\rho) = 0$, then ρ is the unique real root of p in $[\rho, +\infty)$.

Now, we suppose that $\alpha_v < 0$ and that $p'(x)$ has real roots.

Let ρ be the greatest real root of $p'(x)$. We suppose that $p(\rho) > 0$. We consider an arbitrary real number x_0, so that $x_0 > \rho$, $x_0 > 1$, and

$$x_0 > \frac{|\alpha_0| + |\alpha_1| + \cdots + |\alpha_{v-1}|}{-\alpha_v} = \frac{|\alpha_0| + |\alpha_1| + \cdots + |\alpha_{v-1}|}{|\alpha_v|}. \quad (10)$$

By (10), we get (because $|\alpha_v| > 0$ and $x > 0$)

$$|\alpha_v| > \frac{|\alpha_0| + |\alpha_1| + \cdots + |\alpha_{v-1}|}{x_0} \Rightarrow \alpha_v + \frac{|\alpha_0| + |\alpha_1| + \cdots + |\alpha_{v-1}|}{x_0} < 0. \quad (11)$$

Let $x > 1$. Because of $x > 1$, we get

$$x \geq x, x^2 \geq x, \ldots, x^{v-1} \geq x, x^v \geq x \Rightarrow \frac{1}{x^v} \leq \frac{1}{x}, \frac{1}{x^{v-1}} \leq \frac{1}{x}, \ldots, \frac{1}{x} \leq \frac{1}{x} \Rightarrow$$

$$\frac{|\alpha_0|}{x^v} \leq \frac{|\alpha_0|}{x}, \frac{|\alpha_1|}{x^{v-1}} \leq \frac{|\alpha_1|}{x^{v-1}}, \ldots, \frac{|\alpha_{v-1}|}{x} \leq \frac{|\alpha_{v-1}|}{x}.$$

Adding by pairs the previous inequalities, we get

$$\frac{|\alpha_0|}{x^v}+\frac{|\alpha_1|}{x^{v-1}}+\cdots+\frac{|\alpha_{v-1}|}{x}\leq\frac{|\alpha_0|}{x}+\frac{|\alpha_1|}{x}+\cdots+\frac{\alpha_{v-1}|}{x}\Rightarrow$$

$$\alpha_v+\frac{|\alpha_0|}{x^v}+\frac{\alpha_v|}{x^{v-1}}+\cdots+\frac{|\alpha_0-1|}{x}\leq\alpha_v+\frac{|\alpha_0|+|\alpha_1|+\cdots+|\alpha_{v-1}|}{x}. \tag{12}$$

Of course we get

$$\alpha_0\leq|\alpha_0|,\alpha_1\leq|\alpha_1|,\ldots,\alpha_{v-1}\leq|\alpha_{v-1}|\overset{x>0}{\Longrightarrow}\frac{\alpha_0}{x^v}\leq\frac{|\alpha_0|}{x^v},\frac{\alpha_1}{x^{v-1}}$$
$$\leq\frac{|\alpha_1|}{x^{v-1}},\ldots,\frac{\alpha_{v-1}}{x}\leq\frac{|\alpha_{v-1}|}{x}$$

and adding by pairs the previous inequalities, we get

$$\frac{\alpha_0}{x^v}+\frac{\alpha_1}{x^{v-1}}\cdots+\frac{\alpha_{v-1}}{x}\leq\frac{|\alpha_0|}{x^v}+\frac{|\alpha_1|}{x^{v-1}}+\cdots+\frac{|\alpha_{v-1}|}{x}\Rightarrow$$

$$\alpha_v+\frac{\alpha_0}{x^v}+\frac{\alpha_1}{x^{v-1}}+\cdots+\frac{\alpha_{v-1}}{x}\leq\alpha_v+\frac{|\alpha_0|}{x^v}+\frac{|\alpha_1|}{x^{v-1}}+\cdots+\frac{|\alpha_{v-1}|}{x}. \tag{13}$$

Of course as we have seen in basic Lemma 4.8 if $p(\rho)<0$, then p does not have any root in $[\rho,+\infty)$, and if $p(\rho)=0$, then number ρ is the unique real root of p in $[\rho,+\infty)$. So far we have seen how we compute the unique real root of p (if any) in $[\rho,+\infty)$, when ρ is the greatest real root of p'.

In a similar way we compute the unique real root of p in $(-\infty,\rho^*]$ (if any), where ρ^* is the smallest real root of p'. It suffices to observe the following.

We simply write $p(x)=p(-(-x))$ and we find easily a polynomial $q\in\mathbb{R}[x]$, such that $p(x)=q(-x)$ (it is trivial to find such a polynomial q).

Now, for $x_0>1$, $x_0>\rho$, and $x_0>\dfrac{|\alpha_0|+|\alpha_1|+\cdots+|\alpha_{v-1}|}{|\alpha_v|}$ the previous inequalities (11), (12), and (13) hold simultaneously for $x=x_0$; thus, we get

$$\alpha_v+\frac{\alpha_{v-1}}{x_0}+\cdots+\frac{\alpha_1}{x_0^{v-1}}+\frac{\alpha_0}{a_0^v}<0\Rightarrow$$

$$x_0^v\left(\alpha_v+\frac{\alpha_{v-1}}{x_0}+\cdots+\frac{\alpha_1}{x_0^{v-1}}+\frac{\alpha_0}{x_0^v}\right)<0\Rightarrow p(x_0)<0\ \text{ (by the definition of } p).$$

So, we get $p(\rho)\cdot p(x_0)<0$ and p has a root exactly in (ρ,x_0) (say x_0^*), where x_0^* is the unique real root of p in $[\rho,+\infty)$.

We apply the bisection method in the interval $[\rho,x_0]$ and we compute the unique real root $x_0^*\in(\rho,x_0)$ of p in $[\rho,+\infty)$.

Now we suppose that p has real roots, and let ρ be the smallest real root of p. Let r be an arbitrary real root of p. Then $p(r)=0$ and by (14), we take that $q(-r)=0$; that is, $-r$ is a real root of q. This gives that $-\rho$ is the biggest real root of q.

By (14), we also get that $p'(x) = -q'(-x)$, which gives that if α is a real root of p', then $-\alpha$ is a real root of q'. Thus, because by supposition p' has real roots, the same holds for q'. Then, we apply the previous procedure and we compute the greatest real root of q (say $-\rho$), which means that ρ is the smallest real root of p. Thus, we compute the smallest real root of p also in any case. Now, we consider a polynomial $p \in \mathbb{R}[x]$, with $degp(x) = v \in \mathbb{N}$, $v \geq 3$, so that polynomial p' does not have any real root. Because p' does not have any root and $degp'(x) \geq 2$ (because $degp(x) \geq 3$), we conclude that p' is a polynomial of even degree (because any polynomial of odd degree has a real root at least). This means that p is a polynomial of odd degree that has one real root at least. Because $p'(x) \neq 0$ for every $x \in \mathbb{R}$, we take it that $p'(x) > 0$ for every $x \in \mathbb{R}$ or $p'(x) < 0$ for every $x \in \mathbb{R}$. If $\alpha_v > 0$, then $p'(x) > 0$ for every $x \in \mathbb{R}$ and p is a strictly increasing function in $\mathbb{R}$, such that $\lim_{x \to +\infty} p(x) = +\infty$ and $\lim_{x \to -\infty} p(x) = -\infty$. Thus, there exists $x_0 < 0$ so that $p(x_0) < 0$ and $y_0 > 0$ so that $p(y_0) > 0$, which gives that p has a real root in (x_0, y_0), say ρ. Because p is a strictly monotonous function, we take it that the root ρ is the unique real root of p.

We can now compute some numbers x_0, y_0 with the above properties. By inequality $(*)$ in page 40, we take it that if $x \in \mathbb{R}$, so that $x > 1$ and $x > \dfrac{|\alpha_0| + |\alpha_1| + \cdots + |\alpha_{v-1}|}{\alpha_v}$, then we get $\alpha_v + \dfrac{\alpha_v - 1}{x} + \cdots + \dfrac{\alpha_1}{x^{v-1}} + \dfrac{\alpha_0}{x^v} > 0$. We choose some $y_0 > 1$, so that $y_0 > \dfrac{|\alpha_0| + |\alpha_1| + \cdots + |\alpha_v - 1|}{\alpha_v}$, then by inequality $(*)$ in page 40 we get

$$\alpha_v + \frac{\alpha_v - 1}{y_0} + \cdots + \frac{\alpha_1}{y_0^{v-1}} + \frac{\alpha_0}{y_0^v} > 0 \Rightarrow$$

$$y_0^v \left(\alpha_v + \frac{\alpha_v - 1}{y_0} + \cdots + \frac{\alpha_1}{y_0^{v-1}} + \cdots + \frac{\alpha_0}{y_0^v} \right) > 0 \Leftrightarrow p(y_0) > 0. \tag{15}$$

Now, let some $x < -1$. Then we have $(x \neq 0)$

$$\frac{\alpha_0}{x^v} \geq -\left|\frac{\alpha_0}{x^v}\right|, \quad \frac{\alpha_1}{x^{v-1}} \geq -\left|\frac{\alpha_1}{x^{v-1}}\right|, \ldots, \frac{\alpha_{v-1}}{x} \geq -\left|\frac{\alpha_{v-1}}{x}\right|.$$

Adding these inequalities, we get

$$\frac{\alpha_0}{x^v} + \frac{\alpha_1}{x^{v-1}} + \cdots + \frac{\alpha_v - 1}{x} \geq -\left(\left|\frac{\alpha_0}{x^v}\right| + \left|\frac{\alpha_1}{x^{v-1}}\right| + \cdots + \left|\frac{\alpha_{v-1}}{x}\right| \right). \tag{16}$$

We get also

$$|x| > 1 \Rightarrow |x| \geq |x|, |x^2| > |x|, \ldots, |x^{v-1}| > |x|, |x^v| > |x|$$

$$\Rightarrow \frac{1}{|x|} \leq \frac{1}{|x|}, \ldots, \frac{1}{|x^{v-1}|} < \frac{1}{|x|}, \frac{1}{|x^v|} < \frac{1}{|x|}$$

$$\Rightarrow \left|\frac{\alpha_{v-1}}{x}\right| \leq \left|\frac{\alpha_{v-1}|}{|x|}, \ldots, \left|\frac{\alpha_1}{x^{v-1}}\right| < \frac{|\alpha_1|}{|x|}, \left|\frac{\alpha_0}{x^v}\right| < \left|\frac{\alpha_0}{x}\right|$$

$$\Rightarrow -\left|\frac{\alpha_{v-1}}{x}\right| \geq -\left|\frac{\alpha_{v-1}}{x}\right|, \ldots, -\left|\frac{\alpha_1}{x^{v-1}}\right| > -\frac{|\alpha_1|}{|x|}, -\left|\frac{\alpha_0}{x^v}\right| > -\left|\frac{\alpha_0}{x}\right|.$$

Adding by pairs the previous inequalities, we get

$$-\left(\left|\frac{\alpha_0}{x^v}\right| + \left|\frac{\alpha_1}{x^{v-1}}\right| + \cdots + \left|\frac{\alpha_{v-1}}{x}\right|\right) \geq -\frac{|\alpha_0| + |\alpha_1| + \cdots + |\alpha_{v-1}|}{|x|}. \tag{17}$$

By (16) and (17), we get

$$\alpha_v + \frac{\alpha_{v-1}}{x} + \cdots + \frac{\alpha_v}{x^{v-1}} + \frac{\alpha_0}{x^v} \geq \alpha_v + \frac{|\alpha_0| + |\alpha_1| + \cdots + |\alpha_{v-1}|}{|x|} \tag{18}$$

for every $x \in \mathbb{R}$, $x < -1$.

Now we get $x_0 \in \mathbb{R}$, so that:

$$x_0 < -1 \text{ and } x_0 < -\frac{|\alpha_0| + |\alpha_1| + \cdots + |\alpha_{v-1}|}{\alpha_v}. \tag{19}$$

Then by (19), we get

$$\begin{aligned} &-x_0 > \frac{|\alpha_0| + |\alpha_1| + \cdots + |\alpha_{v-1}|}{\alpha_v} > 0 \text{ (because} \alpha_v > 0) \\ &\Rightarrow |x_0| > \frac{|\alpha_0| + |\alpha_1| + \cdots + |\alpha_{v-1}|}{\alpha_v} \\ &\Rightarrow \alpha_v - \frac{|\alpha_0| + |\alpha_1| + \cdots + |\alpha_{v-1}|}{|x_0|} > 0. \end{aligned} \tag{20}$$

So, for $x < -1$, $x_0 < -\frac{|\alpha_0| + |\alpha_1| + \cdots + |\alpha_{v-1}|}{\alpha_v}$ we get from (20) that

$$\alpha_v + \frac{\alpha_{v-1}}{x_0} + \cdots + \frac{\alpha_1}{x_0^{v-1}} + \frac{\alpha_0}{x_0^v} > 0 \Rightarrow$$

(because $x_0 < 0$ and v is odd $x_0^v < 0$)

$$x_0^v \cdots \left(\alpha_v + \frac{\alpha_{v-1}}{x_0} + \cdots + \frac{\alpha_1}{x_0^{v-1}} + \frac{\alpha_0}{x_0^v}\right) < 0 \Leftrightarrow p(x_0) < 0.$$

Thus, for some $x_0 < -1$, $x_0 < -\frac{|\alpha_0| + |\alpha_1| + \cdots + |\alpha_{\upsilon-1}|}{\alpha_\upsilon}$ we get $p(x_0) < 0$ (21).

So with (15) and (12), we get $p(x_0) \cdot p(y_0) < 0$, and by applying the bisection method, we can compute the unique real root ρ of p, so that $\rho \in (x_0, y_0)$.

Finally in the case of $\alpha_\upsilon < 0$, we consider the polynomial $-p(x)$. Then,

$$(-p)'(x) = -p'(x) \neq 0 \text{ for every } x \in \mathbb{R}$$

and the coefficient of the monomial of greater degree of $-p$ is positive now. We apply the previous for $-p$ and we compute the unique real root ρ of $-p$, which is the unique real root of p also.

So in this remark we have covered the gap, we have left from basic Lemma 4.8 and Corollaries 4.9 and 4.10, and we have computed specific numbers x_0, y_0. For the sequel, we also need some tools from real polynomials of two real variables.

Before this, let us give some specific examples for the roots of polynomials.

As it is well known, from elementary calculus, any polynomial of odd degree has a real root at least.

On the contrary, there are many polynomials of any even degree that do not have any real root. For example, let $p(x)$ be any polynomial that is nonconstant, let $k \in \mathbb{N}$, and $\theta > 0$. Then polynomial $q(x) = p(x)^{2k} + \theta$ does not have any real root, as we can easily see, and has degree $2k \cdot \upsilon$, where $\upsilon = deg p(x)$.

Of course for every finite set of real numbers $A = \{\rho_1, \rho_2, \ldots, \rho_\upsilon\}$, $\upsilon \in \mathbb{N}$, $\rho_i \neq \rho_j$, for $i, j \in \{1, 2, \ldots, \upsilon\}$, $i \neq j$, polynomial $p(x) = (x - \rho_1)(x - \rho_2)\ldots(x - \rho_\upsilon)$ has roots the numbers ρ_i, $i = 1, \ldots, \upsilon$, and polynomial

$$q(x) = ((x - \rho_1)(x - \rho_2)\ldots(x - \rho_\upsilon))^{2k} = p(x)^{2k}$$

is a polynomial of even degree $deg q(x) = 2k\upsilon$, with roots the numbers ρ_i, $i = 1, 2, \ldots, \upsilon$, also. Now we consider polynomials of two real variables with real coefficients, that is we consider the set

$$\mathbb{R}^2[x, y] = \{p(x, y) : p(x, y)$$

is a polynomial of two real variables x and y with coefficients in $\mathbb{R}\}$.

Let $p(x, y) \in \mathbb{R}^2[x, y]$. We say that polynomial $p(x, y)$ is a pure polynomial, when $deg_x p(x, y) \geq 1$ and $deg_y p(x, y) \geq 1$, where $deg_x p(x, y)$, $deg_y p(x, y)$ are the greatest degree of its monomials with respect to x (or y, respectively).

The set of roots of $p(x, y)$ is the set

$$L_p(x, y) = \{(x, y) \in \mathbb{R}^2 \mid p(x, y) = 0\}.$$

As in polynomials of one real variable, we can easily see that there are many pure polynomials $p(x, y) \in \mathbb{R}^2[x, y]$, which do not have any roots.

For example, let $p(x, y)$ be any pure polynomial. Then polynomial $q(x, y) = p(x, y)^{2k} + \theta$, where $k \in \mathbb{N}$, $\theta > 0$, is a pure polynomial that does not have any real root, as we easily see. Of course these polynomials are of even degree for x and y.

On the contrary let $A = \{\alpha_1, \alpha_2, \ldots, \alpha_v\}$, $B = \{\beta_1, \beta_2, \ldots, \beta_m\}$, $A \cup B \subseteq \mathbb{R}$, $v, m \in \mathbb{N}$, $\alpha_i \neq \alpha_j$ for every $i, j \in \{1, \ldots, v\}$ $i \neq j$ and $\beta_i \neq \beta_j$ for every $i, j \in \{1, 2, \ldots, m\}$, $i \neq j$. Let also $k \in \mathbb{N}$. We consider the pure polynomial

$$p(x, y) = ((x - \alpha_1)(x - \alpha_2)\ldots(x - \alpha_v))((y - \beta_1)(y - \beta_2)\ldots(y - \beta_m))^{2k}.$$

Then p is a pure polynomial of even degree with respect to x and y, such that

$$L_0 = \{(\alpha_i, \beta_j), i \in \{1, \ldots, v\}, j \in \{1, 2, \ldots, m\}\} \subsetneqq L_p(x, y).$$

We remark also that for ever $y \in \mathbb{R}$, the couple (α_i, y) is a root of $p(x, y)$. This fact differentiates pure polynomials $p(x, y)$ from polynomials of one variable.

That is, there exist uncountable pure polynomials, each one having uncountable set of real roots. Especially, this holds for pure polynomials of an odd degree with respect to x and y. We have the following proposition.

Proposition 4.13 *Let $p(x, y)$ be a pure polynomial such that $deg_x(x, y) = v$ is odd or $degp_y(x, y)$ is odd. Then for every $r \in \mathbb{R}$ the set $L_r = \{(x, y) \in \mathbb{R}^2 : p(x, y) = r\}$ is uncountable.*

Proof We suppose, without loss of generality, that number $v = degp_x(x, y)$ is odd. Then, as we can see easily, we can write polynomial $p(x, y)$ as follows:

$$p(x, y) = \alpha_v(y)x^v + \alpha_{v-1}(y)x^{v-1} + \cdots + \alpha_1(y)x + \alpha_0(y),$$

where $\alpha_i(y) \in \mathbb{R}[y]$ for every $i = 0, 1, \ldots, v$, and $\alpha_v(y) \neq 0$, because $v = degp_x(x, y)$. Because $\alpha_v(y) \neq 0$, polynomial $\alpha_v(y)$ has a finite set of roots. Let A_v be the set of roots of $\alpha_v(y)$; that is, $A_v = \{y \in \mathbb{R} \mid \alpha_v(y) = 0\}$. Let $y_0 \in \mathbb{R} \setminus A_v$. Then $\alpha_v(y_0) \neq 0$. Also let $r \in \mathbb{R}$. We consider the polynomial

$$p_r(x) = \alpha_v(y_0)x^v + \alpha_{v-1}(y_0)x^{v-1} + \cdots + \alpha_1(y_0)x + \alpha_0(y_0) - r.$$

Then, $p_r(x)$ is a polynomial of odd degree $degp_r(x) = v$; thus, polynomial $p_r(x)$ has a real root, say x_0, at least; that is, we have

$$p_r(x_0) = 0 \Rightarrow \alpha_v(y_0)x_0^v + \alpha_{v-1}(y_0)x_0^{v-1} + \cdots + \alpha_1(y_0)x_0 + \alpha_0(y_0) = r \Leftrightarrow$$

$$p(x_0, y_0) = r \Rightarrow (x_0, y_0) \in L_r.$$

That is, we proved that for every $y \in \mathbb{R} \setminus A_v$, we have that there exists some $x \in \mathbb{R}$, such that $(x, y) \in L_r$. Of course, if $y_1, y_2 \in \mathbb{R} \setminus A_v$, $y_1 \neq y_2$ and $(x_1, y_1), (x_2, y_2) \in$

L_r, we have $(x_1, y_1) \neq (x_2, y_2)$ so the set L_r is uncountable, and the proof of this proposition is complete. ■

Corollary 4.14 *Let $p(x, y)$ be a pure polynomial such that $deg_x p(x, y)$ or $deg_y p(x, y)$ is odd. Then the set of real roots of $p(x, y)$ is uncountable.*

Proof It is a simple application of the previous Proposition 4.13 for $r = 0$. ■

As we have noticed previously, there are also such pure polynomials that the numbers $deg_x p(x, y)$ and $deg_y p(x, y)$ are even whose set of real roots is uncountable, as well as there being polynomials that do not have any roots.

Of course here we have the natural question: Are there pure polynomials $p(x, y)$ whose set of real roots $p(x, y)$ is nonempty and finite? Of course, let us give a simple example.

We consider the polynomial: $p(x, y) = (x^2 - 1)^2 + (y^2 - 4)^2$. It is easy to see that

$$L_p(x, y) = \{(1, 2), (1, -2), (-1, 2), (-1, -2)\}.$$

More generally, let $p_1(x)$ be a polynomial with real roots $\alpha_1, \alpha_2, \ldots, \alpha_v$, and $p_2(y)$ be a polynomial with real roots $\beta_1, \beta_2, \ldots, \beta_m$.

We consider the pure polynomial $p(x, y) = p_1(x)^{2k_1} + p_2(y)^{2k_2}$, where $k_1, k_2, \in \mathbb{N}$. Then, it is easy to see that:

$$L_p(x, y) = \{(\alpha_i, \beta_j), i \in \{1, \ldots, v\}, j \in \{1, \ldots, m\}\}.$$

Of course by Corollary 4.14 only pure polynomials whose numbers $deg_x p(x, y)$, $deg_y p(x, y)$ are even can have finite set of roots, as in the previous examples. From the previous results, we also have a significant observation.

These polynomials have the number zero as a global minimum!

For the sequel, we have to concentrate our attention to pure polynomials $p(x, y)$ that have a finite set of roots. So, from the previous observation we are led to ask whether the reverse result holds. That is, does any pure polynomial that has a global minimum have a finite set of real roots? The answer is no, and we can give a simple example. We consider the pure polynomial $p(x, y) = ((x - 1)(y - 2))^2 - 7$. It is easy to check that polynomial $p(x, y)$ has the number -7 as a global minimum. For this polynomial, we get

$$p(x, 3) = (x - 1)^2 - 7 \text{ forevery } x \in \mathbb{R}$$

so $p(1, 3) = -7 < 0$ and $p(4, 1) = 2 > 0$; thus, there exists $x_0 \in (1, 4)$, so that $p(x_0, 3) = 0$. Similarly take any real number $y_0 \in (3, 4)$. That is, $3 < y_0 < 4 \Rightarrow 1 < y_0 - 2 < 2 \Rightarrow 1 < (y_0 - 2)^2 < 4$ (1).

We get

$$p(1, y_0) = -7 < 0$$

$$p(8, y_0) = 7^2(y_0 - 2)^2 > 7^2 > 0$$

by (1), so there exists $x_1 \in (1, 8)$ such that $p(x_1, y_0) = 0$.

Thus, for every $y \in (3, 4)$, there exists $x \in \mathbb{R}$, so that $p(x, y) = 0$, and of course if $y_1, y_2 \in (3, 4)$, $x_1, x_2 \in \mathbb{R}$, $p(x_1, y_1) = p(x_2, y_2) = 0$ and $y_1 \neq y_2$, we have $(x_1, y_1) \neq (x_2, y_2)$, so the set of real roots of p is uncountable even if polynomial $p(x, y)$ has a global minimum. However, the property of a pure polynomial to have a global minimum (or maximum also) is a crucial property that have all pure polynomials that have a finite number of roots, as we will prove now with the following proposition.

Proposition 4.15 (Topological Lemma) *Let $p(x, y)$ be a pure polynomial. We suppose that there exist two couples $(x_1, y_1), (x_2, y_2) \in \mathbb{R}^2$ such that $p(x_1, y_1) \cdot p(x_2, y_2) < 0$. Then, set $L_p(x, y)$ is uncountable.*

Proof We set $A = (x_1, y_1)$, $B = (x_2, y_2)$. We get $A \neq B$, or otherwise we have $A = B$ and $p(x_1, y_1) \cdot p(x_2, y_2) = p(A) \cdot p(B) = p(A)^2 \geq 0$, which is false. So we get $A \neq B$. We consider the mid-perpendicular ℓ of segment $[A, B]$. For every point $\Gamma \in \ell$, we consider the union of two segments $[A, \Gamma] \cup [\Gamma, B]$. We write $A\Gamma B = [A, \Gamma] \cup [\Gamma, B]$ for simplicity. Of course $A\Gamma B \subseteq \mathbb{R}^2$. We consider the restriction $p|_{A\Gamma B}$ for simplicity, and we write $p = p|_{A\Gamma B}$ also for simplicity.

Of course the set $A\Gamma B$ is a compact and connected subset of $\mathbb{R}^2$. So the set $p(A\Gamma B)$ is a closed interval of $\mathbb{R}$.

We suppose that $p(A) < 0$ and $p(B) > 0$, without loss of generality. So $p(A), p(B) \in p(A\Gamma B)$ and gives that $0 \in p(A\Gamma B)$; that is, there exists some point $\Delta \in A\Gamma B$, so that $p(\Delta) = 0$. Of course $\Delta \neq A$ and $\Delta \neq B$. So, for every $\Gamma \in \ell$ and every curve $A\Gamma B$, there exists some $\Delta \in A\Gamma B$, $\Delta \neq A$, $\Delta \neq B$, such that $p(\Delta) = 0$.

Because the set $\mathscr{A} = \{A\Gamma B, \Gamma \in \ell\}$ is an uncountable supset of $P(\mathbb{R}^2)$ (the powerset of $\mathbb{R}^2$), and for every $\Gamma_1, \Gamma_2 \in \ell$, $\Gamma_1 \neq \Gamma_2$, we have that $A\Gamma_1 B \cap A\Gamma_2 B = \{A, B\}$; this means that the set

$$\mathscr{B} = \{\Delta \in A\Gamma B \mid \Gamma \in \ell \text{ and } p(\Delta) = 0\}$$

is uncountable, which gives that the set $Lp(x, y)$ of roots of $p(x, y)$ is uncountable and the proof of this proposition is complete. ■

Corollary 4.16 *Let $p(x, y)$ be a pure polynomial that has a finite set of roots, nonempty. Then, number 0 is the global minimum or maximum of $p(x, y)$, or in other words polynomial $p(x, y)$ has a global maximum or minimum, and when this holds, then this global maximum or minimum is number 0.*

Proof There exist no two points $(x_1, y_1), (x_2, y_2) \in \mathbb{R}^2$, so that:

$$p(x_1, y_1) \cdot p(x_2, y_2) < 0.$$

Or else, if there exist two points $(x_1, y_1), (x_2, y_2) \in \mathbb{R}^2$, so that $p(x_1, y_1) \cdot p(x_2, y_2) < 0$, then set $L_p(x, y)$ is uncountable (by the previous Proposition 4.13), which is false by our supposition.

This means that we have

(i) $p(x, y) \geq 0$ for every $(x, y) \in \mathbb{R}^2$, or
(ii) $p(x, y) \leq 0$ for every $(x, y) \in \mathbb{R}^2$.

We suppose that (i) holds. Because set $Lp(x, y)$ is nonempty, this means that there exists $(x_0, y_0) \in \mathbb{R}^2$ so that $p(x_0, y_0) = 0$, so we get $p(x, y) \geq p(x_0, y_0)$ for every $(x, y) \in \mathbb{R}^2$. So, polynomial $p(x, y)$ has in point (x_0, y_0) its global minimum the number 0, because $p(x_0, y_0) = 0$. If (ii) holds, then we take with a similar way that p has global maximum the number 0 in a point, and the proof of corollary is complete. ■

The above corollary is a basic result that we use in the second stage of our method.

Finally, we refer here the most advanced result that we use in our method.

This result is called many times as Fermat's theorem in calculus of several variables.

Theorem 4.17 *Let $U \subseteq \mathbb{R}^2$, U open, and $f : U \to \mathbb{R}$ be a differentiable function in $x_0 \in U$, where x_0 is a point of local maximum or local minimum of f. Then the following holds: $\nabla f(x_0) = 0$; that is, x_0 is a crucial point of f, where $\nabla f(x_0)$ is the gradient of f in x_0.*

Acknowledgement Many thanks to Vasilli Karali for his contribution in the presentation of this paper.

Appendix

Fundamental Theorem of Algebra is a powerful and basic result in the theory of polynomials, especially in polynomial equations.

Gauss gave the first complete proof of this result in his Ph.D. There are many proofs for this important theorem, but none of them is trivial in order to be presented in books of secondary school.

Its simplest proof comes from complex analysis and uses an advanced theorem of complex analysis, Liouville's theorem. Here we give a proof that uses the most elementary tools that an undergraduate student learns.

We think that it is difficult for an undergraduate student to find this proof in books, so we try to present it with details for educational reasons.

For this reason, we give firstly some elementary lemmas.

Lemma A.1 *Let $p(z) \in \mathbb{C}[z]$ be a complex polynomial, and $z_0 \in \mathbb{C}$. We consider polynomial $Q(z) = p(r + z_0)$, $z \in \mathbb{C}$. If $p(z) \equiv 0$, then of course $Q(z) \equiv 0$. If $p(z) \not\equiv 0$, then $deg Q(z) = deg p(z)$.*

Proof If $deg p(z) = 0$, then the result is obvious. Let $deg p(z) = n \in \mathbb{N}$, $n \geq 1$. We suppose that $n = 1$, so we get $p(z) = az + b$, where $a, b \in \mathbb{C}$, $a \neq 0$. We get $Q(z) = p(z + z_0) = a(r + z_0) + b = az + (az_0 + b)$, and $deg Q(z) = 1$, because $a \neq 0$. So, the result holds for $n = 1$. We prove the result inductively. We suppose $z_0 \neq 0$.

For $n = 1$, the result holds.

We suppose that result holds for $k \in \mathbb{N}$, $k \geq 1$ and for every $j \in \mathbb{N}$, $1 \leq j \leq k$. We prove that result holds for $k + 1$.

We suppose that

$$p(z) = a_0 + a_1 z + \cdots + a_k z + a_{k+1} z^{k+1} \text{ and } a_{k+1} \neq 0, \text{ so } deg p(z) = k + 1.$$

We distinguish two cases:

(i) $q(z) = a_0 + a_1 z + \cdots + a_k z^k \not\equiv 0$.
Then we have $p(z) = q(z) + a_{k+1} z^{k+1}$.
We have

$$Q(z) = p(z + z_0) = q(z + z_0) + a_{k+1}(z + z_0)^{k+1}. \tag{1}$$

We set $r(z) = q(z + z_0)$, $z \in \mathbb{C}$. Because $q(z) \not\equiv 0$, by induction step we have $deg r(z) = deg q(z) \leq k$ (2).
We have by Newton's binomial

$$a_{k+1}(z + z_0)^{k+1} = a_{k+1} \sum_{j=0}^{k+1} z^{k+1-j} z_0^j = \sum_{j=0}^{k+1} a_{n+1} z_0^j z^{k+1-j}. \tag{3}$$

Because $z_0 \neq 0$ (by our supposition) and $a_{k+1} \neq 0$, we have $deg a_{k+1}(z + z_0)^{k+1} = k + 1$ (4), by equality (3).
By (1), (2), and (4), we get $deg Q(z) = k + 1$ and the result holds.

(ii) $q(z) = a_0 + a_1 z + \cdots + a_k z^k \equiv 0$. The proof is similar to case (i), so the result holds by induction.
Of course if $z_0 = 0$, the result is obvious, because $Q(z) = p(z)$, so $deg Q(z) = deg p(z)$.

Lemma A.2 *We consider polynomial $p(z) \in \mathbb{C}[z]$. Of course we have $|p(z)| \geq 0$ for every $z \in \mathbb{C}$. So the set*

$$A = \{x \in \mathbb{R} \mid \exists\, z \in \mathbb{C} : x = |p(z)|\}$$

is low bounded by 0.

We set $m = \inf(A)$. *Then there exists* $R > 0$, *so that:*

$$m = \inf(\{x \in \mathbb{R} \mid \exists z \in \overline{D(0, R)} : x = |p(z)|\}),$$

where $\overline{D(0, R)} = \{z \in \mathbb{C} : |z| \leq R\}$.

Proof We set $B_R = \{x \in \mathbb{R} \mid \exists z \in D(0, R) : x = |p(z)|\}$ for some $R > 0$.

The result is obvious when $p(z) \equiv 0$, so we suppose that $p(z) \not\equiv 0$. It is obvious that $B_R \subseteq A$ by definitions of sets A and B_R for $R > 0$. Let $x \in B_R$ for some $R > 0$. Then $x \in A$. So $m \leq x$, because m is a lower bound of A. So we have $m \leq x$ for every $x \in B_R$. This means that m is a lower bound of B_R, so $m \leq m^*_R$ (1), where $m^*_R = \inf(B_R)$. That is, we get $m \leq m^*_R$ for every $R > 0$.

We suppose that:
$p(z) = a_0 + a_1 z + \cdots + a_n z^n$, where $n \in \mathbb{N} \cup \{0\}$, $a_n \neq 0$. When $n = 0$, we get of course $m^*_R = m = |p(z)|$ for every $z \in \mathbb{C}$ and every $R > 0$, and the result is also obvious. So we suppose that $n \geq 1$.

Then for every $z \in \mathbb{C} \setminus \{0\}$ we get

$$p(z) = z^n \cdot \left(\frac{a_0}{z^n} + \frac{a_1}{z^{n-1}} + \cdots + \frac{a_{n-1}}{z} + a_n\right).$$

By calculus of the elementary limits in complex analysis, we have $\lim\limits_{z\to\infty} \frac{a_0}{z^n} = \lim\limits_{z\to\infty} \frac{a_1}{z^{n-1}} = \cdots = \lim\limits_{z\to\infty} \frac{a_{n-1}}{z} = 0$ and $\lim\limits_{z\to\infty} z^n = \infty$, so we have: $(a_n \neq 0)$ $\lim\limits_{z\to\infty} p(z) = \infty$. By definition of $\lim\limits_{z\to\infty} p(z)$, this means that for $m + 1$, there exists $R_0 > 0$, so that: $|p(z)| > m + 1$ for every $z \in \mathbb{C}$, $|z| > R_0^{(*)}$.

From (1), we have of course $m \leq m^*_{R_0}$ (2). Take $w \in \mathbb{C}$: $|w| > R_0$. Then, by the above we have $|p(w)| > m + 1$ (3).

Now there exists $z_1 \in \mathbb{C}$ so that $|p(z_1)| < m + 1$ (4), or otherwise we have $|p(z)| \geq m + 1$ for every $z \in \mathbb{C}$, so $m + 1$ is a lower bound of A; that is, $m = \inf(A) \geq m + 1$, which is false. Of course $z_1 \in \overline{D(0, R_0)}$ by implication $(*)$, or else $|z_1| > R_0$ that means $|p(z_1)| > m + 1$ (5), which is false by the above inequalities (4) and (5). So we have $m^*_{R_0} \leq |p(z_1)| < m + 1 < |p(w)| \Rightarrow m^*_{R_0} \leq |p(w)|$.

So we get $m^*_{R_0} \leq |p(z)|$ for every $z \in \mathbb{C} : |z| > R_0$. Of course we have also $m^*_{R_0} \leq |p(z)|$ for every $z \in \overline{D(0, R_0)}$ by definition of $m^*_{R_0}$. So we get $m^*_{R_0} \leq |p(z)|$ for every $z \in \mathbb{C}$, which means that $m^*_{R_0}$ is a lower bound of A; that is, $m^*_{R_0} \leq m$ (6). From (2) and (6), we get $m = m^*_{R_0}$; that is, Lemma A.2 has been proven. ∎

Remark A.3 (De Moivre Theorem) We remind here the following result.
Let $n \in \mathbb{N}$, $n \geq 2$. Then every nonzero complex number has exactly n roots; that is, if $w \in \mathbb{C}$, $w \neq 0$, then equation $z^n = w$ has exactly n solutions: This result is proven easily by elementary properties of complex numbers and it is well known as De Moivre's theorem, using properties of functions sine and cosine. We also need a topological theorem.

Theorem A.4 *Let $K \subseteq \mathbb{C}$ be compact and $f : K \to \mathbb{R}$ be continuous. Then f attains its supremum and its infimum and both are finite. For this theorem, see [5]. After the above, we are now ready to give the proof of Fundamental Theorem of Algebra.*

Fundamental Theorem of Algebra A.5
Proof We consider polynomial

$$p(z) = a_0 + a_1 z + \cdots + a_n z^n, \quad a_i \in \mathbb{C} \text{ for every}$$
$$i = 0, 1, \ldots, n, \quad n \in \mathbb{N}, \quad a_n \neq 0, \quad n \geq 1.$$

We prove that $p(z)$ has a root; that is, there exists $z_0 \in \mathbb{C}$, so that $p(z_0) = 0$. First of all, we examine the case of $a_n = 1$.

Of course we have $|p(z)| \geq 0$ for every $z \in \mathbb{C}$. We set

$$A = \{x \in \mathbb{R} \mid \exists z \in \mathbb{C} : x = |p(z)|\}.$$

Set A is low bounded by 0. We set $m = \inf(A)$. Of course $m \geq 0$. For every $R > 0$, we set

$$B_R = \{x \in \mathbb{R} \mid \exists z \in \overline{D(0, R)} : x = |p(z)|\}, \text{ and}$$
$$m_R^* = \inf(B_R), \overline{D(0, R)} = \{z \in \mathbb{C} : |z| \leq R\}.$$

Applying Lemma A.2, we take that there exists $R_0 > 0$, so that: $m = m_{R_0}^*$ (1).

By Theorem 4.3, page 233 [5], Ball $\overline{D(0, R_o)} = \{z \in \mathbb{C} : |z| \leq R_0\}$ is a compact set as a set closed and bounded. Polynomial p is a continuous function in $\mathbb{C}$. This is a well-known result in elementary complex analysis. Usual norm $| \; | : \mathbb{C} \to \mathbb{R}$ is a continuous function also in $\mathbb{C}$, by elementary complex analysis. So, the composition function $F : \mathbb{C} \to \mathbb{R}$, $F = | \cdot | \circ p$, where $p : \mathbb{C} \to \mathbb{C}$, $| \cdot | : \mathbb{C} \to \mathbb{R}$ with formula $F(z) = (| \cdot | \circ p)(z) = |p(z)|$ for every $z \in \mathbb{C}$ is a continuous function as the composition of continuous functions $| \cdot |$ and p. Applying now Theorem A.4 for $K = \overline{D(0, R_0)}$ and $f = F$, we take it that function F attains its infimum in some point $z_0 \in \overline{D(0, R_0)}$. This means that $|p(z_0)| = m_{R_0}^*$ (2). By (1) and (2), we have $m = |p(z_0)|$ (3). We argue that $m = 0$. To take a contradiction, we suppose that $m > 0$. Because $|p(z_0)| = m > 0$, we see that $p(z_0) \neq 0$.

We consider polynomial $Q(z) = \dfrac{p(z + z_0)}{p(z_0)}$, which is well-defined because $p(z_0) \neq 0$.

Applying Lemma A.1, we see that $deg p(z + z_0) = deg p(z) = n$, and by definition of $Q(z)$, we get $deg Q(z) = n$. We have $Q(0) = \dfrac{p(0 + z_0)}{p(z_0)} = 1$, so polynomial $Q(z)$ has constant term equal to 1.

Let

$$Q(z) = 1+c_k z^k+\cdots+c_n z^n,\ \ c_n \neq 0,\ \ \text{for every}\ z \in \mathbb{C},\ \ \text{where}\ k \in \mathbb{N},\ 1 \le k \le n$$

and k be the smallest natural number such that $c_k \neq 0$ (maybe $k = n$ of course).

So, we get $-|c_k|/c_k \neq 0$. From Remark A.3, there exists $j \in \mathbb{C}$, so that $j^k = -|c_k|/c_k$ (4). (Of course there are k different complex numbers such that (4) holds.) By (4), we take $|j^k| = |-|c_k|/c_k| = 1 \Rightarrow |j| = 1$ (5).

By choice of j, we have for $r \in \mathbb{C}$ $|1 + c_k r^k j^k| \overset{(4)}{=} |1 + c_k r^k \cdot (-|c_k|/c_k) = 1 - |c_k| r^k$ (6).

By definition of $Q(z)$, we compute for $z = rj$ for $r \in \mathbb{C}$:

$$\begin{aligned} Q(z) = Q(rj) &= 1 + c_k(rj)^k + \cdots + c_n(rj)^n \\ &= 1 + c_k r^k j^k + c_{k+1} r^{k+1} j^{k+1} + \cdots + c_n r^n j^n. \end{aligned} \tag{7}$$

By (7) and triangle inequality, we get

$$|Q(rj)| \le |1 + c_k r^k j^k| + |c_{k+1} r^{k+1} j^{k+1}| + \cdots + |c_n r^n j^n|. \tag{8}$$

Applying (6), we get by (8)

$$\begin{aligned} |Q(rj)| &\le 1 - |c_k||r^k| + |c_{k+1}||r|^{k+1} + \cdots + |c_n||r^n| \\ &= 1 - |r^k|(|c_k| - |c_{k+1}||r| - \cdots - |c_n||r|^{n-k}),\ \ \text{for every}\ \ r \in \mathbb{C}. \end{aligned} \tag{10}$$

By definition of m, we get

$$m \le |p(z + z_0)|\ \ \text{for every}\ \ z \in \mathbb{C}. \tag{11}$$

By (3) and (11), we get

$$\begin{aligned} &|p(z_0)| \le |p(z + z_0)|\ \ \text{for every}\ \ z \in \mathbb{C} \\ &\Rightarrow \left|\frac{p(z + z_0)}{p(z_0)}\right| \ge 1\ \ \text{for every}\ \ z \in \mathbb{C} \\ &\qquad \Rightarrow |Q(z)| \ge 1\ \ \text{for every}\ \ z \in \mathbb{C}\ \ \text{(by definition of } Q). \end{aligned} \tag{12}$$

Now, we distinguish two cases:

(i) $k = n$. Then, from (10) we get

$$|Q(rj)| \le 1 - |r|^k |c_k|. \tag{11}$$

So, for every $r \neq 0$, we get by (12)

$$|Q(rj)| \geq 1 \text{ and} \tag{13}$$

$$|Q(rj)| \leq 1 - |r^k||c_k| < 1 \tag{14}$$

and we take a contradiction from (13) and (14).

(ii) $k < n$.

By properties of complex limits, we get

$$\lim_{r \to 0} (|c_k| - |c_{k+1}||r| - \cdots - |c_n|r|^{n-k}) = |c_k| > 0.$$

This limit shows us that there exists some small r_0, so that:

$$|c_k| - |c_{k+1}||r_0| - \cdots - |c_n||r_0|^{n-k} > 0. \tag{15}$$

We set $\theta := |c_k| - |c_{k+1}||r_0| - \cdots - |c_n||r_0|^{n-k}$. So we have $\theta_0 > 0$. From (10) and (15), we get

$$|Q(r_0 j)| \leq 1 - |r_0|^k \theta_0 < 1. \tag{16}$$

From (12), we get $|Q(r_0 j)| \geq 1$ (17). By (16) and (17), we get a contradiction. So, our supposition that $m > 0$ is false. So we have $m = 0$, and from (3), we get $0 = p(z_0)$; that is, polynomial p has, as a root number, z_0. If $a_n \neq 1$, we write $\frac{1}{a_n} p(z) = \frac{a_0}{a_n} + \frac{a_1}{a_n} z + \cdots + z^n$, and applying the previous result, we take it that there exists some $w \in \mathbb{C}$ such that $\frac{1}{a_n} p(w) = 0 \Leftrightarrow p(w) = 0$, so polynomial p has a root again. The proof of fundamental theorem has completed now.

■

Remark A.6 Inside our work we have used the well-known binomial equation $x^n = a$, where $a > 0$. We remind how we solve this equation here, for $n \geq 2$, $n \in \mathbb{N}$. We will distinguish two cases:

(i) $a > 1$. We consider function $f : [1, a] \to \mathbb{R}$, with the formula $f(x) = x^n - a$ for every $x \in [1, a]$. We get $f(1) = 1^n - a < 1$, from our supposition and $f(a) = a^n - a = a(a^{n-1} - 1) > 0$. So we have $f(1) \cdot f(a) < 0$, and because f is continuous, we understand from Bolzano theorem that there exists $x_0 \in (1, a)$, so that $f(x_0) = 0 \Leftrightarrow x_0^n - a = 0 \Leftrightarrow x_0 = \sqrt[n]{a}$. Because f is strictly increasing in $[1, a]$ (because $f'(x) = nx^{n-1} > 0$ for every $x \in [1, a]$), equation $f(x) = 0$ has unique root in $[1, a]$, that is number $\sqrt[n]{a}$. Applying bisection method, we approximate number $\sqrt[n]{a}$, or in other words we solve the equation $x^n = a$.

(ii) $a \in (0, 1)$. Then we apply the above procedure similarly to the function $g : [0, 1] \rightarrow \mathbb{R}$ with the formula $g(x) = x^n - a$, for every $x \in [0, 1]$.

References

1. Conte, S. D., and De Boor, C., Elementary Numerical Analysis—An algorithmic approach, 3rd ed., McGraw-Hill, New York 1980.
2. Cox, D., Little, J., O' Shea, D., Ideals Varieties and algorithms.
3. Cox, D., Little, J., O' Shea, D., Using Algebraic Geometry, Springer-Verlag, 1998.
4. Dennis, J. E. and Schnabel, R. B., Numerical methods for unconstrained optimization and nonlinear equations, Prentice Hall, Englewood Cliffs, N.J. 1983.
5. Dugundji, J. Tolopogy, Allyn and Bacon, Boston, 1966.
6. Forsythe, C. E., Malcolm, M. A., Moler, C. B., Computer Methods for Mathematical Computations, Prentice Hall, 1977.
7. Gelfand, I. M., Kapranov, M. M., Zelevinsty, A. V., Discriminants, resultants and multidimensional determinants, Boston, Birkhäuser, 1994.
8. Henrici, P., Essentials of Numerical Analysis with pocket calculator demonstrations, Wiley, New York, 1982.
9. Householder, A. S., The theory of matrices in numerical analysis, Blaisdell, 1964.
10. Milovanovic, G. V., Mitrinovic, D. S. and Rassias, Th. M. Topics in Polynomials Extremal Problems, Inequalities, Zeros, World Scientific Publ. Company, Singapore, New Jersey, London, 1994.
11. Ortega, J. M., and Rheinboldt, W., The numerical solution of nonlinear systems, Academic Press, New York, 1970.
12. Rabinowitz, P., Numerical methods for non-linear algebraic equations, Gordon and Breach, 1970.
13. Rassias, Th. M., Srivastava, H. M. and A. Yanushauskas (eds), Topics in Polynomials of One and Several Variables and their Applications. World Scientific Publishing Company, Singapore, New Jersey, London, 1993.
14. Scheid, F., Schaum's outline series, Numerical Analysis, McGraw-Hill, 1968.
15. Varga, R. S., Matrix iterative analysis, Prentice-Hall, 1962.

Meir–Keeler Sequential Contractions and Pata Fixed Point Results

Mihai Turinici

Abstract The (contractive) maps introduced by Pata [J. Fixed Point Th. Appl., 10 (2011), 299–305] are in fact Meir–Keeler sequential maps. This allows us treating in a unitary manner all fixed point results of this type.

AMS Subject Classification 47H17 (Primary), 54H25 (Secondary)

1 Introduction

Let X be a nonempty set. Call the subset Y of X, *almost-singleton* (in short: *asingleton*) provided $y_1, y_2 \in Y$ implies $y_1 = y_2$; and *singleton* if, in addition, Y is nonempty; note that in this case $Y = \{y\}$, for some $y \in X$. Take a metric $d : X \times X \to R_+ := [0, \infty[$ over X; the couple (X, d) will be then referred to as a *metric space*. Then, let $T \in \mathscr{F}(X)$ be a self-map of X. [Here, for each couple A, B of nonempty sets, $\mathscr{F}(A, B)$ denotes the class of all functions from A to B; when $A = B$, we write $\mathscr{F}(A)$ in place of $\mathscr{F}(A, A)$]. Denote $\text{Fix}(T) = \{x \in X; x = Tx\}$; each point of this set is referred to as *fixed* under T. Concerning the existence and uniqueness of such points, a basic result (referred to as: Banach fixed point theorem; in short: (B-fpt)) may be stated as follows. Call the self-map T, $(d; \alpha)$*-contractive* (where $\alpha \geq 0$), if

(con) $d(Tx, Ty) \leq \alpha d(x, y)$, for all $x, y \in X$.

Theorem 1 *Assume that T is Banach $(d; \alpha)$-contractive, for some $\alpha \in [0, 1[$. In addition, let X be d-complete. Then,*

(11-a) $\text{Fix}(T)$ *is a singleton,* $\{z\}$

(11-b) $T^n x \xrightarrow{d} z$ *as* $n \to \infty$*, for each* $x \in X$.

M. Turinici (✉)
A. Myller Mathematical Seminar, A. I. Cuza University, Iaşi, Romania
e-mail: mturi@uaic.ro

I. N. Parasidis et al. (eds.), *Mathematical Analysis in Interdisciplinary Research*, Springer Optimization and Its Applications 179,
https://doi.org/10.1007/978-3-030-84721-0_36

This result, established in 1922 by Banach [1], found some important applications to the operator equations theory. Consequently, a multitude of extensions for it were proposed. From the perspective of this exposition, the set implicit ones are of interest. These, roughly speaking, may be written as

(i-s-con) $(d(Tx, Ty), d(x, y), d(x, Tx), d(y, Ty), d(x, Ty), d(Tx, y)) \in \mathscr{M}$, for all $x, y \in X$ with $x\mathscr{R}y$;

where $\mathscr{M} \subseteq R_+^6$ is a (nonempty) subset and $\mathscr{R}$ is a relation over X. In particular, when $\mathscr{M}$ is the zero-section of a certain function $F : R_+^6 \to R$; i.e.,

$\mathscr{M} = \{(t_1, \ldots, t_6) \in R_+^6; F(t_1, \ldots, t_6) \leq 0\}$,

the implicit contractive condition above has the functional form:

(i-f-con) $F(d(Tx, Ty), d(x, y), d(x, Tx), d(y, Ty), d(x, Ty), d(Tx, y)) \leq 0$, for all $x, y \in X$, with $x\mathscr{R}y$;

(where $\mathscr{R}$ is taken as before). Note that, when $\mathscr{R} = X \times X$ (the *trivial* relation over X), some basic contributions in the area were obtained, in the explicit case, by Boyd and Wong [4], Reich [36], Matkowski [26], and Rhoades [37]; in addition, for the implicit setting above, certain technical aspects have been considered by Leader [25] and Turinici [43]. Further, when $\mathscr{R}$ is an *order* on X, a couple of 1986 results was established—in the realm of Matkowski type contractions—by Turinici [46, 47]. Two decades later, these fixed point statements have been re-discovered—over the Banach contractive setting—by Ran and Reurings [35]; see also Nieto and Rodriguez-Lopez [32]; and since then, the number of papers devoted to the precise topic increased rapidly. Finally, when $\mathscr{R}$ is an *amorphous* relation over X, some appropriate statements of this type were obtained—in a graph setting—by Jachymski [15], and—in a general context—by Samet and Turinici [40].

Returning to the trivial (modulo $\mathscr{R}$) setting, a basic particular case of the implicit set contractive property above is

(2-i-s-con) $(d(Tx, Ty), d(x, y)) \in \mathscr{M}$, for all $x, y \in X$;

where $\mathscr{M} \subseteq R_+^2$ is a (nonempty) subset. The classical example in this direction is the one due to Meir and Keeler [28]. Further refinements of the method were proposed by Ćirić [7] and Matkowski [27].

Recently, a new contractive condition of the type (2-i-s-con) was introduced by Pata [34]. His methods were appreciated as interesting enough to be used in various fixed point and/or coincidence point problems; see, for example, the survey paper by Choudhury et al. [5]. Having these precise, we may ask about the effectiveness of such methods with respect to the (sketched) old ones. It is our main aim in the present exposition to give a negative answer to this; precisely, to establish that all these results are obtainable via Meir–Keeler sequential contractions. Further aspects will be delineated in a future paper.

2 Dependent Choice Principles

Throughout this exposition, the axiomatic system in use is Zermelo-Fraenkel's (abbreviated: (ZF)), as described by Cohen [8, Ch 2]. The notations and basic facts to be considered are standard; some important ones are discussed below.

(A) Let X be a nonempty set. By a *relation* over X, we mean any (nonempty) part $\mathscr{R} \subseteq X \times X$; then, $(X, \mathscr{R})$ will be referred to as a *relational structure*. Note that $\mathscr{R}$ may be regarded as a mapping between X and $\exp[X]$ (=the class of all subsets in X). In fact, let us simplify the string $(x, y) \in \mathscr{R}$ as $x\mathscr{R}y$; and put

$X(x, \mathscr{R}) = \{y \in X; x\mathscr{R}y\}$ (the *section* of $\mathscr{R}$ through x), $x \in X$;

then, the desired mapping representation is $(\mathscr{R}(x) = X(x, \mathscr{R}); x \in X)$. A basic example of such object is

$\mathscr{I} = \{(x, x); x \in X\}$ [the *identical relation* over X].

Given the relations $\mathscr{R}$, $\mathscr{S}$ over X, define their *product* $\mathscr{R} \circ \mathscr{S}$ as

$(x, z) \in \mathscr{R} \circ \mathscr{S}$, if there exists $y \in X$ with $(x, y) \in \mathscr{R}$, $(y, z) \in \mathscr{S}$.

Also, for each relation $\mathscr{R}$ in X, denote

$\mathscr{R}^{-1} = \{(x, y) \in X \times X; (y, x) \in \mathscr{R}\}$ (the *inverse* of $\mathscr{R}$).

Finally, given the relations $\mathscr{R}$ and $\mathscr{S}$ on X, let us say that $\mathscr{R}$ is *coarser* than $\mathscr{S}$ (or, equivalently: $\mathscr{S}$ is *finer* than $\mathscr{R}$), provided

$\mathscr{R} \subseteq \mathscr{S}$; i.e.: $x\mathscr{R}y$ implies $x\mathscr{S}y$.

Given a relation $\mathscr{R}$ on X, the following properties are to be discussed here:

(P1) $\mathscr{R}$ is *reflexive*: $\mathscr{I} \subseteq \mathscr{R}$
(P2) $\mathscr{R}$ is *irreflexive*: $\mathscr{I} \cap \mathscr{R} = \emptyset$
(P3) $\mathscr{R}$ is *transitive*: $\mathscr{R} \circ \mathscr{R} \subseteq \mathscr{R}$
(P4) $\mathscr{R}$ is *symmetric*: $\mathscr{R}^{-1} = \mathscr{R}$
(P5) $\mathscr{R}$ is *anti-symmetric*: $\mathscr{R}^{-1} \cap \mathscr{R} \subseteq \mathscr{I}$.

This yields the classes of relations to be used; the following ones are important for our developments:

(C0) $\mathscr{R}$ is *amorphous* (i.e. it has no properties at all)
(C1) $\mathscr{R}$ is a *quasi-order* (reflexive and transitive)
(C2) $\mathscr{R}$ is a *strict order* (irreflexive and transitive)
(C3) $\mathscr{R}$ is an *equivalence* (reflexive, transitive, symmetric)
(C4) $\mathscr{R}$ is a *(partial) order* (reflexive, transitive, anti-symmetric)
(C5) $\mathscr{R}$ is the *trivial* relation (i.e.: $\mathscr{R} = X \times X$).

(B) A basic example of relational structure is to be constructed as below. Let

$N = \{0, 1, 2, \ldots\}$, where $(0 = \emptyset, 1 = \{0\}, 2 = \{0, 1\}, \ldots)$

denote the set of *natural* numbers. Technically speaking, the basic (algebraic and order) structures over N may be obtained by means of the *(immediate) successor* function $\text{suc} : N \to N$, and the following Peano properties (deductible in our axiomatic system (ZF)):

(pea-1) ($0 \in N$ and) $0 \notin \text{suc}(N)$
(pea-2) suc(.) is injective ($\text{suc}(n) = \text{suc}(m)$ implies $n = m$)
(pea-3) if $M \subseteq N$ fulfills $[0 \in M]$ and $[\text{suc}(M) \subseteq M]$, then $M = N$.

(Note that, in the absence of our axiomatic setting, these properties become the well known Peano axioms, as described in Halmos [12, Ch 12]; we do not give details). In fact, starting from these properties, one may construct, in a recurrent way, an *addition* $(a, b) \mapsto a + b$ over N, according to

$(\forall m \in N)$: $m + 0 = m$; $m + \text{suc}(n) = \text{suc}(m + n)$.

This, in turn, makes possible the introduction of a (partial) order $(\leq)$ over N, as

$(m, n \in N)$: $m \leq n$ iff $m + p = n$, for some $p \in N$.

Concerning the properties of this structure, the most important one writes

$(N, \leq)$ is well ordered:
any (nonempty) subset of N has a first element.

Denote, for simplicity

$$N(r, \leq) = \{n \in N; r \leq n\} = \{r, r + 1, \ldots, \}, r \geq 0,$$
$$N(r, >) = \{n \in N; r > n\} = \{0, \ldots, r - 1\}, r \geq 1;$$

the latter one is referred to as the *initial interval* (in N) induced by r. Any set P with $N \sim P$ (in the sense: there exists a bijection from N to P) will be referred to as *effectively denumerable*. In addition, given some natural number $n \geq 1$, any (nonempty) set Q with $N(n, >) \sim Q$ will be said to be *n-finite*; when n is generic here, we say that Q is *finite*. As a combination of these, we say that the (nonempty) set Y is (at most) *denumerable* iff it is either effectively denumerable or finite.

Having these precise, let the notion of *sequence* (in X) be used to designate any mapping $x : N \to X$. For simplicity reasons, it will be useful to denote it as $(x(n); n \geq 0)$, or $(x_n; n \geq 0)$; moreover, when no confusion can arise, we further simplify this notation as $(x(n))$ or (x_n), respectively. Also, any sequence $(y_n := x_{i(n)}; n \geq 0)$ with

$(i(n); n \geq 0)$ is *strictly ascending* (hence: $i(n) \to \infty$ as $n \to \infty$)

will be referred to as a *subsequence* of $(x_n; n \geq 0)$. Note that, under such a convention, the relation “subsequence of" is transitive; i.e.

(z_n)=subsequence of (y_n) and (y_n)=subsequence of (x_n)
imply (z_n)=subsequence of (x_n).

(C) Remember that, an outstanding part of (ZF) is the *Axiom of Choice* (abbreviated: (AC)); which, in a convenient manner, may be written as

(AC) For each couple (J, X) of nonempty sets and each function $F : J \to \exp(X)$, there exists a (selective) function $f : J \to X$, with $f(\nu) \in F(\nu)$, for each $\nu \in J$.

(Here, $\exp(X)$ stands for the class of all nonempty elements in $\exp[X]$). Sometimes, when the ambient set X is endowed with denumerable type structures, the existence of such a selective function (over $J = N$) may be determined by using a weaker form of (AC), referred to as: *Dependent Choice* principle (in short: (DC)). Call the relation $\mathscr{R}$ over X, *proper* when

$(X(x, \mathscr{R}) =)\mathscr{R}(x)$ is nonempty, for each $x \in X$.

Then, $\mathscr{R}$ is to be viewed as a mapping between X and $\exp(X)$; and the couple $(X, \mathscr{R})$ will be referred to as a *proper relational structure*. Further, given $a \in X$, let us say that the sequence $(x_n; n \geq 0)$ in X is *$(a; \mathscr{R})$-iterative*, provided

$x_0 = a$, and $x_n \mathscr{R} x_{n+1}$ (i.e. $x_{n+1} \in \mathscr{R}(x_n)$), for all n.

Proposition 1 *Let the relational structure $(X, \mathscr{R})$ be proper. Then, for each $a \in X$ there is at least an $(a; \mathscr{R})$-iterative sequence in X.*

This principle—proposed, independently, by Bernays [3] and Tarski [42]—is deductible from (AC), but not conversely; cf. Wolk [51]. Moreover, by the developments in Moskhovakis [30, Ch 8], and Schechter [41, Ch 6], the *reduced system* (ZF-AC+DC) is comprehensive enough so as to cover the "usual" mathematics; see also Moore [29, Appendix 2].

Let $(\mathscr{R}_n; n \geq 0)$ be a sequence of relations on X. Given $a \in X$, let us say that the sequence $(x_n; n \geq 0)$ in X is *$(a; (\mathscr{R}_n; n \geq 0))$-iterative*, provided

$x_0 = a$, and $x_n \mathscr{R}_n x_{n+1}$ (i.e. $x_{n+1} \in \mathscr{R}_n(x_n)$), for all n.

The following *Diagonal Dependent Choice* principle (in short: (DDC)) is available.

Proposition 2 *Let $(\mathscr{R}_n; n \geq 0)$ be a sequence of proper relations on X. Then, for each $a \in X$ there exists at least one $(a; (\mathscr{R}_n; n \geq 0))$-iterative sequence in X.*

Clearly, (DDC) includes (DC); to which it reduces when $(\mathscr{R}_n; n \geq 0)$ is constant. The reciprocal of this is also true. In fact, letting the premises of (DDC) hold, put $P = N \times X$; and let $\mathscr{S}$ be the relation over P introduced as

$\mathscr{S}(i, x) = \{i + 1\} \times \mathscr{R}_i(x), \ (i, x) \in P$.

It will suffice applying (DC) to $(P, \mathscr{S})$ and $b := (0, a) \in P$ to get the conclusion in our statement; we do not give details.

Summing up, (DDC) is provable in (ZF-AC+DC). This is valid as well for its variant, referred to as: the *Selected Dependent Choice* principle (in short: (SDC)).

Proposition 3 *Let the map $F : N \to \exp(X)$ and the relation $\mathscr{R}$ over X fulfill*

($\forall n \in N$): $\mathscr{R}(x) \cap F(n+1) \neq \emptyset$, for all $x \in F(n)$.

Then, for each $a \in F(0)$ there exists a sequence $(x(n); n \geq 0)$ in X, with

$$x(0) = a, \ x(n) \in F(n), \ x(n+1) \in \mathscr{R}(x(n)), \ \forall n.$$

As before, (SDC) $\Longrightarrow$ (DC) ($\Longleftrightarrow$ (DDC)); just take $(F(n) = X; n \geq 0)$. But, the reciprocal is also true, in the sense: (DDC) $\Longrightarrow$ (SDC). This follows from

Proof *(Proposition 3)* Let the premises of (SDC) be true. Define a sequence of relations $(\mathscr{R}_n; n \geq 0)$ over X as: for each $n \geq 0$,

$\mathscr{R}_n(x) = \mathscr{R}(x) \cap F(n+1)$, if $x \in F(n)$,
$\mathscr{R}_n(x) = \{x\}$, otherwise $(x \in X \setminus F(n))$.

Clearly, $\mathscr{R}_n$ is proper, for all $n \geq 0$. So, by (DDC), it follows that for the starting $a \in F(0)$, there exists an $(a, (R_n; n \geq 0))$-iterative sequence $(x(n); n \geq 0)$ in X. Combining with the very definition above, one derives that conclusion in the statement is holding. □

In particular, when $\mathscr{R} = X \times X$, the regularity condition imposed in (SDC) holds. The corresponding variant of the underlying statement is just (AC(N)) (=the *Denumerable Axiom of Choice*). Precisely, we have

Proposition 4 *Let $F : N \to \exp(X)$ be a function. Then, for each $a \in F(0)$ there exists a function $f : N \to X$ with $f(0) = a$ and $f(n) \in F(n)$, $\forall n \in N$.*

As a consequence of the above facts, (DC) $\Longrightarrow$ (AC(N)) in (ZF-AC). A direct verification of this is obtainable by taking $Q = N \times X$ and introducing the relation $\mathscr{S}$ over it, according to:

$\mathscr{S}(n, x) = \{n+1\} \times F(n+1), \ n \in N, x \in X;$

we do not give details. The reciprocal of the written inclusion is not true; see, for instance, Moskhovakis [30, Ch 8, Sect 8.25].

3 Conv-Cauchy Structures

Let X be a nonempty set; and $\mathscr{S}(X)$ stand for the class of all sequences (x_n) in X. By a (sequential) *convergence structure* on X we mean any part $\mathscr{C}$ of $\mathscr{S}(X) \times X$, with the properties (cf. Kasahara [23]):

(conv-1) $\mathscr{C}$ is *hereditary*:
$((x_n); x) \in \mathscr{C} \Longrightarrow ((y_n); x) \in \mathscr{C}$, for each subsequence (y_n) of (x_n)

(conv-2) $\mathscr{C}$ is *reflexive*: for each $u \in X$,
the constant sequence $(x_n = u; n \geq 0)$ fulfills $((x_n); u) \in \mathscr{C}$.

For each sequence (x_n) in $\mathscr{S}(X)$ and each $x \in X$, we write $((x_n); x) \in \mathscr{C}$ as $x_n \xrightarrow{\mathscr{C}} x$; this reads:

(x_n), $\mathscr{C}$-converges to x (also referred to as: x is the *$\mathscr{C}$-limit* of (x_n)).

The set of all such x is denoted $\mathscr{C} - \lim_n(x_n)$; when it is nonempty, we say that (x_n) is *$\mathscr{C}$-convergent*. The following condition is to be optionally considered here:

(conv-3) $\mathscr{C}$ is *separated*:
$\mathscr{C} - \lim_n(x_n)$ is an asingleton, for each sequence (x_n);

when it holds, $x_n \xrightarrow{\mathscr{C}} z$ will be also written as $\mathscr{C} - \lim_n(x_n) = z$.

Further, by a (sequential) *Cauchy structure* on X we shall mean any part $\mathscr{H}$ of $\mathscr{S}(X)$ with (cf. Turinici [48])

(Cauchy-1) $\mathscr{H}$ is *hereditary*:
$(x_n) \in \mathscr{H} \Longrightarrow (y_n) \in \mathscr{H}$, for each subsequence (y_n) of (x_n)
(Cauchy-2) $\mathscr{H}$ is *reflexive*: for each $u \in X$,
the constant sequence $(x_n = u; n \geq 0)$ fulfills $(x_n) \in \mathscr{H}$.

Each element of $\mathscr{H}$ will be referred to as a *$\mathscr{H}$-Cauchy* sequence in X.

Finally, given the couple $(\mathscr{C}, \mathscr{H})$ as before, we shall say that it is a *conv-Cauchy structure* on X. The optional conditions about the conv-Cauchy structure $(\mathscr{C}, \mathscr{H})$ to be considered here are

(CC-1) $(\mathscr{C}, \mathscr{H})$ is *regular*: each $\mathscr{C}$-convergent sequence is $\mathscr{H}$-Cauchy
(CC-2) $(\mathscr{C}, \mathscr{H})$ is *complete*: each $\mathscr{H}$-Cauchy sequence is $\mathscr{C}$-convergent.

A standard way of introducing such structures is the *(pseudo) metrical* one. By a *pseudometric* over X we shall mean any map $d : X \times X \to R_+$. Fix such an object, with, in addition,

(r-s) d is *reflexive sufficient*: $x = y \iff d(x, y) = 0$;

in this case, (X, d) is called a *rs-pseudometric space*. Given the sequence (x_n) in X and the point $x \in X$, we say that (x_n), *d-converges* to x (written as: $x_n \xrightarrow{d} x$) provided $d(x_n, x) \to 0$ as $n \to \infty$; i.e.,

$\forall \varepsilon > 0, \exists i = i(\varepsilon)$: $i \leq n \Longrightarrow d(x_n, x) < \varepsilon$.

By this very definition, we have the hereditary and reflexive properties:

(d-conv-1) $(\xrightarrow{d})$ is hereditary:
$x_n \xrightarrow{d} x$ implies $y_n \xrightarrow{d} x$, for each subsequence (y_n) of (x_n)
(d-conv-2) $(\xrightarrow{d})$ is reflexive: for each $u \in X$,
the constant sequence $(x_n = u; n \geq 0)$ fulfills $x_n \xrightarrow{d} u$.

As a consequence, $(\xrightarrow{d})$ is a sequential convergence on X. The set of all such limit points of (x_n) will be denoted $\lim_n(x_n)$; if it is nonempty, then (x_n) is called

d*-convergent*. Finally, note that $(\stackrel{d}{\longrightarrow})$ is not separated, in general. However, this property holds, provided (in addition)

(sym) d is *symmetric*: $d(x, y) = d(y, x)$, for all $x, y \in X$
(tri) d is triangular: $d(x, y) \leq d(x, z) + d(z, y), \forall x, y, z \in X$;

i.e. when d is a *metric* on X.

Further, call the sequence (x_n), d*-Cauchy* when $d(x_m, x_n) \to 0$ as $m, n \to \infty$, $m < n$; i.e.,

$\forall \varepsilon > 0, \exists j = j(\varepsilon): \; j \leq m < n \Longrightarrow d(x_m, x_n) < \varepsilon$;

the class of all these will be denoted as $Cauchy(d)$. As before, we have the hereditary and reflexive properties

(d-Cauchy-1) $Cauchy(d)$ is hereditary: (x_n) is d-Cauchy implies (y_n) is d-Cauchy, for each subsequence (y_n) of (x_n)
(d-Cauchy-2) $Cauchy(d)$ is reflexive: for each $u \in X$, the constant sequence $(x_n = u; n \geq 0)$ is d-Cauchy;

hence, $Cauchy(d)$ is a Cauchy structure on X.

Now, the couple $((\stackrel{d}{\longrightarrow}), Cauchy(d))$ will be referred to as a *conv-Cauchy structure* on X generated by d. Note that, by the imposed (upon d) conditions, this conv-Cauchy structure is not (regular or complete), in general. But, when d is symmetric triangular (hence, a metric) the regularity condition holds, as it can be directly seen.

Finally, let us say that $(x_n; n \geq 0)$ is d*-asymptotic*, provided

$d(x_n, x_{n+1}) \to 0$ as $n \to \infty$.

In this case, for each $\gamma > 0$,

$\mathscr{S}((x_n); \gamma) := \{k \in N; n \in N(k, \leq) \Longrightarrow d(x_n, x_{n+1}) < \gamma\}$
is nonempty; hence, $n(\gamma) := \min \mathscr{S}((x_n); \gamma)$ exists;

we then say that $n(\gamma)$ is the *asymptotic rank* attached to γ. Clearly,

$\gamma \mapsto n(\gamma)$ is decreasing: $\gamma_1 \leq \gamma_2 \implies n(\gamma_1) \geq n(\gamma_2)$.

It is immediate that each d-Cauchy sequence appears as d-asymptotic too; the reciprocal of this is not in general true.

We close this section with a few remarks involving convergent real sequences. For each sequence (r_n) in R, and each element $r \in R$, denote

$r_n \to r+$ (resp., $r_n \to r-$), when $r_n \to r$ and $[r_n > r$ (resp., $r_n < r$), $\forall n]$.

Proposition 5 *Let the sequence $(r_n; n \geq 0)$ in R and the number $\varepsilon \in R$ be such that $r_n \to \varepsilon+$. Then, there exists a subsequence $(r_n^* := r_{i(n)}; n \geq 0)$ of $(r_n; n \geq 0)$ with*

$(r_n^*; n \geq 0)$ is strictly descending and $r_n^* \to \varepsilon+$.

Proof Put $i(0) = 0$. As $\varepsilon < r_{i(0)}$ and $r_n \to \varepsilon+$, we have that

$A(i(0)) := \{n > i(0); r_n < r_{i(0)}\}$ is not empty;
hence, $i(1) := \min(A(i(0)))$ is an element of it, and $r_{i(1)} < r_{i(0)}$.

Likewise, as $\varepsilon < r_{i(1)}$ and $r_n \to \varepsilon+$, we have that

$A(i(1)) := \{n > i(1); r_n < r_{i(1)}\}$ is not empty;
hence, $i(2) := \min(A(i(1)))$ is an element of it, and $r_{i(2)} < r_{i(1)}$.

This procedure may continue indefinitely and yields (without any choice technique) a strictly ascending rank sequence $(i(n); n \geq 0)$ (hence, $i(n) \to \infty$ as $n \to \infty$) for which the attached subsequence $(r_n^* := r_{i(n)}; n \geq 0)$ of $(r_n; n \geq 0)$ fulfills

$r_{n+1}^* < r_n^*$, for all n; hence, (r_n^*) is (strictly) descending.

On the other hand, by this very subsequence property,

$(r_n^* > \varepsilon, \forall n)$, and $\lim_n r_n^* = \lim_n r_n = \varepsilon$.

Putting these together, we get the desired conclusion. □

A bi-dimensional counterpart of these facts may be given along the lines below. Let $\pi(t, s)$ (where $t, s \in R$) be a logical property involving pairs or real numbers. Given the couple of real sequences $(t_n; n \geq 0)$ and $(s_n; n \geq 0)$, call the subsequences $(t_n^*; n \geq 0)$ of (t_n) and $(s_n^*; n \geq 0)$ of (s_n), *compatible* when

$(t_n^* = t_{i(n)} n \geq 0)$, and $(s_n^* = s_{i(n)}; n \geq 0)$,
for the same strictly ascending rank sequence $(i(n); n \geq 0)$.

Proposition 6 *Let the couple of real sequences $(t_n; n \geq 0)$, $(s_n; n \geq 0)$, and the pair of real numbers (a, b) be such that*

$t_n \to a+$, $s_n \to b+$ as $n \to \infty$ and ($\pi(t_n, s_n)$ is true, $\forall n$).

There exists then a compatible couple of subsequences $(t_n^*; n \geq 0)$ of $(t_n; n \geq 0)$ and $(s_n^*; n \geq 0)$ of $(s_n; n \geq 0)$ respectively, with

(32-1) $(t_n^*; n \geq 0)$ and $(s_n^*; n \geq 0)$ are strictly descending, compatible
(32-2) ($t_n^* \to a+$, $s_n^* \to b+$, as $n \to \infty$), and ($\pi(t_n^*, s_n^*)$ holds, for all n).

Proof By the preceding statement, $(t_n; n \geq 0)$ admits a subsequence $(T_n := t_{i(n)}; n \geq 0)$, with

$(T_n; n \geq 0)$ is strictly descending, and ($T_n \to a+$, as $n \to \infty$).

Denote $(S_n := s_{i(n)}; n \geq 0)$; clearly,

$(S_n; n \geq 0)$ is a subsequence of $(s_n; n \geq 0)$ with $S_n \to b+$ as $n \to \infty$.

Moreover, by this very construction [$\pi(T_n, S_n)$ holds, for all n]. Again by the statement above, there exists a subsequence $(s_n^* := S_{j(n)} = s_{i(j(n))}; n \geq 0)$ of $(S_n; n \geq 0)$ (hence, of $(s_n; n \geq 0)$ as well), with

$(s_n^*; n \geq 0)$ is strictly descending, and $(s_n^* \to b+$, as $n \to \infty)$.

Denote further $(t_n^* := T_{j(n)} = t_{i(j(n))}; n \geq 0)$; this is a subsequence of $(T_n; n \geq 0)$ (hence, of $(t_n; n \geq 0)$ as well), with

$(t_n^*; n \geq 0)$ is strictly descending, and $(t_n^* \to a+$, as $n \to \infty)$;

Finally, by this very construction (and a previous relation) $[\pi(t_n^*, s_n^*)$ holds, for all $n]$. Summing up, the couple of subsequences $(t_n^*; n \geq 0)$ and $(s_n^*; n \geq 0)$ has all needed properties; and the conclusion follows. □

Note that further extensions of this result are possible, in the framework of quasi-metric spaces, taken as in Hitzler [13, Ch 1, Sect 1.2]; we shall discuss them in a separate paper.

4 Meir–Keeler Relations

Let $\Omega \subseteq R_+^0 \times R_+^0$ be a relation over R_+^0; as a rule, we write $(t, s) \in \Omega$ as $t\Omega s$. The starting global property to be considered upon this object is

(u-diag) Ω is *upper diagonal*: $t\Omega s$ implies $t < s$.

Denote the class of all upper diagonal relations as $\mathrm{udiag}(R_+^0)$. Our exposition below is essentially related to this basic condition.

To begin with, let us consider the global properties over $\mathrm{udiag}(R_+^0)$

(1-decr) Ω is *first variable decreasing*:
$t_1, t_2, s \in R_+^0$, $t_1 \geq t_2$ and $t_1 \Omega s$ imply $t_2 \Omega s$

(2-incr) Ω is *second variable increasing*:
$t, s_1, s_2 \in R_+^0$, $s_1 \leq s_2$ and $t\Omega s_1$ imply $t\Omega s_2$.

Then, define the sequential condition below (for upper diagonal relations)

(M-ad) Ω in *Matkowski admissible*:
$(t_n; n \geq 0)$ in R_+^0 and $(t_{n+1}\Omega t_n, \forall n)$ imply $\lim_n t_n = 0$.

To discuss it, the geometric conditions involving $\mathrm{udiag}(R_+^0)$ are in effect:

(g-mk) Ω has the *geometric Meir–Keeler property*:
$\forall \varepsilon > 0, \exists \delta > 0$: $t\Omega s, \varepsilon < s < \varepsilon + \delta \implies t \leq \varepsilon$

(g-bila-sep) Ω is *geometric bilateral separable*:
$\forall \beta > 0, \exists \gamma \in]0, \beta[, \forall (t, s)$: $t, s \in]\beta - \gamma, \beta + \gamma[\implies (t, s) \notin \Omega$.

The former of these local conditions—related to the developments in Meir and Keeler [28]—is strongly related to the Matkowski admissible property we just introduced. Precisely, the following basic fact is available.

Theorem 2 *Under these conditions, one has in (ZF-AC+DC):*

(41-a) *(for each* $\Omega \in \text{udiag}(R_+^0)$*):*
Ω is geometric Meir–Keeler implies Ω is Matkowski admissible
(41-b) *(for each first variable decreasing* $\Omega \in \text{udiag}(R_+^0)$*):*
Ω is Matkowski admissible implies Ω is geometric Meir–Keeler.

Hence, summing up

(41-c) *(for each first variable decreasing* $\Omega \in \text{udiag}(R_+^0)$*):*
Ω is geometric Meir–Keeler iff Ω is Matkowski admissible.

Proof Three basic stages must be passed.

(i) Suppose that $\Omega \in \text{udiag}(R_+^0)$ is geometric Meir–Keeler; we have to establish that Ω is Matkowski admissible. Let $(t_n; n \geq 0)$ be a sequence in R_+^0, fulfilling $(t_{n+1}\Omega t_n$, for all $n)$. By the upper diagonal property, we get

$(t_{n+1} < t_n$, for all $n)$; i.e. (t_n) is strictly descending.

As a consequence, $\tau = \lim_n t_n$ exists in R_+; with, in addition: $(t_n > \tau, \forall n)$. Assume by contradiction that $\tau > 0$; and let $\sigma > 0$ be the number assured by the geometric Meir–Keeler property. By definition, there exists an index $n(\sigma)$, with

$(t_{n+1}\Omega t_n$ and) $\tau < t_n < \tau + \sigma$, for all $n \geq n(\sigma)$.

This, by the quoted property, gives (for the same ranks)

$\tau < t_{n+1} \leq \tau$; contradiction.

Hence, necessarily, $\tau = 0$; and the conclusion follows.

(ii) Suppose that the first variable decreasing $\Omega \in \text{udiag}(R_+^0)$ is Matkowski admissible; we have to establish that Ω is geometric Meir–Keeler. Suppose by contradiction that this is not true; i.e. (for some $\varepsilon > 0$)

$H(\delta) := \{(t, s) \in \Omega; \varepsilon < s < \varepsilon + \delta, t > \varepsilon\}$ is nonempty, for each $\delta > 0$.

Taking a zero converging sequence $(\delta_n; n \geq 0)$ in R_+^0, we get by the Denumerable Axiom of Choice (AC(N)) [deductible, as precise, in (ZF-AC+DC)], a sequence $((t_n, s_n); n \geq 0)$ in $R_+^0 \times R_+^0$, so as

$(\forall n)$: (t_n, s_n) is an element of $H(\delta_n)$;

or, equivalently (by definition and upper diagonal property)

$(t_n\Omega s_n$ and) $\varepsilon < t_n < s_n < \varepsilon + \delta_n$, for all n.

Note that, as a direct consequence,

$(t_n\Omega s_n$, for all $n)$, and $t_n \to \varepsilon+$, $s_n \to \varepsilon+$, as $n \to \infty$.

Put $i(0) = 0$. As $\varepsilon < t_{i(0)}$ and $s_n \to \varepsilon+$ as $n \to \infty$, we have that

$A(i(0)) := \{n > i(0); s_n < t_{i(0)}\}$ is not empty;
hence, $i(1) := \min(A(i(0)))$ is an element of it, and $s_{i(1)} < t_{i(0)}$;
wherefrom, $s_{i(1)} \Omega s_{i(0)}$ (as Ω is first variable decreasing).

Likewise, as $\varepsilon < t_{i(1)}$ and $s_n \to \varepsilon+$ as $n \to \infty$, we have that

$A(i(1)) := \{n > i(1); s_n < t_{i(1)}\}$ is not empty;
hence, $i(2) := \min(A(i(1)))$ is an element of it, and $s_{i(2)} < t_{i(1)}$;
wherefrom, $s_{i(2)} \Omega s_{i(1)}$ (as Ω is first variable decreasing).

This procedure may continue indefinitely and yields (without any choice technique) a strictly ascending rank sequence $(i(n); n \geq 0)$ in N for which the attached subsequence $(r_n := s_{i(n)}; n \geq 0)$ of $(s_n; n \geq 0)$ fulfills

$r_{n+1} \Omega r_n$, for all n; whence $r_n \to 0$ (as Ω is Matkowski admissible).

On the other hand, by our subsequence property,

$(r_n > \varepsilon, \forall n)$ and $\lim_n r_n = \lim_n s_n = \varepsilon$; that is: $r_n \to \varepsilon+$.

The obtained relation is in contradiction with the previous one. Hence, the working condition cannot be true; and we are done.

(iii) Evident, by the above.

□

In the following, equivalent (sequential) conditions are given for the properties appearing in our (geometric) concepts above. Given the upper diagonal relation Ω over R^0_+, let us introduce the (asymptotic type) conventions

(a-mk) Ω is *asymptotic Meir–Keeler*:
there are no strictly descending sequences (t_n) and (s_n) in R^0_+ and no elements ε in R^0_+, with $((t_n, s_n) \in \Omega, \forall n)$ and $(t_n \to \varepsilon+, s_n \to \varepsilon+)$

(a-bila-sep) Ω is *asymptotic bilateral separable*:
there are no sequences $(t_n; n \geq 0)$ and $(s_n; n \geq 0)$ in R^0_+ and no elements $\beta \in R^0_+$, with $((t_n, s_n) \in \Omega, \forall n)$ and $(t_n \to \beta, s_n \to \beta)$.

Remark 1 The inclusion between these two concepts may be described as

(for each upper diagonal relation $\Omega \subseteq R^0_+ \times R^0_+$):
Ω is asymptotic bilateral separable implies Ω is asymptotic Meir–Keeler.

In fact, let the upper diagonal relation $\Omega \subseteq R^0_+ \times R^0_+$ be asymptotic bilateral separable; and assume by contradiction that Ω is not asymptotic Meir–Keeler:

there exist strictly descending sequences (t_n) and (s_n) in R^0_+ and elements ε in R^0_+, with $((t_n, s_n) \in \Omega, \forall n)$ and $(t_n \to \varepsilon+, s_n \to \varepsilon+)$; hence, $(t_n \to \varepsilon, s_n \to \varepsilon)$.

This tells us that Ω is not asymptotic bilateral separable; in contradiction with the working hypothesis; and the assertion follows.

Concerning the relationships between the introduced asymptotic concepts and their corresponding geometric concepts, the following statement is to be noted.

Theorem 3 *The following generic relationships are valid (for an arbitrary upper diagonal relation $\Omega \subseteq R^0_+ \times R^0_+$), in the reduced system (ZF-AC+DC):*

(42-a) geometric Meir–Keeler equals asymptotic Meir–Keeler
(42-b) geometric bilateral separable equals asymptotic bilateral separable
(42-c) geometric bilateral separable implies geometric Meir–Keeler.

Proof

(i) Firstly, we discuss the case of Meir–Keeler property.

(i-1) Let $\Omega \in \text{udiag}(R^0_+)$ be a geometric Meir–Keeler relation; but—contrary to the conclusion—assume that Ω does not have the asymptotic Meir–Keeler property:

there exist two strictly descending sequences (t_n) and (s_n) in R^0_+ and an element ε in R^0_+, with $((t_n, s_n) \in \Omega, \forall n)$ and $(t_n \to \varepsilon+, s_n \to \varepsilon+)$.

Let $\delta > 0$ be the number given by the geometric Meir–Keeler property of Ω. By definition, there exists a (common) rank $n(\delta)$, such that

$n \geq n(\delta)$ implies $\varepsilon < t_n < \varepsilon + \delta, \varepsilon < s_n < \varepsilon + \delta$.

From the second relation, we must have (by the hypothesis about Ω) $t_n \leq \varepsilon$, for all $n \geq n(\delta)$. This, however, contradicts the first relation above. Hence, Ω is asymptotic Meir–Keeler; as asserted.

(i-2) Let $\Omega \in \text{udiag}(R^0_+)$ be an asymptotic Meir–Keeler relation; but—contrary to the conclusion—assume that Ω does not have the geometric Meir–Keeler property; i.e. (for some $\varepsilon > 0$)

$H(\delta) := \{(t, s) \in \Omega; \varepsilon < s < \varepsilon + \delta, t > \varepsilon\} \neq \emptyset$, for each $\delta > 0$.

Taking a zero converging sequence $(\delta_n; n \geq 0)$ in R^0_+, we get by the Denumerable Axiom of Choice (AC(N)) [deductible, as precise, in (ZF-AC+DC)], a sequence $((t_n, s_n); n \geq 0)$ in $R^0_+ \times R^0_+$, so as

$(\forall n)$: (t_n, s_n) is an element of $H(\delta_n)$;

or, equivalently (by definition and upper diagonal property)

$((t_n, s_n) \in \Omega$ and$)$ $\varepsilon < t_n < s_n < \varepsilon + \delta_n$, for all n.

Note that, as a direct consequence,

$(t_n \Omega s_n$, for all $n)$, and $t_n \to \varepsilon+$, $s_n \to \varepsilon+$, as $n \to \infty$.

By a previous result, there exists a compatible couple of subsequences $(t^*_n := t_{i(n)}; n \geq 0)$ of $(t_n; n \geq 0)$ and $(s^*_n := s_{i(n)}; n \geq 0)$ of $(s_n; n \geq 0)$, with

$(t^*_n \Omega s^*_n, \forall n)$; (t^*_n), (s^*_n) are strictly descending; $t^*_n \to \varepsilon+$ and $s^*_n \to \varepsilon+$.

This, however, is in contradiction with respect to the posed hypothesis upon Ω; wherefrom, our assertion follows.

(ii) Secondly, we discuss the case of bilateral separable property.

(ii-1) Let $\Omega \in \text{udiag}(R^0_+)$ be a geometric bilateral separable relation; we have to establish that Ω is asymptotic bilateral separable. Suppose—contrary to this conclusion—that Ω is not endowed with such a property; that is

there are two sequences $(t_n; n \geq 0)$ and $(s_n; n \geq 0)$ in R^0_+ and an element $\beta \in R^0_+$, with $((t_n, s_n) \in \Omega, \forall n)$ and $(t_n \to \beta, s_n \to \beta)$.

Let $\gamma \in]0, \beta[$ be the number given by the geometric bilateral separable property of Ω. By definition, there exists a (common) rank $n(\gamma)$, such that

$n \geq n(\gamma)$ implies $\beta - \gamma < t_n < \beta + \gamma$ and $\beta - \gamma < s_n < \beta + \gamma$.

This, along with $[t_n \Omega s_n, \forall n \geq n(\gamma)]$ contradicts the geometric bilateral separable property of Ω. Hence, Ω is asymptotic bilateral separable.

(ii-2) Let $\Omega \in \text{udiag}(R^0_+)$ be an asymptotic bilateral separable relation; we have to establish that Ω is geometric bilateral separable. Suppose—contrary to this conclusion—that Ω is not endowed with such a property; that is (for some $\beta > 0$)

$K(\gamma) := \{(t, s) \in \Omega; t, s \in]\beta - \gamma, \beta + \gamma[\} \neq \emptyset$, for each $\gamma \in]0, \beta[$.

Taking a strictly ascending sequence $(\gamma_n; n \geq 0)$ in $]0, \beta[$ with $\gamma_n \to 0$, we get by the Denumerable Axiom of Choice (AC(N)) [deductible, as precise, in (ZF-AC+DC)], a sequence $((t_n, s_n); n \geq 0)$ in Ω, so as

$(\forall n)$: (t_n, s_n) is an element of $K(\gamma_n)$;

or, equivalently (by the very definition above)

$(\forall n)$: $(t_n, s_n) \in \Omega$ and $t_n, s_n \in]\beta - \gamma_n, \beta + \gamma_n[$.

As a consequence of the latter, we must have $(t_n \to \beta, s_n \to \beta)$; and this, along with the former, contradicts the imposed hypothesis. Hence, necessarily, Ω is geometric bilateral separable.

(iii) Finally, it remains to establish the relationships between the underlying geometric properties. This, however, is evident, by the preceding stages and a previous remark; wherefrom, we are done.

□

In the following, some basic examples of (upper diagonal) Matkowski admissible and geometric Meir–Keeler relations are given. The general scheme of constructing these is described along the lines below.

Let $R(\pm\infty) := R \cup \{-\infty, \infty\}$ stand for the set of all *extended real numbers*. For each relation Ω over R^0_+, let us associate a function $\xi : R^0_+ \times R^0_+ \to R(\pm\infty)$, as

$\xi(t, s) = 0$, if $(t, s) \in \Omega$; $\xi(t, s) = -\infty$, if $(t, s) \notin \Omega$.

It will be referred to as the *function* generated by Ω; clearly,

$(t, s) \in \Omega$ iff $\xi(t, s) \geq 0$.

Conversely, given a function $\xi : R^0_+ \times R^0_+ \to R(\pm\infty)$, we may associate it a relation Ω over R^0_+ as

$\Omega = \{(t, s) \in R^0_+ \times R^0_+; \xi(t, s) \geq 0\}$ (in short: $\Omega = [\xi \geq 0]$);
referred to as: the *positive section* of ξ.

Note that the correspondence between the function ξ and its associated relation $[\xi \geq 0]$ is not injective; because, for the function $\eta := \lambda\xi$ (where $\lambda > 0$), its associated relation $[\eta \geq 0]$ is identical with the relation $[\xi \geq 0]$ attached to ξ.

Now, call the function $\xi : R^0_+ \times R^0_+ \to R(\pm\infty)$, *upper diagonal* provided:

(u-diag) $\xi(t, s) \geq 0$ implies $t < s$.

All subsequent constructions are being considered within this setting. This, in particular, includes the sequential condition for upper diagonal functions ξ:

(M-ad) ξ in *Matkowski admissible*:
$(t_n; n \geq 0)$ in R^0_+ and $(\xi(t_{n+1}, t_n) \geq 0, \forall n)$ imply $\lim_n t_n = 0$.

The following geometric conditions involving our functions are—in particular—useful for discussing this property

(g-mk) ξ is *geometric Meir–Keeler*:
$\forall \varepsilon > 0, \exists \delta > 0$: $\xi(t, s) \geq 0, \varepsilon < s < \varepsilon + \delta \implies t \leq \varepsilon$

(g-bila-sep) ξ is *geometric bilateral separable*:
$\forall \beta > 0, \exists \gamma \in]0, \beta[, \forall (t, s)$: $t, s \in]\beta - \gamma, \beta + \gamma[\implies \xi(t, s) < 0$.

The relationships between the geometric Meir–Keeler condition and the Matkowski one attached to upper diagonal functions are nothing else than a simple translation of the previous ones involving upper diagonal relations; we do not give details.

Summing up, the duality principles below are holding:

(DP-1) any concept (like the ones above) about (upper diagonal) relations over R^0_+ may be written as a concept about (upper diagonal) functions in the class $\mathscr{F}(R^0_+ \times R^0_+, R(\pm\infty))$

(DP-2) any concept (like the ones above) about (upper diagonal) functions in the class $\mathscr{F}(R^0_+ \times R^0_+, R(\pm\infty))$ may be written as a concept about (upper diagonal) relations over R^0_+.

For the rest of our exposition, it will be convenient working with relations over R^0_+, and not with functions in $\mathscr{F}(R^0_+ \times R^0_+, R(\pm\infty))$; this, however, is nothing but a methodology question.

We may now pass to the description of some basic objects in this area.

Part-Case I) Let $\mathscr{F}(re)(R^0_+, R)$ stand for the subclass of all $\varphi \in \mathscr{F}(R^0_+, R)$ with

φ is *regressive*: $\varphi(t) < t$, for all $t > 0$.

Call $\varphi \in \mathscr{F}(re)(R^0_+, R)$, *Meir–Keeler admissible* if

(mk-adm) $\forall \gamma > 0, \exists \beta > 0, \forall t$: $\gamma < t < \gamma + \beta \implies \varphi(t) \leq \gamma$.

Some important examples of such functions may be given along the lines below.

For any $\varphi \in \mathscr{F}(re)(R^0_+, R)$ and any $s \in R^0_+$, put

$\Lambda^+\varphi(s) = \inf_{0<\varepsilon<s} \Phi(s+)(\varepsilon)$; where $\Phi(s+)(\varepsilon) = \sup \varphi(]s, s+\varepsilon[)$

$\Lambda^\pm\varphi(s) = \inf_{0<\varepsilon<s} \Phi(s\pm)(\varepsilon)$; where $\Phi(s\pm)(\varepsilon) = \sup \varphi(]s-\varepsilon, s+\varepsilon[)$.

From the regressive property of φ, these limit quantities fulfill

$(-\infty \le)\ \Lambda^+\varphi(s) \le \Lambda^\pm\varphi(s) \le s,\ \forall s \in R^0_+$

but the case of such limits having infinite values cannot be avoided.

The following auxiliary fact will be useful.

Proposition 7 *Let $\varphi \in \mathscr{F}(re)(R^0_+, R)$ and $s \in R^0_+$ be arbitrary fixed. Then,*

(41-1) $\limsup_n(\varphi(t_n)) \le \Lambda^+\varphi(s)$,
for each sequence (t_n) in R^0_+ with $t_n \to s+$

(41-2) $\limsup_n(\varphi(t_n)) \le \Lambda^\pm\varphi(s)$,
for each sequence (t_n) in R^0_+ with $t_n \to s$.

Proof

(i) Given $\varepsilon \in]0, s[$, there exists a rank $p(\varepsilon) \ge 0$ such that $s < t_n < s+\varepsilon$, for all $n \ge p(\varepsilon)$; hence

$$\limsup_n(\varphi(t_n)) \le \sup\{\varphi(t_n); n \ge p(\varepsilon)\} \le \Phi(s+)(\varepsilon).$$

It suffices taking the infimum over ε in this relation to get the desired fact.

(ii) Given $\varepsilon \in]0, s[$, there exists a rank $p(\varepsilon) \ge 0$ such that $s-\varepsilon < t_n < s+\varepsilon$, for all $n \ge p(\varepsilon)$; hence

$$\limsup_n(\varphi(t_n)) \le \sup\{\varphi(t_n); n \ge p(\varepsilon)\} \le \Phi(s\pm)(\varepsilon).$$

Taking the infimum over ε in this relation, we get the desired conclusion.

□

Call $\varphi \in \mathscr{F}(re)(R^0_+, R)$, *Boyd–Wong admissible* [4], if

(bw-adm) $\Lambda^+\varphi(s) < s$, for all $s > 0$.

In particular, $\varphi \in \mathscr{F}(re)(R^0_+, R)$ is Boyd–Wong admissible provided it is *upper semicontinuous at the right* on R^0_+:

$\Lambda^+\varphi(s) \le \varphi(s)$, for each $s \in R^0_+$.

This, e.g., is fulfilled when φ is *continuous at the right* on R^0_+; for, in such a case,

$\Lambda^+\varphi(s) = \varphi(s)$, for each $s \in R^0_+$.

On the other hand, $\varphi \in \mathscr{F}(re)(R^0_+, R)$ is Boyd–Wong admissible when

φ is *strongly Boyd–Wong admissible*: $\Lambda^\pm\varphi(s) < s, \forall s \in R^0_+$.

Further, let $\mathscr{F}(re, in)(R^0_+, R)$ stand for the class of all $\varphi \in \mathscr{F}(re)(R^0_+, R)$, with

φ is increasing on R^0_+ $(0 < t_1 \le t_2$ implies $\varphi(t_1) \le \varphi(t_2))$.

Then, let us say that $\varphi \in \mathscr{F}(re, in)(R^0_+, R)$ is *Matkowski admissible* [26], provided

(m-adm) $(\forall t > 0)$: $\lim_n \varphi^n(t) = 0$, as long as $(\varphi^n(t); n \geq 0)$ exists.

Here, as usual, we denoted for each $t > 0$

$\varphi^0(t) = t, \varphi^1(t) = \varphi(t), \ldots, \varphi^{n+1}(t) = \varphi(\varphi^n(t)), n \geq 1.$

Note that such a construction may be non-effective; e.g.,

$\varphi^2(t) = \varphi(\varphi(t))$ is undefined whenever $\varphi(t) \leq 0$.

Remark 2 Under these conventions,

(BW-mk) each Boyd–Wong admissible function in $\mathscr{F}(re)(R^0_+, R)$ is Meir–Keeler admissible
(M-mk) each Matkowski admissible function in $\mathscr{F}(re, in)(R^0_+, R)$ is Meir–Keeler admissible.

The verification of this is as follows.

(i) (cf. Boyd and Wong [4]). Suppose that $\varphi \in \mathscr{F}(re)(R^0_+, R)$ is Boyd–Wong admissible, and fix $\gamma > 0$; hence $\Lambda^+\varphi(\gamma) < \gamma$. By definition, there exists $\beta = \beta(\gamma) > 0$ with $[\gamma < t < \gamma + \beta$ implies $\varphi(t) < \gamma]$, proving that φ is Meir–Keeler admissible.
(ii) (cf. Jachymski [14]). Assume that $\varphi \in \mathscr{F}(re, in)(R^0_+, R)$ is Matkowski admissible. If the underlying property fails, then (for some $\gamma > 0$):

$\forall \beta > 0, \exists t \in]\gamma, \gamma + \beta[$, such that $\varphi(t) > \gamma$.

Combining with the increasing property of φ, one gets

$(\forall t > \gamma)$: $\varphi(t) > \gamma$ [whence (by induction): $\varphi^n(t) > \gamma$, for each n].

Fixing some $t > \gamma$ and passing to limit as $n \to \infty$, one derives $0 \geq \gamma$; contradiction; hence the claim. Further aspects may be found in Turinici [44].

Having these precise, take a function $\chi \in \mathscr{F}(re)(R^0_+, R)$ and define the associated relation $\Omega := \Omega[\chi]$ over R^0_+, as

$(t, s \in R^0_+)$: $(t, s) \in \Omega$ iff $t \leq \chi(s)$.

Clearly, Ω is upper diagonal. In fact, let $t, s \in R^0_+$ be such that $t\Omega s$; i.e. $t \leq \chi(s)$. As χ is regressive, one has $\chi(s) < s$; and this yields $t < s$; whence the conclusion follows. Further properties of this relation are deductible from

Proposition 8 *Let the function $\chi \in \mathscr{F}(re)(R^0_+, R)$ be given, and $\Omega := \Omega[\chi]$ stand for the associated upper diagonal relation over R^0_+. Then,*

(42-1) Ω *is geometric/asymptotic Meir–Keeler when the starting function* χ *is Meir–Keeler admissible.*

(42-2) Ω *is geometric/asymptotic bilateral separable (hence, necessarily, geometric/asymptotic Meir–Keeler) when* χ *is strongly Boyd–Wong admissible.*

Proof

(i) Let $\varepsilon > 0$ be given; and $\delta > 0$ be the number associated with it, via Meir–Keeler admissible property for χ. Given $t, s \in R_+^0$ with $t\Omega s$, $\varepsilon < s < \varepsilon + \delta$, we have $[t \le \chi(s), \varepsilon < s < \varepsilon + \delta]$. This, according to the underlying property of χ, gives $\chi(s) \le \varepsilon$ [hence, $t \le \varepsilon$]; wherefrom: Ω has the geometric Meir–Keeler property.

(ii) Suppose, by absurd, that Ω is not asymptotic bilateral separable:

there are sequences $(t_n; n \ge 0)$ and $(s_n; n \ge 0)$ in R_+^0 and elements $\beta \in R_+^0$, with $((t_n, s_n) \in \Omega, \forall n)$ and $(t_n \to \beta, s_n \to \beta)$.

By the definition of our relation,

$(t_n \le \chi(s_n), \forall n)$, and $t_n \to \beta, s_n \to \beta$.

Passing to lim sup as $n \to \infty$, yields (by a previous result)

$\beta \le \Lambda^{\pm}\chi(\beta) < \beta$; contradiction;

and this proves our assertion.

□

Part-Case II) Let (ψ, φ) be a couple of functions over $\mathscr{F}(R_+^0, R)$, with

(norm) (ψ, φ) is *normal*:
ψ is increasing and φ is *strictly positive* $[\varphi(t) > 0, \forall t > 0]$.

(This concept may be related to the one introduced by Rhoades [38]; see also Dutta and Choudhury [10]). Then, define the relation $\Omega = \Omega[\psi, \varphi]$ in $\exp(R_+^0 \times R_+^0)$, as

$(t, s) \in \Omega$ iff $\psi(t) \le \psi(s) - \varphi(s)$.

We claim that, necessarily, Ω is upper diagonal. In fact, let $t, s \in R_+^0$ be such that

$(t, s) \in \Omega$; i.e. $\psi(t) \le \psi(s) - \varphi(s)$.

By the strict positivity of φ, one gets $\psi(t) < \psi(s)$; and this, along with the increasing property of ψ, shows that $t < s$; whence the conclusion follows. Further properties of this relation are available under certain supplementary conditions about the normal couple (ψ, φ), like below:

(as-pos) φ is *asymptotic positive*:
for each strictly descending sequence $(t_n; n \ge 0)$ in R_+^0 and each $\varepsilon > 0$ with $t_n \to \varepsilon+$, we must have $\limsup_n(\varphi(t_n)) > 0$.

(bd-osc) (ψ, φ) is *limit-bounded oscillating*:
for each sequence $(t_n; n \geq 0)$ in R_+^0 and each $\beta > 0$ with $t_n \to \beta$,
we have $\limsup_n(\varphi(t_n)) > \psi(\beta + 0) - \psi(\beta - 0)$.

Clearly

(for each normal couple (ψ, φ)):
(ψ, φ) is limit-bounded oscillating implies φ is asymptotic positive.

On the other hand, sufficient conditions under which the first property holds are obtainable (under the same normality setting) via

(φ=increasing or continuous) implies φ=asymptotic positive.

In fact, let the strictly descending sequence $(t_n; n \geq 0)$ in R_+^0 and the number $\varepsilon > 0$ be such that $t_n \to \varepsilon+$. When φ=increasing, we have (by normality)

$\varphi(t_n) \geq \varphi(\varepsilon) > 0, \forall n$; whence $\limsup_n(\varphi(t_n)) \geq \varphi(\varepsilon) > 0$.

On the other hand, when φ=continuous, the same normality condition yields

$\limsup_n(\varphi(t_n)) = \lim_n(\varphi(t_n)) = \varphi(\varepsilon) > 0$, and conclusion follows.

Proposition 9 *Let (ψ, φ) be a normal couple of functions over $\mathscr{F}(R_+^0, R)$; and $\Omega := \Omega[\psi, \varphi]$ be the associated upper diagonal relation. Then,*

(43-1) If φ is asymptotic positive, then the associated relation Ω is asymptotic/geometric Meir–Keeler.
(43-2) If (ψ, φ) is limit-bounded oscillating, then the associated relation Ω is asymptotic/geometric bilateral separable (hence, asymptotic/geometric Meir–Keeler as well).

Proof

(i) Suppose by contradiction that Ω is not asymptotic Meir–Keeler:

There exist strictly descending sequences (t_n) and (s_n) in R_+^0
and elements ε in R_+^0 with $((t_n, s_n) \in \Omega, \forall n)$ and $(t_n \to \varepsilon+, s_n \to \varepsilon+)$.

By the former of these, we get

$(0 <)\varphi(s_n) \leq \psi(s_n) - \psi(t_n), \forall n$.

Passing to limit as $n \to \infty$, and noting that $\lim_n \psi(s_n) = \lim_n \psi(t_n) = \psi(\varepsilon + 0)$, one gets $\lim_n \varphi(t_n) = 0$; in contradiction with the asymptotic positivity of φ. So, necessarily, Ω has the asymptotic Meir–Keeler property; as claimed.

(ii) Suppose by contradiction that Ω is not asymptotic bilateral separable; i.e.

there exist sequences (t_n) and (s_n) in R_+^0 and elements β in R_+^0,
with $((t_n, s_n) \in \Omega, \forall n)$ and $(t_n \to \beta, s_n \to \beta)$.

By the former of these, we get

$(0 <)\varphi(s_n) \leq \psi(s_n) - \psi(t_n), \forall n$.

Passing to lim sup as $n \to \infty$ yields $\limsup_n \varphi(s_n) \le \psi(\beta+0) - \psi(\beta-0)$, in contradiction with (ψ, φ) being limit-bounded oscillating. This tells us that Ω is asymptotic bilateral separable, as claimed.

□

In the following, some basic (and useful) particular choices for the couple (ψ, φ) above are to be discussed.

Part-Case II-a) The construction in the preceding step (involving a certain $\chi \in \mathscr{F}(re)(R_+^0, R))$ is nothing else than a particular case of this one, corresponding to the choice

$$\psi(t) = t, \varphi(t) = t - \chi(t), t \in R_+^0.$$

Since the verification is immediate, we do not give details.

Part-Case II-b) Let $\lambda : R_+^0 \to]1, \infty[$ and $\mu : R_+^0 \to]0, 1[$ be a couple of functions, with λ=increasing. Define a relation $\Omega := \Omega[[\lambda, \mu]]$ over R_+^0 as

$$t\Omega s \text{ iff } \lambda(t) \le [\lambda(s)]^{\mu(s)}.$$

This will be referred to as the *Jleli–Samet relation* attached to $\lambda(.)$ and $\mu(.)$. (The proposed conventions come from the developments in Jleli and Samet [18], corresponding to $\mu(.)$=constant). By a direct calculation, it is evident that

$$t\Omega s \text{ iff } t\Omega[\psi, \varphi]s; \text{ where } \psi(t) = \log[\log(\lambda(t))], \varphi(t) = -\log(\mu(t)), t > 0.$$

Hence, this construction is entirely reducible to the standard one in this series.

Part-Case II-c) Let the couple (ψ, α) over $\mathscr{F}(R_+^0, R)$ be admissible; i.e.

(admi-1) $\psi(.)$ is increasing, right continuous, strictly positive.
(admi-2) $-\alpha(.)$ is right lsc on R_+^0, and $\gamma := \psi - \alpha$ is strictly positive.

Proposition 10 *Let the functions (ψ, α) be as before. Then,*

(44-1) *The couple (ψ, γ) (where $\gamma = \psi - \alpha$) is a normal couple over $\mathscr{F}(R_+^0, R)$, with γ=asymptotic positive.*

(44-2) *The associated with (ψ, γ) relation*

$$t\Omega s \text{ iff } \psi(t) \le \psi(s) - \gamma(s) \text{ (that is: } \psi(t) \le \alpha(s))$$

is upper diagonal and asymptotic (hence, geometric) Meir–Keeler.

Proof

(i) By definition, ψ is increasing and γ is strictly positive.
(ii) Suppose by contradiction that $\gamma(.)$ is not asymptotic positive: There exist $\varepsilon > 0$ and a strictly descending sequence (t_n) in R_+^0, with

$$t_n \to \varepsilon+ \text{ and } \limsup_n(\gamma(t_n)) = 0; \text{ whence, } \lim_n(\gamma(t_n)) = 0.$$

The last relation gives

$\lim_n(-\alpha(t_n)) = -\psi(\varepsilon)$ (as ψ is right continuous).

Combining with $-\alpha(.)$ being right lsc on R^0_+, yields (by this limit process)

$-\alpha(\varepsilon) \leq -\psi(\varepsilon)$; or, equivalently: $\gamma(\varepsilon) \leq 0$;

in contradiction with the strict positivity of γ. Hence, our working assumption is not acceptable, and the claim follows.

(iii) Evident, by our previous facts.

□

Part-Case II-d) Let $\psi \in \mathscr{F}(R^0_+, R)$ and $\Delta \in \mathscr{F}(R)$ be a couple of functions. The following regularity condition involving these objects will be considered here:

(BV-c) (ψ, Δ) is a *Bari–Vetro couple*:
ψ is increasing and Δ is regressive ($\Delta(r) < r$, for all $r \in R$).

In this case, by definition,

$\varphi(t) := \psi(t) - \Delta(\psi(t)) > 0$, for all $t > 0$;

so that, (ψ, φ) is a normal couple of functions over $\mathscr{F}(R^0_+, R)$. Let $\Omega := \Omega[\psi, \Delta]$ be the (associated) *Bari–Vetro relation* over R^0_+, introduced as

$t\Omega s$ iff $\psi(t) \leq \Delta(\psi(s))$.

(This convention is related to the developments in Di Bari and Vetro [9]). From (BV-c), Ω is an upper diagonal relation over R^0_+. It is natural then to ask under which extra assumptions about our data we have that Ω is an asymptotic Meir–Keeler relation. The simplest one may be written as

(a-reg) Δ is *asymptotic regressive*:
for each descending sequence (r_n) in R and each $\alpha \in R$ with $r_n \to \alpha$, we have that $\liminf_n \Delta(r_n) < \alpha$.

Note that, by the non-strict character of the descending property above, one has

Δ is asymptotic regressive implies Δ is regressive.

Proposition 11 *Let the functions ($\psi \in \mathscr{F}(R^0_+, R)$, $\Delta \in \mathscr{F}(R)$) be such that*

(ψ, Δ) is an asymptotic Bari–Vetro couple; i.e.
ψ is increasing and Δ is asymptotic regressive.

Then,

(45-1) *the above defined function φ is asymptotic positive.*
(45-2) *the associated relation Ω is upper diagonal, and asymptotic Meir–Keeler (hence, geometric Meir–Keeler).*

Proof

(i) Let the strictly descending sequence $(t_n; n \geq 0)$ in R^0_+ and the number $\varepsilon > 0$ be such that $t_n \to \varepsilon+$; we must derive that $\limsup_n(\varphi(t_n)) > 0$. Denote

$(r_n = \psi(t_n), n \geq 0); \alpha = \psi(\varepsilon + 0)$.

By the imposed conditions (and ψ=increasing)

(r_n) is descending and $r_n \to \alpha$ as $n \to \infty$.

In this case,

$$\limsup_n \varphi(t_n) = \limsup_n [r_n - \Delta(r_n)] = \alpha - \liminf_n \Delta(r_n) > 0;$$

hence the claim.

(ii) The assertion follows at once from (ψ, φ) being a normal couple with (φ=asymptotic positive), and a previous remark involving these objects. However, for completeness reasons, we provide an argument for this.

(ii-1) Let $t, s > 0$ be such that

$t\Omega s$; i.e. $\psi(t) \leq \Delta(\psi(s))$.

As Δ is regressive,

$\psi(t) < \psi(s)$; whence, $t < s$ (in view of ψ=increasing);

so that, Ω is upper diagonal.

(ii-2) Suppose by contradiction that there exists a couple of strictly descending sequences (t_n) and (s_n) in R^0_+, and a number $\varepsilon > 0$, with

$t_n \to \varepsilon+$, $s_n \to \varepsilon+$, and $t_n \Omega s_n$ [i.e. $\psi(t_n) \leq \Delta(\psi(s_n))$], for each n.

From the increasing property of ψ, one has (under $\alpha := \psi(\varepsilon + 0)$)

$(u_n := \psi(t_n))$ and $(v_n := \psi(s_n))$ are descending sequences in R, with
$u_n \to \alpha$, $v_n \to \alpha$, as $n \to \infty$;

so, passing to $\liminf$ as $n \to \infty$ in the relation above [i.e. $u_n \leq \Delta(v_n)$, $\forall n$], one gets (via Δ=asymptotic regressive)

$$\alpha = \liminf_n u_n \leq \liminf_n \Delta(v_n) < \alpha; \text{ contradiction.}$$

Hence, our working assumption is not acceptable, and the conclusion follows. □

In particular, when ψ and Δ are continuous, our theorem reduces to the one in Jachymski [16].

5 Statement of the Problem

Let (X, d) be a metric space. Further, let $T \in \mathscr{F}(X)$ be a selfmap of X. In the following, sufficient conditions are given for the existence and/or uniqueness of elements in $\text{Fix}(T)$. The way of solving it is by means of local and global (metrical) conditions involving our data.

5-I) Let x_0 be some point in X. By an x_0-*iterative sequence* attached to T, we mean any sequence $X_0 := (x_n; n \geq 0)$ [or, simply, $X_0 := (x_n)$] defined as $(x_n = T^n x_0; n \geq 0)$. Denote also

$U_0 := [X_0] = \{x_n; n \geq 0\}$ (the x_0-*trajectory* attached to $X_0 = (x_n)$)
$V_0 := \text{cl}(U_0)$ (the *complete* x_0-*trajectory* attached to $X_0 = (x_n)$).

The following simple fact is to be noted.

Proposition 12 *Under these conventions,*

$V_0 = U_0 \cup \{z\}$, *whenever* $z := \lim_n (x_n)$ *exists.*

Proof Clearly, $V_0 \supseteq U_0 \cup \{z\}$. Suppose that there exists $v \in V_0$ that is outside $U_0 \cup \{z\}$. By the limit definition, there exists $\sigma > 0$ such that

$X(v, \sigma) := \{x \in X; d(v, x) < \sigma\}$ is disjoint from $U_0 \cup \{z\}$.

In particular, this tells us that v cannot belong to $\text{cl}(U_0) = V_0$; contradiction. Consequently, $V_0 \subseteq U_0 \cup \{z\}$; and we are done. □

5-II) Given the x_0-iterative sequence $X_0 = (x_n)$, two alternatives occur.

Alt-1) The iterative sequence $X_0 = (x_n)$ is *telescopic*, in the sense

(tele) there exists $h \geq 0$, such that $d(x_h, x_{h+1}) = 0$.

By the very definition of our sequence, one derives

$x_h = x_n$, for all $n \geq h$; whence, $z := x_h$ is an element of $\text{Fix}(T)$.

Consequently, this case is completely clarified from the fixed point perspective.

Alt-2) The iterative sequence $X_0 = (x_n)$ is *non-telescopic*, in the sense

(n-tele) $d(x_n, x_{n+1}) > 0, \forall n$.

This is the effective case when the underlying problem is to be solved.

Under the precise framework, let us list the directions under which the proposed problem is to be handled. Given the sequence $(z_n; n \geq 0)$ in X, define the property

(z_n) is *full*: $n \mapsto z_n$ is injective ($i \neq j$ implies $z_i \neq z_j$).

Then, fix some nonempty part W_0 of X with $V_0 \subseteq W_0$.

po-1) We say that the non-telescopic iterative sequence $X_0 = (x_n)$ is *full Picard* (modulo $(d; T)$) when (x_n) is full and d-convergent.

po-2) We say that the non-telescopic iterative sequence $X_0 = (x_n)$ is *strongly full Picard* (modulo $(d; T)$) when (x_n) is full, d-convergent, and $\lim_n (x_n) \in \text{Fix}(T)$.

po-3) We say that the non-telescopic iterative sequence $X_0 = (x_n)$ is W_0-*single strongly full Picard* (modulo $(d; T)$) when (x_n) is full, d-convergent, $\lim_n (x_n) \in \text{Fix}(T)$, and $\text{Fix}(T) \cap W_0$ is an asingleton; whence, $\{\lim_n (x_n)\} = \text{Fix}(T) \cap W_0$.

Clearly, these conventions may be viewed as a local sequential variant of the ones in Rus [39, Ch 2, Sect 2.2].

Sufficient conditions for such properties are being founded on *orbital full* (in short: (o-f)) concepts. Given the sequence $(z_n; n \geq 0)$ in X, define the property

(z_n) is T*-orbital*: $(z_n = T^n x; n \geq 0)$, for some $x \in X$.

Note that the iterative sequence $X_0 = (x_n)$ is an orbital one, but not full in general.

reg-1) Call X, *(o-f,d)-complete* at $X_0 = (x_n)$, provided X_0= full, d-Cauchy implies (x_n) is d-convergent.

reg-2) Let us say that T is *(o-f,d)-continuous* at $X_0 = (x_n)$, if X_0= full and $\lim_n(x_n) = z$ implies $\lim_n(Tx_n) = Tz$. [Note that, in this case, $z = Tz$; because $\lim_n(Tx_n) = \lim_n(x_{n+1}) = z$].

5-III) To solve our problem along the precise directions, the metrical contractive techniques will be used; these are connected with certain Meir–Keeler conditions [28] upon the underlying data. Denote [for $x, y \in X$]

$$Q_1(x, y) = d(x, Tx),\ Q_2(x, y) = d(x, y),$$
$$Q_3(x, y) = d(x, Ty),\ Q_4(x, y) = d(Tx, y),$$
$$Q_5(x, y) = d(Tx, Ty),\ Q_6(x, y) = d(y, Ty),$$
$$\mathcal{Q}(x, y) = (Q_1(x, y), Q_2(x, y), Q_3(x, y), Q_4(x, y), Q_5(x, y), Q_6(x, y)).$$

Further, let us construct the family of functions [for $x, y \in X$]

$$P_0(x, y) = Q_5(x, y),\ P_1(x, y) = (1/2)[Q_3(x, y) + Q_4(x, y)],$$
$$P_2(x, y) = (1/2)[Q_1(x, y) + Q_6(x, y)],$$
$$M_0(x, y) = \min\{Q_1(x, y), Q_2(x, y), Q_5(x, y), Q_6(x, y)\},$$
$$M_0^*(x, y) = \min\{Q_2(x, y), Q_5(x, y)\},$$
$$M_1(x, y) = \max\{Q_1(x, y), Q_6(x, y)\},$$
$$M_2(x, y) = \max\{Q_1(x, y), Q_2(x, y), Q_6(x, y)\},$$
$$M(x, y) = \max \mathcal{Q}(x, y) = \mathrm{diam}\{x, Tx, y, Ty\}.$$

5-III-1) Having this precise, let $P = P(T)$ be a map in $\mathcal{F}(X \times X, R_+)$. For example, one may take

$$P(x, y) = \Theta(\mathcal{Q}(x, y)),\ x, y \in X;$$

where $\Theta : R_+^6 \to R_+$ is a map; but this is not the only possible choice. Let also Y_0 be a nonempty subset of X. (As usual, the case $Y_0 \in \{U_0, V_0, W_0\}$ is considered; but this is not essential for the moment).

We say that T is *Meir–Keeler* $(d; P; Y_0)$*-contractive* if

(mk-1) $x, y \in Y_0$, $P(x, y) > 0$, imply $P_0(x, y) < P(x, y)$;
referred to as: T is *strictly contractive* (modulo $(d; P; Y_0)$)

(mk-2) $\forall \varepsilon > 0, \exists \delta > 0$, such that:
$x, y \in Y_0$, $\varepsilon < P(x, y) < \varepsilon + \delta$ imply $P_0(x, y) \leq \varepsilon$;
referred to as: T has the *Meir–Keeler property* (modulo $(d; P; Y_0)$).

These concepts may be viewed as an extended version of the ones introduced by Meir and Keeler [28]; see also Matkowski [27] and Ćirić [7].

Remark 3 By the former of these conditions, the Meir–Keeler property (modulo $(d; P; Y_0)$) of T writes

(mk-3) $\forall \varepsilon > 0, \exists \delta > 0$, such that:
$x, y \in Y_0, 0 < P(x, y) < \varepsilon + \delta$ imply $P_0(x, y) \leq \varepsilon$.

5-III-2) A geometric version of the above concept may be given along the lines below. Remember that the relation $\Omega \subseteq R^0_+ \times R^0_+$ is called *upper diagonal*, provided

(u-diag) $(t, s) \in \Omega$ implies $t < s$;

the class of all these will be denoted as $\text{udiag}(R^0_+)$. Further, let us introduce the geometric conditions (over the class $\text{udiag}(R^0_+)$)

(g-mk) Ω has the *geometric Meir–Keeler property*:
$\forall \varepsilon > 0, \exists \delta > 0$: $t\Omega s, \varepsilon < s < \varepsilon + \delta \Longrightarrow t \leq \varepsilon$.

(g-bila-sep) Ω is *geometric bilateral separable*:
$\forall \beta > 0, \exists \gamma \in]0, \beta[, \forall (t, s)$: $t, s \in]\beta - \gamma, \beta + \gamma[\Longrightarrow (t, s) \notin \Omega$.

In a close connection with these, a lot of asymptotic conditions are given over the same class of upper diagonal relations $\text{udiag}(R^0_+)$:

(a-mk) Ω is *asymptotic Meir–Keeler*:
there are no strictly descending sequences (t_n) and (s_n) in R^0_+ and no elements
ε in R^0_+, with $((t_n, s_n) \in \Omega, \forall n)$ and $(t_n \to \varepsilon+, s_n \to \varepsilon+)$.

(a-bila-sep) Ω is *asymptotic bilateral separable*:
there are no sequences $(t_n; n \geq 0)$ and $(s_n; n \geq 0)$ in R^0_+ and no elements
$\beta \in R^0_+$, with $((t_n, s_n) \in \Omega, \forall n)$ and $(t_n \to \beta, s_n \to \beta)$.

As precise, any geometric property above is equivalent with its corresponding asymptotic property. In addition (see above) one has

(for each upper diagonal $\Omega \subseteq R^0_+ \times R^0_+$):
Ω is geometric/asymptotic bilateral separable implies
Ω is geometric/asymptotic Meir–Keeler.

Now, given the mapping $P = P(T) : X \times X \to R_+$, the (nonempty) subset Y_0 of X and the relation $\Omega \subseteq R^0_+ \times R^0_+$, let us say that the self-map T is $(d; P; Y_0; \Omega)$-*contractive*, provided

(Om-contr) $(P_0(x, y), P(x, y)) \in \Omega$,
for all $x, y \in Y_0$ with $P_0(x, y), P(x, y) > 0$.

Proposition 13 *Suppose that the self-map T is $(d; P; Y_0; \Omega)$-contractive, where the relation $\Omega \subseteq R_+^0 \times R_+^0$ is upper diagonal and geometric Meir–Keeler. Then, T is Meir–Keeler $(d; P; Y_0)$-contractive.*

Proof

(i) Let $x, y \in Y_0$ be such that $P(x, y) > 0$. If $P_0(x, y) = 0$, all is clear. Suppose now that $P_0(x, y) > 0$. As a consequence of this,

$$(t, s) \in \Omega; \text{ where } t := P_0(x, y), s := P(x, y).$$

Combining with the upper diagonal property of Ω, one gets $t < s$; i.e. $P_0(x, y) < P(x, y)$. Summing up, T is strictly contractive (modulo $(d; P; Y_0)$).

(ii) Let $\varepsilon > 0$ be arbitrary fixed; and $\delta > 0$ be the number assured by the geometric Meir–Keeler property of Ω. Further, let $x, y \in Y_0$ be such that $\varepsilon < s := P(x, y) < \varepsilon + \delta$. As before, if $P_0(x, y) = 0$, all is clear. Suppose now that $P_0(x, y) > 0$. By definition,

$$(t, s) \in \Omega; \text{ where } t := P_0(x, y), s := P(x, y);$$

and this, along with $\varepsilon < s < \varepsilon + \delta$ gives (by the geometric Meir–Keeler property of Ω), $t \leq \varepsilon$; i.e. $P_0(x, y) \leq \varepsilon$. Putting these together, it follows that T has the Meir–Keeler property (modulo $(d; P; Y_0)$). The proof is complete. □

5-III-3) In the following, a converse result is formulated. Given the mapping $P : X \times X \to R_+$ and the subset Y_0 of X, let $\Omega = \Omega[d; P; Y_0; T]$ stand for the associated relation over R_+^0:

$$\Omega = \{(P_0(x, y), P(x, y)); x, y \in Y_0, P_0(x, y), P(x, y) > 0\}.$$

This, by definition, means that

$(t, s) \in R_+^0 \times R_+^0$ belongs to Ω iff the subset $E(t, s)$ of all
 $(x, y) \in Y_0 \times Y_0$ with $(t = P_0(x, y), s = P(x, y))$ is nonempty.

Remark 4 The following sequential aspect of this representation is to be noted, in (ZF-AC+DC). Let $((t_n, s_n); n \geq 0)$ be a sequence in $\Omega[d; P; Y_0; T]$; that is

(seq-1) $((t_n, s_n); n \geq 0)$ is a sequence in $R_+^0 \times R_+^0$.
(seq-2) $E(t_n, s_n) = \{(x, y) \in Y_0 \times Y_0; t = P_0(x, y), s = P(x, y)\} \neq \emptyset$, for all n.

By the Denumerable Axiom of Choice (deductible in (ZF-AC+DC)), we get a sequence $((x_n, y_n); n \geq 0)$ in $Y_0 \times Y_0$, with the property

(seq-3) $(\forall n)$: $(x_n, y_n) \in E(t_n, s_n)$; that is:
 $t_n = P_0(x_n, y_n) > 0, s_n = P(x_n, y_n) > 0$.

Roughly speaking, the upper diagonal and geometric Meir–Keeler properties of the associated to T relation $\Omega[d; P; Y_0; T]$ give, ultimately, a characterization of the Meir–Keeler contractive properties upon T. The following result will certify this.

Proposition 14 *Under these conventions, we have*

(53-1) *If T is Meir–Keeler $(d; P; Y_0)$-contractive, then the attached relation $\Omega := \Omega[d; P; Y_0; T]$ is upper diagonal and geometric Meir–Keeler.*
(53-2) *T is Meir–Keeler $(d; P; Y_0)$-contractive if and only if the attached relation $\Omega := \Omega[d; P; Y_0; T]$ is upper diagonal and geometric Meir–Keeler.*

Proof

(i) Suppose that T is Meir–Keeler $(d; P; Y_0)$-contractive.
(i-1) Let $(t, s) \in R^0_+ \times R^0_+$ be such that $(t, s) \in \Omega$; hence (by definition)

$$t = P_0(x, y),\ s = P(x, y), \text{ where } x, y \in Y_0 \text{ and } [P_0(x, y), P(x, y) > 0].$$

From the strict contractive property of T, we must have $P_0(x, y) < P(x, y)$; or, equivalently, $t < s$; which shows that Ω is upper diagonal.
(i-2) Let $\varepsilon > 0$ be arbitrary fixed; and $\delta > 0$ be the number associated by the Meir–Keeler property of T. Further, let $(t, s) \in R^0_+ \times R^0_+$ be such that $(t, s) \in \Omega$ and $\varepsilon < s < \varepsilon + \delta$; hence (see above)

$$t = P_0(x, y),\ s = P(x, y), \text{ where } x, y \in Y_0 \text{ and } [P_0(x, y), P(x, y) > 0];$$

so that (by definition):

$$P_0(x, y) > 0, \text{ and } \varepsilon < P(x, y) < \varepsilon + \delta.$$

By the underlying Meir–Keeler-property for T, we get

$$P_0(x, y) \le \varepsilon; \text{ i.e. (under our notation): } t \le \varepsilon;$$

so that, Ω has the geometric Meir–Keeler property.
(ii) Suppose that the associated relation $\Omega = \Omega[d; P; Y_0; T]$ over R^0_+ is upper diagonal and has the geometric Meir–Keeler property. By the very definition of this object, T is $(d; P; Y_0; \Omega)$-contractive. Combining with the preceding result, one derives that T appears as Meir–Keeler $(d; P; Y_0)$-contractive.

□

As a consequence of this, it follows that the Meir–Keeler $(d; P; Y_0)$-contractive properties of T are finally reducible to the upper diagonal and geometric Meir–Keeler properties for the associated relation $\Omega[d; P; Y_0; T]$. However, there are some other properties of this relation—like the separable ones—that exceed this Meir–Keeler setting; but, as we will see, these come from certain functional contractions. There are three cases to discuss.

Case I Given $\chi \in \mathscr{F}(R^0_+, R)$, let us say that the self-map T is *Boyd–Wong $(d; P; Y_0; \chi)$-contractive*, if

$$P_0(x, y) \le \chi(P(x, y)), \forall x, y \in Y_0, P_0(x, y), P(x, y) > 0.$$

To discuss it, we need some conditions upon this function. Let $\mathscr{F}(re)(R^0_+, R)$ stand for the subclass of all $\chi \in \mathscr{F}(R^0_+, R)$ with

χ is *regressive*: $\chi(t) < t$, for all $t > 0$.

Then, call $\chi \in \mathscr{F}(re)(R^0_+, R)$,

(mk-adm) *Meir–Keeler admissible* if
$\forall \gamma > 0, \exists \beta > 0, \forall t\colon \gamma < t < \gamma + \beta \Longrightarrow \chi(t) \leq \gamma$
(bw-adm) *Boyd–Wong admissible*, when $\Lambda^+\chi(s) < s, \forall s \in R^0_+$.
(s-bw-adm) *strongly Boyd–Wong admissible*, if $\Lambda^\pm\chi(s) < s, \forall s \in R^0_+$.

Remark 5 We stress that the following relationships are available:

(for each $\chi \in \mathscr{F}(re)(R^0_+, R)$): strongly Boyd–Wong admissible $\Longrightarrow$
Boyd–Wong admissible $\Longrightarrow$ Meir–Keeler admissible.

The verification of these was carried out in a previous place.

Proposition 15 *Suppose that T is Boyd–Wong $(d; P; Y_0; \chi)$-contractive, where $\chi \in \mathscr{F}(re)(R_+)$ is given. Then, the following assertions are true in (ZF-AC+DC):*

(54-1) the associated relation $\Omega[d; P; Y_0; T]$ is upper diagonal.
(54-2) the associated relation $\Omega[d; P; Y_0; T]$ is upper diagonal and geometric Meir–Keeler (whence T is Meir–Keeler $(d; P; Y_0)$-contractive) whenever χ is Meir–Keeler admissible.
(54-3) the associated relation $\Omega[d; P; Y_0; T]$ is upper diagonal, (geometric Meir–Keeler and) geometric bilateral separable if χ is strongly Boyd–Wong admissible.

Proof Let $\Omega := \Omega[\chi]$ be the relation over R^0_+ introduced as

$(t, s \in R^0_+)\colon t\Omega s$ iff $t \leq \chi(s)$.

By definition,

T is Boyd–Wong $(d; P; Y_0; \chi)$-contractive implies T is $(d; P; Y_0; \Omega)$-contractive.

Moreover, by a previous fact,

(p1) Ω is upper diagonal.
(p2) Ω is geometric Meir–Keeler whenever χ is Meir–Keeler admissible.
(p3) Ω is (geometric Meir–Keeler and) geometric bilateral separable whenever χ is strongly Boyd–Wong admissible.

This, along with $\Omega[d; P; Y_0; T] \subseteq \Omega$ yields (see above) the written conclusion. □

Case II Given the functional couple (ψ, φ) over $\mathscr{F}(R^0_+, R)$, let us say that the mapping T is *Rhoades $(d; P; Y_0; \psi, \varphi)$-contractive*, provided

$\psi(P_0(x,y)) \leq \psi(P(x,y)) - \varphi(P(x,y))$, for all $x, y \in Y_0$ with $P_0(x,y), P(x,y) > 0$.

To discuss it, remember that some compatible properties of the couple (ψ, φ) were introduced. First, let us assume that

(norm) (ψ, φ) is *normal*:
ψ is increasing, and φ is *strictly positive* ($\varphi(t) > 0, \forall t > 0$).

Further, let us introduce the conditions upon the normal couple (ψ, φ):

(as-pos) φ is *asymptotic positive*:
for each strictly descending sequence $(t_n; n \geq 0)$ in R^0_+ and each $\varepsilon > 0$ with $t_n \to \varepsilon+$, we must have $\limsup_n(\varphi(t_n)) > 0$.

(bd-osc) (ψ, φ) is *limit-bounded oscillating*:
for each sequence $(t_n; n \geq 0)$ in R^0_+ and each $\beta > 0$ with $t_n \to \beta$, we have $\limsup_n(\varphi(t_n)) > \psi(\beta + 0) - \psi(\beta - 0)$.

Remark 6 The following relationship is available, by this definition

(ψ, φ) is normal and limit-bounded oscillating implies φ=asymptotic positive.

In fact, let the strictly descending sequence (t_n) in R^0_+ and the number $\varepsilon > 0$ be such that $t_n \to \varepsilon+$. By the limit-bounded oscillating property

$\limsup_n(\varphi(t_n)) > \psi(\varepsilon + 0) - \psi(\varepsilon - 0) \geq 0$;

and the assertion follows.

Proposition 16 *Suppose that T is Rhoades $(d; P; Y_0; \psi, \varphi)$-contractive, where the couple (ψ, φ) is normal. Then, the following inclusions are true in (ZF-AC+DC):*

(55-1) the associated relation $\Omega[d; P; Y_0; T]$ is upper diagonal.
(55-2) the associated relation $\Omega[d; P; Y_0; T]$ is upper diagonal and geometric Meir–Keeler (whence T is Meir–Keeler $(d; P; Y_0)$-contractive) whenever φ is asymptotic positive.
(55-3) the associated relation $\Omega[d; P; Y_0; T]$ is upper diagonal, (geometric Meir–Keeler and) geometric bilateral separable if the couple (ψ, φ) has the is limit-bounded oscillating property.

Proof Let $\Omega := \Omega[\psi, \varphi]$ be the associated relation over R^0_+

$(t, s \in R^0_+)$: $t\Omega s$ iff $\psi(t) \leq \psi(s) - \varphi(s)$.

By definition,

T is Rhoades $(d; P; Y_0; \psi, \varphi)$-contractive implies T is $(d; P; Y_0; \Omega)$-contractive.

Moreover, by a previous fact,

(q1) Ω is upper diagonal.
(q2) Ω is geometric Meir–Keeler if φ is asymptotic positive.

(q3) Ω is (geometric Meir–Keeler and) geometric bilateral separable whenever (ψ, φ) is limit-bounded oscillating.

This, along with $\Omega[d; P; Y_0; T] \subseteq \Omega$ yields (see above) the written conclusion. □

Case III Denote, for $x, y \in X$,

$$\mathcal{M}_1(x, y) = (Q_1(x, y), Q_6(x, y)),$$
$$M_1(x, y) = \max \mathcal{M}_1(x, y) = \max\{Q_1(x, y), Q_6(x, y)\},$$
$$\mathcal{M}_2(x, y) = (Q_1(x, y), Q_2(x, y), Q_6(x, y)),$$
$$M_2(x, y) = \max \mathcal{M}_2(x, y) = \max\{Q_1(x, y), Q_2(x, y), Q_6(x, y)\}.$$

Given the couple of maps $g \in \mathcal{F}(R_+)$, $H \in \mathcal{F}(R_+^3, R_+)$, let us say that the mapping T is *Khan $(d; M_2; Y_0; g, H)$-contractive*, provided

(K-con) $g(P_0(x, y)) \leq g(M_2(x, y)) - H(\mathcal{M}_2(x, y)), \forall x, y \in Y_0.$

The class of functions (g, H) appearing here may be described as follows. Let $k \geq 1$ be a natural number. According to Khan et al. [24], we say that $G \in \mathcal{F}(R_+^k, R_+)$ is an *altering function*, in case

(alter-1) G is increasing in each variable
(alter-2) G is reflexive sufficient: $(t_1 = \ldots = t_k = 0)$ iff $G(t_1, \ldots, t_k) = 0$.

The class of all such functions will be denoted $\mathcal{F}(alt)(R_+^k, R_+)$. Note that, given $G \in \mathcal{F}(alt)(R_+^3, R_+)$, the associated function $(g(t) = G(t, t, t); t \in R_+)$ is an element of $\mathcal{F}(alt)(R_+)$. Moreover, by our previous notations,

$$G(\mathcal{M}_2(x, y)) \leq g(M_2(x, y)), \forall x, y \in X.$$

Proposition 17 *Suppose that the mapping T is Khan $(d; M_2; Y_0; g, H)$-contractive, where $g \in \mathcal{F}(alt)(R_+)$ and $H \in \mathcal{F}(alt)(R_+^3, R_+)$. Then, the following inclusions are true in (ZF-AC+DC):*

(56-1) The associated relation $\Omega[d; M_2; Y_0; T]$ is upper diagonal and geometric/asymptotic Meir–Keeler (whence T is Meir–Keeler $(d; M_2; Y_0)$-contractive).
(56-2) The associated relation $\Omega[d; P; Y_0; T]$ is, in addition, geometric/asymptotic bilateral separable (hence, geometric/asymptotic Meir–Keeler) whenever g is continuous.

Proof The verification consists of three stages.

(i) Assume by contradiction that the associated relation $\Omega[d; M_2; Y_0; T]$ is not upper diagonal:

there exist $x, y \in Y_0$ such that $P_0(x, y), M_2(x, y) > 0$ and $P_0(x, y) \geq M_2(x, y)$.

By the contractive condition (and g=increasing),

$g(M_2(x,y)) \leq g(P_0(x,y)) \leq g(M_2(x,y)) - H(\mathscr{M}_2(x,y))$; so, $H(\mathscr{M}_2(x,y)) = 0$.

This, along with $H \in \mathscr{F}(alt)(R_+^3, R_+)$ yields $\mathscr{M}_2(x,y) = 0$; hence $M_2(x,y) = 0$, in contradiction with the posed hypothesis.

(ii) Assume by contradiction that the associated relation $\Omega[d; M_2; Y_0; T]$ does not have the geometric/asymptotic Meir–Keeler property: there exists $\varepsilon > 0$, so that

$C(\delta) := \{(x,y) \in Y_0 \times Y_0; \varepsilon < M_2(x,y) < \varepsilon + \delta,\ P_0(x,y) > \varepsilon\} \neq \emptyset$, for each $\delta > 0$.

Taking a zero converging sequence (δ_n) in R_+^0, we get by the Denumerable Axiom of Choice (deductible in (ZF-AC+DC)), a sequence $((x_n, y_n); n \geq 0)$ in $Y_0 \times Y_0$, with

$(\forall n)$: $(x_n, y_n) \in C(\delta_n)$; that is (by definition and preceding step) $\varepsilon < P_0(x_n, y_n) < M_2(x_n, y_n) < \varepsilon + \delta_n$;

note that, as a direct consequence of this,

$P_0(x_n, y_n) \to \varepsilon+$ and $M_2(x_n, y_n) \to \varepsilon+$, as $n \to \infty$.

By the contractive condition, we get for all n,

$(0 \leq) H(\mathscr{M}_2(x_n, y_n)) \leq g(M_2(x_n, y_n)) - g(P_0(x_n, y_n))$;

and this (via g=increasing) gives (by the above)

$\limsup_n H(\mathscr{M}_2(x_n, y_n)) \leq g(\varepsilon+0) - g(\varepsilon+0) = 0$; so, $\lim_n H(\mathscr{M}_2(x_n, y_n)) = 0$.

On the other hand, by the very construction of our sequence $((x_n, y_n); n \geq 0)$, there must be some index $i \in \{1, 2, 6\}$ such that

$\varepsilon < Q_i(x_n, y_n) < \varepsilon + \delta_n$, for infinitely many n.

Without loss, one may assume that $i = 1$. Combining with (H=increasing in all variables), yields

$H(\mathscr{M}_2(x_n, y_n)) \geq H(\varepsilon, 0, 0)$, for infinitely many n; wherefrom $\lim_n H(\mathscr{M}_2(x_n, y_n)) \geq H(\varepsilon, 0, 0) > 0$;

in contradiction with the limit property above. Consequently, our working assumption is not acceptable; wherefrom, the associated relation $\Omega[d; M_2; Y_0; T]$ does have the geometric/asymptotic Meir–Keeler property.

(iii) Assume by contradiction that the associated relation $\Omega[d; M_2; Y_0; T]$ is not asymptotic bilateral separable: There exists a sequence $((t_n, s_n); n \geq 0)$ in $R_+^0 \times R_+^0$ and an element $\beta > 0$, such that

(r1) $(t_n, s_n) \in \Omega[d; M_2; Y_0; T]$, for all n,
(r2) $t_n \to \beta$ and $s_n \to \beta$ as $n \to \infty$.

By a previous observation involving the underlying object, we get, in the reduced system (ZF-AC+DC), a sequence $((x_n, y_n); n \geq 0)$ in $Y_0 \times Y_0$, with

(r3) $t_n = P_0(x_n, y_n) > 0, s_n = M_2(x_n, y_n) > 0, \forall n.$

This, via convergence property above, gives

(r4) $P_0(x_n, y_n) \to \beta$ and $M_2(x_n, y_n) \to \beta$ as $n \to \infty$.

Moreover, taking the contractive condition into account gives

(r5) $g(P_0(x_n, y_n)) \leq g(M_2(x_n, y_n)) - H(\mathscr{M}_2(x_n, y_n)), \forall n.$

A simple re-arrangement of this last relation yields

$$(0 \leq) H(\mathscr{M}_2(x_n, y_n)) \leq g(M_2(x_n, y_n)) - g(P_0(x_n, y_n)), \forall n;$$

so, passing to lim sup as $n \to \infty$, we get (by the convergence properties)

$$\limsup_n H(\mathscr{M}_2(x_n, y_n)) \leq g(\beta) - g(\beta) = 0; \text{ whence, } \lim_n H(\mathscr{M}_2(x_n, y_n)) = 0.$$

Let (γ_n) be a strictly descending sequence in $]0, \beta[$ with $\lim_n(\gamma_n) = 0$. By the convergence property once again, there must be some index $i \in \{1, 2, 6\}$ such that

$$\beta - \gamma_n < Q_i(x_n, y_n) < \beta + \gamma_n, \text{ for infinitely many } n.$$

Without loss, one may assume that $i = 1$. Combining with (H=increasing in all variables), yields an evaluation like

$$H(\mathscr{M}_2(x_n, y_n)) \geq H(\beta - \gamma_0, 0, 0), \text{ for infinitely many } n;$$
$$\text{wherefrom } \lim_n H(\mathscr{M}_2(x_n, y_n)) \geq H(\beta - \gamma_0, 0, 0) > 0;$$

in contradiction with the limit property above. Consequently, our working assumption is not acceptable; wherefrom, the associated relation $\Omega[d; M_2; Y_0; T]$ is asymptotic (hence, geometric) bilateral separable.

□

Note that similar conclusions may be derived for the pair $(\mathscr{M}_1, M_1)$; we do not give further details.

6 Main Result

Let (X, d) be a metric space. Further, take some self-map $T \in \mathscr{F}(X)$, and put $\text{Fix}(T) = \{z \in X; z = Tz\}$; each point of it will be referred to as *fixed* with respect to T. As precise, we look for existence and uniqueness conditions involving $\text{Fix}(T)$.

Let x_0 be some point in X. By an x_0-*iterative sequence* attached to T, we mean any sequence $X_0 := (x_n)$, where $(x_n = T^n x_0; n \geq 0)$. In the following, we fix such an object, with (cf. a previous discussion)

(non-tele) $X_0 = (x_n)$ is non-telescopic $(d(x_n, x_{n+1}) > 0, \forall n)$.

Denote also

- $U_0 := [X_0] = \{x_n; n \geq 0\}$ (the x_0-*trajectory* attached to $X_0 = (x_n)$).
- $V_0 := \mathrm{cl}(U_0)$ (the *complete* x_0-*trajectory* attached to $X_0 = (x_n)$).

Let also W_0 be some nonempty part of X with $V_0 \subseteq W_0$. The specific directions under which the posed problem is to be solved were already listed; as precise, these are based on regularity conditions involving the non-telescopic x_0-iterative sequence $X_0 = (x_n)$ we just introduced. On the other hand, the metrical tools of our investigations consist in regularity conditions upon certain associated to T relations over R^0_+ constructed over the triple (U_0, V_0, W_0) taken as before.

Precisely, let us introduce the conventions, for $x, y \in X$

$$
\begin{aligned}
&Q_1(x, y) = d(x, Tx),\ Q_2(x, y) = d(x, y), \\
&\quad Q_3(x, y) = d(x, Ty),\ Q_4(x, y) = d(Tx, y), \\
&\quad Q_5(x, y) = d(Tx, Ty),\ Q_6(x, y) = d(y, Ty), \\
&\quad \mathcal{Q}(x, y) = (Q_1(x, y), Q_2(x, y), Q_3(x, y), Q_4(x, y), Q_5(x, y), Q_6(x, y)).
\end{aligned}
$$

Then, let us construct the family of functions [for $x, y \in X$]

$$
\begin{aligned}
&P_0(x, y) = Q_5(x, y),\ P_1(x, y) = (1/2)[Q_3(x, y) + Q_4(x, y)], \\
&\quad P_2(x, y) = (1/2)[Q_1(x, y) + Q_6(x, y)], \\
&\quad M_0(x, y) = \min\{Q_1(x, y), Q_2(x, y), Q_5(x, y), Q_6(x, y)\}, \\
&\quad M_0^*(x, y) = \min\{Q_2(x, y), Q_5(x, y)\}, \\
&\quad M_1(x, y) = \max\{Q_1(x, y), Q_6(x, y)\}, \\
&\quad M_2(x, y) = \max\{Q_1(x, y), Q_2(x, y), Q_6(x, y)\}, \\
&\quad M(x, y) = \max \mathcal{Q}(x, y) = \mathrm{diam}\{x, Tx, y, Ty\}.
\end{aligned}
$$

Further, let $P : X \times X \to R_+$ be a mapping. Usually, this object is of the form

$$P = \Theta(\mathcal{Q});\ \text{i.e. } P(x, y) = \Theta(\mathcal{Q}(x, y)),\ x, y \in X;$$

where $\Theta : R^6_+ \to R_+$ fulfills certain mild conditions. But, this is not the only choice to be considered. Then, let the nonempty set Y_0 of X be given; as a rule, we choose it as $Y_0 \in \{U_0, V_0, W_0\}$. Given the relation $\Omega \subseteq R^0_+ \times R^0_+$, let us say that T is $(d; P; Y_0; \Omega)$-*contractive*, provided

(Om-contr) $(P_0(x, y), P(x, y)) \in \Omega$,
for all $x, y \in Y_0$ with $P_0(x, y), P(x, y) > 0$.

An intrinsic version of this property may be stated as follows. Given the mapping $P : X \times X \to R_+$ and the subset Y_0 of X, let $\Omega[d; P; Y_0; T]$ stand for the associated relation over R^0_+:

$\Omega[d; P; Y_0; T] = \{(P_0(x, y), P(x, y)); x, y \in Y_0, P_0(x, y), P(x, y) > 0\}$;
or, in other words:
$(t, s) \in \Omega[d; P; Y_0; T]$ iff $t = P_0(x, y)$, $s = P(x, y)$, where $x, y \in Y_0$, and $P_0(x, y), P(x, y) > 0$.

It is now clear, by definition, that (with Ω as before)

T is $(d; P; Y_0; \Omega)$-contractive iff $\Omega[d; P; Y_0; T] \subseteq \Omega$.

This, in particular, tells us that all extra properties of Ω to be considered yield corresponding properties of associated to T relation $\Omega[d; P; Y_0; T]$.

As a completion of these, some specific groups of conditions upon our data will be described.

6-I) The first group of such conditions is of starting type; and allows us obtaining upper diagonal properties for each attached to T relation $\Omega[d; P; Y_0; T]$. For each mapping $K : X \times X \to R_+$ define the concept

(posi) $(Y_0; P; K)$ is *positive*: for each $x, y \in Y_0$ we have
$K(x, y) > 0$ implies $P(x, y) > 0$.

The usual choices for our mapping are $K \in \{M_0, M_0^*\}$.

6-II) The second group of conditions has the role of getting a d-asymptotic property and full property for the non-telescopic x_0-iterative sequence $X_0 = (x_n)$ to be considered. It may be formulated as

(o-bd) $(U_0; P; M_1)$ is *orbitally bounded*: $P(x, Tx) \leq M_1(x, Tx)$, $\forall x \in U_0$.

6-III) The third group of conditions allows us determining a d-Cauchy property for the obtained d-asymptotic full x_0-iterative sequence $X_0 = (x_n)$; and writes

if $X_0 = (x_n)$ is d-asymptotic full, then $(X_0; P)$ is *orbitally small*: for each (ε, δ) with $\varepsilon > \delta > 0$, there exists $\gamma \in]0, \delta/4[$ (and the attached asymptotic rank $n(\gamma)$), such that: $j \geq 2$, $k \geq n(\gamma)$, and $d(x_m, x_{m+i}) < \varepsilon + \delta/2$ for ($m \geq k$, $i \in \{1, \ldots, j\}$), imply $P(x_n, x_{n+j}) < \varepsilon + \delta$, whenever ($n \geq k$, $d(x_n, x_{n+j+1}) \geq \varepsilon + \delta/2$).

Concerning this concept, the following practical criteria will be useful for us. For each mapping $K : X \times X \to R_+$ define the concept

$(Y_0; P; K)$ is *bounded*: $P(x, y) \leq K(x, y)$, $\forall x, y \in Y_0$.

Proposition 18 *Suppose that $X_0 = (x_n)$ is d-asymptotic full. Then*

(61-1) if $(U_0; P; M)$ is bounded, then $(X_0; P)$ is orbitally small.
(61-2) if $P_1, P_2 : X \times X \to R_+$ are such that (X_0, P_1) and (X_0, P_2) are orbitally small, then (X_0, P_3) is orbitally small, where $P_3 := \max\{P_1, P_2\}$.

Proof

(i) Let the d-asymptotic full x_0-iterative sequence $X_0 = (x_n)$ and the couple (ε, δ) with $\varepsilon > \delta > 0$ be given. Further, take some $\gamma \in]0, \delta/4[$; and let $n(\gamma)$ stand for the attached asymptotic rank. We claim that

$j \geq 2$, $k \geq n(\gamma)$, and $d(x_m, x_{m+i}) < \varepsilon + \delta/2$ for $(m \geq k, i \in \{1, \ldots, j\})$, imply $P(x_n, x_{n+j}) < \varepsilon + \delta$, for each $n \geq k$;

and this will complete the argument. In fact, let $j \geq 2$, $k \geq n(\gamma)$ be as in premise above; and fix some $n \geq k$. By the very choice of these data

$$d(x_n, x_{n+j}), d(x_{n+1}, x_{n+j}), d(x_{n+1}, x_{n+j+1}) < \varepsilon + \delta/2;$$
$$d(x_n, x_{n+1}), d(x_{n+j}, x_{n+j+1}) < \gamma < \delta/4 < \delta/2.$$

Moreover, taking the triangular inequality into account, one gets

$$d(x_n, x_{n+j+1}) \leq d(x_n, x_{n+1}) + d(x_{n+1}, x_{n+j+1}) < \gamma + \varepsilon + \delta/2 < \varepsilon + \delta.$$

Putting these together yields (by the bounded property)

$$P(x_n, x_{n+j}) \leq M(x_n, x_{n+j}) < \varepsilon + \delta;$$

and our claim follows.

(ii) Given the couple (ε, δ) with $\varepsilon > \delta > 0$, let $\gamma_1 \in]0, \delta/4[$ (with the associated asymptotic rank $n(\gamma_1)$) and $\gamma_2 \in]0, \delta/4[$ (with the associated asymptotic rank $n(\gamma_2)$) be assured by the orbitally small property of (X_0, P_1) and (X_0, P_2), respectively. Then, denote

$$\gamma_3 = \min\{\gamma_1, \gamma_2\} \text{ (hence, } n(\gamma_3) \geq \max\{n(\gamma_1), n(\gamma_2)\});$$

we claim that the desired property of (X_0, P_3) holds with respect to the obtained pair. In fact, let $j \geq 2$, $k \geq n(\gamma_3)$ (hence, $k \geq n(\gamma_s)$, $s \in \{1, 2\}$) be such that

$$d(x_m, x_{m+i}) < \varepsilon + \delta/2 \text{ for } (m \geq k, i \in \{1, \ldots, j\});$$

we have to establish that

$$P_3(x_n, x_{n+j}) < \varepsilon + \delta, \text{ whenever } (n \geq k, d(x_n, x_{n+j+1}) \geq \varepsilon + \delta/2).$$

To verify this, note that, by the imposed hypothesis

$$(\forall s \in \{1, 2\}): d(x_m, x_{m+i}) < \varepsilon + \delta/2 \text{ for } (m \geq k \geq n(\gamma_s), i \in \{1, \ldots, j\}).$$

On the other hand, letting $n \geq k$ be as in the premise above, we have

$$(\forall s \in \{1, 2\}): n \geq k \geq n(\gamma_s) \text{ and } d(x_n, x_{n+j+1}) \geq \varepsilon + \delta/2.$$

Putting these together gives (by the admitted properties of P_1 and P_2)

$$P_s(x_n, x_{n+j}) < \varepsilon + \delta, \forall s \in \{1, 2\}; \text{ whence, } P_3(x_n, x_{n+j}) < \varepsilon + \delta;$$

and the conclusion follows.

□

6-IV) Finally, the fourth group of conditions has, as objective, a deduction of the fixed point property for the full d-convergent iterative sequence $X_0 = (x_n)$. It consists of two conditions

(o-s-asy) P is *orbitally singular asymptotic* at $X_0 = (x_n)$: whenever $\lim_n(x_n) = z$ and $d(z, Tz) > 0$, we have $\liminf_n P(x_n, z) < d(z, Tz)$.
(o-r-asy) P is *orbitally regular asymptotic* at $X_0 = (x_n)$: whenever $\lim_n(x_n) = z$ and $d(z, Tz) > 0$, we have $\lim_n P(x_n, z) = d(z, Tz)$.

Under these preliminaries, we may now state our main (fixed point) result (referred to as *Meir–Keeler sequential contractive principle*; in short: (MK-s-cp)).

Theorem 4 *Suppose that the self-map T, the non-telescopic x_0-iterative sequence $X_0 = (x_n)$, the mapping $P = P(T) : X \times X \to R_+$, and the triple (U_0, V_0, W_0), where $(U_0 \subseteq)V_0 \subseteq W_0$ are such that*

(61-i) the attached to T relation $\Omega[d; P; U_0; T]$ is upper diagonal and geometric/asymptotic Meir–Keeler.
(61-ii) $(V_0; P; M_0)$ is positive.
(61-iii) $(U_0; P; M_1)$ is orbitally bounded.
(61-iv) $(X_0; P)$ is orbitally small, whenever X_0 is d-asymptotic and full.

In addition, let X be (o-f,d)-complete at X_0. Then,

(61-a) *$X_0 = (x_n)$ is full Picard (modulo $(d; T)$).*

(61-b) *$X_0 = (x_n)$ is strongly full Picard (modulo $(d; T)$) provided one of the extra assumptions below is being fulfilled*

(61-b-1) T is (o-f,d)-continuous at X_0.
(61-b-2) P is orbitally singular asymptotic at X_0 and the relation $\Omega[d; P; V_0; T]$ is upper diagonal.
(61-b-3) P is orbitally regular asymptotic at X_0 and the relation $\Omega[d; P; V_0; T]$ is upper diagonal, (geometric/asymptotic Meir–Keeler and) geometric/asymptotic bilateral separable.

(61-c) *$X_0 = (x_n)$ is W_0-single strongly full Picard (modulo $(d; T)$), if $(W_0; P; M_0^*)$ is positive, $(W_0; P; M)$ is bounded, and $\Omega[d; P; W_0; T]$ is upper diagonal.*

Proof There are some steps to be passed.

Step 1 We firstly show that, under these conditions, the non-telescopic x_0-iterative sequence $X_0 = (x_n)$ is full and d-asymptotic. Denote, for simplicity,

$(r_n := d(x_n, x_{n+1}); n \geq 0)$; hence (by hypothesis), $(r_n > 0, \forall n)$.

Let $n \geq 0$ be arbitrary fixed. According to definition,

$$Q_1(x_n, x_{n+1}) = r_n > 0,\ Q_2(x_n, x_{n+1}) = r_n > 0,$$
$$Q_3(x_n, x_{n+1}) = d(x_n, x_{n+2}),\ Q_4(x_n, x_{n+1}) = 0,$$
$$Q_5(x_n, x_{n+1}) = r_{n+1} > 0,\ Q_6(x_n, x_{n+1}) = r_{n+1} > 0;$$

and this yields

$P_0(x_n, x_{n+1}) = r_{n+1} > 0$, $M_0(x_n, x_{n+1}) = \min\{r_n, r_{n+1}\} > 0$,
$M_1(x_n, x_{n+1}) = \max\{r_n, r_{n+1}\}$, $M_2(x_n, x_{n+1}) = \max\{r_n, r_{n+1}\}$.

As a consequence, one has

$P(x_n, x_{n+1}) > 0$ (if we remember that $(V_0; P; M_0)$ is positive).

Denote for simplicity $\Omega_1 := \Omega[d; P; U_0; T]$. By the relations above,

(Om1-contr) $(P_0(x_n, x_{n+1}), P(x_n, x_{n+1})) \in \Omega_1$, for all n.

Two basic consequences of this fact are to be noted.

Conseq 1 By the upper diagonal property of Ω_1, one has

(s-con) $r_{n+1} = P_0(x_n, x_{n+1}) < P(x_n, x_{n+1})$, $\forall n$.

On the other hand, as $(U_0; P; M_1)$ is orbitally bounded,

$P(x_n, x_{n+1}) \leq M_1(x_n, x_{n+1}) = \max\{r_n, r_{n+1}\}$.

This, along with (s-con), gives

(s-contr) $r_{n+1} < P(x_n, x_{n+1}) \leq \max\{r_n, r_{n+1}\}$.

From the inequality between extremal terms, one derives

$(r_{n+1} < r_n, \forall n)$; i.e. (r_n) is strictly descending.

Note that, as a first consequence of this,

(full-is) $X_0 = (x_n)$ is full: $i \neq j$ implies $x_i \neq x_j$.

For, suppose by contradiction that

there exists $i, j \in N$ with $i < j$, $x_i = x_j$.

Then, by definition $x_{i+1} = x_{j+1}$; so that $r_i = r_j$; in contradiction with $r_i > r_j$; and the assertion follows.

Conseq 2 By the Meir–Keeler property of Ω_1, we get

(asy-is) $X_0 = (x_n)$ is d-asymptotic: $d(x_n, x_{n+1}) \to 0$ as $n \to \infty$.

In fact, by the strict decreasing property above,

$r := \lim_n r_n$ exists in R_+; with, in addition, $(r_n > r, \forall n)$.

Suppose by contradiction that $r > 0$; and let $\delta = \delta(r) > 0$ be the number given by the Meir–Keeler property of Ω_1. By definition, there exists $m = m(\delta) \geq 1$ such that

$n \geq m$ implies $r < r_n < r + \delta$.

Let $n \geq m$ be fixed in the sequel. From (s-contr),

$r < r_{n+1} < P(x_n, x_{n+1}) \leq r_n < r + \delta$.

This, along with the contractive property (Om1-contr), gives

$(r <)r_{n+1} \leq r$; contradiction.

Hence, $r = 0$; and this establishes our assertion.

Step 2 As a consequence of this, the iterative sequence $X_0 = (x_n)$ is full and d-asymptotic. We now claim that $X_0 = (x_n)$ is d-Cauchy. Let $\varepsilon > 0$ be given; and $\delta > 0$ be assured by the Meir–Keeler property of Ω_1; clearly, without loss, one may assume that $\delta < \varepsilon$. Further, given the couple (ε, δ) as before, let the number $\gamma \in]0, \delta/4[$ [and the associated asymptotic rank $n(\gamma)$] be assured via $(X_0; P)$=orbitally small. We now establish, via ordinary induction, that, for each $i \geq 1$, the relation below holds

(d-C;i) $d(x_n, x_{n+i}) < \varepsilon + \delta/2$, for each $n \geq n(\gamma)$;

wherefrom, the d-Cauchy property of $X = (x_n)$ follows. The case $i \in \{1, 2\}$ is evident (via d=triangular) in view of $\gamma < \delta/4$ and the very definition of our asymptotic rank $n(\gamma)$. Let $j \geq 2$ be such that

relation (d-C;i) holds for all $i \in \{1, \ldots, j\}$; that is (under the notation $k = n(\gamma)$):
$d(x_m, x_{m+i}) < \varepsilon + \delta/2$, for each $m \geq k$ and each $i \in \{1, \ldots, j\}$;

we claim that our inductive relation holds as well for $i = j + 1$:

(d-C;j+1) $d(x_n, x_{n+j+1}) < \varepsilon + \delta/2$, for all $n \geq k(:= n(\gamma))$.

Suppose by contradiction that this does not hold:

(non;d-C;j+1) $C(\varepsilon, \delta) := \{n \geq k; d(x_n, x_{n+j+1}) \geq \varepsilon + \delta/2\} \neq \emptyset$;

and let $n = \min C(\varepsilon, \delta)$ be the minimal rank in $C(\varepsilon, \delta)$. By the choice of our data

$P(x_n, x_{n+j}) < \varepsilon + \delta$ (as $(X_0; P)$ is orbitally small).

On the other hand, by the full property of $X_0 = (x_n)$ (and $j \geq 2$)

$d(x_n, x_{n+j}) > 0$, $P_0(x_n, x_{n+j}) = d(x_{n+1}, x_{n+j+1}) > 0$;
whence, $M_0(x_n, x_{n+j}) = \min\{r_n, r_{n+j}, d(x_n, x_{n+j}), d(x_{n+1}, x_{n+j+1})\} > 0$;

and this yields

$P(x_n, x_{n+j}) > 0$ (if we remember that $(V_0; P; M_0)$ is positive).

Putting these together gives (by the definition of Ω_1)

$(P_0(x_n, x_{n+j}), P(x_n, x_{n+j})) \in \Omega_1$;

and this, combined with the Meir–Keeler property of Ω_1, yields

$d(x_{n+1}, x_{n+j+1}) = P_0(x_n, x_{n+j}) \leq \varepsilon$.

Taking the triangular inequality into account, one derives

$d(x_n, x_{n+j+1}) \leq d(x_n, x_{n+1}) + d(x_{n+1}, x_{n+j+1}) < \varepsilon + \gamma < \varepsilon + \delta/2$;

in contradiction with the choice of $n \in C(\varepsilon, \delta)$. Hence, the working hypothesis (non;d-C;j+1) does not hold; wherefrom, $(x_n; n \geq 0)$ is d-Cauchy, as claimed.

Step 3 As X is (o-f,d)-complete at X_0, $x_n \xrightarrow{d} z$, for some (uniquely determined) $z \in X$. This, by definition, tells us that X_0 is full Picard (modulo $(d; T)$). In addition, by a previous auxiliary fact, $V_0 = U_0 \cup \{z\}$.

We prove that under the lot of extra conditions above, the obtained limit point is an element of Fix(T). Three alternatives occur.

Alter 1 Suppose that T is (o-f,d)-continuous at X_0. Then, $y_n := Tx_n \xrightarrow{d} Tz$ as $n \to \infty$. On the other hand, $(y_n = x_{n+1}; n \geq 0)$ is a subsequence of $(x_n; n \geq 0)$; whence $y_n \xrightarrow{d} z$; and this gives (as d is separated), $z = Tz$; proving that X_0 is strongly full Picard (modulo $(d; T)$).

Alter 2 Suppose that the remaining alternatives hold. Denote for simplicity $\Omega_2 = \Omega[d; P; V_0; T]$. We claim that $d(z, Tz) > 0$ gives a contradiction. The full property of $X_0 = (x_n)$ assures us that

$E := \{n \in N; Tx_n = Tz\}$ is an asingleton;

so that, the following separation property holds:

(sepa) $\exists h = h(z) \geq 0$: $n \geq h \implies Tx_n \neq Tz$ (hence, $x_n \neq z$).

As a consequence of this, we have for all $n \geq h$

$d(x_n, z) > 0$, $P_0(x_n, z) = d(Tx_n, Tz) > 0$; whence
$M_0(x_n, z) = \min\{d(x_n, x_{n+1}), d(z, Tz), d(x_n, z), d(Tx_n, Tz)\} > 0$;

and this yields (for the same ranks)

$P(x_n, z) > 0$ (if we remember that $(V_0; P; M_0)$ is positive).

Putting these together yields (by the imposed notations)

(Om2-contr) $(P_0(x_n, z), P(x_n, z)) \in \Omega_2$, for all $n \geq h$.

Two possibilities are open before us.

Alter 2-1 Suppose that P is orbitally singular asymptotic at X_0 and Ω_2 is upper diagonal. For the moment,

$\liminf_n P(x_n, z) < d(z, Tz)$ (by the accepted hypotheses upon z and P).

On the other hand, by the upper diagonal property of Ω_2 applied to (Om2-contr)

$d(x_{n+1}, Tz) = P_0(x_n, z) < P(x_n, z), \forall n \geq h$.

Passing to lim inf as $n \to \infty$ yields (by the requirement upon $P(., .)$)

$d(z, Tz) = \liminf_n d(x_{n+1}, Tz) \leq \liminf_n P(x_n, z) < d(z, Tz)$;

a contradiction. Hence, $d(z, Tz) = 0$ (i.e. $z = Tz$); and our assertion follows.

Alter 2-2 Suppose that P is orbitally regular asymptotic at X_0 and Ω_2 is upper diagonal, (geometric/asymptotic Meir–Keeler and) geometric/asymptotic bilateral separable. Denote for simplicity

$(a_n = P_0(x_{n+h}, z); n \geq 0)$, $(b_n = P(x_{n+h}, z); n \geq 0)$.

By the above developments, we have

$(a_n, b_n) \in \Omega_2, \forall n$, and $a_n \to d(z, Tz)$, $b_n \to d(z, Tz)$ as $n \to \infty$.

This, however, is in contradiction with Ω_2 being geometric/asymptotic bilateral separable. Hence, $d(z, Tz) = 0$ (i.e. $z = Tz$); and our assertion follows.

Step 4 We prove that $\text{Fix}(T) \cap W_0$ is an asingleton; and, from this, all is clear. Denote for simplicity $\Omega_3 = \Omega[d; P; W_0; T]$. Let $z_1, z_2 \in \text{Fix}(T) \cap W_0$ be given; and suppose by contradiction that $z_1 \neq z_2$. Clearly,

$P_0(z_1, z_2) = d(z_1, z_2) > 0$; whence $M_0^*(z_1, z_2) = d(z_1, z_2) > 0$;
so that, $P(z_1, z_2) > 0$ (if we remember that $(W_0; P; M_0^*)$ is positive).

This, by definition, yields

$(P_0(z_1, z_2), P(z_1, z_2)) \in \Omega_3$; whence, $d(z_1, z_2) = P_0(z_1, z_2) < P(z_1, z_2)$;

by the upper diagonal property of Ω_3. On the other hand, clearly,

$P(z_1, z_2) \leq M(z_1, z_2) = d(z_1, z_2)$ (in view of $(W_0; P; M)$=bounded).

The contradiction at which we arrived shows that our working assumption is not acceptable; and then, our affirmation follows. The proof is complete.

□

Note that, further enlargements of these facts are possible, over dislocated metric spaces taken as in Hitzler [13, Ch 1, Sect 1.2]. On the other hand, this result admits multivalued type versions, under Nadler's model [31]; see also Turinici [45]. Finally, common fixed point versions of our main result are possible, under the lines in Jachymski [14]. We will discuss all these in a separate paper.

7 Pata Fixed Point Results

In the following, a basic application of our main result is given to a class of fixed point statements over metric spaces involving parametric contractive maps.

Let (X, d) be a metric space. Further, take some self-map $T \in \mathscr{F}(X)$, and put $\text{Fix}(T) = \{z \in X; z = Tz\}$; each point of it will be referred to as *fixed* with respect to T. As already precise, we look for existence and uniqueness conditions involving points of $\text{Fix}(T)$.

Let x_0 be some point in X. By an x_0-*iterative sequence* attached to T, we mean any sequence $X_0 := (x_n)$, where $(x_n = T^n x_0; n \geq 0)$. Denote also

$U_0 := [X_0] = \{x_n; n \geq 0\}$ (the x_0-*trajectory* attached to $X_0 = (x_n)$)
$V_0 := \mathrm{cl}(U_0)$ (the *complete* x_0-*trajectory* attached to $X_0 = (x_n)$).

Remember that the specific directions of solving the posed problem are based on regularity conditions involving the (non-telescopic) x_0-iterative sequence $X_0 = (x_n)$. On the other hand, the metrical tools of our investigations consist in upper diagonal and geometric/asymptotic Meir–Keeler conditions involving certain attached to T relations. It is our aim in the following to show that, in particular, all these are obtainable via parametric contractive conditions upon the considered data.

Let us introduce the conventions [for $x, y \in X$]

$$Q_1(x, y) = d(x, Tx),\ Q_2(x, y) = d(x, y),$$
$$Q_3(x, y) = d(x, Ty),\ Q_4(x, y) = d(Tx, y),$$
$$Q_5(x, y) = d(Tx, Ty),\ Q_6(x, y) = d(y, Ty),$$
$$\mathscr{Q}(x, y) = (Q_1(x, y), Q_2(x, y), Q_3(x, y), Q_4(x, y), Q_5(x, y), Q_6(x, y)).$$

Then, let us construct the family of functions [for $x, y \in X$]

$$P_0(x, y) = Q_5(x, y),\ P_1(x, y) = (1/2)[Q_3(x, y) + Q_4(x, y)],$$
$$P_2(x, y) = (1/2)[Q_1(x, y) + Q_6(x, y)],$$
$$M_0(x, y) = \min\{Q_1(x, y), Q_2(x, y), Q_5(x, y), Q_6(x, y)\},$$
$$M_0^*(x, y) = \min\{Q_2(x, y), Q_5(x, y)\},$$
$$M_1(x, y) = \max\{Q_1(x, y), Q_6(x, y)\},$$
$$M_2(x, y) = \max\{Q_1(x, y), Q_2(x, y), Q_6(x, y)\},$$
$$M(x, y) = \max \mathscr{Q}(x, y) = \mathrm{diam}\{x, Tx, y, Ty\}.$$

Further, let $P : X \times X \to R_+$ be a map. Usually, we may take it as

$$P = \Theta(\mathscr{Q});\ \text{i.e. } P(x, y) = \Theta(\mathscr{Q}(x, y)),\ x, y \in X;$$

where $\Theta : R_+^6 \to R_+$ fulfills certain mild conditions; but this is not the only possible choice. Remember that, given this mapping $P(.,.)$ and the subset Y_0 of X, we introduced the associated with T relation $\Omega := \Omega[d; P; Y_0; T]$ over R_+^0 as

$\Omega = \{(P_0(x, y), P(x, y)); x, y \in Y_0, P_0(x, y), P(x, y) > 0\}$; or, in other words:
$(t, s) \in \Omega$ iff $t = P_0(x, y)$, $s = P(x, y)$, where $x, y \in Y_0$, and $P_0(x, y), P(x, y) > 0$.

As we shall see, the particular cases to be considered are $Y_0 \in \{U_0, V_0, X\}$. Then, let us introduce the mappings $L : X \to R_+$ and $A : X \times X \to R_+$ as

$$L(x) = d(x, x_0),\ x \in X;\ A(x, y) = 1 + L(x) + L(y) + L(Tx) + L(Ty),\ x, y \in X.$$

Further, put $J = [0, 1]$, $J_0 = J \setminus \{0\} =]0, 1]$. Given $\chi \in \mathscr{F}(J, R_+)$, define the concept

(z-cont) χ is *zero-continuous*: $\chi(t_n) \to \chi(0) = 0$ as $t_n \to 0+$.

The class of all these will be denoted as $\mathscr{F}(0 - cont)(J, R_+)$.

Having these precise, let $\varphi \in \mathscr{F}(J, R_+)$ and $\alpha > 0$ be given. We say that T is *Pata $(d; P; \varphi; \alpha)$-contractive*, provided

(Pata-con) $d(Tx, Ty) \leq (1 - \tau)P(x, y) + \tau^{\alpha}\varphi(\tau)[A(x, y)]^{\alpha}, \forall x, y \in X, \forall \tau \in J.$

It is our aim to show that, under $[\varphi \in \mathscr{F}(0-cont)(J, R_+), \alpha \geq 1]$, and certain extra conditions upon $P(.,.)$, the Meir–Keeler sequential contractive principle (MK-s-cp) is applicable here and solves the fixed point question we deal with.

To do this, a lot of preliminary facts is needed. The first one in this series is

Proposition 19 *Suppose that the self-map T is Pata $(d; P; \varphi; \alpha)$-contractive, where $\varphi \in \mathscr{F}(0 - cont)(J, R_+)$ and $\alpha \geq 1$. Then, necessarily,*

(71-1) T is strictly contractive (modulo $(d; P; X)$):
$d(Tx, Ty) < P(x, y)$, for all $x, y \in X$ with $P(x, y) > 0$
(71-2) the attached to T relation $\Omega[d; P; X; T]$ is upper diagonal.

Proof It will suffice verifying the first part. Let $x, y \in X$ be such that $P(x, y) > 0$. Making $\tau = 0$ in (Pata-con), yields $d(Tx, Ty) \leq P(x, y)$. Suppose by absurd that $d(Tx, Ty) = P(x, y)$. Replacing in (Pata-con), we have (under simplification)

$P(x, y) \leq \tau^{\alpha-1}\varphi(\tau)[A(x, y)]^{\alpha}$, for all $\tau \in J_0$.

Taking the limit as $\tau \to 0+$, we derive $0 < P(x, y) \leq 0$; a contradiction. Hence, $d(Tx, Ty) < P(x, y)$; and our claim follows. □

We are now passing to the second auxiliary fact in this series.

Proposition 20 *Suppose that the self-map T is Pata $(d; P; \varphi; \alpha)$-contractive, where $\varphi \in \mathscr{F}(0 - cont)(J, R_+)$, $\alpha \geq 1$, and*

(72-I) $(U_0; P; M_1)$ is orbitally bounded.
(72-II) $(U_0; P; M)$ is bounded.

Then,

(72-1) the iterative sequence $X_n = (x_n)$ is bounded.
(72-2) the x_0-trajectory $U_0 = \{x_n; n \geq 0\}$ and its completion $V_0 = \mathrm{cl}(U_0)$ are bounded subsets in X.
(72-3) $A(.,.)$ is bounded over $U_0 \times U_0$.
(72-4) $A(.,.)$ is bounded on $V_0 \times V_0$, when $X_0 = (x_n)$ is d-convergent.

Proof The case of $X_0 = (x_n)$ being telescopic is clear; so, without loss, one may assume that $X_0 = (x_n)$ is non-telescopic. There are several steps to be passed.

(i) Denote for simplicity $(\rho_n = d(x_n, x_{n+1}); n \geq 0)$. By the orbitally bounded property (and a lot of previous evaluations)

$(\forall n)$: $\rho_{n+1} < P(x_n, x_{n+1}) \leq M_1(x_n, x_{n+1}) = \max\{\rho_n, \rho_{n+1}\}$; whence, $\rho_{n+1} < \rho_n$.

The sequence (ρ_n) is therefore strictly descending; so, $(\rho_n \le \rho_0$, for all $n)$. Denote $(c_n = L(x_n); n \ge 0)$. By the contractive condition, we have

(con-0) $d(x_{n+1}, x_1) = d(Tx_n, Tx_0) \le (1-\tau)P(x_n, x_0) + \tau^\alpha \varphi(\tau)[A(x_n, x_0)]^\alpha$, $\forall n, \forall \tau \in J$.

(ii) Let $n \ge 0$ be arbitrary fixed. We are trying to evaluate the quantities of (con-0), in terms of (ρ_n) and (c_n). For the moment, we have

(eva) $c_n \le d(x_n, x_{n+1}) + d(x_{n+1}, x_1) + d(x_1, x_0) \le 2\rho_0 + d(x_{n+1}, x_1)$.

On the other hand, by the bounded property of $(U_0; P; M)$, we have

$$P(x_n, x_0) \le M(x_n, x_0) = \text{diam}\{x_n, x_{n+1}, x_0, x_1\}, \forall n.$$

But, according to definition, we have for all n

$d(x_n, x_{n+1}) = \rho_n \le \rho_0, d(x_n, x_0) = c_n,$
$d(x_n, x_1) \le d(x_n, x_0) + d(x_0, x_1) = c_n + \rho_0,$
$d(x_{n+1}, x_0) \le d(x_{n+1}, x_n) + d(x_n, x_0) = \rho_n + c_n \le \rho_0 + c_n,$
$d(x_{n+1}, x_1) \le d(x_{n+1}, x_n) + d(x_n, x_0) + d(x_0, x_1) = \rho_n + c_n + \rho_0 \le 2\rho_0 + c_n, d(x_0, x_1) = \rho_0;$

and this, replacing in the preceding inequality, gives $(P(x_n, x_0) \le 2\rho_0 + c_n$, $\forall n)$. Further, again for all n,

$L(x_n) = c_n, L(Tx_n) = d(x_{n+1}, x_0) \le d(x_{n+1}, x_n) + d(x_n, x_0) = \rho_n + c_n \le \rho_0 + c_n, L(x_0) = 0, L(Tx_0) = d(x_1, x_0) = \rho_0;$

and this yields $(A(x_n, x_0) \le 2(1 + \rho_0 + c_n), \forall n)$. Replacing all these in the contractive relation (con-0), one derives [by means of (eva) above]

$$c_n \le 2\rho_0 + (1-\tau)[2\rho_0 + c_n] + 2^\alpha \tau^\alpha \varphi(\tau)[1 + \rho_0 + c_n]^\alpha, \forall n, \forall \tau \in J.$$

This, under the notation $(g_n = 2\rho_0 + c_n; n \ge 0)$, yields the inequality

(ineq) $g_n \le 4\rho_0 + (1-\tau)g_n + 2^\alpha \tau^\alpha \varphi(\tau)(1 + g_n)^\alpha, \forall n, \forall \tau \in J$.

Suppose by contradiction that (c_n) is unbounded. Without loss—passing to a subsequence if necessary—one may take this sequence as divergent:

$\lim_n c_n = \infty$; that is: $g_n \to \infty$, as $n \to \infty$.

Moreover, again without loss, one may assume that

$g_n \ge 2 + 4\rho_0 (\ge 2)$ (whence, $1 + g_n < 2g_n$), for all n.

By a small re-arrangement of (ineq) we then have

(ineq-1) $\tau g_n \le 4\rho_0 + 4^\alpha \varphi(\tau)(\tau g_n)^\alpha, \forall n, \forall \tau \in J$.

Let us now take the sequence $(\tau_n; n \ge 0)$ in J_0 as

$(\tau_n = (4\rho_0 + 1)/g_n; n \ge 0)$; whence, $\tau_n g_n = 4\rho_0 + 1, \forall n$.

Replacing into (ineq-1), we have

(ineq-2) $1 \le 4^\alpha \varphi(\tau_n)(4\rho_0 + 1)^\alpha, \forall n.$

Passing to limit as $n \to \infty$ gives $1 \le 0$; contradiction. Hence, (c_n) is a bounded sequence; and, from this, we are done.

(iii) By the preceding stage,

$$\lambda := \sup\{L(x); x \in U_0\} < \infty.$$

This yields (by the triangular inequality)

$$d(x, y) \le L(x) + L(y) \le 2\lambda, \forall x, y \in U_0; \text{ so, } \mathrm{diam}(U_0) = \mathrm{diam}(V_0) < \infty.$$

(iv) As $T(U_0) \subseteq U_0$, we get

$$A(x, y) = 1 + L(x) + L(y) + L(Tx) + L(Ty) \le 1 + 4\lambda, \forall x, y \in U_0;$$

so that, A is bounded on $U_0 \times U_0$.

(v) By a previous result, $V_0 = U_0 \cup \{z\}$, where $z = \lim_n(x_n)$. In this case, under the convention $\mu := \max\{L(z), L(Tz)\}$,

$$A(x, z) = 1 + L(x) + L(z) + L(Tx) + L(Tz) \le 1 + 2\lambda + 2\mu, \forall x \in U_0;$$
$$A(z, z) = 1 + 2[L(z) + L(Tz)] \le 1 + 4\mu.$$

Putting these together yields

$$\sup\{A(x, y); x, y \in V_0 \times V_0\} \le 1 + 4(\lambda + \mu) < \infty;$$

which tells us that, A is bounded on $V_0 \times V_0$.

□

Given $\psi \in \mathscr{F}(J, R_+)$ and $\alpha > 0$, let $\Omega(\psi, \alpha)$ be the relation over R_+^0

$(t, s) \in \Omega(\psi, \alpha)$ iff $t \le (1 - \tau)s + \tau^\alpha \psi(\tau), \forall \tau \in J.$

Some basic properties of this relation are concentrated in

Proposition 21 *Suppose that $\psi \in \mathscr{F}(0 - cont)(J, R_+)$ and $\alpha \ge 1$. Then,*

(73-1) $\Omega(\psi, \alpha)$ is upper diagonal.
(73-2) $\Omega(\psi, \alpha)$ is geometric/asymptotic Meir–Keeler.
(73-1) $\Omega(\psi, \alpha)$ is geometric/asymptotic bilateral separable.

Proof

(i) Let $t, s > 0$ be such that $(t, s) \in \Omega(\psi, \alpha)$; hence,

(psi-al) $t \le (1 - \tau)s + \tau^\alpha \psi(\tau), \forall \tau \in J.$

Making $\tau = 0$ in (psi-al) yields $t \le s$. Suppose by absurd that $t = s$. Replacing in (psi-al), we have (under simplification)

$$s \le \tau^{\alpha-1}\psi(\tau), \text{ for all } \tau \in J_0.$$

Taking the limit as $\tau \to 0+$, we derive $0 < s \leq 0$; a contradiction. Hence, $t < s$; and our claim follows.

(ii) Suppose by absurd that $\Omega(\psi, \alpha)$ is not asymptotic Meir–Keeler:

there are strictly descending sequences (t_n) and (s_n) in R^0_+ and elements ε in R^0_+, with $((t_n, s_n) \in \Omega(\psi, \alpha), \forall n)$ and $(t_n \to \varepsilon+, s_n \to \varepsilon+)$.

The first half of this relation means

$$t_n \leq (1 - \tau)s_n + \tau^\alpha \psi(\tau), \forall n, \forall \tau \in J;$$

wherefrom, by a limit process (relative to n)

$$\varepsilon \leq (1 - \tau)\varepsilon + \tau^\alpha \psi(\tau), \forall \tau \in J.$$

This yields (under simplification)

$$\varepsilon \leq \tau^{\alpha-1}\psi(\tau), \text{ for all } \tau \in J_0.$$

Taking the limit as $\tau \to 0+$, we derive $0 < \varepsilon \leq 0$; a contradiction. Hence, our working condition cannot be accepted; and the assertion follows.

(iii) Suppose by absurd that $\Omega(\psi, \alpha)$ is not asymptotic bilateral separable:

there are sequences $(t_n; n \geq 0)$ and $(s_n; n \geq 0)$ in R^0_+ and elements $\beta \in R^0_+$, with $((t_n, s_n) \in \Omega(\psi, \alpha), \forall n)$ and $(t_n \to \beta, s_n \to \beta)$.

The first half of this relation means

$$t_n \leq (1 - \tau)s_n + \tau^\alpha \psi(\tau), \forall n, \forall \tau \in J;$$

wherefrom, by a limit process (relative to n)

$$\beta \leq (1 - \tau)\beta + \tau^\alpha \psi(\tau), \forall \tau \in J.$$

This yields (under simplification)

$$\beta \leq \tau^{\alpha-1}\psi(\tau), \text{ for all } \tau \in J_0.$$

Taking the limit as $\tau \to 0+$, we derive $0 < \beta \leq 0$; a contradiction. Hence, our working condition cannot be accepted; and conclusion follows.

□

We are now in position to give an appropriate answer to the posed question. Remember that, for the fixed $x_0 \in X$, we denoted

$$X_0 = (x_n), \text{ where } (x_n = T^n x_0; n \geq 0);$$
$$U_0 = [X_0] = \{x_n; n \geq 0\}, V_0 = \text{cl}(U_0).$$

Fix in the following the iterative sequence $X_0 = (x_n)$ and its attached subsets (U_0, V_0). As precise, there is no loss in generality if we suppose that

(non-tele) $X_0 = (x_n)$ is non-telescopic ($d(x_n, x_{n+1}) > 0, \forall n$).

Let $P : X \times X \to R_+$ be a map, $\varphi \in \mathscr{F}(J, R_+)$ be a function, and $\alpha > 0$ be a number. The following statement (referred to as: Pata fixed point result in metric spaces; in short: (P-fp-ms)) is available.

Theorem 5 *Suppose that the self-map T is Pata $(d; P; \varphi; \alpha)$-contractive, where $\varphi \in \mathscr{F}(0-cont)(J, R_+)$, $\alpha \geq 1$, and*

(71-i) $(V_0; P; M_0)$ is positive.
(71-ii) $(U_0; P; M_1)$ is orbitally bounded.
(71-iii) $(U_0; P; M)$ is bounded, whenever X_0 is d-asymptotic and full.

In addition, let X be (o-f,d)-complete at X_0. Then,

(71-a) *$X_0 = (x_n)$ is full Picard (modulo $(d; T)$).*

(71-b) *$X_0 = (x_n)$ is strongly full Picard (modulo $(d; T)$) provided one of the extra assumptions below is being fulfilled.*

(71-b-1) T is (o-f,d)-continuous at X_0.
(71-b-2) P is orbitally singular asymptotic at X_0.
(71-b-3) P is orbitally regular asymptotic at X_0.

(71-c) *$X_0 = (x_n)$ is X-single strongly full Picard (modulo $(d; T)$), if $(X; P; M_0^*)$ is positive, and $(X; P; M)$ is bounded.*

Proof We show that the Meir–Keeler sequential contractive principle (MK-s-cp) is applicable to our setting. The following steps will clarify this.

(i) By a previous auxiliary fact,

$$\lambda_1 := \sup\{A(x, y); x, y \in U_0\} < \infty.$$

Denote for simplicity $(\varphi_1(t) = \varphi(t)\lambda_1^\alpha; t \geq 0)$; clearly, $\varphi_1 \in \mathscr{F}(0-cont)(J, R_+)$. By another auxiliary fact, the relation $\Omega(\varphi_1, \alpha)$ over R_+^0 introduced as

$$(t, s) \in \Omega(\varphi_1, \alpha) \text{ iff } t \leq (1-\tau)s + \tau^\alpha\varphi_1(\tau), \forall \tau \in J$$

has the properties

$\Omega(\varphi_1, \alpha)$ is upper diagonal, geometric/asymptotic Meir–Keeler, and geometric/asymptotic bilateral separable;

and this, along with $\Omega[d; P; U_0; T] \subseteq \Omega(\varphi_1, \alpha)$ establishes that

$\Omega[d; P; U_0; T]$ is upper diagonal, geometric/asymptotic Meir–Keeler, and geometric/asymptotic bilateral separable.

On the other hand, by an auxiliary fact,

$(X_0; P)$ is orbitally small, whenever X_0 is d-asymptotic and full.

Putting these together, it follows by (MK-s-cp) (the first part) that our first conclusion is available; so that, in particular,

$z = \lim_n(x_n)$ exists; whence, $V_0 = U_0 \cup \{z\}$.

(ii) According to a preceding evaluation, we have that

$\lambda_2 := \sup\{A(x, y); x, y \in V_0\} < \infty$.

Denote for simplicity $(\varphi_2(t) = \varphi(t)\lambda_2^{\alpha}; t \geq 0)$; clearly, $\varphi_2 \in \mathscr{F}(0-cont)(J, R_+)$. By the same argument as before,

$\Omega(\varphi_2, \alpha)$ is upper diagonal, geometric/asymptotic Meir–Keeler, and geometric/asymptotic bilateral separable;

and this, along with $\Omega[d; P; V_0; T] \subseteq \Omega(\varphi_2, \alpha)$ establishes that

$\Omega[d; P; V_0; T]$ is upper diagonal, geometric/asymptotic Meir–Keeler, and geometric/asymptotic bilateral separable.

(iii) Finally, by a previous auxiliary fact,

$\Omega[d; P; X; T]$ is upper diagonal.

Putting these together, it follows via (MK-s-cp) (the second and third part) that the remaining conclusions in our statement follow as well.

□

Now, for an appropriate comparison of this result with the existing ones, we have to derive sufficient conditions upon P under which the positive, orbitally bounded, and bounded properties are holding. To do this, take the mapping $P(.,.)$ as

$P(x, y) = \Theta(\mathscr{Q}(x, y)), x, y \in X$, where $\Theta : R_+^6 \to R_+$ is increasing.

Proposition 22 *Under the precise framework, we have*

(74-1) If $\Theta(\alpha, \alpha, 0, 0, \alpha, \alpha) > 0$ for each $\alpha > 0$, then $(X; P; M_0)$ is positive.
(74-2) If $\Theta(\beta, \beta, 2\beta, 0, \beta, \beta) \leq \beta$, for each $\beta \geq 0$, then $(X; P; M_1)$ is orbitally bounded.
(74-3) If $\Theta(\gamma, \gamma, \gamma, \gamma, \gamma, \gamma) \leq \gamma$, for each $\gamma \geq 0$, then $(X; P; M)$ is bounded.
(74-4) If $\Theta(0, \delta, 0, 0, \delta, 0) > 0$, for each $\delta > 0$, then $(X; P; M_0^*)$ is positive.

Proof

(i) Let $x, y \in X$ be arbitrary fixed; and assume that $\alpha := M_0(x, y) > 0$. By the increasing condition,

$P(x, y) \geq \Theta(\alpha, \alpha, 0, 0, \alpha, \alpha) > 0$; and conclusion follows.

(ii) Let $x \in X$ be arbitrary fixed and denote $\beta := M_1(x, Tx)$. By the increasing condition (and triangular inequality)

$P(x, Tx) \leq \Theta(\beta, \beta, 2\beta, 0, \beta, \beta) \leq \beta$; hence the assertion.

(iii) Let $x, y \in X$ be arbitrary fixed and denote $\gamma := M(x, y)$. As Θ=increasing,

$P(x, y) \leq \Theta(\gamma, \gamma, \gamma, \gamma, \gamma, \gamma) \leq \gamma$; and conclusion follows.

(iv) Let $x, y \in X$ be arbitrary fixed and assume that $\delta := M_0^*(x, y) > 0$. By the increasing condition once again,

$$P(x, y) \geq \Theta(0, \delta, 0, 0, \delta, 0) > 0;$$

and, from this, one gets the desired fact.

□

Having these precise, we may discuss a lot of particular cases of our statement.

Case-1) Suppose that $P = Q_2$. Then, Pata fixed point result in metric spaces (P-fp-ms) is just the basic statement in Pata [34]. But, we must say that our methods are different from the ones in the quoted paper.

Case-2) Suppose that $P = \max\{Q_2, P_1, P_2\}$. Then, Pata fixed point result in metric spaces (P-fp-ms) is just the related statement in Jacob et al. [17].

Case-3) Suppose that $P = \max\{Q_2, M_2\}$. The corresponding version of Pata fixed point result in metric spaces (P-fp-ms) seems to be new.

Finally, it is worth noting that these techniques were applied to a variety of domains, such as (cf. the survey paper in Choudhury et al. [5]).

Dom-1) Coincidence and common fixed points in various structures: Kadelburg and Radenović [21].

Dom-2) Coupled and tripled fixed points: Eshaghi et al. [11], Kadelburg and Radenović [19, 22].

Dom-3) Fixed points of multivalued mappings: Choudhury et al. [6].

Dom-4) Fixed points of cyclic contractions: Kadelburg and Radenović [20].

Dom-5) Fixed points in modular spaces: Paknazar et al. [33].

We close these developments with a methodological question. In a recent paper, Berinde [2] claimed that the original Pata fixed point result [34] is not correct as stated. The inconsistency of his argument was evidentiated in the survey paper by Choudhury et al. [5] we just quoted. Moreover, as proved there, the question of some standard contractive conditions being deductible from the Pata ones has a positive answer. It is our aim in the following to give a small completion of these (sketched) arguments.

Let the general framework be taken as before. Fix some $x_0 \in X$ and denote

$$L(x) = d(x, x_0), x \in X; B(x, y) = 1 + L(x) + L(y), x, y \in X.$$

Given the function $\varphi \in \mathscr{F}(J, R_+)$ and the number $\alpha > 0$, let us say that T is *standard Pata* $(d; \varphi; \alpha)$*-contractive*, provided

(Pata-st) $d(Tx, Ty) \leq (1 - \tau)d(x, y) + \tau^\alpha \varphi(\tau)[B(x, y)]^\alpha, \forall x, y \in X, \forall \tau \in J.$

Clearly, by the immediate relation

$$B(x, y) \leq A(x, y), x, y \in X$$

any such map is Pata $(d; Q_2; \varphi; \alpha)$-contractive.

Proposition 23 *Suppose that* $\lambda \in]0, 1[$ *is such that*

T *is (Banach)* $(d; \lambda)$*-contractive:* $d(Tx, Ty) \leq \lambda d(x, y)$, $\forall x, y \in X$.

Then, fix $\gamma > 0$ *and put*

$C(\lambda, \gamma) = \gamma^\gamma / [(\gamma + 1)^{\gamma+1}(1 - \lambda)^\gamma]$.

In this case, for each $C \geq \max\{1, C(\lambda, \gamma)\}$, *one has*

(Pata-B) $d(Tx, Ty) \leq (1 - \tau)d(x, y) + C\tau^{\gamma+1}B(x, y)$, $\forall x, y \in X$, $\forall \tau \in J$;

that is, T *is standard Pata* $(d; \varphi; 1)$*-contractive, where*

$(\varphi(\tau) = C\tau^\gamma; \tau \in J)$; *hence,* $\varphi \in \mathscr{F}(0 - cont)(J, R_+)$.

Proof Fix a couple of points $x, y \in X$; without loss, one may assume that $x \neq y$ (hence, $d(x, y) > 0$). From the Banach contractive condition and

$d(x, y) \leq L(x) + L(y) \leq B(x, y)$,

a sufficient condition for the desired relation (Pata-B) to hold is

(Pata-B-1) $\lambda d(x, y) \leq (1 - \tau)d(x, y) + C\tau^{\gamma+1}d(x, y)$, for all $\tau \in J$;

or, equivalently (after simplification)

(Pata-B-2) $f(\tau) := \tau - C\tau^{\gamma+1} \leq 1 - \lambda$ for all $\tau \in J$.

To discuss this relation, we start from the derivative

$f'(\tau) = 1 - C(\gamma + 1)\tau^\gamma$, $\tau \in J^0 :=]0, 1[$.

The critical point of f (i.e. the zero of f') is

$\tau_0 = [1/C(\gamma + 1)]^{1/\gamma}$; hence, $\tau_0 \in J^0$ (as $C(\gamma + 1) \geq \gamma + 1 > 1$).

In this case, the maximum of f over J is

$f(\tau_0) = \tau_0[1 - C\tau_0^\gamma] = \tau_0[\gamma/(\gamma + 1)]$.

Consequently, the inequality in (Pata-B-2) is equivalent with

$\tau_0[\gamma/(\gamma + 1)] \leq 1 - \lambda$; or, equivalently, $\tau_0^\gamma[\gamma/(\gamma + 1)]^\gamma \leq (1 - \lambda)^\gamma$.

Combining with the relation that introduces this number yields

$[1/C(\gamma + 1)][\gamma/(\gamma + 1)]^\gamma \leq (1 - \lambda)^\gamma$; or, equivalently,
$C \geq \gamma^\gamma / [(\gamma + 1)^{\gamma+1}(1 - \lambda)^\gamma] = C(\lambda, \gamma)$; evident, by the choice of C.

Summing up, (Pata-B-2) holds; and the proof is complete. □

Concerning the reverse inclusion, remember that given $\varphi \in \mathscr{F}(R_+)$, we say that T is *Boyd–Wong* φ*-contractive*, provided

$d(Tx, Ty) \leq \varphi(d(x, y))$, $\forall x, y \in X$.

By an example in the survey paper by Choudhury et al. [5], it follows that there exists self-maps T that fulfill the standard Pata condition, but not the Boyd–Wong one, even if φ is increasing continuous. Hence, the class of Pata contractions is a strict extension of the Boyd–Wong (hence, Banach) class of contractions. Further extensions of these results may be obtained under the lines in the survey papers by Turinici [49, 50].

References

1. S. Banach, *Sur les opérations dans les ensembles abstraits et leur application aux équations intégrales*. Fund. Math., 3 (1922), 133–181.
2. V. Berinde, *Comments on some fixed point theorems on metric spaces*, Creative Math. Inform., 27 (2018), 15–20.
3. P. Bernays, *A system of axiomatic set theory: Part III. Infinity and enumerability analysis*, J. Symbolic Logic, 7 (1942), 65–89.
4. D. W. Boyd and J. S. W. Wong, *On nonlinear contractions*, Proc. Amer. Math. Soc., 20 (1969), 458–464.
5. B. S. Choudhury, Z. Kadelburg, N. Metiya, and S. Radenović, *A survey of fixed point theorems under Pata-type conditions*, Bull. Malaysian Math. Soc., 43 (2020), 1289–1309.
6. B. S. Choudhury, N. Metiya, C. Bandyopadhyay, and P. Maiti, *Fixed points of multivalued mappings satisfying hybrid rational Pata-type inequalities*, J. Analysis, 27 (2019), 813–828.
7. L. B. Cirić, *A new fixed-point theorem for contractive mappings*, Publ. Inst. Math. 30(44) (1981), 25–27.
8. P. J. Cohen, *Set Theory and the Continuum Hypothesis*, Benjamin, New York, 1966.
9. C. Di Bari and C. Vetro, *Common fixed point theorems for weakly compatible maps satisfying a general contractive condition*, Int. J. Math. Mathematical Sci., Volume 2008, Article ID 891375.
10. P. N. Dutta and B. S. Choudhury, *A generalisation of contraction principle in metric spaces*, Fixed Point Th. Appl., Volume 2008, Article ID 406368.
11. M. Eshaghi, S. Mohsemi, M. R. Delavar, M. De La Sen, G. H. Kim, and A. Arian, *Pata contractions and coupled fixed points*, Fixed Point Th. Appl., 2014, 2014:130.
12. P. R. Halmos, *Naive Set Theory*, Van Nostrand Reinhold Co., New York, 1960.
13. P. Hitzler, *Generalized metrics and topology in logic programming semantics*, PhD Thesis, Natl. Univ. Ireland, Univ. College Cork, 2001.
14. J. Jachymski, *Common fixed point theorems for some families of mappings*, Indian J. Pure Appl. Math., 25 (1994), 925–937.
15. J. Jachymski, *The contraction principle for mappings on a metric space with a graph*, Proc. Amer. Math. Soc., 136 (2008), 1359–1373.
16. J. Jachymski, *Equivalent conditions for generalized contractions on (ordered) metric spaces*, Nonlinear Anal., 74 (2011), 768–774.
17. G. K. Jacob, M. S. Khan, Ch. Park, and S. Yun, *On generalized Pata type contractions*, Mathematics, 6:25 (2018), 1–8.
18. M. Jleli and B. Samet, *A new generalization of the Banach contraction principle*, J. Ineq. Appl., 2014, 2014:38.
19. Z. Kadelburg and S. Radenović, *Fixed point and tripled fixed point theorems under Pata-type conditions in ordered metric space*, Internat. J. Anal. Appl., 6 (2014), 113–122.
20. Z. Kadelburg and S. Radenović, *A note on Pata-type cyclic contractions*, Sarajevo J. Math., 11 (24) (2015), 235–245.
21. Z. Kadelburg and S. Radenović, *Pata-type common fixed point results in b-metric and b-rectangular metric spaces*, J. Nonlinear Sci. Appl., 8 (2015), 944–954.
22. Z. Kadelburg and S. Radenović, *Fixed point theorems under Pata-type conditions in metric spaces*, J. Egypt. Math. Soc., 24 (2016), 77–82.

23. S. Kasahara, *On some generalizations of the Banach contraction theorem*, Publ. Res. Inst. Math. Sci. Kyoto Univ., 12 (1976), 427–437.
24. M. S. Khan, M. Swaleh, and S. Sessa, *Fixed point theorems by altering distances between the points*, Bull. Austral. Math. Soc., 30 (1984), 1–9.
25. S. Leader, *Fixed points for general contractions in metric spaces*, Math. Japonica, 24 (1979), 17–24.
26. J. Matkowski, *Integrable solutions of functional equations*, Dissertationes Math., Vol. 127, Polish Sci. Publ., Warsaw, 1975.
27. J. Matkowski, *Fixed point theorems for contractive mappings in metric spaces*, Časopis Pest. Mat., 105 (1980), 341–344.
28. A. Meir and E. Keeler, *A theorem on contraction mappings*, J. Math. Anal. Appl., 28 (1969), 326–329.
29. G. H. Moore, *Zermelo's Axiom of Choice: its Origin, Development and Influence*, Springer, New York, 1982.
30. Y. Moskhovakis, *Notes on Set Theory*, Springer, New York, 2006.
31. S. B. Nadler Jr., *Multi-valued contraction mappings*, Pacific J. Math., 30 (1969), 475–488.
32. J. J. Nieto and R. Rodriguez-Lopez, *Contractive mapping theorems in partially ordered sets and applications to ordinary differential equations*, Order, 22 (2005), 223–239.
33. M. Paknazar, M. Eshaghi, Y. J. Cho, and S. M. Vaezpour, *A Pata-type fixed point theorem in modular spaces with application*, Fixed Point Th. Appl., 2013, 2013:239.
34. V. Pata, *A fixed point theorem in metric spaces*, J. Fixed Point Th. Appl., 10 (2011), 299–305.
35. A. C. M. Ran and M. C. Reurings, *A fixed point theorem in partially ordered sets and some applications to matrix equations*, Proc. Amer. Math. Soc., 132 (2004), 1435–1443.
36. S. Reich, *Fixed points of contractive functions*, Boll. Un. Mat. Ital., 5 (1972), 26–42.
37. B. E. Rhoades, *A comparison of various definitions of contractive mappings*, Trans. Amer. Math. Soc., 226 (1977), 257–290.
38. B. E. Rhoades, *Some theorems on weakly contractive maps*, Nonlin. Anal., 47 (2001), 2683–2693.
39. I. A. Rus, *Generalized Contractions and Applications*, Cluj University Press, Cluj-Napoca, 2001.
40. B. Samet and M. Turinici, *Fixed point theorems on a metric space endowed with an arbitrary binary relation and applications*, Commun. Math. Anal., 13 (2012), 82–97.
41. E. Schechter, *Handbook of Analysis and its Foundation*, Academic Press, New York, 1997.
42. A. Tarski, *Axiomatic and algebraic aspects of two theorems on sums of cardinals*, Fund. Math., 35 (1948), 79–104.
43. M. Turinici, *Fixed points of implicit contraction mappings*, An. Şt. Univ. "Al. I. Cuza" Iaşi (S I-a, Mat), 22 (1976), 177–180.
44. M. Turinici, *Nonlinear contractions and applications to Volterra functional equations*, An. Şt. Univ. "Al. I. Cuza" Iaşi (S I-a, Mat), 23 (1977), 43–50.
45. M. Turinici, *Multivalued contractions and applications to functional differential equations*, Acta Math. Acad. Sci. Hungaricae, 37 (1981), 147–151.
46. M. Turinici, *Fixed points for monotone iteratively local contractions*, Demonstratio Math., 19 (1986), 171–180.
47. M. Turinici, *Abstract comparison principles and multivariable Gronwall-Bellman inequalities*, J. Math. Anal. Appl., 117 (1986), 100–127.
48. M. Turinici, *Function pseudometric VP and applications*, Bul. Inst. Polit. Iaşi (S. Mat., Mec. Teor., Fiz.), 53(57) (2007), 393–411.
49. M. Turinici, *Implicit contractive maps in ordered metric spaces*, Topics in Mathematical Analysis and Applications (T. M. Rassias and L. Tóth, Eds.), pp. 715–746, Springer Intl. Publ., Switzerland, 2014.
50. M. Turinici, *Contraction maps in pseudometric structures*, Essays in Mathematics and its Applications (T. M. Rassias and P. M. Pardalos, Eds.), pp. 513–562, Springer Intl. Publ., Switzerland, 2016.
51. E. S. Wolk, *On the principle of dependent choices and some forms of Zorn's lemma*, Canad. Math. Bull., 26 (1983), 365–367.

Existence and Stability of Equilibrium Points Under the Influence of Poynting–Robertson and Stokes Drags in the Restricted Three-Body Problem

Aguda Ekele Vincent and Angela E. Perdiou

Abstract In the framework of the circular restricted three-body problem, the dynamical effects of Stokes and Poynting–Robertson (P–R) drag forces on the existence, location, and stability of equilibrium points are investigated. It is found that under constant effects of P–R and/or Stokes drags, collinear equilibrium points cease to exist, but there are in the absence of the perturbing forces. The problem admits five non-collinear equilibrium points, and it is seen that the perturbing forces have significant effects on their positions. The linear stability of the equilibrium points is also studied in certain cases, and it is found that the stability of some of these points significantly depends on the perturbing forces. More precisely, the motion of the infinitesimal body near the non-collinear equilibrium points is unstable under the effect of both kinds of perturbing forces except from the equilibria L_4 and L_5 for which is stable only for Stokes drag effect, namely, the remaining parameter that corresponds to P–R drag is fixed to zero. We may conclude, therefore, that the P–R effect destroys stability of the equilibrium points.

MSC 70F07; 70F15; 70K20; 70K42

1 Introduction

The circular restricted three-body problem (CR3BP) consists of two finite bodies, known as primaries, which rotate in circular orbits around their common center of mass and a massless body that moves in the plane of motion of the primaries under their gravitational attraction and does not affect their motion. The CR3BP

A. E. Vincent
Department of Mathematics, Nigeria Maritime University, Okerenkoko, Delta State, Nigeria

A. E. Perdiou (✉)
Department of Civil Engineering, University of Patras, Patras, Greece
e-mail: aperdiou@upatras.gr

I. N. Parasidis et al. (eds.), *Mathematical Analysis in Interdisciplinary Research*, Springer Optimization and Its Applications 179,
https://doi.org/10.1007/978-3-030-84721-0_37

has been the well-known studied problem in Celestial Mechanics. It possesses five equilibrium points, three of which lie on the x-axis and are called collinear, while the rest two are away from this axis and are called triangular (non-collinear) equilibria. The three collinear points are generally unstable, but the triangular ones are generally stable for values of the mass ratio $\mu \leqslant 0.03850\ldots$ (see, e.g., [1, 2]). The theory and applications for these equilibrium points and the related periodic orbits emanating from them have enabled several space mission explorations, such as ISEE-3, ACE, and PLANCK, among others [3–5], while certain operations are still in progress.

During the past, several more complicated or even simpler modifications of the CR3BP have been proposed in order to make it more realistic for systems of dynamical astronomy, and many scientists and astronomers have studied them (see, for example, [6–12], and the references therein). These modifications involve more bodies and/or include additional forces other than the gravitational one or perturbing forces such as radiation pressure, planetary perturbations, the Poynting–Robertson drag, and the Stokes drag (see, e.g., [13–20]). The case of the CR3BP where at least one of the primary bodies emits radiation is a well-known problem, and it named in the literature as "the photogravitational problem of three bodies." This special case of the CR3BP was firstly studied by Radzievskii [21], and since then it has been extensively investigated [22–26]. The significance of the effect of radiation on natural bodies or artificial satellites has been proved, recognized, and used by many scientists, especially with respect to the solar sail of artificial satellites as well as to the formation of concentrations of interplanetary and interstellar dust in binary star systems [27, 28].

The Doppler shift and absorptions as well as the subsequent re-emission of incident radiation, namely, the Poynting–Robertson drag, are usually neglected in many research works for the approximation of radiation force, although the dissipative effects play a fundamental role in the dynamics of our solar system. The P–R effect is one of the most important mechanisms of dissipation which may be used in the investigation of the stability of zodiacal cloud, asteroidal particles, and dust rings around planets. In this context, Chernikov [29] studied the existence and stability of equilibrium points under the influence of radiation and the P–R effect. He found that despite the absence of a Jacobi integral, six equilibrium points exist at most and pointed out that the collinear points are not positioned on the axis connecting the primaries any more, while the triangular points were not symmetrical with respect to this axis. It was also found that the triangular points are unstable for the P–R effect. Later, Schuerman [30] studied the triangular points of the problem and found that the points are unstable due to the P–R effect. Ragos and Zafiropoulos [31] extended the problem to the case that both main bodies are radiation sources. They studied numerically the equilibrium points lying on the orbital plane of the primaries. Murray [13] discussed the dynamical effect of general drag (nebular drag, gas drag, and P–R drag) in the CR3BP and found that the collinear points are not positioned on the axis joining the two masses, while the displaced triangular points L_4 and L_5 are asymptotically stable for certain classes of drag forces. Researchers like Burns et al. [32], Singh and Simeon [33], Singh and Amuda [34], Umar and Hussain [35], and others studied the CR3BP by taking into account the P–R drag in different views. In a recent study, Vincent and Perdiou [20] investigated the motion of a test particle in the field of Cen-X4 binary system with P–R drag and oblateness together with small perturbations in the Coriolis and centrifugal forces. They asserted that under the constant P–R drag effect, collinear equilibrium points cease to exist numerically and of course analytically. They found that the equilibrium points are unstable for the P–R effect against their conditional stability in the absence of the drag force.

Additional influential perturbing force than that of the gravitational one is Stokes drag. This effect is due to the collisions of particles with the molecules of gas nebula during the formation of a planetary system. In this vein, Celletti et al. [16] performed a dynamical analysis in the framework of the planar CR3BP under different kinds of dissipation (linear, Stokes, or Poynting–Robertson drag). They found the periodic orbit attractors for the case of linear and Stokes drags, while in the case of the P–R effect, no other attractors were found beside the primaries, unless a fourth body is added to counterbalance the dissipative effect. In addition, the stationary points L_4 and L_5 were shown to become unstable for both kinds of dissipation. It was shown in [14] that, in the case of Stokes drag, stationary solutions L_4 and L_5 appear to be stable. Almost recently, Jain and Aggarwal [36] studied the effect of Stokes drag force with oblateness of smaller primary on libration points in the restricted three-body problem. It was established that collinear points cease to exist, while the stability of non-collinear libration points $L_{4,5}$ remains unstable. Later, by taking into consideration the P–R light drag effect, Jain and Aggarwal [18] investigated the existence and stability of equilibrium points of the problem and found that the equilibrium points are unstable due to the effect of the drag.

However, the inclusion of P–R and Stokes drags terms changes the nature of the problem from purely a central force to a dissipative one. In this work, we aim to study numerically the motion of a test particle in the R3BP under the effect of Stokes drag and Poynting–Robertson drag. As it is known [37], the orbital evolution of dust grain in the solar system is affected by the drag forces. The P–R and Stokes drags forces cause dust particle to lose orbital energy and angular momentum and spiral toward the Sun. In the present work, the dissipative forces formulation described by Murray and Dermott [38] and recently applied in the dynamics of the regularized restricted three-body problem with dissipation by Celletti et al. [16] will be used. As a consequence of both kinds of dissipation, the problem becomes a tri-parametric one: a mass ratio, μ, and two constants of dissipation, k_s and k_{pr}, due to Stokes and P–R drags, respectively. Among the different questions that this model may arise, in the current work, we concentrate our study on the existence, locations of equilibrium points, and the corresponding linear stability analysis.

The chapter is organized as follows: in Sect. 2, we present the governing equations of motion for the system in the dissipative framework. In Sect. 3, we determine the existence and locations of the equilibrium points numerically and verify them graphically for various values of the parameters under consideration, while their linear stability is analyzed in Sect. 4. Finally, Sect. 5 summarizes the discussion and conclusion of our study.

2 Equations of Motion with Dissipative Forces

The system we consider is the planar circular restricted three-body problem which is formed by two finite bodies, P_1 and P_2, called the primaries (bigger primary and smaller primary, respectively), with masses $m_1 = 1 - \mu$ and $m_2 = \mu$, correspondingly, where $\mu = m_2/(m_1 + m_2) \leqslant 1/2$ is the mass ratio parameter. We assume that the motion of the three bodies takes place on the same plane under their mutual gravity, but the mass of the third body is so small compared to the masses of the two primaries where its influence on them can be neglected. This system is also dimensionless, i.e., we normalize the units with the supposition such that the sum of the masses and the separation between the primary bodies both be unity and the unit of time is taken as the time period of the rotating frame moving with the angular velocity n of the primaries, where n is normalized to one (for details, see [1]). In this coordinate system, the arising dimensionless equations of motion under both kinds of dissipation take the form [16, 18, 38]:

$$\ddot{x} - 2\dot{y} = \frac{\partial \Omega}{\partial x} = \Omega_x, \qquad \ddot{y} + 2\dot{x} = \frac{\partial \Omega}{\partial y} = \Omega_y, \tag{1}$$

where

$$\begin{aligned} \Omega_x &= \frac{\partial \bar{U}}{\partial x} - k_s\left(\dot{x} - y - \frac{3\alpha y}{2r^{7/2}}\right) - \frac{k_{pr}}{r_1^2}\left[\dot{x} - y + \frac{x}{r_1^2}(x\dot{x} + y\dot{y})\right], \\ \Omega_y &= \frac{\partial \bar{U}}{\partial y} - k_s\left(x + \dot{y} + \frac{3\alpha y}{2r^{7/2}}\right) - \frac{k_{pr}}{r_1^2}\left[\dot{y} + x + \frac{y}{r_1^2}(x\dot{x} + y\dot{y})\right], \end{aligned} \tag{2}$$

while the gravitational effective potential $\bar{U}$ is given by

$$\bar{U} = \frac{1}{2}(x^2 + y^2) + \frac{1-\mu}{r_1} + \frac{\mu}{r_2}, \tag{3}$$

and

$$r_1^2 = (x + \mu)^2 + y^2, \qquad r_2^2 = (x + \mu - 1)^2 + y^2 \tag{4}$$

are the distances of the massless body from the primaries. Also, k_s and $k_{pr} \in [0, 1)$ designate constants of dissipation due to Stokes drag and P–R drag, respectively, while $\alpha \in [0, 1)$ is the ratio between the gas and Keplerian velocities at a given radius $r^2 = x^2 + y^2$ (see, e.g., [14]). We remark here that Stokes dissipation depends on the parameter α.

3 Existence and Locations of the Equilibrium Points

The equilibrium points, or the Lagrangian points for the CR3BP, are obtained when the acceleration $\ddot{x}$, $\ddot{y}$ and velocity $\dot{x}$, $\dot{y}$ components of the infinitesimal body are zero. Therefore, we are able to obtain the coordinates (x_0, y_0) of the equilibrium points as solutions of the following non-linear algebraic equations:

$$
\begin{aligned}
&x - \frac{(1-\mu)(x+\mu)}{r_1^3} - \frac{\mu(x+\mu-1)}{r_2^3} + k_s\left(y + \frac{3\alpha y}{2r^{7/2}}\right) + \frac{k_{pr}y}{r_1^2} = 0, \\
&y - \frac{(1-\mu)y}{r_1^3} - \frac{\mu y}{r_2^3} - k_s\left(x + \frac{3\alpha x}{2r^{7/2}}\right) - \frac{k_{pr}x}{r_1^2} = 0.
\end{aligned} \tag{5}
$$

It is interesting to note that when $k_s = k_{pr} = 0$, the classical model of the restricted three-body problem is recovered, while for $k_s \neq 0$ and $k_{pr} \neq 0$ we have the restricted three-body problem with Stokes drag and P–R drag effects as reported in Jain and Aggarwal [18, 36], correspondingly.

It is well known that in the classical R3BP, there are five equilibrium points. Three of them are on the x-axis and are called collinear points, while the other two are out of the x-axis and are called equilateral equilibrium points. In the present case, we will see that the existence and positions of the equilibria of the problem depend on the drag forces of the primary bodies as well. As indicated by System (5), the drag forces have some effect on the positions of the equilibrium points. On the other hand, the equilibrium solutions (collinear and non-collinear points) under general drag (e.g., Stokes drag, P–R drag) are only poorly studied in the literature by analytical means (with some exception found in [13]). The equations in System (5), containing the constant parameters k_s and k_{pr}, consist of a system of nonlinear algebraic equations. As no simple analytic solution for x and y could be obtained similar to the classical problem [1], we resort to a numerical study of this system. In this study, the equilibrium points are obtained by solving Eqs. (5) simultaneously using the well-known Newton's method. This has also been successfully applied by many authors for the determination of equilibrium points in different model problems of Celestial Mechanics (see, for example, [20, 39–41] and the references therein).

From System (5), it can be seen that the second equation is always not satisfied. With the Stokes (k_s) drag and/or the P–R (k_{pr}) effect, the existence of ordinates of equilibrium points L_1, L_2, and L_3 associated with these equations can be easily verified from the second equation of (5), since the condition $y = 0$ is not fulfilled for them. We note here that the y-components of the equilibrium points $L_{1,2,3}$ are close to zero but not zero, which means that due to the y-components these points are not collinear. This is easy to show geometrically (Fig. 1, bottom frames). The existence of such points was earlier envisaged for the Sun–Jupiter system in [13, 31, 37].

In this case, the second equation of (5) holds and the equilibria are obtained by solving Eqs. (5) simultaneously. In Fig. 1, we illustrate the five non-collinear equilibrium points, L_i, $i = 1, 2, \ldots, 5$, of the problem in the xy-plane along with

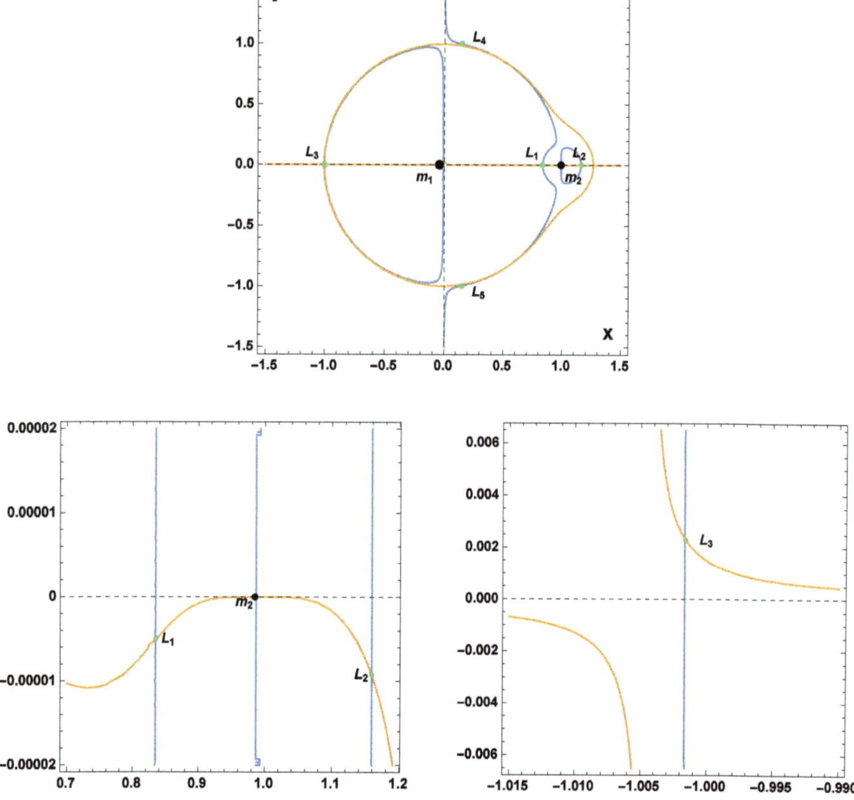

Fig. 1 Top: the five non-collinear equilibria and the positions of the primary bodies for $\mu = 0.01$, $k_s = 10^{-5}$, $k_{pr} = 10^{-5}$, and $\alpha = 0.05$. Bottom: zoomed images of $L_{1,2}$ and L_3, respectively, with intersections of the curves

associated primaries for $\mu = 0.01$, $k_s = 10^{-5}$, $k_{pr} = 10^{-5}$, and $\alpha = 0.05$ for which we have found by solving numerically the above system. We denote here that the equilibria in the xy-plane are given by the mutual intersections of the curves, blue line (first equation) and brown line (second equation). Here we also note that the intersection points of these curves show the coordinates (x_0, y_0) of the equilibria on the xy-plane. It is seen that under the combined effect of the parameters, there exist five non-collinear equilibrium points for which the ordinates of L_1, L_2, and L_3 are close to zero but not zero. We remark that our configuration is similar to the one shown by Murray [13, see figure 4] for the general drag force.

Next, we wish to compute numerically the effects of the perturbing forces on the existence and locations of the equilibrium points. We note that the existence and location of these points depend on the system parameters μ, k_s, and k_{pr} of the problem. In this study, we will keep the value of the mass parameter fixed and equal to $\mu = 0.01$ for all numerical calculations. Figure 2 shows the effects of the

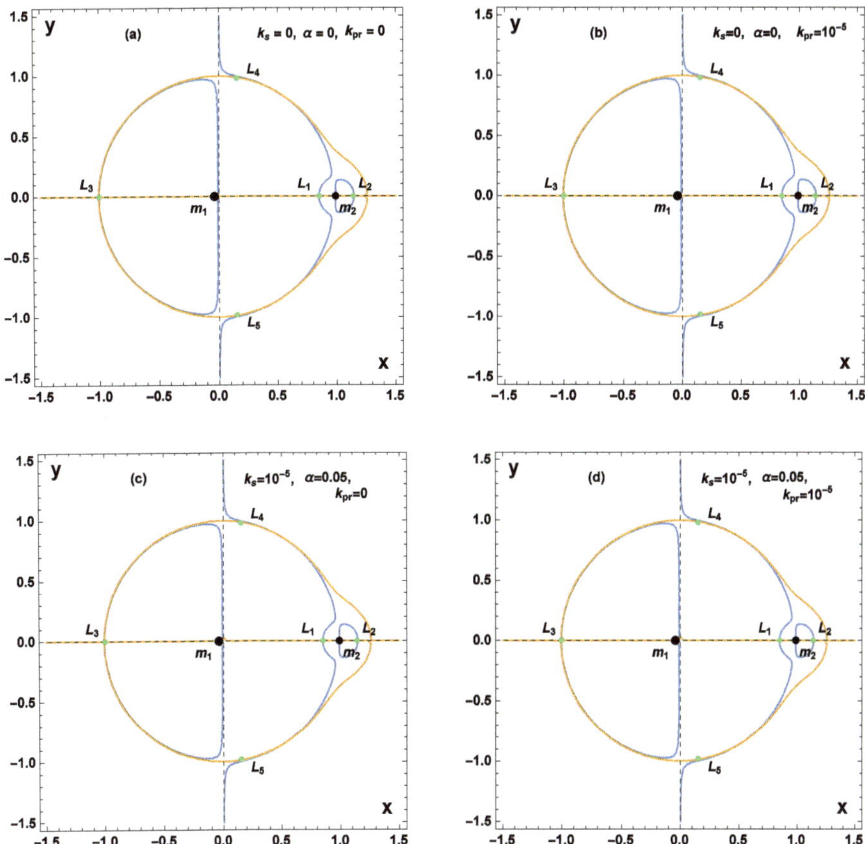

Fig. 2 The positions of the five equilibrium points for (**a**) $k_s = 0,\ \alpha = 0,\ k_{pr} = 0$ (classical case or zero drags), (**b**) $k_s = 0,\ \alpha = 0,\ k_{pr} = 10^{-5}$ (P–R case), (**c**) $k_s = 10^{-5},\ \alpha = 0.05,\ k_{pr} = 0$ (Stokes case), and (**d**) $k_s = 10^{-5},\ \alpha = 0.05,\ k_{pr} = 10^{-5}$ (P–R and Stokes). The value of $\mu = 0.01$ is fixed for all cases

perturbing forces, involved in the problem under consideration, for the classical case and three different cases on the positions of the equilibria. In particular, Fig. 2a is when the effects of the drag forces are neglected (conservative case or zero drag), that is, $k_s = 0 = k_{pr}$, and in this case, there exist five equilibrium points, i.e., three collinear and two triangular points. Additionally, Fig. 2b shows the effect of P–R drag when Stokes drag is absent, while Fig. 2c shows the effect of Stokes drag when P–R drag is absent. Finally, Fig. 2d shows the effect when both Stokes drag and P–R drag are inaction together. In the non-conservative case, we observe that there exist five non-collinear equilibrium points (collinear points cease to exist) and the number of equilibrium points is seen to be independent on the strength and the kind of the involved dissipative forces (see Fig. 2 and Tables 2, 3, 4).

Table 1 Positions (x, y) of the five equilibrium points when the perturbing forces are neglected for $\mu = 0.01$

L_1	L_2	L_3	$L_{4,5}$
(0.848079, 0)	(1.14677, 0)	(−1.00417, 0)	(0.49000, ±0.866025)

Table 2 Positions (x, y) of the five non-collinear equilibrium points under the P–R drag force for $\mu = 0.01$ and $k_s = 0 = \alpha$

k_{pr}	L_1	L_2	L_3	$L_{4,5}$
10^{-5}	(0.848079, -2.83332×10^{-6})	(1.14677, -3.83401×10^{-6})	(−1.00417, 0.00115581)	(0.489615, 0.866245) (0.490385, −0.865805)
10^{-4}	(0.848079, -2.83332×10^{-5})	(1.14677, -3.83401×10^{-5})	(−1.00410, 0.0115581)	(0.486123, 0.868229) (0.493822, −0.863830)
10^{-3}	(0.848079, -2.83332×10^{-4})	(1.14677, -3.83401×10^{-4})	(−0.99744, 0.115658)	(0.448571, 0.888425) (0.525959, −0.844466)

Table 3 Positions (x, y) of the five non-collinear equilibrium points under the Stokes drag force for $\mu = 0.01$ and $k_{pr} = 0$

k_s	α	L_1	L_2	L_3	$L_{4,5}$
10^{-5}	0.05	(0.848079, -2.36470×10^{-6})	(1.14677, -5.36860×10^{-6})	(−1.00417, 0.0012268)	(0.489585, 0.866262) (0.490414, −0.865789)
10^{-4}	0.10	(0.848079, -2.64324×10^{-5})	(1.14677, -5.60682×10^{-5})	(−1.00408, 0.0131125)	(0.485527, 0.868566) (0.494401, −0.863495)
10^{-3}	0.12	(0.848079, -2.75466×10^{-4})	(1.14677, -5.70212×10^{-4})	(−0.99504, 0.134636)	(0.440320, 0.892590) (0.532060, −0.840601)

Table 4 Positions (x, y) of the five non-collinear equilibrium points under the Stokes drag and P–R drag forces for $\mu = 0.01$

$k_s = k_{pr}$	α	L_1	L_2	L_3	$L_{4,5}$
10^{-5}	0.05	(0.848079, -5.19802×10^{-6})	(1.14677, -9.20258×10^{-6})	(−1.00416, 0.0023826)	(0.489200, 0.866482) (0.490798, −0.865569)
10^{-4}	0.10	(0.848079, -5.47656×10^{-5})	(1.14677, -9.44083×10^{-5})	(−1.00386, 0.0246712)	(0.481586, 0.870780) (0.498163, −0.861310)
10^{-3}	0.12	(0.848079, -5.58798×10^{-4})	(1.14677, -9.53612×10^{-4})	(−0.972092, 0.250937)	(0.390558, 0.915742) (0.562872, −0.820118)

As we will show, the effect of the perturbing forces on the non-collinear equilibrium points L_i, $i = 1, 2 \ldots, 5$, is to change their position considerably. The results are shown in Tables 1, 2, 3, 4 for the four cases, namely, when the perturbing forces are neglected, when the P–R drag is dominant, when the Stokes drag is dominant, and when both Stokes drag and P–R drag are inaction together. We found that unlike to the equilibrium points of the classical case, equilibrium points in the presence of the perturbing force or both of them have finite but small y-component for various values of the parameters. The existence of such equilibrium

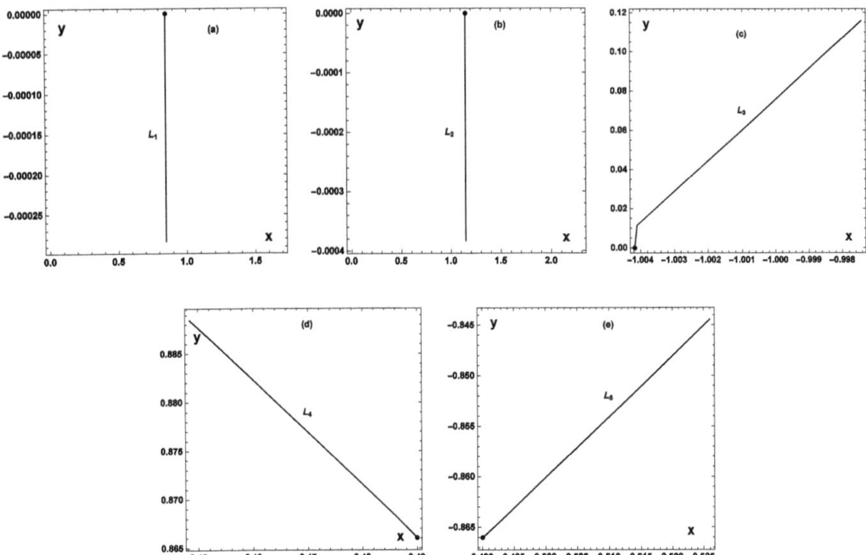

Fig. 3 Location of the non-collinear equilibrium points $L_i,\ i = 1, 2, \ldots, 5$, for $\mu = 0.01$ and various values of the P–R drag. The "•" denotes the position when $k_{pr} = 0$ (classical case), while the thick lines indicate the equivalent paths for various values of P–R (k_{pr})

Table 5 Absolute difference in the coordinates of equilibrium points when $\mu = 0.01$ and $k_{pr} = 10^{-5}$. The coordinates (x, y) denote the position of an equilibrium point $L_i,\ i = 1, 2, \ldots, 5$,, for the classical case, while $(x(k_{pr}), y(k_{pr}))$ stands for the corresponding position when the P–R drag force is included

L_i	x	$x(k_{pr})$	$\lvert x - x(k_{pr})\rvert$	y	$y(k_{pr})$	$\lvert y - y(k_{pr})\rvert$
L_1	0.848079	0.848079	0	0	-2.83332×10^{-6}	2.83332×10^{-6}
L_2	1.14677	1.14677	0	0	-3.83401×10^{-6}	3.83401×10^{-6}
L_3	−1.00417	−1.00417	0	0	0.00115581	0.00115581
L_4	0.49000	0.489615	3.85×10^{-4}	0.866025	0.866245	2.2×10^{-4}
L_5	0.49000	0.490385	3.85×10^{-4}	−0.866025	−0.865805	2.2×10^{-4}

points was shown in [13, 20, 37, and the references quoted therein]. We observe that the positions of these equilibria increase or decrease in a relatively small range with an increase in the respective parameter which already covers a big range from the very low/weak drag coefficient case to the case of high drag coefficient. These effects can be easily seen in the numerical results presented in Tables 2, 3, and 4 and Fig. 3. For instance, in Table 2, the coordinates of the numerically determined non-collinear equilibrium points are shown for various values of the P–R drag. We observe that there is a significant change (difference) in the positions (coordinates) of the equilibrium points.

Table 5 shows the absolute difference in the coordinates of equilibrium points when $k_{pr} = 10^{-5}$ with respect to their classical values. We observe that on the

x-coordinate, L_4 decreases by 0.039 percent while L_5 increases outward by 0.039 percent, while at the same time the y-coordinate of L_1, L_2, L_3, and L_4 increases by 0.00028, 0.00038, 0.12, and 0.022 percents, respectively, and L_5 decreases by 0.022 percent. Moreover, x-coordinates of L_1, L_2, and L_3 are practically unaffected. Therefore, we conclude that positions of the equilibrium points L_3 and L_4 move in opposite direction toward the y-axis, L_5 moves anticlockwise direction toward m_2, while L_1 and L_2 move outward from the x-axis in the negative direction. In the absence of the perturbing forces, we found $L_{4,5}$ to be symmetrically located with respect to the orbital plane. However, the inclusion of the perturbing forces in the system results in a little asymmetry in the location of such equilibria (see Table 2).

The aforementioned discussion can be summarized in Fig. 3, where we plot the positions of the equilibria. Apparently, the P–R drag does change the positions of the equilibria. This can be easily seen by comparing equilibrium point positions with the classical positions. Similar results, though not shown here, are also observed in the absence of P–R drag (Table 3) or when both forces are inaction together (Table 4). Evidently, the variational trend of the corresponding positions in Tables 3 and 4 is similar to the scenario with the variation of P–R drag as described previously. However, by observing the y-coordinate of the equilibrium points L_1, L_3, L_4, and L_5 in Tables 2 and 3, it seems that P–R drag has greater dissipative influence than the Stokes drag. The reason can be found in the fact that the amount of light hitting the particle is proportional to the inverse of the square of the distance from the Sun (see System (5)). All these results are similar with Murray [13] and the references quoted therein.

4 Stability of the Non-collinear Equilibrium Points

To study analytically the solutions in the neighborhood of the non-collinear equilibrium points L_i, $i = 1, 2, \dots, 5$, following Ragos and Zafiropoulos [31] and Singh and Amuda [17], we consider small displacements ξ and η given to the coordinates of an equilibrium point (x_0, y_0) such that

$$\xi = x - x_0, \quad \eta = y - y_0. \tag{6}$$

Then, the variational equations of motion are derived in the following form:

$$\begin{aligned} \ddot{\xi} - 2\dot{\eta} &= \Omega^{(0)}_{x\dot{x}}\dot{\xi} + \Omega^{(0)}_{x\dot{y}}\dot{\eta} + \Omega^{(0)}_{xx}\xi + \Omega^{(0)}_{xy}\eta, \\ \ddot{\eta} + 2\dot{\xi} &= \Omega^{(0)}_{y\dot{x}}\dot{\xi} + \Omega^{(0)}_{y\dot{y}}\dot{\eta} + \Omega^{(0)}_{yx}\xi + \Omega^{(0)}_{yy}\eta, \end{aligned} \tag{7}$$

where only the linear terms in ξ and η have been kept, while dots represent derivatives with respect to time t. The superscript "(0)" means that the corresponding derivatives have been evaluated at an equilibrium point (x_0, y_0). Now, we assume solutions of the variational equations in the form:

$$\xi = A_1 e^{\lambda t}, \quad \eta = A_2 e^{\lambda t}, \tag{8}$$

where A_i, $i = 1, 2$, are arbitrary constants and λ is a complex constant. Substituting (8) in Eqs. (7) and simplifying, we obtain

$$\begin{aligned} (\lambda^2 - \lambda\Omega^{(0)}_{x\dot{x}} - \Omega^{(0)}_{xx})A_1 + (-2\lambda - \lambda\Omega^{(0)}_{x\dot{y}} - \Omega^{(0)}_{xy})A_2 &= 0, \\ (2\lambda - \lambda\Omega^{(0)}_{y\dot{x}} - \Omega^{(0)}_{yx})A_1 + (\lambda^2 - \lambda\Omega^{(0)}_{y\dot{y}} - \Omega^{(0)}_{yy})A_2 &= 0. \end{aligned} \tag{9}$$

For the nontrivial solution, the determinant of the coefficients matrix of system (9) must be zero, namely,

$$\begin{vmatrix} \lambda^2 - \lambda\Omega^{(0)}_{x\dot{x}} - \Omega^{(0)}_{xx} & -2\lambda - \lambda\Omega^{(0)}_{x\dot{y}} - \Omega^{(0)}_{xy} \\ 2\lambda - \lambda\Omega^{(0)}_{y\dot{x}} - \Omega^{(0)}_{yx} & \lambda^2 - \lambda\Omega^{(0)}_{y\dot{y}} - \Omega^{(0)}_{yy} \end{vmatrix} = 0. \tag{10}$$

Simplifying Eq. (10), we obtain the characteristic polynomial corresponding to the system (7) as

$$\lambda^4 + a\lambda^3 + b\lambda^2 + c\lambda + d = 0, \tag{11}$$

where

$$\begin{aligned} a &= -(\Omega^{(0)}_{y\dot{y}} + \Omega^{(0)}_{x\dot{x}}), \\ b &= 4 + \Omega^{(0)}_{x\dot{x}}\Omega^{(0)}_{y\dot{y}} - \Omega^{(0)}_{xx} - \Omega^{(0)}_{yy} - [\Omega^{(0)}_{x\dot{y}}]^2, \\ c &= \Omega^{(0)}_{x\dot{x}}\Omega^{(0)}_{yy} + \Omega^{(0)}_{xx}\Omega^{(0)}_{y\dot{y}} + 2\Omega^{(0)}_{xy} - 2\Omega^{(0)}_{yx} - \Omega^{(0)}_{y\dot{x}}\Omega^{(0)}_{xy} - \Omega^{(0)}_{yx}\Omega^{(0)}_{x\dot{y}}, \\ d &= \Omega^{(0)}_{xx}\Omega^{(0)}_{yy} - \Omega^{(0)}_{yx}\Omega^{(0)}_{xy}. \end{aligned} \tag{12}$$

The obtained eigenvalues determine the stability or instability of the respective equilibrium point. Also, the involved second-order partial derivatives of the modified potential-like function Ω are given by

$$\begin{aligned} \Omega^{(0)}_{xx} = 1 &- \frac{1-\mu}{r_{10}^3} - \frac{\mu}{r_{20}^3} - \frac{2(x_0+\mu)y_0 k_{pr}}{r_{10}^4} + \frac{3(1-\mu)(x_0+\mu)^2}{r_{10}^5} \\ &+ \frac{3\mu(x_0+\mu-1)^2}{r_{20}^5} - \frac{21 k_s \alpha x_0 y_0}{4 r_0^{11/2}}, \end{aligned}$$

$$\Omega^{(0)}_{yy} = 1 - \frac{1-\mu}{r_{10}^3} - \frac{\mu}{r_{20}^3} + \frac{2k_{pr}x_0 y_0}{r_{10}^4} + \frac{3(1-\mu)y_0^2}{r_{10}^5} + \frac{3\mu y_0^2}{r_{20}^5} + \frac{21 k_s \alpha x_0 y_0}{4 r_0^{11/2}},$$

$$\Omega^{(0)}_{xy} = \frac{k_{pr}}{r_{10}^2} - \frac{2k_{pr}y_0^2}{r_{10}^4} + \frac{3(1-\mu)(x_0+\mu)y_0}{r_{10}^5} + \frac{3\mu(x_0+\mu-1)y_0}{r_{20}^5}$$
$$+ k_s(1 - \frac{21\alpha y_0^2}{4r_0^{11/2}} + \frac{3\alpha}{2r_0^{7/2}}),$$

$$\Omega^{(0)}_{yx} = -\frac{k_{pr}}{r_{10}^2} + \frac{2k_{pr}(x_0+\mu)x_0}{r_{10}^4} + \frac{3(1-\mu)(x_0+\mu)y_0}{r_{10}^5} + \frac{3\mu(x_0+\mu-1)y_0}{r_{20}^5}$$
$$- k_s\left(1 - \frac{21\alpha x_0^2}{4r_0^{11/2}} + \frac{3\alpha}{2r_0^{7/2}}\right),$$

$$\Omega^{(0)}_{x\dot{x}} = -\left[k_s + \frac{k_{pr}}{r_{10}^2}(1 + \frac{x_0^2}{r_{10}^2})\right],$$
$$\Omega^{(0)}_{y\dot{y}} = -\left[k_s + \frac{k_{pr}}{r_{10}^2}(1 + \frac{y_0^2}{r_{10}^2})\right],$$
$$\Omega^{(0)}_{x\dot{y}} = -\frac{k_{pr}x_0y_0}{r_{10}^4} = \Omega^{(0)}_{y\dot{x}},$$

with

$$r_0^2 = x_0^2 + y_0^2, \quad r_{10}^2 = (x_0+\mu)^2 + y_0^2, \quad r_{20}^2 = (x_0+\mu-1)^2 + y_0^2.$$

Since we have computed the coordinates (x_0, y_0) of the equilibrium points (presented in Tables 1, 2, 3, 4), we can insert them into the characteristic equation (11) and thus derive their linear stability numerically. An equilibrium point (x_0, y_0) is said to be stable in the sense of Lyapunov if all the four roots of the characteristic polynomial equation (11) are either negative real numbers or distinct imaginary, asymptotically stable if roots are complex with negative real parts, and unstable, otherwise. As a particular example, we compute the characteristic roots λ_i, $i = 1, 2, 3, 4$, which are shown in Tables 6, 7, and 8 for three cases, namely, when the P–R drag is dominant, when the Stokes drag is dominant, and when both forces are inaction together, correspondingly. Our analysis reveals that in all studied cases, all the equilibria are unstable due to a positive real root or a complex root with positive real part except for the equilibria L_4 and L_5 under Stokes drag where we get complex roots with negative real parts, which means that due to the negative real parts these points are stable. From the results, we can conclude that the P–R effect destroys the stability of the equilibrium points (L_4 and L_5) known to be conditionally stable in the classical gravitational restricted three-body problem (see e.g., [1]) or the restricted three-body problem under the force of Stokes drag (see [14]).

Table 6 Stability of the five non-collinear equilibria for $\mu = 0.01$ under P–R drag $k_{pr} = 10^{-5}$

L_i	(x_0, y_0)	$\lambda_{1,2}$	$\lambda_{3,4}$
L_1	$(0.848079, -2.83332 \times 10^{-6})$	$-2.903756,\ 2.903734$	$-0.0000093619 \pm 2.316563i$
L_2	$(1.146770, -3.83401 \times 10^{-6})$	$-2.179447,\ 2.179432$	$-0.0000060981 \pm 1.874813i$
L_3	$(-1.00417, 0.00115581)$	$-0.161350,\ 0.161379$	$-0.000029638 \pm 1.0085933i$
L_4	$(0.489615, 0.866245)$	$-0.0000337395 \pm 0.963376i$	$0.0000187889 \pm 0.2681746i$
L_5	$(0.490385, -0.865805)$	$-0.0000337508 \pm 0.963273i$	$0.0000188004 \pm 0.2685114i$

Table 7 Stability of the five non-collinear equilibria for $\mu = 0.01$ under Stokes drag $k_s = 10^{-5}$ and $\alpha = 0.05$

L_i	(x_0, y_0)	$\lambda_{1,2}$	$\lambda_{3,4}$
L_1	$(0.848079, -2.3647 \times 10^{-6})$	$-2.903749,\ 2.903740$	$-0.0000051451 \pm 2.316563i$
L_2	$(1.146770, -5.36857 \times 10^{-6})$	$-2.179442,\ 2.1794319$	$-0.0000050843 \pm 1.874813i$
L_3	$(-1.00417, 0.0012268)$	$-0.161366,\ 0.1613581$	$-0.0000060627 \pm 1.008593i$
L_4	$(0.489585, 0.866262)$	$-0.0000063370 \pm 0.963374i$	$-0.0000036629 \pm 0.268162i$
L_5	$(0.490414, -0.865789)$	$-0.0000063377 \pm 0.963268i$	$-0.0000036623 \pm 0.268532i$

Table 8 Stability of the five non-collinear equilibria for $\mu = 0.01$ under Stokes and P–R drags $k_s = 10^{-5}$, $\alpha = 0.05$, and $k_{pr} = 10^{-5}$

L_i	(x_0, y_0)	$\lambda_{1,2}$	$\lambda_{3,4}$
L_1	$(0.848079, -5.19802 \times 10^{-6})$	$-2.903761,\ 2.903729$	$-0.000014507 \pm 2.316563i$
L_2	$(1.146770, -9.20258 \times 10^{-6})$	$-2.179447,\ 2.179427$	$-0.000011182 \pm 1.874813i$
L_3	$(-1.00416, 0.0023826)$	$-0.161569,\ 0.161591$	$-0.000035699 \pm 1.008616i$
L_4	$(0.489200, 0.866482)$	$-0.0000400705 \pm 0.963424i$	$0.000015119 \pm 0.267999i$
L_5	$(0.490798, -0.865569)$	$-0.0000400949 \pm 0.963219i$	$0.000015145 \pm 0.268697i$

5 Discussion and Conclusion

The presented work constitutes a numerical study about the effects of dissipative forces on the equilibrium points of the restricted three-body problem. In particular, the positions of the equilibrium points as well as their linear stability were investigated in the framework of this classical problem under the effects of Stokes and Poynting–Robertson drags. In general, it was found that these forces induce considerable changes to the location of all equilibria. This comes directly by the pertinent non-linear algebraic equations, which provide the respective positions, since it was identified analytically that the well-known collinear equilibrium points of the circular restricted three-body problem cease to exist, while the respective triangular Lagrangian points do not form equilateral triangles.

More precisely, we studied the existence, location, and stability of the equilibrium points on the orbital plane (x, y) as the parameters k_{pr} and k_s of the P–R and Stokes drag forces, respectively, vary. To this purpose, we examined four distinct cases: first when the perturbing forces are neglected, second when the P–R drag is dominant, third when the Stokes drag is dominant, and fourth when both Stokes drag and P–R drag are inaction together. In the case where the P–R drag and/or Stokes drag were considered, our numerical investigation for the number and location of the equilibria showed that five non-collinear equilibrium points exist in contrast to the absence of the aforementioned perturbing forces, i.e., the classical case, where five equilibrium points also exist, but three of them lie on the axis joining the primaries, while the rest two form in the plane of motion equilateral triangles with the primaries.

Additionally, it was observed that the involved parameters of the problem not only affect the positions of the corresponding equilibria but also influence their stability as well. For the determination of the stability of the infinitesimal body's motion around the obtained equilibria, we linearized the governing equations of motion around them. For all the considered cases, it was found that all equilibria are unstable due to the existence of a positive real root or a complex root with positive real part except for the equilibria L_4 and L_5 in the case of Stokes drag where we got complex roots with negative real parts, which means that these points are stable. Therefore, we conclude that the stability of the non-collinear equilibrium points L_4 and L_5 is not affected from the Stokes drag force of the primaries, and the motion in their vicinity remains stable for the value of mass parameter considered here. Also, contrary to the classical restricted three-body problem where the three collinear points are generally unstable and the triangular points are linearly stable for sufficiently small ratio of the two masses, we observed that for the problem under consideration all the equilibrium points are unstable due to the presence of the P–R drag effect. The instability of the equilibrium points agrees with the results existing in the literature when the Stokes drag force is not considered.

References

1. V. Szebehely, Theory of orbits: The restricted problem of three bodies, Academic press, New York (1967)
2. C. Marchal, The three body problem. Studies in Astronautics **4**, Elsevier, Amsterdam (1990)
3. R. Farquhar, The flight of ISEE-3/ICE: Origins, mission history, and a legacy. J. Astronaut. Sci. **49**, 23–73 (2001)
4. P. Sharer and T. Harrington, Trajectory Optimization for the Ace Halo Orbit Mission. In AIAA/AAS Astrodynamics Specialist Conference, San Diego, California, July 2931, Paper AAS 96–3601 (1996)
5. M. Giard and L. Montier, Investigating clusters of galaxies with Planck and Herschel. Astrophys. Space Sci. **290**, 159–166 (2004)
6. P. Oberti and A. Vienne, An upgraded theory for Helene, Telesto, and Calypso. Astron. Astrophys. **397,** 353–359 (003)
7. P. Verrier, T. Waters and J. Sieber: Evolution of the L_1 halo family in the radial solar sail circular restricted three–body problem. Celest. Mech. Dyn. Astr., **120**, 373–400 (2014)
8. S.M. Elshaboury, E.I. Abouelmagd, V.S. Kalantonis and E.A. Perdios, The planar restricted three-body problem when both primaries are triaxial rigid bodies: Equilibrium points and periodic orbits. Astrophys. Space Sci. **361**, 315 (2016)

9. N. Pathak, E.I. Abouelmagd and V.O. Thomas, On higher order resonant periodic orbits in the photo-gravitational planar restricted three-body problem with oblateness. J. Astronaut. Sci. **66**, 475–505 (2019)
10. E.E. Zotos and K.E. Papadakis, Orbit classification and networks of periodic orbits in the planar circular restricted five-body problem. Int. J. Nonlin. Mech. **111**, 119–141 (2019)
11. F. Gao and R. Wang, Bifurcation analysis and periodic solutions of the HD 191408 system with triaxial and radiative perturbations. Universe **6**, 35 (2020)
12. V.S. Kalantonis, Numerical Investigation for Periodic Orbits in the Hill Three-Body Problem. Universe, **6**, 72, (2020)
13. C.D. Murray, Dynamical effects of drag in the circular restricted three body problems: 1. Location and stability of the Lagrangian equilibrium points. Icarus **112**, 465–484 (1994)
14. C. Beaugé and S. Ferraz-Mello, Resonance trapping in the primordial solar nebula: the case of a Stokes drag dissipation. Icarus **103**, 301–318 (1993)
15. B. Sicardy, C. Beaugé, S. Ferraz-Mello, D. Lazzaro and F. Roques, Capture of grains into resonances through Poynting–Robertson drag. Celest. Mech. Dyn. Astron. **57**, 373–390 (1993)
16. A. Celletti., L. Stefanelli, E. Lega and C. Froeschlé, Some results on the global dynamics of the regularized restricted three–body problem with dissipation. Celest. Mech. Dyn. Astron. **109**, 265–284 (2011)
17. J. Singh and A. Aminu, Instability of triangular libration points in the perturbed photogravitational R3BP with Poynting–Robertson (P–R) drag. Astrophys. Space Sci. **351**, 473–482 (2014)
18. M. Jain and R. Aggarwal, Existence and stability of non–collinear libration points in restricted three body problem with Poynting Robertson light drag effect. Int. J. Math. Trends and Technol. **19**, 20–33 (2015b)
19. E.E. Zotos and F.L. Dubeibe, Orbital dynamics in the post Newtonian planar circular restricted Sun-Jupiter system. Int. J. Mod. Phys. D **27**, 1850036 (2018)
20. E.A. Vincent and E.A. Perdiou, Poynting–Robertson and oblateness effects on the equilibrium points of the perturbed R3BP: Application on Cen X-4 binary system. In Rassias, Th.M. (Ed.) Nonlinear Analysis, Differential Equations, and Applications, Springer Optim. Its Appl. **173**, Springer, Cham (2021), in press
21. V.V. Radzievskii, The photogravitational restricted problems of three bodies. Astron. J. **27**, 250–256 (1950)
22. J.F.L. Simmons, A.J.C. McDonald and J.C. Brown, The restricted 3-body problem with radiation pressure. Celes. Mech., **35**, 145–187 (1985)
23. A.L. Kunitsyn and E.N. Polyakhova, The restricted photogravitational three–body problem: A modern state, Astron. Astrophys. Trans., **6**, 283–293 (1995)
24. V.S. Kalantonis, E.A. Perdios and O. Ragos, Asymptotic and periodic orbits around L_3 in the photogravitational restricted three–body problem. Astrophys. Space Sci. **301**, 157–165 (2006)
25. E.A. Perdios, V.S. Kalantonis and C.N. Douskos, Straight–line oscillations generating three–dimensional motions in the photogravitational restricted three-body problem. Astrophys. Space Sci. **314**, 199–208 (2008)
26. D.G. Yárnoz, J.P.S Cuartielles and C.R. McInnes, Passive sorting of asteroid material using solar radiation pressure. J. Guid. Control Dyn. **37**, 1223–1235 (2014)
27. H. Baoyin and CR. McInnes, Solar sail halo orbits at the Sun–Earth artificial L_1 point. Celest. Mech. Dyn. Astron. **94**, 155–171 (2006)
28. S. Gong, J. Li and H. Baoyin, Analysis of displaced solar sail orbits with passive control. J. Guid. Control. Dyn. **31**, 782–785 (2008)
29. Yu.A. Chernikov, The photogravitational restricted three–body problem. Soviet Astronomy–AJ **14**, 176–181 (1970)
30. D.W. Schuerman, Influence of the Poynting–Robertson effect on triangular points of the photogravitational restricted three-body problem. Astrophys. J. **238**, 337–342 (1980)
31. O. Ragos and F.A. Zafiropoulos, A numerical study of the influence of the Poynting-Robertson effect on the equilibrium points of the photogravitational restricted three-body problem. I. Coplanar case. Astron. Astrophys. **300**, 568–578 (1995)

32. J. Burns, P. Lamy and S. Soter, Radiation forces on small particles in the Solar system. Icarus **40**, 1–48 (1979)
33. J. Singh and A.M. Simeon, Motion around the triangular equilibrium points in the circular restricted three-body problem under triaxial luminous primaries with Poynting-Robertson drag. Int. Front. Sci. Lett. **12**, 1–21 (2017)
34. J. Singh, J. and T.O. Amuda, Stability analysis of triangular equilibrium points in restricted three-body problem under effects of circumbinary disc, radiation and drag forces. J. Astrophys. Astr. **40**, 5 (2019)
35. A. Umar and A.A. Hussain, Impacts of Poynting–Robertson drag and dynamical flattening parameters on motion around the triangular equilibrium points of the photogravitational ER3BP. Adv. Astron. **vol. 2021**, Article ID 6657500 (2021)
36. M. Jain and R. Aggarwal, A study of non–collinear libration points in restricted three body problem with Stokes drag effect when smaller primary is an oblate spheroid. Astrophys. Space Sci. **358**, 51–58 (2015a)
37. J.C. Liou, H.A. Zook, and A.A. Jackson, Radiation pressure, Poynting–Robertson drag, and solar wind drag in the restricted three–body problem. Icarus **116**, 186–201 (1995)
38. C.D. Murray and S.F. Dermott, Solar System Dynamics. Cambridge University Press, Cambridge (1999)
39. A.N. Baltagiannis and K.E. Papadakis, Equilibrium points and their stability in the restricted four-body problem. Int. J. Bifurc. Chaos **21**, 2179–2193 (2011)
40. M. Arribas, A. Abad, A. Elipe and M. Palacios, Out–of–plane equilibria in the symmetric collinear restricted four-body problem with radiation pressure. Astrophys. Space Sci. **361**, 210–280 (2016)
41. A.E. Vincent, J.J. Taura and S.O. Omale, Existence and stability of equilibrium points in the photogravitational restricted four-body problem with Stokes drag effect. Astrophys. Space Sci. **364**, 183 (2019)

Nearest Neighbor Forecasting Using Sparse Data Representation

Dimitrios Vlachos and Dimitrios Thomakos

Abstract The method of the nearest neighbors as well as its variants have proven to be very powerful tools in the non-parametric prediction and categorization of experimental measurements. On the other hand, the number of data available today as well as their dimensionality and complexity is growing rapidly in many scientific fields, such as economics, biology, chemistry, medicine, and others. Usually, the data and their characteristics have semantic dependence and a lot of noise. At this point, the sparse data representation that deals with these problems with great success is involved. In this paper we present the application of these two tried and tested techniques for prediction in various fields related to economics. New techniques are presented as well as exhaustive tests for the evaluation of the proposed methods. The results are encouraging to continue research into the possibilities of sparse representation combined with good proven machine learning techniques.

1 Introduction

In recent years, deep learning networks have made tremendous advances in machine learning but still face obstacles when it comes to implementing more complex applications. The usual solution to this problem is to create larger and more complex models that in turn bring new problems such as over-fitting and the excessive need for computing power [1, 2]. On the other hand, as is usually the case in the evolution of machine learning algorithms, it is useful to focus on the ways in which nature and the human brain deal with such problems. It is obvious that the human brain solves

D. Vlachos
University of Peloponnese, Tripoli, Greece
e-mail: dvlachos@uop.gr

D. Thomakos (✉)
National and Kapodistrian University of Athens, Athens, Greece
e-mail: dthomakos@ba.uoa.gr

I. N. Parasidis et al. (eds.), *Mathematical Analysis in Interdisciplinary Research*,
Springer Optimization and Its Applications 179,
https://doi.org/10.1007/978-3-030-84721-0_38

problems much more efficiently than a deep learning network. The brain is estimated to require only 20 watts of power to perform a wide range of tasks, from logic to language, processing visual and audio inputs, and performing complex behaviors. On the contrary, today's networks for deep Learning have huge energy requirements and often require large amounts of data at the stage of training running on multiple servers for many days. How is the brain so smart with such amazing performance?

One reason is that most of it is sparse. The brain stores and processes information in the context of extremely sparse neural activity and sparse connectivity [3]. Concerning the neo-cortex, one of the most important and remarkable observations is that throughout it, neuronal activity is sparse. Only a small percentage of neurons send signals at any given time. Activity can range from less than one percent to quite a few percent, but it is always extremely sparse. In addition, unlike deep learning networks, the connection between neurons in the brain is also sparse. Deep learning traditionally uses dense representations in which almost all neurons are connected and most of them are constantly active. To calculate the output for each neuron, the contribution of each connected neuron must be considered. This calculation is usually expressed as matrix multiplication, in which each row vector must be multiplied by each column vector. With a sparse representation, we can create sparse versions of these networks [4–7]. In this sparse network, as a result of both limited connectivity and limited activation, the matrix calculations required are performed in arrays for which the majority of matrix values are zero. When these sparse rows and columns are multiplied together, a large fraction of the products can be eliminated. If an application is able to "skip" the calculation of zero products, significant performance benefits can arise.

But beyond the acceleration in the calculations, the sparse representation as we will explain in detail below, allows the conceptual transformation of the data. Consider, for example, a trajectory of the stock price following a stock. Even if this snapshot is identical to another, the price level compared to previous ones, the time of day or even the day of the week can dramatically differentiate the price evolution in both cases.

In the present work, we will develop a classification-based technique based on the nearest neighbors to sparsely represented data. First we will give one a detailed description of both the classical method and the methodology we follow to represent the data. After an exhaustive experimental process we will present and comment on the improvement that results with the proposed technique.

2 A Modification of Classical Nearest Neighbors Forecasting

The nearest neighbors (NN) approach for classification and prediction has a venerable history and an active research present, and obviously future as well. We sample several recent papers from across a number of fields to illustrate the use and usefulness of the NN methods in prediction and forecasting. We emphasize that the references that follow are far from a complete list; rather they clearly show the vast expanse of methods and applications to the NN approach across different disciplines.

The original articles on NN dates back to Cover and Hart [8], Cover [9] and Hart [10], followed by many other papers in the statistics literature, see, for example, Devroye [11], Devroye [12], Stute [13], Bhattacharya and Mack [14], Devroye et al. [15] and Gadat et al. [16]—all these papers dealing with various theoretical aspects of NN specification and consistent estimation and classification in a non-parametric regression context.

Turning to the use of NN on prediction and forecasting we start with a recent monograph of Chen and Devavrat [17] that is entirely devoted to the use of NN in the context of prediction—see also the many references therein. Stroup and Mulitze [18] present results on best linear unbiased prediction using NN, and is another early reference on the topic, Ghosh [19] discusses NN classification with adaptive choice of the number of NN to use and Jensen and Cornelis [20] present results on fuzzy-based NN classification and prediction. Going closer to today, we find proposals for new uses of NN such as in Zhang et al. [21] and Talavera-Llames et al. [22, 23] on the use of multidimensional NN approaches for prediction and forecasting, also in the context of big data.

A number of other papers, with either new NN methods applications or both, include the following. Nikolopoulos et al. [24] and Kück and Freitang [25] use NN in the context of demand forecasting while Li et al. [26], Andrada-Félix et al. [27], Zhang et al. [21], Chen and Hao [28], Cheng et al. [29] and Kyriazi and Thomakos [30] all use NN approaches for prediction in an economics/finance context. The use of NN is essentially everywhere, for EEG monitoring in Erla et al. [31], for RNA prediction in Chou et al. [32], for battery life prediction in Ma et al. [33], for sand liquefaction prediction in Huang et al. [34], while in Yesilbudak et al. [35], Pedro and Coimbra [36], Wood [37] and Dong et al. [38] the use of NN methods is used for prediction in various energy-related problems.

We next turn to the use of the NN methodology for this chapter, where we consider a standard (or classical) approach to the forecasting problem with the use of NN and offer small modification—the latter makes possible sense in the context of "sparsity" discussed earlier in this chapter. Consider thus a bivariate time series of interest say $Z_t = [Y_t, X_t]$, as a (1×2) row vector, and assume that there are n available observations, $t = 1, 2, \ldots, n$. The variable Y_t is the target variable and the variable X_t is an auxiliary (or explanatory variable) to aid in the prediction of Y_t. The aim is to forecast the out-of-sample value Y_{n+h} by the forecast $\hat{Y}_{n+h|n}$ for some $h \geq 1$, the forecasting horizon.

We first construct the trajectory matrix $\mathcal{T}_{m,n}$ where m is the embedding dimension, $m < n/2$. This matrix, obtained by overlapping stacking of the elements of Z_t, has the following form:

$$\mathcal{T}_{m,n} = \begin{bmatrix} Z_1 & Z_2 & \dots & Z_m \\ Z_2 & Z_3 & \dots & Z_{m+1} \\ \vdots & \vdots & \vdots & \vdots \\ Z_{n-m+1} & Z_{n-m+2} & \dots & Z_n \end{bmatrix} \tag{1}$$

and naturally defines the target vector as its last row, i.e., the most recent m values of the time series. Thus, let us define $N = n - m + 1$ the row dimension of the trajectory matrix, as $\tau_N = [Z_{n-m+1}, Z_{n-m+2}, \dots, Z_n]$ the $(1 \times m)$ target vector and as $\mathcal{T}_{N-1,n}$ the trajectory matrix $\mathcal{T}_{m,n}$ with its last row removed.

We can now compute the distance vector between the elements of the target vector τ_N and the rows of the $\mathcal{T}_{N-1,n}$ trajectory matrix as in:

$$d_{N-1,n} = \mathcal{D}\left(\mathcal{T}_{N-1,n} - e_N \otimes \tau_N\right) \tag{2}$$

where $\mathcal{D}$ is a distance function applied row-wise and where $e_N = [1, 1, \dots, 1]^\top$ is the $(N \times 1)$ vector of ones. Standard distances, like the Euclidean or Manhattan, were used in our computations. If the time series in question were binary (as in the discussion of the sparse representation) the plain matching distance function was also used.

The indices of the distance vector define the NN which are to be used in forming the forecast. Suppose thus that we require a percentage of α NN to be included in our computation. Then, if $d^{(j)}_{N-1,n}$ denotes the jth ordered NN, $j = 1, 2, \dots N - 1$ and where higher exponents indicate lower proximity (more distance), we consider only the set of indices satisfying:

$$\mathcal{S}_\alpha = \left\{ j^* | j^* \le \alpha N \right\} \tag{3}$$

Once this set of indices is available we then move them h-periods forward, where h is the forecasting horizon, adjusting for the shape and positioning of the trajectory matrix:

$$\mathcal{S}^f_\alpha = \left\{ j^* | j^* = j + h + m - 1, \forall j \in \mathcal{S}_\alpha \right\} \tag{4}$$

and the (global) NN forecast is defined as the average of the NN values for the target variable Y_t as in:

$$\hat{Y}^g_{n+1|n} = \frac{1}{\mathcal{N}(\mathcal{S}^f_\alpha)} \sum_{t \in \mathcal{S}^f_\alpha} Y_t \tag{5}$$

where $\mathcal{S}^f_\alpha)$ is the cardinality of the set $\mathcal{S}^f_\alpha$ (note that if it happens than an index exceeds n this is naturally dropped from the set). It should be clear from the above context that the use of the auxiliary variable X_t is for obtaining a, possibly more accurate, estimate of the distances to be used in forming the final forecast.

We next add another layer of smoothing in the computation of our NN forecast, the idea being very simple: each NN index is in itself a sparse representation of the original time series and therefore one could look into the neighborhood of each such index and repeat the smoothing; now we have local smoothing before the final global averaging. The procedure by which we do this is as follows. For each index $t \in \mathcal{S}_{\alpha}^{f}$ we look into a radius of size β around it, either on both sides of the index, forward only or backward only. That is, for each $t \in \mathcal{S}_{\alpha}^{f}$ we consider the observations in the set:

$$S_{t,\beta}^{c} = \left\{t^{*}|t - \beta n \leq t^{*} \leq t + \beta n\right\} \tag{6}$$

for a symmetric (centered) neighborhood or, for example, into the set:

$$S_{t,\beta}^{f} = \left\{t^{*}|t \leq t^{*} \leq t + \beta n\right\} \tag{7}$$

for a forward neighborhood. Once these sets are defined for all $t \in \mathcal{S}_{\alpha}^{f}$ we then compute the average of averages (local) forecast as in:

$$\hat{Y}_{n+1|n}^{\ell} = \frac{1}{\mathcal{N}(\mathcal{S}_{\alpha}^{f})} \sum_{t \in \mathcal{S}_{\alpha}^{f}} \left\{ \frac{1}{\mathcal{N}(S_{t,\beta}^{f})} \sum_{t^{*} \in S_{t,\beta}^{f}} Y_{t^{*}} \right\} \tag{8}$$

3 Sparse Data Representation

One of the most important problems one faces when designing a physical model is the representation of knowledge. There are various techniques by which we try to represent our data, always making sure that they have a form that we can process with the mathematical and computational tools we have in our hands. As an example we can bring here the breakthrough brought to the area of natural language processing by the representation of words or sentences or paragraphs in the form of vectors with the Skip Gramm model introduced by Mikolov et al. [39]. This problem is exacerbated by the fact that it is not always easy or obvious to represent our data relationships in a way that computers can work. The main problem here is that our knowledge is not limited to distinct events with well-defined relationships. Our knowledge is limited and the relationships that arise, are too many and so very difficult to capture in the traditional variables recognized by algorithms and computers.

Fortunately, the human brain does not have this problem. Information in the brain is represented by a series of neurons, among which only a small percentage are activated. The flexibility and creativity observed by the human intelligence are intertwined with this method of representation. Why has nature chosen this way?

Let us start by defining as Sparse Distributed Representations or SDR a set of bits of which only a small percentage is 1 and all the rest is 0. Each bit in an SDR corresponds to a neuron in the brain (with the obvious correspondence , 1 corresponds to an active neuron and 0 to an inactive one). The most important property of SDRs is that every bit makes sense. Thus, the set of active bits in any particular representation encodes the set of semantic features of what is represented. The meaning of the bits is not predefined but arises with the successive transformations they undergo in a learning system. It is obvious that by determining the overlap of bits between two SDRs (by overlap we refer to the number of active bits in the same positions in both SDRs) we can conclude if the two expressions are semantically similar or not (if we go back to the example we gave for the representation of words, it means that two words are semantically close to each other if the angle formed by the two vectors corresponding to them is small).

Based on the semantic meaning of the overlap of two SDRs, this representation acquires enormous possibilities for encoding the concepts and relationships that are reflected in our data. The current value of each bit in a given representation changes depending on the context in which it is located (i.e., the combination of the other active bits). So, for example, in a sequence of symbols, the same bit at one point in time may indicate that the particular symbol is a vowel while at another time it may indicate the exact opposite!

To better understand the properties of SDRs, let us look at a simple example which is borrowed from [40]: In computers, we represent information in bytes and words using 8, 32, or 64 bit words. The ASCII code for the letter "m" is represented by: 01101101. Notice that the combination of all eight bits encodes the "m" and the individual bits in this representation mean nothing by themselves. There is no specific meaning in the second or the third bit but the combination of all eight bits is required. Also note that such a representation is extremely vulnerable. If you change only one bit in the ASCII code for "m" as follows: 01100101 you get the representation for a completely different letter, "e". An error or the presence of a small noise bit and the meaning changes completely. Imagine the effect this can have on a forecasting system.

Unlike an SDR, every bit makes sense. For example, representing the letters of the alphabet using SDR, it is expected that there will be bits showing if the letter is consonant or vowel, bits that represent how the letter sounds, bits that represent where the letter appears in the alphabet, bits that represent how the letter is drawn, etc. To represent a particular letter, we can, for example, select 40 bits (characteristics that best describe this letter) in a sequence of 2 or even 3 thousand bits. It is obvious here that even if some of these bits change for some reason (either from 1 to 0 or vice versa) then the letter will probably remain unchanged and can be successfully identified (we will see later the amazing tolerance of these representations in noise).

The use of Sparse Data Representations has proved very efficient alternative in several problems (Wielgosz et al. [41], Kirtay et al. [42], Alshammari [43], Zhou [44], Ibrayev et al. [45], Osegi [46], Pilinszki-Nagy et al. [47], Dauletkhanuly et al. [48], Dobric et al. [49]). A very recent theoretical paper has shown that simple linear sparse networks may be more resistant to adversary attacks (Guo et al. [50]). A number of papers have shown that it is possible to effectively introduce sparsity through pruning and retraining (Han et al. [51], Frankle et al. [52], Lee et al. [53]). We present first a short review of the definitions and properties of Sparse Data Representations.

3.1 The Mathematics of SDR

A detailed presentation of the Sparse Data Representation can be found in Ahmad and Hawkins [54].
Definitions and notation:

1. A **binary vector** is a vector of the form b^i for $i = 0, 1, \ldots, N-1$ with values in $\{0, 1\}$ and size N.
2. The **sparsity** of a binary vector is the ration of 1s.
3. The **cardinality** of a binary vector is the total number of 1s.
4. The **overlap** score of two binary vectors x, y is their inner product $x \cdot y = x_i y^i$.
5. The **match** of two binary vector given a threshold θ is $x \sim_\theta y \iff x \cdot y \geq \theta$.

Note here that if $\theta = w$ where w is the cardinality, then we have the case of **exact match**. If $\theta < w$ we have the case of **inexact match**. In the case where $\theta > w$ there is no match between the two vectors.

Now let us assume we are given a vector with size N and cardinality w. There will be

$$\binom{N}{w} = \frac{N!}{w!(N-w)!} \tag{9}$$

unique representations. This number is by far smaller than the total representations (2^N) that we can obtain with N bits but in the common case that we have 2048 bits and cardinality 40, we can represent 2.37×10^{84} different patterns (compare this number with the number of atoms in the universe which is $\sim 10^{80}$).

Given two patterns, the probability that they have the same SDR representation is $\binom{N}{w}^{-1}$ which in the previous example is almost 0. Thus, choosing a large size, even if we keep the sparsity low, we can have an enormous amount of capacity keeping the probability that two patterns coincide almost 0.

Let us focus now in the crucial operation in binary vectors, the matching. If we have a vector x with cardinality w_x and size N, the number of vectors with the same size and cardinality w which have exactly b active bits in common is given by

$$|\Omega_x(N, w, b)| = \binom{w_x}{b}\binom{N - w_x}{w - b} \tag{10}$$

where the first term is the combinations of b bits among the w_x active ones in the vector x and the second term gives the combinations (selecting the w_x bits and the b active ones from the vector y) of the rest $w - b$ active bits in y. Using this result we can calculate both the cardinality and the threshold of matching in order to have an efficient representation. More specifically, assuming that both vectors have the same cardinality, by setting the threshold equal to w_x, even the change of one bit can cancel the matching. On the other hand, if we lower the threshold to $w_x/2$, then the representation is tolerant to a noise of 50% but the probability of false positive (false matching of two vectors) is significant. There is a trade-off here between the noise tolerance and false positive probability which can be easily resolved:

The probability of false positive match is given by

$$fp_w^N = \frac{\sum_{b=\theta}^{w} |\Omega_x(N, w, b)|}{\binom{N}{w}} \tag{11}$$

Figure 1 illustrates the effect of the matching threshold and the SDR size on the false positive probability. A typical selection of an SDR size of 2000 bits with cardinality around 40 and threshold around 30 gives a probability of false positive match in the order of 10^{-30}.

Let us now consider the case where we want to reliably compare against a sub-sampled version of a vector. This is the case where we want to recognize a large

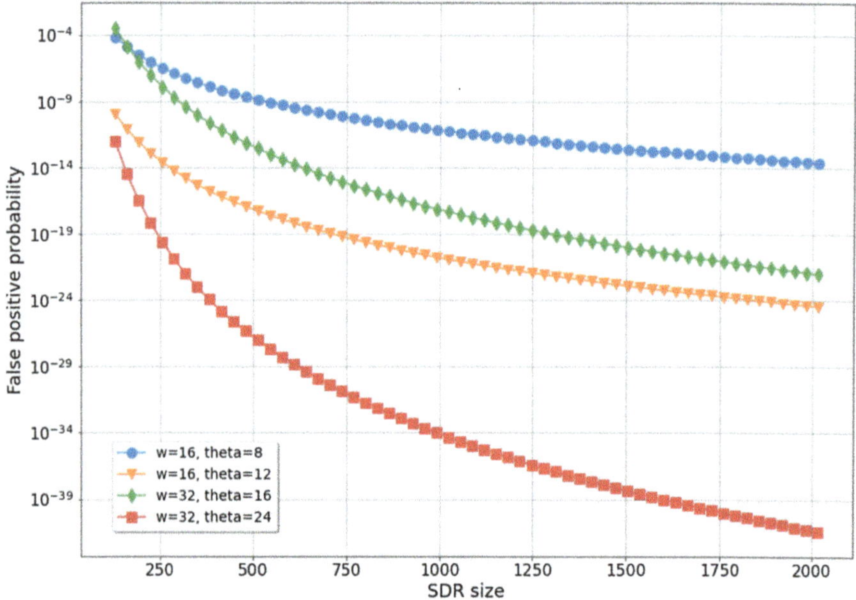

Fig. 1 The probability of false matching as a function of the matching threshold and the size of the SDR

pattern given only a subset of the active bits in it. Let x and x' the original and the sub-sampled vectors with cardinalities $w_{x'} \leq w_x$. Of course, if the matching threshold is such that $\theta \leq w_{x'}$ then there be always a match. But the probability of a false matching of the vector x and y given only the sub-sampled version x' can be easily calculated as

$$fp_{w_y}^N(\theta) = \frac{\sum_{b=\theta}^{w_{x'}} |\Omega_{x'}(n, w_y, b)|}{\binom{N}{w_y}} \tag{12}$$

where the symbols are easily interpreted and $\Omega_{x'}(n, w_y, b)$ is the subset of all representations with n bits, cardinality w_y that have exactly b active bits in common with the sub-sampled version x'. Again, with a proper selection of size and sparsity, the probability of false positive matching is practically zero.

3.2 Classification

One of the most useful operation with SDR is their ability to classify representations. Let $X = (x_1, x_2, \ldots, x_M)$ be a set of M vectors. Given a vector y, we can classify if y belongs to the set X if

$$y \in X \iff \exists_{x_i \in X} : y \sim_\theta x_i \tag{13}$$

Let us see now how reliably can we classify a vector in the presence of noise. The existence of noise means that there is a chance that t bits out of n (the size of the vector) change their state from 1 to 0 and vice versa. The probability now for a false positive classification of the vector y in the set X is given by

$$fp_{y,X}(\theta) = 1 - \left(1 - fp_w^n(\theta)\right)^M \tag{14}$$

where for practical reasons we can use the inequality

$$fp_{y,X}(\theta) \leq M fp_w^n(\theta) \tag{15}$$

You can see in Fig. 2, that while we keep the threshold ratio the same (2/3), by increasing the size of the SDR we obtain a drastic decrease in the false positive classification. As we will see later, this exceptional property of SDRs has led us to design a very effective classifier that in combination with a correlative memory can lead to a predictor based on the principle of the nearest neighbor.

Another property of SDR is their union. The union of SRDs x, y is defined as

$$x + y = x \vee y \tag{16}$$

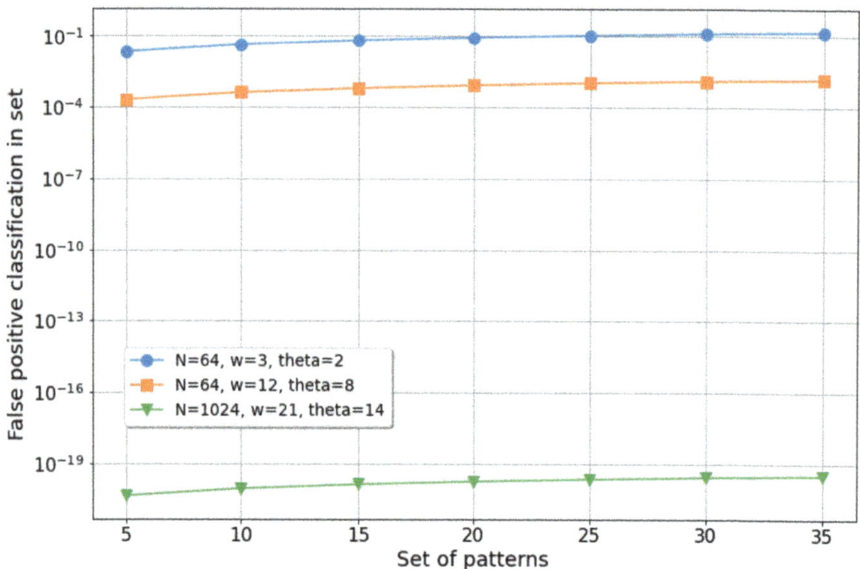

Fig. 2 The probability of false positive classification of a vector in a set as a function of the number of vectors in the set for different values of the size and cardinality of the vectors and the matching threshold

which means that a bit becomes active if at least one of the corresponding bits in x and y is active. The union gives us a way to store a batch of SDRs in a single one. Suppose that we have M vectors and we take their union. Then each one of the initial vectors will have a match with the union, but as the number of vectors increase, the union saturates (almost all bits become active) and the probability of a false classification is increased.

Let us examine the case of exact match. If we have only one vector in the set X then the probability of a bit to be zero is w/n where w is the cardinality and n the size of the vector. In case of M vectors in the set X, the probability of a bit to be zero is:

$$p_0 = \left(1 - \frac{w}{n}\right)^M = (1-s)^M \tag{17}$$

where $s = w/n$. The probability now of a bit to be active in the set X is $1 - p_0$. We can calculate now the probability of a positive matching of a random vector Y with the set X:

$$fp_{w,M} = (1 - p_0)^w = (1 - s^M)^w \tag{18}$$

As we can see, after a proper selection of size and cardinality, we can efficiently classify a vector y in a set of vectors X (Fig. 3).

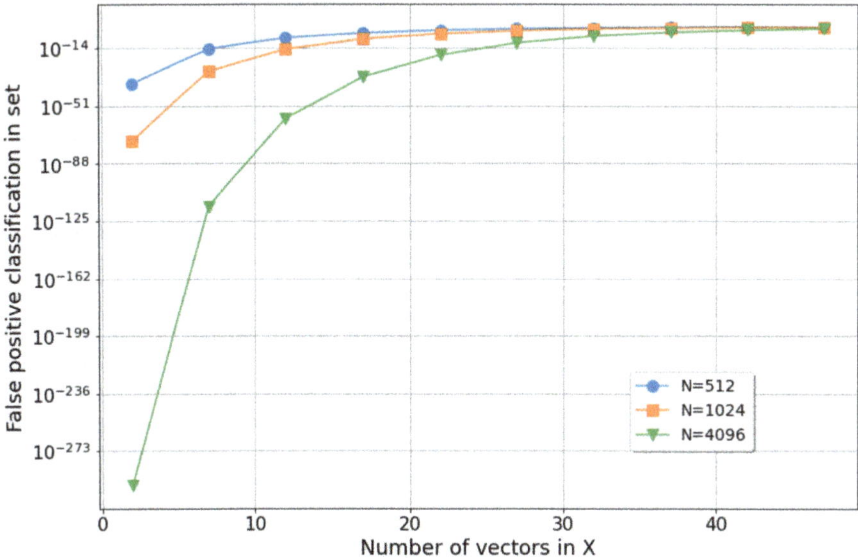

Fig. 3 The probability of false positive classification as a function of the number of patterns in the set X and the cardinality. In all cases, the sparsity is constant (0.1)

3.3 Encoding

The process by which we assign discrete or continuous variables to an SDR is called encoding. For encoding to be effective, the following basic principles must apply:

1. The values of the variables corresponding to conceptually related observations must correspond to sufficiently overlapping SDRs.
2. The same values must always correspond to the same SDRs.
3. The encoding of all values should lead to SDRs with the same dimension and similar sparsity.

Let us look at two examples of coding: Let us first consider a continuous variable that takes all values from 0 to 1. We also want to have an SDR with 1000 bits and sparsity 0.05, ie 50 active bits. The obvious solution is for each SDR to have 50 consecutive active bits at some point in the representation and all other bits to be 0. Thus, assigning 0 to

$$111\ldots111000000\ldots0000$$

and 1 to

$$0000...000001111\ldots111$$

we can calculate that we will have $1000 - 50 = 950$ different SDRs and so we will have a quantization error in the representation of the order of 10^{-3}. Also, if we assume that 2 SDRs match when 75% of their active bits match, then each representation will match about 25 other ones and so values that are less than 0.025 apart will match. If we want to change that, we can double the SDR dimension and the cardinality so, keeping the sparsity constant, we will have a half distance match.

In the second example, suppose we want to code the day of the week. Here, we must keep in mind that the last day (Sunday) is the previous of the first (Monday). For this reason we will adopt a circular encoding.

Let us look at the general case: Suppose we have m values that we want to encode in an SDR and we want two adjacent values to have an overlap with ratio a (that is, if we have cardinality equal to w then there will be $a \cdot w$ overlapping bits in two adjacent representations). If the first representation started with bit 0, then, for the last one to have the required number of overlapping bits with the first, it must start at $n - (1 - a)w$ bit so that the remaining aw bits are located at the beginning of the SDR. Because each consecutive representation shifts $(1 - a)w$ bits, we have:

$$(m - 1) * (1 - a)w = n - (1 - a)w \implies$$
$$\frac{w}{n} = \frac{1}{m(1 - a)}$$

where w/n is the sparsity of the representation.

One of the most important properties of SDRs is that we can combine them and include many variables and measurements in a single representation. The obvious way to do this is to stack the individual representations into a larger representation. So if we have the representations (n_1, w_1) and (n_2, w_2) where n, w correspond to size and cardinality, we can make the representation $(n_1 + n_2, w_1 + w_2)$ simply by placing the first representation after second. For example, we can encode in the same SDR the day of the year (from 1 to 365), the hour of the day, the temperature, and the wind speed. The only problem here is that the process of matching becomes complicated. To see this, assume that we have 40 active bits for the temperature and 5 active bits for the hour of the day. Then matching two SDRs is determined almost exclusively by the temperature.

To overcome this difficulty we introduce a different mechanism to join SDRs. Let a^i and b^i two vectors (SDRs). The Kronecker product of these two vectors is

$$c^i = (a \otimes b)^i = a^{i//n_b} b^{i\%n_b} \tag{19}$$

where n_b is the size of the vector b^i, $//$ is the integer division and $\%$ is the modulo. Notice now how well this coupling can lead to a conceptually compatible representation of the information. Two representations, in order to have a match, must have sufficient matching in both vectors from which they originate, regardless of their cardinality. Thus, one can assume that the second variable is represented in the context of the first (and this applies recursively if we have more variables). The flaw here is that every time we take the Kronecker product, the resulting vector has length the product of lengths and sparsity the product of sparsities. In other words, we are gradually leading to larger and more sparse representations. However, this difficulty can be overcome with the appropriate transformations.

3.4 Transformations of SDR

Given that we want the representation of our data to be done in such a way in an SDR, so that the various bits are rendered conceptually, we must consider transformations of the representations that lead to this result. As mentioned before this will be done through a learning process. We start with an initial SDR x and create a new y as follows:

1. For each bit of the vector y we create a perception field that contains a number of bits of x. This y-bit can only sense the bits that belong to this set.
2. Between each bit of the vector y we assume that there is a pairing with each of the bits belonging to the field of perception. This connection may or may not be active and this depends on the value of a parameter. If the value of this parameter is greater than a given threshold, then we consider the connection active.
3. Each y bit adds the active bits from the perception field (taking into account only those that have an active connection). If this sum is above a threshold, the bit is activated.
4. To maintain a desired level of sparsity, if the bits in the vector y are more than we want, then we get the most active, that is, those that have resulted from a larger number of active bits of their perception fields.
5. For each bit that was activated, we increase the strength of the connections with the active bits of the field of perception and, respectively, decrease the strength of connections with the inactive bits.

The conceptual nature of the above transformation stems from the fact that a bit in the vector y will only be activated when an appropriate combination of input bits are enabled. Figure 4 shows the perception field and the active connections of an output bit. This transformer is known as **Spatial Pooler**.

Let us see now how we can calculate the various parameters related to a Spatial Pooler in order to achieve both the desired size of the representation and the sparsity. Let us start with a representation that has length n_i and sparsity s_i. That is, $n_i \times s_i$

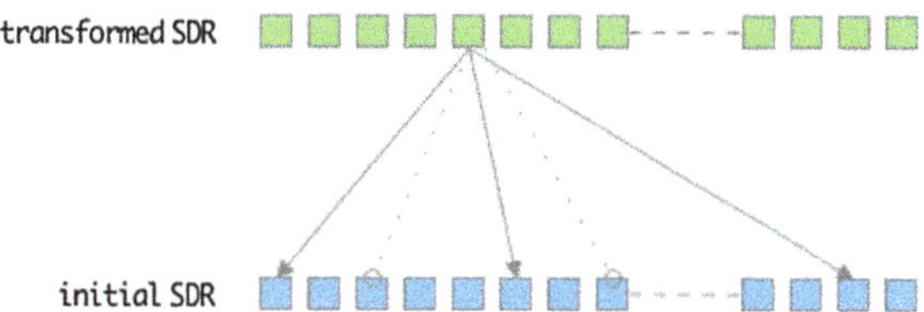

Fig. 4 The perception field and the active connections of an output bit

bits of the representation are active while all the rest are 0. Next, let us consider a bit y_j of the output and let d be the percentage of the bits of the input that belong to the field of perception of y_j. Our purpose is to calculate the threshold for the activation of bit y_j if we want the output to have n_o digits and sparsity s_o.

We select $n = d \cdot n_i$ bits from the input which has a mixture of $K = s_i \cdot n_i$ active bits and $n_i - K = (1 - s_i) \cdot n_i$ zeros. The probability of sampling k active bits follows the Hypergeometric distribution and its equal to:

$$p(x = k) = \frac{\binom{K}{k}\binom{n_i - K}{n-k}}{\binom{n_i}{n}} \tag{20}$$

where $\binom{K}{k}$ is the number of ways that we can draw k active bits for the set of K total active ones, $\binom{n_i-K}{n-k}$ is the number of ways that we can draw $n - k$ inactive bits from the set of $n_i - K$ total inactive ones and finally the denominator is the total number of ways that we can draw n bits out of n_i. Replacing the given values we have

$$p(x = k) = \frac{\binom{s_i \cdot n_i}{k}\binom{n_i - s_i \cdot n_i}{d \cdot n_i - k}}{\binom{n_i}{d \cdot n_i}} \tag{21}$$

Let now $F(k) = \sum_{j=0}^{k} p(x = k)$ be the cumulative distribution function. Then, in order to have the desired sparsity s_o at the output, the probability of an output bit to become active is s_o, which guide us to fix the threshold parameter θ such that $1 - F(\theta) = s_o$ which means that the probability to sample more than θ active bits from the input is s_o. In Fig. 5 the dependence of the output sparsity on the activation threshold is shown.

It is obvious that output sparsity is very sensitive to the threshold. But this is not a problem because we can always put the active bits of the output to compete with each other and always select the desired number of active bits. This is a process that also occurs in nature, as we know very well today that a significant percentage of neurons act prohibitively for others, that is, when they are activated, they prevent neighboring neurons from being activated. In this case, all we need is to select a threshold number that is guaranteed to give at least as many active bits as we need and then deactivate the unnecessary bits that have the least active bits in their receptive field.

More details about Spatial Pooler (with pseudo-code include) can be found in Hawking et al. [40], *BIOLOGICAL AND MACHINE INTELLIGENCE*.

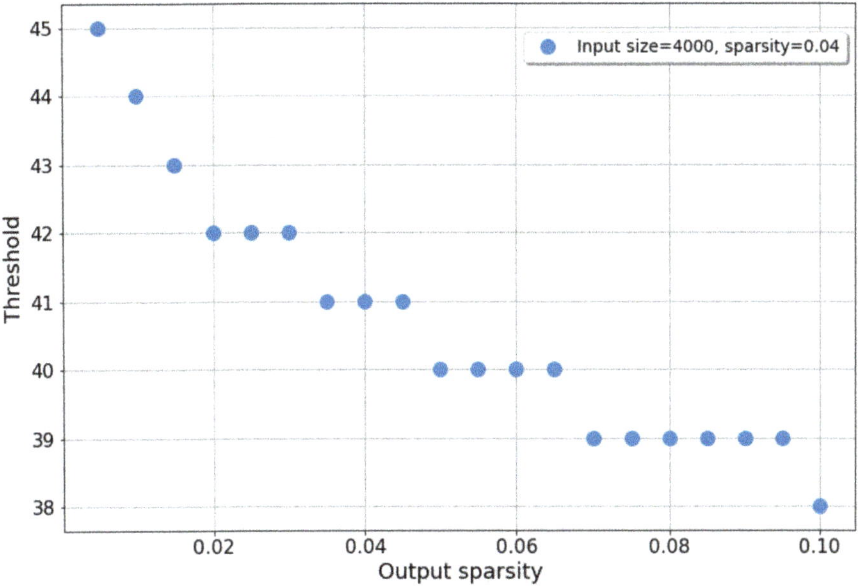

Fig. 5 The maximum threshold required to obtain a desired output sparsity. The size of the input representation is 4000 and its sparsity is 0.04 which means that 160 bits are active. The receptive field of each output bit is 20% of the input bits

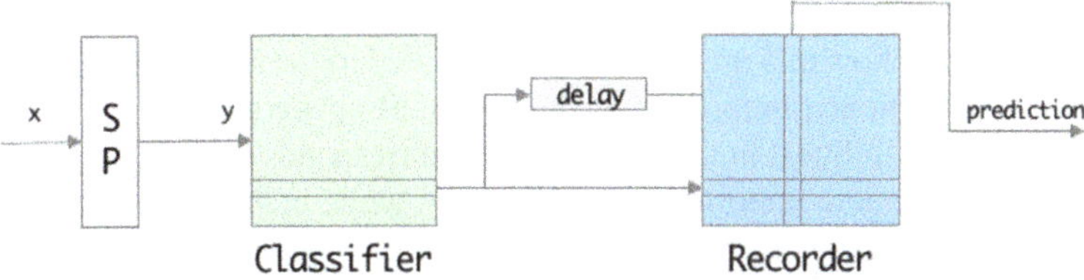

Fig. 6 The architecture of the classifier-predictor system

3.5 *Classification and Prediction Using SDR*

Let us know see how by combining the aforementioned properties of SDRs, we can create an efficient non-supervised classifier. Figure 6 shows the architecture of the system.

The steps of the classification and prediction system are:

1. Initially, a vector x is transformed using a Spatial Pooler to a vector y.
2. The transformed vector y is compared with the stored vectors in the classifier. If there is a match, then the output is the matched vector stored in the classifier. If there is not a match, then the vector y is added in the classifier and becomes the output.

a. For each vector in the classifier there is a vector with weights, one for every bit.
b. The weights of the bits that correspond to the active bits of the input are increased.
c. Only the bits with the higher weights become active (in this way we can keep the sparsity constant).
d. In order to include a "forgetting" mechanism, with each representation in the classifier we keep a measure of its "duty cycle" which is just an exponential moving average of the times this representation is matched. If the duty cycle of a representation falls below a predefined threshold, the representation is removed from the classifier.

3. The recorder is a 2-dimensional array in which rows represent the previous vector produced by the classifier and the columns the present ones. The prediction is the column with the higher entry in the row that corresponds to the present vector.
4. The entry in the recorder that corresponds to the previous vector and the present one is increased. The whole row is normalized because it can be interpreted as a probability distribution of the next vector given the present one.

The role of the classifier is to find the nearest neighbors of the present state. A vector finds a match in the classifier only if a similar (nearest neighbor) vector has appeared in the past. Moreover, in the classifier is stored only a representative member of the class of vectors that give the same match and this member changes in an adaptive manner with time in order to follow the conceptual changes of the input class.

Let us see now how this implementation is related to the technique of the nearest neighbor we mentioned before. Each time an SDR appears in the memory input, there are two possibilities: in the first this SDR corresponds to an input that has not reappeared, so the new vector enters the memory, while in the second case, there is a pattern stored in the memory which fits quite well with the incoming SDR. In the second case, a similarity is identified with one or more inputs that have appeared in the past and the representative of the class of these snapshots that has appeared most often is selected (this is the pattern that is stored in the memory). This results from the mechanism by which the stored pattern is informed, in which the bits that appear most often are ultimately those that are stored (here we must keep in mind that passing the input through the Spatial Pooler, every bit of the representation acquires a conceptual meaning). This process is perfectly compatible with the technique of the nearest neighbor, since if a previous state that is close to the current one, has occurred many times, this state will gain more weight as it will appear more times in the sum that calculates the average. Of course, the difference here is that in this implementation we do not take averages but only the winner of the comparison. But in the end, the average exists because the pattern selected has resulted from an averaging adaptive process as we explained earlier.

Once a pattern matching the input has been found, the prediction can be made using the Recorder in two ways: we can see the probability distribution of the patterns that have followed the current state and get this state that has the highest probability (Winner takes all) or to take as an output the probability distribution itself. The first case can be useful if we use many such identical units in a boosted chain of predictors while the second case is more suitable for stand-alone forecasting systems that use only one unit.

4 Experimental Results

The aim of the experimental process is to compare the application of the modification of classical nearest neighbors forecasting with that of sparse data representation. The tests include:

- The modification of the classical method (NN)
- The application of the classical method but the vector of the features is an SDR (NN-SDR)
- The application of the Classifier-Predictor system without the use of SP (NN-CP)
- The application of the Classifier-Predictor system with the use of SP (NN-CP-SP)

In addition, we will test one-dimensional and multidimensional input vectors to study the effect of auxiliary features on conceptually compatible sparse data representation using the Kronecker product.

For our empirical analysis we use two datasets, one financial and one economic. The financial dataset is the monthly returns and volume changes of the exchange traded fund (ETF) with ticker name "SPY" which tracks the temporal evolution of the S&P500 index. For this dataset the target variable is the monthly return while the auxiliary variable is the percent change in volume of transactions and the range of observations is from 1993 to 2020. The economic dataset is the monthly US unemployment rate along with two series of leading indicators. For this dataset the target variable is the percent change in the unemployment rate and the auxiliary variable is the percent change of one of the two leading indicators while the range of observations is from 1959 to 2020.

The hyper-parameters of the applied methods are:

- **r**: The rolling window size (applies to NN and NN-SDR methods)
- **p**: The p-norm used for distances in the classical method
- **k**: The trajectory size k
- **h**: The forecast step
- **a**: The ration of nearest neighbors to keep
- **steps**: The sub-sampling step
- **method**: The method used for forecast. Can be `center`, `forward`, `backward` or `regress`

The result of the simulations is summarized in Table 1.

The simulation was performed in a huge combination of hyper-parameters as follows (total 2592 combinations):

Table 1 Mean Absolute Error (MAE) and Mean Square Error (MSE) compared to the benchmark method

									Unemployment data							
									NN		NN-SDR		NN-CP		NN-CP-SP	
Roll	p	k	h	a	b	Steps	Type	id	MSE	MAE	MSE	MAE	MSE	MAE	MSE	MAE
60	2	4	1	0.5	0.1	1	for	UN-UN	0.983	0.951	0.981	0.950	0.981	0.949	0.976	0.945
120	0	6	1	0.1	0.1	1	for	UN-MU	0.976	0.950	0.944	0.938	0.938	0.939	0.922	0.919
60	0	2	1	0.5	0.9	1	for	UN-UN			0.980	0.950	0.981	0.949	0.976	0.945
120	0	2	1	0.1	0.9	1	for	UN-MU			0.938	0.931	0.938	0.930	0.936	0.922
								UN-UN					0.975	0.942	0.970	0.935
								UN-MU					0.928	0.936	0.928	0.920
								UN-UN							0.920	0.919
								UN-MU							0.915	0.918
									*S&P*500 data							
90	2	6	1	0.5	0.1	1	for	SPY-UN	0.988	0.961	0.983	0.952	0.983	0.951	0.978	0.955
120	0	6	1	0.5	0.1	1	for	SPY-MU	0.982	0.965	0.979	0.958	0.978	0.959	0.975	0.955
90	0	2	1	0.5	0.9	1	for	SPY-UN			0.981	0.950	0.981	0.949	0.976	0.945
90	2	2	1	0.1	0.5	1	for	SPY-MU			0.982	0.935	0.944	0.932	0.940	0.928
								SPY-UN					0.970	0.940	0.965	0.943
								SPY-MU					0.965	0.937	0.959	0.940
								SPY-UN							0.940	0.929
								SPY-MU							0.935	0.922

- roll: [60, 90, 120, 150]
- p: [0, 1, 2]
- k: [2, 4, 6, 8, 10, 12]
- a: [0.1, 0.5, 0.9]
- b: [0.1, 0.5, 0.9
- steps: 1
- type: [forward, center, backward, regress]

In Table 1 only the best combinations are shown. In each row, the best combination of the first non-empty entry is shown. Summarizing some of the results:

- In almost all cases, the sparse data representation outperforms the classical method. The reason for this is that the system is more noise-tolerant and does not spend its resources and capabilities to learn the noise.
- The use of Spatial Pooler improves further the performance of the technique since it codes basic relations between the input features.
- The inclusion of auxiliary variables outperforms in both techniques.
- Although the improvement of the classical technique using multivariate data is marginal, this is not the case in the sparse representation. The reason is that, with the use of the Kronecker product, an increase in the feature space dimension is obtained and thus patterns can be more easily separated and identified.

5 Conclusions

According to Chen et al. [17], there are 4 main reasons that can justify the success of the techniques based on the nearest neighbors:

1. The flexibility in choosing the metric relationship that determines the proximity and consequently the characterization of neighbors
2. The low computational cost of these methods that makes it possible to apply them to large problems
3. These methods are non-parametric
4. They are easy to evaluate because they provide complete data on the neighbors on whom their conclusions are based.

The use of Sparse Data Representation contributes to both the flexibility and the ability to handle large problems. Combined with the use of Spatial Pooler, further increases the flexibility of the methods by significantly reducing the effect of noise without the need of averaging techniques that dramatically reduce the variety of information that can be extracted from the data. In addition, the computational cost is even lower because all data can be represented in sparse arrays.

In addition, difficult-to-manage and configurable variables, which, however, contribute significantly to the formation of a conceptual framework in which the phenomenon we observe takes place, can be very easily integrated into the input data. This is shown by the experiments, where the inclusions of such variables contribute decisively to the improvement of the predictions. In our estimation, the significant improvement occurs exactly when the technique is enriched by this conceptual coding of input characteristics achieved by sparse representation and when it is combined with the effect that Spatial Pooler has on data pre-processing. Many auxiliary variables, like the time or the season, can be easily integrated in the presentation data and boost the performance of the method.

Finally, it is almost certain that data representation can significantly determine the performance of a machine learning technique. Under this prism, the great flexibility and simplicity of sparse representation needs to be further explored in relation to well-tried techniques and algorithms.

References

1. Jiawei Su, Danilo Vasconcellos Vargas, and Kouichi Sakurai. One pixel attack for fooling deep neural networks. *IEEE Transactions on Evolutionary Computation*, 23, 2019.
2. Yuwei Cui, Subutai Ahmad, and Jeff Hawkins. The htm spatial pooler—a neocortical algorithm for online sparse distributed coding. *Frontiers in Computational Neuroscience*, 11, 2017.
3. Jeff Hawkins, Marcus Lewis, Mirko Klukas, Scott Purdy, and Subutai Ahmad. A framework for intelligence and cortical function based on grid cells in the neocortex. *Frontiers in Neural Circuits*, 12, 2019.
4. Emma Strubell, Ananya Ganesh, and Andrew McCallum. Energy and policy considerations for deep learning in NLP. In *ACL 2019 - 57th Annual Meeting of the Association for Computational Linguistics, Proceedings of the Conference*, 2020.
5. DEAN MOUHTAROPOULOS. Training a single ai model can emit as much carbon as five cars in their lifetimes | MIT technology review. *MIT Technology Review*, 2020.
6. Dario Amodei and Danny Hernandez. Ai and compute. *Blog Open AI*, 2018.
7. Neil C. Thompson, Kristjan Greenewald, Keeheon Lee, and Gabriel F. Manso. The computational limits of deep learning. *arXiv*, 2020.
8. T. M. Cover and P. E. Hart. Nearest neighbor pattern classification. *IEEE Transactions on Information Theory*, 13, 1967.
9. Thomas M. Cover. Estimation by the nearest neighbor rule. *IEEE Transactions on Information Theory*, 14, 1968.
10. Peter E. Hart. The condensed nearest neighbor rule. *IEEE Transactions on Information Theory*, 14, 1968.
11. Luc P. Devroye. The uniform convergence of nearest neighbor regression function estimators and their application in optimization. *IEEE Transactions on Information Theory*, 24, 1978.
12. Luc Devroye. Necessary and sufficient conditions for the pointwise convergence of nearest neighbor regression function estimates. *Zeitschrift für Wahrscheinlichkeitstheorie und Verwandte Gebiete*, 61, 1982.
13. Winfried Stute. On a class of stopping times for m-estimators. *Journal of Multivariate Analysis*, 14, 1984.
14. P. K. Bhattacharya and Y. P. Mack. Weak convergence of k-nn density and regression estimators with varying k and applications. *The Annals of Statistics*, 15, 2007.
15. Luc Devroye. On the non-consistency of an estimate of Chiu. *Statistics and Probability Letters*, 20, 1994.
16. Sébastien Gadat, Thierry Klein, and Clément Marteau. Classification in general finite dimensional spaces with the k-nearest neighbor rule. *Annals of Statistics*, 44, 2016.
17. George H. Chen and Devavrat Shah. Explaining the success of nearest neighbor methods in prediction. *Foundations and Trends in Machine Learning*, 10, 2018.

18. W. W. Stroup and D. K. Mulitze. Nearest neighbor adjusted best linear unbiased prediction. *The American Statistician*, 45, 1991.
19. Anil K. Ghosh. On nearest neighbor classification using adaptive choice of k. *Journal of Computational and Graphical Statistics*, 16, 2007.
20. Richard Jensen and Chris Cornelis. Fuzzy-rough nearest neighbour classification and prediction. *Theoretical Computer Science*, 412, 2011.
21. Ningning Zhang, Aijing Lin, and Pengjian Shang. Multidimensional k-nearest neighbor model based on EEMD for financial time series forecasting. *Physica A: Statistical Mechanics and its Applications*, 477, 2017.
22. R. Talavera-Llames, R. Pérez-Chacón, A. Troncoso, and F. Martínez-Álvarez. Big data time series forecasting based on nearest neighbours distributed computing with spark. *Knowledge-Based Systems*, 161, 2018.
23. R. Talavera-Llames, R. Pérez-Chacón, A. Troncoso, and F. Martínez-Álvarez. Mv-kwnn: A novel multivariate and multi-output weighted nearest neighbours algorithm for big data time series forecasting. *Neurocomputing*, 353, 2019.
24. Konstantinos I. Nikolopoulos, M. Zied Babai, and Konstantinos Bozos. Forecasting supply chain sporadic demand with nearest neighbor approaches. *International Journal of Production Economics*, 177, 2016.
25. Mirko Kück and Michael Freitag. Forecasting of customer demands for production planning by local k-nearest neighbor models. *International Journal of Production Economics*, 231, 2021.
26. Hui Li, Jie Sun, and Bo Liang Sun. Financial distress prediction based on OR-CBR in the principle of k-nearest neighbors. *Expert Systems with Applications*, 36, 2009.
27. Julián Andrada-Félix, Fernando Fernández-Rodríguez, and Ana Maria Fuertes. Combining nearest neighbor predictions and model-based predictions of realized variance: Does it pay? *International Journal of Forecasting*, 32, 2016.
28. Yingjun Chen and Yongtao Hao. A feature weighted support vector machine and k-nearest neighbor algorithm for stock market indices prediction. *Expert Systems with Applications*, 80, 2017.
29. Ching Hsue Cheng, Chia Pang Chan, and Yu Jheng Sheu. A novel purity-based k nearest neighbors imputation method and its application in financial distress prediction. *Engineering Applications of Artificial Intelligence*, 81, 2019.
30. Foteini Kyriazi and Dimitrios D. Thomakos. Distance-based nearest neighbour forecasting with application to exchange rate predictability. *IMA Journal of Management Mathematics*, 31, 2020.
31. Silvia Erla, Luca Faes, Enzo Tranquillini, Daniele Orrico, and Giandomenico Nollo. K-nearest neighbour local linear prediction of scalp EEG activity during intermittent photic stimulation. *Medical Engineering and Physics*, 33, 2011.
32. Fang Chieh Chou, Wipapat Kladwang, Kalli Kappel, and Rhiju Das. Blind tests of RNA nearest-neighbor energy prediction. *Proceedings of the National Academy of Sciences of the United States of America*, 113, 2016.
33. Guijun Ma, Yong Zhang, C. Cheng, Beitong Zhou, Pengchao Hu, and Ye Yuan. Remaining useful life prediction of lithium-ion batteries based on false nearest neighbors and a hybrid neural network. *Applied Energy*, 253, 2019.
34. Shuai Huang, Mingming Huang, and Yuejun Lyu. A novel approach for sand liquefaction prediction via local mean-based pseudo nearest neighbor algorithm and its engineering application. *Advanced Engineering Informatics*, 41, 2019.
35. Mehmet Yesilbudak, Seref Sagiroglu, and Ilhami Colak. A new approach to very short term wind speed prediction using k-nearest neighbor classification. *Energy Conversion and Management*, 69, 2013.
36. Hugo T.C. Pedro and Carlos F.M. Coimbra. Nearest-neighbor methodology for prediction of intra-hour global horizontal and direct normal irradiances. *Renewable Energy*, 80, 2015.
37. David A. Wood. Lithofacies and stratigraphy prediction methodology exploiting an optimized nearest-neighbour algorithm to mine well-log data. *Marine and Petroleum Geology*, 110, 2019.

38. Yunxuan Dong, Xuejiao Ma, and Tonglin Fu. Electrical load forecasting: A deep learning approach based on k-nearest neighbors. *Applied Soft Computing*, 99, 2021.
39. Tomas Mikolov, Kai Chen, Greg Corrado, and Jeffrey Dean. Efficient estimation of word representations in vector space. In *1st International Conference on Learning Representations, ICLR 2013 - Workshop Track Proceedings*, 2013.
40. J. Hawkins, S. Ahmad, S. Purdy, and A. Lavin. Biological and machine intelligence (BAMI), 2016. Initial online release 0.4.
41. Maciej Wielgosz and Marcin Pietroń. Using spatial pooler of hierarchical temporal memory to classify noisy videos with predefined complexity. *Neurocomputing*, 240, 2017.
42. Murat Kirtay, Lorenzo Vannucci, Ugo Albanese, Alessandro Ambrosano, Egidio Falotico, and Cecilia Laschi. Spatial pooling as feature selection method for object recognition. In *ESANN 2018 - Proceedings, European Symposium on Artificial Neural Networks, Computational Intelligence and Machine Learning*, 2018.
43. Nasser Owaid Alshammari. Anomaly detection using hierarchical temporal memory in smart homes. *PQDT - UK & Ireland*, 2018.
44. TAO ZHOU, ZHEN-ZHEN ZHANG, and YUAN-YUAN CHEN. Hierarchical temporal memory network for medical image processing. *DEStech Transactions on Computer Science and Engineering*, 2018.
45. Timur Ibrayev, Ulan Myrzakhan, Olga Krestinskaya, Aidana Irmanova, and Alex Pappachen James. On-chip face recognition system design with memristive hierarchical temporal memory. In *Journal of Intelligent and Fuzzy Systems*, volume 34, 2018.
46. E. N. Osegi. Using the hierarchical temporal memory spatial pooler for short-term forecasting of electrical load time series. *Applied Computing and Informatics*, 2018.
47. Csongor Pilinszki-Nagy and Bálint Gyires-Tóth. Performance analysis of sparse matrix representation in hierarchical temporal memory for sequence modeling. *Infocommunications Journal*, 12, 2020.
48. Yeldos Dauletkhanuly, Olga Krestinskaya, and Alex Pappachen James. *HTM theory*, volume 14. Springer, 2020.
49. Damir Dobric, Andreas Pech, Bogdan Ghita, and Thomas Wennekers. Scaling the htm spatial pooler. *International Journal of Artificial Intelligence & Applications*, 11, 2020.
50. Yiwen Guo, Chao Zhang, Changshui Zhang, and Yurong Chen. Sparse DNNs with improved adversarial robustness. In *Advances in Neural Information Processing Systems*, volume 2018-December, 2018.
51. Song Han, Jeff Pool, John Tran, and William J. Dally. Learning both weights and connections for efficient neural networks. In *Advances in Neural Information Processing Systems*, volume 2015-January, 2015.
52. Jonathan Frankle and Michael Carbin. The lottery ticket hypothesis: Finding sparse, trainable neural networks. In *7th International Conference on Learning Representations, ICLR 2019*, 2019.
53. Namhoon Lee, Thalaiyasingam Ajanthan, and Philip H.S. Torr. Snip: Single-shot network pruning based on connection sensitivity. In *7th International Conference on Learning Representations, ICLR 2019*, 2019.
54. Subutai Ahmad and Jeff Hawkins. How do neurons operate on sparse distributed representations? a mathematical theory of sparsity, neurons and active dendrites, 2016. arXiv:1601.00720.

On Two Kinds of the Hardy-Type Integral Inequalities in the Whole Plane with the Equivalent Forms

Bicheng Yang, Dorin Andrica, Ovidiu Bagdasar, and Michael Th. Rassias

Abstract By the use of weight functions, a few equivalent conditions of two kinds of Hardy-type integral inequalities with multi-parameters in the whole plane are obtained. The constant factors related to the extended Riemann-zeta function are proved to be the best possible. Applying our results, we deduce a few equivalent conditions of two kinds of Hardy-type integral inequalities in the whole plane and some particular cases.

2000 Mathematics Subject Classification 26D15, 65B10

1 Introduction

If $f(x), g(y) \geq 0$,

B. Yang
Department of Mathematics, Guangdong University of Education, Guangzhou, Guangdong, P. R. China
e-mail: bcyang@gdei.edu.cn

D. Andrica
Faculty of Mathematics and Computer Science, Babeş-Bolyai University, Cluj-Napoca, Romania
e-mail: dandrica@math.ubbcluj.ro

O. Bagdasar
School of Computing and Engineering, University of Derby, Derby, UK
e-mail: o.bagdasar@derby.ac.uk

M. Th. Rassias (✉)
Department of Mathematics and Engineering Sciences, Hellenic Military Academy, Athens, Greece

Moscow Institute of Physics and Technology, Dolgoprudny, Russia

Institute for Advanced Study, Program in Interdisciplinary Studies, Princeton, NJ, USA
e-mail: michail.rassias@math.uzh.ch

I. N. Parasidis et al. (eds.), *Mathematical Analysis in Interdisciplinary Research*,
Springer Optimization and Its Applications 179,
https://doi.org/10.1007/978-3-030-84721-0_39

$$0 < \int_0^\infty f^2(x)dx < \infty \text{ and } 0 < \int_0^\infty g^2(y)dy < \infty,$$

then we have the following Hilbert integral inequality with the best possible constant factor π (see [1]):

$$\int_0^\infty \int_0^\infty \frac{f(x)g(y)}{x+y} dxdy < \pi \left(\int_0^\infty f^2(x)dx \int_0^\infty g^2(y)dy \right)^{\frac{1}{2}}, \tag{1}$$

By the use of weight functions, several extensions of (1) were established in [2] and [3]. Some Hilbert-type and Hardy-type inequalities were introduced and proved in [4–9]. In 2017, Hong [10] also considered an equivalent condition between a Hilbert-type inequality with homogenous kernel and a few parameters. Some other kinds of Hilbert-type as well as Hardy-type inequalities were proved in [11–19, 30–36]. Most of these inequalities are built on the first quadrant of the whole plane.

In 2007, Yang [20] established the following Hilbert-type integral inequality in the whole plane:

$$\int_{-\infty}^\infty \int_{-\infty}^\infty \frac{f(x)g(y)}{(1+e^{x+y})^\lambda} dxdy$$
$$< B\left(\frac{\lambda}{2}, \frac{\lambda}{2}\right) \left(\int_{-\infty}^\infty e^{-\lambda x} f^2(x)dx \int_{-\infty}^\infty e^{-\lambda y} g^2(y)dy \right)^{\frac{1}{2}}, \tag{2}$$

with the best possible constant factor $B(\frac{\lambda}{2}, \frac{\lambda}{2})$ ($\lambda > 0$, $B(u, v)$ stands for the beta function) (see [21]). He et al. [22–35] also deduced some Hilbert-type integral inequalities in the whole plane.

In this chapter, by the use of weight functions, a few equivalent conditions of two kinds of the Hardy-type integral inequalities with multi-parameters in the whole plane are obtained. The constant factors related to the extended Riemann zeta function are proved to be the best possible. In the form of applications, we deduce a few equivalent conditions of two kinds of Hardy-type integral inequalities in the whole plane and some particular cases.

2 An Example and Two Lemmas

Example 1 We set

$$h(u) := \frac{(\min\{|u|, 1\})^{1-\lambda}}{|u-1|} |\ln|u||^\beta \quad (u \in \mathbf{R}).$$

(i) For $\beta > 0, \sigma + \mu = \lambda, \mu < 1$, we obtain that

$$\begin{aligned} k^{(1)}(\sigma) &:= \int_{-1}^{1} h(u)|u|^{\sigma-1}du = \int_{0}^{1} (h(-u)+h(u))u^{\sigma-1}du \\ &= \int_{0}^{1} (\min\{u,1\})^{1-\lambda}(-\ln u)^{\beta}u^{\sigma-1}\left(\frac{1}{u+1}+\frac{1}{|u-1|}\right)du \\ &= \int_{0}^{1} (-\ln u)^{\beta}\left(\frac{1}{u+1}+\frac{1}{1-u}\right)u^{-\mu}du \\ &= 2\int_{0}^{1} (-\ln u)^{\beta}\frac{u^{-\mu}}{1-u^{2}}du = 2\int_{0}^{1} (-\ln u)^{\beta}\sum_{k=0}^{\infty} u^{2k-\mu}du. \end{aligned}$$

By the Lebesgue term by term integration theorem (cf. [38]), we derive that

$$\begin{aligned} k^{(1)}(\sigma) &= 2\sum_{k=0}^{\infty}\int_{0}^{1} (-\ln u)^{\beta}u^{2k-\mu}du \\ &= 2\sum_{k=0}^{\infty}\frac{1}{(2k-\mu+1)^{\beta+1}}\int_{0}^{\infty} v^{\beta}e^{-v}dv \\ &= \frac{\Gamma(\beta+1)}{2^{\beta}}\zeta\left(\beta+1,\frac{1-\mu}{2}\right)\in\mathbf{R}_{+}, \end{aligned} \tag{3}$$

where

$$\zeta(s,a) = \sum_{k=0}^{\infty}\frac{1}{(k+a)^{s}} \quad (Res>1;\, a>0)$$

is the extended Riemann-zeta function (where $\zeta(s,1)=\zeta(s):=\sum_{k=1}^{\infty}\frac{1}{k^{s}}$ $(Res>1)$ is the Riemann-zeta function) (cf. [21]).
In particular, for $\beta>0, \sigma=\lambda+1, \mu=-1\ (<1)$, we have

$$\begin{aligned} k^{(1)}(\lambda+1) &= \int_{-1}^{1}\frac{(\min\{|u|,1\})^{1-\lambda}}{|u-1|}|\ln|u||^{\beta}|u|^{\lambda}du \\ &= \frac{\Gamma(\beta+1)}{2^{\beta}}\zeta(\beta+1)\in\mathbf{R}_{+}. \end{aligned}$$

(ii) For $\beta>0, \sigma+\mu=\lambda, \sigma<1$, we also obtain that

$$\begin{aligned} k^{(2)}(\sigma) &:= \int_{\{u;|u|\geq 1\}} h(u)|u|^{\sigma-1}du = \int_{1}^{\infty}(h(-u)+h(u))|u|^{\sigma-1}du \\ &= \int_{-1}^{1}\frac{(\min\{|v|,1\})^{1-\lambda}}{|v-1|}|\ln|v||^{\beta}|v|^{\mu-1}dv \end{aligned}$$

$$= \frac{\Gamma(\beta+1)}{2^{\beta}}\zeta\left(\beta+1, \frac{1-\sigma}{2}\right) = k^{(1)}(\mu) \in \mathbf{R}_{+}, \tag{4}$$

$$k^{(2)}(-1) = \frac{\Gamma(\beta+1)}{2^{\beta}}\zeta(\beta+1).$$

Remark 1 It is obvious that for $\sigma + \mu = \lambda$, $k^{(1)}(\sigma) < \infty$ if and only if $\mu < 1$ (or $\sigma > \lambda - 1$) with $\beta > 0$; $k^{(2)}(\sigma) < \infty$ if and only if $\sigma < 1$ (or $\mu > \lambda - 1$) with $\beta > 0$.

In the sequel, we shall always assume that $p > 1, \frac{1}{p} + \frac{1}{q} = 1, \sigma + \mu = \lambda$.

Lemma 1 *If $\sigma_1 \in \mathbf{R}$, and there exists a constant M_1 such that for any $f(x) \geq 0$ and $g(y) \geq 0$ in $\mathbf{R}$ the following inequality*

$$\int_{-\infty}^{\infty} g(y)\left[\int_{-\frac{1}{|y|}}^{\frac{1}{|y|}} \frac{(\min\{|xy|, 1\})^{1-\lambda}}{|xy-1|}|\ln|xy||^{\beta} f(x)dx\right]dy$$

$$\leq M_1\left[\int_{-\infty}^{\infty} |x|^{p(1-\sigma)-1} f^p(x)dx\right]^{\frac{1}{p}}\left[\int_{-\infty}^{\infty} |y|^{q(1-\sigma_1)-1} g^q(y)dy\right]^{\frac{1}{q}} \tag{5}$$

holds true, then we have $\sigma_1 = \sigma > \lambda - 1$ with $\beta > 0$ and $k^{(1)}(\sigma) \leq M_1$.

Proof If $\sigma_1 > \sigma$, then for

$$n \geq \frac{1}{\sigma_1 - \sigma} \ (n \in \mathbf{N}),$$

we set

$$f_n(x) := \begin{cases} |x|^{\sigma+\frac{1}{pn}-1}, & 0 < |x| \leq 1 \\ 0, & |x| > 1 \end{cases}, \quad g_n(y) := \begin{cases} 0, & 0 < |y| < 1 \\ |y|^{\sigma_1-\frac{1}{qn}-1}, & y \geq 1 \end{cases},$$

and derive that

$$J_1 := \left[\int_{-\infty}^{\infty} |x|^{p(1-\sigma)-1} f_n^p(x)dx\right]^{\frac{1}{p}}\left[\int_{-\infty}^{\infty} |y|^{q(1-\sigma_1)-1} g_n^q(y)dy\right]^{\frac{1}{q}}$$

$$= \left(2\int_0^1 x^{\frac{1}{n}-1}dx\right)^{\frac{1}{p}}\left(2\int_1^{\infty} y^{-\frac{1}{n}-1}dy\right)^{\frac{1}{q}} = 2n.$$

We obtain

$$I_1 := \int_{-\infty}^{\infty} g_n(y)\left[\int_{-\frac{1}{|y|}}^{\frac{1}{|y|}} \frac{(\min\{|xy|, 1\})^{1-\lambda}}{|xy-1|}|\ln|xy||^{\beta} f_n(x)dx\right]dy$$

$$= \int_{-\infty}^{-1} \left[\int_{\frac{1}{y}}^{\frac{-1}{y}} \frac{(\min\{|xy|, 1\})^{1-\lambda}}{|xy-1|} |\ln|xy||^{\beta} |x|^{\sigma+\frac{1}{pn}-1} dx \right] (-y)^{\sigma_1-\frac{1}{qn}-1} dy$$

$$+ \int_{1}^{\infty} \left[\int_{\frac{-1}{y}}^{\frac{1}{y}} \frac{(\min\{|xy|, 1\})^{1-\lambda}}{|xy-1|} |\ln|xy||^{\beta} |x|^{\sigma+\frac{1}{pn}-1} dx \right] y^{\sigma_1-\frac{1}{qn}-1} dy$$

$$= \int_{1}^{\infty} \left[\int_{\frac{-1}{y}}^{\frac{1}{y}} (h(-xy) + h(xy)) |x|^{\sigma+\frac{1}{pn}-1} dx \right] y^{\sigma_1-\frac{1}{qn}-1} dy$$

$$= 2 \int_{1}^{\infty} \left[\int_{0}^{1} (h(-u) + h(u)) u^{(\sigma+\frac{1}{pn})-1} du \right] y^{(\sigma_1-\sigma)-\frac{1}{n}-1} dy, \quad (6)$$

and then by (5), it follows that

$$2k^{(1)} \left(\sigma + \frac{1}{pn}\right) \int_{1}^{\infty} y^{(\sigma_1-\sigma)-\frac{1}{n}-1} dy = I_1 \le M_1 J_1 = 2M_1 n. \quad (7)$$

Since $(\sigma_1 - \sigma) - \frac{1}{n} \ge 0$, it follows that

$$\int_{1}^{\infty} y^{(\sigma_1-\sigma)-\frac{1}{n}-1} dy = \infty.$$

By (7), for

$$k^{(1)} \left(\sigma + \frac{1}{pn}\right) > 0,$$

we have

$$\infty \le 2M_1 n < \infty,$$

which is a contradiction.

If $\sigma_1 < \sigma$, then for

$$n \ge \frac{1}{\sigma - \sigma_1} \ (n \in \mathbf{N}),$$

we set

$$\widetilde{f}_n(x) := \begin{cases} 0, & 0 < |x| < 1 \\ |x|^{\sigma-\frac{1}{pn}-1}, & |x| \ge 1 \end{cases}, \quad \widetilde{g}_n(y) := \begin{cases} |y|^{\sigma_1+\frac{1}{qn}-1}, & 0 < |y| \le 1 \\ 0, & |y| > 1 \end{cases},$$

and get

$$\widetilde{J}_1 := \left[\int_{-\infty}^{\infty} |x|^{p(1-\sigma)-1} \widetilde{f}_n^p(x)dx\right]^{\frac{1}{p}} \left[\int_{-\infty}^{\infty} |y|^{q(1-\sigma_1)-1} \widetilde{g}_n^q(y)dy\right]^{\frac{1}{q}}$$

$$= \left(2\int_1^{\infty} x^{-\frac{1}{n}-1}dx\right)^{\frac{1}{p}} \left(2\int_0^1 y^{\frac{1}{n}-1}dy\right)^{\frac{1}{q}} = 2n.$$

We obtain

$$\widetilde{I}_1 := \int_{-\infty}^{\infty} \widetilde{f}_n(x)\left[\int_{-\frac{1}{|x|}}^{\frac{1}{|x|}} \frac{(\min\{|xy|, 1\})^{1-\lambda}}{|xy-1|} |\ln|xy||^{\beta} \widetilde{g}_n(y)dy\right]dx$$

$$= \int_{-\infty}^{-1}\left[\int_{\frac{1}{x}}^{\frac{-1}{x}} \frac{(\min\{|xy|, 1\})^{1-\lambda}}{|xy-1|} |\ln|xy||^{\beta} |y|^{\sigma_1+\frac{1}{qn}-1}dy\right](-x)^{\sigma-\frac{1}{pn}-1}dx$$

$$+\int_1^{\infty}\left[\int_{-\frac{1}{x}}^{\frac{1}{x}} \frac{(\min\{|xy|, 1\})^{1-\lambda}}{|xy-1|} |\ln|xy||^{\beta} |y|^{\sigma_1+\frac{1}{qn}-1}dy\right]x^{\sigma-\frac{1}{pn}-1}dx$$

$$= \int_1^{\infty}\left[\int_{-\frac{1}{x}}^{\frac{1}{x}} (h(-xy)+h(xy))|y|^{\sigma_1+\frac{1}{qn}-1}dy\right]x^{\sigma-\frac{1}{pn}-1}dx$$

$$= 2\int_1^{\infty}\left[\int_0^1 (h(-u)+h(u))u^{\sigma_1+\frac{1}{qn}-1}du\right]x^{(\sigma-\sigma_1)-\frac{1}{n}-1}dx, \tag{8}$$

and then by the Fubini theorem (cf. [38]) and (5), it follows that

$$2k^{(1)}\left(\sigma_1+\frac{1}{qn}\right)\int_1^{\infty} x^{(\sigma-\sigma_1)-\frac{1}{n}-1}dx$$

$$= \widetilde{I}_1 = \int_{-\infty}^{\infty} \widetilde{g}_n(y)\left(\int_{\frac{-1}{|y|}}^{\frac{1}{|y|}} h(xy)\widetilde{f}_n(x)dx\right)dy \le M_1\widetilde{J}_1 = 2M_1 n. \tag{9}$$

Since $(\sigma-\sigma_1)-\frac{1}{n} \ge 0$, it follows that

$$\int_1^{\infty} x^{(\sigma-\sigma_1)-\frac{1}{n}-1}dx = \infty.$$

By (9), for

$$k^{(1)}\left(\sigma_1+\frac{1}{qn}\right) > 0,$$

we deduce that

$$\infty \le 2M_1 n < \infty,$$

which is a contradiction.

Hence, we conclude that $\sigma_1 = \sigma$. For $\sigma_1 = \sigma$, we reduce (9) as follows:

$$k^{(1)}\left(\sigma_1 + \frac{1}{qn}\right) = \int_0^1 (h(-u) + h(u))u^{\sigma_1 + \frac{1}{qn} - 1} du \le M_1.$$

Since $\{(h(-u) + h(u))u^{\sigma + \frac{1}{qn} - 1}\}_{n=1}^{\infty}$ is increasing in $(0, 1)$, by Levi's theorem (cf. [38]), we obtain that

$$k^{(1)}(\sigma) = \int_0^1 \lim_{n\to\infty} (h(-u) + h(u))u^{\sigma + \frac{1}{qn} - 1} du$$
$$= \lim_{n\to\infty} \int_0^1 (h(-u) + h(u))u^{\sigma + \frac{1}{qn} - 1} du \le M_1 < \infty.$$

By Remark 1, it follows that $\sigma > \lambda - 1$ with $\beta > 0$.

This completes the proof of the lemma. □

Lemma 2 *If $\sigma_1 \in \mathbf{R}$, and there exists a constant M_2 such that for any $f(x) \ge 0$ and $g(y) \ge 0$ in $\mathbf{R}$ the following inequality*

$$\int_{-\infty}^{\infty} g(y) \left[\int_{\{x;|x| \ge \frac{1}{|y|}\}} \frac{(\min\{|xy|, 1\})^{1-\lambda}}{|xy - 1|} |\ln |xy||^{\beta} f(x) dx\right] dy$$
$$\le M_2 \left[\int_{-\infty}^{\infty} |x|^{p(1-\sigma)-1} f^p(x) dx\right]^{\frac{1}{p}} \left[\int_{-\infty}^{\infty} |y|^{q(1-\sigma_1)-1} g^q(y) dy\right]^{\frac{1}{q}} \quad (10)$$

holds true, then we have $\sigma_1 = \sigma < 1$ with $\beta > 0$ and $k^{(2)}(\sigma) \le M_2$.

Proof If $\sigma_1 < \sigma$, then for $n \ge \frac{1}{\sigma - \sigma_1}$ $(n \in \mathbf{N})$, we set the functions $\widetilde{f}_n(x)$ and $\widetilde{g}_n(y)$ as in Lemma 1 and obtain that

$$\widetilde{J}_1 = \left[\int_{-\infty}^{\infty} |x|^{p(1-\sigma)-1} \widetilde{f}_n^p(x) dx\right]^{\frac{1}{p}} \left[\int_{-\infty}^{\infty} |y|^{q(1-\sigma_1)-1} \widetilde{g}_n^q(y) dy\right]^{\frac{1}{q}} = 2n.$$

We get

$$\widetilde{I}_2 := \int_{-\infty}^{\infty} \widetilde{g}_n(y) \left[\int_{\{x;|x| \ge \frac{1}{|y|}\}} \frac{(\min\{|xy|, 1\})^{1-\lambda}}{|xy - 1|} |\ln |xy||^{\beta} \widetilde{f}_n(x) dx\right] dy$$
$$= \int_{-1}^{0} \left[\int_{\{x;|x| \ge \frac{-1}{y}\}} \frac{(\min\{|xy|, 1\})^{1-\lambda}}{|xy - 1|} |\ln |xy||^{\beta} |x|^{\sigma - \frac{1}{pn} - 1} dx\right] (-y)^{\sigma_1 + \frac{1}{qn} - 1} dy$$

$$+\int_0^1\left[\int_{\{x;|x|\geq\frac{1}{y}\}}\frac{(\min\{|xy|,1\})^{1-\lambda}}{|xy-1|}|\ln|xy||^{\beta}|x|^{\sigma-\frac{1}{pn}-1}dx\right]y^{\sigma_1+\frac{1}{qn}-1}dy$$

$$=\int_0^1\left[\int_{\{x;|x|\geq\frac{1}{y}\}}(h(-xy)+h(xy))|x|^{\sigma-\frac{1}{pn}-1}dx\right]y^{\sigma_1+\frac{1}{qn}-1}dy$$

$$=2\int_0^1\left[\int_1^{\infty}(h(-u)+h(u))u^{\sigma-\frac{1}{pn}-1}du\right]y^{(\sigma_1-\sigma)+\frac{1}{n}-1}dy,$$

and thus by (10), we have

$$2k^{(2)}\left(\sigma-\frac{1}{pn}\right)\int_0^1 y^{(\sigma_1-\sigma)+\frac{1}{n}-1}dy=\widetilde{I}_2\leq M_2\widetilde{J}_1=2M_2n. \tag{11}$$

Since $(\sigma_1-\sigma)+\frac{1}{n}\leq 0$, it follows that

$$\int_0^1 y^{(\sigma_1-\sigma)+\frac{1}{n}-1}dy=\infty.$$

By (11), for

$$k^{(2)}\left(\sigma-\frac{1}{pn}\right)>0,$$

we have

$$\infty\leq 2M_2n<\infty,$$

which is a contradiction.

If $\sigma_1>\sigma$, then for $n\geq\frac{1}{\sigma_1-\sigma}$ $(n\in\mathbf{N})$, we set the functions $f_n(x)$ and $g_n(y)$ as in Lemma 1 and deduce that

$$J_1=\left[\int_{-\infty}^{\infty}|x|^{p(1-\sigma)-1}f_n^p(x)dx\right]^{\frac{1}{p}}\left[\int_{-\infty}^{\infty}|y|^{q(1-\sigma_1)-1}g_n^q(y)dy\right]^{\frac{1}{q}}=2n.$$

We obtain

$$I_2:=\int_{-\infty}^{\infty}f_n(x)\left[\int_{\{y;|y|\geq\frac{1}{|x|}\}}\frac{(\min\{|xy|,1\})^{1-\lambda}}{|xy-1|}|\ln|xy||^{\beta}g_n(y)dy\right]dx$$

$$=\int_{-1}^{0}\left[\int_{\{y;|y|\geq\frac{-1}{x}\}}\frac{(\min\{|xy|,1\})^{1-\lambda}}{|xy-1|}|\ln|xy||^{\beta}|y|^{\sigma_1-\frac{1}{qn}-1}dy\right](-x)^{\sigma+\frac{1}{pn}-1}dx$$

$$+\int_0^1\left[\int_{\{y;|y|\geq\frac{1}{x}\}}\frac{(\min\{|xy|,1\})^{1-\lambda}}{|xy-1|}|\ln|xy||^{\beta}|y|^{\sigma_1-\frac{1}{qn}-1}dy\right]x^{\sigma+\frac{1}{pn}-1}dx$$

$$=\int_0^1\left[\int_{\{y;|y|\geq\frac{1}{x}\}}(h(-xy)+h(xy))|y|^{\sigma_1-\frac{1}{qn}-1}dy\right]x^{\sigma+\frac{1}{pn}-1}dx$$

$$=\int_{\{u;|u|\geq 1\}}(h(-u)+h(u))|u|^{(\sigma_1-\frac{1}{qn})-1}du\int_0^1 x^{(\sigma-\sigma_1)+\frac{1}{n}-1}dx,$$

and then by the Fubini theorem (cf. [38]) and (8), we have

$$\begin{aligned}&2k_2\left(\sigma_1-\frac{1}{qn}\right)\int_0^1 x^{(\sigma-\sigma_1)+\frac{1}{n}-1}dx\\&\quad=I_2=\int_0^{\infty}g_n(y)\left(\int_{\{x;|x|\geq\frac{1}{|y|}\}}h(xy)f_n(x)dx\right)dy\leq M_2J_1=2M_2n.\end{aligned}\tag{12}$$

Since $(\sigma-\sigma_1)+\frac{1}{n}\leq 0$, it follows that

$$\int_0^1 x^{(\sigma-\sigma_1)+\frac{1}{n}-1}dx=\infty.$$

By (12), for

$$k^{(2)}\left(\sigma_1-\frac{1}{qn}\right)>0,$$

we obtain that

$$\infty\leq 2M_2n<\infty,$$

which is a contradiction.

Hence, we conclude that $\sigma_1=\sigma$. For $\sigma_1=\sigma$, we reduce (12) as follows:

$$k^{(2)}\left(\sigma-\frac{1}{qn}\right)=\int_1^{\infty}(h(-u)+h(u))u^{\sigma-\frac{1}{qn}-1}du\leq M_2.\tag{13}$$

Since $\{(h(-u)+h(u))u^{\sigma-\frac{1}{qn}-1}\}_{n=1}^{\infty}$ is increasing in $[1,\infty)$, by Levi's theorem (cf. [38]), we obtain that

$$\begin{aligned}k^{(2)}(\sigma)&=\int_1^{\infty}\lim_{n\to\infty}(h(-u)+h(u))u^{\sigma-\frac{1}{qn}-1}du\\&=\lim_{n\to\infty}\int_1^{\infty}(h(-u)+h(u))u^{\sigma-\frac{1}{qn}-1}du\leq M_2<\infty.\end{aligned}$$

By Remark 1, we have $\sigma < 1$ with $\beta > 0$.

This completes the proof of the lemma. □

3 Main Results and Particular Cases

Theorem 1 *If $\sigma_1 \in \mathbf{R}$, then the following conditions are equivalent:*

(i) There exists a constant M_1, such that for any $f(x) \geq 0$, satisfying

$$0 < \int_{-\infty}^{\infty} |x|^{p(1-\sigma)-1} f^p(x)dx < \infty,$$

we have the following Hardy-type integral inequality of the first kind, with the nonhomogeneous kernel:

$$J := \left\{\int_{-\infty}^{\infty} |y|^{p\sigma_1 - 1} \left[\int_{\frac{-1}{|y|}}^{\frac{1}{|y|}} \frac{(\min\{|xy|, 1\})^{1-\lambda}}{|xy-1|} |\ln|xy||^{\beta} f(x)dx\right]^p dy\right\}^{\frac{1}{p}}$$
$$< M_1 \left[\int_{-\infty}^{\infty} |x|^{p(1-\sigma)-1} f^p(x)dx\right]^{\frac{1}{p}}. \tag{14}$$

(ii) There exists a constant M_1, such that for any $f(x), g(y) \geq 0$,

$$0 < \int_{-\infty}^{\infty} |x|^{p(1-\sigma)-1} f^p(x)dx < \infty,$$

and

$$0 < \int_{-\infty}^{\infty} |y|^{q(1-\sigma_1)-1} g^q(y)dy < \infty,$$

we have the following inequality:

$$I := \int_{-\infty}^{\infty} g(y) \left[\int_{\frac{-1}{|y|}}^{\frac{1}{|y|}} \frac{(\min\{|xy|, 1\})^{1-\lambda}}{|xy-1|} |\ln|xy||^{\beta} f(x)dx\right] dy$$
$$< M_1 \left[\int_{-\infty}^{\infty} |x|^{p(1-\sigma)-1} f^p(x)dx\right]^{\frac{1}{p}} \left[\int_{-\infty}^{\infty} |y|^{q(1-\sigma_1)-1} g^q(y)dy\right]^{\frac{1}{q}}. \tag{15}$$

(iii) $\sigma_1 = \sigma > \lambda - 1$ and $\beta > 0$.

If Condition (iii) is satisfied, then the constant factor $M_1 = k^{(1)}(\sigma)(\in \mathbf{R}_+)$ in (14) and (15) (for $\sigma_1 = \sigma$) is the best possible.

Proof $(i) \Rightarrow (ii)$. By Hölder's inequality (cf. [37]), we have

$$I = \int_{-\infty}^{\infty} \left(|y|^{\sigma_1 - \frac{1}{p}} \int_{\frac{-1}{|y|}}^{\frac{1}{|y|}} \frac{(\min\{|xy|, 1\})^{1-\lambda}}{|xy-1|} |\ln |xy||^{\beta} f(x) dx \right) \left(|y|^{\frac{1}{p} - \sigma_1} g(y) \right) dy$$

$$\leq J \left[\int_{-\infty}^{\infty} |y|^{q(1-\sigma_1)-1} g^q(y) dy \right]^{\frac{1}{q}}. \tag{16}$$

Then by (14), we deduce (15).

$(ii) \Rightarrow (iii)$. By Lemma 1, we have $\sigma_1 = \sigma > \lambda - 1$ with $\beta > 0$.

$(iii) \Rightarrow (i)$. We obtain the following weight function:

For $y \neq 0$,

$$\begin{aligned}
\omega_1(\sigma, y) &:= |y|^{\sigma} \int_{\frac{-1}{|y|}}^{\frac{1}{|y|}} \frac{(\min\{|xy|, 1\})^{1-\lambda}}{|xy-1|} |\ln |xy||^{\beta} |x|^{\sigma-1} dx \\
&= |y|^{\sigma} \int_{\frac{-1}{|y|}}^{0} h(xy)(-x)^{\sigma-1} dx + |y|^{\sigma} \int_{0}^{\frac{1}{|y|}} h(xy) x^{\sigma-1} dx \\
&= |y|^{\sigma} \int_{0}^{\frac{1}{|y|}} h(-xy) x^{\sigma-1} dx + |y|^{\sigma} \int_{0}^{\frac{1}{|y|}} h(xy) x^{\sigma-1} dx \\
&= |y|^{\sigma} \int_{0}^{\frac{1}{|y|}} (h(-x|y|) + h(x|y|) x^{\sigma-1} dx \\
&= \int_{0}^{1} (h(-u) + h(u)) u^{\sigma-1} du = k^{(1)}(\sigma).
\end{aligned} \tag{17}$$

Then by Hölder's inequality with weight and (17), we obtain that

$$\begin{aligned}
&\left[\int_{\frac{-1}{|y|}}^{\frac{1}{|y|}} \frac{(\min\{|xy|, 1\})^{1-\lambda}}{|xy-1|} |\ln |xy||^{\beta} f(x) dx \right]^p \\
&= \left\{ \int_{\frac{-1}{|y|}}^{\frac{1}{|y|}} h(xy) \left[\frac{|y|^{(\sigma-1)/p}}{|x|^{(\sigma-1)/q}} f(x) \right] \left[\frac{|x|^{(\sigma-1)/q}}{|y|^{(\sigma-1)/p}} \right] dx \right\}^p \\
&\leq \int_{\frac{-1}{|y|}}^{\frac{1}{|y|}} h(xy) \frac{|y|^{\sigma-1} f^p(x)}{|x|^{(\sigma-1)p/q}} dx \left[\int_{\frac{-1}{|y|}}^{\frac{1}{|y|}} h(xy) \frac{|x|^{\sigma-1}}{|y|^{(\sigma-1)q/p}} dx \right]^{p-1} \\
&= \int_{\frac{-1}{|y|}}^{\frac{1}{|y|}} h(xy) \frac{|y|^{\sigma-1}}{|x|^{(\sigma-1)p/q}} f^p(x) dx \cdot \left[\omega_1(\sigma, y) |y|^{q(1-\sigma)-1} \right]^{p-1}
\end{aligned}$$

$$= (k^{(1)}(\sigma))^{p-1}|y|^{-p\sigma+1}\int_{\frac{-1}{|y|}}^{\frac{1}{|y|}} h(xy)\frac{|y|^{\sigma-1}}{|x|^{(\sigma-1)p/q}}f^p(x)dx. \tag{18}$$

If (18) assumes the form of equality for a $y \in \mathbf{R}\backslash\{0\}$, then (cf. [37]) there exist constants A and B, such that they are not all zero, and

$$A\frac{|y|^{\sigma-1}}{|x|^{(\sigma-1)p/q}}f^p(x) = B\frac{|x|^{\sigma-1}}{|y|^{(\sigma-1)q/p}} \text{ a.e. in } \mathbf{R}.$$

Let us suppose that $A \neq 0$ (otherwise $B = A = 0$). It follows that

$$|x|^{p(1-\sigma)-1}f^p(x) = |y|^{q(1-\sigma)}\frac{B}{A|x|} \text{ a.e. in } \mathbf{R},$$

which contradicts the fact that

$$0 < \int_{-\infty}^{\infty} |x|^{p(1-\sigma)-1}f^p(x)dx < \infty.$$

Hence, (18) assumes the form of strict inequality.

For $\sigma_1 = \sigma > \lambda - 1$ with $\beta > 0$, by Remark 1, we have $k^{(1)}(\sigma) \in \mathbf{R}_+$. In view of the above results and Fubini's theorem (cf. [38]), we deduce that

$$J < (k^{(1)}(\sigma))^{\frac{1}{q}}\left\{\int_{-\infty}^{\infty}\left[\int_{\frac{-1}{|y|}}^{\frac{1}{|y|}} H(xy)\frac{|y|^{\sigma-1}}{|x|^{(\sigma-1)p/q}}f^p(x)dx\right]dy\right\}^{\frac{1}{p}}$$

$$= (k^{(1)}(\sigma))^{\frac{1}{q}}\left\{\int_{-\infty}^{\infty}\left[\int_{\frac{-1}{|x|}}^{\frac{1}{|x|}} H(xy)\frac{|y|^{\sigma-1}}{|x|^{(\sigma-1)(p-1)}}dy\right]f^p(x)dx\right\}^{\frac{1}{p}}$$

$$= (k^{(1)}(\sigma))^{\frac{1}{q}}\left[\int_{-\infty}^{\infty} \omega_1(\sigma, x)|x|^{p(1-\sigma)-1}f^p(x)dx\right]^{\frac{1}{p}}$$

$$= k^{(1)}(\sigma)\left[\int_{-\infty}^{\infty} |x|^{p(1-\sigma)-1}f^p(x)dx\right]^{\frac{1}{p}}.$$

Setting $M_1 \geq k^{(1)}(\sigma)$, we have

$$J < k^{(1)}(\sigma)\left[\int_{-\infty}^{\infty} |x|^{p(1-\sigma)-1}f^p(x)dx\right]^{\frac{1}{p}} \leq M_1\left[\int_{-\infty}^{\infty} |x|^{p(1-\sigma)-1}f^p(x)dx\right]^{\frac{1}{p}},$$

namely, (14) follows.

Therefore, conditions (i), (ii) and (iii) are equivalent.

When Condition (iii) is satisfied, if there exists a constant $M_1 \leq k^{(1)}(\sigma)$, such that (15) is valid, then by Lemma 1, we still have $k^{(1)}(\sigma) \leq M_1$. It follows that the constant factor $M_1 = k^{(1)}(\sigma)$ in (15) is the best possible. The constant factor $M_1 = k^{(1)}(\sigma)$ in (14) is also the best possible. Otherwise, by (16) (for $\sigma_1 = \sigma$), we would conclude that the constant factor $M_1 = k^{(1)}(\sigma)$ in (15) is not the best possible.

This completes the proof of the theorem. □

In particular, for $\sigma = \sigma_1 = \frac{1}{p}$ in Theorem 1, we have:

Corollary 1 *The following conditions are equivalent:*

(i) There exists a constant M_1, such that for any $f(x) \geq 0$, satisfying

$$0 < \int_{-\infty}^{\infty} |x|^{p-2} f^p(x)dx < \infty,$$

we have the following inequality:

$$\left\{\int_{-\infty}^{\infty}\left[\int_{\frac{-1}{|y|}}^{\frac{1}{|y|}} \frac{(\min\{|xy|, 1\})^{1-\lambda}}{|xy-1|} |\ln|xy||^{\beta} f(x)dx\right]^p dy\right\}^{\frac{1}{p}}$$
$$< M_1\left(\int_{-\infty}^{\infty} |x|^{p-2} f^p(x)dx\right)^{\frac{1}{p}}. \tag{19}$$

(ii) There exists a constant M_1, such that for any $f(x), g(y) \geq 0$,

$$0 < \int_{-\infty}^{\infty} |x|^{p-2} f^p(x)dx < \infty,$$

and

$$0 < \int_{-\infty}^{\infty} g^q(y)dy < \infty,$$

we have the following inequality:

$$\int_{-\infty}^{\infty} g(y)\left[\int_{\frac{-1}{|y|}}^{\frac{1}{|y|}} \frac{(\min\{|xy|, 1\})^{1-\lambda}}{|xy-1|} |\ln|xy||^{\beta} f(x)dx\right] dy$$
$$< M_1\left(\int_{-\infty}^{\infty} |x|^{p-2} f^p(x)dx\right)^{\frac{1}{p}}\left(\int_{-\infty}^{\infty} g^q(y)dy\right)^{\frac{1}{q}}. \tag{20}$$

(iii) $\lambda < \frac{1}{p} + 1$ and $\beta > 0$.

If Condition (iii) is satisfied, then the constant factor $M_1 = k^{(1)}(\frac{1}{p})\ (\in \mathbf{R}_+)$ *in (19) and (20) is the best possible.*

Setting

$$y = \frac{1}{Y},\quad G(Y) = g\left(\frac{1}{Y}\right)\frac{1}{Y^2}$$

in Theorem 1, and then replacing Y by y, we have

Corollary 2 *If* $\sigma_1 \in \mathbf{R}$, *then the following conditions are equivalent:*

(i) There exists a constant M_1, *such that for any* $f(x) \geq 0$, *satisfying*

$$0 < \int_{-\infty}^{\infty} |x|^{p(1-\sigma)-1} f^p(x)dx < \infty,$$

we have the following inequality:

$$\left\{\int_{-\infty}^{\infty} |y|^{-p\sigma_1-1}\left[\int_{-|y|}^{|y|} \frac{(\min\{|x/y|, 1\})^{1-\lambda}}{|x/y-1|}|\ln|x/y||^{\beta} f(x)dx\right]^p dy\right\}^{\frac{1}{p}}$$

$$< M_1\left[\int_{-\infty}^{\infty} |x|^{p(1-\sigma)-1} f^p(x)dx\right]^{\frac{1}{p}}. \quad (21)$$

(ii) There exists a constant M_1, *such that for any* $f(x), G(y) \geq 0$,

$$0 < \int_{-\infty}^{\infty} |x|^{p(1-\sigma)-1} f^p(x)dx < \infty,$$

and

$$0 < \int_{-\infty}^{\infty} |y|^{q(1+\sigma_1)-1} G^q(y)dy < \infty,$$

we have the following inequality:

$$\int_{-\infty}^{\infty} G(y)\left[\int_{-|y|}^{|y|} \frac{(\min\{|x/y|, 1\})^{1-\lambda}}{|x/y-1|}|\ln|x/y||^{\beta} f(x)dx\right]dy$$

$$< M_1\left[\int_{-\infty}^{\infty} |x|^{p(1-\sigma)-1} f^p(x)dx\right]^{\frac{1}{p}}\left[\int_{-\infty}^{\infty} |y|^{q(1+\sigma_1)-1} G^q(y)dy\right]^{\frac{1}{q}}. \quad (22)$$

(iii) $\sigma_1 = \sigma > \mu - 1$ *and* $\beta > 0$.
If Condition (iii) holds, then the constant factor $M_1 = k^{(1)}(\sigma)\ (\in \mathbf{R}_+)$ *in (21) and (22) (for* $\sigma_1 = \sigma$*) is the best possible.*

For $g(y) = y^{\lambda}G(y)$ and $\mu_1 = \lambda - \sigma_1$ in Corollary 2, we have

Corollary 3 *If $\mu_1 \in \mathbf{R}$, then the following conditions are equivalent:*

(i) There exists a constant M_1, such that for any $f(x) \geq 0$, satisfying

$$0 < \int_{-\infty}^{\infty} |x|^{p(1-\sigma)-1} f^p(x)dx < \infty,$$

we have the following Hardy-type integral inequality of the first kind with the homogeneous kernel:

$$\left\{\int_{-\infty}^{\infty} y^{p\mu_1-1}\left[\int_{-|y|}^{|y|} \frac{(\min\{|x|,|y|\})^{1-\lambda}}{|x-y|}|\ln|\frac{x}{y}||^{\beta} f(x)dx\right]^p dy\right\}^{\frac{1}{p}}$$
$$< M_1\left[\int_{-\infty}^{\infty} |x|^{p(1-\sigma)-1} f^p(x)dx\right]^{\frac{1}{p}}. \quad (23)$$

(ii) There exists a constant M_1, such that for any $f(x), g(y) \geq 0$,

$$0 < \int_{-\infty}^{\infty} |x|^{p(1-\sigma)-1} f^p(x)dx < \infty,$$

and

$$0 < \int_{-\infty}^{\infty} |y|^{q(1-\mu_1)-1} g^q(y)dy < \infty,$$

we have the following inequality:

$$\int_{-\infty}^{\infty} g(y)\left[\int_{-|y|}^{|y|} \frac{(\min\{|x|,|y|\})^{1-\lambda}}{|x-y|}|\ln|\frac{x}{y}||^{\beta} f(x)dx\right]dy$$
$$< M_1\left[\int_{-\infty}^{\infty} |x|^{p(1-\sigma)-1} f^p(x)dx\right]^{\frac{1}{p}}\left[\int_{-\infty}^{\infty} |y|^{q(1-\mu_1)-1} g^q(y)dy\right]^{\frac{1}{q}}; \quad (24)$$

(iii) $\mu_1 = \mu < 1$ and $\beta > 0$.
If Condition (iii) holds, then the constant factor $M_1 = k^{(1)}(\sigma)\ (\in \mathbf{R}_+)$ in (23) and (24) (for $\mu_1 = \mu$) is the best possible.

In particular, for $\lambda = 1, \sigma = \frac{1}{q}, \mu = \frac{1}{p} < 1$ in Corollary 3, we have

Corollary 4 *The following conditions are equivalent:*

(i) There exists a constant M_1, such that for any $f(x) \geq 0$, satisfying

$$0 < \int_{-\infty}^{\infty} f^p(x)dx < \infty,$$

we have the following inequality:

$$\left\{\int_{-\infty}^{\infty}\left[\int_{-|y|}^{|y|}\frac{|\ln|x/y||^{\beta}}{|x-y|}f(x)dx\right]^p dy\right\}^{\frac{1}{p}} < M_1\left(\int_{-\infty}^{\infty} f^p(x)dx\right)^{\frac{1}{p}}. \tag{25}$$

(ii) There exists a constant M_1, *such that for any* $f(x), g(y) \geq 0$,

$$0 < \int_{-\infty}^{\infty} f^p(x)dx < \infty,$$

and

$$0 < \int_{-\infty}^{\infty} g^q(y)dy < \infty,$$

we have the following inequality:

$$\int_{-\infty}^{\infty} g(y)\left[\int_{-|y|}^{|y|}\frac{|\ln|x/y||^{\beta}}{|x-y|}f(x)dx\right]dy$$
$$< M_1\left(\int_{-\infty}^{\infty} f^p(x)dx\right)^{\frac{1}{p}}\left(\int_{-\infty}^{\infty} g^q(y)dy\right)^{\frac{1}{q}}. \tag{26}$$

(iii) $\beta > 0$.
If Condition (iii) is satisfied, then the constant factor $M_1 = k^{(1)}(\frac{1}{q}) (\in \mathbf{R}_+)$ *in (25) and (26) is the best possible.*

Remark 2

(i) For $\sigma_1 = \sigma = \lambda + 1$ in (14), we have the following inequality with the best possible constant factor $\frac{\Gamma(\beta+1)}{2^{\beta}}\zeta(\beta+1)\ (\beta > 0)$:

$$\left\{\int_{-\infty}^{\infty}|y|^{p(\lambda+1)-1}\left[\int_{\frac{-1}{|y|}}^{\frac{1}{|y|}}\frac{(\min\{|xy|,1\})^{1-\lambda}}{|xy-1|}|\ln|xy||^{\beta}f(x)dx\right]^p dy\right\}^{\frac{1}{p}}$$
$$< \frac{\Gamma(\beta+1)}{2^{\beta}}\zeta(\beta+1)\left[\int_{-\infty}^{\infty}|x|^{-p\lambda-1}f^p(x)dx\right]^{\frac{1}{p}}. \tag{27}$$

(ii) For $\mu_1 = \mu = -1$ in (23), we have the following inequality with the best possible constant factor $\frac{\Gamma(\beta+1)}{2^{\beta}}\zeta(\beta+1)\ (\beta > 0)$:

$$\left\{\int_{-\infty}^{\infty} y^{-p-1}\left[\int_{-|y|}^{|y|} \frac{(\min\{|x|,|y|\})^{1-\lambda}}{|x-y|}|\ln|\frac{x}{y}||^{\beta} f(x)dx\right]^p dy\right\}^{\frac{1}{p}}$$

$$< \frac{\Gamma(\beta+1)}{2^{\beta}}\zeta(\beta+1)\left[\int_{-\infty}^{\infty} |x|^{-p\lambda-1} f^p(x)dx\right]^{\frac{1}{p}}. \tag{28}$$

(iii) For $\beta = 1$ in (25), we have the following inequality with the best possible constant factor $\frac{1}{2}\zeta(2, \frac{1}{2q})$:

$$\left\{\int_{-\infty}^{\infty}\left[\int_{-|y|}^{|y|} \frac{|\ln|x/y||}{|x-y|} f(x)dx\right]^p dy\right\}^{\frac{1}{p}} < \frac{1}{2}\zeta\left(2, \frac{1}{2q}\right)\left(\int_{-\infty}^{\infty} f^p(x)dx\right)^{\frac{1}{p}}. \tag{29}$$

Similarly, in view of Lemma 2, we obtain the following weight function: For $y \neq 0$,

$$\omega_2(\sigma, y) := |y|^{\sigma} \int_{\{x;|x|\geq \frac{1}{|y|}\}} \frac{(\min\{|xy|, 1\})^{1-\lambda}}{|xy-1|}|\ln|xy||^{\beta}|x|^{\sigma-1}dx$$

$$= \int_1^{\infty} (H(-u) + H(u))u^{\sigma-1}du = k^{(2)}(\sigma).$$

Similarly, we also have:

Theorem 2 *If $\sigma_1 \in \mathbf{R}$, then the following conditions are equivalent:*

(i) There exists a constant M_2, such that for any $f(x) \geq 0$, satisfying

$$0 < \int_{-\infty}^{\infty} |x|^{p(1-\sigma)-1} f^p(x)dx < \infty,$$

we have the following Hardy-type integral inequality of the second kind with the nonhomogeneous kernel:

$$\left\{\int_{-\infty}^{\infty} y^{p\sigma_1-1}\left[\int_{\{x;|x|\geq \frac{1}{|y|}\}} \frac{(\min\{|xy|, 1\})^{1-\lambda}}{|xy-1|}|\ln|xy||^{\beta} f(x)dx\right]^p dy\right\}^{\frac{1}{p}}$$

$$< M_2\left[\int_{-\infty}^{\infty} |x|^{p(1-\sigma)-1} f^p(x)dx\right]^{\frac{1}{p}}. \tag{30}$$

(ii) There exists a constant M_2, such that for any $f(x), g(y) \geq 0$,

$$0 < \int_{-\infty}^{\infty} |x|^{p(1-\sigma)-1} f^p(x)dx < \infty,$$

and

$$0<\int_{-\infty}^{\infty}|y|^{q(1-\sigma_1)-1}g^q(y)dy<\infty,$$

we have the following inequality:

$$\int_{-\infty}^{\infty}g(y)\left[\int_{\{x;|x|\geq\frac{1}{|y|}\}}\frac{(\min\{|xy|,1\})^{1-\lambda}}{|xy-1|}|\ln|xy||^{\beta}f(x)dx\right]dy$$

$$<M_2\left[\int_{-\infty}^{\infty}|x|^{p(1-\sigma)-1}f^p(x)dx\right]^{\frac{1}{p}}\left[\int_{-\infty}^{\infty}|y|^{q(1-\sigma_1)-1}g^q(y)dy\right]^{\frac{1}{q}}. \quad (31)$$

(*iii*) $\sigma_1=\sigma<1$ *and* $\beta>0$.
If Condition (iii) is satisfied, then the constant factor $M_2=k^{(2)}(\sigma)$ $(\in\mathbf{R}_+)$ *in (30) and (31) (for* $\sigma_1=\sigma$*) is the best possible.*

In particular, for $\sigma=\sigma_1=\frac{1}{p}<1$ in Theorem 2, we have the following:

Corollary 5 *The following conditions are equivalent:*

(*i*) *There exists a constant* M_2*, such that for any* $f(x)\geq 0$*, satisfying*

$$0<\int_{-\infty}^{\infty}|x|^{p-2}f^p(x)dx<\infty,$$

we have the following inequality:

$$\left\{\int_{-\infty}^{\infty}\left[\int_{\{x;|x|\geq\frac{1}{|y|}\}}\frac{(\min\{|xy|,1\})^{1-\lambda}}{|xy-1|}|\ln|xy||^{\beta}f(x)dx\right]^p dy\right\}^{\frac{1}{p}}$$

$$<M_2\left(\int_{-\infty}^{\infty}|x|^{p-2}f^p(x)dx\right)^{\frac{1}{p}}. \quad (32)$$

(*ii*) *There exists a constant* M_2*, such that for any* $f(x),g(y)\geq 0$*,*

$$0<\int_{-\infty}^{\infty}|x|^{p-2}f^p(x)dx<\infty,$$

and

$$0<\int_{-\infty}^{\infty}g^q(y)dy<\infty,$$

we have the following inequality:

$$\int_{-\infty}^{\infty} g(y) \left[\int_{\{x;|x| \geq \frac{1}{|y|}\}} \frac{(\min\{|xy|, 1\})^{1-\lambda}}{|xy-1|} |\ln|xy||^{\beta} f(x)dx \right] dy$$

$$< M_2 \left(\int_{-\infty}^{\infty} |x|^{p-2} f^p(x)dx \right)^{\frac{1}{p}} \left(\int_{-\infty}^{\infty} g^q(y)dy \right)^{\frac{1}{q}}. \tag{33}$$

(iii) $\beta > 0$.
If Condition (iii) is satisfied, then the constant factor $M_2 = k^{(2)}(\frac{1}{p}) (\in \mathbf{R}_+)$ *in (32) and (33) is the best possible.*

Setting

$$y = \frac{1}{Y}, \ G(Y) = g\left(\frac{1}{Y}\right)\frac{1}{Y^2}$$

in Theorem 2, and then replacing Y by y, we have:

Corollary 6 *If* $\sigma_1 \in \mathbf{R}$, *then the following conditions are equivalent:*

(i) There exists a constant M_2, *such that for any* $f(x) \geq 0$, *satisfying*

$$0 < \int_{-\infty}^{\infty} |x|^{p(1-\sigma)-1} f^p(x)dx < \infty,$$

we have the following inequality:

$$\left\{ \int_{-\infty}^{\infty} y^{-p\sigma_1 - 1} \left[\int_{\{x;|x| \geq |y|\}} \frac{(\min\{|x/y|, 1\})^{1-\lambda}}{|x/y-1|} |\ln|x/y||^{\beta} f(x)dx \right]^p dy \right\}^{\frac{1}{p}}$$

$$< M_2 \left[\int_{-\infty}^{\infty} |x|^{p(1-\sigma)-1} f^p(x)dx \right]^{\frac{1}{p}}. \tag{34}$$

(ii) There exists a constant M_2, *such that for any* $f(x), G(y) \geq 0$,

$$0 < \int_{-\infty}^{\infty} |x|^{p(1-\sigma)-1} f^p(x)dx < \infty,$$

and

$$0 < \int_{-\infty}^{\infty} |y|^{q(1+\sigma_1)-1} G^q(y)dy < \infty,$$

we have the following inequality:

$$\int_{-\infty}^{\infty} G(y) \left[\int_{\{x;|x| \geq |y|\}} \frac{(\min\{|x/y|, 1\})^{1-\lambda}}{|x/y-1|} |\ln|x/y||^{\beta} f(x)dx \right] dy$$

$$< M_2 \left[\int_{-\infty}^{\infty} |x|^{p(1-\sigma)-1} f^p(x) dx \right]^{\frac{1}{p}} \left[\int_{-\infty}^{\infty} |y|^{q(1+\sigma_1)-1} G^q(y) dy \right]^{\frac{1}{q}}. \quad (35)$$

(iii) $\sigma_1 = \sigma < 1$ *and* $\beta > 0$.
If Condition (iii) is satisfied, then the constant factor $M_2 = k^{(2)}(\sigma)$ $(\in \mathbf{R}_+)$ *in (34) and (35) (for* $\sigma_1 = \sigma$*) is the best possible.*

For $g(y) = y^{\lambda} G(y)$ and $\mu_1 = \lambda - \sigma_1$ in Corollary 6, we have:

Corollary 7 *If* $\mu_1 \in \mathbf{R}$, *then the following conditions are equivalent:*

(i) There exists a constant M_2, *such that for any* $f(x) \geq 0$, *satisfying*

$$0 < \int_{-\infty}^{\infty} |x|^{p(1-\sigma)-1} f^p(x) dx < \infty,$$

we have the following inequality:

$$\left\{ \int_{-\infty}^{\infty} y^{p\mu_1 - 1} \left[\int_{\{x;|x| \geq |y|\}} \frac{(\min\{|x|, |y|\})^{1-\lambda}}{|x-y|} |\ln |x/y||^{\beta} f(x) dx \right]^p dy \right\}^{\frac{1}{p}}$$

$$< M_2 \left[\int_{-\infty}^{\infty} |x|^{p(1-\sigma)-1} f^p(x) dx \right]^{\frac{1}{p}}. \quad (36)$$

(ii) There exists a constant M_2, *such that for any* $f(x), g(y) \geq 0$,

$$0 < \int_{-\infty}^{\infty} |x|^{p(1-\sigma)-1} f^p(x) dx < \infty,$$

and

$$0 < \int_{-\infty}^{\infty} |y|^{q(1-\mu_1)-1} g^q(y) dy < \infty,$$

we have the following inequality:

$$\int_{-\infty}^{\infty} g(y) \left[\int_{\{x;|x| \geq |y|\}} \frac{(\min\{|x|, |y|\})^{1-\lambda}}{|x-y|} |\ln |x/y||^{\beta} f(x) dx \right] dy$$

$$< M_2 \left[\int_{-\infty}^{\infty} |x|^{p(1-\sigma)-1} f^p(x) dx \right]^{\frac{1}{p}} \left[\int_{-\infty}^{\infty} |y|^{q(1-\mu_1)-1} g^q(y) dy \right]^{\frac{1}{q}}. \quad (37)$$

(iii) $\mu_1 = \mu > \lambda - 1$ *and* $\beta > 0$.
If Condition (iii) holds true, then the constant factor $M_2 = k^{(2)}(\sigma)$ $(\in \mathbf{R}_+)$ *in (36) and (37) (for* $\mu_1 = \mu$*) is the best possible.*

In particular, for $\lambda = 1, \sigma = \frac{1}{q}(< 1), \mu = \frac{1}{p}$ in Corollary 7, we have:

Corollary 8 *The following conditions are equivalent:*

(i) There exists a constant M_2, such that for any $f(x) \geq 0$, satisfying

$$0 < \int_{-\infty}^{\infty} f^p(x)dx < \infty,$$

we have the following inequality:

$$\left\{\int_{-\infty}^{\infty} \left[\int_{\{x;|x|\geq|y|\}} \frac{|\ln|x/y||^\beta}{|x-y|} f(x)dx\right]^p dy\right\}^{\frac{1}{p}} < M_2 \left(\int_{-\infty}^{\infty} f^p(x)dx\right)^{\frac{1}{p}}. \tag{38}$$

(ii) There exists a constant M_2, such that for any $f(x), g(y) \geq 0$,

$$0 < \int_{-\infty}^{\infty} f^p(x)dx < \infty,$$

and

$$0 < \int_{-\infty}^{\infty} g^q(y)dy < \infty,$$

we have the following inequality:

$$\int_{-\infty}^{\infty} g(y) \left[\int_{\{x;|x|\geq|y|\}} \frac{|\ln|x/y||^\beta}{|x-y|} f(x)dx\right] dy < M_2 \left(\int_{-\infty}^{\infty} f^p(x)dx\right)^{\frac{1}{p}} \left(\int_{-\infty}^{\infty} g^q(y)dy\right)^{\frac{1}{q}}. \tag{39}$$

(iii) $\beta > 0$.

If Condition (iii) holds true, then the constant factor $M_2 = k^{(2)}(\frac{1}{q}) (\in \mathbf{R}_+)$ in (38) and (39) is the best possible.

Remark 3

(i) For $\sigma_1 = \sigma = -1$ in (30), we have the following inequality with the best possible constant factor $\frac{\Gamma(\beta+1)}{2^\beta}\zeta(\beta+1)$ $(\beta > 0)$:

$$\left\{\int_{-\infty}^{\infty} y^{-p-1} \left[\int_{\{x;|x|\geq\frac{1}{|y|}\}} \frac{(\min\{|x/y|, 1\})^{1-\lambda}}{|x/y-1|} |\ln|x/y||^\beta f(x)dx\right]^p dy\right\}^{\frac{1}{p}}$$

$$< \frac{\Gamma(\beta+1)}{2^{\beta}}\zeta(\beta+1)\left[\int_{-\infty}^{\infty}|x|^{2p-1}f^{p}(x)dx\right]^{\frac{1}{p}}. \tag{40}$$

(ii) For $\mu_1 = \mu = \lambda + 1$ in (36), we have the following inequality with the best possible constant factor $\frac{\Gamma(\beta+1)}{2^{\beta}}\zeta(\beta+1)$ $(\beta > 0)$:

$$\left\{\int_{-\infty}^{\infty} y^{p(\lambda+1)-1}\left[\int_{\{x;|x|\geq|y|\}}\frac{(\min\{|x|,|y|\})^{1-\lambda}}{|x-y|}|\ln|x/y||^{\beta}f(x)dx\right]^{p}dy\right\}^{\frac{1}{p}}$$

$$< \frac{\Gamma(\beta+1)}{2^{\beta}}\zeta(\beta+1)\left[\int_{-\infty}^{\infty}|x|^{2p-1}f^{p}(x)dx\right]^{\frac{1}{p}}. \tag{41}$$

(iii) For $\beta = 1$ in (38), we have the following inequality with the best possible constant factor $\frac{1}{2}\zeta(2, \frac{1}{2p})$:

$$\left\{\int_{-\infty}^{\infty}\left[\int_{\{x;|x|\geq|y|\}}\frac{|\ln|x/y||}{|x-y|}f(x)dx\right]^{p}dy\right\}^{\frac{1}{p}}$$

$$< \frac{1}{2}\zeta\left(2,\frac{1}{2p}\right)\left(\int_{-\infty}^{\infty}f^{p}(x)dx\right)^{\frac{1}{p}}. \tag{42}$$

4 Conclusions

In this chapter, by the use of weight functions, a few equivalent conditions of two kinds of Hardy-type integral inequalities with multi-parameters in the whole plane are obtained in Theorems 1 and 2. The constant factors related to the extended Riemann-zeta function are proved to be the best possible. In the form of applications, a few equivalent conditions of two kinds of Hardy-type integral inequalities in the whole plane are deduced in Corollaries 3 and 7. We also consider some particular cases in Corollaries 1, 4, 5, 8, Remarks 2 and 3. In our investigation, methods of real analysis are essential and play a key role for the proof of the equivalent inequalities with the corresponding best possible constant factors. The lemmas and theorems provide an extensive account of these types of inequalities.

Acknowledgments B. Yang: This work was supported by the National Natural Science Foundation (No. 61772140) and the Science and Technology Planning Project Item of Guangzhou City (No. 201707010229). We are grateful for their support.

References

1. G.H. Hardy, J.E. Littlewood, G. Pólya, Inequalities, Cambridge University Press, Cambridge, USA, 1934.
2. B.C. Yang, The Norm of Operator and Hilbert-type Inequalities, Science Press, Beijing, China, 2009.
3. B.C. Yang, Hilbert-Type Integral Inequalities, Bentham Science Publishers Ltd., The United Arab Emirates, 2009
4. B.C. Yang, On the norm of an integral operator and applications, J. Math. Anal. Appl., **321**(2006), 182–192.
5. J.S. Xu, Hardy-Hilbert's inequalities with two parameters, Advances in Mathematics, **36**(2), 63–76.
6. B.C. Yang, On the norm of a Hilbert's type linear operator and applications, J. Math. Anal. Appl., **325**(2007), 529–541.
7. D.M. Xin, A Hilbert-type integral inequality with the homogeneous kernel of zero degree, Mathematical Theory and Applications, **30**(2)(2010), 70–74.
8. B.C. Yang, A Hilbert-type integral inequality with the homogenous kernel of degree 0, Journal of Shandong University (Natural Science), **45**(2)(2010), 103–106.
9. L. Debnath, B.C. Yang, Recent developments of Hilbert-type discrete and integral inequalities with applications, International Journal of Mathematics and Mathematical Sciences, Volume 2012, Article ID 871845, 29 pages.
10. Y. Hong, On the structure character of Hilbert's type integral inequality with homogeneous kernal and applications, Journal of Jilin University (Science Edition), 2017, 55(2), 189–194.
11. M.Th. Rassias, B.C. Yang, On half-discrete Hilbert's inequality. Applied Mathematics and Computation, **220**(2013), 75–93.
12. B.C. Yang, M. Krnic, A half-discrete Hilbert-type inequality with a general homogeneous kernel of degree 0. Journal of Methematical Inequalities, **6**(3)(2012), 401–417.
13. B.C. Yang, An Extended Multidimensional Half-Discrete Hardy-Hilbert Type Inequality with a General Homogeneous Kernel, in "Differential and Integral Inequalities", D. Andrica, Th.M. Rassias Eds., Springer, 2020, 831–854.
14. Th.M. Rassias, B.C. Yang, A multidimensional half - discrete Hilbert - type inequality and the Riemann zeta function, Applied Mathematics and Computation, **225**(2013), 263–277.
15. M.Th. Rassias, B.C. Yang, On a multidimensional half - discrete Hilbert - type inequality related to the hyperbolic cotangent function. Applied Mathematics and Computation, **242**(2013), 800–813.
16. M.Th. Rassias, B.C. Yang, A multidimensional Hilbert - type integral inequality related to the Riemann zeta function, Applications of Mathematics and Informatics in Science and Engineering (N. J. Daras, ed.), Springer, New York, 417–433, 2014.
17. Q. Chen, B.C. Yang, A survey on the study of Hilbert-type inequalities. Journal of Inequalities and Applications (2015), 2015:302.
18. M.Th. Rassias, B.C. Yang, A. Raigorodskii, On the Reverse Hardy-Type Integral Inequalities in the Whole Plane with the Extended Riemann-Zeta Function, Journal of Mathematical Inequalities, **14**(2)(2020), 525–546.
19. M.Th. Rassias, B.C. Yang, A. Raigorodskii, Two kinds of the reverse Hardy-type integral inequalities with the equivalent forms related to the extended Riemann zeta function. Applicable Analysis and Discrete Mathematics, **12**(2) (2018),273–296.
20. B.C. Yang, A new Hilbert-type integral inequality, Soochow Journal of Mathematics, **33**(4)(2007), 849–859.
21. Z.Q. Wang, D.R. Guo, Introduction to Special Functions, Science Press, Beijing, China, 1979.
22. B. He, B.C. Yang, On a Hilbert-type integral inequality with the homogeneous kernel of 0-degree and the hypergeometrc function, Mathematics in Practice and Theory, **40**(18)(2010), 105–211.

23. B.C. Yang, A new Hilbert-type integral inequality with some parameters, Journal of Jilin University (Science Edition), **46**(6)(2008), 1085–1090.
24. B.C. Yang, A Hilbert-type integral inequality with a non-homogeneous kernel, Journal of Xiamen University (Natural Science), **48**(2)(2008), 165–169.
25. Z. Zeng, Z.T. Xie, On a new Hilbert-type integral inequality with the homogeneous kernel of degree 0 and the integral in whole plane, Journal of Inequalities and Applications, Vol. 2010, Article ID 256796, 9 pages.
26. A.Z. Wang, B.C. Yang, A new Hilbert-type integral inequality in whole plane with the non-homogeneous kernel, Journal of Inequalities and Applications, Vol. 2011, 2011: 123.
27. D.M. Xin, B.C. Yang, A Hilbert-type integral inequality in whole plane with the homogeneous kernel of degree -2, Journal of Inequalities and Applications, Vol. 2011, Article ID 401428, 11 pages.
28. B. He, B.C. Yang, On an inequality concerning a non-homogeneous kernel and the hypergeometric function, Tamsul Oxford Journal of Information and Mathematical Sciences, **27**(1)(2011), 75–88.
29. B.C. Yang, A reverse Hilbert-type integral inequality with a non-homogeneous kernel, Journal of Jilin University (Science Edition), **49**(3)(2011), 437–441.
30. Z.T. Xie, Z. Zeng, Y.F. Sun, A new Hilbert-type inequality with the homogeneous kernel of degree -2, Advances and Applications in Mathematical Sciences, **12**(7)(2013), 391–401.
31. Q.L. Huang, S.H. Wu, B.C. Yang, Parameterized Hilbert-type integral inequalities in the whole plane, The Scientific World Journal, Volume 2014, Article ID 169061, 8 pages.
32. Z. Zhen, K. Raja Rama Gandhi, Z.T. Xie, A new Hilbert-type inequality with the homogeneous kernel of degree -2 and with the integral, Bulletin of Mathematical Sciences & Applications, **3**(1)(2014), 11–20.
33. M.Th. Rassias, B.C. Yang, A Hilbert - type integral inequality in the whole plane related to the hyper geometric function and the beta function, Journal of Mathematical Analysis and Applications, 428(2): 1286 - 1308 (2015).
34. X.Y. Huang, J.F. Cao, B. He, B.C. Yang, Hilbert-type and Hardy-type integral inequalities with operator expressions and the best constants in the whole plane. Journal of Inequalities and Applications (2015), 2015:129.
35. Z.H Gu, B.C. Yang, A Hilbert-type integral inequality in the whole plane with a non-homogeneous kernel and a few parameters. Journal of Inequalities and Applications (2015), 2015:314.
36. M.Th. Rassias, B.C. Yang, A reverse Mulholland-type inequality in the whole plane with multi-parameters. Applicable Analysis and Discrete Mathematics, **13**(1)(2019),290–308.
37. J.C. Kuang, Applied inequalities. Shangdong Science and Technology Press, Jinan, China, 2004.
38. J.C. Kuang, Introduction to Real Analysis. Hunan Educiton Press, Changsha, China, 1996.

Product Formulae for Non-Autonomous Gibbs Semigroups

Valentin A. Zagrebnov

Abstract We consider linear evolution corresponding to non-autonomous Gibbs semigroup on a separable Hilbert space $\mathfrak{H}$. It is shown that evolution family $\{U(t,s)\}_{0\leq s\leq t\leq T}$ solving the non-autonomous Cauchy problem can be approximated in the *trace-norm* topology by product formulae. The rate of convergence of product formulae approximants $\{U_n(t,s)\}_{\{0\leq s<t\leq T,\, n\geq 1\}}$ to the solution operator $\{U(t,s)\}_{\{0\leq s<t\leq T\}}$ is also established.

1 Introduction and Main Result

We study linear dynamics of a *non-autonomous* perturbation of Gibbs semigroups. Recall that they are strongly continuous semigroups (that is, C_0-semigroups) on a separable Hilbert space $\mathfrak{H}$ with values in the trace-class operators $\mathcal{C}_1(\mathfrak{H})$, [20].

The aim of this note is to prove the convergence of the *product formulae approximants* to the corresponding to this dynamics *solution operator* $\{U(t,s)\}_{\{0\leq s\leq t\}}$, known also as evolution family, fundamental solution, or propagator, see [1] Ch.VI, Sec.9, in topology of the trace-class ideal $\mathcal{C}_1(\mathfrak{H})$.

To this end we consider a linear non-autonomous dynamics given on a separable Hilbert space $\mathfrak{H}$ by evolution equation of the type:

$$\begin{aligned} &\frac{\partial u(t)}{\partial t} = -\,C(t)u(t), \quad u(s) = u_s, \quad s \in [0,t) \subset \mathbb{R}_0^+, \qquad t \in \mathcal{I} := [0,T], \\ &C(t) := A + B(t), \quad u_s \in \mathfrak{H}, \end{aligned} \tag{1.1}$$

where $\mathbb{R}_0^+ = \{0\} \cup \mathbb{R}^+$ and linear operator A is generator of the Gibbs semigroup. Note that for the autonomous Cauchy problem (ACP), when $B(t) = B$ in (1.1), the

V. A. Zagrebnov (✉)
Institut de Mathématiques de Marseille (UMR 7373)—AMU, Centre de Mathématiques et Informatique—Technopôle Château-Gombert, Marseille, France
e-mail: Valentin.Zagrebnov@univ-amu.fr

I. N. Parasidis et al. (eds.), *Mathematical Analysis in Interdisciplinary Research*, Springer Optimization and Its Applications 179,
https://doi.org/10.1007/978-3-030-84721-0_40

outlined programme corresponds to the Trotter product formula approximation of the Gibbs semigroup generated by a closure of operator $A + B$, [20, Ch.5].

The main result of the present note concerns the non-autonomous Cauchy problem (nACP) (1.1) under the following:

Assumptions

(A1) The operator $A \geq \mathbb{1}$ in a separable Hilbert space $\mathfrak{H}$ is self-adjoint, and $\{B(t)\}_{t\in\mathcal{I}}$ is a family of non-negative self-adjoint operators in $\mathfrak{H}$.

(A2) There exists $\alpha \in [0, 1)$ such that inclusion: $\mathrm{dom}(A^{\alpha}) \subseteq \mathrm{dom}(B(t))$, holds for a.a. $t \in \mathcal{I}$. Moreover, the function $t \mapsto B(t)A^{-\alpha} \in \mathcal{L}(\mathfrak{H})$, $t \in \mathcal{I}$ is strongly measurable and essentially bounded in the operator norm:

$$C_{\alpha} := \operatorname*{ess\,sup}_{t\in\mathcal{I}} \|B(t)A^{-\alpha}\| < \infty. \tag{1.2}$$

(A3) The map $t \mapsto A^{-\alpha}B(t)A^{-\alpha} \in \mathcal{L}(\mathfrak{H})$, $t \in \mathcal{I}$, is Hölder continuous in the operator norm: that is, for some $\beta \in (0, 1]$, there is a constant $L_{\alpha,\beta} > 0$ such that

$$\|A^{-\alpha}(B(t) - B(s))A^{-\alpha}\| \leq L_{\alpha,\beta}|t - s|^{\beta}, \quad (t, s) \in \mathcal{I} \times \mathcal{I}. \tag{1.3}$$

(A4) The operator A is a generator of the Gibbs semigroup $\{G(t) = e^{-tA}\}_{t\geq 0}$, which is a C_0-semigroup such that $G(t)|_{t>0} \in \mathcal{C}_1(\mathfrak{H})$. Here $\mathcal{C}_1(\mathfrak{H})$ denotes the $*$-ideal of *trace-class* operators of bounded operators $\mathcal{L}(\mathfrak{H})$.

Remark 1.1 Assumptions *(A1)–(A3)* are introduced in *[4]* to prove the *operator-norm* convergence of the product formula approximants: $\{U_n(t, s)\}_{0\leq s\leq t}$, to solution operator $\{U(t, s)\}_{0\leq s\leq t}$. Then they were widely used for product formula approximations in [10–15] in the context of the *evolution semigroup* approach to the nACP, see [6–9].

Remark 1.2 The following main facts were established *(e.g., [4, 7, 18, 19])* about the nACP for perturbed evolution equation of the type (1.1)*:*

(a) Because of assumptions *(A1)–(A2)* the operators $\{C(t) = A + B(t)\}_{t\in\mathcal{I}}$ have a common $\mathrm{dom}(C(t)) = \mathrm{dom}(A)$ and they are generators of contraction holomorphic semigroups. Hence, the nACP (1.1) is of *parabolic* type *[5, 16].*

(b) Since domains $\mathrm{dom}(C(t)) = \mathrm{dom}(A)$, $t \geq 0$, are dense, the nACP is *well-posed* with time-independent *regularity* subspace $\mathrm{dom}(A)$.

(c) Assumptions *(A1)–(A3)* provide the existence of *evolution family* solving nACP (1.1) which we call the *solution operator.*

It is a strongly continuous, uniformly bounded family of operators $\{U(t, s)\}_{(t,s)\in\Delta}$, $\Delta := \{(t, s) \in \mathcal{I} \times \mathcal{I} : 0 \leq s \leq t \leq T\}$, such that the conditions

$$\begin{aligned} U(t, t) &= \mathbb{1} \quad \text{for} \quad t \in \mathcal{I}, \\ U(t, r)U(r, s) &= U(t, s), \quad \text{for}, \quad t, r, s \in \mathcal{I} \quad \text{for} \quad s \leq r \leq t, \end{aligned} \tag{1.4}$$

are satisfied and $u(t) = U(t,s)u_s$ for any $u_s \in \mathfrak{H}_s$ is in a certain sense (e.g., *classical, strict, mild*) solution of the nACP (1.1).

(*d*) Here $\mathfrak{H}_s \subseteq \mathfrak{H}$ is an appropriate *regularity subspace* of initial data. Assumptions *(A1)–(A3)* provide that $\mathfrak{H}_s = \mathrm{dom}(A)$ and $U(t,s)\mathfrak{H} \subseteq \mathrm{dom}(A)$ for $t > s$.

In the present note we essentially focus on convergence of the product *approximants* $\{U_n(t,s)\}_{(t,s)\in\Delta, n\geq 1}$ to solution operator $\{U(t,s)\}_{(t,s)\in\Delta}$. Let

$$s = t_1 < t_2 < \ldots < t_{n-1} < t_n < t, \quad t_k := s + (k-1)\tfrac{t-s}{n}, \tag{1.5}$$

for $k \in \{1,2,\ldots,n\}$, $n \in \mathbb{N}$, be partition of the interval $[s,t]$. Then the corresponding approximants may be defined as follows:

$$\begin{aligned} W_k^{(n)}(t,s) :=& e^{-\frac{t-s}{n}A} e^{-\frac{t-s}{n}B(t_k)}, \quad k = 1,2,\ldots,n, \\ U_n(t,s) :=& W_n^{(n)}(t,s) W_{n-1}^{(n)}(t,s) \times \cdots \times W_2^{(n)}(t,s) W_1^{(n)}(t,s). \end{aligned} \tag{1.6}$$

It turns out that if the assumptions (A1)–(A3), *adapted* to a Banach space $\mathfrak{X}$, are satisfied for $\alpha \in (0,1)$, $\beta \in (0,1)$ and in addition the condition $\alpha < \beta$ holds, then solution operator $\{U(t,s)\}_{(t,s)\in\Delta}$ admits the operator-norm approximation

$$\operatorname*{ess\,sup}_{(t,s)\in\Delta} \|U_n(t,s) - U(t,s)\| \leq \frac{R_{\beta,\alpha}}{n^{\beta-\alpha}}, \quad n \in \mathbb{N}, \tag{1.7}$$

for some constant $R_{\beta,\alpha} > 0$. This result shows that convergence of the approximants $\{U_n(t,s)\}_{(t,s)\in\Delta,n\geq 1}$ is determined by smoothness of the perturbation $B(\cdot)$ in (A3) and by the parameter of inclusion in (A2), see [13].

In [10] the Lipschitz case $\beta = 1$ was examined in Banach space $\mathfrak{X}$. There it was shown that if $\alpha \in (1/2, 1)$, then one gets estimate

$$\operatorname*{ess\,sup}_{t\in\mathcal{I}} \|U_n(t,s) - U(t,s)\| \leq \frac{R_{1,\alpha}}{n^{1-\alpha}}, \quad n = 2,3,\ldots. \tag{1.8}$$

For the Lipschitz case in Hilbert space $\mathfrak{H}$, the assumptions (A1)–(A3) yield a stronger result [4]:

$$\operatorname*{ess\,sup}_{(t,s)\in\Delta} \|U_n(t,s) - U(t,s)\| \leq R\,\frac{\log(n)}{n}, \quad n = 2,3,\ldots. \tag{1.9}$$

Note that actually it is the best of known estimates for operator-norm rates of convergence under conditions (A1)–(A3).

The estimate (1.7) was improved in [12] for $\alpha \in (1/2, 1)$ in a Hilbert space using the *evolution semigroup* approach [2, 3, 9]. This approach is quite different from technique used for (1.9) in [4], but it is the same as that employed in [10].

Proposition 1.3 ([12]) *Let assumptions* (A1)–(A3) *be satisfied for* $\beta \in (0,1)$. *If* $\beta > 2\alpha - 1 > 0$, *then estimate*

$$\operatorname*{ess\,sup}_{(t,s)\in\Delta} \|U_n(t,s) - U(t,s)\| \leq \frac{R_\beta}{n^\beta} \tag{1.10}$$

holds for $n \in \mathbb{N}$ *and for some constant* $R_\beta > 0$.

Note that the condition $\beta > 2\alpha - 1$ is weaker than $\beta > \alpha$ (1.7), but it does not cover the Lipschitz case (1.8) because of condition $\beta < 1$.

The main result of the paper is the *raising* of the known *operator-norm* bounds (1.7)–(1.10) (we denote them by $R_{\alpha,\beta}\varepsilon_{\alpha,\beta}(n)$) to estimates in the *trace-norm* topology $\|\cdot\|_1$. This is a subtle matter even for ACP, see [20] Ch.5.4:

- The first step is construction for nACP (1.1) a *trace-norm* continuous solution operator $\{U(t,s)\}_{(t,s)\in\Delta}$, see Theorem 2.3 and Corollary 2.4.
- Then in Sect. 3 for assumptions (A1)–(A4) we prove (Theorem 1.4) the corresponding *trace-norm* estimate $R_{\alpha,\beta}(t,s)\varepsilon_{\alpha,\beta}(n)$ for difference $\|U_n(t,s) - U(t,s)\|_1$.

Theorem 1.4 *Let assumptions* (A1)–(A4) *be satisfied. Then the trace-norm estimate*

$$\|U_n(t,s) - U(t,s)\|_1 \leq R_{\alpha,\beta}(t,s)\varepsilon_{\alpha,\beta}(n) \tag{1.11}$$

holds for $n \in \mathbb{N}$ *and* $0 \leq s < t \leq T$ *for some* $R_{\alpha,\beta}(t,s) > 0$.

2 Preliminaries

Besides Remark 1.2(a)–(d), we also remind the following assertion, and we refer to [16] Theorem 1, and [17] Theorem 5.2.1.

Proposition 2.1 *Let assumptions* (A1)–(A3) *be satisfied:*

(a) *Then solution operator* $\{U(t,s)\}_{(t,s)\in\Delta}$ *is a family of strongly continuously differentiable contractions for* $0 \leq s < t \leq T$ *and*

$$\partial_t U(t,s) = -(A + B(t))U(t,s). \tag{2.1}$$

(b) *Moreover, the unique function* $t \mapsto u(t) = U(t,s)\,u_s$ *is a* classical *solution of nACP* (1.1) *for initial data* $\mathfrak{H}_s = \mathrm{dom}(A)$.

Note that solution of nACP (1.1) is called *classical* if $u(t) \in C([0,T],\mathfrak{H}) \cap C^1([0,T],\mathfrak{H})$, $u(t) \in \mathrm{dom}(C(t))$, $u(s) = u_s$, and $C(t)u(t) \in C([0,T],\mathfrak{H})$ for all $t \geq s$, with convention that $(\partial_t u)(s)$ is the right derivative, see [16] Theorem 1, or [1] Ch.VI.9.

Since involved into (A1), (A2) operators are non-negative and self-adjoint, Eq. (2.1) implies that the solution operator consists of *contractions*:

$$\partial_t \|U(t,s)u\|^2 = -2(C(t)U(t,s)u, U(t,s)u) \leq 0, \quad \text{for } u \in \mathfrak{H}. \tag{2.2}$$

On account of (A1) the contraction semigroup generated by A is holomorphic. Therefore, $G(t)|_{t>0} = e^{-tA} : \mathfrak{H} \to \text{dom}(A)$, which by Remark 1.2(d) is a regularity subspace for solution operator $\{U(t,s)\}_{(t,s)\in\Delta}$, see Proposition 2.1(b). Then applying to (2.1) the *variation of parameter* arguments we obtain for contractions $U(t,s)$ (2.2) the integral equation:

$$U(t,s) = G(t-s) - \int_s^t d\tau\, G(t-\tau)\, B(\tau)\, U(\tau,s), \quad U(s,s) = \mathbb{1}\,. \tag{2.3}$$

As a consequence the evolution family $\{U(t,s)\}_{(t,s)\in\Delta}$, which is defined by Eq. (2.3), can be considered as a *mild* solution of the operator-valued nACP (2.1) for $0 \leq s \leq t \leq T$ on the Banach space $\mathcal{L}(\mathfrak{H})$ of bounded operators, cf. [1], Ch.VI.7.

Owing to assumptions (A1)–(A2) one gets for holomorphic (Gibbs) semigroup $\{G(t)\}_{t\geq 0}$ and for $\{B(\tau)\}_{\tau\in(s,t)}$:

$$\|A^\alpha G(t-s)\| \leq \frac{M_\alpha}{(t-s)^\alpha} \quad \text{and} \quad \|B(\tau)\, A^{-\alpha}\| \leq C_\alpha\,, \tag{2.4}$$

where $0 \leq s < t \leq T$. Then (2.4) yields estimate

$$\left\| \int_s^t d\tau\, B(\tau)\, G(t-\tau) \right\| \leq \frac{M_\alpha C_\alpha}{1-\alpha}(t-s)^{1-\alpha}\,, \quad \alpha \in [0,1)\,. \tag{2.5}$$

For that reason, we can construct solution operator $\{U(t,s)\}_{(t,s)\in\Delta}$ as uniformly *operator-norm* convergent Dyson–Phillips series $\sum_{n=0}^{\infty} S_n(t,s)$ by iteration of the integral formula (2.3) for $t > s$.

To this aim we define the recurrence relation

$$\begin{aligned} S_0(t,s) &= G(t-s), \\ S_n(t,s) &= -\int_s^t \mathrm{d}s\, G(t-\tau)\, B(\tau)\, S_{n-1}(\tau,s), \quad n \geq 1. \end{aligned} \tag{2.6}$$

Seeing that by (A2) operators $S_{n\geq 1}(t,s)$ in (2.6) are the n-fold strongly convergent Bochner integrals for $n \geq 1$ (with convention that $\tau_0 = s$ and $\tau_{n+1} = t$):

$$\begin{aligned} S_n(t,s) = \int_s^t \mathrm{d}\tau_n \int_s^{\tau_n} \mathrm{d}\tau_{n-1} \ldots \int_s^{\tau_2} \mathrm{d}\tau_1 \\ G(t-\tau_n)(-B(\tau_n))G(\tau_n-\tau_{n-1}) \cdots G(\tau_2-\tau_1)(-B(\tau_1))G(\tau_1-s)\,, \end{aligned} \tag{2.7}$$

and that semigroup $\{G(t)\}_{t\geq 0}$ is contraction, by estimate (2.5) there exists interval $[s,t]$ such that $\xi_{t,s}(n) := M_\alpha(n)C_\alpha(t-s)^{1-\alpha}/n^{(1-\alpha)} < 1$. On that account (2.7) provides estimate

$$\|S_n(t,s)\| \leq \xi^n_{t,s}(n)\,, \quad n \geq 1. \tag{2.8}$$

Consequently, the Dyson–Phillips series $\sum_{n=0}^{\infty} S_n(t,s)$ converges in the operator-norm topology uniformly in $[\mu,\nu] \subset [s,t]$ and satisfies the integral equation (2.3). Thus, we obtain bounded operator

$$U(t,s) = \sum_{n=0}^{\infty} S_n(t,s)\,, \tag{2.9}$$

which is the mild solution of nACP (2.1). It is operator-norm continuous for $\mu < \nu$ and strongly continuous at $\mu = \nu$. This result can be extended to any $0 \leq s < t \leq T$ using (1.4) and to the operator-norm differentiability for $s < t$ by making use of condition (A3), see [16] Theorem 1, or [17] Theorem 5.2.1. Then on account of (1.4) and (2.2) the solution operator (2.9) is contraction: $\|U(t,s)\| \leq 1$.

For extension of this result to the *trace-norm* topology we need to use assumption (A4) and the following preparatory lemma.

Lemma 2.2 *Let self-adjoint positive operator A be such that $e^{-tA} \in \mathcal{C}_1(\mathfrak{H})$ for $t > 0$, and let $V_1, V_2, \ldots, V_n$ be bounded operators from $\mathcal{L}(\mathfrak{H})$. Then*

$$\Big\|\prod_{j=1}^{n} V_j e^{-t_j A}\Big\|_1 \leq \prod_{j=1}^{n} \|V_j\| \|e^{-(t_1+t_2+\ldots+t_n)A/4}\|_1\,, \tag{2.10}$$

for any set $\{t_1, t_2, \ldots, t_n\}$ of positive numbers.

Proof First we prove this assertion for a set of *compact* operators: $V_j \in \mathcal{C}_\infty(\mathfrak{H})$, $j = 1, 2, \ldots, n$.

Let $t_m := \min\{t_j\}_{j=1}^n > 0$ and $T := \sum_{j=1}^n t_j > 0$. For any $1 \leq j \leq n$, we define an integer $\ell_j \in \mathbb{N}$ by condition: $2^{\ell_j} t_m \leq t_j \leq 2^{\ell_j+1} t_m$. Then we get $\sum_{j=1}^n 2^{\ell_j} t_m > T/2$ and

$$\prod_{j=1}^{n} V_j e^{-t_j A} = \prod_{j=1}^{n} V_j e^{-(t_j - 2^{\ell_j} t_m)A} (e^{-t_m A})^{\ell_j}. \tag{2.11}$$

By the definition of the $\|\cdot\|_1$-norm and by inequalities for singular values $\{s_k(\cdot)\}_{k\geq 1}$ of compact operator

$$\Big\|\prod_{j=1}^{n} V_j e^{-t_j A}\Big\|_1 = \sum_{k=1}^{\infty} s_k\Big(\prod_{j=1}^{n} V_j e^{-(t_j - 2^{\ell_j} t_m)A} (e^{-t_m A})^{2^{\ell_j}}\Big)$$

$$\leq \sum_{k=1}^{\infty} \prod_{j=1}^{n} s_k \left(e^{-(t_j - 2^{\ell_j} t_m)A}\right) \left[s_k(e^{-t_m a})\right]^{2^{\ell_j}} s_k(V_j)$$

$$\leq \sum_{k=1}^{\infty} s_k(e^{-t_m A})^{\sum_{j=1}^{n} 2^{\ell_j}} \prod_{j=1}^{n} \|V_j\| . \tag{2.12}$$

Here we used that $s_k(e^{-(t_j - 2^{\ell_j} t_m)A}) \leq \|e^{-(t_j - 2^{\ell_j} t_m)A}\| \leq 1$ and that $s_k(V_j) \leq \|V_j\|$. Let $N := \sum_{j=1}^{n} 2^{\ell_j}$ and $T_m := N t_m > T/2$. Given that $A = A^*$ and $e^{-tA} \in \mathcal{C}_1(\mathfrak{H})$ for $t > 0$, we obtain, by definition of the $\| \cdot \|_q$-norm on the von Neumann–Schatten ideal $\mathcal{C}_{q \geq 1}(\mathfrak{H})$, that

$$\|e^{-tA}\|_1 = \sum_{k=1}^{\infty} s_k(e^{-tA/q})^q = (\|e^{-tA/q}\|_q)^q . \tag{2.13}$$

Then inequality (2.12) yields for $q = N$:

$$\left\| \prod_{j=1}^{n} V_j e^{-t_j A} \right\|_1 \leq \left(\|e^{-T_m A/N}\|_N \right)^N \prod_{j=1}^{n} \|V_j\|. \tag{2.14}$$

Now we consider integer $p \in \mathbb{N}$ such that $2^p \leq N < 2^{p+1}$. Then, $T/4 < T_m/2 < 2^p T_m/N$, and consequently,

$$\left(\|e^{-T_m A/N}\|_N \right)^N = \sum_{k=1}^{\infty} s_k^N (e^{-T_m A/N}) \tag{2.15}$$

$$\leq \sum_{k=1}^{\infty} s_k^{2^p} (e^{-2^p T_m A/2^p N}) \leq \sum_{k=1}^{\infty} s_k^{2^p} (e^{-TA/2^{p+2}}) = \|e^{-TA/2^2}\|_1 ,$$

where we used (2.13). Therefore, the estimates (2.14), (2.15) give the bound (2.10).

Now, let $V_j \in \mathcal{L}(\mathfrak{H})$, $j = 1, 2, \ldots, n$, and set $\tilde{V}_j := V_j e^{-\varepsilon A}$ for $0 < \varepsilon < t_m$. Hence, $\tilde{V}_j \in \mathcal{C}_1(\mathfrak{H}) \subset \mathcal{C}_\infty(\mathfrak{H})$ and $s_k(\tilde{V}_j) \leq \|\tilde{V}_j\| \leq \|V_j\|$. If we set $\tilde{t}_j := t_j - \varepsilon$, then

$$\left\| \prod_{j=1}^{n} V_j e^{-t_j A} \right\|_1 \leq \prod_{j=1}^{n} \|V_j\| \|e^{-(\tilde{t}_1 + \tilde{t}_2 + \cdots + \tilde{t}_n)A/4}\|_1 , \tag{2.16}$$

by the preceding arguments for compact case. Since the semigroup $\{e^{-tA}\}_{t \geq 0}$ is $\| \cdot \|_1$-continuous for $t > 0$, we can take in (2.16) the limit $\varepsilon \downarrow 0$. This gives the result (2.10) in general case. □

Theorem 2.3 *Let assumptions* (A1)–(A4) *be satisfied. Then strongly continuous solution operator* $\{U(t,s)\}_{(t,s)\in\Delta}$ (2.9) *yields for* $t > s$ *a trace-norm continuous mild solution of nACP* (2.1) *on Banach space* $\mathcal{C}_1(\mathfrak{H})$.

Proof To this aim we again use the Dyson–Phillips series $\sum_{n=0}^{\infty} S_n(t,s)$. Then to estimate (2.7) we define $V(\tau_j)$ (cf.(2.10) by

$$V(\tau_j) := (-B(\tau_j))G((\tau_j - \tau_{j-1})/2)\,, \quad j = 1, 2, \ldots, n\,, \quad \tau_0 = s\,. \tag{2.17}$$

As a consequence, one gets for (2.7) representation

$$\begin{aligned} S_n(t,s) = \int_s^t d\tau_n \int_s^{\tau_n} d\tau_{n-1} \ldots \int_s^{\tau_2} d\tau_1\, G(t-\tau_n)\, V(\tau_n)\, G((\tau_n - \tau_{n-1})/2)\cdot \\ \cdot V(\tau_{n-1})\, G((\tau_{n-1} - \tau_{n-2})/2) \cdots G((\tau_2 - \tau_1)/2)\, V(\tau_1)\, G((\tau_1 - s)/2)\,. \end{aligned} \tag{2.18}$$

Let $V_0 := G((t-\tau_n)/2)$. Note that $\|V_0\| \le 1$. Then on account of inequality (2.10)

$$\begin{aligned} \|V_0 G((t-\tau_n)/2) \prod_{j=n}^{1} V(\tau_j)\, G((\tau_j - \tau_{j-1})/2)\|_1 \le \\ \le \prod_{j=1}^{n} \|V(\tau_j)\|\, \|G((t-s)/8)\|_1\,. \end{aligned} \tag{2.19}$$

Because of (2.5), (2.17) and due to (2.19) we infer from representation (2.18) the trace-norm estimate

$$\|S_n(t,s)\|_1 \le \left\{ \frac{2^{\alpha}\, M_{\alpha}(n) C_{\alpha}}{n^{(1-\alpha)}} (t-s)^{1-\alpha} \right\}^n \|G((t-s)/8)\|_1\,, \quad \alpha \in [0,1)\,, \tag{2.20}$$

for $s < t$ and $n \ge 1$, where $M_{\alpha}(n) := 2\sqrt{2\pi}\, M_{\alpha}\Gamma(1-\alpha)((1-\alpha)\sqrt{2\pi n})^{-1/n}$.

Owing to (2.20) the Dyson–Phillips series (2.9) converges for $2^{\alpha}\,\xi_{t,s} < 1$ in the trace-norm topology uniformly in $[\mu, \nu] \subset [s,t]$ and satisfies the integral equation (2.3). It can be extended to any $0 \le s < t \le T$ using (1.4). Thus, we obtain operator

$$U(t,s) = \sum_{n=0}^{\infty} S_n(t,s) \in \mathcal{C}_1(\mathfrak{H})\,, \quad \|U(t,s)\|_1 \le M_T\, \|G((t-s)/8)\|_1\,, \tag{2.21}$$

where $M_T > 0$. Family $\{U(t,s)\}_{0 \le s < t \le T}$ is a mild solution of the Gibbs nACP (2.1). This solution is trace-norm continuous for $0 \le s < t \le T$ and strongly continuous at $s = t$. □

Corollary 2.4 *For $t > s$ the evolution family $\{U(t,s)\}_{(t,s)\in\Delta}$ (2.21) is a strict solution of the Gibbs nACP* :

$$\begin{aligned} &\partial_t U(t,s) = -\, C(t)U(t,s), \;\; t \in (s,T) \quad \text{and} \quad U(s,s) = \mathbb{1}, \\ &C(t) := A + B(t), \end{aligned} \qquad (s,T) \subset [0,T], \tag{2.22}$$

on Banach space $\mathcal{C}_1(\mathfrak{H})$ and $\|U(t,s)\| \leq 1$.

Proof Since by Remark 1.2(c),(d) the function $t \mapsto U(t,s)$ for $t \geq s$ is strongly continuous and since $U(t,s) \in \mathcal{C}_1(\mathfrak{H})$ for $t > s$, the product $U(t+\delta,t)U(t,s)$ is continuous in the trace-norm topology for $|\delta| < t-s$. Moreover, since $\{u(t)\}_{s\leq t\leq T}$ is a classical solution of nACP (1.1), Eq. (2.1) implies that $U(t,s)$ has strong derivative for any $t > s$. Then again by Remark 1.2(d) the trace-norm continuity of $\delta \mapsto U(t+\delta,t)U(t,s)$ and by inclusion of ranges: $\mathrm{ran}(U(t,s)) \subseteq \mathrm{dom}(A)$ for $t > s$, the trace-norm derivative $\partial_t U(t,s)$ at $t(> s)$ exists and belongs to $\mathcal{C}_1(\mathfrak{H})$.

Therefore, $U(t,s) \in C((s,T],\mathcal{C}_1(\mathfrak{H})) \cap C^1((s,T],\mathcal{C}_1(\mathfrak{H}))$ with $U(s,s) = \mathbb{1}$ and $U(t,s) \in \mathcal{C}_1(\mathfrak{H})$, $C(t)U(t,s) \in \mathcal{C}_1(\mathfrak{H})$ for $t > s$, which means that solution $U(t,s)$ of (2.22) is *strict*, cf. [19] Definition 1.1. On account of (1.4) and (2.2) the strongly continuous solution operator (2.21) is contraction: $\|U(t,s)\| \leq 1$. □

We note that these results for ACP on Banach space $\mathcal{C}_1(\mathfrak{H})$ are well known for Gibbs semigroups, see [20], Chapter 4.

Now, to proceed with the proof of Theorem 1.4 about trace-norm convergence of the solution operator approximants (1.6), we need the following preparatory lemma.

3 Proof of Theorem 1.4

We follow the line of reasoning of the *lifting* lemma developed in [20], Ch.5.4.1:

1. By virtue of (1.4) and (1.6) we obtain for difference in (1.11) formula:

$$U_n(t,s) - U(t,s) = \prod_{k=n}^{1} W_k^{(n)}(t,s) - \prod_{l=n}^{1} U(t_{l+1},t_l). \tag{3.1}$$

Let integer $k_n \in (1,n)$. Then (3.1) yields the representation:

$$\begin{aligned} U_n(t,s) - U(t,s) &= \left(\prod_{k=n}^{k_n+1} W_k^{(n)}(t,s) - \prod_{l=n}^{k_n+1} U(t_{l+1},t_l)\right) \prod_{k=k_n}^{1} W_k^{(n)}(t,s) \\ &+ \prod_{l=n}^{k_n+1} U(t_{l+1},t_l) \left(\prod_{k=k_n}^{1} W_k^{(n)}(t,s) - \prod_{l=k_n}^{1} U(t_{l+1},t_l)\right), \end{aligned}$$

which entails the trace-norm estimate

$$\|U_n(t,s) - U(t,s)\|_1 \le \left\| \prod_{k=n}^{k_n+1} W_k^{(n)}(t,s) - \prod_{l=n}^{k_n+1} U(t_{l+1},t_l) \right\| \left\| \prod_{k=k_n}^{1} W_k^{(n)}(t,s) \right\|_1$$

$$+ \left\| \prod_{l=n}^{k_n+1} U(t_{l+1},t_l) \right\|_1 \left\| \prod_{k=k_n}^{1} W_k^{(n)}(t,s) - \prod_{l=k_n}^{1} U(t_{l+1},t_l) \right\| . \tag{3.2}$$

2. Now we assume that $\lim_{n\to\infty} k_n/n = 1/2$. Then (1.5) yields $\lim_{n\to\infty} t_{k_n} = (t+s)/2$, $\lim_{n\to\infty} t_n = t$, and uniform estimates (1.7)–(1.10) with the bound $R_{\alpha,\beta}\varepsilon_{\alpha,\beta}(n)$ provide

$$\operatorname*{ess\,sup}_{(t,s)\in\Delta} \left\| \prod_{k=n}^{k_n+1} W_k^{(n)}(t,s) - U(t,(t+s)/2) \right\| \le R^{(1)}_{\alpha,\beta}\varepsilon_{\alpha,\beta}(n), \tag{3.3}$$

$$\operatorname*{ess\,sup}_{(t,s)\in\Delta} \left\| \prod_{k=k_n}^{1} W_k^{(n)}(t,s) - U((t+s)/2,s) \right\| \le R^{(2)}_{\alpha,\beta}\varepsilon_{\alpha,\beta}(n), \tag{3.4}$$

for $n \in \mathbb{N}$ and for some constants $R^{(1,2)}_{\alpha,\beta} > 0$.

3. Since $\lim_{n\to\infty} k_n/n = 1/2$ and $t > s$, by definition (1.6) and by Lemma 2.2 for contractions $\{V_k = e^{-\frac{t-s}{n}B(t_k)}\}_{k=1}^{n}$, we obtain

$$\left\| \prod_{k=k_n}^{1} W_k^{(n)}(t,s) \right\|_1 = \left\| \prod_{k=k_n}^{1} e^{-\frac{t-s}{n}A} e^{-\frac{t-s}{n}B(t_k)} \right\|_1 \le \|e^{-\frac{t-s}{8}A}\|_1 . \tag{3.5}$$

On account of (2.21), one gets

$$\left\| \prod_{l=n}^{k_n+1} U(t_{l+1},t_l) \right\|_1 \le M_T \, \|e^{-\frac{t-s}{16}A}\|_1 . \tag{3.6}$$

4. Seeing that $M_T > 1$ and $\|e^{-\frac{t-s}{8}A} \le \|e^{-\frac{t-s}{16}A}\|_1$, for $t > s$, by (3.2)–(3.6), we conclude the proof of estimate (1.11) for

$$R_{\alpha,\beta}(t,s) := M_T(R^{(1)}_{\alpha,\beta} + R^{(2)}_{\alpha,\beta})\, c(t-s) , \tag{3.7}$$

where $c(t-s) := \|e^{-\frac{t-s}{16}A}\|_1 < \infty$ and $0 \le s < t \le T$. □

Corollary 3.1 *By virtue of Lemma 2.2, the proof of Theorem 1.4 can be carried over almost verbatim for approximants* $\{\widehat{U}_n(t,s)\}_{(t,s)\in\Delta,n\ge 1}$:

$$\begin{aligned}\widehat{W}_k^{(n)}(t,s) &:= e^{-\frac{t-s}{n}B(t_k)}e^{-\frac{t-s}{n}A}, \quad k = 1, 2, \ldots, n, \\ \widehat{U}_n(t,s) &:= \widehat{W}_n^{(n)}(t,s)\widehat{W}_{n-1}^{(n)}(t,s) \times \cdots \times \widehat{W}_2^{(n)}(t,s)\widehat{W}_1^{(n)}(t,s),\end{aligned} \tag{3.8}$$

as well as for self-adjoint approximants $\{\widetilde{U}_n(t,s)\}_{(t,s)\in\Delta,n\geq 1}$:

$$\begin{aligned}\widetilde{W}_k^{(n)}(t,s) &:= e^{-\frac{t-s}{n}A/2}e^{-\frac{t-s}{n}B(t_k)}e^{-\frac{t-s}{n}A/2}, \quad k = 1, 2, \ldots, n, \\ \widetilde{U}_n(t,s) &:= \widetilde{W}_n^{(n)}(t,s)\widetilde{W}_{n-1}^{(n)}(t,s) \times \cdots \times \widetilde{W}_2^{(n)}(t,s)\widetilde{W}_1^{(n)}(t,s).\end{aligned} \tag{3.9}$$

For the both cases the rate of convergence $\varepsilon_{\alpha,\beta}(n)$ *for approximants* (3.8), (3.9) *is the same as in* (1.11).

Note that extension of Theorem 1.4 to Gibbs semigroups generated by a family of non-negative self-adjoint operators $\{A(t)\}_{t\in\mathcal{I}}$ can be done along the arguments outlined in Section 2 of [18]. To this end one needs to add more conditions to (A1)–(A4) that allow to control the family $\{A(t)\}_{t\in\mathcal{I}}$.

Acknowledgments The paper was motivated by a conference meeting conversation with Fëdor Sukochev. I have greatly benefited from discussions with him on the early stage of this project. I am grateful to Valeri Frolov for a useful discussion of Theorem 2.3.

I am thankful to Professor Themistocles M. Rassias for his kind invitation to write this note for the present SPRINGER volume on Mathematical Analysis in Interdisciplinary Research.

References

1. K.-J. Engel and R. Nagel. *One-parameter semigroups for linear evolution equations*. Springer-Verlag, New York, 2000.
2. D. E. Evans. Time dependent perturbations and scattering of strongly continuous groups on Banach spaces. *Math. Ann.*, 221(3):275–290, 1976.
3. J. S. Howland. Stationary scattering theory for time-dependent Hamiltonians. *Math. Ann.*, 207:315–335, 1974.
4. T. Ichinose and H. Tamura. Error estimate in operator norm of exponential product formulas for propagators of parabolic evolution equations. *Osaka J. Math.*, 35(4):751–770, 1998.
5. T. Kato. Abstract evolution equation of parabolic type in Banach and Hilbert spaces. *Nagoya Math. J.*, 19:93–125, 1961.
6. S. Monniaux and A. Rhandi. Semigroup method to solve non-autonomous evolution equations. *Semigroup Forum*, 60:122–134, 2000
7. R. Nagel and G. Nickel. Well-posedness of nonautonomous abstract Cauchy problems. *Progr. Nonlinear Diff. Eqn. and Their Appl.*, vol.50:279–293, 2002
8. H. Neidhardt. *Integration of Evolutionsgleichungen mit Hilfe von Evolutionshalbgruppen*. Dissertation, AdW der DDR. Berlin 1979.
9. H. Neidhardt. On abstract linear evolution equations. I. *Math. Nachr.*, 103:283–298, 1981.
10. H. Neidhardt, A. Stephan, and V.A. Zagrebnov. On convergence rate estimates for approximations of solution operators for linear non-autonomous evolution equations. *Nanosyst., Phys. Chem. Math.*, 8(2):202–215, 2017.

11. H. Neidhardt, A. Stephan, and V. A. Zagrebnov. Remarks on the operator-norm convergence of the Trotter product formula. *Int.Eqn.Oper.Theory*, 90:1–15, 2018.
12. H. Neidhardt, A. Stephan, and V. A. Zagrebnov. Trotter Product Formula and Linear Evolution Equations on Hilbert Spaces. *Analysis and Operator Theory.* Dedicated in Memory of Tosio Kato's 100th Birthday, Springer vol.146, Berlin 2019, pp. 271–299.
13. H. Neidhardt, A. Stephan, and V. A. Zagrebnov. Convergence rate estimates for Trotter product approximations of solution operators for non-autonomous Cauchy problems. Publ. RIMS Kyoto Univ., 56:83–135, 2020.
14. H. Neidhardt and V. A. Zagrebnov. Linear non-autonomous Cauchy problems and evolution semigroups. *Adv. Differential Equations*, 14(3–4):289–340, 2009.
15. G. Nickel. Evolution semigroups and product formulas for nonautonomous Cauchy problems. *Math.Nachr*, 212:101–116, 2000.
16. P. E. Sobolevskii. Parabolic equations in a Banach space with an unbounded variable operator, a fractional power of which has a constant domain of definition. *Dokl. Akad. Nauk USSR*, 138(1): 59–62, 1961.
17. H. Tanabe. *Equations of evolution.* Pitman (Advanced Publishing Program), Boston, Mass.-London, 1979.
18. P.-A. Vuillermot, W. F. Wreszinski, and V. A. Zagrebnov. A general Trotter–Kato formula for a class of evolution operators. *Journal of Functional Analysis*, 257(7): 2246–2290, 2009.
19. A. Yagi. Parabolic evolution equation in which the coefficients are the generators of infinitely differentiable semigroups, I. *Funkcialaj Ekvacioj*, 32:107–124, 1989.
20. V. A. Zagrebnov, *Gibbs Semigroups*, Operator Theory Series: Advances and Applications, Vol. 273, Bikhäuser - Springer, Basel 2019.

www.ingramcontent.com/pod-product-compliance
Ingram Content Group UK Ltd.
Pitfield, Milton Keynes, MK11 3LW, UK
UKHW021832270726
14058UKWH00001B/99
* 9 7 8 3 0 3 0 8 4 7 2 3 4 *